ADVANCED ACCELERATOR CONCEPTS

ADVANCED ACCELERATOR CONCEPTS
Seventh Workshop

Lake Tahoe, California October 1996

EDITORS
Scientific Editor:
Swapan Chattopadhyay
Lawrence Berkeley National Laboratory

Technical Editors:
Julie McCullough
Per Dahl
Lawrence Berkeley National Laboratory

◎ CD-ROM INCLUDED

AIP CONFERENCE
PROCEEDINGS 398

American Institute of Physics   Woodbury, New York

Authorization to photocopy items for internal or personal use, beyond the free copying permitted under the 1978 U.S. Copyright Law (see statement below), is granted by the American Institute of Physics for users registered with the Copyright Clearance Center (CCC) Transactional Reporting Service, provided that the base fee of $10.00 per copy is paid directly to CCC, 222 Rosewood Drive, Danvers, MA 01923. For those organizations that have been granted a photocopy license by CCC, a separate system of payment has been arranged. The fee code for users of the Transactional Reporting Service is: 1-56396-697-2/ 97 /$10.00.

© 1997 American Institute of Physics

Individual readers of this volume and nonprofit libraries, acting for them, are permitted to make fair use of the material in it, such as copying an article for use in teaching or research. Permission is granted to quote from this volume in scientific work with the customary acknowledgment of the source. To reprint a figure, table, or other excerpt requires the consent of one of the original authors and notification to AIP. Republication or systematic or multiple reproduction of any material in this volume is permitted only under license from AIP. Address inquiries to Office of Rights and Permissions, 500 Sunnyside Boulevard, Woodbury, NY 11797-2999; phone: 516-576-2268; fax: 516-576-2499; e-mail: rights@aip.org.

L.C. Catalog Card No. 97-72788
ISBN 1-56396-697-2–set
ISBN 1-56396-727-8–cloth
ISBN 1-56396-726-X–CD-ROM
ISSN 0094-243X
DOE CONF- 9610210

Printed in the United States of America

Contents

Preface xiii
Dedication xv
AAC '96 Scientific, Organizing, and Program Committee xvii

I. INVITED PAPERS

Status and Future Directions for Advanced Accelerator
Research—Conventional and Non-Conventional Collider Concepts 3
    R. H. Siemann
Plasma-Based Acceleration Concepts 13
    C. E. Clayton
Electron Beam Monitoring at High Frequencies and Ultra-Fast
Time Scales 23
    W. P. Leemans
Recent Progress in Photoinjectors 40
    I. Ben-Zvi
Microwave Sources 55
    W. Lawson
Ultrahigh Intensity Laser for Laser Wakefield Acceleration 68
    G. Mourou, J. Nees, and S. Biswal
Guiding of Sub-100 Femtosecond Pulses in Preformed Plasma Channels 76
    S. P. Nikitin, T. R. Clark, and H. M. Milchberg
Recent Results of Laser Wakefield Acceleration in KEK/U.
Tokyo/JAERI 83
    K. Nakajima, M. Kando, H. Ahn, H. Kotaki, T. Watanabe,
    T. Ueda, M. Uesaka, H. Nakanishi, A. Ogata, T. Kawakubo, and K. Tani
Laser Driven Acceleration in Vacuum and Gases 96
    P. Sprangle, E. Esarey, B. Hafizi, R. Hubbard, J. Krall, and A. Ting
Ultrashort-Pulse Relativistic Electron Gun/Accelerator 106
    E. Dodd, J. K. Kim, and D. Umstadter
Performance of the Argonne Wakefield Accelerator Facility
and Initial Experimental Results 116
    W. Gai, M. Conde, G. Cox, R. Konecny, J. Power, P. Schoessow,
    J. Simpson, and N. Barov
The Drive Beam Generation for Two Beam Accelerators 126
    R. Corsini
Developments in Relativistic Channeling 146
    R. A. Carrigan, Jr.

II. WORKING GROUP SUMMARIES

Summary Report: Working Group 1 on "Nonconventional Collider Concepts
and Luminosity Paradigms" 167
    J. S. Wurtele

Summary Report: Working Group 2 on "Plasma-Based Acceleration
Concepts" 175
    T. Katsouleas
Summary Report: Working Group 3 on "Structure-Based Accelerator
Concepts" 181
    J. B. Rosenzweig
Summary Report: Working Group 4 on "Beam Monitoring, Conditioning, and
Control at High Frequencies and Ultrafast Timescales" 187
    X. J. Wang
Summary Report: Working Group 5 on "Particle Beam Sources" 196
    L. Serafini and A. D. Yeremian
Summary Report: Working Group 6 on "Microwave Sources" 207
    G. A. Westenskow
Summary Report: Working Group 7 on "Laser Sources for Particle Accelera-
tion" 214
    M. C. Downer and C. W. Siders

III. WORKING GROUP PAPERS

Group 1: Nonconventional Collider Concepts and Luminosity Paradigms

**Studies of Laser-Driven 5 TeV e^+e^- Colliders in Strong Quantum
Beamstrahlung Regime** 233
    M. Xie, T. Tajima, K. Yokoya, and S. Chattopadhyay
Emittance in Particle and Radiation Beam Techniques 243
    K.-J. Kim
Suppression of Radiation Excitation in Focusing Environment 254
    Z. Huang and R. D. Ruth
Luminosity-Limiting Coherent Beam-Beam Phenomena 263
    S. Krishnagopal
**Crystal Channel Collider: Ultra-High Energy and Luminosity
in the Next Century** 273
    P. Chen and R. J. Noble
**Final Crystal Cooling to Reduce the Beam Current and Detector
Backgrounds for a $\mu^+\mu^-$ Collider** 286
    S. A. Bogacz, D. B. Cline, and D. A. Sanders
**On the Physical Limitations to the Lowest Emittance (Toward Colliding
Electron-Positron Crystalline Beams)** 294
    A. A. Mikhailichenko
Nonlinear Dynamics of Semicrystalline Beams 301
    V. A. Skvortsov and P. V. Milyutin
**Nonlinear Interactions between Relativistic Electrons and Ultrahigh
Intensity Laser Pulses in Vacuum** 309
    F. V. Hartemann, A. L. Troha, J. R. Van Meter, N. C. Luhmann, Jr.,
    A. K. Kerman, and T. S. Chu
Particle Dynamics in an Active Medium 326
    L. Schächter

Group 2: Plasma-Based Acceleration Concepts

Plasma Channel Formation in the Wake of a Short Laser Pulse 337
    E. Esarey, A. Ting, K. Krushelnick, C. Moore, M. Baine, and P. Sprangle

Beam-Generated Plasma Channels for Laser Wakefield Acceleration 344
    G. Shvets and N. J. Fisch

Laser Instabilities and Coupling Efficiency in Hollow Channel Plasma Wakefield Accelerators. 357
    T. C. Chiou, T. Katsouleas, W. B. Mori, and G. Shvets

First Measurement of Laser Wakefield Oscillations by Longitudinal Interferometry. 372
    C. W. Siders, S. P. LeBlanc, B. Rau, D. Fisher, T. Tajima, M. C. Downer,
    A. Babine, A. Stepanov, and A. Sergeev

Optical Guiding of High Intensity Laser Pulses in Plasma Channel: Interferometrical Investigations 378
    N. Vogel

Formation of Self-Channeling and Electron Jet in an Underdense Plasma Excited by Ultrashort High Intensity Laser Pulses 390
    M. Kando, H. Ahn, H. Kotaki, K. Tani, T. Watanabe, T. Ueda, M. Uesaka,
    Y. Kishimoto, J. Koga, H. Watanabe, K. Nakajima, M. Arinaga,
    T. Kawakubo, H. Nakanishi, and A. Ogata

Accelerated Electron Measurements in the Self-Modulated Laser Wakefield Accelerator 400
    C. I. Moore, K. Krushelnick, A. Ting, C. Manka, H. R. Burris, R. Fischer,
    M. Baine, E. Esarey, P. Sprangle, and R. Hubbard

Electron Acceleration by Self-Modulated Laser Wakefield in a Relativistically Self-Guided Channel 408
    S.-Y. Chen, R. Wagner, A. Maksimchuk, D. Gustaphson, and D. Umstadter

Laser Injection of Electrons into Wakefields: A Qualitative Understanding 417
    J. L. Bobin

Charged Particle Acceleration in Nonuniform Plasmas 422
    S. V. Bulanov, G. I. Dudnikova, N. M. Naumova, F. Pegoraro,
    I. V. Pogorelsky, and V. A. Vshivkov

Laser-Driven Acceleration with Bessel and Gaussian Beams. 433
    B. Hafizi, E. Esarey, and P. Sprangle

Proof-of-Principle Experiment of the Vacuum Beat Wave Accelerator 443
    A. Ting, B. Hafizi, R. Burris, R. Fischer, C. I. Moore, K. Krushelnick,
    E. Esarey, and P. Sprangle

Group 3: Structure-Based Accelerator Concepts

Recent Results & Plans for the Future on SLAC Damped Detuned Structures (DDS) 455
    N. M. Kroll, R. M. Jones, C. Adolphsen, K. L. F. Bane, W. R. Fowkes,
    K. Ko, R. H. Miller, R. D. Ruth, M. Seidel, and J. W. Wang

Analysis and Application of Manifold Radiation in DDS 1:
First Experiences 465
    R. M. Jones, N. M. Kroll, M. Seidel, C. Adolphsen, K. L. F. Bane,
    W. R. Fowkes, K. Ko, R. H. Miller, R. D. Ruth, and J. W. Wang

Cold Test Results of a Standing Wave Muffin-Tin Structure
at X-Band 473
    P. J. Chou, S. M. Hanna, H. Henke, A. Menegat, R. H. Siemann,
    and D. Whittum

Planar Millimeter-Wave RF Structures............................. 485
    H. Henke

The Fabrication of Millimeter-Wavelength Accelerating Structures........... 501
    P. J. Chou, G. B. Bowden, M. R. Copeland, A. Farvid,
    R. E. Kirby, A. Menegat, C. Pearson, L. Shere, R. H. Siemann,
    J. E. Spencer, and D. H. Whittum

Recent Progress on Photonic Band Gap Accelerator Cavities................ 518
    D. R. Smith, D. Li, D. C. Vier, N. Kroll, S. Schultz, and H. Wang

Wake-Field Studies on Photonic Band Gap Accelerator Cavities 528
    D. Li, N. Kroll, D. R. Smith, and S. Schultz

A Proposed Dielectric-Based Laser-Driven Electron Accelerator
Using Crossed Cylindrical Laser Focusing............................. 538
    Y. C. Huang and R. L. Byer

The Concept of a Linac Driven by a Traveling Laser Focus................. 547
    A. A. Mikhailichenko

An Inverse Cherenkov Accelerator Using a Dielectric Channelled
Waveguide 564
    W. Gai and J. Simpson

Theory and Simulation of an Inverse Free-Electron Laser Experiment 575
    S. K. Gou, A. Bhattacharjee, J.-M. Fang, and T. C. Marshall

Status of the BNL IFEL Accelerator 591
    A. van Steenbergen, J. Gallardo, M. Babzien, K. Kusche, R. Malone,
    I. Pogorelsky, T. Romano, J. Sheehan, J. Skaritka, X.-J. Wang,
    J. Sandweiss, J.-M. Fang, and X. Qiu

Observation of 10 μm Smith-Purcell Radiation from 45 MeV/c Electrons..... 601
    R. C. Fernow, H. G. Kirk, S. H. Robertson, J. H. Brownell, and J. E. Walsh

Inverse Cerenkov Acceleration Using an IFEL Prebuncher 608
    W. D. Kimura, I. V. Pogorelsky, Y. Liu, K. P. Kusche,
    A. van Steenbergen, J. C. Gallardo, J. Sandweiss, D. B. Cline,
    D. C. Quimby, and M. Babzien

Microwave Inverse Cerenkov Accelerator 618
    T. B. Zhang, T. C. Marshall, M. A. LaPointe, and J. L. Hirshfield

Simulation Results and Experimental Design for the Microwave
Inverse FEL Accelerator 629
    R. B. Yoder, T. B. Zhang, T. C. Marshall, and J. L. Hirshfield

Ionization Effects in Inverse Cherenkov Accelerators 638
    P. Sprangle, B. Hafizi, and R. F. Hubbard

Group 4: Beam Monitoring, Conditioning, and Control at High Frequencies and Ultrafast Timescales

Temporal Characterization of a Self-Modulated Laser Wakefield 651
    S. P. Le Blanc, M. C. Downer, R. Wagner, S.-Y. Chen, A. Maksimchuk,
    G. Mourou, and D. Umstadter

Micro-Bunching Diagnostics for the IFEL by Coherent Transition Radiation 664
    Y. Liu, D. B. Cline, X. J. Wang, M. Babzien, J. M. Fang,
    and V. Yakimenko

Inverse Transition Radiation 673
    L. C. Steinhauer, R. D. Romea, and W. D. Kimura

Generation, Measurement, and Application of Synchronized 700 fs Electron, 100 fs T^3 Laser, and Picoseconds X-Ray Single Pulses 687
    M. Uesaka, T. Ueda, T. Watanabe, M. Kando, K. Nakajima, H. Kotaki,
    and A. Ogata

Group 5: Particle Beam Sources

Initial Commissioning Results of the Next Generation Photoinjector 695
    D. T. Palmer, X. J. Wang, R. H. Miller, M. Babzien, I. Ben-Zvi,
    C. Pellegrini, J. Sheehan, J. Skaritka, H. Winick, M. Woodle, and V. Yakimenko

The Design and Fabrication of an X-Band RF Gun 705
    C. H. Ho, W. K. Lau, T. T. Yang, J. Y. Hwang, S. Y. Hsu, Y. C. Liu,
    G. P. Le Sage, F. V. Hartemann, and N. C. Luhmann, Jr.

High Power Operation of a 17 GHz Photocathode RF Gun 717
    S. Trotz, W. J. Brown, B. G. Danly, J.-P. Hogge, M. Khusid,
    K. E. Kreischer, M. Shapiro, and R. J. Temkin

Table Top, Pulsed, Relativistic Electron Gun with GV/m Gradient 730
    T. Srinivasan-Rao and J. Smedley

Acceleration of Kiloampere Current at 2.65 GV/m 739
    F. Villa

Thin Film Photoemission Experiments 747
    R. Brogle, P. Muggli, and C. Joshi

Measurements of the Argonne Wakefield Accelerator's Low Charge, 4 MeV RF Photocathode Witness Beam 757
    J. Power, E. Chojnacki, M. Conde, W. Gai, R. Konecny, P. Schoessow,
    and J. Simpson

Microbunching and Coherent Acceleration of Electrons by Subcycle Laser Pulses 766
    B. Rau, T. Tajima, and H. Hojo

Nonlinear Space-Charge Effects in High-Brightness Beams 782
    Y. Fink, C. Chen, and W. P. Marable

The Space Charge Limits of Longitudinal Emittance in RF Photoinjectors 793
    D. H. Dowell, S. Joly, and A. Loulergue

Group 6: Microwave Sources

The Next Linear Collider Test Accelerator's RF Pulse Compression and Transmission Systems 805
    S. G. Tantawi, A. E. Vlieks, K. Fant, T. Lavine, R. J. Loewen,
    C. Pearson, R. Pope, J. Rifkin, and R. D. Ruth

Active High Power RF Pulse Compression Using Optically Switched Resonant Delay Lines 813
    S. G. Tantawi, R. D. Ruth, A. E. Vlieks, and M. Zolotorev

Pulse Compressor Based on Electrically Switched Bragg Reflectors 822
    M. I. Petelin, A. L. Vikharev, and J. L. Hirshfield

Development of an X-Band Magnicon Amplifier for the Next Linear Collider 832
    S. H. Gold, A. W. Fliflet, A. K. Kinkead, B. Hafizi, O. A. Nezhevenko,
    V. P. Yakovlev, J. L. Hirshfield, and R. True

RK-TBA Studies at the RTA Test Facility 842
    S. Lidia, D. Anderson, S. Eylon, E. Henestroza, T. Houck, L. Reginato,
    D. Vanecek, G. Westenskow, and S. Yu

Progress in High Power, High Efficiency Relativistic Traveling Wave Tube Amplifiers 852
    J. A. Nation, S. A. Naqvi, G. S. Kerslick, and L. Schächter

High Power Gyroklystron Development for Advanced Accelerator Applications 865
    W. Lawson, J. Anderson, J. P. Calame, J. Cheng, M. Castle,
    V. L. Granatstein, B. Hogan, M. Reiser, and G. P. Saraph

Gyroklystrons for Driving Linear Colliders at 35 GHz 874
    V. L. Granatstein, G. P. Saraph, G. S. Nusinovich, A. Singh,
    and W. Lawson

High-Power Microwave Production by Gyroharmonic Conversion and Co-Generation 887
    M. A. LaPointe, R. B. Yoder, M. Wang, A. K. Ganguly, C. Wang,
    B. Hafizi, and J. L. Hirshfield

A Racetrack Geometry to Avoid Undesirable Azimuthal Variations of the Electric Field Gradient in High Power Coupling Cavities for TW Structures 898
    J. Haimson, B. Mecklenburg, and E. L. Wright

The Results of the 7 GHz Pulsed Magnicon Investigation 912
    O. A. Nezhevenko, E. V. Kozyrev, I. G. Makarov, A. A. Nikiforov,
    G. N. Ostreiko, B. Z. Persov, G. V. Serdobintsev, V. V. Tarnetsky,
    S. V. Shchelkunoff, V. P. Yakovlev, and I. A. Zapryagaev

Group 7: Laser Sources for Particle Acceleration

CO_2 Laser Technology for Advanced Particle Accelerators 923
    I. V. Pogorelsky, A. Van Steenbergen, R. Fernow, W. D. Kimura,
    and S. V. Bulanov

The First Terawatt Picosecond CO_2 Laser for Advanced Accelerator Study at the Brookhaven ATF 937
      I. V. Pogorelsky, I. Ben-Zvi, J. Skaritka, Z. Segalov, M. Babzien,
      K. Kusche, I. K. Meskovsky, V. A. Lekomtsev, A. A. Dublov,
      Y.-A. Boloshin, and G. A. Baranov

Prospects for Compact High-Intensity Laser Synchrotron X-Ray and Gamma Sources 951
      I. V. Pogorelsky

List of Participants. 967
Photographs. 971
Author Index. 975

PREFACE

The Seventh Workshop on Advanced Accelerator Concepts (AAC) was held from October 12–18, 1996, in the Granlibakken Conference Center at Lake Tahoe, California, USA. This series started in the early 1980s. The last two workshops were held at the Abbey at Lake Geneva, Wisconsin, USA, in 1994 and in Port Jefferson on Long Island, New York, USA, in 1992. AAC '96 was sponsored, as per tradition, by the US Department of Energy (High Energy Physics Division, Advanced Technology Branch). Also sponsoring the workshop was the Center for Beam Physics at Ernest Orlando Lawrence Berkeley National Laboratory (Berkeley Lab) of the University of California.

The Advanced Accelerator Concepts workshop is the only acknowledged and fully sponsored forum that provides a platform for inter- and cross-disciplinary discussions on various aspects of advanced accelerator and beam physics/technology concepts. A wide range of applications is covered—from High Energy Colliders to Synchrotron Radiation Sources. This wide scope has included new methods of particle acceleration to high energies, techniques for production of ultra-high gradient electromagnetic fields in the laboratory, diagnostics and control of particle/photon beams in ultrashort dimensions and ultrafast time scales, and various energy and beam sources. Of special note is the fact that this was the first AAC workshop following the demise of the Super-conducting Super Collider (SSC). Therefore, the workshop achievements and proceedings will be crucial in directing the future of accelerator- and beam-based science world wide.

The AAC '96 program consisted of a set of invited talks throughout the week, solicited from international experts and recommended by the Advisory and Program Committee. These talks filled the plenary sessions—taking the entire first day of the workshop and consisting of one or two workshops per day thereafter. The last day of the workshop was devoted to the presentation of the Working Group Summaries. The rest of the workshop time was spent in working-group discussions (seven in all), spontaneous group discussions, and poster presentations. The posters were exhibited throughout the entire duration of the workshop. There was an informal reception upon arrival, a free afternoon of tours, a Boating Excursion, and a Workshop Banquet.

The workshop was attended by a total of 165 participants. The participation was by invitation only, and a significant number of students and junior colleagues active in the fields of accelerators, lasers, plasmas, and particle/photon beam physics and technology participated. There was significant participation by international experts as well. The composition of workshop participants reflected a balance among various fields, institutions, students, colleagues, experts, national laboratories, and key international figures.

I would like to take this opportunity to acknowledge the unwavering and continuous support of our field over the last two decades by the US Department of Energy's High Energy Physics Division and in particular, by Dr. David F. Sutter, Chief of its Advanced Technology Branch. I would also like to thank the Scientific Advisory Program and Organizing Committee for their hard work and support. The scientific and social program they put together was enhanced by everyone's zealous participation. The workshop would not have been possible without the continual support, dedication, and care by the administrative staff of Berkeley Lab's Center for Beam Physics and Conference Coordination Center—in particular Mollie Field, Joy Kono, and "Sam" Vanecek. Many thanks also go to Al Early, Keith Groves, Del Thomas, and Olivia Wong for numerous administrative, logistics, and computer support. A final note of appreciation is given to the staff of Granlibakken for their hospitality and graceful hosting of many events during the Workshop (including lending sailboats for the now-famous AAC boat race, whose winner this year was ...!).

I feel privileged to have chaired this Workshop, and I take much pleasure in presenting to you now its ultimate product—these proceedings, which also appear on CD-ROM. Thanks go to Berkeley Lab's technical and copy editor, Julie McCullough, for carefully and delicately putting together both the electronic and hard-copy versions. I also wish to thank Per Dahl and the Editorial

Advisory Committee for their significant help. This book is full of exciting new ideas and a record of achieved results!

Do enjoy these proceedings, and happy reading!

Swapan Chattopadhyay
Berkeley, California

To the Memory of Raymond G. Herb—physicist, teacher, and colleague

Banquet Speech, October 16, 1996

by Fred Mills

I want to comment on the passing of a pioneer in our field. This workshop in its many meetings has been a celebration of creativity and ingenuity. It is exactly in the sense of this celebration that I want to speak. Ray Herb, if he were here, would fit in perfectly.

Let me start out by reviewing a few of Ray's accomplishments in Nuclear Physics. He, together with Kerst and Parkinson, made the first precision measurements of p-p scattering. He developed the precision electrostatic analyzer which allowed measurements of the properties of nuclear states. Ray was principally responsible for the establishment of the long, successful Nuclear Physics program at the University of Wisconsin.

Ray made many contributions to the development of particle accelerators and related technology, particularly electrostatic accelerators and vacuum. He developed pressurized accelerators, leading to multi-MeV acceleration. He pioneered the use of cantilevered horizontal pressurized accelerator structures. He invented the "Pelletron" charging chain, yielding higher currents and increased stability. Eventually, most existing electrostatic accelerators utilized them. The use of baked accelerating columns simplified voltage conditioning. Ray was an early user of ion pumping in his systems. He invented the Evaporion and Orbion pumps, used on many accelerators, including the AGS at BNL. Ray understood the importance of beam cooling and through his company, the National Electrostatics Corporation, and with his test accelerator, collaborated with Delbert Larson, David Cline, and myself on the first tests of multi-MeV electron beam recirculation and measurements of beam emittance. This technology is now being put into use at Fermilab in the Recycler Antiproton Accumulation Ring.

During World War II, Ray placed his efforts at the MIT Radiation Laboratory developing Radar, while he sent his accelerator, the "long tank," and his student, Joseph McKibben, to Los Alamos. McKibben long played an important role at Los Alamos.

In the late sixties, Ray retired from the University of Wisconsin and formed the National Electrostatics Corporation, where many of the above activities took place. He built the 25 MV tandem accelerator for Oak Ridge National Laboratory, the highest energy electrostatic accelerator so far built. Ray had plans for the construction of a 50 MV accelerator when the world was ready for it.

Ray trained many students, some of whom also played important roles in accelerator development. Don Kerst invented the Betatron accelerator, the boiling water reactor, and made key contributions to the development of colliding beams. Dave Parkinson built cyclotrons and used them for Nuclear Physics at the University of Michigan. Clarence Turner built the injector for the Cosmotron at Brookhaven National Laboratory. Jim McGruer developed ultra high vacuum and helped build the 45 MeV Beam Stacking Accelerator at MURA as well as doing Nuclear Physics research at the University of Pittsburgh for many years. Per Dahl, who is here tonight, went to Brookhaven National Laboratory and became an expert in superconducting magnet technology there and at the SSC. Jack Bittner went to Brookhaven National Laboratory and helped build the AGS and develop the program there. There were many more students who learned their fields from Ray.

Ray was also a wonderful husband and father. Among the diverse careers of his and Ann's children, their son Steve Herb played a key role in the discovery of the B quark at Fermilab, developed micro-beta for CLEO, and now works at Cornell and DESY. My family enjoyed canoeing and camping with the Herbs over a period of thirty years.

I propose to the Organizing Committee that this Workshop be dedicated to the memory of Raymond G. Herb. Thank you.

AAC '96 Scientific, Organizing, and Program Committee

Swapan Chattopadhyay (Berkeley), Chair
Ilan Ben-Zvi (Brookhaven)
David L. Burke (Stanford)
Patrick L. Colestock (Fermilab)
Jean-Pierre Delahaye (CERN)
Heino H. Henke (Berlin)
Chan Joshi (UCLA)
Howard Milchberg (Maryland)

Gerard G. Mourou (Michigan)
Kazuhisa Nakajima (KEK)
Claudio C. Pellegrini (UCLA)
Andrew Sessler (Berkeley)
James D. Simpson (Argonne)
Phillip Sprangle (NRL)
David Sutter (DOE)
Anahid Dian Yeremian (Stanford)

Local Organizers

Swapan Chattopadhyay
Al Early
Mollie Field
Keith Groves
Joy Kono
Julie McCullough
Romy Perry
Del Thomas
"Sam" Vanecek
Olivia Wong

Hosted By
Center for Beam Physics
Ernest Orlando Lawrence Berkeley
National Laboratory

Sponsored By
The U.S. Department of Energy
Ernest Orlando Lawrence Berkeley
National Laboratory

Editorial Advisory Committee

Jonathan Wurtele (UC Berkeley)
Thomas Katsouleas (USC)
James Rosenzweig (UCLA)
X. J. Wang (BNL)
Luca Serafini (UCLA/INFN-Milano)

Anahid Dian Yeremian (SLAC)
Glen Westenskow (LLNL)
Michael Downer (UT Austin)
Craig Siders (UT Austin)
David Sutter (DOE)

I. INVITED PAPERS

Status and Future Directions for Advanced Accelerator Research - Conventional and Non-Conventional Collider Concepts

R. H. Siemann*

Stanford Linear Accelerator Center,
Stanford University, Stanford, CA 94309

ABSTRACT: The relationship between advanced accelerator research and future directions for particle physics is discussed. Comments are made about accelerator research trends in hadron colliders, muon colliders, and e^+e^- linear colliders.

COLLIDERS AND HIGH ENERGY PHYSICS

The mass scale of interest to particle physics is the range of ~ 0.5 to 2 TeV where electroweak symmetry is broken. Experiments at colliders with high enough energy are expected to detect evidence of electroweak symmetry breaking and to shed light on the symmetry breaking mechanism. Is it the classic Higgs phenomena, supersymmetry, strong coupling, or something else? History suggests that discovering the origin of electroweak symmetry breaking will also raise questions about subjects unknown today.

The Large Hadron Collider (LHC) is a technically proven and funded project that could reach high enough energy and luminosity for the study of electroweak symmetry breaking, and the NLC, JLC and TESLA, linear colliders being designed for center-of-mass energies E_{CM} = 0.5 to 1.5 TeV, promise an unrivaled environment for the study of this phenomenon. The sizes and costs of these colliders raise questions that are at the heart of the future of particle physics

1. Are the colliders and detectors needed for the study of electroweak symmetry breaking affordable? The costs of these facilities are modest on the scale of many governmental activities, so the issue is whether our elected representatives decide that high energy physics pursued at this scale is or is not in the national interest. The SSC was started when they decided it was, but that project was terminated when their opinion changed. CERN and the LHC may be facing problems of the same nature with the discussion of budget cuts initiated by the German government.

International collaboration on the design, construction and operation of large colliders is the proposed solution to the high cost of these facilities. The cost per country is reduced, but the involvement and commitment of each country is reduced also. Will one or two large colliders located somewhere in the world meet the needs of the governments that support particle physics, the institutions

---

* Work supported by the Department of Energy, contract DE-AC03-76SF00515.

that commit faculty and staff to this scholarly field, and the physicists who perform the research? If these needs are not met, support for and interest in the field could drop precipitously. The discussion of international collaboration has concentrated on the cost reductions without much consideration of these needs and the consequences of not meeting them.

2. Do we have the technology and accelerator physics to move on to the next energy scale? The colliders of today are based on a combination of principles, technologies, and accomplishments that has led to many of the past discoveries in particle physics and has placed the field on the verge of studying the Higgs phenomenon. However, these accomplishments are not enough for the future. We are at the limit of affordability, and an extension of present techniques is not a way to reach the next energy scale.

High energy physics based on an extrapolation of present trends will be a field posing exciting scientific questions but with few opportunities to explore them and with high costs. Reduced opportunities and remoteness from universities, laboratories and nations could reduce institutional and national commitment to particle physics, and the LHC and the next generation of linear collider could be the last major facilities constructed for this science.

This demise of particle physics seems inevitable unless there is a revolutionary change in particle accelerators that reduces costs. This must be a revolution comparable to that which replaced vacuum tubes with integrated circuits and telephone wires with fiber optics and cellular facilities. These are examples of inventions that were so dramatic that new, previously undreamed of ideas became possible. Particle physics must have inventions of comparable impact.

COLLIDERS

Characteristics

Colliders have characteristics that describe the particle physics potential: luminosity, center-of-mass energy, lepton or hadron beams, backgrounds, interactions per crossing, energy spread, collision spot size, etc. Some of these such as luminosity and center-of-mass energy are the raison d'être, and these should be the goal of accelerator development.

Others have major impact on experiments, and accelerator physicists try to make those impacts as favorable as possible. Backgrounds and interactions per crossing are examples. These can be given less emphasis, perhaps even ignored, when revolutionary changes in accelerators are required. Comparable changes in experimentation are going to be necessary also. The choice is as stark as it is for future colliders - work on innovations in experimentation or the survival of particle physics is in question.

Topology

The general topology of a collider has a particle source, accelerator, storage system, and collision system. The two examples given in Table 1 show

Table 1. General Collider Topology and Two Examples

General Topology	SLC	Tevatron
Particle Source	Polarized Gun, Damping Rings	Ion Source, $\bar{p}$ Cooler and Accumulator
Accelerator	SLAC S-Band Linac	Booster, Main Ring, Tevatron
Storage System	------	Tevatron
Collision System	Final Focus	High β Quadrupoles and Interaction Region

that *i)* the SLC has three of four of these systems and *ii)* the functions are combined or closely connected in the Tevatron.

A collider must have most of these systems, and they must work together, complement each other, and the properties of one system can strongly influence other systems. Two examples of that the dominant role of $\bar{p}$ production and cooling in all of the other Tevatron systems, and need for flat beams at the collision point of a linear collider determining many of the parameters of the damping rings and accelerator. While much of this is obvious, it is often ignored in the advanced accelerator community which can become fascinated with an aspect of performance without considering possible functioning as a collider.

OLD AND NEW INVENTIONS

The accelerators and colliders of today are based on:
1) *Great principles* of accelerator physics: phase stability, strong focusing, and colliding beam storage rings;
2) *Dominant technologies*: superconducting magnets, high power RF production, and normal and superconducting RF acceleration;
3) Many other *substantial accomplishments* in accelerator physics and technology: non-linear dynamics, collective effects, beam diagnostics, etc.;
4) Years of *experience* with operating colliders. This is closely related to the previous element. Overcoming performance limits has often required development of sophisticated theories, experiments, or instrumentation.

A change in the future of high energy physics will require inventions and new ideas of comparable importance to the great principles and dominant technologies. These must encompass both accelerator physics and technology to have the needed impact.

Particle physics is only a small part of science, and these critical ideas may arise in other contexts and have other driving forces including market forces. The accelerator community needs to be aware of developments throughout science and technology and constantly be considering the application of new developments to particle physics. High peak power lasers are a clear example. These devices are being developed for a wide range of scientific and commercial applications, and in the process devices with enormous potential for producing high acceleration gradients are becoming available.

HADRON COLLIDERS

This is the first of three sections that deal with the colliders that could have a role in the future and with issues related to them.

High energy hadron colliders are a proven way to reach the energy scales of interest to high energy physics. Unfortunately the costs of today's technology are prohibitive for thinking about future extrapolations, and the focus of hadron collider development has to be cost reduction.[1] The SSC can be used to understand costs and to identify areas with potentially significant savings. The Appendix shows that the superconducting magnets of the collider ring were almost half of the SSC cost. This is the area where there must be significant savings.

There is extensive experience at the Tevatron, HERA, RHIC, SSC, and LHC with 4 - 8 T $\cos\theta$ magnets. This is the technology determining the present energy frontier. However, since this type of magnet is well developed, it is unlikely to be the basis for the qualitative changes needed in the future. Directions that hold promise for such changes are low-field, superferric magnets and high temperature superconductors.

The low-field superferric magnet[3] addresses many of the costly aspects of higher field magnets. The geometry is simple with a single conductor placed in a low magnetic field region. The principle disadvantage is that the field is low, $B \leq 2T$, because iron is used to shape it. As a result the collider must be large, several hundred km in circumference, and that has consequences for beam stability, stored beam energy, etc.[2] Magnet development together with further work on the consequences of low field should indicate whether this is a viable and cost effective idea.

Table 2 is a comparison of superconductors which shows the high critical magnetic fields and critical temperatures of the high T_c superconductors BSCCO and YBCO. These intrinsic properties make the materials attractive, but the superconductor volume fraction, the mechanical properties, and the production of material must be improved. There will be help from outside high energy physics because of potential commercial applications. In addition to improving the

Table 2. Comparison of Superconductors (Ref. 2)

Property	NbTi	Nb$_3$Sn	BSCCO-2223	YBCO
Upper Critical Magnetic Field (T)	15	25	~ 100	~ 100
Critical Temperature (K)	9.5	18	110	92
Critical Current Density* (kA/mm^2)	2 - 2.3	1 - 2.4	< 0.9	< 2.4
Superconductor Volume Fraction (%)	40 - 50	35 - 40	35 - 40	~4
Conductor Type	multifila-ment wire	multifila-ment wire	multifila-ment tape	micro-bridge
Mechanical Property	Ductile	Brittle	Brittle	Brittle
Longest Piece Made	~ 10 km	> 1 km	~ 1 km	~ 10 mm

* The magnetic fields and temperatures for the critical current densities are: NbTi - 7 T & 4.2 K or 10 T & 1.8 K; Nb$_3$Sn - 10 T & 4.2 K; BSCCO - 20 T & 20 K; YBCO - 20 T & 77K.

materials, there need to be ideas about how high T_c superconductors might be used in an accelerator magnet. High field magnets are not attractive at the present time, and the superferric magnet appears to be the only possibility.

MUON COLLIDERS

There are two premises leading to the interest in muon colliders for ultra-high energies. The first is that lepton-lepton collisions are necessary because the radiation damage to detectors at hadron colliders will be unacceptable, and the second is that beam-beam effects are a critical flaw of linear e+e- colliders. These are strong criticisms of hadron and linear e+e- colliders, and they deserve being addressed. Possible answers could include *i)* novel experimental techniques, *ii)* changes to the linear collider paradigm, and *iii)* the muon collider.

The muon collider consists of a high intensity proton synchrotron, a muon production system, ionization cooling stages, accelerators capable of bringing the beams to collision energy rapidly, and a collider ring.[4] A system approach has been taken to the design of a muon collider with all of the elements of the general topology of Table 1 being considered at the same time. Since each of the major component systems has significant technological and/or beam dynamics issues, this approach optimizes the collider concept and focuses research on critical issues.

Some people believe that since a complete collider concept is being discussed, the muon collider has moved from the realm of advanced accelerator research to that of project oriented research. This is not the case. The muon collider poses research questions in many fundamental areas of accelerator physics and technology. Beam current limits in proton synchrotrons and ionization cooling are two examples. The muon collider provides a context for the study of this accelerator physics just as an e+e- linear collider and a hadron collider provide ones for research in high gradient acceleration and high T_c superconductors, respectively.

ELECTRON-POSITRON LINEAR COLLIDERS

There is no complete concept for a 5 - 10 TeV e+e- linear collider, but there are several issues of clear importance.

Limitations of the Beam-Beam Interaction

The expressions for luminosity, $\mathcal{L}$, beam power, P_B, and the number of beamstrahlung photons per incident particle, n_γ, can be combined to give

$$\mathcal{L} \approx \frac{1}{8\pi\alpha r_e} \frac{P_B n_\gamma}{E \sigma_y} . \tag{1}$$

The beam energy is denoted by E, and the vertical beam size, σ_y, is assumed much smaller than the horizontal beam size, σ_x. The other quantities in the equation are: α = fine structure constant; and r_e = electron classical radius. This equation shows the well-known trade-offs between beam power, vertical spot size and beamstrahlung. The factor n_γ in the numerator is taken as a measure of backgrounds produced by the beam-beam interaction. Increasing the collision point electromagnetic fields increases beamstrahlung and luminosity. If there is a limit on beamstrahlung from detector backgrounds, there is a limit on luminosity.

This expression is valid when the collision point electromagnetic fields are much less than the critical magnetic field, $B_C = 4.4 \times 10^{13}$ G. When the fields are comparable to B_C, phenomena such as coherent pair production increase backgrounds dramatically.[5] The parameter Y,

$$Y = \frac{\gamma B}{B_C} \approx \frac{r_e^2 \gamma N}{\alpha \sigma_z (\sigma_x + \sigma_y)} \qquad (2)$$

($\gamma = E/mc^2$; B = collision point magnetic field; N = number of particles per bunch; σ_z = bunch length), is usually kept Y < 0.3 in linear collider designs. This becomes increasingly difficult at high energies because of *i)* the direct proportionality to γ, *ii)* high gradient structures have short wavelengths and the bunch length must be a small fraction of the wavelength, and *iii)* the need for small σ_y together with limits on σ_x/σ_y from beam optics.[6] If Y < 0.3 is necessary, this could be the critical flaw of e^+e^- linear colliders mentioned earlier in the muon collider section. However, there are several possible ways to deal with the limitations of the beam-beam interaction within the linear collider concept.

The first is to *ignore it*. This may be wishful thinking, but perhaps it isn't. High field Quantum Electrodynamics with Y ~ 1 has been studied experimentally in laser - electron beam interactions,[8] but there is no experience with beam-beam related backgrounds in a linear collider. Real life will be different than the Monte Carlos studied to date which have considered backgrounds in an extrapolation of today's high energy collider detectors. A compelling multi-TeV linear collider concept will spark creativity in the experimental physics community, and innovative approaches to experimentation could emerge.

The second approach to the limitations of the beam-beam interaction are to *avoid them* with a different collision paradigm. One possibility is photon-photon rather than e^+e^- collisions.[9] There are no issues of beamstrahlung or coherent pair production in a photon-photon collider, and the dominant problem is the configuration near the collision point. Accelerated electrons have to be converted to photons by Compton scattering with an intense laser, and this conversion point must be close to the collision point for high luminosity.

The other possibility of a different collision paradigm is plasma[10] or beam compensation where fields at the collision point are reduced by neutralization. There would be substantial backgrounds from interactions in a plasma if one were used to neutralize the collision. The creativity of experimentalists would be required to deal with them. Compensation with beams would require overlapping electron and positron beams. Efficient generation and control of such beams together with the stability of the compensated configuration are all problems to be solved. There are ideas for this.[11]

Harnessing the Potential of the Laser

High peak power lasers are a breakthrough technology, and exploiting their enormous potential for particle acceleration is one of the major challenges for accelerator physics research. They have found use already for the generation of low emittance beams in laser driven RF guns, and they could have a role in generation of power at high frequencies.[12] However, the primary interest has to be with the high gradients possible in a laser driven accelerator.

Different laser driven accelerators have been studied both theoretically and experimentally. Far field accelerators (of which the Inverse Free Electron Laser (IFEL) is the most prominent) couple to the transverse electric field of the laser by giving particles a transverse component of motion. This motion generates synchrotron radiation which limits the beam energy. Far field accelerators could find application as injectors or bunchers, but the energy limit makes them relatively uninteresting for high energy physics.

There have been many ideas for direct acceleration of a beam with a laser by using structures to give a longitudinal component to the laser field. Structures with features comparable to the laser wavelength are similar to RF driven linacs. Lithographic techniques could be used for fabrication, but there will be stringent limitations on accelerated charge from wakefields. These limitations are so severe that interest in this type of structure has dropped substantially. Current interest is focused on structures with the features in at least one dimension large compared to the laser wavelength. Crossed laser beams[13] and a structure similar to the open optical waveguide are being considered.[14] Both promise gradients ~ 1 GeV/m with substantially lower wakefields than optical renditions of RF linacs.

The highest acceleration gradients achieved to date have been with laser driven plasma accelerators. Plasma waves can be excited resonantly in the laser beatwave accelerator or by the excitation of a wakefield with a short, high intensity laser. The laser pulse is self-modulated when the pulse is long compared to the plasma wavelength. Gradients of ~ 100 GeV/m have been observed in the latter configuration.[15] This type of result has attracted widespread interest, and the field of laser driven plasma accelerators is moving on to achieving this acceleration over long distances, staging of multiple accelerators, and beam quality and stability. When these have been successfully addressed the plasma accelerator will attract the interest of the mainstream accelerator community.

Short Wavelength & High Gradient Limits of Metallic Structures

The SLC has an RF wavelength of 10.5 cm and an accelerating gradient of $G \sim 20$ MeV/m. While there is a variety of RF technologies being considered for a next generation of linear collider, the tendency is towards shorter wavelengths and higher gradients. A 5 - 10 TeV collider could be possible by going even further in this direction to mm wavelengths and GeV/m gradients.

The arguments for this include energy efficiency, which for a fixed gradient and number of particles is proportional to λ^{-2}, and the dependence of gradient on wavelength. The dominant phenomena limiting gradient at 1 - 10 cm wavelengths are *i)* capture and acceleration of dark current and *ii)* RF breakdown. Dark current capture depends on wavelength as $1/\lambda$.[16] Loew and Wang[17] have measured RF breakdown at a fixed pulse length of 1 µs and different frequencies. They find that the breakdown gradient is proportional to $\lambda^{-1/2}$. Correcting for reduced pulse length at shorter wavelengths, Wilson estimates that the gradient limit from RF breakdown is proportional to $\lambda^{7/8}$.[16] These are empirical results, and, while further research is needed to clarify underlying mechanisms, they argue for short wavelengths.

There are several disadvantages of short wavelengths. Longitudinal and transverse wakefields scale as $1/\lambda$ and $1/\lambda^3$, respectively. New ideas for aligning and stabilizing accelerating structures and beams are needed. Recent work on structure alignment based on detecting RF induced in deflecting modes may provide a basis.[18] There is a possible gradient limitation from pulsed heating. This is thought to scale as $1/\lambda^{1/8}$,[11] but the experimental information about pulsed heating in RF systems is contradictory. An experiment studying pulsed heating in RF systems is planned.[19] Structures and filling times get shorter with shorter wavelength, and the peak power per meter depends on gradient and wavelength as $G^2\lambda^{1/3}$.[16] The consequences are that new RF power sources and pulse compression techniques are sure to be required. These problems must be solved for short wavelength, high gradient RF to be viable.

CONCLUDING REMARK: ACCELERATOR IR&D

The future of high energy physics and successful accelerator *Invention, Research and Development (IR&D)* are one and the same. The last three sections have discussed and commented on some of the current directions for advanced accelerator research in hadron, muon, and linear colliders for future generations of high energy physics colliders. Most of the ideas are not the revolutionary ones that are needed. However, my hope is that the combination of motivated, intelligent people and a supportive atmosphere will produce the critical insight that is so badly needed.

REFERENCES

1. Much of the current thinking about high energy hadron colliders can be found in ref. 2.
2. G. Dugan, P. Limon and M. Syphers, "Really Large Hadron Collider Working Group Summary", <u>Proc. of 1996 Snowmass Workshop</u>.
3. G. W. Foster and E. Malamud, "Low-Cost Hadron Colliders at Fermilab", Fermilab TM-1976 (1996).
4. R. B. Palmer, A. Tollestrup and A. Sessler, "Status Report of a High Luminosity Muon Collider and Future Research and Development Plans", <u>Proc. of 1996 Snowmass Workshop</u>.
5. P. Chen, AIP Conf Proc **184**, 633(1989).
6. These general statements are supported with more detail in ref. 7.
7. S. Chattopadhyay, D. Whittum, and J. Wurtele, "Advanced Accelerator Technologies A Snowmass '96 Subgroup Summary", <u>Proc. of 1996 Snowmass Workshop</u>.
8. C. Bula *et al*, PRL **76**, 3116 (1996).
9. S. Chattopadhyay and A. Sessler editors, NIM **A355**, 1 (1995).
10. A. M. Sessler and D. Whittum, AIP Conf Proc **279**, 939 (1993).
11. D. Whittum presentation at 1996 Snowmass Workshop. See ref. 7.

12. W. Budiarto *et al*, "High Intensity THz Pulses at 1 kHz Repetition Rate", submitted to JQE (1996).
13. Y.-C. Huang presentation at 1996 Snowmass Workshop. See ref. 7.
14. R. Pantell, presentation at the 1996 Free Electron Laser Conference.
15. A. Modena *et al*, IEEE Trans Plasma Sci **24**, 289 (1996).
16. P. Wilson, SLAC-PUB-7256(1996).
17. G. A. Loew and J. W. Wang, SLAC-PUB-5320 (1990).
18. M. Seidel *et al*, to be submitted to PRL.
19. D. Pritzkau *et al*, private communication.
20. T. Elioff, "A Chronicle of Costs", SSCL-SR-1242 (April, 1994).
21. Table 6-2 of ref. 20.
22. Tables 6-3 and 6-4 of ref. 20.

APPENDIX: SSC COST ANALYSIS

While a detailed SSC cost analysis is complicated because project evolution and schedule changes had large impacts the cost,[20] the "Site Specific Conceptual Design" can be used to show relative costs. From Tables A-1 and A-2 one sees that 52% of the Total Project Cost (TPC) was in the accelerator system with the collider accounting for 42% of the TPC. When it is assumed that project management, contingency, R&D, and administrative and technical support should be apportioned according to system costs rather than appearing as separate items in the budget these percentages become 75% and 61%.

The accelerator systems (not including the magnets), superconducting magnets, and conventional systems of the collider are 17%, 44% and 10% of the TPC, respectively. Almost one-half of the cost is associated with the collider ring superconducting magnets.

Table A-1. SSC Site Specific Conceptual Design Costs* [21]

Category	SCDR Costs FY90$
Construction	
1.0 Technical Systems	2,986,400,000
2.0 Conventional Systems	1,051,500,000
3.0 Project Management	48,700,000
Contingency	753,000,000
Construction Subtotal (TEC)	*4,839,600,000*
Other Program Costs	
4.0 R&D, Pre-Operations, Administrative and Technical Support	975,900,000
5.0 Experimental Systems	752,100,000
Other Subtotal	*1,728,000,000*
Total Project Cost (TPC)	**6,567,600,000**

* These numbers correspond to a proposed actual year cost of $7,836,600,000 which was increased to $8,249,000,000 after reviews by the Department of Energy.

Table A-2. SSC Accelerator Technical and Conventional Systems[22] (1)

System	Accelerator Systems (2)	Conventional Systems	System Cost (3)		% of TPC (3)	
Linac	37	3	40	(58)	0.6	(0.9)
LEB	42	5	47	(68)	0.7	(1.0)
MEB	113	35	147	(212)	2.2	(3.2)
HEB	326	74	400	(576)	6.1	(8.8)
Injector	*518*	*117*	*635*	*(915)*	*9.7*	*(13.9)*
Collider	2,304	464	2,768	(3987)	42.1	(60.7)
Accelerators	*2,822*	*581*	*3,403*	*(4901)*	*51.8*	*(74.6)*

Notes: 1. Costs in FY90 M$. 2. Including superconducting magnets which are $1,668M$ of the collider cost. 3. The numbers in ()'s indicate costs and percentages with project management, contingency, R&D etc. allocated in proportion; (cost) = cost ¥ [1 + (48.7+753.0+975.9)/(2986.4 + 1051.5)].

Plasma-Based Acceleration Concepts

C. E. Clayton
Department of Electrical Engineering
University of California at Los Angeles
Los Angeles, CA 90095

Abstract

This paper reviews the experimental progress made towards plasma-based accelerators since the last Advanced Accelerator Concepts Workshop at Lake Geneva in June, 1994. It has been a very productive two years, with many new experimental results from all over the world. The latest results from the Plasma Beat Wave, Laser Wakefield, and Plasma Wakefield Accelerator concepts will be reviewed. The explosion of results from the instability-driven or self-modulated Laser Wakefield Accelerator experiments will be reviewed as well as such enabling technologies as plasma channel formation which is necessary to guide laser pulses over long distances.

INTRODUCTION

In this seventh edition of the workshop on Advanced Accelerator Concepts (called Laser Acceleration of Particles for the first two workshops) dating back to 1982, we find the field mature yet still vital, generating new ideas and experimental results at each meeting. This paper reviews the experimental progress made towards plasma-based accelerators [1] since the last Advanced Accelerator Concepts Workshop at Lake Geneva in June, 1994. [2] This year proved to be the most exciting as far as plasma-based acceleration concepts were concerned. This general overview attempts simply to list the experimental research at the various institutions around the world. The reader is referred to the Summary of the Working Group on Plasma-Based Accelerators in these proceedings [3] and to a recent Special Issue on Second Generation Plasma Accelerators [4] for more review material on the progress of plasmas in accelerator applications.

This article will be divided along the following lines of research: Plasma Beat Wave Accelerators (PBWA), Laser Wakefield Accelerators (LWFA), self-modulated Laser Wakefield Accelerators (sm-LWFA), Plasma Wakefield Accelerators (PWFA), Plasma Channels, and finally, a brief look at the future experiments in this field.

PLASMA BEAT WAVE ACCELERATOR

The PBWA concept [5] is the most mature of all the plasma acceleration concepts due, in a large part, to the modest laser requirements needed to drive substantially large electron plasma waves (EPW), thus giving this scheme about a 10 year experimental head-start [6] over some of the other laser-based schemes.

Recent PBWA experiments at UCLA [7] which time-resolved the forward-scattered Stokes and anti-Stokes light, along with supporting computer simulations, were performed to elucidate the dynamics of the process and it was confirmed that the poderomotive force of the EPW itself terminates the EPW growth by radial expulsion of plasma through its own ponderomotive force. The frequency matching condition that the beat frequency of the two laser beams equal the plasma frequency is subsequently destroyed, stopping the growth. Thermalization of the highly energetic and non-maxwellian plasma were also studied via a 2 MeV electron probe beam which confirmed the existence of magnetic fields due to a probable combination of the Weibel instability due to relaxation of the coherent electron motion in the EPW (prompt B-fields) and long-time-scale (5 ns) hydrodynamic fields later in the evolution of the plasma.

Another resent addition to the succession of PBWA experiments is one performed at AECL in Chalk River, Canada. This experiment [8], completed but not presented or published as of the last Workshop, differed from others in the use of Argon rather than Hydrogen or Deuterium as the working gas. Electrons were injected by a linac at 12 MeV and on several shots, 29 MeV electrons were detected. Argon, having a lower ionization threshold and larger mass than Hydrogen may have limited the role of ponderomotive-force-driven hydrodynamics but increased the role of multiply-ionized atoms, making the dynamics of the experiment difficult to interpret.

LASER WAKEFIELD ACCELERATOR

In the last two years, the availabiliyt of ≈ 0.5 TW, 100 fs class lasers has made it possible to drive up large amplitude EPW's, albeit in a tightly focused, nonlinear 3-D regime unsuitable for sustained electron acceleration as of yet. It is unusual for laser groups to also have access to a quality relativistic electron beam for probing the electric fields of the EPW's. A breakthrough came when researchers at the University of Texas applied "longitudinal interferometry" to directly probe the electron density fluctuations associated with the wave [9]. A "witness" optical probe beam was injected into the EPW (the "wake") created by an intense driving pulse (0.2 TW, 100 fs). the phase shift experienced by the probe beam was thus a measure of the perturbed refractive index caused by the wave. This was no easy task as, due to the tight focusing, the interaction length and thus the accumulated phase shift was extremely small–only about 10 mrad of an optical cycle! Fortunately, the technique had a phase resolution of a few mrad and the EPW behind the driving pulse could be mapped out over many periods.

A group at the LULI laser facility in Ecole Polytechnique pulse (0.2 TW, 72 fs) improved on this idea by adding a transverse component to the diagnostic and were able to image a transverse slice of the phase shift as a function of delay [10]. The result was an extraordinarily clear picture of the longitudinal and transverse components of the density perturbations of this 3-D wave.

Preliminary reports from a group at KEK which *does* have both a 0.5 TW, 90 fs laser and a 15 MeV electron probe beam showed that the electron energy "spectra" due to the electrons interacting with the EPW fields driven by the laser pulse extending beyond 30 MeV. To view higher energies, they were force to place detectors in the near-forward direction where energy resolution is poor. The highest energy observed was 80 ± 40 MeV. Images of the interaction region (static fill of Helium gas at 10^{18} cm^{-3}) showed plasma emission over a distance much greater than the Rayleigh range of the f/10 focused laser–about 2 cm in length. This apparent self-guiding contradicts the theoretical predictions for known short-pulse guiding mechanisms [11] and needs further study.

SELF-MODULATED LASER WAKEFIELD ACCELERATOR

By far, the easiest experimental route to accelerating electrons with laser-driven EPW's is to operate at high enough densities until the number of e-foldings of the Forward Raman Scattering (FRS) instability, over a Rayleigh range, is >> 1. This requires the use of a gas jet with a sharp gas-vacuum interface to avoid ionization-induced refraction which is especially troublesome at high densities. This setup was common to many recent experiments (all since the last workshop) which will briefly be described in historical order in this section. Typical minimum parameters for these sm-LWFA experiments are a few TW of laser power over about 600 fs and a few hundred microns of interaction length and a density of $1.5-3 \times 10^{19}$ cm^{-3}.

These were the parameters of a LLNL/UCLA collaboration experiment [12] which was the first unambiguous demonstration of the correlation of the number of FRS e-foldings and the acceleration of *background* plasma electrons to relativistic (2 MeV) energies. These results came several weeks after the Lake Geneva workshop in 1994.

Subsequently, an Imperial College (London)/UCLA/LLNL/LULI collaboration, using the powerful (25 TW) Vulcan Laser at Rutherford Appleton Laboratory (RAL) in the UK was able to drive the EPW's from FRS to wavebreaking levels resulting in a copious emission of broadband electrons up to the electron-spectrometer limit of 44 MeV [13]. Later, with an upgraded spectrometer, 100 MeV electrons were recorded under similar conditions, an energy higher than the linear dephasing limit for the density used [14].

The next results were from the NSF Center for Ultrafast Sciences at the University of Michigan (U of M) where high fluxes of 1.5-6 MeV electrons were observed with a few TW at 400 fs [15]. Angularly-resolved electron spectral measurements suggested the surprising characteristic that the electron spectrum, contained solely within the cone angle of the pump laser, was roughly independent of direction. The 1.5 MeV cutoff at the low end was a diagnostic artifact.

Researchers at the Navel Research Laboratory (NRL) were next on board with a forward-scattering pump-probe determination [16] of the lifetime of the EPW (< 2 ps FWHM) in a similar laser parameter regime as the U of M. Thomson scattering at larger angles revealed the late-time (10–100 ps) presense of ion waves associated with the hydrodynamics of plasma expansion. Electrons out to about 30 MeV were also observed in this experiment [17].

A U of M/University of Texas (UT) collaboration later performed a very carefully timed pump-probe experiment similar in nature to the NRL experiment. The goal was to view the onset, peak, and fall-off of the EPW. With their pump/probe timing calibrated, they found that the onset of substantial EPW amplitude was nearly 1 ps ahead of the peak of the 0.2 ps HWHM driving pulse [18]. They attributed this early onset to an enhanced EPW noise level due to an ionization-front-driven EPW [19] followed by growth from FRS. This explanation seems to fit their data quite well. Their measured lifetime ($\approx$ 2ps FWHM) agrees with that of the NRL group.

Both the U of M/UT group and the NRL group are looking at their data for energy partitioning; that is, can the acceleration of electrons explain the observed decay rate of the EPW or are there other channels to which the EPW couples.

It should be noted that this instability-driven class of accelerators may be difficult to control with accuracy necessary for staging. Ideally, one would operate at a somewhat lower density to increase the maximum energy per stage (a dephasing issue and a wavelength issue) and produce a large noise source to minimize the number of e-foldings required for a desired EPW amplitude (a control issue) [20]. Nevertheless, such accelerators may one day be considered for "low tech" production of high spectral-brightness, broadband electrons in the 0.1–1.0 GeV range [13], in "competition" with linacs and storage rings. Although the LWFA and PBWA concepts are more amenable to controllable, high-gradient structures for high-energy physics applications, the sm-LWFA experiments, where the laser and plasma frequencies are not too disparate, are more amenable to today's computer capability to simulate these experiments and are therefore useful for developing and benchmarking massively-parallel particle-in-cell (PIC) codes for stepping into true 3-D computational physics in the next couple of years.

PLASMA WAKEFIELD ACCELERATORS

The PWFA scheme is the electron-driver analogue to the laser-driven LWFA. Current research in this area is focusing on working in the so-called underdense regime. That is, where the density of the electron bunch that drives the EPW is higher than the background plasma density. This is also referred to as the "blowout" regime because the plasma is too dilute to completely charge-neutralize the driver bunch, thus leaving an ion channel in its wake.

A UCLA/Argonne National Laboratory (ANL) collaboration showed preliminary data on the first PWFA experiments in this regime. A witness pulse of electrons experienced a EPW accelerating gradient of 10–20 MeV/m over about 10 cm of plasma. Measurements of the beam profile at the end of the plasma indicated ion-channel focusing with a beam-to-plasma density ratio of about 2.

In an experiment reported on from LANL, an 8 MeV L-band RF gun plus chicane compressor produced 10–20 microbunches of kA (1 ps, 1 nC) currents [21]. This beam was focused into a jet of Neon with the goal of converting the beam energy into inner-shell soft xrays. In this case, the beam density of about 10^{16} cm^{-3} was about 10–100 times lower than the Neon neutral density. The first three bunches brought ionized the Neon until 22 nm radiation form Ne^{+3} was observed for the remainder of the train. This unique electron beam has the capability to drive EPW's to gradients near 1 GeV/m.

PLASMA CHANNELS

Plasma channels or plasma fibers, as the name suggests, is a way to guide laser light over long distances (10's of cm), in analogy to optical fibers made from glass or plastic. As discussed in respect to EPW's, plasma channels can either be created in a controlled manner or through an instability. To date, the impressive results from the University of Maryland [22] stand alone as the only controlled channel formation experiments. These were made by an axicon lens which brings the channel-forming laser pulse in from the side, thus avoiding ionization-induced refraction. Subsequent hydrodynamic expansion of this plasma creates a density depression on-axis. This in turn constitutes an index of refraction maximum on-axis which guides light injected longitudinally into this channel. However, in these experiments, there is some concern as to the maximum intensity that can be channeled due to the fact that the ions making the channel were not fully stripped. Recently Maryland has synchronized a short pulse Ti:sapphire laser with there plasma fiber and is in the process of boosting the guided power as their laser system grows.

In a related experiment [23], a refractive index profile suitable for guiding was achieved by discharging a capacitor across a polypropylene capillary tube (350 µm inner diameter, 1 cm long). In this case, the plasma produced by the capillary discharge acts like the plasma formed by the axicon lens in the Maryland experiment except that the actual plasma properties are much less well known and controllable than in the axicon case. With proper matching at the entrance to the capillary, high transmission fractions of a moderately-high intensity laser pulse were achieved, even when the capillary was bent to a 10 cm radius of curvature.

Other experimental attempts at channel formation relied on ponderomotive or relativistic effects to alter the on-axis refractive index profile into a guiding structure. Filamentary structures [24] or linear chains of "sparks" [25] have been reported in the past but with inadequate diagnostics to confirm the nature of the refractive index changes.

A recent UCLA/IC/LULI experiment at RAL [26], presented as part of this review talk, used five imaging diagnostic, four of which contained spectral information, to study self-focusing of a laser of power $P \approx 20$ TW with a 1.2 ps pulse at 1.5×10^{19} cm^{-3} where $P/P_{cr} \approx 18$. Here, P_{cr} is the threshold power needed for relativistic self-focusing for a gaussian beam. The key diagnostic was the spatial- and frequency-resolved collective Thomson scattering of an external probe beam off EPW's *within* the channel. Using this same external probe beam, this experiment showed the refractive index gradients associated with the channel via Schlieren photography as well as the local plasma frequency inside the channel. Large dynamic frequency changes at one point in the channel suggest that propagating intensities were $> 5 \times 10^{19}$ W/cm^2. This in turn implies that the originally 5×10^{18} W/cm^2 beam self-focused inside the channel. A simple measurement of the channel diameter indicated that the initial 20 µm laser spot did indeed collapse, consistent with the higher propagating intensities. Having an estimate of the laser intensity and spot size in the channel, one can estimate the power trapped in the channel. This turned out to be about 1/3 of the initial 20 TW giving $P/P_{cr} > 5$ inside the channel of > 3 mm (> 10 Rayleigh lengths) in length.

FUTURE WORK

On the PBWA front, only UCLA continues their research and is in the progress of combining their (upgraded) ≈ 1 TW CO_2 laser with a state-of-the-art RF photoinjector. The intent is to drive large *diameter* EPW's of about 3 GeV/m (the same gradient of the earlier experiments of Ref. 8) and focus the low emittance, 16 MeV electron beam into the core of the EPW. The phase velocity of the EPW will match the velocity of the electrons minimizing capture dynamics and every effort will be made to "match" the electron beam into the PBWA structure. Finally, the diagnostics will focus on the beam-properties of the > 100 MeV output beam and try to understand any emittance blowup and longitudinal energy spread.

In addition to Maryland, plans are underway at LBL to produce controlled plasma fibers and inject high intensity laser pulses to demonstrate the extended-interaction-length version of the LWFA [27].

Imperial College is planning on frequency-doubling the upgraded (150 TW) Vulcan laser to increase the phase velocity of the EPW's and thereby increase the ultimate energy in the sm-LWFA scheme to well beyond 100 MeV [28].

NRL [29] and KEK/JAERI [30] are commissioning RF guns similar to the one at UCLA to be used as an electron injector for their LWFA experiments.

The LANL group has proposed to use their ps electron pulses to perform high-density, and therefore high gradient, PWFA experiments. They estimate that 1 GeV/m is possible with a transformer ratio > 2.

Finally, Michigan plans on trying out a very clever, all-optical electron injection scheme [31]. Here a second, tightly-focused laser pulse, derived from the main LWFA driving pulse with a $200 beamsplitter, knocks background electrons into the sepratrix of the LWFA EPW. This eliminates the need for an external device to deliver synchronous, sub-50 fs electron bunches to these high frequency plasma structures.

Should this or other ideas prove successful, some future experiments (5-10 year scale) may very well produce high repetition rate, narrowband, high-charge single electron bunches– perhaps in a channel-guided LWFA or PBWA configuration–and the suitability of these schemes to high energy physics (staging, for example) can be further studied. The next Advanced Accelerator Concepts

Workshop should be exciting as more and more of the enabling technologies for Plasma Accelerators are put to the experimental test.

ACKNOWLEDGMENTS

The author would like to thank W. Leemans, D. Umstadter, E. Eserey, A. Ting, K. Nakajima, C. Siders, M. Downer, A. Modena, A. E. Dangor, H. Milchberg, T. Katsouleas, W. Mori, and C. Joshi for useful conversations. This work is supported by DOE contract number DE-FG03-92ER40727.

REFERENCES

1. T. Tajima and J. M. Dawson, Phys. Rev. Lett. 43, 267 (1979).
2. *Advanced Accelerator Concepts, AIP Conference Proceedings* 335, P. Schoessow, ed. (AIP Press, New York).
3. T. Katsouleas, "Summary of the Working Group on Plasma-Based Accelerators", these Proceedings.
4. E. Esarey et al., "Overview of Plasma-Based Accelerator Concepts", IEEE Trans. on Plasma Science 24, 252 (1996).
5. C. Joshi et al., Nature 311, 525 (1984).
6. C. E. Clayton, et al., Phys. Rev. Lett. 54, 2343 (1985).
7. A. Lal et al., Physics of Plasmas 4, April (1997).
8. N. A. Ebrahim, J. Appl. Phys. 76, 7645 (1994).
9. C. W. Siders et al., IEEE Trans. on Plasma Science 24, 301 (1996); C. W. Siders et al., Phys. Rev. Lett. 76, 3570 (1996).
10. J. R. Marques et al., Phys. Rev. Lett. 76, 3566 (1996).
11. P. Sprangle et al., Phys. Rev. Lett. 64, 2011 (1990).
12. C. Coverdale et al., Phys. Rev. Lett. 74, 4659 (1995).
13. A. Modena et al., Nature 337, 606 (1995).
14. D. Gordon et al., submitted to Phys. of Plasmas.
15. D. Umstadter et al., Science 273, 472 (1996).
16. A. Ting et al., Phys. Rev. Lett. 77, 5377 (1996).
17. C. I. Moore et al., Bull. Am. Phys. Soc. 41, 1602 (1996).
18. S. P. Le Blanc et al., Phys. Rev. Lett. 77, 5381 (1996).
19. W. B. Mori and T. Katsouleas, Phys. Rev. Lett. 69, 3495 (1992).

20. V. V. Goloviznin et al., Phys. Rev. E. 52, 5327 (1995).
21. B. E. Carlsten and S. J. Russell, Phys. Rev. E 53, R2072 (1996).
22. C. G Durfee,III and H. M. Milchberg, Phys. Rev. Lett. 71, 2409 (1993).
23. Y. Ehrlich et al., Phys. Rev. Lett. 77, 4186 (1996).
24. P. Monot et al., Phys. Rev. Lett. 74, 2953 (1995).
25. A. B. Borisov et al., Phys. Rev. Lett. 68, 2309 (1992).
26. C. E. Clayton et al., submitted to Phys. Rev. Lett..
27. W. P. Leemans, private communication.
28. A. E. Dangor, private communication.
29. A. Ting, private communication.
30. K. Nakajima, private communication.
31. D. Umstadter et al., Phys. Rev. Lett. 76, 2073 (1996).

Electron Beam Monitoring at High Frequencies and Ultra-fast Time Scales

W.P. Leemans

Ernest Orlando Lawrence Berkeley National Laboratory
University of California, Berkeley, California 94720

Abstract. Radiation based methods for electron beam monitoring that are capable of providing information on ultra-fast time and micron spatial scales are discussed. The radiation is produced through interaction of the electron beam with a medium (transition and Cerenkov radiation), or through interaction with fields from a bending magnet or a high intensity laser pulse. Examples of techniques based on optical transition radiation (OTR) as well as on laser Thomson scattering are presented, which allow measurement of both transverse and longitudinal phase space distributions.

I. INTRODUCTION

Electron accelerators using photo-cathode radio-frequency guns are being designed and built with unprecedented low normalized emittances on the order of 1 π mm-mrad [1]; sub-micron beam spot sizes have been achieved [2] in specially designed magnetic focusing lattices; sub-picosecond, high peak current electron bunches have been produced through magnetic compression of electron bunches with a temporally correlated energy spread (chirp) [3]; femtosecond and even sub-femtosecond electron bunches might be produced in FEL-based bunchers [4] or in laser driven accelerators [5].

Measurement of the transverse and longitudinal phase space properties of electron bunches produced in these high performance devices, requires development of beam diagnostics with high spatial (micron or sub-micron) and temporal (femtosecond) resolution requirements. In this paper electron beam monitoring techniques will be reviewed, with emphasis on ultra-fast timescales. The techniques to be considered will rely mainly on the use of radiation produced when the beam propagates through a medium (in specific optical transition radiation [6]), and when the beam interacts with a high intensity ultra-short pulse laser [7]. OTR is produced promptly when an electron beam encounters a discontinuity in dielectric properties of the medium and allows measurement of both transverse (spot size and divergence) and longitudinal properties. Using laser beams as microprobes has allowed measurement of spot sizes with unmatched resolution [2], and has also recently been used to measure divergence and longitudinal beam profile with sub-picosecond time resolution [7].

Sec. II begins with a discussion of coherent and incoherent radiation from charged particles, followed by an overview of some of the properties of optical transition radiation, with emphasis on spatial and temporal resolution issues. Beam profiling techniques based on OTR are then reviewed which use the coherent as well as the incoherent contribution to the radiation. The concept of fluctuational interferometry is introduced and a bunch profiling method based on an optical gating technique is proposed.

In Sec. III, scaling laws describing the interaction of a relativistic electron beam with a high intensity laser beam will be presented. Using results from experiments conducted at the Center for Beam Physics' Beam Test Facility (BTF) it will be shown that measurement of the properties of the scattered radiation provides an effective technique for obtaining information on the transverse and longitudinal phase space distribution. A summary and conclusions are given in Sec. IV.

II. OPTICAL TRANSITION RADIATION BASED DIAGNOSTICS

A. Transverse beam distribution

Optical transition radiation is generated when a relativistic particle of charge e crosses an interface with a discontinuity in dielectric properties [6]. The radially polarized radiation is promptly emitted in a cone with half opening angle $1/\gamma$ (γ is the usual Lorentz factor), and contains information about the energy, divergence and spot size of the particle beam. The general equations describing the radiation process for oblique angles of incidence are described in Ref. [8]. In practice, metallized foils are used at a 45° angle of incidence and the backward emitted radiation cone is collected onto an imaging system. The intensity of the backward emitted radiation W, in a frequency range $d\omega$ and solid angle $d\Omega$, has been derived using an image positron approach and is given by [9]

$$\frac{d^2W}{d\omega d\Omega} = \frac{\mu_0 c}{16\pi^3} \left| \frac{e\sin\theta'}{(1-\beta\cos\theta')} + \frac{-e\sin\theta}{(1-\beta\cos\theta)} \right|^2 \tag{1}$$

Here θ is the angle with respect to the normal of the interface, μ_0 is the vacuum permeability, and β is the particle velocity normalized to the speed of light c. Oblique incidence gives rise to an asymmetry in the OTR cone. For highly relativistic particles ($\gamma \gg 1$) and $\theta \ll \pi/2$, Eq.(1) reduces to

$$\frac{d^2W}{d\omega d\Omega} = \frac{\mu_0 c e^2}{4\pi^3} \frac{\theta^2}{(\gamma^{-2}+\theta^2)^2} \tag{2}$$

The number of radiated photons within the wavelength range (λ_{end}, λ_{start}) is then

$$\frac{dN(\theta)}{d\Omega} = \frac{\mu_0 c e^2}{4\pi^3 \hbar} \ln(\frac{\lambda_{end}}{\lambda_{start}}) \frac{\theta^2}{(\gamma^{-2}+\theta^2)^2} \tag{3}$$

where $d\omega$ is integrated from λ_{start} to λ_{end}, and $\hbar$ is Planck's constant.

The number of photons emitted during the radiation process can be obtained by integrating Eq. [3] over the collecting solid angle. Figure 1 shows the relative radiated total intensity as a function of opening angle of the detector for $\gamma = 10^2$ and $\gamma = 10^5$. At angles large compared to $1/\gamma$, a substantial number of photons is radiated, which will be discussed in the next section. The total number of photons emitted per incident electron in the forward direction is then given by integrating over a solid angle extending from 0 to $\pi/2$:

$$N(\pi/2) = -\frac{\mu_0 c e^2}{4\pi^3} \frac{1}{\hbar} \ln(\frac{\lambda_{end}}{\lambda_{start}}) \left(\frac{2\beta + \beta^2 + 2\log(1-\beta)}{\beta^3} \right) \tag{4}$$

which, for γ greater than 10 is well approximated by

$$N(\pi/2) \approx \frac{\alpha}{\pi} \ln(\frac{\lambda_{end}}{\lambda_{start}})(2\log\gamma - 0.81). \tag{5}$$

The photon yield roughly scales as the fine structure constant $\alpha \approx 1/137$ and is logarithmically dependent on the beam energy. For typical electron beam parameters, i.e. 1 nC charge per bunch, energy 50 MeV about $6 - 8 \times 10^7$ photons are generated.

From Fig. 1, for a collection angle on the order of 40 - 100 mrad (limited by aberrations in the collection optics) about 2 - 4 x 10^7 can be imaged onto a detection system.

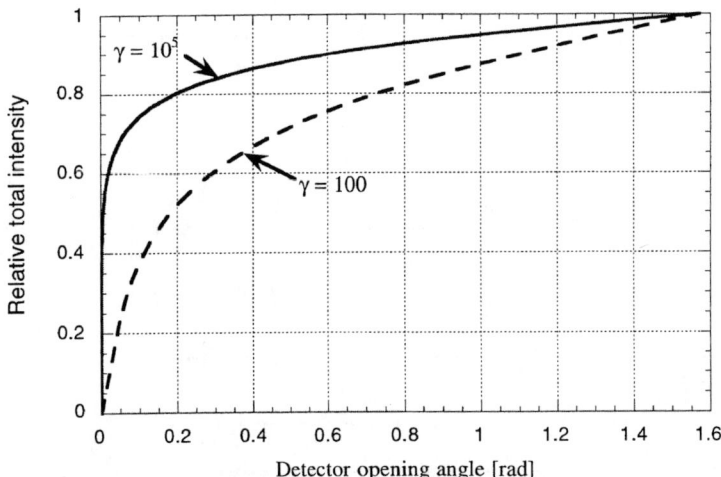

Figure 1: Relative total radiated intensity as a function of opening angle of the detector for transition radiation. The solid(dashed) curve is for a beam energy of 50 GeV (MeV).

OTR has become one of the essential diagnostics [12] at the Beam Test Facility (BTF) [10], which is operated by the Center for Beam Physics at Lawrence Berkeley National Laboratory. At the BTF, a magnetic transport line can bring the 50 MeV electron beam from the Advanced Light Source [11] linac injector to a dedicated experimental area where a variety of experiments on the interaction of relativistic electron beams with lasers and plasmas can be carried out. . Radiation is generated using either 5 µm thick aluminized nitrocellulose pellicles or aluminum coated, 3 mm thick quartz substrates. The beam is imaged 8.5 m away using an optical telescope system with a factor 3 magnification, onto a Peltier cooled (-30 °C), 16 bit CCD camera (Photometrics Series 200). The CCD camera has a chip with 512x512 square pixels with a pixel size of 19 µm.

Typical experimental results of measuring the vertical and horizontal beam size for different longitudinal positions of an OTR foil, mounted at 45° with respect to the direction of propagation of the electron beam are shown in Fig. 2. From such measurement, the time-integrated (i.e. integrated over the electron beam temporal profile) beta-function and beam emittance can directly be obtained. Alternatively, a conventional quadrupole scan technique can be used with a fixed OTR foil.

Information on the electron beam divergence can be obtained independently by measuring the radiation cone profile. This is accomplished using a lens positioned a focal distance away from the camera sensing area. The experimentally measured OTR cone is than fitted with a convolution of Eq. (3) and a Gaussian distribution (with rms. width α) for the angular spread in the electron beam:

$$\frac{dN(\theta,\alpha)}{d\Omega} = \frac{Q\mu_o ce^2}{4\pi^3} \frac{1}{\hbar} \ln(\frac{\lambda_{end}}{\lambda_{start}}) \int_{-\infty}^{+\infty} \frac{1}{\sqrt{2\pi}\alpha} e^{-\frac{1}{2}\left(\frac{\theta'}{\alpha}\right)^2} \frac{(\theta-\theta')^2}{(\gamma^{-2}+(\theta-\theta')^2)^2} d\theta' \quad (6)$$

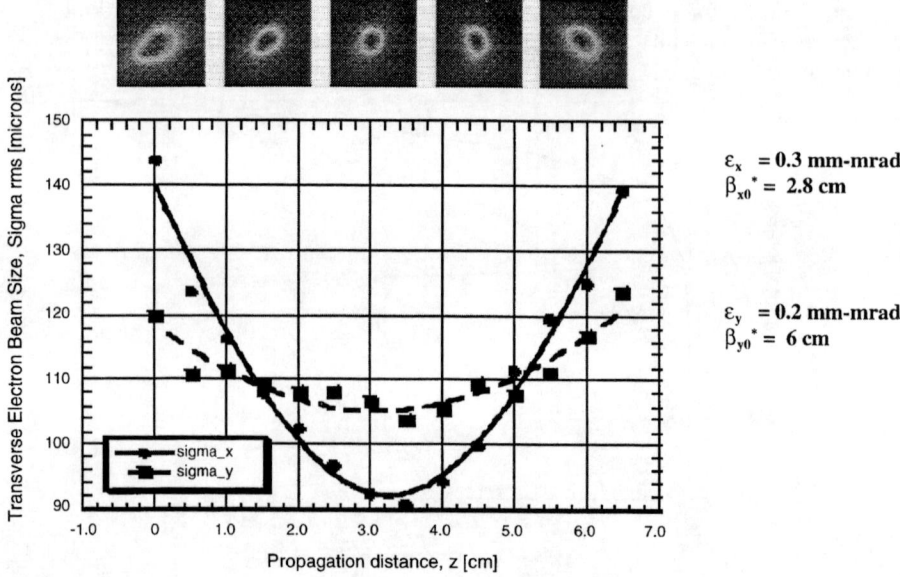

$\varepsilon_x = 0.3$ mm-mrad
$\beta_{x0}^* = 2.8$ cm

$\varepsilon_y = 0.2$ mm-mrad
$\beta_{y0}^* = 6$ cm

Figure 2: time-integrated electron beam spots (top) and beam size as a function of propagation distance (graph). The images were obtained by imaging OTR from an aluminized foil onto a CCD camera with a small f-number telescope. Spatial resolution was measured to be 14 μm. The solid (dashed) line is a quadratic fit through the horizontal (vertical) beam size measurements. From these fits, the horizontal and vertical beta-functions and beam emittance is obtained.

The effect of increasing the beam divergence α is to fill in the center of the radiation cone. A rough estimate for the beam divergence can be obtained by calculating the ratio of the minimum and maximum of the radiation distribution [12], which is shown in Fig. 3.

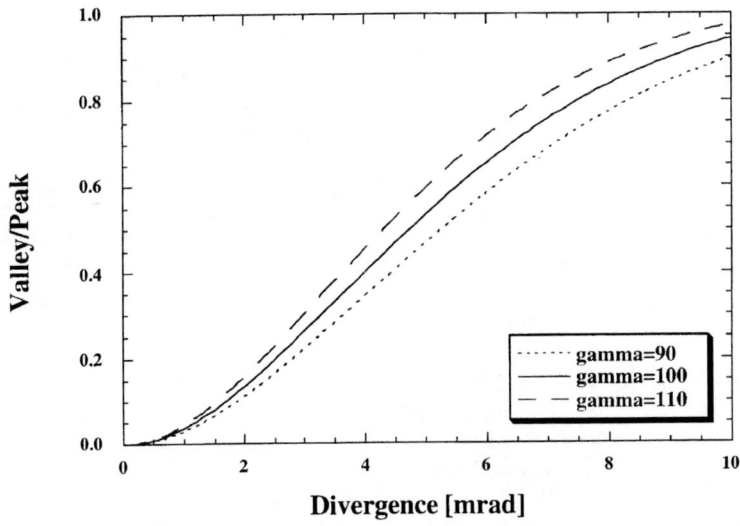

Figure 3: Valley/Peak characteristic for various γ's.

In the case of $\gamma \approx 100$, this ratio varies only slightly for a divergence angle less than 1 mrad and hence becomes an insensitive measure of beam divergence. A substantial increase in sensitivity can be obtained using an OTR interferometer as demonstrated by Wartski et al. [9].

B. Diffraction limited resolution

Optical diffraction can impose a severe limit on the use of conventional imaging for relativistic beam spot size measurements. Consider two radiating charged particles separated transversely by a distance d (see Fig. 4).

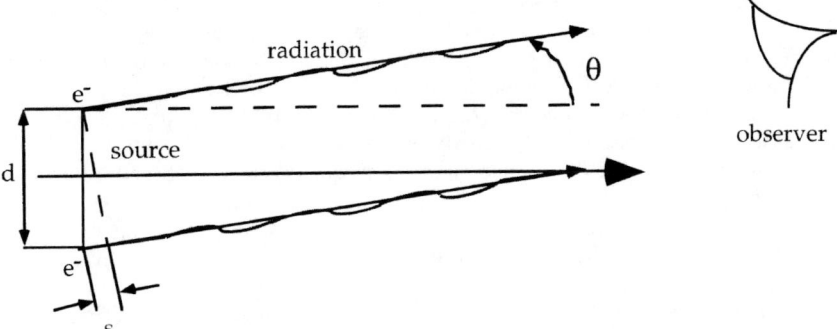

Figure 4: Schematic of an observer looking at two point sources separated by a distance d.

For an observer looking at an angle θ with respect to the propagation direction of the particles, these two particles will be indistinguishable if $\theta < \frac{\lambda}{2\pi d}$. In other words, the source will be transversely coherent for spot sizes smaller than d. For a Gaussian radiation distribution with opening angle σ_θ the diffraction limited spot size is given by $\sigma_{dif} = \frac{\lambda}{4\pi\sigma_\theta}$. In the case of synchrotron radiation the radiated intensity per unit solid angle is described by an Airy function and reduces exponentially with angle. An equivalent Gaussian opening angle of the radiation can be expressed as [13]

$$\sigma_\theta \approx \frac{0.6}{\gamma}\left(\frac{E_c}{E_{ph}}\right)^{1/3} \text{ for } E_{ph} \ll E_c \qquad (7)$$

or, using $E_c[keV] \approx 0.665 E_{beam}^2[GeV]B[T]$ with E_{beam} the electron beam energy and B the magnetic field strength of the bending magnet,

$$\sigma_{dif} = 10\left(\frac{\gamma}{E_{ph}^2[eV]B[T]}\right)^{1/3} [\mu m] \qquad (8)$$

As an example, the diffraction limited spot size for imaging a 50 GeV electron beam spot, using visible radiation (E_{ph} = 2 eV) produced with a 1 T bending magnet, is from Eq.(8) about 290 µm which is several orders of magnitude larger than the minimum produced spot size.

In the case of OTR, Eq.(3) indicates that the radiated intensity per unit solid angle decreases much slower than for synchrotron radiation. As is evident from Fig. 1, the fraction of radiated intensity at large collection angles could allow larger observation angles to be used without suffering substantial detection sensitivity loss. This in turn results in transverse resolution approaching the wavelength of the radiation.

Techniques with fractional wavelength resolution will be discussed under laser based probing methods.

C. Longitudinal bunch profile measurements and time resolved emittance measurement.

Since OTR is prompt radiation, it can be used for longitudinal profile measurements. Bunch profiling in the picosecond and sub-picosecond regime has been done mainly using detectors based on the streak camera principle. A detailed overview of the state-of-the-art for such devices can be found in Ref. [14]. Streak cameras have been used to look at OTR, Cerenkov radiation, spontaneous emission, optical synchrotron radiation, and FEL radiation and are equally applicable to both incoherent and coherent radiation sources. State-of-the-art time resolution for single sweep cameras has been reported at 68 fs (rms). As an example of time-resolved OTR, Fig.(5 a,b) shows streak images of the electron beam spot size and electron beam divergence.

To obtain the beam divergence, vertically linear polarized light was first selected from the radiation cone using a polarizing cube and then as projected onto the slit of a picosecond time resolution streak camera (Soliton Ltd. K-001). From these streak measurements, the time resolved beam divergence can be obtained using Eq.(6), and combined with spot size measurements, a slice emittance scan can be carried out.

A different slice emittance technique has been developed by Qiu et al. [15]. Picosecond slices were selected from a 10 ps long energy chirped electron bunch by a slit in a dispersive region. For each slice a quadrupole scan technique was then used to measure the slice emittance allowing detailed studies of emittance compensation in a photocathode RF gun.

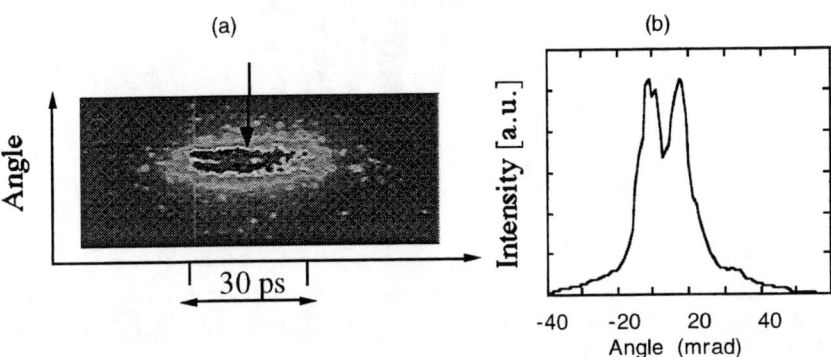

Figure 5. a) Streak image of an OTR radiation cone with picosecond time resolution and b) angular radiation profile for a given temporal slice (marked by the arrow) through the streak image.

Recently, a fluctuational interferometry technique relying on the incoherent contribution to the radiation has been proposed [16]. In this technique, it is assumed that the number of photons in the coherence volume is much greater than unity, allowing classical descriptions to be used. The concept behind the proposed method can be understood as follows. For a radiation pulse to be longitudinally incoherent, the spectral bandwidth $\Delta\omega$ must be much larger than the inverse of the pulse duration τ_p, i.e. $\Delta\omega\tau_p \gg 1$. Using a bandpass filter, centered around ω_0 and with spectral width $\delta\omega$, temporal coherence can be imposed on the emitted radiation pulse with an associated coherence time $\tau_{coh} \propto \delta\omega^{-1}$, effectively breaking the pulse up in N shorter pulses where $N = \tau_p / \tau_{coh}$ (Fig. (6)).

From shot-to-shot, the electric field amplitude will vary on the order of $\sqrt{N}$. Measurement of the variance of the intensity fluctuations will then give a measure for N and hence $\tau_p \approx N/\delta\omega$. It is demonstrated in Ref. [16] that the variance of the Fourier transform $S(\tau) = \int P(\omega)e^{-i\omega\tau}d\omega$ of the radiation spectrum $P(\omega)$ is proportional to the convolution function of the beam current, i.e.

$$\langle |S(\tau)|^2 \rangle - |\langle S(\tau) \rangle|^2 \propto \int I_b(t)I_b(t-\tau)dt \tag{9}$$

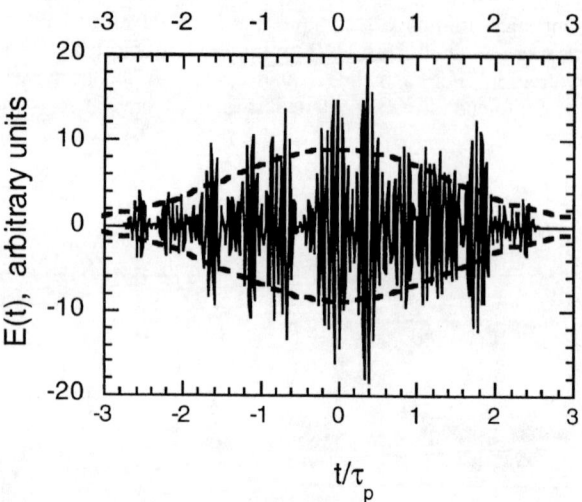

Figure 6: Example of an incoherent radiated electric field, within a Gaussian envelope, which has been spectrally filtered with a bandpass filter. The effect of the bandpass filter is to induce temporal coherence.

By measuring this variance and using phase retrieval techniques [17], bunch shapes can be obtained for many practical cases. In a practical set-up, beam radiation is spectrally dispersed in a high resolution spectrometer. Spectra, recorded with a high dynamic range CCD camera, are then collected on a shot-to-shot basis, and inverse Fourier transformed digitally. Each pixel of the CCD camera essentially detects light going through a specific bandpass filter similar to the previous discussion. The spectral resolution determines the temporal resolution, which implies the use of a high resolution spectrum and a CCD camera with small pixel size. Statistics on the fluctuation in the obtained quantity are then accumulated over many shots.

A method relying on optical gating could also be used in which optical-field or laser induced birefringence [18] changes the polarization of a radiation beam which propagates through the laser activated medium. Such techniques are widely used for temporal laser pulse characterization and are the building blocks of FROG systems [19]. First, the beam radiation is sent through a high contrast polarizer to determine a specific polarization direction. After the polarizer, the radiation travels through a Kerr cell (e.g. quartz or glass), followed by an analyzing polarizer. Using a CCD camera, the polarizer and analyzer are adjusted for maximum extinction. An ultra-short femtosecond laser pulse, synchronized with the electron beam and intense enough to cause polarization rotation in the medium, can then effectively gate out femtosecond slices of the radiation, either in a single shot or multi-shot scanning correlator configuration. Whereas the laser intensity requirement is easily met, the synchronization requirement will essentially be the limiting factor in time resolution of this method.

The use of coherent transition radiation (CTR) has also been proposed [20] and successfully applied for bunch profiling of sub-picosecond electron bunches [3]. In general, the radiation from a bunch of particles has a temporally coherent and incoherent contribution. The radiated intensity at frequency ω emitted by a bunch of particles can, in the far field limit and under the assumption of transverse coherence be written as

$$I_{tot}(\omega) = I(\omega)[N + N(N-1)F(\omega)] \tag{10}$$

with

$$F(\omega) = \left| \int dz S(z) e^{i\frac{\omega}{c}z} \right|^2 \tag{11}$$

and S(z) the longitudinal particle density distribution. As can be seen from Eq. (10), the incoherent (coherent) part scales linearly (quadratically) with the number of particles in the bunch. As can be seen from Eq.(11), measurement of the coherent radiation spectrum can provide information on the longitudinal bunch profile.

This frequency resolved method was implemented by Kung et al. [3] using a far-infra-red Michelson interferometer. The multi-bunch scanning auto-correlator measured electron bunches with a duration as short as 330 fs [21].

Bunch shape recovery is complicated due to the fact that, in principle, the complete radiation spectrum needs to be measured, and that both intensity and phase are needed for Fourier transformation. In practice however, experimentally measured intensity spectra are measured in a finite spectral band. Under the proper assumptions, the frequency range can be extended towards high and low frequency using theoretical predictions. In addition, recently a method was proposed for asymmetric bunch shape retrieval [17] by applying a Kramers-Kronig relation to the spectral form factor to find the minimal phase. The authors showed that for most reasonable bunch shapes, the minimal phase agrees with the actual phase. The technique has not been widely used or tested yet but has been applied to analyzing the shapes of the submillimeter-long electron bunches at the Cornell linear accelerator [22].

CTR has recently been suggested [23] as a diagnostic for measuring FEL bunching. The longitudinal bunching occurring in an FEL with periodicity near the resonant radiation wavelength, could become a useful approach for the generation of femtosecond, or even sub-femtosecond electron bunches. The effect of the periodic bunch train is to modify the OTR spectrum (Eq.(2)): the single electron spectrum is multiplied by a form factor which is the Fourier transform of the electron bunch distribution h(r,z)

$$F(\omega,\theta) = \left| \iiint h(r,z) \exp(-i\mathbf{k}.\mathbf{x}) d^3 x \right|^2 \tag{12}$$

This results in a reduction in opening angle of the radiation cone and an increase in brightness.

III. LASER BASED DIAGNOSTICS

A different approach to generating radiation from particle beams for beam monitoring is to use the interaction of the beam with high intensity laser fields. This radiation process can either be viewed classically as the emission of radiation by a particle subjected to acceleration in the electro-magnetic fields of the laser beam or as a scattering process between an electron and a photon, with energy and momentum determined by conservation of energy and momentum. In effect, the laser acts as an electro-magnetic undulator and the properties of the emitted radiation can be accurately predicted using an equivalent undulator model [24]

The use of lasers for particle beam characterization was proposed early after the first lasers became operational [25], [26] and used to diagnose low energy electron beams [27] in the early sixties. The scattered radiation contains information on energy as well as on transverse and (for short laser pulses) longitudinal distributions of the electron beam.

The early scattering experiments mentioned above yielded a low flux of scattered photons due to limits in available laser power, and low current electron beams, combined with the fact that the number of scattered photons is proportional to the well known Thomson cross-section, $\sigma_\tau = \frac{8}{3}\pi r_e^2$, which is only about 6.65×10^{25} cm^{-2}. Here r_e is the classical electron radius. Advances in sub-picosecond high peak power laser systems based on chirped pulse amplification [28], have rekindled the interest in scattering between laser light and relativistic electrons. Laser systems which deliver multi-terawatt peak power, 100 fs long pulses at a repetition rate of 10 Hz are now readily available.

At the Final Focus Test Beam (FFTB) at SLAC, transverse e-beam sizes as small as 70 nm were measured, by scanning a 50 GeV e-beam across the intensity fringes of an optical standing wave [2] produced by crossing two laser beams. Such resolution is beyond OTR or synchrotron radiation based methods. At the BTF, transverse and longitudinal phase space distribution were measured [7] for a 50 MeV electron beam using a focused terawatt, 100 fs long laser pulse. The laser beam was orthogonally incident on the electron beam. The 90° geometry used in this Thomson scattering experiment allows beam size and divergence to be obtained for a 200 - 300 fs slice of the electron beam, which will be discussed next.

The experiments were conducted at the BTF and a schematic of the experiments is shown in Fig. (7). Electron bunches with energy of 50 MeV (energy spread 0.2 - 0.4 %) containing typically 1.3 nC of charge within a 10-15 ps (rms) bunch length, were produced by the linac from the Advanced Light Source. The linac consists of a thermionic gun operated at 120 kV which produces about a 2 ns long electron bucnh, three RF buncher cavities (125 MHz, 500 MHz and 3 GHz) which compress the pulse to about 30 ps and two 3 GHz accelerator structures which accelerate the electrons to an energy of 50 MeV. transported Bend magnets and quadrupoles (BTF line) transport the beam to an interaction chamber where the beam was focused and scattered against the laser beam. After the interaction chamber, a 60° bend magnet deflected the electron beam onto a heavily shielded beam dump, away from the forward scattered x-rays.

The terawatt laser system, with center wavelength at 800 nm, was based on chirped pulse amplification in Ti:Al$_2$O$_3$ [28, 29]. The Kerr lens modelocked oscillator operated at the fourth sub-harmonic (125 MHz) of the 500 MHz master oscillator source for the linac. Individual laser oscillator pulses, with an energy of a few nJ were extracted at a 10 Hz repetition rate, and were temporally stretched using a grating and a telescope based on a parabolic reflector. Amplification to the 100 - 200 mJ range was achieved in an 8-pass pre-amplifier and a 3-pass power amplifier pumped by second harmonic radiation from a Q-switched Nd:YAG laser. The amplified pulses were temporally compressed in a vacuum chamber using a grating pair and were propagated to the e-beam/laser interaction chamber through an evacuated beam line. Amplified laser pulses as short as 50 fs (2 TW peak power) have been produced with this system, but typical operating parameters for the experiment were 100 fs long pulses containing about 40 mJ energy. A 75 cm radius of curvature mirror was used to focus the S-polarized amplified laser pulses to about a 30 µm diameter spot at the IP (measured by a charge coupled device (CCD) camera at an equivalent image plane outside the vacuum chamber).

To measure the spot size (and position) of the electron beam at the interaction point (IP), a 2 µm thick Al-coated (35 nm) nitrocellulose foil, mounted on a retractable plunger at 45° with respect to the beam, was installed in the chamber. An image of the electron beam was obtained by relaying OTR onto a 16 bit CCD camera or optical streak camera using a small f-number telescope. The spatial resolution of the imaging system was 14 µm. Electron beam spot sizes as small as 35 µm rms have been measured.

Synchronization between the laser oscillator and linac was accomplished by using a phase-locked loop which dynamically adjusts the oscillator cavity length [30]. The phase error signal was generated by mixing the fourth harmonic (500 MHz) of the oscillator repetition frequency (generated from a photodiode) with the master oscillator

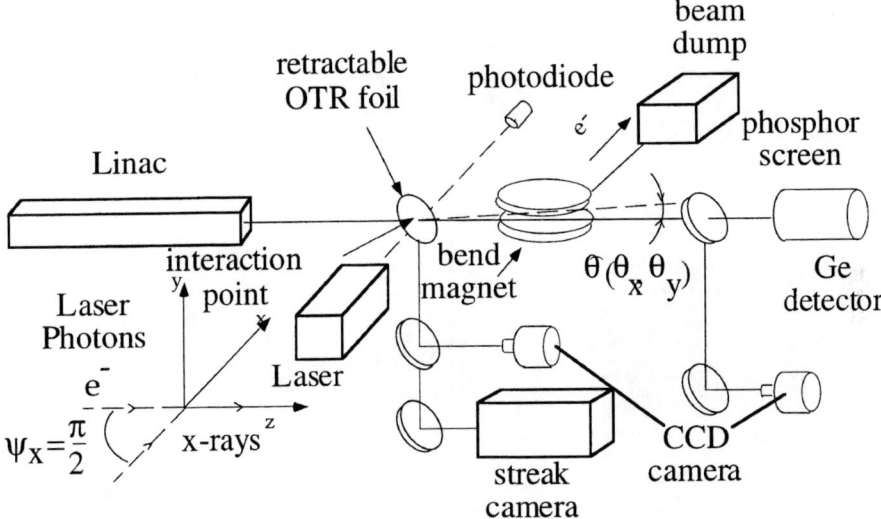

Figure 7: Experimental set-up for the 90° Thomson scattering experiment of laser based beam probing and femtosecond x-ray generation.

source for the linac. Timing jitter measurements (using a streak camera with an instrument response of 1.5 - 2 ps) which simultaneously detected a laser pulse and OTR from the electron bunch indicated an rms jitter of 1-2 ps.

To monitor the arrival time of the electron bunch and laser pulse at the IP, a button-type radio-frequency pick-up and a 6 ps rise time photoconductive switch were used to provide a signal from the electron beam and laser pulse, respectively. The signals were measured on a transient digitizer with 4.5 GHz bandwidth and an optical delay line allowed changing the arrival time of the laser pulse. Initial spatial alignment of the beams was accomplished by optimizing the e-beam and laser transmission through a cube with two 250 µm diameter intersecting orthogonal holes.

During the interaction of an electron beam and laser beam, scattered x-ray photons are produced with energy U_x, given by (for $\gamma \gg 1$)

$$U_x = \frac{2\gamma^2 \hbar \omega_0}{1+\gamma^2 \theta^2}(1-\cos\psi) \qquad (13),$$

where ω_0 is the frequency of the incident photons, ψ is the interaction angle between the electron and laser beam ($\psi=\pi/2$ in our experiments), and θ is the angle at which the radiation is observed and assumed to satisfy $\theta \ll 1$ (see Fig. 7). Equation (13) implies that a one-to-one correlation exists between the observation angle and the energy of the scattered photon. When the 50 MeV electron beam ($\gamma = 98$) collided at 90° with the laser pulse (center wavelength = 800 nm), x-rays with a maximum energy of 30 keV (0.4 Å) were generated.

To measure the transverse electron beam distribution for a given slice of the electron beam, we scanned the laser beam transversely across the electron beam and monitored the x-ray yield on the phosphor screen. The laser beam position was changed vertically at the interaction point, in steps of 10 μm, by changing the tilt of the focusing mirror. It was found (Fig. 8) that the laser based technique and the results from OTR were in good agreement and give a half-width half maximum (HWHM) vertical size of 66 μm. However, whereas the beam core overlaps, the tails are different and both non-Gaussian. This result was reproducible and may indicate real differences in the time structure of the tails of the spatial distribution of the actual beams that were being sampled, which will be discussed below. From the OTR data an HWHM horizontal size of 47 μm was obtained.

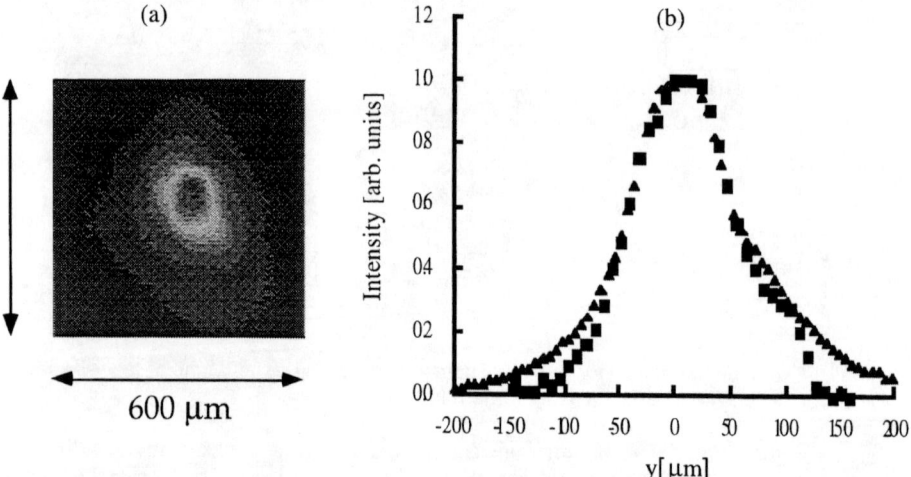

Figure 8: a) OTR image of the focused electron beam and b) ▲ (triangle) - vertical line-profile through the OTR image of the electron beam; ■ (square)- x-ray yield vs. vertical laser beam position.

Measurement of the electron beam divergence for a fixed longitudinal location (i.e. fixed delay time between the laser and electron beam) of a time slice of the electron beam, with a duration equal to the convolution of the transit time of the laser pulse and the laser pulse duration, was done by monitoring the spatial x-ray beam profile on the phosphor screen using the CCD camera (see Fig.(9)). The scattered x-ray energy flux contains information of the angular distribution of the electron beam. By convoluting the single electron spectrum [24] with a Gaussian distribution for the horizontal and vertical angles ($\sigma_{\theta x}$ and $\sigma_{\theta y}$ are the rms. widths of the angular distribution of the electron beam in the horizontal and vertical direction respectively) and integrating over all energies and solid angle[7, 31] the energy flux can be written as:

$$\frac{dP}{d\theta_x d\theta_y} \propto \int_0^{2\pi} d\phi \int_0^1 d\kappa\, F(\kappa)\kappa\left[1-4\kappa(1-\kappa)\cos^2\phi\right]$$

$$\exp\left[-\frac{(\theta_x - \gamma^{-1}\sqrt{\frac{1}{\kappa}-1}\sin\phi)^2}{2\sigma_{\theta x}^2}\right]\exp\left[-\frac{(\theta_y - \gamma^{-1}\sqrt{\frac{1}{\kappa}-1}\cos\phi)^2}{2\sigma_{\theta y}^2}\right]$$

(14).

Here dP is the radiated x-rays intensity in a solid angle $d\theta_x d\theta_y$, ϕ is the azimuthal angle and $F(\kappa)$ is an x-ray energy dependent function which takes into account overall detector sensitivity and x-ray vacuum window transmission, $\kappa = U/U_{max} = (1+\gamma^2\theta^2)^{-1}$ with $U_{max} = 2\gamma^2\hbar\omega$ and we have assumed a single incident laser frequency.

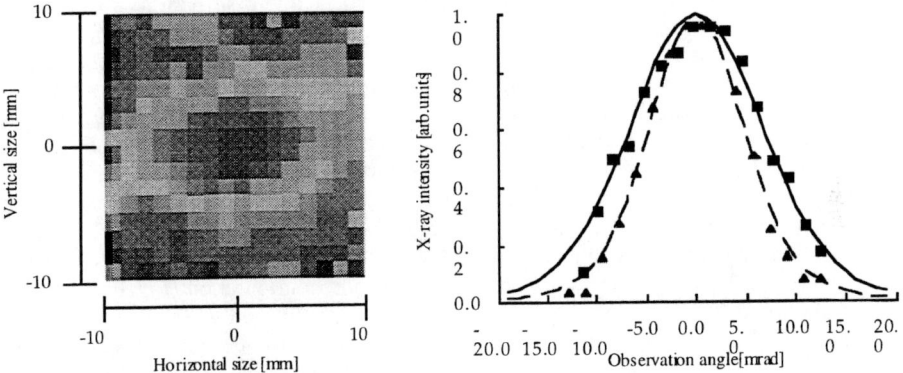

Figure 9. a) False color CCD image of the spatial profile of a 30 keV x-ray pulse on the phosphor screen, which is located 80 cm from the IP; b) ■ (square)- horizontal line-profile and fitting curve (solid line), ▲ (triangle) -vertical line-profile and fitting curve (dashed line) from Fig. 8 (a). The scale has been converted into angular units.

By fitting the data (see Fig. 9) using Eq.(14), an electron beam divergence of $\sigma_{\theta x}$ ($\sigma_{\theta y}$) = 6.3 ± 0.2 (3.9 ± 0.2) mrad was found. $F(\kappa)$ was adjusted to account for the spectral dependence of the x-ray window transmission. The difference between $\sigma_{\theta x}$ and $\sigma_{\theta y}$, is due to a combination of, the electron beam being focused astigmatically at the IP, resulting in a tilted phase space ellipse (y, y′), and a laser spot size much smaller than the vertical electron beam size. As the laser beam crosses the focal volume of the electron beam, the complete horizontal (direction of propagation of the laser) phase space (x, x′) is sampled by the laser beam. However, only electrons occupying the region in the vertical phase space defined by the spatial overlap with the laser beam will contribute to the x-ray flux. As opposed to the transition radiation based detector, the laser beam therefore acts as an optical microprobe of a finite region of the transverse phase space. Also note that this value of the electron beam divergence is consistent with an effective angular divergence of the electron beam of 3.5 - 4 mrad obtained from analyzing the x-ray spectra. Of course, the main difference is that measurement of the spatial profile is a single shot technique as opposed to measuring the x-ray spectra which requires accumulation of thousands of shots.

The spot size measurements and beam divergence measurements imply a horizontal geometric slice emittance $\sigma_x \sigma_{\theta x}$ for the electron beam of 0.25±0.03 mm-mrad. A linac beam emittance of 0.32±0.02 mm-mrad was measured using a quadrupole scan technique [32], which is in reasonable agreement with the x-ray slice measurements.

Finally, since the x-ray yield is sensitive to both the longitudinal bunch profile and the degree of transverse overlap between the laser and electron beam, time-correlated phase space properties of the electron beam can be studied. As an example, when an electron bunch, which exhibits a finite time-correlated energy spread (chirp), is focused at the IP with a magnetic lattice which has large chromatic aberrations, different temporal slices of the bunch will be focused at different longitudinal locations. The transverse overlap between e-beam and laser will therefore strongly depend on which time slice the laser interacts with. This in turn will lead to a time dependence of the x-ray yield varying faster than the actual longitudinal charge distribution. To illustrate this, we next present results of measuring the x-ray flux as a function of the delay between laser and e-beam, for two different magnetic transport lattices. In both lattices, the magnet settings were optimized to obtain a minimum electron beam spot size in the horizontal and vertical plane (as well as zero dispersion at the IP), but chromatic aberrations were about 5 times larger in the second lattice. An example of a 60 ps long scan (time step of 1 ps) and time-resolved OTR from the streak camera for the lattice with low and high chromatic aberrations are shown in Fig. 10(a, b) and (c,d) respectively. Whereas the temporal scan (Fig.10a) for the lattice with low chromatic aberrations is in good agreement with the time-resolved OTR (Fig. 10b) measured with a visible streak camera, the scans taken for the second configuration (Fig.10c) typically showed a 2-3 times larger amplitude 5 ps wide peak sitting on a 20 ps wide pedestal. This is to be compared to the time resolved OTR from the streak camera which typically showed a 25-30 ps wide electron beam without any sharp time structure (Fig. 10d).

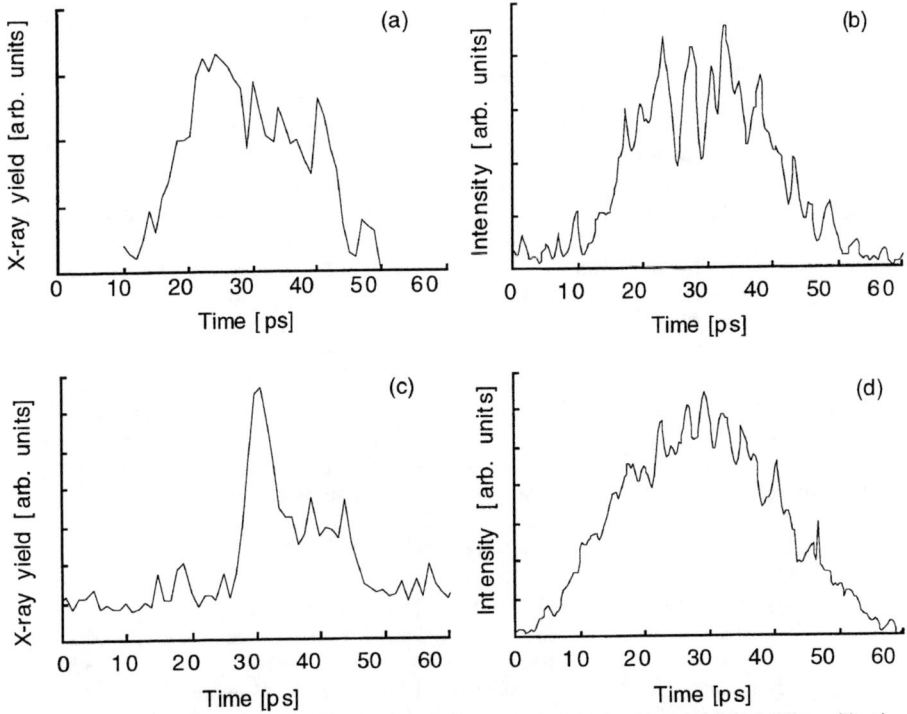

Figure 10. x-ray yield vs. delay time between laser and electron beam for a lattice with a) large and c) small chromatic aberrations; profile of time resolved OTR image from a streak camera for a magnetic lattice with b) large and d) small chromatic aberrations.

Lattice calculations for our experiment with MAD [33] indicated that an energy change of 0.25 % would increase the vertical spot size by a factor two at the IP, compared to best focus, resulting in a proportional reduction in vertical overlap between the laser and electron beam, and hence in x-ray yield. The measurements indicate the potential of the laser based Thomson diagnostic to measure time-correlated energy changes of less than a percent, with sub-picosecond time resolution.

IV. CONCLUSION

Various beam monitoring techniques have been discussed for transverse and longitudinal bunch distribution measurements. In particular, the properties of optical transition radiation were briefly reviewed, followed by a discussion of transverse and longitudinal bunch profiling using OTR. For transverse profiling, the larger opening angle of the OTR cone compared to synchrotron radiation can provide better spatial resolution for ultra-relativistic electron beams. For longitudinal profiling, streak camera based methods continue to be powerful tools, even for sub-picosecond bunches. Slice emittance measurement techniques were discussed which use chirped electron beams combined with dispersive filter sections. A fluctuational interferometry technique was discussed for bunch length and shape measurement using incoherent radiation. An optical gating technique using laser induced birefringence in an optical Kerr medium was proposed for measurement of bunch profiles with femtosecond time resolution. Using coherent transition radiation, electron bunches as short as 300 fs have been detected. Laser based probing of electron beams has been used for measurement of longitudinal and transverse bunch distributions of picosecond and sub-picosecond slices. Results from an experiment in which a 90° geometry was used were discussed. From a study of x-ray beam images and total flux, the transverse electron beam phase space distribution of essentially a 300 fs slice of the electron beam was obtained. By scanning the laser beam in time along the electron bunch, not only the longitudinal density distribution was measured, but it was also found that the Thomson scattering technique can become a powerful and sensitive tool to measure detailed longitudinal phase space properties. The main limitation on the slice duration arose from the finite transit time of the laser pulse across the electron beam. Shorter slices and higher photon yields are expected when using RF photocathode guns which produce beams with normalized emittance on the order of a few π mm-mrad. Synchronization jitter between the laser and the linac were experimentally found to be on the order of a picosecond. The laser based methods are better suited for photocathode driven RF guns, where lower jitter between laser and electron beam is expected.

V. ACKNOWLEDGMENTS

This work would not have been possible without the help and support of many people. In particular, my collaborators on the laser based probing work R. Schoenlein, P. Volfbeyn, A. Chin, P. Balling, T.E. Glover, M. Zolotorev, K.J. Kim, S. Chattopadhyay and C.V. Shank; the OTR work: R. Govil, S. Wheeler, B. vander Geer, M. de Loos, and M. Conde. I want to thank Leon Archambault, and Jim Dougherty for engineering and technical support and T. Byrne for help with the linac and BTF operations. I would also like to acknowledge useful discussions with R. Fiorito and many enlightening conversations with Max Zolotorev and A. Zholents.

VI. REFERENCES

1. D.T. Palmer et al., IEEE Proc. of the 1995 Particle Accelerator Conference, Dallas, TX, p. 2432-4 vol.4. (1995).
2. T. Shintake, Nucl. Inst. & Methods A**311**, pp. 453 (1992), Balakin, V. et al., Phys. Rev. Lett. **74**, p. 2479-82 (1995).
3. P. Kung, H. Lihn and H. Wiedemann, Phys. Rev. Lett. **73**, 967 (1994); M. Uesaka, K. Tauchi, T. Kozawa, T. Kobayashi, and others., Phys. Rev. E. **50**, 3068 (1994); B. E. Carlsten and S.J. Russell, Phys. Rev. E **53**, R2072 (1996).
4. J. Gardelle, J. Labrouche, and J.L. Rullier, Phys. Rev. Lett. **76**, 4532 (1996); J.S. Wurtele et al., Phys. Fluids B **2**, p. 401 (1990); X.J. Wang et al., Phys. Rev. E 54, R3121-4 (1996).
5. An excellent overview with many references is given by E. Esarey et al., IEEE Trans. on Plasma Science **24**, pp. 252 (1996).
6. I. Frank and V. Ginzburg, J. Phys. USSR, Vol. 9, pp. 353 (1945), L. Wartski et al., J. Appl. Phys., vol. 46, pp. 3644 (1975); D. W. Rule, Nucl. Inst. and Meth., Vol. B**24/25**, pp. 901 (1987).
7. W.P. Leemans et al., Phys. Rev. Lett. **77**, pp. 4182 (1996).
8. M.L. Ter-Mikaelian, "High Energy electromagnetic processes in condensed media", Interscience tracts on physics and astronomy, Wiley-Interscience, New-York, pp. 212 (1972).
9. L. Wartski, J. Marcou and S. Roland, IEEE Trans. in Nucl. Science, vol. 20, pp. 544 (1973).
10. W. P. Leemans, G. Behrsing, K.J. Kim, J. Krupnick, C. Matuk, F. Selph, and S. Chattopadhyay, Proc. 1993 Part. Accel. Conf., 83 (1993).

11. "1-2 GeV Synchrotron Radiation Source", Conceptual Design Report, LBL-PUB 5172 Rev., 1986.
12. M.J. de Loos, S. B. van der Geer, and W.P. Leemans, Proc. Fourth European Part. Accel. Conf, pp. 1679, London 27 June- 1 July (1994).
13. Max Zolotorev, private communication.
14. A. Lumpkin, AIP Proceedings **367**, pp. 327 (1995).
15. X. Qiu et al., Phys. Rev. Lett. **76**, pp. 3723 (1996).
16. M. Zolotorev and G. Stupakov, SLAC PUB-7132, (1996).
17. R. Lai and J. Sievers, Phys. Rev. E **50**, R3342 (1994), R. Lai, U. Happek, and A.J. Sievers, Phys. Rev. E **50**, R4294 (1994)
18. Y.R. Shen, "The Principles of Nonlinear Optics", Wiley-Interscience (1984).
19. D.J. Kane and R. Trebino, IEEE J. of Quant. Electron. **29**, pp. 571 (1993); K.W. DeLong, D.N. Fittinghoff, R. Trebino, IEEE J. of Quant. Electron. **32**, pp. 1253 (1996).
20. W. Barry, Proc. Workshop on Advanced Beam Instrumentation, pp. 224, KEK Tsukuba (1991).
21. H.C. Lihn et al., "Measurement of Subpicosecond electron bunch lengths", AIP Conf. Proc. **367**, pp. 435 (1996).
22. R. Lai and A.J. Sievers, AIP Conf. Proc. **367**, pp. 312 (1996).
23. J. Rosenzweig, G. Travish, and A. Tremaine, Nucl. Instr. Meth. A **365**, pp. 255 (1995).
24. K.J. Kim, S. Chattopadhyay, and C.V. Shank, Nucl. Instr. Meth. A **341**, 351 (1994).
25. F.R. Arutyumian and V.A. Tumanian, Phys. Lett. **4**, 176 (1963).
26. R. H. Milburn, Phys. Rev. Lett. **10**, 75 (1963).
27. G. Fiocco and E. Thompson, Phys. Rev. Lett. **10**, 89 (1963).
28. D. Strickland and G. Mourou, Opt. Comm. **56**, 219 (1985).
29. C. LeBlanc, G. Grillon, J.P. Chambaret, A. Migus, and A. Antonetti, Opt. Lett. **18**, 140 (1993).
30. M.J.W. Rodwell, D.M. bloom, K.J. Weingarten, IEEE J. Quant. Electron. **25**, 817 (1989).
31. W.P. Leemans et al., J. Quant. Electron., to be published.
32. J. Bengtson, W. Leemans, and T. Byrne, Proc. 1993 Particle Accelerator Conf., pp. 567(1993).
33. H. Groter and F. Iselin, "The MAD program", CERN/SL/90-13 (1993).

Recent Progress in Photoinjectors[*]

Ilan Ben-Zvi
Brookhaven National Laboratory

INTRODUCTION

In photoinjector electron guns, electrons are emitted from a photocathode by a short laser pulse and then accelerated by intense RF fields in a resonant cavity. Photoinjectors are very versatile tools. Normally we think of them in terms of the production of high electron density in 6-D phase space, for reasons such as injection to laser accelerators, generation of x-rays by Compton scattering and short wavelength FELs. Another example of the use of photoinjectors is the production of a high charge in a short time, for wake-field acceleration, two-beam accelerators and high-power long-wavelength FELs. There are other potential uses, such as the generation of polarized electrons, compact accelerators for industrial applications, and more.

Photoinjectors are in operation in many electron accelerator facilities, and a large number of new guns are under construction. The purpose of this work is to present some trend-setting recent results that have been obtained in some of these laboratories. In particular, the subjects of high density in 6-D phase space, new diagnostic tools, photocathode advances, and high-charge production will be discussed.

SLICE EMITTANCE

Any future improvement in the quality of the electron beam of a photoinjector will require sophisticated diagnostics to study in detail the phase space of the beam and provide guidance for multi-parameter adjustment of the photoinjector. Study of the phase space of a short longitudinal slice of the electron beam is particularly important. The slice-emittance technique [1] is such a diagnostic. In this technique, a short slice is selected out of an energy chirped beam by a slit in a dispersive region. The emittance of this slice is measured downstream of the slit using the quadrupole scan technique (or any other method). The measurement is repeated for different slices and for different beam conditions, like a new value for the current in the emittance compensating solenoid.

In the emittance compensation technique, proposed by Bruce Carlsten [2], the observed emittance growth due to linear components of space charge forces in the

---

[*] Work supported by the US Department of Energy under contract No. DE-AC02-76CH00016

photoinjector is compensated for by passing the electron beam through a laminar-flow beam waist. The space charge interaction in this beam waist results in a differential rotation of the slice ellipses to bring them into alignment at some point downstream of this waist. To understand this emittance growth and compensation, one has to look at the slice emittance of a number of slices as they evolve.

Figure 1 shows the schematic layout of the slice-emittance measurement diagnostic at the Brookhaven National Laboratory (BNL) Accelerator Test Facility. It is important to note that the beam is going through the laminar-flow beam waist while it is accelerated by the linac. Thus, it is possible to achieve the perfect compensation and freeze the phase space distribution (space charge forces are practically eliminated by the acceleration) so as to avoid subsequent deterioration. The measured phase-space ellipses of three slices for three values of the solenoid current

This measurement can be used under conditions in which the linear compensation technique of Carlsten is inadequate. These conditions are expected to occur with higher beam peak currents. It has been analyzed by Gallardo and Palmer [3] that there is a large degree of correlation, including nonlinear terms, in the radial and londitudinal phase-spaces. They suggested that nonlinear emittance compensation techniques can be applied in such cases to produce higher electron beam brightness. The application of nonlinear correction schemes requires the adjustment of more than one parameter. The slice emittance measurement demonstrated here provides the diagnostic tool to observe and control a nonlinear emittance correction scheme. We plan to carry out such an experiment, using the longitudinal power distribution of the laser pulse to control the currents of individual slices. The method suggested by Gallardo [4] is control of the laser intensity in 3-D to produce an improved electron beam distribution. Gallardo's discussion did not include the Carlsten laminar beam waist linear correction. If one uses the laminar beam waist technique together with a programmed longitudinal and transverse modulation of the laser's intensity, a much better correction can be expected with convenient and precisely controlled variables.

MEASUREMENT OF EMITTANCE COMPENSATED BEAMS

A systematic program of measuring the emittance of compensated beams is in progress, using the new BNL/SLAC/UCLA collaboration photoinjector (see gun in Figure 3 and detailed throughout). Emittance measurement, in particular that of a very small emittance, is difficult. One problem is the handling of beam halo. In a parametric measurement of emittance compensated beams, one may encounter difficulty due to a limited dynamic range of the frame grabber.

As an example, consider the quad scan measurement of the total emittance as a function of solenoid current. As seen in Figure 4, the emittance initially decreases as expected as a result of emittance compensation. However, the increase in emittance is not as rapid as one would expect. This is the result of a crossover

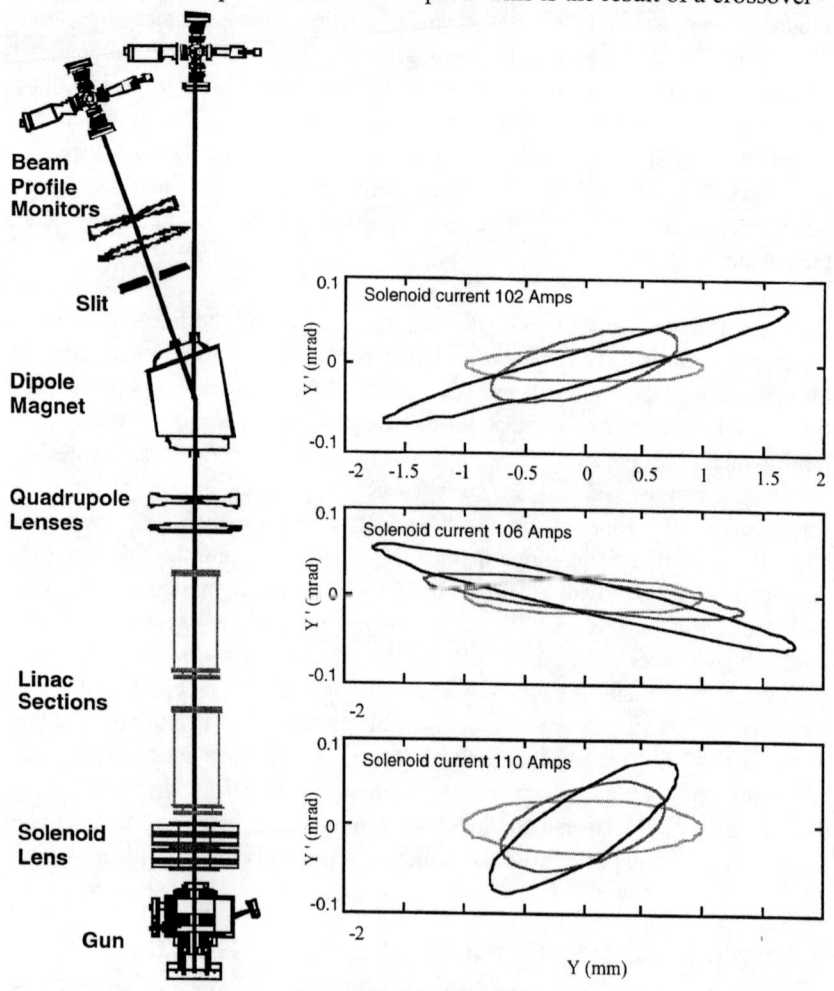

Figure 1. Setup for the measurement of slice emittance at the BNL ATF.

Figure 2. The fitted beam ellipses for the slice emittance measurement. The slice at the end of the beam bunch is taken as a beam waist, showing the relative rotation of the front and middle slices as the solenoid current is increased.

(non-laminar) of some part of the beam. Crossover results in a large divergence and the creation of a low intensity beam halo that is not registered by a low dynamic range camera and frame-grabber system. The loss of the halo electrons due to the dynamic range results in a better emittance than should be observed otherwise.

The reason for the crossover on some low current slices at a large solenoid field and the resultant beam halo can be shown by integrating the envelope equation of the beam in the linac. The relativistic envelope equation of a beam with space charge and emittance is given by:

$$R'' + R'\frac{\gamma'}{\gamma} = \frac{2I}{\gamma^3 I_A}\frac{1}{R} + \frac{\varepsilon_n^2}{\gamma^2}\frac{1}{R^3}$$

R is the local envelope, γ is the energy, I_A is the van Alfven current, I is the local current and ε_n is the normalized local emittance.

The Carlsten emittance correction relies on a laminar flow beam waist, which is described by the envelope equation without the emittance term. We can define a critical beam waist envelope size, R_c by equating the two terms on the right hand side, resulting in:

$$R_c = \sqrt{\frac{\gamma \varepsilon_n^2 I_A}{2I}}$$

Figure 3. The BNL/SLAC/UCLA s-band photoinjector. The waveguide is at 3 o'clock, symmetrizer and vacuum port at 9 o'clock, laser ports at 1 and 7 O'cl o'clock, rf pickup at 12 o'clock.

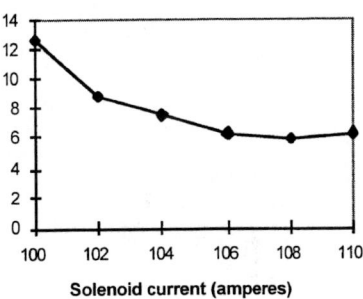

Solenoid current (amperes)

Figure 4. Normalized total vertical rms emittance as a function of solenoid current.

A beam waist that is larger (smaller) than the critical size, R_c, is dominated by space charge (emittance). The emittance dominated beam waist leads to a crossover of electrons at that particular slice. Since R_c is a function of the current, the low current slices will cross over while high current ones will maintain a laminar flow.

Now we numerically integrate the envelope equation through the linac. We integrate the equation both with and without the emittance term. For comparison we also plot the critical envelope R_c. We can show that at a low solenoid field the beam waist is larger than R_c, even for a small slice current, but at a larger solenoid field low current slices will cross over while high current slices will not (that is will have an envelope size larger than R_c.) The typical values for which we integrate the equation are: local (slice) normalized emittance $\varepsilon_n = 2 \times 10^{-6}$ meter radians; initial energy at the entrance to the linac $\gamma = 9$; initial beam envelope at the entrance to the linac $R_0 = 2.5 \times 10^{-3}$ meters; accelerating gradient in the linac $\gamma' = 14$ meter $^{-1}$.

In Figure 5 we use an initial beam envelope derivative of $R'_0 = 3.7 \times 10^{-3}$ radians (corresponding to a large solenoid field) and a slice current of $I = 30$ amperes. Under these conditions we get a cross over, since the envelope gets smaller than the critical size (at about $z=2$ meters). Note the large divergence of the beam at the linac exit. Figure 6 shows that with the same R_0' (solenoid field), a higher current of 50 amperes does not cross over, and the divergence of the beam at the end of the linac is markedly smaller. Also the effect of the emittance term is much smaller, as expected for a laminar flow beam waist.

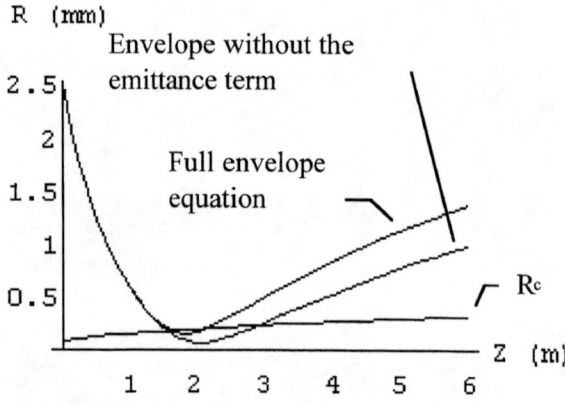

Figure 5. Beam envelope as a function of position for a large solenoid current and a small slice current.

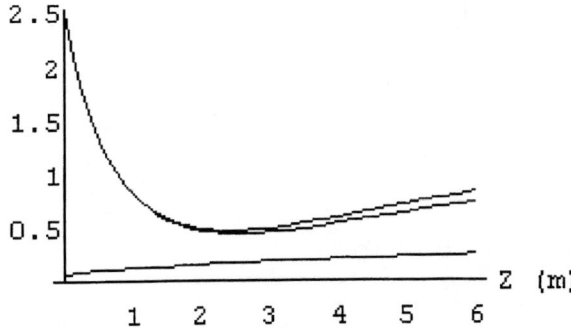

Figure 6. Beam envelope as a function of position for a large solenoid current and a large slice current.

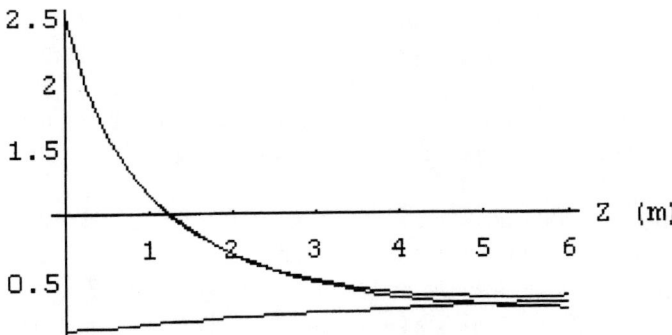

Figure 7. The beam envelope as a function of position for a small solenoid current and a small slice current

Figures 7 and 8 have the same beam slice currents, but a weaker solenoid field (initial divergence of 2.8 mrad). Figure 7 illustrates a slice at 30 amps, and Fig. 8 a slice at 50 amps at the same solenoid setting. Both cases are characterized by a laminar flow beam waist and a small divergence of the beam.

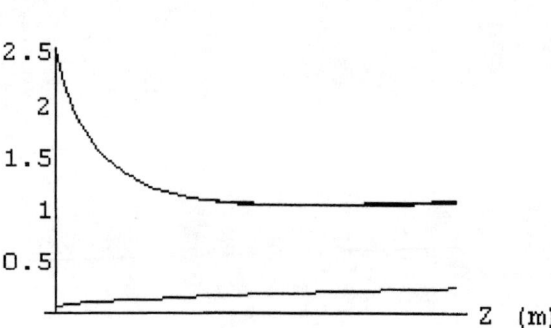

Figure 8. The beam envelope as a function of position for a small solenoid current and a large slice current.

TOMOGRAPHY

We lose a lot of information by discussing beam ellipses or by making *a priori* assumptions concerning the distribution of the beam in phase space in order to fit a quadrupole scan data (or other measurement techniques) to some beam parameters. As has been shown by McKee, O'Shea, and Madey [5], one can apply tomographic techniques to a quadrupole scan and derive the full phase space distribution. This is a powerful technique that, together with the slice emittance measurement technique, will provide a greater understanding of the beams of photoinjectors and help improve the performance.

Tomography is being applied at the BNL ATF to gain an understanding of the transverse phase space of the photoinjector. As an example [6], in Figure 9 below we have the recovered x-x' phase space at the end of the ATF linac for a 7 ps (full width) electron bunch with a charge of 350 pC.

MICROBUNCHING

Femtosecond electron bunches that contain a large number of electrons and have small dimensions in the six-dimensional phase space have emerged as a powerful tool for the investigation of transient effects. Extremely short, low emittance electron pulses are of interest to a number of disciplines. Femtosecond pulses of X-ray can be produced by a head-on Thomson scattering geometry of laser light from a femtosecond electron bunch. Such radiation is the most effective probe of structural dynamics of materials on the time scale of the motion of atoms. The best coherence and intensity of the produced X-rays require electrons with a high density in six-dimensional phase space. Other uses are injection into future linear colliders or laser accelerators and the generation of powerful broad band radiation in the millimeter to Far-Infrared (FIR) wavelength, as well as industrial uses, such as lithography.

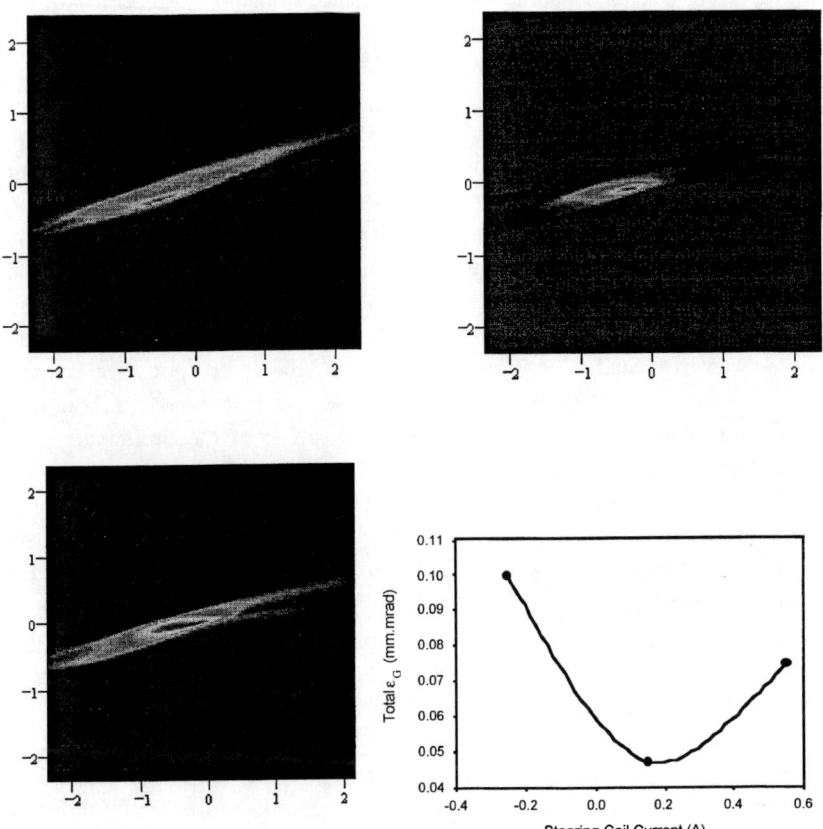

Figure 9. Tomographic measurement of the transverse phase space and a conventional emittance scan for three values of the steering coil in the linac's entrance. In the tomographic phase space plots, electron density is color coded, the vertical axis is divergence in mrad, and the horizontal axis is in mm.

At Los Alamos, bunch compression has achieved sub-picosecond pulses with charges in the range of 0.1 to 1 nC at 8 MeV [7]. The compression has been done with a chicane magnet. A compression in excess of 40 and peak currents greater than 1 kA have been measured. The method of measurement of the short bunch lengths was spot size increase due to a transversely deflecting rf cavity. The typical normalized rms emittance of the Los Alamos photoinjector is 2.5 mm mrad; however, a measurement of emittance under the compression conditions was not reported. One expects that the low emittance of the photoinjector will be, if not actually preserved, not degraded too much. The measured bunch length as a function of the phase for the Los Alamos result is shown in Figure 10.

Figure 11 shows the measured longitudinal peak current distribution as a function of longitudinal position in picoseconds (which is nearly 2856 Mhz in degrees) for a number of initial laser to rf phases at the BNL ATF gun. The salient feature of this demonstration is that the bunching was done without magnetic compression (no chicane or alpha magnet). For this reason, the emittance can be very good even after compression, leading to a high bunch density. In a recent measurement [8], a short, high brightness electron bunch with a 40 pC charge ($2.5 \cdot 10^8$ electrons), was produced. For this measurement, the laser spot size on the cathode was reduced to 0.4 mm diameter. For the layout of the system, one may consult Fig. 1. The calculated β function was set to 3 m at the momentum slit. The slit opening was set to the equivalent of 500 fs. With this setting, better than 95% of the charge, or 40 pC, passed through the slit. An emittance measurement was done for this bunch yielding 0.5 π mm mrad normalized rms and the intrinsic energy spread of the beam was $\Delta \gamma_i / \gamma = 0.15\%$ full width. The emittance contribution to the beam size can be shown to be negligible, but the intrinsic energy spread contributes nearly half of the beam size on the momentum slit. Therefore, the estimated bunch length (95% charge) is 370 fs. The corresponding peak current is 170 amperes. This is a record 6-D phase-space density.

The 6-D phase-space density can be improved by operating the photoinjector in a linear regime of the energy vs. phase curve. The length of the first cell, (normally the half cell in the BNL gun) has an effect on the linearity of the longitudinal phase space [9]; thus, the 1.6 cell (and later the 3.6 cell) design was adopted by the Grumman-BNL collaboration for its photoinjector (termed "Gun II" at BNL). In this work, it was established that the increased length does not affect the emittance and, incidentally, it leads to a reduction of the peak surface field in the gun so that the highest electric field is attained on the cathode rather than on the iris. Therefore, this type of gun can be operated at a higher field before breakdown occurs.

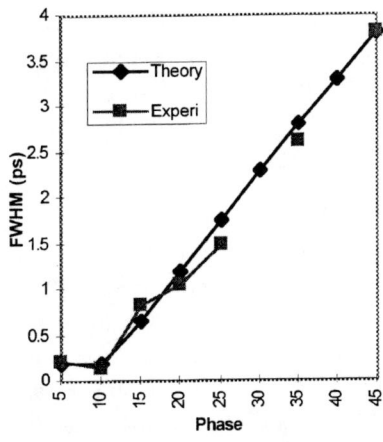 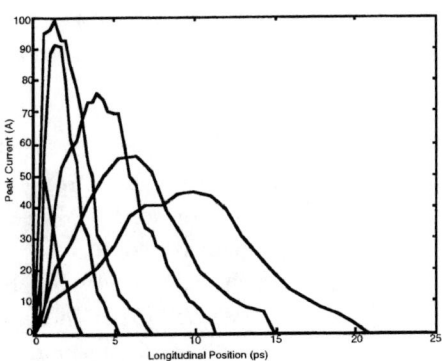

Figure 10. Electron bunch length as a function of the laser to rf phase for the Los Alamos gun.

Figure 11. Measured electron bunch current distribution as a function of time for a few laser to rf phases in the Brookhaven gun.

HIGH CHARGE PRODUCTION

A high surface field on the cathode and the high yield of electrons possible by photo emission, permit a very large current density, $J \sim 10^4$ to 10^5 A/cm^2 [10]. This is much larger than thermionic emission (about 10 A/cm^2). To achieve this, it is necessary to have a good quantum efficiency and a large damage threshold for the cathode material. The progress that has been made recently in cesium telluride in various laboratories is very impressive, and this is the material of choice for very large charge production. This material is reasonably stable and provides quantum efficiencies of 2-3% in a routine manner. Good reviews on this subject are given by Guy Suberlucq and Paolo Michelato [11]. However, this is not necessarily the case for single bunch production.

A very high-charge photoinjector is the Argonne Advanced Wake Accelerator [12], where charges in a single pulse of up to 100 nC have been achieved using magnesium cathodes. The Schottky effect can enhance the quantum efficiency tremendously if the cathode is operated at a high electric field. A high field operation is a must in order to avoid space charge limits. The Child-Langmuir Law (in an approximate expression for two infinite plane parallel electrodes and a short pulse) relates the charge density q/A in nC per cm^2 to the electric field E in MV/m by q/A=0.885E. Therefore it is natural to design a high field photoinjector for high charge operation. Indeed, the ANL gun operates at 100 MV/m even though it is an L-band gun. The emittance of this gun is quite good for this high charge. Figure 12 shows that at a 70 nC charge the pulse length is under 50 ps (more than 1.4 kA current!). The measured emittance at this charge is 20 mm mrad.

The CLIC photoinjector has a somewhat different objective than the ANL device—that is, to generate high frequency rf (transferring energy in a resonant structure). For this purpose, a high charge in a bunch train is required. The 1 _ cell S-band CLIC photoinjector together with a four cell booster have successfully accelerated a bunch train of 48 bunches containing 450 nC (!) to a beam momentum of 9.5 MeV/c [13] . The maximum charge for a single bunch was 35 nC with a bunch length of 16 ps—more than 2 kA! An intense bunch train beam loading can lead to a bunch timing error as a result of the change in field before the electrons become relativistic. The CLIC team is designing a special multi-mode gun that will minimize this timing error.

A novel system, based on a pulse transformer, is in operation at BNL. This table-top device has demonstrated electric fields of up to 1000 MV/m over a gap of 1 mm with a flat top of the order of 1 ns (see scope trace in Figure 13) and very little dark current. This is a unique ultra high electric field test bed. In addition, it holds the promise of ultra-high charge extraction and a good emittance (no rf effects). Photoemission of 3 nC has already been demonstrated [14] at a reduced gradient of 150 MV/m, and the synchronization of the high voltage pulse with the laser is already better than 0.5 ns. The rather compact device is shown in Figure 14.

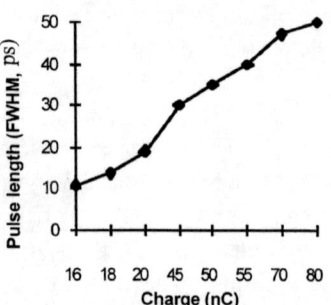

Figure 12. Pulse length as a function of charge for the Argonne gun.

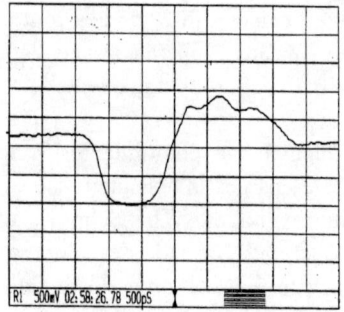

Figure 13. A 1 MV over 1 mm gap pulse, 1 ns flat top, 150 ps rise time for the BNL pulsed gun.

THE BNL/SLAC/UCLA COLLABORATION PHOTOINJECTOR

Using the expertise and resources of a number of laboratories can facilitate the development of a new photoinjector at low cost and improved performance. Such a collaboration was started a few years ago [15] and led to the construction and testing of one prototype, with three production guns nearing completion. The salient features of this gun (shown in Figure 3) are 1.6 cell, 2.856 GHz, high degree of symmetry, replaceable cathode with the joint at the low electric field rim of the fractional cell; increased aperture; simplified machining; and precision mount. The solenoid also has been extensively modified as compared to previous BNL guns. It is a single, iron enclosed solenoid with a careful conductor design. It includes fields straightners, has a high degree of symmetry, and is made to exacting tolerances.

The design and handling of the gun have been successful in obtaining a very high electric field on the cathode with a relatively short conditioning. The gun has been operated stably as high as 150 MV/m peak field on the cathode, and operation at fields of 120 to 130 MV/m is routine.

Following the gun production, a User's Experiment was started at the ATF to characterize the gun in great detail and compare its performance and many parameters with numerical simulations. Initial results of this program were reported at this workshop [16]. One of the systematic measurements is the emittance as a function of charge, shown in Figure 15. This curve was taken at a constant solenoid current of 106 amperes (the solenoid was not optimized for each charge), phase of 45 degrees, gun field of 120 MV/m, laser spot size of 2 mm diameter, laser pulse length of 10 ps, and at a final linac energy of 40 MeV. The charge could not be taken above 450 pC at the time due to a laser power limitation. The emittance is in agreement with what is expected from simulations given that the longitudinal intensity of the laser is Gaussian and not uniform. With a flat-top laser distribution one would expect an emittance of 1 mm mrad at 1 nC charge.

FUTURE DIRECTIONS

Where should we go from here? Below is a list (not prioritized) of needed developments:

- Nonlinear emittance correction, with the objective of maximizing the transverse brightness.
- Compression studies, with the objective of maximizing the longitudinal brightness.

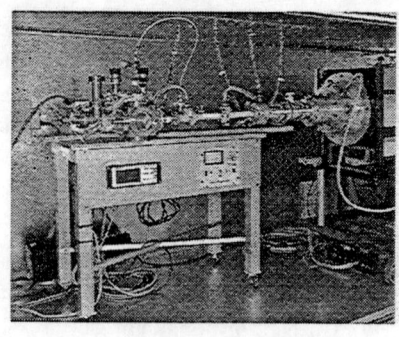
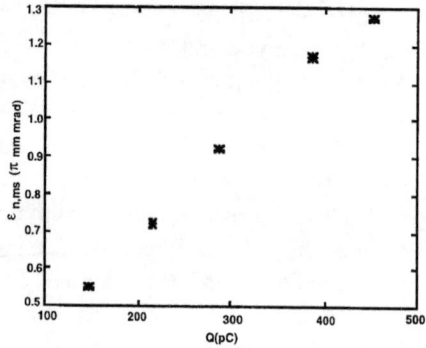

Figure 14. A photograph of the compact GV/m pulsed photoinjector at BNL.

Figure 15. Normalized rms emittance as a function of the bunch charge for the BNL/SLAC/UCLA gun.

- Systematic parametric measurement of photoinjectors and comparison to simulations.
- Measurement of the thermal emittance for various emitters to establish the lower limit of emittance.
- Tomographic analysis of the six dimensional phase space of the electron bunch.
- Research of the emission properties of the photocathode.
- Development of superconducting photoinjectors.
- Development of polarized electron photoinjectors.
- Multiple RF frequency photoinjectors for improved brightness.
- Improved lasers for photoinjectors with the objective of improvement in stability.

We need an exhaustive set of measurements to expose the dependence of emittance on the relevant gun and laser parameters, such as phase, electric fields, laser spot size, laser pulse length, solenoid setting, and charge. These measurements should be carried out for both integrated and slice emittance to gain a complete understanding of the photoinjector. Experimental work aimed at this goal is now in progress as a user experiment at the BNL ATF [16]. Such a parametric study would not be complete or meaningful unless the complete six dimensional phase space of the gun is measured (four dimensions if one can assure cylindrical symmetry) using tomographic techniques. This would be the only way to gain true understanding of the various emittance growth mechanisms and possible trade-off among parameters.

Research into non-linear emittance compensation would be most rewarding in terms of potential improvement of the beam brightness. This direction has been made possible by the advent of the slice-emittance diagnostic technique. Nonlinear compensation can be done longitudinally by shaping the longitudinal power distribution of the photocathode laser. Transverse compensation can be done by shaping the transverse distribution of the laser power. Ultimately, both longitudinal and transverse correction must be made simultaneously.

Bunch compression in and immediately following the photoinjector was recently achieved. Partial compression of the electron bunch at this early stage can be important because simulations show that the optimal bunch length under emittance compensation conditions is relatively long—enough so that energy spread due to the curvature of the linac RF wave-form and transverse wakefield emittance growth may be a problem. A reduction of the bunch length immediately following the gun will help. It is very desirable to avoid magnetic compression at the low energy of the photoinjector. Emittance compensation under such compression has to be studied.

The emission properties of the photocathode cover many areas requiring extensive R&D. The thermal emittance of photocathodes has to be measured and its dependence on the photocathode material explored. We must continue the study of various photocathodes to improve the quantum efficiency and robustness. A better understanding of the Schottky effect, the field enhancement coefficient, the work function and quantum efficiency for the common cathode material such as copper, magnesium, cesium telluride is necessary. It is possible that materials that combine a better combination of high quantum efficiency and robustness than what we have now will be discovered.

A collaboration between SLAC and CERN has started to develop galium arsenide photoinjectors [17]. The motivation is to make a high brightness source of polarized electrons for future linear colliders The status of this work is that a cesiated GaAs crystal was mounted and tested in a CLIC S-band photoinjector and taken up to an electric field of 50 MV/m. The gun conditioned very rapidly to 50 MV/m with surprisingly little dark current and some photoemission was observed. This is a very promising direction for photoinjector R&D.

REFERENCES

[1] X. Qiu, K. Batchelor, I. Ben-Zvi and X.J. Wang, Demonstration of emittance compensation through the measurement of the slice emittance of a 10 picosecond electron bunch, Phys. Rev. Let. 76 No. 20, 3723, (1996) BNL 62386

[2] B.E. Carlsten, New Photoelectric Injector Design for the LANL XUV FEL Accelerator, Nucl. Instr. And Meth. In Phys. Res. **A285**, 313 (1989)

[3] J.C. Gallardo and R.B. Palmer. Preliminary Study of Gun Emittance Correction. IEEE Journal **QE-26** 1328-1331 (1990)
[4] J.C. Gallardo, Control of non-linear space-charge emittance growth, Brookhaven National Laboratory Report BNL-522246, BNL Center for Accelerator Physics Report CAP 58 -90R
[5] C.B. McKee, P.G. O'Shea and J.M.J. Madey, Phase Space Tomography of Relativistic Electron beams, Nucl. Inst. And Meth. In Phys. Res. **A358**, 264 (1995)
[6] X. Qiu, private communication.
[7] Bruce E. Carlsten and Steven E. Russell, Phys. Rev. **E53**, Rapid Communication, R2072,(1996)
[8] X.J. Wang, X. Qiu and I. Ben-Zvi, Experimental Observation of High-Brightness Micro-Bunching in a Photocathode RF Gun To be published in Phys. Rev. E Rapid Communications
[9] I.S. Lehrman, I.A. Birnbaum, S.Z. Fixler, R.L. Heuer, S. Siddiqi, I. Ben-Zvi, K. Batchelor, J.C. Gallardo, H.G. Kirk, T. Srinivasan-Rao, Design of a High Brightness, High Duty Factor Photocathode RF Gun, Nuclear Instruments & Methods in Physics Research A318,247 (1992)
[10] T. Srinivasan-Rao, J. Fischer and T. Tsang, Journal Appl. Phys. **69**, 3291 (1991).
[11] G. Suberlucq, Development and Production of Photocathodes for the CLIC Test Facility. P. Michelato, Photocathodes for RF Photoinjectors. Both papers presented in the 1996 International FEL Conference, Rome August 26-31, 1996. To be published in Nucl. Instr. And Meth. In Phys. Res. **A**.
[12] W. Gai, M. Conde, G. Cox, R. Konecny, J. Power, P. Schoessow, J. Simpson and N. Barov, Performance of the Argonne Wakefield Accelerator Facility and Initial Experimental Results, presented in the 1996 International Linac Conference, Geneva August 26-31, 1996.
[13] R. Bossart, H. Braun, M. Dehler and J.-C. Godot, A 3 GHz Photoelectron Gun for High Beam Intensity, CLIC Note 29, PS/RF Note 95-25, CERN, Switzerland.
[14] T. Srinivasan-Rao and J. Smedley, Tabletop, Pulsed Relativistic Electron Gun with GV/m Gradient, these proceedings.
[15] The collaboration was initialized between I. Ben-Zvi, (BNL), C. Pellegrini, (UCLA) and H. Winick, (SLAC). X.-J. Wang (BNL) is leading the gun development and measurements. The production and measurement of the gun are part of the Ph.D. thesis of D. Palmer advised by R. Miller (SLAC). The engineering at BNL was done by J. Sheehan K. Batchelor (microwave) M. Woodle (gun-mechanical) and J. Skaritka (solenoid and instrumentation). K. Halbach's help in the solenoid magnetic design is gratefully acknowledged.
[16] D.T. Palmer, X.-J. Wang, R.H. Miller, I. Ben-Zvi, C. Pellegrini, J. Sheehan, J. Skaritka, H. Winick, M. Woodle and V. Yakimenko, Commissioning Results of the Next Generation Photoinjector, these proceedings.
[17] K. Aulenbacher, R. Bossart, H. Braun, J. Clendenin, J.P. Delahaye, J. Madsen, G. Mulhollan, J. Sheppard, G. Suberlucq and H. Tang, CLIC Note 303, CERN, Switzerland, and NLC Note 20, SLAC-Stanford, CA, May 1, 1996.

Microwave Sources

W. Lawson

Electrical Engineering Department and Institute for Plasma Research
University of Maryland, College Park, MD 20742, USA

Abstract

The availability of microwave sources which satisfy the requirements for peak power, total efficiency, frequency, gain, reliability, cost, phase stability, and pulse energy, duration, and shape is one of the critical limiting factors in most of the current scenarios for future linear colliders. In this paper, some of the key parameter requirements are discussed before the current efforts in microwave source development for advanced accelerator applications are summarized. Recent efforts on klystrons, traveling-wave tubes, cyclotron resonance maser devices, and free electron lasers for normal-conducting accelerators are described. A brief discussion of the technical issues that remain to be resolved is also given.

INTRODUCTION

As the design studies for the next generation of linear colliders have matured, so have the performance requirements for the RF drivers (1,2). With the exception of the 3 GHz DESY design and the S- and C-Band KEK designs for the next generation of linear colliders, the consensus is that colliders at the 1 TeV CM energy level and beyond will require RF sources at X-Band frequencies or higher. This requirement is based on practical limits to collider lengths and on the well-known frequency scaling of peak field gradients based on thresholds for breakdown, dark current capture, melting, and fatigue in accelerator structures (3). A recent estimate of a 5 TeV collider system (4), for example, estimates that over 16,000 discrete sources with a peak power of 80 MW (each) will be required at 34 GHz. This estimate assumes that a 16x pulse compression system can be realized with 80% efficiency, otherwise the peak power requirement is significantly higher.

While the details of the above design will likely change with future progress, several conclusions regarding the RF drivers are unlikely to change. In addition to the usual requirements for gain, phase stability, etc., overall system efficiency, cost, and reliability are of paramount importance due to the large number of tubes required. Tube lifetimes will have to at least match those of the SLAC 5045 klystrons (~50,000 hours). The efficiency of every subsystem will have to pushed to new levels so that the overall wall plug-to-RF output efficiency can be near

50%. This will require advances in modulator and pulse compression technologies and require the use of either permanent or superconducting magnets. Electronic efficiencies will need to approach 70% or energy recovery techniques for high voltage systems will need to be further developed.

The simultaneous attainment of all requirements on the RF system for any collider design (≥1 TeV) have yet to be realized experimentally. Nonetheless, considerable progress on high power microwave sources for future linear colliders has been made since the last workshop on advanced accelerator concepts (5). Perhaps the most significant result is the successful development of an X-Band PPM klystron at SLAC which satisfies the expected requirements for a 0.5 TeV CM electron-positron collider. This effort is summarized in the next section, along with the results of the 150 MW S-Band tubes for DESY. Given these results, the focus is now on discrete tubes that can produce more than 50 MW of power at X-Band frequencies and above and on RF sources for two-beam accelerators. A section on conventional round beam klystron development at other institutions follows the description of the SLAC klystrons. These tubes include the approach of the VLEPP, KEK and HRC relativistic klystrons, and the LBL/LLNL relativistic klystron two-beam accelerator. Various novel high perveance designs with sheet, annular, and multiple beam klystrons are then discussed. A description of recent traveling wave tube research follows. Fast wave approaches, which are reported in subsequent sections, include the low-harmonic gyroklystron and magnicon research, the high harmonic cyclotron resonance maser (CRM) devices, and the current work on annular beam free electron lasers (FELs). Auxiliary systems are touched upon briefly before the results are summarized and suggestions on the future direction of RF source work are made.

The survey of RF research activities that this paper reports is meant to give an indication of the progress in the variety of the approaches that are being explored in multi-megawatt pulsed tubes operating at frequencies above several gigahertz. It is by no means comprehensive, in that many institutions are working on similar devices and space limitations do not allow a discussion of every effort.

KLYSTRON DEVELOPMENT AT SLAC

At the Stanford Linear Accelerator Center, the on-going work on klystrons over the past few years has produced a number of excellent results including 150 MW at S-Band and 75 MW at X-Band (6-8). The S-Band tube was built for DESY and operates at 3 GHz with a 3 μs pulse width. The nominal beam voltage is 535 kV, the perveance is 1.8 μK, the efficiency is about 45%, and the gain is 57 dB. Stainless steel drift tubes were used to enhance stability. Based on the 150 MW tube and the usual frequency scaling, the output power of a scaled klystron at X-Band should have been at best 10 MW. The improvements in tube design that have enabled SLAC to exceed the 50 MW power level at 11.424 GHz include an enhancement in beam area compression and the use of tapered traveling-wave output circuits. The XL4 tube produced 75 MW at 11.424 GHz with an efficiency of 48% via the interaction with a 450 kV, 354 A (1.17 µK) beam. Both the S-Band and X-Band tubes utilized solenoidal fields furnished by electromagnets.

A comparison of the theoretical predictions and the experimental results for SLAC's latest X-Band klystron (9) is given in Table 1. The beam voltage and current for this tube are 470 kV and 193 A, respectively, resulting in a low perveance design of 0.6 µK. The pulse width was 1.2 µs. The beam interception listed in the table is given at the maximum RF output level. There are a total of 6 cavities (one input, two gain, and three buncher cavities) and the output traveling-wave section has 5 cells. Power is extracted through two TE_{01} (circular) windows. The focusing system uses a periodic permanent magnet arrangement with 38 cells. The peak axial field amplitude is about 3 kG on axis and the periodicity is about 1.25 cm. The only power required for the magnetic system is used to energize two small solenoidal coils which are needed for beam compression and matching. The peak power of 56 MW is in excellent agreement with the CONDOR simulations and the realized electronic efficiency of 60% is unmatched by any other TeV linear collider RF candidate to date.

TABLE 1. SLAC X-Band PPM Klystron Results

Item	Experiment	Simulation
Peak RF power (MW)	56	59
RF efficiency (%)	60	65
Beam voltage (kV)	459	470
Beam Current (A)	205	193
Beam interception (%)	1.0	0.0

OTHER ROUND-BEAM KLYSTRON DESIGNS

Discrete Tubes

Several groups are continuing to develop klystrons that are designed to either exceed the 100 MW level at X-Band or to operate above X-Band. The VLEPP 14 GHz klystron (10) has achieved a peak power of 100 MW with a pulse width of 250 ns, an efficiency of 40% and a gain of about 80 dB. The nominal beam voltage and current are 1 MV and 250 A (0.25 µK), respectively. The tube features PPM focusing and a gridded cathode. The maximum magnetic field is 4.5 kG and the period is 6.4 cm. Beam transmission is improved with a wide aperture tube and stability is enhanced by placing microwave-absorbing material between the 6th and 7th cavities.

The XB72k klystron at KEK (11) has undergone a number of design changes and is currently expected to produce 120 - 130 MW of power at 11.424 GHz with an efficiency of 45%. The nominal beam voltage and current are 550 kV and 490 A, (1.2 µK) respectively. The tube uses a 5 cell traveling-wave output section, which was designed at the Budker Institute for Nuclear Physics. Two TE_{11} output windows are used to extract the power. An earlier version of this tube, XB72k#2 (12), produced 95 MW of peak power with an efficiency less than 35% and a pulse width of 600 ns.

The Haimson Research Corporation has developed a Ku-Band klystron in support of high gradient accelerator research in that frequency range (13). The X7100 tube was designed to be compatible with existing hardware at the MIT laboratory where it was originally tested. The nominal beam parameters include a voltage of 560 kV, a current of 95 A (0.23 µK), and a pulse width of about 1 µs (though the RF drive width was only 150 ns). The RF circuit was about 42 cm in length and consisted of one drive cavity, two gain cavities, two prebuncher cavities, and an eleven-cell traveling-wave output structure. Higher-order modes were suppressed in the output structure via the incorporation of terminated, stainless steel HOM extraction channels that reduced HOM quality factors to below 50. Power is extracted in two WR62 waveguides. The tube produced 26 MW of output power at 52% electronic efficiency (49% at the load) and a gain of 67 dB. About 22 dB of the gain was produced by the output structure. Currently, an effort is underway to increase the efficiency level to 60% and the corresponding output power to 30 MW (14). The modified tube incorporates several new features, including a "race-track" output cavity designed to reduce surface field gradient variations.

The Relativistic Klystron Two-Beam Accelerator

There is currently a major collaboration between LLNL and LBL to utilize the klystron in a two beam approach to colliders. This effort, originally begun at LLNL is quite comprehensive and includes preliminary relativistic klystron experiments (15), re-acceleration experiments, a total system design of a two-beam accelerator drive unit, complete with cost estimate (16), and an on-going effort to build a scaled test facility for the RF drive (17). The re-acceleration experiment, for example, extracted a total peak power of 172 MW from three traveling-wave output structures which were separated by two induction cells.

The TBA approach generates a 300 ns bunched electron beam via a 2.5 MeV, 1.2 kA induction accelerator-based injector and a transverse beam modulation technique known as a choppertron (18). The beam energy will be increased to 10

MeV in an adiabatic compressor. This beam then passes through an alternating series of klystron cavities and induction (re-acceleration) units. The 3-cell traveling-wave extraction cavities remove about 6% of the beam's kinetic energy via the extraction of 360 MW of power at 11.424 GHz. Magnetic focusing is achieved via PPM quadrapoles with a peak field of 800 G and a lattice periodicity of 3.33 cm. The RF feeds are separated by about 2 meters and the overall TBA unit length is 300 meters. Approximately 50 of these units would be required to power a 1 TeV electron-positron collider. Overall wall-plug to RF feed efficiency is predicted to exceed 40%.

The group has taken a two-step approach to answering questions of long-term beam quality and stability. First they have simulated beam stability with a variety of codes. Transverse displacement due to HOM is a concern which seems to be minimized by HOM loading, Landau damping, and the "betatron node" scheme. This scheme matches the betatron wavelength to the RF cavity spacing so that the displaced beam returns to its original radial location in the next cell, thereby minimizing the effects of off-centering. A second issue is the ability to produce nearly constant power per output cavity, regardless of the cavity's relative position in the TBA unit. Simulations show that inductively detuned cells can be used effectively to equalize the output power.

The second step in proving the concept involves the construction and operation of a scaled-down prototype TBA unit (called the RTA). The RTA, having a pulse duration of 200 ns, a peak energy of 4 MeV and an overall length of 24 m, will be able to test many of the beam quality and stability issues. This facility is currently in the preliminary construction stage.

HIGH PERVEANCE KLYSTRONS

For a given power level in any tube, there are clearly only two options that can be pursued to increase the peak power. Increasing the beam voltage much beyond 500 kV requires a change in the power supply technology since modulator transformers become quite large and high efficiency becomes subsequently more difficult to achieve. The dc approach at VLEPP and the induction approach of LBL/LLNL are two viable possibilities in that direction. The other possible path is to increase beam current, but because of the strong dependence of electronic efficiency on perveance in round-beam systems, new geometries are required. Three different beam cross-sections that have been investigated for collider applications have given rise to annular-beam, sheet beam, and cluster klystrons. Each of these devices has a novel approach to high power production, but each has concerns regarding the high-quality formation of the electron beam and the stability of the circuit to HOM that need to be addressed via experiment. Annular beam klystrons have been investigated experimentally, but only in low-frequency tubes (19,20). The LANL tube produced 500 MW at 1.3 GHz in a 500 ns pulse with a 600 kV, 5 kA beam. Simulations indicate that higher efficiencies are possible and that powers in excess of 100 MW can be produced at X-Band, but experimental verification has not yet been achieved. The sheet-beam klystron is another concept that has been explored theoretically (21) but has not yet had experimental results with accelerator driver-relevant numbers. The cluster klystron approach has also received considerable design effort at BNL (22), but has not yet been tested experimentally. This approach avoids the problem of highly overmoded cavities by placing a large number of discrete low perveance tubes in a single vacuum jacket. The tubes are driven by a single dc supply, and the beam formation is handled by individual, gridded magnetron injection guns with reservoir-type dispenser cathodes.

TRAVELING WAVE TUBES

The possible application of traveling-wave tubes as advanced accelerator drivers has been under investigation at Cornell University for several years. The best results in terms of peak power were achieved a few years ago with a two-stage, rippled-wall circuit (23). A total power of 410 MW at 8.76 GHz was achieved in 70 ns pulses with an efficiency of 45%, and a gain of 37 dB. The beam voltage was 850 kV and the current was near 1 kA. The principle drawback to operation was that nearly half the power was obtained in sidebands of the drive frequency.

The researchers at Cornell have since looked at a number of novel circuit designs aimed at eliminating the sidebands and improving performance. One approach was to use a circuit with a very slow group velocity so that the signal would be transit-time isolated (24). The method eliminated the sidebands, but did not result in increased power at the drive frequency. Another approach involved the use of a coaxial extraction circuit (25). The current focus is to push the electronic efficiency above 50% by using a two stage circuit in which the phase velocity of the second stage is lower than that of the first stage (26). The transition between stages occurs near the point of maximum beam bunching. Simulations have indicated that this approach, combined with coaxial energy extraction can result in the target efficiency. As seen in klystrons, further improvements may also be possible with lower perveance beams.

FAST WAVE DEVICES

Fast wave devices are able to utilize overmoded microwave circuits and offer the possibility of high power operation at higher frequencies than slow wave tubes can achieve. A number of fast wave devices have been tested in the past few years in order to ascertain their potential as RF drivers for linear colliders. Some of the experimental parameters for a few of the more promising candidates are given in Table 2. The FEL result was obtained a few years ago at MIT on a reversed-field configuration (27). The other results are described in the following sections, where the current work on gyroklystrons, magnicons, and FELs are also summarized. Furthermore, some novel concepts which have yet to be fully characterized experimentally are also described below.

TABLE 2. Recent Fast Wave Experimental Results

Experiment Value	Magnicon	Gyroklystron	FEL
Frequency (GHz)	7.0 / 11.1	19.7	33.3
Peak RF power (MW)	30 / 14	32	61
RF efficiency (%)	35 / 10	29	27
Beam voltage (kV)	401 / 650	460	750
Beam Current (A)	210 / 225	245	300
Pulse length (ns)	700 / 100	800	30

Gyroklystrons

The University of Maryland has been running a comprehensive program to study the suitability of gyroklystrons as drivers for linear collider applications (28). The best experimental result to date achieved 32 MW at 29% efficiency in a two-cavity second harmonic gyroklystron near 20 GHz (as indicated in Table 2). An effort is currently underway to develop a tube at 17.136 GHz capable of producing about 160 MW of peak power in 1 µs with an efficiency of at least 41% and a gain of 49 dB (29). The beam voltage and current are 500 kV and 770A, respectively. The microwave circuit consists of one 8.57 GHz TE_{011} drive cavity, one 17.136 GHz TE_{021} buncher cavity and a 17.136 GHz TE_{021} output cavity. The magnetic field is solenoidal with a peak axial value of about 5 kG. The coaxial circuit utilizes microwave-absorber-lined drift tubes to suppress instabilities. The single-anode magnetron injection gun has a peak dc electric field below 100 kV/cm. The peak RF fields in the output cavity are also significantly below their klystron counterparts. Preliminary experimental testing with a TE_{011} output cavity at 8.57 GHz is expected to begin in the next few months.

High Harmonic Devices

Devices which operate at high harmonics of the cyclotron frequency substantially reduce the magnetic field requirement, increase the nominal beam and cavity dimensions, and are promising candidates for high frequency operation. The principle drawback to these devices is the tendency for efficiency to drop off rapidly with increasing harmonic number.

One device which theoretically overcomes this limitation (30) utilizes gyroharmonic conversion and is being investigated at Yale/Omega-P. The basic idea of this device is to energize a pencil beam by cyclotron autoresonance acceleration (CARA) and to then extract most of its energy into an output structure at an harmonic of the CARA frequency. For this process to be of interest for collider applications, both the conversion efficiency and the extraction efficiency must be quite high. To date, a good deal of theory and simulation results for a number of specific converter configurations have been produced. An experiment that appears to establish CARA as a viable means of preparing an energetic gyrating beam has also been performed (31). The experiment utilized a 95 kV, 25 A injected beam and added about 7.2 MW of RF power with an efficiency of up to 96%. Experiments to produce gyroharmonic power from beams energized with CARA have been in progress for nearly a year. So far, peak powers in excess of 1 MW have been observed at the 2nd and 3rd harmonic, but a quantitative comparison with theory has not yet been performed.

An idea for efficient extraction of 7th harmonic power from a CARA beam has also been developed (32). This idea is based on a near degeneracy in cylindrical waveguide between the TE_{11} mode at the fundamental frequency, and the TE_{72} mode at the 7th harmonic, i.e., both modes have nearly the same group velocity. As a result, one can show theoretically that efficient power transfer from the first to the 7th harmonic can occur in CARA, when the group velocity is low enough. This has appeal for accelerator applications since most of the injected RF power emerges from the output as RF, some at the 7th harmonic and some at the first. The latter can be recirculated to drive another stage, while the former can be sent to the load.

A second high-harmonic device is under investigation at Physical Sciences, Inc. This device utilizes a klystron-like cavity and a drift region to ballistically bunch a linearly-streaming annular beam (33). A non-adiabatic magnetic transition is subsequently used to transform most of the beam's linear momentum into perpendicular momentum. A gyrotron-like output cavity then converts much of the beam power to microwaves. The design features a nominal beam voltage and current of 500 kV and 500 A, respectively. An efficiency of 43.5% is expected at the 8th harmonic of the cyclotron frequency, resulting in an output of 109 MW at 17.136 GHz in a magnetic field of only 1.18 kG.

Magnicons

The magnicon is a scanning beam device which has demonstrated ideal-beam simulated efficiencies well in excess of 60% at accelerator-relevant frequencies and a low-frequency experimental electronic efficiency of over 80% (34). Efficiency is often reduced by velocity spread, finite beam thickness, and other non-ideal effects. These devices impart perpendicular energy to a pencil beam in a series of TM_{m10} deflection cavities (the first one driven at a sub-harmonic of the output frequency). The beam "spins up" in the applied magnetic field and energy is extracted in a final TM_{n10} output cavity (n/m is the ratio of the output to the drive frequency). The best experimental result to date at higher frequencies produced 30 MW of peak power at 7 GHz in a 700 ns pulse with an efficiency of 35% and a gain of 55 dB (35). The beam voltage and current were 401 kV and 210 A, respectively.

An X-Band magnicon was designed, built, and tested recently at NRL (36). The beam was produced by a cold-cathode gun and had a voltage and current of 650 kV and 225 A, respectively. The circuit was driven at 5.56 GHz, had three additional deflection cavities, and power was extracted at 11.12 GHz. The axial magnetic field was in the range 6.7 - 8.2 kG. Efficiencies of about 10% were achieved in 14 MW, 100 ns pulses. Operation was hampered by gain saturation that was caused by plasma loading which resulted from poor vacuum conditions and from RF breakdown in the penultimate cavity. A thermionic, high-vacuum version of the magnicon (with slightly modified beam parameters and an additional deflection cavity) is currently under construction. Predicted power and efficiency are 66 MW and 63%, respectively. A design of a fourth-harmonic magnicon, which would operate at 1/2 the applied field of the existing tube, was simulated to have and efficiency of nearly 60% with an ideal beam (37).

The LANL 17 GHz FEM

A novel free electron maser has been analyzed and is under construction at the LANL (38). This device utilizes a rippled wall structure instead of the conventional magnetic wiggler and operates with a TM_{02} mode. The beam is annular and has an average radius which coincides with the nominal location at the null in E_z. As with a conventional FEL, the beam slips in phase by 180° with respect to the EM wave in 1/2 period, but the rippled wall causes the field to reverse direction, so that efficient energy extraction can occur. The nominal beam voltage and current are 500 kV and 5 kA, respectively, and the beam radius is 6 cm. The tube is expected to produce 500 MW of power at 17 GHz with 20% efficiency.

The X-Band Coaxial PPM Ubitron

A somewhat more conventional approach is being undertaken at UC Davis, where they have simulated performance of a TE_{01} Ubitron which uses a coaxial PPM magnetic wiggler (39). The beam voltage and current are 500 kV and 1 kA, respectively. The initial wiggler period is 2.9 cm and the wiggler strength at the beam is 1.84 kG. With an untapered wiggler, efficiencies of 37% and gains of 53 dB are expected. If the wiggler period is increased to nearly 65 cm over a 132 cm length after the beam is optimally bunched, an efficiency of 56% is expected with a peak power of 280 MW and a gain of 55 dB. The tapered system is quite long and stability may be an issue. Variations on the Ubitron microwave circuit are currently under development.

AUXILIARY SYSTEMS

In addition to RF source development, substantial improvements in related systems will be necessary if colliders with energies well above 1 TeV CM are to be realized. Efficient pulse compressors with compression factors greater than 10 may be required. Gridded electron guns may be necessary to improve power supply efficiency sufficiently. It will have to be determined if power supplies can be built to take advantage of beam energy recovery. The future direction of some RF development might be substantially impacted by future developments achieved or limitations discovered in these auxiliary systems.

SUMMARY AND FUTURE DIRECTIONS

The PPM klystron result at SLAC has clearly demonstrated many of the RF tube requirements for a 0.5 TeV collider and the focus in X-Band should now shift to the issues of overall system efficiency, cost, and reliability. Because the field stresses and beam power densities are much larger in the PPM tube compared to the 5045 tube, lifetime must be evaluated. New concepts in X-Band should perhaps be explored only if they appear to offer considerable advantages in efficiency, cost, or reliability, or if the approach obviates the use of some auxiliary system (e.g. pulse compressors, discrete modulators, etc.).

At higher frequencies, the need still exists to evaluate novel RF sources in terms of peak power capability, gain, electronic efficiency, phase stability, etc. Much of this active research appears to be focused in Ku-Band and no device has yet demonstrated substantial advantages over the other approaches. While fast wave devices appear to have the best peak power capability due to their overmoded microwave circuits, the essentially untested klystron tubes with non-circular beam cross-sections may be competitive in this frequency range. The next two years promise to resolve some of the uncertainty in performance capability as several of these devices are scheduled to undergo experimental evaluation. A summary of the simulated parameters of these designs is given in Table 3.

Considerably less effort has been expended on advanced accelerator-compatible RF tube designs in K and Ka bands. This range is likely to be dominated by the fast wave devices, with high-energy free-electron lasers and high-harmonic CRM devices being particularly viable candidates. The frequency range above 40 GHz has received virtually no attention in terms of accelerator-compatible RF sources and the field appears to be wide open for new ideas and devices.

TABLE 3. A Few Impending High Power Experiments

Experiment Value	FEL (LANL)	Gyroklystron (Maryland)	Magnicon (NRL)	TWT (Cornell)
Frequency (GHz)	17.1	17.1	11.4	9
Peak RF power (MW)	500	158	66	200 / 100
RF efficiency (%)	20	41	63	52
Beam voltage (kV)	500	500	500	800 / 500
Beam Current (A)	5000	770	210	500 / 400
Pulse length (μs)	1.0	1.2	1.5	0.10 / 0.35

ACKNOWLEDGMENTS

The author would like to thank G. Caryotakis, M. Fazio, S. Gold, J. Haimson, J. Hirshfield, J. Nation, O. Nezhevenko, D. Sprehn, and G. Westenskow for their input and comments.

REFERENCES

[1] "Zeroth-Order Design Report for the Next Linear Collider" SLAC Report 474 (1995).
[2] Siemann, R. H., "Overview of Linear Collider Designs," In Proc. 1993 Particle Accelerator Conf., 532 - 539, (1993).
[3] Palmer, R. B., "Linear Collider RF: Introduction and Summary," in *Pulsed RF Sources for Linear Colliders*, AIP Conf. Proc. 337, 1 - 15 (1994).
[4] Wilson. P., "RF Power Sources for 5-15 TeV Linear Colliders," presented at the RF 96 Workshop, April 8, 1996.
[5] Danly, B.G., "RF Sources for Linear Colliders," In *Advanced Accelerator Concepts*, AIP Conf. Proc. 335, 25 - 38 (1994).
[6] Caryotakis, G., "High-Power Microwave Tubes: In the Laboratory and On-Line," *IEEE Trans. Plasma Sci.* 22, 683 - 691 (1994).
[7] Sprehn, D., R. M. Phillips, and G. Caryotakis, "Performance of a 150-MW S-Band Klystron," in *Pulsed RF Sources for Linear Colliders*, AIP Conf. Proc. 335, 43 - 49 (1994).
[8] Caryotakis, G., "X-Band Source Development at SLAC", presented at the RF 96 Workshop, April 8, 1996.
[9] Sprehn, D., "Development of a 50-MW PPM Focused X-Band Klystron," presented at the RF 96 Workshop, April 8, 1996.
[10] Dolbilov, G. V., "Development Status of Wide Aperture 14 GHz VLEPP Klystron with Distributed Suppression of Parasitic Modes," ," presented at the RF 96 Workshop, April 8, 1996.
[11] Mizuno, H., "X-Band RF System for JLC," ," presented at the RF 96 Workshop, April 8, 1996.
[12] Mizuno, H. "A prototype RF power source system for the X-Band Linear Collider," ," in *Pulsed RF Sources for Linear Colliders*, AIP Conf. Proc. 335, 89 - 93 (1994).
[13] Haimson, J., B. Mecklenburg, and B. G. Danly, "Initial Performance of a High Gain, High Efficiency 17 GHz Traveling Wave Relativistic Klystron for High Gradient Accelerator

Research," in *Pulsed RF Sources for Linear Colliders*, AIP Conf. Proc. 335, 146 - 159 (1994).
[14]   Haimson, J., and B. Mecklenburg, in these proceedings.
[15]   Westenskow, G. A. and T. Houck, "Relativistic Klystron Two-Beam Accelerator," *IEEE Trans. Plasma Sci.* 22, 750 - 755 (1994).
[16]   Yu, S., et al., "RK-TBA based power source for a 1 TeV NLC," LBID-2085/UCRL-ID-119906, LBNL Internal Document (1995).
[17]   Houck, T., F. Deadrick, G. Giordano, E. Henestroza, S. Lidia, L. Reginato, D. Vanacek, G. Westenskow, and S. Yu, "Prototype Microwave Source for a Relativistic Klystron Two-Beam Accelerator," *IEEE Trans. Plasma Sci.* 24, 938 - 946 (1996).
[18]   Haimson, J. and B. Mecklenburg, "Design and Construction of a Chopper-driven 11.4 GHz traveling-wave RF Generator," in Proc. 1989 IEEE Part. Acc. Conf., 243-245 (1989).
[19]   Haynes, W. B., M. V. Fazio, B. E. Carlston, and R. M. Stringfield, "Experimental and Theoretical Development Towards a 500 MW, One-Microsecond, L-Band Relativistic Klystron Amplifier," in *Pulsed RF Sources for Linear Colliders*, AIP Conf. Proc. 335, 36 - 42 (1994).
[20]   Serlin, V. and M. Friedman, "Development and Optimization of the Relativistic Klystron Amplifier," *IEEE Trans. Plasma Sci.* 22, 692 - 700 (1994).
[21]   Yu, D. U. L., J. S. Kim, and P. B. Wilson, "Design of a high-power sheet-beam klystron," In *Advanced Accelerator Concepts*, AIP Conf. Proc. 279, 85 - 102 (1992).
[22]   Palmer, R. B., et al., "Status of the BNL-MIT-SLAC Cluster Klystron Project," in *Pulsed RF Sources for Linear Colliders*, AIP Conf. Proc. 337, 94 - 102 (1994).
[23]   Shiffler, D., J. A. Nation, L. Schächter, J. D. Ivers, and G. S. Kerslick, "A High-Power Two-Stage, Traveling-Wave Tube Amplifier," *J. Appl. Phys.* 70, 106-113 (1991).
[24]   Kuang, E., T. J. Davis, G. Kerslick, J. A. Nation, and L. Schächter, "Transit Time Isolation of a High-Power Microwave Amplifier," *Phys. Rev. Lett.* 71, 2666 - 2669 (1993).
[25]   Naqvi, S. A., G. S. Kerslick, and J. A. Nation, "Axial Extraction of High-Power Microwaves from Relativistic Traveling-Wave Amplifiers," *Appl. Phys. Lett.* 69, 1550 - 1552 (1996).
[26]   Nation, J. A., S. A. Naqvi, G. S. Kerslick, and L. Schächter, "Progress in High Power, High Efficiency Relativistic Traveling-Wave Tube Amplifiers," in these proceedings.
[27]   Conde, M. E. and G. Bekefi, ""Amplification and Superradient Emission from a 33.3 GHz Free Electron Laser with a Reversed Guide Magnetic Field," *IEEE Trans. Plasma Sci.* 20, 240 - 246 (1992).
[28]   Granatstein, V. L. and W. Lawson, "Gyro-Amplifiers as Candidate RF Drivers for TeV Linear Colliders," *IEEE Trans. Plasma Sci.* 24, 648 - 665 (1996).
[29]   Saraph, G. P., W. Lawson, M. Castle, J. Cheng, J. P. Calame, and G. S. Nusinovich, "100-150 MW Designs of Two and Three Cavity Gyroklystron Amplifiers Operating at Fundamental and Second Harmonics in X- and Ku-Band," *IEEE Trans. Plasma Sci.* 24, 671 - 677 (1996).
[30]   LaPointe, M. A., R. B. Yoder, C. Wang, A. K. Ganguly, and J. L. Hirshfield, *Phys. Rev. Lett.* 76, 2718 - 2721 (1996).
[31]   Hirshfield, J. L., in these proceedings.
[32]   Ganguly, A. K. and J. L. Hirshfield, "Linear and Non-Linear Theory of Gyroharmonic Radiation into Modes of a Cylindrical Waveguide from Spatiotemporally Modulated Electron Beams," *Phys. Rev. E* 47, 4364 - 4380 (1993).
[33]   W. Lawson and W. W. Destler, "The Axially Modulated, Cusp-Injected, Large-Orbit Gyrotron Amplifier," IEEE Trans. Plasma Sci. 22, 895-901, (1994).
[34]   Karliner, M., E. V. Kozyrev, I. G. Makarov, O. A. Nezhevenko, G. N. Ostreiko, B. Z. Persov, and G. V. Serdobinstev, "The Magnicon - an Advanced Version of the Gyrocon," *Nucl. Instrum. Methods Phys. Res.* **A269**, 459 - 473 (1988).
[35]   Kozyrev, E. V., I. G. Makarov, O. A. Zezhevenko, B. Z. Persov, G. V. Serdobintsev, S. V. Shchelkunoff, V. V. Tarnetsky, V. P. Yakovlev, and I. A. Zapryagaev, "Performance of the High Power 7 GHz Magnicon Amplifier," *Particle Accelerators* **52**, 55-64 (1996).

[36] Gold, S. H., A. K. Kinkead, A. W. Fliflet, B. Hafizi, and W. M. Manheimer, "Initial Operation of a High-Power frequency-Doubling X-Band Magnicon Amplifier," *IEEE Trans. Plasma Sci.* **24**, 947 - 956 (1996).

[37] Fliflet, A. W. and S. H. Gold, "Mode Competition in Fourth Harmonic Magnicon Amplifiers," *IEEE Trans. Plasma Sci.* **24**, 957 - 963 (1996).

[38] Fazio, M. V., in LINAC 96 conference proceedings.

[39] Balkcum, A. J., D. B. McDermott, R. M. Phillips, A. T. Lin, and N. C. Luhmann, Jr, "250-MW X-Band TE_{01} Ubitron Using a Coaxial PPM Wiggler," *IEEE Trans. Plasma Sci.* **24**, 802 - 807 (1996).

Ultrahigh Intensity Laser for Laser Wakefield Acceleration

Gerard Mourou, John Nees, and Subrat Biswal

Center for Ultrafast Optical Science, University of Michigan
2200 Bonisteel Blvd, IST 1006, Ann Arbor, Michigan 48109-2099

The next generation of high peak power CPA laser systems will be improved in compactness, simplicity, and cost. Yb:glass is a suitable choice for the amplifier medium in such a laser system necessary for an all-optical GeV accelerator.

The concept of Chirped Pulse Amplification (CPA) has revolutionized the generation of high peak power pulses as shown in Fig. 1 and made possible the generation of optical pulses of several terawatts by table top systems that are 10^3-10^4 above what was obtained by the same size system (1, 2). The concept is very general and has been applied from the microjoule to the kilojoule using amplifying systems as diverse in size as doped fibers or building size amplifier built in national laser fusion facility like Lawrence Livermore National Laboratory, CEA, Osaka, etc... Although these recent achievements have been very impressive, there is still much room for improvement. The next generation of CPA laser systems will have vastly improved characteristics in terms of peak power near petawatt, compactness, simplicity and cost. These improvements will come from a more judicious choice of materials and pumping sources.

Right now one of the most important applications of CPA has been in the area of wakefield acceleration, a concept proposed in 1979 by Tajima and Dawson (3). Their concept has been recently verified by the important work of Clayton et al. (4), Umstadter et al. (5), Downer et al. (6), and Amiranoff et al. (7). For this particular application it is important that we get the peak intensity in the 10^{18} W/cm^2 range. In order to get the best efficiency between the laser and the accelerated electrons, it has been proposed to excite the plasma with a succession of pulses with a changing, pulse duration and interval between pulses. This concept is known as the RLPA (8) for Resonantly Driven Plasma Accelerator. Here, CPA is also ideal to produce a pulse with a prescribed time structure that will optimize the energy transfer between the laser pulse. The pulse duration is one of the most fundamental parameter. For single pulse excitation, (not RLPA), the optimum duration is between 50 - 100 fs. The focused pulse will create a plasma wave with a period about equal to the optical pulse duration. Associated with this plasma wave will be a very large field

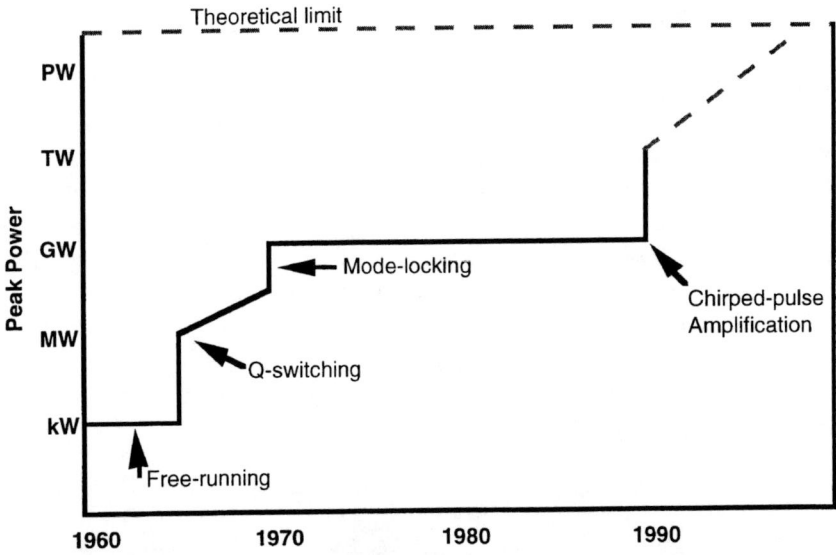

Figure 1. Evolution of laser peak power.

gradient of the order of 100 GeV/m that will form an acceleration bucket of about 50 fs in duration. The very large field existing in the acceleration buckets is at the expense of their short duration. In order to have no energy spread we need to inject the electrons at one particular point of the acceleration bucket. This requirement becomes impossible to fill with present technology like, photogun injection where the shortest pulse with a significant amount of charge (of the order of one nC) is of the order of 1ps, that is 20 times larger than the bucket duration. This apparent insurmountable difficulty was solved by Umstadter et al. (5) that proposed that the injection could be done with a femtosecond auxiliary pulse, synchronized with the main pulse that would kick the electrons by ponderomotive force into the acceleration bucket. Because of the formidable gradients, the simulations show that a femtosecond low energy spread electron pulse with a charge of several nC could be very simply produced. This concept has been dubbed LILAC for Laser Injected Laser Accelerator. The laser wakefield combined with the LILAC opened the possibility to make an all optical GeV electron accelerators.

We are going to describe some of the CPA embodiments well suited in terms of high peak and average power, compactness, simplicity and efficiency for an All Optical GeV Accelerator.

CPA Laser for Wakefield Acceleration.

The size of a CPA system scales with the saturation fluence $F_s = h\nu/\sigma$. The saturation fluence is inversely proportional to the cross section σ. The cross section of different laser materials can vary over many orders of magnitude. What defines a good energy storage materials is its ability to store optical energy before it is depleted by amplified spontaneous emission (ASE). The ASE increases exponentially with the product σNl, where σ is the stimulated emission cross section, N is the atom or molecule density, and l is the amplifier length. We see that for a given σNl, a good storage medium will have the smallest emission cross section so the largest amount of atoms or molecules can be present in the smallest volume keeping the ASE to a low level.

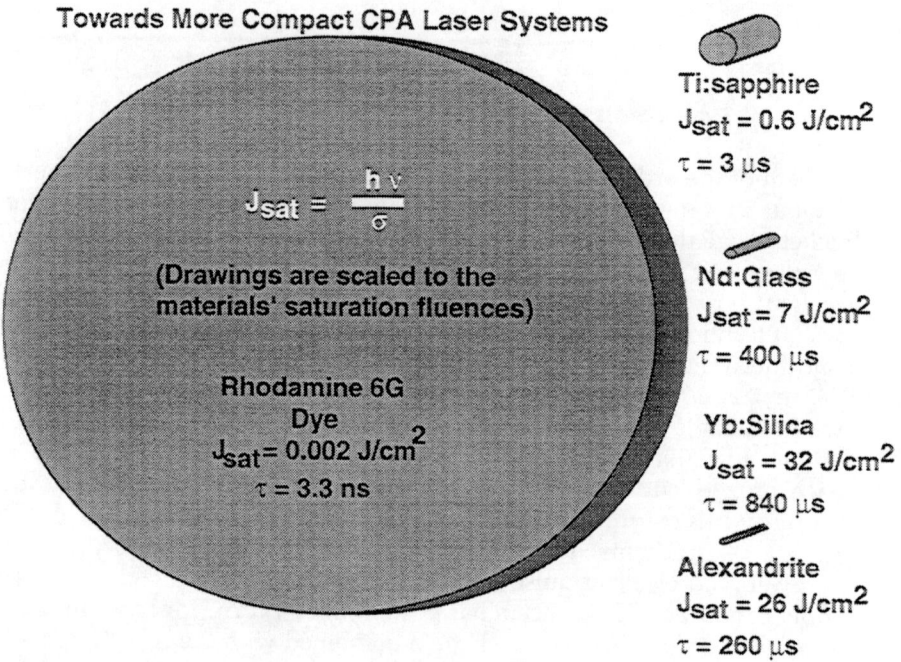

Figure 2. Relative area of various amplifier media for a given amount of energy

The lasing emission cross section can vary widely for different amplifying media from 10^{-16} cm^2 for dye to 10^{-20} cm^2 for Yb:glass, that is over 4 orders of magnitude. The fundamental problem which the CPA technique solved was the ability to extract a large amount of energy while keeping the nonlinear effects at a manageable level. This was demonstrated 10 years ago by stretching the pulse in time before the amplification.[1] After energy extraction the pulse is compressed back to its initial value. Before this technique, only dye amplifier was used in a straight amplification configuration. With CPA, low emission cross section materials, that is good energy storage solid-state materials, can be used such as Ti:sapphire, Nd:glass, Cr:LiSAF, etc... Fig. 2 shows the relative physical cross sectional area of different amplifying media for a given amount of stored energy. It clearly expresses the vast difference between the different materials. Of course all these materials have a broad gain bandwidth, necessary to amplify ultrashort pulses in the sub-picosecond domain.

Besides the good energy storage characteristics, one additional property that is particularly important is the storage time τ_S. This characteristic can also vary widely over many orders of magnitude, between the different amplifier materials. For instance, dyes have storage time in the nanoseconds, while for Ti:sapphire it is 3 µs. For Nd:glasses it is 300 µs and up to 2 ms for Yb:glass. This time imposes a very different constraint on the pump power for different materials. If we want to store one joule into an amplifier, the pump power has to be in the gigawatts for the dye, megawatt for Ti:sapphire, and kilowatt for Nd:glass and Yb:glass. The size, cost, and complexity of the pump source will here again be a very important function of the pump peak power.

Laser acceleration will need high peak power, but also high average power. Since the inception of CPA, the average power of femtosecond systems has increased by at least two orders of magnitude. Fig. 3 shows the average power for different repetition rate femtosecond CPA systems. Average power is a function of the quantum defect between the absorption and emission wavelengths and the thermal conductivity of the amplifying media. The average power can be enhanced by rotating or translating the amplifier (10), that would be made easier if the physical dimensions of the amplifier are small and not birefringent.

Over the past few year at the University of Michigan we looked at alternatives to enhance the compactness and simplicity of CPA systems by using a judicious combination of amplifier materials and pump sources. Our search has been centered on long storage time materials with low cross section (good energy storage) that could be pumped with an inexpensive pump source working in free running mode or with laser diodes. Our first choice has been the combination of Nd:glass and free running alexandrite (11). This had the advantage to use an inexpensive material, Nd:glass, with excellent storage energy characteristics due to the low cross section corresponding to a stored energy density greater than 1 J/cm^3. A free running alexandrite laser is an inexpensive tunable source of photons and can deliver of the order of 5 J at 10 Hz. Its tunability over the absorption bandwidth of Nd:glass between 700-800nm allows for the adjustment of the absorption length. Also the free running pump pulse duration time is of the order of 100 µs, which is a fraction of the Nd:glass storage time of 300 µs. The long pump pulse seriously reduces the damage threshold problem in the optical components and permits pumping of the Nd:glass with a high energy fluence on the order of 100 J/cm². Recently an alexandrite-pumped Nd:glass 10 Hz, 50 mJ system with bandwidth to support a sub-picosecond pulse was demonstrated (12). The pulse duration due to the relatively narrow gain bandwidth of Nd:silicate of 25 nm was of the order of 400 fs after compression. We have also shown that this pulse duration could be further reduced to the sub-100 fs range, by frequency doubling the compressed pulse using a scheme based on two KDP crystals, one for predelay and one for harmonic generation (13). The conversion efficiency was 75%. The pulse duration was measured to be only of the order of 100 fs. We think that the observed temporal broadening was due to nonlinear effects in the crystal due small scale self focusing.

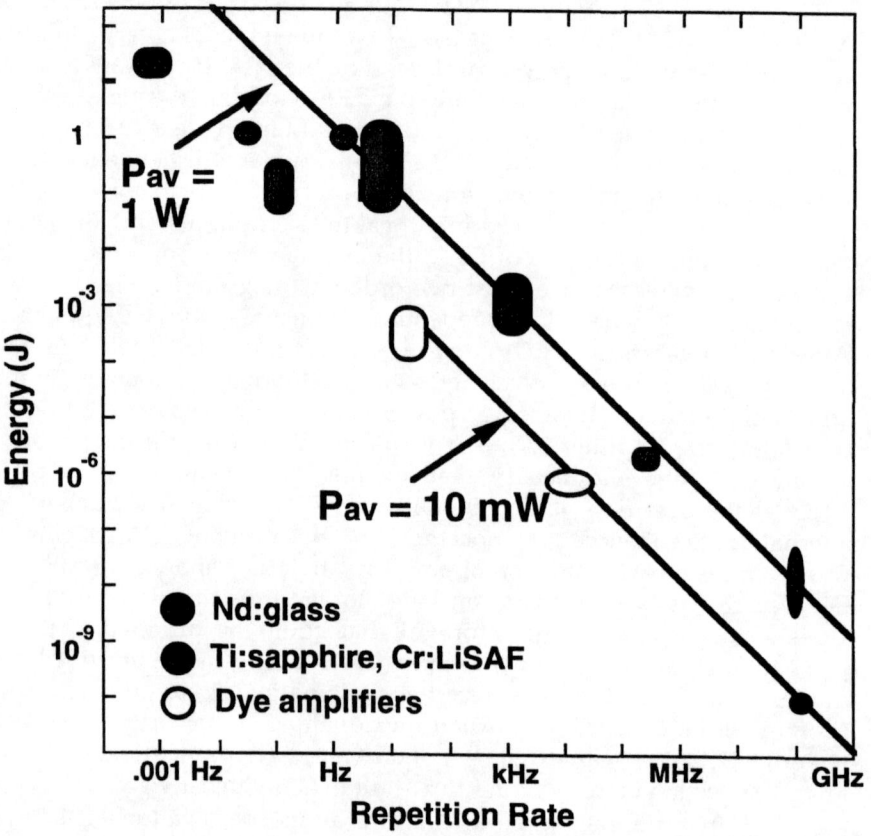

Figure 3. Average power for various high peak power laser systems.

Lately our efforts have concentrated on Yb: glass pumped with a free running flashlamp pumped Ti:sapphire and Cr:LiSAF. As we will see later, Yb:glass is ideal for diode pumping. Yb:glass has a very large bandwidth to accommodate 30 fs pulse duration (14, 15) and a very good storage efficiency of the order of 10 J/cm^3. Also, the upper level lifetime of Yb:glass of up to 2 ms, makes this material ideal for laser diode pumping. The simple two electronic energy level structure of Yb^{3+} allows for a high doping concentration of > 10^{21}/cm^3 in glass. If all the ions could be inverted and no ASE was present, the 10^{21}/cm^3 concentration corresponds to a stored energy density of 200 J/cm^3. The small quantum defect between the absorption band at 900 - 980 nm and emission at 1000 - 1030 nm is good for high average power operation. Free running Ti:sapphire and Cr:LiSAF are excellent laser diode simulators, but are also rugged, cheap and compact source of photons. For instance with a quarter inch diameter Cr:LiSAF rod, a free-running, multi-mode output of 10 J at 900 nm can be obtained.

In order to extract efficiently the energy from the amplifier, one technique is to have an input fluence F_{in} of the order of the saturation fluence of the material $F_s=h\nu/\sigma$. For superior energy storage materials, F_s is very large in the range of 5 J/cm^2 for Nd:glass to 50 J/cm^2 for Yb:glass. This saturation fluence corresponds to the bulk damage threshold value for the input stretched pulse duration of a few nanoseconds. The solution to the problem of damage threshold will be a combination of large pulse stretching, to several nanoseconds - it has been determined empirically that the damage threshold of dielectrics increases with the square root of the pulse duration - and techniques to enhance damage thresholds. Superpolishing techniques can significantly enhance the damage threshold and rely on chemical etching, ion-beam polishing, or laser conditioning (16). Of course because of their large saturation fluences, the amplifier physical cross section will be small. For a 1 J system, the physical cross section will be of the order of one millimeter so the damage conditioning has to be only done over a small area of 1 mm. Another solution to extract the energy is to use a low gain regenerative amplifier in which the population inversion is gradually depleted resulting in a peak fluence below the saturation fluence (17).

Theoretical Peak Power and Rabi Intensity.

The highest peak power can be estimated, by recognizing that the maximum extractable energy from an optical amplifier is dictated by the saturation fluence $F_s=h\nu/\sigma$. This energy, can be extracted by a pulse, limited by the gain bandwidth $\Delta\omega$ that will ultimately limit the pulse duration to approximately $1/\Delta\omega$. We end up with a very simple expression for the theoretical peak power $P_{th}=h\nu\,\Delta\omega/\sigma$. Note, that of course this is the maximum extractable power per unit area.

It is easy to recognize that this power corresponds in fact to the Rabi intensity necessary to take the excited population to the ground state in a medium. The theoretical peak powers for the most commonly used CPA materials are listed in table 1.

Table 1. Theoretical peak power per cm^2 of different laser materials.

Laser material	σ (10^{-20}cm^2)	$\Delta\lambda$ (nm)	Δt (fs)	τ_s (µs)	P_{th} (TW/cm^2)
Ti:sapphire	40	200	5	3	130
Nd:silicate	2.4	30	60	400	140
Cr:LiSAF	5	150	7	70	650
Alexandrite	1	100	10	250	2500
Yb:silicate	0.5	100	15	800	2500

Size of CPA wake-field System.

An All Optical GeV Accelerator using the RLPA and the LILAC concepts will require few terawatt peak power pulses with sub-100 fs duration. A quick look at the table 1 indicates that this peak power could be obtained by using a CPA based Yb:glass system with a 1 mm diameter. The fact the rod can be very small, opens the possibility for efficient cooling by rotating the glass rod for instance. Or by changing the rod for each shot. The advantage of glass here is that it is very cheap and can be grown to a large size. The alignment could be done for each shot using adaptive optic techniques at 10 Hz for instance.

Conclusion.

We have shown that CPA is entering a new phase where we will intentionally take advantage of the exceptional energy storage, long lifetime and laser diode pumping, of materials such as Yb:glass. Yb:glass can store more than 10 J/cm^3, with a long storage lifetime of 2 ms, and can be easily and directly pumped with free running lasers like Ti:sapphire, Cr:LiSAF, or laser diodes. The obtainable pulse duration should be of the order of 30 fs. The system will be very compact with a mm diameter rod and could be ideally suited for producing pulses at the terawatt level that will be necessary to inject and accelerate electrons to the GeV level using the LILAC and the RLPA at a 10 Hz repetition rate.

Acknowledgments

We gratefully thank M. Myers and S. Jiang of Kigre Inc. for their fruitful discussions and for providing the QX/Yb glass samples. We also thank ELIGHT Laser Systems and Polytec for providing the flashlamp-pumped Ti:sapphire laser.

This research was partially supported by the National Science Foundation through the Center for Ultrafast Optical Science under STC PHY 8920108.

References

1. Strickland, D. and Mourou, G., Opt. Commun. **56**, 219 (1985).
2. Perry, M. D. and Mourou, G., Science **264**, 917-24 (1994).
3. Tajima, T. and Dawson, J. M., Phys. Rev. Lett. **43**, 267 (1979).
4. Modena, A., *et al.* Nature **377**, 606 (1995).
5. Umstadter, D., Chen, S. Y., Maksimchuk, A., Mourou, G., and Wagner, R., Science **273**, 472 (1996).
6. Downer, M., *et al.* Bullet. Amer. Phys. Soc. **40**, 1892 (1995).
7. Amiranoff, F., *et al.* Adv. Accel. Conc., AIP Conf. Proc. 335, 612-34 (1995).
8. Umstadter, D., Esarey, E., and Kim, J. Phys. Rev. Lett. **72**, 1224 (1994).
9. Nakajima, K., et al. Phys. Rev. Lett. 74, 4428 (1995).
10. Biswal, S., Nees, J., and Mourou, G., submitted to Applied Opt.
11. Squier, J., Coe, S., Clay, K., and Mourou, G., Opt. Commun. **92**, 73 (1992).
12. Biswal, S., Coe, J. S., and Mourou, G., "High repetition rate, subpicosecond alexandrite-pumped Nd:glass amplifier system," presented at the Conference on Lasers and Electro-Optics, Baltimore, MD, May 1995.
13. Chien, C. Y., Korn, G., Coe, J. S., Squier, J., Mourou, G., and Craxton, R. S., Optics Letters **20**, 353 (1995).
14. Walton, D., Nees, J., and Mourou, G., Opt. Lett. **21**, 1061 (1996).
15. Pask, H. M., Carman, R. J., Hanna, D. C., Tropper, A. C., Mackechnie, C. J., Barber, P. R., and Dawes, J. M., IEEE JQE 1, 2-13 (1995).
16. Koechner, W., Solid-State Laser Engineering, Berlin: Springer-Verlag, 1992, ch. 11, pp. 579-89.
17. Biswal, S., Druon, F., Nishimura, A., Nees, J., and Mourou, G., submitted to the Conference on Lasers and Electro-Optics, Baltimore, MD (1997).

Guiding of sub-100 femtosecond pulses in preformed plasma channels.

S.P. Nikitin, T.R. Clark and H.M. Milchberg

Institute for Physical Science and Technology, University of Maryland at College Park
College Park, MD 20742
tel: (301) 405-4816, fax: (301) 314-9404, e-mail: milch@ipst.umd.edu

Abstract. The guiding of 80 fs pulses over a 1 cm distance in a preformed plasma waveguide has been demonstrated. The guided pulses were produced by a Ti:sapphire laser system while the plasma waveguide has been created in a neutral gas by using a separate Nd:YAG laser system. In order to achieve stable guiding these two laser systems were synchronized so that the time jitter between their pulses does not exceed 1 ns. In addition, time-resolved electron density distributions have been obtained by using a single shot interferometer.

RESULTS

Our femtosecond laser system is based on the chirped pulse amplification (CPA) scheme (1) and consists of a Ti:sapphire oscillator, an all-reflective broadband stretcher, a Ti:sapphire regenerative amplifier and a diffraction grating compressor.

The oscillator contains a 10 mm long Ti:sapphire crystal pumped by 4 Watts of an Ar+ ion laser in the 'all lines' regime. The typical output of the oscillator in the self mode-locked regime is a train of pulses shorter than 50 fs with a spectrum maximum at 780 nm and width of 25 nm FWHM. The average power of the output is about 300 mW.

Femtosecond pulses from the oscillator are stretched to ~ 200 ps in a single grating all-reflective stretcher, similar to the one described by the authors of (2). The stretcher has a dispersion of 10 ps/nm, an 80 nm bandpass and a 30% efficiency.

The stretched pulses are seeded into the regenerative amplifier, which is pumped by 35 mJ from a frequency doubled Q-switched Nd:YAG laser. Typical output of the amplifier is above 1 mJ at 10 Hz repetition rate with a spectrum width of

~20 nm FWHM. Finally, the amplified 200 ps pulses compress to ~80 fs duration FWHM in the diffraction grating compressor, which has a 50% efficiency.

The Ti:sapphire oscillator is currently located in a room separate from the regenerative amplifier and the plasma waveguide producing Nd:YAG system. The resulting distance between the regenerative amplifier and oscillator is of the order of 10 m. Though this requires more careful alignment of the stretcher it was found to improve optical isolation between the regenerative amplifier and oscillator, thus reducing the probability of leakage from the regenerative amplifier terminating the self mode-locked regime of the oscillator.

In order to measure the output pulse duration and chirp we use a frequency resolved optical gating (FROG) technique based on the polarization gating nonlinearity (5). A polarization gating FROG trace of the output pulse is shown in Fig 1. The corresponding pulse duration is 80 fs FWHM.

To inject the femtosecond pulse into the preformed plasma waveguide the output of the femtosecond system must be synchronized with a Nd:YAG laser system which is used for the waveguide production. This Nd:YAG system consists of an actively mode locked oscillator. Single oscillator pulses of 100 ps duration are amplified by a flashlamp pumped Nd:YAG regenerative amplifier/power amplifier system up to 500 mJ with a 10 Hz repetition rate. More detailed description of the Nd:YAG can be found in (3)

Typical time scales for the plasma waveguide evolution are a few hundred picoseconds (4). This fact was verified experimentally by measuring time and space resolved electron density profiles. These measurements have been made by using a folded wave interferometer in a pump-probe configuration. A synchronous probe pulse ($0.532\mu m$, 70ps width, ~$100\mu J$, ~1cm diameter) with an adjustable delay (-1 to 11ns) is directed transversely through the plasma waveguide. The probe beam picks up a phase shift as a function of vertical position in the plasma, whose overall diameter is no greater than ~200 μm for the delays used here. Most of the probe beam is not phase shifted and can be used as a phase reference in the interferometer, which consists of an optical quality BK7 glass wedge, a matched two lens imaging system, and a microscope objective, producing a net magnification of 23X.

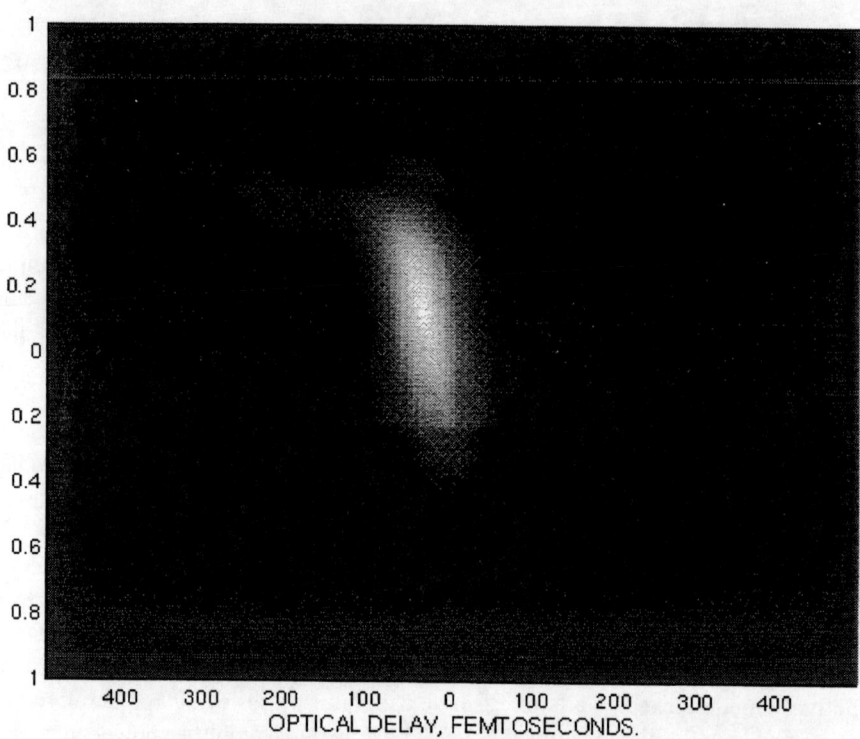

FIGURE 1. PG FROG trace of a femotosecond Ti:sapphire system. Corresponding pulse duration is 80 fs FWHM.

The high degree of axial and radial symmetry of the waveguides produced is evident in the sample interferogram of Fig. 2. The optical phase shift is extracted from the interferograms using fast Fourier transform techniques (6). Obtaining the electron density is then reduced to the well known Abel problem for a cylindrically symmetric object from which chordally integrated information is known (7). Fig. 3 shows the electron density development for a plasma produced in a background gas mixture of 230 torr Ar and 20 torr N_2O. The experimental uncertainty is due mostly to the calibration of the magnification of our lens system and is measured to be ~4%. The data confirms the general features of channel formation predicted by our calculations and suggested by our guiding experiments (3,4) and leads to the conclusion that Nd:YAG and Ti:sapphire laser systems should be synchronized with ~ 100 picosecond time accuracy.

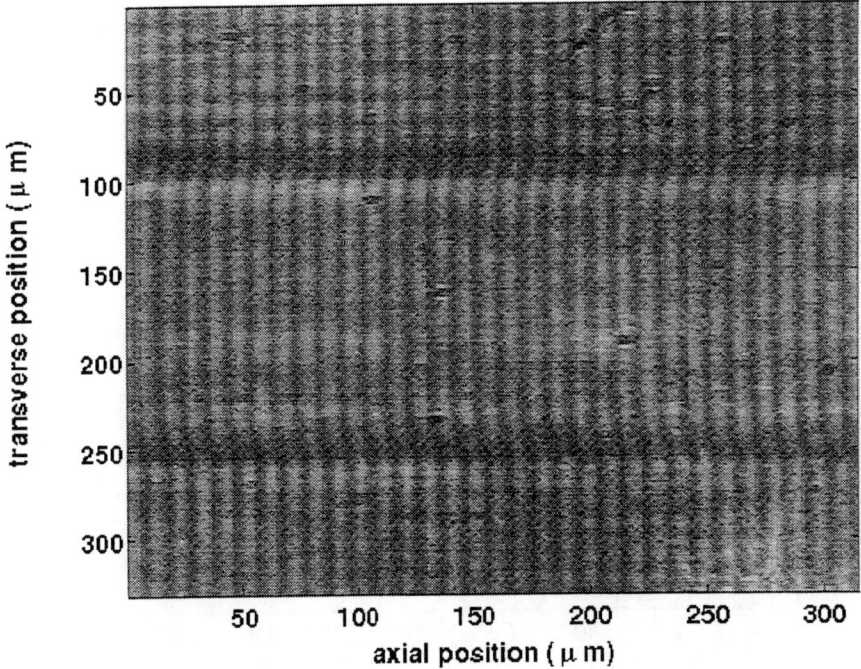

FIGURE 2. Interferogram data using 532 nm probe at a delay of 5 ns, showing the central 300 μm of a waveguide formed in a 230/20 torr Ar/N$_2$ gas mixture.

The ~100 ps synchronization has been realized by using a feedback loop which adjusts the cavity length of the Ti:sapphire oscillator so that the repetition rate, or to be more exact, the phase of the pulse train generated by this oscillator is exactly the same as one from the Nd:YAG oscillator. Fine cavity length adjustment is done by putting one of the folding mirrors on a piezo-driven translator. A voltage applied to the PZT is produced by a phase comparator which compares phases of an electrical signal produced by the Ti:sapphire pulse train on a photodiode and the frequency doubled 38 MHz reference signal from the modelocker of the Nd:YAG oscillator. Since long term cavity changes due to thermal variations may be larger than the range of a PZT, the end HR mirror is positioned on a motorized drive which allows the preliminary equalization of the cavity lengths 'manually' until the difference between them is within a PZT working range.

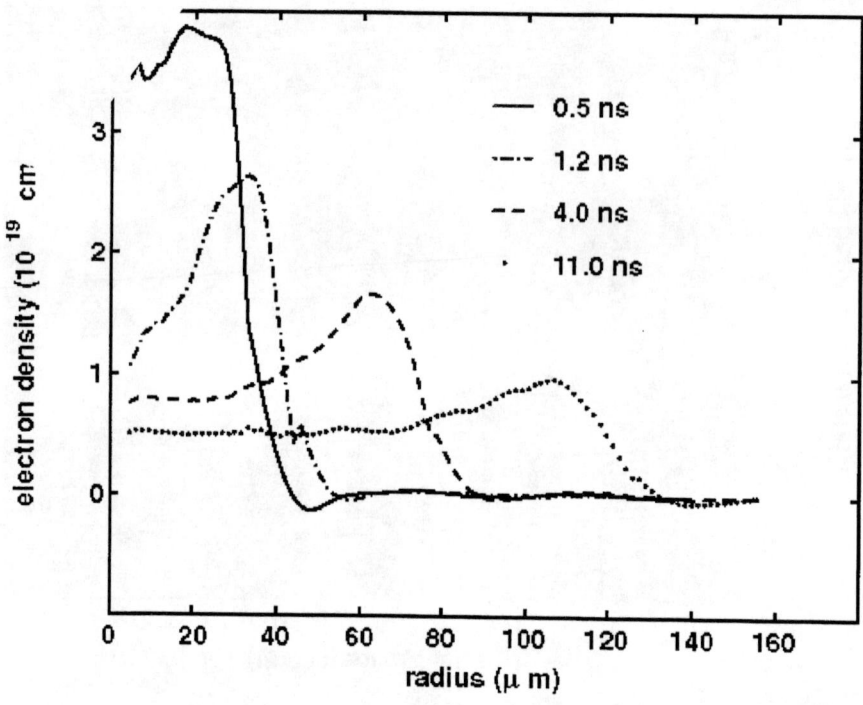

FIGURE 3. Density (10^{19} cm^{-3}) vs. Radius (μm): Electron density profiles for 230/20 torr Ar/N$_2$O gas mixture.

The guiding experiment has been set as follows. A pulse (1.064μm, 100ps width, up to 500mJ) from the Nd:YAG regenerative/ power amplifier system generates the plasma waveguide at the ~1 cm long focus of a 35° base angle axicon, with intensities of ~5 x 10^{13} W/cm^2, as described in (4). Approximately 1 ns after the waveguide forming pulse, a femtosecond pulse at 0.1 mJ energy level generated by the Ti:sapphire system has been focused into the plasma waveguide with f/25 optics. The output plane of the ~1cm long plasma waveguide is imaged to a CCD camera located outside of the chamber. A typical output mode of an 80 fs pulse guided by the plasma waveguide is shown in Fig. 4. In case the plasma waveguide formation is suppressed or not synchronized the spot size in the plane of observation is bigger than the surface of the CCD camera and is practically indistinguishable from the background.

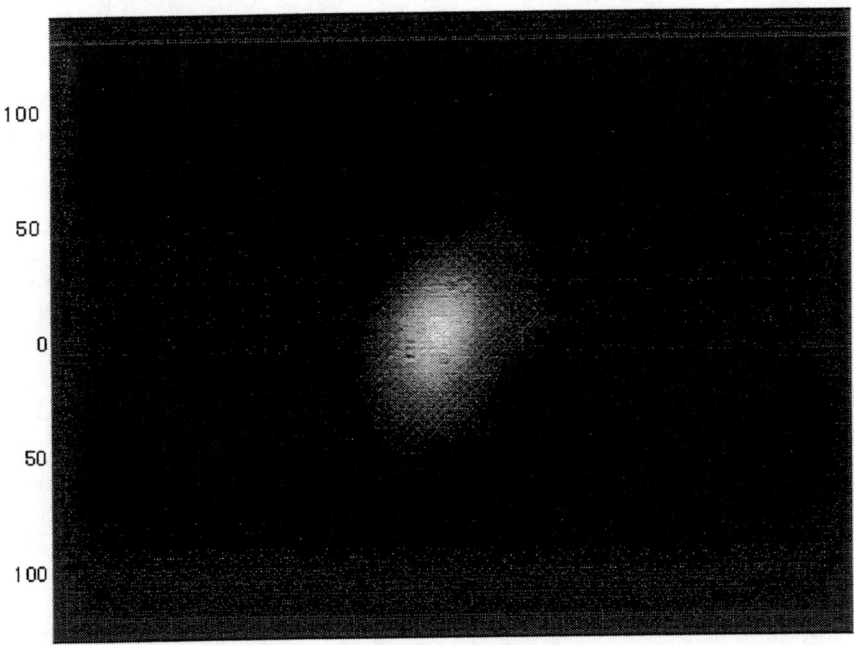

FIGURE 4. Spatial intensity profile of the femtosecond pulse at the output of the plasma waveguide

CONCLUSIONS

In conclusion, these experiments demonstrate the ability to inject and propagate an intense femtosecond pulse in a synchronously preformed plasma waveguide over several centimeters. Single shot interferometer measurements have produced high resolution electron density profiles which are important for experiments using higher intensity guided pulses. Among the other possible applications of these techniques, the use of preformed plasma waveguides in plasma electron accelerators is one of the most promising.

ACKNOWLEDGMENTS

The authors acknowledge the support of the National Science Foundation (ECS - 9224520 and PHY-9515509) and the Air Force of Scientific Research (F49620-96-10095)

REFERENCES

1. Strickland D. and Mourou G., *Opt. Commun.* **56**, 219 (1985)
2. Cheriaux G., Rousseau P., Salin F., Shambaret J.P., Walker B. and Dimauro L., *Opt. Lett.* **21**, 414 (1996)
3. Durfee III C.G., Clark T.R. and Milchberg H.M., *J. Opt. Soc. Am.* **B13**, 59 (1996)
4. Durfee III C.G, Lynch J. and Milchberg H.M., *Phys. Rev.* **E51**, 2368 (1995)
5. Kane D.J. and Trebino R., *Opt.Letters* **18**, 825 (1993)
6. Takeda M., Ina H., Kobayashi S., *J. Opt. Soc. Am. B* **72**, 156 (1981).
7. Hutchinson I.H., *Principles of plasma diagnostics*, Cambridge : Cambridge University Press, 1987.

Recent Results of Laser Wakefield Acceleration in KEK/U. Tokyo/JAERI

K. Nakajima[1,3], M. Kando[4], H. Ahn[3], H. Kotaki[3],
T. Watanabe[2], T. Ueda[2], M. Uesaka[2], H. Nakanishi[1],
A. Ogata[1], T. Kawakubo[1], and K. Tani[3]

[1] National Laboratory for High Energy Physics (KEK),
Tsukuba, Ibaraki, Japan
[2] Nuclear Engineering Research Laboratory (NERL),
The University of Tokyo, Tokai, Ibaraki, Japan
[3] Japan Atomic Energy Research Institute (JAERI),
Tokai, Ibaraki, Japan
[4] Institute for Chemical Research, Kyoto University,
Kyoto, Japan

Abstract

We report recent results of the laser wakefield acceleration (LWFA) project in KEK/U. Tokyo/JAERI. The project aims at achieving high energy particle acceleration to energies more than 1 GeV on a table-top scale owing to a channel-guided LWFA scheme by the use of 100 fs, 2 TW table-top-terawatt (T^3) laser system. We have demonstrated the self-channeling of ultrashort laser pulses with a relativistic intensity over a few cm. We have achieved synchronization of a 10 ps electron beam with a 100 fs laser pulse within a few ps in order to accelerate injected electrons firmly due to wakefields induced by laser pulses in the rate of 10 Hz. Recently we have observed laser wakefield acceleration of an electron beam, injected at 17 MeV, up to 100 MeV.

1 Introduction

Laser-driven particle accelerators have been conceived over the past decade to be the next-generation particle accelerators, promising super-high field particle acceleration and a compact size compared with conventional accelerators[1]. Among a number of laser accelerator concepts, laser wakefield accelerators have great potential for producing ultra-high-field gradients of plasma waves excited by intense ultrashort laser pulses[2]. Recently wakefield excitation of the order of $\sim$ 10 GeV/m in a plasma has been directly confirmed by the use of a compact terawatt laser system; the so called T^3 lasers[3]. In a homogeneous plasma, however, diffraction of the laser propagation limits the laser-plasma interaction distance to the extent of the vacuum Rayleigh length. This effect deducts the advantage of ultrahigh gradient acceleration from laser-driven accelerators. Therefore it is essential for laser wakefield accelerators to achieve a long distance interaction of an intense ultrashort laser pulse with an underdense plasma in order to bridge the energy gain from tens of MeV to the order of o n e G e V .

In this context we have conceived the channel-guided LWFA scheme as the second step for the laser-driven accelerator development. In the first step we demonstrated the proof-of-principle experiments of laser wakefield acceleration[2]. It will be quite important for practical application of the laser wakefield accelerator concept to be able to generate a high energy gain as it keeps the high gradient acceleration. In order to exceed the limit of acceleration length determined by diffraction, optical guiding in a plasma has been proposed as a promising way of propagating a high-power laser pulse over many Rayleigh lengths[4]. A laser beam may be guided through a plasma of which the refractive index along the optical axis is sufficiently high to compensate diffraction. It is known that there are two mechanisms which cause refractive guiding of intense laser pulses. For low focused laser intensities below the relativistic regime, a preformed plasma channel with density minima on axis guides a laser pulse in a linear propagation regime similar to that in solid optical fibers. A plasma waveguide with radially increasing density may be formed due to a shock wave created by a first pulse to guide a second pulse after some delay[5]. In relativistic regimes self-focusing and self-channeling in a homogeneous plasma have been predicted to occur above the critical power, given by $P_c^{th} = 17(\omega^2/\omega_p^2)$ GW where ω is the laser frequency and ω_p is the plasma frequency[6]. Relativistic self-focusing arises from the increase of the electron mass in a plasma due to relativistic effects because the refractive index of the plasma is peaked on the axis where the intensity has a maximum. It is believed that relativistic self-channeling is ineffective in preventing diffraction of ultrashort pulses shorter than the plasma wavelength[7]. In the case where the pulse length and width are comparable with the plasma wavelength, a strong wakefield excited by the ponderomotive force may cause channel guiding of ultrashort laser pulses. The 2D simulation shows the pulse is trapped in a pocket of the electron density depletion traveling with the laser pulse. As the pulse length is increased to be longer than the plasma wavelength, wakefields excited by the stimulated Raman scattering instability affects the focusing properties of the pulse leading to its self-modulation[8]. These effects are the third mechanism to cause the self-channeling of the laser pulse induced by wakefields. This mechanism implies that the relativistic self-focusing can occur at the lower laser power than the critical power P_c^{th}.

In this paper we present our current project of the LWFA test facility constructed at the Nuclear Engineering Research Laboratory (NERL) of The University of Tokyo. The project has been carried out under the collaboration of KEK, The University of Tokyo and JAERI, aiming at an energy gain of the order of GeV in a table-top size using a 100 fs, 2 TW T^3 laser system. The channel-guided LWFA is pursued by means of the self-channeling mechanism of intense short laser pulses in plasmas. Experimental results on propagation of ultrashort laser pulses in plasmas will be reported in a separate paper[9]. Here we report the first experimental results of laser wakefield acceleration of an externally injected electron beam synchronized with T^3 laser pulses.

2 Channel-guided LWFA

The axial and radial wakefields are calculated from the wake potential resulting from the density oscillation with the plasma frequency $\omega_p = \sqrt{4\pi e^2 n_e/m_e}$ for the ambient density n_e of plasma electrons. The maximum amplitude of the axial wakefield is achieved at the plasma wavelength $\lambda_p = \pi\sigma_z$: $(eE_z)_{max} \simeq 1.3 m_e c^2 a_0^2/\sigma_z$, where σ_z is the rms pulse length and a_0 is the normalized vector potential of the laser field given by $a_0^2 = 0.73 \times 10^{-18} I \lambda_0^2$ for the peak intensity I in units of W/cm^2, the laser wavelength $\lambda_0 = 2\pi c/\omega_0$ in units of μm, and the laser frequency ω_0.

Assuming a Gaussian beam propagation of the laser pulse at the peak power (P) in an underdense plasma ($\omega_0 \gg \omega_p$), the peak amplitude of the accelerating wakefield is

$$eE_z = \frac{\Omega_0 P}{\sqrt{\pi} m_e c^2} \left(\frac{\lambda_0}{\lambda_p}\right)\left(\frac{k_p \sigma_z}{Z_R}\right) \exp\left(-\frac{k_p^2 \sigma_z^2}{4}\right), \tag{1}$$

where Ω_0 is the vacuum resistivity (377Ω), λ_0 is the laser wavelength, $k_p = 2\pi/\lambda_p$, σ_z is the temporal $1/e$ half-width of the pulse and Z_R is the Rayleigh length, i.e. $Z_R = \pi R_0^2/\lambda_0$, where R_0 is the spot radius at the focus. Diffraction limits the laser-plasma interaction distance to $\simeq \pi Z_R$. Thus, the maximum energy gain of relativistic electrons is obtained as $\Delta W = eE_z \cdot \pi Z_R$. For the optimum plasma density, $n_e = 1/\pi r_e \sigma_z^2$, where r_e is the classical electron radius;

$$\Delta W_{max}[\text{MeV}] \simeq 1.4 P[\text{TW}]\lambda_0[\mu m]/\tau_0[\text{ps}], \qquad (2)$$

where τ_0 is the pulse duration in FWHM, $c\tau_0 = (2\ln 2)\sigma_z$. For the diffraction-limited case, the energy gain is at most 22 MeV for a 100 fs, 2 TW laser pulse.

In order to increase the energy gain, it is essential to propagate a short laser pulse in a plasma beyond the vacuum Rayleigh length limited by diffraction. Optical guiding of a Gaussian laser pulse with a focal spot radius of R_0 can be made through the plasma density channel with a parabolic electron-density profile given by $n(r) = n(0) + \Delta n r^2/R_0^2$. If the channel density depth satisfies $\Delta n = 1/(\pi r_e R_0^2)$, propagation of a laser pulse occurs with a uniform spot size (R_0). When the optical guiding can be accomplished through the plasma density channel, the acceleration distance is limited due to detuning of accelerated particles from a correct acceleration phase of plasma waves. As a phase detuning distance is limited to be $L_\phi \simeq \lambda_p(\lambda_p/\lambda_0)^2$, the maximum energy gain is given by $\Delta W = (2/\pi)eE_z L_\phi$. For the plasma density at the channel axis, $n(0) = 1/(\pi r_e \sigma_z^2)$, the energy gain is given by

$$\Delta W[\text{GeV}] \simeq 0.6 P[\text{TW}](\sigma_z/R_0)^2. \qquad (3)$$

In the channel-guided LWFA, the maximum energy gain exceeds 5.5 GeV at the dephasing length of $\sim$ 50 cm for a 100 fs, 2 TW laser pulse in the optimized plasma density channel ($n(0) = 2.4 \times 10^{17}$ W/cm^2). We can obtain the energy gain exceeding 1 GeV with propagation distance of $\sim$ 10 cm in the plasma channel.

3   LWFA Test Facility

We have developed the LWFA test facility to achieve GeV energies based on the channel-guided LWFA scheme. A schematic of the test facility is shown in Fig. 1. The T^3 laser system is installed in the adjacent room of the RF linac. The laser beam is transported through a 30 m long vacuum pipe to the pulse compressor chamber located in the linac room. Intense ultrashort laser pulses produced from the compressor chamber are focused in the acceleration chamber to induce

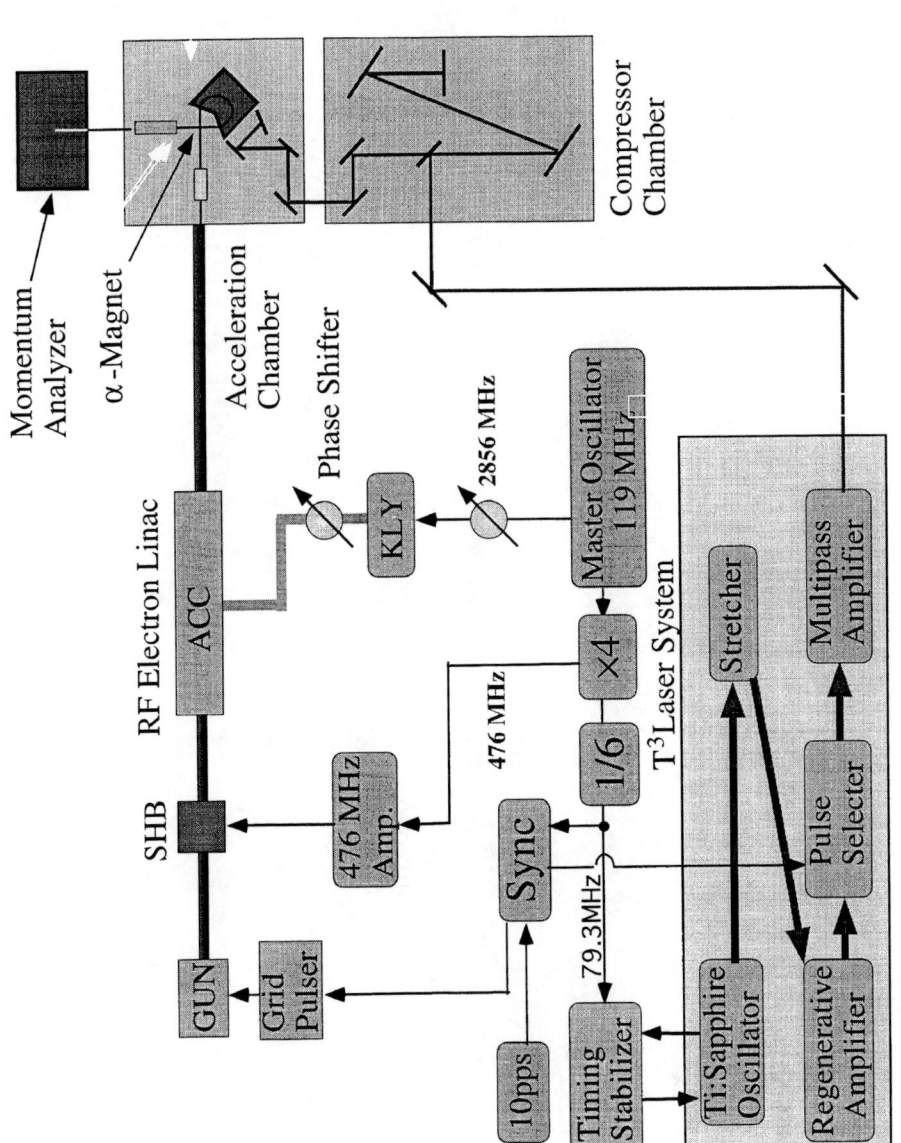

Figure 1. Schematic of the LWFA test facility.

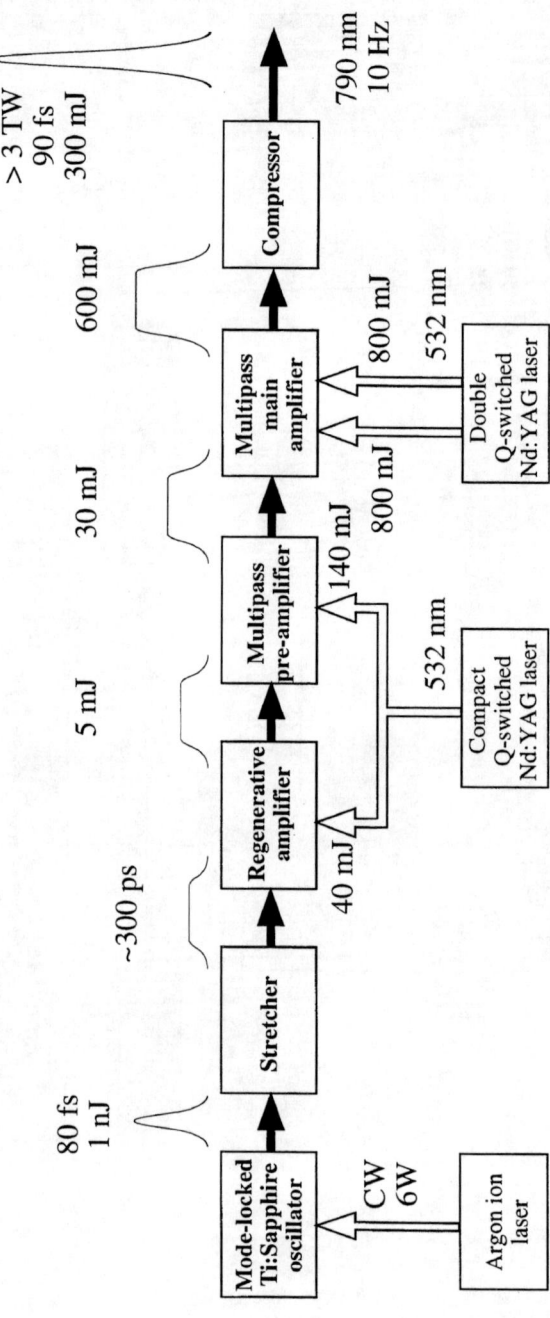

Figure 2. Schematic of the T3 laser system.

wakefields accelerating an electron beam delivered from the RF linac. The accelerated electrons are detected with the energy analyzing spectrometer. Using this facility, we plan to execute experiments on laser-plasma interaction for observing wakefields and a subpico-second X-ray pulse generation through Thomson scattering as well as laser wakefield acceleration.

4 T^3 Laser System

We have constructed the Ti:sapphire T^3 laser system on a 2 × 4 m^2 table based on the chirped-pulse amplification (CPA) technique at 790 nm. The oscillator is a mode-locked Ti:sapphire laser pumped by a cw-argon-ion laser at a power of 6 W. It produces pulses of 60 fs duration at a repetition rate of 79.33 MHz to deliver an output power of 0.5 W at 790 nm. The seed pulse from the oscillator is stretched to 320 ps in a four-pass grating arrangement with a reflective telescope. A stretched pulse is amplified to ~ 5 mJ in the Ti:sapphire regenerative amplifier (RGA) pumped at 10 Hz by 35 mJ, 6 ns pulses of a Q-switched Nd:YAG laser at 532 nm. The output from the regenerative amplifier is further amplified to > 400 mJ through a multipass pre-amplifier and a multipass main amplifier. Both faces of a Ti:sapphire crystal are pumped with two frequency-doubled pulses of 100 mJ for a pre-amplifier and 1.3 J for a main amplifier from a Q-switched Nd:YAG laser which produces a total energy of 1.6 J at 532 nm. The amplified pulse is compressed in a two-pass grating configuration to 90 fs with an energy of > 200 mJ, corresponding to a peak power of 2 TW. Since we have succeeded in producing the maximum output energy of 600 mJ at the main amplifier, we can generate the maximum peak power of 3 TW at 10 Hz with the transmission efficiency of 50% in the compressor. A schematic of the T^3 laser system is shown in Fig. 2.

5 Electron Beam Injector

It is necessary for the LWFA to inject the electron beam with an appropriate initial energy so that electrons can be trapped and accelerated by relativistic wakefields. The minimum energy is 0.2 MeV for an accelerating gradient of 18 GeV/m at a plasma density of $n_e = 2.4 \times 10^{17}$ cm^{-3}. We use the RF linac at NERL as an electron injector. This linac, driven at 2856 MHz RF frequency, produces a 17 MeV elctron beam with a bunch length of 10 ps containing ~ 1.0 nC at the repetition rate of 10 Hz. We made a good single bunch beam with energy spread of 0.17 MeV in FWHM.

An injected electron beam must spatially overlap with wakefields of which amplitudes are distributed inside the laser radial profile. Since the focusing force of the radial wakefield exists at $r < R_0/2$, the electron beam should be focused to the diameter less than a half laser spot size. An electron beam from the injector must be brought to a focus in the chamber with the rms beam size of $\sim 10\mu m$ through a beamline consisting of a permanent quadrupole doublet, triplet and a triple focusing magnet. A design of the injection beam envelope is shown in Fig. 3 for the normalized injection beam emittance of 10π mm-mrad. The RF linac and the beamline are separated with a $20\mu m$ thick titanium window from the interaction chamber to maintain ultrahigh vacuum in the electron injector. Since this window causes emittance blow-up due to multiple scattering of electrons, the collimator slit is installed at the downstream of the window to reduce the beam emittance.

An electron bunch should be synchronized to wakefields excited by a 100 fs laser pulse with the phase locked control of the mode-locked oscillator. The phase locked loop maintains synchronization of the oscillator repetition period (79.33 MHz) with every 36th RF period of the linac (2856 MHz). The synchronization and timing system of the RF linac with T^3 laser system at 10 Hz will be reported in a separate paper[10]. We have measured a timing jitter between the laser pulse and Cherenkov radiation from the electron beam with the streak camera with the time resolution of 200 fs. Synchronization between two pulses was achieved within 4 ps as shown in Fig. 4.

6 Diagnostic System

The electron beam is aligned with the laser beam by monitoring a beam image on the fluorescence screen with a CCD camera. In order to measure a timing between the laser pulse and the electron pulse, a laser light and fluorescence emitted by electrons from the screen at the focus are collected on the streak camera. The energies of accelerated electrons are measured with the magnetic spectrometer consisting of a dipole magnet and an array of 32 scintillation detectors. The pulse heights of the detectors were recorded with ADC triggered by a gate signal synchronized with the linac electron pulse. We made the energy calibration of the spectrometer, using a 17 MeV electron beam from the RF linac. The energy gain of E (MeV) is obtained from the magnet excitation voltage of V (V) and the de-

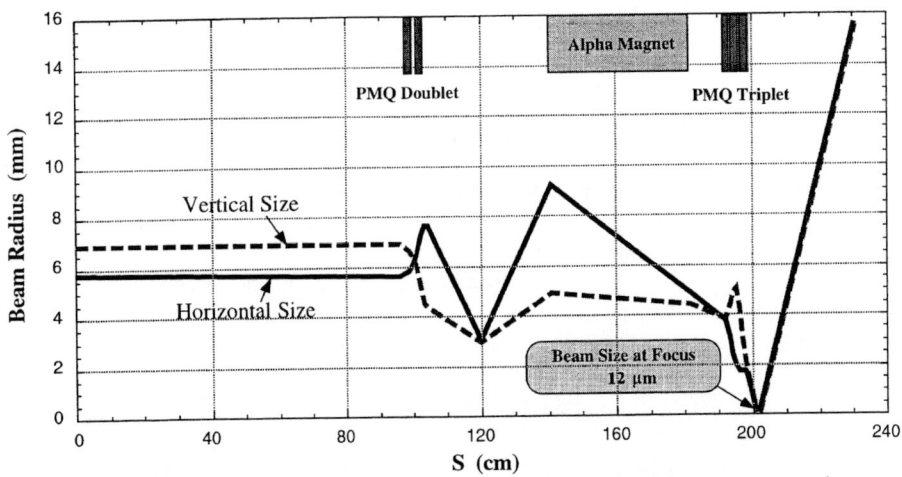

Figure 3. Envelopes of the injection beam for the emittance of 10 πmm-mrad.

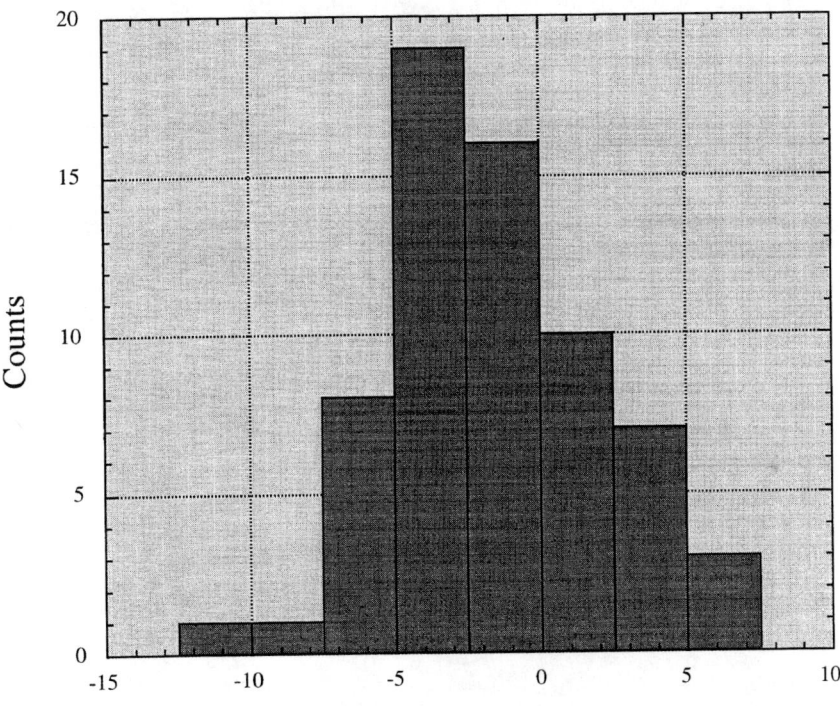

Figure 4. Measurement of timing jitter between laser and electron pulses. The rms jitter of 3.7 ps was obtained.

tector position of X (cm) by $E = 117.6V/(30.6-X) - 17$. The energy resolution is given by $\Delta E/E = 1/(30.6-X)$.

7 Experimental Results

At the beginning of the experiment the acceleration chamber was evacuated to $< 10^{-3}$ Torr with a turbo molecular pump. The laser transport pipe and the pulse compressor chamber were separately pumped out to the order of 10^{-3} Torr. Afterward the laser beam was aligned with the remote-controlled mirror so that the laser propagation axis should pass through the center of the output beam duct. Then the electron beam was roughly overlapped to the laser beam. The timing between laser and electron pulses was adjusted by changing a phase delay of the reference RF to the phase locked loop so that streak images of two pulses were overlapped. After a He gas was filled in the acceleration chamber, fine adjustment of overlapping two spots of laser and electron beams was carried out within 50 μm.

Two sets of pulse height data of the scintillator array were taken with pump laser pulses and without them as a background. A net pulse height corresponding to the number of electrons accelerated was obtained by subtracting the data without the pump from the data with the pump. The number of electrons was estimated to range from 2 to 4 per ADC count for all detectors.

As a beginning of the acceleration experiments, we made a pressure scan ranging from 4 mTorr to 760 Torr for the laser peak power of 0.5 TW. As the experimental run and its data analysis are still in progress, we can present only the preliminary results of this series of the experiments. Fig. 5 shows the distribution of the energy gain of accelerated electrons for 20 Torr. The maximum energy gain of $\sim$ 100 MeV was obtained from this figure. Fig. 6 shows the maximum energy gain as a function of the pressure. We have still obeserved accelerated electrons for a high pressure corresponding to the plasma density much higher than the resonant density. We have measured off-timing interaction of electrons with respect to the pump laser pulse in order to investigate a duration of wakefields after the pump pulse. When the electron pulse preceded the laser pulse, we have observed no acceleration of electrons. As the delay time was increased from 0 to 1 ns, we have observed accelerated electrons up to the delay of 500 ps.

We have investigated the self-trapping of plasma electrons due to wakefields without the electron beam injection as the gas pressure was increased up to 760 Torr. We have observed no electrons accelerated to energies higher than $\sim$ 1 MeV.

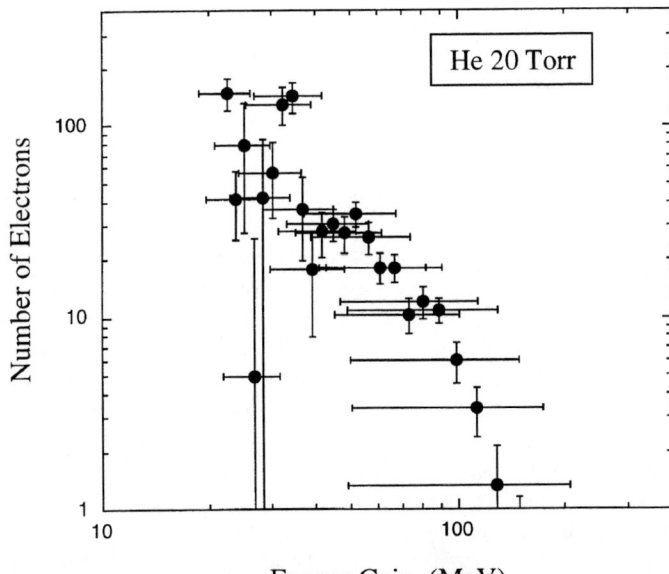

Figure 5. Distribution of the energy gain of accelerated electrons at the He pressure of 20 Torr for the laser peak power of 0.5 TW.

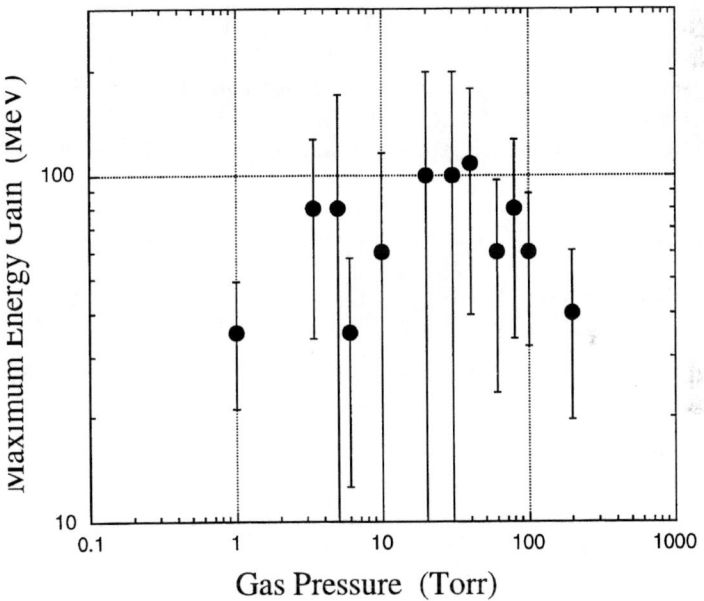

Figure 6. The maximum energy gain as a function of the He pressure of the laser peak power of 0.5 TW.

It implies that a large amplitude wakefield excitation caused by the stimulated Raman instability may be suppressed for ultrashort laser pulses of 100 fs.

8 Conclusions

The laser wakefield acceleration has been in progress under the project group of KEK, The University of Tokyo and JAERI. We made the first experiment for demonstrating electron acceleration in wakefields excited by intense ultrashort laser pulses, using a 2 TW, 100 fs T^3 laser system as a driving power source and a 17 MeV RF linac as an electron beam injector. We have successfully carried out acceleration of electrons injected from the RF linac synchronized with the laser at the repetition rate of 10 Hz. We have measured a pressure scan and a delay scan for 0.5 TW. Although the data analysis is still under way, it is found that acceleration occurs even at higher pressures of He gas than the resonant plasma density. This indicates that wakefields are excited in the nonlinear regime. We have measured the maximum energy gain that is higher than a theoretically expected value for a diffraction limited case. It implies that enhancement of the acceleration length may occur due to the self-channeling of laser pulses in a plasma. Results of the delay scan showed a considerably longer duration of wakefield than we expected.

Acceleration experiments for a higher peak power than 2 TW will be carried out to approach an energy gain of 1 GeV. Afterward we will measure the frequency and amplitude of excited plasma wakefields in detail with the frequency domain interferometry technique[3] using frequency-doubled probe pulses generated by a BBO crystal from a fraction of the incident pulse. In order to increase the intensity of accelerated electrons, we will install the RF driven photoinjector and a chicane beam line for electron pulse compression to improve a beam emittance and an efficiency of electrons captured into an acceleration phase of the wakefield.

References

[1] T. Tajima and J. M. Dawson, Phy. Rev. Lett., **43**, 267 (1979); P. Sprangle et al., Appl. Phys. Lett. **53**, 2146 (1988).

[2] K. Nakajima et al., Phys. Rev. Lett., **74**, 4428 (1995); A. Modena et al., Nature (London) **377**, 606 (1995).

[3] J.R. Marquès et al., phys. Rev. Lett., **76**, 3566 (1996); C.W. Siders et al., Phys. Rev. Lett., **76**, 3570 (1996).

[4] E. Esarey, P. Sprangle, J. Krall, A. Ting, and G. Joyce, Phys. Fluids B, **5**, 2690 (1993).

[5] C. G. Durfee III and H. M. Milchberg, Phys. Rev. Lett., **71**, 2409 (1993).

[6] G. Z. Sun, E. Ott, Y. C. Lee, and P. Guzdar, Phys. Fluids **30**, 526 (1987). D. C. Barnes, T. Kurki-Suonio, and T. Tajima, IEEE Trans. Plasma Sci. **PS-15**, 154 (1987). P. Sprangle, C. M. Tang, and E. Esarey, IEEE Trans. Plasma Sci. **PS-15**, 145 (1987).

[7] P. Sprangle, E. Esarey, and A. Ting, Phys. Rev. Lett., **64**, 2011 (1990).

[8] S. V. Bulanov, F. Pegoraro and A. M. Pukhov, Phys. Rev. Lett., **74**, 710 (1995).

[9] K. Nakajima et al., submitted to Phys. Rev. Lett.; M. Kando et al., contributed to this conference.

[10] M. Uesaka et al., contributed to this conference.

Laser Driven Acceleration in Vacuum and Gases

P. Sprangle, E. Esarey, B. Hafizi,[+] R. Hubbard, J. Krall, and A. Ting

Beam Physics Branch, Plasma Physics Division
Naval Research Laboratory, Washington, DC 20375

[+] Icarus Research, Inc., P.O. Box 30780, Bethesda, MD 20824-0780.

ABSTRACT

Several important issues pertaining to particle acceleration in vacuum and gases are discussed. The limitations of laser vacuum acceleration as they relate to electron slippage, laser diffraction, material damage, and electron aperture effects are presented. Limitations on the laser intensity and particle self-fields due to material breakdown are quantified. In addition, the reflection of the self-fields associated with the accelerated particles places a limit on the number of particles. Two configurations for the inverse Cherenkov accelerator (ICA) are considered, in which the electromagnetic driver is propagated in a waveguide that is i) lined with a dielectric material or ii) filled with a neutral gas. The acceleration gradient in the ICA is limited by tunneling and collisional ionization in the dielectric liner or gas. Ionization can lead to significant modification of the optical properties of the waveguide, altering the phase velocity and causing particle slippage, thus disrupting the acceleration process. Maximum accelerating gradients and pulse durations are presented for a 10 μm and a 1 mm wavelength driver. We show that the use of an unguided Bessel (axicon) beam can enhance the energy gain compared to a higher order Gaussian beam. The enhancement factor is $N^{1/2}$, where N is the number of lobes in the Bessel beam.

I. INTRODUCTION

Advances in laser technology (1-3) have made possible compact terawatt laser systems with high intensities ($\geq 10^{18}$ W/cm^2), modest energies (≤ 100 J), and short pulses (≤ 1 psec). The peak amplitude of the transverse electric field of a linearly polarized laser pulse is given by $E_L [TV/m] = 2.7 \times 10^{-9} I^{1/2}$ [W/cm^2], where I is the laser intensity. There are three fundamental issues that must be addressed in any type of high gradient laser driven acceleration mechanism. These are: i) laser beam guiding over extended distances, ii) phase coherence (slippage) between the accelerated particles and laser field, and iii) material breakdown arising from either the laser fields or from the self-fields of the accelerated particles.

II. LASER ACCELERATION IN VACUUM

The acceleration of electrons in vacuum (4-12) by optical fields is limited by diffraction effects and by electron slippage. The phase velocity of the optical field co-propagating in the direction of the accelerated electrons is greater than the speed of light c and given approximately by $v_{ph}/c \cong 1 + 1/(kZ_R)$, where $\omega = ck = 2\pi c/\lambda$ is the laser frequency, $Z_R = kr_0^2/2$ is the Rayleigh length and r_0 is the spot size. Consequently electrons will phase slip with respect to the field and decelerate. This will occur over a slippage distance Z_s, which for highly relativistic electrons is $\sim Z_R$.

Maxwell's equations show that under certain conditions no net energy gain is possible using optical fields in vacuum. This has become known as the Lawson-Woodward (LW) theorem (4,5). The LW theorem assumes: (i) the region of interaction is infinite, (ii) the laser fields are in vacuum with no walls or boundaries, (iii) the electron is highly relativistic along the acceleration path, (iv) no static fields are present, and (v) nonlinear effects (e.g., ponderomotive, v x B, and radiation reaction forces) are neglected. One or more of the assumptions of LW theorem must be violated in order to achieve a net energy gain. For example, finite energy gain can be achieved by introducing a background gas or dielectric boundary as in the inverse Cherenkov accelerator (ICA) (13-17). In vacuum, a nonzero energy gain can also be achieved by the introduction of boundaries which limit the interaction distance to a region ($Z_s \sim Z_R$) about the laser focus (12).

A. Acceleration Using Higher Order Gaussian Beams

Higher order Gaussian modes can provide an axial electric field component E_z on axis for electron acceleration in vacuum (5-12). Consider a radially polarized higher order Gaussian mode propagating along the positive z-direction, after having been reflected off a mirror located at some negative z-position, as shown in Fig. 1. Near the axis, the fields are given by $E_r = E_1(rr_0/r_s^2)\sin\psi$ and $E_z = E_1(2r_0/kr_s^2)\cos\psi$, where $\Psi = kz - \omega t - 2\tan^{-1}(z/Z_R)$, $r_s = r_0(1 + z^2/Z_R^2)^{1/2}$ is the laser spot size, and E_1 is a constant. The phase velocity along the axis (r = 0) and near the focal point, $|z| \leq Z_R$, is given by $v_{ph}/c \cong 1 + 1/(2\gamma_c^2)$, where $\gamma_c = \pi r_0/\sqrt{2}\lambda$ defines a critical energy. The slippage distance (9-11) Z_s, defined as the distance required for the electron to phase slip by π, is given by $\omega Z_s |v_{ph}^{-1} - v^{-1}| \cong \pi$, i.e., $Z_s \cong (\pi Z_R/2)\left(1 + \gamma_c^2/\gamma^2\right)^{-1}$, where $v/c \cong 1 - 1/2\gamma^2$. In the high energy limit ($\gamma \gg \gamma_c$), $Z_s \cong \pi Z_R/2$ and in the low energy limit ($\gamma \ll \gamma_c$), $Z_s \cong \lambda\gamma^2 \ll Z_R$.

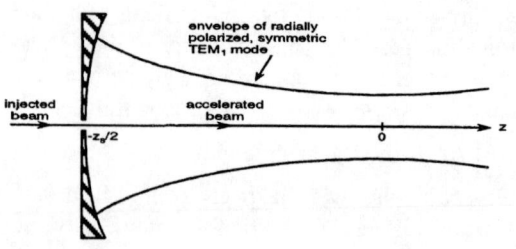

FIGURE 1. Schematic of electron acceleration in vacuum by reflecting a higher order Gaussian beam from a mirror placed near focus.

Figure 2 shows the accelerating field as a function of axial distance for high (solid curve) and low (dashed curve) energy injection. In both cases, the total area under the curve is zero, i.e., there is no net energy gain from $-\infty < z < \infty$. For a finite interaction region centered about the origin, the maximum energy gain (9-11) ΔW is given approximately by the peak amplitude of the axial electric field, $2E_0/kr_0$, multiplied by the slippage distance, Z_s, i.e.,

$$\Delta W[\text{MeV}] \cong 31 P^{1/2}[\text{TW}]\left(1 + \gamma_c^2/\gamma^2\right)^{-1}, \qquad (1)$$

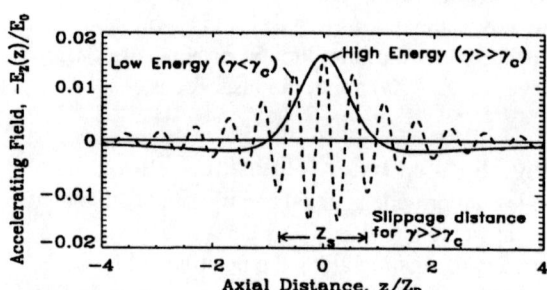

FIGURE 2. Acceleration axial field $-E_z$ plotted versus position along the z-axis for the higher order Gaussian mode. The solid curve is the high energy limit $\gamma \gg \gamma_c$, the dashed curve is the low energy limit $\gamma \ll \gamma_c$.

where $P = cE_1^2 r_0^2/32$ is the laser power. The critical energy $W_c \cong mc^2 \gamma_c$, can be written as $W_c[\text{MeV}] \cong 1.1(r_0/\lambda)$. In the high energy limit ($\gamma \gg \gamma_c$), the energy gain can be substantial, i.e., $\Delta W \cong 100$ MeV for $P = 10$ TW. In the low energy limit, however, this energy gain is reduced by the factor $(\gamma/\gamma_c)^2 \ll 1$. When damage thresholds are considered, the low energy limit $\gamma \ll \gamma_c$ appears to be the relevant regime for typical parameters of interest.

B. Limitations Due to Material Damage

In principle, limiting the interaction distance to a small region near the focus can lead to substantial energy gains. In practice, however, the energy gain can be limited by the intensity damage threshold of the reflecting surface material

(10,11). As an example, consider placing a mirror at a distance $-Z_s/2$ from the focus ($z = 0$) and using a higher order Gaussian mode, see Fig. 1. The electron energy gain is one-half the value given by Eq. (1).

The laser intensity on the surface of the mirror I_s must be less than the damage threshold limit, $I_s \cong P/(\pi r_m^2) < I_d$, where r_m is the radiation spot size on the mirror surface and I_d is the mirror damage threshold intensity. Typically (18,19), for a 1 ps laser pulse, $I_d \leq 5$ TW/cm^2. Since $Z_s \cong Z_R$ and $I_s \leq I_d$, we find that the spot size $r_0 \geq (P/\pi I_d)^{1/2}$, where we assumed $r_0 = r_m$. Increasing r_0 increases the critical energy. The condition $I_s < I_d$ implies $\gamma_c^2 > \pi P/2I_d\lambda^2$, which corresponds to a critical energy $W_c[\text{GeV}] = (6.4/\lambda[\mu m])(P[\text{TW}]/I_d[\text{TW/cm}^2])^{1/2}$. For $P = 10$ TW, $\lambda = 1$ μm, and $I_d = 5$ TW/cm^2, the injected beam energy should be greater than $W_c = 9$ GeV to be in the high energy limit. On the other hand, if the injected energy is below the critical value, the energy gain is one-half that given by Eq. (3) with $\gamma \ll \gamma_c$. For $\lambda = 1$ μm, $P = 10$ TW, $I_d = 5$ TW/cm^2 and $W_I = 1$ GeV, the energy gain is small, $\Delta W < 0.6$ MeV.

C. Limitations Due to Apertures

To limit the electron and optical beam interaction distance within an acceleration stage, the electron bunch must propagate through an aperture and the optical beam reflected, see Fig. 1. A portion of the self-fields associated with the electron bunch extends beyond the aperture and will essentially be reflected by the boundary (dielectric mirror). A limit on the maximum accelerated charge per bunch is reached when the reflected self-field energy equals the energy gain per stage or when the self-fields on the reflecting boundary become comparable to the damage threshold value (11). The self-fields for a single electron of charge q, at position $r = 0$, and $z = z_0 + vt$, are $E_{r,1} = q\gamma r(r^2 + \gamma^2(z-z_0)^2)^{-3/2}$ and $B_{\theta 1} = (v/c)E_{r,1}$. The radial field of an electron bunch of length ℓ_b, radius r_b and consisting of N electrons, evaluated at $z = 0$ (midplane of bunch) and $r > r_b \ll \gamma\ell_b$, is

$$E_r = (N/\ell_b) \int_{-\ell_b/2}^{\ell_b/2} dz_0 E_{r,1} = qN(\gamma/r)\left(r^2 + \gamma^2\ell_b^2/4\right)^{-1/2}. \quad (2)$$

The total self-field energy which intercepts the boundary is given approximately by

$$W_{self} \cong \int_a^\infty r\,dr \int_{-z_b/2}^{z_b/2} dz\, E_r^2/2$$

$$= \left(q^2 N^2/\ell_b\right)\left(2\ell n\left(\left(1+x_a^2\right)^{1/2}/x_a\right) + \pi/2 - \tan^{-1} x_a\right), \quad (3)$$

where $z_b = \ell_b + r/\gamma$, a is the aperture radius, $a \gg r_b$, and $x_a = 2a/(\gamma \ell_b)$. The self-field energy intercepting the aperture in the long electron bunch limit, $\ell_b \gg 2a/\gamma$, is

$$W_{self} \cong (q^2 N^2 / \ell_b)(2\ell n(\gamma \ell_b / 2a) + \pi / 2), \quad (4)$$

and in the short electron bunch limit, $\ell_b \ll 2a/\gamma$, is $W_{self} \cong q^2 N^2 \gamma / 2a$.

Equating the total energy gain per stage ($N\Delta W$) with the total reflected self-field energy (RW_{self}) places a limit on the amount of charge which can be accelerated, where ΔW is the single electron energy gain per stage and R is the self-field energy reflection coefficient. The reflection coefficient is approximately given by $R = |(1-\sqrt{\varepsilon})/(1+\sqrt{\varepsilon})|^2$, where ε is the dielectric constant of the reflecting surface and is taken to be independent of frequency even though the self-fields have a frequency spectrum which peaks in frequency around $\cong \pi c / \ell_b$. In the long electron bunch limit the maximum number of electrons which can be accelerated to energy γmc^2 is

$$N_{max} = (\ell_b / r_e)(\Delta\gamma / 2R)(\ell n(\ell_b \gamma / 2a) + \pi / 4)^{-1}, \quad (5)$$

and in the localized electron bunch limit is $N_{max} = (2a/r_e)(\Delta\gamma/\gamma)/R$ where $r_e = q^2/mc^2$ is the classical electron radius and $\Delta\gamma = \Delta W/mc^2$. For highly relativistic beams, the long bunch limit is appropriate, $\ell_b \gg a/\gamma$. As an example, consider a 10 GeV electron bunch ($\gamma = 2 \times 10^4$) with $\ell_b = 0.1$ μm, $\Delta\gamma = 1$, R = 0.5, and a = 5 μm. Equation (5) yields $N_{max} \cong 6 \times 10^6$, whereas the peak self-field at the edge of the aperture is $\cong 30$ GV/m, as obtained from Eq. (2) with $N = 6 \times 10^6$. This self-field level far exceeds typical damage threshold values, which are ~ 1 GV/m.

III. INVERSE CHERENKOV ACCELERATION

A. Dielectrically Lined Waveguide

In the ICA (13-16) the phase velocity can be reduced by lining the interior of a waveguide with a dielectric material. Using a dielectrically lined waveguide in an ICA i) avoids diffraction of the driving laser beam, ii) overcomes electron slippage, and iii) the acceleration is accomplished in a vacuum. The dielectric liner, however, is susceptible to electrical breakdown, i.e., ionization. The large electric fields associated with short, intense electromagnetic pulses can readily ionize the dielectric material. A small amount of ionization can result in the disruption of the acceleration process since the phase velocity of the electromagnetic fields will be altered, resulting in phase slippage (16).

The cross-sectional view of a dielectrically lined waveguide is shown in Fig. 3, where the dielectric liner occupies the region $a \leq r \leq b$, with a conducting surface at r = b. This configuration can support an axially symmetric transverse-magnetic (TM) mode which consists of a radial and an axial electric field, along with an azimuthal magnetic field. Denoting the dielectric constant of the liner by ε and the wavelength by λ, in the limit $2\pi(\varepsilon - 1)^{1/2}a/\lambda \gg 1$ the dispersion relation for the TM mode is given by (16)

$$\omega^2/c^2 - k^2 = 8/a^2 - (4\alpha_0/\varepsilon a)\tan[\alpha_0(b-a)], \qquad (6)$$

where $\alpha_0 = \omega(\varepsilon - 1)^{1/2}/c$, k is the axial wavenumber, and $\lambda = 2\pi c/\omega$. For a phase velocity equal to c the axial electric field is uniform while the radial electric field increases linearly with the radius in the central (i.e., vacuum) region. Inside the dielectric both components of the field fall-off with r, the radial field being large compared to the axial field.

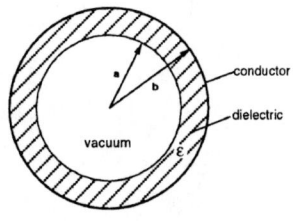

FIGURE 3. Schematic of a dielectrically-lined optical waveguide. The dielectric material with dielectric constant ε lies between the vacuum ($0 \le r \le a$) region and perfect conductor at r = b.

The large electric fields associated with intense electromagnetic pulses can ionize the dielectric. The generation of a plasma in the dielectric, due to tunneling and collisional ionization, will modify the dispersive properties of the waveguide. When the average electron density reaches a critical value, n_{crit}, the optical properties of the waveguide and the phase velocity of the electromagnetic pulse are significantly modified and electron slippage takes place, disrupting the acceleration process. The critical electron density is (16)

$$n_{crit} = \varepsilon/(\pi r_e a(b-a)). \qquad (7)$$

To avoid disruption of the acceleration process the electromagnetic pulse duration τ_L must be short compared to the critical time τ_{crit} required for the plasma density to reach the value n_{crit}, i.e.,

$$\tau_L \ll \tau_{crit} = W_m^{-1} \log_e\left[(1+\tau_0 W_m)/(1+fn_{p0}W_m/\langle S \rangle)\right], \qquad (8)$$

where $\tau_0 = n_{crit}/\langle S \rangle$, S is the photo-ionization (i.e., electron tunneling) source term, W_m is the collisional ionization source term in the electron continuity equation, n_{p0} is the initial electron density, f is a filling factor and $\langle ... \rangle$ denotes a radial average over the dielectric region (16).

We illustrate the analysis by describing the results for the dielectrically lined ICA configuration with two examples, one with a laser driver at $\lambda = 10$ μm, the other with a millimeter wave driver at $\lambda = 1$ mm (16). Figure 4 is a plot of the critical time as a function of the accelerating gradient E_z, for $\lambda = 10$ μm, as the inner radius of the waveguide is varied. The dielectric layer thickness $b - a = 50$ μm in all three cases, with $a = 30$, 50, and 100 μm for the three curves. The scaling of $\tau_{crit}(E_z)$ is primarily due to the relationship $E_z = \lambda E_0 / \pi a$ between the axial component of the field and the amplitude of the radial component of the field E_0.

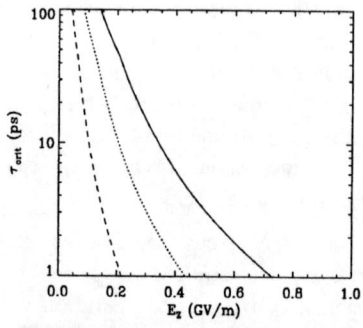 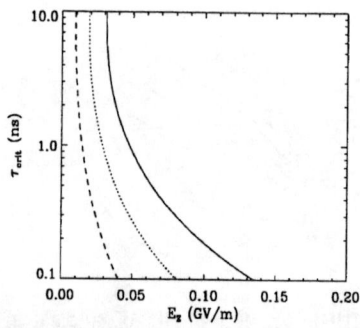

FIGURE 4. Plot of critical pulse time τ_{crit} versus the accelerating gradient E_z for a laser of wavelength $\lambda = 10$ μm, inner radius $a = 30$ μm (solid curve), 50 μm (dotted curve) and 100 μm (dashed curve, $b - a = 50$ μm. In addition, $\varepsilon = 3$, $U_i = 8$ eV, $n_{p0} = 0$, $v_m = 10^{15}$ sec^{-1} and $v_0 = 10^{11}$ sec^{-1}.

FIGURE 5. Plot of critical pulse time τ_{crit} versus the accelerating gradient E_z for a millimeter wave driver, $\lambda = 1$ mm. The inner radius $a = 3$ mm (solid curve), 5 mm (dotted curve) and 10 mm (dashed curve), where $b - a = 5$ mm. In addition, $\varepsilon = 3$, $U_i = 8$ eV, $n_{p0} = 0$, $v_m = 10^{15}$ sec^{-1} and $v_0 = 10^{11}$ sec^{-1}.

Figure 5 shows the variation of τ_{crit} with E_z, for $\lambda = 1$ mm, as the radius of the inner wall is varied. The scaling here is similar to the $\lambda = 10$ μm case shown in Figure 4, but the allowable accelerating gradient is lower. The longer pulse lengths, compared to the 10 μm wavelength example, is the principle reason for the lower accelerating gradients. For a wide range of parameters, the accelerating gradient is limited to less than ~ 1 GV/m (100 MV/m) for a driver wavelength of $\lambda = 10$ μm ($\lambda = 1$ mm) (16).

B. Gas-Filled Waveguide

Another configuration for the inverse Cherenkov accelerator consists of a gas-filled waveguide. The presence of the neutral gas allows for phase velocity $v_{ph} = c$ of the accelerating field. The critical average plasma density in the gas-filled ICA is given by (16)

$$n_{crit} = \pi\Delta\varepsilon / (r_e \lambda^2), \tag{9}$$

where $\Delta\varepsilon > 0$ is the contribution of the neutral gas to the dielectric constant. Typically $\Delta\varepsilon \sim 10^{-4}$ for a gas at atmospheric pressure and for a $\lambda = 10$ μm laser-driven gas-filled ICA the critical density is found to be $n_{crit} \cong 10^{15}$ cm^{-3}. At 10 μm the gas-filled case has a higher accelerating gradient than the typical dielectric liner case because the lower densities and collision frequencies result in lower collisional ionization rates and hence less plasma (16).

C. Energy Gain in ICA Using Bessel (Axicon) Beams

The ICA could also be driven by a first order Bessel (axicon) beam (14,17) which is axially-symmetric, radially-polarized and has an axial field peaked along the z-axis. Nonideal Bessel beams (finite in transverse extent) can be formed using axicon mirrors (14). Both the nonideal Bessel beam and the higher-order Gaussian beam diffract, limiting the acceleration distance. For a fixed total optical beam power, however, the energy gain in an ICA can be substantially higher when driven by a Bessel beam as opposed to a higher-order Gaussian beam.

A solution of the wave equation in the paraxial approximation for a radially-polarized field E_r propagating in a medium with linear refractive index n_0 is

$$E_r = E_0 J_1(k_\perp r) \exp(i(k - \Delta k)z - i\omega t) + c.c., \tag{10}$$

where J_1 is the Bessel function of the first kind of order unity, $\Delta k = k_\perp^2 / 2k$, $k_\perp$ is the transverse wavenumber, $k = \omega/v$, $v = c/n_0$, $\omega = 2\pi c/\lambda$ is the frequency, λ is the vacuum wavelength, and E_0 is the radial field amplitude. The ideal Bessel field in Eq. (10) (infinite in transverse extent) is nondiffracting (20,21) in the sense that the transverse profile remains constant. The power, however, contained within an ideal Bessel beam is infinite. Associated with the radially-polarized electric field in Eq. (10) is the axial electric field $E_z = ik_\perp (E_0 / (k - \Delta k)) J_0(k_\perp r) \exp(i(k - \Delta k)z - i\omega t) + c.c.$ The axial accelerating field is peaked along the z-axis and has axial phase velocity $v_{ph} = \omega/(k - \Delta k) \cong v(1 + (k_\perp v/\omega)^2 / 2)$, which can be less than c for $n_0 > 1 + (k_\perp c/\omega)/2$, where it is assumed that $k_\perp c/\omega \ll 1$.

The ideal Bessel beam consists of an infinite number of rings (lobes) extending radially to infinity and having a radial width of $r_0 \cong \pi/k_\perp$. Since the asymptotic form ($k_\perp r \gg 1$) for the Bessel function is $J_1(k_\perp r) \sim (\pi k_\perp r)^{-1/2} \cos(k_\perp r - 3\pi/4)$. Then the total power contained in a nonideal Bessel beam of a finite radial extent R_{max} is $P \cong NP_0$, where $N = R_{max}/r_0$ is the number of rings and P_0 is the power in each ring. In principle, the number of rings can be large, $N \gg 1$. A nonideal Bessel beam consisting of N rings diffracts away sequentially starting with the outermost ring (21). The outermost ring diffracts after a distance $\sim \pi r_0^2 / \lambda$, the next ring diffracts after a distance $\sim \pi r_0^2 / \lambda$ and so on. Hence, the maximum propagation distance of a nonideal Bessel beam consisting of N rings of width r_0 is (21)

$$L_{max} \cong N z_{R0}, \qquad (11)$$

where $Z_{R0} = \pi r_0^2 / \lambda$ is the Rayleigh length associated with the individual rings, assuming $n_0 \cong 1$.

The maximum energy gain in the ICA driven by a nonideal Bessel beam is $W_{max} = -q E_{z0} L_{max}$, assuming that the axial phase velocity is matched to the electron velocity, where $E_{z0} = (k_\perp/k) E_0$ is the axial accelerating field along the z-axis. The radial field amplitude in terms of the power within a ring is $E_0 \cong (2\pi/r_0)(P_0/cn_0)^{1/2}$. The maximum energy gain, in terms of the total optical power is

$$W_{max}(MeV) = 1.7 N^{1/2} [P(GW)]^{1/2}, \qquad (12)$$

for an ICA driven by a nonideal Bessel beam. If a higher-order Gaussian optical beam, of the same total power P, were used instead of the nonideal Bessel, the maximum energy gain would be $W_{max}(MeV) = 2.3 [P(GW)]^{1/2}$. That is, the energy gain in the ICA is $\sim N^{1/2}$ times greater for a nonideal Bessel beam as compared to a higher-order Gaussian beam of the same total power.

IV. SUMMARY

In this paper we have discussed some of the key issues that arise in the context of laser driven acceleration in vacuum and gases. Two configurations for the inverse Cherenkov accelerator (ICA) have been considered, in which the electromagnetic driver is propagated in a waveguide that is i) lined with a dielectric material or ii) filled with a neutral gas. The acceleration gradient in the ICA was shown to be limited by tunneling and collisional ionization in the dielectric liner or gas. Partial ionization of the dielectric liner or gas leads to significant modifications of the optical properties of the waveguide, altering the phase velocity of the accelerating field and causing particle slippage, thus disrupting the acceleration process. Maximum accelerating gradients and pulse durations have been presented for a 10 μm and a 1 mm wavelength driver. We have also shown that the use of an unguided Bessel (axicon) beam can enhance the energy gain compared to a higher order Gaussian beam. The enhancement factor is $N^{1/2}$ where N is the number of lobes in the Bessel beam.

ACKNOWLEDGMENTS

This work was supported by DOE and ONR.

REFERENCES

1. Maine, P. Strickland, D., Bado, P., Pessot, M., and Mourou, G., *IEEE J. Quantum Electron.* **24**, 398 (1988); Mourou, G. and Umstadter, D. *Phys. Fluids B* **4**, 2315 (1992).
2. Perry, M.D., and Mourou, G., *Science* **264**, 917 (1994).
3. See, e.g., *Advanced Accelerator Concepts,* edited by Schoessow, P. *AIP Conf. Proc.* **335** (Amer. Inst. Phys., NY, 1995); *Advanced Accelerator Concepts,* edited by Wurtele, J. *AIP Conf. Proc. No. 279* (AIP, New York, 1993).
4. Lawson, J.D. *Rutherford Laboratory Report No. RL-75-043* (1975); *IEEE Trans. Nucl. Sci.* **NS-26**, 4217 (1979); Woodward, P. *J. IEE* **93**, 1554 (1947).
5. Palmer, R.B., in *Frontiers of Particle Beams*, Lecture Notes in *Physics* **296**, edited by Month, M. and Turner. S, (Springer-Verlag, Berlin, 1988), p. 607; *Part. Accel.* **11**, 81 (1980).
6. Edighoffer, J.A., and Pantell, R.H., *J. Appl. Phys.* **50,** 6120 (1979).
7. Bochove, E.J., Moore, G.T., and Scully, M.O., *Phys. Rev. A* **46**, 6640 (1992); Scully, M.O. and Zubairy, M.S., *Phys. Rev. A* **44**, 2656 (1991).
8. Steinhauer L.C. and Kimura, W.D., *J. Appl. Phys.* **72, (1992).**
9. Sprangle, P., Esarey, E., Krall, J., and A. Ting, "Vacuum Laser Acceleration," *Optics Communications* **124**, 69-73, (1996).
10. Esarey, E., Sprangle, P., and Krall, J., *Phys. Rev. E.* **52**, 5443-5453 (1995).
11. Sprangle, P., Esarey, E., Krall, J., "Laser Driven Electron Acceleration in Vacuum, Gases and Plasmas," *Phys. Plasmas* **3**, (5) 2183-2190 (1996).
12. Huang, Y.C., Zheng, D., Tulloch, W.M., and Byer, R.L., *Appl. Phys. Lett.,* **68**, (6) 753-755 (1996).
13. Fontana, J.R. and Pantell, R.H., *J. Appl. Phys.* **54** 4285 (1993).
14. Kimura, W.D., Kim, G.H., Romea, R.D., Steinhauer, L.C., Pogorelsky, I.V., Kusche, K.P., Fernow, R.C., Wang, X., and Liu, Y., *Phys. Rev. Lett*, **74**, 546 (1995).
15. Sprangle, P. Esarey, E. And Krall, J., "Self-Guiding and Stability of Intense Beams in Gases Undergoing Ionization," *Phys. Rev. E*, **54** (4) Oct (1996).
16. Sprangle, P., Hafizi, B., and Hubbard, R., "Ionization Effects in Inverse Cherenkov Accelerators," submitted to *Phys. Rev. E* (1996), also in these proceedings.
17. Hafizi, B., Esarey, E., and Sprangle, P., "Laser-Driven Acceleration with Bessel Beams," submitted to *Phys. Rev. E*; also in these proceedings.
18. Du, D., Liu, X., Korn, G., Squier, J., and Mourou, G., *Appl. Phys. Lett.*, **64**, 3071 (1994).
19. Stuart, B.C., Feit, M.D., Rubenchik, A.M., Shore, B.W., and Perry, M.D., *Phys. Rev. Lett*, **74**, 2248 (1995).
20. Durnin, J., Miceli, J.J., and Eberly, J.H., *Phys. Rev. Lett.* **58**, 1499 (1987); Durnin, J., *J. Opt. Soc. Am. A* **4**, 651 (1987).
21. Sprangle, P., and Hafizi, B., *Phys. Rev. Lett.* **66**, 837 (1991); Hafizi, B., and Sprangle, P., *J. Opt. Soc. Am. A* **8**, 705 (1991).

Ultrashort-pulse relativistic electron gun/accelerator

E. Dodd, J. K. Kim and D. Umstadter
Center for Ultrafast Optical Science
University of Michigan, Ann Arbor, MI 48109

Abstract

Laser driven plasma waves have up to now been considered exclusively as second stage accelerators. Conventional linacs are used in this case as the first stage of acceleration to inject MeV electrons into the plasma. This paper shows it to be advantageous to instead use laser wake fields in the first stage for greater simplicity and better emittance. The concept presented makes this possible with all-optical generation and acceleration of electrons. It is tested using two dimensional particle-in-cell simulations.

1 INTRODUCTION

In a recent publication [1] we proposed an electron injection scheme for plasma-wave-based accelerators. These types of accelerators are possible because of CPA based lasers [2], which provide the necessary power level, and pulse length. Laser-wakefield plasma waves have a very short wavelength compared to conventional electron beam bunch lengths. Therefore a novel method of injection is called for to inject beams into plasma waves. The actual scheme uses two laser pulses, one for a pump and the other for injecting electrons. LILAC, or Laser Injected Laser ACcelerator, is purely optical, using no external RF electron source. New computer simulation work on LILAC is presented here.

Besides the short bunch lengths, there exist other important characteristics of LILAC. First, the solid state lasers are compact and w i l l fit on a table top, so by purely optical injection this electron gun will also be a table top device. Besides removing the need for an externally triggered RF electron source to make the electron bunch, it also is inherently synchronized with femtosecond accuracy to the laser pulses. The third result of all optical injection is cost. The same laser creates both pulses and no money is spent on a RF injector. LILAC's advantages are not limited to cost and simplicity, the bunch quality created is better than currently available by other means. Both emittance and bunch length are smaller than conventional electron guns, as will be demonstrated in the paper.

2 METHODS OF INJECTION

As mentioned previously, LILAC uses two laser pulses to create an ultrashort electron bunch. The first pulse acts as the pump used to create the wave for acceleration. The second, or injection pulse, intersects the wave and alters the electron's motion in such a way as to cause some of them to become trapped and accelerated. In the particular method analyzed here, the ponderomotive force of the injection pulse gives an impulse to electrons in the background so they may be injected. The particles with a large enough velocity in the direction parallel to the pump pulses' propagation fall into the wave's potential and then are trapped. These electrons form the desired beam, without the necessity for an external source.

The method detailed so far is general. In fact, several geometries have been considered. The first is the orthogonal orientation analyzed in the previous paper [1], where the transverse drift of electrons out of the injection pulse causes trapping, Fig. 1a. The most obvious change is to the orientation of the two pulses, with the injection pulse parallel to the pump pulse, Fig. 1b, and its longitudinal ponderomotive force acting on the electrons [3]. This method is currently under study. One can, actually, envision orienting the pulses at any angle between the two extremes

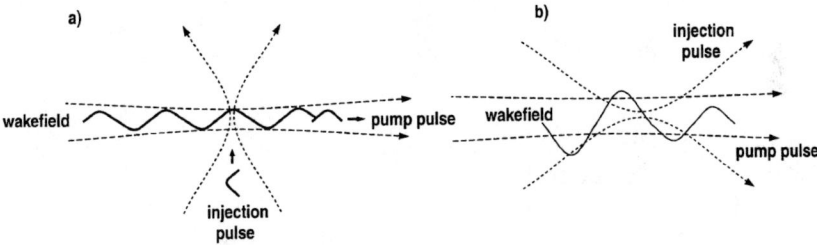

Figure 1: a) Schematic of the transverse LILAC accelerator concept. b) Schematic diagram of the colinear LILAC. Please note that in b) only the contours of intensity are shown.

talked about. In this paper we will consider the action of the injection pulse to give an impulse to the electrons. At large amplitudes, electrons quivering in the wake have velocities larger than needed for injection. However, these velocities are $\pi/2$ rad out of phase so will not inject until the wave breaking limit is reached. The injection pulse can push electrons into the proper phase for trapping. In the first model, dephasing was not dealt with for simplicity. Another aspect ignored was the fact that the injection pulse creates its own wake, and will modify that of the pump pulse. The motion of electrons in the injection pulses' wake, and alterations to the pump's wake can cause injection. Both cases will cause a modification to the seperatrix of the wake field, capturing and accelerating electrons. The commonality of all versions, is that we are using a laser pulse's ponderomotive force to affect electrons in the wake, to enhance trapping in a small spatial region. The injection pulse essentially acts as a controllable switch turning on injection at the desired time. Photoionization can also act as a switch for injection [1], if the appearance intensity is between the intensities of the pump and injection pulses. If electrons are produced at the proper phase in the wave from this effect, they will also be trapped and accelerated.

One last aspect that should be addressed is more numerical in nature. The simulation uses very sharp boundaries between vacuum and plasma which will create strong fields at this edge. These f i elds will oscillate out of phase with the plasma wave and may cause injection. This effect needs to be avoided in simulation so as not to be confused with ponderomotive injection. Though if we could create very sharp boundaries, this would be a physical process.

3 MODEL

In our previous work [1] we approached this problem in a number of ways. First we developed a simple analytic model in order determine the minimum intensity of the injection pulse. The authors were only able to achieve a closed form for an approximate and idealized case. Nevertheless, this model has the benefit of defining the order of magnitude of the problem and providing useful information about laser injection in general, as discussed in Sec. 2 . To fully develop LILAC, we turned to numerical methods for a solution. The simulations will be addressed in a later section.

The wake field of the pump laser moving through the plasma defines an electric potential that can be used to accelerated electrons and is the basis for all laser plasma based accelerator concepts [4]. The problem is that the potential is moving at near the speed of light, and the electrons must start with a velocity in order to become trapped and accelerated. With an impulsive kick the electrons move into the seperatrix defined by the wake field's potential well, and then they may interact with the wave and draw energy from it. Our idea uses the ponderomotive force of the injection pulse to give an electron the necessary velocity. We start by calculating the imparted drift velocity,

$$\Delta(\frac{p_z}{m_e c}) = \frac{b_0^2}{(1+b_0^2/2)^{1/2}} \sqrt{\frac{\pi}{8}} \exp(-1/2). \tag{1}$$

The value, b_0, is the normalized intensity of the injection pulse normally called a_0. Next we calculate the velocity needed by the electron to fall into the wake. This is a previously solved problem [5], analogous to other potential wells, such as the Kepler problem or atomic structure. To be trapped, the electron must be moving in a potential well, such that the well's depth is greater than or equal to the particle's kinetic energy. With the difference that this is true in the moving frame of the wave, so a Lorentz transformation, is needed to define the minimum trapping energy in the lab frame, $\Gamma = \gamma_\phi^2 \{\varepsilon + 1/\gamma_\phi - \beta_\phi | (\varepsilon + 2/\gamma_\phi)\varepsilon|^{1/2}\}$, with $\varepsilon = \phi_{max} - \phi_{min}$. The quantity Γ is the minimum trapping energy for an electron in a wake field potential ϕ, and peak to trough value of ε. Trapping now also depends on the phase velocity of the wave γ_ϕ, due to the transformation.

Finally we can say that LILAC injects electrons into the plasma wave when the drift velocity of the electron is larger than the trapping velocity,

$$\Delta(\frac{p_z}{m_e c}) \geq (\Gamma^2 - 1)^{1/2}. \quad (2)$$

Using Eq. 2 to combine Eq. 1 with the minimum trapping energy, we arrive at the intensity b_{th}, needed to trap. This is represented graphically in Fig. 2. The dashed line is the plotted trapping condition. So our simple model says that assuming we

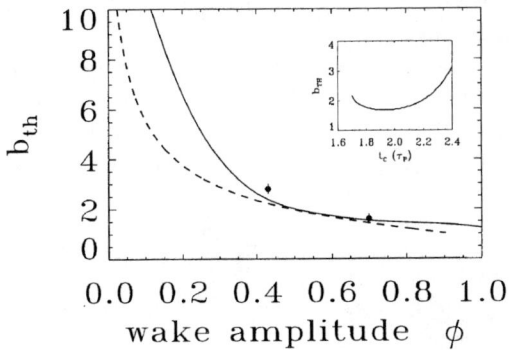

Figure 2: The trapping threshold, b_{th}, plotted versus the plasma-wave amplitude ϕ. The dashed line represents the results of Eq. 2. The trapping region is above of the curves. Inset: $(\gamma\beta)_z$ vs t_c, valid only along $y = 0$.

have correctly matched an electron's drift velocity to the phase of the wave, it may now be accelerated by the wave and form a beam. Given an a_0 of a laser pulse, and the amplitude ϕ of the created wake field, we now know the minimum b_0 of the injection pulse for electron to be injected. The problem with our calculation is that it does not take into account that electrons are moving in the wake field at the same time they feel the injection pulse. The more complete problem of calculating the drift velocity at the same time the electrons move in the background has no easy analytic answer. It was necessary to actually do a numeric integration to find a better trapping condition. Also we could then deal with the problem of matching the injection pulse to the wave's phase. This is the solid line in the plot. The inset show the fact that there is one optimal phase at which to inject electrons. The two points on the plot represent 1D particle simulations previously presented [1], to test the trapping.

4 TRANSVERSE LILAC

To test the previous model, before setting up an experiment, we used particle-in-cell simulations to actually try out LILAC. As previously mentioned, we first worked in one dimension. However the problem is inherently multidimensional due the orthogonal orientation of the two laser pulses. So now we have run simulations with two spatial and three velocity dimensions, to better model the physics. They inherently include all electromagnetic, and space charge effects. By performing a series of runs we can find the threshold intensity for the injection pulse. For the beam we may find its final energy, spread of the energy, and the beam emittance. The specific parameters used are $a_0 = 1.6$, $b_0 = 1.6$, and $\tau_{pe} = 5\lambda_l/c$. As before we use the LWFA with the pump pulse resonant with the plasma frequency. For the injection pulse we use $\tau = 2\tau_{pe}$ to reduce ponderomotive forces perpendicular to the direction of acceleration.

The following figures then represent one sample simulation with the excellent characteristics of a LILAC electron beam. In Fig. 3 we can see the electron pulse, highlighted, riding in the plasma wave. Note the separation of the beam electrons from the back ground. Fig. 4 shows the actual volume of phase space inhabited by

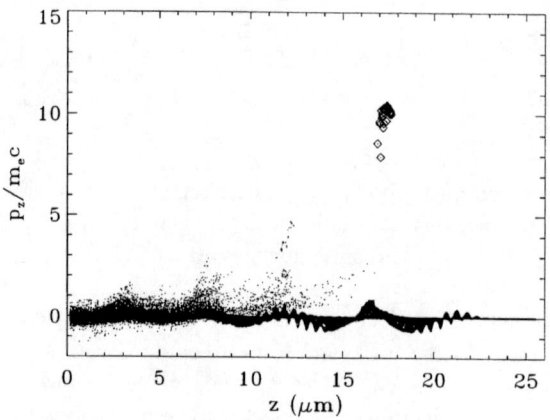

Figure 3: A PIC simulation showing a trapped electron bunch due to laser injection. The trapped electrons are highlighted with diamonds

the beam. As stated before the electron distribution is localized to a small area giving the beam excellent qualities. Empirically we note that during the acceleration

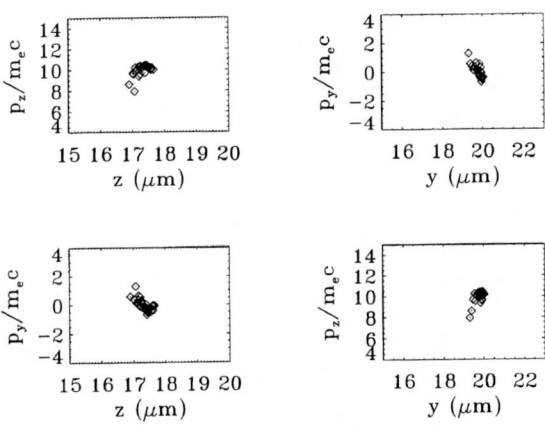

Figure 4: The same PIC simulation as Fig. 3. Plotted is the volume of the bunch for various planes in phase space.

process this volume remains constant, only being altered when the electrons finally outrun the wave. Table 1 summarizes the characteristics found from this particular simulation. Now that it has been shown that an electron beam can be accelerated, it

$\varepsilon_{\perp n}$	$.16\pi$ mm·mrad
$\varepsilon_{\parallel}$	$.3 \times 10^{-9}$ eV·sec
τ_b	1.3 fs (.4 μm)
$\Delta E/E$ @ 100 MeV	0.5%
n_b	1.5×10^7 per bunch (2 pC)

Table 1: A summary of the results for LILAC

is relevant to compare it with existing ones. To do this we use the emittance, a common quantity used with particle beams to examine their quality. In two dimensions there will be both longitudinal and transverse emittances. Basically they represent the volume of phase space occupied by the beam. In the transverse case we shall calculate the normalized emittance by $\varepsilon_{\perp n} = \pi\gamma\beta 2r_0 \frac{p_y}{p_z}$, where r_0 is the spot size of the beam, p_y and p_z are the transverse and longitudinal normalized momenta, p/mc. Comparing these results with some recent work on electron guns [6] we see that LILAC is potential as good or better these newer devices. Typical values of newer electron guns are 1 - .5 π mm·mrad. The number of electrons produced by LILAC is 1.5×10^7, smaller than reported for new electron guns. However it should be pointed out that the length of this electron bunch is so short that to achieve high particle numbers would cause the bunch to blow apart due to space charge effects.

Increasing the plasma wavelength allows more particles to be trapped, thereby increasing the total number in the bunch. At this point in time the electrons have an average energy of 10 MeV, and a relative energy spread of 5%. Since ΔE appears to be invariant with increasing energy, if the electrons reach 100 MeV the relative energy spread reduces to .5%. This spread in the energies is consistent with the change in accelerating gradients over the bunch length. For the purposes of this paper we will represent the longitudinal emittance by the integral $\varepsilon_{\|} = \oint dp_z \, dz$. It is also observed to be a constant of the motion, with a value from the simulation of $.3 \times 10^{-9}$ eV·sec. This small value is partly to do with the bunch length on the order of 1.3 fs. Even with a large energy spread, the area in phase space will be small with such a short bunch length. The two dimensional simulations again show that LILAC works in theory, and produces a beam of excellent quality.

Fig. 4 shows a number of plots, including $p_y/m_e c$ by y, or the transverse dimension. The positive y-axis is the direction in which the injection pulse travels. From inspection it can be seen that the particles have a much smaller transverse velocity than in the longitudinal direction. By making the injection pulse longer, the ponderomotive force longitudinal to the injection pulse (and transverse to the pump's wake), is reduced. Enough so that particles do not drift out of the wake, decreasing the number of electrons in the accelerated bunch. Also, the radial wake is necessary for the injection process. During injection, electrons are simultaneously kicked sideways and forward. With a large enough radial wake, electrons are unable to drift transversely out of the accelerating region. One can calculate the velocity needed to escape the wake in the radially given the depth of the potential, in this case about $\gamma\beta = 1.7$.

5 COLINEAR LILAC

Now we consider the other variation of LILAC which employs a different geometry where the injection pulse shares the same axis of propagation of the pump pulse. More tightly focused than the pump, it can act as an injection "switch" by turning on and off the wakefield enhancement. The schematic diagram of this idea appears in Fig. 1b. Injection occurs due to the ponderomotive force interacting with electrons through collective effects. Contribution from the longitudinal and transverse wakefield components will lead to wave breaking, or trapping of background electrons into the wave. Longitudinally, the accelerating phase of the wakefield has a larger value of the gradient for the time interval during which the injection pulse is at its focus. Accordingly, the separatrix, which determines the trapping of electrons, is expanded in phase space. It allows a certain group of electrons to be captured if those electrons satisfy the trapping condition. Inward transverse wakefield breaking by the injection pulse plays a role, in that it actually drags the electrons from outside the channel into the central region near the axis. The longitudinal fields were well below the one dimensional wavebreaking limit, nevertheless we still observe significant amounts of electron trapping. We believe this is inherent in two dimensional wavebreaking. Since this relatively small fraction of electrons, if injected at the optimal acceleration phase, satisfy the trapping condition, they can be accelerated along a narrow region on axis. The optimization of the injection process was done by adjusting the pulse delay between the injection and pump pulses. This wavebreaking process is observed in 2D and 3D PIC simulations, where the constant-phase wave front develops a horseshoe-like shape. In the 1D limit of pump pulse and plasma wave, an injection pulse could introduce transverse wave-breaking in the plasma channel, eventually leading to the generation of fast electrons on axis.

Fig. 5 and Fig. 6 show the longitudinal and transverse momentum of trapped electrons respectively. These are simulations with similar plasma and pulse pa-

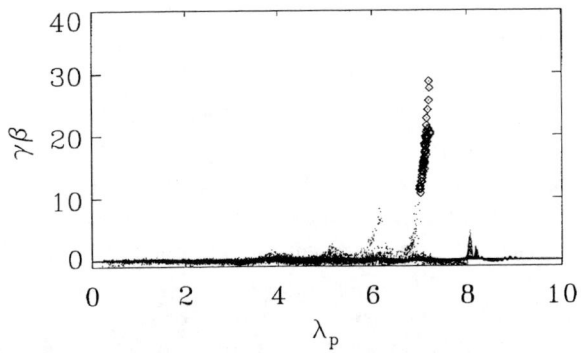

Figure 5: Longitudinal momentum of electrons: $a_0 = 1.5$, $b_0 = 3.0$

rameters as in the previous section, except the injection pulse's spot size is smaller, $a_0 = 1.5$, and $b_0 = 3.0$. The injection pulse travels on axis with the pump, and delayed behind it. The large transverse momentum spread is believed to be coming from the steep transverse profile of the injection pulse at its focus. In Fig. 5, the electrons, after trapping, are accelerated up to several MeVs, filling the multiple "buckets" 5 plasma periods (80 fs) behind the pump. Each bunch has a well-defined linear chirp in the momentum over the bunch length, which makes possible the ultrashort compression of these electrons. The problem of filling up multiple buckets might be resolved by using a second pump pulse and driving down the wakefield after the first pulse. An additional improvement can be achieved if the injection pulse uses a smaller laser wavelength than the pump so that the on-off time for the injection switch, from diffraction, is reduced. The beam characteristics of colinear LILAC may be improved if the Rayleigh range of the injection pulse is small compared to the plasma wavelength. We suspect that a lower density is easiest

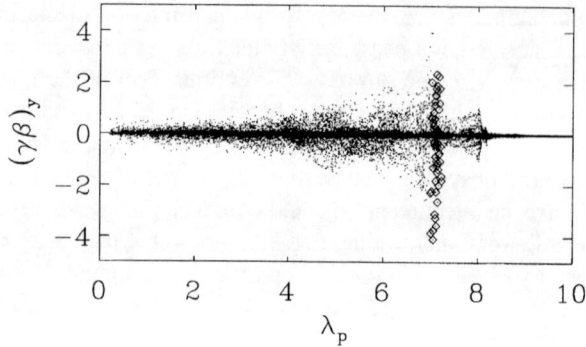

Figure 6: Transverse momentum of electrons from same simulation in Fig. 5

for trying this idea. Our next set of simulations will explore these changes to the LILAC parameters.

6 CONCLUSION

We have defined and analyzed a method for injecting electrons into plasma waves. As demonstrated in this paper, the initial analysis produces a beam of excellent quality, achieving characteristics equivalent or better than presently possible by other means. Starting with the general idea of using ponderomotive forces to inject electrons into plasma waves, we have investigated two particular methods of interest. Particle-in-cell simulations were used to test a model of transverse LILAC, and see if it provides accurate parameters. Normalized intensities of $a_0 = 1 - 2$ are predicted by the model, and do appear to cause injection in the simulations. These numbers are correct within an order of magnitude for transverse LILAC. There are more possible variations to LILAC, and they may be varied to find the best possible way to implement an all optical electron gun. The two particular versions reported in this paper are under study in preparation for experiment.

We would like to acknowledge the support of the NSF and to thank Torsten Neubert and Gerard Mourou for many valuable discussions. Computing services were provided by the University of Michigan Center for Parallel Computing, which is partially funded by NSF grant CDA-92-14296.

References

[1] D. Umstadter, J. K. Kim and E. Dodd, Phys. Rev. Lett. **76** 2073 (1996); E. Dodd, J. K. Kim and D. Umstadter, "An ultrashort-pulse relativistic electron

gun/accelerator," 37th Annual Meeting of the American Physical Society, Division of Plasma Physics, Louisville, KY, November 6-10, 1995.

[2] P. Maine et al., IEEE J. Quantum Electron. **24**, 398 (1988); G. Mourou and D. Umstadter, Phys. Fluids B **4**, 2315 (1992); M.D. Perry and G. Mourou, Science **264**, 917 (1994).

[3] D. Umstadter, "Laser Acceleration", The Accelerator Working Group at Snowmass, Snowmass, CO, June 25 - July 12, 1996. J. K. Kim, E. Dodd, and D. Umstadter, "Laser injection of electrons into Laser Wakefields", 38th Annual Meeting of the American Physical Society, Division of Plasma Physics, Denver, CO, November 11-15, 1996.

[4] T. Tajima and J. M. Dawson, Phys. Rev. Lett. **43**, 267 (1979).

[5] T. Katsouleas et al.,in *Advanced Accelerator Concepts,* AIP Conf. Proc. 130 (1985).

[6] X. J. Wang et al., Nucl. Instr. and Meth. in Phys. Res. A **375**, 82 (1996); C. Pellegrini et al., Nucl. Instr. and Meth. in Phys. Res. A **341**, 326 (1994); R. L. Sheffield et al., Nucl. Instr. and Meth. in Phys. A **341**, 371 (1994).

Performance of the Argonne Wakefield Accelerator Facility and Initial Experimental Results

W. Gai, M. Conde, G. Cox, R. Konecny, J. Power, P. Schoessow and
J. Simpson

High Energy Physics Division, Argonne National Laboratory
Argonne, Illinois 60439, USA

N. Barov
Physics Department, UCLA
Los Angeles, CA 90024, USA

Abstract: The Argonne Wakefield Accelerator (AWA) facility has begun its experimental program. This unique facility is designed to address advanced acceleration research which requires very short, intense electron bunches. The facility incorporates two photo-cathode based electron sources. One produces up to 100 nC, multi-kiloamp 'drive' bunches which are used to excite wakefields in dielectric loaded structures and in plasma. The second source produces much lower intensity 'witness' pulses which are used to probe the fields produced by the drive. The drive and witness pulses can be precisely timed as well as laterally positioned with respect to each other. We discuss commissioning, initial experiments, and outline plans for a proposed 1 GeV demonstration accelerator.

1. Overview of the AWA Facility

The generation of high gradients (> 100 MV/m) in wakefield structures requires a short pulse, high intensity electron drive beam. The main technological challenge of the AWA program is the development of a photo injector capable of fulfilling these requirements. The goal of the AWA is to demonstrate high gradient and sustained acceleration of charged particle beam by using wakefield method. In the past year we have made considerable progress towards attaining the design goals of the AWA.

Fig.1 shows the schematic diagram of the AWA facility, consisting of 3 major components: 1) an L-band rf photocathode and Linac capable of producing a 100 nC electron drive beam; 2) a second L-Band photocathode gun generates a low emittance and low charge beam which probes the wakefield produced by the intense drive beam and 3) An experimental test section for wakefield experiments.

AWA Facility Schematic Layout

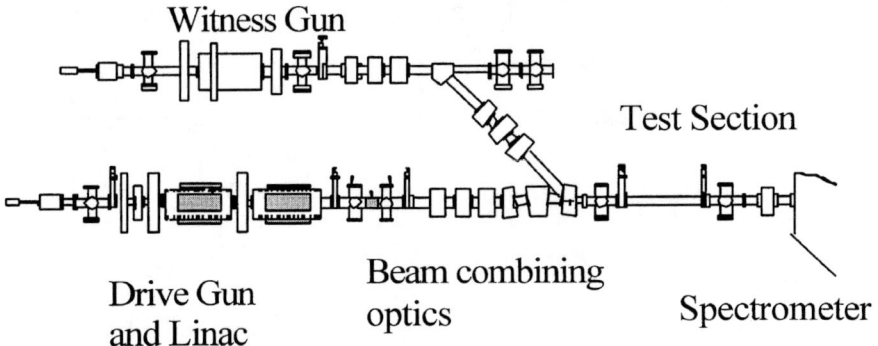

Figure 1. Schematic layout of the AWA facility.

A picosecond UV laser with up to 8 mJ/pulse output is used to illuminate the photocathode for both guns.

In this paper we present detailed descriptions of the facility and initial characterization of its performance. The preliminary results of dielectric and plasma wakefield experiments are discussed. The near and long term plans for experiments and facility upgrades will be described.

2. **Photocathode Gun and Drive Linac System**

The gun and drive linac are shown in Fig. 2. The laser photocathode sources was designed to deliver 100 nC bunches at 2 MeV to the drive linac. The photocathode gun is a single cell standing wave cavity with designed peak field of 90 MV/m on the cathode [1]. Some of the novel features incorporated into the gun to attain high intensities include a large (2 cm diameter) cathode, the use of a curved laser wave front and nonlinear focusing solenoids matched to the angle-energy correlation computed for the 100 nC bunch. So far, only flat laser pulses have been used for the experiment. However, for most AWA experiments, only 40 ~ 60 nC pulses are needed as discussed below.

The AWA drive linac [2] consists of two sections of $\pi/2$ standing wave structures. Each section is about a meter long. The linac is designed to deliver 18 MeV electron beam with 5 ~ 10 % of energy spread at 100 nC.

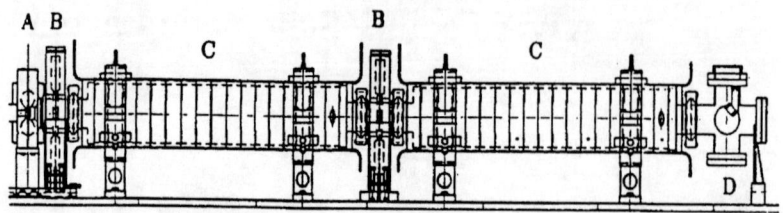

Figure 2. Drive Linac. A: High current photocathdoe gun. B: Focusing Solenoids (bucking solenoid not shown), C: Linac Cavities, D: Laser and diagnostic port.

3. **Witness Gun**

The witness gun is a six-cell, copper, iris loaded, rf photocathode operating at 1.3 GHz in a p/2 standing wave mode. A low charge, low emittance witness beam (0.1 nC charge, 1 p mm-mrad 90% physical emittance) is produced to probe (i.e. witness) the wakefields left behind by the drive beam . The witness gun is a scaled down version of the s-band Mark IV accelerator that was used at SLAC, as described in reference [3]. Since the Mark IV Accelerator was a linac, some adjustments were made to turn it into a photocathode gun using the rf design code URMEL. The witness gun has a photocathode in the first 1/2 cell, a coupling iris in the fourth full cell and a beam exit hole in the last half cell.

In order to probe the test devices properly, the witness beam must have a kinetic energy of 4 MeV, a physical emittance of 1 p mm-mrad, an energy spread of less than 1% and a bunch length of about 5 ps. Extensive simulations with PARMELA have shown the Mark IV type gun to be capable of achieving the design parameters. Using a 1.5 mm spot size and a phase launch of 65 degrees we obtain the following results

Energy	90% Emittance	Energy Spread	Bunch Length
4.53 MeV	0.76 p mmmrad	0.5% FW	5.6 psec

4. **Lasers and Control**

1The picosecond KrF laser system

The laser consists of a front end that produces picosecond pulses at 248 nm and a final KrF amplifier. The central component of the front end is a synchronously pumped mode locked dye oscillator (Coherent 702). The dye laser is tuned to the desired wavelength of 497 nm by a single-plate birefringence filter. Coumarin 102 dissolved in benzyl alcohol and ethylene glycol is the lasing medium, and DOCI dissolved in benzyl alcohol and ethylene glycol is the saturable absorber. A harmonic tripled mode locked Nd:YAG laser is used to pump the dye laser. The frequency of the mode locker is 40.625 MHz of which the 32nd harmonic is exactly 1.3 GHz.

A single pulse from the dye laser output train is amplified to 300 µJ through a three-stage amplifier. The dye amplifier is Lambda-Physik FL2003 pumped by 100 mJ, 308 nm pulses from a Lambda-Phyisk LPX105i excimer laser. The duration of the pump pulse is shortened to 10 ns so only one pulse from the dye oscillator can be amplified. The output from the dye amplifier is frequency doubled in a 3x3x7mm angle matched BBO crystal. Output at 248 nm is typically 25 - 30 µJ. Because the length of this doubling crystal, temporal broadening of the input pulse is expected.

Amplification of the ultra-short UV pulses is done in a single stage KrF excimer laser (Lambda-Physik LPX105i). The input pulses pass through the amplifier twice in order to fully utilize its stored energy. Typical output of 8 - 10 mJ is obtained routinely. The length of the final pulse is measured by Hamamatsu streak camera (model C1587) which has resolution of 2 ps. The typical measured pulse length (FWHM) is 3 - 4 ps. No satellite pulses observed. Repetition rate of the of the laser can be as high as 35 Hz.

In order to have certain flexibility of the experiment, we can run the Coherent 702 dye laser in a single jet mode. In the single jet mode, the laser is capable of producing pulse length from 5 ps to 30 ps. We have verified the laser pules length by using the autocorrealtor and streak camera. The laser energy is from 5 - 7 mJ/pulse with nominal fluctuation of 10% for the long laser pulses..

Controls and data acquisition

The design of the AWA control system[6] is based in part on the experience gained at the Advanced Accelerator Test facility (AATF), and also on more extensive data acquisition systems used for high energy physics experiments. The goal of the AWA system is to provide easy selection and adjustment of accelerator and beamline parameters, as well as the online analysis of diagnostic and physics data.

At the core of the system is an HP-750 RISC workstation using the UNIX operating system. The workstation is interfaced to VMEbus via a high speed adapter with dual port RAM. A 68060 CPU board on the VMEbus handles command requests from the workstation and provides auxiliary processing capabilities. Most of the control and monitoring functions are handled through a VME-CAMAC parallel bus interface. Video signals from beam position monitors and from the streak camera, comprising the actual physics data from the experiment, are acquired using a high resolution VME-based frame grabber. The AWA control software was developed in house and is based on the Tcl/Tk scripting language. The various codes comprising the system are written in C and FORTRAN77.

5. Initial Characterizations of the drive and witness beam

Detailed characterization of the both AWA drive and witness beam is currently underway. We have made an initial measurement of the beam properties at the exit of the Linac. Attempts were made to measure the pulse length and emittance vs the charge.

One unexpected problem encountered during the experiment was the low observed quantum efficiency of Magnesium photocathode, compared to measurements reported in the literature [4]. The QE found for Mg is $1 \sim 1.5 \times 10^{-4}$. Hence almost all the available laser energy is required to generate a 100 nC beam. However, a higher intensity laser pulse generally induces the photocathode to emit electron continuously ("explosive mode") [5]. Therefore, our initial measurements were made with charges generally less than 100 nC.

A diagnostic port at the exit of the Linac consists of an insertable pepper pot and a phosphor screen for emittance measurements. A calibrated integrated current transformer (ICT) device is used here for online nondestructive charge monitoring and a thin quartz (1 mm thick) plate is used as a Cherenkov radiator for pulse length measurement.

High Charge Generation

A 20-27 nC beam can be produced by 1 mJ laser pulse regardless the laser pulse length. It appears that we run into the space charge limit when we increase the laser power to 2 - 3mJ for short laser (5 ps). The maximum charge produced is 55 nC with 5 mJ of laser power. Increasing the laser pulse length resulted in higher charges as expected. A 100nC per pulse were observed, and 90nC pulses can be reached consistently with 5mJ laser power

Pulse Length Measurement

The electron pulse length is measured by using a streak camera situated in the laser room. The Cherenkov light from the quartz plate in the diagnostic port is collected and transported to the laser room. The Cherenkov light transport line was carefully built to ensure that no electron beam information can be lost.

Electron Pulse Length vs Charge

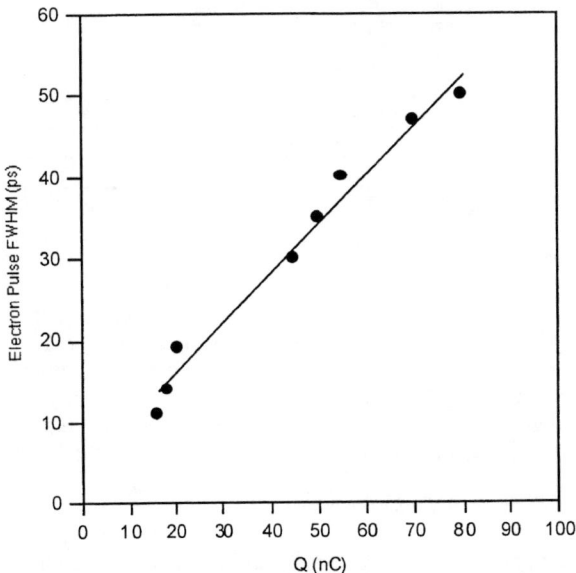

Fig 3 Electron pulse length vs charge

Results of the measurement are summarized in Figure 3. Because the electron pulse is not a gaussian, all the data were characterized by the full width half maximum (FWHM). The bunch length has a strong dependence on the charge. The shortest bunch length is 11 ps for 18 nC beam. At each charge, we average several data points to minimize the "random error" due to the pulse to pelse charge fluctuations. For 80 nC beam (with long laser pulse length), the measured electron pulse length is 48 ps, longer than the design goal (100 nC, 30ps). We are in the process of setting up an experiment for further investigation to attempt to reduce the bunch length. Note that while the Linac focusing is optimized for the curved laser bunch, only planar wavefront laser beam have been used

Emittance Measurement

A "pepper pot" with 0.5 mm holes and 2.5 mm spacing with a phosphor plate placed 40 cm downstream used for emittance measurement. Because of the small electron beam spot and relatively large holes of the pepper pot, and the resolution is about 10 mm mrad. Therefore, we have only estimated an upper limit on the emittance.

The following table summarizes the results of several measurements.

Charge	Measured rms physical emittance
20nC	10 mm mrad
55nC	13 mm mrad
70nC	20 mm mrad

The Witness Beam

The witness gun and its associated beam lines were recently installed and commissioned. Properties of the witness beam are being studied. The charge produced in the witness gun ranges from 0.1 ~ 3 nC. The beam energy is 4 MeV. The beam has been used for the initial dielectric wakefield measurements. Emittance measurements using a quadrupole scan technique and bunch length measurements using Cherenkov radiation are underway.

Synchronization of the drive and witness beam

Once the drive beam and the witness beam are generated, both beams are transported to the experimental section and combined. Since both the drive and witness beam are generated using the same laser pulse, a laser beam splitter is used to reflect a small amount of the laser beam through an adjustable delayed. Time delay between the two beams can be adjusted precisely using a mirror mounted on a movable stage for the witness laser beam line, while at the sametime, adjusting the rf phase to the witness gun to maintain a constant laser injection phase. The typical delay range used in the wakefield experiments is -50 ps to 400 ps. Delays up to 10 ns are possible using this system limited only by the adjustable stage.

6. Initial Wakefield Experiment Results

We have performed several collinear wakefield experiments to verify the performance of the AWA facility. Initial choice of the wakefield device were dielectric structure fabricated from Borosilicate glass. This material has a sufficiently large DC conductivity to minimize charging effects during beam tuning when scraping of the drive beam is worst.

Dielectric Wakefield Experiment

We have measured the wake field in two different dielectric structures (7 and 15 GHz). The results for the 7 GHz structure are shown in Figure 4. The wake amplitude is 1.5 MV/m for 20 nC drive beam. The structure has an inner radius of 1.25 cm and an outer radius of 1.6 cm with dielectric constant of 4. The measured wakefield amplitude and frequencies agree well the theory. This directly tested all the components of the AWA facility, and the results are satisfactory.

Another dielectric tube with inner radius 5 mm and outer radius 7.7 mm was also studied in the wake field experiments. The resonant frequency for this tube is 15 GHz. A wake field amplitude of > 5 MV/m was observed. Further tuning of the drive beam (more charge and shorter pulse length) should produce a wake field in the excess of 15 MV/m in this structure.

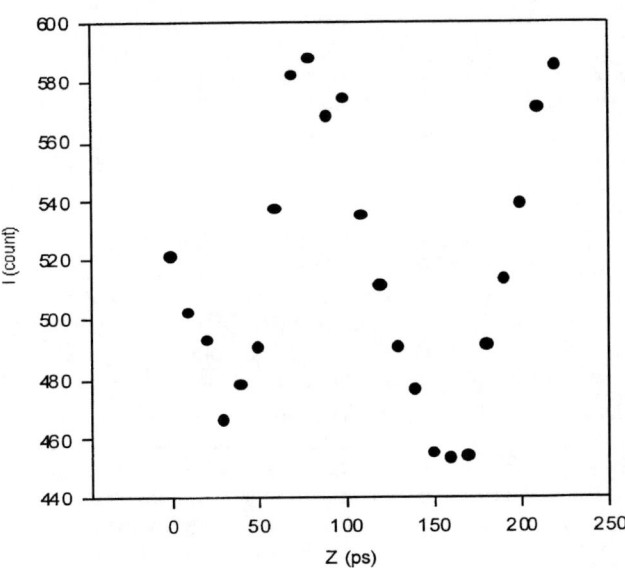

Figure 4. Measured longitudinal wake field for the 7 GHz dielectric wake field structure. The peak corresponds to 1.5 MV/m.

Plasma Wake field Acceleration and Focusing Experiment

In collaboration with an UCLA group, we have performed several preliminary experiments to study the plasma wakefield acceleration in the blowout regime. The first set of experiments demonstrated acceleration of a witness beam as a result of the plasma wave excitation caused by the drive beam. There is a current effort to study the self focusing of the drive beam. In order for the drive beam energy to be optimally coupled to the plasma wave, the drive beam must be focused to a very small spot, and the radius of a significant part of the beam must be kept nearly constant by the plasma's focusing force for the length of the plasma. The aim of the current experiment is to quantify this focusing and propagation, which depends greatly on the beam's emittance, charge and initial matching, as well as on the plasma properties.

7. Future Planned Experiment

Near term plans

a). Fully characterize the AWA beams, particularly the drive beam. Studying the beam properties (bunch length and emittance vs charge) dependence on the machine parameters.
b). High gradient collinear wakefield experiments using a dielectric structures. Generation of a electron pulse train and test the step-up transformer concepts. Ultimately to test the dielectric breakdown of the dielectric materials.
c). Continuation of nonlinear plasma focusing and acceleration experiments.
d). Colinear Wakefield Plasma Experiment. This experiment will be very similar to the AATF experiment [6]. Since the drive beam charge from the AWA is much higher than the charge from AATF, one should expect much more intense wakefield. Although this experiment will be in the non-blowout regime, we still believe it is very interesting. We can scan the charge from 2 nC ~ 40 nC in the range of plasma densities of 10^{12} ~ 5×10^{13}. The justification of this experiment is that although PWFA has been a subject of the intense theoretical investigations, no one has experimentally studied PWFA in detail. Since we have the capability of mapping out the wakefields, this experiment should be straightforward to carry out. The expected acceleration gradient produced in the plasma would be in the range of 10 - 50 MV/m.

Long term plan

Its well know that a major constraint of collinear wakefield acceleration is the transformer ratio. To overcome this difficulty, an accelerating field step-up transformer scheme of the dielectric wakefield accelerator was proposed[7]. The approach is to extract rf power from an intense drive beam traveling in a relatively large diameter dielectric wake field tube (stage I). This power is then transferred to a smaller diameter dielectric loaded guide (stage II) where the enhanced axial electric field is used to accelerate electrons. Field enhancement results both from a lower group velocity in stage II than in stage I (longitudinal compression), and from geometrical effects made possible by the use of the dielectric loaded guide (transverse compression). High net acceleration can be realized if one uses a train of large number (10 - 20) electron pulses. The spacing of the drive pulses can be arranged in such way that a long rf pulse is generated to fill stage II. This also permits us to identify less stringent parameters for the drive beam than previously described. Using this new procedure we predict that Phase-I of the AWA (20 MeV drive beam) can accelerate a witness beam to over 100 MeV in a meter or less.

The current plans for the AWA (phase I) is to generate 40 nC, 20 ps long electron pulse train consisting of 10 -20 pulses. Further upgrade of the drive beam energy in excess of 100 MeV (phase II) without changing any of other parameters would enable us to achieve net acceleration of the witness beam to 1 GeV energy in a less of 10 meters. Therefore, successful demonstration of the multiple pulse driven step-up transformer is critical.

8. Summary

Installation of the AWA Phase I facility has been completed. The facility was successfully commissioned. The drive gun and Linac has produced up to 100 nC beam with maximum pulse length of 50 ps (FWHM). The witness gun has produced high quality beams being used for the wakefield experiments. More detailed characterizations of both beams are currently underway. Initial collinear dielectric wakefield experiments verified the new wakefield measurement system. High gradient wakefield acceleration experiments in dielectric structures and in plasma are being pursued.

We would like to thank L. Balka, A. Caired, C. Keyser, B. Taylor and K. Wood for their technical support. This work is supported by the Department of Energy, Division of the High Energy Physics, Contact No. W-31-109-ENG-38.

References

[1]. C. H. Ho, PhD Thesis, UCLA, 1992
[2]. E. Chojnacki *et al.*,Proceedings of the 1993 IEEE Paricle Accelerator Conference, pp 815-817.
[3] The Stanford Two-Mile Accelerator, (Chapter 6)
[4] T. Srinivasan-Rao *et al.*, J. Appl. Phys **69** (5), 1991 p.3291
[5] X. J. Wang, PhD Thesis, UCLA, 1992
[6] J. Rosenzweig *et al*, Phys. Rev. Lett. **61**, p 98, 1988
[7] E. Chojnacki, Proceedings of Particle Accelerator Conference, San Francisco, May 6-9, 1991, p2557-2559.

The Drive Beam Generation for Two Beam Accelerators

R. Corsini
CERN, CH-1211 Geneva 23, Switzerland

Abstract. This paper reviews the schemes of two beam accelerators proposed for electron-positron colliders in the TeV energy range. Attention is given to the different methods of generating and handling the high-charge drive beam. Different options are discussed, and in particular three different possibilities for the generation of the drive beam for the Compact Linear Collider (CLIC) project are described.

INTRODUCTION

One of the main issues of the proposed electron-positron linear colliders is the generation of the RF power used for acceleration. While conventional designs make use of klystron tubes for this scope, a good alternative is the two-beam accelerator scheme, or TBA. In such a scheme, the RF power is extracted from a high-current relativistic electron beam. This beam is passed through a number of either extraction RF cavities or free-electron laser (FEL) sections, where it generates the microwave power. The power is applied to adjacent high-gradient acceleration structures, which accelerate the lower charge main electron beam to high energy.

Several options are possible in the choice of the drive beam charge, energy and time structure, and the technology used for its acceleration and the subsequent power extraction, but one of the main difficulties is the generation of the drive beam itself and its acceleration. Good energy efficiency is paramount, since it allows us to use the lowest power level possible (for a given high-energy beam power), facilitating the above tasks and increasing the overall collider efficiency, in itself one of the main positive aspects of linear colliders.

In Fig. 1 the conceptual scheme of a TBA is shown. In the drive beam line, a bunching section to provide a beam time structure capable of exciting the extraction units may or may not be present. The same is true for the re-accelerating units distributed along the drive linac. Two main lines of investigation are at present followed fro the drive beam acceleration, namely the use of superconducting RF structures and the use of induction modules. both have the advantage of efficient acceleration of the drive beam to the desired energy. The sources considered include RF photoinjectors as well as induction injectors. Bunching can be provided via a beam chopper, standard RF plus magnetic compression, bunching cavities or FEL interaction, while the extraction units can be travelling or standing wave RF cavities as well as FEL sections.

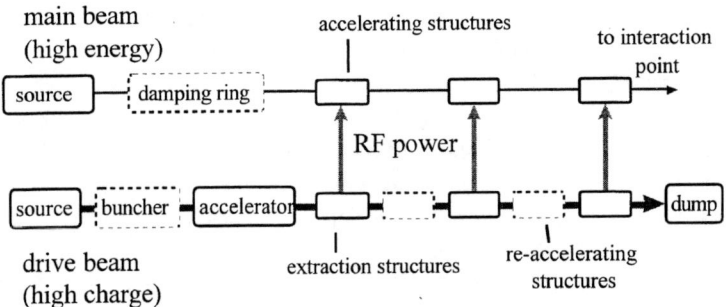

Figure 1. The TBA conceptual scheme

THE RK-TBA APPROACH

For several years a Lawrence Berkeley National Laboratory (LBNL) and Lawrence Livermore National Laboratory (LLNL) collaboration has performed both experimental work and theoretical studies on the Relativistic Klystron (RK) version of the TBA, based on the use of induction cells for beam acceleration and re-acceleration and RF cavities for power extraction. The effort included a preliminary design study for an RK power source suitable for an NLC energy upgrade in the TeV c.m. region [1], referred to as the TBNLC.

To provide the RF power, the TBNLC design requires several independent RK units. Each RK unit is about 300 m long. A 1.5 kA beam is generated and accelerated to 2.5 MeV in an induction injector. A 5.7 GHz chopper is used to obtain a beam bunched at 11.4 Ghz. Chopping reduces the dc current from 1.5 kA to 600 A. The bunched beam is then accelerated to 10 MeV in an 'adiabatic compressor' section, where bunching cavities are used to increase the bunching (from 240° to 70°). RF cavities located every 2 m extract power from the beam, which is periodically re-accelerated in induction modules at 300 kV/m. The RK is divided into identical 2 m modules with six 100 kV induction cells and one extraction cavity each. Both the drive beam current and re-acceleration voltage have a rise time of 125 ns and a 200 ns flat-top, with a fall time that is comparable to the rise time. At the end of each RK unit, an 'afterburner' section extracts more RF power from the beam before it is collected at low energy (2 to 3 MeV) in a beam dump. The adiabatic compressor is 26 m long, while the main RK is 276 m long (138 extraction cavities). The afterburner consists of 12 cavities, making a total of 150 extraction cavities per unit.

The drive beam to RF power efficiency is estimated to be 90%. The 10% loss is shared among the beam loss on the chopper (3.7%), beam dump (2.8%), and RF into induction cavities (3.6%). The efficiency for wall plug to drive beam is ~ 55%. Hence the total efficiency from wall plug to RF is ~ 50%.

A test facility (RTA) at LBNL has been established to verify the analysis used in the design study. All major components of a RK unit will be tested. However, the prototype will have only 8 RF extraction structures. The pulsed power system and induction cells in the extraction section will be very similar for both machines. Other common features include transverse chopping for initial beam modulation, adiabatic compression to increase the RF current component while accelerating the beam, a PPM quadrupole focusing system, and detuned RF extraction structures.

The main difficulties in realizing an RK-TBA module are linked to the drive-beam propagation over a long distance. In particular, the beam-break-up (BBU) instability through an RK unit is known to be severe. While BBU suppression techniques have been successfully used for a few cavities [2,3], and simulation results show that their use in a long RK-TBA should control the instability, an experimental demonstration on a bigger scale would be important. The same is true for the longitudinal stability of the RF bunches. Other problems include the RF phase and amplitude stability, which depends on the current and energy stability over the electron pulse. Finally, the RTA will assess the questions of efficiency and cost.

THE CLIC PROJECT STUDY

The CLIC Study Team examined the possibility of building an electron-positron linear collider in the TeV energy range, whose main characteristics are the use of a TBA scheme and the high frequency (30 GHz) chosen for the acceleration of the main beam. The scheme plan will use superconducting RF cavities for the acceleration of the drive beam to ultrarelativistic energies (above 1 GeV), and low impedance travelling wave structures (CLIC Transfer Structure - CTS) for 30 GHz power production and possibly no re-acceleration, in order not to have active elements in the main tunnel. The main goal parameters are listed in Table 1. The characteristics of the CLIC accelerating structure (CAS) are as follows [4]:

R/Q	=	3920 W/structure	shunt impedance (circuit convention)
l	=	0.32 m	length
Q	=	2800	quality factor
v_G	=	0.066 c	group velocity
E_{ACC}	=	100 MV/m	gradient (loaded)

In the following we will consider the 1 TeV cm upgrade of the CLIC, since such an upgrade is considered to be not renounceble. In this case 15200 CASs will be used.

Table 1. CLIC main parameters for 1 TeV operation.*

RF frequency	ν_{RF}	30	GHz
Accelerating Field	E	100	MV/m
Number of bunches / RF pulse	n	30	
Distance between bunches	Δ	30	cm
RF pulse repetition frequency	f_{rep}	700	Hz
Total Two-Linac AC power	P_{AC}	248 (222)	MW
Beam power/beam	P_b	13.5	MW
AC power efficiency	η_b	10.5 (12)	%
Number of Particles per bunch	N_b	$8 \; 10^9$	
Normalized emittances (hor. x vert.)	$\gamma \; \varepsilon_{x,y}$	$457 \times 12.5 \; 10^{-8}$	m rad
RMS bunch length	σ_s	160	µm
Relative energy loss	d_B	7.5	%
Luminosity with pinch	$\mathcal{L}$	$1.5 \; 10^{34}$	$cm^{-2} s^{-1}$

*The numbers in parentheses refer to the collector ring drive beam generation scheme (explained later), while the others correspond to the reference scheme.

THE CLIC REFERENCE SCHEME FOR DRIVE BEAM GENERATION

At present, the CLIC study group is investigating different options for the generation of the drive beam. The so-called "reference" scheme is the more conservative approach, making use of existing techniques of acceleration and beam handling. The main disadvantage of such a scheme is the high capital cost due to the large number of superconducting cavities needed for acceleration. Lately, a version of this scheme was presented at Linac '96 [5] that makes use of 250 MHz cavities instead of the 352 MHz LEP type cavities considered up to now, in order to decrease the total installed voltage (by reducing the beam loading due to the higher stored energy). This version should be considered the "state of the art" and is the one described below and depicted in Fig. 2. With respect to the parameters reported in Table 1, only 26 main beam bunches instead of 30 are foreseen in this scheme for the moment. The $1.5 \; 10^{34}$ luminosity level is reached by increasing the repetition rate to 800 Hz. One drive beam consists of 11 bunchlet train pairs. Each train is composed of 18 to 30 bunchlets. They are generated in 10 photocathode rf guns followed by short S-band linacs, and combined in a magnetic switchyard [6] at 1 cm distance. The switchyard can provide some bunch compression as well, bringing the bunchlet rms length down to the desired 0.6 mm. They are then accelerated in the two 250 MHz SC linacs as shown in Fig. 2.

The use of two independent linac pairs in push-pull (antiphase) configuration has the following advantages.

a) Twice as many drive bunchlets (60 instead of 30) per 250 MHz period can be accelerated, halving the required charge per bunchlet (from 100 to 50 nC in our case).
b) With two trains of 30 bunchlets per period (instead of one) the bunchlet deceleration variation inside one train is reduced (from ±76% to ±23%, see Fig. 3).

To guarantee a good efficiency it is necessary to accelerate the drive beam bunchlets inside each train matching the above decelerating wake curve, such that at the end of the CLIC drive linac, the bunchlets are dumped at an almost equal and low average energy, e.g., 0.3 GeV, but above 0.2 GeV to avoid electron losses.

The train energy profile can be pre-shaped by the combination of off-crest acceleration in the 250 MHz SC cavities and additional 4th harmonic cavities, as shown in Fig. 3. An almost linear energy ramp can thus be generated that matches the longitudinal wake behaviour shown in Fig. 2.

The acceleration optimisation has been carried out on the flat-top trains (which have the most severe deceleration). This is also true for the optimisation of the beam-loading compensation in the 250 MHz SC structures.

Each of the two CASs are fed by one CTS. To deliver two flat-top 103 MW CAS input power, the main CTS parameters can be chosen as follows:

R/Q = 1.4 Ω/structure   shunt impedance (circuit convention)
l = 0.71 cm   length
d = 4 ns = 1/250 MHz   drain time
v_G = 0.37 c   g group velocity

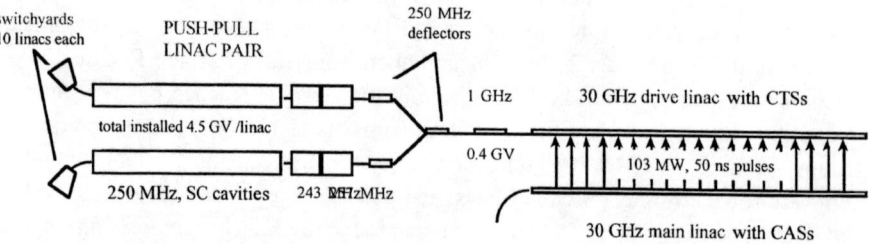

Figure 2. Generation of one drive beam upstream of one drive linac.

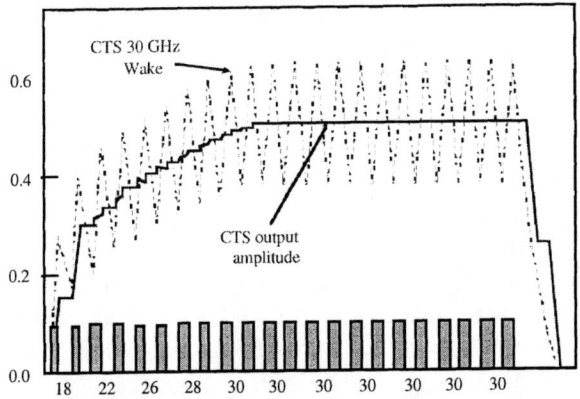

Figure 3. Computer optimised bunchlet numbers, relative intensities, normalised 30 GHz CTS wake amplitudes (dashed) and the normalized CTS output amplitude.

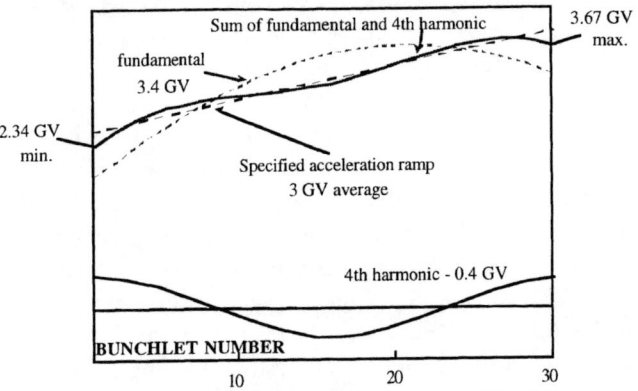

Figure 4. Acceleration ramp synthesis.

The internal CTS losses are 5 %, while CTS to CAS transmission losses are 10%. The flat-top bunchlet intensity is 50 nC. The RF output pulse has an initial ramp shaped to compensate the beam loading in the main beam [7], and to obtain constant energy along the main beam multibunch pulse (an energy spread of ±0.2% being the CLIC final focus acceptance). The flat top power is 103 MW per CLIC Accelerating Structure (CAS).

In order to obtain the necessary RF pulse profile, the 5 pre-fill train pairs have increasing numbers of bunchlets (18, 22, 26, 28 and 30 bunchlets/train) [8], combined with small variations of bunchlet intensities [9]. This particular time structure of the drive beam also allows the self-excitation of the 4th harmonic correction structures.

The energy spread of the accelerated 26 multibunches of the main beam is 0.1% rms.

The voltage decrease of the fundamental frequency cavities (R/Q = 96.6 W/m, circuit convention) has been compensated by using two groups of cavities with slightly different frequencies (243 and 257 MHz). They produce a fractional beat during the passage of the drive beam, yielding low deceleration for the first pre-fill trains and high acceleration for the flat-top ones [9]. A beneficial (for drive beam to RF efficiency) voltage increase for the pre-filling trains resulted from optimisation (see Fig. 5, curve D).

Figure 6 shows the self-excitation of the 4th harmonic cavities by the first 4 pre-filling train pairs (excitation occurs when the trains last less than one oscillation period), as calculated using R/Q of 386 W/m at 1 GHz. The flat-top trains lasting exactly one period cause no net excitation. The resulting field amplitude is 6.1 MV/m and the total active length is 67 m. The geometry is scaled from the 250 MHz structures.

From Fig. 5 it can be seen that an installed voltage of 3.5 GV per linac is needed at the fundamental frequency. Beam loading compensation requires an additional installed voltage in the fundamental frequency range of 33%. Thus, a total fundamental installed voltage of 18.5 GV is needed. To evaluate the needed cryogenic power, we follow investigations by K. Hübner [10] and I. Wilson [11] for LEP type structures at 6 MV/m, 352 MHz and Q = 4 10^9 with static losses of 29.5 W/m and dynamic losses of 32.3 W/m. For the 250 MHz structures static losses are estimated to be 15% and the Q-value to be 75% higher than for the 352 MHz ones. Taking into account by a factor of 3/4 the yo-yoing stored RF energy level between pulses (the cavities yielding half their energy to the passing drive beam), the dynamic losses are 19.7 W/m. Applying a cryo-factor of 250, the total cryogenics mains power becomes 41.4 MW for the 2 push-pull linac pairs.

The most important assumptions, leading to a wall plug to main beam efficiency of 10.5%, are illustrated in Fig. 6. RF, and main power level indications are for both main beams. Using a greater number of drive beam train pairs would increase the overall efficiency. Unfortunately, because of beam-loading in the 250 MHz SC cavities, it is difficult to accelerate more than 11 drive train pairs corresponding to 26 main beam bunches. In this compact double push-pull linac proposal, most of the capital investment would be for 250 MHz cavities (with their klystrons and cryostats at 4.2°K) providing usable stored energy for acceleration.

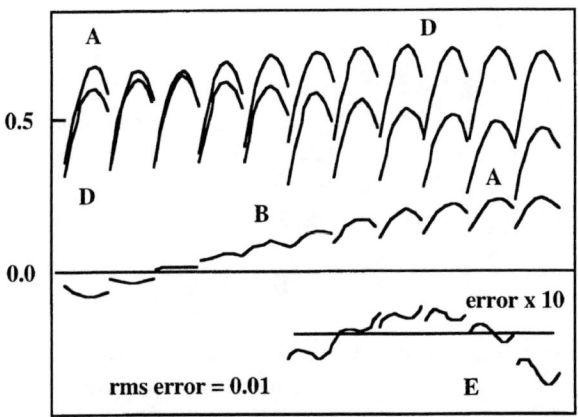

Figure 5. The decrease of normalised accelerating voltage (fundamental only) due to beam loading over 11 trains is shown (A). Curve B shows the compensating voltage. Curve D is the sum of A and B and corresponds to the total voltage seen by the trains. In E the remaining voltage error is shown. The unpopulated 270° in each of the 11 periods have been cut out.

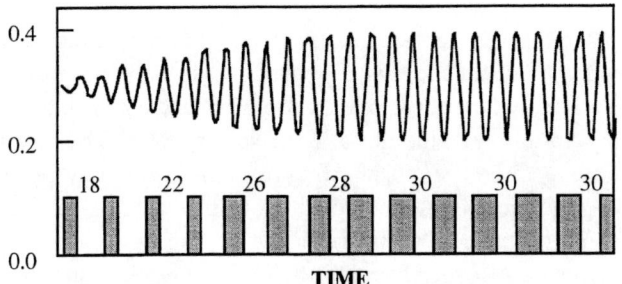

Figure 6. H = 4 excitation by the prefill train pairs(upper trace). The drive train pairs (bottom trace) are also shown. Only a small fraction (600J, 0.7%) of one drive beam energy (90 kJ) is used for energizing. Furthermore it is taken from the first trains which are underdecelerated in the drive linac. The voltage obtained is 6.1 MV/m.

The main disadvantages of the scheme seem to be:

a) A significant amount of RF and cryogenics hardware for the complete drive-beam generation complex, corresponding to about 8 times the LEP2 upgrade.
b) A high bunchlet charge of 50 nC.
c) An overall efficiency of only 10.5%, essentially because of the limited stored RF energy in the 250 MHz structures, which limits the number of drive bunchlet trains per pulse.

The main advantages appear to be:

a) Simplicity: no long drive-beam transport lines, no 180° and 90° arcs, and no fast kickers as in previous concepts. The drive-beam generation process thus occurs over a limited and almost straight length of about 1.4 km (including SC cavities, magnets, and straight sections).
b) Acceleration in mainly large-aperture (33 cm) 250 MHz SC structures causing low wakes.
c) Only 41 kW/m input power for the SC cavities.

The forty 3 GHz linacs, upstream of the switchyards delivering up to 33 bunchlets of 50 nC per pulse, seem to be a difficult challenge. The development work can, however, be attempted within the modest framework of the CLIC Test Facility (CTF). Dark currents in the 250 MHz SC cavities should be investigated. Cavity tune changes due to ponderomotive forces, periodic with the 800 Hz pulsing, may be acceptable thanks to the shortness (< 50 ns) of the drive pulse.

CLIC DRIVE BEAM GENERATION USING A COLLECTOR RING

In the reference scheme the strong beam loading prevents the use of a single, short linac. In order to overcome this problem, it has been proposed [12] to first accelerate each individual bunch of the drive beam and combine them later with the right spacing. In this way the global beam loading would be distributed in time over the entire repetition period, allowing its compensation by power refill of the accelerating structures. On top of that, a continuous train can be produced, in which most of the bunches will be decelerated in the drive linac in the same way. A good efficiency can therefore be obtained (only the first bunches will be dumped at higher energies), without the need to use the linac cavities to pre-compensate the difference in energy loss along the trains.

The main problem is how to store the bunches until all of them are produced. The natural solution is to use a ring, both for storing and combining them. However, having very high-charge, extremely short bunches circulating in a ring for a relatively long time poses a number of problems (collective instabilities, synchrotron radiation losses, debunching due to non-isochronicity, etc.). In order to minimise these problems, a number of expedients have been devised:

a) Divide the drive linac into a number of sections N_D, each one powered by a different drive beam pulse. In this way the charge per pulse and the charge per bunch can be decreased by a factor N_D.

b) Compress the bunches to their final length only after the storage time in the ring. In this way, the isochronicity requirements are relaxed, and the resistive wall effect in the ring is minimised.
c) Separate the functions of storage and combination into two different rings.

Both rings can be separately optimised, and the bunch compression can be made before the combination, since the time spent in the combiner ring is very short. Relatively low frequency cavities can therefore be used for bunch compression. The final drive beam temporal structure to be obtained with the present scheme is depicted in Fig. 7, together with the temporal structure of the main beam and of the RF pulse. As explained before for the reference scheme, the RF pulse is composed of a pre-fill that is used to establish steady state conditions in the accelerating structure (CAS), followed by a constant power part corresponding to the duration of the main beam pulse. The drive beam pulse is composed of 496 bunches of variable charge, in order to generate the needed RF pulse shape. The main differences with respect to the reference scheme are: a) the drive beam pulse is continuous rather than composed of several trains; b) the individual bunchlets are 3 cm apart, instead of 1 cm.

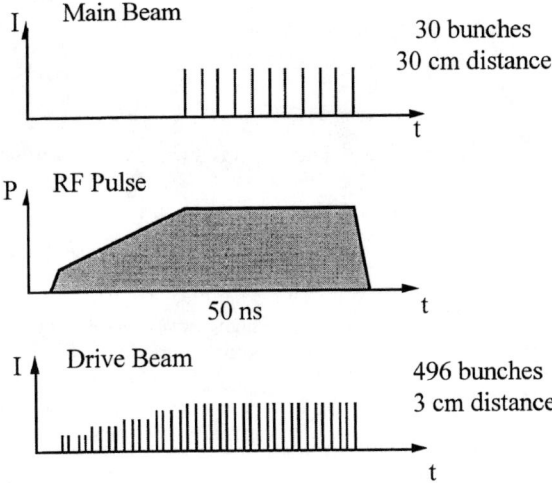

Figure 7. Main Beam pulse, RF pulse and Drive Beam pulse temporal structure, 1 TeV operation.

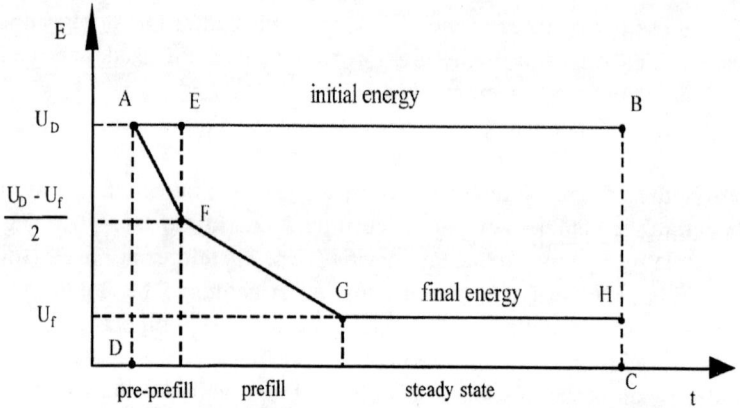

Figure 8. Energy distribution along the drive beam pulse at the beginning and at the end of a drive linac section. The drive beam extraction efficiency is the ratio between area ABCD and area EFGHB.

The power fed to the CAS depends on the parameters of the drive beam pulse and of the CTS in the following way:

$$P_{CAS} = \eta_T \, \eta_{CTS} \frac{W_{CTS}}{\tau_{CTS}} = \frac{\eta_T \, \eta_{CTS} \, \omega \, r'_{CTS} \, v_g \, q_{ss}^2 \, F^2(\sigma) \, \tau_{CTS}^2}{4 \quad 1 - v_g/c \quad \Delta t_{ss}} \quad ,$$

where (assuming a different CTS design with respect to the reference scheme):

P_{CAS}	= 2 × 10³	MW		power to feed into CAS
h_{CTS}	= 0.95			extraction efficiency from CTS
h_T	= 0.9			transfer efficiency from CTS to CAS
t_{CTS}	= 3.78	ns		CTS drain time
r'_{CTS}	= 50	W/m		CTS shunt impedance (circuit convention)
v_g	= 0.38	c		CTS group velocity
q_{ss}	= 4350	nC		drive beam charge in flat top
Dt_{ss}	= 29	ns		drive beam pulse duration (flat top)
$F(s)$	= 0.93			form factor (for s = 0.6 mm)

Given the above CTS parameters, the efficiency losses, and the needed steady state pulse duration, the flat top charge is evaluated as q_{ss} = 4.35 mC and the charge needed during pre-fill is 2.15 mC. Actually the pre-fill is composed of two parts: the real pre-fill (during one filling time of the accelerating structure and with 1860 nC charge), and an initial part ("priming," during one drain time of the transfer strucure and with 290 nC) used to optimise the energy extraction efficiency. The "priming" part of the RF pulse is not seen by any of the main beam bunches. The total charge of the drive beam pulse is $q_D = 2.83$ mC. The total energy to be given to the main beams is $W_{beam} = 33$ kJ. For 30 bunch-operation the efficiency from RF to beam is $h_{RF} = 30$ %, therefore the total RF energy needed for one main beam pulse is $W_{RF} = W_{beam} \times h_{RF} = 117$ kJ.

This energy is linked to the main beam parameters as:

$$W_{RF} = \eta_T \eta_{CTS} \eta_D F(\sigma) q_D U_D N_D,$$

where N_D is the number of drive beam lines, U_D is the drive beam injection energy, and hD is the extraction efficiency from the drive beam. This last is defined as (see Fig. 8):

$$\eta_D = 1 - \frac{q_P U_D + q_F \left(\frac{U_D + U_f}{4}\right) + q_{ss} U_f}{q_D U_D},$$

where q_P is the charge during "priming," q_F is the charge during pre-fill, and U_f is the minimum energy of the drive beam bunches when they are dumped at the end of every drive beam line.

Using the above equations, one can calculate the initial drive beam energy needed and the extraction efficiency as a function of the number of drive beam lines, as shown in Figs. 9 and 10. The choice of the operation point ($N_D = 20$, $U_D = 1.53$ GeV) is a trade-off between efficiency, installed voltage, and considerations based on longitudinal beam dynamics during the generation process.

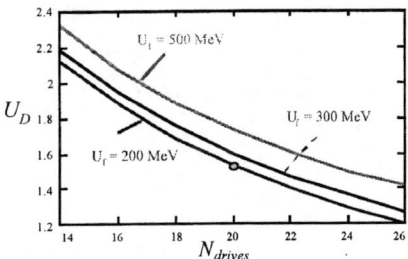

Figure 9. Drive Beam energy (GeV) as a function of the number of drive beam lines for different final energies. The circle indicates the proposed operating point ($N_D = 20$, $U_D = 1.53$ GeV).

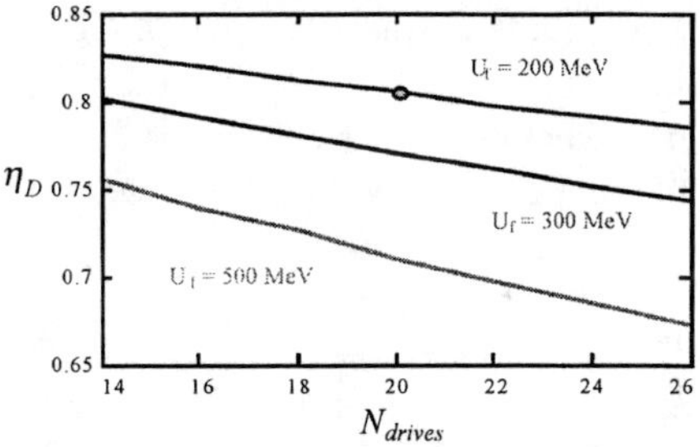

Figure 10. Energy extraction efficiency from Drive Beam to RF (not including TRS and transfer losses) as a function of the number of drive beam lines. The circle indicates the proposed operating point ($N_D = 20$, $h_D = 0.805$).

As shown in Fig. 11, each main beam pulse is accelerated to the 500 GeV final energy by 10 drive linac sections (50 GV each). Every section is powered by an individual drive beam pulse. Therefore, 20 drive beam pulses must be generated for each e^+ / e^- pulse in the main linac. Initially, 2 ×160 trains composed of 31 bunches each are generated in two 612.5 MHz photoinjectors. The bunches in a train are spaced by 48 cm, corresponding to the above frequency, and have an rms length of 5 mm. This is repeated with the repetition rate of the main linac (700 Hz). The trains are then accelerated up to 1.53 GeV in two 612.5 MHz superconducting linacs in push-pull configuration. A total of 3.1 GV of cavities at fundamental plus ~ 90 MV cavities at 612.5 MHz ± e to provide beam loading compensation along the train will be installed.

The beam loading in the injector linac can be calculated as follows:

$$\frac{dU}{U} = \frac{1}{2}\frac{dW}{W_{stored}} = \frac{r'_I \omega_I q}{2 E_I}$$

where:

ω_I	=	$2\pi \times 612.5$	MHz	pulsation of injector linac
r'_I	=	490	W/m	shunt impedance per meter
E_I	=	10	MV/m	accelerating field in injector linac
q	=	405	nC	charge per train

The charge per train is obtained allowing for a 2.5 % charge loss in the accelerator chain up to the drive beam. The resulting energy spread is of the order of 4 %. This will be compensated by additional cavities at 612.5 MHz ± e to the level of $6 \cdot 10^{-4}$. The single bunch energy spread is due to the combined effect of the longitudinal wakefields and of the RF curvature, and is of the level of 0.26 %. The beam loading from train to train is compensated by RF refilling of the injector linac structures inbetween the passage of the 160 trains at a repetition frequency of 700 Hz. The necessary RF power at 1.25 GHz is:

$$P = q_D N_D U_I f_{rep\, r}$$

where U_I is the total installed voltage of the injector linac and f_{rep} is the repetition rate. The resulting power is 70 MW. The power per meter of structure will be 440 kW/m. The trains are then combined two by two using a transverse RF deflector at the same frequency of the linacs, halving the distance between bunchlets and doubling their number per train. The distance between trains at this stage will be 8.9 ms ~ 2.7 km.

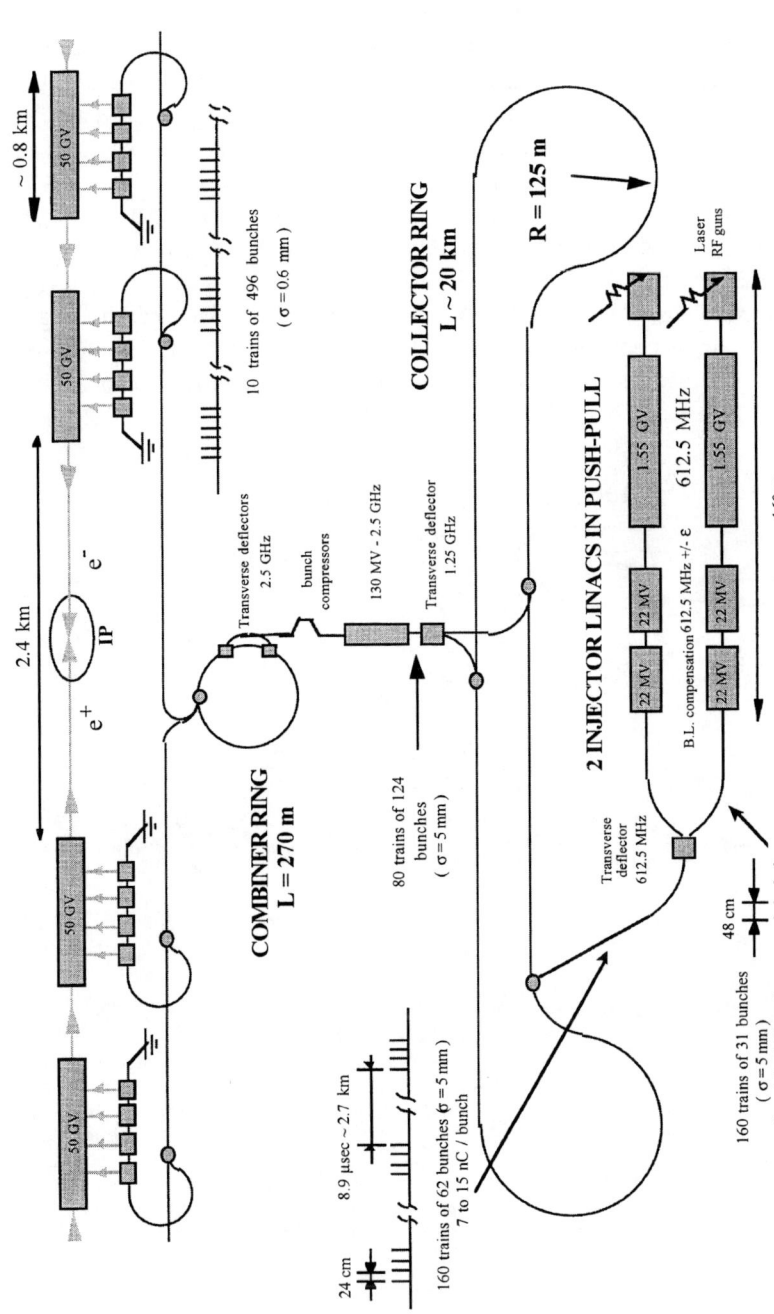

Figure 11. Layout of drive beam generation using a Collector Ring.

The trains are then injected into the collector ring using simple magnetic kickers. The ring is of dog-bone shape, and the two long straight sections can be located in the same tunnel as the main and drive beams. The end bends are composed of two 360° arcs with the same radius (125 m) needed for the main beam bends, and could be in the same tunnels.

The ring length (~ 21 km) corresponds to 8 times the distance between trains, plus 125 m. Therefore, after the second turn the 5th train will be injected 125 m after the first, the 6th will be injected 125 after the 2nd, and so forth. After 1.4 ms, the whole ring will be filled. When the last train is injected, the ring is emptied. Every second train is ejected by magnetic kickers in two different locations of the ring, at a 10-km distance, and each train couple is recombined in a transverse RF deflector operating at 1.25 GHz.

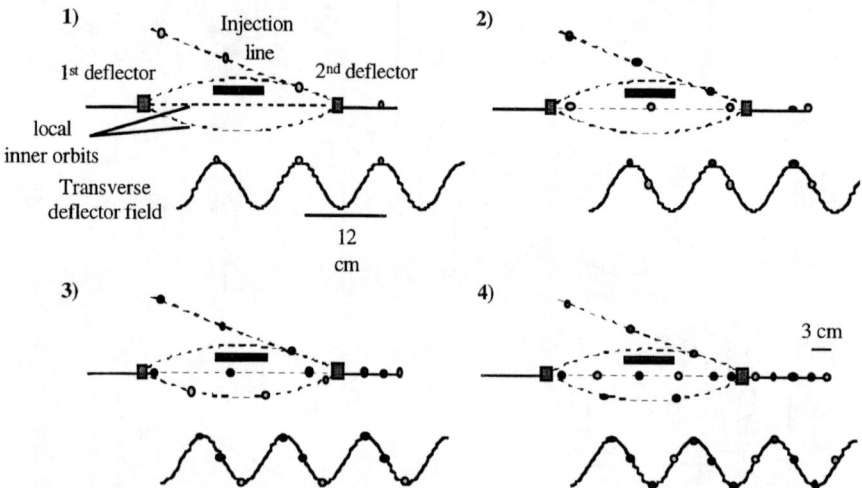

Figure 12. Schematic of the four turns injection into the combiner ring. 1) When the first train arrives all of its bunchlets are displaced by the 2nd transverse kicker on the equilibrium orbit. 2) When the first train comes back, its bunchlets arrive at the kickers at zero crossing of the RF field, hence it stays on an internal orbit. The second train arrives 90° later, and its bunches are displaced by the 2nd kicker on the equilibrium orbit. 3) Now the 1st train bunchlets are kicked inside the orbit, the 2nd train bunchlets arrive at the zero crossing, and the 3rd train bunchlets are injected. 4) The 1st train bunchlets arrive again at the zero crossing, the 2nd train bunchlets are in the inner orbit, the 3rd train bunchlets are also at the zero crossing, and the 4th train bunchlets are injected. After the 2nd kicker, the four trains are combined in a continuous train with 3 cm spacing between bunches.

The bunches are then compressed in length by RF (120 MV, 2.5 GHz, superconducting) + magnetic chicanes down to an rms length < 0.6 mm. At this stage the pulse is composed of 80 trains of 124 bunches each, with 12 cm distance between bunches and 270 m distance between trains.

The trains are then combined four by four in a combiner ring (270 m long), in order to obtain 20 trains of 496 bunches at 3 cm distance. The injection into the combiner ring is made using 2 transverse RF deflectors at 2.5 GHz, which creates a time dependent local deformation of the equilibrium orbit. The revolution time is not exactly a multiple of the kicker frequency (while the distance between incoming trains is) so that for each revolution time the RF phase increases by 90°.

When all of the four trains are inside the ring, they are extracted by a magnetic kicker, and the whole cycle is repeated for the next four. The long trains so obtained are alternatively switched by a magnetic kicker in the two drive linacs.

The collector ring is composed of two long straight sections (~ 10 km each), located in the main tunnel, and two 360° bend (270° + 90°) at the ends (785 m each). The last train will be ejected during its first turn, while the first train will stay in the ring for quite a long time (1.4 ms, corresponding to ~ 20 turns). It is therefore important to avoid deterioration of the quality of the first train bunchlets during this time. Possible deleterious effects in the ring include: synchrotron radiation losses, resistive wall effect., non-isochronicity, collective effects. The first two effects will produce an energy spread within the bunches, different from train to train. Non-isochronicity in the ring will cause a bunch lengthening, dependent on the incoming energy spread and on the energy spread developed by the bunch in the ring itself. Collective instabilities could result in the loss of some or all the bunches or could spoil the transverse emittance. The average current in the ring will be (at the end of the cycle) < 2 A, which should be acceptable with respect to collective instabilities. In any event, the peak current (in each bunch or along a bunch train) is much higher, and further investigations are necessary in this respect.

Figure 13 shows the longitudinal phase space for the flat top bunchlets at injection and extraction from the collector ring. The longitudinal wakes and RF curvature in the injector linac, the radiation losses, the resistive wall effect, the space charge, and the non-isochronicity in the collector ring are included in the calculation. Figure 14 shows the longitudinal phase space for bunchlets after compression.

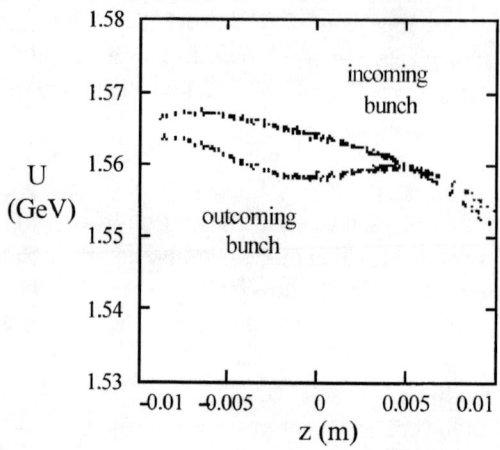

Figure 13. Longitudinal phase space for bunchlets at injection and extraction (20 turns) from the collector ring.

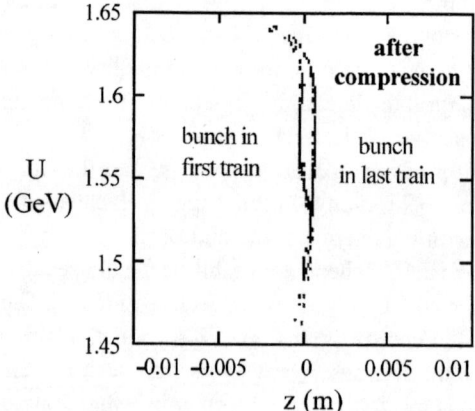

Figure 14. Longitudinal phase space after compression for flat top pulses belonging to the first train (20 turns in the ring) and the last train (half a turn).

In Fig. 15 a comparison is made at normal luminosity between the power flow (wall plug to main beam) and efficiencies for the reference scheme and the collector ring scheme, for the nominal luminosity. The efficiencies are quite similar, but the advantage of the second scheme lies in the lower installed voltage (capital cost). In addition, the second scheme should be able to handle more total charge than at present, such that an increase in the number of multibunches in the main beam can be considered, leading to a higher efficiency.

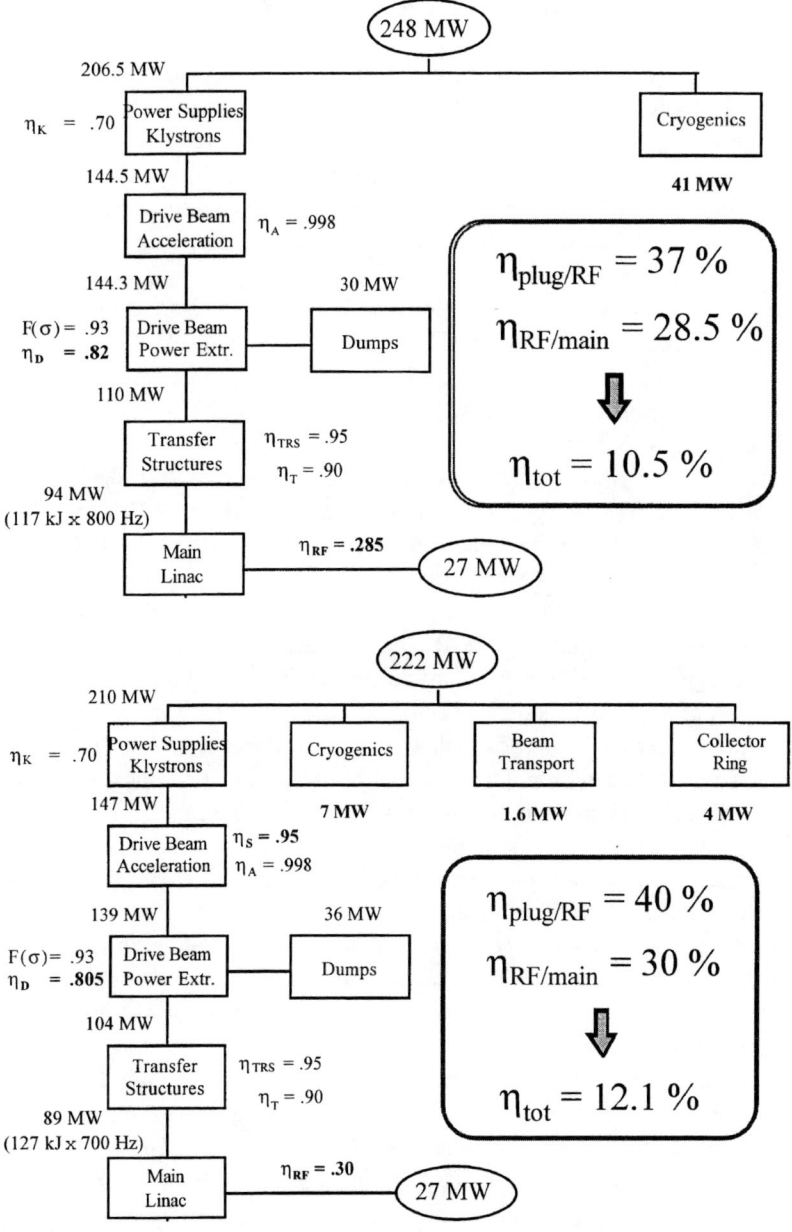

Figure 15. Power flow in the present CLIC scheme, from wall plug to main beam, for the reference scheme (top) and the collector ring scheme (bottom).

CLIC DRIVE BEAM GENERATION USING THE RK-TBA SCHEME

Following the discussions arising at the "CLIC/RK-TBA Collaboration Meeting," in Geneva in January 1996, an analysis was made of the use of RK-TBA units as an RF power source for a CLIC type linac at 30 GHz. This analysis is called RK-CLIC, and was made by the LLNL-LBNL collaboration [13]. Although the analysis was performed using a CLIC parameter list considered at the time, which is different from the one discussed above, its conclusions are fully valid with the new parameters. In particular, the pulse duration is equal to the present one (50 ns), while the repetition rate is only slightly different (600 Hz). The main changes have been in the CAS length (32 cm instead of the earlier 42 cm), the main beam time spacing (30 cm instead of 20 cm) and the number of main beam bunches (30 instead of 50), and the RF pulse power (2 ¥ 103 MW instead of 2 ¥ 95, implying 80 MV/m in CAS).

Many elements of the TBNLC are the same for a collider designed at a higher frequency. We propose to power the main linac using 50 RK units. Each unit provides RF power for 300 meters of main linac. Similar to the TBNLC, each unit consists of a 1.2 kA, 2.5 MeV induction injector, a beam modulation unit, an adiabatic capture section followed by the main power extraction unit and an afterburner. At the entrance to the RF power extraction unit the beam has an average energy of 10 MeV and 630 A of dc current. Each RF power extraction unit is comprised of about 300 modules. Each module is one meter long. It is composed of 3 induction cells and one extraction cavity, and powers two CASs. The proposed beam modulation unit is a chopper as in TBNLC, but other methods are possible. The drive beam dynamics issues are similar to those in the TBNLC case. The longitudinal stability should improve with respect to the TBNLC design due to the shorter synchrotron wavelength and closer spacing of the output structures.

The smaller aperture of the 30 GHz extraction cavities stresses the importance of generation and conservation of a small emittance beam. Alternatively, an increase of the average drive beam energy would relax the requirements. The beam dynamics for the low frequency BBU due to beam interaction with the induction acceleration gaps (2–4 GHz) should be the same as in the TBNLC. This instability is controlled by Landau damping from energy spread. However, the lower emittance needed may put higher requirements on and necessitate reducing transverse impedance of the induction gap. The highfrequency BBU due to higher order modes in the RF extraction cavity (~ 40 GHz) appears to be the main problem because of the higher impedance of the 30 GHz structures and their number (twice as many as compared to the TBNLC case). To counter this effect in the TBNLC, the structures were placed at a betatron wavelength distance. The concept is valid also for the RK-CLIC case, where they are spaced at half a betatron wavelength. The total effect should lead to an instability growth similar to that for the TBNLC. However, a higher tolerance on focusing field errors might be needed. The Ceramic Magnetics CMD-5005 ferrite is used for the accelerator cores, 50 ns being the limit where the use of the Allied-Signal METGLAS® alloy 2714AS is more efficient. Magnetic pulse compression and switching are proposed instead of the thyratron switching used in the TBNLC, which cannot reach the short rise time (17 ns) needed and is not suitable for a 600 Hz repetition rate. The evaluated efficiency from wall plug to RF is 43%, the drive beam efficiency 90%, plus an additional 94% due to the pulse fall time, not used for acceleration. It should be noted that the RK can handle longer pulses than the one considered here, with a higher efficiency. As already noted, an increased pulse length would also improve the RF-to-beam conversion.

ACKNOWLEDGEMENTS

The author would like to thank G. Westenskow and S. Yu for all the information and the useful discussions on the RK-TBA concept. Many thanks go also to L. Thorndahl who did all the work on the CLIC reference scheme and provided a lot of useful material and assistance, and J.P. Delahaye who invented the collector ring drive beam generation method and kept it alive with his suggestions and ideas.

REFERENCES

[1] "An RF Power Source Upgrade to the NLC Based on the Relativistic-Klystron Two-Beam-Accelerator Concept," Appendix A of the Zeroth-Order Design Report for the Next Linear
[2] J. Haimson, and B. Mecklenburg, "Suppression of beam induced pulse shortening modes in high power RF generator and TW output structures," SPIE Symposium on Intense Microwave and Particle Beams III, proceedings Vol. 1629-71, 209 (1992).
[3] T.L. Houck, and G.A. Westenskow, "Status of the Choppertron Experiments," Proc. 16th International Linac Conference, Ottawa, Ontario, Canada, Aug. 23–28, pp. 498-450 (1992).
[4] L. Thorndahl, "Push-Pull Pairs to Generate Two Drive Beams for CLIC Multibunch Operation," 18th International Linac Conference Proceedings, Geneva, Switzerland (1996), CERN/PS 96-38 and CLIC Note 317.
[5] M. Dehler, I. Wilson, and W. Wuensch, "CLIC waveguide damped accelerating structure studies," 18th International Linac Conference Proceedings, Geneva, Switzerland (1996), CERN/PS 96-39 (LP) and CLIC Note 318.
[6] B. Autin, "Design of a Magnetic SwitchYard for the CLIC Drive Beam," CLIC Note 220.
[7] K.A. Thompson and R.D. Ruth, "Simulation of multibunch energy variation in 0.5 and 1 TeV linear collider designs," SLAC Pub. 5882.
[8] I. Syrachev, private communication.
[9] L. Thorndahl, "Drive Beam Bunchlet Trains for Multibunching," CLIC Note 291.
[10] K. Hübner, "Wall-plug power of 2*0.25 TeV CLIC," CLIC Note 167.
[11] I. Wilson, "RF power requirements for CLIC," CLIC Note 270.
[12] B. Autin, K. Bane, R. Bossart, R. Corsini, J.P. Delahaye, G. Guignard, C.D. Johnson, A. Mikhailichenko, J. Nation, L. Thorndahl, and I. Wilson, "CLIC Drive Beam Generation, a Feasibility Study," 1994 European Particle Accelaration Conference, London, United Kingdom (1994), CERN/PS 94-16 and CLIC Note 228.
[13] "CLIC/RK-TBA Collaboration Meeting," Novotel-Geneva Airport, 15-16 Jan. 1996. CLIC Note 295.

Developments In Relativistic Channeling

Richard A. Carrigan, Jr.
Fermi National Accelerator Laboratory
Batavia, IL 60510

Abstract

The possibility of using channeling as a tool for high energy accelerator applications and particle physics has now been extensively investigated. Bent crystals have been used for accelerator extraction and for particle deflection. Applications as accelerating devices have been discussed but have not yet been tried.

1. INTRODUCTION

Over the last decade there has been significant progress on the application of channeling to accelerators and high energy physics. Channeling extraction has turned out to be a remarkably interesting technique. Channeling has been used for beams at Serpukhov, CERN, and Fermilab and more possibilities continue to arise. Spin precession has been demonstrated at Fermilab but looks challenging for charm spin measurements. Development of simulation tools has continued, particularly for extraction. Channeling acceleration remains a dream but some interesting theoretical work and the advent of terawatt chirped lasers have drawn the curtain back further on an interesting possibility.

Recently there have been several excellent reviews of channeling by Biryukov et al. (BKC) and Møller. Many papers from a recent meeting on relativistic channeling will be available soon. The following sections sketch basic channeling parameters, summarize applications to high energy beams and extraction, review proposals for acceleration, and consider exotic extensions.

2. BASIC CHANNELING PARAMETERS

Critical Angle: A particle no longer channels when the so-called critical transverse kinetic energy is equal to the channeling maximum potential energy U^m, that is:

$$E_\perp^c = \frac{1}{2} p \beta c \psi_c^2 = U^m \qquad (1)$$

where ψ_c is the critical angle, p is the momentum, βc is the velocity, and U^m is the maximum potential. The Lindhard critical angle for planar channeling for a singly charged particle is:

$$\psi_p = \sqrt{\frac{4Ze^2 Nd\, Ca_{TF}}{p\beta c}} \tag{2}$$

where Z is the atomic number of the crystal, N is the atomic density, d_p is the planar spacing, a_0 is the Bohr radius, C is approximately $\sqrt{3}$, and a_{TF}, the Thomas-Fermi screening parameter, is equal to $0.8853a_0(Z^{2/3}+1)^{-1/2}$. At 1 TeV, $\psi_p = 5$ μrad for the (110) plane in silicon. Note that a channeling dip or peak is far from Gaussian. From the standpoint of applications a more useful quantity is $\psi_{1/2}$, the half angle at half width. This is roughly equal to the critical angle. Another factor that needs to be considered for some applications is the impact of lattice vibrations. Since these are temperature dependent the critical angle will vary with temperature. This is particularly important for advanced accelerator applications where crystals might even vaporize.

Bent Crystal Channeling: The first serious suggestion that channeling in bent crystals could be used to steer charged particles was made by Tsyganov. When a crystal is bent the potential well is modified by a linear centrifugal barrier. At a small enough bending radius, the Tsyganov radius,

$$R_T = \frac{E}{eE_c} \tag{3}$$

the centrifugal barrier exactly equals the depth of the normal potential and particles are no longer captured into channeling orbits and deflected. Here E is the total energy and E_c is the interatomic field at a distance from the plane of the crystal lattice where the trajectory of the particle no longer remains stable due to its interaction with individual atoms. More complete treatments of bending dechanneling have been developed by Ellison, Kudo, and Kaplin and Vorobiev.

The equivalent magnetic field for channeling for the case $\beta \approx 1$ in a crystal bent with a uniform radius R is:

$$B = \frac{p}{0.3R} \tag{4}$$

(here B is in tesla, p is in GeV/c, and R is in m). Equivalent fields up to 1000 tesla are feasible.

Dechanneling: Particles dechannel because of multiple scattering in the channel and defects such as dislocations. Although dechanneling is a diffusion process practically it can be described with a dechanneling length. For planar dechanneling BKC give:

$$\lambda_D = \frac{256}{9\pi^2} \frac{p\beta c}{\ln(2m_e c^2 \gamma / I) - 1} \frac{a_{TF} d_p}{r_e m_e c^2} \quad (5)$$

where γ is the Lorentz factor, I is the ionization potential, d_p is the interplanar spacing, m_e is the mass of the electron, and r_e is the classical electron radius.

In a bent crystal the channeling potential is shallower and narrower. To account for this the dechanneling length for the straight crystal case should be multiplied by $(1-R_T/R_m)^2$ (see BKC). R_m is the minimum radius of curvature for the crystal.

Materials Issues: Materials affect every aspect of channeling. The planar critical angle is proportional to the cube root of the atomic number so that a high Z material like tungsten has a critical angle almost twice as large as silicon. The planar dechanneling length is less sensitive to Z but is proportional to the interplanar spacing, d_p, so that wide planes like $(111)_w$ increase the dechanneling length. The Tsyganov bending radius is inversely proportional to the critical field, that is Z^{-1}, so that tungsten should be better than silicon for bending.

A regular crystal structure, that is a lattice free of many dislocations, is imperative for high energy channeling. By far the best material in this regard is silicon. Presently available silicon single crystals should easily handle the energy of the LHC. Good germanium crystals are also available and have been used up to several hundred GeV. Tungsten is another matter. Available zone-refined crystals tend to be smaller and have more dislocations. A program is now underway at Dubna to develop better tungsten crystals.

A fourth material property that is important for some applications is the behavior in a high radiation environments. Tests with MeV-level channeling found little significant degradation due to radiation damage up to fluences of $4 \cdot 10^{20}/cm^2$ in a 28 GeV proton beam. Since MeV channeling is more sensitive to induced point defects, high energy applications may be even less affected. A silicon crystal in a 70 GeV beam at Serpukhov continued to channel with high efficiency after a fluence of more than $10^{19}/cm^2$ at an operating temperature of 150 °C.

Channeling Radiation: Particles moving in a crystal channel radiate much like they would in an undulator. The theory of channeling radiation was originally formulated by Kumakhov. The channeling radiation maximum energy goes as $E^{3/2}$ for positrons in the 10 MeV to GeV regime. While this is a much higher energy than corresponding undulator radiation it is small compared to bremsstrahlung so there is no natural application of channeling radiation in high energy physics. However, recent work has emphasized the importance of channeling radiation as a damping mechanism for exotic accelerator applications (see later section on cooling).

The possibility of stimulated emission of channeling radiation has also been considered. There has been no experimental progress on the subject. High but not un-realizable current densities would be required.

3. APPLICATIONS OF CRYSTALS AT HIGH ENERGY ACCELERATORS

Channeling in bent crystals has now been widely used at accelerators for both extraction and for secondary beam deflection. Transmission is the critical factor in applying a bent crystal. If the beam emittance is substantially greater than the crystal acceptance the transmission efficiency of a bent crystal in an external beam can be approximated by:

$$E = E_c \left(\frac{\phi_b^{50}}{\phi} \right) \left(1 - \frac{R_T}{R_m} \right) e^{(-sp_o/\lambda_{no}p)} \qquad (6)$$

In the formula ϕ_b^{50} is the phase space acceptance of the bent crystal (proportional to the thickness times the channeling critical angle), Φ is the 50% phase space emittance of the particle beam for the crystal bending direction, E_c is the surface acceptance of the crystal, p is the beam momentum, s is the length of the crystal, and λ_{b0} is the bent crystal dechanneling length at p_0. As noted earlier, to account for the decreased channeling length in a bent crystal the dechanneling length for the straight crystal case should be multiplied by $(1-R_T/R_m)^2$. Both bending and temperature effects should be included in calculating the critical angle in the crystal acceptance so the critical angle (including lattice vibration effects) should be multiplied by $(1-R_T/R_m)$ as discussed by BKC. For a harmonic potential in a straight crystal when Θ_b (the beam divergence half width) $\gg \psi_c$, BKC gives for E_c:

$$E_c = \frac{\pi x_c}{2d_p} \qquad (7)$$

where x_c, the critical transverse distance or effective half width of the channel, depends on the bend, screening, and lattice vibrations.

Up until now bent crystals have been used or considered for four major types of applications; external beam deflection, focusing, extraction from accelerators, and deflection and spin measurements of short-lived particles. These are discussed below.

External Beam Deflection: The first application of a crystal as a beam element appears to have been in 1984 in the Meson Bottom beam at Fermilab. The peak energy of the beam had been 225 GeV limited by two 3 m long septum magnets. In a demonstration, these magnets were replaced by a 2.7 cm long crystal enabling the beam to go to 400 GeV. Use was constrained by safety considerations. The radiation safety experts were concerned that the crystal would work all together too well!

A bent crystal was also used at Fermilab as a beam throttle in NE operating at 800 GeV. The crystal reduced the beam after a high intensity experiment so it could be used downstream in a low intensity emulsion experiment. The observed beam transmission was 0.05%, about a factor of six less than expected. The difference might have been due to some combination of misalignment of the body of the crystal relative to the planar direction, the onset of problems with dislocations or interstitial imperfections, overestimation of the surface acceptance, or an improper understanding of the crystal beam optics.

The most ambitious application of crystals to external beams is at Serpukhov where they have been applied in a variety of applications including beam splitting and beam diagnostics. Serpukhov (see BKC) holds the record for the largest deflection (150 mrad) and the longest crystal used (15 cm).

More recently a bent crystal has been used as a beam splitter to produce the K_S beam for CERN NA48, an experiment to measure CP-violation with high precision. A much-attenuated proton beam is required to produce the K_S near the detection apparatus since the K_S decays quickly. Indeed, one challenge is reducing the beam by a factor of $0.5*10^{-4}$, well below a typical ratio of crystal channel phase space to secondary beam phase space. A second important consideration for the application has been the high flux on the crystal, on the order of 10^{12} protons every 14.4 s, leading to a fluence of $10^{18}/cm^2$ per year.

Focusing: Crystals can also be used as focusing elements. In the first multi-hundred GeV experiments at Fermilab losses were noted halfway around the bend in a three point bending jig. These losses were traced to additional local curvature due to the pressure of the middle bending pin. Sun suggested that this type of compression could be exploited to construct an element with a focal length of several meters. A more straight-forward approach is to bevel the end faces of a bent crystal so that some rays are deflected more. In a beautiful series of experiments at Serpukhov, Smirnov and his colleagues focused a 70 GeV proton beam down to a 40 micron line with a crystal with a 0.5 m focal length.

Spin Precession: The spin of a channeled particle moving in a bent crystal should precess through an angle φ given by:

$$\phi = \frac{1}{2}\gamma\omega\,(g-2) \qquad (8)$$

for $\gamma \gg 1$, where γ is the Lorentz factor, g is the gyromagnetic ratio, and ω is the deflection angle of the channeled particle. This occurs because the crystal bend leads to an average electric field that points in to the center of curvature, resulting in a net effective magnetic field perpendicular to the plane of curvature. The spin of a particle moving in the channel precesses around that effective magnetic field.

In a recent Fermilab spin precession demonstration polarized Σ^+ hyperons from the Fermilab charged hyperon beam were channeled in two silicon crystals with bends of ±1.65 mrad (with an equivalent field of 45 tesla). These bends resulted in a spin precession of 60±17°, in agreement with the predicted value of 62° based on the world average of the measurements of the Σ^+ magnetic moment. Improvements such as the use of crystals with more active area and five to ten times the bending angle would have permitted this experiment to match precision experiments done in the eighties.

Because they produce large deflections in a short length of crystal, the high effective magnetic fields associated with bent crystal channeling offer a unique possibility for the measurement of charm particle magnetic moments. On the other hand, the small angular acceptance for planar channeling is a significant limitation.

An experiment for a charm particle magnetic moment measurement would look quite different than the channeling Σ^+ measurement. Since charm lifetimes are short, there is not a beam of charm baryons in the conventional sense. In an experiment charm baryons would be produced in a thin, high Z amorphous target upstream of the bent crystal. An amorphous target is necessary since particles produced on nuclei in the channeling planes of a crystal cannot channel.

Daniels and Lach, Carrigan and Smith, and Samsonov have studied the rates for a charm particle magnetic moment measurement. The conclusions from these studies are sobering. The physics is such that only two charm baryons, Λ_c^+ and Ξ_c^+, are likely to be measurable and both of them would have small precession angles. There is very little hope to measure beauty baryon magnetic moments since there are no relatively stable positively charged states. Conclusions about experiment running time depend sensitively on assumptions about how successfully one can trigger with a high intensity beam. The crystal bend could enrich the trigger, since the charm particles are deflected from the forward cone. The long-lived channeled particles go further around the bend so they will not be as large a background on the charm fraction of the beam. Samsonov has estimated that an experiment could be done in several hundred hours using a 10^9/s proton beam.

Extraction: Almost from the first suggestion of bent crystal channeling there has been interest in exploiting it for extraction from particle accelerators. The first demonstration of channeling extraction occurred at the JINR synchrophasotron at Dubna in 1984. Since relatively large particle beam deflections can be achieved with short crystals, an interesting feature of channeling extraction is to applications where the potential extraction path is limited. This feature has been exploited successfully at the 70 GeV Serpukhov accelerator where the available straight sections are short.

As usual the critical angle for channeling is a limitation. This is less of a problem for extraction than it would first seem since many unchanneled particles multiple scatter in the crystal and remain in the accelerator to channel on a later pass. This "multi-turn" extraction was first studied in simulations and confirmed in experiments at 120 GeV at CERN.

For channeling extraction a crystal can be placed at the edge of the accelerator beam where it can extract a limited portion of the beam. This is particularly interesting for colliders where there may be enough halo to create significant external beams with little impact on the integrated luminosity. During the planning stages for the SSC such a technique was proposed to produce a 20 TeV proton beam suitable for beauty production. A recent Fermilab experiment, E853, was undertaken at the Tevatron to investigate that possibility at 900 GeV. E853 is the highest energy channeling experiment yet performed. No difficulties were experienced from such potential problems as dislocations or radiation damage.

Several mechanisms were used to pump halo beam onto the crystal. Beam-gas scattering and power supply modulation produced some natural beam growth. A fast kicker could provide transverse kicks of 0.5 mm at the crystal on an individual bunch. Beam diffusion on to the crystal was also stimulated with a fast horizontal damper. Finally, beam-beam collisions at the collider detector interaction regions stimulated halo beam growth and diffusion out to the crystal.

To study the effect of luminosity-driven extraction during E853 the circulating beams were prepared so that there were 36 proton bunches and three antiproton bunches in the Tevatron. This gave rise to 6 colliding bunches and thirty proton-only bunches. The E853 measurements found that the rate was roughly proportional to luminosity after the background was subtracted. The rates for colliding bunches were about 6 times higher than the proton-only bunches at a bunch luminosity of $0.4*10^{30}$/cm^{2}s[1]. The effect of that luminosity was equivalent to moving the crystal into the beam on the order of 1 σ.

E853 made several different measurements of efficiency. One practical measurement of efficiency is the amount of beam extracted down the beam line divided by the beam lost in the accelerator. This measurement is not easy since determining the loss rate from the accelerator characteristically involves a difference between two large fitted numbers for beam lifetime in the accelerator. A second way to measure the efficiency is to determine the number of particles that interact with the crystal when it is not aligned versus the number that interact when it is correctly positioned for channeling. Calculations by Biryukov predicted a practical efficiency of 30-45% for E853. The data for E853 not yet been completely analyzed but is slightly below the prediction. Efficiencies up to 15.4% were measured in a recent CERN 120 GeV run.

The beam that can be extracted from an accelerator using a crystal is limited by damage to the crystal, by the losses that can be incurred elsewhere in the accelerator and at the colliding experiments, and by the tolerable rate of beam current attenuation with time. For E853 beams of up to 0.5 - 1 M/s were obtained without significantly disturbing other operations and experiments.

Based on the E853 experience a design for a 1000 GeV, 100 KHz parasitic test beam for use during collider operations has been developed. The beam, extracted at A0, would feed into the Fermilab fixed target areas. It has been designed for minimum impact on the existing complex rather than to optimize beam flux. The design makes use of two bent crystals, one for extraction with a bend angle of 16.4 mrad and another one with a bend of 7.5 mrad for redirecting the beam in to the switchyard. Because the angles are large, the transmission of the extraction crystal is 9% while it is 40% for the redirection crystal. The overall flux down the beam could be improved by using germanium crystals and perhaps bending less and adding more conventional magnetic elements to replace the second crystal and cut the bend required for the first one.

One other exotic application of extraction is the possibility of creating long base-line neutrino beams aimed toward large neutrino cosmic ray detectors. While intriguing, dechanneling and small acceptance remain significant problems. Interestingly, unless physics dictates otherwise, there is no advantage in going to high energy to exploit linearly-rising neutrino cross sections because it takes longer to accelerate. Since the required deflection angles are large, there is a premium on high Z crystals. As with any neutrino beam, an expensive infrastructure of a meson decay pipe and a beam dump is required. In addition, any neutrino experiment requires a very large fluence of protons.

4. POSSIBLE APPLICATIONS TO ADVANCED ACCELERATORS

Over the past decades a number of suggestions have been made to apply channeling to advanced accelerator applications. The large electromagnetic fields, hundreds of times higher than laboratory fields, have the right feel for what is needed for an acceleration breakthrough. To be useful, channeling actually has to solve some problem like beam cooling. But channeling also comes with several penalties-the channeling critical angle is small, electrons cause dechanneling, negative particles don't channel well, and really high energy densities vaporize crystals.

Cooling. Beam cooling was one of first accelerator channeling applications to be discussed. A. Kanofsky and E. Tsyganov, collaborators in the original Fermilab channeling experiment in the late seventies, raised the possibility of cooling particle beams with channeling. Radiative processes are significant for light particles and will be discussed in more detail later. At present energies radiative processes that can produce transverse cooling for heavy particles are small so that beam heating due to multiple scattering from the electrons in channels is the important process. In the early Fermilab investigations of axial channeling of heavy particles in a germanium crystal heating was observed rather than cooling, as expected.

Channeling extraction raises the possibility of an alternative approach to cooling. Multiple-pass channeling extraction is like a cross between a Maxwell demon and the famous "Energizer Bunny" that just keeps going. The extracted beam has the low emittance of a crystal channel, potentially much smaller than an internal accelerator beam. Many particles that don't channel return and eventually channel. Of course some particles are lost to nuclear interactions and dechanneling. However if one could cut the beam size by 4-10 while losing half the particles there would still be a gain in luminosity if the beam was reused.

As an example, one could extract beam from the Fermilab Tevatron operating at 150 GeV into the Main Injector. The phase space would be reduced by the ratio of the critical angle to the beam divergence times the effective septum width divided by the beam size. This could be a factor on the order or 10. The extraction efficiency might be 25-50% so that the later luminosity gain would be several fold. There are several problems. One is that multi-pass channeling extraction requires multi-turn injection into the following accelerator or storage device, a challenging problem. A second is that this only works for positive particles. The third, and most challenging problem, is that handling the full accelerator beam does violence to the crystal and the accelerator.

It is possible to envision a system where the crystal "extracts" into a different part of the accelerator phase space. The crystal could kick the beam from one dip in a magnetic potential similar to a Higgs' potential in the Standard Model across an electrostatic septum into the other valley of the potential. After the entire beam had been "extracted" the electrostatic septum would be physically removed and the Higgs' potential would be adiabatically erased. Whether something like this is possible is debatable. The loss problem would be significant. For the E853 run and a 10^{12} proton beam the loss in the crystal due to ionization would be several Joules. Unless something was done such as moving the crystal or increasing the effective crystal septum width (which diminishes the effective beam cooling) the energy would be deposited in a small region so the temperature rise would be enormous. Much worse, 10^8 Joules would be lost somewhere else in the accelerator if the nuclear interactions losses were O(10%). This could easily quench a superconducting accelerator if it was lost in one place.

The possibilities for cooling positrons are more interesting. Recently Huang, Chen, and Ruth, (HCR) have studied radiation damping in a continuously focusing planar channel. They find the particle damps down to a transverse ground state with a very small emittance of:

$$\gamma^\varepsilon \min = \hbar/2mc \qquad (9)$$

where m is the mass of the particle. Once the particle is in the ground state it can be accelerated without any radiative energy loss. The damping constant for the process is:

$$\Gamma_c = 2r_e K/mc \qquad (10)$$

where K is the focusing strength of the channel, that is $V(x) = Kx^2/2$. HCR note that for a typical case $K = 10^{11}$ GeV/m^2 so that $1/\Gamma_c = 10$ ns. A 100 Mev particle in a crystal channel has an initial quantum number of about 500 so it requires 6 e-folding times to reach the ground state or 60 ns. This corresponds to a channel length of 18 m. At first blush, this is discouraging since the dechanneling length is much shorter. However HCR argue that the channeling radiative damping rate suppresses both bremsstrahlung and transverse growth due to multiple scattering. In addition, if this process is used in conjunction with some advanced acceleration scheme it might be possible to accelerate fast enough to boost into a very long dechanneling length regime.

HCR illustrate the potential of this approach with an example of a 5 TeV on 5 TeV crystal collider. For 10^9 particles/bunch, 10 bunches in a train, and a repetition rate of 180 Hz they suggest the luminosity could be $L = 3*10^{36}$ cm^{-2}s^{-1} without the need for a final focus. The total beam power would be only 3 MW.

With the recent interest in cooling muons for muon colliders Bogacz, Cline, and Sanders suggested channeling could be used for that purpose. Unfortunately the damping time for the HCR picture goes as the mass of the particle so that it is in the microsecond regime for muons. To overcome this problem, they suggest using an acoustic wave to set up a micro-undulator and get stimulated emission. In another approach, they propose exploiting ionization cooling in crystal channeling for muon cooling.

Acceleration: In the late seventies several suggestions were put forward for possible channeling accelerators using lasers. Some early proponents argued dielectric materials were needed to avoid electrical breakdown with the high laser accelerating fields. Kanofsky, apparently the first person to publish a concept for the application of channeling to acceleration, investigated a laser acceleration scheme suggested by Csonka where a laser beam strikes a dielectric from the side. As noted by Kanofsky, the use of a channeling medium with acceleration along an axis minimizes scattering and energy loss and also exploits the transverse focusing of the channel. In the Csonka approach the crystal is masked so that the beam particles see only the in-phase electric field. Csonka estimated that purely laser acceleration could give accelerating gradients in the neighborhood of 0.1 GV/cm and would require laser power densities on the order of 10^9 W/cm^2. This approach has problems with diffraction and near-field effects. Kanofsky implicitly discussed dechanneling and seemed to recognize that acceleration would help to ameliorate the problem. However, since his laser acceleration gradient is "small" his estimates of transmission are extremely pessimistic. He did not consider the possibility of adiabatic damping of beam emittance later discussed by Chen and Noble.

Another early technique proposed by Grishaev and Nasonov suggested a system of two coupled lasers to produce a longitudinal wave in a crystal with a non-linear optical susceptibility. They estimated the accelerating gradient would be 0.01-0.1 GV/cm. Both the lasers as well as the accelerated particle beam would have to be phase matched, a difficult problem over an extended acceleration region. Pisarev discussed the use of a transverse laser swept along a crystal which would generate a static longitudinal electric polarization and "light rectification". Pisarev argued that this could produce phase matching. He estimated the accelerating gradient to be 0.04 GV/cm. The required power density was in the neighborhood of 10^{12} W/cm^2. Nasonov suggested setting up a static charge distribution in alkali-halide crystals by driving a charge wave through atomic displacements with optical phonons. This delivers a relatively-modest gradient of 0.01 GV/cm. None of these people discuss dechanneling. Indeed, channeling considerations are a minor aspect of the three schemes.

Another variant of a laser-type accelerator in a modified charge distribution lattice has been proposed by Bogacz and his collaborators. This scheme visualizes a strain-modulated lattice, either from a super-lattice or an acoustic wave. The modulated channel serves as an undulator in an inverse free electron accelerator. Bogacz notes that too much undulator gain might cause rapid dechanneling. A typical acceleration gradient might be 0.03 GV/cm.

Belotshitskii and Kumakhov studied both the physical limitations of the laser process as well as the efficacy of channeling for ameliorating the problems. They appear to have been the first to fully appreciate that rapid acceleration could overcome some of the problems with dechanneling. They also noted that in view of the short acceleration time the acceleration of unstable particles was possible. For the increase of transverse energy in a distance dx Belotshitskii and Kumakhov give:

$$\Delta E_1 = e\kappa V_0 dx / E \tag{11}$$

where κ is a coefficient that depends on crystal structure and is about 1 GV/cm for positive particles channeled axially, V_0 is the potential barrier for the channel, and E is the particle energy. To overcome the dechanneling it is necessary that the energy gain in dx, ΔE, be of the same magnitude as E. Chen and Noble treated this problem by using a dechanneling length that scaled with the total energy. They also introduced a normalized rms acceptance from accelerator theory:

$$\varepsilon_{cn} = \frac{1}{2}\gamma a \psi_c \tag{12}$$

where a is the axial channel radius and ψ_c is the critical angle. Note that this acceptance will scale as $\sqrt{E}$. They show that particles will remain channeled as they are accelerated provided the accelerating gradient $G \geq \Lambda^{-1}$, where $\Lambda = e\lambda_d/E$ is a normalized dechanneling length.

Belotshitskii and Kumakhov estimated the required power density for laser acceleration as 10^{15} W/cm^2 but also observed that crystal destruction occurs for power densities of 10^{11} W/g. They suggested that one way to address this problem was to look for crystals with absorption coefficients less than 10^{-4} cm^{-1}.

With power densities this high the process is moving into the plasma regime. In a plasma accelerator a longitudinal electric field is established with a traveling wave in an electron plasma. The maximum gradient for a plasma can be found by using Poisson's equation and taking the case where all the plasma electrons are removed at points of rarefaction for the plasma wave. Substituting the plasma frequency $\omega_p = (4\pi n_0 e^2/m_e)^{1/2}$ into Poisson's equation gives:

$$eE_{max} \approx 0.97\sqrt{n_0} \qquad (13)$$

where E_{max} is the so-called cold wave-breaking field and the force is eV/cm. Here n_0 is the equilibrium electron density. In a gas n_0 could be up to $10^{18}/cm^3$ giving rise to a gradient of 1 GV/cm. Clearly this is a significant step forward beyond the earlier schemes.

Chen and Noble appear to have been the first to look at the accelerating process in a channeling crystal in terms of a plasma. With that recognition they could separate the nature of the driver source used to create the plasma from the plasma acceleration mechanism. The plasma could be stimulated by either a laser or a particle beam. For a particle beam there is an additional advantage in channeling since energy loss due to the driver may also be lowered. For a crystal with a plasma density of $n_0 = 10^{22}$ electrons/cm^3 the energy gradient will approach 100 GV/cm. Of course, the required power drive densities will be very high, in the range of 10^{15} to 10^{19} W/cm^2.

Another approach to plasma acceleration was proposed by Tajima and Cavenago. They suggested driving the plasma using Bormann anomalous transmission of x-rays. For Bormann transmission the crystal geometry is arranged so that the x-rays go into a channel at the Bragg angle. They proposed use of 40 keV x-rays and stated that the scheme could give gradients of 1 GV/cm. Tajima and his collaborators have investigated the beam transport in the crystal, the x-ray optics, and the crystal survivability. Survivability issues are constrained by the required power densities of 10^{19} W/cm^2 or a power of 10^9 Watts.

As noted earlier, there are several very severe problems with all of these schemes. One is the extremely high power density required. Belotshitskii and Kumakhov appear to have been the first to consider crystal damage in detail. They note that crystal destruction takes place at a power density of 10^{12} W/cm^3 for nanosecond-long pulses. This corresponds to current densities of 10^5 A/cm^2. This is roughly related to the fracture threshold for thermal shock. The exact fate of the crystal for a given energy density will depend on such things as the relaxation time for converting plasmon energy to phonons (a pico-second range process). Clearly many of these schemes rely on power densities well beyond the crystal breaking limit. Of course one can ask if the acceleration process can be completed before the damage occurs. If it does, it might be possible to use a new crystal for each acceleration cycle.

5. EXOTICS

The basic problem with channeling applications is the small phase space of a typical channel. If nature had been kinder there might have been wider channels or higher Z atoms. Indeed nature has provided two of these. A third option is to alter the natural state.

Nature has already provided a range of elements with atomic number higher than silicon. As noted in the materials section, some work is underway to produce better tungsten crystals for channeling applications.

The conditions occurring in a solid state accelerator are an illustration of altering the natural state. What happens to channeling when a crystal is struck by a beam from a powerful laser or particle beam? To answer this it is useful to examine the behavior of the critical angle for axial channeling as the electron screening and the temperature are changed. A useful form for investigating this

Macro-channeling using hollow nano-scale tubes has been discussed by Kumakhov. In Kumakhov's scheme all surfaces bend. However extremely good surfaces are needed. Carbon nanotubes with apertures in the nanometer range and micron lengths are available. Recently schemes have been developed for making almost millimeter lengths of 20 to 200 nm caliber tubes using bio-membrane techniques. Channeling experiments have already been carried out in highly ordered pyrolytic graphite. There are several potential roadblocks to using hollow-bore nanoscale structures for channeling. These include surface irregularity and channel alignment. These are similar to problems with dislocations and bending beyond the Tsyganov radius. A second type of problem relates to the surface acceptance.

has been suggested by Andersen:

$$\Psi_{1/2} = \frac{\Psi_L}{\sqrt{2}} \sqrt{\ln\left(\frac{r_0^2}{u_2^2 \ln 2}\right) + \ln\left(\frac{(Ca_{TF})^2 + u_2^2 \ln 2}{(Ca_{TF})^2 + r_0^2}\right)} \qquad (14)$$

where $C = \sqrt{3}$, u_2 is the rms two dimensional lattice vibration amplitude, and r_0 is some channel radius. Removing most of the electrons is equivalent to a large screening length or letting a_{TF} become large. For practical purposes the screening length reaches its limiting values when $a_{TF} = r_0$. For silicon at high temperatures $u_2 = 0.006\sqrt{T}$ where u_2 is in Å and T is in °K. The critical angle behavior as a function of a_{TF} and T are illustrated in Figure 1. Perhaps surprisingly, the changes in the critical angle are not large.

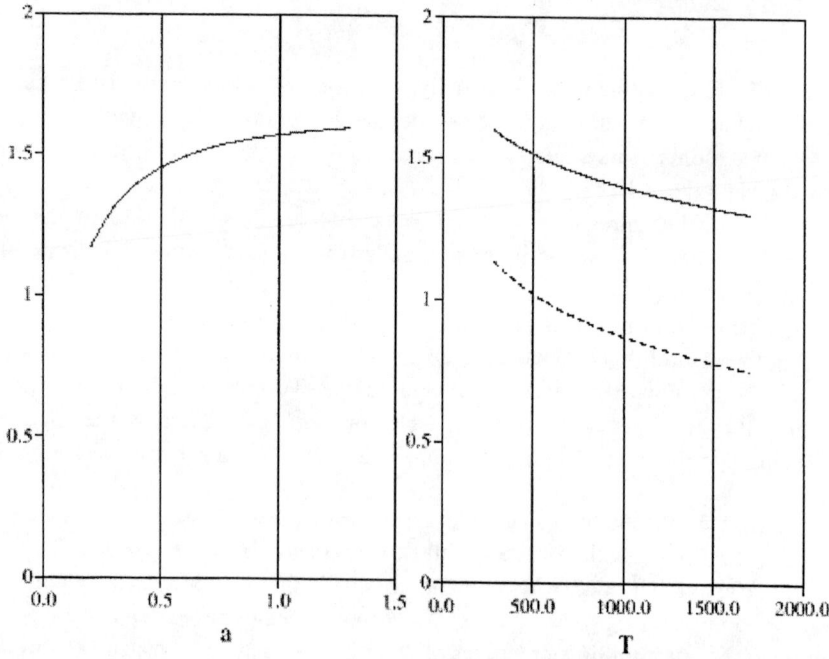

Figure 1. Axial critical angle for silicon in units of the Lindhard angle as a function of a variable screening length and temperature. The solid line for temperature is the unscreened case.

To associate these numbers with the time history of the crystal order it is necessary to understand the evolution of the electron density and the lattice vibrations. How these evolve is a complicated problem. Many of the materials issues have been addressed in connection with laser inertial confinement. In a typical case, a substantial fraction of the electrons might be swept away in much less than a picosecond and the critical angle would grow by 50%. The expanding plasma of ionized electrons might impede the passage of x-rays through the crystal. If the electron plasma was heated too fast the electron density would become too low for some exotic accelerator applications. As the process continued the critical angle would then shrink as ionized electrons generated phonons.

These are the same conditions that naturally prevail for any solid state accelerator. The reasons for considering channeling are beam radiative cooling and the possibility that there may be some channeling gains such as higher effective fields and longer dechanneling lengths. These will be transient effects that will disappear quickly as the crystal vaporizes. Interestingly, beam currents are high enough here that it may be important to consider coherent channeling effects similar to those seen in conventional particle accelerators.

With the advent of "tabletop" terrawatt lasers it may be possible to study channeling dynamically under such conditions. Relatively low energy (MeV regime) channeling back-scattering techniques (RBS) may be satisfactory since one is probably talking about thin crystals. However picosecond channeling RBS studies will require instantaneous currents of many kiloamps. A candidate facility is the Karlsruhe light ion accelerator (KALIF). Another possibility might be to observe bending of intense multi-GeV beams. Relaxation of the bend due to heating could be a complication. Detecting particles with sub-picosecond time resolution would require innovative tools like Kerr cells. The Fermilab wake-field facility may also offer another venue for tests. At Fermilab a high-powered chirped laser will be used to produce electrons by photo-ionization. This arrangement will serve as the source for a 20 MeV linac. Part of the laser beam can be split off to illuminate a crystal placed in the electron beam. With this arrangement one might hope to study the time evolution or degradation of channeling radiation in the picosecond regime as the crystal was heated by the laser.

ACKNOWLEDGMENTS

The author would like to acknowledge useful discussions with J. Andersen, J. Ellison, T. Murphy, R. Noble, and A. Sorensen. Operated by Universities Research Associations, Inc. under Contract No. DE-AC02-76CHO3000 with the United States Department of Energy.

REFERENCES

1. V. M. Biryukov, V. I Kotov, Yu. A. Chesnokov, Physics - Uspekhi **37**, 937 (1994).
2. S. P. Møller, Nucl. Instr. and Meth. **A361**, 403 (1995).
3. 1995 Aarhus Workshop on Channeling and Other Coherent Crystal Effects at Relativistic Energies, eds H. Andersen, R. Carrigan, and E. Uggerhøj, Nucl. Instr. and Meth. B (to be published).
4. D. S. Gemmell, Rev. Mod. Phys. **46**, 129 (1974).
5. E. N. Tsyganov, Fermilab TM-682, TM-684, Batavia (1976).
6. J. A. Ellison, Nucl. Phys. **B206**, 205 (1982).
7. H. Kudo, Nucl. Instr. and Meth., **189**, 609 (1981).
8. V. V. Kaplin and S. A. Vorobiev, Phys. Lett., **67A**, 135 (1978).
9. A. Taratin, private communication. A. D. Kovalenko, V. A. Mikhailov, A. M. Taratin, V. V. Boiko, S. I. Kozlov, and E. N. Tsyganov, JINR Rapid Communications, No. 4 (72), p. 9 (1995).
10. S. I. Baker, R. A. Carrigan, Jr., V. R. Cupps II, J. S. Forster, W. M. Gibson, and C. R. Sun, Nucl. Instr. and Meth. **B90**, 119 (1994).
11. Yu. Chesnokov, et al., Proc. of the 15th Inter. Conf. on High Energy Acc., J. Rossbach,

ed, Hamburg, **1**, 173 (1992).
12. For a summary see M. A. Kumakhov and R. Wedell, **Radiation of Relativistic Light Particles During Interaction with Single Crystals**, Spektrum Physics-Heidelberg (1991).
13. See, for example, G. Kurizki, p. 505 in **Relativistic Channeling**, eds. R. A. Carrigan, Jr., J. A. Ellison (Plenum, 1987). See also Kumakhov and Wedell, p. 190.
14. R. A. Carrigan, Jr., p. 339 in **Relativistic Channeling**, eds. R. A. Carrigan, Jr., J. A. Ellison (Plenum, 1987).
15. S. I. Baker, et al., Nucl. Instr. and Meth. **A234**, 602 (1985).
16. S. I. Baker, R. A. Carrigan, Jr., R. L. Dixon, H. C. Fenker, R. J. Stefanski, J. S. Forster, R. L. Wijayawardana, and S. Reucroft, Nucl. Instr. and Meth. **A248**, 301 (1986).
17. N. Doble, L. Gatignon, and P. Grafström, 1995 Aarhus Workshop on Channeling and Other Coherent Crystal Effects at Relativistic Energies, eds H. Andersen, R. Carrigan, and E. Uggerhøj, Nucl. Instr. and Meth. B (to be published).
18. C. R. Sun, p. 379 in **Relativistic Channeling**, eds. R. A. Carrigan, Jr. and J. A. Ellison (Plenum, 1987).
19. See, for example, A. S. Denisov, et al., Nucl. Instr. and Meth. **B69**, 382 (1992).
20. V. G. Baryshevskii, Pis'ma Zh. Tekh. Fiz. **5**, 182 (1979), Sov. Tech. Phys. Lett. **5**, 73 (1979). L. Pondrom, private communication and **Proc. of the 1982 DPF Summer School on Elementary Particle Physics and Future Facilities**, p. 98, eds. R. Donaldson, R. Gustafson, and F. Paige, Snowmass, CO (1982). V. L. Lyuboshits, Yad. Fiz. **31**, 986 (1980) [Sov. J. Nucl. Phys. **31**, 509 (1980)]. I. J. Kim, Nucl. Phys. **B229**, 251 (1983).
21. D. Chen et al., Phys. Rev. Lett. **69**, 3286 (1992).
22. D. Daniels, Fermilab Hyperon note 569 (1992).
23. R. A. Carrigan, Jr. and V. J. Smith, p. 123 in CHARM 2000 Workshop, eds. D. Kaplan and S. Kwan, Fermilab Conf.-94/190 (1994).
24. V. M. Samsonov, contribution to the 1995 Aarhus Workshop on Channeling and Other Coherent Crystal Effects at Relativistic Energies, eds H. Andersen, R. Carrigan, and E. Uggerhøj, Nucl. Instr. and Meth. B (to be published).
25. V. V. Avdeichikov et al., JINR Communication (1984). English translation: Fermilab FN-429 (1986).
26. A. A. Asseev, et al., Nucl. Instr. and Methods, **A309**, 1 (1991).
27. V. Biryukov, Nucl. Instr. and Meth. **B53**, 202 (1991). A. Taratin, S. Vorobiev, M. Bavizhev, and I. Yazynin, Nucl. Instr. and Meth. **B58**, 103 (1991).
28. X. Altuna et al., Phys. Lett. **B357**, 671 (1995).
29. S. E. Anassontzis, et al., Nucl. Phys. (Proc. Sup.) **B27**, 352 (1992).
30. T. Murphy, 1995 Aarhus Workshop on Channeling and Other Coherent Crystal Effects at Relativistic Energies, eds H. Andersen, R. Carrigan, and E. Uggerhøj, Nucl. Instr. and Meth. B (to be published).
31. V. Biryukov, Phys. Phys. Rev. **E52**, 6818 (1995).
32. R. Carrigan, Fermilab TM-1978 (1996).
33. R. Carrigan, p. 199, **Non-Accelerator Particle Astrophysics**, eds. E. Bellotti, R. A. Carrigan, Jr., G. Giacomelli, and N. Paver, World Publishing, Singapore (1996).
34. A. Kanofsky, Let. al Nuovo Cimento **17**, 191 (1976).
35. C. R. Sun, et al., Nucl. Phys. **B203**, 40 (1982).
36. Z. Huang, P. Chen, and R. Ruth, Phys. Rev. Lett., **74**, 1759 (1995).
37. Z. Huang, P. Chen, and R. Ruth, SLAC-PUB-95-7071, and 1995 Aarhus Workshop on Channeling and Other Coherent Crystal Effects at Relativistic Energies, eds H. Andersen, R. Carrigan, and E. Uggerhøj, Nucl. Instr. and Meth. B (to be published).
38. P. Chen, Z. Huang, and R. Ruth, SLAC-PUB-95-6814, contribution to the Fourth Tamura Symposium on Accelerator Physics, Austin, Texas (1994).

39. S. A. Bogacz and D. B. Cline, Int. Jour. of Mod. Phys., **A11**, 2613 (1996).
40. S. A. Bogacz, D. B. Cline, and D. B. Sanders, 1995 Aarhus Workshop on Channeling and Other Coherent Crystal Effects at Relativistic Energies, eds H. Andersen, R. Carrigan, and E. Uggerhøj, Nucl. Instr. and Meth. B (to be published).
41. A. Kanofsky, Rev. Sci. Instrum., **48**, 34 (1977).
42. P. L. Csonka, Particle Accelerators **5**, 129 (1973).
43. P. Chen and R. J. Noble, p. 517 in **Relativistic Channeling**, eds. R. A. Carrigan, Jr. and J. A. Ellison (Plenum, 1987).
44. I. A. Grishaev and N. N. Nasonov, Sov. Tech. Phys. Lett. **3**,446 (1977).
45. A. F. Pisarev, Sov. Phys. Tech. Phys., **24**, 456 (1979).
46. N. N. Nasonov, Sov. Tech. Phys. Lett., **6**, 214 (1980).
47. S. Bogacz, Particle Accelerators, **42**, 181 (1993).
48. For a summary see R. Palmer, Part. Accel. **11**, 81 (1980).
49. V. V. Belotshitskii and M. A. Kumakhov, Sov. Phys. Dokl., **24**, 916 (1979).
50. T. Katsouleas, C. Joshi, J. M. Dawson, F. F. Chen, C. Clayton, W. B. Mori, C. Darrow, and D. Umstadter, p. 63 in **Laser Acceleration of Particles**, C. Joshi and T. Katsouleas, eds., A. I. P. Conf. Proc. No. 130, New York (1985).
51. T. Tajima and M. Cavenago, Phys. Rev. Lett., **59**, 1440 (1987).
52. T. Tajima, B. Newberger, F. Huson, W. MacKay, B. Covington, J. Payne, N. Mahale, and S. Ohnuma, Part. Acc. **32**, 235 (1990).
53. B. S. Newberger and T. Tajima, Phys. Rev. A, **40**, 6897 (1989).
54. M. Kumakhov, Sov. Tech Phys. Lett. **5** 283 (79).
55. T. W. Ebbesen, Ann. Rev. Mater. Sci., **24**, 235 (1994).
56. E. Evans, H. Bowman, A. Leung, D. Needham, and D. Tirrell, Science, **273**, 933 (1996).
57. B. S. Elman, G, Braunstein, M. S. Dresselhaus, G. Dresselhaus, T. Venkatesan, and B. Wilkens, J. Appl. Phys. **56**, 2114 (1984), D. Schroyen, M. Bruggerman, I. Dezsi, and G. Langouche, Nucl. Instr. and Meth. **B15**, 341 (1986).
58. J. U. Andersen, private communication.
59. See, for example, the recent special section on laser and particle induced shock waves in Laser and Particle Beams, **14** (1996).
60. K. Baumung, et al., Laser and Particle Beams, **14**, 181 (1996).
61. **Proposal for Staged Plasma Wake-field Accelerator Experiment at the Fermilab Test Facility**, J. Rosenzweig, et al. (1996).

II. WORKING GROUP SUMMARIES

Summary Report: Working Group 1 on "Nonconventional Collider Concepts and Luminosity Paradigms"

R. Carrigan, S. Chattopadhyay, P. Chen, D.B. Cline, N. J. Fisch,
R. Fernow, J. Gallardo, Z. Huang, K-J. Kim, H. Kirk,
N. Kroll, N. Marquardt, A.A. Mikhailichenko, F. Mills, R. Noble,
A. M. Sessler, R. Siemann, A. Spitkovsky, D. Sutter, D. H. Whittum,
J.S. Wurtele, M. Xie, A. Zholents, and M. Zolotorev

Reported by J.S. Wurtele

University of California, Berkeley, and
Center for Beam Physics
Lawrence Berkeley National Laboratory

Abstract. This working group was charged with examining new luminosity paradigms and nonconventional collider concepts. The most significant new luminosity paradigm considered was the operation of 5 TeV center-of-mass colliders in a regime that relies strongly on quantum mechanical suppression of beamstrahlung. The nonconventional accelerator concepts—those that did not fall properly into one of the other working groups—were muon colliders and crystal accelerators. We also examined beam cooling and diagnostics.

Introduction

Three main areas were discussed in the working group on new luminosity paradigms and nonconventional collider concepts: high Υ collider design, beam cooling, and crystal accelerators. An outline of the topics we considered is presented below. The following sections of this working group summary report our discussions of each of these areas.

1. High Υ
   (a) Motivation
   (b) Beam Combining and Compensation
   (c) Model and Simulation
   (d) $\gamma-\gamma$ collider

2. Cooling
   (a) Optical Stochastic Cooling
   (b) Ionization Cooling for muon colliders
   (c) Suppression of Quantum Excitation

3. Crystal Acceleration
   (a) Experiments on extraction and deflection
   (b) Theory and possible acceleration experiments

High Υ Colliders

The group reviewed the basic scaling laws for 5 TeV center-of-mass colliders and the rationale [1] for the operation at high values of the beam-beam parameter Υ (and, as well, high-frequency accelerating wavelength). This parameter measures the ratio of the energy of a typical radiated photon (as calculated classically) to the particle energy. Thus, if $\Upsilon \gg 1$ the radiation is no longer classical and a proper QED description must be used. These calculations have been in the literature for some time. Nevertheless, the recent consideration of 5 TeV linear colliders has led many of us to believe that such colliders will likely have to operate with $\Upsilon \approx 10^3$, and we are motivated to examine the limits of our theoretical understanding and our simulation capabilities. We heard talks on the scaling laws and collider parameters and the modeling in the codes such as CAINII [2]. Perhaps the most important regime not yet explored in detail is that wherein the formation length becomes important. In particular, the simulations do not yet handle this regime. At $\Upsilon \approx (207)^2$, muon pair creation occurs, but this is a regime well beyond present consideration. While reviews of the beam-beam interaction, including disruption, radiation and pair production, are available, the details of beam-beam phenomena with parameters appropriate to a 5TeV machine have not been examined in detail. Some work on 5 TeV designs can be found elsewhere in the proceedings [3].

One consequence of operation at high Υ may be increased backgrounds. We examined possible ways to deal with the background, and considered it important that some thought be given to detectors which can operate in this environment. One method for avoiding the background associated with high Υ collider design is to operate with neutralized beams. This can be achieved, for example, by accelerating multiple beams, electrons and positrons, in each linac, and then combining them at the IP. An innovative "matrix" linac concept with beam combining at the final focus was presented by Whittum [4]. Another idea is to collide bunch trains–each bunch collides with many bunches of the opposing beam. In this model one would need a continuous focus (at least long compared to the length of the bunch trains). We noted that with laser acceleration concepts bunch spacing and bunch lengths may be quite short, and disruptions decrease with bunch length.

We discussed the stability issues with neutralized beams. It is well known that there is an instability with four beams. It was felt that the area of multiple beam collisions is ripe for further investigation. A careful study of the physics of multiple beams at the interaction point should be performed using codes. Various phenomena which may reduce the growth rate of the instability, such as tapered density profiles, should be included in a theoretical analysis.

Another method to avoid the undesirable aspects of the beam-beam interaction is to collide photons rather the electrons. An overview of this concept (which is part of the NLC ZDR [5]), the $\gamma-\gamma$ collider, was presented [6]. A third potentially viable method for avoiding beamstrahlung and disruption is plasma compensation. This was discussed [7] in the context of suppressing the beam-beam interaction in the muon collider ring.

Our discussions can best be summarized with a few key points:

1. A 5 TeV center-of-mass energy is likely to require $\Upsilon \approx 10^3$.

2. Such a value of υ may produce significant backgrounds. The level of the background and its consequences should be carefully studied in collabora-

tion with a detector design group. Innovation in detector design may prove to be very important.

3. There was no obvious problem with the model of the beam-beam interaction until bunches are either very short or very dilute, when formation length phenomena and field fluctuations may need to be implemented.

4. The innovative collider designs that implement ideas such as beam combination and multibunch collisions should be explored.

Cooling and manipulation of phase space

A second thrust of the group was in the area of beam cooling techniques, in particular stochastic cooling, laser cooling and ionization cooling for muons. Other topics examined included new applications of lasers, such as beam slicing and crystal-based accelerator schemes.

Cooling of Muon Beams

The muons collider concept (interest in which grew, in part, out of prior workshops on advanced accelerator concepts), has developed to the point where it now warrants an entirely separate set of meetings. There seemed little point in devoting much time to the muon collider, given the numerous other opportunities to attend workshops on the topic. We did, however, consider new ideas for muon cooling, perhaps the most challenging and least developed aspect of the muon collider complex. The discussion of muon cooling comprised an overview [8] of cooling requirements and present scenarios for their realization. This was followed by a detailed discussion [9] of a helical lithium cooling rod to efficiently provide both transverse and longitudinal cooling. Laser cooling for electron linacs was presented by Kim [10]. Finally, the use of a crystal to cool the muon beam was suggested [11].

Emittance limits in rings

Huang [12] presented work on the minimal emittance that can be achieved in a ring. This work unifies the previous treatment of emittance limits in rings developed over two decades ago (see Ref. [13]) with more recent work on the emittance of a beam in a linear focusing channel [14]. A major point in the theoretical analysis is the identification of the ratio of the coherence length to the local radius of curvature as the parameter whose value distinguishes the linear result from the "classical" ring result. Work is in progress on a set of practical (realizable) parameters for such a ring. A discussion [15] on fundamental limits on emittance and applications to colliders can be found in the proceedings.

We heard a report [16] on recent progress in the development of stochastic cooling at optical wavelengths. Experiments are planned and, if, successful, could lead to greatly reduced cooling times for anti-protons at FNAL. Cooling schemes such as stochastic cooling rely on measuring and feedback on fluctuations, from the finite number of particles, in the bunch distribution that arise. The synchrotron radiation of a bunch has statistical properties that, with appropriate diagnostic systems, can be used to infer its macroscopic properties [17]. In Ref. [17] it is proposed that fluctuations in the spontaneous emission spectra be used to measure the length of short bunches.

Applications of lasers: Beam Slicing

Many users of light sources would prefer a shorter photon beam, but with similar brightness to the longer beams now provided by storage rings. Practical considerations, having to due with lattice design and collective instabilities, do not permit rings to operate with high current and very short bunches. However, if a slice of a bunch could be removed it could be used to generate short pulses of xrays. A method for achieving this, suggested by Zholents and Zolotorev [18], uses the inverse free-electron laser interaction and a very short laser pulse. The idea is for the short pulse laser to overlap part of the beam as it propagates through a wiggler. An energy spread on the beam will be induced by the IFEL interaction. Magnets can then select those particles that gain or loose energy–i.e., some of those that happened to overlap with the laser pulse–thereby creating a short electron bunch.

Channeling of particle beams and crystal acceleration

We heard a summary of experiments in which bent crystals were used for extraction and particle deflection [19]. The potential of crystal accelerators to reach very high energy was presented [20]. Some of the inherent difficulties were discussed, along with similarities in the approach to the laser-plasma schemes that use plasma channels.

Acknowledgments

This work was supported by the department of Energy, Division of High Energy and Nuclear Physics.

References

[1] S. Chattopadhyay, D. H. Whittum and J. S. Wurtele, "Report of the Working group on advanced technologies," Proc. Snowmass 96 workshop.

[2] K. Yokoya, CAINII Simulation Code; for a review of beam-beam phenomena see K. Yokoya and P. Chen, "Beam-Beam Phenomena in Linear Colliders," Frontiers of Particle Beams: Intensity Limits, Lecture Notes in Physics, Editors: M. Dienes, M. Month and S. Turner, Vol.400, Berlin: Springer-Verlag, 1992, pp.415-445.

[3] M. Xie, et al., "Design Considerations for 5 TeV Laser Driven Linear Collider," these proceedings.

[4] D. Whittum, "Multiple beam collisions and recombination," and "The matrix accelerator," presentations to the working group.

[5] "Zeroth-Order Design Report for the Next Linear Collider," SLAC Report 474, 1996.

[6] Appendix B in ibid, and K.J. Kim, "$\gamma - \gamma$ colliders," presentation to the working group.

[7] J. Gallardo, "Plasma Compensation at the IP in the Muon Collider Ring," presentation to the working group.

[8] R Fernow, "Muon Cooling Scenarios," presentation to the working group.

[9] F. Mills, "Helical Lithium Rod for Muon Cooling," presentation to the working group.

[10] K.-J. Kim, "Laser Cooling in Linacs," presentation to the working group and "Emittance in Particle and Radiation Beam Techniques," these proceedings. This paper contains a broad discussion of cooling and emittance concepts. The linac laser cooling was suggested by V. Telnov (1996).

[11] S.A. Bogacz, D.B. Cline, and D.A. Saunders, "Final Crystal Cooling to Reduce the Beam Current and Detector Backgrounds for a $\mu^+ - \mu^-$ Collider," these Proceedings.

[12] Z. Huang, P. Chen and R.D. Ruth, *Phys. Rev. Lett.* **74**, 1759 (1995).

[13] M. Sands, "The Physics of Electron Storage Rings," SLAC Report 121 (1970).

[14] Z. Huang and R. D. Ruth, "Suppression of Radiation Excitation in a Focusing Environment," these proceedings.

[15] A.A. Mikhailichenko, "On the Physical Limitations to the Lowest Emittance (Toward Colliding Crystalline Beams)," these proceedings.

[16] A. Zholents, presentation to the working group, and A. Zholents and M. Zolotorev, "Transit-time Method of Optical Stochastic Cooling," Phys. Rev. E, V50, p. 3087, (1994).

[17] M. Zolotorev, presentation to the working group, and M.S. Zolotorev and G.V. Stupakov, "Fluctuational Interferometry for Measurement of Short Pulses of Incoherent Radiation," SLAC-PUB-7132, 1996.

[18] A.A. Zholents and M.S. Zolotorev, "Femtosecond X-Ray Pulses of Synchrotron Radiation," *Phys. Rev. Lett.* **76**, 912 (1996).

[19] R. Carrigan, "Developments in Relativistic Channeling," these proceedings.

[20] P. Chen and R. Noble, "Crystal Channel Collider: Ultra-High Energy and Luminosity in the Next Century," these proceedings.

Table 1: Talks in the Working Group arranged by Topic

Scaling Laws

Wurtele	Consequences of Scaling Laws for High Energy Collider Design
Xie	Cain-II and 5 TeV colliders
Kim	$\gamma-\gamma$ colliders

High Υ Beam-Beam Interaction

Chen	Overview of the beam-beam interaction at high Υ
Kim	Formation length considerations

Beam Combining and Compensation

Whittum	Multiple beam collider concept: acceleration, combining and IP dynamics
Cline	Plasma compensation
Gallardo	Plasma compensation at the IP in a muon collider ring

Cooling

Zholents	Optical stochastic cooling
Fernow	Overview of cooling concepts for a muon collider
Mills	Ionization cooling of Muons
Huang	Suppression of quantum excitation

Diagnostics

Zolotorev	A bunch length diagnostic based on synchrotron radiation

Crystal Acceleration

Noble	Perspectives on accelerators in the 21st century
Chen	A proposed crystal channel experiment
Carrigan	Experimental results on beam deflection and extraction with channels

Summary Report: Working Group 2 on "Plasma-Based Acceleration Concepts"

Thomas Katsouleas
University of Southern California
Department of Electrical Engineering-Electrophysics
Los Angeles, CA 90089-0484

Abstract. There has been tremendous progress in plasma-based accelerator research in the last two years, and this is reflected in the papers in this section and in Table 1 below.

TABLE OF CURRENT EXPERIMENTS

Table 1 is a summary of self-reported results of current experiments represented at this workshop.[1] Of the ten experiments in the table, all but three (UCLA, LLNL, France-BWA) are new since the previous Advanced Accelerator Concepts Workshop in Lake Geneva. This table offers a quick snapshot of how far the field has come and how fast. The units in the table themselves tell the story: *nanoCoulombs, mm-mrads, 100 MeVs, 100 GeV/m, 10Hz*. These are more remarkable in light of the fact that five years ago (just before the Port Jefferson Workshop), not **one** electron had been accelerated by even **one MeV** in a high-gradient plasma acceleration experiment.

Digging deeper into the Table, we see that the first nine experiments are laser-driven and are grouped by the three principal mechanisms of plasma wave excitation by lasers[2]: (1) Laser Wakefield Accelerators (LWFA), (2) Self-modulated (SM) or Forward Raman Scattering (FRS) and (3) Beatwave Accelerators (PBWA). The tenth experiment is driven by a particle beam (Plasma Wakefield Accelerator - PWFA). In essence all are plasma wakefield accelerators -- the LWFA wake is produced by a single short laser pulse that is shorter than a plasma period; the SM or RFS wake is produced by a single laser pulse that is longer than a plasma period and that becomes modulated at the plasma frequency by an instability; the PBWA wake is produced by a laser that is pre-modulated or beating at near the plasma frequency. These broad divisions formed the basis for four of the Working Group's sub-session presentations. The remaining sub-session presentations were on the topics of Injectors and Bunchers, Channels for Laser Guiding, Plasma Lenses, New Concepts and Applications. Each is briefly highlighted next.

TABLE 1. Current Experiments

EXPERI-MENT	Mech.	P(TW)	t(ps)/ $w_p t$	l (mm)	r(mm)/ $k_p r$	Guide?	Dg_{max} (MeV)	N	n_0	a_0	w/w_p
JAPAN	LWFA	0.5	0.09, 5	0.79	10/ 1.9	self-channeling	100	10^3	10^{18}	0.5	42
FRANCE wakefield	LWF	0.2	0.072, 0.3~1.6	0.8	6/ .08~.4	No	-	-	5×10^{15} ~ 1.5×10^{17}	0.3	590 ~108
UT	LFW	0.2	0.1, 3	0.8	3.6/ 0.4	No, but possible modification	-	-	3×10^{17}	0.3	78
RAL	self-mod (FRS)	20	0.8, 180	1.055	10/ 73	not necessary but often observed	100	~nC	1.5×10^{19}	2.0	8.1
UM/UT	self-mod	1-6	0.4, 120	1	9/ 9	Yes	>20MeV	10^9	3×10^{19}	1-2	5
NRL	self-mod	3	0.4, 85	1.053	5/ 3.5	Yes	30MeV	10^9	1.4×10^{19}	2.2	9
LLNL	self-mod (FRS)	3	0.6, 150	1.055	10/ 83	Not diagnosed	>2MeV	~pC	2×10^{19}	1.0	7.1
UCLA	PBWA	0.2	150, 860	1.06+ 10.3	150/ 2.9	No	30	~fC	5.8×10^{16}	0.2	34
FRANCE	PBWA	0.022 0.13	200, 90 3800,1700	1.064 1.053	60/ 3.8	No	1.3	$>10^3$	1.1×10^{17}	0.01 .025	94.5
AWA/ UCLA	PWFA	20nc. 1kA peak e^- beam	20,3.6	-	450/ 0.27	self-formed ion channel	0.5 measured 5 projected	-	10^{13}	-	-

Column abbreviations: Mech. is excitation mechanism, P is laser power, t is laser pulse length, (2nd number is t normalized to ω_p^{-1}), l is laser wavelength, r is spot radius, Dg_{max} is energy gain in MeV, N is number of accelerated electrons, n_0 is plasma density in cm^{-3}, a_0 is normalized laser amplitude, L is the interaction length, g_{inj} is the injected beam energy, f is the rep. rate, # cycles is the number of wake oscillations observed, E_{acc} is the accelerating gradient and dn/n is the normalized amplitude of the plasma density perturbation associated with the wake.

TABLE 1 (CONTINUED). Current Experiments

EXPERI-MENT	Target/Gas	L(mm) L/L$_R$	g$_{inj}$	f (HZ)	Number of Cycles	Limiting Mechanism	E$_{acc}$ (GeV/m)	dn/n (%)
JAPAN	-	20, 50	33	10	~1000	channel length	~5	5
FRANCE wakefield	back filled He	He^{2+}:0.05,5 epw:0.16, p/2	-	10	2~50$^+$	diffraction	-	~100
UT	He	0.052, 1	-	10	6	diffraction	-	~100
RAL	4mm jet He	4, 11	self-trapped (1)	10^{-3}	~30	dephasing	>100	50~100
UM/UT	He 0.75 mm	0.75, 4	-	1/200	~100	gas jet distance	50~150	10~30
NRL	H$_2$/He gas jet	2, 27	-	1/180	67	dephasing	~60	10
LLNL	0.8mm jet He	0.8, 3.2	self-trapped (1)	3×10^{-3}	~30	number of FRS e-flodings	>2.5	>1
UCLA	static fill H$_2$	2.5, 2.5	2MeV	3×10^{-3}	~100	diffraction	3	35
FRANCE	static D$_2$ 2.27mb	4.4, p	5.9	1/1000	>57	dephasing	0.7	2.4
AWA/UCLA	Ar	120, 12	26	1~30	-	-	0.05	~100

LASER WAKEFIELD, PHOTON AND BEATWAVE ACCELERATORS

One of the major highlights of the Workshop was the detailed measurement of the plasma wake excited by the LWFA mechanism at Ecole Polytechnique (J.R. Marques et al.) and at UT Austin (C. Siders et al.). Both groups used interferometry to measure the "photon acceleration" of secondary probe lasers. By varying the delay time between the LWFA laser and probe lasers, the groups mapped out the wakefield profile in r and z. To obtain their single color image of the wake the Ecole collaboration required a plasma wake that was stable and reproducible for over 2000 shots.

The UCLA Beatwave experiment is being redesigned for a 100 MeV experiment with an injected electron beam. K. C. Tzeng, et al. presented PIC simulations illustrating the importance of pre-bunching for maintaining beam quality.

The most surprising results were those from the LWFA experiment in Japan by K. Nakajima et al. The self-guiding of their short laser pulse over 2 cm (50 Rayleigh lengths) is not predicted by simulations and theory; identification of the mechanism for this behavior becomes one of the important unsolved problems for the field.

INJECTORS AND BUNCHERS

The recent proposal of D. Umstadter et al. for a LILAC or Laser-injected-laser-accelerator spawned a rich discussion of cathodeless injection possibilities. E. Dodd and R. Hemker presented PIC simulations of the LILAC in which one laser excited the accelerating wakefield and a second laser crossing or tightly focused behind the first kicks electrons from the background plasma into a region of phase space where they can be trapped and accelerated. An intriguing variation employing two colliding secondary lasers was proposed by E. Esarey. B. Rau et al. and D. Gordon et al. gave interesting talks on the use of sub-cycle or beating lasers in vacuum to bunch beams.

CHANNELS

With the exception of the Japanese results mentioned above, the laser accelerator experiments in Table 1 are on the scale of millimeters. The limiting mechanism in the LWFA and PBWA experiments is generally diffraction of the laser. Thus there is a considerable effort to explore the use of channels to guide the laser and increase the interaction length. Experimental work was presented by H. Milchberg et al., N. Vogel et al. and E. Esarey et al. Theoretical work by P. Volfbeyn, B. Shadwick, J. Wurtele and collaborators illustrated the subtleties in modeling this problem analytically. T. C. Chiou et al. showed PIC simulations of wakes in channels indicating lasers could be propagated over 100 Rayleigh lengths stably with at least 10% extraction of laser energy into plasma wakes. G. Shvets gave an interesting talk on the use of a dense relativistic particle beam (rather than a pre-formed channel) to guide a laser.

SELF-MODULATED AND RFS WAKEFIELD ACCELERATION

A main highlight of the Working Group was the success of the several experiments around the world (RAL, UM, NRL and LLNL) on the acceleration of electrons in self-modulated or RFS laser experiments. Acceleration gradients of 10 -100 GeV/m were typical of these experiments. C. Clayton et al. reported a maximum energy gain of 100

MeV in the RAL-UCLA-Imperial College experiment. There was a great deal of consistency in the data between the separate experiments as reported by A. Ting, S. P. Le Blanc, C. Moore, S. Y. Chen and D. Umstadter. W. B. Mori and E. Esarey gave theoretical interpretations of the phenomena and C. Hubbard and K. C. Tzeng gave numerical analysis and PIC simulations, respectively.

Self-modulated laser accelerators rely on the growth of an instability and are not candidates for future high energy physics accelerator. Nevertheless, they are exceedingly simple (involving only a gas jet and a focused laser -- no injectors, preformed plasmas, etc.) and make robust sources of intense relativistic beams. Bunch densities may exceed 10^{18} cm^{-3}, and this led to considerable discussion of space charge issues with the Sources Working Group. More importantly, the self-modulated experiments provide a valuable testbed for simulation and theory models as well as for developing diagnostic and experimental techniques. These techniques will be needed for experiments on LWFA and PBWA accelerators which **are** collider candidates.

PLASMA WAKEFIELD ACCELERATORS (PWFA)

Plasma wakefield accelerators driven by particle beams will be of increasing interest in the next few years. Preliminary experimental results were reported by N. Barov (ANL/UCLA) and B. Carlsten (LANL) indicating gradients of order 50 MeV/m. But the real excitement may be in new experiments proposed by these and other groups in the next two to five years. J. Rosenzweig described a proposal for a two-stage experiment at Fermi Lab and R. Assmann described plans for proposing a 1 GeV experiment at SLAC. The latter experiment involves USC, UCLA, LBL and SLAC and would accelerate the tail of a 30 GeV SLAC beam in mm-scale plasma wakes ($n_o \sim 2 - 10 \times 10^{14}$ cm^3) excited by the beam itself.

NEW CONCEPTS, PLAMSA LENSES AND APPLICATIONS

Plasma accelerator work is generating an ever enriching array of spin-off applications, including "energy spread compensators", "ionization collimators", plasma lenses and photon accelerator light sources. R. Govil described new plasma lens work beginning at LBL. T. Katsouleas reported preliminary calculations on two new ideas from T. Raubenheimer and S. Heifets at SLAC. The first is the energy spread compensator in which a few meters of plasma are used to cancel the 5 GeV correlated energy spread at the end of the 500 GeV NLC design beam. The proposed GeV experiment at SLAC could serve as a partial proof-of-principle of the concept. The ionization collimator uses a meter-scale gas section to clip the halo from a high energy NLC-class beam. The central beam is intense enough to tunnel ionize the gas, producing a plasma lens effect that strongly focuses the central core but not the halo. S. Chattopadhyay described the use of unique temporal and spatial structure of plasma accelerated micro-beams for applications in ultra-fast chemical dynamics and other basic research. P. Chen explored the possibility of using plasma acceleration to produce detectable "Unruh radiation".

WORKING SUB-GROUP TOPICS

The plasma group broke into smaller sub-groups to work on specific problems. Several will undoubtedly become the subject of the next AAC Workshop. The sub-groups were tasked with the following:

1. Calculate the optimized longitudinal laser profile to maximize laser-to-plasma coupling efficiency. A simple estimate by W. Mori, P. Chen and T. Katsouleas suggested a linear ramp profile may enable pump depletion with minimal laser distortion.

2. Design an experiment to study explicitly laser coupling efficiency in LWFAs. The group of W. Mori, K. Marsh, J. R. Marques and J. Wurtele identified the key parameter to be laser power/(# of laser cycles)^3, thus favoring short high-power lasers. A 30 TW in 30 fs laser at $\lambda = .8$ μm could show 30% pump depletion in a Rayleigh range.

The remainder of the group worked on various topics including identifying mechanisms of LILAC injection (J. Bobin, W. Leemans et al) and interpreting recent experimental results.

CONCLUSION

There has been tremendous progress since the Lake Geneva meeting. Electron beams have been accelerated in laboratories around the world with prodigious amounts of charge, energy and gradients, and with reasonable emittances. However, there are also significant limitations in these current experiments. All of them have essentially 100% energy spread and exist over typically mm-scales. These limitations point to the most urgent research directions; namely, (1) pre-bunchers and injectors capable of synchronously injecting an electron bunch shorter than about 60 fs and (2) channel guiding schemes capable of extending the interaction lengths to 1 to 10 cm. Many interesting possibilities for achieving these goals were discussed at this Workshop and the near term prospects for success are promising. In addition, the new proposals on particle wakefield accelerators and applications of these are very exciting. Finally, the rapid advances in laser technology are bringing issues such as high repetition rate and multi-pulse stability onto the horizon far sooner than expected.

ACKNOWLEDGEMENTS

I wish to thank the members of the Working Group for making this one of the most scientifically and socially enjoyable meetings I have experienced. Special thanks to the working group sub-chairs: E. Esarey, W. Leemans, A. Ting, J-L. Bobin, C. Clayton, R. Assmann, R. Hubbard, W. B. Mori and K. Nakajima; and to those who stayed up late into Thursday night helping to prepare the Working Group Summary talk: J-L. Hsu, T. C. Chiou, K. Marsh, R. Assmann, E. Esarey, J-R. Marques and D. Bernard. Finally, thanks to S. Mistry for manuscript preparation, Sam Vanacek for technical support at Lake Tahoe, Swapan Chattopadhyay for organizing a great Workshop and Dave Sutter and DOE for supporting it.

Work supported in part by US DOE grant # DE-FG03-92ER40745.

REFERENCES

1. Private communication from C. E. Clayton (UCLA, RAL, LLNL), J. R. Marques (Ecole Polytechnique), A. Ting and C. Moore (NRL), C. Siders and M. Downer (UT), D. Umstadter (UM), N. Barov (AWA/UCLA), K. Nakajima (Japan).
2. Tajima, T. and Dawson, J. M., *Phys. Rev. Lett.* **43**, 267 (1979).

Summary Report: Working Group 3 on "Structure-Based Acceleration"

J.B. Rosenzweig
UCLA Department of Physics and Astronomy
405 Hilgard Ave., Los Angeles CA, 90095-1547

Abstract

The working group on structure-based accelerators had an expansive mission, consisting of externally powered structures of high frequency, novel material and/or novel structure design, wake-field (excluding plasma) accelerators, and inverse radiative process accelerators. Within the context of these themes, discussions were held on the subjects of field gradient, efficiency, and accelerated beam quality, with an eye on technical and experimental issues brought on by the need to experimentally develop these schemes. We discuss here the progress presented in the working group, as well as recommendations for future directions in this area of advanced accelerator techniques.

INTRODUCTION

The notion of structure-based acceleration techniques was broadly defined in the 1996 Advanced Accelerator Concepts (AAC) Workshop held at Lake Tahoe. The discussions encompassed all externally powered structures which are sufficiently short wavelength to be considered novel, novel material and/or novel structure design), all wake-field (excepting plasma) and other two-beam accelerators, and all inverse radiative process accelerators — inverse Cerenkov, inverse transition (vacuum), and inverse FEL. These fields have shown considerable progress since the 1994 AAC, with many more participants taking part in the discussions in the working group — more than forty in all.

Within the context of these schemes, particular emphasis was placed on experimental issues, both in the results of present experiments and in future plans. Attention was also paid to the viability of the schemes in application to linear colliders with discussions of power and particle sources (particularly for near term experimental needs), relevant aspects of transverse motion — accelerated beam quality and collective instabilities — and projected accelerator efficiency.

INVERSE RADIATIVE PROCESS ACCELERATORS

While all acceleration techniques can be rigorously viewed as inverse radiative processes, for the purpose of the working group discussion this class of accelerator was somewhat arbitrarily (but in an intuitively and aesthetically well accepted way) was limited to all far-field processes. These processes, which in practice are taken to be those employing infrared or optical radiation with conducting boundaries more than a wavelength away from the beam path, include the so-called vacuum acceleration, which is more properly termed inverse transition acceleration (ITA), the inverse Cerenkov acceleration (ICA), and the inverse free-electron laser (IFEL).

Vacuum acceleration, which is most like standard acceleration techniques in that both employ metallic structures with vacuum beam channels, was the subject of a plenary presentation by P. Sprangle[1] as well as a working group discussion led by L. Steinhauer[2]. In these schemes, the nominally transverse-polarized vacuum field pattern is focused by means of metallic or dielectric obstacles (lenses, masks or irises) to a symmetry axis or plane, yielding a longitudinal fields giving a localized acceleration. One of the main issues associated with these schemes concerns the minimization of the electric field on the obstacles, as excessive fields on these solid objects limit the acceleration gradient due to concern over breakdown. From this point of view, it should be noted that for wavelengths of far IR or shorter that dielectrics are preferable, with limits on field gradient and pulse length set by avalanche breakdown[3].

The IFEL results obtained at the Brookhaven ATF, showing MeV acceleration and excellent agreement with theory and simulation, were discussed by A. van Steenbergen[4]. Perhaps the most notable aspect of the IFEL program at BNL, however, is the proposal to use an IFEL section to pre-microbunch the beam at the 10.6 microns in order to inject into a next-generation ICA experiment. W. D. Kimura presented this proposal[5], including a summary of ICA progress to date, and emphasizing the role of coherent transition radiation diagnosis of the microbunching[6] in the planned experiments. The IFEL is also the subject of experiments in the microwave regime (MIFELA) planned at Yale[7], and the simulation results and design of this experiment were presented in the working group by R.B. Yoder. The Yale group also is investigating the ICA in the microwave limit, and an analysis of the beam dynamics for this accelerator, as well as dielectric breakdown studies in this wavelength regime were discussed by J.L. Hirshfield[8].

The microwave ICA (MICA) is not a gas-loaded device, but a dielectric-loaded wave guide with a vacuum beam hole, a pure travelling wave analogue of the common disk-loaded accelerator, which was first discussed over 45 years ago. Discovery of new, low loss materials for use at relatively short wavelengths in these devices has led to renewed interest in the MICA concept. The main problem with extending this concept to shorter wavelengths is the existence of power sources. One of the candidates for this source is the wake-field transformer under investigation at ANL. Computer studies of the travelling wave propagation in X-band MICA structures were presented by M. Conde.

The MICA concept originated at cm to mm wavelengths, but has been proposed, in order to take advantage of power source availability, at IR or optical wavelengths by W. Gai[9]. In this laser-excited device, one must couple radially polarized optical pulse into a travelling wave mode in a hollow optical fiber with an axicon lens. It was pointed out in discussion that this scheme suffers from excessively large wake-field problems, as well as potential Raman scattering degradation of the pump laser at "interesting" fields (1 GV/m).

Neither the IFEL nor the gas ICA are analogous to rf linacs, as they do not have paraxial particle velocity in vacuum. The IFEL, in particular, is different in that it is a second order acceleration, proportional to product of the laser and wiggler field amplitudes. A similar phenomena was reported on by J.-L. Hsu[10], who discussed the acceleration possible by ponderomotive effects in terawatt-to-petawatt laser focus. This acceleration can be considered as the inverse process of laser wake-field excitation of plasma electrons.

SLAB SYMMETRIC STRUCTURES

The luxury afforded by those who wish to accelerate particles in short wavelength structures is that the power source (*e.g.* laser) considerations are not the limiting factor in accelerating gradient optimization, as they are in microwave-based accelerators. On the other hand, very serious limitations arise due to the decreased size of the structures: geometric tolerances and power coupling become more challenging, and beam-loading and transverse wake-field effects become much more important.

It is now recognized that these limitations and challenges can be partially addressed by use of slab-symmetric structures. A precursor of the most recent ideas was discussed by H. Kirk of BNL, who discussed Smith-Purcell radiation experiments at BNL[11] and plans for related acceleration experiments. The main attraction of the inverse Smith-Purcell effect is ease of radiation coupling, as it entails use of a one-sided grating structure. However, this asymmetric system produces strongly non-uniform acceleration (as a function of distance from the grating), and a related, second order deflecting force which pushes particles away from the grating.

In order to remove the dependence of laser accelerators on peak power, as well as to symmetrize the system, a proposal for a side-injected, slab-symmetric, resonant dielectric-loaded laser accelerator[12] was presented by J. Rosenzweig. In this scheme, a dielectric-lined Fabry-Perot resonator with a small longitudinal variation (periodic at the radiation vacuum wavelength) in the input reflectivity allows a resonant standing wave accelerating field to be built up during several picosecond infrared laser illumination. This illumination can be made synchronous with the beam by used of an electro-optic sweeping technique proposed by A.A.

Mikhailichenko[13]. P. Schoessow showed 2-D time-dependent electromagnetic simulations verifying the mode patterns and filling dynamics of the structure[14].

In these slab-symmetric structures, there are several advantages obtained over cylindrically symmetric structures or open structures — there is a strong second order vertical focusing force[12,15], as well as a natural suppression of the transverse (deflecting) wake-fields *if* the beam is also very wide compared to the vertical dimensions of the structure. Analytical wake-field calculations[16] were discussed by A. Tremaine and Rosenzweig, and computational results on the problem were shown by Schoessow[14].

A scheme which is part way between the slab-symmetric resonator and the vacuum accelerator, a dielectric-loaded, slab-symmetric laser-driven accelerator operating in vacuum, was discussed by Y.C. Huang[17]. In this scheme nearby cylindrical lenses and prisms bring the laser light into a focus only in the vertical dimension. This scheme was shown to be well optimized from the point of view of media breakdown and maximum acceleration, which was estimated to be 0.5 GeV/m. Unaddressed concerns include wake-fields and a full treatment of particle dynamics.

These concepts are also under investigation in the mm-wave regime by groups at SLAC, ANL and TU-Berlin. H.Henke[18] gave a presentation covering cavity design results, 3-D computational electromagnetic modelling of structures, and fabrication techniques such as LIGA. P.J. Chou also discussed work ongoing at SLAC on fabrication and low-power testing of muffin tin structures at X-band and W-band.

An interesting result on the general question of structure coupling was presented by J. Haimson[19], who discussed a racetrack geometry which azimuthally symmetrizes the coupling cells in nominally cylindrically symmetric structures. While these results were presented for travelling wave linac sections, they are perhaps most usefully applied to standing wave rf gun designs.

EXOTIC ACCELERATORS

The attraction of the slab-symmetric systems with sheet beams is in the mitigation of short range transverse wake-field effects. Other structures have been proposed to allow damping of long-range wake-field effects. These proposals range from those already implemented in a sophisticated manner, like the damped detuned structure (DDS) at SLAC, discussed by N. Kroll and R. Jones[20], and the more exotic, like the photonic band gap accelerator (PBG). The PBG is a structure composed of many periodic (in the transverse plane) electromagnetic obstructions, with one removed at the beam hole position. This forms a defect mode which is well confined spatially, while other higher order modes are deconfined, leading to long-range transverse wake-field suppression. Recent progress on the PBG work done at UCSD was reviewed by D.R. Smith[21], and computational studies of the wake-fields studies the PBG were discussed by Derun Li[22].

The electromagnetic interaction of a charged particle with its surrounding medium — wake-field generation— is usually considered to be a linear, dissipative process. L. Schachter has proposed and analyzed cases where this is not true, when the medium is active, not reactive. In this case, particles can be directly accelerated by energy stored at the atomic level in the medium. This breaking of assumptions about energy transfer in accelerators is potentially of great importance; the usefulness of the notions proposed by Schachter await further experimental investigation.

CONCLUSIONS

The field of structure-based accelerators is going through a period of rapid growth and maturation, with some penetrating insights now being obtained from theory, computation and experiment. The level of activity is on the rise in this field, both in number of researchers and number of laboratories engaged in the work. Although in comparison to plasma accelerators, the field is still in start-up — developing concepts and technology, one can reasonably expect new and important results before the next AAC workshop.

ACKNOWLEDGMENTS

The author would like to thank all the members of the working group for a stimulating week at Granlibakken. This work performed with partial support from U.S. Dept. of Energy grants DE-FG03-90ER40796 and DE-FG03-92ER40693.

REFERENCES

1. "Vacuum Acceleration", P. Sprangle, these proceedings.
2. "Inverse Transition Acceleration" L. C. Steinhauer, these proceedings.
3. D. Du, *et al., Appl. Phys. Lett.* **64**, 3073 (1994).
4. A. van Steenbergen, et al., *Phys. Rev. Lett.* **76**, (1996).
5. "Inverse Cerenkov Acceleration With an IFEL Prebuncher", W. Kimura, *et al.*, these proceedings.
6. "Coherent Transition Radiation Diagnosis of Electron Beam Microbunching", J. Rosenzweig, *et al., Nucl. Instr. Meth. A* **365**, 255 (1995); Y. Liu, these proceedings.
7. "Simulation results and experimental design for the microwave inverse FEL accelerator", R.B. Yoder, these proceedings.
8. "Analysis and dielectric breakdown studies for a microwave inverse Cerenkov accelerator", J.L. Hirshfield, these proceedings.
9. "Cylindrically symmetric laser excited fiber accelerator" W. Gai, these proceedings.
10. "Vacuum Acceleration with an Intense Laser" Jui-Lung Hsu, these proceedings.
11. "Smith-Purcell Radiation and Acceleration", Harold Kirk, these proceedings.
12. J. Rosenzweig, et al., *Phys. Rev. Lett.* **74**, 2467 (1995).
13. "A Linac Driven by a Traveling Laser Focus", A.A.Mikhailichenko, these proceedings.
14. "Computational modelling of slab-symmetric, dielectric loaded structures" P. Schoessow, these proceedings.
15. "Experimental Determination of the Transverse Trace Space Map of a Standing Wave Linear Accelerator", S. Reiche, *et al.*, submitted to *Phys. Rev. E.*

16. "Electromagnetic Wake-fields in Slab-symmetric Dielectric Structures", A. Tremaine, *et al.*, to be published in Proceedings of Snowmass '96; also A. Chao, same proceedings.
17. "Design for a dielectric-loaded laser-driven accelerator operating in vacuum". Y.C. Huang, these proceedings.
18. "Slab-symmetric structures at mm wavelength" H.Henke, these proceedings.
19. "A Racetrack Geometry to Avoid Undersirable Azimuthal Variations of the Electric Field Gradient in Coupling Cavities for TW Structures", J. Haimson, these proceedings.
20. "The Damped Detuned Accelerator Concept and its First Implementation", N. Kroll, *et al.;* "Recent Results & Plans For The Future On SLAC Damped Detuned Structures (DDS): Theory & ASSET Measurements" , R. M. Jones, these proceedings.
21."Progress on photonic band gap accelerator cavities" D.R. Smith, these proceedings.
22. "Wake-field studies on photonic band gap (PBG) Accelerator cavities" Derun Li, these proceedings.
23. "PASER: particle acceleration by stimulated emission of radiation" L. Schachter, *Physics Letters* **205,** 355 (1995).

Summary Report: Working Group 4 on "Beam Monitoring, Conditioning, and Control at High Frequencies and Ultrafast Timescales"

X.J. Wang

Brookhaven Accelerator Test Facility
National Synchrotron Light Source
Upton, NY 11973

Abstract. The highlights of Seventh Advanced Accelerator Concepts (AAC) working group IV (Beam Monitoring, Conditioning and Control at High Frequencies and Ultrafast Timescales) are presented in this report. The talks given at the working group covered a wide range of beam monitoring subjects. They include a new technique for measuring sub-picosecond electron beam bunch length, an optical stochastic cooling experiment, timing jitter measurement of a photocathode injector, and a proposed experiment to measure micro-bunching of an IFEL accelerator. Working group IV also carried out extensive discussions on the longitudinal and transverse emittance characterization of short (sub-picosecond) low emittance (normalized rms emittance < 1 mm-mrad) electron beams, and beam diagnostic requirements for a Muon collider.

I. INTRODUCTION

With successful experimental demonstrations of the first generation of laser plasma [1], Inverse Cerenkov [2], and Inverse Free Electron Laser (IFEL) [3] accelerators, the generation and characterization of ultra-short, low emittance electron beams for future laser accelerators now become one of the major challenges for advanced accelerator research. The proposed X-ray FEL by SLAC and DESY also has similar requirements. The charge of working group IV at the Seventh Advanced Accelerator Concepts (AAC) Workshop was to discuss the recent developments in beam monitoring, conditioning, and control, and to identify critical areas of future research.

The works reported in working group IV are summarized here. Three discussion sessions were held on longitudinal emittance measurement, transverse emittance characterization, and Muon collider beam diagnostics. The following sections reflect my personal view of the discussions. There are many important developments in beam diagnostics that were not discussed in our working group, such as nanometer beam size measurement [4] and sub-micrometer beam position monitors [5], due to both the time limitation and participants' interests.

II. SUMMARY OF WORKS REPORTED

Table 1 summarizes the subjects covered by the talks presented in working group IV. The three talks on the transverse emittance measurement discuss three different techniques of emittance measurement. J. Power of ANL [6] discusses using the quad scan technique for measuring the emittance of a space-charge dominated beam. By including space charge effect in the envelope equation instead of the traditional matrix formula, the measurement agreed reasonably well with the beam emittance.

Table 1: Summary of the talks presented at working group IV.

Subject	Number of talks
Transverse Emittance Measurement	3
Pulse length and longitudinal emittance measurement and technique	4
Beam Conditioning- Optical Stochastic Cooling	1
Timing Jitters Measurement	1
Muon Collider Beam Diagnostics	1

The use of transition and diffraction radiations for electron beam diagnostics was reviewed by R. Fiorito [7]. The basic properties of TR and DR were first discussed. Comparison of TR and DR with synchrotron radiation reveals that the angular distribution of TR and DR are very broad, so the spatial resolution of TR and DR are mainly determined by the collection angle of the imaging optics. Fiorito presented preliminary data of an OTR image from the CEBAF 3.25 GeV beam. Data presented are consistent with wire scanner measurement and limited optical resolution, and smaller than self diffraction limits.

Four talks presented covered techniques of measuring a wider spectrum of pulse length and longitudinal emittance. Y. Liu of UCLA presented an experimental scheme for measuring micro-bunches (<15 fs) produced by the Inverse Free Electron Laser (IFEL) accelerator [8]. M. Uesaka of the University of Tokyo discussed a technique for utilizing an S-band linac and magnet compression system to produce sub-picosecond electron pulse and pulse length measurement using a streak camera [9]. He pointed out the advantage of the streak camera for short pulse length measurement, such as the pulse structure and direct longitudinal emittance measurement in the dispersion region. D.H. Dowell of Boeing presented his experiment in which he measured longitudinal phase space of a 144 MHz photocathode RF gun at Bruyeres-le-Chatel [10]. He reported the first observation of a longitudinal beam breakup caused by the space charge effect in an RF gun. M. S. Zolotorev of LBNL discussed a new technique for measuring short electron beam pulse length [11]. This technique is based on measuring the visibility fluctuation of incoherent radiation interference fringe and statistical analysis to reconstruct the pulse shape. The main advantage of this technique is that it allows the measurement to be carried out in the visible and x-ray region, where sensitive detectors are readily available.

The only paper discussing beam conditioning was given by A. A. Zholents of LBNL. He first discussed the basic principle of the transit-time method of optical stochastic cooling [12] and its advantages. Both transverse and longitudinal phase space can be cooled simultaneously with a wider bandwidth, as compared with microwave stochastic cooling. An experiment is under construction at LBNL to test the tolerance of optical stochastic cooling on the synchronization and phase mixing.

Other talks presented in the working group were timing jitter measurement in the RF gun injector and beam diagnostics requirements for a Muon collider. R. Fernow of BNL gave an outline of the future Muon collider and basic beam parameters. The discussion following the talk focused on the emittance measurement and background produced by the Muon decays. Emittance on-line monitoring is critical for the success of the Muon collider. The main challenge in Muon beam diagnostics is the large number of electrons produced by the Muon decay (10^9) and the background associated with both beam position and profile measurement. The author discussed four different techniques for measuring timing jitters of a photocathode RF gun injector. The timing jitter resolution of 0.5 ps has been experimentally demonstrated using a quarter-wave-length beam position monitor [13].

III. SURVEY OF SUB-PICOSECOND BUNCHLENGTH MEASUREMENT TECHNIQUES

Our discussion about sub-picosecond electron bunch length measurement is delineated in Table 2, which summarizes the electron bunchlength requirements for various frequency linacs and laser accelerators based on the requirement that the energy spread be smaller than a few percents. For FEL, wakefield control, and other beam dynamics studies, the bunch shape and longitudinal emittance are also important issues for consideration.

Table 2: Typical pulse length for advanced accelerator applications.

S-band Linac	500fs - 10 ps
30 GHz Linac	100fs-1ps
90 GHZ Linac	<200 fs
X-Ray FEL	≈100 fs
CO_2 Based IFEL, ICA	<15 fs
Plasma Based Laser Accelerators	10 fs - 1 ps
Number of Particles	$10^7 - 10^9$

As seen in Table 2, bunch lengths required for most advanced accelerators and X-ray FEL applications are in the range a few femtoseconds to a sub-picosecond. The techniques that either demonstrate or have the potential to measure the sub-picosecond bunch length are discussed in the remaining part of this section.

Time Domain: The only technique in the time currently able to measure sub-picosecond pulse length is the streak camera. The state-of-art streak camera is the FCS-200 streak camera manufactured by Hammamatsu. It has a resolution of 200 fs (FWHM). Using a streak camera for charged particle beam measurement first involves production of radiation, then imaging the radiation onto the streak camera. Many forms of radiation, such as synchrotron radiation, Cerenkov radiation, and transition radiation, can be used to produced the bunch length information of the beam. Care must be taken for sub-picosecond measurement, such as dispersion of the optics and the source effect must be minimized. Transition radiation is the choice for very short bunch length measurement because of its promptness. For sub-picosecond measurement, a synch-scan streak camera is required. The streak camera has played an important role in short bunch length characterization in both circular [15] and linear accelerators [16]. Its measurement is direct and single shot, and it is capable of measuring the charge distribution within the bunch. It can also measure the longitudinal phase space of the radiation located in the dispersion region. The main disadvantages of the streak camera are that it is expensive, and requires a large amount of charge and a small dynamics range. Using RF deflector technology, it is believed that streak camera resolution will reach 50 fs (FWHM) in a few years.

Frequency Domain: Techniques considered in the frequency domain can be further divided into coherent and incoherent radiation.

Coherent Radiation: The total radiation power generated by a bunch of N charged particles can be expressed as,

$$I_{tot}(\omega) = I(\omega)[N + N(N-1) F(\omega)] \tag{1}$$

where $I(\omega)$ is the single electron radiation power and

$$F(\omega) = |\int d\mathbf{r}\, S(\mathbf{r})\, e^{i\mathbf{k}\cdot\mathbf{r}}|^2 \tag{2}$$

is the Fourier transform of three-dimension beam distribution $S(\mathbf{r})$. The first part of Eq. (1) is the incoherent radiation and the second part is the coherent radiation. It is coherent radiation that contains the bunch length information, and coherent radiation wavelength is comparable with the bunch length. Two techniques are used with coherent radiation to extract electron beam bunch length information.

The first one utilizes the auto-correlation method proposed by W. Barry of CEBAF to measure the bunch length. A Michelson interferometer with one arm on the movable translation stage was developed at Stanford. The interferogram is the direct measurement of the bunch length [16]. It was reported that bunch lengths as short as 100 fs (rms) were measured with a resolution of 30 fs [16].

Another way to extract bunch length information from coherent radiation is by frequency analysis using a Kramers-Kronig relation [17]. By measuring the coherent radiation over a wide frequency range, it is possible to extract both the amplitude and the phase information of the radiation sources applying a Kramers-Kronig relation to the spectrum form factor to find the minimum phase. The reported resolution of frequency analysis techniques is on the order of 50 fs [18]. Using coherent radiation to obtain the bunch length information has many advantages, such as its simplicity and demonstrated capability in the sub-picosecond to femtosecond resolution. Since the radiation usually ranges from mm to IR, the detector will be major challenge for advanced accelerator applications due to the small number of electrons (10^7).

Incoherent Radiation: The fluctuational interferometer technique discussed in the previous section [11] is used to measure the incoherent radiation interference fringe visibility fluctuation to extract the bunch information. Since the radiation must be incoherent, the working spectrum will be in visible and x-ray regions, and measurement could be single shot. This technique will soon be tested at the LBNL beam test facility.

All frequency domain techniques are limited by the bandwidth of the detector and optical system. The reconstruction bunch information is incomplete, and hence the detailed structure of the bunch distribution may be lost. It will be very important in future experiments to compare the frequency domain techniques with streak camera measurement in the sub-picosecond scale.

RF deflector cavity and Linac based techniques: RF cavities operating in the TM_{110} mode have been used for electron beam bunch length measurement in many facilities [19,20]. The principle of using the RF deflector cavity for bunch length measurement is similar to that for the streak camera. The fields of the cavity can be described by

$$E_z = E_0 \cos(\frac{\pi x}{a}) \sin(\frac{2\pi y}{d}) \cos\omega t$$

$$H_x = -\frac{\lambda}{d} E_0 \cos(\frac{\pi x}{d}) \cos(\frac{2\pi y}{d}) \sin\omega t \qquad (3)$$

$$H_y = -\frac{\lambda}{2d} E_0 \sin(\frac{\pi x}{d}) \sin(\frac{2\pi y}{d}) \sin\omega t$$

where a and d are x and y dimensions of the cavity, on axis particles passing through the cavity suffer a vertical (y) kicker, and the amplitude of the kick depends on the relative phase between the cavity RF fields and the particles. The angular kick will translate into vertical displacement after a drift distance, hence by measuring the beam size increase in vertical dimension we can obtain the bunch information of the particles. One of the advantages of using the RF kicker cavity is that it can be very accurately calibrated by varying the phase of the RF and measuring the centroid variation of the beam [20]. The resolution of the RF kicker cavity depends on the location (beam optics) of the cavity and the emittance of the beam. Sub-picosecond resolutions were reported for both L-band and S-band cavities [19,20].

Several techniques can be used to obtain electron beam bunch information with the RF linac. The energy spread of the beam near the beam crest can described by

$$\frac{\Delta E}{E} = \frac{1}{8}\Delta\phi^2 \qquad (4)$$

where $\Delta\phi$ is the electron beam bunch length measured in the RF phase spread. A technique developed at the BNL accelerator test facility using the linac can measure the charge distribution within the bunch with half picosecond resolution [21]. The linear energy chirp was first produced by dephasing the linac about 30° from the crest, and the chirped electron was passed through a dipole magnet and energy selection slit, the opening of the slit and the beam emittance determining the resolution. The distribution of charge was measured by a beam position monitor after the electron was passed through the energy selection slit. The different sections of the beam were passed through the slit by varying the linac phase. This technique has the advantage of a large dynamics range, hence one can measure the detailed structure of the bunch.

The tomographic technique is another method of using the linac to measure the electron beam longitudinal phase space [22]. The energy spectra of the electron beam were taken as the function of the relative phase between the beam and accelerating field. The longitudinal phase space of the electron beam was reconstructed using inverse Radon Transform technique. With an energy resolution of 0.1 %, the resolution of the bunch length can be 250 fs.

Laser based techniques. With the dramatic increase in peak laser power, electron pulse length measurement based on the interaction between the electron beam and laser field, i.e., Compton scattering, becomes more feasible [23]. The two techniques proposed are the laser micro-probe [24] and Shintake laser heterodyne method [25]. A femtosecond laser with a delay line was used to probe the electron pulse structure. If the laser pulse is much smaller than the electron beam pulse length, the detected x-ray flux variation as the laser relative to the beam timing was varied gives the longitudinal distribution of the electron beam. An experiment has demonstrated that this technique has the sub-picosecond resolution [24]. The timing jitter between the laser and electron beam and transverse electron beam size are the main limitation of resolution.

Another technique proposed by Shintake uses two laser beams with different frequencies to generate the intensity modulation (beam wave). A normal incident electron beam will produce x-ray flux due to Compton scattering. The fluctuation depth of the x-ray gives the Furies spectrum of the injected bunch. Scanning the frequency of the one laser allows measurement of the wide spectrum of the bunch. All laser-based techniques not only require a powerful laser, but also laser number of electrons due to the small cross section of the Compton scattering process.

IV. BEAM PROFILE MONITORING AND EMITTANCE MEASUREMENT

Emittance is one of the fundamental parameters characterizing charged particle beams. It determines the final spot size in the linear collider and the wave length at which the X-ray FEL will be operating. Emittance characterization always plays an important role in all accelerator development. In the advanced accelerator research community, emittance measurement will become more critical for second generation laser acceleration experiments; for example, emittance preservation in the plasma-based accelerators will determine its potential applications, and ultra-low emittance (normalized rms emittance $< 10^{-8}$ mm-mrad) is required for structure-based laser accelerators. There are many excellent reviews and lectures on emittance concepts and measurement techniques [26,27].

I will devote the rest of this section to the importance of the accuracy of emittance measurement and a brief discussion on the several commonly used emittance measurement techniques. The importance of emittance and accurately determining emittance can be best illustrated using an x-ray FEL application. To operate an FEL, the geometric emittance of the electron beam must satisfy the following conditions

$$\lambda = \frac{\lambda_W}{2\gamma^2}(1 + K^2 + \gamma^2\theta^2)$$

$$\lambda \approx \frac{\varepsilon}{4\pi}$$

where λ is the x-ray wavelength, and ε is the geometric emittance. And using $\varepsilon = \varepsilon_n/\gamma$ and FEL resonance condition, we have

$$\varepsilon_n = \frac{4\pi\lambda_W}{2\gamma}. \qquad (5)$$

The gain of single pass FEL G,

$$G \propto B^{1/3} \propto 1/\varepsilon_n^{2/3}, \qquad (6)$$

where B is the beam brightness. Equations (5) and (6) show that the beam energy and gain of FEL are almost inversely proportional to the normalized beam emittance. Since the beam energy is linearly proportional to the linac length, and the wiggler length is inversely proportional to the gain of FEL, the cost uncertainty of linac and wiggler is directly proportional to the uncertainty of the emittance measurement.

All transverse emittance measurement techniques involve determining both the beam spot size and its divergence angle. So accurately measuring emittance requires precision in measuring both the spot size and angle. We can classify emittance measuring techniques into two categories. The first one usually involves only measuring the beam profile, such as the multi-screen method, quadrupole magnet scan, and pinhole method. The accuracy of emittance measurement is directly proportional to the resolution of the beam profile monitor, typical on the order of 50 μm. Using transition radiation, wire scanner, and YAG crystal [28], we may improve the beam profile resolution to the order of 10 μm. The accuracy of emittance measurement based on beam profile measurement also depends on a special arrangement because the beam divergence is derived from the beam profile information. For example, in three beam profile monitor techniques, the emittance measurement accuracy is directly determined by the separation between the beam profile monitors and the beam waist location [29]. Another good example is the quadrupole magnet scan techniques—the minimum spot size measured is directly proportional to the emittance, and the divergence of the beam is basically determined by quadrupole current far from the minimum spot size. To measure the small divergence, the spot size of the beam image must be varied significantly. This requires not only good beam profile monitor resolution, but also a large dynamic range. A 10- to 12-bit CCD camera with pixel size on the order of a few microns is needed for the quadrupole magnet scan technique.

The second type of emittance measurement technique can measure both the beam profile and angular divergence simultaneously. This includes transition radiation, diffraction radiation, synchrotron radiation, and laser scattering. One of the most successful techniques so far developed for emittance measurement is the two-foil interometer. It uses optical transition radiation (OTR) to measure the beam profile while utilizing an interference pattern to measure the angular distribution [7]. The angular resolution of 100 μrad was measured experimentally.

V. ACKNOWLEDGMENTS

I would like to thank all participants of working group IV: R. M. Alvis, K. Bane, D.H. Dowell, R. Fernow, R. Fiorito, Y.Liu, N.Marquardt, D.T. Palmer, J.Power, M. Uesaka , A. Zholents and M. Zolotorev. This work is supported by the U.S. Department of Energy under contract No. DE-AC02-76CH00016.

VI. REFERENCES

1. K. Nakajima et al, Phys. Rev. Lett. Vol. 74, No.22, 4428 (1995).
2. W.D. Kimura, G.H. Kim, R.D. Romea, L.C. Steinhauer, I. V. Pogorelsky, K.P. Kusche, R. Fernow, X.J. Wang and Y. Liu, Phys. Rev. Lett., Vol.74, No.4, 546-549 (1995).
3. A. van Steenbergen, J. Gallardo, J. Sandweiss, M. Babzien, J.M. Fang, X. Qiu, J. Skaritka and X.J. Wang, Phys. Rev. Lett. Vol.77, No. 20, 4280 (1996).
4. T. Shintake, Nucl. Inst. Meth. A311 (1992) 453.
5. J.P. H. Sladen, I. Wilson and W. Wuensch, Proc. of EPAC 96, 1609 (1996).
6. J. Power, these proceedings.
7. R.B. Fiorito and D.W. Rule, AIP Conf. Proc. 319 (1991) 21-37.
8. Y. Liu, D.B. Cline, X.J. Wang, M. Babzien, J.M. Fang and V. Yakimenko, these proceedings.
9. M. Uesaka et al, these proceedings.
10. D.H. Dowell, S. Joly and A. Loulergue, these proceedings.
11. M.S. Zolotorev and G.V. Stupakov, Submitted to Phys. Rev. Lett., SLAC-PUB-7132 (1996).
12. M.S. Zolotorev and A. A. Zholents, Phys. Rev. E, Vol. 50, No. 4, 3087-3091 (1994).
13. X.J. Wang, I. Ben-Zvi and Z. Segalov, Proc. of EPAC 96, 1576 (1996).
14. E. Rossa, AIP Conf. Proc. 333 (1993), 148-159.
15. R. L. Holtzapple, SLAC-487 (1996).
16. H. Wiedemann, AIP Conf. Proc. 367 (1995) 293-306.
17. R. Lai and A.J. Sievers, AIP Conf. Proc. 367, 312-326 (1995).
18. F. Amirmadhi, C.A. Brau, M. Mendenhall, J.R. Engholm and U. Happek, Nucl. Inst. and Meth. In Phys Resea. A375 (1996) 95.
19. B.E. Carlsten et al, AIP Conf. Proc. 367 (1995) 21-35.
20. X.J. Wang , T. Srinivasan-Rao, K. Batchelor, I. Ben-Zvi and J. Fischer, Nucl. Inst. and Meth. In Phys. Resea. A 356, 159-166 (1995).
21. X.J. Wang, X. Qiu and I. Ben-Zvi, Phys. Rev. E Vol. 54, No. 4, R3121-R3124 (1996).
22. E.R. Cross et al, Nucl. Inst. and Meth. A 375 87-90(1996).
23. J.L. Bobin, AIP Proc. 367, 371-380 (1996).
24. W.P. Leemans, R.W. Schoenlein, P. Volfbeyn, A.H. Chin, T.E. Glover, P. Balling, M. Zolotorev, K.J. Kim, S. Chattopadhyay and C.V. Shank, Phys. Rev. Lett. Vol. 77, No.20, 4182-4185 (1996).
25. T. Shintake, KEK Preprint 96-81 (1996), to be published in the Proc. of 7th Beam Instrumentation Workshop.
26. C. Lejeune and A. Aubert, Adv. In Electronics and Electron Physics, Supplement 13 A, 128 (1980).
27. O. R. Sander, AIP Proc. 212, 127-155 (1989).
28. W. S. Graves, private communication.
29. K. D. Jacobs, J.B. Flanz and T. Russ, Proc. Of 1989 PAC, 1529 (1989).

Summary Report: Working Group 5 on "Particle Beam Sources"

L. Serafini
INFN-Milano and UCLA Dept. of Physics and Astronomy
405 Hilgard Ave., Los Angeles, CA 90095-1547

A.D. Yeremian
SLAC - Stanford
P.O. Box 4349, M/S 26, Stanford, CA 94309

Abstract

We report here on the activity carried in the working group on "Particle Beam Sources", which was actually focused on electron beams and mostly covered the progress and future perspectives of laser driven RF photo-injectors. Recent experimental work on other types of electron sources were also presented: the pulsed power guns in our working group, while the plasma-based electron injectors in a joint session held with the working group on "Plasma-Based Accelerator Concepts". Several beam dynamics issues of general interest in the field of high brightness beams production and manipulation have also been addressed by discussions and a number of communications in our working group.

I. Introduction

Although Laser Driven RF Photoinjectors have moved from the first generation scenario, where the performances in terms of beam quality were limited by Kim's laws[1], and they have definitely entered the second generation era, which is basically dominated by the understanding and wide application of the emittance correction process[2], there is still space open for further improvements of the performances as well as for exotic applications like flat beams and polarized beams production.

At the same time other type of sources are emerging, some based on old ideas revisited, like the pulsed power guns, other completely new, like the plasma-based electron injectors.

The common denominator in the beam dynamics of these devices is the highly collective and non-linear behavior of space charge forces which chiefly characterize the beam quality of the quasi-laminar relativistic beams typically produced at the exit of these sources.

In order to push the emittance rush, which is pointing toward lower and lower emittances and has now entered the sub mm·mrad range (in simulations), we need both a complete understanding of the dynamics of cold plasma oscillations which are the basis of emittance correction process, and the development of new challenging beam diagnostic tools supposedly able to measure such unprecedented ultra-low emittances.

Eventually, the extrapolation of present experimental data into the low bunch charge domain, which is foreseen to be of interest for future colliders and optical accelerators, indicates the chance to reach $5 \cdot 10^{-8}$ π m·rad at 10 pC bunch charge, setting a challenging goal for future developments.

I. BEAM DYNAMICS

The theory and explanation of the emittance correction technique, originally introduced by Carlsten[2], has been now fully integrated into the theoretical framework based on the rms envelope equation treatment of surface plasma oscillations in beams as formerly developed by Lapostolle and Sacherer[3].

This task has been accomplished by Serafini and Rosenzweig[4], who found a new equilibrium mode for a quasi-laminar accelerated beam called *invariant envelope*, characterized by vanishing longitudinal correlations in the transverse space charge field, namely the basic condition to achieve emittance correction. The interesting output of this analysis is that we can perform emittance correction in many steps, connecting each other different equilibria, i.e. Brillouin flow in drifts and invariant envelopes in accelerating sections: the matching between these equilibria is achieved automatically once the accelerating sections are set at the proper accelerating gradient and the input beam in the first section is properly matched onto the invariant envelope. The process is halted, and the residual emittance is set at the value achieved up to that point, when the parameter ρ_{INV}, given by $\rho_{INV} = \dfrac{2 \cdot 10^6}{3 T_{cat}[eV]} \left[\dfrac{4(I_p/I_0)}{\gamma \sigma_{cat} \gamma'} \right]^2$ (where T_{cat} is the equivalent photo-cathode temperature, I_p is the beam peak current ($I_0 = 17$ kA), σ_{cat} is the cathode laser spot size and γ is the beam normalized energy growing like $\gamma = \gamma_i + \gamma' z$), reaches values close to 1, which represents the threshold between the laminar space charge dominated regime and the emittance thermal flow regime. For this reason, sources producing beams characterized by $\rho_{INV} \gg 1$ (*i.e.* laminar) are still subject to possible either emittance degradation or further emittance correction.

There is in any case a fundamental limitation to the emittance correction process: this is set up by the thermal emittance generated by the emission processes at the cathode and it represents a real Liouvillian invariant. Although we are missing so far precise measurements of a photo-cathode thermal emittance, which would be very welcome, in a talk by Mikhailichenko[5] we learned how much a beam of electrons, treated as a Fermi gas, can be compressed in phase space before it becomes degenerate due to quantum effects. Although aimed at establishing fundamental limits in the damping ring performances in terms of minimum attainable emittances, Mikhailichenko's work can be qualitatively adapted to the case of a cathode emission of N electrons over a thickness δ, obtaining a minimum normalized thermal emittance $\varepsilon_{n,th}$, due to degeneration, given by $\varepsilon_{n,th} = \lambda_c \sqrt[3]{\dfrac{N \sigma_{cat}}{\delta}}$ (λ_c is the electron Compton wavelength).

Moving further from the photo-injector and pointing toward typical issues of magnetic compressors, we had a comprehensive talk from B. Carlsten[6] reporting on impressive results obtained at Los Alamos in compressing 1 nC bunches down to sub-picosecond lengths and discussing the emittance growth in bends due to noninertial space charge forces (near field effect) and coherent synchrotron radiation forces (far field). The agreement in the predicted energy loss, due to synchrotron radiation, between numerical simulations with a special version of Parmela and an analytical formula by J. Schwinger[7] has been reported to be within a factor 2. Because of this uncertainty, we certainly acknowledge the need of experimental measurements on these predicted effects: the LANL experiment seems to be the most promising so far in pursuing such investigation.

The space charge non linearities and their impact on halo production and chaotic phase space behavior were discussed in a presentation by C. Chen[8], showing that space charge dominated beams in the non-laminar regime can develop halos even in a very few lattice periods when rms matched into periodic focusing channels. The halo growth rate tends to scale like the parameter $\frac{SK}{4\sigma_0\varepsilon}$ (where S is the lattice period, K the normalized beam perveance, σ_0 the phase advance in vacuum and ε the beam emittance). This effect is potentially harmful in the transport of high quality beams and surely deserves further studies.

II. PRESENT EXPERIMENTS AND NEW DESIGNS

While the general overview about recent progress on photoinjectors has been given in the plenary session by I. Ben-Zvi[9], more detailed reports on the activity at various laboratories were presented in the working group, starting with the Next Generation Photoinjector program[10], a project based on a BNL/SLAC/UCLA collaboration: D.T. Palmer[11] presented emittance measurements performed after the booster linac at 40 MeV showing an rms normalized emittance of 0.7 π mm·mrad at 0.3 nC bunch charge, which represents so far the first evidence of emittance correction performed at S-band after the first demonstration achieved at Los Alamos[12] at L-band. It is remarkable noticing the agreement between the measured beam rms spot at the same location where the emittance were measured, namely 0.7 mm, and the predicted value from the invariant envelope theory, i.e. $rms_\sigma = \sqrt{(2I_p)/(I_0\gamma'^2\gamma)}$ [4], which gives 0.75 mm at 50 A peak current and 6 MeV/m accelerating gradient in the S-band travelling wave sections used as booster linac. The beam at 40 MeV is yet quite a bit in the quasi-laminar regime, since $\rho_{INV} \approx 100$, implying that the emittance can be in principle even further reduced.

J. Power[13] reported about the beam measurements of the Witness Beam injector at the Argonne Wakefield Accelerator: this RF gun, using a standing wave structure operated in the $\pi/2$ mode, has so far produced 0.2 nC bunches with an emittance of 4 π mm·mrad. An activity report from the UCLA Saturnus lab[14] was also presented in the working group, showing significant progress in the beam quality delivered by the RF gun + PWT linac apparatus, as well as some preliminary data, still under investigation, of SASE-FEL infrared radiation emission from the wiggler downstream.

The limitation in the longitudinal emittance attainable when operating a RF gun close to the maximum charge limit (and even above) was discussed in a presentation by D. Dowell[15], where experimental data taken at Bruyere Le Chatel (CEA) have been analyzed to show the degradation of the longitudinal emittance due to beam breakup effects in the longitudinal phase space distribution.

At Fermilab the advanced accelerator R&D activity is currently focused on the commissioning of the L-band symmetric emittance injector for TTF as we heard from E. Colby[16], while the construction of a high duty cycle gun is in progress, as well as the design work on the flat beam asymmetric emittance injector.

High frequency RF guns, those at X-band or above, still wait the first commissioning: two different initiatives were presented in the working group, dealing with a 8.548 GHz 1.5 cell RF gun, the one based on a UC-Davis+LLNL+SRRC collaboration, and with a 17 GHz gun, the other one carried on at MIT. The X-band experiment was extensively presented by G. Le Sage[17] (RF structure design and characterisation and general lay-out), C.H. Ho (RF structure and solenoid fabrication and testing) and F. Harteman[18] with a dedicated presentation of the 1 ps pulse UV laser system able to run mode-locked to the 8.548 GHz RF system with an impressive rep rate of 2 GHz and a jitter, measured so far, of 400 fs. The progress activity report on the MIT 17 GHz gun by S. Trotz[19] showed a measured filling of the gun cavity with 7.5 MW of power coupled into it, giving about 250 MV/m peak electric field at the cathode.

Due to the rising needs of ultra-fast responding photo-cathodes, for applications like 2nd generation plasma accelerators[20] or ultra-short X-ray pulse production, a dedicated R&D program has been launched at UCLA[21]. C. Joshi reported[22] on the preliminary experimental measurements performed on thin-films of Mg and Cu, showing that response times down to 15 fs can be deduced by the analysis of front-side and back-side photo-emission for film thickness varying between 5 and 100 nm.

Alternative sources to RF guns for the production of high brightness beams can be the pulsed power electron guns, as those under investigation now at Brookhaven and at Frantel Inc. . T. Srinivasan-Rao[23] presented both some preliminary results and the predicted capabilities of a high gradient (>1 GV/m) table-top pulsed photo-diode which should be able to produce a 1 MeV beam at 1 nC, 10 ps bunch length with an emittance lower than 0.5 π mm·mrad. F. Villa[24] illustrated the measurements done on a pulsed plasma diode gun, able to reach a 2.7 GV/m electric field in the cathode-anode area, producing a beam of 2.3 MeV and 1 π mm·mrad emittance at 50 nC (emittance measurements performed with a pepper-pot 2.5 cm far from the anode). These devices basically attain nicely low emittances because of the high peak field applied at the cathode, in full agreement with Kim's laws on space charge emittance growth: nevertheless, the coupling of such guns with a booster accelerator and the related issues of beam emittance preservation should be addressed, since the beams they produce are still largely in the high ρ_{INV} regime (typically ρ_{INV} is in excess of a few hundreds), hence they should be properly accelerated to avoid emittance growth.

The initial beam performances for NLCTA were also presented by A.D. Yeremian[25], showing that the design goals have been almost achieved in terms of beam fraction captured in the X-band booster and energy spread measured at 60 MeV, with still some improvements needed on the requested charge per bunch and emittance (not yet measured).

III. PLASMA-BASED ELECTRON SOURCES

During a joint session with the "Plasma-Based Accelerator Concepts" working group, other two electron sources of completely new concept have been presented. The analysis of the LILAC scheme, presently under investigation at the Univ. of Michigan, was illustrated by E. Dodd[26], while the microbunching and coherent acceleration of electrons by subcycle laser pulses is being studied at the Univ. of Texas at Austin[27]. Both represent elegant solutions for the production of a beam directly by a plasma-based accelerator, avoiding all the not yet solved problems related to the injection, matching and capture of a beam into a plasma accelerator (see for instance ref.20 for an overview).

The results shown, based on numerical simulations, show quite promising performances in terms of bunch charge, length and transverse emittance obtained. Nevertheless, some issues should be further investigated, mainly concerning the thermal emittance of the electrons captured by the plasma wave (i.e. the inherent Liouvillian phase space occupied by the particles when the emission process is terminated) and the beam quality preservation when the beam exits the plasma channel by crossing the plasma-vacuum interface (is still the beam in the laminar high ρ_{INV} regime?). Also, the statistics of the phase space distribution of the captured electrons should be comparable to typical multi-particle simulations performed for RF photo-injectors, to account for possible space charge correlations which have a space-time scale length quite smaller than the bunch sizes.

IV. FUTURE PERSPECTIVES

The final session of our working group was occupied by a general discussion on future perspectives of electron beam sources and the anticipated demands from future high energy accelerators.

In a joint session with working group on "Structure-Based Accelerator Concepts" we learned that optical accelerators require at injection normalized transverse emittances scaling like the acceleration wavelength[28] because of the matching conditions to the second order (ponderomotive) focusing inherent in such devices: since the beta function scales like $\beta \propto \gamma/\gamma'$ and the beam spot like $\sigma \propto \lambda$, hence $\varepsilon_n \propto \lambda^2 \gamma'$, from which, since the acceleration gradient scales like $\gamma' \propto \lambda^{-1}$, we obtain $\varepsilon_n \propto \lambda$. Furthermore, wake-field constraints set up a limitation in the number N of particles per bunch which can be accelerated without significant beam quality degradation, giving qualitatively $N = 10^8$ at $\lambda = 10\,\mu\mathrm{m}$ and $N = 10^6$ at $\lambda = 1\,\mu\mathrm{m}$. The actual requested emittance comes out to be $\varepsilon_n = 10^{-8}\,\pi$ m·rad at $\lambda = 10\,\mu\mathrm{m}$ and $\varepsilon_n = 10^{-9}\,\pi$ m·rad at $\lambda = 1\,\mu\mathrm{m}$. Such low emittances are in principle needed only in one plane, so that the use of flat sheet beams should mitigate the requirements on the injectors. Therefore, we think that the experimental as well theoretical investigations of asymmetric emittance beam generation should be further encouraged.

In order to address such a challenging demand we performed a scaling of the emittances by starting from the values measured or modeled in the region of nC bunches and going down to lower charges. From previous work on round beams[29] we know that the residual emittance growth $\Delta\varepsilon_n$ at the exit of a photoinjector operated in emittance correction regime scales like $\Delta\varepsilon_n \propto Q_b^{2/3}$ in the range of sub-nC bunch charges Q_b. The total normalized rms emittance ε_n will be therefore given by the quadratic sum of $\Delta\varepsilon_n$ with the thermal normalized emittance $\varepsilon_{n,th}$, which typically scales as $\varepsilon_{n,th} \propto \sigma_{cat}\sqrt{T_{cat}}$. Recalling that the laser spot size at the cathode σ_{cat} must scale like $\sigma_{cat} \propto Q_b^{1/3}$ in order to achieve properly the emittance correction [see ref. 29 and 4, which agree completely on this prediction], we may cast the scaling law for ε_n just in terms of the bunch charge Q_b, once the cathode temperature is specified, as $\varepsilon_n \propto \sqrt{c_1 Q_b^{2/3} + c_2 Q_b^{4/3}}$.

The determination of the coefficients is done for the two cases of available numerical simulations and actual beam measurements: the main choice consists in taking an optimum cathode spot given by $\sigma_{cat}[mm] = 0.8 \cdot Q_b^{1/3}[nC]$ and a cathode temperature $T_{cat} = 0.1$ eV, implying a thermal emittance $\varepsilon_{n,th} = 2.5 \cdot 10^{-7} Q_b^{1/3}$. The reference measurement is taken to be the Next-Generation (ng) measurement[11], while the simulations are taken by ref.30 (jr) and 4 (ls). The predicted scaling is shown in Figure 1.

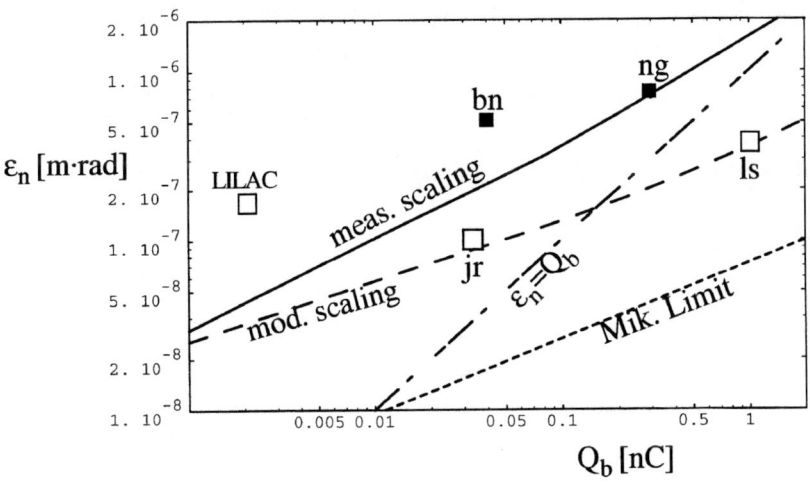

Figure 1. - Scaling of the normalized rms emittance in the sub-pC range

Black square dots in the diagram indicate measured data while white dots report simulation results: the black dots marked "bn" reports measurements done at Brookhaven[31] on a 40 pC bunch with RF gun set-up not optimized for emittance correction but for sub-ps bunch production. The dotted-dashed line reports the linear scaling $\varepsilon_n = 10^{-6} Q_b$, just for sake of comparison, as well as the dotted line gives the Mikhailichenko limit (discussed above) for a cathode thickness of 30 nm. Part of the difference between experimental values and simulations (solid and dashed line, respectively) in the 0.1-1 nC range can be explained by the different laser pulse intensity time profile in use: simulations are carried on with optimal uniform time profiles, while measurements are so far performed with gaussian distributions (see ref.4 for a detailed discussion on this issue). The bunch length at the injector exit scales also like $\sigma_z \propto Q_b^{1/3}$, with a typical $\sigma_z \cong 1-2$ mm at 1 nC.

It is clear from Figure 1 that the predicted scaling for the rms normalized emittance at 16 pC bunch charge ($N = 10^8$) misses by a factor 5 the requirements of optical accelerators if round beams are employed: since the total emittance seems to be dominated at low charges by the thermal contribution, we really need more experimental investigation on this quantity.

In conclusion, the working group identified the following areas of investigation and R&D to be recommended as of higher priority:

    i) diagnostics for ultra-low emittance measurements
    ii) photo-cathode thermal emittance characterisation
    iii) 3D self consistent simulations (*i.e.* code development) for flat beam generation and emittance growth in magnetic compressors
    iv) polarized electron beam production in RF guns

As a last remark, we recall that the working group also updated a comprehensive table[32] which collects the main characteristics and performances of electron beam sources in operation and/or development around the world. The table was originally set up by R. Sheffield[33].

ACKNOWLEDGEMENTS

We thank all the participants of the working group for the enthusiasm and collaboration in the working group activity: in particular we are grateful to David Reis for updating and editing the source table. This work was supported in part by U.S. Dept. of Energy grants DE-FG03-93ER40796 and DE-FG03-92ER40693, the Alfred P. Sloan Foundation grant BR-3225.

REFERENCES

1. K.J. Kim, *Nucl. Instr. Methods A* **275**, 201 (1988)
2. B.E. Carlsten, *Nucl. Instr. Methods A* **285**, 313 (1989)
3. P. Lapostolle, *Proton Linear Accelerators* (Los Alamos, 1980)
4. L. Serafini and J.B. Rosenzweig, Envelope Analysis of Intense Relativistic Quasi-Laminar Beams in RF Photoinjectors: a Theory of Emittance Compensation, submitted to *Phys. Rev. E*
   L. Serafini, Overview on Production and Dynamics of High Brightness Beams, presented at ICFA Workshop on Nonlinear and Collective Phenomena in Beam Physics, Arcidosso, Italy, September 1-6, 1996
5. A.A. Mikhailichenko, these proceedings
6. B. Carlsten et al., *Phys. Rev. E* **53**, 124 (1996)
7. J. Schwinger, *Phys. Rev.* **75**, 1912 (1949)
8. C. Chen, these proceedings
9. I. Ben-Zvi, Invited Paper on Recent Progress on Photoinj., these proceedings
10. D.T. Palmer, *et al.* in *Proc. 1995 Particle Acc. Conf.* 2432 (IEEE, 1996)
11. D.T. Palmer, these proceedings
12. R.L. Sheffield, *Nucl. Instr. Methods A* **341**, 371 (1994)
13. J. Power, these proceedings
14. G. Travish, *et al.*, *Nucl. Instr. Methods A* **365**, 255 (1995)
    A. Tremaine, these proceedings
15. D. Dowell, these proceedings
16. E. Colby, these proceedings
17. G. Le Sage, these proceedings

18. F. Harteman, these proceedings
19. S. Trotz, these proceedings
20. C. Clayton and L. Serafini, *IEEE Trans. on Plasma Sc.* **24**, No.2, 400 (1996)
21. P. Muggli, *et. al.*, *IEEE Trans. on Plasma Sc.* **24**, No.2, 428 (1996)
22. C. Joshi, these proceedings
23. T. Srinivasan-Rao, these proceedings
24. F. Villa, these proceedings
25. A.D. Yeremian, these proceedings
26. D. Umstadter, *et al.*, *Phys. Rev. Letters* **76**, 2073 (1996)
27. B. Rau, T. Tajima and H. Hojo, these proceedings
28. J.B. Rosenzweig, private communication
29. J.B. Rosenzweig and E. Colby, AIP-CP **335**, 724 (1995)
    L.C.L. Lin, S.C. Chen, J.S. Wurtele, AIP-CP **335**, 704 (1995)
30. J. B. Rosenzweig, N. Barov, and E. Colby, *IEEE Trans. on Plasma Sc.* **24**, No.2, 409 (1996)
31. X.J. Wang, *et al.*, Proc. Micro Bunches Workshop, Brookhaven, Sept. 1995
32. A postscript file named SOURCETAB.PS containing the table can be found in the Web page of UCLA-PBPL at address http://pbpl.physics.ucla.edu/
33. R. Sheffield, Proc. of ICFA Workshop on Nonlinear and Collective Phenomena in Beam Physics, Arcidosso, Italy, September 1-6, 1996

Summary Report: Working Group 6 on "Microwave Sources"

Glen A. Westenskow

Lawrence Livermore National Laboratory
P.O. Box 808, Livermore, CA 94550 USA

Abstract

This report summarizes the discussions of the Microwave Source Working Group during the Advanced Accelerator Concepts Workshop held October 13-19, 1996 in the Granlibakken Conference Center at Lake Tahoe, California. Progress on rf sources being developed for linear colliders is reviewed. Possible choices for high-power rf sources at 34 GHz and 94 GHz for future colliders are examined.

INTRODUCTION

In the past few workshops much of the work by this group focused on providing a X-band power source for a 1-TeV electron-positron linear collider system. SLAC started a serious effort on high-peak power X-band sources in the mid 1980s. The successful demonstration of an efficient 50-MW X-band tube this summer by SLAC shifts the emphasis of this group to the power source for future accelerators beyond the NLC.

These proceedings include an excellent review of the microwave source development for colliders by W. Lawson [1] and many contributed papers on individual research efforts by members of the working group. This summary will concentrate more on the issues that were discussed by the working group.

RESEARCH EFFORTS ON DISCRETE RF POWER SOURCES

SLAC's PPM klystron tube is the best choice for the NLC rf power source. However, a number of the X-band rf sources being studied have the potential to improve the efficiency of future collider systems or to lower their cost. The following is a brief report of the status of the research efforts on discrete sources by members of the group with references to where additional information can be found.

CP398, *Advanced Accelerator Concepts,* edited by S. Chattopadhyay, J. McCullough, and P. Dahl
AIP Press, New York © 1997

PPM klystron tube - SLAC in July 1996 successfully demonstrated [2] that the PPM klystron tube could produce 56 MW of 11.4 GHz power in a 1.2 μsec pulse. Work on the beam transport using periodic permanent magnets (PPM) went smoothly after several years of working on the design of the output structure. The tube shows 60% electronic efficiency, close to that predicted by CONDOR. The beam loss in the tube is small over an acceptable voltage spread. There are still a number of "engineering" issues for the tube that will be studied. Since almost 10,000 tubes are planned for 1-TeV NLC operation construction techniques are being considered to reduce the tube cost. It is expected that this type of tube will satisfactorily provide the rf power for the NLC.

Cluster klystron - Experiments during the past year at BNL [3] on the cluster klystron concept have focused on the creation of the annular electron beam which will be needed in the device. The close packing of the tubes does not allow for use of a high cathode-to-beam compression ratio as in many of the other devices. This places a difficult requirement on the MIG guns being proposed for the system. The pulsed power for the test device has been prepared and cathode testing should start soon.

Sheet-beam klystron - None of the workshop participants were working on the sheet beam klystron approach, however this device [4] is seen as a possible efficient rf source in the higher frequency collider schemes.

Magnicon - The work at NRL [5] and INP [6] on magnicons was reported during the workshop. Magnicons have the advantage of lowering the energy density from that of a high power X-band klystron. They have shown impressive results at low frequency, but experiments at X-band have shown that better vacuum and microwave techniques are required to reduce pulse shortening. NRL is working on a new magnicon with a thermionic cathode for which simulations have shown good performance.

TWT - Cornell University's recent simulations [7] have shown that the efficiency of their device can be increased to about 50% by an appropriate phase velocity transition in their interaction structure and the use of a coaxial extraction circuit. The group has also made good progress on the control of side-bands in their device.

Gyroklystron - Experimental efforts at the University of Maryland have focused on a 2nd harmonic gyroklystron [8] for which simulations have shown 158 MW at 17.1 GHz for a 1.2 μsec pulse. The expected increase in rf output power is from careful attention paid to the gun and transport systems.

Gyroharmonic Converters - Yale/Omega-P is working on a cyclotron autoresonance acceleration (CARA) [9] device. In this device rf power is used to "energize" an electron beam. RF power can then be extracted from the beam at a higher harmonic. They have examined a system where a 2.856 GHz drive source could be used to provide 30 GHz rf power (7th harmonic).

Cyclotron Resonance Klystron - UC Davis is proposing a device that uses an FEL interaction to bunch the electron beam, and a klystron like output cavity for power extraction.

Ubitron - Work around 1960 on ubitrons produced tubes that ranged up to 50 GHz. Promising conceptual work [10] on a 280-MW, 11.4-GHz system has been done at SLAC and UC Davis. A thin annular beam is required in this device. The 2.3 meter design length is more compact than standard FEL devices. The 11.4-GHz device uses a 500-kV 1-kA thin annular beam.

FEMs - LANL has started construction of a novel 17-GHz free-electron maser (FEM) [11]. The beam interacts with rippled walls, instead of the normal wiggler magnetic field. Simulation has shown rf power production of 500 MW using a 5-kA, 600-kV beam source.

TWO-BEAM ACCELERATORS

To demonstrate the efficacy of the two-beam accelerator concept the two major efforts are constructing test facilities which include prototype modules for their larger systems. In these schemes high conversion efficiencies of wall power to beam power will only be seen in the larger systems. Work on the two-beam accelerator schemes is summarized [12] by an invited talk by R. Corsini.

Several new options for the creation of the drive beam for CERN's two-beam accelerator scheme were discussed. The effects of short field wakes from the transfer structures in the drive beam are better understood, and a solution is being demonstrated in the CTF test facility [13]. The test facility will address many of the issues of rf power generation using the CLIC approach.

The major effort of LBL/LLNL's two beam accelerator approach is to provide an upgrade to the NLC power sources at 11.4 GHz. Work on establishing the test facility[14] and rf cavity design [15] was reported on during the workshop. With minor variation, the test facility could provide power up to 34 GHz. A study for a 5-TeV collider at 30 GHz using induction-driven two-beam accelerator was presented [16]. Additional studies of the wake fields in the rf output structures and in the main accelerator at these higher frequency scheme is needed for this study. Two-beam accelerator schemes using a FEL interaction could operate at even higher frequencies.

WORK AT 34 GHZ

There are compelling reasons [17,18] that push the designs of high-energy room-temperature electron-positron linear colliders to higher frequencies. Several studies have suggested that for a 5-TeV center-of-mass energy the optimal linac operating frequency is in the 30 to 35 GHz range. However, in choosing the frequency, each of the studies has made some questionable assumptions with regard to the rf source. A challenge for the working group was to examine possible rf power sources at 34.3 GHz (12 times SLAC's operating frequency). Many of the fast wave devices (i.e. gyroklystron, FELs) and the two-beam accelerator systems are already designed for frequencies above 11.4 GHz and could be scaled to the 34 GHz. A design for 35-GHz gyroklystron was presented [19].

During the workshop SLAC presented an interesting 50-MW klystron design at 34 GHz with 40% efficiency. It uses a 620-kV, 240-A electron beam. The required guide field of 2.7 kGauss would be too high for PPM focusing. The only high power (>10MW) klystrons above 3 GHz are those being developed for linear colliders. Besides SLAC and KEK tubes at 11.4 GHz, there is the VLEPP tube at 14 GHz and the HRC tube at 17.1 GHz.

Some initial thoughts on scaling TWTs and magnicons and other devices were also done by the workshop participants. Overall, there is no clear choice on which device will give the best performance at 34 GHz. We expect that additional simulation work will be performed before the next workshop on 34 GHz rf sources.

W-BAND SOURCES

During the last year there has been growing interest in the development of 94 GHz (W-band) accelerator structures. Many components exist for operation in this band for radar applications, however there is a shortage of high peak power sources. Critical issues to be studied involve surface heating effects in the accelerating structures for these short wavelengths at high gradients. Initial experiments do not require that the rf source be efficient. However, the final choice of optimal frequency for a future collider depends on suitable efficient rf sources. In the higher energy machine overall system efficiency of wall-plug to beam will become even more important.

Gyrotrons have already produced high average power in this frequency range. For 500-ms pulses they have produced 0.68 MW of rf power. The goal would be to increase the peak power to about 100 MW for shorter pulse widths (~1µs). The highest power produced in the gyroklystron experiment at University of Maryland has been 32 MW at 20 GHz. A rough estimate indicates that 50 MW of rf power at 100 GHz could be produced using gyroklystrons. The efficiency was estimated at 15% for this device without energy recovery, and 37% for a single depressed-collector system.

Free-electron lasers (FELs) are another possible candidate for a laboratory high peak power source. LLNL produced about 2 GW of rf power at 140 GHz with about 20 ns duration [20]. A smaller system at MIT produced 61 MW for 30 ns pulses at 33 GHz [21]. A recommendation is that the issues of phase stability and cost be reviewed for using a FEL as a driver in this frequency range. A fast-forward feedback circuit [22] could reduce some of the earlier concerns with systematic phase variations.

TECHNOLOGIES

During the workshop several technologies were discussed that could improve the performance of rf sources for accelerator applications in general. The following is brief summary of a few of these ideas.

RF Pulse Compression

An area of exciting research involves schemes to improve the efficiency of pulse compressors with high compression ratios. This would allow a low-power tube to deliver the high peak powers which are needed for high-gradient accelerators. SLAC has recently demonstrated efficient high-power pulse compression [23] for NLC. They are attempting to increase the efficiency of SLED by introducing an appropriate phase shift in the delay lines [24]. OMEGA-P also presented a scheme based on closing switches [25].

Electron Source Development

Advances in cathode technology could enable new rf power sources. The generation of the required annular beams in many of the devices discussed (i.e. cluster klystron, gyroklystron, ubitron, FEMs) are viewed as a major issue in their development. Even in the "conventional" klystron tube the cathode is a major consideration for the tube lifetime.

Work on several electron sources was discussed at the workshop. Cornell is investigating ferroelectric cathodes [26]. They have obtained long pulse operation with their sources. However, the electron from the longer pulses may be from plasma emission. Work on an electron sources using secondary electron emission was reported on by FMT [27]. These have shown stable operation and have produced good low-emittance beams. Work has started on long-life high-current oxide cathodes at SLAC.

Cavity wall coatings

Another technology that promises to have a major impact on rf sources is cavity coating to increase the breakdown voltage level and/or lower multipactoring. Coating have been found that have good DC breakdown holdoff, low rf absorption, and are physically rugged. Hot testing in a klystron output structure should take place within the next few months. The coating in DC tests had a substantial effect on reducing dark-current emission.

ACKNOWLEDGMENTS

Wish to thank W. Lawson for help in preparing summary of the working group. Also to T. Houck and S. Lidia for reviewing the article.

REFERENCES

1. Lawson, W., "Microwave Sources," in the Proceedings of the 1996 Advanced Accelerator Concepts Workshop, Lake Tahoe, CA, 1996.
2. Wilson, P. B., "Advanced RF Power Sources for Linacs," in the Proceedings of the XVIII International Linac Conference, Geneva, Switzerland, 1996.
3. Palmer, R. B., et al., "Status of the BNL-MIT-SLAC Cluster Klystron Project," in the Proceedings of the 1994 Pulsed RF Sources for Linear Colliders Workshop, Montauk, NY, (*AIP Conference Proceedings 337*), p. 94.
4. Yu, D.U.L., et al, "Design of a High Power Sheet Beam Klystron," Fifth Advanced Accelerator Concepts Workshop, Port Jefferson, NY, (*AIP Conference Proceedings 279, 1993*), p. 85.
5. Gold, S.H., et al., "Development of an X-Band Magnicon Amplifier for the Next Linear Collider," in the Proceedings of the 1996 Advanced Accelerator Concepts Workshop, Lake Tahoe, CA, 1996.
6. Nezhevenko, O., et al, "The Results of the 7 GHz Pulsed Magnicon Investigation," in the Proceedings of the 1996 Advanced Accelerator Concepts Workshop, Lake Tahoe, CA, 1996.
7. Nation, J.A., et al, "Progress in High Power, High Efficiency Relativistic Traveling Wave Tube Amplifiers," in the Proceedings of the 1996 Advanced Accelerator Concepts Workshop, Lake Tahoe, CA, 1996.
8. Lawson, W., et al., "High Power Gyroklystron Development for Advanced Accelerator Applications," in Proceedings of the 1996 Advanced Accelerator Concepts Workshop, Lake Tahoe, CA, 1996.
9. LaPointe, M.A., et al., "High-Power Microwave Production by Gyroharmonic Conversion and Co-Generation,' in the Proceedings of the 1996 Advanced Accelerator Concepts Workshop, Lake Tahoe, CA, 1996.
10. McDermott, D.B., et al., "Periodic Permanent Magnet Focusing of an Annular Electron Beam and its Application to a 250 MW Ubitron FEL," submitted to Physics of Plasmas (1995).
11. Fazio, M. V., et al., "The development of an Annular-Beam, High Power Free-Electron Maser for Future Linear Colliders," in the Proceedings of the XVIII International Linac Conference, Geneva, Switzerland, 1996.
12. Corsini, R., "The Drive Beam Generation for Two Beam Accelerators," in the Proceedings of the 1996 Advanced Accelerator Concepts Workshop, Lake Tahoe, CA, 1996.
13. Wilson, I., et al, "CLIC Test Beam Facilities – Status and Results," in the Proceedings of the XVIII International Linac Conference, Geneva, Switzerland, 1996.
14. Lidia, S., et al., "RK-TBA Prototype RF Power Source," in the Proceedings of the 1996 Advanced Accelerator Concepts Workshop, Lake Tahoe, CA, 1996.
15. Kim, J.S., et al., "RF Structures Design for the TBNLC," in the Proceedings of the XVIII International Linac Conference, Geneva, Switzerland, 1996.
16. Lidia, S.M., et al, "Relativistic Klystron Two-Beam Accelerator Approach to Multi-TeV e$^+$e$^-$ Linear Colliders," Contributed to the APS New Directions for High Energy Physics Workshop, Snowmass, CO, USA, July 1996.
17. Palmer, R.B., Ann. Rev. Nucl. Sci. 40, 529-92 (1990).
18. Gaponov-Grekhov, A.V., and Granatstein, V.L., eds., *Applications of High Power Microwaves*, Boston: Artech House, 1994.
19. Granatstein, V.L., et al., "Gyroklystron for Driving Linear Colliders at 35 GHz," in Proceedings of the 1996 Advanced Accelerator Concepts Workshop, Lake Tahoe, CA, 1996.
20. Allen, S. L., et al., "Generation of High Power 140 GHz Microwaves with an FEL for the MTX Experiment," in the Proceedings of the 1993 Particle Accelerator Conference, p. 1551.
21. Conde, M. E. and G Bekefi, "Amplification and Superradient Emission from a 33.3 GHz Free Electron Laser with a Reversed Guide Magnetic Field," IEEE Trans. Plasma Sci **20**, 240-246 (1992).

22. Hopkins, D., et al, "Phase and Amplitude Stabilization of Short-Pulsed, High-Power Microwave Amplifiers," in the Proceedings of the 1991 Particle Accelerator Conference, p. 1335.
23. Tantawi, S. G., et al., "The Next Linear Collider Test Accelerator's RF Pulse Compression and Transmission Systems," in the Proceedings of the 1996 Advanced Accelerator Concepts Workshop, Lake Tahoe, CA, 1996.
24. Tantawi, S. G., et al., "Active High Power RF Pulse Compression Using Optically Switched Resonant Delay Lines," in the Proceedings of the 1996 Advanced Accelerator Concepts Workshop, Lake Tahoe, CA, 1996.
25. Petelin, M. I., et al., "Pulse Compressor Based on Electrically Switched Bragg Reflectors," in the Proceedings of the 1996 Advanced Accelerator Concepts Workshop, Lake Tahoe, CA, 1996.
26. Flechtner, D., et al, "Electron Emission from Ferroelectric Cermanic," August submission to Journal of Applied Physics.
27. Mako, F.M., Peter, P., "A High Current Micro-Pulse Electron Gun," in the Proceedings of the 1993 Particle Accelerator Conference, p. 2702.

Summary Report: Working Group 7 on "Laser Sources for Particle Acceleration"

M. C. Downer
The University of Texas at Austin
Department of Physics

C. W. Siders
Los Alamos National Laboratory

INTRODUCTION AND OVERVIEW

The Laser Sources working group (a.k.a. working group 7) concerned itself with recent advances in and future requirements for the development of laser sources relevant to high-energy physics (HEP) colliders, small scale accelerators, and the generation of short wavelength radiation. We heavily emphasized pulsed terawatt peak power laser sources for several reasons. First, their development over the past five years has been rapid and multi-faceted, and has made relativistic light intensity available to the advanced accelerator community, as well as the wider physics community, for the first time. Secondly, they have strongly impacted plasma-based accelerator research over the past two years, producing the first experimental demonstrations of the laser wakefield accelerator (LWFA) in both its resonantly-driven [1] and self-modulated [2] forms. Thirdly, their average power and wall-plug efficiency currently fall well short of projected requirements for future accelerators and other high average power applications, but show considerable promise for improving substantially over the next few years. A review of this rapidly emerging laser technology in the context of advanced accelerator research is therefore timely.

With the help of T. Tajima (U. Texas), our working group began by estimating the laser peak power, average power, efficiency, and channeling length requirements for its most ambitious, long-term projected application: a laser-driven plasma-based electron-positron linear collider beyond the Next Linear Collider (NLC). C. Clayton (UCLA) and R. Freeman (LLNL) then reviewed terawatt laser requirements for a nearer term application: γ-ray generation via Compton scattering from the proposed NLC beam for a projected γ-γ collider. W. Leemans (LBL) and P. Chen (Stanford) discussed terawatt laser requirements, respectively, for x-ray generation near $\lambda = 1$ angstrom, and for a proposed fundamental study of Unruh radiation generated by relativistically accelerated electrons in the 100 eV range. The laser requirements for short-wavelength generation experiments were both more modest and more predictable than those for an e^+e^- collider, but motivate laser development in the same general direction. We then examined recent, current, and planned development of terawatt laser sources at several institutions, including both compact solid state table-top-terawatt (T^3) laser systems operating at wavelengths near $\lambda = 1\ \mu m$ [U. Michigan (G. Mourou), U. California-San Diego (C.P.J. Barty), and the Japan Atomic Energy Research Institute (JAERI- K. Tani)], and a terawatt CO_2 laser system operating near 10 μm under development at Brookhaven National Laboratory (BNL, I.V. Pogorelsky). Potential advantages of 10 μm radiation for the inverse Cerenkov accelerator (ICA) and the inverse free electron laser (IFEL) were emphasized by W. D. Kimura (STI Optronics) and A. van Steenbergen (BNL). Laser pulse shaping and sequencing schemes for optimizing laser-plasma coupling efficiency for the LWFA were presented by P. Chen (Stanford), I. V. Pogorelsky (BNL) and D. Umstadter (Michigan). The current status and near term research problems facing optical guiding of high intensity pulses in preformed plasma channels were presented by S. Nikitin and H. M. Milchberg (U. Maryland).

These presentations showed that current technologies for ultrashort pulse, peak power generation and pulse shaping already meet the requirements of laser-driven accelerators, but that average power and efficiency fall short of projected requirements by several orders of magnitude. Several speakers discussed key enabling technologies which are opening prospects for multi-kilowatt average power, high efficiency terawatt lasers, including laser diode array pump sources, high energy storage laser crystals and glasses amenable to diode pumping, phase conjugate mirrors and adaptive optics to correct thermal distortions, new cooling technologies, beam multiplexing and combining techniques, wide bandwidth CO_2 amplifiers, and high average power compressor gratings. Rapid development of these technologies, coupled with simultaneous driving by applications in materials processing and atmospheric sensing, justifies cautious optimism that terawatt laser sources which meet the average power and efficiency requirements of future accelerators will become available within the next decade.

TERAWATT LASER REQUIREMENTS FOR HEP COLLIDERS

The 1996 Snowmass meeting projected that an e^+e^- collider beyond the NLC must provide a center of mass energy of 5 TeV and luminosity at the interaction point (IP) of $10^{35}\ cm^{-2}s^{-1}$. In order to help estimate the laser requirements for such a collider, T. Tajima presented a "strawman" scenario, described in more detail in a sep-

Parameter	5 TeV LWFA	NLC γ-γ	Solid St.		CO_2	Units
λ	1–10	1	0.8	1.06	10	μm
E	1	1	0.1	20	15	J
τ	100	1000	20	1000	3000	fs
P_{peak}	10	1	5	20	5	TW
f_{rep}	60	15	50×10^{-3}	10^{-5}	10^{-4}	kHz
P_{avg}	60	15	5×10^{-3}	2×10^{-4}	1.5×10^{-3}	kW
$\eta_{wall-plug}$	0.1	—	$< 10^{-4}$	$< 10^{-4}$	~ 0.01	

Table 1: Laser Requirements for a 500 stage, 5 TeV LWFA and NLC-based γ-γ collider (85 bunches/macropulse at 180 Hz macropulse rate) and current state of the art for solid-state, i.e. Ti:S (0.8 μm, [3, 4]) or Nd:x (1.06μm, [5]), and CO_2 systems [6]. For the NLC γ-γ collider, current efficiencies are adequate since only one laser system is needed.

arate paper in this volume, which envisioned that such a collider would be a LWFA driven resonantly in a well-controlled linear regime of laser-plasma interaction to a plasma wave amplitude approaching the cold wavebreaking (WB) limit $\delta n_e/n_e \sim 1$. Collider parameters were chosen to satisfy known collider physics requirements. In particular, bunch size was limited to 10^8 e^- in order to hold IP beamstrahlung losses and collective instabilities to acceptable levels. Preference was given to small individual stage acceleration lengths ($L \sim 1$ m) by assuming acceleration gradients (~ 10 GeV/m) well beyond the limit of conventional rf accelerators, thus dictating a plasma density $n_e \sim 10^{17}$ cm^{-3}, for which $L_{dephasing} = L_{pump\ depletion} = 1$ m and $E_{WB} = 10$ GeV/m. The plasma wave was assumed to transfer its stored electrostatic energy to beam energy with an efficiency of 10%. Small beam size and therefore emittance ($e_{yn} = e_{xn} = 2 \times 10^{-7}$ m-rad) were assumed achievable and appropriate to a small radius acceleration channel driven by a short wavelength ($\lambda \sim 1$ μm) laser source. Finally 500 MW of electrical power were assumed to be available for the lasers. See the paper by M. Xie et al. in this volume for further discussion and justification of the assumed parameters. From the above scenario, one can straightforwardly derive the terawatt laser requirements shown in the second column of Table 1 and compare them with the capabilities of existing solid state and CO_2 laser technology shown in the fourth and fifth columns. Because of uncertainty in the underlying assumptions, the figures in the second column should be treated as, at best, order of magnitude estimates.

A parallel set of laser requirements can be derived, with much less uncertainty, for γ-ray generation at NLC for a γ-γ collider. In this case the laser requirements are dictated directly by known e^- beam characteristics. First, the pulse format of the laser must follow that of the e^- bunches, which at NLC consists of macropulses of 120 ns duration at 180 Hz repetition rate, each consisting of 85 micropulses separated by 1.5 ns. Second, the lower limit of laser wavelength is set by the need to avoid e^+e^- pair production by secondary rescattering of γ-ray photons with incident laser photons ($\gamma + \omega_0 = e^+ + e^-$), as dictated by the inequality $\omega_0\omega_\gamma < m^2c^4$, or equivalently $\lambda > 4.2\ E_0[\text{TeV}]\ \mu$m. For NLC ($E_0 = 0.25$ TeV), this implies $\lambda > 1\ \mu$m, an appropriate wavelength for solid state TW laser technology. For a future higher energy collider (e.g. $E_0 = 2.5$ TeV), the same inequality implies $\lambda > 10\ \mu$m, appropriate for CO_2 technology. Thirdly, the required pulse duration $\tau \sim 1$ ps is determined by matching the longitudinal e^- bunch profile, while the pulse energy $E \sim 1$ J is set by requiring unity conversion efficiency of optical to γ-ray photons. These requirements are summarized (peak power 1 TW, average power 15 kW, at 1 μm wavelength) in the third column of Table 1.

Several broad conclusions emerge from these intellectual exercises. First, starting with the good news, the individual laser pulse requirements ($E_{\text{pulse}} \sim 1$ J, $\tau_{\text{pulse}} \sim 0.1$–1 ps, $P_{\text{peak}} \sim 1$–10 TW) for both the e^+e^- and γ-γ collider scenarios have already been met by existing 1 μm solid state T^3 laser technology, as described in more detail in the next section. Subpicosecond terawatt CO_2 sources appear poised to meet such requirements in the near future. Secondly, the estimated number of acceleration stages (500), and therefore laser systems, for the e^+e^- collider appears to be within the budget ($\sim$\$200 million) of a large national or international collider facility, assuming current prices ($\sim$\$400K) for terawatt laser systems. For the γ-γ collider, only two such laser systems would be required. Thirdly, the laser-plasma coupling efficiency $\eta_{\text{laser-plasma}} \sim 10\%$ required by the LWFA-based e^+e^- collider scenario appears reasonable provided the required channeling lengths can be achieved, and that pulse shaping techniques discussed further in the next section are fully exploited.

At the same time, the collider scenarios reveal several aspects of terawatt laser technology which are in need of substantial further development. First, the IP luminosity requirement for the e^+e^- collider dictates that 1 J, 100 fs pulses must be delivered at a repetition rate $f \sim 60$ kHz, corresponding to an average power of $P_{\text{avg}} \sim 60$ kW. This, as well as the 15 kW requirement of the γ-γ collider, both exceed the current capability of solid state terawatt laser systems by over three orders of magnitude. CO_2 technology has demonstrated multi-kilowatt average powers, but so far only for pulses of nanosecond or longer duration. The physical causes of the current limitations and efforts in progress to overcome them are described in the next section. Secondly, the available electrical power for a e^+e^- collider dictates wallplug-laser efficiency near 10%, again several orders of magnitude beyond the capability of current solid state laser technology. High power CO_2 lasers achieve efficiency near the required 10% level, but this needs to be demonstrated for subpicosecond pulse duration. For the γ-γ collider, and other short-wavelength generation applications, current wall-plug efficiency levels are tolerable because of the much smaller number of laser systems.

TERAWATT LASER TECHNOLOGIES IN HAND

Over the past ten years, the dual emergence of Ti:sapphire and other solid-state vibronic media as high-quality, ultrabroadband, high energy storage amplification materials and of chirped pulse amplification (CPA) techniques [7] have spurred the generation of multi-terawatt ultrashort pulses near 1 μm wavelength in table-top laser systems around the world. In CPA, low energy short duration seed pulses are expanded several thousand-fold in duration by chirping, thus enabling amplification at high fluences without intensity-dependent damage of the solid-state amplifier components. Early CPA systems achieved peak powers near 1 TW in the form of 0.1 to 1 J pulses of 100 to 500 fs duration. Later refinements to CPA included the development of seed pulse oscillators at the 10 fs level, development of high order dispersion compensation techniques, and the development of passive spectral shaping elements to compensate spectral gain narrowing in the amplifier. These refinements have resulted in generation of ultrashort laser pulses with peak power higher than 100 TW. These single pulse properties exceed projected requirements for plasma-based particle accelerators and gamma ray sources of interest to the high energy physics community. The dramatic recent developments in solid state TW laser technology have been reviewed in detail by Barty [4] and by Mourou and Perry[8].

Participants in the Laser Sources working group pointed out emerging solid state technologies likely to lead to still further improvements in multi-TW peak power generation. G. Mourou described several potential advantages of Yb:glass as an amplifying medium for CPA. The glass host is cheaper and scalable to larger sizes than crystalline hosts. Moreover, Yb:glass, which has a gain bandwidth sufficient for 30 fs pulse amplification, offers much higher energy storage capacity (30 J/cm^3) than most ion-doped crystals, although limitations to usable fluence set by optical damage must be carefully evaluated. Other possible advantages of Yb:glass are discussed in the next section. C.P.J. Barty, on the other hand, emphasized that CPA based on now standard, wider bandwidth Ti:S crystals, which support amplification of pulses as short as 10 fs, could be scaled to pulse energies as high as 40 J at repetition rates of 6 Hz by exploiting recent advances in phase-conjugated Nd:glass pump lasers. For example, eight such lasers each providing 25 J at 532 nm, simultaneously pumping a Ti:S final amplifier stage would be adequate to produce a 40 J pulse. K. Tani described plans to construct a system similar to this at the Japan Atomic Energy Research Institute, where generation of pulses approaching the 1 PW level at 5 to 10 Hz repetition rate is anticipated by the year 2000.

While solid state CPA systems are currently the most advanced sources of TW laser pulses, I. V. Pogorelsky described a novel TW picosecond CO_2 laser system now under development at the Brookhaven Accelerator Test Facility (ATF). Conventional subnanosecond, multi-GW CO_2 lasers have been, and continue to be, used successfully in laser beatwave accelerator (LWBA) experiments, where externally-injected electrons have been accelerated to energies as high as 30 MeV [9, 10]. The projected generation of ps, multi-TW CO_2 pulses will not only upgrade ongoing far field (inverse Cerenkov and inverse FEL) experiments, but will open application of CO_2 lasers to plasma-based LWFA schemes for the first time. Pogorelsky, A. Van Steenbergen, and W. D. Kimura emphasized several advantages which the 10 μm wavelength of the CO_2 laser can offer in these applications. In far field accelerators, the long wavelength lengthens the phase slippage distance between particles and oscillating fields, reduces sensitivity to bunch smearing and gas scattering, and increases the achievable longitudinal accelerating field compared to a shorter wavelength driver. For the LWFA, a 10 μm driver pulse produces a hundred-fold higher ponderomotive potential in the plasma, resulting in a higher axial accelerating field E_z than for a 1 μm pulse of the same intensity and duration. For plasmas of the same density this advantage is offset by the shorter phase slippage distance ($L_{\text{dephasing}} \sim \lambda_p^3/\lambda^2$) for the 10 μm driver, caused by its smaller group velocity in the plasma. However, a practical 10 μm driven LWFA would probably operate with lower plasma density and looser focus than its 1 μm counterpart, in which case the $L_{\text{dephasing}}$ disadvantage is reduced or eliminated. For the self-modulated LWFA, drivers of equivalent relativistic self-focusing power $P[\text{GW}] = 17(\omega/\omega_p)^2$ may provide the most relevant comparison. From this point of view, the CO_2 driven accelerator operates at hundred-fold lower plasma density than, and $L_{\text{dephasing}}$ equivalent to, the short wavelength driven accelerator, and thus provides greater total acceleration $E_z L_p \sim \lambda$ over a dephasing length [12]. Emittance degradation from electron beam scattering should also be reduced at lower plasma density. The paper by Pogorelsky *et al.* in this volume discusses these comparisons further.

A 10 GW (1 J, 100 ps) CO_2 laser is currently operational at the ATF. The basic technologies for upgrading the current system to multi-TW (15 J, 3 ps) power exist, and are being combined and implemented at BNL. Semiconductor switching techniques [11] will be used to generate picosecond 10 μm seed pulses. Isotopic mixing and high pressure ($\sim$ 10 atm) collisional broadening are being used to broaden and smooth the CO_2 gain spectrum into a 1 THz bandwidth continuum which supports 0.5 ps pulse amplification. A high volume- penetrating x-ray pre-ionizer and fast energy loading from a 1 MV Marx generator will be used to achieve a spatially uniform discharge in the large active volume ($\sim$ 10 liter) required to reach TW power levels in the small energy storage density ($\sim 10^{-2}$ J/cm^3) CO_2 medium. Initial repetition rate of 0.1 Hz is anticipated. The ATF laser is expected to be available as a user facility by 1998. Further details of the upgrade parameters and schedule are presented in the paper by Pogorelsky *et al.* in this volume.

In addition to high peak power generation, several working group participants emphasized the importance of temporal shaping and sequencing of TW pulses for optimizing laser-plasma coupling efficiency in plasma-based advanced accelerators. P. Chen predicted analytically that an asymmetric temporal pulse profile consisting of a slow rising edge and a sharp falling edge maximizes efficiency for driving the

LWFA with a single pulse in the linear regime. A similar conclusion based on PIC simulations [12] was presented by I.V. Pogorelsky, who noted that nearly optimized pulse shapes can be produced by the semiconductor switching system used to seed picosecond CO_2 amplifiers. For a more strongly driven LWFA, D. Umstadter presented 1D simulations showing that short pulse trains with pulses spaced by approximately a plasma period improve driving efficiency over a single pulse [13]. In an optimized pulse train, the pulse spacing increases and pulsewidth decreases as the plasma amplitude grows. These optimized LWFA driving schemes have not yet been demonstrated in the laboratory. Such experiments are needed to determine their sensitivity to jitter in pulse shape and spacing, to plasma density nonuniformity, and to radial pulse structure. Nevertheless the technology for shaping and optimizing ultrashort pulses and pulse sequences has been developed in other contexts. Programmable liquid crystal displays or acousto-optic modulators placed in the Fourier plane of the stretcher of a CPA system modulate the spatially dispersed frequency components of a short pulse to produce the desired temporal shape and/or sequence after the amplifier and compressor [14, 15]. Temporal features on a 20 fs times scale has been demonstrated [15], and appears adequate for LWFA applications. Pulse sequencing challenges beyond the single- or several-pulse regime which arise in multi-kilohertz repetition rate applications such as the γ-γ collider are discussed in the next section.

For resonantly-driven LWFAs based in plasmas of density less than several 10^{18} cm^{-3}, the driving pulse or pulse sequence must be optically guided at peak intensities of 10^{18} W/cm^2 over distances much longer than its vacuum Rayleigh length in order to convert plasma electrostatic energy efficiently into beam energy. Leemans et al. [16] have carried out a detailed design study of guiding requirements for next generation LWFA experiments. The guide length should be approximately equal to the phase slippage length λ_p^3/λ^2, which ranges from several cm to several meters for typical resonantly-driven LWFA schemes. The U. Maryland group of H. Milchberg has pioneered the development of preformed plasma waveguides which have guided pulses as intense as 5×10^{15} W/cm^2 over several hundred Rayleigh lengths (3 cm) [17]. The waveguide is created by bringing a 100 ps, 1 μm, 0.5 J pulse to a cylindrically symmetric focus using an axicon lens. A density cavity with wavelength-independent optical guiding properties evolves on a nanosecond time scale as a shock wave driven by axial electron heating propagates radially outward. The Maryland group has characterized the radial mode structure of guided pulses [18] and has analyzed the temporal evolution of the waveguide structure using folded wavefront interferometry [19]. The experimental approach and theory of a guiding technology relevant to plasma-based accelerators is thus already well-developed. Nevertheless several problems still need to be addressed. Guiding properties must be experimentally tested at the intensity and pulsewidth (10^{18} W/cm^2, 100 fs) required by the LWFA. Distortions induced on both the waveguide and the intense driving pulse by relativistic nonlinearities may become significant and must be carefully evaluated. Moreover the input and output coupling efficiency, and shot-to-shot pointing stability, will have to be evaluated at high intensity. Waveguide-induced slowing of the group velocity will degrade phase slippage distance, and will have to be figured into the accelerator design. Finally, waveguides of 10 cm to 1 m length will require laser pulses of 1 to 20 J energy to form the channel. Such lasers become a major fraction of the power consumption of any accelerator scheme, and will face the same difficulty in scaling to higher repetition rates as the lasers which power CPA systems. These issues will be key subjects of near-term research in plasma channeling for accelerator applications.

Several short wavelength sources driven by visible and near infrared TW or multi-GW laser pulses have been developed in the past few years. Incoherent picosecond soft x-ray pulses have been produced by the interaction of intense high-contrast pulses with solid targets [20, 21, 22]. Coherent XUV radiation has been generated by TW-pulse-driven laser action [23] and by high order harmonic generation [24] in gases. Of particular relevance to the advanced accelerator community is the recent generation of fs x-ray pulses by 90 degree Thomson scattering of a TW laser pulse (100 fs, 40 mJ, 800 nm., 10^{15} W/cm^2) from electron pulses (1.5 nC, 15 ps, 50 MeV) in the injector linac at the Berkeley Advanced Light Source [25]. Initial experiments have produced 10^5, 30 keV x-ray photons in a pulse of 200 fs duration with an angular divergence of 10 mrad. A similar physical interaction scaled to higher electron energy (.25 to 2.5 TeV) and laser pulse energy (1 J), involving similar laser-electron pulse synchronization issues, underlies proposed γ-ray generation schemes for future γ-γ colliders [26], as discussed further in the next section. P. Chen and T. Tajima also proposed to the working group that existing or near-future TW lasers (10–100 TW, 10 fs, 10 μm focus guided over 1 cm) could generate detectable (∼ 100 photons) Unruh radiation [27] in the 100 eV range because of perturbation of the vacuum by relativistically accelerated electrons in the laser pulse front. Thus within the last few years TW laser technology has introduced not only a variety of promising new short wavelength sources, but the promise of new table-top physics experiments at the most fundamental level.

Most of the laser technologies and applications outlined above have developed since 1991, demonstrating that progress can be very fast. We can thus approach remaining problems in terawatt laser technology with some confidence that they, too, can be overcome on a reasonable time scale.

TW LASER TECHNOLOGIES NEEDING MAJOR DEVELOPMENT

The principal shortcomings of current TW laser technology for projected collider applications are low average power and, for solid-state lasers, low wall-plug efficiency. These limitations originate from several sources. Current solid state TW laser systems are very inefficiently pumped. Typical Ti:sapphire oscillators are pumped by an argon laser which consumes > 10 kW of electrical power to produce a 5 W output beam before pumping the oscillator with 15% efficiency. Flashlamp-powered Nd:YAG lasers, which often pump the Ti:sapphire amplifier crystals and the plasma channelling system, typically operate with about 1% wall-plug efficiency. In such lasers the flashlamps convert about half of the electrical input power to a blackbody output spectrum of which only 10–20% falls within the absorption bands of the laser rod. Of this useful fraction, typically 10 to 20% will appear as laser output, depending on material parameters of the gain medium, geometrical factors, and optical losses, while the rest is dissipated as heat [28]. Similar ratios apply for flashlamp-pumped Nd:glass amplifiers, where the poor thermal conductivity of glass results in slower heat removal. Thermal stressing and thermo-optic beam distortion from cumulative heating limit the repetition rate to tens of Hz for typical (10 W average power, 1 cm beam diameter, Q-switched) Nd:YAG pump lasers. In fact the average power of nearly all high peak-power solid-state laser systems, from Nova-scale down to short-pulse oscillators does not exceed approximately 10 Watts. The currently highest power CPA systems (5–50 J/pulse) use cylindrical Nd: glass rods of several cm diameter in the final stage, where heat removal limits the repetition rate to as little as two to three pulses per hour.

For a simple pulsed amplifier this "10 Watt limit" is straightforwardly rationalized. The repetition rate f is inversely proportional to the cooling time $\tau_{th} = (x_0^2 c\gamma)/\kappa$, where c is the heat capacity, γ the mass density, κ the heat conductivity, and x_0 is a scale length for heat flow. For a simple cylindrical rod amplifier cooled about its sides, x_0 would be proportional to the radius r_0. The damage fluence, $F_{dam} \sim 1$ J/cm^2, for both the amplifier media and associated optical elements (mirrors, lenses, etc.). In order to accommodate large pulse energies E with a cylindrical amplifier, the beam must have an area $A \geq E/F_{dam}$, and therefore the average power $P_{avrg} = E \cdot f \sim F_{dam}\kappa/c\gamma$ is constant. For typical solid-state laser hosts, e.g. sapphire and YAG, this constant is on the order of a few watts. For glass hosts, which can be made in large slab geometries, the significant increase in extracted energy is offset by the $\sim 50\times$ smaller heat conductivity, leading coincidentally to the same few watt limit. Similar results hold for quasi-CW operation, in which a more careful analysis of the heat flow must be performed. The figures used above represent conservative estimates.

CO_2 lasers can operate with 15 to 20% efficiency with 50 kilowatt average power for CW discharges and nearly 1 MW for gas-dynamic lasers [29]. For long pulses, they can operate at 1 kHz with several kW average power. The TW peak power, subpicosecond pulsed laser now under construction at BNL will initially operate with unoptimized repetition rate (0.1 Hz), average power (1.5 W), and wall-plug efficiency (0.01), which are adequate for its intended use for proof-of-principle high field physics experiments. However, in the future it should be feasible to scale these numbers significantly upward, using known technologies, while maintaining the 10 liter volume of the gain medium. Pogorelsky estimates that repetition rates of 1 kHz, limited by the power supply and the speed of gas exchange through the discharge region, should be feasible with 50 TW peak power, limited by window damage, 10 kW average power and 0.5 to 0.8 ps pulse duration. The initial wall-plug efficiency is limited by the weakess of the x-ray (compared to e-beam) preionization, which necessitates a higher than optimum discharge voltage, and by the window damage threshold and an unoptimized optical configuration. Efficiency of stored energy extraction can be improved several-fold by amplifying a 100 ns train of 1 ps pulses instead of a single 1 ps pulse. Pogorelsky estimates that efficiencies approaching 20% are achievable in ps TW mode.

DRIVING APPLICATIONS

Shortcomings in efficiency and repetition rate are not unique to TW lasers or to advanced accelerator applications, but currently limit many high average power laser applications [26]. For example, TW lasers scaled to multi-kilohertz repetition rates could provide a compact, bright, inexpensive source of soft x-rays for photolithographic manufacture of future submicron integrated circuits [30]. Atmospheric sensing applications, including global wind measurements for weather modelling and local diagnostics of wind shear near airports and vorticity in tornado-prone areas[31], generation of pulsed laser guidestars [32], and laser ranging and targeting applications are driving development of high average power, nanosecond-pulse solid state lasers in the near infrared. Since such lasers often pump solid state TW lasers, scaling of their repetition rate and average power will directly benefit TW technology. Nanosecond CO_2 lasers, which can operate at high efficiency (10–20%) and average power (multi-kilowatt) compared to their long-pulse solid-state counterparts, are also becoming increasing attractive for remote sensing applications as new nonlinear optical materials are developed [33] for efficiently frequency doubling their output into the atmospheric propagation window at 4–5 μm. Finally, materials processing applications such as welding, cutting and annealing also increasingly require kilowatt average power lasers of various wavelengths and pulse durations. Femtosecond laser micromachining [34] and 3D optical storage inside transparent materials [35] are emerging applications which will also benefit from higher average power sources. Since the driving applications are numerous, the community-wide motivation for overcoming current limitations to TW technology is high.

ENABLING TECHNOLOGIES

Group 7 participants outlined the key enabling technologies which are now being developed to achieve efficient high repetition rate laser operation. The first two of the following technologies – efficient pumping and improved thermal management

– apply primarily to solid state lasers. The third – beam multiplexing – applies equally to solid state and CO_2 lasers.

1. Diode array pumping. The replacement of flashlamps and argon lasers by laser diode arrays promises to have an impact on solid state laser technology in general, and TW laser technology in particular, as profound as the replacement of vacuum tubes by solid state integrated circuits. The ability to fabricate linear bars and 2D arrays with tight control of material composition, layer thickness and device geometry has progressed significantly during the past decade with advances in epitaxial semiconductor growth techniques such as metal organic chemical vapor deposition (MOCVD). The excellent spectral match between their output and the absorption bands of solid state gain media renders essentially all of the radiation useful. Consequently pumping efficiency is greatly increased, while thermal loading is reduced compared to flashlamp pumping, thus permitting an increase in repetition rate and average power. For example, diode-array pumped Nd:YAG lasers with cylindrical rods have already been operated with wall-plug efficiencies of 8% at 0.8 J/pulse, average powers 160 W, repetition rate 200 Hz [28]. Longer lifetime (10^9 shots) and reduced system maintenance compared to flashlamps (10^7 shots) also result from the reduced thermal loading and negligible UV output. The principal current drawback of diode arrays is their high cost resulting from customized, labor-intensive manufacturing. However, R. Freeman estimated during group 7 discussions that high demand would reduce their cost 10- to 100-fold within the next several years.

Group 7 participants described several ways in which laser diode arrays could impact TW laser systems. C.P.J. Barty emphasized their role in scaling Nd:YAG and glass lasers for use as high average power pump sources for wide-bandwidth Ti:sapphire gain crystals, which because of short fluorescence lifetime cannot be directly diode-pumped. In a similar vein, K. Tani described extensive use of diode-pumped Nd:YAG lasers for pumping a planned high repetition rate TW Ti:sapphire system at JAERI while R. Freeman and C. Clayton emphasized the central role of diode-pumped Nd lasers in planned γ-γ collider development. G. Mourou, on the other hand, emphasized the potential role of laser diodes as direct pumps of wide bandwidth gain media such as Yb:glass, which have storage times as long as 2 ms. This approach favors overall system compactness, although the problem of heat removal from glass must be addressed, as discussed further below. The emergence of femtosecond oscillators based on Nd:glass and Cr:LiSaF which can be directly laser-diode pumped [36], though not presented at AAC, is also enabling more efficient and compact laser sources for advanced accelerators by replacing inefficient argon lasers.

2. Thermal management. Even with efficient diode pumping, the scalability of solid state amplifiers to higher average power is still limited by residual thermal distortions in the solid state gain medium. Methods for management of and compensation for these residual thermal effects therefore complement diode pumping as a key enabling technology in high average power generation. Face-pumped slab gain medium architectures, employing a zig-zag optical path to suppress thermal distortions present in cylindrical rod geometries, have been a subject of active research and development since their original proposal in 1972 [37]. Residual thermal distortions from edge and end effects set the practical performance limits of such systems, and place a premium on uniform pumping deposition, uniform cooling and optimized slab design [28]. Scalability of such systems to kilowatt average power has been demonstrated for long pulse operation. For example, a diode-array-pumped Nd:YAG slab laser has generated 1 kW average power in 150 microsecond pulses at 2 kHz repetition rate, although with poor beam divergence [38]. Improvements to this design are expected to achieve 10% wall-plug efficiency [28]. To improve output beam quality, several groups have supplemented slab amplifiers with active correction of thermal wavefront distortions by optical phase conjugation [39]. For example, a diode-pumped Nd:YAG slab laser using phase conjugate mirrors based on stimulated Brillouin scattering (SBS) has produced 100 W average power in the form of 1 J pulses at 100 Hz [40]. Even in flashlamp-pumped Nd:glass amplifiers, SBS phase conjugation has enabled unprecedented generation of 30 J, 14 ns pulses at > 150 W average power (6 Hz repetition rate) [41]. Although the powers and repetition rates achieved to date are lower than in uncorrected lasers, these phase conjugated systems have produced nearly diffraction-limited output beams. The bandwidth of available phase conjugate nonlinear media may limit their direct use in TW CPA systems, where they must respond to extremely broadband, sub-nanosecond stretched pulses whose re-compressibility is very sensitive to material dispersion. In these cases alternative adaptive optic technologies such as deformable mirrors may provide the required wavefront correction. According to K. Tani, extensive use of adaptive optic technologies is planned in the high average power TW systems at JAERI.

G. Mourou described an alternative thermal management approach appropriate for media such as Yb:glass, which have poor thermal conductivity but can be scaled to large sizes. In this scheme, diode-pumping and amplification are restricted to a region near the edge of an Yb:glass cylinder, which is then continuously rotated during amplification so that a cooled region of the gain medium is presented to each pulse. The center of the cylinder can be hollowed to provide a channel for coolant flow.

3. Beam multiplexing. Efficient pumping and good thermal management may push solid state TW technology one to two orders of magnitude beyond the "10 watt limit". However, further progress will require beam multiplexing. G. Mourou, R. Freeman, and C. Clayton each independently described one generic approach to beam multiplexing. In their scheme, a single pulse from a low repetition rate preamplifier is divided into numerous individual beam lines, each containing one thermal- and fluence-limited ~ 10–100 W T^3 amplifier. After recombination, the copropagating but temporally separated pulses form a "brigade" of pulses with a net average power in the multi-kW regime. C. Clayton described a variety of specific methods for dividing and recombining the pulse train. The simplest approach uses a passive cascade of beam splitters and optical delay lines. An alternative active approach traps the low repetition rate input pulses in an optical storage ring in which the pulses can be regeneratively amplified and shaped while the pulse frequency is multiplied. Frequency-domain multiplexing is also possible. These multiplexing schemes can be used not only to increase the repetition rate, but also to produce a desired laser pulse format. For example, to produce gamma rays by Compton backscattering from the NLC beam, the laser pulses must be synchronized with an electron pulse format consisting of 120 ns. macrobunches at 180 Hz repetition rate, each containing approximately 85 pulses at 1.5 ns separation. These various approaches to multiplexing are described in more detail in [26].

A long-range proposal for a more ambitious multiplexing scheme, discussed by C.W. Siders, utilizes the r_0^2 dependence of the cooling time, combined with a monolithic amplifier consisting of, for example, 10 μm diameter × 1 cm long amplifier rods in a 5000 × 5000 (10 cm × 10 cm) array, to achieve up to MHz repetition rates with MW average powers. Rather than multiplexing in the time domain, this scheme would utilize a microlens array to coherently separate, amplify, and recombine tens of millions of individual "microbeams". Optical pumping could be achieved either via dichroic longitudinal pumping with a large array of fiber-coupled diode lasers or, perhaps, via diode lasers embedded in the monolithic structure and with transverse gradient-index coupling to the amplifier rods. Though this scheme has considerable attraction as a compact (100 cm^3) "single-stage" amplifier, considerable engineering would be required in the area of coolant flow within the structure and subsequent heat dissipation. Nevertheless, similar designs have been successfully used for ∼ 10 MW operation in commercial nuclear power station fuel bundles and design work has been performed on similarly compact MW-class fission-fragment-pumped lasers[42].

CONCLUDING REMARKS AND PROSPECTUS

Rapid developments in solid-state terawatt laser technology within the past five years have already met the single pulse requirements for a variety of laser-driven advanced accelerator schemes. A new CO_2 terawatt technology is now emerging, and promises to complement the capabilities of solid state sources in many advanced accelerator applications. The basic technologies for improving average power and (for solid state lasers) efficiency by as much as two or three orders of magnitude are now rapidly being developed in numerous laboratories. It appears likely that kilowatt average power, terawatt peak power lasers which meet the basic requirements for developing a γ-γ collider in conjunction with the NLC or other linear collider can be developed within the next decade. A future high luminosity, laser-driven

e^+e^- collider will require hundreds of such laser systems. Wall-plug efficiencies of at least 10% will therefore be essential to contain power consumption costs. There are grounds for optimism that this efficiency requirement can also be met. Nevertheless simultaneous unprecedented requirements for multi-stage alignment and beam emittance in such a collider will raise new technological challenges which have yet to be seriously addressed.

Acknowledgements

This work was supported by U.S. Department of Energy Grant No. DEFG05-92-ER-40739 and NSF Grant No. PHY-9417558. C. W. Siders acknowledges support of a Postdoctoral Fellowship from Los Alamos National Laboratory.

References

[1] K. Nakajima et al., in Proc. AIP Conf. Advanced Accelerator Concepts, vol. 335, P. Schoessow, ed., New York: Am. Inst. Phys. 1995, pp. 145-155; C. W. Siders et al., Phys. Rev. Lett. **76**, 3570 (1996); J. R. Marques et al., Phys. Rev. Lett. **76**, 3566 (1996).

[2] K. Nakajima et al., Phys. Rev. Lett. **74**, 4428 (1995); A. Modena et al., Nature **377**, 606 (1995); C. A. Coverdale et al., Phys. Rev. Lett. **74**, 4659 (1995); D. Umstadter et al., Science **273**, 472 (1996).

[3] C. P. J. Barty et al., Opt. Lett. **21**, 668 (1996).

[4] C. P. J. Barty, Laser Focus World **32**, 93 (1996).

[5] C. N. Danson et al., Optics Commun. **103**, 392 (1993).

[6] I. V. Pogorelky et al., this volume.

[7] D. Strickland and G. Mourou, Opt. Commun. **56**, 219 (1985); P. Maine et al., IEEE J. Quantum Electron. **24**, 398 (1988).

[8] M. D. Perry and G. Mourou, Science **264**, 917 (1994).

[9] M. Everett et al., Nature **368**, 527 (1994).

[10] N. A. Ebrahin, J. Appl. Phys. **76**, 7645 (1994).

[11] P. B. Corkum and D. Keith, J. Opt. Soc. B. **2**, 1873 (1985).

[12] S. V. Bulanov, et al., IEEE Trans. Plasma Sci. **24**, 393 (1996).

[13] D. Umstadter, *et al.*, Phys. Rev. Lett. **72**, 1224 (1994).

[14] A. M. Weiner, *et al.*, J. Opt. Soc. Am. B **5**, 1563 (1988); A. M. Weiner *et al.*, IEEE J. Quantum Electron. **28**, 908 (1992); A. M. Weiner, *et al.*, Progr. Quantum Electron. **19**, 161 (1995).

[15] D. H. Reitze, A. M. Weiner, and D. E. Laird, Appl. Phys. Lett. **61**, 1260 (1992).

[16] W. P. Leemans *et al.*, IEEE Trans. Plasma Sci. **24**, 331 (1996).

[17] C. G. Durfee III and H. M. Milchberg, Phys. Rev. Lett. **71**, 2409 (1993); C. G. Durfee III, J. Lynch, and H. M. Milchberg, Phys. Rev. E **51**, 2368 (1995).

[18] C. G. Durfee III, J. Lynch and H. M. Milchberg, Opt. Lett. **19**, 1937 (1994).

[19] M. Takeda *et al.*, J. Opt. Soc. Am. B **72**, 156 (1982); see also the paper by H. M. Milchberg *et al.* in this volume.

[20] M. M. Murnane *et al.*, Science **251**, 531 (1991).

[21] J. D. Kmetec *et al.*, Phys. Rev. Lett. **68**, 1527 (1992).

[22] J. Workman *et al.*, Phys. Rev. Lett. **75**, 2324 (1995).

[23] B. E. Lemoff *et al.*, Phys. Rev. Lett. **74**, 1574 (1995).

[24] J. J. Macklin *et al.*, Phys. Rev. Lett. **70**, 766 (1993).

[25] W. Schoenlein *et al.*, Science **274**, 236 (1996).

[26] C. E. Clayton *et al.*, Nucl. Inst. and Meth. A **355**, 121 (1995).

[27] W. G. Unruh, Phys. Rev. D **14**, 870 (1976).

[28] W. Koechner, Solid State Laser Engineering, 4th ed. (Springer Verlag, Berlin, 1996).

[29] W. J. Witteman, The CO_2 laser (Springer Verlag, Berlin 1987).

[30] L. A. Hackel *et al.*, Appl. Opt. **32**, 6914 (1993).

[31] Solid State Lasers II, ed. G. Dube, Proc. SPIE **1871** (1993).

[32] D. G. Sandler *et al.*, J. Opt. Soc. Am. A **11**, 858 (1994).

[33] L. Gordon *et al.*, Electron. Lett. **29**, 1942 (1993); M. W. McGeoch, in Intense Laser Beams and Applications, Proc. SPIE **1871**, 62 (1993).

[34] X. Liu, private communication.

[35] E. N. Glezer, *et al.*, Ultrafast Phenomena X, Springer Series in Chemical Physics **62**, Eds: P.F. Barbara, J.G. Fujimoto, W.H. Knox, and W. Zinth, pg. 157.

[36] D. Kopf *et al.*, Opt. Lett. **20**, 1169 (1995); D. Kopf *et al.*, Opt. Lett. **20**, 1782 (1995); R. Mellish *et al.*, Opt. Lett. **20**, 2312 (1995).

[37] W. S. Martin and J. P. Chernoch, U.S. Patent 3,633,126 (1972).

[38] B. J. Comaskey *et al.*, IEEE J. Quantum Electron. **28**, 992 (1992).

[39] B. Ya. Zel'dovich *et al.*, Principles of Phase Conjugation, Springer Ser. Opt. Sci., Vol. **42** (Springer, Berlin, Heidelberg 1985).

[40] R. J. St. Pierre *et al.*, in Conf. Lasers and Electro-Optics, OSA Tech. Dig. Ser. (OSA, Washington, 1994), Vol. **8**, p. 283.

[41] C. B. Dane *et al.*, IEEE J. Quantum Electron. **31**, 148 (1995).

[42] G. A. Hebner and G. N. Hays, Proc. SPIE **2121**, 10 (1994); G. A. Hebner, private communication.

III. WORKING GROUP PAPERS

GROUP 1

"Nonconventional Collider Concepts and Luminosity Paradigms"—
Jonathan Wurtele (LBNL/UCB)

Studies of Laser-Driven 5 TeV e^+e^- Colliders in Strong Quantum Beamstrahlung Regime

M. Xie[1], T. Tajima[2], K. Yokoya[3]
and S. Chattopadhyay[1]

[1] *Lawrence Berkeley National Laboratory, USA*
[2] *University of Texas at Austin, USA*
[3] *KEK, Japan*

Abstract.
We explore the multidimensional space of beam parameters, looking for preferred regions of operation for a e^+e^- linear collider at 5 TeV center of mass energy. Due to several major constraints such a collider is pushed into certain regime of high beamstrahlung parameter, Υ, where beamstrahlung can be suppressed by quantum effect. The collider performance at high Υ regime is examined with IP simulations using the code CAIN. Given the required beam parameters we then discuss the feasibility of laser-driven accelerations. In particular, we will discuss the capabilities of laser wakefield acceleration and comment on the difficulties and uncertainties associated with the approach. It is hoped that such an exercise will offer valuable guidelines for and insights into the current development of advanced accelerator technologies oriented towards future collider applications.

INTRODUCTION

It is believed that a linear collider at around 1 TeV center of mass energy can be built more or less with existing technologies. But it is practically impossible to go much beyond that energy without employing a new, yet largely unknown method of acceleration. However, apart from knowing the details of the future technologies, certain collider constraints on electron and positron beam parameters are considered to be quite general and have to be satisfied, e.g. available wall plug power and the constraints imposed by collision processes: beamstrahlung, disruption, backgrounds, etc. Therefore it is appropriate to explore and chart out the preferred region in parameter space based on these constraints, and with that hopefully to offer valuable guidelines for and insights into the current development of advanced accelerator technologies oriented towards future collider applications.

Taking such a point of view, we examine collider performance at the final interaction point (IP) of a e^+e^- collider over a large space of beam parameters. We show that it becomes increasingly necessary at higher energy to operate colliders in high Υ regime and use to our advantage the quantum effect to suppress beamstrahlung. Although the quantum suppression effect was known and studied before with simple models [1–4], it has not been checked with full-blown simulation at high Υ regime that we are considering in this paper. As will be shown later, there are indeed several surprising features revealed by our simulations, in particular in the differential luminosity spectrum, which is a crucial factor for colliders.

Given beam parameters that are confirmed by simulation to be within acceptable level of beamstrahlung, we then discuss its implications for laser-driven acceleration. In particular we examine general characteristics and capabilities of laser wakefield acceleration and comment on the difficulties and uncertainties associated with the approach.

COLLIDER CONSIDERATIONS

In this section we will first discuss major collider requirements and constraints and organize the beam parameters in a way more convenient for exploration. We then scan the parameter space to find optimal regime of operation, and discuss its characteristics, as well as design options and trade-offs. These optimal designs are shown to be in high Υ regime. The collider performance at high Υ regime is examined with CAIN [5] simulations.

IP Requirements

The primary drive for developing ever more advanced accelerators is to expand both energy and luminosity frontiers for high energy physics applications. An important collider performance parameter is the geometrical luminosity given by $\mathcal{L}_g = f_c N^2/4\pi\sigma_x\sigma_y$ where f_c is the collision frequency, N is the number of particles per bunch, σ_x and σ_y are, respectively, the horizontal and vertical rms beam sizes at the IP. The real luminosity, however, depends on various dynamic processes at collision. Among them the most important ones are beamstrahlung and disruption [6]. These two processes are characterized by the beamstrahlung parameter $\Upsilon = 5r_e^2\gamma N/6\alpha\sigma_z(\sigma_x + \sigma_y)$, and the disruption parameter $D_y = 2r_eN\sigma_z/\gamma\sigma_y(\sigma_x + \sigma_y)$, where γ is the Lorentz factor, r_e the classical electron radius, α the fine structure constant, and σ_z the rms bunch length. Beamstrahlung is in classical regime if $\Upsilon \ll 1$, and strong quantum regime if $\Upsilon \gg 1$. The physical effect of beamstrahlung is not directly reflected in the magnitude of Υ, but rather it is more conveniently monitored through the average number of emitted photons per electron $n_\gamma = 2.54\,(\alpha\sigma_z\Upsilon/\lambda_c\gamma)\,U_0(\Upsilon)$ and relative electron energy loss $\delta_E = 1.24\,(\alpha\sigma_z\Upsilon/\lambda_c\gamma)\,\Upsilon U_1(\Upsilon)$. where $\lambda_c = \hbar/mc$ is the Compton wavelength, $U_0(\Upsilon) \approx 1/(1+\Upsilon^{2/3})^{1/2}$, and $U_1(\Upsilon) \approx 1/(1+(1.5\Upsilon)^{2/3})^2$.

So far we have given the major constraints imposed at the collision, which require n_γ and δ_E not be too large to cause luminosity degradation. Generally speaking, when these requirements are satisfied, other deteriorating effects such as pair creation and hardronic background will also be small [6]. Another major constraint for collider design is the available wall plug power which limits the beam power, given accelerator efficiency. We define the average power of both colliding beams $P_b = 2E_b N f_c$, the center of mass energy $E_{cm} = 2E_b$, and the beam energy $E_b = \gamma mc^2$.

It is noted from all the formulas given above that there are only six independent parameters and they are chosen for convenience to be $\{E_{cm},\,\mathcal{L}_g,\,P_b,\,R,\,N,\,\sigma_z\}$, where R is the aspect ratio σ_x/σ_y. For collider design considerations we are interested in monitoring six quantities $\{f_c, \sigma_y, \Upsilon, D_y, n_\gamma, \delta_E\}$, and they are expressed in terms of the six independent parameters as follows

$$f_c = \left(\frac{P_b}{E_{cm}}\right)\left(\frac{1}{N}\right) \tag{1}$$

$$\sigma_y = \left(\frac{1}{\sqrt{4\pi}}\right)\left(\frac{1}{\sqrt{R}}\right)\left(\sqrt{\frac{P_b}{E_{cm}\mathcal{L}_g}}\right)(\sqrt{N}) \tag{2}$$

$$\Upsilon = \left(\frac{5\sqrt{\pi}r_e^2}{6\alpha mc^2}\right)\left(\frac{\sqrt{R}}{1+R}\right)\left(\sqrt{\frac{E_{cm}^3\mathcal{L}_g}{P_b}}\right)\left(\frac{\sqrt{N}}{\sigma_z}\right) \tag{3}$$

$$D_y = \left(16\pi mc^2 r_e\right)\left(\frac{R}{1+R}\right)\left(\frac{\mathcal{L}_g}{P_b}\right)(\sigma_z) \tag{4}$$

$$n_\gamma = 2.54 U_0(\Upsilon)F,\qquad \delta_E = 1.24\Upsilon U_1(\Upsilon)F \tag{5}$$

$$F = \left(\frac{5\sqrt{\pi}r_e^2}{3\lambda_c}\right)\left(\frac{\sqrt{R}}{1+R}\right)\left(\sqrt{\frac{E_{cm}\mathcal{L}_g}{P_b}}\right)(\sqrt{N}). \tag{6}$$

The advantage of organizing the independent and dependent parameters in such a way lies in its convenience for design optimization in the multidimensional parameter space, since in most situations many of the independent parameters can be fixed. For example, in this paper, we set $E_{cm} = 5\text{TeV}$

and $\mathcal{L}_g = 10^{35} \text{cm}^{-2}\text{s}^{-1}$ as our goal in energy and luminosity frontiers. For laser-driven acceleration, we assume $R = 1$ for reasons that will be explained later in this section. Furthermore, given maximum wall plug power, it is often adequate to consider P_b at a few discrete values corresponding to different accelerator efficiencies. Then for each fixed value of P_b we are left with only two independent parameters $\{N, \sigma_z\}$ to vary, and all the dependent parameters can thus be conveniently visualized in a surface or contour plot, as will be shown in the next section.

The design approach given here can be extended to integrate more collider parameters and the associated boundary conditions into the process of constrained optimization. For example, the beam size σ_y is related to two other important parameters: the normalized rms emittance ε_y and the betafunction at IP β_y by $\sigma_y = \sqrt{\beta_y \varepsilon_y / \gamma}$. Once σ_y is determined, ε_y and β_y can be chosen according to other constraints, and vice versa. One constraint that is of immediate importance for the IP is the Oide limit [7], which sets the minimum achievable beam size: $\sigma_{min}[\text{m}] = 1.7 \times 10^{-4} \varepsilon_y[\text{m}]^{5/7}$. Here we have used in the Oide limit a smaller numerical factor proposed by Irwin [8]. For later use, we define $F_{oide} = \sigma_y / \sigma_{min}$, the Oide limit is violated if $F_{oide} < 1$.

Before going to the exploration of parameter space using Eqs.(1-6), it is instructive to look at the more transparent scaling laws in two dimensional parameter space $\{N, \sigma_z\}$ when $\{E_{cm}, \mathcal{L}_g, P_b, R\}$ are considered fixed

$$f_c \sim 1/N, \quad \sigma_y \sim \sqrt{N}, \quad D_y \sim \sigma_z, \quad \Upsilon \sim \sqrt{N}/\sigma_z \qquad (7)$$

$$n_\gamma \sim U_0(\Upsilon)\sqrt{N}, \quad \delta_E \sim \Upsilon U_1(\Upsilon)\sqrt{N}. \qquad (8)$$

In the limit $\Upsilon \gg 1$, $U_0(\Upsilon) \to 1/\Upsilon^{1/3}, \Upsilon U_1(\Upsilon) \to 1/\Upsilon^{1/3}$. Eq.(8) becomes

$$n_\gamma \sim (N\sigma_z)^{1/3}, \quad \delta_E \sim (N\sigma_z)^{1/3}. \qquad (9)$$

We see from Eqs.(7,9) that once in the high Υ regime there are two approaches to reduce the effects of beamstrahlung: either by reducing N or by reducing σ_z. The consequences on the collider design and the implied restrictions on the approaches, however, can be quite different. Reducing N requires f_c to be increased and σ_y decreased, thus the approach is limited by the constraints on f_c and σ_y. Reducing σ_z, on the other hand, is not directly restricted in this regard. Also the dependencies of Υ on the two approaches are quite the opposite. The second approach clearly demonstrates the case that beamstrahlung can indeed be suppressed by having larger Υ.

We now come to explain why it is reasonable to assume round beam $R = 1$. The current designs of linear colliders at 0.5 TeV are all based on damping ring technology which provides much smaller emittance in the vertical dimension. Taking advantage of this feature, beam distribution at the IP has been made very flat, $R \gg 1$, to suppress beamstrahlung. However, first of all, it is not clear at this point what would be the injector of choice for future laser-driven accelerator, if emittances can be made as asymmetrical as in the damping ring, or if possible, would it be compatible with, for example, transverse focusing channel of the acceleration scheme. Secondly, as will be shown in the next section for round beam, the required beam size is already in the Å level. A flat beam requires the beam size in one dimension be made even smaller, thus pushing the limit for tight beam positioning control. Nonetheless, one should keep in mind that making $R \gg 1$ is still a knob for further suppression of beamstahlung, even in strong quantum regime as can be seen from Eqs.(1-6).

Parameter Optimization

Using the formulas provided in the previous section: Eqs.(1-6), we are now ready to explore the parameter space. As mentioned before we will consider the situation with $\{E_{cm} = 5\text{TeV}, \mathcal{L}_g = 10^{35}\text{cm}^{-2}\text{s}^{-1}, R = 1\}$. Assuming wall plug power for such a collider is limited to 2 GW [8], and the overall "wall plug to beam" efficiency is within the range of 0.1% to 10%, we will look at three cases with P_b being 2 MW, 20 MW and 200MW, respectively.

Figure 1 shows the contour plots of parametric scans for the cases with $P_b = 2$MW (left column) and 20MW (right column). Due to page limitation, we show only a few out of many quantities that can be monitored in $\{N(10^8), \sigma_z(\mu\text{m})\}$ space, they are, starting from the top row: n_γ, Υ and σ_y(nm). From these scans one may chose optimal operation point $\{N, \sigma_z\}$ based on various constraints imposed on the independent as well as dependent quantities. Using the plots in the bottom row one can also determine ε_y(nm) and $\beta_y(\mu\text{m})$ at different values of σ_y, and from there to check F_{oide}. The type of parametric scans shown here are used as a guide to pick specific parameter sets given in Table 1 for three values of beam power. Several performance parameters computed from the formulas are given in Table 2, some of them can be directly compared with simulations. It is noted here we have chosen to make n_γ significantly less than 1 and same for all three cases, and violate the Oide limit by about 10% on purpose to relax other parameters.

High Υ IP Simulation

Although the simple formula, Eq.(5), takes into account strong quantum beamstrahlung with high Υ, some important effects are nonetheless neglected, for example, disruption and multiphoton processes [6]. It is therefore necessary to examine its predictions with full-blown simulations. We use a Monte-Carlo simulation code recently developed by Yokoya [5] to study QED processes at the IP for e^+e^- and $\gamma\gamma$ colliders. This code is a superset of the well-known code ABEL by the same author. Care has been taken to ensure that there is

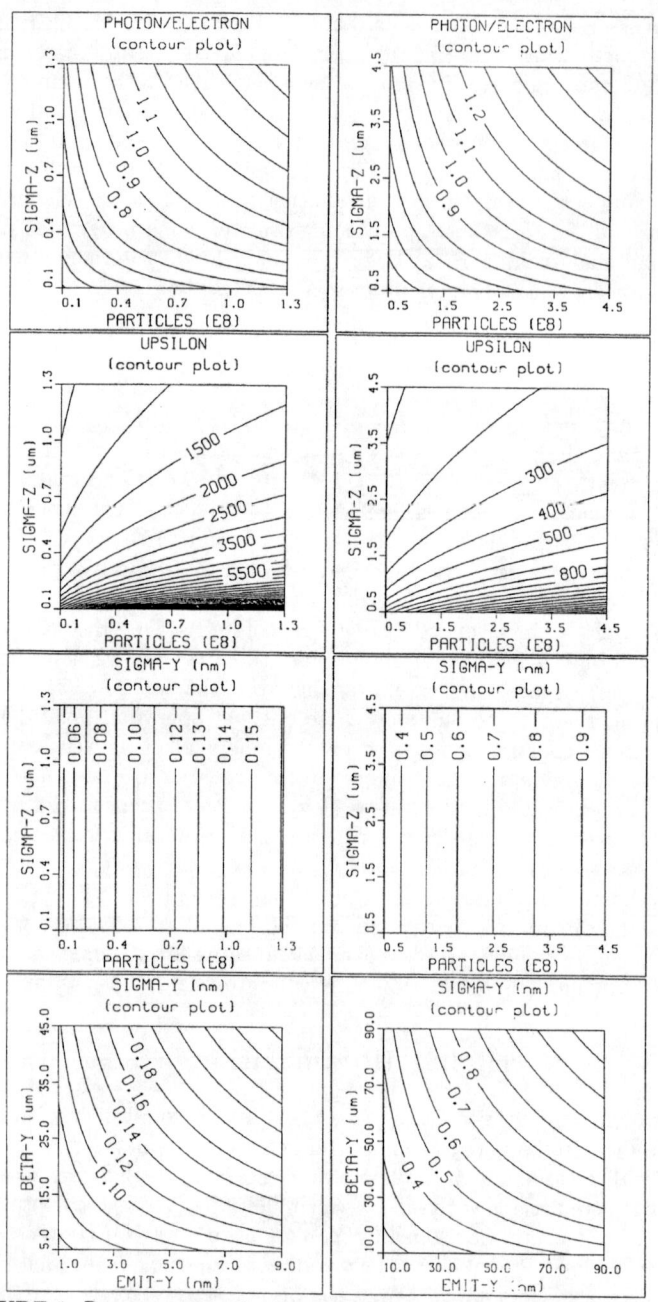

FIGURE 1. Parameter scans for $P_b = 2\text{MW}$ (column 1) and 20MW (column 2).

enough resolution in the simulation at such high Υ values to yield reliable QED prediction. This is established by verifying that results changes insignificantly by changes of resolution grids.

Figure 2 gives the differential e^+e^- luminosities for the case I, II, III in Table 1. It is noted that the luminosity spectrum is characterized by an outstanding core at the full energy and a very broad, nearly flat halo. One see from Table 3, taking case II for example, although on average the beam loses 26% of its energy and has a rms energy spread of 36%, the core itself within 1% of full energy still accounts for 65% of the geometrical luminosity. The outstanding core is more than two orders of magnitude above the halo. The sharpness and the high luminosity of the core is rather surprising but pleasantly so. Comparing simulation results in Table 3 for n_γ and δ_E with that calculated from the formulas in Table 2, one see the agreement varies from being reasonably good at lower Υ to rather poor at higher Υ. It seems to indicate that the formulas can be used only as a rough guideline for collider design at high Υ. It is interesting to note that the core luminosity is somewhat larger for the case with higher beamstrahlung loss, which is probably due to disruption enhancement as indicated by the larger value of D_y in Table 2.

Another major deteriorating process at high Υ is coherent pair creation. The number of pairs created per primary electron, n_p, is given in Table 2 by formulas [6] and in Table 3 by simulations. According to our simulations the incoherent pair creation is 2 to 3 orders of magnitude smaller than that of the coherent pairs, thus negligible. Finally we point out that such a differential luminosity spectrum should be rigorously assessed together with the background of beamstrahlung photons and coherent pairs from the point of view of particle physics and detector considerations. Only then, one may judge if operation of colliders at high Υ regime is indeed a viable approach for high energy physics applications.

ACCELERATOR CONSIDERATIONS

As seen from Eq.(9), an effective way to suppress beamstrahlung is to reduce σ_z, which naturally favors laser acceleration as it offers much shorter acceleration wavelength than that of conventional microwaves. For laser wakefield acceleration, typical wavelength of accelerating wakefield is ~ 100 μm, which is in the right range for the required bunch length in Table 1. Laser wakefield acceleration [9,10] has been an active area of research in recent years primarily due to the major technological advance in short pulse TW lasers [11]. The most recent experiment at RAL has demonstrated an acceleration gradient of 100 GV/m and produced beam-like properties with 10^7 accelerated electrons at 40MeV $\pm$ 10% and a normalized emittance of $\varepsilon < 5\pi$ mm-mrad [12].

For beam parameters similar to that in Table 1, we consider a laser wakefield accelerator system consisting of multiple stages with a gradient of 10 GeV/m.

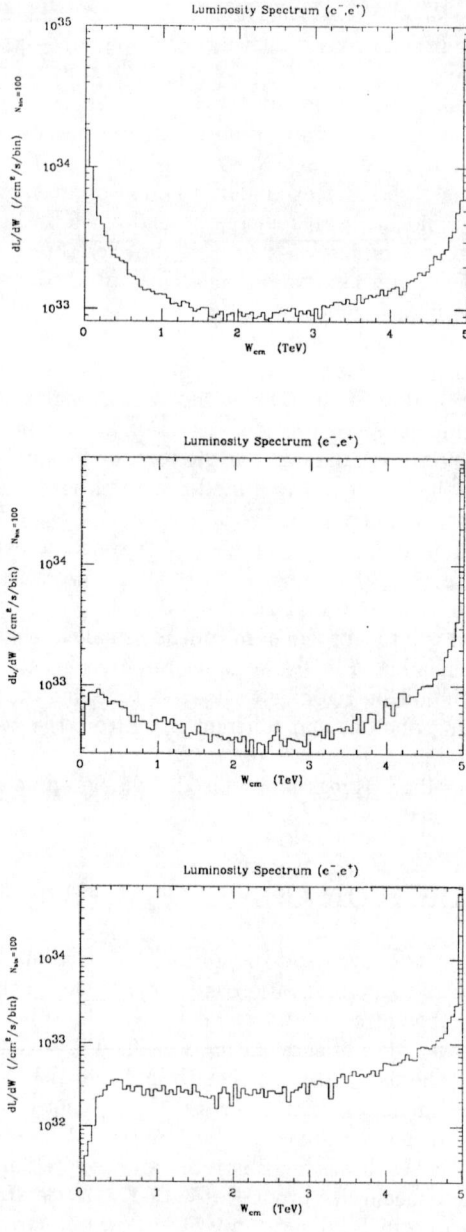

FIGURE 2. e^+e^- luminosity spectrum for case I (top), II (middle), III (bottom).

With a plasma density of 10^{17}cm^{-3}, such a gradient can be produced in the linear regime with more or less existing T^3 laser, giving a plasma dephasing length of about 1 m [13]. If we assume a plasma channel tens of μm in width can be formed at a length equals to the dephasing length, we would have a 10 GeV acceleration module with an active length of 1 m. Of course, creating and maintaining a plasma channel of the required quality is no simple matter. To date, propagation in a plasma channel over a distance of up to 70 Rayleigh lengths (about 2.2 cm) of moderately intense pulse ($\sim 10^{15}\text{W}/\text{cm}^2$) has been demonstrated [14]. New experiment aiming at propagating pulses with intensities on the order of $10^{18}\text{W}/\text{cm}^2$ (required for a gradient of 10 GeV/m) is underway [13].

Table 1. Beam Parameters at Three Values of Beam Power

CASE	P_b(MW)	$N(10^8)$	f_c(kHz)	ε_y(nm)	$\beta_y(\mu\text{m})$	σ_y(nm)	$\sigma_z(\mu\text{m})$
I	2	0.5	50	2.2	22	0.1	0.32
II	20	1.6	156	25	62	0.56	1
III	200	6	416	310	188	3.5	2.8

Table 2. Results Given By the Formulas

CASE	Υ	D_y	F_{oide}	n_γ	δ_E	n_p	$\mathcal{L}_g(10^{35}\text{cm}^{-2}\text{s}^{-1})$
I	3485	0.93	0.89	0.72	0.2	0.19	1
II	631	0.29	0.89	0.72	0.2	0.12	1
III	138	0.081	0.91	0.72	0.2	0.072	1

Table 3. Results Given By CAIN Simulations

CASE	n_γ	δ_E	σ_e/E_0	n_p	$\mathcal{L}/\mathcal{L}_g(W_{cm} \in 1\%)$	$\mathcal{L}/\mathcal{L}_g(W_{cm} \in 10\%)$
I	1.9	0.38	0.42	0.28	0.83	1.1
II	0.97	0.26	0.36	0.12	0.65	0.80
III	0.84	0.21	0.32	0.06	0.62	0.75

Although a state-of-the-art T^3 laser, capable of generating sub-ps pulses with 10s of TW peak power and a few Js of energy per pulse [11], could almost serve the need for the required acceleration, the average power or the rep rate of a single unit is still quite low, and wall-plug efficiency inadequate. In addition, injection scheme and synchronization of laser and electron pulse from stage-to-stage to good accuracy have to be worked out. Yet another important consideration is how to generate and maintain the small beam emittance in the transverse focusing channel provided by plasma wakefield throughout the accelerator leading to the final focus. There are various sources causing emittance growth, multiple scattering [15], plasma fluctuations [16] and mismatching between acceleration stages, to name just a few. Should the issues of guiding, staging, controllability, emittance preservation, etc. be worked out, there is hope that wakefields excited in plasmas will have the necessary characteristics for particle acceleration to ultrahigh energies.

CONCLUSIONS

We have explored the possibilities of operating a 5 TeV linear collider in the strong quantum beamstrahlung regime. To take the full advantage of quantum suppression of beamstrahlung, we have searched a large space of multidimensional collider parameters for the preferred regime of operation. By making collider scaling laws transparent, we found that reducing bunch length is an effective approach to suppress beamstrahlung, which naturally favors laser-driven acceleration. The prediction of scaling laws has been checked with full-blown IP simulations, and the results are quite encouraging. We have discussed the implied requirements for laser wakefield acceleration. The parameters of a 10 GeV module in a 5 TeV collider vision demonstrates both encouraging and sobering features that calls for further developments and innovations.

REFERENCES

1. Himel T., Siegrist J., AIP Conf. Proc., **130**, 602 (1985).
2. Chen P., Yokoya K., *Phys. Rev. Lett.*, **61**, 1101 (1988).
3. Blankenbecler R., Drell S., *Phys. Rev. D*, **37**, 3308 (1988).
4. Jacob M., Wu T. T., *Nucl. Phys.*, **B318**, 53 (1989).
5. For code and manual, see http://jlcux1.kek.jp/subg/ir/Program-e.html.
6. Yokoya K., Chen P., *Frontiers of Particle Beams*, **400**, 415 (1992).
7. Oide K., *Phys. Rev. Lett.*, **61**, 1713 (1988).
8. Irwin J., AIP Conf. Proc., **335**, 3 (1995).
9. Tajima T., Dawson J., *Phys. Rev. Lett.*, **43**, 267 (1979).
10. For recent experiment see Nakajima K., et. al., these proceedings.
11. For a review, see Downer M. C., Siders C. W., these proceedings.
12. For reference see Chattopadhyay S., et. al., Snowmass'96, LBL-39655, (1996).
13. Leemans W. P., et.al., *IEEE Trans. on Plasma Science*, **24**, 331 (1996).
14. Durfee III C. G., Milchberg H. M., *Phys Rev. Lett.* **71**, 2409 (1993).
15. Montague B. W., Schnell W., AIP Conf. Proc., **130**, 146 (1985).
16. Horton W., Tajima T., et.al., *Phys. Rev. A*, **31**, 3937 (1985).

Emittance in Particle and Radiation Beam Techniques

Kwang-Je Kim*

Lawrence Berkeley National Laboratory, One Cyclotron Road, MS 71-259; Berkeley, CA 94720, USA

Abstract. We discuss the important and diverse role of the phase space area—the emittance—in the advanced techniques involving interaction of particle and radiation beams. For undulator radiation from unbunched beams, the radiation phase space is diluted from the coherent phase space of the single electron radiation. When the undulator radiation is used as a light source, it is important to minimize the dilution by decreasing the beam emittance and matching the phase space distributions of the particle and the radiation beams. For optical stochastic cooling, on the other hand, the phase space should be maximally mismatched for efficient cooling. In the case particles are bunched to a length much shorter than the radiation wavelength, the emittance appears as an intensity enhancement factor. In the operation of free electron lasers, the phase space matching becomes doubly important, once as the dilution factor in the initial stage of energy modulation and then as the radiation efficiency factor at the end where the beam is density modulated. We then discuss some of the beam cooling techniques producing smaller emittances, especially the recent suggestions for relativistic heavy ions in storage rings or electron beams in linacs. These are based on the radiative cooling that occurs when particle beams backscatter powerful laser beams.

1. INTRODUCTION

Particle beams propagating in free space are characterized by the rms emittance ε_x and the Twiss parameter β_x at the beam waist. The beam size and angle at the waist are given respectively by

$$\Delta x_e = \sqrt{2\pi \varepsilon_x \beta_x} \text{ and } \Delta x'_e = \sqrt{2\pi \varepsilon_x / \beta_x}. \tag{1}$$

Here Δx_e is roughly the FWHM value, defined as the rms size times $\sqrt{2\pi}$, and similarly for the angular divergence. The letter x refers to the x-direction, there being the corresponding quantities in the y-direction.

---

* This work was supported by the Director, Office of Energy Research, Office of High Energy and Nuclear Physics, Division of High Energy Physics, of the U.S. Department of Energy under Contract No. DE-AC03-76SF00098.

Coherent radiation beams can be described in a similar way; the quantity corresponding to β_x is the Rayleigh length z_R. Thus the spot size and angle at the waist are respectively given by

$$\Delta x_r = \sqrt{\lambda z_R / 2} \text{ and } \Delta x_r' = \sqrt{\lambda / 2 z_R} , \qquad (2)$$

where λ is the radiation wavelength. The radiation phase space area (corresponding to $2\pi\varepsilon_x$ for the particle beam) is

$$\Delta x_r \Delta x_{r'} = \lambda / 2 . \qquad (3)$$

The analogy goes further. Thus, the beam envelope at a distance z from the waist for a particle beam is given by

$$\Delta x_e(z) = \sqrt{2\pi\varepsilon_x (\beta_x + z^2 / \beta_x)} , \qquad (4)$$

and for a radiation beam,

$$\Delta x_r(z) = \sqrt{(\lambda/2)(z_R + z^2 / z_R)} . \qquad (5)$$

We will often consider the spontaneous radiation generated by electrons in passing through an undulator (1). A single electron passing through an undulator generates a laser-like beam with a waist in the middle of the undulator with

$$\Delta x_r \sim \sqrt{\lambda L / 8\pi}, \quad \Delta x_{r'} \sim \sqrt{\pi\lambda / L} . \qquad (6)$$

Here L is the length of the undulator. The Rayleigh length is therefore $z_R \sim L/\sqrt{8\pi}$. A cautionary remark: Eq. (6) was derived under the assumption that the single-electron undulator radiation can be approximated by the Gaussian TEM$_{00}$ mode, which is not very accurate (2).

The radiation pulse length and bandwidth are, respectively:

$$\ell_r \sim \frac{1}{2} N_u \lambda, \quad \frac{\Delta\omega}{\omega} \sim \frac{1}{N_u} . \qquad (7)$$

Note that the product

$$\ell_r \frac{\Delta\omega}{\omega} \sim \lambda / 2 \qquad (8)$$

is the same as in the case of the transverse phase space.

2. PHASE SPACE DILUTION OF RADIATION FROM PARTICLE BEAMS

In considering radiation by electron beams, it is convenient to distinguish the unbunched case from that of the bunched case. To define these cases, we introduce the so-called bunching parameter as follows:

$$b = \left|\left\langle e^{ikz_i} \right\rangle\right|, \quad (9)$$

where z_i is the electron location, $k = 2\pi / \lambda$, and λ the radiation wavelength. In this section we consider the case where the bunch length is much longer than the radiation wavelength, and the particles are distributed randomly over the bunch. This is normally the case of particle beams in accelerators and storage rings. This will be referred to as the unbunched case, for which

$$b = 0. \quad (10)$$

The phase space distribution of radiation from an unbunched electron beam is given by a convolution of the distribution of a single electron radiation and the electron beam distribution. Thus the spot size and angular divergence is given by

$$\Delta x = \sqrt{(\Delta x_r)^2 + (\Delta x_e)^2} \;,\; \Delta \phi = \sqrt{(\Delta \phi_r)^2 + (\Delta \phi_e)^2} \;. \quad (11)$$

This is schematically illustrated in Figure 1.

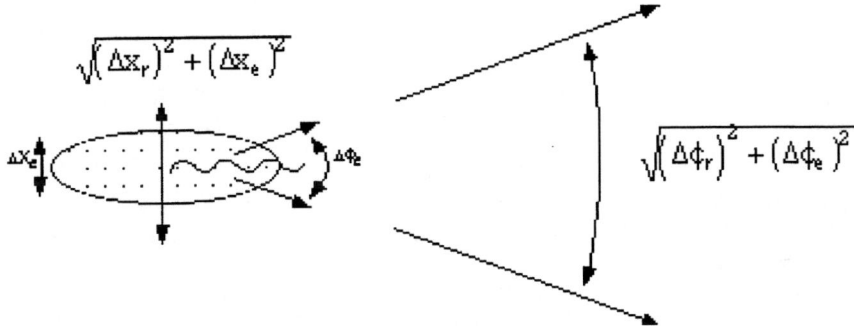

FIGURE 1. The spatial-angular characteristics of undulator radiation by a beam of electrons (unbunched case).

Thus the transverse phase space density is diluted due to the electron beam. Actually, the phase space dilution takes place also in the temporal-spectral domain: the length of the optical pulse and the spectral bandwidth are both broadened from the expression given in Eq. (7) due to the electron beam effect:

$$\ell = \sqrt{\ell_r^2 + \ell_e^2}, \quad \frac{\Delta\omega}{\omega} = \sqrt{\left(\frac{\Delta\omega}{\omega}\right)_r^2 + \left(\frac{\Delta\omega}{\omega}\right)_e^2}, \tag{12}$$

where ℓ_e is the e-beam bunch length, and

$$\left(\frac{\Delta\omega}{\omega}\right)_e = 2\frac{\Delta E_e}{E_e}. \tag{13}$$

The factor 2 is due to the fact that the frequency of the undulator radiation is proportional to E_e^2.

The phase space dilution means that the radiation phase space area in each dimension becomes larger than $\lambda/2$, i.e., the radiation beam becomes partially or completely incoherent depending on the degree of the dilution. More quantitatively, the radiation density in the 6-dimensional phase space, known as the *brightness*, becomes

$$B = N_e B_o R_\perp R_\parallel, \tag{14}$$

where N_e is the total number of the electrons in the bunch, B_o is the brightness due to a single electron, and

$$R_\perp = \frac{\lambda/2}{\sqrt{\left((\Delta x_r)^2 + (\Delta x_e)^2\right)\left((\Delta\phi_r)^2 + (\Delta\phi_e)^2\right)}}, \tag{15}$$

$$R_\parallel = \frac{\lambda/2}{\sqrt{(\ell_r^2 + \ell_e^2)}\sqrt{\left(\frac{\Delta\omega}{\omega}\right)_r^2 + \left(\frac{\Delta\omega}{\omega}\right)_e^2}}. \tag{16}$$

We have discussed the phase space properties of the radiation from a beam of particles as if the radiation beam could be treated just as in geometric optics. When the wave nature of the electro-magnetic field is taken into account, the concept of the radiation phase space does not have the usual physical significance. However, the phase space concept, as a mathematical entity behaving similarly to the real phase space of geometric optics, can be justified based on the Wigner distribution function (3),(4).

We now consider two examples in particle-radiation technique, one in which the factor R needs to be maximized and one in which the factor R needs to be minimized.

2.1 Brightness of Synchrotron Radiation Sources

The brightness is an invariant field strength of a radiation source. Considerable progress has been made in recent years in increasing the brightness of synchrotron radiation sources by increasing each factor in Eq (14); the factor N_e by increasing the stored current, the factor $R_\perp$ by reducing the electron beam emittance and choosing the β-function so that the radiation beam phase space is "matched" to that of the electron beam, and the factor $R_{\parallel}$ by reducing the energy spread. For advanced synchrotron radiation sources operating or being built currently at several laboratories around the world, the stored average current is 0.5 to 1 Amperes; the horizontal and the vertical rms emittances are about 0.5×10^{-8} m-rad and 10^{-10} m-rad, respectively; and the e-beam energy spread about a fraction of 10^{-3}. With these values, insertion devices can be built with spectral brightness of about $10^{19} - 10^{20}$ photons/(sec)(0.1% BW)(mm^2)(mrad^2), permitting experiments with remarkable spectral, angular and spatial resolutions.

2.2 Optical Stochastic Cooling

It is well known that the key parameter in microwave stochastic cooling (5) is the number of particles in a *sample* N_s, i.e., those particles that can be affected by the signal of a given particle. It is given by

$$N_s = \frac{Nc}{\ell \Delta W}, \qquad (17)$$

where N is the total number of particles, c is the speed of light, ℓ is the length of the beam, and ΔW is the bandwidth of the cooling system. The signal from other particles in a sample generate noise to the cooling signal of a given particle. Therefore N_s needs to be minimized for faster cooling. This is achieved by employing a broad band microwave system consisting of the pick-up, amplifier and kicker. The entire beam is cooled through the *mixing* process in which particles are continuously reshuffled into different samples. In normal stochastic cooling, the samples are characterized by the longitudinal coordinates, i.e., by the time axis.

Recently it was pointed out that the stochastic cooling may be extended to the optical region (6) using the undulators as pick-up and kicker and taking advantage of the recent development in broad band optical amplifiers. In particular, the transit time method appears to be very promising (7).

New effects occur in optical stochastic cooling, since the transverse phase space area of optical beams could be comparable to that of the particle beam. Therefore the sample associated with a given particle must be those particles within the phase space volume of the optical signal of the original particle. Thus Eq. (17) needs to be generalized as (8)

$$N_s = N R_\perp R_\parallel. \quad (18)$$

We may recover Eq. (17) is a special case of Eq. (18) when

$$R_\perp = 1, \; \ell_e \gg \ell_r, \; \text{and} \; (\Delta\omega/\omega)_r \gg (\Delta\omega/\omega)_e. \quad (19)$$

In general, the factor $R_\perp R_\parallel$ needs to be minimized in optical stochastic cooling. Therefore, the phase space distributions of the radiation beam and the particle beam need to be mismatched as much as possible. This is exactly opposite to the case of radiation sources considered in the previous subsection.

Realizing that N_s is given by the general expression in Eq. (18), one can contemplate a new way of implementing stochastic cooling. Thus, one may choose the parameters so that

$$\ell_e \ll \ell_r, \; (\Delta\omega/\omega)_r \ll (\Delta\omega/\omega)_e. \quad (20)$$

Equation (18) then becomes

$$N_s = NR_\perp \frac{(\Delta\omega/\omega)_r}{(\Delta\omega/\omega)_e}. \quad (21)$$

In this case, the cooling is optimized by choosing the bandwidth of the amplifier as narrow as possible, and the mixing should occur in the energy axis. This regime of stochastic cooling may be referred to as the "energy sampling" as opposed to the "time sampling" in which Eq. (19) is applicable. The energy sampling may be useful for cooling of bunched beams.

3. RADIATION FROM BUNCHED BEAMS

Consider a disc of an electron beam generating undulator radiation. We assume that the disc is so short that the bunching parameter b, defined by Eq. (9), does not vanish, namely, that the beam is bunched. The radiation pattern in this case is schematically illustrated in Figure 2.

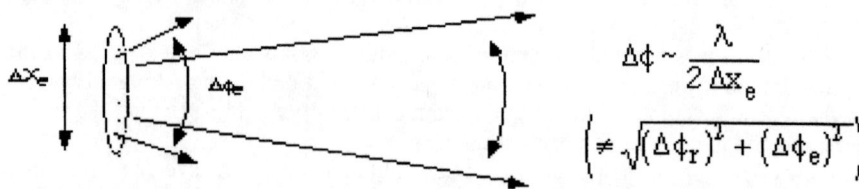

FIGURE 2. Radiation pattern from a bunched electron beam.

It is markedly different from the case of the unbunched beam; the angular divergence is given by the diffraction limit of the transverse size of the beam,

$$\Delta\phi \sim \frac{\lambda}{2\Delta x_e}, \qquad (22)$$

rather than the convolution of the angular divergences of the electron beam and the radiation beam. Does the phase space ratio R become irrelevant in this case? No! The phase space ratio is still important, but it enters in a different way; the total energy radiated is given by

$$N_e^2 b^2 U_o R_\perp, \qquad (23)$$

where U_o is the energy radiated by a single electron. Therefore, the factor $R_\perp$ appears as the radiation efficiency rather than as a phase space dilution factor.

Electron beams in free electron lasers develop a periodic density modulation as schematically shown in Figure 3.

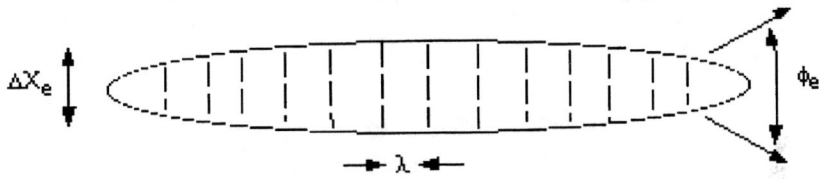

FIGURE 3. A particle bunch with periodic density modulation.

In this case the temporal bandwidth of the radiation is given by that corresponding to the Fourier transform of the bunch length, i.e.,

$$\frac{\Delta\omega}{\omega} \approx \frac{\lambda}{2\ell_e}. \qquad (24)$$

Once again, this is quite different from the un-bunched beam case given by the second member of Eq. (12). However, the temporal phase space ratio $R_\parallel$ is still important because it enters in the radiation efficiency. Thus, for a periodically modulated beam, the factor $R_\perp$ in Eq. (23) is replaced by a more general expression $R = R_\perp R_\parallel$.

In a free electron laser, the electron beam receives a periodic energy modulation with period λ in the beginning part of the undulator. The energy modulation becomes the density modulation after passing through some length of the undulator. The density modulated beam then emits amplified radiation in the final section of the undulator. The ratio R therefore enters twice in the expression of the FEL gain; once as the phase space overlap factor at the beginning and then as the radiation efficiency factor at the end of the undulator. Thus the importance of the beam quality is much more crucial in FELs than in synchrotron radiation devices.

The shortest wavelength record of FELs is 2400 Å achieved at BINP (9). With the recent development of the RF photocathode gun (10), and the advances in the linac technology (11), the beam qualities in the linac have progressed significantly so that it is now possible to contemplate the operation of a self-amplified spontaneous emission for intense, coherent x-rays for 1 Å or shorter wavelengths (12).

4. FUNDAMENTAL LIMIT AND BEAM COOLING TECHNIQUES

4.1 Fundamental Limits

As previously discussed, the minimum photon beam phase space area is $\lambda/2$. There is a similar fundamental limit for the particle beam arising from quantum mechanics

$$\gamma \Delta x \Delta x' \geq \lambda_c / 2. \tag{25}$$

Here $\lambda_c = \hbar/mc$ is the Compton wavelength corresponding to the particle mass m. For an electron, $\lambda_c \approx 4 \times 10^{-13}$ m, and the fundamental limit given by Eq. (25) is many orders of magnitude smaller than that obtained by the RF photocathode techniques, which is about

$$\gamma \Delta x \Delta x' \sim 1\text{-}2 \times 10^{-6} \text{ m} - \text{rad} / \text{nC}. \tag{26}$$

The particle beam phase space area can be reduced by various non-Liouvillain beam cooling techniques, such as radiative cooling in electron storage rings, electron cooling of heavy ions, ionization cooling, microwave or optical stochastic cooling, channeling, laser cooling of non-relativistic ions, etc. (13). We have already discussed the stochastic cooling in the previous section. The radiative cooling is well-known (14), and is essential for operation of high-brightness electron storage rings in advanced synchrotron radiation facilities.

Recently the radiative cooling method was extended to the case of relativistic heavy ions (15) and to the case of electron linacs (16) by making the particle beams radiate in the presence of high-field laser beams, as shown schematically in Figure 4.

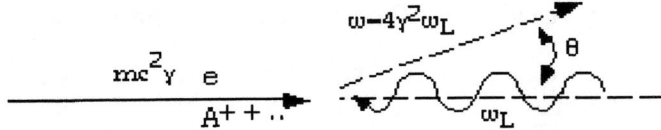

FIGURE 4. Backscattering of a laser beam by a charged particle.

4.2 Radiative Cooling

Radiative cooling occurs when particle beams are forced to emit spontaneous radiation. The equilibrium normalized emittance is

$$\gamma \Delta y \Delta y' \sim \beta_y \frac{u}{mc^2 \gamma^2}, \tag{27}$$

where β_y is the average β-function in the y-direction and u is the characteristic energy of the radiated photons. Equation (27) is valid when there is no dispersion and the excitation of the betatron oscillation is solely due to the angular deviation of the emitted photons from the direction of the particle momentum. In case there is a dispersion of an average magnitude η_x, the equilibrium emittance is

$$\gamma \Delta x \Delta x' \sim \frac{\eta_x^2}{\beta_x} \frac{u}{mc^2}. \tag{28}$$

In an electron storage ring where the emission is dominated by the radiation from bending magnets, $u \sim \hbar c \gamma^3 / \rho$. The dispersion is normally non vanishing in the horizontal (x) direction but vanishes in the vertical (y) direction. Therefore

$$\gamma \Delta x \Delta x' \sim \lambda_e \gamma^3 \eta_x^2 / \beta_x \rho, \tag{29}$$

$$\gamma \Delta y \Delta y' \sim \lambda_e \beta_y \gamma / \rho. \tag{30}$$

Here λ_e is the electron's Compton wavelength. Equations (29) and (30) are applicable for the horizontal direction (the bending plane) and the vertical direction, respectively. The horizontal emittance is many orders of magnitude larger than the fundamental limit. The vertical emittance is much smaller, but the limit is hard to achieve due to the coupling of the horizontal motion to the vertical direction.

4.3 Radiative Laser Cooling in Relativistic Heavy Ion Rings

Here the laser beam is backscattered by relativistic heavy ions in the dispersion-free straight sections (15). The average energy of the scattered photons is

$$u \sim 2\gamma^2 \hbar\omega_L, \qquad (31)$$

where ω_L is the laser frequency. By choosing the laser frequency to be in resonance with one of the transition frequencies of the heavy ion, the interaction cross section is increased from the Thomson cross section $\sim r_e^2$ by a factor $(\lambda^*/r_e)(\omega_L/\Delta\omega_L)$. Here λ^* is the transition wavelength in the ion's rest frame, and $\Delta\omega_L$ is the bandwidth of the laser which is chosen to cover the full Doppler broadening of the ion beam. This scheme is different from the usual laser cooling, which depends on the line shape of the individual ions, and thus employs narrow bandwidth lasers. The limiting equilibrium emittance in this scheme is obtained by inserting Eq. (31) in Eq. (27):

$$\gamma \Delta y \Delta y' \approx \lambda_M k_L \beta_y. \qquad (32)$$

Here λ_M is the Compton wavelength of the heavy ion. This is a very small emittance.

Explicit examples for RHIC and LHC indicate that the heavy ion beams can be cooled efficiently in this way, and in the process generate diffraction limited x-rays or γ-rays.

4.4 Radiative Laser Cooling in Electron Linacs

With a sufficiently intense laser pulse, a high energy electron beam can be cooled significantly during a single collision with the laser pulse. The radiative process here may be viewed either as the Thomson backscattering or as the undulator radiation. The process here is similar to the one discussed in Section 4.2 except that the beam parameters change substantially during the laser-electron beam collision. It can be applied to the cooling of the electron beams in the linac (16).

The required laser pulse energy W_L to reduce the electron energy from E_o to $E \ll E_o$ is

$$W_L = \frac{mc^2}{4E} \frac{\Sigma}{\sigma_{Th}}. \qquad (33)$$

Here Σ is the cross-sectional area of the laser beam and σ_{Th} is the Thomson cross section. The achievable emittance is similar to Eq. (32):

$$\gamma \Delta x \Delta x' = \lambdabar_e \frac{K}{4} k_L \beta_x . \tag{34}$$

Here K is the deflection parameter of the electron trajectory in the laser field. [In Eq. (34), K is assumed to be of order unity or less. If $K \gg 1$, then Eq. (34) needs to be multiplied by K^2 to account for the dispersion generated by the laser field itself.] Explicit examples show that this could be an attractive cooling method either for the TeV linear colliders or for high brightness electron beams for x-ray SASE.

5. SUMMARY

In this paper we have summarized the important role of emittance and phase space in particle and radiation beam techniques, emphasizing the physical pictures rather than mathematical rigor. We have also discussed some cooling techniques made feasible with the recent advent in the high power laser technology (17).

REFERENCES

1. For a review, Kim, K.-J., "Characteristics of Synchrotron Radiation," *AIP Proceedings* **184**, M. Month and M. Dienes, eds., pp. 565–632, AIP (1989).
2. Kim, K.-J., *Proc. 1987 Particle Accel. Conf., Washington, DC* (March 1987), IEEE Catalog **No. 87**, CH2387–9, 194 (1987).
3. Kim, K.-J., *Nucl. Instr. Meth.* **A246**, 71 (1986).
4. Coisson, R., and Walker, R. P., *SPIE Proceedings* **582**, 24 (1986).
5. van der Meer, S., CERN/ISR PO/72–31(1972). For a review, Mohl, D., *CERN Accelerator School Report*, No. CERN 87–03, page 453 (1987).
6. Mikhailichenko, A., and Zolotorev, M., *Phys. Rev. Lett.* **71**, 4146 (1993).
7. Zolotorev, M. S., and Zholents, A. A., *Phys. Rev.* **E 50**, 3087 (1994).
8. Kim, K.-J., *Proceedings of 1995 Part. Acc. Conference, Dallas, Texas* (1995).
9. Drobyazka, I. B., et al., *Nucl. Instr. Meth.* **A331**, 98 (1988).
10. Sheffield, R. L., *AIP Proceedings* **184**, M. Month and M. Dienes, eds., p. 1500 (1989).
11. For a review, Raubenheimer, T. O., *Nucl. Instr. Meth.* **A358**, 40 (1995).
12. For a review, Winick, H., et al., *Nucl. Instr. Meth.* **A 347**, 199 (1994).
13. For a review, Sessler, A. M., LBL–38278 (1996).
14. Kolomenski, A., and Lebedev, A., CERN Symposium, 447(1956); Robinson, K., *Phys. Rev.* **111**, 373 (1958); Sands, M., SLAC Pub–121 (1970).
15. Bessonov, E.G., and Kim, K.-J., *Phys. Rev. Lett.* **76**, 431 (1996).
16. Telnov, V., NSF–ITP–96–142, Symposium at Santa Barbara, October, 1996.
17. Strickland, D., and Mourou, G., *Opt. Comm.* **56**, 219 (1985).

Suppression of Radiation Excitation in Focusing Environment*

Zhirong Huang and Ronald D. Ruth

Stanford Linear Accelerator Center
Stanford University, CA 94309 USA

Abstract. Radiation damping and quantum excitation in an electron damping ring and a straight focusing channel are reviewed. They are found to be the two limiting cases in the study of a general bending and focusing combined system. In the intermediate regime where the radiation formation length is comparable to the betatron wavelength, quantum excitation can be exponentially suppressed by focusing field. This new regime may have interesting applications in the generation of ultra-low emittance beams.

INTRODUCTION

Many applications of particle accelerators require very low emittance beams. In an electron damping ring, synchrotron radiation created by bending magnets is utilized to damp the beam emittance in all three degrees of freedom. It is well known [1, 2] that the damping effect is counteracted by quantum excitation due to random photon emissions, which leads to an equilibrium beam emittance when the damping and excitation rates balance.

On the other hand, we have shown [3, 4] that in a straight, continuous focusing channel, the transverse damping rate is independent of the particle energy, and that no quantum excitation is induced. In fact, the final normalized emittance in a generic focusing system is limited only by the uncertain principle and is equal to one half of the Compton wavelength of the electron, which is much smaller than the equilibrium emittance achieved in a normal damping ring.

In this paper, we first review these distinct results. In order to illustrate the transition between bending systems and focusing ones, we study the radiation effects on particle beams in a system where both bending and focusing are present. We show that, in general, quantum excitation can be suppressed by the focusing environment because a photon emission does not take place instantaneously. Finally, we investigate the possibility of an ultra-low emittance damping ring based on this effect.

---

* Work supported by Department of Energy contract DE–AC03–76SF00515.

AN ELECTRON DAMPING RING

In an electron damping ring, the transverse focusing quadruples are present to confine the beam. Their contribution to the radiation effects is secondary relative to the bending dipoles. The typical length associated with a photon emission (the radiation formation length) is on the order of ρ/γ [1, 2], where ρ is the bending radius and γ is the electron energy in units of its rest energy. The standard treatment of quantum excitation can be quasi-classical because the radiation formation length is much shorter than the average beta function β. Thus, one can model the radiation to be instantaneous with a continuous spectrum of frequencies and treat the quantum nature of radiation as fluctuations about the average rate.

On the average, the radiation from the bending dipoles takes away the electron's momenta in all three degrees of freedom, while the rf acceleration only replenishes the longitudinal momentum of the electron. Thus, the transverse damping rate is comparable to the energy damping rate which is given by the characteristic damping constant

$$\Gamma_b = \frac{1}{E}\left|\frac{dE}{dt}\right| = \frac{2}{3}\frac{r_e c \gamma^3}{\rho^2}, \tag{1}$$

where $r_e = e^2/mc^2$ is the classical electron radius.

Although the position of the electron does not change instantaneously right after a photon emission, the horizontal betatron displacement is suddenly changed as a result of the equilibrium orbit shift

$$\delta x_\beta = -\eta \frac{u}{E}, \tag{2}$$

where u is the photon energy, η is the dispersion function, and we have assumed that $d\eta/ds = 0$ for simplicity. Because of the random nature of this sudden change, the transverse normalized emittance diffuses at a rate [1, 2]

$$\left(\frac{d\varepsilon_N}{dt}\right)_{QE} = \gamma\left\langle\frac{(\delta x_\beta)^2}{2\beta}\frac{1}{u}\left|\frac{dE}{dt}\right|\right\rangle_{\text{averaged over the radiation spectrum}}, \tag{3}$$

$$= \frac{55}{48\sqrt{3}}\frac{r_e \hbar \gamma^6}{m}\frac{\eta^2}{\beta\rho^3} = \frac{55\sqrt{3}}{96}\Gamma_b \frac{\lambda_c \gamma^3 \eta^2}{\beta\rho}$$

where $\lambda_c = \hbar/mc$ is the Compton wavelength for electron. The equilibrium is reached when the damping rate is equal to the quantum excitation rate. Thus, we obtain

$$(\varepsilon_N)_{\min} \sim \frac{\lambda_c \gamma^3 \eta^2}{\beta \rho} \sim \lambda_c \frac{\gamma^3}{\upsilon^3}. \tag{4}$$

In the last step of Eq. (4), we have used the smooth approximation so that the dispersion function is $\eta \sim \beta^2/\rho$ and the betatron tune is $\upsilon \sim \rho/\beta$. Eq. (4) indicates that the equilibrium emittance increases with higher particle energy, and decreases with higher betatron tune. It already suggests that stronger focusing (i.e., increasing the tune) can allow for the lower emittance. Before we study the radiation effects due to focusing in a general system, we first look at the simpler situation of a straight focusing channel.

A STRAIGHT FOCUSING CHANNEL

Following Ref. 3 and 4, we consider a planar focusing system that provides a continuous parabolic potential $Kx^2/2$, where K is the focusing strength. An electron of energy E oscillates in the transverse x direction while moving freely in the longitudinal z direction with a constant longitudinal momentum p_z in the absence of radiation, i.e.

$$\begin{aligned} E &= \sqrt{m^2 c^4 + p_z^2 c^2 + p_x^2 c^2} + \frac{1}{2} K x^2 \\ &\approx \underbrace{\sqrt{m^2 c^4 + p_z^2 c^2}}_{E_z} + \frac{p_x^2 c^2}{2 E_z} + \frac{1}{2} K x^2, \end{aligned} \tag{5}$$

Defining the transverse frequency as $\omega_z = \sqrt{Kc^2/E_z}$, we obtain from a simple quantum mechanical analysis that

$$E(n, p_z) = E_z + \hbar \omega_z \left(n + \frac{1}{2} \right), \tag{6}$$

where $n = 0, 1, 2,...$ is the transverse quantum level and is related to the normalized beam emittance by

$$\varepsilon_N = \lambda_c \left\langle n + \frac{1}{2} \right\rangle_{\text{beam}} \approx \lambda_c \langle n \rangle_{\text{beam}} \qquad \text{for large } n. \tag{7}$$

Eq. (6) indicates that n is another independent constant of motion besides p_z in the absence of radiation. Instead of building a semi-classical model for the photon emission process, we can calculate the change of the transverse quantum level (ultimately related to the evolution of the normalized beam emittance) directly by conservation laws before and after a photon emission (namely, the conservation of total energy and total longitudinal momentum). A simple kinematical argument shows [3, 4] that n must drop after an arbitrary photon emission. The existence of the focusing field suppresses the direct transverse recoil and absorbs the excess transverse momentum. Therefore, no quantum excitation is induced to the transverse emittance in this focusing system.

When the transverse oscillation amplitude is very small, the transverse motion looks like a one-dimensional harmonic oscillator in the co-moving frame of the electron. It is straightforward to obtain the damping rate in that frame and transform back to the lab frame, then we have

$$\frac{d\varepsilon_N}{dt} = -\frac{2}{3}\frac{r_e K}{mc}\varepsilon_N \equiv -\Gamma_c \varepsilon_N, \tag{8}$$

where $\Gamma_c = 2r_e K/(3mc)$ is the energy-independent damping constant given in Ref. 3 and 4.

In the case of arbitrary transverse oscillation, the transverse motion exhibits a figure of eight motion in the co-moving frame. The damping rate can be calculated using a semiclassical method or using the Lorentz-Dirac radiation damping force [4] and can be written in the form

$$\frac{d\varepsilon_N}{dt} = -\Gamma_c \varepsilon_N - \frac{3}{4}\frac{1}{E}\left|\frac{dE}{dt}\right|\varepsilon_N. \tag{9}$$

The second term of Eq. (9) comes directly from the energy loss, similar to the radiation damping in a normal damping ring. It is the dominant term when the oscillation amplitude is large. However, the direct momentum suppression due to the transverse focusing gives rise to an additional term for the damping (the first term in Eq. (9)), which becomes more significant as the oscillation amplitude becomes smaller.

In the absence of quantum excitation, the electron damps to the transverse ground state ($n = 0$) that corresponds to a theoretical minimum emittance for the beam:

$$(\varepsilon_N)_{min} = \lambda_c/2 \sim 10^{-13}\,\text{m}. \tag{10}$$

This ultimate emittance is limited only by the uncertainly principle, and is analogous to the diffraction limited photon beam emittance because the Compton wavelength here plays the role of natural wavelength for the electron.

We notice that the damping constant Γ_c is independent of energy and is proportional to the focusing strength K. The damping effect is usually negligible for any practical straight focusing device. Thus, we extend this effect to a bent focusing system where the electron beam can be recirculated for a long period of time.

A BENT FOCUSING SYSTEM

In a system where the radiation effects due to focusing are as important as those from bending, the quasi-classical picture of instantaneous photon emissions may not be valid because the oscillation wavelength can be the same order as the radiation formation length. In this case we can follow the treatment of the above section and calculate the evolution of constants of motion when the radiation is turned on. Let us consider a simple model with continuous focusing superimposed by a global bending field. Suppose a reference electron with momentum p_0 has a circular trajectory with radius ρ, then the vector potential for the uniform bending field in a curvilinear coordinates system (x, s, y) is [5]

$$A_s \equiv (\vec{A}\cdot\hat{s})\left(1+\frac{x}{\rho}\right) = -\frac{cp_0}{e}\left(\frac{x}{\rho}+\frac{x^2}{2\rho^2}\right). \tag{11}$$

Let the continuous focusing force $(-Kx)$ be in the transverse x direction and neglect the dynamics in the other transverse y direction, the total energy of the electron can be decomposed as

$$E = \sqrt{m^2c^4 + p_x^2c^2 + \frac{(p_s - eA_s/c)^2 c^2}{(1+x/\rho)^2} + \frac{1}{2}Kx^2}$$

$$\approx \underbrace{\sqrt{m^2c^4 + p_s^2c^2}}_{E_s} + \frac{p_x^2 c^2}{2E_s} + \frac{1}{2}\underbrace{\left(K + \frac{p_0^2 c^2}{E_s \rho^2}\right)}_{K_e} x^2 - (p_s - p_0)c\frac{x}{\rho}, \tag{12}$$

$$\approx E_s + \frac{p_x^2 c^2}{2E_s} + \frac{1}{2}K_e \underbrace{(x-x_\varepsilon)^2}_{\tilde{x}} - \frac{1}{2}K_e x_\varepsilon^2$$

where the equilibrium orbit displacement $x_\varepsilon = (p_s - p_0)c/(K_e \rho)$ and the betatron oscillation frequency $\omega_s = \sqrt{K_e c^2/E_s} \equiv c/\beta$ are both functions of p_s. Similar to the straight channel analysis, the total energy of the electron

$$E(n, p_s) = E_s + \hbar\omega_s\left(n+\frac{1}{2}\right) - \frac{1}{2}K_e x_\varepsilon^2 \tag{13}$$

is a function of n and p_s, with $n = 0, 1, 2,...$ being the transverse quantum level. Both n and p_s are constants of motion in the absence of radiation.

The change of the transverse quantum level n due to spontaneous radiation can be calculated with first-order, time-dependent perturbation theory. Since we are interested in the total radiation effects, we can integrate over the angular and frequency distribution of the radiated photons to obtain the total transition rate W_{fi}, it can be shown that [6]

$$\frac{dn}{dt} = \sum_{f(n',p_s)}(n'-n)W_{fi}$$
$$= -\frac{2}{3}\frac{e^2\gamma^3}{\rho^2 mc}(\chi^2-1)n + \frac{e^2\gamma^3}{\rho^2 mc}\frac{\exp(-2\sqrt{3}\chi)}{144\chi^3}F(\chi) , \quad (14)$$

where

$$F(\chi) = 55\sqrt{3} + 330\chi + 262\sqrt{3}\chi^2 + 300\chi^3 + 48\sqrt{3}\chi^4 ,$$
$$\text{and } \chi \equiv \frac{\rho/\gamma}{\beta} = \frac{\text{radiation formation length}}{\text{reduced betatron wavelength}} . \quad (15)$$

From Eq. (7), the evolution of the normalized emittance is then given by

$$\frac{d\varepsilon_N}{dt} = -\Gamma_b\left\{(\chi^2-1)\varepsilon_N - \lambda_c\frac{\exp(-2\sqrt{3}\chi)}{96\chi^3}F(\chi)\right\}, \quad (16)$$

where Γ_b is the damping constant defined in Eq. (1). Equation (16) describes the general result of radiation (anti-)damping (the first term) and quantum excitation (the second term) in this combined function system. We can now take various limits for different situations. For example, when $\chi \ll 1$ or $\rho/\gamma \ll \beta$, Eq. (16) reduces to

$$\frac{d\varepsilon_N}{dt} = \Gamma_b\left\{\varepsilon_N + \lambda_c\frac{55\sqrt{3}}{96\chi^3}\right\} = \Gamma_b\left\{\varepsilon_N + \lambda_c\frac{55\sqrt{3}\gamma^3}{96\upsilon^3}\right\}, \quad (17)$$

where $\upsilon = \rho/\beta$ is the betatron tune in this smooth system. The first term of Eq. (17) is anti-damping instead of damping because the combined function system studied here has a negative horizontal damping partition number ($J_x = -1$) [1]. However, the second term of Eq. (17) gives the same quantum excitation rate as using the quasi-classical model in an electron damping ring (see Eq. (3) with $\eta \sim \beta^2/\rho$).

When $\chi \gg 1$ or $\rho/\gamma \gg \beta$, Eq. (16) also predicts the correct result for a straight focusing channel ($\rho \to \infty$), i.e.,

$$\frac{d\varepsilon_N}{dt} = -\Gamma_b\chi^2\varepsilon_N = -\Gamma_c\varepsilon_N . \quad (18)$$

As expected, no quantum excitation is induced in the straight focusing channel.

In the intermediate regime where the radiation formation length is on the order of reduced betatron wavelength ($\rho/\gamma \sim \beta$), the quantum excitation is exponentially suppressed according to Eq. (16) and starts to depart from Eq. (3) based on the quasi-classical model (see Figure 1). The transverse energy spectrum of the electron is highly discrete due to the strong transverse focusing force, and excitation (jumping up transverse levels) becomes impossible for almost all photon emissions. Therefore, the betatron oscillation is adiabatically suppressed to the new ideal orbit during the radiation process.

Finally, we note that all of the above results can be extended to alternating-gradient and separated function systems when longitudinal variations of both bending and focusing fields are short compared with the radiation formation length [6]. Thus, the beam in such lattices will damp instead of anti-damp. We consider a realistic lattice design in the following section.

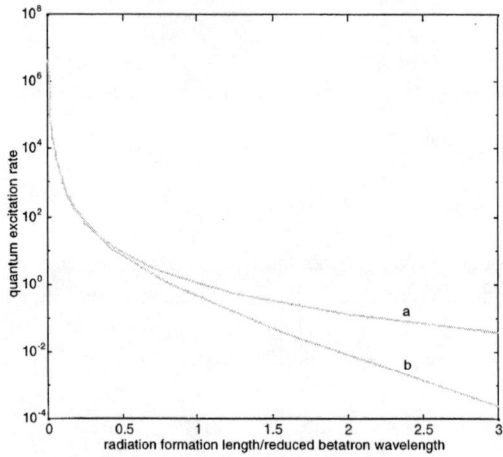

FIGURE 1. Quantum excitation rate in units of $\Gamma_b \lambda_c$, predicted by (a) quasi-classical model, i.e., Equation (3) and (b) quantum mechanical calculation, i.e., the second term of Equation (16).

A FOCUSING-DOMINATED DAMPING RING

In this section, we study the parameters of a focusing-dominated damping ring where quantum excitation can be strongly suppressed. Suppose that the ring is composed of many repetitive cells. Each cell of length $4L$ consists of four basic elements of equal length L: focusing quad, bend, defocusing quad, and another identical bend. Both quads have the same field gradient g. Furthermore, we assume that the phase advance per cell is 60 degrees. If we treat the bending as gradual and the cell as a basic FODO cell with drift space $2L$, we obtain

$$L[\text{cm}] \approx \left[\frac{E[\text{MeV}]}{6g[\text{Tesla/cm}]} \right]^{1/2}. \tag{19}$$

The averaged beta function (reduced betatron wavelength) for the 60 degrees cell is

$$\beta = \frac{24L}{2\pi} = \frac{12L}{\pi}. \tag{20}$$

By choosing $\chi \approx 1$ or the averaged ring radius $\rho \approx \gamma\beta = 12\gamma L / \beta$, quantum excitation is kept at the minimum level and the equilibrium emittance is on the order of the Compton wavelength.

These simple lattice scaling formulas suggest that in order to design a compact ring, it is favorable to use high-gradient focusing quads and low-energy electron beams. As an example, we assume that permanent magnet quads have a field gradient $g = 4$ Tesla/cm, and we take the electron energy to be 25 MeV, we then arrive at

$$L = 1.0 \text{ cm}, \quad \beta = 3.9 \text{ cm}, \quad \rho = 1.9 \text{ m}. \tag{21}$$

The transverse damping rate is about the same for both the focusing effect and the bending effect since $\rho/\gamma \approx \beta$. The two damping constants are

$$\Gamma_b = \Gamma_c = 0.11 \text{ sec}^{-1}. \tag{22}$$

The transverse size that corresponds to the Compton wavelength is

$$\sigma_x = \sqrt{\frac{\lambda_c}{\gamma} \beta} = 1.8 \times 10^{-6} \text{ cm}. \tag{23}$$

The energy loss per turn is mainly due to the bends, as long as the betatron amplitude is not too large. Thus, we have

$$(\Delta E)_{\text{per turn}} = \frac{2\pi\rho}{c} \Gamma_b E \approx 0.11 \text{ eV}. \tag{24}$$

It can be replenished by either radio-frequency or betatron-type acceleration. The equilibrium energy spread is determined by the effect of discrete photon emissions, and is given by

$$\frac{\sigma_E}{E} = \sqrt{\lambda_c \frac{\gamma^2}{2\rho}} = 1.6 \times 10^{-5}. \tag{25}$$

However, at such low energy, space charge and intra-beam scattering effects are significant. It might be conceivable to operate the ring below the transition energy when $\rho/\beta = \nu \equiv \gamma_t > \gamma$ is satisfied, then the Coulomb interaction between electrons, together with the external focusing environment, tend to stabilize the beam by the crystallization effect [7]. Other collective effects such as wakefields and beam-gas scattering can also influence the stability of the system and may determine the final beam emittance. These effects have yet to be studied in this new regime of operation.

ACKNOWLEDGMENT

We would like to thank Max Zolotorev for many useful discussions.

REFERENCES

1. Sands, M., The Physics of Electron Storage Rings, SLAC Report-121, 1970.
2. Sokolov, A. A., and Telnov, I. M., *Synchrotron Radiation*, Pergamon Press 1968.
3. Huang, Z., Chen, P., and Ruth, R. D., *Phys. Rev. Lett*. **74**, 1759-1762 (1995).
4. Huang, Z., Chen, P., and Ruth, R. D., *Nucl. Instrum. Methods Phys. Res., Sect. B* **119**, 192-198 (1996).
5. Ruth, R. D., Single Particle Dynamics in Circular Accelerators, in *Physics of Particle Accelerators*, AIP Conference Proceedings **153**, New York, 1987, pp. 150-235.
6. Huang, Z., and Ruth, R. D., in preparation.
7. Wei, J., Li, X., and Sessler, A. M., *Phys. Rev. Lett*. **73**, 3089-3092 (1994).

Luminosity-limiting coherent beam-beam phenomena

Srinivas Krishnagopal

FEL Section, Centre for Advanced Technology, Indore 452013, India

Abstract. We discuss details and results of a new simulation program that correctly models the transverse beam-beam dynamics in lepton colliders. We find evidence for both kinds of coherent beam-beam phenomena – flip-flop and period-n beam-size oscillations – that have been observed in operating colliders. We study the dependence of these phenomena on beam-size ratio, tune and damping, and comment on their potential to limit the luminosity of colliders. Some of these results have been published earlier in Ref. 1.

INTRODUCTION

The beam-beam interaction is widely believed to be responsible for the observed tune-shift and luminosity limitations in operating electron-positron colliders, though the precise underlying mechanism is not fully understood. There is growing evidence, however, that *coherent* beam-beam effects could play an important role in the observed performance limitations. For example, it is thought that in the case of VEPP-2M coherent (possibly synchrobetatron) resonances may be the cause of performance limitations [2]. It is also widely believed that coherent effects were responsible for the failure of the DCI space-charge compensation experiments [3]. More recently, periodic beam-size oscillations were observed at LEP [4] that were identical to coherent beam-beam effects predicted for round beams [5,6]. Finally, there is the ubiquitous *flip-flop instability*, widely observed in operating colliders, that could have its origin in coherent beam-beam phenomena.

Dipole effects, that displace the beam centroids, are easily cured by feedback systems; the potential for performance limitations comes from *quadrupole* effects, that act to distort the beam shape. The study of such coherent quadrupolar phenomena requires a fully self-consistent calculation, in which both beams are allowed to influence each other. Then, a beam that starts out Gaussian cannot retain that shape after experiencing the nonlinear force produced by the opposing Gaussian beam, and hence becomes non-Gaussian. The analytic formula of Bassetti & Erskine [7] can no longer be used, and one must take recourse to numerical simulations. Earlier we have developed an algorithm to study coherent beam-beam effects in beams with arbitrary (non-Gaussian) distributions but nearly-round profiles [6]. However that algorithm fails for the flat beams that coast in nearly all operating e^+-e^- colliders. We have therefore developed a new field-calculation algorithm that does away with the constraint of nearly-round beams, and is valid for beams of arbitrary ellipticity.

NUMERICAL ALGORITHM

We assume that the beams collide only once per turn at the interaction point. We only model dynamics in the two transverse dimensions; longitudinal dynamics is not included. Particles comprising both beams are initialized in a Gaussian distribution in all four phase-space dimensions, with any chosen ellipticity κ (defined as the ratio of the horizontal to the vertical beam size). Typically 10,000 particles are used per beam. They are then transported once around the ring using maps for the betatron transport, radiation damping and fluctuations, and the beam-beam interaction. The maps are then iterated, typically for 10 damping times, until the beams achieve equilibrium.

The betatron transport assumes that the magnetic lattice is linear and the horizontal and vertical motions are decoupled, so that the (transverse) transport of particles around the ring can be described by two 2×2 rotation matrices. The treatment of the radiation damping and fluctuations calculates the average effect, over one turn, of synchrotron radiation and energy gain in an RF cavity, and puts this in at a single point in the ring [6,8].

The model for the beam-beam interaction assumes that the beams are ultrarelativistic. In this case the force due to the magnetic field has the same magnitude and direction as that due to the electric field. One can therefore ignore the magnetic field and solve the corresponding electrostatic problem (by Lorentz transforming to the rest frame of the bunch and solving numerically for the electrostatic field from the coordinates of the test particles comprising the beam). The actual force on the particle is then twice that given by the electrostatic calculation.

The electrostatic field calculation is done on a two-dimensional Cartesian grid [9]. The primary references for the general techniques of particle and field simulation are the books by Hockney & Eastwood [10] and by Birdsall & Langdon [11]. There are four steps involved in going from a distribution of particles to the field generated by those particles: (a) casting the particles onto the grid to calculate the charge density; (b) from the charge density, solving Poisson's equation to calculate the potential at the grid points; (c) from the potential, calculating the field at the grid-points; (d) interpolating from the field at the grid-points to the field at any arbitrary point.

Density calculation. The technique used is second-order weighting or quadratic spline, assigning charge to the five nearest grid-points. The weighting is done independently in the two dimensions. The advantage of this technique is that the particle shape and its first derivative are both continuous. This localizes statistical errors and reduces long-range fluctuations. Details are in Ref. 10, p. 136, and Ref. 11, p. 314-5.

Potential calculation. The heart of the technique is the Poisson solver for the electrostatic potential, which is based on the Fourier Analysis by Cyclic Reduction (FACR) method developed by Hockney and implemented in the code DELSQPHI [12]. In this method, which is one of the general class of 'Rapid Elliptic Solvers',

Fourier analysis in one dimension is combined with cyclic reduction (of the difference equations for Poisson's equation at the various grid-points) in the other, to evolve a numerical method that is both fast and accurate, and best suited for Coloumbic problems in rectangular coordinates. For further discussion of the advantages of this method the reader is directed to Chapter 6 of Ref. 10, and Section 6-5-3 in particular.

Field calculation. To calculate the electric field at the grid-points, from the potential at these points, one needs to solve Gauss's Law numerically on the grid. To do this we use a six-point differencing scheme for the gradient operator.

Field interpolation. For interpolating from the grid points to any arbitrary point we must, to conserve momentum, use the same scheme as the density calculation (Ref. 11, p. 162). We therefore use a quadratic spline, separately in each dimension.

Diagnostics

We have made extensive efforts to test the program. In particular we have performed the following checks:
(1) Starting with a random potential distribution, we differentiated it twice to calculate the charge density. This charge density was then used to regenerate the potential using the FACR algorithm. The original and recalculated potentials should agree at every grid-point; they do. This checks the Poisson solver.
(2) We initialized particles in a two-dimensional Gaussian distribution and used our numerical algorithm to calculate the fields. These were compared with the analytic formulas for the fields produced by Gaussian charge distributions [7]. The agreement was excellent.
(3) We checked the sensitivity of the algorithm to grid-spacing, grid-size and number of test particles. Based on these checks, typically in the simulations the step-size was $\sigma/5$ (where σ is the transverse beam size), the grid extended out to around 25σ, and there were 10,000 test particles.
(4) We used this program in a beam-beam simulation in which particles were initialized with a round beam-profile (i.e. were initially axisymmetric) and Gaussian distribution, but were not constrained to remain so. We have performed such simulations earlier [6], using a completely different field-calculation algorithm. The present simulations are able to reproduce the earlier results; in particular the coherent simulations seen in the earlier simulation are seen here too, at the same values of the parameters (especially the current). This checks the algorithm for a beam-size ratio of 1:1.
(5) In addition, some of the results reported below (for $\kappa = 2$) have been confirmed independently, using our earlier field-calculation algorithm [13].
(6) Finally, we have developed a new diagnostic that operates pass-by-pass and is built into the code. Every so many turns (100 at present) the code takes the calculated potential, differentiates it to get the density along the X and Y axes, and compares this derived density with the original density. They are required to agree to within a specified tolerance (10% presently), at every grid-point at which the density is greater than a certain fraction (1% presently) of the maximum density --

if the density is less than 1% the number is too close to zero for accurate comparison. In the results presented below this diagnostic did not turn up any problems with the algorithm. (Note that this diagnostic is not sensitive to any systematic discrepancy in the potential that is linear in the coordinates, since only a second differential of the potential is being compared. Any such linear trend is ruled out by comparison with the analytic formula for Gaussian beams.)

In one particular case, when interesting coherent beam-beam physics was observed within the first couple of hundred turns, the diagnostic was turned on *every* turn; it revealed no discrepancy. For the same run X & Y slices of the density, potential and electric field were written out for both beams, and their plots were observed manually. In all cases the plots were smooth, with no discontinuities or abrupt features that could be numerical in origin.

RESULTS OF THE SIMULATION

Dependence on beam-size ratio

We chose storage-ring parameters of our test collider to correspond to those of the Cornell Electron Storage Ring (CESR). We studied the dynamics at a tune of Q_β (= Q_x = Q_y) = 0.79, with a damping decrement of $\delta = 1 \times 10^{-3}$. We have seen in our earlier work that period-3 coherent oscillations appear in this region [6]. We studied the change in the nature of the dynamics as the ellipticity of the beam was varied from $\kappa = 1$ (round) to $\kappa = 6$, as a function of current (or tune-shift parameter). The parameters for the flat beams were derived from those of the round by requiring that the nominal luminosity and tune-shift parameters be identical in the two cases.

Figure 1 shows the nature of the equilibrium state, after running for 10 transverse damping times (20,000 turns). For round beams only period-n oscillations are seen, and the region of coherent activity is restricted to the range 25 - 35 mA. At currents above and below that there is no coherent activity. As soon as one gets away from round beams, however, flip-flop solutions appear. In fact, for lower κ these are the only solutions, but at higher values of κ both phenomena are seen. However the flip-flop always occurs at a lower current and is therefore more likely to be the luminosity-limiting factor in operating colliders. Note that the threshold for the onset of coherent motion seems to be largely independent of the beam-size ratio (at least over the range of ratios considered here).

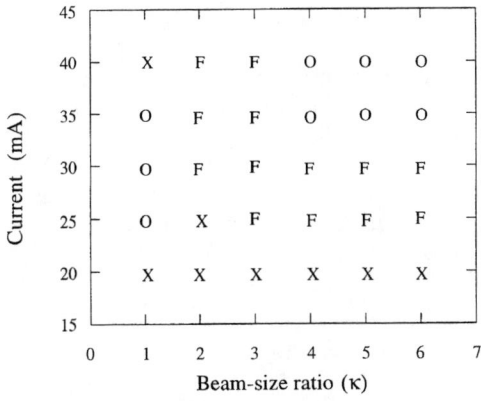

FIGURE 1. Dependence of the equilibrium coherent solution as a function of beam-size ratio. k = 1 corresponds to round beams. The 'X' indicates lack of any coherent motion, 'F' indicates flip-flop motion, and 'O' indicates period-n oscillations.

Figure 2 shows typical plots of the evolution of the beam sizes for two different sets of parameters, when the equilibrium solution is a flip-flop and a period-n oscillation respectively. From Fig. 2 it is clear that in the case when the flip-flop is the equilibrium solution, there are initially period-n beam-size oscillations, so that both possible effects are competing. Ultimately, however, it is the flip-flop that is dominant and emerges as the equilibrium solution. The initial onset of coherent oscillations that later die out to leave behind a flip-flop is common, but we have never observed the reverse situation, i.e., the beams finding a transient flip-flop solution which dies out to leave behind a period-n oscillation as the equilibrium state. This suggests that the flip-flop is typically stronger than the period-n oscillation.

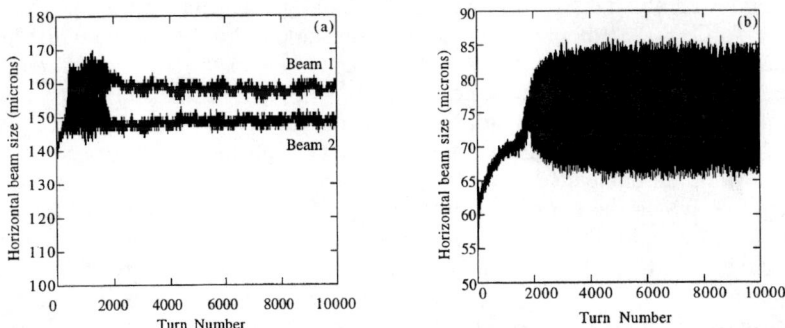

FIGURE 2. Plots of beam size as a function of turn number, for different parameters, showing (a) a flip-flop solution, and (b) a period-n solution.

Dependence on tune

In our earlier work on round beams [6] we had found that the period-n coherent resonances observed span a characteristic tubular region in current vs. tune (I vs. Q_β) space. Coherent motion was observed only within this region. Thus, there were clear thresholds for the onset and offset of coherent motion. We therefore looked for a similar structure in the case of flat beams; here chosen as beams with $\kappa = 4$. Figure 3 shows this plot in two different tune regions corresponding to tunes just below the 5/6 (= 0.83) and the 4/6 (= 0.67) resonances.

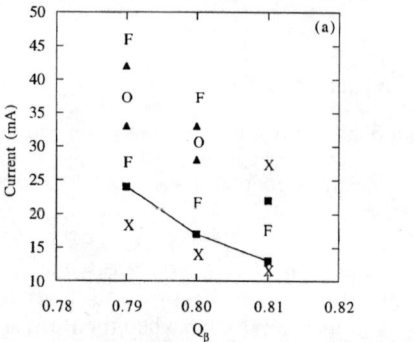
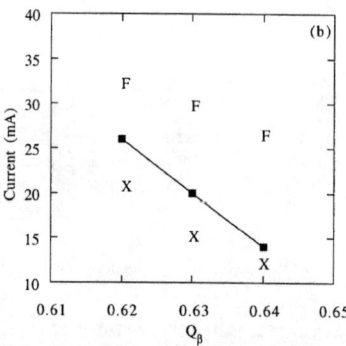

FIGURE 3. Stability diagram at two different tune regimes, (a) below the 5/6 (0.83) resonance, and (b) below the 4/6 (0.67) resonance. The line shows the threshold for the onset of coherent motion. The 'X' indicates lack of any coherent motion, 'F' indicates flip-flop motion, and 'O' indicates period-n oscillations.

In Fig. 3(a) the line joining the solid rectangles marks the lower threshold current below which no coherent activity is seen. It can be seen that as one moves away from the resonance, the threshold current increases. At all three tunes it is the flip-flop instability that sets in above this threshold (indicated by the letter 'F' in the figure). At $Q_\beta = 0.81$ there is a finite *offset* threshold (22 mA) beyond which there is, once again, no coherent motion. At the other two tunes the story is different and more complex. There is no upper threshold beyond which coherent motion disappears – at least up to the maximum current of 50 mA that we investigated. Just above the onset threshold and at very large currents the equilibrium solution is the flip-flop, but embedded in that region is a region, with clear onset and offset currents, in which the period-n oscillation is the equilibrium solution. For $Q_\beta = 0.80$ the onset and offset currents for the period-n oscillations are 28 and 34 mA respectively, and for $Q_\beta = 0.79$ they are 33 and 43 mA respectively. Thus, as one goes away from the resonance line the width of the region of dominance of the period-n oscillations increases, but it always remains embedded within the region of the flip-flop solutions.

The situation is quite different at tunes near the 4/6 resonance. Figure 3(b) shows that there is still a lower threshold for the onset of coherent motion, but there is no upper threshold at all three tunes, and more interestingly, *there is no period-n behaviour seen.* By this we mean that period-n oscillations were never the equilibrium solution, though they did often appear transiently, along the lines of Fig. 2(b).

Figure 3 shows that the dependence of the two types of coherent motion on tune is interesting and exceeding complex. Figure 3(a) suggests that close to the resonance line the flip-flop solutions are very strong and therefore there are no equilibrium period-n solutions. As one moves away the flip-flop solutions weaken and it is possible to find equilibrium period-n solutions. It is not clear why there is an offset current for the period-n behaviour, but it clear suggest a complicated, non-monotonic dependence of the strengths of these coherent phenomena on the current. Close to the resonance, at higher currents the particles are perhaps quickly tune-shifted beyond the resonance, whereas far away from the resonance the flip-flop is strong enough to lock in the tunes. The absence of period-n oscillations below the 4/6 resonance suggests that the strengths of the coherent resonances depend strongly on the nature of the resonance involved. Overall, it seems that the flip-flop solutions are 'stronger' than the period-n solutions, and are therefore more ubiquitous.

Dependence on damping

Clearly, the damping decrement chosen in the above simulations ($\delta = 1 \times 10^{-3}$) is unrealistically large; more typical numbers are around an order of magnitude smaller. We therefore decided to study the dependence on damping for some fixed tune; we chose $Q_\beta = 0.82$. We then looked at how the four features seen in Fig. (3) – the thresholds for the onset & offset of coherent motion, and the thresholds for the onset & offset of period-n behaviour – changed as a function of damping. For this study we chose only two additional values of the damping: $\delta = 2.5 \times 10^{-4}$ and $\delta = 1 \times 10^{-4}$; at these lower values of the damping decrement the computational time needed becomes exorbitant.

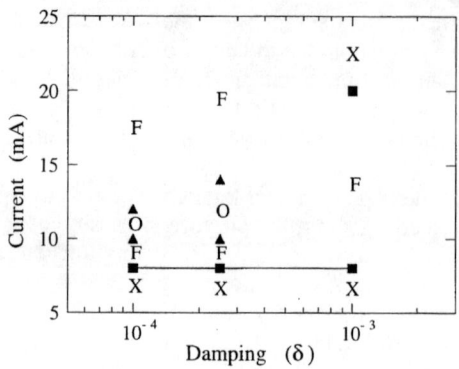

FIGURE 4. Plot of the variation of the various coherent thresholds as a function of damping, at a tune of $Q_b = 0.82$, for a flat beam with a beam-size ratio of 4:1. The line indicates the onset of coherent motion, the 'X' indicates lack of any coherent motion, 'F' indicates flip-flop motion, and 'O' indicates period-n oscillations.

Results are summarized in Fig. 4. The line connecting the solid rectangles indicates the threshold for the onset of coherent motion. Clearly, that threshold is independent of the damping. At this tune there are no period-n oscillations for $\delta = 1 \times 10^{-3}$, but there is a clear threshold for the offset of the flip-flop (at 20 mA), indicated by the solitary solid rectangle. At the lower values of damping period-n oscillations are seen, and the threshold for their onset and offset are indicated in the figure by solid triangles. Above the region of period-n oscillations, the flip-flop takes over again, and persists up to the maximum current of 40 mA that was investigated.

DISCUSSION AND CONCLUSIONS

We have presented here some results from a new beam-beam code that handles the beam-beam dynamics self-consistently, and makes no approximations regarding the nature of the beam-profiles or the beam-distributions. We find that the simulations are able to reproduce the two coherent phenomena that have been seen in experiments: the flip-flop effect and period-n beam-size oscillations. This very strongly suggests that these effects have their origin in the coherent beam-beam interaction.

The results show that coherent effects can set in at currents as low as 10 mA which, for our parameters, corresponds to a tune-shift parameter of 0.03. Since the B factories presently under construction have been designed assuming a tune-shift parameter of 0.03–0.05, it is possible that these coherent phenomena can limit their luminosity. Of course, the precise thresholds depend strongly on the tune, and it should be possible, by a suitable choice of tune, to walk around these resonances.

However that would require more careful and detailed studies of the coherent beam-beam effects for the *B* factory parameters.

It should be emphasized that the results presented above depend crucially on the self-consistent nature of the calculation and the non-Gaussian nature of the beam distribution; strong-strong simulations that assume Gaussian distributions for the beams do not give the same results.

It is also interesting to note that for round beams the dynamics seems much more benign. In particular the flip-flop instability is not seen, and the region of coherent instability is finite for tunes and damping at which the corresponding flat beams have no offset threshold.

The nature of the dependence of the various thresholds on tune and damping also suggest that what is going on is pretty involved and complex. Nonetheless, at least some order emerges from amongst this chaos. First, we note from Fig. 4 the striking fact that the threshold for coherent motion seems independent of the damping. The details of what happens beyond that threshold, especially in terms of what the equilibrium state will be, is clearly complex, but it seems safe to argue that the position of the threshold should be derivable from a simpler analysis which need not include the effects of radiation. Similarly, we note from Fig. 3 that while the details of the stability diagram are complex, the threshold for the onset of coherent motion seems to have an uncomplicated structure. These two observations suggest that the Vlasov-equation models of Dikansky and Pestrikov [14] and of Chao and Ruth [15] should be able to make *quantitative* predictions of the onset of coherent beam-beam effects. This in itself should be extremely useful information at the design stage, since one would be more concerned about avoiding luminosity-limiting coherent beam-beam effects, than about investigating the precise nature of these effects.

Second, we consider for a moment Fig. 3. In Fig. 3(a) the highest tune, $Q_\beta = 0.81$, is only 0.02 away from the resonance line (at 0.83). In Fig. 3(b) however, the highest tune, $Q_\beta = 0.64$, is 0.03 away from the resonance line (at 0.67). Since in the former case there is a clear offset threshold for the onset of coherent oscillations, and in the latter case there isn't, the existence of that threshold is a function of distance away from the resonance line. Presumably, is one is close enough to the resonance line then particles are quickly tune-shifted over the line, and therefore are no longer affected by the coherent resonance. As you go away from the resonance line, the particles need to be tune-shifted by a greater amount to get over the resonance, which means larger currents; but these larger currents also make the resonance stronger, so that particles are now locked onto the resonance and there is therefore no offset threshold.

Third consider Fig. 4, which is at a tune of $Q_\beta = 0.82$. Again the figure can be understood as arising from an interplay between the resonance strength and the damping mechanism. At the rather large damping of $\delta = 1 \times 10^{-3}$ there is a definite offset threshold, but as the damping is decreased the effective strength of the resonance increases and at $\delta = 2.5 \times 10^{-4}$ and below the particles are trapped by the resonance and there is no offset threshold.

In summary we have shown that a fully self-consistent calculation of the beam-beam interaction shows both the coherent effects that have been observed experimentally – the flop-flop instability and the period-n oscillations – and therefore unambiguously identifies the underlying mechanism behind these phenomena to be the beam-beam interaction. The details of the dependence of these coherent effects on beam-size ratio, tune and damping is complex, but some simple trends emerge which can form the basis for analytic calculations. Since these coherent phenomena are seen at tune-shifts as small as 0.03 they have the potential to limit the luminosity of B-factory colliders.

REFERENCES

1. S. Krishnagopal, Phys. Rev. Lett. **76**, 235 (1996).

2. P. Ivanov et al., in *Proceedings of the Third Advanced International Committee for Future Accelerators Beam Dynamics Workshop on Beam-Beam Effects in Circular Colliders, Novosibirsk, U. S. S. R., 1989*, edited by I. Koop and G. Tumaikin (Institute of Nuclear Physics, Novosibirsk, U. S. S. R., 1990), p. 26.

3. J. Le Duff et al., in *Proceedings of the Eleventh International Conference on High-Energy Accelerators, Geneva, 1989*, edited by W. S. Newnam (Birkhauser, Geneva, 1980), p. 707.

4. CERN Courier, May 1994, p. 20.

5. S. Krishnagopal, Ph. D. dissertation, Cornell University, 1991.

6. S. Krishnagopal and R. Siemann, Phys. Rev. Lett. **67**, 2461 (1991).

7. M. Bassetti and G. A. Erskine, Report No. CERN-ISR-TH/80-06 (unpublished).

8. M. Sands, Stanford Linear Accelerator Center Report No. 121, 1970 (unpublished).

9. S. Krishnagopal, Centre for Advanced Technology Internal Report No. CAT/95-5, 1995 (unpublished).

10. *Computer Simulation Using Particles*, R. W. Hockney and J. W. Eastwood, McGraw-Hill (USA, 1981).

11. *Plasma Physics Via Computer Simulation*, C. K. Birdsall and A. B. Langdon, Adam Hilger (Bristol, Great Britain, 1991).

12. J. P. Christiansen and R. W. Hockney, Computer Phys. Commun. **2**, 139 (1971).

13. R. Siemann, private communication.

14. N. S. Dikansky and D. V. Pestrikov, Part. Accel. **12**, 27 (1982).

15. A. W. Chao and R. D. Ruth, Part. Accel. **16**, 201 (1985).

CRYSTAL CHANNEL COLLIDER: ULTRA-HIGH ENERGY AND LUMINOSITY IN THE NEXT CENTURY

Pisin Chen
Stanford Linear Accelerator Center [*]
Stanford, California 94309 USA

Robert J. Noble
Fermi National Accelerator Laboratory [†]
Batavia, Illinois 60510 USA

ABSTRACT

We assume that, independent of any near-term discoveries, the continuing goal of experimental high-energy physics (HEP) will be to achieve ultra-high center-of-mass energies, possibly approaching the Planck scale (10^{28} eV), in the next century. To progress to these energies in such a brief span of time will require a radical change in accelerator and collider technology. High-gradient acceleration of charged particles along crystal channels and the possibility of colliding them in these same strong-focusing atomic channels have been separately investigated in earlier proposals. Here we expand further upon the concepts of emittance damping and plasma wave generation to explore a new paradigm for HEP machines early in the next century: the crystal channel collider. Energy and emittance limitations in natural crystal accelerators are determined. The technologies needed to begin experimental research on this accelerator concept are now emerging. The excitation of 1 to 100 GV/cm plasma waves in semiconductor and metal crystals by either the laser wakefield or side-injected laser techniques appears experimentally feasible with near-term lasers.

---

[*]Work supported by the U.S. Department of Energy under contract No. DE-AC03-76SF00515.
[†]Work supported by the U.S. Department of Energy under contract No. DE-AC02-76CH03000.

INTRODUCTION

High-energy physics has progressed twelve orders of magnitude in energy during the last one-hundred years (1 eV to 10^{12} eV or 1 TeV). Modern high-energy colliders are both microscopes and time machines allowing us to probe fundamental physics at distances of 10^{-16} centimeters and hence understand the relevant phenomena 10^{-10} seconds after the Big Bang. It is thought by some that by advancing only one or two orders of magnitude higher in energy, experiments will place enough constraints on unified field theories to yield one consistent "Theory of Everything" including gravity. Machine builders instead assume that regardless of any intermediate discoveries, the continuing goal of experimental high-energy physics will be to achieve ultra-high center-of-mass energies, possibly approaching the Planck scale (10^{28} eV), in the next century. Electromagnetic acceleration is limited to about 10^{16} V/cm (the critical field for pair production), and a Planck-scale linear accelerator would then have a length of about one-tenth the Earth-Sun separation - not an inconceivable task for an advanced technological society. Still to reach these energies with their attendant high luminosity in such a brief span of time will require a radical change in accelerator and collider technology. In all likelihood more than one paradigm shift in accelerators will be needed. In this paper we investigate a concept that may enable us to reach energies of order 10^{18} eV early in the next century: the crystal channel collider.

For the next few decades, our sphere of technological influence will likely be limited to the near-Earth neighborhood. Accelerators with lengths of order 10^3 to 10^4 kilometers will probably become feasible. Very high gradients are the only avenue available then to attain truly "cosmic" energies in the immediate future. The energy density in an accelerator increases with the square of the acceleration gradient. To keep the total deposited energy manageable and maintain a small beam for high luminosity, an accelerator with small transverse dimensions is called for. The idea of using atomic structures to accelerate charged particles to high energy in a short distance was expounded early, notably by Hofstadter [1]. That paper remarkably contains the germ of several ideas which were independently discovered and developed by various workers over the following twenty years, including channeling to guide accelerated particles.

Ten years ago the present authors made a cursory study of a concept to accelerate positively-charged particles along crystal channels by the electron plasma waves in metals [2,3]. The maximum electric field of a plasma wave is of order $\sqrt{n_o}$ V/cm, where n_o is the electron number density in units of cm^{-3}. Acceleration gradients of 100 GV/cm or more were implied based on the electron densities in solids. The strong electrostatic focusing of the atomic channels combined with the high gradients were found to maintain low beam emittance in spite of multiple scattering on channel electrons. The technological demands to excite such large amplitude plasma waves with lasers or particle beams appeared daunting then, and crystal behavior at picosecond to femtosecond time scales and high power densities was uncertain at best. Some of the early estimates made for crystal survival were certainly too optimistic.

The development of ultra-short pulse-length lasers, nano-fabrication technology and a better experimental and theoretical understanding of high energy density effects in solids motivate us to return to the topic of a crystal channel accelerator. Recent work on radiative damping of channeled particle emittance has also opened up the possibility of achieving both high luminosity and high energy in a crystal collider [4,5]. An improved picture of a crystal accelerator emerges which allows us to further illucidate the advantages of crystals for acceleration and emittance control as well as point out the constraints imposed by the use of natural crystals as high-energy particle accelerators. Limits on high luminosity may ultimately be more difficult to overcome than achieving ultra-high energy.

CHANNELING ACCELERATION AND EMITTANCE DAMPING

The basic concept of crystal channel acceleration combines plasma wave acceleration [6] with the well known channeling phenomenon [7] to allow positively charged particles to be accelerated over long distances without colliding with nuclei in the accelerating medium. Positively charged particles are guided by the average electric fields produced by the atomic rows or planes in a crystal. The particles make a series of glancing collisions with many atoms and execute classical oscillatory motion along the interatomic channels. The condition for classical motion is that the transverse de Broglie wavelength $\hbar/p\psi$, where p is the total momentum and ψ is the channeling angle relative to an atomic row or plane, be much less than the typical atomic screening length (~ 0.1 Å) where single atomic collisions become important. In contrast negatively charged particles oscillate about the atomic nuclei and rapidly suffer large-angle Coulomb scattering [8].

Acceleration in the crystal is provided by an electron plasma oscillation [9] with phase velocity near the speed of light. The maximum electric field of a relativistic plasma wave is the wave-breaking field [10,11]

$$\mathcal{E}_{WB} = \sqrt{2(\gamma_p - 1)}\,\mathcal{E}_0, \tag{1}$$

where $\mathcal{E}_0 = m_e\omega_p c/e$, m_e is the electron rest mass, $\omega_p = (4\pi n_o e^2/m_e)^{1/2}$ is the electron plasma frequency, and γ_p is the usual relativistic factor for a wave with phase velocity v_p. In convenient units, $\mathcal{E}_0(V/cm) \simeq 0.96(n_o(cm^{-3}))^{1/2}$ and $\omega_p/2\pi(sec^{-1}) \simeq 9000(n_o(cm^{-3}))^{1/2}$. For phase velocities approaching the speed of light (needed for relativistic acceleration), extremely high electric fields could be achieved if the necessary power densities can be applied to the plasma in the correct geometry. Doped semiconductors typically have carrier densities of 10^{14} to 10^{18} cm^{-3} corresponding to $\mathcal{E}_0 = 10$ MV/cm to 1 GV/cm, the same as typical laboratory gas plasmas. Conduction electrons in metals have densities of 10^{22} to 10^{23} cm^{-3} while the total electron density of solids is of order 10^{24} cm^{-3} implying gradients of order 1 TV/cm.

A basic obstacle to accelerating particles over long distances in crystals is beam loss from dechanneling. The transverse momentum of channeled particles increases due to collisions with electrons in the interatomic channels. Dechanneling occurs when a particle's transverse kinetic energy $E\psi^2/2$, where E is the total particle energy, allows it to overcome the channel's potential energy barrier V_c ($\sim 10 - 1000\, ze$ volts for a particle of charge ze). At this point close encounters with atomic cores quickly scatter particles out of the channel. This defines the critical channeling angle $\psi_c = (2V_c/E)^{1/2}$. In many crystals the electron density n over most of the channel is roughly constant. From Poisson's equation the channel potential energy function in either plane is simply $V = K_c x^2/2$, where $K_c = 4\pi ze^2 n$ is the focusing strength. The channel half-width a corresponds to the point where $V = V_c = K_c a^2/2$.

The increase in the *rms* angular divergence per unit length (projected onto a plane) of a channeled particle due to electron multiple scattering can be written as

$$\frac{d\langle\psi^2\rangle_{ms}}{ds} = \frac{4\pi(ze^2)^2 n}{E^2}\ln(b_{max}/b_{min}) \equiv \frac{\psi_c^2}{2\ell_d}, \qquad (2)$$

where n is the channel electron density (which is typically less than the average electron density n_o in the crystal), the impact parameters in the Coulomb logarithm are $b_{max} \simeq c/\omega_p$ and $b_{min} = \hbar/\gamma m_e c$, and γ is the relativistic factor for the channeled particle [12]. The characteristic dechanneling length is $\ell_d = \Lambda E/ze$, and the dechanneling constant $\Lambda = a^2/2e\ln(b_{max}/b_{min})$ is essentially only a function of the channel width. In natural crystals where $a \simeq 1$ to 3 Å, Λ is typically of order 1 to 10 μm/MV, consistent with experimental dechanneling lengths for MeV to 100 GeV particles [13]. Note that ℓ_d and Λ are independent of the electron density in the channel to first order.

In the harmonic potential approximation, each crystal channel acts like a smooth focusing accelerator with betatron focusing function (wavelength/2π of transverse oscillations)

$$\beta_F = (E/K_c)^{1/2} = a/\psi_c. \qquad (3)$$

The normalized *rms* channel acceptance, $A_n \equiv \gamma a^2/2\beta_F = \gamma a\psi_c/2$, defines the available transverse phase space for a channeled particle. Multiple scattering in a transverse focusing system randomly excites betatron oscillations leading to growth in the normalized *rms* emittance $\varepsilon_n = \gamma\varepsilon = \gamma\sigma^2/\beta_F$, where ε is the geometric emittance, and σ^2 is the *rms* amplitude of the particle [14]. In this terminology, dechanneling occurs when the particle emittance exceeds the channel acceptance.

Particle dechanneling in a crystal accelerator is modified by several effects. Acceleration reduces multiple scattering with increasing energy as is evident from Eqn. (2). The presence of any transverse fields in addition to the natural channel forces will change the betatron focusing function and channel acceptance. For example a plasma wave of amplitude A and transverse size b has longitudinal and transverse fields near the central axis ($x \ll b$) $\mathcal{E}_z = A\cos\phi$ and $\mathcal{E}_x = -2A\sin\phi x/k_p b^2$, respectively, where k_p is the plasma wavenumber and ϕ is the particle phase with respect to the wave crest [15]. There is a phase region in which both acceleration and transverse focusing are possible. If this wave is centered on the crystal channel axis, then particles experience a total focusing strength K given by the sum of K_c and $K_p = 2zeA\sin\phi/k_p b^2$ of the plasma wave. Charged particles oscillating in a transverse focusing system radiate and make transitions to lower energy levels of the potential with an energy-independent decay constant $\Gamma_c = 2r_{cl}K/3mc$, where $r_{cl} = (ze)^2/mc^2$ is the classical particle radius [4,5]. The channeled particle is assumed to be in the so-called undulator radiation regime where $\gamma\psi \ll 1$ and dipole radiation dominates. These radiative transitions act to damp the particle's normalized emittance. Collisional energy loss to electrons in the channel can also damp emittance (ionization cooling) but with an energy-dependent decay parameter $E^{-1}(dE_{coll}/ds) = (E/m_e c^2)d\langle\psi^2\rangle_{ms}/ds$ for a relativistic particle.

Combining these effects, the evolution equation for the normalized emittance in a crystal channel accelerator is

$$\begin{aligned}\frac{d\varepsilon_n}{ds} &= -\frac{\Gamma_c}{c}(\varepsilon_n - \hbar/2mc) - \frac{1}{E}\frac{dE_{coll}}{ds}\varepsilon_n + \frac{\gamma\beta_F}{2}\frac{d\langle\psi^2\rangle_{ms}}{ds} \\ &= -\frac{\Gamma_c}{c}(\varepsilon_n - \hbar/2mc) - \frac{zeK_c a^2}{2m_e c^2 \Lambda E}\varepsilon_n + \frac{zeK_c a^2}{4mc^2\Lambda(KE)^{1/2}},\end{aligned} \quad (4)$$

where $E = E_i + zeGs$, E_i is the initial particle energy, and G is the (net) acceleration gradient which is assumed to be a constant. The term $\hbar/2mc$ is the minimum quantum emittance of a particle in the ground state of the transverse potential. In a natural crystal channel ($K = K_c$) with electron density $n = 10^{23}$ cm^{-3}, the focusing strength is $K_c \simeq 20$ eV/Å^2, and the radiative damping distance c/Γ_c is 15 cm for positrons, 6 km for muons and 500 km for protons.

Because of the different energy dependencies in the three terms of Eqn. (4), not all terms are equally important in an arbitrary energy regime. The effectiveness of ionization cooling clearly falls off rapidly with increasing energy. Ionization cooling dominates over radiation damping for energies $E < 3m^2 c^2 K_c a^2/4zem_e K\Lambda$. For example, this corresponds to roughly 10 MeV positrons, 600 GeV muons and 50 TeV protons in a natural crystal when $K = K_c$. The solution to Eqn. (4) when radiation damping can be neglected is

$$\varepsilon_n = \left(\varepsilon_{ni} - \frac{(K_c/K)A_{ni}}{\Lambda G + K_c a^2/m_e c^2}\right)(\gamma_i/\gamma)^{K_c a^2/2m_e c^2\Lambda G} + \frac{(K_c/K)A_n}{\Lambda G + K_c a^2/m_e c^2}, \quad (5)$$

where $A_n = \gamma(K/E)^{1/2}a^2/2$ is the normalized channel acceptance including all transverse focusing forces, and A_{ni} is the channel acceptance at energy E_i. In the special case of ionization cooling with no net acceleration, the first term in Eqn. (5) becomes a damped exponential, and the emittance approaches the equilibrium value $\varepsilon_n = (m_e c^2/Ka^2)A_n$. Only if $Ka^2 > m_e c^2 \simeq 511$ keV is the particle's equilibrium emittance within the channel acceptance. This value far exceeds the channel potential energy barrier found in natural crystals and could only be obtained in an artificially wide channel (10 to 100 Å) or by added focusing such that $K \gg K_c$. In the opposite limit of high gradients, $G \gg K_c a^2/\Lambda m_e c^2 \sim 1$ MV/cm, ionization cooling has a negligible effect on the emittance evolution in the channel compared to acceleration, and the emittance becomes

$$\varepsilon_n = \varepsilon_{ni} + \frac{(K_c/K)A_n}{\Lambda G}(1-(\gamma_i/\gamma)^{1/2}). \tag{6}$$

Accelerated particles remain indefinitely channeled provided $G \geq K_c/K\Lambda$. This corresponds to 1 to 10 GV/cm in natural crystal channels when $K = K_c$. Note that the equilibrium *rms* amplitude is $\sigma^2 = (K_c/K)a^2/2\Lambda G$.

Neglecting ionization cooling for very high-energy channeled particles, the solution to the differential equation (4) is

$$\begin{aligned}\varepsilon_n(s) &= \varepsilon_{ni}\exp(-\Gamma_c s/c) + (\hbar/2mc)(1-\exp(-\Gamma_c s/c)) \\ &+ \exp(-\Gamma_c s/c)\int_0^s \exp(\Gamma_c s'/c)\frac{zeK_c a^2}{4mc^2\Lambda K^{1/2}}\frac{ds'}{(E_i+zeGs')^{1/2}}.\end{aligned} \tag{7}$$

The integral can be rewritten in terms of Dawson's integral [16]

$$D(\chi) \equiv \exp(-\chi^2)\int_0^\chi \exp(t^2)dt, \tag{8}$$

by the change of variable $\chi(s) = [(\Gamma_c/c)(s+E_i/zeG)]^{1/2}$. The solution becomes

$$\begin{aligned}\varepsilon_n(s) &= \varepsilon_{ni}\exp(-\Gamma_c s/c) + (\hbar/2mc)(1-\exp(-\Gamma_c s/c)) \\ &+ \frac{zeK_c a^2}{2\Lambda mc^2}\sqrt{\frac{c}{zeGK\Gamma_c}}[D(\chi(s))-\exp(-\Gamma_c s/c)D(\chi(0))].\end{aligned} \tag{9}$$

For $\Gamma_c s/c \ll 1$, this solution reduces to Eqn. (6) as expected. The function $D(\chi)$ reaches a maximum value of approximately 0.541 at $\chi \simeq 0.924$, and asymptotically approaches $1/2\chi$ for $\chi > 1$. For distances $s \gg c/\Gamma_c$ and $E \gg E_i$, the normalized emittance can be written approximately as

$$\varepsilon_n(s) = \frac{\hbar}{2mc} + \frac{3mc^2(K_c/K)a^2}{8ze\Lambda(zeGKs)^{1/2}}, \tag{10}$$

which damps like $\gamma^{-1/2}$ provided the net gradient G can be maintained constant with the increasing radiative energy loss. Note that the *rms* amplitude σ^2 of the channeled particle damps like γ^{-1} in this regime. The presence of K_c in Eqn. (10) reflects the deleterious effect of electron multiple scattering, and prevents one from realizing the ideal quantum emittance in such a collective accelerator.

To obtain small emittances and high luminosity in a channeling accelerator then, it is advantageous to have a high acceleration gradient and strong transverse focusing such that $K \gg K_c$. In practice the available technology will limit the plasma wave amplitude G_0 that can be generated in a crystal channel accelerator. When the magnitude of the radiative energy loss rate $(dE/ds)_{rad} = -\Gamma_c \gamma^2 K \sigma^2$ becomes comparable to zeG_0, a limiting energy is reached. In the regime $\Gamma_c s/c < 1$ where Eqn. (6) is valid, the radiation rate is proportional to γ^2, and the limit is

$$E_{max} \simeq \sqrt{\frac{3\Lambda}{zea^2 K K_c}} m^2 c^4 G_0 . \tag{11}$$

The presence of K and K_c in Eqn. (11) reflects the competing effects of strong focusing and multiple scattering in the channel. This places a fundamental energy limit on natural crystal accelerators with $K = K_c$ because the electron density ($\sim K_c$) is fixed by the atomic structure. For example if $G_0 = 100$ GV/cm and $K = K_c = 20$ eV/Å^2, then the maximum energy is about 300 GeV for positrons, 10^4 TeV for muons and 10^6 TeV for protons. On the other hand if one can artificially arrange that $K_c < (4ze\Lambda/3a^2)K$, accelerated particles will enter the regime $s > c/\Gamma_c$ before the limit (11) is reached. Here the radiation rate is only proportional to γ, and the energy limit is $E_{max} \simeq 4m^2 c^4 \Lambda G_0/K_c a^2$. The channeled particle is assumed to be in the undulator radiation regime [4,5] which is true if $K_c < (4ze\Lambda/3a^2)K$. If this is not the case, the radiation rate is higher than in Eqn. (4), and particles will stop being accelerated earlier or be quickly damped back into the undulator regime.

Only for acceleration gradients $G \geq \Lambda^{-1} \simeq 1 - 10$ GV/cm will particles remain channeled over long distances in a natural crystal accelerator. Since Λ is proportional to a^2, it may be useful to consider artificially wide channels ($a > 3$ Å) to reduce the gradient demand, at least for early experiments where gradients may be limited. Still, a large amplitude plasma wave with a field of 100 GV/cm or more is ultimately desirable to shorten the accelerator and keep the emittance as low as possible.

Two regimes of the crystal accelerator can be distinguished based on whether the plasma wave amplitude or the fields used to excite the wave are greater than or less than $I/r_a \simeq 1 - 10$ V/Å, where I is the ionization energy of electrons in an atom of size r_a. For fields greater than this, the Coulomb potential of an atom is sufficiently deformed to induce significant tunneling ionization. For an oscillating electric field $\mathcal{E}$, electrons tunnel from atoms within a time $v_e/c\varepsilon\omega$, where $v_e = (2I/m_e)^{1/2}$, $\varepsilon = e\mathcal{E}/m_e\omega c$ is the normalized field strength, and ω is the frequency [17]. Typically v_e/c is of order the fine-structure constant $\alpha \simeq 1/137$, so for field strengths $\varepsilon > 10^{-2}$, electrons escape the atom within an oscillation period. In this high field regime, the lattice ionizes, but does not yet dissociate, on this time scale. If an intense laser ($> 10^{14} - 10^{15}$ W/cm^2) is used to build up the plasma wave, the lattice will already be in this ionized state prior to plasma-wave formation.

For laser and plasma fields below 0.1 to 1 GV/cm, reusable crystal accelerators can probably be built which might survive multiple pulses, and many of our conclusions on crystal survival in References 2 and 3 probably still hold. Plasma wave decay is determined by interband transitions with a timescale of 10 to 100 ω_p^{-1} in this regime [18]. For fields above a GV/cm, only disposable accelerators, perhaps in the form of fibers or films, are possible. The lattice is highly ionized by the laser driver used to excite the plasma wave in a few optical periods, and the free electron density immediately increases to 10^{23} cm^{-3} or more for any solid. Plasma wave build-up and channeled particle acceleration must occur before the ionized lattice disrupts due to ion motion. The lattice dissociates by absorbing plasmon energy on a timescale determined by the inverse ion plasma frequency $\omega_{pi}^{-1} = (m_i/m_e)^{1/2}\omega_p^{-1} \sim 10^{-14}$ sec, where m_i is the ion rest mass. Within this time, the ions have not moved appreciably, and the lattice remains sufficiently regular to allow channeling.

The generation of large-amplitude plasma waves in a crystal requires an intense power source to supply the plasma-wave energy before the lattice dissociates. A gradient of 100 GV/cm corresponds to an energy density of 3×10^7 J/cm^3, and this must be created and used within ω_{pi}^{-1}. Because of the increased availability of high peak-power lasers, two excitation methods which may have promise, at least for initial experiments, are considered here: the side-injected laser [19] and laser wakefield [20] techniques.

In the side-injected laser method, a laser with frequency $\omega_o \simeq \omega_p$ impinges (perpendicular to the acceleration axis) on a plasma containing a spatially periodic density perturbation, either formed by an acoustic wave ($\omega_{ac} \ll \omega_p$) or a grating. The initial electron density follows this pattern. The period of the density perturbation is set at the desired plasma wavelength λ_p and defines a wavevector $\vec{k}_p$ in the plasma such that the phase velocity $v_{ph} = \omega_p/k_p \simeq c$. The laser is linearly polarized parallel to this wavevector and is of course near cutoff ($\vec{k}_o \simeq 0$) upon entry. The laser ($\omega_o, \vec{k}_o \simeq 0$) and the density modulation ($\omega_{ac} \simeq 0, \vec{k}_p$) quasiresonantly excite forward and backward traveling plasma waves with $\omega = \omega_o \simeq \omega_p$ and $\vec{k} \simeq \pm \vec{k}_p$. For a crystal accelerator, the density modulation is probably easiest to form by epitaxially growing a superlattice (in the longitudinal direction) with period λ_p consisting of two alternating materials with different electron densities. The initial spatial electron density modulation $\delta n/n_o$ is then automatically formed with the desired periodicity.

For initial side-injected laser experiments in metal crystals, a low gradient of 1 GV/cm may be adequate to demonstrate some channeling acceleration. For example a 10 MeV positron would channel about 10 μm in a metal crystal gaining about 1 MeV in energy from the plasma wave. Assuming that little tunnel-ionization occurs, the free electron density is just the conduction value 10^{22} cm^{-3} typical of metals. The plasma and laser wavelengths are 0.3 μm (near UV). The plasma wave decays via interband transitions, and the plasma-wave amplitude ε_p saturates according to the usual damped oscillator expression $\varepsilon_p = \varepsilon_o(\delta n/n_o)\hbar\omega_p/2\Gamma_p$, where ε_o is the normalized laser strength, $\delta n/n_o$ is the initial electron density modulation, and Γ_p is the plasmon decay width. The gradient of 1 GV/cm corresponds to $\varepsilon_p = 10^{-2}$. Taking $\delta n/n_o = 10^{-1}$ and $\Gamma_p \simeq \hbar\omega_p/20$, the required laser strength is $\varepsilon_o = 10^{-2}$ giving an intensity of 10^{14} W/cm^2, consistent with minimal tunnel-ionization. In contrast, for later experiments with high gradients of 100 GV/cm, the crystal will become highly tunnel-ionized to a density of about 10^{23} cm^{-3}. Plasma wave saturation will probably be determined by relativistic frequency detuning according to $\varepsilon_p \simeq (\varepsilon_o \delta n/n_o)^{1/3}$ since there will be few interband transitions available for damping. For $\varepsilon_p = 0.3$ (100 GV/cm if $n_o = 10^{23}$ cm^{-3}), the required laser strength is $\varepsilon_o = 0.3$ corresponding to an intensity of 10^{18} W/cm^2 at a wavelength of 0.1 μm.

In the laser wakefield method, a series of short laser pulses with frequency $\omega_o \gg \omega_p$, each separated by about a plasma period, are directed onto a plasma collinear with the desired acceleration direction. The longitudinal ponderomotive force arising from each pulse's intensity gradient excites a plasma oscillation. The laser frequency must be much greater than the plasma frequency so that the laser group velocity is near the speed of light. This imprints the driven plasma wave with a phase velocity near c. Simulations are typically required to determine the optimal increase in pulse spacing as ω_p detunes relativistically as well as the change in pulse shape as the nonlinear plasma wave steepens. For estimating purposes we simply use the analytic square-pulse result for the plasma-wave amplitude arising from a series of N identical laser pulses of strength ε_o [21],

$$\varepsilon_p = (1+\varepsilon_o^2)^{N/2} - (1+\varepsilon_o^2)^{-N/2} \simeq N\varepsilon_o^2, \tag{12}$$

where the last equality applies when $\varepsilon_o^2 \ll 1$. A semiconductor crystal with carrier density 10^{18} cm^{-3} may be suitable for use in an initial laser wakefield experiment to attain a gradient of 1 GV/cm. This corresponds to $\varepsilon_p \simeq 1$ and $\lambda_p = 30$ μm at this density. We take the laser wavelength as 1 μm (near IR) and assume that the number of laser pulses (each about 50 femtosec long) is $N = 30$. The required laser amplitude is $\varepsilon_o \simeq 0.17$ corresponding to an intensity of 3×10^{15} W/cm^2, so some tunnel-ionization of the crystal occurs.

THE CRYSTAL CHANNEL COLLIDER

Conceivably side-injected laser, laser wakefield or another driving mechanism (e.g. electron beam-plasma wakefield) could be used to excite plasma waves in a future crystal channel collider. For low gradients (< 1 GV/cm) reusable accelerators probably would take the form of crystal slabs on some alignable substrate. For higher gradients replaceable films or fibers are more appropriate since these are expected to be vaporized on each pulse. Alignment is certainly problematic here, and awaits the invention of fast, repeatable atomic-scale positioning. This is needed to permit staging of crystal accelerator sections with atomic precision and maintain a straight accelerator. Dislocations, unintended crystal curvature, and misalignment between sections will likely be the practical limits to long crystal accelerators.

The emittance solutions above suggest that small beamlets can be maintained with a high acceleration gradient and strong transverse focusing in crystal channels. As noted in Ref. 5, the small beamlets can in principle be brought into collision with a high probability if the crystals of each collider arm can be aligned channel to channel. This improves the luminosity, but limitations are still reached because the bunch population cannot be made arbitrarily high, as is true in all accelerators with small transverse dimensions and short wavelengths. The crystal lattice disrupts after about 10^{-14} sec, or a hundred plasma oscillations, so the number of accelerated bunches in each channel is limited to $n_b \simeq 100$. The number of particles in each bunch is denoted by N. The bunches pass through all bunches of the oncoming train so the luminosity is proportional to $n_b^2 N^2$. Of course the accelerating crystal contains a huge number of parallel atomic channels, n_{ch}, each accelerating its own n_b bunches. The luminosity of this parallel array of accelerators is then

$$L = f_{rep} n_{ch} n_b^2 N^2 \gamma / 4\pi \beta^* \varepsilon_n. \tag{13}$$

Here f_{rep} is the repetition rate of the accelerator, and β^* is the channel beta function $(E/K)^{1/2}$ since no additional focusing at the crossing is assumed.

For the sake of discussion, let us assume a natural crystal with $K = K_c$, $a \simeq 1$ Å, and that the emittance is given by Eqn. (6) with an acceleration gradient

$G = 10\Lambda^{-1} \simeq 100$ GV/cm. The number of accelerated particles in each plasma oscillation bucket is limited by beam loading [15] to a value $n_{ch}N \simeq n_{ch}A_{ch}G/8\pi e$, where $A_{ch} \simeq \pi$ Å^2 is the area of an atomic channel. This yields $N \simeq 10$, and the luminosity becomes $L(cm^{-2}sec^{-1}) \simeq 2 \times 10^{22} f_{rep} n_{ch}$. To use a proton collider for discovering new physics at a center-of-mass energy E_{cm} may require a luminosity $L(cm^{-2}sec^{-1}) \simeq 10^{29}(E_{cm}(TeV))^2$, although this may be an overestimate. This implies $f_{rep}n_{ch} \simeq 5 \times 10^{12}$ at 10^3 TeV and 5×10^{18} at 10^6 TeV. The average beam powers at these energies are 800 GW and 8×10^8 TW, respectively. These high powers result from the inherent disadvantage of having many parallel accelerators each with a small number of particles per bunch. The situation can be improved by having low electron density and/or strong focusing ($K \gg K_c$) in each channel so that particles would enter the radiation damping regime where σ^2 damps like γ^{-1}, thus increasing the luminosity. The method for doing this for each channel independently is unclear. Alternatively it may be simpler to add final focusing and combine the beamlets from the channels into a single high-density beam spot for collision, giving a large luminosity enhancement, as is done in conventional linear colliders.

CONCLUSION

Although further study is needed on this concept, the chief advantages of collective acceleration in crystal channels remain the avoidance of emittance growth due to multiple scattering on atomic nuclei and the potential for very high acceleration gradients. The crystal naturally provides a confined, uniform electron plasma for acceleration and a strong focusing system to maintain a small beam size and increase luminosity. In natural crystal accelerators, multiple scattering on channel electrons competes strongly with radiative emittance damping, and keeps the transverse particle amplitudes from being reduced to the quantum mechanical limit. The resulting radiative energy loss limits the maximum attainable energy which is then proportional to the acceleration gradient that can be generated. For a gradient of 100 GV/cm, proton energies of order 10^{18} eV are possible. Channels with low electron density and/or strong additional focusing are suggested to raise the energy limit. Some form of final focusing of the crystal beamlets will probably be required to increase the luminosity. Independent of the acceleration mechanism, the quest for higher luminosity may ultimately prove more difficult than that of reaching ultrahigh energy in the next century.

ACKNOWLEDGEMENTS

The authors wish to thank R. Carrigan, M. Downer, T. Tajima and D. Umstadter for helpful discussions.

REFERENCES

1. R. Hofstadter, "The Atomic Accelerator", HEPL Report 560, April 23, 1968 (Stanford University, Stanford, California).

2. P. Chen and R. J. Noble, "A Solid State Accelerator" in *Advanced Accelerator Concepts*, AIP Conf. Proc. **156**, ed. F.E. Mills (AIP, New York, 1987), p. 222.

3. P. Chen and R. J. Noble, "Channeled Particle Acceleration by Plasma Waves in Metals" in *Relativistic Channeling*, ed. R. A. Carrigan and J. A. Ellison (Plenum, New York, 1987), p. 517.

4. Z. Huang, P. Chen and R. Ruth, Phys. Rev. Lett. **74**, 1759 (1995).

5. P. Chen, Z. Huang and R. Ruth, "Channeling Acceleration: A Path to Ultrahigh Energy Colliders", in Proc. of the Fourth Tamura Symposium on Accelerator Physics, AIP Conf. Proc. **356**, ed. T. Tajima (AIP, New York, 1995).

6. For reviews, see for example, *Advanced Accelerator Concepts*, AIP Conf. Proc. **91, 130, 156, 193, 279** and **335** (Amer. Inst. of Physics, New York).

7. For a review, see for example D. S. Gemmel, Rev. Mod. Phys. **46**, 129 (1974).

8. The exception occurs when negatively-charged particles spiral around atomic rows with sufficient angular momentum to stay outside the atomic screening radius. The channeling motion in this case is a classical helical motion about the atomic row.

9. C.J. Powell and J.B. Swan, Phys. Rev. **115**, 869 (1959).

10. A. I. Akhiezer and R. V. Polovin, Zh. Eksp. Teor. Fiz. **30**, 915 (1956) [Soviet Physics JETP **3**, 696 (1956)].

11. A. I. Akhiezer, I. A. Akhiezer, R. V. Polovin, A. G. Sitenko and K. N. Stepanov, *Plasma Electrodynamics, Vol. 2* (Pergamon, New York, 1975).

12. J.D. Jackson, *Classical Electrodynamics*, 2nd edition (Wiley, New York, 1975), Chapter 13. For positrons scattering on channel electrons, the minimum impact parameter becomes $(2/(\gamma-1))^{1/2}\hbar/m_e c$.

13. R.A. Carrigan, "The Application of Channeling in Bent Crystals to Charged Particle Beams", in *Relativistic Channeling*, ed. R.A. Carrigan and J.A. Ellison (Plenum, New York, 1987), p. 339.

14. B.W. Montague and W. Schnell, "Multiple Scattering and Synchrotron Radiation in the Plasma Beat-Wave Accelerator", in *Laser Acceleration of Particles*, ed. C. Joshi and T. Katsouleas, AIP Conf. Proc **130** (AIP, New York, 1985),p. 146.

15. P. Chen and R. Ruth, "A Comparison of the Plasma Beat-Wave Accelerator and the Plasma Wakefield Accelerator", in *Laser Acceleration of Particles*, ed. C. Joshi and T. Katsouleas, AIP Conf. Proc. **130** (AIP, New York, 1985), p. 213.

16. *Handbook of Mathematical Functions*, ed. M. Abromowitz and I. A. Stegun (National Bureau of Standards, Applied Mathematics Series 55, Tenth Printing, Dec. 1972).

17. L.V. Keldysh, Zh. Eksp. Teor. Fiz. **47**, 1945 (1964) [Soviet Physics JETP **20**, No. 5, 1307 (1965)].

18. P.C. Gibbons *et al*, Phys. Rev. **B13**, 2451 (1976). P.C. Gibbons, Phys. Rev. **B23**, 2536 (1981).

19. T. Katsouleas *et al*, IEEE Trans. Nucl. Sci. **NS-32**, 3554 (1985).

20. T. Tajima and J.M. Dawson, Phys. Rev. Lett. **43**, 267 (1979).

21. D. Umstadter *et al*, "Resonantly Driven Laser-Plasma Accelerators", in *Advanced Accelerator Concepts*, ed. P. Schoessow, AIP Conf. Proc. **335** (AIP, New York, 1995), p. 551.

Final Crystal Cooling to Reduce the Beam Current and Detector Backgrounds for a $\mu^+\mu^-$ Collider

S.A. Bogacz, D.B. Cline, and D.A. Sanders

Center for Advanced Accelerators, Department of Physics and Astronomy,
Box 951547, University of California, Los Angeles, CA 90025-1547

Abstract. There are a few major outstanding problems for a high luminosity $\mu^+\mu^-$ collider. Here we address two of them: the high current of $\mu^\pm$ in the storage ring and the backgrounds due to the decays of these $\mu^\pm$. Both problems may be solved if a final stage $\mu^\pm$ cooling employing a bent crystal cooling system, previously proposed by the authors, is used. We show how the transverse emittance can be reduced by a factor of at least 10^2 with such a cooling system. The cost of the collider would also be substantially reduced.

INTRODUCTION

Over the past four years the possibility of a real $\mu^+\mu^-$ collider has gained interest (1–3). With the proceedings of the recent San Francisco conference (4) and the "Snowmass Book" (5), the design goals have gained even more credibility. However, there are several serious problems still remaining, which are:

1) the high backgrounds in the detector from $\mu^\pm$ decay products (1–3),
2) the very high $\mu^\pm$ content in the final collider ($\sim 5 \times 10^{12}$ $\mu^\pm$ per bunch),
3) the relatively poor reduction of the phase-space using medium energy ionization cooling (6), and
4) the current high cost of the source due to the large yield of $\mu^\pm$ required to reach high luminosity.

Most of these problems can be partially cured if the number of $\mu^\pm$ required to produce high luminosity can be reduced.

Over the past few years we have studied the use of crystal channels for cooling (7,8) and even for a µ+µ− collider that would use bent crystal beam confinement. In this note we show how crystal cooling for the high energy µ± beams could result in a dramatic decrease in the beam emittance and therefore provide high luminosity with a substantially reduced beam intensity and backgrounds of µ± in the storage ring (1,2). Also, reducing the beam emittance at high energy helps the relatively poor low energy beam cooling (3). Finally, the lower yield of µ± reduces the required proton current in the µ± source, thus possibly reducing considerably the cost of the overall µ+µ− collider (4).

We now turn to a discussion of the cooling concept. Our model calculation, presented here, shows that one can decrease the normalized emittance to less than $\varepsilon_N = 10^{-8}$ mrad by passing the muon beam through a cascade of many cooling modules.

CRYSTAL CHANNEL BEAM COOLING

We consider motion of planar channeled particles in a crystal, which is bent elasticity in a direction perpendicular to the particle velocity and to the channeling planes. The effect of bending introduces a centripetal force to the equation of transverse motion (9) (by adding a linear piece to the crystal potential), which is equivalent to lowering one side of the continuum potential well and raising the other. The equilibrium planar trajectory moves away from the midpoint of the planar channel toward the plane on the convex side of the curved planar channel. However, such shift would cause some fraction of the channeled particles to leave the potential well (dechannel) (10). The curvature at which no particle can remain channeled is reached when the equilibrium point of planar channeled motion is shifted to the position of the planar wall on the outside of the curve. This critical radius of curvature, known as the Tsyganov radius (11), ρ_T, is

$$\rho_T = \frac{2E_\mu}{\phi a}, \tag{1}$$

where $\phi = 6 \times 10^{12}$ GeV m^{-2} is a material constant, related to the curvature of the potential well (12), and "a" is the distance between adjacent atomic planes (≈ 2 Å).

Using a simple formula linking equivalent magnetic bending field, B, with the trajectory's curvature, ρ, namely

$$B[\text{Tesla}] \times \rho_T[\text{m}] = 3.34 \times E_\mu[\text{GeV}], \tag{2}$$

one can calculate the maximum available equivalent bending field corresponding to the Tsyganov curvature. From Eqs. (1) and (2), this field is given by

$$B_T[\text{Tesla}] = 3.34 \times \frac{1}{2} \phi\, a. \tag{3}$$

Its numerical value for silicon is evaluated as: $B_T = 2 \times 10^3$ Tesla. We note in passing, that the maximum bending field is energy independent.

Here we propose a fast muon cooling scheme based on the ionization energy loss (13) experienced by high energy muons (25 GeV) channeling through a silicon crystal. Applying classical theory of ionization energy loss (14), a relativistic (γ) charged particle passing through a silicon crystal of length, ΔL, loses total energy of $\Delta E[\text{MeV}] = 4 \times 10^2 \times \Delta L[\text{m}]$. One can introduce a characteristic damping length, Λ,

$$\frac{1}{\Lambda} = \frac{1}{E_\mu} \frac{\Delta E}{\Delta L}, \tag{4}$$

over which the particle loses all its energy. Relativistic muons passing though the crystal lose energy uniformly in both the transverse and longitudinal directions according to Eq. (4). After passing through a short section of a crystal ($\Delta L \ll \Lambda$) muons are re-accelerated longitudinally to compensate for the lost longitudinal energy. This leads to the transverse emittance shrinkage.

Introducing normalized transverse emittance, $\varepsilon_N = \gamma\, \sigma_x\, \sigma_{x'}$, one can write the normalized emittance budget in the form of the following cooling/heating equation:

$$\frac{d\varepsilon_N}{dL} = -\frac{\varepsilon_N}{\Lambda} + \left(\frac{\Delta \varepsilon_N}{\Delta L}\right)_{\text{scatt}}. \tag{5}$$

The last term in the above equation accounts for transverse heating processes contributing to the beam divergence increase according to the following relationship (12):

$$\left(\frac{\Delta \varepsilon_N}{\Delta L}\right)_{\text{scatt}} = \frac{1}{2} \gamma \beta \frac{\Delta \langle \theta \rangle^2_{\text{scatt}}}{\Delta L}. \tag{6}$$

Here β is the beta function of a focusing crystal channel, which has an enormously small value ($\beta = 2 \times 10^{-6}$ m, for 25 GeV muons channeling through a silicon crystal).

For muon channeling in a dielectric crystal, the dominant scattering process comes from elastic (Rutherford) muon scattering off the conduction electrons present in the channel. One can integrate the Rutherford cross section over the solid angle, which yields the following formula:

$$\alpha = \left(\frac{\Delta\varepsilon_N}{\Delta L}\right)_{scatt} = 40\pi n \frac{r_\mu^2}{\gamma} \beta. \qquad (7)$$

Here $r_\mu = 1.4 \times 10^{-17}$ m is the classical muon radius and $n = 6 \times 10^{29}$ m^{-3} is the average concentration of the conduction electron gas in silicon crystal.

Integrating the cooling equation, Eq. (5), one obtains the following compact solution in terms of the normalized transverse emittance evolution:

$$\varepsilon_N = \varepsilon_N^0 \, e^{-(L/\Lambda)} + \Lambda\alpha \left[1 - e^{-(L/\Lambda)}\right]. \qquad (8)$$

The last term in Eq. (8) sets the equilibrium cooling limit of

$$\varepsilon_N^{min} = \Lambda\alpha \quad , \quad L \to \infty. \qquad (9)$$

Assuming 25 GeV muons, one gets $\Lambda = 62.5$ m and the equilibrium limit of the normalized emittance of

$$\varepsilon_N^{min} = 8 \times 10^{-9} \text{ mrad}. \qquad (10)$$

This value of the normalized emittance will be used in our achievable luminosity estimate.

BENT-CRYSTAL COOLING RING

Here we employ previously discussed properties of the planar channeling of high energy muons in silicon to design components of a storage ring. Particularly, we are interested in a section of bent crystal followed by two straight pieces providing alternating horizontal–vertical focusing. A basic guiding cell is depicted schematically in Fig.1. One can notice that the induced configuration of guiding fields in this element is equivalent to a powerful alternating gradient achromat.

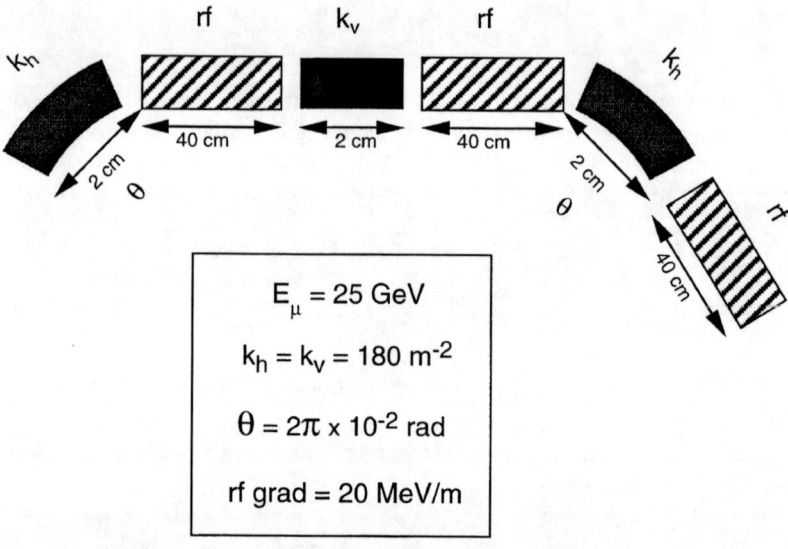

FIGURE 1. Layout of a 'cooling ring' consisting of 50 bending–focusing–acceleration multi-functional cells. A straight piece of silicon crystal rotated by 90° separating two sections of bent crystals provides vertical focusing, which maintains betatron phase stability in the proposed lattice. Conventional rf, 40-cm–long inserts (20 MeV/m) follow every 2-cm–long section of silicon crystal absorber.

Relativistic muons channeling through an Si crystal are confined between two neighboring atomic planes – they experience strong focusing electrostatic crystal-potential in the direction perpendicular to these planes, while there is virtually no confinement in the direction parallel to the planes (no focusing or defocusing). The focusing gradient $k = 1/\beta$ is equivalent to the magnetic quadrupole strength, k_1 (magnetic gradient), where

$$k_1 = \frac{1}{B\rho}\frac{\partial B_y}{\partial x}, \qquad (11)$$

and

$$k = \frac{\phi}{E_\mu}. \qquad (12)$$

E_μ is the total muon energy and $\phi = 6 \times 10^{12}$ GeV m^{-2} is a material constant, related to the curvature of the potential well (16). Assuming 25 GeV muons, crystal focusing gradient, k, yields an enormous value of 180 m^{-2} exceeding conventional quadrupole strength by four or five orders of magnitude.

As discussed in "Crystal Channel Beam Cooling" (Sec. 2), one can bend the crystal slightly, so that channeling muons follow the curvature of the guiding field, which results in bending of muon trajectories similar to the effect of a bending magnetic field. Projecting experimental results for proton channeling in a bent silicon crystal, one can assume that 25 GeV muons channeling through a 2-cm–long crystal should follow (without significant dechanneling effects) a bend of $\theta = 2\pi \times 10^{-2}$ rad (compared with the critical bending angle $\theta_T = 2 \times 10^{-1}$ rad). The lattice design presented here is based on these two numbers, k and θ.

Figure 1 illustrates a functional bending–focusing cell, where alternating sections of the horizontal and vertical continuous focusing channels are combined with sections of horizontally bent Si crystals. The following sequence of crystal elements: a horizontally focusing bent crystal (2-cm long) – a short drift space – a conventional rf re-acceleration section (40-cm long) – a short (2-cm long) vertically focusing straight crystal – a short drift space – another conventional rf re-acceleration section (40-cm long) – finally, a horizontally focusing bent crystal (2-cm long) followed by a conventional rf re-acceleration section (40-cm long) completes the proposed elementary cell. At 25 GeV, one could close the entire collider ring using 50 of the above $F_hOF_vOF_hO$ cells. For a sequence of the above described cells, one can find periodic betatron trajectories in both the horizontal and vertical planes – the betatron phase stability is provided by the proposed lattice configuration (alternating horizontal/vertical focusing). By virtue of [110] planar channeling, discussed in detail in the previous section, a crystal channel provides an ultra-strong electrostatic focusing gradient in the [110] direction with practically no confinement or defocusing in the plane perpendicular to the [110] direction. This fact guarantees both local and global decoupling of the horizontal and vertical betatron motions for the proposed collider lattice.

Practical realization of muon cooling at 25 GeV could be done in a compact 'cooling ring'. Assuming characteristic damping length, Λ, of 62.5 meters, the energy loss suffered by the muon beam after passing through a 2-cm–long section of a silicon crystal, is equal to 8 MeV. In principle, a conventional high gradient (20 MeV/m) acceleration inserts (40-cm long rf insert following every 2-cm–long crystal absorber) could be used to replenish the suffered energy loss (0.4 m × 20 MeV/m = 8 MeV). The proposed cooling ring of 50-fold symmetry would have a nominal circumference of 63 meters!

Our goal is to start with the initial muon phase-space of the normalized emittance of 2.5×10^{-7} mrad and cool it down to the final emittance of 2.5×10^{-9}

mrad. One can see from Eq. (8) that to achieve this goal, muons have to pass through the total silicon crystal length of

$$L = 2 \log 10 \times \Lambda = 280 \text{ m}. \tag{13}$$

In the proposed cooling cell architecture, the total cooling medium (silicon) length of L = 280 m is equivalent to about 90 turns of the beam circulation in the ring. The lost energy is replenished every ΔL = 2 cm, which satisfies the adiabatic re-acceleration condition ($\Delta L \ll \Lambda$ = 62.5 m).

To go beyond the above simple analytic calculation, we are planning to carry out realistic computer simulations of planar channeling in bent crystals. One should track a charged particle through the distorted crystal lattice with the use of a realistic continuous-potential approximation and take into account the processes of both single and multiple scattering of electrons and nuclei, as well as on various defects and imperfections of the crystal lattice.

APPLICATION TO HIGH ENERGY COOLING: CONCLUSIONS

In this paper we show how crystal cooling for the high energy $\mu^\pm$ beams could result in a dramatic decrease in the beam emittance and therefore would provide high luminosity with a substantially reduced beam intensity and backgrounds of $\mu^\pm$ in the storage ring. Also, reducing the beam emittance at high energy helps the relatively poor low energy beam cooling. Finally, the lower yield of $\mu^\pm$ reduces the required proton current in the $\mu^\pm$ source, thus possibly considerably reducing the cost of the overall $\mu^+\mu^-$ collider.

We suggest employing ionization energy loss in an alternating focusing crystal channel as a cooling mechanism, since initially small muon phase space allows for efficient channeling through long sections of silicon crystal. Ultra-strong focusing in a crystal channel combined with alternating bending makes it a powerful focusing cell with ultra-small beta function. The cooling equation derived here shows that it is quite feasible to reduce transverse emittance by two orders of magnitude. Our model calculation done for 25 GeV muons shows that final emittances as low as 10^{-9} mrad are readily achievable, limited only by multiple scattering off the valence electrons in the crystal.

We conclude our study with the following observation: The proposed ionization crystal cooling could be used at some later stages of the collider scheme (e.g., for the final cooling), because of 'favorable' energy scaling of the relevant cooling characteristics, α, Λ, and ε_{min}^N. They can be summarized as follows:

$$\alpha \sim \gamma^{-3/2}, \tag{14}$$

$$\Lambda \sim \log \gamma, \qquad (15)$$

and

$$\varepsilon_N^{min} = \Lambda\alpha \sim (\gamma^{-3/2})\log \gamma. \qquad (16)$$

Therefore, the proposed cooling mechanism scaled to higher energies looks even more attractive.

ACKNOWLEDGMENTS

We wish to thank J. Wurtle and R. Fernow for discussions concerning the work reported here. Stimulating comments from S. Geer and P. Chen are also acknowledged.

REFERENCES

1. Cline, D.B, *Advanced Accelerator Concepts*, edited by P. Schoessow (AIP Conference Proceedings 335), New York: AIP Press, 1994, p. 659.
2. *Physics Potential and Development of $\mu^+\mu^-$ Colliders*, edited by D.B. Cline (AIP Conference Proceedings 352), New York: AIP Press, 1994.
3. *Beam Dynamics and Technology Issues for $\mu^+\mu^-$ Colliders*, edited by J.C. Gallardo (AIP Conference Proceedings 372), New York: AIP Press, 1995.
4. *Physics Potential and Development of $\mu^+\mu^-$ Colliders*, edited by D.B. Cline (Proc., San Fransisco, 1995) *Nucl. Phys. B* (PS), **51A** (1996) in press.
5. "$\mu^+\mu^-$ Collider a Feasibility Study" (Snowmass Book), Report Nos. BNL-52503, FNAL–Conf.–96/092, LBNL–38946, 1996.
6. Fernow, R., (private communication).
7. Bogacz, S.A., and Cline, D.B, *Int. J. Mod. Phys. A*, **11**, 2613 (1996).
8. Bogacz, S.A., Cline, D.B., and Sanders, D.A., (Proc., Relativistic Channeling Wksp., Aarhus, Denmark, July 1995), edited by H.H. Andersen, *NIMB*, **119** (1,2), 206-209 (1996).
9. Gibson, W.M., in: *Relativistic Channeling*, NATO ASI Series, vol. 165B, edited by R.A. Carrigan, Jr. and J.A. Ellison (Plenum, New York, 1986) p. 27.
10. Biryukov, V.M., *Phys. Rev. Let. E.* **51**, 3522 (1995).
11. Tsyganov, E.N., FERMILAB Internal Report No. TM-682, 1976.
12. Bogacz, S.A., and Cline, D.B., "A Bend Crystal Undulator to Provide Gold Positron Beams," NIMB (1996) submitted.
13. Skrinsky, A.N., and Parakhomchuk, V.V., *Sov. J. Part. Nucl.*, **12**, 223 (1981).
14. Parkhomchuk, V.V., and Skrinksy, A.N., in: *Proceedings of the 12th International Conference on High Energy Accelerators*, edited by F.T. Cole and R. Donaldson (1983) p. 485.

On the Physical Limitations to the Lowest Emittance
(Toward Colliding Electron-Positron Crystalline Beams)

A.A. Mikhailichenko
Cornell University, Wilson Laboratory, Ithaca, NY, 14850

Abstract. In this paper we explore the conditions under which a relativistic beam of electrons or positrons, treated as a Fermi gas, becomes degenerate. The quantum effects give an absolute limit to the emittance of the beam. We discuss here some methods for preparing this ultra-cold electron-positron gas in a framework of colliding beams at high energy.

1. INTRODUCTION

It is well known that electron gas has a tendency to degenerate when its temperature is lowered. In the fully degenerated state, each particle occupies the volume in a phase space of the order $\Delta p \Delta x \approx \hbar$. A typical requirement for degeneration is that the temperature of the electron gas must be low compared to Fermi energy ε_F. The Fermi energy is [1]

$$\varepsilon_F \cong (3\pi^2)^{2/3} \frac{\hbar^2}{2m} \rho^{2/3} \qquad (1)$$

or

$$\varepsilon_F = (3\pi^2)^{1/3} \hbar c \rho^{1/3},$$

if the particles in the rest frame are relativistic. Here $\rho \approx \left(\frac{N}{V}\right)$ is a density, N is the number of states in volume V, m is a rest mass of electron, and $\hbar$ is Planck's constant. The condition for degeneration becomes

$$k_B T \leq \frac{\hbar^2}{m} \rho^{2/3} = mc^2 \lambda_C^2 \rho^{2/3},$$

where $k_B \cong 1.38 \cdot 10^{-23}\, J/°K$ is Boltzmann's constant, $\lambda_C = \hbar/mc$. By keeping the temperature constant, one can see that the gas becomes more degenerate as the density ρ increases. Typical density in metals is $\rho \approx 5 \cdot 10^{22}\, 1/cm^3$, which allows the electron gas to degenerate at a temperature of about $1\, eV$.

Normally in metals the charge of electron gas is neutralized by the charge of the nuclei -- eZ. To satisfy the condition that the Coulomb interaction energy of the electrons and nuclei is small compared with Fermi energy, one requires $\rho \geq \left(\frac{e^2 m}{\hbar^2}\right)^3 Z^2$ [1]. Again, by increasing the density, this condition is better satisfied.

An ultrarelativistic electron beam is a unique object of statistical physics due to cancellation of the forces between charged particles $\approx 1/\gamma^2$, where $\gamma = E/mc^2$, E being the energy of the electron (or positron). The gas becomes more ideal with increasing energy. For intermediate energy, that is, say, 10 GeV, the factor of cancellation is of the order $1/\gamma^2 \approx 2.5 \cdot 10^{-9}$. In principle, some amount of particles with an opposite charge can be added to the linearly moving beam.

2. CONDITION FOR DEGENERATION

For an electron gas in a volume V, the number of states with absolute magnitude of momentum in the interval from $p_{x,y,s}$ to $p_{x,y,s} + dp_{x,y,s}$ is [1]

$$dn \cong 2 \frac{dp_x dp_y dp_s \cdot V}{(2\pi\hbar)^3},$$

where factor 2 reflects two possibilities for spin orientation. The total number of electrons in these states can be estimated for uniform distribution as

$$N = \int dn \cong 2 \frac{p_x p_y \Delta p_\| S_\perp l_b \gamma}{(2\pi\hbar)^3} \cong 2 \frac{\gamma\varepsilon_x \gamma\varepsilon_y \gamma l_b (\Delta p/p_0)}{(2\pi\lambda_c)^3}, \qquad (2)$$

where $\gamma l_b (\Delta p/p_0)$ is an invariant longitudinal emittance, l_b is the bunch length, and $\gamma\varepsilon_x$ and $\gamma\varepsilon_y$ are the transverse horizontal and vertical emittances. Again, if N is close to the number of the particles in the bunch, then the particles in the bunch are close to the degeneration condition. We can also say that the beam with the number of particles N cannot have emittances lower than those defined by Equation (2). The physical sense is clear: all lower states are occupied.

3. THERMALIZATION

The electron gas temperature T in a moving frame can be represented as the following (see for example [2])

$$\frac{3}{2} k_B T \cong \frac{1}{2} mc^2 \gamma \left[\frac{\gamma\varepsilon_x}{\beta_x} + \frac{\gamma\varepsilon_y}{\beta_y} + \frac{1}{\gamma}\left(\frac{\Delta p}{p_0}\right)^2 \right], \qquad (3)$$

where $(\Delta p/p_0)$ is the energy spread in the beam, and $\beta_{x,y}$ is the envelope function in the damping ring. For damping rings developed as injectors for future linear colliders, the typical values are $\gamma\varepsilon_x \cong 3\cdot 10^{-4} cm \cdot rad$, $\gamma\varepsilon_y \cong 3\cdot 10^{-6} cm \cdot rad$, $\beta_{x,y} \cong 10 m$, $\Delta p/p \cong 10^{-3}$, $\gamma \cong 6\cdot 10^3$ (3 GeV). This allows for the temperature in the beam $\frac{3}{2} k_B T \cong \frac{1}{2} mc^2 \gamma [3\cdot 10^{-7} + 3\cdot 10^{-9} + 1.5\cdot 10^{-10}]$.

One can see that the longitudinal temperature is the lowest. This allows the possibility of temperature redistribution from the transverse to the longitudinal direction due to intra-beam scattering or due to the help of the mechanism, as described in [10]. This will be discussed later. Let's investigate the condition for degeneration due to temperature. By comparing equation (3) with (1), one obtains

$$\frac{1}{2}mc^2\gamma\left[\frac{\gamma\varepsilon_x}{\beta_x}+\frac{\gamma\varepsilon_y}{\beta_y}+\frac{1}{\gamma}\left(\frac{\Delta p}{p_0}\right)^2\right]\leq(3\pi^2)^{1/3}\hbar c\tilde{\rho}^{1/3},$$

where $\tilde{\rho}=\rho/\gamma$ is the density in the moving frame. Neglecting the longitudinal temperature and supposing that $\gamma\varepsilon_x \approx \gamma\varepsilon_y$, one can determine the condition $(\gamma\varepsilon_x)^4 \leq \frac{\beta_x^2}{l_b}\lambda_c^3 N$. Supposing that $\beta_x \approx l_b \approx 1cm$, $N \approx 10^{10}$, one can estimate $\gamma\varepsilon_x \leq (5.5\cdot 10^{-22})^{1/4} \cong 4.8\cdot 10^{-6} cm\cdot rad$. This is an estimation for the maximal possible transverse emittance required for degeneration. One can see that the radial emittance needs to be lowered about two orders of magnitude.

The expression for temperature yields $v_\perp^2 \cong c^2\gamma\cdot\left[\frac{\gamma\varepsilon_x}{\beta_x}+\frac{\gamma\varepsilon_y}{\beta_y}+\frac{1}{\gamma}\left(\frac{\Delta p}{p_0}\right)^2\right]$.

The frequency of collisions $f_{x,y}$ is $\frac{1}{Nf_{x,y}} \cong \frac{a_{x,y}}{v_{x,y}} \cong \sqrt{\frac{\gamma\varepsilon_{x,y}\beta_{x,y}}{\gamma}} / c\sqrt{\gamma(\frac{\gamma\varepsilon_{x,y}}{\beta_{x,y}})} = \frac{\beta_{x,y}}{c\gamma}$

and is not dependent on emittance. For longitudinal motion, one can estimate $f_s^{-1} \cong \gamma_b / v_\| \cong \frac{\gamma_b}{c(\Delta p/p)}$. Longitudinal mass is $\mu^{-1}=\frac{1}{m\gamma}\left(\frac{1}{\gamma^2}-\alpha\right)=\frac{1}{m\gamma}\left(\frac{1}{\gamma^2}-\frac{1}{\gamma_{tr}^2}\right)$. It becomes negative above the transition energy, when $\gamma > \gamma_{tr} \cong \alpha^{-1/2}$. Here $\alpha = \frac{p}{C}\frac{\partial C}{\partial p}$, where C is the circumference, is a momentum compaction factor.

Considerations in [6,7] show that thermalization due to intrabeam scattering (not only direct Coulomb interaction between particles, but through a third agent such as a resistive wall or a parasitic cavity) can happen only *below* the critical energy (the longitudinal mass is positive here). So, for thermalization, the damping ring must have a value of *zero* or *negative* α. The kayak-paddle cooler and LDS system, considered below, satisfy this condition. Some investigations have been made recently to verify the conditions for operation with positive longitudinal mass [9]. This circumstance is out of the scope of designers working with damping rings for linear colliders, however.

The thermalization also can be used in the OSC method for cooling *all degrees of freedom* of the beam, although cooling is applied only to *one* degree of freedom. Here thermalization will equalize the temperature over all degrees of freedom. This can reduce emittance requirements arising from simultaneous cooling in transverse and longitudinal directions with the Hereward cooling method [8]. One method also considered recently [10] allows effective redistribution of the modes of damping in a focusing-dominated system, a quadrupole wiggler, for example.

Particle spacing. The distance between particles δ in the laboratory frame can be estimated as

$$\delta^3 \cong \frac{V}{N} \cong \frac{\sqrt{\varepsilon_x \beta_x \varepsilon_y \beta_y} \cdot l_b}{N}.$$

By substituting here expressions for emittances from the previous relation and the estimate $\beta_{x,y} \cong l_b$, one can determine that in the case of degeneration

$$\delta^6 \cong (2\pi\lambda_C)^3 \frac{l_b^3}{2\gamma^3 N(\Delta p / p_0)}.$$

Let us estimate possible values for emittances that satisfy the condition of degeneration (1). We shall consider ordinary synchrotron damping with the minimal possible invariant value, obtained by the Linear Damping System (in an LDS [3] type damping ring), kayak-paddle cooler [4], and with the help of the OSC method [5].

4. THE COOLERS

4.1. Linear damping system (LDS) [3] is a sequence of wigglers and accelerating structures installed along a straight line, basically at the beginning of the linear collider main accelerating structure. The losses in the wigglers are the same as the gain of the energy given by the accelerating structures, so the particles are moving with almost constant energy. The formulas are the same as for kayak-paddle ring, so we will consider them in the following section.

4.2. Kayak-paddle cooler (KPC) [4] is a damping ring, which consists of a sequence of wigglers and accelerating structures, installed along a straight line and having bends at the end. The bends only give a small input into the cooling dynamics, so we can overlook their influence. In that sense the cooler is similar to the LDS. However, here there is no requirement that the linear rate of losses be the same as the energy gain per unit length. In the LDS, as in any damping ring, a particle needs to re-radiate its full energy a few times. The length, which characterizes the speed of damping, is $l_s = -\frac{\gamma}{d\gamma/ds}$, $d\gamma/ds \cong -\frac{2}{3}r_0\frac{K^2}{\lambda^2}\gamma^2$, where $K = eH_\perp \lambda / mc^2$, $H_\perp$ is a magnetic field value in the wiggler, $2\pi\lambda$ is the wiggler period, and $r_0 = e^2/mc^2$. Notice here that $d\gamma/ds$ is not a function of the wiggler period. By substituting here for estimation $\gamma \cong 8 \cdot 10^3$ (4 GeV), $K \approx 5$

($H_\perp \cong 0.36T$, $2\pi\lambda \cong 30\,cm$), one can obtain $\frac{d\gamma}{ds} \cong -\frac{2}{3} 2.8 \cdot 10^{-13} \frac{25}{25} 6.4 \cdot 10^7 \cong 1.2 \cdot 10^{-5}$ [$1/cm$] or 0.0012[$1/m$]. For the characteristic damping length, one can obtain $l_s = -\frac{\gamma}{d\gamma/ds} \cong \frac{8 \cdot 10^3}{1.3 \cdot 10^{-3}} \cong 6 \cdot 10^6\,m$.

The fraction of the wigglers is about 1, due to relatively slow damping (in contrast to the LDS, where the fraction is about 1/2—half the space is occupied by the accelerating structures). If we suppose that the circumference is about 3 km, the number of revolutions will be 2000, which takes about 20 ms. Emittance dynamics are defined by

$$\frac{d\varepsilon_x}{ds} = \left\langle \left(H_x + \frac{\beta_x}{\gamma^2}\right) \frac{d(\Delta E/E)^2_{tot}}{ds} \right\rangle - 2\alpha_x \varepsilon_x,$$

with a similar equation for vertical motion, where

$$H_{x,y} = \frac{1}{\beta_{x,y}}\left(\eta^2_{x,y} + (\beta_{x,y}\eta'_{x,y} - \frac{1}{2}\beta'_{x,y}\eta_{x,y})^2\right),$$

$\eta_{x,y}$ are the dispersion functions. Derivatives are taken over the longitudinal direction. Decrements $\alpha_{x,y,s}$ are defined in the usual way $\alpha_i = \frac{J_i}{2l_s}$, where $J_x \cong 1$, $J_y = 1$, $J_s \cong 2$. Notice here that the decrement for the energy spread is the same as for the emittance decrease,

$$\frac{d(\Delta E/E)^2_{tot}}{ds} = \frac{d(\Delta E/E)^2_{IBS}}{ds} + \frac{d(\Delta E/E)^2_{QE}}{ds} - \alpha_s\left(\frac{\Delta E}{E}\right)^2,$$

where $\frac{d(\Delta E/E)^2_{QE}}{ds} \cong \frac{55}{48\sqrt{3}} \frac{r_0^2 \cdot \gamma^5}{\alpha|\rho|^3}$, $\frac{d(\Delta E/E)^2_{IBS}}{ds} \cong \frac{N r_0^2 \ln_C}{\gamma^3 \varepsilon_x \sqrt{\varepsilon_y \beta_y l_b}}$, and $\ln_C$ is Coulomb's logarithm. For the dipole wiggler, the periodic solution for η_x can be expressed as

$$\eta_x = \frac{K\lambda}{\gamma}Sin(s/\lambda) = \frac{\lambda^2}{\rho}Sin(s/\lambda),$$

where $\rho = \lambda\gamma/K$ is the bending radius in the magnetic field of the wiggler. For the function H_x, we can estimate $H_x \cong \beta_x\eta'^2_x$ and for $\eta'_x \cong K/\gamma$. For the vertical emittance, the term β_y/γ^2 dominates. This gives the equilibrium emittances under condition $d\varepsilon_{x,y}/ds = 0$ as:

$$(\gamma\varepsilon_x) \cong \frac{1}{2}\lambda\bar{\beta}_x K^2\gamma/\rho \cong \frac{1}{2}\lambda\bar{\beta}_x K^3/\lambda$$

$$(\gamma\varepsilon_y) \cong \frac{1}{4}\lambda_c \overline{\beta}_y \gamma / \rho \cong \frac{1}{2}\lambda_c \overline{\beta}_y K / \lambda,$$

where $\overline{\beta}_{x,y}$ is the averaged values of the envelope functions in the wiggler. Notice here that quantum equilibrium vertical emittance is not dependent on the wiggler period and neither equilibrium invariant emittances is dependent on energy. Substitute here for estimation $\overline{\beta}_{x,y} \approx 1 m$, $\lambda \cong 5\, cm$, $K \cong 5$,

$$(\gamma\varepsilon_x) \cong \frac{1}{2} 3.86 \cdot 10^{-11} \cdot 100 \cdot 125 / 5 \cong 5 \cdot 10^{-8}\, cm \cdot rad$$

$$(\gamma\varepsilon_x) \cong \frac{1}{2} 3.86 \cdot 10^{-11} \cdot 100 \cdot 5 / 5 \cong 2 \cdot 10^{-9}\, cm \cdot rad.$$

According to equation (2), the number of states for these emittances will be $N \cong 2 \dfrac{5 \cdot 10^{-8} \cdot 2 \cdot 10^{-9} \cdot 8 \cdot 10^{3} \cdot 10^{-1} \cdot 10^{-4}}{(2\pi \cdot 3.86 \cdot 10^{-11})^3} \cong 1.1 \cdot 10^{12}$, which indicates that these emittances are still too great to raise a degeneration. Here we substitute, optimistically, $l_b \cong 0.1 cm$, $(\Delta p / p_0) \cong 10^{-4}$ at $\gamma \cong 8 \cdot 10^{3}$ (4 GeV).

4.3. OSC method. The appropriate way to obtain a beam with an extremely small emittance is the Optical Stochastic Cooling (OSC) method [4]. The amplification coefficient κ of an optical amplifier can be expressed as follows

$$\kappa \cong \frac{\varepsilon_\parallel}{r_0} \frac{1}{N} \frac{\Delta f}{f},$$

where $\gamma\varepsilon_\parallel = \gamma l_b \Delta E / E$ is an invariant longitudinal emittance, $\Delta f / f$ is a relative bandwidth, and N is the number of particles. One can see that only phase density in the longitudinal direction $\rho_\varepsilon = \gamma\varepsilon_\parallel / N$ is important for the level of amplification. The decrease of emittance is equal to

$$\frac{\varepsilon_f}{\varepsilon_0} \cong \frac{1}{\alpha N_u}.$$

Here $\alpha = e^2 / \hbar c \cong 1/137$, N_u is the number of the particles in the bandwidth, and $N_u \cong \dfrac{f}{\Delta f} N \dfrac{\lambda_u}{l_b} \cong \dfrac{f}{\Delta f} N \dfrac{L}{l_b \gamma^2}$, where $L = 2\pi\lambda$ is the wiggler period, l_b is the bunch length, and λ_u is the central wavelength of radiation amplified. For $N \cong 10^{10}$, $\lambda_u \cong 10^{-4}\, cm\, (1\, \mu m)$, $l_b \cong 0.5 cm$, $\Delta f / f \cong 20\%$, one can expect $N_u \cong 5 \cdot 10^{10} \cdot 2 \cdot 10^{-4} \cong 10^{7}$ and $\varepsilon_f / \varepsilon_0 \cong 10^{-5}$. This is of course the maximal possible value of cooling. Real value will be within 0.1:0.01. This procedure can be applied after the LDS or KPC, so one can expect the emittance $\varepsilon_y \leq 10^{-12}\, cm\, rad$ at 1 GeV (or invariant emittance $\gamma\varepsilon_y \leq 2 \cdot 10^{-11} m\, rad$). According to equation (2), the number of states for these emittances will be $N \cong 5.6 \cdot 10^{11}$, where we substitute $(\Delta p / p_0) \cong 10^{-3}$. As we only have the number of the particles in the beam $N \cong 10^{10}$, this method also does not give the yield for full degeneration. Nevertheless, the calculations are not drastically out of range and some optimization is possible. However, this method needs to be tested experimentally.

5. CONCLUSION

These considerations indicate that for present day coolers the condition for degeneration could not be satisfied. At the same time, *optimistic* estimations indicate that the real condition for degeneration is not out of scale for coolers in the near future. Meanwhile, one can try to investigate experimentally the condition for degeneration in routinely working machines. At the same time, the physics of this quantum crystal state in a relativistic beam may be of general interest. Behavior of the degenerate beam may modify both the characteristics of synchrotron radiation and the scattering process in the IP of electron-positron colliders.

For protons the situation is not so optimistic. Despite the smaller emittance of the proton machines, the Compton wavelength of the proton is about 1836 times smaller, thereby excluding satisfaction of condition (1).

6. ACKNOWLEDGMENT

In conclusion, the author thanks M.S. Zolotorev for numerous discussions and criticism.

7. REFERENCES

[1] L.D. Landau, E.M. Lifshitz, "Statistical Physics," Pergamon Press, 1985.
[2] M. Conte, W. MacKay, "An Introduction to the physics of particle accelerators," Word Scientific, 1991.
[3] N.S. Dikansky, A.A. Mikhailichenko, "A linear damping system for obtaining high energy $e^{\pm}$ with extremely low emittance," EPAC 92, Berlin, 1992, Proc., p. 898.
[4] A.A. Mikhailichenko, "Damping Ring for VLEPP linear collider," III Int'l Workshop on Linear Colliders LC91, Protvino, September 17-27, 1991. Proceedings, Edited by V.E. Balakin, S. Lepshokov, N.A. Solyak, Serpukhov, (IFVE). 1991. Serpukhov, USSR: BINP (1991).
[5] A.A. Mikhailichenko, M.S. Zolotorev, "Optical Stochastic Cooling," Phys. Rev. Let. 71, 4146 (1993).
[6] A. Piwinsky, "IBS," Proc. 9th Intern. Conf. on High Energy Acc., Stanford, CA, 2-7 May, 1974, SLAC, 1974. p. 405.
[7] Ya. S. Derbenev, "Collision relaxation of the heavy particles in a damping ring," All Union Conference, Dubna, 1979, vol. 1, p.119-121.
[8] H.G. Hereward, in *Theoretical Aspects of the Behavior of Beams in Accelerators and Storage Rings* (CERN 77-13, Geneva, 1977), p. 281.
[9] A. Nagji *et al.*, "Experiments with low and negative momentum compaction factor with SUPER-ACO," Super-ACO/96-03, presented at the 5th European Particle Accelerator Conference (EPAC'96), Barcelona, Spain, 1996.
[10] Z. Huang, P. Chen, R. Ruth, "Radiation Damping in Focusing-Dominated Systems," PAC'95, Dallas, TX, May 1-5, 1995, Proceedings, Vol. 5, p. 3326.

Nonlinear dynamics of semicrystalline beams

V.A. Skvortsov, P.V. Milyutin

*High Energy Density Research Center of the Russian Academy of Sciences,
127412, Moscow, Izhorskaya 13/19, IVTAN*

Abstract. The Boussinesq equation and equation of Korteveg-de-Vries are obtained to describe the nonlinear dynamics of a semicrystalline beam of charge particles (SBCP) in continuum approximation. Analysis of its soliton solutions shows that the mathematical problem of SBCP evolution is similar to that of Fermi-Pasta-Ulam. It is established that the SBCP possesses a positive dispersion. The obtained equations are generalized in classical approximation in the case of multiparticle Coulomb interaction. It is shown that nonlinear disturbances in SBCP can modify the spectrum and intensity of electromagnetic (in the case of free electron laser) and gravitational (in the case of free ion laser) radiation as a result of the interaction of SBCP with laser beams. Thus, one can diagnose nonlinear waves of charge density by using the characteristic of such radiation.

1. Introduction

This work is devoted to the study of the nonlinear dynamics of a semicrystalline beam of charged particles. One of the main properties of such systems is excitation and distribution to them of nonlinear waves of density charge. As was first shown in the work of Rivlin [1], by virtue of discrete elementary charge in a beam of charged particles in vacuum and their particle interaction, it is possible to compare such beams to semicrystalline structures. The semicrystalline beams of charged particles, unlike usual structures, are stable only in motion and are not electrically neutral.

First we construct a theoretical model and investigate the processes of excitation and development of nonlinear waves of density charge in SBCP. Let the beam with a current be distributed in a vacuum in a strong magnetic field, so that the cross section of beam $d_b << a = ev/I$, where a is the average longitudinal distance between neighboring charges, e is the charge of a particle, v is the speed along the axis z, and length of beam $L >> a$. Such a thin beam is a kind of linear discrete electronic chain, resembling known models of one-dimensional crystals. Under the action of longitudinal Q-forces, there is the tendency to ordering, i.e., to limiting of potential energy of a chain to a minimum. Thermal movement hinders this tendency, imposing on temperature T of a beam the known restriction [1]

$$\left|\frac{z}{a}\right| \approx \left[\frac{ak_bT_b}{4.8e^2}\right]^{1/2} = 0.46\left[\frac{ak_bT_b}{r_omc^2}\right] << 1 \quad , \tag{1}$$

where z is the longitudinal deviation from balance, a is the average longitudinal distance between particles of a chain, k is the constant of B, m and r are mass and classical radii of electron, and e is the speed of light. The strict consideration of fluctuations in nonlinear chains started to be seriously studied in the beginning of 1950, when Fermi, Pasta, and Ulam (FPU) numerically studied a problem of equipartition of energy [2].

The estimates that were made in [3] show the possibility of effective laser cooling of relativistic ion beams in the storage rings of modern accelerators. The high-intensity and low-emittance ion beams in turn can be used in fundamental and applied physics—for example, in elementary particle physics, for ion thermonuclear fission, free-ion lasers, quantum generators on moving ions, and high-power hard quasi-monochromatic spontaneous incoherent backward Thomson scattering sources, as well as in electromagnetic and gravitational Free Particle lasers (FPLs) [3].

2. The Boussinesq equation for a semicrystalline beam of charged particles

We shall consider a beam of charged particles with charge e in a magnetic field directed along an axis of a beam. The amplitude of the magnetic field needed to make this beam one-dimensional, i.e., $d/a \ll 1$ (where d is the diameter of a beam, and a is the average distance between particles in a beam) is:

$$H \geq 3.1 \left[m_e c^2 a^{-3} \right]^{1/2} \tag{2}$$

Now, when we have found the conditions for existence of an SBCP, we shall write equations of motion of beam particles. We shall consider longitudinal movement along the z axis. To start, we shall take into account only forces acting between the nearest particles. Let us take a system of coordinates in the beam frame. In this system of coordinates, we have:

$$m \frac{d^2 Q_n}{dt^2} = F(Q_{n+1} - Q_n) - F(Q_n - Q_{n-1}) \quad , \tag{3}$$

where Q_n is displacement of the n-th particle, and $F(Q)$ is some function which includes, as usual, linear interaction as well as some small nonlinearity. For electrostatic interaction:

$$F(Q_{n+1} - Q_n) = -\frac{e^2}{\left(a + (Q_{n+1} - Q_n)\right)^2} \tag{4}$$

Assuming that $Q/a \ll 1$ in equation (4), it is possible to write:

$$F(Q_{n+1} - Q_n) \simeq -\frac{e^2}{a^2} \left\{ 1 - 2\left[\frac{Q_{n+1} - Q_n}{a}\right] + 3\left[\frac{Q_{n+1} - Q_n}{a}\right]^2 - \cdots \right\} \tag{5}$$

$$F(Q) = const + f(Q) \quad , \quad \text{where } f(Q) = \gamma Q + \alpha Q^2 \tag{6}$$

The dimensionless equation of motion of the n-th particle can be recorded as:

$$\frac{d^2 Q_n}{d\tau^2} = f(Q_{n+1} - Q_n) - f(Q_n - Q_{n-1}) \tag{7}$$

where

$$f(Q) = \gamma Q + \alpha Q^2 \quad , \quad \gamma = \frac{2e^2 t_o^2}{m_e a^3} \quad , \quad \alpha = -\frac{3e^2 t_o^2}{m_e a^3}\left[\frac{a_o}{a}\right] \tag{8}$$

It is obvious that the mathematical problem of the evolution of linear SBCP is similar to that in FPU [2]. In our case, we have a system with square-law nonlinearity, with a very large ($N = L/a \gg 100$, $L = \upsilon \tau_b$ is length a beam, τ_b its duration) number of particles. Therefore, for the description of nonlinear dynamics of SBCP, it is necessary to use continuum approximation [5].

$$\frac{d^2 Q_n}{d\tau^2} = \gamma\left[(Q_{n+1} - Q_n) - (Q_n - Q_{n-1})\right] + \alpha\left[(Q_{n+1} - Q_n)^2 - (Q_n - Q_{n-1})^2\right] \tag{9}$$

$$\frac{d^2 Q_n}{d\tau^2} = \gamma(Q_{n+1} + Q_{n-1} - 2Q_n) + \alpha(Q_{n+1} - Q_{n-1})(Q_{n+1} + Q_{n-1} - 2Q_n) \tag{10}$$

$$\frac{d^2 Q_n}{d\tau^2} = \gamma(Q_{n+1} + Q_{n-1} - 2Q_n)\left[1 + \frac{\alpha}{\gamma}(Q_{n+1} - Q_{n-1})\right] \tag{11}$$

Continuum approximation replaces the discretic Q_n by continuous $Q = Q(x)$.

$$Q_{n+1} = Q_n \pm h\frac{\partial Q_n}{\partial x} + \frac{h^2}{2}\frac{\partial^2 Q_n}{\partial x^2} \pm \frac{h^3}{6}\frac{\partial^3 Q_n}{\partial x^3} + \frac{h^4}{24}\frac{\partial^4 Q_n}{\partial x^4} \pm \cdots \tag{12}$$

Substituting equation (12) in the equation of motion and taking into account the fourth order of on h, we get

$$c_o^{-2}\frac{\partial Q}{\partial \tau} = \left(1 + \beta\frac{\partial Q}{\partial x}\right)\frac{\partial^2 Q}{\partial x^2} + \frac{h^2}{12}\frac{\partial^4 Q}{\partial x^4} \tag{13}$$

where $c_o = h\gamma^{1/2}$, $\beta = 2\alpha h/\gamma$.

Making the replacement $U = \partial Q/\partial x$, equation (13) can be rewritten as follows:

$$c_o^{-2}\frac{\partial^2 U}{\partial \tau^2} = \frac{\partial^2}{\partial x^2}\left(U + \frac{\beta}{2}U^2 + \frac{h^2}{12}\frac{\partial^2 U}{\partial x^2}\right) \tag{14}$$

This equation is known as the Boussinesq equation and describes waves that can moved to the left or to the right. Its soliton solution is:

$$U = U_o \operatorname{sech}^2\left(\frac{(x-ct)}{\Delta}\right) \tag{15}$$

$$c^2 = c_o^2\left(1 + \frac{\beta U_o}{3}\right) \quad , \quad \Delta^2 = \frac{h^2}{\beta U_o} \tag{16}$$

3. The KdV equation for a system SBCP

The structure of the Boussinesq equation looks like that of the KdV equation, and it is possible to expect that it will be reduced to a KdV equation for waves, with motion only in one direction. Initial disturbance will be seen only in one direction along a beam. We shall use scale transformation variables x, t, u. First, we will enter the small parameter e and choose new space and temporal variables:

$$\eta = e^p(x - c\tau) \quad , \quad \hat{\tau} = e^q \tau \tag{17}$$

Then we will expand U as:

$$U = eU^{(1)} + e^2 U^{(2)} + \cdots \tag{18}$$

Taking all members with the fourth derivative, deriving one at a time, as well as nonlinear members of the same order, put: $c = c_o$, $p = 1/2$, $q = 3/2$. Thus, the equation for U is the KdV equation.

$$\frac{\partial U}{\partial \hat{\tau}} + \frac{h^2 c_o}{24} \frac{\partial^3 U}{\partial \eta^3} - \frac{3c_o}{2} U \frac{\partial U}{\partial \eta} = 0 \tag{19}$$

Its soliton solution is:

$$U = U_o \operatorname{sech}^2\left(\frac{(\eta - v\hat{\tau})}{\Delta}\right) \quad , \quad \text{where} \quad v = \frac{U_o c_o}{2} \quad , \quad \Delta^2 = \frac{h^2}{3U_o} \tag{20}$$

4. The case of multipartial interactions

For Q force to essentially take into account influence not only of the nearest particles, we write an equation of motion in which we consider the action of all particle systems:

$$\frac{d^2 Q_n}{d\tau^2} = f(Q_{n+k} - Q_n) + \cdots + f(Q_{n+1} - Q_n) - f(Q_n - Q_{n-1}) - \cdots - f(Q_n - Q_{n-k}) \tag{21}$$

$$\frac{d^2 Q_n}{d\tau^2} = \{f(Q_{n+1} - Q_n) - f(Q_n - Q_{n-1})\} + \cdots + \{f(Q_{n+k} - Q_n) - f(Q_n - Q_{n-k})\}$$

The difference in force, acting on the part of the subsequent pairs, depends on the distance as:

$$\gamma = \frac{\gamma_o}{k^3} \quad , \quad \alpha = \frac{\alpha_o}{k^4} \tag{22}$$

where $k = 1,2,3...$

Therefore, equation (2.14) can be replaced as follows:

$$\frac{d^2 Q_n}{d\tau^2} = \hat{f}(Q_{n+1} - Q_n) - \hat{f}(Q_n - Q_{n-1}) \tag{23}$$

where:

$$\hat{f}(Q) = \hat{\gamma}Q + \hat{\alpha}Q^2 \; ; \; \hat{\gamma} = \gamma_o \sum_{k=1}^{k=\infty} \frac{1}{k^3} \; ; \; \hat{\alpha} = \alpha_o \sum_{k=1}^{k=\infty} \frac{1}{k^4} \tag{24}$$

$$\hat{\gamma} = 2.36 \frac{e^2 t_o^2}{m_e a^3} \; ; \; \hat{\alpha} = -3.24 \frac{e^2 t_o^2}{m_e a^3} \left(\frac{a_o}{a}\right) \tag{25}$$

Account of multipartial interaction results in an increase of speed of distribution $c \approx 1.1 c_0$.

5. Conclusions

The Boussinesq and Korteveg-de-Vries equations describe the nonlinear dynamics of a semicrystalline beam of charge particles in continuum approximation. The analysis of its soliton solutions is completed.

It is established that the SBCP possesses a positive dispersion. The obtained equations are generalized in classical approximation in the case of multiparticle Coulomb interaction. In this case, the velocity of disturbances increases.

It is easy to show that nonlinear disturbances in SBCP can modify the spectrum and intensity of electromagnetic radiation as a result of the interaction of SBCP with laser beams. We can therefore diagnose nonlinear waves of charge density by using the characteristics of such radiation.

A. Equation of beam electrical fields

$$E_p = \sum_{v=1}^{\infty} \left(-\frac{e\rho}{\left((va-z)^2 + \rho^2\right)^{3/2}} - \frac{e\rho}{\left((va+z)^2 + \rho^2\right)^{3/2}} \right) =$$

$$-\frac{e}{a^2} \frac{\rho}{a} \sum_{v=1}^{\infty} v^{-3} \left\{ \left[\left(1 - \frac{1}{v}\frac{z}{a}\right)^2 + v^{-2}\left(\frac{\rho}{a}\right)^2\right]^{-3/2} + \left[\left(1 + \frac{1}{v}\frac{z}{a}\right)^2 + v^{-2}\left(\frac{\rho}{a}\right)^2\right]^{-3/2} \right\}$$

Expand electrical field E_p around point $z = 0$ as

$$E_\rho \approx -\frac{2e}{a^2}\frac{\rho}{a}\sum_{v=1}^{\infty} v^{-3}\left[1 - \frac{3}{2}v^{-2}\left(\frac{\rho}{a}\right)^2\right] = -\frac{2e}{a^2}\frac{\rho}{a}\left[\sum_{v=1}^{\infty} v^{-3} - \frac{3}{2}\sum_{v=1}^{\infty} v^{-5}\left(\frac{\rho}{a}\right)^2\right] \approx$$

$$-2.4\frac{e}{a^2}\frac{\rho}{a}\left[1 - 1.3\left(\frac{\rho}{a}\right)^2\right]$$

B. Deduction of the Boussinesq equation

$$\frac{d^2 Q}{d\tau^2} = \gamma\left(h^2 \frac{\partial^2 Q}{\partial x^2} + \frac{h^4}{12}\frac{\partial^4 Q}{\partial x^4}\right)\left(1 + \frac{\alpha}{\gamma}\left(2h\frac{\partial Q}{\partial x} + \frac{h^3}{3}\frac{\partial^3 Q}{\partial x^3}\right)\right)$$

$$\frac{d^2 Q}{d\tau^2} = \gamma h^2 \frac{\partial^2 Q}{\partial x^2} + \frac{\gamma h^4}{12}\frac{\partial^4 Q}{\partial x^4} + 2\alpha h^3 \frac{\partial Q}{\partial x}\frac{\partial^3 Q}{\partial x^3}$$

$$c_o^{-2}\frac{d^2 Q}{d\tau^2} = \left(1 + \beta\frac{\partial Q}{\partial x}\right)\frac{\partial^2 Q}{\partial x^2} + \frac{h^2}{12}\frac{\partial^4 Q}{\partial x^4}$$

$$c_o = h\sqrt{\gamma} \quad , \quad \beta = \frac{2\alpha h}{\gamma} = -3 \; .$$

Introducing $U = \dfrac{\partial Q}{\partial x}$ we can find that

$$c_0^{-2}\frac{\partial^2 U}{\partial \tau^2} = \frac{\partial^2}{\partial x^2}\left(U + \frac{\beta}{2}U^2 + \frac{h^2}{12}\frac{\partial^2 U}{\partial x^2}\right)$$

Let us seek a solution for $U = U(x - c\tau)$.

$$c_0^{-2}c^2 U'' = U'' + \frac{\beta}{2}(U^2)'' + \frac{h^2}{12}U''''$$

Integrate twice and take constants of integration equal to zero.

$$\left(\frac{c^2}{c_0^2} - 1\right)U = \frac{\beta}{2}U^2 + \frac{h^2}{12}U''$$

$$U = U_0 \operatorname{sech}^2\left(\frac{x - c\tau}{\Delta}\right)$$

$$\left(\frac{c^2}{c_0^2} - 1\right)U_0 \operatorname{sech}^2 y = \frac{\beta}{2}U_0^2 \operatorname{sech}^2 y + \frac{h^2}{12}\left(-\frac{2U_0}{\Delta^2}\right)(-2\operatorname{sech}^2 y + 3\operatorname{sech}^4 y)$$

$$\left(\frac{c^2}{c_0^2} - 1\right)U_0 = \frac{4U_0 h^2}{12\Delta^2} \quad \Rightarrow \quad c^2 = c_0^2\left(1 + \frac{h2}{3\Delta^2}\right)$$

$$\frac{\beta U_0^2}{2} = \frac{h^2}{2}\frac{U_0}{\Delta^2} \Rightarrow \Delta^2 = \frac{h^2}{\beta U_0}$$

$$c^2 = c_0^2\left(1 + \frac{\beta U_0}{3}\right)$$

$$\beta = \frac{2\alpha h}{\gamma} = \frac{-6e^2 t_0^2}{m_e a^3}\frac{a_0}{a}\frac{a}{a_0}\frac{m_e a^3}{2e^2 t_0^2} = -3$$

$$c^2 = c_0^2(1 - U_0)$$

$$\Delta^2 = \frac{h^2}{\beta U_0} = -\frac{h^2}{3U_0}$$

C. Deduction of the KdV equation

$$c_0^{-2}\frac{\partial^2 U}{\partial \tau^2} = \frac{\partial^2}{\partial x^2}\left(U + \frac{\beta}{2}U^2 + \frac{h^2}{12}\frac{\partial^2 U}{\partial x^2}\right)$$

$$\eta = e^p(x - c\tau) \quad \hat{\tau} = e^q \tau$$

$$\frac{\partial^2}{\partial \tau^2} = c^2 e^{2p}\frac{\partial^2}{\partial \eta^2} + e^{2q}\frac{\partial^2}{\partial \hat{\tau}^2} - 2ce^{p+q}\frac{\partial^2}{\partial \eta \partial \hat{\tau}}$$

$$\frac{\partial^2}{\partial x^2} = e^{2p}\frac{\partial^2}{\partial \eta^2}$$

$$U = eU^{(1)} + e^2 U^{(2)} + \cdots$$

$$c^2 c_0^{-2} e^{2p+1}\frac{\partial^2 U^{(1)}}{\partial \eta^2} + c_0^{-2} e^{2q+1}\frac{\partial^2 U^{(1)}}{\partial \hat{\tau}^2} - 2cc_0^{-2} e^{p+q+1}\frac{\partial^2 U^{(1)}}{\partial \eta \partial \hat{\tau}} =$$

$$= e^{2p}\frac{\partial^2}{\partial \eta^2}\left(eU^{(1)} + \frac{\beta}{2}e^2\left(U^{(1)}\right)^2 + \frac{h^2}{12}e^{2p+1}\frac{\partial^2 U^{(1)}}{\partial \eta^2}\right)$$

$$c^2 c_0^{-2} = 1 \Rightarrow c = c_0$$

$$2p + 2 = p + q + 1, \; 4p + 1 = 2p + 2 \Rightarrow p = \tfrac{1}{2}, q = \tfrac{3}{2}$$

$$\frac{2}{c_0}\frac{\partial U^{(1)}}{\partial \hat{\tau}} + \frac{h^2}{12}\frac{\partial^3 U^{(1)}}{\partial \eta^3} + \beta U^{(1)}\frac{\partial U^{(1)}}{\partial \eta} = 0$$

$$\frac{\partial U^{(1)}}{\partial \hat{\tau}} + \frac{c_0 h^2}{24}\frac{\partial^3 U^{(1)}}{\partial \eta^3} + \frac{\beta c_0}{2}U^{(1)}\frac{\partial U^{(1)}}{\partial \eta} = 0$$

$$\frac{\partial U^{(1)}}{\partial \hat{\tau}} + \frac{c_0 h^2}{24}\frac{\partial^3 U^{(1)}}{\partial \eta^3} + \frac{3}{2}c_0 U^{(1)}\frac{\partial U^{(1)}}{\partial \eta} = 0$$

$$U = U(\eta - v\tau)$$

$$-vU' + \frac{c_0 h^2}{24}U''' - \frac{3c_0}{2}UU' = 0$$

$$\frac{c_0 h^2}{24}U'' = vU + \frac{3c_0 U^2}{4}$$

$$U = U_0 \operatorname{sech}^2\left(\frac{\eta - v\tau}{\Delta}\right)$$

$$\frac{c_0 h^2}{24\Delta^2}\left(4U_0 \operatorname{sech}^2 y - 6U_0 \operatorname{sech}^4 y\right) = vU_0 \operatorname{sech}^2 y + \frac{3c_0 U_0^2}{4}\operatorname{sech}^4 y$$

$$\frac{c_0 h^2 U_0}{6\Delta^2} = vU_0 \quad \Rightarrow \quad v = \frac{c_0 h^2}{6\Delta^2}$$

$$\Delta^2 = -\frac{h^2}{3U_0} \quad \Rightarrow \quad v = -\frac{c_0 U_0}{2}$$

References

[1] L.A. Rivlin. *Collective effects in discrete-electrons beams.* Pis'ma in Zh.E.T.F., 1971, V. 13, pp. 362–365.
[2} E. Fermi. *Scientific proceedings.* V. 2.M.:"Nauka" 1971.
[3] E.G. Bessonov. *Free-ion lasers with radiative ion cooling.* Journal of Russian Laser Research, 1994, v. 15, N. 5 pp. 403–416.
[4] R. Dodd et al. *Solitons and nonlinear wave equations.* M.: "Mir" 1988.
[5] M. Toda. *The theory of non-linear lattice.* M.:"Mir" 1984.

Nonlinear Interactions between Relativistic Electrons and Ultrahigh Intensity Laser Pulses in Vacuum

F.V. Hartemann*, A.L. Troha, J.R. Van Meter and N.C. Luhmann, Jr.
Department of Applied Science, University of California, Davis, CA 95616

A.K. Kerman
Physics Department and Center for Theoretical Physics,
Massachusetts Institute of Technology, Cambridge, MA 02139

T.S. Chu
*Nonlinear Technologies, Oakland, CA 94602

Abstract. Different nonlinear interactions between relativistic electrons and ultrahigh intensity laser pulses in vacuum are considered. We first briefly review the wave equation in vacuum to describe the three-dimensional laser field distribution at focus, including longitudinal field components, and discuss the accuracy of the paraxial wave equation and its Gaussian spherical solutions. The effect of radiation pressure on the Doppler-shifted radiation scattered by the electrons is then discussed for plane waves, and the corresponding nonlinear spectra are presented. We also introduce the concept of temporal laser beam shaping, and show how it can alleviate the nonlinear Doppler shift problem. At high energies, radiation reaction plays an important role in the electron dynamics, and is briefly discussed within the context of the classical Dirac-Lorentz equation.

1. INTRODUCTION

The nonlinear interaction between relativistic electrons and ultrahigh intensity lasers is a subject of considerable theoretical and experimental interest, ranging from vacuum laser acceleration, to ultrahigh intensity Compton backscattering. For example, the generation of tunable, focused X-ray pulses via ultrahigh intensity Compton backscattering has numerous potential applications ranging from biology and medicine to basic research and microchip technology. The practical realization of such a revolutionary light source, however, has been hampered by three serious problems: the lack of ultrahigh intensity sources at optical wavelengths, the absence of focusing optics in the X-ray range, and the nonlinear Doppler shift associated with ultrahigh intensity Compton scattering, which distributes the X-ray energy over many spectral lines [1]. The first issue has now been resolved through the pioneering of laser chirped pulse amplification (CPA) [2]. The remaining issues can be addressed by two novel ideas: electron beam lensing of the X-rays, as proposed for the future γ–γ collider, and spectral filtering at the Fourier plane of a CPA laser to generate flat-top laser pulses with constant Doppler shift during the interaction. Combined together, these ideas alleviate the major potential problems for the production of focused X-rays. Such devices would have a wide range of applications in biology and medicine including cancer research and biomolecular imaging.

An accurate description of the three-dimensional focus of a laser wave both in the near-field and far-field regions is required to properly describe the interaction of the electromagnetic field with charged leptons. In particular, the validity of the paraxial ray approximation [3], when used to model problems involving relativistic electrons co-propagating with a laser wave over many Rayleigh ranges, must be firmly established. For applications involving ultrahigh intensity [1] and nonlinear [4] Compton scattering, such as the γ–γ collider or focused X-ray sources, detailed knowledge of the three-dimensional electromagnetic field distribution in the focal region is of paramount importance, and the axial component of the fields may play a major role in the electron dynamics.

In recent years, considerable interest has been given to the detailed properties of laser focusing, partly because of potential novel applications such as plasma [5] and vacuum-based laser acceleration schemes [6]. For example, supergaussian rings have been thoroughly studied in an analysis [7] which shares some similarities with our work. More in line with our motivation, the effect of the ponderomotive potential [8] associated with an ultrahigh intensity laser wave on the radial confinement of relativistic electrons co-propagating with the pulse has been investigated by Moore [9]. This analysis indicates that higher-order Gaussian modes can indeed confine the electrons through the focus, because of the inward radiation pressure gradient. In this particular case, an exact three-dimensional field distribution, satisfying both the vacuum wave equation and the gauge condition, is needed to conclusively demonstrate the validity of this approach. In Section 2, we study exact solutions to the wave equation in vacuum for a three-dimensional laser focus in both rectangular and cylindrical geometries. In rectangular coordinates, the electromagnetic field is Fourier transformed into transverse and longitudinal wavepackets, and diffraction is described by the different phase shifts accumulated by the various Fourier components, as constrained by the dispersion relation. In the case of cylindrical geometry, the starting point of our derivation is the expansion of the 4-vector potential in terms of the well-known transverse electric (TE) and magnetic (TM) eigenmodes of the cylindrical wave equation. Within this context, the axial and temporal dependence of the propagating wave is represented by a double Fourier transform, while its azimuthal dependence corresponds to harmonic functions. To define the boundary conditions for this problem, the beam profile can be matched to a Gaussian-Hermite distribution at focus where the wavefront is planar. Because the radial boundary condition extends out to infinity, a continuous spectrum of Bessel functions is required, in contrast with the case of a cylindrical waveguide structure, where the finite radial boundary yields a discrete spectrum of eigenmodes [10]. The radial dependence of the focusing wave is thus described as a continuous spectrum of Bessel functions, and is obtained by using Hankel's integral theorem [11].

The coherence and spectral characteristics of Compton backscattered radiation are very strong functions of the electron dynamics during the interaction because of the relativistic Doppler effect. In particular, it has been predicted that at ultrahigh intensities, the laser radiation pressure modulates the electron axial velocity, resulting in a inhomogeneous Doppler shift, yielding frequency modulation effects [1] and nonlinear spectra. The detailed study of these nonlinear effects is strongly motivated by the recent advent of extremely high power, ultrafast lasers using fiber compression, Kerr-lens modelocking, and, more recently, chirped pulse amplification (CPA) [2], which makes it possible to study Compton scattering experimentally at ultrahigh intensities, where the normalized vector potential associated with the laser wave exceeds unity. For a wavelength of 1 $\mu\mu$, this translates into a focused intensity larger than 0.055 TW/mm^2. This type of effect is studied here by considering the covariant dynamics of an electron subjected to a classical electromagnetic field of arbitrary strength. The framework of classical electrodynamics should be approximately valid for an intense laser when the number of photons scattered by the electron is sufficiently high and the coherence of the laser allows the indefinite number of photons to be treated as a classical, continuous electromagnetic field.

Starting from general considerations on the electron dynamics *in vacuo*, including the conservation of the transverse and axial canonical momenta, and the Lorentz invariance of the electron phase, an expression for the Compton backscattered radiation spectrum is derived for arbitrary laser field intensities, in Section 3. This derivation indicates that, in general, nonlinear spectra are obtained at very high laser intensities, where the scattered light is distributed over a large number of spectral lines, even in the case of circular polarization. To alleviate this problem, we show how spectral filtering [12] can be used at the Fourier plane of the chirped pulse laser amplifier to optimize both the temporal pulse shape and the spectrum of the backscattered radiation by increasing the contrast ratio between the transient lines radiated during the rise and fall of the laser pulse, and the main spectral line radiated during the flat-top part of the optimized laser pulse.

Finally, the problem of radiation reaction effects, which becomes important at high energies, is discussed in Section 4. The electron self-interaction problem is central to the foundations of both classical and quantum electrodynamics. In the study of classical electrodynamics at high field strengths, the Dirac-Lorentz equation [13] describes the covariant dynamics of a point charge, including radiative corrections representing the recoil momentum of the fields interacting with the particle. This effect is assumed to be equivalent to a reaction force connected to the self-interaction of the charge with its electromagnetic field. Although the quantum electrodynamical nature of the electron-photon interaction must be taken into account for a full description of such phenomena, it is hoped that a large class of interactions may be appropriately studied within the context of high field strength classical electrodynamics. In addition, a thorough understanding of that topic is required for a comprehensive approach to high field QED. A number of conceptual problems arise within the classical framework, including mass renormalization [13,14], runaway solutions and preacceleration or acausal effects, and must be carefully addressed. In QED, the Dirac equation describes the temporal evolution of the wavefunction of a relativistic spin 1/2 particle. At high field strengths, the Dirac-Coulomb problem can be solved exactly, owing in part to the hidden supersymmetry of this problem, but a general treatment of QED in time-dependent external fields remains to be defined. In particular, multiphoton (nonlinear) Compton scattering has not yet been fully described in terms of the Dirac equation and the classical relativistic particle limit remains elusive. Finally, one might quote Dirac's comment concerning the electron self-interaction: "...it seems more reasonable to suppose that the electron is too simple a thing for the question of the laws governing its structure to arise, and thus quantum mechanics should not be needed for the solution of the difficulty" [14].

2. THREE-DIMENSIONAL FIELD DISTRIBUTION IN A LASER FOCUS

In vacuum, the wave equation takes the form

$$\left[\nabla^2 - \frac{1}{c^2}\partial_t^2\right]A_\mu = \left[\partial_\mu \partial^\mu\right]A_\mu = 0 . \tag{1}$$

It is well known that a general solution to the vacuum wave equation can be constructed as a Fourier superposition of wavepackets of the form

$$A_\mu(x_\nu) = \frac{1}{(2\pi)^2} \iiint \tilde{A}_\mu(k_\nu) \exp(i\, k_\nu\, x^\nu)\, d^4k_\nu , \tag{2}$$

where the 4-wavenumber $k_\mu \equiv \left(\frac{\omega}{c}, \mathbf{k}\right)$ satisfies the vacuum dispersion relation $\frac{\omega^2}{c^2} - \mathbf{k}^2 = k_\mu k^\mu = 0$.

In the case where the laser pulse characteristics are defined at focus ($z = 0$), we can obtain the electromagnetic field distribution on any given z-plane by performing the following integral

$$A_\mu(x,y,z,t) = \frac{1}{(2\pi)^{3/2}} \iiint \tilde{A}_\mu(\mathbf{k}_\perp, \omega, z=0) \exp\left[i\left(\omega t - k_x x - k_y y - \sqrt{\frac{\omega^2}{c^2} - \mathbf{k}_\perp^2}\, z\right)\right] d^2\mathbf{k}_\perp d\omega. \tag{3}$$

The physics of this solution can be understood as follows: the temporal evolution of each wavepacket is described by the frequency spectrum, while the transverse profile of the laser wave is described by an integral over a continuous spectrum of transverse vacuum eigenmodes. The dispersion relation indicates how each transverse and temporal component of the wavepacket propagates, thus yielding wavefront curvature and transverse spreading (diffraction) of the wavepacket. It should also be noted that the axial wavenumber can become purely imaginary, in which case the corresponding waves become evanescent modes.

We now turn our attention to the specific case of a linearly polarized Gaussian laser pulse with a Gaussian focal distribution. This example can then be used to derive the paraxial ray approximation [3], and to compare the results obtained in both cases. The transverse component of the 4-vector is given by

$$A_x(x,y,z=0,t) = \Re\left\{A_0 \exp\left[i\omega_0 t - \left(\frac{t}{\Delta t}\right)^2 - \left(\frac{r}{w_0}\right)^2\right]\right\}, \quad (4)$$

where we have introduced the pulse duration Δt, the focal beam waist w_0, and the laser frequency ω_0. The radius is defined by $r^2 = x^2 + y^2$. Since A_x is a function of x, and since the 4-vector has no other transverse component, the gauge condition requires the existence of an axial component, such that $\partial_x A_x + \partial_z A_z = 0$. In Fourier space, this requirement yields

$$\tilde{A}_z = -k_x\left(\frac{\omega^2}{c^2} - \mathbf{k}_\perp^2\right)^{-1/2} \tilde{A}_x. \quad (5)$$

At this point, it is interesting to show that one can recover the so-called paraxial ray approximation by performing a Taylor expansion in the transverse wavenumber. We start by deriving the transverse wave spectrum at focus for the focal field distribution given in Eq. (4); in this case, we obtain

$$\tilde{A}_x(\mathbf{k}_\perp,\omega,z=0) = \frac{A_0 w_0^2 \Delta t}{2\sqrt{2}} \exp\left\{-\left[\frac{(\omega-\omega_0)\Delta t}{2}\right]^2\right\} \exp\left[-\left(\frac{w_0 \mathbf{k}_\perp}{2}\right)^2\right]. \quad (6)$$

The expression of the transverse field at any axial position is thus

$$A_x = \frac{A_0 w_0^2 \Delta t}{(4\pi)^{3/2}} \iiint \exp\left\{-\left(\frac{w_0 \mathbf{k}_\perp}{2}\right)^2 - \left[\frac{(\omega-\omega_0)\Delta t}{2}\right]^2 \right.$$
$$\left. + i\left(\omega t - \mathbf{k}_\perp \bullet \mathbf{x} - \sqrt{\frac{\omega^2}{c^2} - \mathbf{k}_\perp^2}\, z\right)\right\} d^2\mathbf{k}_\perp d\omega. \quad (7)$$

The integral over the transverse eigenwavenumber spectrum can easily be performed analytically if we Taylor expand the square root in the complex exponential. For small values of the transverse wavenumber, we have

$$\sqrt{\frac{\omega^2}{c^2} - k_\perp^2} = \frac{\omega}{c}\sqrt{1 - \left(\frac{ck_\perp}{\omega}\right)^2} \cong \frac{\omega}{c}\left[1 - \frac{1}{2}\left(\frac{ck_\perp}{\omega}\right)^2\right] = k\left[1 - \frac{1}{2}\left(\frac{ck_\perp}{\omega}\right)^2\right]. \quad (8)$$

This expansion corresponds exactly to the paraxial ray approximation. Indeed, we now have

$$A_x = \frac{A_0 w_0^2 \Delta t}{(4\pi)^{3/2}} \int_{-\infty}^{+\infty} \exp[i(\omega t - kz)] \exp\left\{-\left[\frac{(\omega - \omega_0)\Delta t}{2}\right]^2\right\} d\omega$$

$$\times \int_{-\infty}^{+\infty} \exp\left\{-\left(\frac{w_0 k_x}{2}\right)^2 - i\left(k_x x - \frac{1}{2}\left(\frac{k_x}{k}\right)^2 kz\right)\right\} dk_x \quad (9)$$

$$\int_{-\infty}^{+\infty} \exp\left\{-\left(\frac{w_0 k_y}{2}\right)^2 - i\left(k_y y - \frac{1}{2}\left(\frac{k_y}{k}\right)^2 kz\right)\right\} dk_y .$$

A general form for the integrals over k_x and k_y can now be found in reference [15]. In our particular case, they reduce to

$$\int_{-\infty}^{+\infty} e^{-ax^2} \exp[i(px^2 + 2qx)]dx = \frac{\sqrt{\pi}}{\sqrt[4]{a^2 + p^2}} \exp\left[-\frac{aq^2}{a^2 + p^2}\right]$$

$$\exp\left\{i\left[\frac{\tan^{-1}\left(\frac{p}{a}\right)}{2} - \frac{pq^2}{a^2 + p^2}\right]\right\}. \quad (10)$$

Identifying the different parameters with the physical quantities in Eq. (9), $a = \left(\frac{w_0}{2}\right)^2$, $p = -\frac{z}{2k}$, $q = -\frac{x}{2}$ or $-\frac{y}{2}$, we can now introduce the Rayleigh range $z_0 = \pi\frac{w_0^2}{\lambda} = \frac{1}{2}k w_0^2$, defined such that $\frac{p}{a} = \frac{z}{z_0}$.

Performing the k_x and k_y integrals, multiplying the results as prescribed in Eq. (9), and recalling that $r^2 = x^2 + y^2$, we obtain

$$A_x(x_\mu) = \frac{\Delta t}{\sqrt{4\pi}} \int_{-\infty}^{+\infty} \exp[i(\omega t - kz)] \exp\left\{-\left[\frac{(\omega-\omega_0)\Delta t}{2}\right]^2\right\} \frac{A_0}{\sqrt{1+\frac{z^2}{z_0^2}}}$$

(11)

$$\exp\left\{-\left[\frac{r}{w(z)}\right]^2\right\} \exp\left\{i\left[k\frac{r^2}{2R(z)} - \Psi(z)\right]\right\} d\omega ,$$

where we have defined the wavefront curvature $R(z) = z + \frac{z_0^2}{z}$, the variable beam waist $w(z) = w_0\sqrt{1+\left(\frac{z}{z_0}\right)^2}$, and the Guoy phase $\Psi(z) = \tan^{-1}\left(\frac{z}{z_0}\right)$.

Equation (11) corresponds exactly to the Gaussian spherical waves obtained by solving the paraxial wave equation [3]. Note that both the beam waist, wavefront curvature, and Guoy phase [3] are implicit functions of the frequency, since the Rayleigh range is inversely proportional to the wavelength. Therefore, when we describe an ultrashort laser pulse, we must include these variations to describe the physical curvature of the focusing pulse; as a result, the time-dependent field must be calculated by summing over the laser frequency spectrum. The physical content of the paraxial ray approximation is illustrated in Fig. 1. The Gaussian transverse wavenumber spectrum is shown for $w_0 k_0 = 20$, where $k_0 = 2\pi/\lambda_0$. Evanescent modes correspond to $k_\perp/k_0 > 1$. The axial wavenumber is also shown, both for the exact dispersion relation, and in the paraxial ray approximation. It is clear that for physically realizable foci, where the beam waist is significantly larger than the wavelength, the region of transverse wavenumber space where the paraxial phase differs significantly from the exact value corresponds to very small spectral amplitudes.

We can now derive the axial component of the 4-potential by using the gauge condition given in Eq. (5):

$$A_z(x_\mu) = \frac{1}{(2\pi)^{3/2}} \iiint \frac{-k_x}{\sqrt{\frac{\omega^2}{c^2} - k_\perp^2}} \tilde{A}_x(\mathbf{k}_\perp, \omega)$$

(12)

$$\exp\left[i\left(\omega t - \mathbf{k}_\perp \cdot \mathbf{x} - \sqrt{\frac{\omega^2}{c^2} - k_\perp^2} z\right)\right] d^2\mathbf{k}_\perp d\omega .$$

We first expand the square root according to Eq. (8), to obtain

$$\frac{k_x}{\sqrt{\frac{\omega^2}{c^2} - k_\perp^2}} \cong \frac{k_x}{k}\left[1 + \left(\frac{k_\perp}{k}\right)^2\right] .$$

(13)

Keeping only the linear and quadratic terms, the integral in Eq. (12) reduces to

$$A_z = -\frac{A_0 w_0^2 \Delta t}{(4\pi)^{3/2}} \int_{-\infty}^{+\infty} \exp[i(\omega t - kz)] \exp\left\{-\left[\frac{(\omega - \omega_0)\Delta t}{2}\right]^2\right\} d\omega$$

$$\times \int_{-\infty}^{+\infty} \exp\left\{-\left(\frac{w_0 k_x}{2}\right)^2 - i\left(k_x x - \frac{1}{2}\left(\frac{k_x}{k}\right)^2 kz\right)\right\} dk_x \quad (14)$$

$$\int_{-\infty}^{+\infty} \exp\left\{-\left(\frac{w_0 k_y}{2}\right)^2 - i\left(k_y y - \frac{1}{2}\left(\frac{k_y}{k}\right)^2 kz\right)\right\} dk_y .$$

The integral in k_y is given by Eq. (10), and the integral over k_x can be obtained by using the fact that $i\, d_x[e^{-ik_x x}] = k_x e^{-ik_x x}$. We can thus replace the integral over k_x by

$$\frac{i}{k} d_x \left[\int_{-\infty}^{+\infty} \exp\left\{-\left(\frac{w_0 k_x}{2}\right)^2 - i\left(k_x x - \frac{1}{2}\left(\frac{k_x}{k}\right)^2 kz\right)\right\} dk_z\right]; \quad (15)$$

because the other integrals are independent of x, we can now identify

$$A_z(x_\mu) = -\frac{i}{k} d_x A_x(x_\mu) . \quad (16)$$

This equation can be regarded as the paraxial approximation to the gauge condition. Explicitly taking the derivative of Eq. (11) with respect to x, we obtain the sought-after expression for the axial field component

$$A_z(x_\mu) = \left[\frac{2i}{k w^2(z)} + \frac{1}{R(z)}\right] x A_x(x_\mu) = \frac{x}{q(z)} A_x(x_\mu) . \quad (17)$$

We have thus derived a general solution to the wave equation in vacuum, which reduces to the well-known paraxial ray approximation in the limit of small transverse wavenumbers. In addition, the derivation of an analytical expression of the axial field component in the case of linear polarization, within the paraxial approximation, may prove quite useful to study the relativistic dynamics of electrons in ultrahigh intensity laser fields. In particular, this derivation can be extended to higher order Gaussian modes, which are believed to yield the particle confinement [9] required for vacuum laser acceleration applications [6]. A comparison between the exact solutions and Gaussian spherical waves, shown in Fig. 2 for $\lambda_0 = 1$ μm, $w_0 = 5$ μm, and $z/z_0 = 10$ demonstrates their excellent agreement.

3. NONLINEAR DOPPLER SHIFT

In this Section, we first briefly review the nonlinear dynamics of a relativistic electron subjected to the time-dependent field of a plane wave of arbitrary intensity propagating in vacuum. The electron normalized 4-velocity and 4-momentum are defined as

$$u_\mu = \frac{1}{c}\frac{dx_\mu}{d\tau} = \gamma(1,\beta) = (\gamma, \mathbf{u}) \quad , \quad p_\mu = m_0 c u_\mu,$$

where τ is the proper time along the electron world line, $x_\mu(\tau)$. The energy-momentum transfer equations are given by the Lorentz force

$$d_\tau u_\mu = -\frac{e}{m_0 c}\left(\partial_\mu A_\nu - \partial_\nu A_\mu\right) u^\nu, \tag{18}$$

where we have introduced the 4-vector potential of the laser wave

$$A_\mu = \left(\frac{\varphi}{c}, \mathbf{A}\right) \quad , \quad \mathbf{A} = \hat{x} A_x(\phi) + \hat{y} A_y(\phi) \quad , \quad \varphi = 0, \tag{19}$$

and defined the invariant phase of the traveling wave $\phi = \omega_0\left(t - \frac{z}{c}\right) = k^\mu x_\mu(\tau)$, as a function of the characteristic laser frequency ω_0. This form of the potential corresponds to a transverse plane wave propagating in vacuum.

Along the electron trajectory, we have the important relation: $d_\tau \phi = \omega_0(\gamma - u_z)$.

The energy-momentum transfer equations now read

$$d_\tau u_{x,y} = \omega_0 (\gamma - u_z) d_\phi a_{x,y}(\phi), \tag{20}$$

$$d_\tau u_z = d_\tau \gamma = \omega_0 \left[u_x d_\phi a_x(\phi) + u_y d_\phi a_y(\phi)\right], \tag{21}$$

where $\mathbf{a} = e \mathbf{A}/m_0 c$ is the invariant normalized vector potential of the laser wave. Equation (21) yields the well-known canonical momentum invariant $\gamma - u_z = \gamma_0(1 - \beta_0)$ [6]. An important consequence is that the electron phase and proper time are proportional to within a constant: $d\phi/d\tau = \omega_0 \gamma_0 (1 - \beta_0)$.

Equation (20) can now readily be integrated to obtain the transverse canonical invariant $u_{x,y}(\tau) = a_{x,y}(\phi)$

Using the conservation of canonical momentum, the normalized electron energy and axial momentum are derived, with the result that

$$u_z(\tau) = \gamma_0 \left[\beta_0 + \mathbf{a}^2(\phi)\left(\frac{1+\beta_0}{2}\right)\right], \tag{22}$$

$$\gamma(\tau) = \gamma_0 \left[1 + \mathbf{a}^2(\phi)\left(\frac{1+\beta_0}{2}\right) \right]. \qquad (23)$$

These results are quite general and hold as long as plane waves are considered [6]. An important difference between polarization states immediately appears: $\mathbf{a}^2(\phi)$ varies adiabatically as the pulse envelope for circular polarization, while there is an extra modulation at $2\omega_0$ for linear polarization. The transverse electron momentum depends linearly on the laser field, but the axial momentum modulation is a quadratic function of that field, as it results from the coupling of the transverse velocity to the laser magnetic field through the ponderomotive force. At low intensities, the radiation pressure of the laser pulse is negligible, and the electron basically propagates through the laser pulse at constant axial velocity, while its transverse momentum is modulated at the Doppler-shifted laser frequency. This results in the radiation of the well-known FEL spectral lines on-axis [16]. The situation is very different when the normalized vector potential of the laser wave exceeds unity: the strong modulation of the electron axial momentum results in a nonlinear Doppler shift which translates into frequency modulation effects.

To derive the spectrum of the scattered radiation, we shall need an expression of the electron's position as a function of its nonlinear phase, ϕ. We change variables:

$$d_\phi \mathbf{x} + \frac{d_t \mathbf{x}}{d_t \phi} = \frac{c\boldsymbol{\beta}}{\omega_0(1-\beta_z)} = \frac{c\gamma\boldsymbol{\beta}}{\omega_0\gamma(1-\beta_z)} = \frac{c\mathbf{u}}{\omega_0\gamma_0(1-\beta_0)}. \qquad (24)$$

Eq. (24) can be formally integrated to obtain the electron's axial position

$$z(\phi) = \frac{c}{\omega_0(1-\beta_0)} \int_{-\infty}^{\phi} \left[\beta_0 + \mathbf{a}^2(\psi)\left(\frac{1+\beta_0}{2}\right) \right] d\psi. \qquad (25)$$

The distribution of energy radiated per unit solid angle per unit frequency can be derived by considering the instantaneous radiated power, as described by the Larmor formula, and applying Parsival's theorem. Using the nonlinear electron phase ϕ as the independent variable, we obtain

$$\frac{d^2 I(\omega,\mathbf{n})}{d\omega\, d\Omega} = \frac{e^2\omega^2}{16\pi^3\varepsilon_0 c} \left| \int_{-\infty}^{+\infty} \frac{\mathbf{n}\times[\mathbf{n}\times\mathbf{u}(\phi)]}{\omega_0\gamma_0(1-\beta_0)} \exp\left[i\omega\left(\frac{\phi}{\omega_0} + \frac{z(\phi)-\mathbf{n}\bullet\mathbf{x}(\phi)}{c} \right) \right] d\phi \right|^2. \qquad (26)$$

The most interesting case is that of the radiation emitted on-axis, where most of the power is radiated, and where we obtain the maximum relativistic Doppler upshift. We first consider the forward scattered wave, which is the classical equivalent to stimulated emission. Equation (26) then reduces to

$$\frac{d^2I(\omega,\hat{z})}{d\omega\,d\Omega} = \frac{e^2}{16\pi^3\varepsilon_0 c}\left(\frac{1+\beta_0}{1-\beta_0}\right)\varpi^2\left|\int_{-\infty}^{+\infty}\mathbf{a}(\phi)\exp[i\varpi\phi]d\phi\right|^2, \qquad (27)$$

where we have introduced the normalized frequency $\varpi = \omega/\omega_0$.

Equation (27) shows that the spectrum of the forward scattered wave is always similar to that of the laser wave. This is due to the fact that the relativistic Doppler shift on the laser frequency in the electron frame is always exactly compensated by the opposite shift on the forward scattered radiation. For example, in the case of a linearly polarized Gaussian wavepacket of normalized width $\Delta\phi$,

$$\mathbf{a}(\phi) = \hat{x}\,a_0\exp\left[-\left(\frac{\phi}{\Delta\phi}\right)^2\right]\exp(-i\phi), \qquad (28)$$

the Fourier transform can be evaluated analytically [15], and we find

$$\frac{d^2I(\omega,\hat{z})}{d\omega\,d\Omega} = \frac{e^2 a_0^2 \Delta\phi^2}{16\pi^3\varepsilon_0 c}\left(\frac{1+\beta_0}{1-\beta_0}\right)\varpi^2\exp\left[-\frac{\Delta\phi^2}{2}(\varpi-1)^2\right], \qquad (29)$$

which describes a quadratic-Gaussian spectrum centered around the normalized frequency $\varpi = 1$.

For the backscattered radiation spectrum, we use the result concerning the electron's axial position [Equation (25)], and recast the expression obtained so that its properties under Lorentz transformation are manifest:

$$\frac{d^2I(\omega,-\hat{z})}{d\omega\,d\Omega} = \frac{e^2}{16\pi^3\varepsilon_0 c}\left(\frac{1-\beta_0}{1+\beta_0}\right)\chi^2\left|\int_{-\infty}^{+\infty}\mathbf{a}(\phi)\exp\left\{i\chi\left[\phi + \int_{-\infty}^{\phi}\mathbf{a}^2(\psi)d\psi\right]\right\}d\phi\right|^2. \qquad (30)$$

Here, $\chi = \frac{\omega}{\omega_0}\left(\frac{1+\beta_0}{1-\beta_0}\right)$ is the normalized Doppler-shifted frequency.

Before studying the nonlinear spectrum, it is interesting to briefly review the case of a Gaussian wavepacket [Equation (28)] of small amplitude, where one can neglect $\mathbf{a}^2$ in the argument of the exponential. Equation (30) can then be integrated analytically [15]:

$$\frac{d^2I(\omega,-\hat{z})}{d\omega\,d\Omega} = \frac{e^2 a_0^2 \Delta\phi^2}{16\pi^3\varepsilon_0 c}\left(\frac{1+\beta_0}{1-\beta_0}\right)\varpi^2\exp\left[-\frac{\Delta\phi^2}{2}\left[\varpi\left(\frac{1+\beta_0}{1-\beta_0}\right)-1\right]^2\right]. \qquad (31)$$

We obtain the usual Doppler-shifted Gaussian radiation spectrum centered around the normalized frequency $\varpi = \left(\frac{1-\beta_0}{1+\beta_0}\right)$. In the FEL case, $\beta_0 \to -1$, and we recover the well-known formula $\varpi \approx 4\gamma_0^2$ [16], for an electromagnetic wiggler.

We note that the duration of the backscattered pulse is also Doppler compressed, because $\Delta\phi$, which is related to the number of optical oscillations in the pulse, is also a relativistic invariant.

We now focus on the nonlinear effects induced by the variation of the electron Doppler factor along its trajectory. As seen in Eq. (30), the functional dependence of the spectrum is now independent from β_0, which only sets the frequency scale. This fact is not surprising, as it results directly from relativistic invariance: by changing the reference frame in which the scattering process is viewed, one can vary the sign of β_0 and continuously go from the FEL [16] geometry to the laser acceleration [5,6] geometry. For the FEL, the laser frequency is Doppler-upshifted in the electron frame, while it is downshifted in the second case. In both cases, the normalized vector potential and the average photon number are conserved as they are Lorentz invariants. In particular, one can choose a frame where the electron is initially at rest.

We now consider the case of circular polarization, where we have $\mathbf{a}(\phi) = g(\phi)[\hat{x}\sin\phi + \hat{y}\cos\phi]$, and $\mathbf{a}^2(\phi) = g^2(\phi)$. Here, we choose a simple physical model of the pulse envelope, namely a hyperbolic secant

$$g(\phi) = a_0 \cosh^{-1}\left(\frac{\phi}{\Delta\phi}\right). \tag{32}$$

The electron's axial position can then be determined analytically [15], and we have

$$\int_{-\infty}^{\phi} \mathbf{a}^2(\psi)d\psi = \int_{-\infty}^{\phi} \frac{a_0^2 d\psi}{\cosh^2(\psi/\Delta\phi)} = a_0^2 \Delta\phi\left[1 + \tanh\left(\frac{\phi}{\Delta\phi}\right)\right]. \tag{33}$$

The nonlinear backscattered spectrum is now proportional to

$$\chi^2 \left| a_0 e^{i\chi a_0^2 \Delta\phi} \int_{-\infty}^{+\infty} \frac{\hat{x}\sin\phi + \hat{y}\cos\phi}{\cosh(\phi/\Delta\phi)} \exp\left\{i\chi\Delta\phi\left[\frac{\phi}{\Delta\phi} + a_0^2 \tanh\left(\frac{\phi}{\Delta\phi}\right)\right]\right\} d\phi \right|^2. \tag{34}$$

To evaluate this Fourier transform, we make a first change of variable, and introduce $y = e^{\phi/\Delta\phi}$. The integral in Eq. (34) now reads

$$\Delta\phi \int_0^{+\infty} \frac{y^{i\Delta\phi(\chi\pm 1)}}{y^2+1}(\hat{y} \mp i\hat{x}) \exp\left[i a_0^2 \chi\Delta\phi\left(\frac{y^2-1}{y^2+1}\right)\right] dy, \tag{35}$$

where the plus and minus signs simply indicate the different contributions of the sine and cosine functions to the argument of the exponential. We now make a second change of variable, namely we let $x = (y^2 - 1)/(y^2 + 1)$. Equation (35) reduces to

$$\int_{-1}^{+1}(1+x)^{-\frac{1}{2}-\frac{1}{2}\Delta\phi(\chi\pm1)}(1-x)^{-\frac{1}{2}+\frac{1}{2}\Delta\phi(\chi\pm1)}(\hat{y}\mp i\hat{x})\exp\left[ia_0^2\chi\Delta\phi x\right]dx \,, \quad (36)$$

which has an exact analytical expression [15]. The expression for the nonlinear Compton backscattered spectrum can then be given in terms of B, the beta function (Euler's integral of the first kind), and Φ, the degenerate (confluent) hypergeometric function [18]. Using the properties of the beta function, and defining $\mu\pm = \frac{1}{2}[1 + i\Delta\phi(\chi\pm 1)]$, the nonlinear spectrum further reduces to

$$\frac{d^2 I(\omega,-\hat{z})}{d\omega\,d\Omega} = \frac{e^2 a_0^2 \Delta\phi^2}{8\pi\varepsilon_0 c}\left(\frac{1-\beta_0}{1+\beta_0}\right)\chi^2 \left\{ \frac{\left|\Phi(\mu_-,1,2ia_0^2\chi\Delta\phi)\right|^2}{\left|\cosh\left[\frac{\pi}{2}\Delta\phi(\chi-1)\right]\right|} \right.$$

$$\left. + \frac{\left|\Phi(\mu_+,1,2ia_0^2\chi\Delta\phi)\right|^2}{\left|\cosh\left[\frac{\pi}{2}\Delta\phi(\chi+1)\right]\right|} \right\}. \quad (37)$$

At low intensity ($a_0 \ll 1$), $\Phi \sim 1$, and we recover the Doppler-shifted $\cosh^{-2}$ spectral line.

Although the derivation presented here is purely classical, it is interesting to translate Equation (37) into an average number of photon radiated per unit frequency per unit solid angle by dividing this equation by $\hbar\omega$. The resulting nonlinear spectral and angular distribution of the photon probability density is presented in Fig. 3, and the spectral lines can now be tentatively identified with nonlinear 1, 2, ..., n-photon processes. In that case, the effect of a multiphoton interaction can be described as a combination of ponderomotive recoil and harmonic generation.

The most important consequence of the nonlinear Doppler effect, however, resides in the fact that, at ultrahigh intensities, the peak photon number density in each line is approximately constant across the spectrum. This indicates that for ultrashort laser pulses, even in the case of circularly polarized light, the backscattered energy is redistributed over a wide spectral range instead of contributing to a single, narrow Compton backscattered line. This is a potentially serious difficulty for applications, such as the γ–γ collider, which require the generation of a single, intense, highly collimated, narrow γ-ray line. Such a problem can be partially alleviated by shaping the temporal envelope of the pump laser pulse in order to minimize the nonlinear Doppler shift during the interaction. In such a scheme, as illustrated in Fig. 4 (top), the main part of the laser pulse is flat, thereby yielding constant axial electron velocity during most of the interaction. The associated Doppler shift thus remains nearly constant, resulting in the radiation of a narrow spectral line, as indicated in Fig. 4 (bottom). During the rise and fall of the laser pulse envelope, transient lines are radiated, but they are kept to a minimum by using this technique, which is rather analogous to the use of a tapered wiggler entrance for a FEL [16].

The beneficial effects of square optical pulses, which can be generated by holographic filtering at the Fourier plane of a CPA laser, as demonstrated by Weiner *et al.* [12], are illustrated in Fig. 5. Here, the ratio of the energy in the main line to the total backscattered energy is evaluated as a function of the laser pulse shape, for different values of the normalized vector potential and circular polarization. The pulse shape is parametrized by the ratio of the flat-top to the FWHM: for zero, the pulse is a hyperbolic secant, and for unity, the pulse is square. At low intensity, the hyperbolic secant pulse generates a single Doppler-upshifted line, which contains all the backscattered energy, while the square pulse contrast ratio is limited to approximately 95% by the corresponding *sinc* spectrum. At high intensities, however, the contrast ratio remains high for the square pulse, while the backscattered energy is distributed over a wide spectrum for the hyperbolic secant pulse, thus demonstrating the interest of operating with square optical pulses.

4. DIRAC-LORENTZ ELECTRODYNAMICS

The Dirac-Lorentz equation [13] describes the covariant dynamics of a classical point electron, including the radiation reaction effects due to the electron self-interaction. In the case of nonlinear Compton scattering, the Dirac-Lorentz equation can be given as

$$\frac{du_\mu}{d\tau} = a_\mu = L_\mu + \tau_0 \left[\frac{da_\mu}{d\tau} - u_\mu(a_v\, a^v) \right], \quad \mathbf{L}_\perp = w\mathbf{E}_\perp, \quad L_z = L_0 = \mathbf{u}_\perp \bullet \mathbf{E}_\perp, \quad (38)$$

where we recognize the light-cone variable, $w = \gamma - u_z$, and the laser transverse electric field; $\tau_0 = 2\, r_0/3\, c = 0.626\ 10^{-23}$ s is the Compton time-scale. Subtracting the axial component of Eq. (38) from the temporal component, we obtain an equation governing the evolution of the light-cone variable, which is no longer invariant

$$\frac{dw}{d\tau} = \tau_0 \left[\frac{d^2w}{d\tau^2} - w(a_v\, a^v) \right]. \quad (39)$$

Noting that $\mathbf{E}_\perp = w \frac{d\mathbf{A}_\perp}{d\phi}$, we also obtain an equation governing the evolution of the canonical momentum,

$$\frac{d}{d\tau}\left(\mathbf{u}_\perp - \frac{e\mathbf{A}_\perp}{m_0 c}\right) = \tau_0 \left[\frac{d^2 \mathbf{u}_\perp}{d\tau^2} - \mathbf{u}_\perp(a_v\, a^v) \right], \quad (40)$$

which is no longer invariant. If we introduce the expansion parameter $\varepsilon = \omega_0 \tau_0$, and use the nonlinear electron phase as the independent variable, Eq. (39) now reads

$$\frac{dw}{d\phi} = \varepsilon \left[\frac{d^2}{d\phi^2}\left(\frac{w^2}{2}\right) - w^2 \left(\frac{du_\mu}{d\phi}\frac{du^\mu}{d\phi}\right) \right]. \tag{41}$$

Since the right-hand side of Eq. (41) is at least of order ε, we can replace the terms in the brackets by their zeroth-order (Lorentz dynamics) approximation; in this case, we obtain a simple differential equation for the light-cone variable

$$\frac{d}{d\phi}\left[\frac{1}{w(\phi)}\right] \cong \varepsilon g^2(\phi), \tag{42}$$

where we recognize the envelope of the circularly polarized laser pulse. Equation (42) can easily be integrated to yield

$$\frac{1}{w(\phi)} = \frac{1}{w_0} + \varepsilon \int_{-\infty}^{\phi} g^2(\Psi) d\Psi. \tag{43}$$

This equation describes the electron recoil, as illustrated in Fig. 6, for beam parameters similar to those of SLAC. It is clear that at sufficient intensities, the relative radiative energy loss becomes quite significant.

REFERENCES

1. F.V. Hartemann et al., Phys. Rev. **E54**, 2956 (1996).
2. M.D. Perry and G. Mourou, Science **264**, 917 (1994).
3. A.E. Siegman, "Lasers" (University Science Books, Mill Valley, CA 1986), Chaps. 16-21.
4. F.V. Hartemann and A.K. Kerman, Phys. Rev. Lett. **76**, 624 (1996).
5. C.E. Clayton et al., Phys. Rev. Lett. **70**, 36 (1993).
6. F.V. Hartemann et al., Phys. Rev. **E51**, 4833 (1995).
7. J. Ojeda-Castaneda, J.C. Escalera and M.J. Yzuel, Opt. Comm. **114**, 189 (1995).
8. L.S. Brown and T.W.B. Kibble, Phys. Rev. **133**, A705 (1963).
9. C.I. Moore, J. Mod. Optics **39**, 2171 (1992).
10. F.V. Hartemann, Phys. Rev. **A42**, 2906 (1990).
11. N.N. Lebedev, "Special Functions and their Applications" (Dover, New York, NY 1972),
12. A.M. Weiner, J.P. Heritage, and E.M. Kirschner, J. Opt. Soc. Am. **B5**, 1563 (1988).
13. P.A.M. Dirac, Proc. R. Soc. London Ser. **A167**, 148 (1938).
14. S.S. Schweber, "QED and the Men who Made it" (Princeton Univ. Press, Princeton, NJ, 1994).
15. L.S. Gradshteyn and I.M. Ryzhik, Table of Integrals, Series and Products (Academic, Orlando, FL, 1980).
16. C.W. Roberson and P. Sprangle, Phys. Fluids **B1**, 3 (1989).

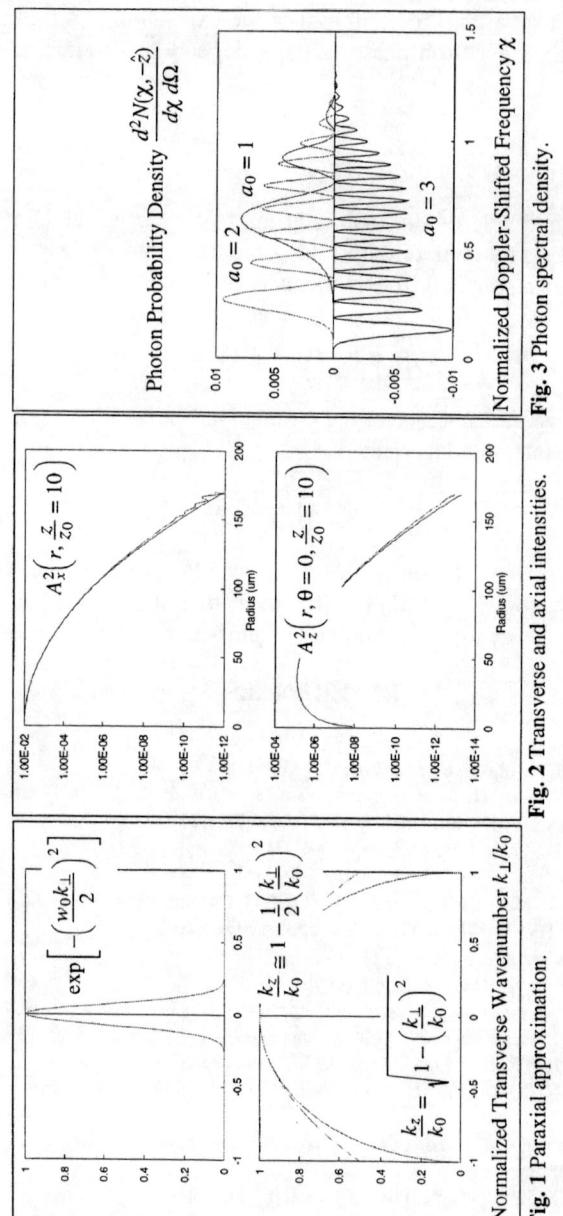

Fig. 1 Paraxial approximation. **Fig. 2** Transverse and axial intensities. **Fig. 3** Photon spectral density.

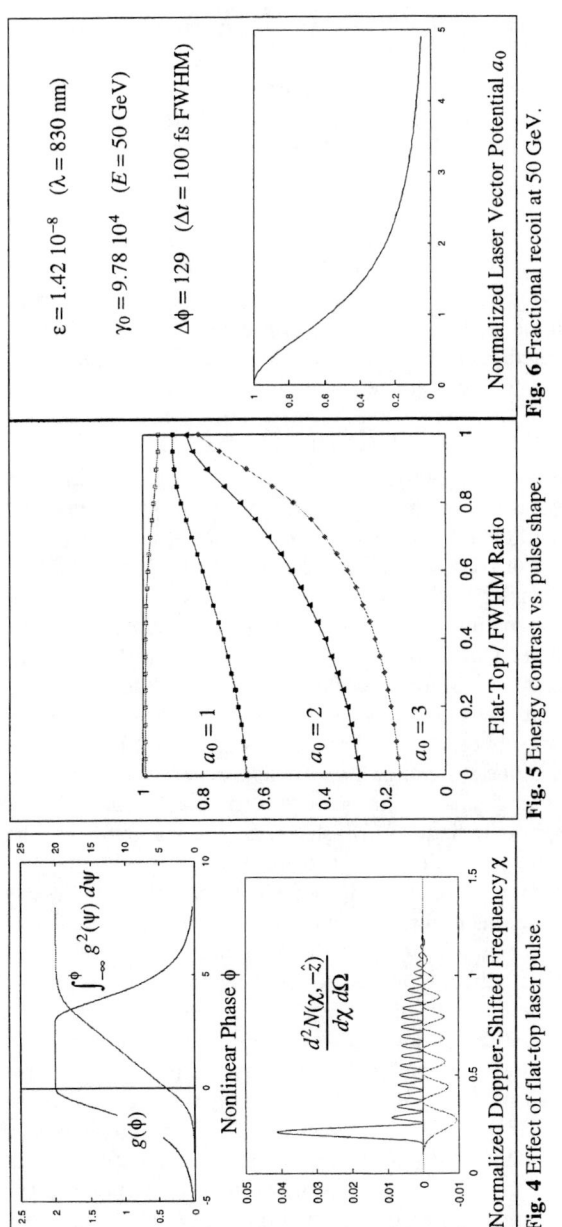

Fig. 4 Effect of flat-top laser pulse. **Fig. 5** Energy contrast vs. pulse shape. **Fig. 6** Fractional recoil at 50 GeV.

Particle Dynamics in an Active Medium

Levi Schächter
Department of Electrical Engineering
Technion- Israel Institute of Technology
Haifa 32000, Israel

ABSTRACT

When a point-charge moves in an active medium it can gain energy at the expense of that stored in the medium. The maximum gradient is evaluated and its relation to the energy stored in the medium is established. The dynamics of a distribution of electrons was also examined and it is reported here.

INTRODUCTION

When an electron moves along a vacuum channel in a dielectric material it may cause Cerenkov radiation to be emitted. The electromagnetic energy comes at the expense of the particle's kinetic energy or in other words, the particle is decelerated. If instead of a passive dielectric medium, an active medium is used, the action of this secondary field may cause the particle to accelerate [1-2]. Thus energy stored in the medium can be transferred to the moving electron. Since the mechanism resembles the inverse of a laser, we call it PASER, which stands for Particle Acceleration by Stimulated Emission of Radiation.

In the past, there were two schemes in which an active medium was suggested in order to support the acceleration process [3, 4]. In both cases the active medium facilitates the generation of solitons extending into the interaction region. Fisher and Tajima [4] have shown that an active medium can be used in order to preserve the radiation pulse-shape, energy and velocity. The scheme relies partially on the self-induced transparency theory which was first developed by McCall and Hahn [5] in 1969. Fisher and Tajima envision their system to use the outer shell electrons to form the plasma and the inner shell electron resonant transition as the constituent of the active medium.

Usually the radiative processes in atoms decay rapidly, typically on a time scale of tens of nanoseconds; therefore, one can wonder how can substantial energy be stored in the medium in order to facilitate the acceleration of an entire bunch. The potential of storing energy in an ensemble of atoms was demonstrated as early as 1927 when Zemansky [6] observed that the decay time of mercury vapors of densities larger than 10^{15} cm^3 is on the order of 10^{-4} seconds, namely more than 3 orders of magnitude larger than the characteristic life-time of a single excited atom. This result can be understood in terms of multiple absorption and emission processes of the resonant atoms; in fact the effect was called at the time *imprisonment of resonance radiation* in gases. This result indicates that electromagnetic energy can be stored in atoms for relatively long periods of times, even in non-metastable states. Thus proper choice of conditions may allow us to have sufficient electromagnetic energy, for a sufficient duration of time, in order to accelerate a bunch of electrons.

FORCE ON A SINGLE PARTICLE

Consider a point-charge q moving at a constant velocity, v. This charge moves in a vacuum channel bored in an otherwise infinite medium characterized by a dielectric coefficient ε_r,

$$\varepsilon_r = 1 - \chi \frac{(\omega - \omega_0)T_2 + j}{1 + \zeta^2 + (\omega - \omega_0)^2 T_2^2}. \tag{1}$$

This expression represents a two-state system, characterized by angular frequency ω_0 and a relaxation time T_2; ζ represents the saturation effect and is given by E/E_{cr} where the critical field is $E_{cr} = (h/2\pi)/\mu(\tau T_2)^{1/2}$ and τ is the characteristic time the ensemble returns to equilibrium. χ is the normalized population (inversion) $\chi = \mu^2(2\pi)(N_1 - N_2)T_2/\varepsilon_0 h$. The transition intensity is represented here by the dipole moment μ and N_1, N_2 represent the atoms density in the lower and upper energy state, respectively.

It has been shown [1-2] that in the case of a medium described by (1), the reaction force on a point-charge is given by

$$E = \frac{q}{2\pi\varepsilon_0 R^2} \frac{\chi}{1 + (E/E_{cr})^2} \frac{\omega_0 R}{c} \frac{R}{cT_2}; \tag{2}$$

this expression is valid at the limit when R, the radius of the channel, is much smaller than the resonance wavelength - which is reasonable if we take R as the average distance between atoms. Throughout this study we shall ignore non-linear effects ($E_{cr} \gg E$) therefore

$$E = \frac{q}{2\pi\varepsilon_0} \frac{\mu^2(N_1 - N_2)(2\pi)}{\varepsilon_0 h} \frac{\omega_0}{c^2}. \tag{3}$$

In order to have a measure for the typical electric field we assume the point-charge is $q = 1.6 \times 10^{-19} \times 10^{10}$, the resonance wavelength of the medium is $\lambda = 10^{-6}$m, the dipole moment $\mu = (1.6 \times 10^{-19}) \times (1 \times 10^{-10})$Cm and the population inversion $|N_1 - N_2| \sim 10^{21}$m^{-3}. With these parameters, the electric field is of the order of 0.1GV/m.

ACCELERATION IN MULTI-STATES SYSTEM

The main constraint of the previous model is that the medium is idealized to be a two states system. In this section we shall address this problem by examining a multi states system.

Exciting an ensemble of atoms of a gas (such as Argon, Krypton, etc) means that there are many possible states. However, before we consider the excited states let us recapitulate the characteristics of an ensemble in equilibrium. The density of atoms in a given state is given by

$$N_i = N \frac{\exp(-\varepsilon_i/k_B T)}{\Sigma_j \exp(-\varepsilon_j/k_B T)} \tag{4}$$

where N is the total density of atoms, ε_i is the energy of the i'th state, T is the temperature in the equilibrium state and the summation is over all possible states. The internal (average) energy stored in the medium in equilibrium is

$$\langle \varepsilon \rangle = \sum_i \varepsilon_i \frac{N_i}{N} . \tag{5}$$

It is evident from (4) and (3) that, in equilibrium, the gas can only decelerate the particle, since always $N_1 > N_2$. Let us assume now that the gas has been excited such that the probability of finding an atom in the i'th state is P_i and therefore the additional energy stored in the medium is

$$\delta\varepsilon = \sum_i \varepsilon_i P_i - \langle \varepsilon \rangle. \tag{6}$$

We shall examine now, how this energy is related to the acceleration force which acts on a point-charge as it traverses the medium.

For sake of simplicity we shall now express the dipole moment in terms of the energy states of the atom using a quasi-phenomenological approach. Since the energy of the i'th state can be approximated by virtue of Coulomb force by

$$\varepsilon_i = \frac{e^2 Z}{4\pi\varepsilon_0 r_i} , \tag{7}$$

where Z is the gas atomic number and r_i is the average radius of the internal electron's trajectory, we can approximate the dipole moment between two states with

$$\mu_{2,1} \sim e|r_2 - r_1| \sim \frac{\varepsilon_2 - \varepsilon_1}{\varepsilon_2 \varepsilon_1} \frac{Z e^3}{4\pi\varepsilon_0}. \tag{8}$$

Substituting this expression in (3) we obtain

$$\delta E = \frac{qN}{2\pi\varepsilon_0^2 (hc/2\pi)^2} \sum_\upsilon \left[\frac{\varepsilon_{\upsilon,2} - \varepsilon_{1,\upsilon}}{\varepsilon_{\upsilon,1} \varepsilon_{\upsilon,2}} \frac{Ze^3}{4\pi\varepsilon_0} \right]^2 (\varepsilon_{\upsilon,2} - \varepsilon_{\upsilon,1})$$
$$\times \left[\left(P_{\upsilon,2} - \frac{N_{\upsilon,2}}{N} \right) - \left(P_{\upsilon,1} - \frac{N_{\upsilon,1}}{N} \right) \right] \quad (9)$$

where υ represents the transition whereas 2 and 1 represent the upper and lower state respectively. It should be emphasized that only single transitions are considered here. For Argon (Z=18), and expressing the energies in eV's, (9) reads

$$\delta E = 10^{-18} N[m^{-3}] N_e \sum_\upsilon \frac{(\varepsilon_{\upsilon,2} - \varepsilon_{\upsilon,1})^3}{(\varepsilon_{\upsilon,2} \varepsilon_{\upsilon,1})^2}$$
$$\times \left[\left(P_{\upsilon,2} - \frac{N_{\upsilon,2}}{N} \right) - \left(P_{\upsilon,1} - \frac{N_{\upsilon,1}}{N} \right) \right] \quad (10)$$

where N_e is the number of electrons in the bunch. Note that in the framework of this notation $\sum_\upsilon (P_{\upsilon,1} + P_{\upsilon,2}) = 1$; in a similar way $\sum_\upsilon (N_{\upsilon,1} + N_{\upsilon,2}) = N$.

There is no simple expression (we are aware of) which can describe the probability of finding an atom in a state i after the whole ensemble was excited. In fact, according to Ref. 7, for a given energy of the impinging electron, the probability of excitation varies dramatically from one state to another. Therefore, we shall proceed by assuming that P is uniformly distributed between 0 and 1, and all the results presented below rely on an averaging over 1000 different uniform distributions. We clearly can not expect this approach to provide us with an accurate result regarding the variation of the gradient as a function of the total energy stored in the medium, but we can anticipate obtaining the correct trend and limits.

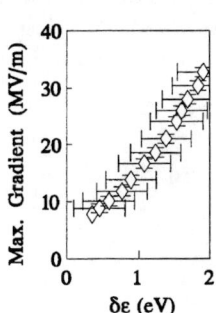

Fig. 1: Maximum gradient which acts on a point-charge as a function of the energy stored in the medium; $NN_e \sim 10^{28} m^{-3}$.

The energy levels considered were those of Argon, as described in Ref. 8; 234 such transitions were examined ranging from Ar up to Ar^{+8}. We found that the energy stored in the medium due to excitation decreases monotonically as a function of the equilibrium temperature from about 2.7eV at zero temperature, to 0.3eV when the temperature is $k_B T \sim 2eV$. According to this simulation, see

Fig. 1, the maximum gradient increases as a function of the stored energy; this result is valid for values smaller than 2eV. The error bars represent the standard deviation associated with the 1000 "experiments" used in the simulations. In this calculation it was assumed that the final state of the ensemble is the equilibrium state; for this reason we refer to this gradient as the maximum gradient. For a stored energy larger than 1eV the gradient is within a good approximation linear with $\delta\varepsilon$ and it is given by $\delta E[\text{MV/m}] \approx 21\,\delta\varepsilon[\text{eV}] - 7$. Beyond the 2eV level the gradient drops very rapidly to zero (not shown). This is because the probability that all the high energy-states are excited is slim and as a result, the gradient drops to zero.

BUNCH DYNAMICS: TWO STATES SYSTEM

Now, after we examined the force which acts on a point-charge as it moves in a multi-state excited medium, we can go back to the simple two energy-state model and investigate the dynamics of a bunch of electrons which are not glued together as in the previous section. Furthermore, the size of the bunch is not much smaller than the resonance wavelength of the medium.

With the expression presented in (3) we can establish the electric field which acts on a single electron (E_1). On axis and not very far (<10 λ) behind the particle we can approximate the electric field generated by the i'th electron as

$$E_i(t,z) = E_1 \cos\left[\omega_0\left(t - \frac{z - z_i(0)}{v_i}\right)\right] h\left(t - \frac{z - z_i(0)}{v_i}\right) \quad (11)$$

where, $h(x)$ is the step function, v_i is the velocity of this electron and $z_i(0)$ is its location at $t=0$; it is tacitly assumed that the electron is relativistic such that the change in the velocity is small but the change in the relativistic factor γ can be large. Assuming that $\gamma \gg 1$, then the electric field which acts on the j'th electron is given by

$$\begin{aligned} E_j(t) = \; & E_1 \sum_i \cos\left[\omega_0 t\left(\frac{1}{2\gamma_i^2} - \frac{1}{2\gamma_j^2}\right) + \frac{\omega_0}{c}\left(z_i(0) - z_j(0)\right)\right] \\ & \times h\left[\omega_0 t\left(\frac{1}{2\gamma_i^2} - \frac{1}{2\gamma_j^2}\right) + \frac{\omega_0}{c}\left(z_i(0) - z_j(0)\right)\right]. \end{aligned} \quad (12)$$

It is convenient to define the phase term

$$\psi_i \equiv \omega_0 t \frac{1}{2\gamma_i^2} + \frac{\omega_0}{c} z_i(0) \;, \quad (13)$$

which enables us to write the following two equations for the dynamics of the electrons in the bunch:

$$\frac{d}{dt}\gamma_j = \frac{e}{mc}|E_1|N_e \left\langle \cos\left(\psi_i - \psi_j\right) h\left(\psi_i - \psi_j\right) \right\rangle_i,$$
$$\frac{d}{dt}\psi_j = \frac{\omega_0}{2\gamma_j^2}.$$

(14)

Note that in this expression it was assumed that the medium is active ($E_1 < 0$); furthermore, the quantity $|E_1|N_e$ is the gradient which acts on the bunch if this had a point-like distribution. Before we illustrate the result of simulations of this set of equations, it is important to emphasize their characteristics comparing to the case of a regular bunch subject to a wave generated externally as is the case in a regular accelerator: (i) each particle is affected by the field generated by the particles in front of it (and itself). In a relatively narrow bunch this effect can have a very positive effect since it *compresses* the particles together. (ii) The phase of the wave is not set by an external source but by the bunch itself - and this is the correct phase for acceleration!! (iii) As in regular accelerators, particles can be trapped since if $\gamma \to \infty$ then the phase remains constant.

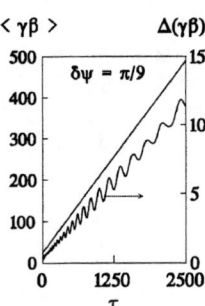

Figure 2 illustrates the results of a simulations of the system's equations [$\lambda_0 = 1\mu m$, $\langle\gamma(0)\rangle = 24.5$, $\langle\psi\rangle = 0$, $\Delta\psi(0) = \delta\psi/2$, $\Delta\gamma(0)/\langle\gamma(0)\rangle \sim 0.005$ and $|E_1|N_e = 1GV/m$]. The electrons in the bunch are divided into 300 macro- particles which are (initially) uniformly distributed in a phase domain $\delta\psi$.

Fig. 2: The variation in time, of the average momentum and the momentum spread.

Specifically, for $\delta\psi = 20°$ we observe that average momentum and the momentum spread increase in time, $\tau \equiv \omega_0 t/2\langle\gamma(0)\rangle^2$; the interaction length is 3m. The energy of the bunch increases linearly from 12MeV to 250MeV which corresponds to an average gradient of 80MV/m. Figure 3 illustrates the characteristics of the phase. Specifically, we observe that the phase approaches a "saturation" value which is indicative of a trapping process.

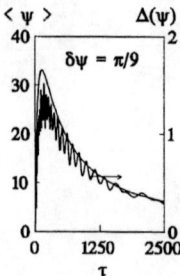

Fig. 3: The average phase and the phase spread as a function of time.

Note that the average ψ increases in the first part of the interaction. It reaches a peak value at $\tau \sim 135$ and decreases beyond this point. The reduction in the average phase can be understood when examining (14), since when averaged we obtain

$$\frac{d}{dt}\langle\psi\rangle = \frac{\omega_0}{2}\left\langle\frac{1}{\gamma^2}\right\rangle. \qquad (15)$$

Bearing in mind that γ increases linearly in time the average phase $\langle\psi\rangle \propto 1/t$; the spread in phase, for long periods of time, behaves as $\Delta\psi \propto 1/\sqrt{t}$.

An even better illustration of the trapping process is presented in Figure 4 where the average momentum is plotted vs. the average phase of the bunch (the spread in both quantities is smaller than the width of the points). At the beginning of the interaction both the average momentum and phase increase. Beyond the maximum phase point the momentum increases rapidly, whereas the average phase reaches an asymptotic value.

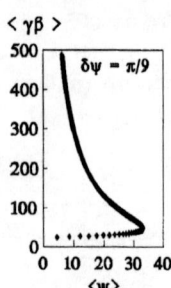

Fig. 4: The average momentum as a function of the average phase

SUMMARY

The dynamics of particles in an active medium was considered. Two cases of interest were reported: in the first part the medium was represented by 234 transitions (lines) - which correspond to the dominant transitions in Argon. A simple model was employed to show the relation between the maximum gradient achievable from an excited ensemble of atoms, as a function of the energy stored in the gas. In the second part the dynamics of a bunch of electrons was examined. We shown that electrons can be trapped and asymptotically the bunch is compressed.

Acknowledgement This study was supported by the United States Department of Energy, the Bi-National United States - Israel Science Foundation and the Center for Absorption in Science, Ministry of Immigrant Absorption State of Israel.

REFERENCES

1. L. Schächter; Phys. Lett. A Vol. **205**, p.355. (1995)
2. L. Schächter; Phys. Rev. E. **53**, p.6427 (1996).
3. K. Mima *et. al.* ; PRL **57**, 142 (1986).
4. D. Fisher and T. Tajima; PRL **71**, p.4338 (1993).
5. S. McCall and E. Hahn; Phys. Rev. p.457 (1969).
6. M.W. Zemansky; Phys. Rev. **29**, 513(1927).
7. F.L. Arnot;
    Collision Process in Gases,
    John Wiley & Sons, New York 1955 p.33.
8. A.R. Striganov and N.V. Svetitskii;
    Tables of Spectral Lines of Neutral and Ionized Atom,
    IFI/Plenum, New York 1968, p.34 .

GROUP 2

"Plasma-Based Acceleration Concepts"—Thomas Katsouleas (USC)

Plasma Channel Formation in the Wake of a Short Laser Pulse

E. Esarey, A. Ting, K. Krushelnick,[1] C. Moore,[2] M. Baine,[3] and P. Sprangle

Beam Physics Branch, Plasma Physics Division
Naval Research Laboratory, Washington DC 20375-5346

[1] *Laboratory for Plasma Studies, Cornell University, Ithaca NY 14853*
[2] *National Research Council/NRL Postdoctoral Fellow*
[3] *Department of Physics, University of California San Diego*
La Jolla CA 92093

Abstract. The process by which a short ($\lesssim 1$ ps), intense ($\gtrsim 10^{18}$ W/cm^2) laser pulse creates a plasma density channel in its wake is analyzed. A relativistically guided laser pulse expels electrons as it propagates. This leads to a radial space charge force which initiates ion motion. After the passage of the pulse, the ions drift radially and form a plasma channel. This channel can guide a trailing probe pulse, as has been observed in pump-probe experiments.

INTRODUCTION

The propagation of intense ($\gtrsim 10^{18}$ W/cm^2), short ($\lesssim 1$ ps) laser pulses in plasmas has a number of possible applications including laser-driven accelerators [1,2], x-ray lasers [3], harmonic generators [4,5], and laser-fusion [6]. These applications could benefit by enhancing the propagation distance beyond the vacuum diffraction range (Rayleigh length) Z_R. Methods for optically guiding a laser pulse in a plasma include relativistic self-focusing [7-9], which requires that the laser power exceed the critical power $P \geq P_c$, and preformed plasma channels [9-11], which requires that the channel depth exceed a critical depth $\Delta n \geq \Delta n_c$. Plasma channel guiding of laser pulses has been recently demonstrated using channels created either by an axicon focusing geometry [10] or by a slow capillary discharge [11]. In this paper the production of a plasma channel in the wake of a short intense laser pulse is analyzed. An extended channel ($> Z_R$) can be generated [12] provided that the pulse power exceeds the critical power for relativistic self-focusing $P \geq P_c$ and that the pulse length exceeds the electron plasma wavelength $L > \lambda_{pe}$.

The regime in which $P > P_c$ and $L > \lambda_{pe}$ corresponds to the self-modulated regime of the laser wakefield accelerator (LWFA) [13-16]. In the self-modulated LWFA, the laser pulse is guided over several Z_R through an underdense plasma by a combination of relativistic and ponderomotive self-focusing. As the pulse propagates, it undergoes a rapid self-modulation and/or forward Raman instability and becomes highly modulated at λ_{pe} [13-19]. Associated with this self-modulation is the growth of a large amplitude plasma wave wakefield. This plasma wave can trap and accelerate a fraction of the plasma electrons, as has been observed in recent experiments [20-24]. The majority of theoretical and numerical modelling of the self-modulated LWFA assumes that the ions remain stationary [13-19]. The motion of the plasma ions, however, can be significant in these experiments. In particular, a large density channel can be formed in the plasma region trailing the intense laser pulse. Furthermore, this density channel has been observed to guide a low power ($P \ll P_c$) probe laser pulse in recent pump-probe experiments [12].

PLASMA ION MOTION

Consider a pulse which is long compared to a electron plasma wavelength but short compared to an ion plasma wavelength, $\lambda_{pe} < L < \lambda_{pi}$, where L is the laser pulse length, $\lambda_{pe,i} = 2\pi c/\omega_{pe,i}$, $\omega_{pe,i} = (4\pi n_0 q^2/m)_{e,i}^{1/2}$ is the plasma frequency, n_0 is the ambient density, q is the charge, m is the mass, and the subscripts e and i refer to the electrons and ions, respectively. The laser pulse is assumed to have a power near the critical power P_c such that it will be guided for many Rayleigh lengths through the plasma, where $P_c \simeq 17(\lambda_{pe}/\lambda)^2$ GW is the critical power for relativistic self-focusing [7,8], $Z_R = \pi r_0^2/\lambda$ is the Rayleigh length, r_0 is the focal spot size of the laser pulse, and λ is the laser wavelength. In the following the envelope of the laser pulse is assumed to be nonevolving as it is guided through the plasma. The evolution of the laser pulse, including the effects of diffraction, self-focusing, self-modulation, and instabilities, has been treated in numerous publications under the assumption of nonevolving plasma ions [1,2,13-19]. In this paper, the laser pulse is assumed to be nonevolving and the evolution of the plasma ions is considered.

The ponderomotive force associated with an intense laser pulse leads to the creation of a plasma density channel as follows. The laser pulses exerts a ponderomotive force on the plasma electrons and expels them radially, as is the case in ponderomotive self-channeling [1,8,13-16]. This sets up a large space charge force which subsequently drags the ions outward from the axis. After the passage of the intense laser pulse, the ions continue to drift radially at approximately the ion acoustic speed $C_s = (ZT_e/m_i)^{1/2}$, thus creating a plasma channel, where Z is the number of electrons per ion and T_e (T_i) is the electron (ion) temperature.

The plasma electron motion is described by the radial force balance [8,14]

$$e\nabla_\perp \phi = m_e c^2 \nabla_\perp (1 + a^2/2)^{1/2} + n_e^{-1} \nabla_\perp P_e, \qquad (1)$$

where ϕ is the space charge potential and $\mathbf{a} = e\mathbf{A}_\perp/m_e c^2$ is the normalized vector potential of the laser field. For a linearly polarized field, $a^2 \simeq 7.2 \times 10^{-19} \lambda^2 I$, where λ is the laser wavelength in μm and I is the laser intensity in W/cm^2. In Eq. (1), $e\nabla_\perp \phi$ is the space charge force, $m_e c^2 \nabla_\perp (1 + a^2/2)^{1/2}$ is the ponderomotive force, $P_e = T_e n_e$ is the electron pressure, and n_e is the electron density. This expression neglects the generation of plasma waves and assumes that the axial length of the laser pulse is large compared to the laser spot size r_0 and the electron plasma wavelength λ_{pe}.

In the linear regime, the ion motion is described by the continuity equation

$$\partial \delta n_i / \partial t = -n_{i0} \nabla_\perp \cdot \mathbf{v}_\perp, \qquad (2)$$

and the momentum equation

$$m_i \partial \mathbf{v}_\perp / \partial t = -Ze\nabla \phi, \qquad (3)$$

where δn_i and n_{i0} are the perturbed and ambient ion densities, $\mathbf{v}_\perp$ is the radial ion velocity, and $T_i \ll T_e$ is assumed. Combining the electron radial force balance, the ion continuity equation, and the ion momentum equation yields

$$\left(\frac{\partial^2}{\partial t^2} - C_s^2 \nabla_\perp^2 \right) \frac{\delta n_i}{n_{i0}} \simeq \frac{Zm_e}{4m_i} \nabla_\perp^2 a^2, \qquad (4)$$

assuming $\delta n_i^2/n_{i0}^2 \ll 1$, $a^2 \ll 1$, and an isothermal equation of state. Equation (4) is similar to that used to describe stimulated Brillouin scattering and ponderomotive self-focusing of long ($L > \lambda_{pi}$) laser pulses [25].

Assuming a nonevolving laser intensity profile $a^2 = a^2(\zeta, r)$, which is function of only the variables r and $\zeta = z - ct$, Eq. (4) can be solved using the 2D Green's function for the wave equation,

$$\frac{\delta n_i}{n_{i0}} = \frac{Zm_e}{8\pi\beta_s m_i} \int_{-\infty}^{\infty} d\zeta' \int_{-\infty}^{\infty} dx' \int_{-\infty}^{\infty} dy' \frac{{\nabla'_\perp}^2 a^2(\zeta', r')}{[\beta_s^2(\zeta - \zeta')^2 - |\mathbf{r} - \mathbf{r}'|^2]^{1/2}}, \quad (5)$$

where $\beta_s = C_s/c$, $\mathbf{r} = x\mathbf{e}_x + y\mathbf{e}_y$, and the integrand is only nonzero in the region $\beta_s(\zeta - \zeta') > |\mathbf{r} - \mathbf{r}'|$.

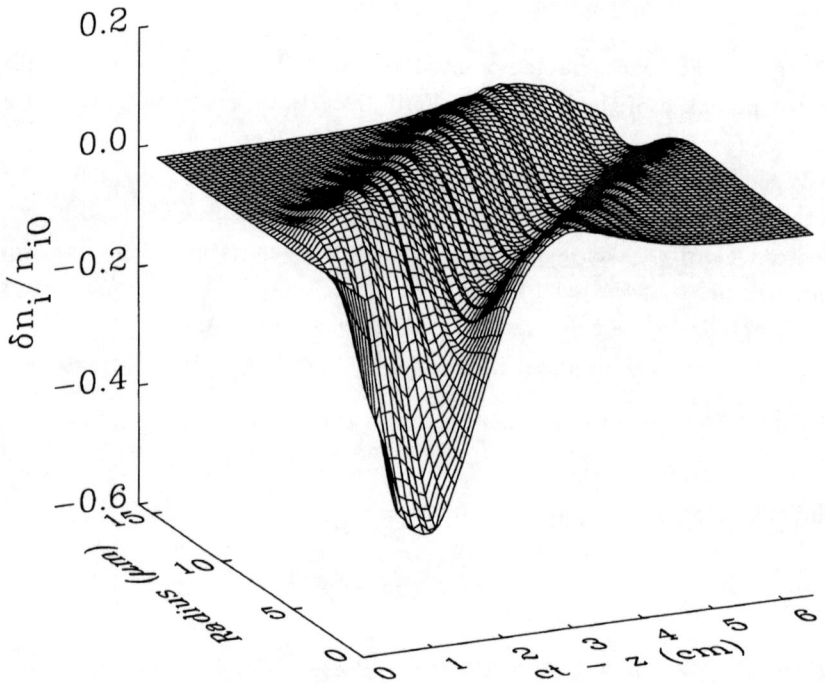

FIG. 1: The evolution of the density channel in the region trailing the laser pulse as obtain from Eq. (5) for the parameters $a_0 = 0.25$, $r_0 = 10\ \mu\text{m}$, $L = 120\ \mu\text{m}$, $T_e = 100\ \text{eV}$, $Z = 2$, $m_i/m_e = 7300$, and $\beta_s = 2.3 \times 10^{-4}$. The laser pulse is moving to the left.

Equation (5) has been evaluated for a nonevolving laser pulse of the form $a^2 = a_0^2 f(\zeta) \exp(-2r^2/r_0^2)$, with $f = \sin^2(\pi\zeta/L)$ for $0 \leq \zeta \leq L$ and $f = 0$ otherwise. The evolution of the density channel is shown in Fig. 1 for the parameters $a_0 = 0.25$, $r_0 = 10$ μm, $L = 120$ μm, $T_e = 100$ eV, $Z = 2$, $m_i/m_e = 7300$, and $\beta_s = 2.3 \times 10^{-4}$. Analysis indicates that the density channel reaches a maximum depth of $\delta n_{i0}/n_{i0} \simeq -\alpha_1(Zm_e/\beta_s m_i)a_0^2 L/r_0$ after a distance of $\zeta_c = \alpha_2 r_0/\beta_s$, where $\alpha_{1,2}$ are constants which depend on the shape of the laser pulse profile (for the example in Fig. 1, $\alpha_1 = 0.46$ and $\alpha_2 = 0.78$). For $\zeta^2 \ll \zeta_c^2$, the channel depth increases roughly linearly, $\delta n_i/n_{i0} \simeq \alpha_3(\zeta/\zeta_c)\delta n_{i0}/n_{i0}$, and for $\zeta^2 \gg \zeta_c^2$, the channel depth decreases roughly as $\delta n_i/n_{i0} \simeq \alpha_4(\zeta_c/\zeta)\delta n_{i0}/n_{i0}$, where $\alpha_{3,4}$ are constants ($\alpha_3 = 0.77$ and $\alpha_4 = 0.40$ in Fig. 1).

DISCUSSION

If the laser pulse is of sufficiently high power $P > P_c$ and long compared to an electron plasma wavelength $L > \lambda_{pe}$, it can be guided through the plasma over a distance of many Rayleigh lengths by a combination of relativistic self-focusing and ponderomotive self-channeling. As the pulse propagates, its ponderomotive force expels electrons which sets up a radial space charge force. This space charge force is exerted on the plasma ions for the duration of the laser pulse and initiates a radial ion motion. After the passage of the laser pulse, the ions continue to drift radially at the ion acoustic speed, thus forming a large plasma density channel behind the pulse. Such a channel is capable of guiding a low power probe pulse, injected some distance behind the intense pump pulse. This effect has been observed in pump-probe experiments in the self-modulated LWFA regime [12].

In experiments at the Naval Research Laboratory [12], a 2 TW, 400 fs, 1 μm pump pulse was injected into a gas jet (Hydrogen or Helium) of density 10^{19} cm^{-3} ($\lambda_{pe} \simeq 10$ μm and $P_c \simeq 1.7$ TW). In vacuum a focal spot size of $r_0 \simeq 5$ μm was measured corresponding to a focused intensity of 5×10^{18} W/cm^2 and a vacuum Rayleigh length of $Z_R \simeq 80$ μm. Prior to injection, a small fraction (10%) of the pulse energy was split off and frequency doubled to form a 0.5 μm probe pulse. The probe pulse was run through an adjustable delay line and injected some distance behind the pump pulse. By imagining the transversely scattered light, the propagation characteristics of both the 1 μm pump and the 0.5 μm probe pulses could be measured. The pump pulse

($P > P_c$) was observed to be guided through the entire 2.5 mm ($\simeq 30 Z_R$) width of the gas jet. The propagation characteristics of the probe pulse ($P \ll P_c$) varied as a function of the delay behind the pump pulse. For small delays (a few ps) the probe pulse diffracted approximately as it would in vacuum. For larger delays (tens of ps), the probe pulse was observed to be guided through the entire 2.5 mm of plasma. This strongly suggests the formation of a plasma density channel in the wake of the pump laser pulse. Furthermore, this experiment demonstrates the relativistic guiding of the 1 μm pump pulse over a distance of $30 Z_{R,pump}$ and the channel guiding of the 0.5 μm probe pulse over a distance of $15 Z_{R,probe}$.

For the parameters of the experiment, the linear theory presented in the previous section would predict complete evacuation of the plasma channel. This violates the assumption of a linear ion response. A more accurate description of the experiment requires the development of a nonlinear model of channel formation.

ACKNOWLEDGMENTS

The authors thank H.R. Burris for technical assistance. This work was support by the Office of Naval Research and the Department of Energy.

REFERENCES

[1] E. Esarey, P. Sprangle, J. Krall, and A. Ting, IEEE Trans. Plasma Sci. **PS-24**, 252 (1996).
[2] P. Sprangle, E. Esarey, and J. Krall, Phys. Plasmas **3**, 2183 (1996).
[3] H.M. Milchberg, C.G. Durfee, and T.J. McIlrath, Phys. Rev. Lett. **75**, 2494 (1995).
[4] I.P. Christov, J. Zhou, J. Peatross, A. Rundquist, M.M. Murnane, and H.C. Kapteyn, Phys. Rev. Lett. **77**, 1743 (1996).
[5] P. Eisenberger and S. Suckewer, Science **273** (in press).
[6] M. Tabak, J. Hammer, M. Glinsky, W.L. Kruer, S. Wilks, J. Woodworth, E.M. Campbell, M. Perry, and R. Mason, Phys. Plasmas **3**, 2183 (1996).
[7] P. Sprangle, C.M. Tang, and E. Esarey, IEEE Trans. Plasma Sci. **PS-15**, 145 (1987).
[8] G.Z. Sun, E. Ott, Y.C. Lee, and P. Guzdar, Phys. Fluids **30**, 526 (1987).
[9] P. Sprangle and E. Esarey, Phys. Fluids 4, 2241 (1992).
[10] C. Durfee and H. Milchberg, Phys. Rev. Lett. **71**, 2409 (1993); H.M. Milchberg, T.R. Clark, C.G. Durfee, T.M. Antonsen, and P. Mora, Phys. Plasmas **3**, 2149 (1996).

[11] A. Zigler, Y. Ehrlich, C. Cohen, J. Krall, and P. Sprangle, J. Opt. Soc. Am. B **13**, 68 (1996); Y. Ehrlich, C. Cohen, A. Zigler, J. Krall, P. Sprangle, and E. Esarey, Phys. Rev. Lett. **77**, (1996).

[12] K. Krushelnick, A. Ting, C.I. Moore, H.R. Burris, E. Esarey, P. Sprangle, and M. Baine, submitted to Phys. Rev. Lett. (1996).

[13] N.E. Andreev, L.M. Gorbunov, V.I. Kirsanov, A.A. Pogosova, and R.R. Ramazashvili, Pis'ma Zh. Eksp. Teor. Fiz. **55**, 551 (1992).

[14] P. Sprangle, E. Esarey, J. Krall, and G. Joyce, Phys. Rev. Lett. 69, 2200 (1992).

[15] T.M. Antonsen, Jr. and P. Mora, Phys. Rev. Lett. **69**, 2204 (1992); Phys. Fluids B **5**, 1440 (1993).

[16] E. Esarey, P. Sprangle, J. Krall, A. Ting, and G. Joyce, Phys. Fluids B **5**, 2690 (1993); J. Krall, A. Ting, E. Esarey, and P. Sprangle, Phys. Rev. E **48**, 2157 (1993).

[17] W.B. Mori, C.D. Decker, D.E. Hinkel, and T. Katsouleas, Phys. Rev. Lett. **72**, 1482 (1994); C.D. Decker, W.B. Mori, and T. Katsouleas, Phys. Rev. E **50**, 3338 (1994).

[18] E. Esarey, J. Krall, and P. Sprangle, Phys. Rev. Lett. **72**, 2887 (1994).

[19] N.E. Andreev, V.I. Kirsanov, and L.M. Gorbunov, Phys. Plasmas **2**, 2573 (1995).

[20] K. Nakajima, D. Fisher, T. Kawakubo, H. Nakanishi, A. Ogata, Y. Kato, Y. Kitagawa, R. Kodama, K. Mima, H. Shiraga, K. Suzuki, K. Yamakawa, T. Zhang, Y. Sakawa, T. Shoji, Y. Nishida, N. Yugami, M. Downer, and T. Tajima, Phys. Rev. Lett. **74**, 4428 (1995).

[21] C.A. Coverdale, C.B. Darrow, C.D. Decker, W.B. Mori, K.C. Tzeng, K.A. Marsh, C.E. Clayton, and C. Joshi, Phys. Rev. Lett. **74**, 4659 (1995).

[22] A. Modena, Z. Najmudin, A.E. Dangor, C.E. Clayton, K.A. Marsh, C. Joshi, V. Malka, C.B. Darrow, C. Danson, D. Neely, and F.N. Walsh, Nature **337**, 606 (1995).

[23] D. Umstadter, S.Y. Chen, A. Maksimchuk, G. Mourou, and R. Wagner, Science **273**, 472 (1996).

[24] C. Moore, A. Ting, K. Krushelnick, H.R. Burris, E. Esarey, P. Sprangle, and M. Baine, submitted to Phys. Rev. Lett. (1996).

[25] W.L. Kruer, *The Physics of Laser Plasma Interactions* (Addison-Wesley, New York, 1988).

Beam-Generated Plasma Channels for Laser Wakefield Acceleration

G. Shvets and N. J. Fisch
Princeton Plasma Physics Laboratory

Abstract

A hybrid of laser-wakefield and plasma wakefield accelerators is proposed, in which an intense laser pulse is optically guided by a plasma channel created by the leading portion of a high-current low-energy electron beam. This accelerator configuration appears to combine to advantage features of both laser wakefield and plasma wakefield accelerators.

Limitations of Traditional Acceleration Schemes

Because of the very high electric field it can sustain, plasma has been proposed [1, 2, 3, 4, 5, 6, 7] as a medium for high-gradient particle acceleration. Moreover, in a plasma it is possible to convert strong transverse fields, characteristic of lasers and fast-moving charged particles, into longitudinal waves, which are suitable for accelerating particles to high energy. The plasma waves can be excited by relativistic electron bunches [2, 3, 4] in a plasma wakefield accelerator or by intense laser pulses [1, 8, 9, 10, 11, 12] in a laser wakefield accelerator.

Plasma wakefield accelerator (PWFA) is a very promising advanced accelerator concept because of the very high (in excess of 1GeV/m) accelerating gradient and a potentially long interaction distance. The long interaction distance is due to the effect of self-focusing that the driving bunch experiences in the plasma. In the absence of the plasma the space-charge radial electric field of the bunch is almost exactly (to order $1/\gamma^2$) offset by the self-pinching effect of the azimuthal magnetic field. In the presence of the plasma the space-charge of the bunch is neutralized by the plasma electrons, resulting in an overall magnetic self-focusing. One of the most apparent limitations of plasma wakefield accelerator is a low transformer ratio. The fundamental wake theorem [2, 13] limits the transformer ratio – the peak accelerating field E_+ experienced by the accelerated electrons over the average decelerating field $\langle E_-\rangle$ acting on the driving electron bunch – to less than 2 for symmetric bunches. Thus, accelerating electrons to 1 GeV energy would require a 0.5 GeV driving beam.

The transformer ratio might be increased to $T \approx \pi N_p$ by shaped bunches[3, 4], with a slow linear rise in density over N_p plasma periods and an abrupt termination at $\xi = N_p \lambda_p$, where $\lambda_p = 2\pi c/\omega_p$ and $\omega_p = \sqrt{4\pi e^2 n_0/m}$ are the plasma wavelength and plasma frequency, respectively, and $\xi = ct-z/\beta_g$ is a co-moving with a bunch coordinate. However, an abrupt termination over a distance smaller than a collisionless skin depth may be difficult to accomplish in practice.

In contrast, short laser pulses, which are used to drive a plasma wave in a laser wakefield accelerator (LWFA), carry much more energy than an electron driver of the same duration, capable of exciting an identical plasma wake. Energy content of a circularly polarized laser pulse of duration τ_p and radius σ, with a normalized vector-potential $a_0 = eA/mc^2$ and frequency ω_0, is given by $U_L = a_0^2 \sigma^2 c \tau_p (m\omega_0 c/2e)^2$. Energy content of a beam with the dimensions is given by $U_b = \pi \sigma^2 \tau_p \gamma_b mc^3$. Taking the ratio of the two, and assuming that $a_0^2 = 2n_b/n_0$, obtain

$$U_L/U_b = \gamma_L^2/(2\gamma_b), \tag{1}$$

where $\gamma_g = \omega_0/\omega_p$. If the excited wakes have identical phase velocities, then $\gamma_b = \gamma_g$, and $U_L/U_b = \gamma_b/2$. Thus, in comparison with PWFA, which can only accelerate electrons by $\Delta E = 2\gamma_b mc^2$, LWFA is capable of accelerating injected electrons by by $\Delta E = \gamma_b^2 mc^2$ before pulse depletion becomes an issue.

On the other hand, laser acceleration is limited by diffraction of the laser pulse, which reduces the propagation distance to about a Rayleigh length $Z_R = \pi r_L^2/\lambda_0$, where λ_0 is the laser wavelength. Laser pulses might be guided by plasma channels [10, 11, 12], which have been formed [15] by a hydrodynamic expansion of a laser breakdown spark in an ambient gas. Differential "leakage" rates [15] for different quasi-bound transverse laser modes allows for some control over the modal content and Raman-type instabilities of the laser pulse[16].

We suggest a novel method of creating a narrow plasma channel which utilizes the leading portion of a high-current electron beam. An intense laser pulse, timed with the electron bunch so that it is placed several plasma wavelengths behind the head of the pulse, is guided by the plasma channel, generated by the beam, and excites a strong plasma wake. This wake accelerates the portion of the electron beam that trails behind the laser pulse. Since the width and density depression of the plasma channel are determined by the electron beam, they can be accurately controlled by the beam optics. Creating the channel by the same beam, a slice of which is accelerated, aligns the injected electrons and the plasma channel.

It is essential that the leading portion of the electron beam, which creates the plasma channel, is several plasma periods long, and that the beam density is smaller than the plasma density. This results in a high transformer ratio, with the decelerating field experienced by the leading beam much smaller than the accelerating field experienced by the trailing portion of the beam. The transformer ratio for the present accelerator scheme does not have the meaning of transferring energy from the driving beam to the trailing beam (or the wake left behind), since almost all of the energy for acceleration comes from the laser pulse.

It appears that a small portion of the beam can then be accelerated to GeV energies if: (i) the electron beam is stable and guides the laser pulse (ii) the plasma density provides sufficient focusing to the beam and a sufficient accelerating gradient (iii) the beam emittance is practically attainable consistent with reasonable laser power required for acceleration. We find that all these requirements might be met in, for example, an apparently feasible compact accelerator, with the parameters given in Table 1.

In the next section, we analyze in detail the requirements (i)–(iii). We show that the parameters that simultaneously meet these requirements lead also to rather interesting behavior of the electron beam, which is subjected to the self-consistent wakefields generated by both the laser and the beam itself.

Requirements for Laser Guiding

For the purpose of this calculation we assume that the electron beam has a top-hat axisymmetric radial density profile of radius r_b. The electron beam is assumed to be much longer than a plasma period:

$$n_b(\xi, r) = n_{b0} H(r_b - r) \exp{-\xi^2/(2\sigma_z^2)}, \qquad (2)$$

with $\sigma_z \gg c/\omega_p$, so that the density depression in the plasma is exactly equal to the beam density, n_b. Injecting a laser pulse near $\xi = 0$ insures that the laser pulse experiences the largest density depression, n_{b0}. Since beam electrons are assumed highly relativistic, they do not affect the electromagnetic wave because of their increased mass. Neglecting the relativistic mass increase of the plasma electrons reduces the guiding problem to that of an optical fiber with a radially dependent refractive index:

$$\begin{aligned} n_{ref} &= 1 - \frac{n_0 - n_b}{n_{crit}} \quad \text{for} \quad r < r_b \\ n_{ref} &= 1 - \frac{n_0}{n_{crit}} \quad \text{for} \quad r > r_b, \end{aligned} \qquad (3)$$

where $n_{crit} = m\omega_0^2/(4\pi e^2)$ is the critical density for the laser. For $\lambda_0 = 1\mu$ $n_{crit} = 10^{21}\,\text{cm}^{-3}$. Assuming circularly polarized laser light, the electric field of the laser is given by [11]

$$\vec{A} = (\vec{e}_x + i\vec{e}_y) \begin{bmatrix} A_0 J_0(hr) e^{i(k_0 z - \omega_0 t)}, & r < r_b \\ A_0 \frac{J_0(hr_b)}{K_0(qr_b)} K_0(qr) e^{i(k_0 z - \omega_0 t)} & r > r_b, \end{bmatrix}$$

with the characteristic equation given by

$$h \frac{J_1(hr_b)}{J_0(hr_b)} = q \frac{K_1(qr_b)}{K_0(qr_b)}$$
$$h^2 + q^2 = k_{pb}^2, \qquad (4)$$

where $k_{pb}^2 = 4\pi n_{b0} e^2 / mc^2$.

It can be demonstrated [19] that the wavelength independent product $V = k_{pb} r_b = \sqrt{4I/I_A}$, where $k_{pb}^2 = 4\pi n_{b0} e^2/mc^2$, and $I_A = 17\text{kA}$ is Alfven current, plays the role of a fiber parameter for the laser pulse. To confine one mode only, take $V < 1.8$. The appearance of a second mode could lead to unwanted hosing instabilities [17, 18]. Numerical studies of leaky channels, which have a significantly higher attenuation coefficient for higher-order modes than for the fundamental mode [16], demonstrated that various Raman instabilities (such as hosing [17, 18]) are strongly suppressed. On the other hand, choosing the fiber parameter to be too small leads to excessive transverse spreading and unwanted loss of laser power. Thus we assume fiber parameter to be unity. For $k_{pb} r_b = 1$ we find that the amplitude of the laser pulse drops to half of its amplitude at $r \approx 2r_b$ [11]. The effective spot size of the laser pulse is also about $r_L = 2r_b$. Peak current of $I_b = I_A/4 \approx 4.25$ kA is thus required for effective guiding of the laser pulse.

(ii) In choosing the plasma density, it is important to insure that the beam self-focusing is not strongly reduced by the return current in the plasma. Hence, the ambient plasma density has to be chosen such that the beam radius does not much exceed the collisionless skin-depth c/ω_p. Non-relativistically, we estimate the return current density on axis as $J_- = F_R(r=0) e n_b c$, which reduces the on-axis magnetic field pinching force by a factor $\eta = 1 - F_R(r=0)$, where the reduction factor F_R is associated with the finite radius of the electron bunch and is given by [14]:

$$F_R(r) = \begin{bmatrix} 1 - k_p r_b K_1(k_p r_b) I_0(k_p r), & \text{for } r < r_b \\ k_p r_b I_1(k_p r_b) K_0(k_p r), & \text{for } r > r_b \end{bmatrix}. \qquad (5)$$

E-beam energy	γmc^2	25 MeV
E-beam current	I_b	4.25 kAmp
E-beam radius	r_b	17 μ
E-beam emittance	ε_n	30 πmm mrad
E-beam density	n_b	10^{17} cm^{-3}
Bunch length	σ_z	1.0 psec
Laser radius	r_L	34 μ
Vector potential	a_0	0.9
Laser intensity	I	$2.3 \cdot 10^{18}$ W/cm^2
Laser power	P	80 TW
Pulse duration	τ_L	85 fs
Plasma density	n_p	$4 \cdot 10^{17}$ cm^{-3}
Accelerating gradient	W_z	16.3 GeV/m
Final energy	$\gamma_f mc^2$	1.0 GeV
Total length	L_t	6.3 cm
Rayleigh length	Z_r	0.3 cm

Table 1: List of parameters for beam-channeled laser wakefield accelerator

Choosing $k_p r_b = 2$, we find the reduction on axis $F_R(r=0) \equiv F_R = 0.72$.

For a self-focused beam normalized emittance, beam radius, and beam energy are related through [19]

$$\varepsilon_n = r_b \left(\frac{\gamma \eta I_b}{2 I_A} \right)^{1/2}. \qquad (6)$$

(iii) In choosing the electron beam spot size several criteria should be met: (1) the radius should not be prohibitively large so as to make the required laser power exceed the capabilities of the present and soon-to-come laser systems (2) the radius should not be too small as to make the required beam emittance too stringent. Laser power and emittance given in Table 1 are achievable with today's technologies.

Using these parameters as guidelines, we show in the next section how a GeV electron accelerator can be designed and demonstrate the advantages of a BCLWA over conventional laser wakefield and plasma wakefield accelerators.

Wakefield Generation

To simplify the calculation we assume that the laser pulse has a flat-top longitudinal profile of width equal to half plasma wavelength:

$$a_0^2(\xi, r) = a_0^2 \left[H(\xi) - H(L - \xi) \right] \psi(r), \tag{7}$$

where $\psi(r)$ is a normalized transverse laser profile, with $\psi(0) = 1.0$, and $L = \lambda_p/2$. In the limit of $a_0^2 \ll 1$ the accelerating field at $r = 0$ is approximately given by

$$E_z(\xi, r = 0) = 4\pi e \int_{-\infty}^{\xi} d\xi' \frac{\sin k_p(\xi - \xi')}{k_p} \left[n_0 \frac{\partial}{\partial \xi'} \left(\frac{a_0^2}{2} \right) + F_R(r = 0) \frac{\partial n_b(\xi')}{\partial \xi'} \right], \tag{8}$$

where we have neglected the slight depletion in the plasma density inside the channel (since $n_{b0}/n_p = 0.25$). A more accurate description should also take into account an additional accelerating field from the surface charges at the channel boundary [11, 12]. For linear polarization a_0^2 is replaced by $a_0^2/2$.

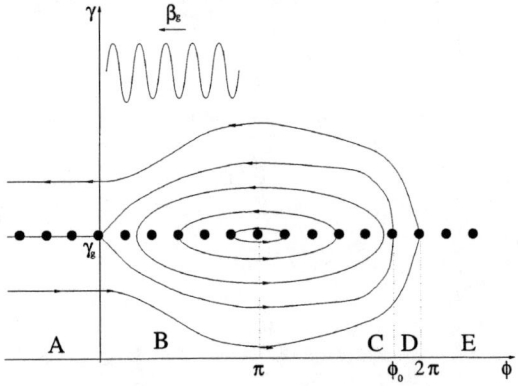

Figure 1: Schematic of particle phase space in a beam channeled laser wakefield accelerator. A – guiding beam; B – laser pulse location, region of beam erosion; C – trapped accelerated electrons; D – untrapped accelerated electrons; E – trailing beam. Initially monoenergetic electrons are schematically shown as shaded circles.

The self-consistent nature of the accelerating wake is clear from Eq.(8) since the beam profile $n_b(\xi)$ is affected by the wake. In obtaining Eq.(8) a flat-top radial profile for the electron beam was assumed. Figure 1 presents a schematic phase space of beam electrons, which are assumed to have the initial velocity equal to the group velocity of the laser $c\beta_g$. Electrons in region A are, by causality, unaffected by the laser, and are only subject to weak self-generated decelerating field. Their function is to create a plasma channel and provide guiding to the laser pulse. On the other hand, electrons in the region B experience a strong decelerating wake, rapidly slip forward in phase, and are transfered into the accelerating region C. In the context of a GeV-scale acceleration we are mainly concerned with the wake generation in region C. Thus we do not study the dynamics of the beam electrons and the wake for $\xi > \lambda_p$ (region E).

The wake generation process can now be separated into two stages: the initial stage, when erosion of the decelerated portion of the electron beam takes place, and the final stage, when beam electrons are accelerated to high energy in a self-consistent wakefield, generated by both the laser and the beam. This separation of the erosion and acceleration stages is possible due to separation of time scales for particle deceleration and acceleration. During the erosion stage beam electrons initially residing in region slow down and slips forward in ξ. Depending on laser intensity, some, or all of the initially decelerated electrons, injected with velocity equal to phase velocity of the wake, are trapped. Only the case of high laser intensity is schematically shown in Fig.1, indicating that all beam electrons between $\phi = 0$ and $\phi = 2\pi$, where $\phi = k_p\xi$, are trapped. In both cases, electrons, which are initially at a decelerating phase, rapidly execute half a synchrotron oscillation, thus arriving at an accelerating phase. Their departure generates a sharp edge in the electron density, which modifies the wake. Untrapped electrons (if any) slip further in ξ, leaving the region of interest, while the trapped electrons are accelerated to high energy.

- **Erosion stage**

Here we demonstrate that electrons in region A, which are responsible for generation of the plasma channel, are only weakly decelerated by the self-generated wake. Indeed, in the limit of $k_p\sigma_z \gg 1$, for $\xi < 0$, Eq.(8) can be shown to reduce to

$$E_- \approx \frac{4\pi e n_0}{k_p^2} F_R(r=0) \frac{\partial}{\partial \xi'} \left(\frac{n_b(\xi')}{n_0} \right). \qquad (9)$$

Using the beam profile from Eq.(2) and calculating the maximum decelerating field on the beam results in

$$E_-^{max} \approx \frac{4\pi e n_0}{k_p} \frac{0.6 n_b F_R(r=0)}{n_0(k_p\sigma_z)}, \qquad (10)$$

which has to be compared with the peak accelerating field generated by the laser pulse:

$$E_+^{max} \approx \frac{4\pi e n_0}{k_p} a_0^2. \quad (11)$$

An effective transformer ratio can be phenomenologically introduced as $T = E_+^{max}/E_-^{max}$, yielding

$$T \approx \frac{1.7 a_0^2 (k_p \sigma_z) n_0}{n_b F_R(r=0)}. \quad (12)$$

For the parameters from Table 1 we obtain $T \approx 300$. Even though in the after-erosion phase the peak accelerating gradient will be significantly reduced by the wake generated by the sharply-edged guiding beam (which happens to be out of phase with the wake generated by the laser), the resulting transformer ratio will still significantly exceed the required $T_{crit} = \gamma_f/\gamma = 40$, necessary to accelerate the witness beam to the final energy $\gamma_f mc^2 = 1\text{GeV}$ using the low-energy $\gamma_b mc^2 = 25\text{MeV}$ guiding beam.

- **After-erosion stage**

Below we calculate a plasma wake, based on a simplifying assumptions that beam density is piece-wise constant and vanishes in the regions of $E_z > 0$.

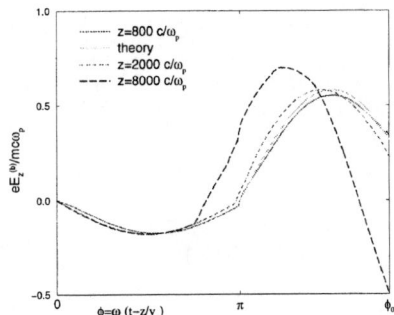

Figure 2: Wake generated by the bunched beam; $a_0 = 0.84$, $\phi_0 = 1.82\pi$. Final energy 1GeV is reached at $z_{acc} = 8000 c/\omega_p$.

The relativistic electron dynamics in a fixed wakefield can be described by a Hamiltonian [14]

$$H = \sqrt{\gamma^2 - 1} - \frac{\gamma}{\beta_g} + V(\phi), \qquad (13)$$

where $\phi = k_p \xi$, and $E_z = (mc\omega_p/e)\partial V/\partial \phi$. Equations of motion are then given by $\phi' = \partial H/\partial \gamma$, and $\gamma' = -\partial H/\partial \phi$, where prime denotes differentiation with respect to $\bar{z} = k_p z$. The wake potential $V(\phi)$ is determined self-consistently from Eq.(8). Depending on the laser intensity, two different regimes are possible: $V(2\pi) < V(0)$ (high-intensity regime, schematically shown in Fig.1, $a_0^2 > 4F_R n_{b0}/n_0$) and $V(2\pi) > V(0)$ (low-intensity regime, $2F_R n_{b0}/n_0 < a_0^2 < 4F_R n_{b0}/n_0$). We consider these two regimes separately.

In the high-intensity regime electrons in region $0 < \phi < \phi_0$, where $V(0) = V(\phi_0)$, are trapped. Electrons, from region B are rapidly transferred after a half-synchrotron oscillation to region C, generating sharp density gradients at $\phi = 0$ and $\phi = \phi_0$. Both the trapped ($\pi < \phi < \phi_0$) and untrapped ($\phi_0 < \phi < 2\pi$) accelerated electrons slip back in ϕ at a much slower rate, and can be assumed fixed in phase. Assuming that decelerated electrons from region B are uniformly spread out through the region C, we estimate the electron density in region C as $n'_b = n_{b0}\phi_0/(\phi_0 - \pi)$, and the electron density in region D as unchanged at n_{b0}. Introducing $u = (a_0^2/2 - n_{b0}/n_0 F_R)$ and $g = [a_0^2 - (n_{b0}/n_0 + n'_b/n_0)F_R]/u$, one can demonstrate that the wake potential is given by

$$V(\phi) = -u\cos\phi \qquad \text{for } 0 < \phi < \pi \qquad (14)$$
$$= -u(g\cos\phi + g - 1) \quad \text{for } \pi < \phi < \phi_0, \qquad (15)$$

where $1 < g < 2$. The resulting beam density n'_b, consistent with such a wake, becomes an implicit function of the laser intensity (characterized by the quantity u), and can be expressed in terms of u and g. The quantity g, itself an implicit function of u, can now be determined from a nonlinear equation:

$$\sin^2\left(\frac{\pi}{2}\frac{F_R n_{b0}/n_0}{u(2-g)}\right) = \frac{1}{g}, \qquad (16)$$

where $C = F_R n_{b0}/(n_0 u) < 1$.

In the low-intensity regime, the self-consistent wake potential is also described by Eqs.(14,15), except that the domain in Eq.(15) is extended from to 2π. The difference between the two regimes is that here $g < 1$, so that electrons initially at $0 < \phi < \phi_1$, (where $V(\phi_1) = V(2\pi)$), are untrapped, and rapidly leave the

$0 < \phi < 2\pi$ region of interest. At the same time, electrons initially occupying $\phi_1 < \phi < \pi$ are trapped, and are transferred to the region $\pi < \phi < 2\pi$, resulting in an overall density there $n'_b = n_{b0}(2\pi - \phi_1)/\pi$. A nonlinear equation for g, analogous to Eq.(16) can be derived for this regime:

$$\sin^2\left(\frac{\pi}{2}\frac{u(2-g)}{F_R n_{b0}/n_0}\right) = g, \qquad (17)$$

Equations (16,17) can be solved numerically for various values of C, corresponding to different laser intensities. The required laser intensity and peak accelerating gradient achieved at $\phi = 3\pi/2$ are given by $a_0^2 = 2n_{b0}/n_0 F_R(1+C)/C$ and $W_z = 4\pi e n_0 g F_R n_b/(k_p C n_0)$.

As an example, consider a high-intensity regime $C = 0.75$, and, using the plasma and beam parameters from Table 1, obtain $a_0^2 \approx 0.84$, $\phi_0 \approx 1.82\pi$, and $W_z \approx 15.6 \text{GeV/m}$. An important caveat here is that the laser intensity needed to achieve such an accelerating gradient may be somewhat higher since the present calculation assumed $a_0^2 \ll 1$, which is only marginally satisfied for the suggested parameters.

For a non-evolving potential given by Eq.(15) equations of motion for beam particles were integrated numerically to verify that the plasma wake, generated by the evolving electron beam are indeed close to the ones assumed in deriving Eq.(15). Comparison between the theoretically assumed and actual beam wakefields (computed for different propagation distances) is presented in Fig.2. As evident from Fig.2, the actual beam wakefields become almost indistinguishable from the theoretically assumed after a propagation distance $z < 0.1 z_{acc}$, where $z_{acc} = 8000 c/\omega_p \approx 6.3$cm is the distance required for 1 GeV acceleration. Beam electrons occupying the region around $\xi = 3\lambda_p/2$ are accelerated by being trapped in the plasma wakefield. The maximum energy the trapped electrons can acquire is limited by the height of the separatrix, and for $u \ll 1$ is given by [14] $\gamma_{max} = 4\gamma_g^2 gu = 1.3 \text{GeV}/mc^2$. As it is clear from Fig.2, beam wakefield starts deviating from the theoretically assumed for $z \approx z_{acc}$. The physical reason for this is bounce motion of beam electrons nearby the equilibrium point $\phi = \pi$. Self-consistent particle simulations are needed to accurately include the two-dimensional effects, relativistic effects, and to refine the calculation of beam wake for long acceleration distances.

For comparison, a laser-wakefield accelerator with identical laser and plasma parameters, limited by diffraction, would only be able to accelerate electrons to $2eE_+^{max} Z_R = 300$MeV. A plasma wakefield accelerator would only be effective in accelerating the witness beam to twice the energy of the driving beam, unless special shaping, with slow rise and abrupt termination, could be achieved.

An interesting way of reducing the beam-generated wake (which interferes destructively with the laser-generated wake) was suggested by E. Esarey [20]. Choosing a shaped laser pulse, with a long rise (over several plasma periods) and a sharp termination at $\xi = 0$ would result in a very small *deceleration* field in the region A and a large *acceleration* field at $\xi = \pi/2\omega_p$. In this configuration, electron beam does not develop sharp density gradients for $\xi < \pi/2\omega_p$, and the peak accelerating gradient is approximately given by $W_z = 2\pi e n_0 a_0^2 / k_p$. Hence, the same acceleration gradient of 15.6 GeV/m can be achieved with lower laser intensity $a_0^2 \approx 0.5$.

In summary, we show how an intense (over 4kA peak current) electron beam can create a plasma channel which guides a short laser pulse, generating a plasma wake for electron acceleration. Among the advantages of this scheme over the conventional laser-wakefield designs is the long acceleration distance (not limited by diffraction); perfect alignment between the channel and the accelerated electrons; absence of potentially dangerous transverse instabilities of the laser pulse (such as hosing); control over the profile of a plasma channel through the control over the electron beam profile.

Acknowledgments

The authors acknowledge insightful discussions with J. S. Wurtele, T. C. Katsouleas, and E. Esarey. One of us (GS) acknowledges the support of a US DOE Postdoctoral Fellowship. This work was supported in part by the US DOE under contract number DE-AC02-76-CHO3073.

References

[1] T. Tajima and J. M. Dawson, *Phys. Rev. Lett.* **43**, 267 (1979); L. M. Gorbunov and V. I. Kirsanov, *Zh. Exp. Teor. Fiz.* **93**, 509 (1987) [*Sov. Phys. JETP* **66**, 290 (1987)]; C. Joshi, W. B. Mori, T. Katsouleas, *et al.*, *Nature* **311**, 525 (1984).

[2] R. D. Ruth, A. Chao, P. L. Morton, and P. B. Wilson, *Part. Accel.* **17**, 171 (1985).

[3] K. L. F. Bane, P. Chen, P. B. Wilson, *IEEE Trans. Nucl. Sci.* **32,** 3524 (1985).

[4] T. Katsouleas, *Phys. Rev. A* **33,** 2056 (1986).

[5] J. S. Wurtele, *Phys. Fluids B* **5,** 2363 (1993); J. S. Wurtele, *Physics Today* **47** (no. 7), 33 (1994), and references therein.

[6] P. Sprangle, E. Esarey, *Phys. Fluids B* **4,** 2241 (1992).

[7] C. E. Clayton, *et al.*, *Phys. Rev. Lett.* **30,**37 (1993).

[8] P. Sprangle, E. Esarey, A. Ting, and G. Joyce, *Appl. Phys. Lett.* **53,** 2146 (1988).

[9] J. M. Rax and N. J. Fisch, *Phys. Fluids B* **4,** 1323 (1992); J. M. Rax and N. J. Fisch, *Phys. Fluids B* **5,** 2680 (1993).

[10] E. Esarey, P. Sprangle, J. Krall, A. Ting, and G. Joyce, *Phys. Fluids B* **5,** 2578 (1993).

[11] T. C. Chiou, T. Katsouleas, C. Decker, W. B. Mori, J. S. Wurtele, G. Shvets, and J. J. Su, *Phys. Plasmas* **2,** 310 (1995).

[12] G. Shvets, J. S. Wurtele, T. C. Chiou, and T. Katsouleas, *IEEE Trans. Plasma Sci.* **24,** 351 (1996).

[13] A. W. Chao, in *Physics of High Energy Particle Accelerators (SLAC Summer School, 1982)*, AIP Conf. Proc. No. 105, ed. by M. Month (AIP, New York, 1983).

[14] E. Esarey, P. Sprangle, J. Krall, and A. Ting, *IEEE Trans. Plasma Sci.* **24,** 252 (1996).

[15] C. G. Durfee and H. M. Milchberg, *Phys. Rev. Lett.* **71,** 2409 (1993); C. G. Durfee, J. Lynch, and H. M. Milchberg, *Phys. Rev. E* **51,** 2368 (1995); H. M. Milchberg, T. R. Clark, C. G. Durfee, T. M. Antonsen, P. Mora, *Phys. Plasmas* **3,** 2149 (1996).

[16] T. M. Antonsen, Jr. and P. Mora, Phys. Rev. Lett. **744,** 4440 (1995).

[17] G. Shvets and J. S. Wurtele, *Phys. Rev. Lett.* **73,** 3540 (1994).

[18] P. Sprangle, J. Krall, and E. Esarey, *Phys. Rev. Lett.* **73,** 3544 (1994).

[19] D. H. Whittum, PhD thesis, UC Berkeley (1990).

[20] E. Esarey, private communication (1996).

Laser Instabilities and Coupling Efficiency in Hollow Channel Plasma Wakefield Accelerators

T. C. Chiou, T. Katsouleas
Department of Electrical Engineering–Electrophysics,
University of Southern California
Los Angeles, CA 90089-0484

W. B. Mori
University of California, Los Angeles, CA 90024

G. Shvets
Princeton Plasma Physics Laboratory, Princeton, NJ 08543

Abstract

The stability and distortion of intense laser pulses ($I \lesssim 10^{18} W/cm^2$) that are propagating in and guided by underdense plasma channels is investigated both analytically and through Particle-In-Cell (PIC) simulations. In a hollow channel where the plasma is totally evacuated from the channel center Raman Forward Scattering (RFS) can be largely suppressed if the channel radius is wide enough. The laser hosing can be suppressed by reducing the channel radius such that the channel can guide only one mode($a < \pi/2$). The pump depletion and dispersive spread lengths are also calculated. The PIC simulations are performed and shown to agree with these analytical predictions. These results suggest that lasers are more stable in hollow channels than in other plasmas and that high energy gain and high beam quality may be achievable in hollow plasma channels.

I INTRODUCTION

Recently it has been shown that the wakefield structure in hollow plasma channels has advantages for maintaining high beam quality compared to parabolic channels or homogeneous plasmas [1]. To achieve high energy gain and high coupling efficiency, in any of these cases the laser is required to propagate many Rayleigh lengths. Thus laser instabilities and laser distortion become important issues. In both parabolic channels and homogeneous plasmas the laser was shown to have Raman instabilities and hose instability [2,3].

In hollow channel plasmas, the laser hosing can be avoided by making a single mode channel. However, it may still undergo Raman instabilities. In Ref.[1] it was shown that a hollow plasma channel can support both a body mode (mode oscillating at frequency ω_p) and a surface mode (mode with frequency $\omega_{ch} = \omega_p/\sqrt{1 + k_p a}$).

These modes must co-exist in order to satisfy the boundary conditions. In Ref. [4] the coupling between the laser and the body mode through Raman Forward Scattering (RFS) was estimated. Here in Sec. II we revisit this calculation including additional contributions from the surface mode. The growth rates for both the body mode and the surface mode are found. In Sec. III, we turn to the question of how efficiently laser energy can be coupled to the plasma wake in a hollow channel (i.e., we are concerned with how much laser pump depletion can occur and with that effect on the quality of the wake). The zeroth order expressions of pump depletion and dispersion lengths are derived. In Sec.IV, PIC simulations are performed to test the theories given in previous sections. Sec. V is a brief discussion and conclusion.

II Raman Forward Scattering in Hollow plasma channels

Recently we have analysized RFS in hollow plasma channels [4] in space-time domain where we have neglected the contribution from the plasma/vacuum boundaries. In this paper we use Fourier analysis to recover the terms missing in the previous analysis.

Again our simplified analytic model consists of an infinite homogeneous, cold plasma of uniform density n_0 with an evacuated region for $-a < y < a$ in slab geometry. The laser propagates in the channel in the z-direction and is initially given by the TE_{00} optical fiber mode structure :

$$\vec{E}_0 = \hat{x} \left[\begin{array}{ll} E_0 \cos(hy) e^{i(k_0 z - \omega_0 t)} & |y| \leq a \\ E_0 \cos(ha) e^{-p(|y|-a)} e^{i(k_0 z - \omega_0 t)} & |y| \geq a \end{array} \right] \quad (1)$$

where p is defined implicitly through $h^2 + p^2 = k_p^2$ and $p = h\tan(ha)$.

The wave equation that governs both the plasma modes and the laser wave can be derived by combining Maxwell's equations and the fluid momentun equation as [5]

$$c^2 \nabla \times \nabla \times \vec{E} + \omega_p^2 \vec{E} + \frac{\partial^2 \vec{E}}{\partial t^2} = \vec{Q}_{NL} \quad (2)$$

where $\vec{Q}_{NL} = \frac{-m\omega_p^2}{2e} \nabla(\vec{v}\cdot\vec{v}) - \frac{\partial}{\partial t}(\vec{v}(\nabla\cdot\vec{E}))$. Notice that $\vec{Q}_{NL}$ consists of two terms. The $\vec{v}\cdot\vec{v}$ term comes from the combination of the convective derivative $(\vec{v}\cdot\nabla)$ and the magnetic force $(\frac{\vec{v}}{c}\times\vec{B})$ terms in the fluid momentum equation while the $\vec{v}(\nabla\cdot\vec{E})$ term originates from the current source in the Ampere's law.

Let's consider 3-wave process first, we can then write the total electric field as the superposition of the pump wave ($\vec{E}_0$), the decayed light wave ($\vec{E}_1$) and the plasma wave ($\vec{E}_p$). To first order, the electron velocity can be approximated as $\partial \vec{v}_i/\partial t = -e\vec{E}_i/m$ where $i = 0, 1, p$. Since plasma waves have both y and z components while the laser light wave has only an x component (TE mode), thus, to first order, $\vec{v_{0,1}} \cdot \vec{v_p} = 0$. Also we can assume $\nabla \cdot \vec{E}_{0,1} = 0$ and $\nabla \cdot \vec{E}_p = -4\pi e \delta n$ Eq.(2) can then be decomposed into

$$\nabla \times \nabla \times \vec{E}_1 + \frac{\omega_p^2}{c^2}\vec{E}_1 + \frac{1}{c^2}\frac{\partial^2 \vec{E}_1}{\partial t^2} = \frac{4\pi e}{c^2}\frac{\partial}{\partial t}(\delta n \vec{v}_0) \quad (3)$$

for the decayed light wave and

$$\nabla \times \nabla \times \vec{E}_p + \frac{\omega_p^2}{c^2}\vec{E}_p + \frac{1}{c^2}\frac{\partial^2 \vec{E}_p}{\partial t^2} = \frac{\omega_p^2}{c^2}\nabla(-\frac{m}{e}(\vec{v}_0 \cdot \vec{v}_1)) \quad (4)$$

for the plasma wave where we have assumed the pump wave amplitude is a constant (i.e., no pump depletion). Let $\vec{E}_p = E_{py}\hat{y} + E_{pz}\hat{z}$, $\vec{E}_{0,1} = E_{0,1}\hat{x}$ and assume all the components have the following propagation wave solution forms

$$E_{py,z} = \frac{1}{2}[E_{y,z}(y)e^{i(kz-\omega t)} + c.c.]$$

$$E_{0,1} = \frac{1}{2}[E_{0,1}(y)e^{i(k_{0,1}z-\omega_{0,1}t)} + c.c.]$$

$$-4\pi e \delta n = \frac{1}{2}[n_p(y)e^{i(kz-\omega t)} + c.c.]$$

Here we have assumed that ω_0 and all k's are real but allowed ω and ω_1 to be complex numbers. The dispersion relation for (ω_0, k_0) is $\omega_0^2 = \omega_p^2 + c^2 k_0^2 - c^2 p^2$ and in order to match the frequencies we require $k = k_0 - k_1$ and $\omega = \omega_0 - \omega_1^*$.

Substituting the above expressions into the wave equations (3) and (4) we can get

$$ik\frac{\partial E_z}{\partial y} + (k^2 + \frac{\omega_p^2}{c^2} - \frac{\omega^2}{c^2})E_y = \frac{-1}{2}\frac{\omega_p^2}{c^2}\frac{e}{m\omega_0\omega_1^*}\frac{\partial(E_0 E_1^*)}{\partial y} \quad (5)$$

$$ik\frac{\partial E_y}{\partial y} - \frac{\partial^2 E_z}{\partial y^2} + (\frac{\omega_p^2}{c^2} - \frac{\omega^2}{c^2})E_z = \frac{-1}{2}\frac{\omega_p^2}{c^2}\frac{e}{m\omega_0\omega_1^*}(ik)(E_0 E_1^*) \quad (6)$$

$$-\frac{\partial^2 E_1}{\partial y^2} + \Delta_1^2 E_1 = \frac{1}{2}\frac{eE_0\omega_1}{m\omega_0 c^2}n_p^* \quad (7)$$

where $\Delta_1^2 = k_1^2 + \frac{\omega_p^2}{c^2} - \frac{\omega_1^2}{c^2}$. The solutions of Eqs. (5) and (6) can be found by following the same precedures as in Ref.[1] which are

$$E_z = \begin{bmatrix} A\cosh(\Delta y) & |y|<a \\ Be^{-s(|y|-a)} + \frac{ik\omega_p^2 F}{\omega_p^2-\omega^2} & |y|>a \end{bmatrix} \quad (8)$$

$$E_y = \begin{bmatrix} \frac{-ik}{\Delta}A\sinh(\Delta y) & |y|<a \\ \frac{ik}{s}Be^{-s(|y|-a)} + \frac{\omega_p^2}{\omega_p^2-\omega^2}\frac{\partial F}{\partial y} & |y|>a \end{bmatrix} \quad (9)$$

where $\Delta = \sqrt{k^2 - \frac{\omega^2}{c^2}}$, $s = \sqrt{\Delta^2 + \frac{\omega_p^2}{c^2}}$ and $F = -\frac{1}{2}\frac{eE_0 E_1^*}{m\omega_0 \omega_1^*}$. The constants A and B can be determined by matching the boundary conditions : (1) E_z continuous, (2) $\omega_p^2 \partial F/\partial y + (\omega^2 - \omega_p^2)E_y$ continuous. Notice that the second boundary condition is slightly different from that found in Ref.[1]. This is due to the fact that $\partial E_1/\partial y$ is discontinuous at the boundaries which can be checked ,a posteriori, from Eq. (7). We can then obtain

$$A = \frac{-ik\omega_p^2 F(a)}{(\omega^2 - \omega_p^2)\cosh(\Delta a) + \omega^2(sa)\frac{\sinh(\Delta a)}{\Delta a}} \quad (10)$$

$$B = \frac{\omega^2 s}{\omega_p^2 - \omega^2} \frac{\sinh(\Delta a)}{\Delta} A \quad (11)$$

We also notice that E_y is discontinuous at the channel edges. This implies that there are surface charges accumulated at the channel boundaries. From Poisson's equation it is easy to show $n_p = \frac{\partial E_y}{\partial y} + ikE_z$. Thus n_p can be found as

$$n_p = \begin{bmatrix} 0 & |y|<a \\ \frac{\omega_p^2}{\omega_p^2 - \omega^2}[\frac{\partial^2 F}{\partial y^2} - k^2 F] & |y|>a \\ \frac{\omega_p^2}{\omega_p^2 - \omega^2}[\frac{\partial F}{\partial y} + \frac{ikA}{\Delta}\sinh(\Delta a)]\delta(y \pm a) & |y|=a \end{bmatrix} \quad (12)$$

Substituting into Eq. (7) we can get

$$\pounds E_1 = \begin{bmatrix} 0 & |y|<a \\ \frac{1}{4}(\frac{V_{osc}}{c})^2 \frac{\omega_p^2 \psi}{\omega_p^2 - (\omega^*)^2}[\frac{\partial^2 (E_1 \psi)}{\partial y^2} - k^2(E_1 \psi)] & |y|>a \\ \frac{1}{4}(\frac{V_{osc}}{c})^2 \frac{\omega_p^2 \psi}{\omega_p^2 - (\omega^*)^2}[\frac{\partial (E_1 \psi)}{\partial y} + k_s(E_1 \psi)]\delta(y \pm a) & |y|=a \end{bmatrix} \quad (13)$$

where $\pounds = \partial^2/\partial y^2 - \Delta_1^2$, $V_{osc} = \frac{eE_0}{m\omega_0}\cos(ha)$, $\psi(y) = e^{-p(|y|-a)}$ and

$$k_s = \frac{(k^2 a)\omega_p^2 \frac{\sinh(\Delta^* a)}{\Delta^* a}}{[(\omega^*)^2 - \omega_p^2]\cosh(\Delta^* a) + (\omega^*)^2 (s^* a)\frac{\sinh(\Delta^* a)}{\Delta^* a}}$$

For the study of RFS, it is reasonable to assume $\omega/k \approx c$ because all the waves should propagate at speeds close to c. That is $\Delta a \approx 0$. Under this assumption, Eq. (13) can be reduced to

$$\pounds E_1 = \begin{bmatrix} 0 & |y|<a \\ \frac{1}{4}(\frac{V_{osc}}{c})^2 \frac{\omega_p^2 \psi}{\omega_p^2 - (\omega^*)^2}[\frac{\partial^2 (E_1 \psi)}{\partial y^2} - k^2(E_1 \psi)] & |y|>a \\ \frac{1}{4}(\frac{V_{osc}}{c})^2 \frac{\omega_p^2 \psi}{\omega_p^2 - (\omega^*)^2}[\frac{\partial (E_1 \psi)}{\partial y} + \frac{\omega_{ch}^2 k^2 a}{(\omega^*)^2 - \omega_{ch}^2}(E_1 \psi)]\delta(y \pm a) & |y|=a \end{bmatrix} \quad (14)$$

Eq. (14) can be solved by using the mode expansion method. That is, assuming $E_1(y) = \sum_n a_n \phi_n(y)$ where $\phi_n(y)$ is the normal mode of the operator $\pounds$ at frequency

ω_1 which is just the optical fiber mode. Substituting into Eq. (14), multiplying both sides by $\phi_n^*(y)$, integrating across the transverse coordinate y, and, for a single mode channel, retaining only the $n = 0$ term, we can obtain

$$p^2 - k_1^2 - \frac{\omega_p^2}{c^2} + \frac{\omega_1^2}{c^2} = \frac{1}{4}(\frac{V_{osc}}{c})^2 \frac{\cos(ha)^2}{2 + 2pa} \frac{\omega_p^2}{\omega_p^2 - (\omega^*)^2}(4p^2 - k^2) +$$

$$\frac{1}{4}(\frac{V_{osc}}{c})^2 \frac{\cos(ha)^2}{2 + 2pa} \frac{\omega_p^2}{\omega_p^2 - (\omega^*)^2}[-8p^2 + \frac{\omega_{ch}^2 k^2 (4pa)}{(\omega^*)^2 - \omega_{ch}^2}] \quad (15)$$

To study the growth rate of the body mode, we may assume $\omega^* = \omega_p + i\omega_i$ where $\omega_p \gg \omega_i \gg \omega(\omega_p/\omega_0)$ and choose k such that $(\omega_0 - \omega_p)^2 = \omega_p^2 + c^2(k_0 - k)^2 - c^2 p^2$. Then it is easy to show that $ck \approx \omega_p$ and $ck_0 \approx \omega_0$. Under these assumption, the growth rate ω_i can be found as

$$\omega_i^2 = \frac{1}{16}(\frac{V_{osc}}{c})^2 \frac{\omega_p^3}{\omega_0 - \omega_p} \frac{\cos(ha)^2}{2 + 2pa}[(1 - \frac{4p^2}{k_p^2}) + (\frac{8p^2}{k_p^2} - \frac{4p}{k_p})] \quad (16)$$

The first parenthesis in the square bracket is the contribution from the plasma outside the channel. The second parenthesis contains two terms. The first term is due to the discontinuity of the ambient plasma density (or the plasma/vacuum boundaries). The second term is the contribution from the surface mode induced surface charges. It is easy to see that as $2p \to k_p$, $\omega_i \to 0$ (i.e., as the channel width a increases from 0 to $\pi/3\sqrt{3}$, the growth rate drops from the homogeneous RFS growth rate to zero). However, as $k_p a$ increases further, the square bracket increases again and a nonzero growth rate is recovered. This inconsistency between physical intuition and the theory may be due to the fact that the single mode assumption is only good at $y \gg a$ as can be seen from Eq. (14) where the right-hand-side (RHS) approaches zero and the solution approaches the normal mode solution. However, as y approch a, the RHS is non-negligible and the single mode solution is no longer valid.

For 4-wave process, we may assume $E_1 = \frac{1}{2}\{E_+(y)e^{i(k+z-\omega_+ t)} + E_-(y)e^{i(k-z-\omega_- t)} + c.c.\}$ where $k_\pm = k_0 \pm k$, $\omega_+ = \omega_0 + \omega$ and $\omega_- = \omega_0 - \omega_*$. The pondermotive potential F is now replaced by $F = \frac{-1}{2}\frac{eE_0(y)}{m\omega_0}(\frac{E_-^*}{\omega_-^*} + \frac{E_+}{\omega_+})$. The density pertubation has the same form as Eq. (12). The equations for E_+ and E_- become

$$-\frac{\partial^2 E_+}{\partial y^2} + \Delta_+^2 E_+ = \frac{1}{2}\frac{eE_0\omega_+}{m\omega_0 c^2}n_p \quad (17)$$

$$-\frac{\partial^2 E_-}{\partial y^2} + \Delta_-^2 E_- = \frac{1}{2}\frac{eE_0\omega_-}{m\omega_0 c^2}n_p^* \quad (18)$$

where $\Delta_\pm^2 = k_\pm^2 + \omega_p^2/c^2 - \omega_\pm^2/c^2$. Using the single mode approximation, the dispersion relation for (ω, k) can be found as

$$D_+ D_-^* = \alpha\beta(D_+ + D_-^*) \quad (19)$$

where
$$D_\pm = \omega_\pm^2/c^2 - \omega_p^2/c^2 - k_\pm^2 + p^2,$$

$$\alpha = \frac{1}{4}\left(\frac{V_{osc}}{c}\right)^2 \frac{\cos(ha)^2}{2+2pa} \frac{\omega_p^2}{\omega_p^2 - \omega^2},$$

and
$$\beta = -4p^2 - k^2 + \frac{\omega_{ch}^2 k^2}{\omega^2 - \omega_{ch}^2}(4pa).$$

Again by assuming $\omega^* = \omega_p + i\omega_i$ and $(\omega_0 - \omega_p)^2 = \omega_p^2 + c^2(k_0 - k)^2 - c^2 p^2$, it is easy to show that

$$\omega_i^2 = \frac{1}{8}\left(\frac{V_{osc}}{c}\right)^2 \frac{\omega_p^4}{\omega_0^2} \frac{\cos(ha)^2}{2+2pa}\left[(1 - \frac{4p^2}{k_p^2}) + (\frac{8p^2}{k_p^2} - \frac{4p}{k_p})\right] \quad (20)$$

It is also easy to show that at $|\omega| \approx \omega_{ch}$ we have a surface instability which can excite the surface electromagnetic mode. However, instead of dealing with discrete frequencies it is instructive to solve Eq. (19) numerically for each given k. The result is shown in Fig. 1 for $V_{osc}/c = 0.8$ and $k_p a = 1$. In solving the dispersion relation, we have only kept solutions with $Re(\omega) \approx ck$ consistent with the assumption $\triangle a \approx 0$. Fig. 1 clearly shows that the surface mode instability dominates the

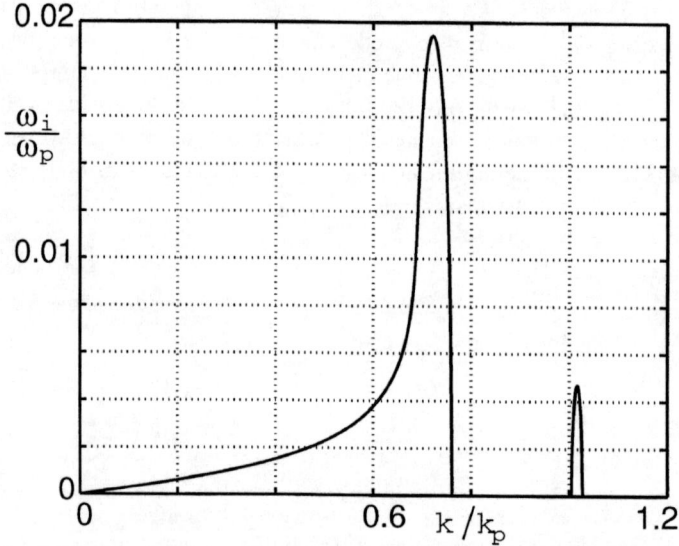

Figure 1: 4-wave RFS growth rate vs. k for $k_p a = 1$ and $V_{osc}/c = 0.8$.

body mode instability.

III Pump Depletion and Dispersion Spread of the Laser

In the previous section, we have neglected the evolution of the pump wave $\vec{E}_0$. There are at least two factors that can affect the pump laser field. First, due to energy conservation, the laser amplitude must decrease, i.e. pump deplete. Second, finite length pulses (especially short pulses) have finite bandwidth and since the group velocity of the pulse is a function of frequency, the pulse will have dispersion. Thus as the pulse propagates down the channel its length begins to spread.

In the following analysis, we consider these two effects separately. That is, we assume these two effects won't affect each other. Under this assumption, the pump depletion length (L_p) can be estimated by simply equating the energy in the laser pulse to the wavefield energy plus the particle kinetic energy. The result is

$$L_p = 24 l_\tau G_m \left(\frac{V_{osc}}{c}\right)^{-2} \left(\frac{\omega_0}{\omega_p}\right)^2 \tag{21}$$

where l_τ is the pulse length and G_m is the geometry factor given by

$$G_m = \frac{k_p a (1 + \frac{1}{pa})}{\frac{4+k_p a}{1+k_p a} + \frac{6+9k_p a + 6(k_p a)^2 + 4(k_p a)^3}{3(1+k_p a)^2}} \cdot \frac{1}{\cos(ha)^4} \tag{22}$$

ω_0 is the laser frequency and ω_p is the plasma frequency. ha and pa are determined by the equations given below Eq. (1).

For a short pulse, the spread in group velocity can be estimated by

$$\Delta v_g = \frac{\partial v_g}{\partial \omega_0} \Delta \omega_0 \tag{23}$$

In a hollow channel with $\omega_0 \gg \omega_p$, the group velocity of the pulse can be found as $v_g/c \approx 1 - \omega_p^2/2\omega_0^2$ (a more preccise result is given in Ref.[1]). For a pulse with profile proportional to $e^{-z^2/2l_\tau^2}$, i.e. a Gaussion pulse, the bandwidth of the pulse is $\Delta \omega_0 \approx c/l_\tau$. If we define the dispersion length as the distance where the pulse length has doubled, then the dispersion length (L_d) can be found as

$$L_d = \left(\frac{\omega_0}{\omega_p}\right)^3 l_\tau^2 \frac{c}{\omega_p} \tag{24}$$

For typical experimental parameters the laser frequency is much higher than the plasma frequency. Thus L_d is much longer than L_p. For example, if we choose the plasma density to be $n_0 = 6.7 \times 10^{16} cm^{-3}$, channel radius $a = 20.5 \mu m$ and the laser parameters as : $\lambda_0 = 1 \mu m$, pulse length $l_\tau = 0.25 ps \cdot c$, energy $E = 7.5 joules$, spot size $\sigma = 41 \mu m$. The corresponding normalized parameters are $k_p a = 1$, $V_{osc}/c = 0.64$ and $\omega_0/\omega_p = 128.8$. In this case $L_p = 133.5m$ and $L_d = 586.3m$. For completeness, we note that the Rayleigh length of an unguided laser with these parameters would be only $0.5cm$. Also the dephasing length for very relativistic particles is $L_\phi \approx 2\pi c/\omega_p (\frac{\omega_0}{\omega_p})^2 \approx 2m$.

IV PIC simulations

To test the above theories, we perform two-dimensional PIC simulations using the code ISIS. First we study the stability of "long" pulses (i.e., pulse length much longer than c/ω_p; Figs. 2 - 6). Then we study the evolution of short pulses over long propagating distances (Figs. 7 - 9), with particular attention to pump depletion, dispersion and their effect on the wake amplitude and phase velocity. For the long pulse cases, the grid size and time step are $\Delta z = 0.1 c/\omega_p$, $\Delta y = 0.117 c/\omega_p$ and $\Delta t = 0.25 \omega_0^{-1}$. The pulse length is $37 c/\omega_p$ (FWHM), $V_{osc}/c = 0.8$ and 6 particles per cell. Figs. 2 show the initial laser profile on the axis and its k-spectrum. Figs. 3 and 4 show

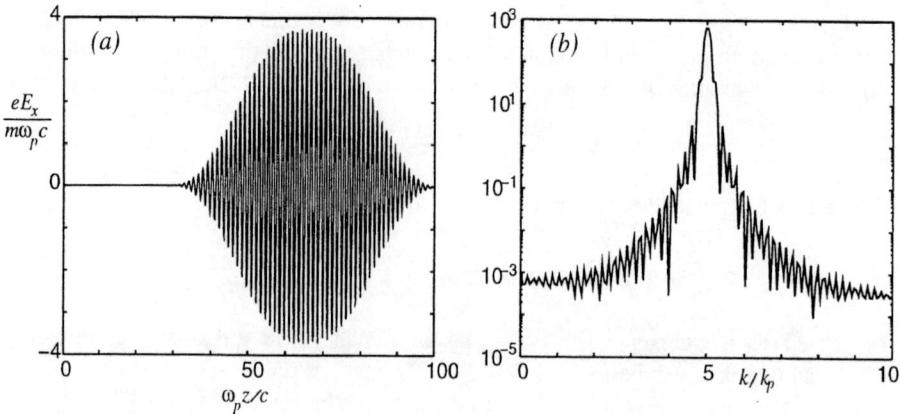

Figure 2: (a) Initial laser field along the axis for each case and (b) its k-spectrum.

the laser profiles in homogeneous plasma and parabolic channel plasma, respectively. In both cases, either the beam breakup in real space [Figs. (3 a) and (4 a)] or the spectral cascading [Figs. (3 b) and (4 b)] show evidence of Raman scattering. Figs. 5 show the laser profile in a hollow channel plasma with $k_p a = 1$. Evidence for suppression of RFS is in Fig. (5 b) which shows no sidebands after the same propagation distance. Figs. 6 show the results in a narrow hollow channel plasma ($k_p a = 0.25$). Because the channel is very narrow, we choose $V_{osc}/c = 0.3$ to avoid distorting the channel boundaries. We also seed 1% of the $\omega_0 - \omega_p$ component in the pulse to speed up the growth of the instability. The clear growth of E_z and $\delta n/n_0$ in Figs. (6 c) and (6 d) show evidence of RFS of the body mode. However, none of these two simulations has shown the growth of the surface mode. There are two factors which can cause this. First, the growth of the surface mode relies on the charges accumulated at the boundaries. An intense laser can distort the channel edge significantly and thus destroy the accumulation of the surface charges. Second, due to the

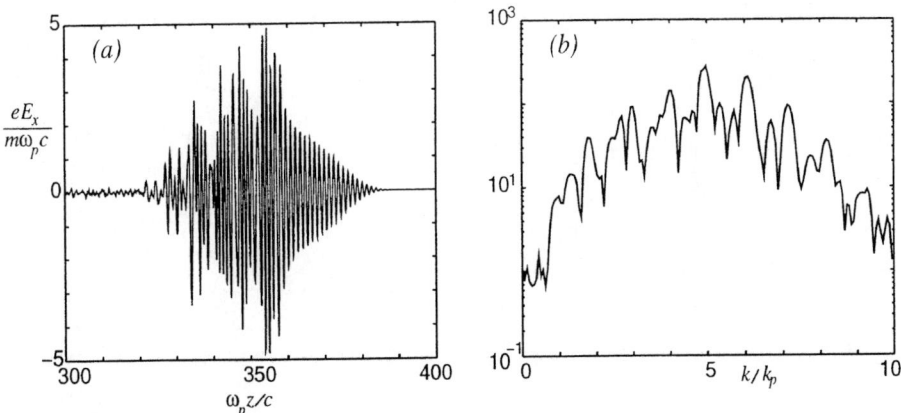

Figure 3: (a) Laser field in homogeneous plasma at $\omega_p t = 300$ and (b) its k-spectrum.

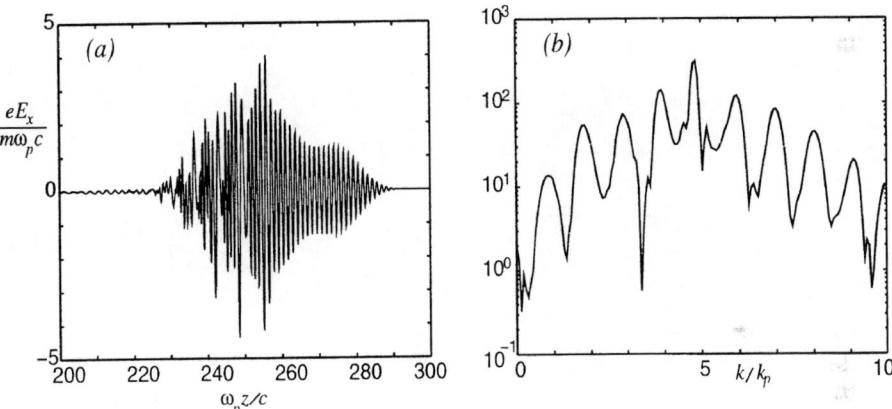

Figure 4: (a) Laser field in parabolic channel plasma at $\omega_p t = 200$ and (b) its k-spectrum.

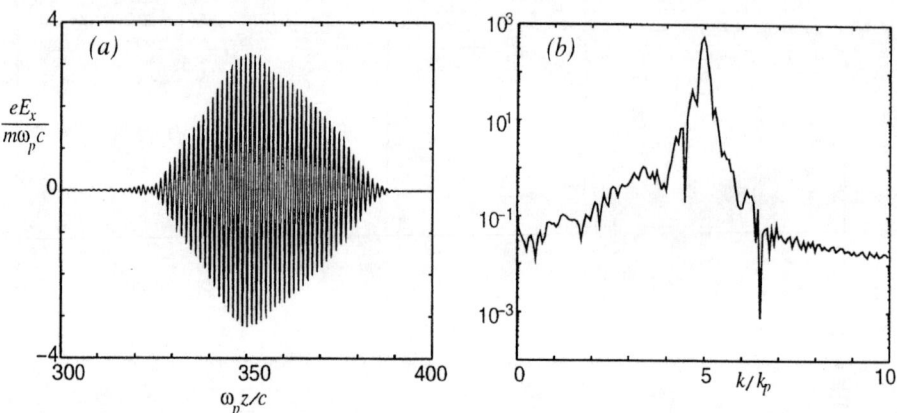

Figure 5: (a) Laser field in hollow channel plasma at $\omega_p t = 300$ ($k_p a = 1.0$) and (b) its k-spectrum.

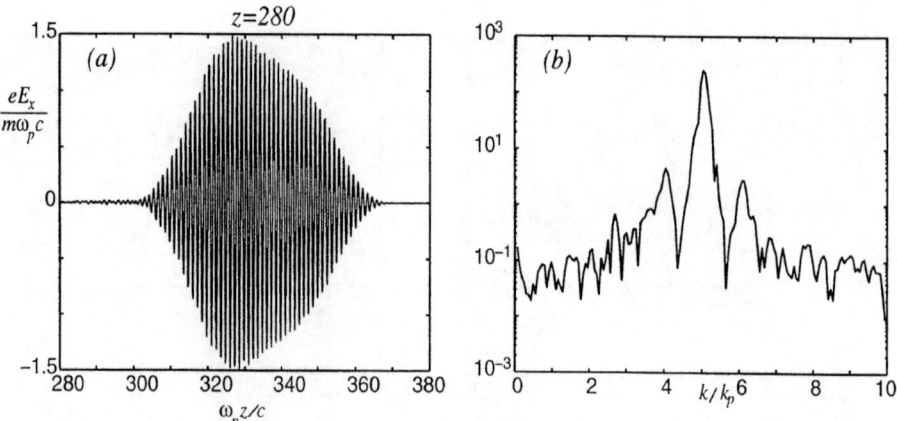

Figure 6: (a) Laser field in hollow channel plasma at $\omega_p t = 280$ ($k_p a = 0.25$) and (b) its k-spectrum.

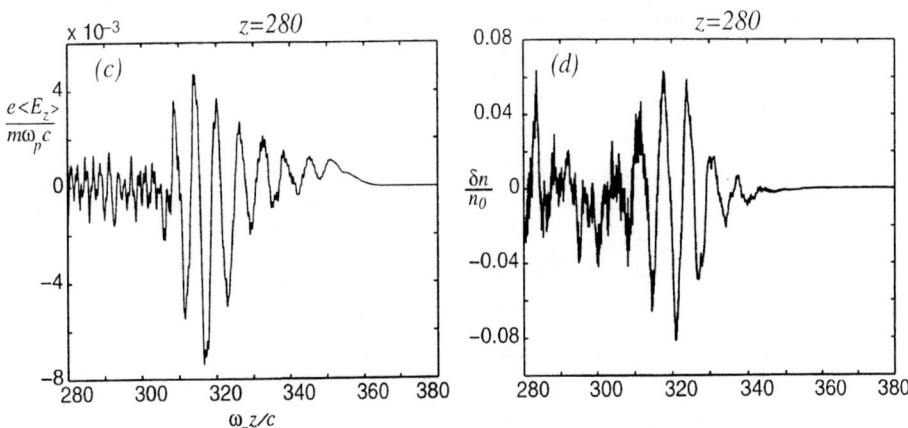

Figure 6: (c) The averaged longitudinal wake along the axis at $z = 280c/\omega_p$ and (d) the averaged density perturbation $\delta n/n_0$ near the channel edge.

finite grid size in the simulation ($\triangle y = 0.117c/\omega_p$), the plasma density will ramp from 0 to n_0 over a distance of 2 grids. This finite thickness of the channel wall will damp the surface mode by resonant absorption [6]. The damping rate can be estimated as

$$\omega_{im} \approx 0.2\omega_{ch}\delta$$

where δ is the ratio of the channel wall thickness to the channel radius. For $k_p a = 1$ and $\delta = 0.234$, then $\omega_{ch} = 1/\sqrt{2}\omega_p$ and the damping rate is $\omega_{im} \approx 0.033\omega_p$ which is bigger than the growth rate found from Fig. 1. On the other hand, there is no corresponding resonant absorption damping of the body mode, and the body mode is seen in the simulations.

To test the linear distortions, i.e. pump depletion and dispersive spread, it is hard to simulate the experimental parameters given in the last section because of the high frequency ratio. So instead we choose $k_p a = 1$, $V_{osc}/c = 0.5$, $\omega_0/\omega_p = 4$ and pulse length $l_\tau = \sqrt{2}\pi(c/\omega_p) = \lambda_{ch}/2$. This yields $L_p = 11286(c/\omega_p)$ and $L_d = 1263(c/\omega_p)$ which makes the dispersion spread much more severe than the pump depletion. The corresponding Rayleigh length for the laser is about $2c/\omega_p$. Figs. 7 show the laser after propagating $120c/\omega_p$ and the wakefield left behind. Both the laser amplitude and the pulse width have negligible changes as compared to the initial laser profile (not shown here). However, after propagating $1000c/\omega_p$ as shown in Figs. 8 ,the amplitude of the pulse has dropped from 2 to 1.25 and the pulse length has almost doubled. The slip-back of the pulse with respect to the window is because the window moves at the speed of light while the pulse moves at group velocity $v_g/c \approx 1 - \omega_p^2/2\omega_0^2$. The detailed evolution of the pulse as it propagating

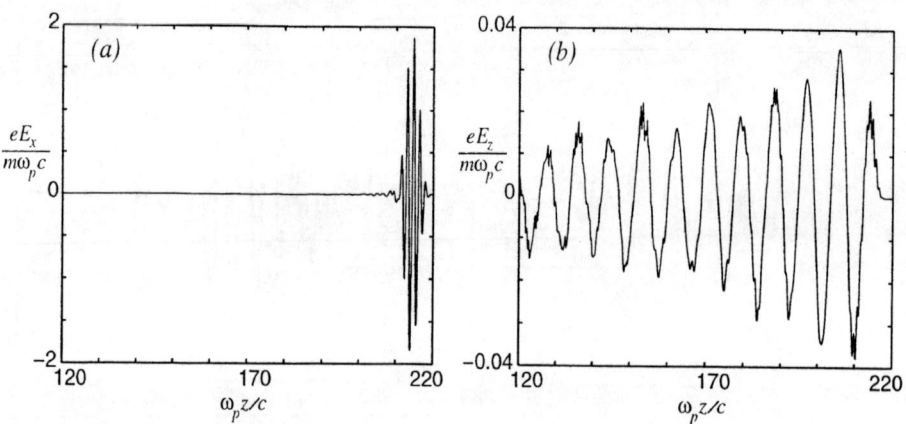

Figure 7: (a) Laser field and (b) wakefield on the channel axis at $z = 0.03 L_p$ where L_p is the pump depletion length.

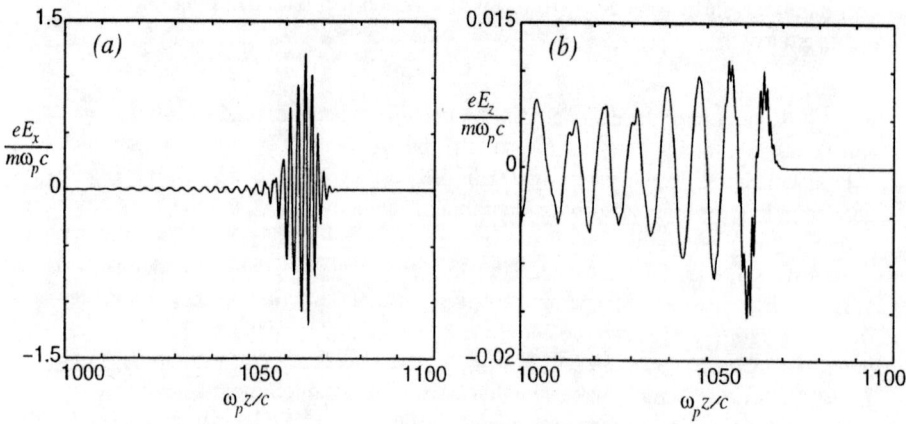

Figure 8: (a) Laser field and (b) wakefield on the channel axis at $z = 0.09 L_p$.

down the channel is shown in Fig. 9. The circles indicate the measured phase velocity of the wakefield v.s. time which shows no apparent decreasing or increasing tendency. The dashed line shows the decrease of the wake amplitude with time and the solid line shows evolution of the laser energy all normalized to the initial values. As mentioned earlier, because of the parameters we chose for this simulation both pump depletion and dispersive spread have significant effects on the laser pulse. Since the wakefield amplitude depends on the ponderomotive force which is proportional to E_0^2/ω_0 or laser energy over ω_0, we expect the wake amplitude to decrease more slowly than the laser energy. However, in our example dispersion further reduces the ponderomotive force and the wake decreases more quickly than the laser. This was shown in Fig. 9. We also notice that at $\omega_p t = 1000 (or z = 0.09 L_p)$ the

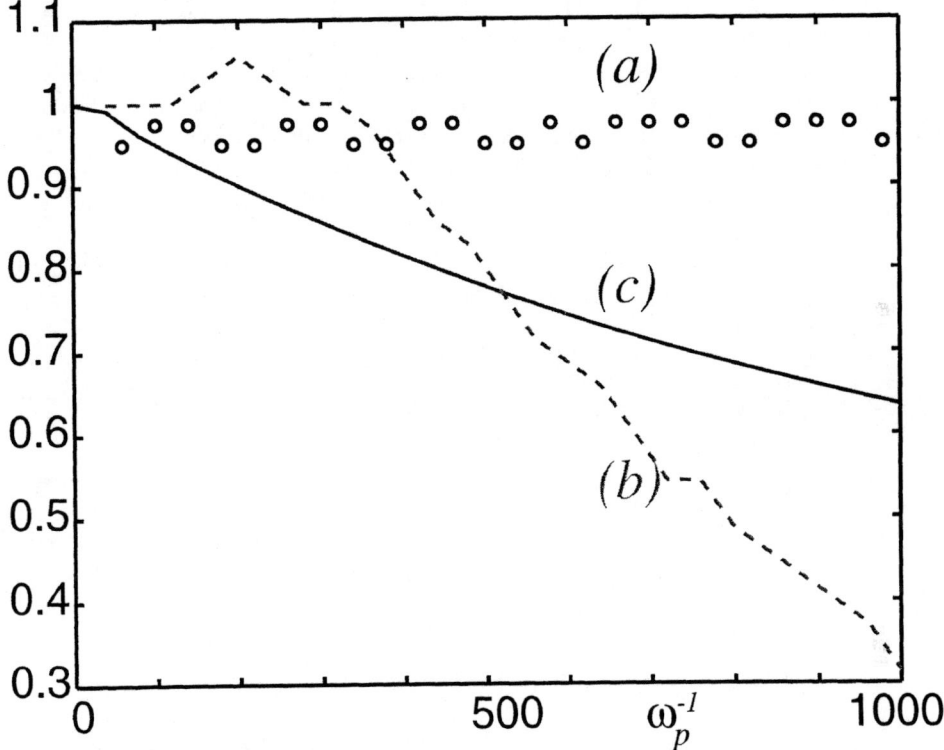

Figure 9: (a) Measured phase velocity of the wakefield v_φ/c vs. time. (b) Peak wakefield amplitude normalized to the initial wake amplitude at $t = 40\omega_p^{-1}$ vs. time. (c) Laser energy normalized to the initial energy vs. time.

laser energy has dropped to 63% of its initial value. At this point we estimated that

9% of the laser energy has been transferred to the plasma wake. Thus the coupling effficiency is approximately 25% of the depleted laser energy.

V Conclusion

We have examined the propagation of intense laser pulses through various types of plasmas. The RFS instabilities in both homegeneous and parabolic channel plasmas is shown in PIC simulations. The pulse in parabolic channel plasmas seems to be more unstable than in homogeneous plasmas. This may be due to the addition guiding of the laser pulse provided by the plasma channel. The laser pulse was shown both analytically and through PIC simulations to be more stable than in the other plasmas. The growth of the body mode is largely suppressed in hollow plasma channels. The group velocity of the laser and thus the phase velocity of the wakefield is relative constant. This maybe due to the fact that most of the laser energy is confined in the channel where there is no plasma density perturbations. For ultrashort (on the order of plasma wavelength) pulses, PIC simulations show that hollow plasma channels can guide the laser over 500 Rayleigh lengths without suffering any instabilities. Although 9% coupling efficiencyes from laser to plasma wake was measured in one PIC simulation. Further work is needed to make predictions about cases with parameters relevant to current experiments. In further work, we plan to include beam loading and detailed studies of the overall coupling efficiencies from laser to particles.

REFERENCES

1. T.C. Chiou, T. Katsouleas, C. Decker, W. B. Mori, J.S. Wurtele, G. Shvets and J. J. Su, Phys. of Plasmas **2**, 310 (1995).
2. W.B. Mori, C. D. Decker, D. E. Hinkel and T. Katsouleas, Phys. Rev. Lett., **72**, 1482 (1994). T. Antonsen and P. Mora, Phys. Rev. Lett. ,**74**, 4440 (1995)
3. G. Shvets and J. S. Wurtele, Phys. Rev. Lett., **73**, 3540 (1994). P. Sprangle, J. Krall and E. Esarey, Phys. Rev. Lett., **73**, 3544 (1994)
4. T.C. Chiou, T. Katsouleas and W. B. Mori, Phys. of Plasmas **3**, 1700 (1996).
5. Klaus Baumgartel and Konrad Sauer,Topics on Nonlinear Wave-Plasma Interaction, Birkhauser Verlag, Basel, 1987 chapter 2.
6. G. Shvets, J. S. Wurtele, T.C. Chiou and T. Katsouleas, IEEE Trans. on Plasma Science, Vol. 24, 2, 351 (1996).

First Measurement of Laser Wakefield Oscillations by Longitudinal Interferometry

C. W. Siders
Los Alamos National Laboratory

S. P. Le Blanc, B. Rau, D. Fisher, T. Tajima, M. C. Downer
The University of Texas at Austin, Department of Physics

A. Babine, A. Stepanov, A. Sergeev
The Institute of Applied Physics, Nizhny Novgorod, Russia

Because the electrostatic fields present in plasma waves can exceed those achievable in conventional accelerators and approach atomic scale values ($E_a \sim 500$ GV/m), plasma based accelerators have received considerable attention as compact sources of high-energy electron pulses [1]. Although stimulated Raman scattering [2] or terahertz radiation at ω_p [3] provided spatially averaged optical signatures of the plasma wave's existence, new diagnostic techniques are required to map the the temporal and spatial structure of the plasma wave directly since such information is vital for addressing fundamental issues of wakefield generation and propagation. In this paper, we report femtosecond time resolved measurements of the longitudinal and radial structure of laser wakefield oscillations using an all optical technique known as interferometric "photon acceleration" [4], or Longitudinal Interferometry [5].

In a simple version of the experiment, a probe pulse co-propagates behind an intense pump pulse ($I = 3 \times 10^{17} W/cm^2$, $\lambda = 0.8 \mu m$, $\tau = 100 fs$) tightly focused ($f^\# = 4.2$) in helium gas. As the pump pulse ionizes the gas and exerts ponderomotive pressure on the resulting plasma, the probe pulses experiences electron density gradients behind the pump pulse which cause both DC phase shifts as well as blue/red shifting of the probe pulse frequency spectrum. In order to detect the small changes in frequency ($\Delta\omega/\omega \sim 10^{-4} - 10^{-5}$) and phase with femtosecond resolution, our photon accelerator diagnostic uses multiple, temporally separated probe pulses which produce frequency domain interferograms [5].

Two types of experiments were conducted to temporally resolve the wakefield oscillations. In the first, probe pulses propagated in front of and behind the pump

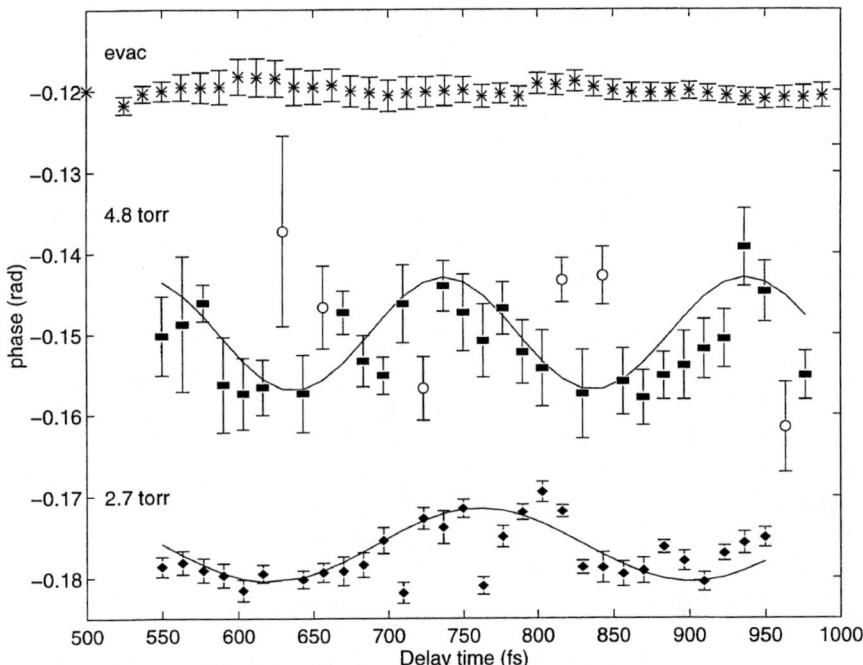

Figure 1: Measured wakefield oscillations in helium. For the 4.8 Torr data, the two probe pulses are separated by 2.2 ps about the pump, while in the 2.7 Torr data (offset from zero and shifted by -400 fs) the probes trail the pump with 415 fs separation. For the 4.8 (2.7) Torr data, 10 (9) mJ of energy was focused with an e^{-1} radius of 3.6 (5.0) μm. The solid lines show the calculated phase shift due to the wakefield oscillations, while the top line of data shows the noise level for a scan in an evacuated chamber

pulse and the delay of the pump pulse was varied relative to the two probe pulses. Fig. 1 shows measured phase shifts in 4.8 (2.7) Torr helium oscillating with a period of $220 \pm 25 fs$ ($270 \pm 10 fs$) and an amplitude of 0.007 rad (0.005 rad). Under these conditions, we detect wakefield oscillations 3-5 (4-5) cycles behind the pump pulse. From the amplitude of the phase modulation in Fig. 1, we estimate that the amplitude of the wakefield oscillation is at least $\delta n_e / n_e = 0.8$. This amplitude is much larger than a simple one dimensional estimation of the laser plasma interaction due to the fact that the radial component of the ponderomotive force is order ten times larger than the axial component. The peak longitudinal electric field is estimated as ~ 10 GV/m.

A second set of experiments was conducted by fixing the pump and probe pulse delays while varying the helium gas pressure. Such a pressure scan allows the wakefield to be scanned across the second probe pulse. Fig. 2a shows the measured phase shift between probe pulse 1 and 2 for two different pump pulse intensities as the helium pressure varied from 2-12 Torr He. Resonant excitation of the wakefield is obtained when the plasma wave period ($2\pi/\omega_p$) is approximately twice the pump pulse duration. Longitudinal and radial averaging cause the measured phase shifts to be similar for the two different pump pulse intensities. To help reduce the effect of radial averaging, a second pressure scan (Fig. 2b) was performed with a smaller spectrometer entrance slit.

Though our use of dual-beam spectroscopy eliminates most systematic contribution to our data on a 2000:1 (long term) level, approximately 10% of the data points fall significantly away from the calculated pressure scan curves in Fig. 1 and 2. Uncorrected drifts in beam pointing, center wavelength, and spectral shape on the time scale of the data collection (~ 40 sec for each data point) have been found to contribute significantly to such noise in the data. Even so, nonlinear effects such as radial density peaking, radial dephasing and wave breaking [6] may also contribute to the data in ways which are not well understood at present.

In an effort to evaluate nonlinear contributions quantitatively, numerical simulations were performed with a 2D, multi-grid, fully relativistic, cold fluid model in which a Gaussian laser pulse propagates through a preformed plasma. The $\mathbf{v} \times \mathbf{B}$ term of the Lorentz force was not included; thus only relativistic and electrostatic influences on ω_p were modeled. Figure 3 shows the calculated wakefield structures for near-resonant excitation for the same focal geometry and pulse energies (2.5 and 10 mJ) as used in the experiment. The higher energy simulation (Fig. 3a) clearly shows the excitation of nonlinear plasma waves with significant density peaking and a maximum $\delta n/n \sim 5$. Even in the intense focus, these plasma waves oscillate for at least five cycles after the pump pulse. The lower energy simulation (Fig. 3b) shows significantly reduced peaking with $\delta n/n \sim 1$, as expected from an analytic solution. Careful examination showed that the higher energy simulation has a period longer than either the lower energy simulation or the linear result in the focus, thus suggesting that relativistic period lengthening dominates over electrostatic period

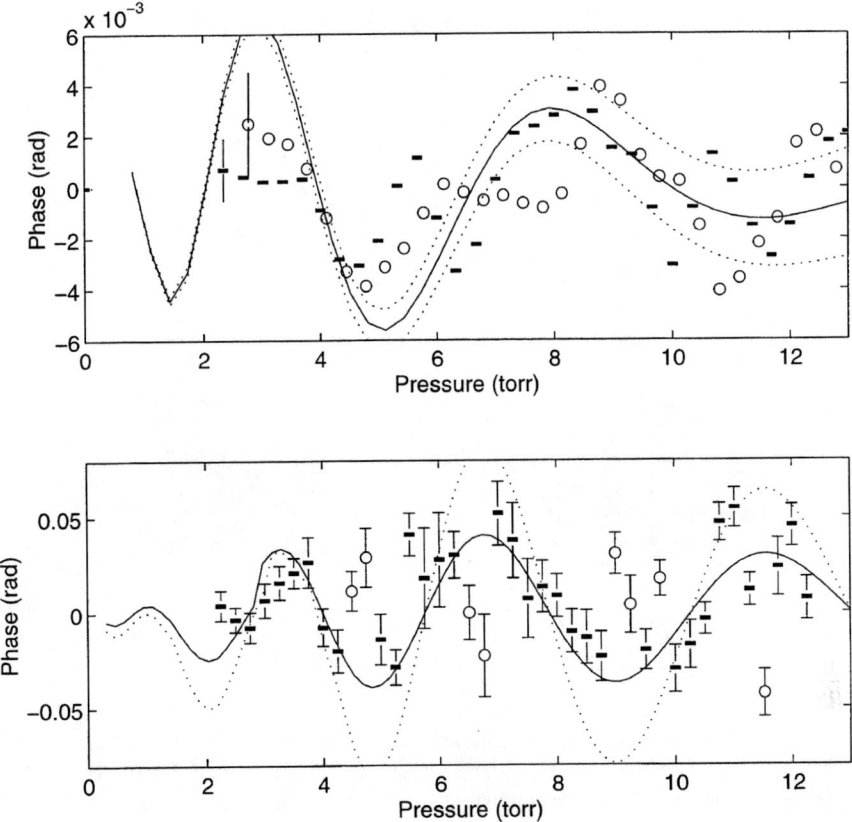

Figure 2: Phase shift as a function of the He gas pressure. (a) For two pulse energies: 10 mJ (filled square) and 2.5 mJ (circle). The solid line indicates a theoretical calculation of the phase shift for the higher energy. Representative error bars shown. (b) Pressure scan (10 mJ) with narrow slit. Curves are theoretical calculations of the phase shift without (dotted line) and with (solid line) radial averaging.

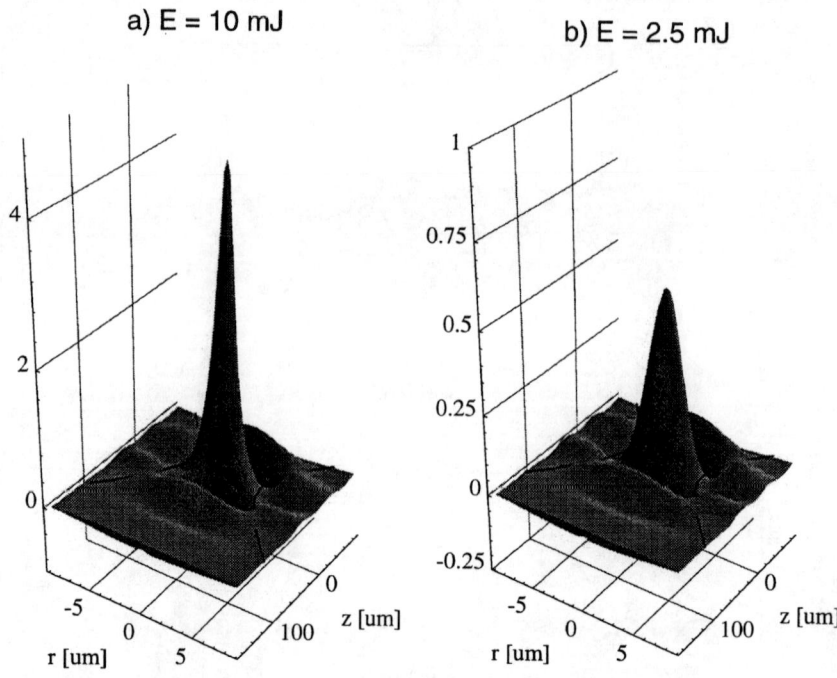

Figure 3: Two dimensional (r,z) numerical simulation of wakefield oscillations $\delta n_e/n_e$ corresponding to $E = 10(2.5)$ mJ, 3.6μm spot radius, $\tau = 100$ fs, $n_e = 3 \times 10^{17}$cm^{-3}. The figure shows the electron density oscillations within the confocal parameter of the tightly focused beam and in the moving frame of the pump pulse (centered at $z = 111\mu$m and moving in the positive z direction, but not shown). The heavy line represents the e^{-1} contour of the laser focus.

shortening. Thus for our parameters only a slight (~ few percent) period lengthening is expected and then only in the most intense portion of the focus, consistent with the observed wakefield periods.

Numerical integration of the data in Fig. 3 confirms that $\Delta\phi \sim 10$ mrad is expected for our focal geometry and probe pulse widths for both 2.5 mJ and 10 mJ pump energy, consistent with the data in Fig. 2a and with predictions ($\delta n/n \sim 1, \Delta\phi \sim 10$ mrad) for the 2.5 mJ pump. The sharply peaked density perturbation in Fig. 3a does not result in a larger measured $\Delta\phi$ than the broader, lower peak in Fig. 3b because the high electron density is concentrated in a volume smaller than the probe pulse, and thus is not spatially resolved in our experiment.

In summary, we have used longitudinal pump-probe interferometry to excite and measure laser wakefield oscillations with femtosecond resolution in both timedelay and pressure scan configurations. From the data, we estimate density perturbations of order unity and longitudinal fields of order 10 GV/m, consistent with the predictions of both an analytic 2D linear nonrelativistic fluid analysis and a fully relativistic nonlinear 2D self-consistent numerical model. By using tightly focused laser pulses, nonlinear wakefield oscillations were driven with subrelativistic laser intensity ($I < 10^{18}$ W/cm^2). As this technique utilizes a necessary component of any laser-based plasma accelerator, i.e. the intense driving pulse, it promises to be a powerful tool for on-line monitoring and control of future plasma based particle accelerators.

Acknowledgements

This work was performed under U.S. Department of Energy Grant No. DEFG05-92-ER-40739, NSF Grant No. PHY-9417558, and Russian Basic Science Foundation Grants No. 93-02-03571 and No. 94-02-03849. C. W. Siders acknowledges support in the form of a Postdoctoral Fellowship from Los Alamos National Laboratory.

References

1. T. Tajima and J. M. Dawson, Phys. Rev. Lett. **43**, 267 (1979); P. Sprangle and E. Esaray,Phys. Fluids B **4**, 2241 (1992).
2. C. E. Clayton et al., Phys. Rev. Lett. **54**, 2343 (1985).
3. H. Hamster et al., Phys. Rev. E **49**, 671 (1994).
4. S. C. Wilks et al., Phys. Rev. Lett. **62**, 2600 (1989); W. M. Wood et al., Phys. Rev. Lett. **67**, 3523 (1991).
5. Reynaud et al., Opt. Lett. **14**, 275 (1989); E. Tokunaga et al., Opt. Lett. **17**, 1131 (1992); J. P. Geindre et al., Opt. Lett. **19**, 1997 (1994); C. W. Siders et al., IEEE Trans. Plasma Sci.**24**, 301 (1996); C. W. Siders et al., Phys. Rev. Lett.**76**, 3570 (1996).
6. J. M. Dawson, Phys. Rev.**113**, 383 (1959); A. R. Bell et al., Plasma Phys. Controlled Fusion**30**, 1319 (1988).

Optical Guiding of High Intensity Laser Pulses in Plasma Channel: Interferometrical Investigations

Nadja Vogel

University of Technology Chemnitz, Department of Physics, Optical Spectroscopy and Molecule Physics, 09107 Chemnitz, Germany,
presently visiting scientist at Lawrence Berkeley Laboratory, University of California,
1 Cyclotron Road, Berkeley, California 94720 USA

Abstract. The exitation of the electric and self-generated magnetic field by pondermotive force during propagation of 100 ps laser pulse in air are investigated experimentally. Measurements of electron density distribution with high temporal (100 ps) and spatial resolution (< 1 µm) by interferometry and absorption photography are presented. It is shown that under certain conditions a hollow current channel can be generated. The azimuthal magnetic field in the micro-channel was determined by Faraday rotation of a probing laser beam to 7.6 MG. The charged partical densities in channel exceed $6 \cdot 10^{20}$ cm^{-3}. Ion acceleration in a pinched annular current channel up to 6 MeV analogous to a micro -"plasma focus" conditions may be realized just at length of 100 µm.

INTRODUCTION

For particle accelerations to high energy by laser accelerator, optical guiding schemes in a plasma have been proposed as a way to increase the propagating distance to many Rayleigh lengths [1-7]. The optical guiding mechanisms are based on the principle of refractive guiding, which can be induced by relativistic effects (relativistic optical guiding,) or by a preformed plasma density channel (for example, with parabolic density distribution or by hollow channels with zero density on axis) [8-9]. In relativistic self-focusing [10], the relativistic quiver motion of plasma electrons in the laser field causes a reduction of the index refraction on axis

$$\eta_R \cong 1 - \frac{\omega_p^2}{2\omega^2} \frac{N_e(r)}{N_0 \gamma(r)} \qquad (1),$$

through the relativistic factor $\gamma(r)$ such that $\frac{\partial \eta_R}{\partial r} < 0$. Here $\omega_p = \left(\frac{4\pi N_0 e^2}{m_e} \right)^{1/2}$ is the electron plasma frequency, N_0 is the ambient electron density, e is electron charge, m_e is electron mass, N(r) is radial electron density profile and ω is the radian frequency of laser radiation. Above a threshold power

$P_c[GW] \cong 17(\omega/\omega_p)^2$, Ref. [10-13] this mechanism provides optical guiding of the main body of the pulse.

The optical guiding due to a preformed plasma density channel was proposed several years ago by Tajima [14] and has been demonstrated for modest laser intensity (up to 10^{14} W/cm^2) by Durfee *et al.* [15-16]. Experiments on relativistic guiding and pondermotive self-channeling have been performed [17-18], in which was shown that the radial pondermotive force of a laser pulse during its propagation in an initially uniform plasma expels electrons from the axis and thus cteates a channel with a hollow density distribution.

In order to create optimal conditions for particle acceleration in self-modulated laser accelerators, information about plasma parameter, such as the radial electron density distribution and its temporal evolution is necessary. Stimulated Raman scattering or Thomson scattering from the plasma channel cannot resolve a fine structure of a steep density profile in plasma. This, however, is one of the most important parameters in the theoretical description of the self-focusing process. . There is a need for a diagnostic technique with high temporal and spatial resolution.

We report here on experimental investigations of self-focusing and self-channeling processes of 100 ps laser pulse in air. The temporal evolution of the plasma density , the generation of a longitudinal electric field and very strong lowfrequency magnretic field due to the pondermotive force have been studied during laser beam propagation through focus area.

EXPERIMENTAL

Experimental setup

A high intensity IR pulse (60 mJ, 100 ps, 1064 nm) from Nd: YAG laser was focused to a 40 μm, diameter spot by 25 cm focal length lens to a maximum intensity of 3 x 10^{13} W/cm^2. The laser beam (s-polarised) was incident normal on the planar Al target in air. The target was placed in one arm of a Michelson interferometer. The probing laser beam at the second harmonic (λ= 532 nm) was optically delayed by - 1.3 ns to 10 ns with respect to the ignition beam. The focus area is located in front of the target, with the focal plane 200 μm away from the surface. This distance was found to be optimum for the effects discribed later. The target and the focus area of the laser pulse were imaged with a magnification optics (x125). An interference filter (centred at λ = 532 nm with full width at half-maximum (FWHM) = 10 nm) was used to reduce the light emitted by the laser-produced plasmas. A charge-coupled-device (CCD) camera images was used to capture the images (512x512 pixel frame, 8 bit resolution). The spatial magnification was as high to have 0.47 μm per pixel. Holographic interferometry and absorption photography were used as main diagnostic technique. The distribution of the interference phase was determinated by a linear FFT algorithm. When using only one mirror of the Michelson interfermeter, this arrangement was used to obtain absorption pictures. With the Nd:YAG laser pulse of 100 ps

FWHM this laser diagnostic give a "snapshot" of the density profile in laser-produced plasma. By turning the delay of the probing laser besam, the evolution of the plasma density can be monitored.

Time-resolved measurements of the electron density distribution

Optical interferometry was used for measuring the refractivity of laser-produced plasmas by directly comparing the phase of a laser wavefront passing through the plasma with a reference. In highly ionized plasmas, the refractive index is related to the free-electron density. We assume that (i) the plasma density is less than the critical density at which the plasma frequency equals the laser frequency, (ii) the wavelength of the probe beam is far from the characteristic absorption lines of atoms or ions. Under these conditions we can determine electron densities N_e from changes in the wavefront caused by the plasma electrons in units of fringe shifts K in interferogram with

$$K = \frac{e^2 \lambda}{2\pi mc^2} \int_0^l N_e(r) dr \qquad (2)$$

where λ is the wavelength of the probing light, r is measured along the optical path of the probing beam, l is path length through the plasma and c is the light velocity. The signal received by a detector is an intergration along a path length through the plasma. The electron density distribution was obtained by Abel inversion of the observed shifts, with assumption of cylindrical symmetry. To observe a specific electron density maximum with a steep profile, large aperture optics (long distance microscope) was used to accommodate the strong deviation imposed on the probe beam. Absorption photography were used as a complementary plasma diagnostics technique to obtain an overview over the plasma development.

EXPERIMENTAL RESULTS

Plasma channel development

Using the interferometric system we study the electron density distribution prodused by 1.064 µm laser radiation on Al-target in air. The diffraction-limited spatial resolution was ≤ 1 µm. The probe pulse had a duration of 100 ps at a wavelength of 0.532 µm. To study the temporal evolution of a plasma the probe pulse was delayed relative to the ignition pulse between -100 ps and 10 ns.

Askaryan in Ref. [19] has shown, that the heating and ionisation of a gas medium by laser beam can produce a low density channel. Following, charged particles and quanta without strong scattering can be guided in a channel.

In our experiments we investigated the plasma production during the first phase of laser propagation in the focus area. Fig. 1 gives an example of radial electron density profiles derived from holographic interferogram and absorption image for a delay time t =-20 ps.

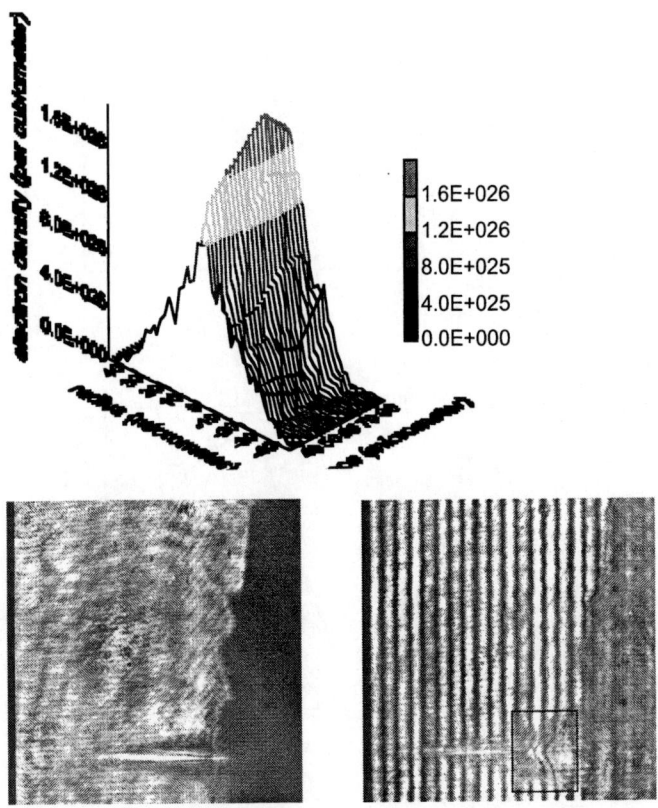

Figure 1. Radial electron density profile derived from holographic interferogram and absorption image for delay time t = - 20 ps

This implies that the probe pulse illuminates the focal region in the vicinity of the Al-target - 20 ps before a pulse envelope with 100 ps FWHM reaches the focuse point. It is evident tthat a sharp electron density front propagats in the direction of the incident laser beam. The electron density at the front reaches the value of $6 \cdot 10^{20}$ cm^{-3}. A short plasma channel with a refractive index as shown in Fig. 2 has already formed at the rising edge of the shape.

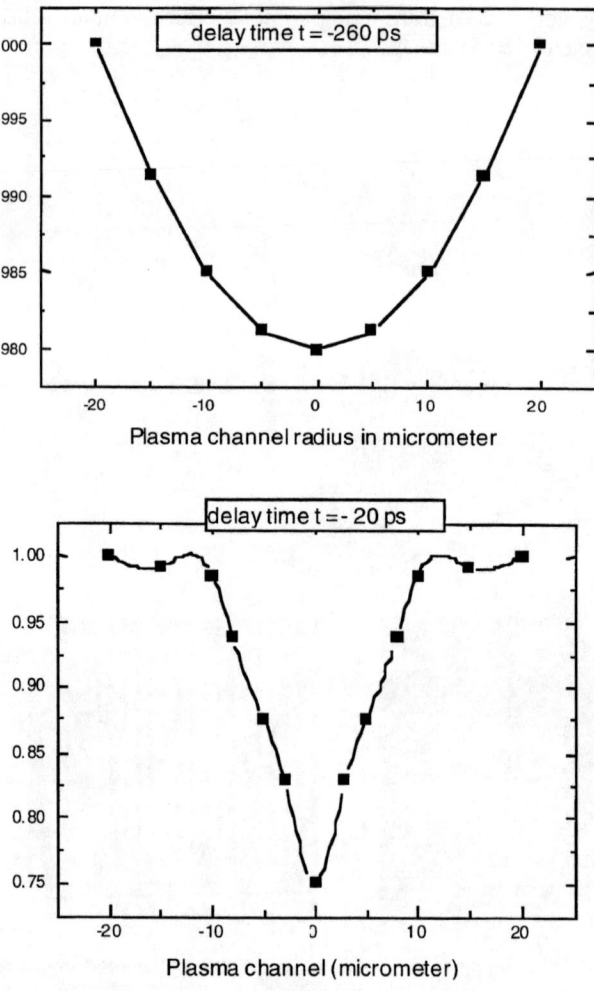

Figure 2. Refractive index in plasma channel for a). -delay time t = -260 ps and b). -delay time t = -20 ps

It is necessary to note that a long prepulse stage with a significant amplified spontaneous emission (ASE) exists in our laser system (a prepulse peak intensity is about 10^{11} W/cm^2). A full ionized plasma channel can be formed during propagation of prepulse radiation through the focus area in gas since the breakdown threshold for plasma generation in air is relatively small ($8 \cdot 10^{10}$ W/cm^2), Ref. [20-21].

Several processes are responsable for the creation of primary electrons in focus area. At laser intensities $10^{12} - 10^{13}$ W/cm^2, tunneling as well as a multiphoton ionisation can occur. The tunneling process tends to prevail at strong irradiation. In contrast, at relatively low levels of laser prepulse intensity cascade ionisation is more probable. In this case, free electrons acquire energy from the laser electromagnetic wave via elastic collisions with neutral molecules. After several collisions they may obtain enough energy to ionize the molecule and create a second generation of free electrons. From interferograms and shadowgrams it is evident that very intensive absorption of laser radiation occurs already during the prepulse for t = - 260 ps before the main peak of laser pulse reaches the focus area. The size of the absorption region can be estimated as 40 μm in diameter and 50 μm in length. Taking into account the fringe displacement on the order of 0.5 at 40 μm we can estimate the maximum of the electron density to $5.2 \cdot 10^{19}$ cm^{-3} with a parabolic like distribution along the channel radius. The preformed plasma channel is very effecient for the following guiding and focussing of the main pulse to high power density. It had been shown [22-25], that a channel with parabolic density can guide a Gaussian laser pulse, provided that the channel depth satisfies the condition $\Delta n > \Delta n_c$. Here $\Delta n_c = \left(\pi r_e r_0^2 \right)^{-1}$ is the critical depth and $r_e = \dfrac{e^2}{m_e c^2}$ is the classical electron radius. For a laser spot size $r_0 = 20$ μm, the relation $\Delta n > \Delta n_c > 2.6 \cdot 10^{17}$ cm^{-3} is fullfilled.

The corresponding refractive index profiles in plasma channel for two different delay time is shown in Fig. 2. A rapid increase in the channel depth connected with an intensive expelling of electrons in focus area is evident. The induced plasma channel leads to self-focusing of the laser beam in narrow channel. Following, the laser radiation of the main peak will be" locked" in a small volume, increasing the power density up to $5 \cdot 10^{15}$ W/cm^2. This value is already above relativistic threshold for self-focusing at the neodymium glass laser wavelength in a plasma [26]. Sun et. al [27] have shown that inclusion of both types of nonlinearities (ponderomotive and relativistic self-focusing) can lead to final states for which there is complete expulsion of electrons from the beam region for power levels which are only 10% above the standard self-focusing threshold. Numerical modeling performed with coupled electrodynamic and plasma equations by Hora et. al [28] for propagation 5 ps Nd laser pulse of peak power 10^{13} W/cm^2 in plasma have demonstrated a rapid relativistic self-focusing down to a beam diameter of one micron at length of the order of the original beam diameter, as well as the production of GeV ions moving against the laser light.

As it is obvious from the interferogram in Fig. 1, that an intensive electron beam propagates toward the laser beam. A strong density gradient in the front of the electron beam have been formed at the incident wave's rise time t = - 20 ps. We can estimate the density gradient length $L = \left(\dfrac{\partial \ln N_e}{\partial r} \right)^{-1} = 2.5 \cdot 10^{-7}$ m and the electron displacement in laser wave electric field $r = \dfrac{V_{ocs}}{\omega} = \dfrac{eE_0}{m\omega^2} = 7 \cdot 10^{-7}$ m, where E_0 is the amplitude of electric field in laser wave

$E_0 = 1.2 \cdot 10^{11}$ V / m *for the laser power density* $5 \cdot 10^{15}$ W / cm². We obtained that the minimum length necessary to shield the electrostatic field for a plasma at the critical density is larger than the density discontinuity length . Under this conditions, a "vacuum heating" mechanism for electron acceleration may become more efficient than the usual resonant absorption [29-32]. At an ascent slope of laser pulse, electrons can efficiently accumulate energy from laser wave as long as incident-wave frequency is above the plasma frequency $\omega > 2\pi v_{eff}$, $\omega = 1.8 \cdot 10^{15}$ s^{-1}. The electrons absorb most energy of laser radiation and may be accelerated to keV energies [33]. Electrons pulled out from the focus area will tend to drag and accelerate ions with them. Numerical simulations with particle-in-cell (PIC) codes [34] have shown that in the process of "vacuum heating", ions can be accelerated to energies corresponding to equality of ion and electron velocities, for all that without the assumption that the plasma is quasi-neutral. The mechanism for the generation of directional ion jets in our case seems to be similar to the acceleration mechanism of free plasma expansion in vacuum in the presence of supra-thermal electrons. If the pressure of electrons inside the channel exceeds the pressure out the channel, the jet of superhot electrons from plasma surface cannot be compensated by the return current of slow electrons [35]. Obviously this is the case under the experimental conditions.

In Fig. 3 the velocity evolution of a sharp electron density front during laser wave propagation through focus region is presented. The maximum value of 1.4×10^8 cm/s was reached in peak of the laser pulse. The velocity decreased down to 2×10^7 cm/s in the falling slope of the pulse.

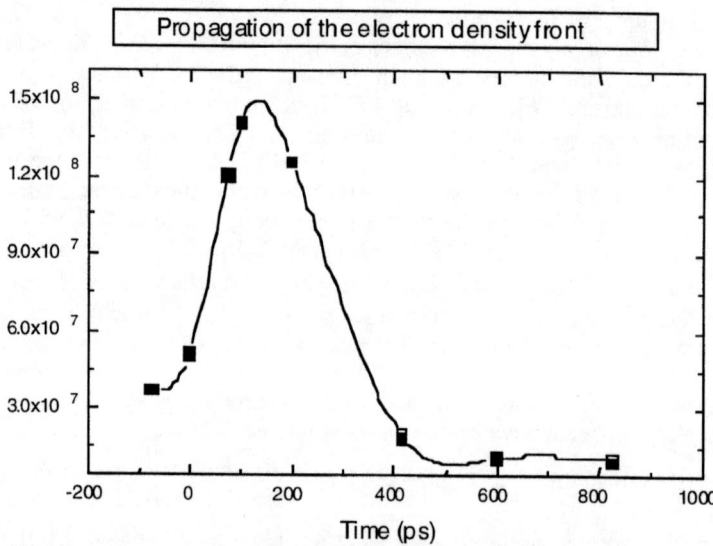

Figure 3. Velosity of electron density front in cm/s during laser pulse.

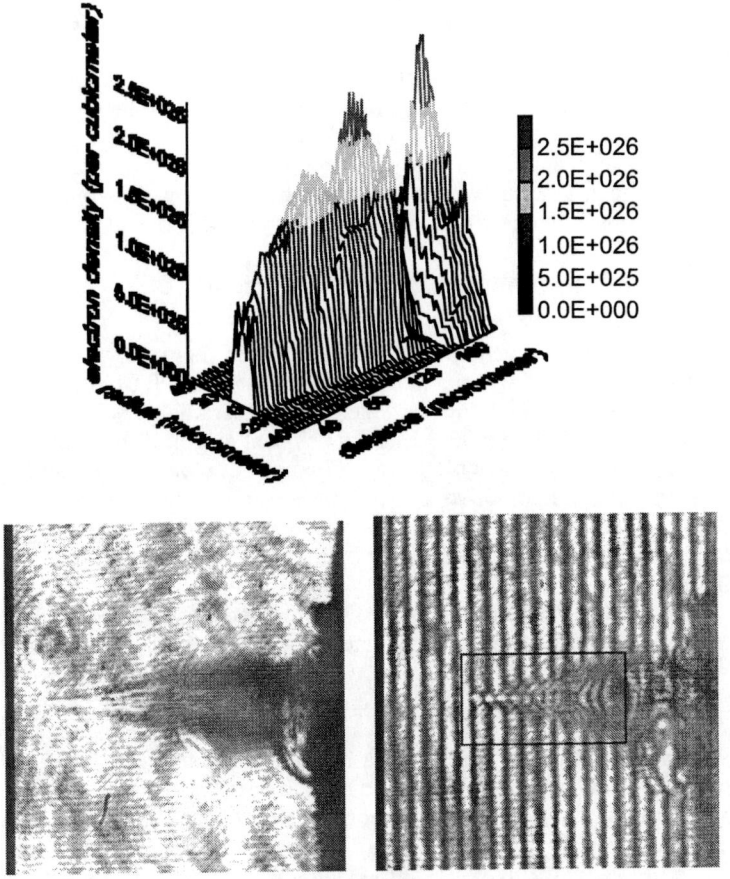

Figure 4. Radial electron density profile derived from holographic interferogram and absorption image for delay time t = 130 ps

Fig. 4 demonstrates the evolution of electron density profile for delay time t = 130 ps between pump and probe pulses. The electric field along the axis of the laser beam produced by the relativistic self-focusing can be determined from the density modifications in Fig. 1, 4. It causes a strong nonlinear force which accelerates ions to high energies against the laser beam. From the interferogram represented in Fig. 4, which was recorded at the falling edge of the laser pulse, we can evaluate axial electric field drawing ions from focus area of order of $8 \cdot 10^{10}$ V/m. The number of electrons leaving the focus area can be estimated from the interferogram to be $2 \cdot 10^{13}$. Assuming that the flow velocity equals the velocity of the density front as shown Fig. 3, we obtain a current density of order of $J = 6 \cdot 10^{13} \ A/m^2$.

The discontinuity of interference fringes in interferogram in Fig.4, in the radial direction indicates the presence of a fine structure in the electron density distribution. This kind of spreading in image of interference fringes verifies rapid electron motion in the channel, as well as a possible existence of interior tubular channels which enclose one another with distances of the order of 1 μm. The spatial and temporal resolution was not sufficient to resolve these structures.

With a delay time in the range from 1.45 ns to 2 ns, it can be seen that sharp directional channel begins to disintegrate and the strong plasma turbulence exists. For longer delay times (t > 2 ns) the sharp front vanished completely and only laser corona plasma with expansion region of 140 μm could be ascertained.

Magnetic field measurements

The first experimental observation of a dc magnetic field generated in laser-plasma interaction was explained by Stamper *et al*. [36] in terms of thermo electric currents associated with large temperature gradients near the laser target. It had been shown by Sudan [37] that the physical origin of the dc magnetic field must be strongly coupled with electron dynamics, which can be expelled in Z-direction by the ponderomotive force of the laser pulse. So long as the laser pulse increases in amplitude the pondermotive force generates an average electron current in the Z direction. The sign of B is such that it is created by a current flowing in the negative Z direction.

From interferograms and shadowgraphs recorded for delay times t > 50 psec we measure self-generated magnetic field in narrow electron jet channel. The probe pulse at second harmonic λ= 532 nm illuminated the focus region at delay time t = 100 psec related to the ignition point by IR laser beam. The plane of polarisation for probe laser beam was determinated by a polarizer just before the target and electric field vector in laser wave was vertical polarized. A second polarizer at the other side of the plasma have been used to change polarisation plane for transmitted radiation from horizontal (it means, that two polarizer are crossed) to vertical position. In this way absorption images have been recorded as a function of the analyzing polarization angle. A peak of intensity ratio at φ = 10° proves a linearly polarised radiation in narrow channel at λ = 532 nm with rotation of the polarisation plane for probing beam of order of 0.17 rad. Assuming that rotation of the plane of polarzation is the result of the magnetic field generation in narrow channel, we can estimate the magnetic field magnitude required for this Faraday rotation according to

$$\phi = 2.62*10^{-17} \lambda^2 \int_0^l N_e(r) B(r) dr \qquad (3)$$

where ϕ is rotation angle in radians, λ is the wavelength of the light in centimeters and the magnetic field B(r) in Gauss is the component of magnetic field in the direction of the propagation vector. From measurement of the electron density along the path of the beam ($l = 5$ μm) and rotation angle $\phi = 0.17$ rad we calculate the magnetic field value on the order of 7.6 MG. It shows a double-humped structure corresponding spatially to a region where the density gradient reverses. The toroidal nature of the field is evident from two opposite direction for orientation of the second polarizer. The probe beam polarisation must be rotated to the right, viewing along the beam, when travelling parallel to **B**, and to the left, when the magnetic field changes its direction.

The magnetic field is caused in the channel by the intense current flux of runaway electrons moving away from the focus area. From the interferogram represented in Fig. 4 the magnitude of the electron density at the steep front is $N_e \cong 10^{21}$ cm^{-3} and the directional velocity $V \approx 2 \cdot 10^8$ cm/s. We obtain an upper limit for current density $J = eN_e V = 3.2 \cdot 10^{10}$ A/cm^2 and magnetic field of $B = 0.2$ J • r $\approx 6 \cdot 10^6$ G where r is the radial coordinate. This estimate is in satisfactory agreement with the optical measurements of the Faraday rotation. Similar to an electric conductor, the magnetic field distribution within the channel has a tubular structure, with the maximum magnitude at the channel radius $R \approx 5$ μm and $B = 0$ on the axis. Intense ion acceleration occurs just at the ascent slope of the laser pulse by a collective mechanism (electrons pull ions) as well by the interaction with cylindrical magnetic field. Interaction between the laser radiation and the electron beam with a Gaussian radial profile must be result in modification of the laser wave front.The laser beam focusing reproduces a tubular structure of the wave front and amplifies by this way ring-shaped structure for electron density distribution in the channel. The magnetic field configuration has a cylindrical form, resulting in an effective ion acceleration within the micro-channel.

CONCLUSIONS

A plasma profile diagnostic has been represented for the investigation of self-focusing and self-channeling process of a relatively long laser pulse (100 ps) in gas. Direct measurements of electron density profiles with high spatial resolution have shown that a rapid change in the refractive index on the axis caused an increase in the laser power density in the focus area above the critical power for relativistic optical guiding. The spatial gradients in the strongly modified plasma channel and nonstationary character of the pondermotive force lead to the generation of a lowfrequency azimuthal magnetic field and an axial electric field. The electric field along the axis accelerates ions which are space-charge coupled with electrons. As a result, the particle achieve a high portion of energy, moving toward the laser beam. Futhermore, ions acquire energy in collisions with contracting magnetic wall in annular current channel and move along the channel axis over long distance.

In conclusion, we have presented experimental results for direct measurements of electron density distribution in optically guided picosecond laser beam with high spatial resolution. We have observed an interesting effect of an electron micro jet generation with current density on the order of 10^{10} A/cm^2 causing the generation of a self-magnetic field in the narrow channel with a magnitude of 7.6 MG.Ions acceleration up to 6 MeV appears according to following mechanisms:

- ion acceleration in electric field at the sharp electron density front, as well as
- an acceleration by interaction with cylindrical wall of magnetic field in annular electron beam channel.

It is evident, that a continuous regime of microchanneling and anomalous ion acceleration with an acceleration rate of order of $6 \cdot 10^8$ eV/cm can be achieved in a repetitively operated intense short laser pulse.

ACKNOWLEDGMENTS

This work was supported by the Deutsche Forschungsgemeinschaft, project No. Vo 527/1-2.

REFERENCES

1. D. C. Barnes, T. Kurki-Suoinio, and T. Tajima, "Laser self-trapping for the plasma fiber accelerator", IEEE Trans. Plasma Sci., vol. PS-15, pp.154-160, 1987
2. E. Esarey and A. Ting,"Comment on cascade focusing in the beat-wave accelerator", Phys. Rev. Lett., vol 65, p.1961, 1990
3. P. Sprangle and E. Esarey, "Interaction of ultrahigh laser fields with beams and plasmas", Phys. Fluids B, vol. 4, pp.2241-2248, 1992.
4. E. Esarey, P. Sprangle, J. Krall, A. Ting, and G. Joyce, "Propagation and guiding of intense laser pulses in plasmas", Phys. Rev. Lett., vol. 69, pp. 2200-2203, 1992.
5. E. Esarey, P. Sprangle, J. Krall, A. Ting, and G. Joyce, "Optically guided laser wakefield acceleration", Phys. Fluids B, vol. 5, pp. 2690-2697, 1993.
6. J. S. Wurtele, "The role of plasma in advanced accelerators", Phys. Fluids B, vol. 5, pp.2363-2370, 1993.
7. S. V. Bulanov and F. Pegoraro, "Acceleration of charged particles and photons in the wake of a short laser pulse in a thin channel", Laser Phys., vol. 4, pp.1120-1131, 1994.
8. T. C. Chiou, T. Katsouleas, C. Decker, W. B. Mori, J. S. Wurtele, G. Shvets, and J. J. Su, "Laser wakefield acceleration and optical guiding in hollow plasma channel", Phys. Plasmas, vol. 2, pp. 310-318, 1995.
9. T. M. Antonsen, Jr. and P. Mora, "Leaky channel stabilisation of intense laser pulses in tenuous plasmas", Phys. Rev. Lett., vol. 74, pp. 4440-4443, 1995.
10. T. C. Chiou and T. Katsouleas, Phys. Plasmas, 2, (1995), pp.310-318.
11. C. Max, J. Arons, and A. B. Langdon, Phys. Rev. Lett., 33, (1974), p.209.
12. A. G. Litvak, Sov. Phys. JETP, vol. 30, (1969), p.344.
13. G. Schmidt and W. Horton, *Comments Plasma Phys. Controlled Fusion*, vol. 9, (1985), p.85-90.
14 P. Sprangle, C. M. Tang, and E. Esarey, IEEE Trans. Plasma Sci., vol. PS-15, (1987), pp.145-153.
15. D. C. Barnes, T. Kurki-Suonio, and T. Tajima, IEEE Trans. Plasma Sci., PS-15, (1987), p.154.
16. C. G. Durfee III and H. M. Milchberg, 71, (1993), p. 2409

17. C. G. Durfee, III, J. Lynch, and H. M. Milchberg, Phys. Rev. E, vol. 51, (1995), pp. 2368-2389.
18. A. B. Borisov, A. V. Borovskiy, V. V. Korobkin, A. M. Prochorov, O. B. Shiryev, X. M. Shi, T. S. Luk, McPherson, J. C. Solem, K. Boyer, and C. K. Phodes, Phys. Rev. Lett., vol. 68, (1992), pp. 2309-2312.
19. P. Monot, T. Auguste, P. Gibbon, F. Jakober, G. Mainfray, A. Dulieu, M. Lous-Jacquet, G. Malka, and J. L. Miquel, Phys. Rev. Lett., vol. 74, (1995), pp.2953-2957.
20. G. A. Askaryan, N. M. Tarasova, Letter in JETF, 20, pp. 277-280, 1974 (in Russian)
21. P. Nelson, P. Veyrie, M. Berry, Y. Durand, Phys. Rev. Lett., 13, (1964), p.226
22. S. A. Achmanov, A. I. Kovrshin, M. M. Strukov, R. V. Chochlov, JETP Lett., 42, (1965), p.1 (in Russian)
23. E. Esarey and A. Ting, Phys. rev. Lett., 65, (1990), p.1961.
24. P. Sprangle ands E. Esarey, Phys. Fluids B, 4, (1992), p.2241.
25. P. Sprangle, E. Esarey, J. Krall, and G. Joyce, Phys. Rev. Lett., 69, (1992), p.2200
26. E. Esarey, J. Krall, and P. Sprangle, Phys. Rev. Lett., 72, (1994), p. 2887.
27. D. A. Jones, E. L. Kane, P. Lalousis, P. R. Wiles, and H. Hora, Appl. Phys. B 27, (1982), pp.157-159
28. G. Z. Sun, E. Ott, Y. C. Lee, and P. Guzdar, Phys. Fluids, 30, (1987), p.526-532.
29. D. A. Jones, E. L. Kane, P. Lapousis, P. Wiles, and H. Hora, Phys. Fluids, 25, (1982), pp. 2295-2301.
30. F. Brunel, Phys. Rev. Lett., 59, (1987), p.52
31. F. Brunel, Phys. Fluids, 31, (1988), 2714
32. P. Gibbon and R. Bell, Phys. Rev. Lett., 68, (1992), p. 1535
33. J. Delettrez et. al, Bull. Am. Phys. Soc., 10, (1993), p. 1987
34. R. L. Keck, L. M. Goldman, M. C. Richardson, W. Seka, and K. Tanaka, Phys. Fluids, 27, (1984), p. 2762
35. S. V. Bulanov, N. M. Naumova, F. Pegoraro, Plasma Physics, 20, (1994), p. 640 (in Russian)
36. A. V. Gurevich, A.P. Mescherkin, JETP, 80, (1981), p. 1810-1826 (in Russian)
37. J. Stamper, K. Papadapoulos, R. N. Sudan, E. McLean, S. Dean, and J. Dawson, Phys. Rev. Lett., 26, (1971), p. 1012.
38. R. N. Sudan, AIP Conf. Proceedings 318, 11th Int. Workshop, Monterey, CA, 1993, p. 91-96.

Formation of Self-Channeling and Electron Jet in an Underdense Plasma Excited by Ultrashort High Intensity Laser Pulses

M. Kando[1], H. Ahn, H. Kotaki, K. Tani, T. Watanabe[*], T. Ueda[*],
M. Uesaka[*], Y. Kishimoto[†], J. Koga[†], H. Watanabe[†], K. Nakajima[**],
M. Arinaga[**], T. Kawakubo[**], H. Nakanishi[**], and A. Ogata[**]

Advanced Photon Research Center, Japan Atomic Energy Research Institute
2-4 Shirakata shirane, Tokai, Naka, Ibaraki 319-11, Japan

[*]*Nuclear Engineering Research Laboratory, The University of Tokyo*
2-22 Shirakata shirane, Tokai, Naka, Ibaraki 319-11, Japan

[†]*Naka Fusion Research Establishment*
Japan Atomic Energy Research Institute
Naka, Naka, Ibaraki 311-01, Japan

[**]*National Laboratory for High Energy Physics*
Oho, Tsukuba, Ibaraki 305, Japan

Abstract. Experimental investigations on interactions of a tera-watt laser (0.5-2.5 TW) with an underdense plasma are reported. Formation of self-channeling longer than the vacuum Rayleigh length is observed when ultrashort (100 fs in FWHM) high power (~2 TW) laser pulses are focused into a gas (He, Ar, N_2) filled chamber.

In addition to the formation of self-channeling along the laser axis, a bright radiation in the transverse direction of the channel was observed for the highly excited dense gases. This sideway radiation was observed for all three gases and especially for N_2, sideway jet-like radiation (we call it sideway electron jet) was monitored. The energy of the emitted electrons was estimated to be ~10 keV from the range of electron trajectories in N_2.

INTRODUCTION

Self-focusing and self-channeling of intense laser pulses through a plasma have attracted a great deal of interest for laser-plasma-based accelerators[1] and laser-pumped x-ray lasers[2]. In particular laser wakefield accelerator (LWFA), one of laser-plasma-based particle accelerators, recently made great progress in generation of ultrahigh accelerating field[3]. However, their energy gains are still below 100 MeV, because their acceleration length is limited around the vacuum Rayleigh length. This effect deducts the advantage of ultrahigh gradient acceleration from la-

---

[1] On leave from Institute for Chemical Research, Kyoto University

ser-plasma accelerators. Therefore it is essential for LWFA to cause a long distance interaction of an intense ultrashort laser pulse with an underdense plasma in order to jump the energy gain from tens of MeV to the order of GeV.

If the refractive index of a plasma along the optical axis is sufficiently high, a laser pulse may be guided without suffering from the diffraction. There are two mechanisms to cause refractive guiding of intense laser pulses. For low focused laser intensities below relativistic regime, a preformed plasma channel with density minima on axis guides a laser pulse in a linear propagation regime similar to that in solid optical fibers. A plasma waveguide with radially increasing density may be formed due to a shock wave created by a first pulse to guide a second pulse after some delay[4]. In relativistic regimes self-focusing and self-channeling in a homogeneous plasma have been predicted to occur above the critical power, given by $P_c=17(\omega^2/\omega_p^2)$ [GW] where ω is the laser frequency and ω_p is the plasma frequency [5]. Relativistic self-focusing arises from increase of electron mass in a plasma due to relativistic effects because the refractive index of the plasma is peaked on the axis where the intensity of the laser has a maximum. For the ultrashort pulses shorter than the plasma wavelength, however, it is believed that relativistic self-channeling is substantially reduced[6]. In case a pulse length and width are comparable with the plasma wavelength, a strong wakefield excited by a ponderomotive force may cause channel guiding of ultrashort laser pulses. The 2D simulation shows the pulse is trapped in a pocket of the electron density depletion traveling with the laser pulse. As increasing the pulse length to be longer than the plasma wavelength, wakefields excited stimulated Raman scattering instability affect focusing properties of the pulse to lead to its self-modulation[7]. These effects are the third mechanism to raise the self-channeling of the laser pulse induced by wakefields. This mechanism implies that the relativistic self-focusing can occur at a lower laser power than the critical power.

LASER-PLASMA EXPERIMENT

T^3 Laser System

We have constructed the T^3 laser system on a 2×4 m^2 table based on the chirped-pulse amplification (CPA) technique at 790 nm. The schematic of the system is depicted in Fig.1. The oscillator is a mode-locked Ti:sapphire laser pumped by a cw-argon-ion laser at a power of 6 W. It produces pulses of 70 fs duration at a repetition rate of 80 MHz to deliver an output power of 0.75 W at 790 nm. The seed pulse from the oscillator is stretched to 320 ps in a four-pass grating arrangement with a reflective telescope. A stretched pulse is amplified to ~5 mJ in the Ti:sapphire regenerative amplifier (RGA) pumped at 10 Hz by 35 mJ, 6 ns pulses of a Q-switched Nd:YAG laser at 532 nm.

The output from the regenerative amplifier is further amplified to > 400 mJ through a multipass pre-amplifier and a multipass main amplifier. Both faces of a Ti:sapphire crystal are pumped with two frequency-doubled pulses of 100 mJ for a pre-amplifier and 1.3 J for a main amplifier from a Q-switched Nd:YAG laser that produces a total energy of 1.6 J at 532 nm. The amplified pulse is compressed in a two-pass grating configuration to 100 fs with an energy of >200 mJ, corresponding to a peak power of 2 TW. Since we have succeeded in producing the maximum output energy of 600 mJ at the main amplifier, we can generate the maximum peak power of 3 TW at 10 Hz with the transmission efficiency of 50 % in the compressor.

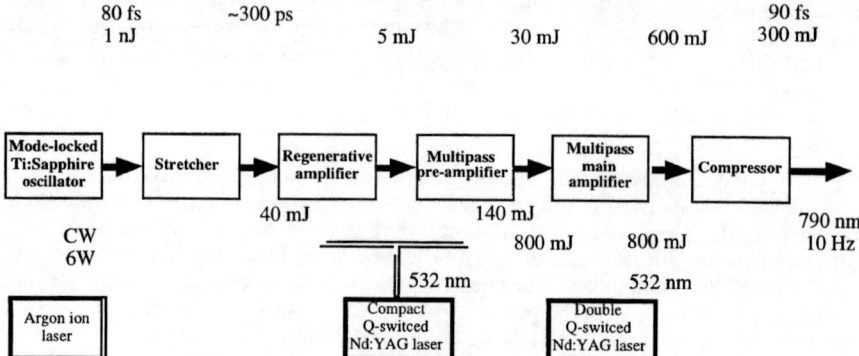

FIGURE 1. Schematic of our T^3 laser system.

Plasma Production

A plasma is produced through tunneling ionization when an intense laser pulse is focused in an experimental chamber which is filled with a gas such as helium(He), argon(Ar) and nitrogen(N_2) at pressures of 10^{-3} - 760 Torr. For example, the threshold intensities of tunneling ionization are 2×10^{15} W/cm^2 for He$^+$ and 9×10^{15} W/cm^2 for He^{2+}. In vacuum the laser pulse focuses to a peak intensity of 1.5×10^{18} W/cm^2 for a peak power of 2.4 TW using a f/10 off-axis parabolic mirror with a focal length of 480 mm.

Laser Propagation Experiment

We measured the spot size of a laser pulse along the propagation axis. The diagnostic system is shown in Fig.2. The forward scattered laser light was imaged onto a charge-coupled device (CCD) camera coupled to a microscope objective through a 10 nm FWHM interferential filter. The exposure time of the CCD camera was 100 ms synchronized with the 10 Hz laser trigger. Many neutral-density (ND) filters and reflecting mirrors were used to reduce the intensity of the laser beam.

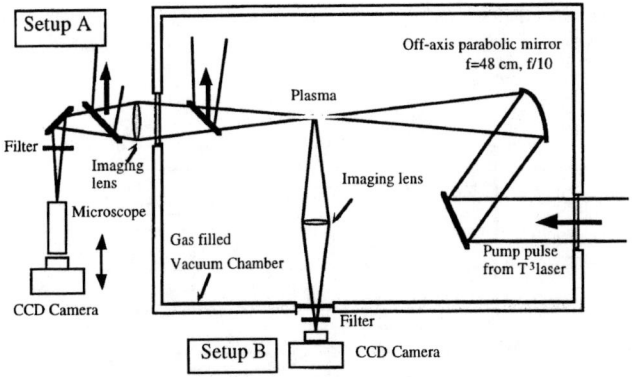

FIGURE 2. Setup for the laser-plasma experiments. Setup A is used for the spot size measurement of the laser, while Setup B is for the measurement of fluorescence from a plasma.

Plasma Fluorescence Measurement

Plasma fluorescence was taken by a CCD camera through a blue-pass filter or a 10 nm FWHM around 780 nm band-pass interferential filter. In case of the interferential filter, we can observe the Thomson scattered image of the incident laser pulses. The diagnostic setup is indicated setup B in Fig.2. The exposure time of the camera was selected to 100 ms so as to take a single shot of the plasma/Thomson scattered image.

Spectrum Measurement

Forward scattered (90°) spectra of the laser lights were measured by a spectrometer. The spectrometer consists of three gratings (300, 600, and 1200 lines/mm) which have different resolutions and cover the wavelength over 500 - 1000 nm and a CCD camera where images reflected by the grating are projected. The experimental setup was similar to the setup A/B case in Fig.2, except that the optical fiber probe was used instead of the CCD camera.

EXPERIMENTAL RESULTS

Laser Propagation Experiment

The measured FWHM spot sizes along the propagation axis in He are shown in Fig.3. In vacuum (3×10^{-4} Torr), the spot size along the axis was in good agreement with the theoretical beam envelope for the Gaussian beam propagation without regard to the laser power. When the gas pressure increases toward more than 100

Torr, the spot size decreases down to the smaller size than the expected Gaussian propagation over the range of a few cm as the laser power increases toward more than 1 TW.

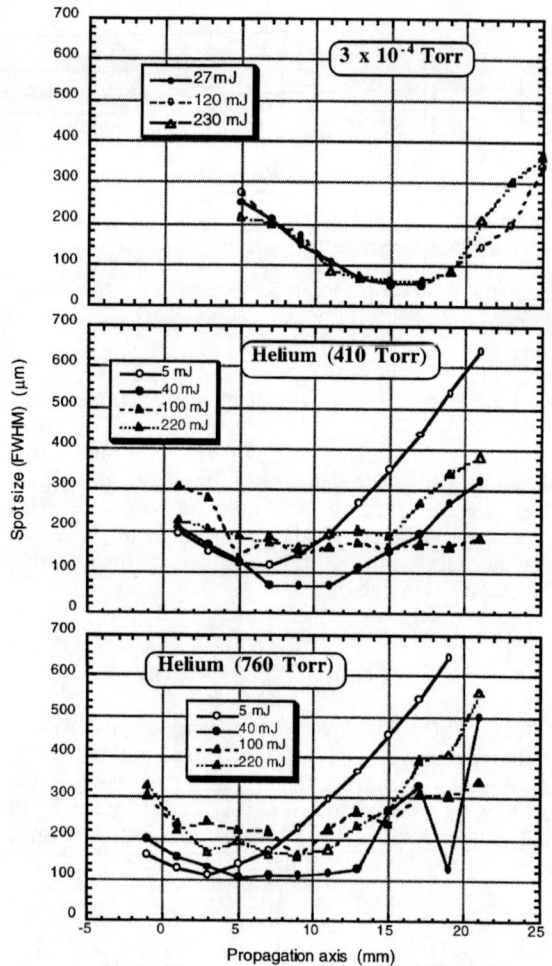

FIGURE 3. Measured FWHM spot size of the laser beam along the propagation axis in vacuum (top), He 410 Torr (middle), and He 760 (bottom).

Plasma Observation through Blue-pass Filter

Figure 4 shows the images of N_2 plasma fluorescence emitted in the process of recombination. In the figure the laser beam came from the right side. Relatively, at a low pressure (< 1 Torr), we can observe long plasma fluorescence as expected from the Gaussian beam propagation. As a pressure and a laser power increase, the fluorescence become brighter. In particular, bright radiations in the transverse direction can be seen when a pressure is 50 Torr and a laser power was more than 1 Torr. These radiation can be observed for all three gases (He, Ar and N_2) and especially for N_2. Since a shape of this bright radiation was deformed with a magnetic field and they are axially symmetric, it was inferred that high energy plasma electrons were accelerated by wakefields left trajectories due to gas ionization. Therefore we call this bright radiation sideway electron jet. For the lower incident power, these are ejected perpendicularly outward. A length of the jet becomes shorter as the gas pressure increased since the range of electrons is inversely proportional to the gas pressure. The energy of emitted electrons is estimated to be ~10 KeV from the range of electron trajectories due to ionization energy loss in N_2 gas.

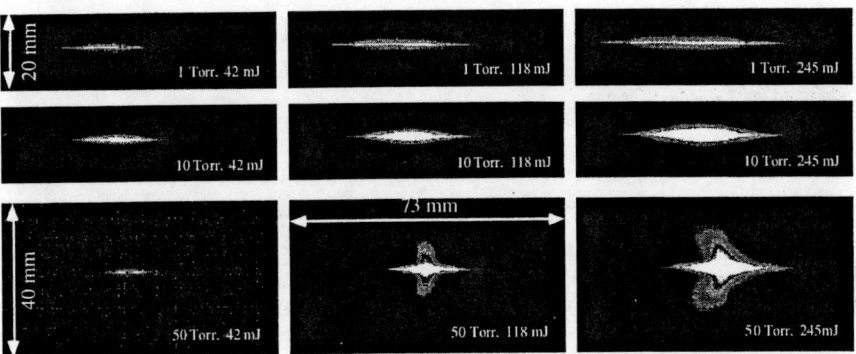

FIGURE 4. N_2 Plasma fluorescence images taken through a blue-pass (340 - 560 nm) filter. The laser beam came from the right side.

Thomson Scattered Image through Interferential Filter

When we use an interferential filter around the incident laser wavelength, an image of the scattered laser light depicts the laser intensity and the plasma electron density distributions because the Thomson scattering from the plasma region depends on the product of them[8] . Figure 5 shows the measured image and the Thomson axial profiles. When relativistic self-focusing occurs, a flat broad peak appears on the Thomson scattered profile along
the laser propagation axis while the profile shows a single narrow peak at the focus below the self-focusing threshold. This feature of the Thomson profile results from the electron density depletion in the laser focal volume due to the ponderomotive force and the self-focused laser intensity. The Thomson axial profiles shown in Fig.5 indicate features arising from self-focusing and self-channeling above 1 TW in 87 Torr He. Relativistic self-focusing should take place at $P > P_c = 5$ TW at 87 Torr fully ionized He plasma ($n_e = 6.1 \times 10^{18}$ cm^{-3}) and at $P > P_c = 0.56$ TW at 760 Torr fully ionized He plasma ($n_e = 5.4 \times 10^{19}$ cm^{-3}).

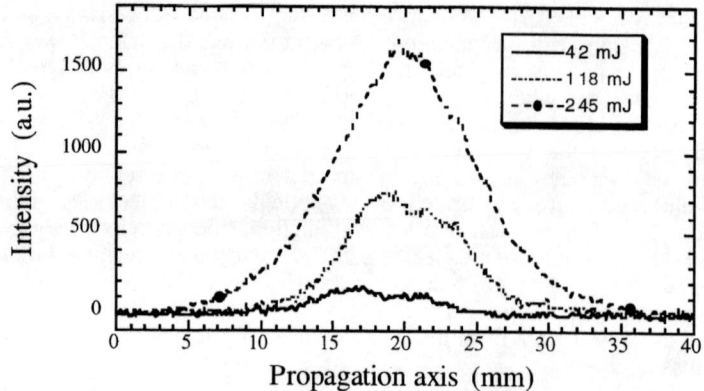

FIGURE 5. Axial Thomson profiles for 87 Torr He. The laser beam came from the right side.

Spectrum Measurement

Figure 6 shows the spectra of the incident laser lights and the forward scattered lights at pressures of 1 and 22 Torr He for various laser powers. The wavelength of the incident laser (after compressor) was centered at 790 nm and the bandwidth was ~10 nm (FWHM). The pulse duration was measured to be ~100 fs by autocorrelation method. At low pressure (below 20 Torr), the forward scattered spectrum was almost keep the incident distribution even if the laser power was increased to 2 TW. However, at high pressure (over 20 Torr), frequency up-shift of the laser beam took place gradually if the laser power was increased. The shifted peak was 56 nm below the initial peak (792 nm) and the degree of the up-shift was independent of the laser power. In case the laser power was increased to 2 TW, almost all laser photons were blue-shifted.

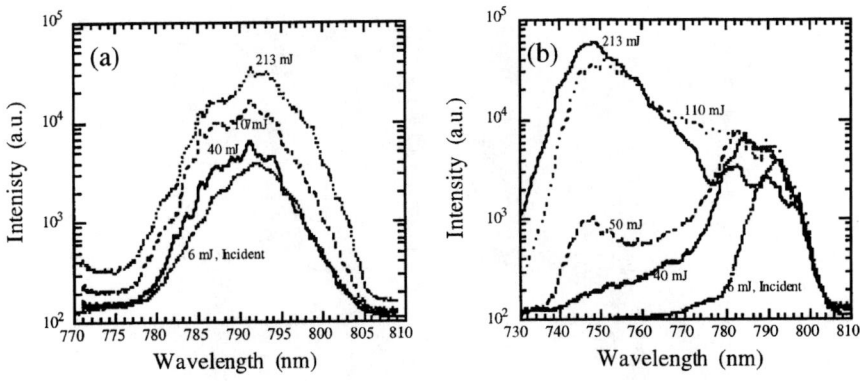

FIGURE 6. Measured spectra of the forward scattered laser lights. (a) He 1 Torr. (b) He 22 Torr. The spectrum of the incident laser is measured in air.

DISCUSSIONS

In order to analyze the experimental results we carry out a series of 2D PIC (Particle-in-Cell) simulations. Typical parameters of our simulation system are: Laser wavelength $\lambda_0/\lambda_p = 0.672$, pulse length $\omega_{pc}\tau_{FWHM} = 8.55$, laser waist at focus $w_0/\lambda_p = 2.53$, pulse energy $E_l/m_ec^2 = 1.13 \times 10^{10}$, and Rayleigh length $Z_R/\lambda_p = 30$ where $\lambda_p = c/\omega_{pc}$ is the plasma skin depth. The simulation size was $4096\Delta x \times 256\Delta y$ where $\Delta x = \lambda_p/13$, $\Delta y = 2\lambda_p/13$. The laser pulse is a linearly polarized Gaussian beam. These parameters are close to those of the experiment. However, some of the parameters have not faithfully been reproduced, although their effects are believed to play little role in the present investigation. In the present simulation the gas is assumed to have been ionized by the very front skirt of the laser. Figure 7 shows evidence of self-channeling of the laser pulse.

Figure 7 indicates that (i) a small well collimated self-channeling; (ii) a small fraction of laser energy is diffracted to form lobes; (iii) the self-channeling happened substantially below the relativistic self-focusing theoretical threshold P_c. The reason for behavior (iii) may be ascribed to the transverse ponderomotive force that creates a favorable index of refraction for self-focusing.

Simultaneous to this phenomenon are the production of sideway jets of electrons (and associated current formation and ω_p radiation (THz radiation)) and the sudden onset of total blue-shift of the laser pulse. The latter phenomenon, once again believed to be related to the onset of self-channeling, will be reported in a separate publication. The sideway jet formation is shown in Fig.8 from a simulation with a short laser pulse satisfying the condition for wake field excitation. The observed experimental properties of jet formation, i.e., its pressure dependence and laser power dependence, as well as its morphology, seem to be reproduced by this simulation.

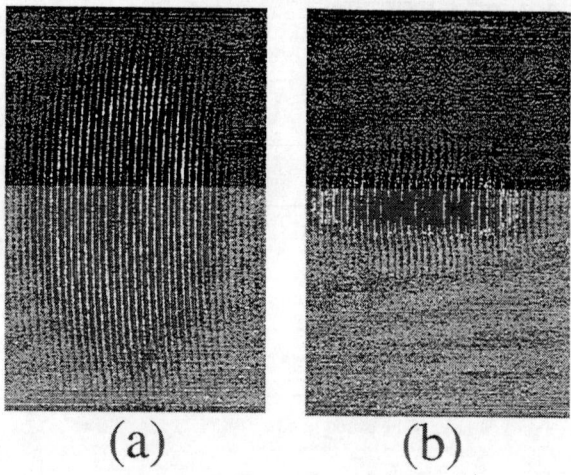

FIGURE 7. Contour plot of laser field E_z. (a) At initial time $\sim -4.5 Z_r$, near the vacuum focus point, (b) after propagating $\sim 3.5 Z_r$ past the focus point. Z_r is the vacuum Rayleigh length.

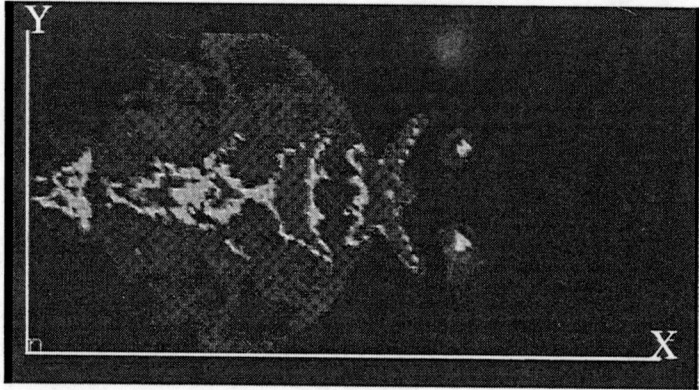

FIGURE 8. Density contour plot of high energy particles for a short pulse simulation where the laser pulse is moving to the right.

CONCLUSION

We have first demonstrated that the self-focusing and self-channeling of ultrashort multi-terawatt laser pulses occur over the order of a few cm in an underdense plasma below the theoretical threshold intensity of the relativistic self-focusing. We have observed that side-way electron jets are formed. It is inferred that electron jets are induced due to wakefields generated by ultrashort laser pulses. We have observed that the almost fully blue-shift of the incident laser lights to a degree of 56 when the gas pressure is increased to more than 20 Torr and the power of the laser is 2 TW.

References

1. T. Tajima and J. M. Dawson, Phy. Rev. Lett., **43**, 267 (1979); M. Gorbunov and V. I. Kirsanov, Zh. Eksp. Teor. Fiz. **93**, 509 (1987) [Sov.Phys. JETP **66**, 290 (1987)]; P. Sprangle et al., Appl. Phys. Lett. **53**, 2146 (1988).
2. N. H. Burnett and P. B. Corkum, J. Opt. Soc. Am. **B 6**, 119 (1989).
3. K. Nakajima et al., Phys. Rev. Lett., **74**, 4428 (1995); A. Modena, et al., Letts. Nature **377**, 606 (1995).
4. C. G. Durfee III and H. M. Milchberg, Phys. Rev. Lett., **71**, 2409 (1993).
5. G. Z. Sun, E. Ott, Y. C. Lee, and P. Guzdar, Phys. Fluids **30**, 526 (1987). D. C. Barnes, T. Kurki-Suonio, and T. Tajima, IEEE Trans. Plasma Sci. **PS-15**, 154 (1987). P. Sprangle, C. M. Tang, and E. Esarey, IEEE Trans. Plasma Sci. **PS-15**, 145 (1987)
6. P. Sprangle, E. Esarey, and A. Ting, Phys. Rev. Lett., **64**, 2011 (1990).
7. S. V. Bulanov, F. Pegoraro and A. M. Pukhov, Phys. Rev. Lett., **74**, 710 (1995).
8. P. Monot et al., Phys. Rev. Lett., **74**, 2953 (1995)

Accelerated electron measurements in the self-modulated laser wakefield accelerator

C.I. Moore,[1] K. Krushelnick,[2] A. Ting,[1] C. Manka,[1] H.R. Burris,[1] R. Fischer,[1] M. Baine,[3] E. Esarey,[1] P. Sprangle,[1] and R. Hubbard[1]

[1] Plasma Physics Division, Naval Research Laboratory, Washington DC, 20375
[2] Laboratory of Plasma Studies, Cornell University, Ithaca NY, 14853
[3] University of California, San Diego, La Jolla CA, 92093

High energy electrons (up to 30 MeV) have been measured in the self-modulated laser wakefield accelerator using a 2.5 TW laser pulse and a high sensitivity detector, a scintillator coupled to a photo-multiplier tube (PMT). Highly non-linear plasma waves have been detected using forward Raman scattering as a plasma diagnostic and a correlation between the non-linear plasma waves and electron signal has been observed.

Acceleration of charged particles using laser produced plasma waves has been the focus of much research in the past few years (1). We have been performing experiments using the self-modulated laser wakefield accelerator (SM-LWFA) technique (2—6). The SM-LWFA generates large amplitude relativistic plasma waves suitable for particle acceleration through a combination of the forward Raman scattering (FRS) instability (7), the self-modulation instability (2), and the ponderomotive force of the laser pulse. The ponderomotive force of a high-intensity laser pulse passing through a plasma generates a copropagating relativistic plasma wave. When the laser power exceeds the critical power for relativistic self focusing,

$$P_c = 17(\omega_0/\omega_p)^2 \text{ GW} \qquad (1)$$

(where ω_0 is the laser frequency and ω_p is the plasma frequency), and the pulse length is greater than the plasma period, $1/\omega_p$, the laser pulse envelope undergoes self modulation at the plasma frequency which resonantly enhances the ponderomotive force and the resulting plasma wave amplitude. This process can generate extremely high gradient electric fields (~100 GeV/m) traveling at the phase velocity of the laser pulse (3,6). Since the laser pulse phase velocity is close to the speed of light, these plasma waves are ideal for high energy particle acceleration.

Recent SM-LWFA experiments by Modena *et al.* used a 25 TW laser pulse to drive nonlinear wakefield plasma waves in a helium plasma resulting in the capture and acceleration of background plasma electrons to 44 MeV (6). Their experiments showed a broadening of the anti-Stokes lines in the forward Raman scattering (FRS) spectrum concurrent with the onset of high energy electron production at approximately 7 TW. This broadening and high energy electron production was attributed to the onset of wavebreaking (8). Our experiments use an order of magnitude lower power laser pulse (2.5 TW). In this lower power regime, we see the production of high energy electrons (up to 30 MeV) and highly non-linear wakefields (5,6,9,10) with no evidence of wave breaking.

In earlier experiments, we measured relativistic guiding of a high power pulse (~2 TW). These measurements showed optical guiding of the laser for >30 Rayleigh lengths which was the length of the gas jet used in the experiments (~3 mm long, n_e ~10^{19} cm^{-3}, and P_c~1.7 TW). We also measured the wakefield lifetime using a pump-probe setup. Coherent Thomson scattering of a variable delay probe (relative to pump timing) showed a wakefield lifetime of 5 ps. In addition, 90° imaging of the scattered probe light showed that the pump formed a plasma channel in its wake which guided the probe for ~3 mm, i.e., over the same distance as the pump was guided. The channel was formed by ponderomotive expulsion of electrons followed by space charge expulsion of the ions. The channel formation was well modeled by a simple thermal expansion model of the plasma. These results (the relativistic optical guiding, the wakefield lifetime measurements and channel formation observations) have been published elsewhere (11,12). This paper will focus on high energy electron measurements, forward Raman scattering measurements, and the correlation between the two.

The schematic of the experiment is shown in figure 1. The laser system used in the experiments was a CPA system (13) consisting of a Ti:Sapphire oscillator, a Ti:Sapphire regenerative amplifier, and three Nd:Glass single pass amplifiers. The output laser pulse had a wavelength of 1.053 mm and a typical pulse length of 400 fs. The final energy was approximately 1 joule resulting in a peak power of 2.5 TW. The laser pulse was focused in a 3 mm diameter helium gas jet with an f/4 off axis parabola (peak intensity ~5×10^{18} W/cm^2). The laser pulse completely ionized the helium resulting in a plasma density of n_e~1.5×10^{19} cm^{-3}. Since the laser power was approximately twice the critical power for relativistic self-focusing (P_c~1.2 TW for n_e~1.5×10^{19} cm^{-3}), the laser pulse was self-modulated through the forward Raman and self-modulation instabilities and in the process excited large amplitude plasma waves. Forward Raman scattered laser light at 40° to the laser axis was imaged on the entrance slit of a spectrometer to measure the relative wakefield amplitude and linearity. An inline spectrometer configuration measured the energy distribution of electrons accelerated from the background plasma, i.e., with no external injection of electrons.

The electron spectrometer consisted of an electromagnet for electron deflection and a plastic scintillator directly coupled to a PMT for electron detection. The electromagnet used a 0 to 2500 Gauss magnetic field in a field region 5.5 cm long. Graphite and carbon shielding were arranged with a small gap centered on the laser axis which allowed only high energy electrons with less than an 8° deflection in the magnet to strike the scintillator. Electrons with lower energies were deflected more than 8° and dumped in a graphite block to minimize x-ray production. This inline spectrometer configuration therefore detects all electrons above a cutoff energy which is determined by the magnetic field strength and the maximum acceptance angle of the gap in the shielding (8°).

The gas jet density was determined through the frequency shift of the forward Raman scattered anti-Stokes lines (14,15). This frequency shift is given by the plasma frequency which is related to the plasma density by

$$\omega_p = (4\pi n_e e^2/m_e)^{1/2}. \qquad (2)$$

The density found in this way was in good agreement with previous plasma density measurements using backward Raman scattering at low intensities (15).

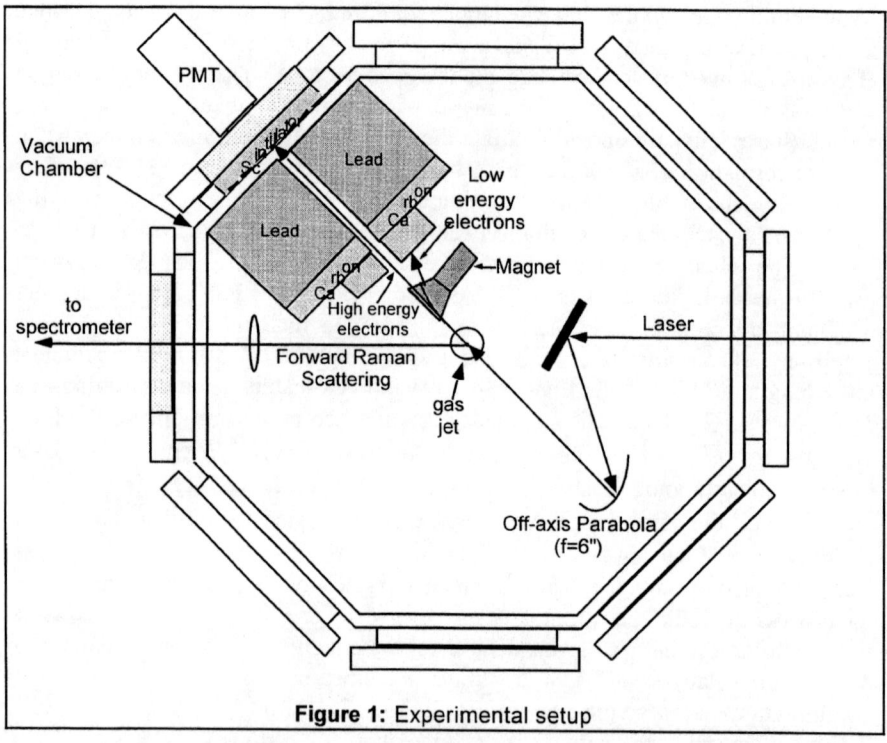

Figure 1: Experimental setup

An energy scan is shown in figure 2. This is raw data showing the electron signal at a range of energy settings (1-30 MeV). The data points are the total relative number of electrons above the energy specified on the x-axis. The large fluctuations are known to be shot-to-shot fluctuations since low energy (200 keV - 5 MeV) electron spectrum measurements using film have shown a smooth monotonically decreasing energy spectrum.. The fluctuations were not due to changes in laser parameters since no laser parameters were changed between shots except for small fluctuations in energy (~10%) which did not correlate with electron signal fluctuations. One possible explanation is the growth of the Raman instability. Since this instability is seeded from noise, plasma wave growth is expected to fluctuate from shot-to-shot. The source of the background accelerated electrons may also contribute to the large fluctuations. A possible source of the background accelerated electrons other than from wave breaking is the beating of backward Raman scattered light with the laser pulse (16). This beating generates a low phase velocity plasma wave capable of accelerating low energy electrons to the energy necessary for wakefield acceleration. Numerical simulations using this model suggest that small changes in wakefield amplitude (~10%) can result in orders of magnitude fluctuations in electron production, which is consistent with our experimental results. This theoretical work is presented in another paper from these proceedings (16).

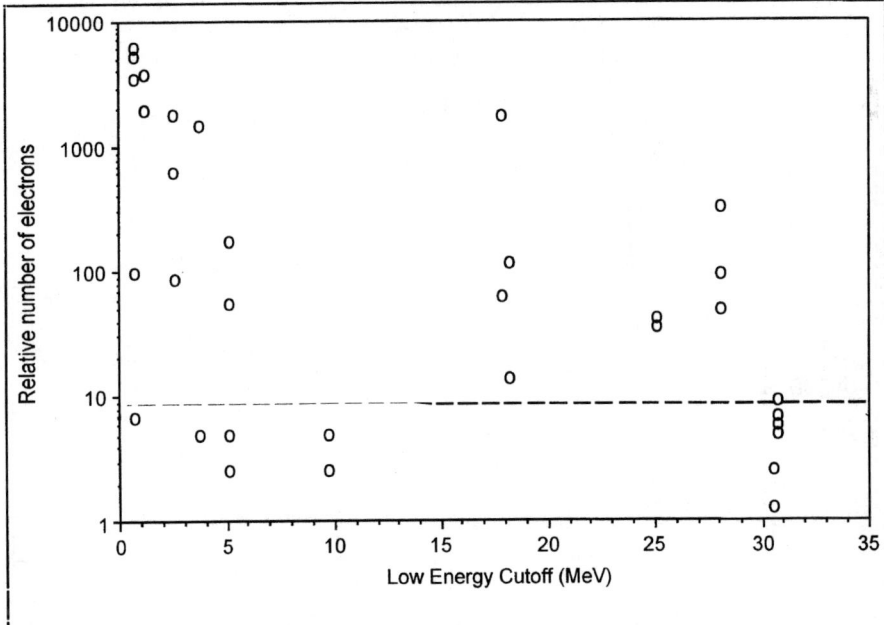

Figure 2: The relative number of electrons above the cutoff energy of the spectrometer. The maximum noise level is the background noise level of the PMT. A signal above this level represents a clearly discernible signal from electrons striking the scintillator.

Concurrent with electron measurements we examined the spectrum of forward Raman scattered light. Typical spectra are shown in figure 3. These spectra show the strong non-linearity of the plasma wave. As the plasma wave amplitude approaches or exceeds $\Delta n/n_0 = 1$, the standard linear plasma wave theory

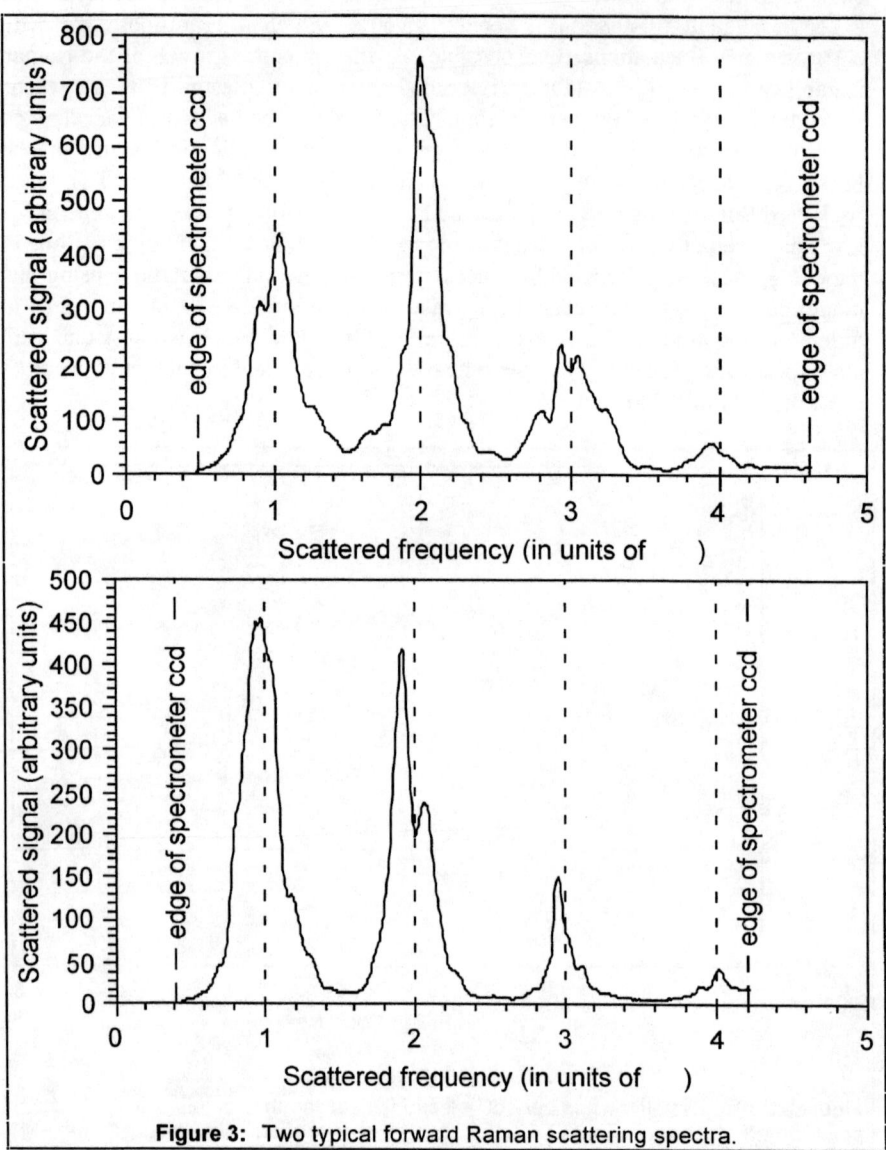

Figure 3: Two typical forward Raman scattering spectra.

breaks down. At this point the plasma wave "steepens" and harmonics of the plasma wave frequency became apparent (5,6,9,10). The presence of these plasma wave harmonics is readily apparent in the scattered spectra as higher orders of the anti-Stokes line. The relative amplitudes of the different order anti-Stokes lines (~1) show the presence of large plasma density modulations (~1) (10). Also, there is no indication of line broadening in the FRS spectra. The width of the 1st order anti-Stokes lines shown in figure 3 are the same as those we observed in the linear plasma wave regime (where only the 1st order anti-Stokes line is observed). This indicates that although the wave has steepened, it has not begun to break as was observed by Modena *et al.* at higher powers (6).

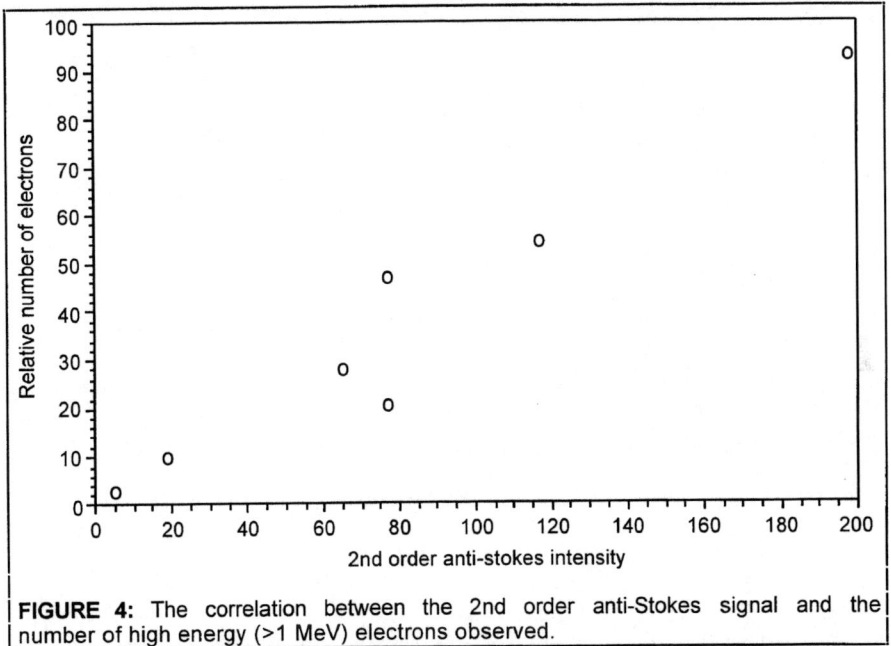

FIGURE 4: The correlation between the 2nd order anti-Stokes signal and the number of high energy (>1 MeV) electrons observed.

The correlation of 2nd order anti-Stokes signal to electron production has been studied. For these experiments, the electron spectrometer measured the number of electrons with greater than 1 MeV of energy. For each laser shot the electron number and forward Raman scattering spectrum were recorded. The 2nd order anti-Stokes signal vs. electron signal is shown in figure 4. A strong correlation between 2nd order anti-Stokes and electron signal was observed. This perhaps indicates that electrons are only accelerated from the background plasma as the plasma wave becomes non-linear.

In conclusion, we have observed the wakefield acceleration of background plasma electrons to high energy, ~30 MeV. Highly non-linear plasma waves were detected through forward Raman scattering diagnostics with no evidence of wave breaking. A correlation between 2nd order anti-Stokes signal and high-energy electron signal was observed.

ACKNOWLEDGMENTS

This work was supported by the Office of Naval Research and the U. S. Department of Energy. This work was performed while the principal author held a National Research Council-Naval Research Laboratory Research Associateship.

REFERENCES

1 T. Tajima and J. M. Dawson, *Phys. Rev. Lett.* **43**, 267 (1979); E. Esarey, P. Sprangle, J. Krall, and A. Ting, *IEEE Trans. Plasma Sci.* **24**, 252 (1996) and references contained within.

2 P. Sprangle, E. Esarey, J. Krall, and G. Joyce, *Phys. Rev. Lett.* **69**, 2200 (1992); T. M. Antonsen and P. Mora, *Phys. Rev. Lett.* **69**, 2204 (1992); N. E. Andreev, L. M. Gorbunov, V. I. Kirsanov, A. A. Pogosova, and R. R. Ramazashvili, *JETP Lett.* **55**, 571 (1992).

3 J. Krall, A. Ting, E. Esarey, and P. Sprangle, *Phys. Rev. E* **48**, 2157 (1993).

4 K. Nakajima *et al.*, *Phys. Rev. Lett.* **74**, 4428 (1995); R. Wagner, S. Y. Chen, A. Maksimchuk, E. Dodd, J. K. Kim, and D. Umstadter, *Bull. Am. Phys. Soc.* **40**, 1799 (1995); C. I. Moore, K. Krushelnick, A. Ting, R. Burris, A. Fisher, P. Sprangle, E. Esarey, and D. Umstadter, *Bull. Am. Phys. Soc.* **40**, 1797 (1995).

5 C. Coverdale, C. B. Darrow, C. D. Decker, W. B. Mori, K. C. Tzeng, K. A. Marsh, C. E. Clayton, and C. Joshi, *Phys. Rev. Lett.* **74**, 4659 (1995).

6 A. Modena, Z. Najmudin, A. E. Dangor, C. E. Clayton, K. A. Marsh, C. Joshi, V. Malka, C. B. Darrow, C. Danson, D. Neely, and F. N. Walsh, *Nature* **337**, 606 (1996).

7 D. W. Forslund, J. M. Kindel, and E. L. Lindman, *Phys. Fluids* **18**, 1002 (1975); R. E. Turner, K. Estabrook, R. P. Drake, E. A. Williams, H. N. Kornblum, W. L. Kruer, and E. M. Campbell, *Phys. Rev. Lett.* **57**, 1725 (1986); W. B. Mori, C. D. Decker, D. E. Hinkel, and T. Katsouleas, *Phys. Rev. Lett.* **72**, 1482 (1994).

8 J. M. Dawson, *Phys. Rev.* **113**, 383 (1959); A. I. Azhiezer and R. V. Polovin, Sov. Phys. JETP **3**, 696 (1956); J. B. Rosenzweig, *Phys. Rev. A* **38**, 3634 (1988); T. Katsouleas and W. B. Mori, *Phys. Rev. Lett.* **61**, 90 (1988).

9 P. Sprangle, E. Esarey, and A. Ting, Phys. Rev. Lett. **64**, 2011 (1990); D. Umstadter, R. Williams, C. Clayton, and C. Joshi, *Phys. Rev. Lett.* **59**, 292 (1987).

10 D. Umstadter, S.-Y. Chen, A. Maksimchuk, G. Mourou, and R. Wagner, *Science* **273**, 472 (1996).

11 A. Ting, K. Krushelnick, C. I. Moore, H. R. Burris, E. Esarey, J. Krall, and P. Sprangle, *NRL int. memo report*, NRL/MR/6790-96-7876 (also submitted to *Phys. Rev. Lett.*).

12 K. Krushelnick, A. Ting, C. I. Moore, H. R. Burris, E. Esarey, P. Sprangle, and M. Baine, *NRL int. memo report*, NRL/MR/6790-96-7879 (also submitted to *Phys. Rev. Lett.*).

13 P. Maine, D. Strickland, P. Bado, M. Pessot, and G. Mourou, *IEEE J. Quantum Electron.* **QE-24**, 398 (1988).

14 M. D. Perry, C. Darrow, C. Coverdale, and J. K. Crane, *Opt. Lett.* **17**, 523 (1992).

15 A. Ting, K. Krushelnick, H. R. Burris, A. Fisher, C. Manka, and C. I. Moore, *Opt. Lett.* **21**, 1096 (1996).

16 R. F. Hubbard, P. Sprangle, E. Esarey, A. Ting, H. R. Burris, C. I. Moore, and K. Krushelnick, *Proc. AIP Conf.*, 7th Workshop on Advanced Accelerator Concepts, Oct. 12-18 1996.

Electron Acceleration by Self-Modulated Laser Wakefield in A Relativistically Self-Guided Channel

S.-Y. Chen, R.Wagner, A. Maksimchuk, D. Umstadter
Center for Ultrafast Optical Science, University of Michigan, Ann Arbor, MI
48109

Abstract. The relativistic self-focusing of an intense laser pulse ($I \sim 4\times10^{18}$ W/cm^2, λ = 1µm, τ = 400 fs) in a gas jet 750 µm in length was observed using sidescattering imaging. A self-modulated laser wakefield was generated to accelerate self-trapped electrons. The energy distribution and transverse emittance of the electron beam changed due to the onset of the relativistic self-guiding.

Introduction

Plasma-based accelerators are of great interests for electron acceleration [1,2] because they have the ability to support much larger acceleration gradient (E > 100 GV/m) than conventional RF accelerators. Several methods can be used to drive large amplitude plasma wave. Since the size of terawatt laser systems has been reduced by chirped pulse amplification, the laser wakefield accelerator (LWFA) [1] and the self-modulated LWFA [1,3-9], have recently received considerable attention and show promising results. In order to drive a plasma wave to large amplitudes, the laser pulse has to be focused to small spot size (on the order of tens of microns) to achieve high intensity. Nonetheless, it also limits the length of the plasma wave channel to be less than twice the Rayleigh length (confocal parameter) due to the laser beam diffraction. In order to extend this plasma wave channel, two methods have been proposed. One way is to use a prepulse to create a concave plasma channel which can then guide the main laser pulse over the length of this channel [3,10-11]. Another way is to make the laser pulse self-guided by relativistic self-focusing (self-guiding) so that it can propagate over many Rayleigh lengths while maintaining high peak intensity [12-14]. In this paper we, for the first time, report the interplay between relativistic self-guiding and acceleration of electrons in self-modulated laser wakefield.

Experiments

When a terawatt-peak-power laser is focused to high intensity into a gas, a plasma is created by tunneling ionization, and the free electrons begin to quiver at velocities close to the speed of light (c) in the laser's transverse oscillating electromagnetic field. The laser light group velocity v_g depends on the index of refraction $v_g = cn$, where the latter is given by

$$n = \left[1 - \left(\omega_p/\omega\right)^2\right]^{1/2}, \qquad (1)$$

in which ω is the laser frequency and the plasma frequency is given by

$$\omega_p = \left(4\pi n_e e^2 / \gamma m_0\right)^{1/2}, \qquad (2)$$

where e is the electron charge, m_0 is the electron rest mass, and n_e is the plasma electron density. The relativistic factor γ associated with the motion perpendicular to the direction of laser propagation depends on the laser field through the electron quiver velocity v_{os}, $\gamma = [1 + a^2]^{1/2}$, where $a = \gamma v_{os}/c = eE/m_0\omega c = 8.5\times10^{-10}\lambda[\mu m]I^{1/2}[W/cm^2]$ is the normalized vector potential, E is the laser electric field, and $I = cE^2/8\pi$ is the laser intensity. For a terawatt laser pulse, with $\gamma > 1$, several nonlinear effects arise. First, the plasma frequency decreases when the laser intensity increases, due to the increase in γ. Second, the rapid temporal change in γ for a short laser pulse broadens and modulates the transmitted light spectrum (relativistic self-phase modulation). Finally, the relativistic self-focusing effect, which occurs because the laser intensity varies transversely, can focus the laser to a smaller spot size. This increases the peak laser intensity and extends the length of the high intensity region, forming a channel. Usually, when self-focusing occurs, electron cavitation, which is the expelling of electrons out of laser axis by the transverse ponderomotive force, will also occur, enhancing the self-guiding of the laser beam [12,15] and lowering the electron density. Theoretical work shows that this self-focusing (self-guiding) occurs at a critical power given approximately by $P_c = 17(\omega/\omega_{p0})^2$ GW, where ω_{p0} is ω_p in eqn.2 with $\gamma = 1$ [16].

Laser pressure (ponderomotive force)—combined with ion inertia, which provide an electrostatic restoring force—can drive a high amplitude electron plasma wave (epw). By this process, some of the laser energy is converted to a longitudinal electrostatic laser wakefield, propagating at near the speed of light, which can continuously accelerate electrons in the direction of laser propagation. If the laser pulse width τ is much larger than a plasma period, $\tau \gg 2\pi/\omega_p$, the resulting local density changes will have time to feedback on the light pulse. This case is referred to as the stimulated Raman forward scattering (RFS) instability [3-5]. Modulations in the index of refraction break the pulse into a sequence of shorter pulses with a periodicity equal to a plasma period, which resonantly drive the electron plasma wave to high amplitude. Once the plasma wave is excited, it can trap hot background electrons which may be heated by above-threshold-ionization (ATI) [17] or Raman backscattering, sidescattering, and small angle forward scattering [18], and accelerate them to relativistic energies. The acceleration of trapped electrons will in turn damp the plasma wave, limiting the plasma wave amplitude and its temporal duration [19].

In the present experiments, a hybrid Ti:Sapphire - Nd:Glass laser system based on chirped pulse amplification and capable of delivering 3 J, 400 fs laser pulses at $\lambda = 1.053$ μm was used to drive the self-modulated LWFA. The 43 mm diameter beam was focused by an f/4 off-axis parabolic mirror to a spot size (e^{-2} intensity) of $r_0 = 8.5$ μm, giving a maximum vacuum intensity of $I = 6.2 \times 10^{18}$ W/cm^2 ($a_0 = 2.2$). The laser was focused onto a supersonic helium gas jet with a sharp gradient (250 μm) and a flat-top width of 750 μm. The neutral gas density varied linearly with backing pressure and reaches 1.8×10^{19} cm^{-3} at a backing pressure of 1000 psi. An underdense plasma is created when the gas is tunneling-ionized by the foot of the laser pulse.

The existence of a large amplitude epw is inferred from the presence of up to the 4th anti-Stokes sideband in the Raman forward scattering spectra obtained with a prism spectrometer [9]. In order to determine the absolute plasma wave amplitude and its temporal distribution, a copropagating green probe pulse generated by frequency-doubling 20% of the pump pulse is used to probe the plasma wave by using collective Thomson scattering [19]. The plasma wave amplitude is observed to be about 1.5 ps in duration and the peak amplitude changes from 10 % of n_0 to 40 % of n_0 when the laser peak power is increased from 2 TW to 3 TW, where n_0 is the background electron density.

A sidescattering imaging system composed of collecting lens, filter, and 12-bit CCD is used to measure to spatial distribution of laser-plasma interaction. Fig. 1 shows side-scattering intensity distribution along the laser axis as a function of laser power for fixed gas density. At low laser power (< 1 TW), the channel length is fixed around 125 μm. When laser power exceeds 1.8 TW (3.9 P_c), the channel length extends to 250 μm and stays fixed with increasing laser power. When the laser power is increased above 3.3 TW (7.2 P_c), the length extends futher to 750 μm, which seems to be limited by the gas jet used in this experiment. The peaks of the images show the position of multi-foci, which seem to stay fixed in space regardless of laser power. Spectral analysis of the sidescattered light using imaging spectrometer shows that > 96 % of the emission comes from the Thomson scattering of the blue-shifted pump light. Imaging of the Thomson scattering by using narrow bandpass filter centered at λ shows an emission channel of the exactly same spatial distribution, only much weaker.

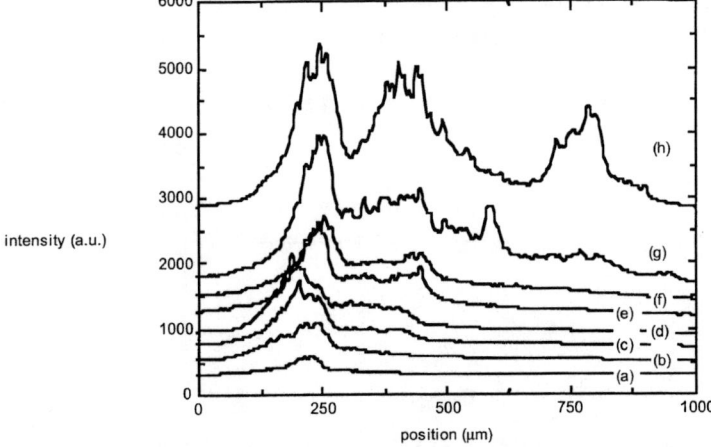

Fig.1 On-axis lineouts of the sidescattering images as a function of peak laser power for fixed backing pressure 1000 psi (corresponding to a gas density of 1.8 x 10 cm^{-3}). (a) 0.8 TW (b) 1.2 TW (c) 1.5 TW (d) 1.9 TW (e) 2.5 TW (f) 2.8 TW (g) 3.4 TW (h) 3.8 TW

A recent simulation and experiment [12] shows that when $P_0/P_c = 3 - 4$, a second focus is clearly visible (defined as the criterion for self-focusing), and when $P_0 \approx 6.4\, P_c$, guiding occurs (appearance of four foci or a flat extended channel after two foci). The consistency between our result and theirs indicates that the extension of the plasma wave channel in this case is actually due to the relativistic self-guiding. A similar channel extension occurs if the gas density is varied at fixed laser power. For a 3.9 TW laser pulse, the channel extends to 250 μm at 400 psi backing pressure (3.2 P_c) and 750 μm at 800 psi (7.0 P_c). This consistent behavior at specific values of P_c for varying laser power or plasma density indicates again that the channeling mechanism is relativistic self-focusing. Because of the ionization defocusing effect, the laser beam divergence increases by a factor of two with respect to the beam divergence in vacuum at all laser powers. The fact that the laser beam divergence doesn't increase when self-guiding occurs indicates that the cross section of the channel is close to the size without self-guiding.

The copropagating collective Thomson scattering is also used to monitor the change of plasma frequency when self-guiding occurs. As seen in Fig.2, the plasma wave frequency decreases gradually from 3.3×10^{14} rad/s to 2.5×10^{14} rad/s with increasing laser power. Three effects could cause the decrease of plasma wave frequency in this case. First, the plasma frequency decreases with increasing laser intensity, as can be seen in eqn.2. Second, the plasma wave frequency will decrease when the plasma wave amplitude grows larger and plasma wave becomes nonlinear. Third, when self-focusing occurs, the increase of laser intensity can expel the electrons out of the axis (electron cavitation), reducing the electron density and thus the plasma frequency. It is believed that the first mechanism is dominating here [21].

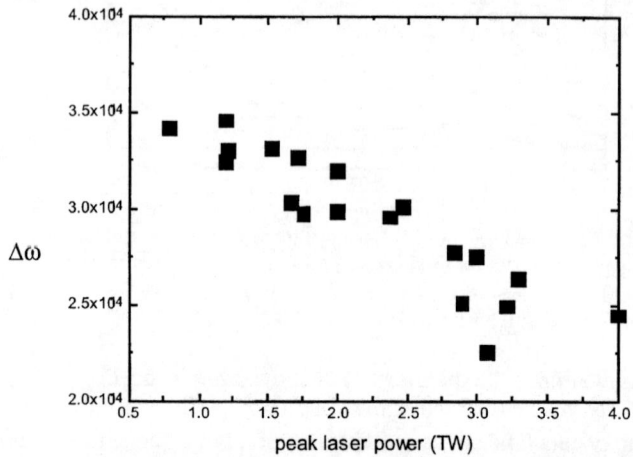

Fig. 2 Plasma wave frequency as a function of peak laser power.

The energy spectrum of the accelerated electron beam is measured by a collimator and a 60° sector dipole magnet for low energy range (< 3 MeV), a rectangular dipole magnet for medium energy range (2 MeV - 15 MeV) and aluminum absorbers for high energy range. The detector is a LANEX film (scintillating film) imaged by a CCD. The highest electron energy is known to be larger than 15 MeV, but the maximum value has not been determined yet. Fig.3 shows the electron kinetic energy distribution in the low energy range with varying laser power or plasma density. The number of electrons decreases exponentially with higher electron energy. In higher energy range, the electron number decreases much more slowly with higher electron energy (the slope decreases by a factor of 5 or more) compared to the low energy range. The slope of this energy distribution on the semilog scale basically decreases with higher laser power or higher gas density due to the fact that a larger plasma wave amplitude is achieved. In addition, a sudden jump in the slope is observed when self-guiding occurs by changing laser power or gas density. The extension of plasma channel not only provides longer acceleration distance but also allows the plasma wave to grow to larger amplitude. Recent PIC simulations show that such exponential distribution is typical for acceleration of self-trapped electrons and the slope of electron energy distribution in the low energy range becomes less steep with longer propagation distance [22]. In addition, the sudden jump of channel length should result in a sudden jump of electron energy distribution, as in our observation. About 50% of the accelerated electrons have kinetic energy higher than 1 MeV and the energy transfered to the electron beam is about 0.5 mJ.

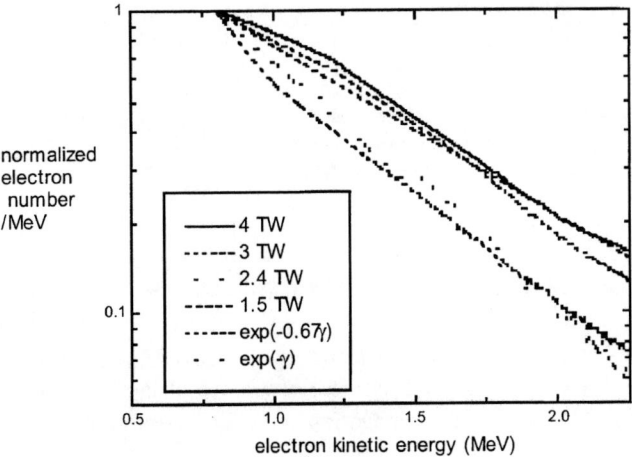

Fig. 3(a) Normalized electron energy distribution as a function of laser power for fixed backing pressure (1000 psi). Also shown are the fits to exponential function.

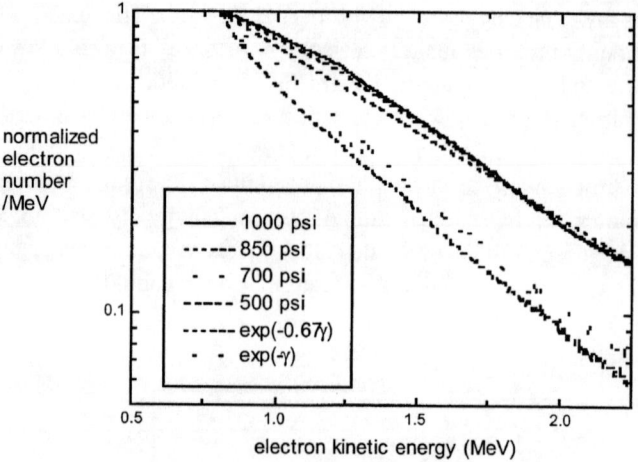

Fig. 3(b) Normalized electron energy distribution as a function of gas backing pressure (electron density) for fixed laser power (3 TW). Also shown are the fits to exponential function.

The number of accelerated electrons are measured using both a Faraday cup and a plastic scintillator coupled to a photo-multiplier tube. The electron acceleration occurs with a sharp threshold at ~ 750 GW (1.5 Pc) and then the number increases exponentially with laser power and gradually saturates after 2 TW. At 3 TW (6 P_c), 6×10^9 electrons are accelerated out of the plasma toward the forward direction (i.e., the laser propagation direction). Although the plasma wave should pick up a factor of three more electrons when self-guiding occurs (longer channel), it is difficult to observe this change since the electron number is increasing exponentially and the shot-to-shot fluctuation is large.

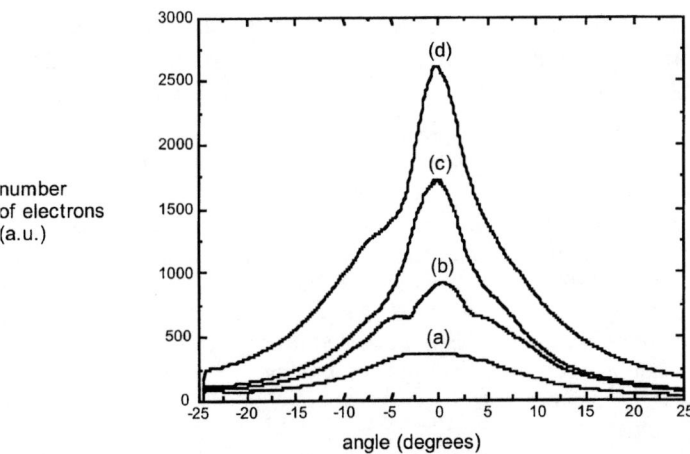

Fig. 4 Electron beam divergence relative to the laser propagation direction as a function of peak laser power. P/P_c = (a) 3.4, (b) 5.0, (c) 6.0, and (d) 7.5.

The accelerated electrons are well confined in the forward direction since the accelerating field is in the forward direction and the transverse field can confine the beam transversely during acceleration.. Fig.4 shows the angular distribution of accelerated electron beam with changing laser power. Because the number of electrons in the high energy range is very small compared to electrons at low energy (< 3 MeV), this image is basically the distribution of electrons in the range of 100 keV - 3 MeV, where 100 keV is the energy electrons need in order to pass through the plastic and aluminum piece in front of the LANEX. Using aluminum absorbers, it is found that the electron beam divergence is independent of electron energy in this energy range. For the high energy range, it is expected that the beam divergence should decrease with higher electron energy. At low laser power, the electron beam divergence is measured to be about 10° in radius at half maximum and the beam profile is Gaussian-like. When the laser power is increased and the channel length extends to ~ 250 μm, a new distribution of smaller divergence starts to grow out of the original distribution. Above 2 TW, the electron beam divergence becomes 5° in radius and the beam profile is Lorentzian-like. This minimum electron beam divergence possibly comes from the space charge effect after leaving the plasma wave, which can be large due to the fact that most of the electrons are in the few MeV range (γ small) and the current is high (large electron number and short electron bunch). Assuming 10^9 electrons of 1 MeV kinetic energy and 1 ps bunch length, the electron beam divergence due to the space charge effect is calculated to be 6°. The decrease of electron beam divergence with laser power is caused by the increase of electron energy (lower space charge effect and relatively low transverse momemtum). The electron beam emittance defined by the radius of the source multiplied by the angle of beam divergence is 0.4 π-mm-mrad in the best case and is same regardless of electron energy in the low energy range.

In summary, a collimated electron beam is produced by self-modulated laser wakefield, and the effect of relativistic self-focusing on electron acceleration was studied. The existence of relativistic self-guiding is observed by sidescattering imaging. The electron beam energy distribution and divergence changes when self-guiding occurs, leading to lower beam emittance and higher energy.

Acknowledgements

The authors would like to thank E. Dodd, D. Gustafson, G. Mourou, C. Keppel, and W. Buck for their support and useful discussions. This work was supported by DOE/LLNL subcontract B307953 and NSF STC PHY 8920108.

References

[1] T. Tajima and J. Dawson, Phys. Rev. Lett. **43**, 267 (1979).
[2] E. Esarey *et al.*, IEEE Trans. Plasma Science **24**, 252 (1996).
[3] P. Sprangle *et al.*, Phys. Rev. Lett. **69**, 2200 (1992).
[4] T. M. Antonsen and P. Mora, Phys. Rev. Lett. **69**, 2204 (1992).
[5] N. E. Andreev *et al.*, JETP Lett. **55**, 571 (1992).
[6] C. A. Coverdale *et al.*, Phys. Rev. Lett. **74**, 4659 (1995).
[7] K. Nakajima *et al.*, Phys. Rev. Lett. **74**, 4428 (1995).
[8] A. Modena *et al.*, Nature (London) **377**, 606 (1995).
[9] D. Umstadter *et al.*, Science **273**, 472 (1996).
[10] C. G. Durfee III *et al.*, Phys. Rev. E **51**, 2368 (1995).
[11]] P. E. Young *et al.*, Phys. Rev. Lett. **75**, 1082 (1995).
[12] A. Chiron *et al.*, Phys. Plasmas **3**, 1373 (1996).
[13] P. Monot *et al.*, Phys. Rev. Lett. **74**, 2953 (1995).
[14] A. B. Borisov *et al.*, J. Opt. Soc. Am. B **11**, 1941 (1994).
[15] A. B. Borisov *et al.*, Phy. Rev. A **45**, 5830 (1992).
[16] C. Max *et al.*, Phys. Rev. Lett. **33**, 209 (1974).
[17] B. M. Penetrante *et al.*, Phys. Rev. A **43**, 3100 (1991).
[18] Y. Kishimoto *et al.*, to be published.
[19] S. P. Le Blanc *et al.*, to be published.
[20] S. C. Wilks *et al.*, Phys. Plasmas **2**, 274 (1995).
[21] A. Modena *et al.*, IEEE Trans. Plasma Sci. 24, 289 (1996).
[22] W. B. Mori (private communication).

Laser Injection of Particles into Wakefields: a Qualitative Understanding

Jean Louis Bobin

Université Pierre et Marie Curie, 4 place Jussieu,
75252 Paris CEDEX 05, France

Abstract: The "phase space" portrait of the laser driven plasma wakefield is modified by the ponderomotive potential of an auxiliary laser pulse. Background plasma electrons are thus allowed to channel through regions in phase space where acceleration is possible.

INTRODUCTION

Plane longitudinal electron plasma waves with electric fields of order 100 GV/m are considered for electron acceleration. Among the methods proposed to produce such waves, laser induced wakefields look promising. When a high power subpicosecond laser pulse propagates through an underdense plasma, the electron response is a high amplitude wake [1] whose properties have been thoroughly investigated. Providing a sizable increase of the accelerating field, the resonant plasma acceleration concept [2] is a specially attractive improvement. Now, electron bunches to be accelerated should have a longitudinal size much shorter than the plasma wave length. They also ought to be properly phased. These requirements can be met thanks to a recently invented laser injection scheme. An auxiliary laser pulse is shined onto the wake either orthogonally or longitudinally [3]. The associated ponderomotive force gives the background electrons the necessary kick to catch up with the wave. The present note intends to give a qualitative understanding of this effect.

PHASE SPACES

Consider a plane electron plasma wave propagating in the z direction. Denoting by E_0 the field amplitude, by k_p and ω_p the wave number and the circular frequency respectively, and by φ_0 the initial phase, the relativistic equation of motion of electrons is

$$\frac{dv}{dt} = -\frac{e}{m_0 \gamma^3} E_0 \cos(k_p z - \omega_p t + \varphi_0) \quad , \tag{1}$$

where

$$\gamma = \frac{1}{\sqrt{1-\frac{v^2}{c^2}}}$$

$$\xi = k_p z - \omega_p t \quad . \tag{2}$$

Then the equation of motion (1) is better rewritten in terms of γ and ξ:

$$\frac{d\gamma}{dt} = -\frac{eE_0}{m_0 c}\sqrt{1-\frac{1}{\gamma^2}} \cos(\xi+\varphi_0) \quad ,$$

$$\frac{d\xi}{dt} = ck_p\sqrt{1-\frac{1}{\gamma^2}} - \omega_p \quad , \tag{3}$$

from which a "phase portrait" can be drawn (Figure 1b). Separatrices (thick curves) limit 3 domains. Assuming the wave propagate from left to right, low γ passing trajectories represent the plasma background electrons moving from right to left and to which high γ passing and intermediate γ trapped trajectories are inaccessible. Thus, except in case of wavebreaking, background electrons are not to be accelerated by the wave.

As shown in [6], in resonant plasma acceleration, wavebreaking is expected after only a few pulses (typically 4). However, this nonlinear mechanism associated with dramatic distortions of the phase trajectories cannot be considered a reliable method of injecting electrons to be accelerated by the wave.

On the contrary, the laser injection [3] changes the phase portrait in a controllable way. This can readily be seen by adding in the right hand side of the equation of motion, a space dependent ponderomotive force. Take for instance a gaussian ponderomotive potential. Then equation (3) is replaced by

$$\frac{d\gamma}{dt} = -\frac{eE_0}{m_0 c}\sqrt{1-\frac{1}{\gamma^2}}\left[\cos(\xi+\varphi_0) - \frac{\partial}{\partial \xi}e^{-(\xi-3+\varphi_0)^2}\right] \quad ,$$

$$\frac{d\xi}{dt} = ck_p\sqrt{1-\frac{1}{\gamma^2}} - \omega_p \quad . \tag{4}$$

The corresponding phase portrait is displayed on Figure 2b. The split separatrix allows a channel of phase space trajectories (arrow) to link low to high γ regions of passing trajectories. Through the channel, electrons from the background plasma with a negative velocity with respect to a reference frame R moving with the phase velocity of the wave, can be driven to that domain of the phase space in which they have a positive velocity.

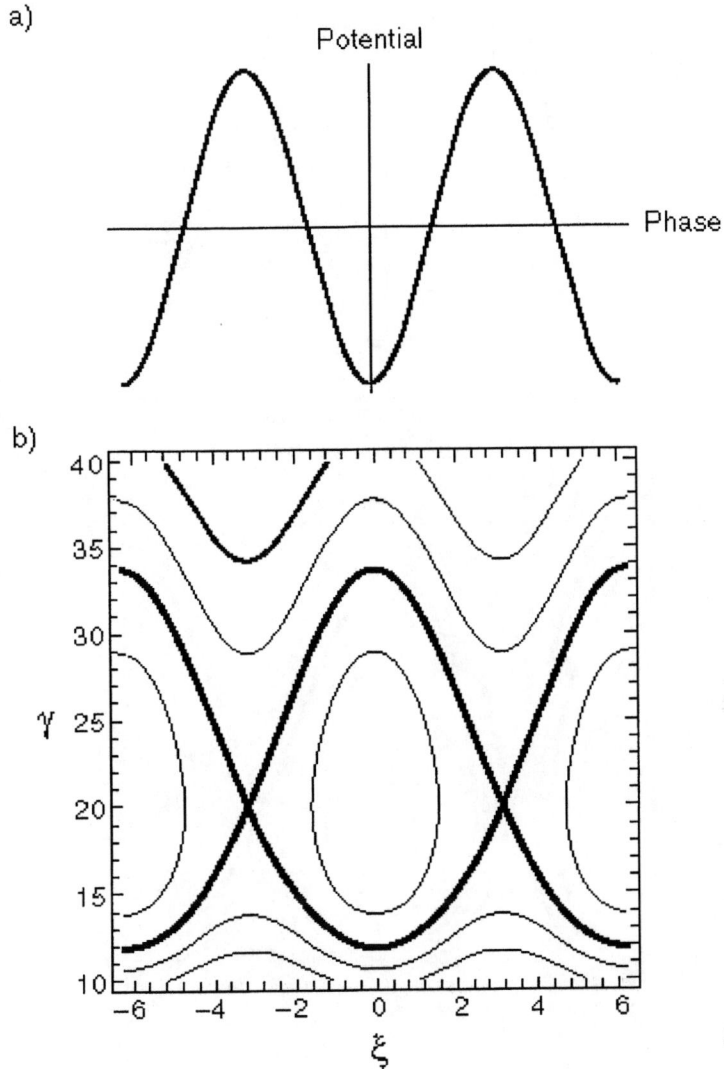

Figure 1. Electron plasma wave potential profile a) and "phase portrait" b).

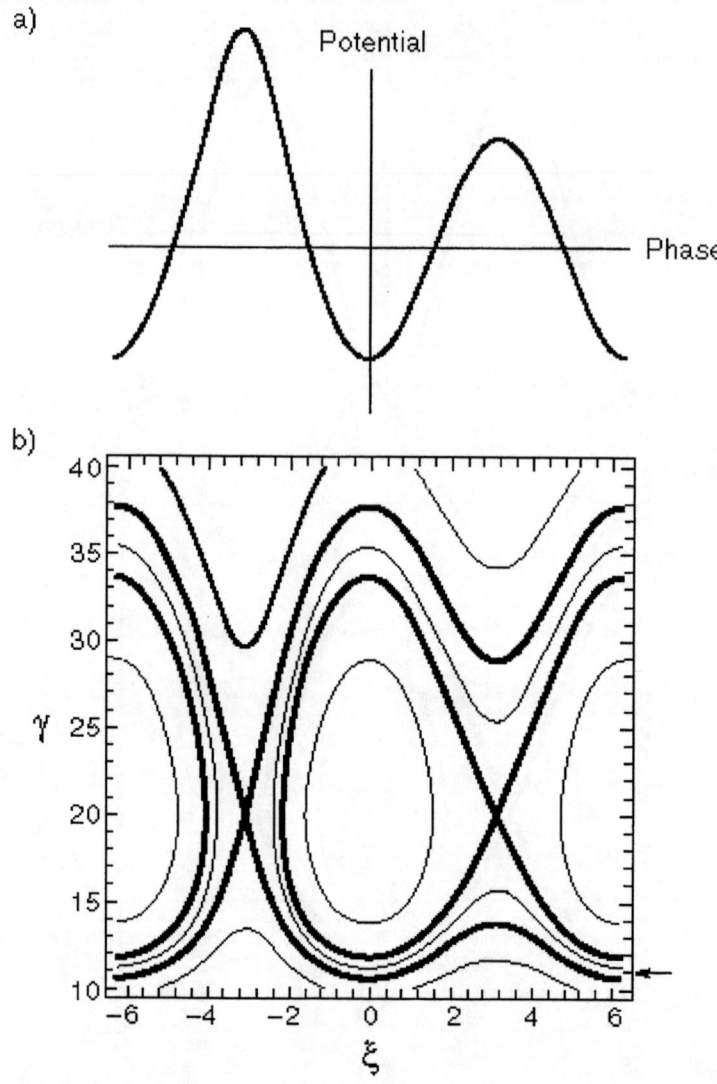

Figure 2. Potential profile a) and "phase portrait" b) of a short pulse ponderomotive potential superimposed to a plasma wave potential.

ELECTRON INJECTION

Through the channel, electrons from the background plasma with a negative velocity with respect to a reference frame R moving with the phase velocity of the wave, can be driven to that domain of the phase space in which they have a positive velocity.

Now, when due to another laser pulse, the ponderomotive force is actually time dependent. Consequently, the phase space configuration of Figure 2b exists only for a short time interval. Furthermore, if the auxiliary pulse propagates perpendicularly to the plasma wave, the corresponding ponderomotive potential is receding in R. In any case, channeling electrons are to do so for a limited time.

It is then expected that a bunch of electrons from the background plasma will be injected mainly in the domain of trapped phase space trajectories. Hopefully, the particles are to follow on trajectories close enough to the separatrix of Figure 1b so that they might still undergo a significant acceleration.

REFERENCES

1. Gorbunov, L.M. and Kirsanov, V.I. *Soviet Phys. J.E.T.P.* **66**. 290 (1987).
2. Umstadter, D. Esarey, E. and Kim, J. *Phys. Rev. Lett.* **72**. 1224 (1994).
3. Umstadter, D. Kim, J. and Dodd, E. *Phys. REv. Lett.* **76**. 2073 (1996); Umstadter, D. these proceedings.
4. Bobin, J.L. Proc. of the Workshop on New Developments in Particle Acceleration Techniques, CERN 87-11, ECFA 87-110, Orsay S. Turner, ed. (1987).
5. Williams, R.L. Clayton, C.E. Joshi, C. Katsouleas, T. and Mori, W. *Laser and Particle Beams*, **8** 427 (1990).
6. Bobin, J.L. Bonnaud, G. and Teychenné, D. A.I.P. Proc. no. 335, Advanced Accelerator Concepts, P. Schoessow ed. (1995).

CHARGED PARTICLE ACCELERATION IN NONUNIFORM PLASMAS

S.V. Bulanov[1], G.I. Dudnikova[2], N.M. Naumova[1],
F. Pegoraro[3], I.V. Pogorelsky[4], V.A. Vshivkov[2]

[1]*General Physics Institute, Russian Academy of Sciences, Moscow, Russia*
[2]*Institute of Computation Technology, Novosibirsk, Russia*
[3]*Department of Theoretical Physics, University of Turin, Turin, Italy*
[4]*Brookhaven National Laboratory, 725C, Upton, NY 11973-5000, USA*

Abstract. We consider the interaction of high-intensity laser pulses with nonuniform underdense plasmas and address the problem of the excitation of strong wake plasma waves with regular electric fields that can provide a high acceleration rate of charged particles.

I. INTRODUCTION

The high-gradient electron acceleration schemes that have been demonstrated using LWFA [1,2] appear promising for the development of plasma-based laser accelerators into practical devices. However, a question still exists: how to avoid the wake field deterioration and the loss of the phase synchronism between the plasma wave and the electrons that prevent them from being accelerated up to the theoretical limit that corresponds to $\gamma_{max} \approx (\omega/\omega_p)^2$ [3] and can be even as high as $a(\omega/\omega_p)^3$ [4,5]. Here γ is the Lorentz factor, ω is the laser frequency, ω_p is the Langmuir frequency, and $a \equiv eE_\perp/m\omega c$ is the dimensionless amplitude of the laser radiation. In order to obtain the highest possible values of the wake electric field, we must use as intense laser pulses as possible (i.e., pulses with dimensionless amplitudes $a \gg 1$).

Pulses with $a \gg 1$ tend to be subject to a host of instabilities, such as relativistic self-focusing, self modulation and stimulated Raman scattering, that affect their propagation in the plasma. Such processes could be beneficial, in so far as they increase the pulse energy density, enhance the wake field generation, and provide the mechanism for transporting the laser radiation over several Rayleigh lengths without diffraction spreading. However, it is still far from certain that these processes can be exploited in a controlled form and can lead to regular, stationary wake fields.

It is known that, in order to create good quality wake fields, it would be preferable to use laser pulses with steep fronts of order λ_p [5,6]. The present paper aims at further analyzing the influence of the laser pulse shape and of the plasma nonuniformity on the charged particle acceleration. This study is based on the results obtained with one dimensional Particle in Cell (PIC) simulations. In these simulations circularly polarized laser pulses with amplitude $a \approx 1$ and different profiles interact with nonuniform plasmas. Test particles with random momentum distribution in the interval $0 \div 5mc$ are used to determine the maximum acceleration rate and the energy of the accelerated particles.

II. PRODUCTION OF PULSES WITH SHARP FRONTS

The feature of the pulse that is specifically responsible for providing good conditions for particle acceleration is the sharp rise of the pulse front. Sharpening of relativistically intense pulses propagating in an underdense plasma via the backward stimulated Raman scattering process has been observed in [7] and [8].

Here we consider another method for producing laser pulses with sharp fronts which consists of transmitting the pulse through a thin foil. The results obtained with the UMKA2D3V code simulations for this case are illustrated in Fig. 1. Sharpening of relativistically intense pulses interacting with a thin slab of the overdense plasma has been observed in the 1D PIC computer simulations presented in Ref. [5].

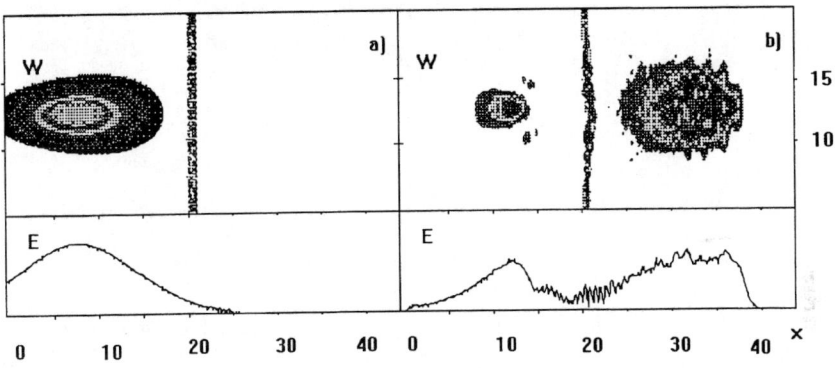

FIGURE 1. 2-D simulation of the formation of a sharp leading front in a pulse with dimensionless amplitude a = 1 irradiated over a thin foil where ω/ω_p = 1.1: a) pulse shape before interaction with the foil; b) transmitted pulse with a typical step-like shape.

In Fig. 1 the results of simulations for a pulse with initial amplitude $a=1$, pulse length 10λ, and width 5λ, with foil thickness equal λ, and plasma density corresponding to $\omega/\omega_p = 1.1$ are presented. As a result of the interaction, a sharp,

≈ 3λ long, leading edge is produced. In addition we can see a sharp, ≈ 3λ long, rear edge of the reflected part of the laser pulse.

III. CHARGED PARTICLE ACCELERATION BY WAKE FIELD

III.1. Acceleration in nonuniform plasma without phase-slippage limit

To be effectively accelerated, a charged particle must be in the proper phase of a wake plasma wave. In a plasma with uniform density, the phase velocity of the wave does not change along the propagation while the particle velocity increases in the course of the acceleration. This breaks the wave-particle resonance conditions and hence restricts the final particle energy. In nonuniform plasmas the group velocity and the amplitude of the electromagnetic wave packet depend on the coordinate along the way of the pulse propagation, and thus the plasma wake wave phase velocity and amplitude vary. With an appropriate choice of the plasma density profile one can increase the acceleration length [9,10] considerably. The solution of this problem gives an example of the application of the well known phase stability principle by V. I. Veksler and E. McMillan to the problems of charged particle acceleration by laser radiation.

The equations of the electron motion in the electric field of a one dimensional wake plasma wave can be written in the form

$$\frac{d}{dx}\left(\frac{\psi}{\omega_p}\right) = \frac{(m^2c^2 + p^2)^{1/2} - p}{cp} - \frac{\omega_p^2}{2\omega^2 c}, \qquad (1)$$

$$\frac{d}{dx}(m^2c^4 + p^2c^2)^{1/2} = -eE, \qquad (2)$$

where
$$\psi = \omega_p(t - t_0) = \omega_p\left(t - \int_0^x \frac{dx'}{v_g}\right) \qquad (3)$$

is the wave phase, t_0 is the time at which the pulse reaches the point x, p is the particle momentum, and E is the wake field.

The electric field depends on the coordinate x and the phase ψ. In a uniform plasma, an ultrarelativistic particle in a moderately strong plasma wave acquires an energy of the order of $\Delta E \approx eE_m l_{acc}$, with l_{acc} the acceleration length which is given by $l_{acc} \approx (2c/\omega_p)\gamma_{ph}^2 = c/\pi\omega_p(\omega/\omega_p)^2$ [3,11]. This length is larger than the plasma wave length by the factor $(\omega/\omega_p)^2$. In an inhomogeneous plasma with a density that depends on the coordinate as $n(x) = n_0(L/x)^{2/3}$, with $L \approx (c/3\omega_p)(\omega/\omega_p)^2$, a laser pulse with amplitude, $a < 1$, and length l_p excites a wake plasma wave with electric field $E(x,t) = -\omega_p^2(x)(ml_p a^2/4e)\cos\psi$. In this wave the acceleration length becomes formally infinite and the particle energy growth is unlimited

$$E(x) \approx mc^2 \left(\frac{\omega}{\omega_p}\right)^2 \left(\frac{x}{L}\right)^{1/3}. \qquad (4)$$

III.2. Acceleration of charged particles at the wave-breaking regime

A second possibility of enhancing the efficiency of the acceleration of charged particles is to increase the value of the electric field in a nonstationary plasma wave.

In the course of the laser pulse propagation in an underdense plasma, if the pulse scale nonuniformity is shorter than λ_p for nonrelativistic amplitude, $a < 1$, and shorter than λ_p/a for $a > 1$, the amplitude of the wake field produced behind the pulse corresponds to the maximum value of the electrostatic potential $\varphi_m = (mc^2/2e)a^2$. Taking into account the relativistic dependence of the relativistic Langmuir wave frequency on the wave amplitude [12], the longitudinal component of the electric field in the wake is equal to $E_m \approx (m\omega_p c/e)a^2$ when $a < 1$, and $E_m \approx (m\omega_p c/e)a$ when $a > 1$, respectively.

As was demonstrated in [12,13] the maximum field in a stationary plasma wave is given by $eE_m/mc\omega_p = (2(\gamma_{ph}-1))^{1/2}$, where $\gamma_{ph} = (1-\beta_{ph}^2)^{-1/2}$. For larger values of the electric field, the wave breaks and its structure becomes distorted and transient in time. However the region where the wave has a regular structure can exist for a rather long time due to the relatively low wake velocity in the frame comoving with the laser pulse. An estimate of the maximum wake field amplitude can be obtained by requiring that no wave-breaking occurs inside the laser pulse before the electrostatic potential in the pulse reaches its maximum amplitude. In the pulse the electrons of the background plasma are accelerated in the direction of the pulse motion up to the velocity $v_e \approx c(1-2/a^2)$. Thus the requirement $v_e < v_{ph}$ gives the maximum value of the laser pulse amplitude $a < 2\gamma_{ph} = 2(1-\beta_{ph}^2)^{-1/2}$. When this condition is satisfied, background electrons inside the pulse are not trapped in the plasma wave. Then, in the case of a long pulse, the maximum value of the electric field in the wake behind the pulse is given by [14]

$$\frac{eE_m}{mc\omega_p} = \gamma_{ph} = \frac{1}{\left(1-\beta_{ph}^2\right)^{1/2}}, \quad (5)$$

which is much bigger than the value in the case of stationary plasma wake waves. The particle energy gain in a plasma wave in a uniform background plasma is given by $\Delta E \approx \varphi_m \gamma_{ph}^2$. For a stationary plasma wave φ_m cannot be greater than γ_{ph} [12], so that ΔE is limited to γ_{ph}^3. In a breaking wave it can be of order γ_{ph}^4.

IV. SIMULATIONS WITH DIFFERENT PULSE PROFILES

To address this problem we made use of numerical simulations with a 1-D PIC code. In this code, physical variables depend on x only and a $1000\delta x$ grid is used with approximately 10^5 particles.

In order to compare the particle acceleration gain, we performed 1D PIC simulations of the interaction with a uniform plasma (n_0=0.004 n_{cr}) of two triangular pulses with length 40λ, one with a sharp rise (Fig.2) and one with a sharp rear edge (Fig.3).

We see that the triangular pulses excite regular plasma waves. Due to the development of the forward stimulated Raman scattering, self-modulation of the pulse amplitude takes place. The maximum of the pulse amplitude, as is seen in Fig. 2a, propagates with a velocity smaller than the linear e.m. wave group velocity. This can be explained by the local downshift of the electromagnetic wave frequency [7,15].

This frequency downshift leads to the decrease of the wave phase velocity of the wake plasma seen in Fig. 2b. Nevertheless, electrons injected into the plasma with initial temperature equal 5 mc^2, acquire an energy of the order of 500-600 mc^2.

In the first case (Fig.2) in the wake field produced by the pulse with a sharp rise, beam electrons are linearly accelerated up to the energy 400 mc^2 before the wake wave-breaking and up to 700 mc^2 after the wake wave-breaking.

In the second case (Fig.3), when the laser pulse has a sharp rear edge, the beam electrons are accelerated up to the energy 700 mc^2 before the wake wave-breaking and up to the energy 1200 mc^2 after the wave-breaking.

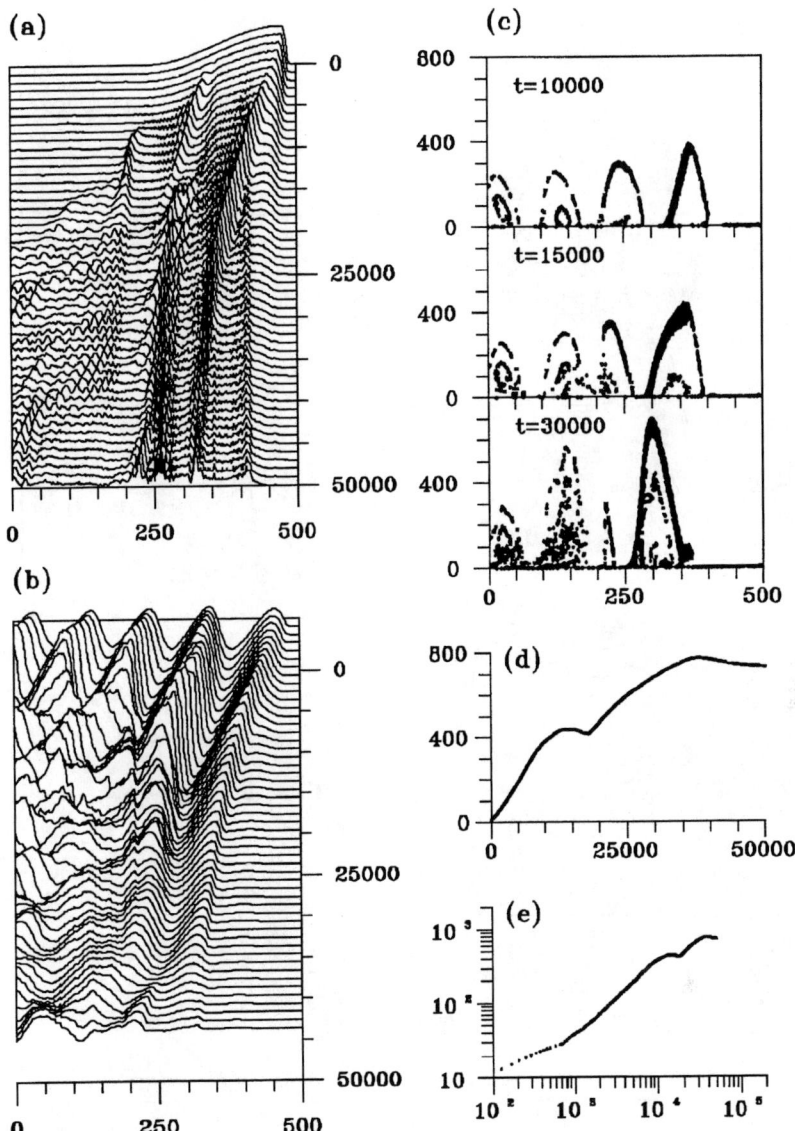

FIGURE 2. 1-D simulation of the nonlinear evolution of a triangular laser pulse with sharp leading edge in an underdense plasma: a) $E_\perp$ distribution as a function of time; b) electric field in wake plasma wave; c) time evolution of the longitudinal phase plane of the beam electrons; dependence of the maximum energy of the accelerated electrons versus time in d) linear and e) logarithmic scale; ($a=1$, $(\omega/\omega_p)^2=250$).

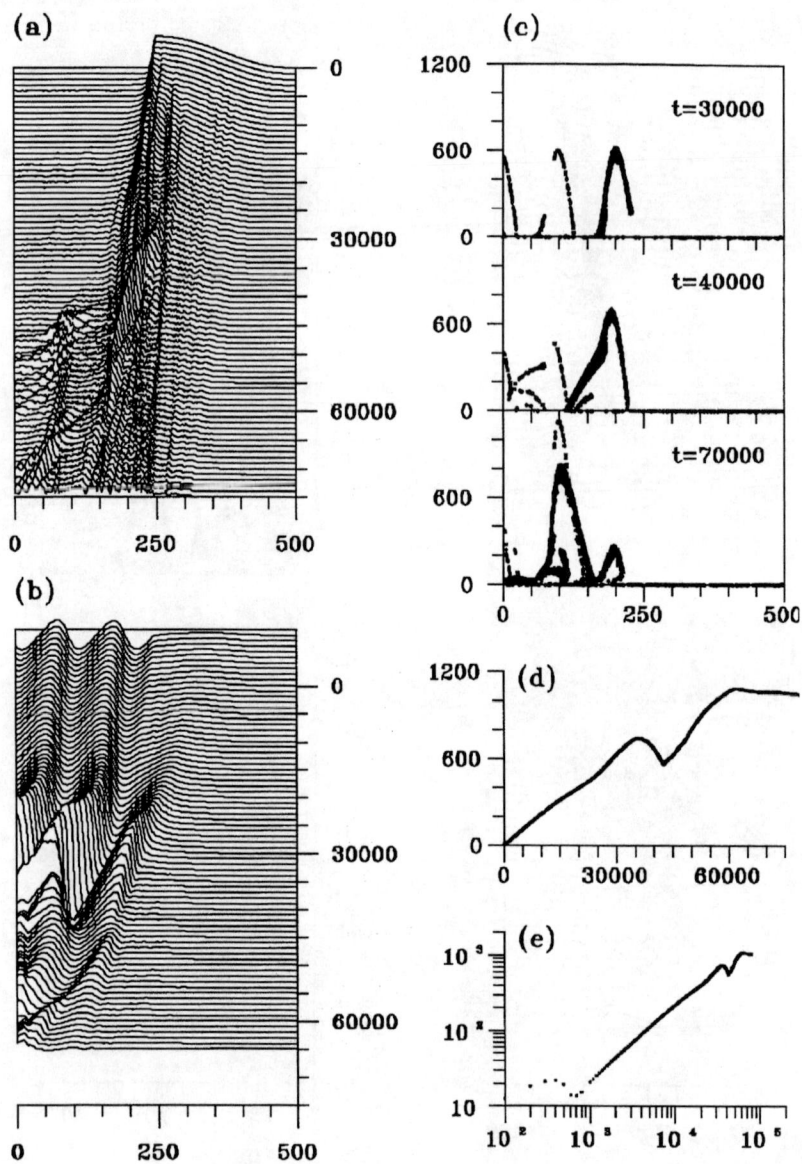

FIGURE 3. 1-D simulation of the nonlinear evolution of a triangular laser pulse with sharp rear edge in an underdense plasma: a) $E_\perp$ distribution as a function of time; b) electric field in wake plasma wave; c) time evolution of the longitudinal phase plane of the beam electrons; dependence of the maximum energy of the accelerated electrons versus time in d) linear and e) logarithmic scale; ($a = 1$, $(\omega/\omega_p)^2 = 250$).

In a nonuniform plasma with a properly chosen plasma profile where the Langmuir frequency depends on the coordinate as $\omega_p=\omega_{p0}(L/x)^{1/3}$, with $L=(1/3)(\omega/\omega_p)^3(c/\omega)$, the energy of accelerated particles is predicted to grow as $E=E_0(x/L)^{1/3}$. PIC simulations performed for a short pulse have demonstrated such acceleration rate (see Fig.4a). The temporal evolution of the pulse demonstrates the pulse stability without self-modulation or significant change of the phase velocity of the wake wave.

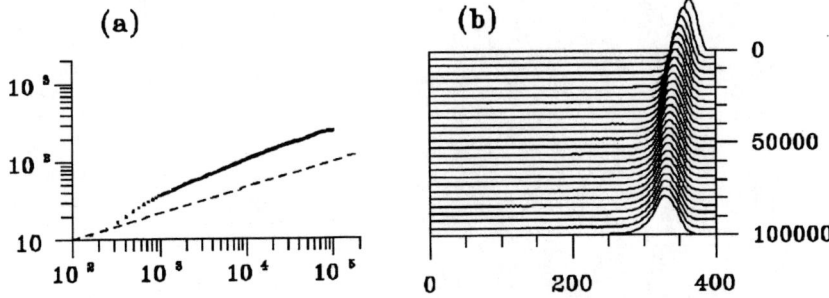

FIGURE 4. The short laser pulse with the length equal 8λ propagates in the plasma with the density $n=n_0(x/L)^{-2/3}$. a) Dependence of the maximum energy of fast electrons on time and b) the evolution of the electric field in the laser pulse.

To compare acceleration rates in nonuniform plasmas we performed PIC simulations. The pulse length is $l_p=30\lambda$ ($l_p>\lambda_{p0}$) and it has a sharp rise ($l_r=2\lambda$). The plasma density is equal to $\omega_p=\omega_{p0}(L/x)^\alpha$, with ω_{p0} corresponding to the density equal $n_0=0.004\,n_{cr}$ and α is a parameter.

For a plasma with increasing density, $\omega_p=\omega_{p0}(x/L)^{1/3}$, (Fig.5) the energy of the accelerated particles grows as $E=E_0(x/L)^{3/2}$. In this case, the pulse modulation and depletion appear earlier than in a uniform plasma. The charged particles are accelerated up the energy $300\,mc^2$ during $500\omega^{-1}$ and, after the wake wave-breaking, they acquire an energy of order $400mc^2$.

In a plasma with the decreasing density profile with $\alpha=1/3$ (Fig.6), the charged particles are accelerated up to the energy $900mc^2$ at $\omega t=200000$. In this case their energy increases as $E=E_0(x/L)^{1/2}$.

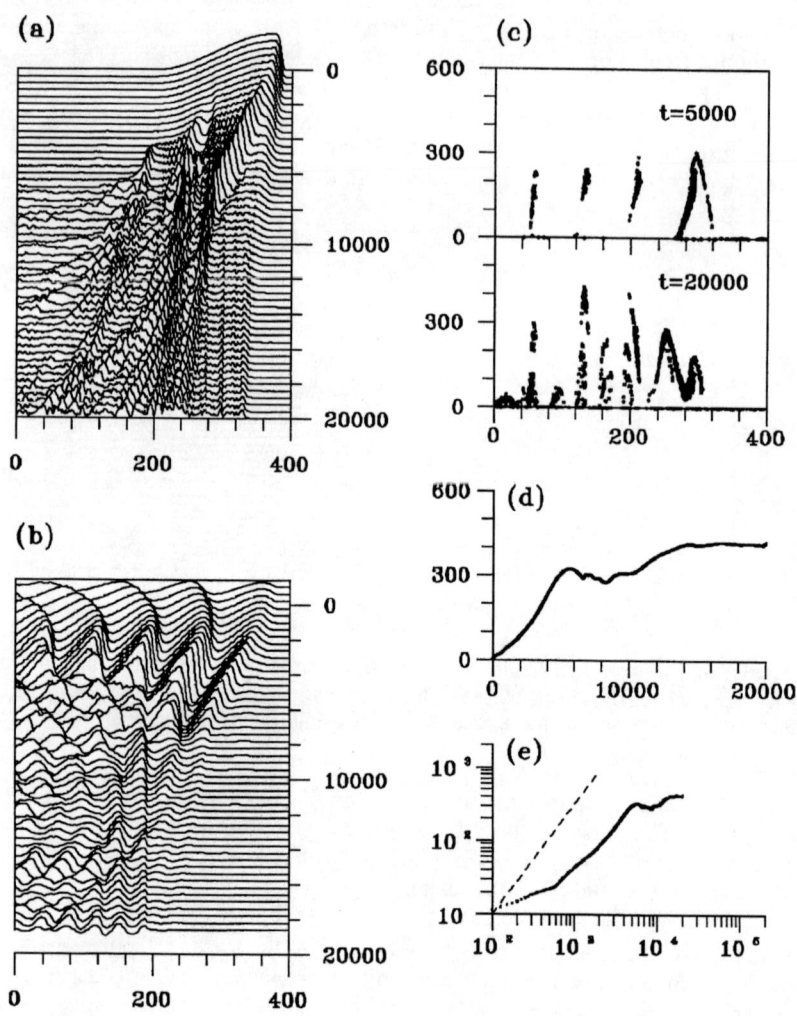

FIGURE 5. 1-D simulation of the nonlinear evolution of a triangular laser pulse with sharp rise in the plasma with the density $n = n_0(x/L)^{2/3}$: a) $E_\perp$ distribution as a function of time; b) electric field in wake plasma wave; c) time evolution of the longitudinal phase plane of the beam electrons; d), e) dependence of the maximum energy of the accelerated electrons versus time, the dashed line corresponds to $y = kx^{3/2}$.

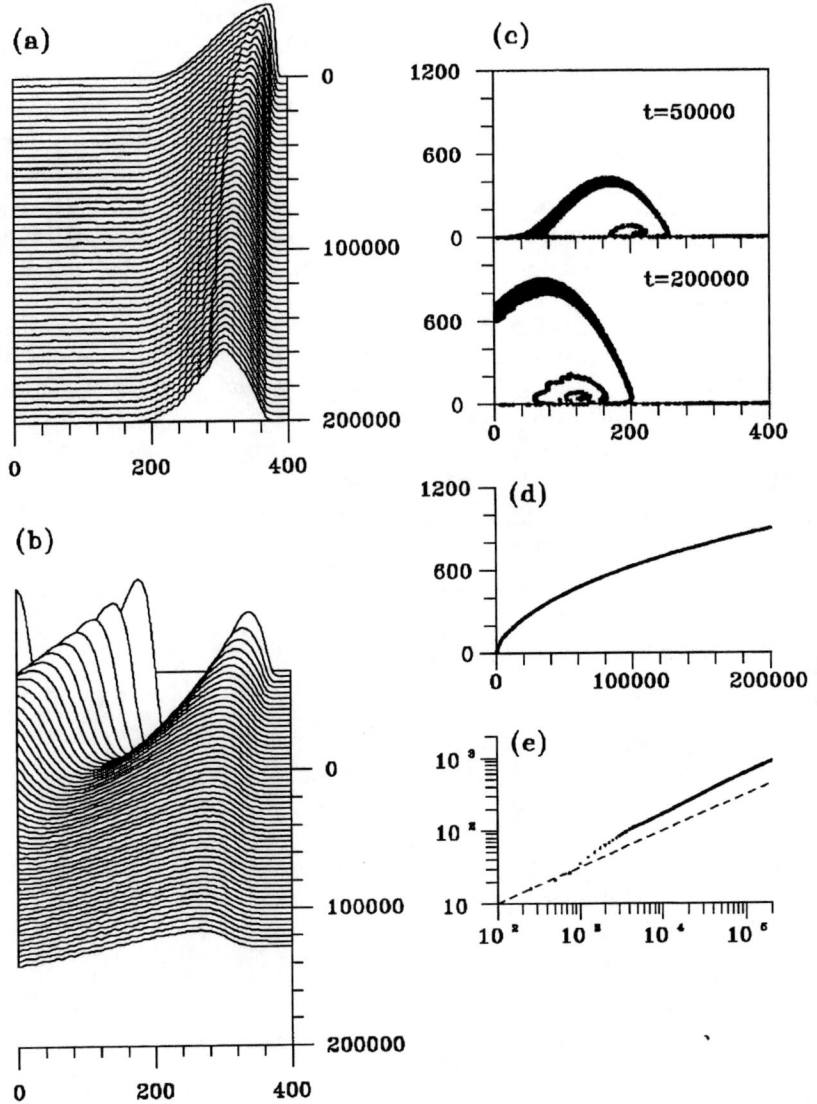

FIGURE 6. 1-D simulation of the nonlinear evolution of a triangular laser pulse with sharp rise in the plasma with the density $n=n_0(x/L)^{-2/3}$: a) $E_\perp$ distribution as a function of time; b) electric field in wake plasma wave; c) time evolution of the longitudinal phase plane of the beam electrons; d), e) dependence of the maximum energy of the accelerated electrons versus time, the dashed line corresponds to $y=kx^{1/2}$.

V. DISCUSSIONS

The present study is stimulated by the ongoing development at the ATF of a new high-power laser source, a 5-ps 4-TW CO_2 laser [16]. Let us estimate the value of achievable energy for a pulse produced by a CO_2 laser with $\lambda=10$ μm, power ~4 TW, pulse length 0.15 cm, i.e., duration 5ps corresponding to $N=155$ periods, and a width 100 μm, or 10λ. The maximum of the pulse amplitude corresponds to $a\approx0.8$. In an underdense plasma with density such that $\omega/\omega_p<15$ the pulse amplitude can increase up to a value of the order of 2-3 due to relativistic self-focusing. In this regime the pulse energy is transported over several Rayleigh lengths. Efficient transport of the electromagnetic radiation can be achieved when the pulse propagates inside a narrow channel of width $R_\perp$. The plasma density inside and outside the channel corresponds to ω_{p1} and ω_{p2}, respectively. In the channeling regime we have $R_\perp > c/(\omega_{p1}^2 - \omega_{p2}^2)^{1/2}$. Further we assume that $\omega_{p1} \approx \omega_{p2} \approx \omega_p$. A laser pulse propagating inside such a channel produces a wake field where the limiting value of the longitudinal electric field, corresponding to the wave break limit, is $eE_m/mc\omega_{p2}=[2(\gamma_{ph}-1)]^{1/2}\approx\omega/[\omega_{p2}^2+(2\pi c/R_\perp)^2]^{1/2} \approx(\omega/\omega_{p1})$, i.e., $eE_m/mc\omega\approx1$. The maximum value of the electrostatic potential is $\varphi_m = (mc^2 a/e)(\omega/\omega_p)$. The maximum energy of the accelerated charged particles can be estimated as $\Delta E \approx amc^2(\omega/\omega_p)^3$. In a plasma with $(\omega/\omega_p)=12$, the wake field produced by such a pulse is $E_{max} \approx 1000 mc^2 = 500 MeV$.

REFERENCES

[1] K. Nakajima et al., Phys. Rev. Lett. 74, 4428 (1996).
[2] A. Modena, et al. IEEE Transaction on Plasma Science 24, 289 (1996).
[3] T. Tajima, Laser and Particle Beams 3, 351 (1985).
[4] E. Esarey, and M.Piloff, Phys. Plasmas 2, 1432 (1995).
[5] S.V. Bulanov, T.Zh. Esirkepov, N.M. Naumova, F. Pegoraro, I.V. Pogorelsky, and A.M.Pukhov, IEEE Transaction on Plasma Science 24, (1996).
[6] S.V. Bulanov, F. Pegoraro, and A.M. Pukhov, Phys. Rev. Lett. 74, 710 (1995).
[7] S.V. Bulanov, I.N. Inovenkov, V.I. Kirsanov, N.M. Naumova, and A.S. Sakharov, Phys. Fluids B 4, 1935, (1992).
[8] C.D. Decker, W.B. Mori, and T. Katsouleas, Phys. Rev. E 50, R3338 (1994).
[9] T. Katsouleas, Phys. Rev. A 33, 2062 (1986).
[10] S.V. Bulanov, V.I. Kirsanov, F. Pegoraro, and A.S. Sakharov, Laser Physics 6, 1078 (1993).
[11] T. Tajima, and J.M. Dawson, Phys. Rev. Lett. 43, 262, (1979).
[12] A.I. Akhiezer, R.V. Polovin, Sov. Phys. JETP 3, 696 (1956).
[13] T. Katsouleas, W.B. Mori, Phys. Rev. Lett. 61, 90 (1988).
[14]. S.V. Bulanov, V.I. Kirsanov, and A.S. Sakharov, JETP Lett. 53, 540 (1991).
[15] S.V. Bulanov, V.I. Kirsanov, and A.S. Sakharov, Sov. J. Plasma Phys. 16, 543 (1990).
[16] I.V. Pogorelsky, W.D. Kimura, C. Fisher, F. Kannari, and N. Kurnit, in 6th Workshop on Advanced Accelerator Concepts, June 12-18, 1994, Fontana, WI, AIP Conf. Proc. 335, 405 (1995).

Laser-Driven Acceleration with Bessel and Gaussian Beams

B. Hafizi,[†] E. Esarey, and P. Sprangle

Beam Physics Branch, Plasma Physics Division
Naval Research Laboratory, Washington DC 20375-5346

[†]*Icarus Research, Inc., P.O. Box 30780, Bethesda, MD 20824-0780*

Abstract. The possibility of enhancing the energy gain in laser-driven accelerators by using Bessel laser beams is examined. Scaling laws are derived for the propagation length, acceleration gradient, and energy gain in various accelerators for both Gaussian and Bessel beam drivers. For equal beam powers, the energy gain can be increased by a factor of $N^{1/2}$ by utilizing a Bessel beam with N lobes, provided that the acceleration gradient is linearly proportional to the laser field. This is the case in the inverse free electron laser and the inverse Cherenkov accelerators. If the acceleration gradient is proportional to the square of the laser field (e.g., the laser wakefield, plasma beat wave, and vacuum beat wave accelerators), the energy gain is comparable with either beam profile.

INTRODUCTION

Laser-driven accelerators rely on the large intensities obtained when laser beams are focused down to spot sizes on the order of several wavelengths [1-15]. A shortcoming of many of these schemes is that the interaction length over which the high intensity can be sustained is relatively short due to transverse spreading (diffraction). For a Gaussian beam the Rayleigh length, i.e., the free-space scale length for diffraction, is given by $Z_{RG} = kr_0^2/2$, where r_0 is the minimum spot size of the beam at the focal point and $\lambda = 2\pi/k$ is the free-space wavelength [16].

This paper addresses the scaling of and the maximization of the energy gain in various accelerators driven by lasers with various transverse mode profiles. In particular, laser accelerators driven by Gaussian beams will be compared to those driven by Bessel beams [17-27]. It is shown that a Bessel beam can enhance the energy gain by a factor of $N^{1/2}$ compared to a Gaussian beam of the same power, provided that (i) the acceleration gradient is linearly proportional to the laser field, and (ii) the acceleration distance is limited by diffraction and not by phase detuning (or some other mechanism), where N is the number of transverse rings (lobes) in the Bessel beam.

BESSEL AND GAUSSIAN LASER BEAMS

A. Ideal Bessel Beams

In vacuum, the cartesian components of the laser electric field, E_i ($i = x, y, z$), satisfy the paraxial wave equation [16] $(\nabla_\perp^2 + 2ik\partial/\partial z)\hat{E}_i = 0$, where $E_i = \text{Re}\hat{E}_i \exp(ikz - i\omega t)$, $\hat{E}_i$ is the slowly varying laser field envelope, $|\partial \hat{E}_i/\partial z| \ll |k\hat{E}_i|$, $\omega = ck = 2\pi c/\lambda$ is the laser frequency, and c is the speed of light. An exact solution of the paraxial wave equation is the fundamental Bessel mode [17-20]

$$\hat{E}_x = E_0 J_0(k_\perp r) \exp(-ik_\perp^2 z/2k), \tag{1}$$

where J_0 is the zeroth order Bessel function, E_0 is the peak field amplitude, and $k_\perp \ll k$ is the transverse wavenumber. The radius of the central lobe of the fundamental Bessel mode is given by $p_{01}/k_\perp$, where p_{01} is the first zero of J_0. Associated with the transverse field is an axial field component E_z such that $\nabla \cdot \mathbf{E} = 0$. For $E_x \sim J_0$, however, E_z is zero along $r = 0$.

For laser acceleration of particles in vacuum or in gas [4-11], a more useful laser field is a radially polarized, first-order Bessel mode of the form

$$\hat{E}_r = E_0 J_1(k_\perp r) \exp(-ik_\perp^2 z/2k), \tag{2}$$

$$\hat{E}_z = ik_\perp E_0 (k - k_\perp^2/2k)^{-1} J_0(k_\perp r) \exp(-ik_\perp^2 z/2k). \tag{3}$$

For the first order Bessel beam, E_z is maximum along $r = 0$ whereas E_r is zero. Notice that the axial wavenumber for the above Bessel beams is $k_z \simeq k - k_\perp^2/2k$. This implies an axial phase velocity $v_p = \omega/k_z$ given by $v_p/c = 1 + k_\perp^2/2k^2$. That is, the phase velocity exceeds c and particle slippage prevents acceleration to high energies.

Although ideal Bessel beams do not diffract, they have infinite power, since $J_n(k_\perp r) \sim (2/\pi k_\perp r)^{1/2} \cos[k_\perp r - (2n+1)\pi/4]$ for $k_\perp r \gg 1$. An ideal

Bessel beam consists of an infinite number of rings (lobes) each having a radial width of $r_b \simeq \pi/k_\perp$. Since the asymptotic width of each ring is the same, the power contained in each ring $P_b \simeq (c/4)E_0^2/k_\perp^2$ is nearly equal.

B. Gaussian Beams

It is useful to compare the Bessel beam solutions to the well-know Gaussian beam solutions [16]. For example, a radially-polarized, first-order Gaussian mode is given by

$$\hat{E}_r = E_0(rr_0/r_s^2)\exp\left[-(1-i\alpha)r^2/r_s^2 - 2i\tan^{-1}\alpha\right], \qquad (4)$$

where $r_s = r_0(1+\alpha^2)^{1/2}$ is the spot size, $\alpha = (z-z_0)/Z_{RG}$, r_0 is the minimum spot size at the focal point $z = z_0$, and $Z_{RG} = kr_0^2/2$ is the Rayleigh length. The axial field component associated with Eq. (4) is

$$\hat{E}_z \simeq \frac{2ir_0 E_0}{kr_s^2}\left[1 - (1-i\alpha)\frac{r^2}{r_s^2}\right]\exp\left[-(1-i\alpha)\frac{r^2}{r_s^2} - 2i\tan^{-1}\alpha\right]. \qquad (5)$$

Near the focal point $\alpha \simeq 0$ and along the z axis the axial wavenumber associated with this field is given by $k_z \simeq k - 2/Z_{RG}$. This corresponds to an axial phase velocity $v_p = \omega/k_z$ given by $v_p/c = 1 + 2/kZ_{RG}$. In vacuum $v_p > c$ and particle slippage prevents acceleration to high energies, as is the case for a Bessel beam. The scale length over which the Gaussian beams diffract is the Rayleigh length Z_{RG}. The total power associated with a Gaussian beam is $P_G = cE_0^2 r_0^2 f_g/16$, where $f_g = 1$ for the fundamental and $f_g = 1/2$ for the first-order Gaussian beam.

C. Nonideal Bessel Beams

In principle, finite power nonideal Bessel beams can be created by truncating the ideal solutions at a finite aperture radius $r = a$ [17-20]. A nonideal Bessel beam consists of $N \simeq a/r_b = ak_\perp/\pi$ rings, with a total power given by N times the power in a single ring, $P_B \simeq NP_b \simeq (c/4)NE_0^2/k_\perp^2$. Roughly speaking, a nonideal Bessel beam consisting of N rings diffracts away sequentially starting with the outermost ring [18]. The outermost ring diffracts after a distance $\sim \pi r_b/\lambda$, the next ring diffracts after a distance $2\pi r_b^2/\lambda$, and so on until the innermost ring diffracts away after a distance $\sim N\pi r_b^2/\lambda$. Hence, the maximum propagation distance of a nonideal Bessel beam consisting of

N rings is [17,18] $L_{max} \simeq NZ_{RB}$, where $Z_{RB} = kr_b^2/2 = (\pi^2/2)k/k_\perp^2$ is the Rayleigh length associated with the asymptotic width of an individual ring. More accurately, the analysis presented in the following section gives $L_{max} = (2/\pi)NZ_{RB} = ak/k_\perp$.

Axicon lenses can be used to create nonideal Bessel beams [11,27]. A schematic for creating a radially polarized, first order Bessel beam is shown in Fig. 1 [10,11,25,26]. Here a radially polarized beam is focused by an axicon lens of radius R_c, such that it crosses the z-axis at an angle θ_c and forms a focal region of length L_c. A circularly symmetric interference pattern develops along the focal region with a radial field component given by $\hat{E}_r = E_c(z)J_1(k_\perp r)$, where the field amplitude $E_c(z)$ depends on the laser intensity at the surface of the axicon. Assuming the field components have the form given by Eqs. (2)-(3), the axicon focal parameters (R_c, L_c, and $\theta_c \ll 1$) are related to the Bessel beam parameters ($k_\perp = \pi/r_b$, $L_{max} = 2NZ_{RB}/\pi$, and $N = a/r_b$) by $R_c \simeq L_c\theta_c \simeq 2a$, $k_\perp \simeq k\theta_c$, $L_c \simeq 2L_{max}$, and $N \simeq (R_c/\lambda)\theta_c$.

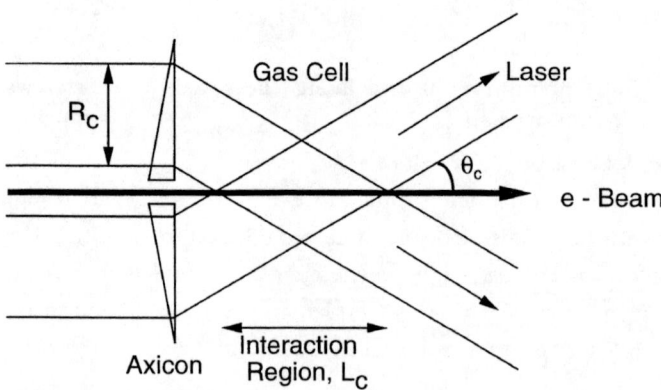

Fig. 1: Schematic of the axicon focusing geometry used to create a line focus. At the midplane of the focus, the transverse profile of the laser field is a nonideal Bessel beam. The axicon focal parameters are related to the Bessel beam parameters by $k_\perp \simeq k\theta_c$, $L_c \simeq 2L_{max}$, and $N \simeq (R_c/\lambda)\theta_c$.

BESSEL BEAM DIFFRACTION

The scale length for diffraction of a nonideal Bessel beam can be determined analytically using scalar diffraction theory based on Huygen's principle [16]. In this method the transverse beam profile is specified in the plane of an aperture and the beam is propagated forward via an integral formulation. The solution to the wave equation for a laser beam of frequency $\omega = ck$ is given by the Kirchhoff integral [16]

$$E_i(\mathbf{r}) = \frac{k}{2\pi i} \int_A dS' \left(1 + \frac{i}{kR}\right) \frac{\mathbf{n} \cdot \mathbf{R}}{R^2} \exp(ikR) E_i(\mathbf{r}'), \tag{6}$$

where $\mathbf{R} = \mathbf{x} - \mathbf{x}'$ is the position vector from the element of surface integration dS' at $\mathbf{x}'$ to the point of observation $\mathbf{x}$, $\mathbf{n}$ is a unit vector that is normal to the plane of the aperture (located at $z' = 0$) and directed towards the observation point, and A denotes the surface area of the aperture. In the limit $(r^2 + r'^2)/z^2 \ll 1$, the integral Eq. (6) can be evaluated by the method of stationary phase. Assuming a transverse Bessel profile at the surface of the aperture, to leading order, the amplitude of the laser field along the axis of propagation is given by

$$\hat{E}_z(r = 0) \simeq E_0 \exp\left(-ik_\perp^2 z/2k\right) \left[1 - H\left(z - ka/k_\perp\right)\right], \tag{7}$$

where $H(z - ka/k_\perp)$ is the Heaviside step function and a circular aperture of radius a has been assumed. Hence, the maximum propagation distance for a Bessel beam passing through an aperture of radius $r = a$ is

$$L_{max} = ka/k_\perp = (2/\pi) N Z_{RB}, \tag{8}$$

where $r_b = \pi/k_\perp$ is the asymptotic width of a Bessel ring, $N = a/r_b$ is the total number of rings, and $Z_{RB} = kr_b/2$ is the Rayleigh length associated with an individual Bessel ring. This value for L_{max} is in agreement with previous estimates [17,18].

LASER ACCELERATION OF PARTICLES

A. Ponderomotive Accelerators

In the laser wakefield accelerator, the plasma beat wave accelerator, and the vacuum beat wave accelerator, particle acceleration is the result of the ponderomotive force of the laser fields, i.e, the accelerating gradient is proportional to the square of the laser field amplitude, $E_{acc} \sim E_0^2$ [1-5,8,9]. In this case, there is no enhancement in the energy gain when a Bessel beam is used, as is evident by the following scaling laws. For a Gaussian beam undergoing vacuum diffraction, the acceleration distance is on the order of a Rayleigh length, $L \sim Z_{RG}$. In terms of the total power $P_G \sim E_0^2 r_0^2$, the energy gain scales as $W_G \sim LE_{acc} \sim P_G$. For a J_0 Bessel beam, the total propagation length is $L \sim NZ_{RB}$ and the total beam power is $P_B \simeq NP_b \sim NE_0^2 r_b^2$. Hence, $W_B \sim NZ_{RB} E_0^2 \sim P_B$ and

$$W_B/W_G \sim P_B/P_G, \qquad (9)$$

where the subscripts B and G refer to the Bessel and Gaussian beams, respectively. For equal beam powers, there is no enhancement in the energy gain, assuming that the acceleration distance is limited by diffraction.

B. Inverse Free Electron Laser

Consider the case of an accelerating gradient that is linearly proportional to the amplitude of the transverse laser field, $E_{acc} \sim E_0$. This is the case in the inverse free electron laser [12-15], in which acceleration results from the ponderomotive force of the beat wave of the laser field with the wiggler magnetic field. For a Gaussian driver, $W_G \sim E_0 Z_{RG} \sim r_0 P_G^{1/2}$. For a J_0 Bessel driver, $W_B \sim E_0 NZ_{RB} \sim r_b(NP_B)^{1/2}$. Hence,

$$W_B/W_G \sim (r_b/r_0)(NP_B/P_G)^{1/2}. \qquad (10)$$

Uniform acceleration of the electron beam requires both r_0 and r_b to be greater than the electron beam radius. Assuming $r_b \simeq r_0$ and $P_G \simeq P_B$, the energy gain can be enhanced by the use of a Bessel beam by a factor of $N^{1/2}$. This assumes that the electron remains synchronous and that the acceleration distance is limited by diffraction.

C. Direct Acceleration in Vacuum

For direct acceleration in vacuum [4-8], in which the particles are accelerated by the axial component of the laser field ($E_{acc} \simeq E_z$), phase detuning can limit the acceleration length. The detuning length is defined as the distance required for a particle to phase slip by one-half wavelength with respect to the axial electric field, i.e., $L_d(v_p - v_e) \simeq \lambda/2$, where v_e is the axial particle velocity. For a Gaussian beam, the phase velocity is $v_p > c$ and the detuning length for a highly relativistic electron ($v_e \simeq c$) is $L_d \simeq (\pi/2)Z_{RG}$. For a Bessel beam, $v_p > c$ as well and $L_d \simeq (4/\pi)Z_{RB}$. Hence, in vacuum the effective acceleration length will be limited to $L_{acc} = L_d$, even if the beam propagates a distance longer than L_d. For a Gaussian beam $E_z \simeq 2E_0/kr_0$ and $W_G \sim (2E_0/kr_0)Z_{RG} \sim P_G^{1/2}$. For a J_1 Bessel beam $E_z \simeq kE_0/k_\perp$ and $W_B \sim (kE_0/k_\perp)Z_{RB} \sim (P_B/N)^{1/2}$. Thus

$$W_B/W_G \sim (P_B/NP_G)^{1/2}, \qquad (11)$$

and the energy gain for direct acceleration in vacuum will be reduced by a factor of $N^{1/2}$ for the Bessel beam, assuming equal beam powers.

D. Inverse Cherenkov Accelerator

In the inverse Cherenkov accelerator (ICA) [4,10,11], a background of neutral gas is introduced to control the phase velocity of the laser field. Acceleration is the result of the axial component of a radially polarized first-order laser field. In a gas the dispersion relation for a plane wave is $\omega = nck$, where n is the index of refraction. Typically, $\Delta n \equiv n-1$ is positive, much less than unity, and proportional to the gas density. The phase velocity can be tuned by adjusting the gas density and phase synchronism can be achieved.

1. Gaussian Beam ICA

For a Gaussian beam of the form given by Eqs. (4)-(5), the energy gain in the ICA can be calculated by integrating the axial electric field, Eq. (5), over $-\infty < z < \infty$. Assuming that the particle moves at constant velocity $v_e \simeq c(1 - 1/2\gamma^2)$ ($\gamma \gg 1$) along the z axis, the energy gain is

$$W_G = 2\pi q E_{z0} \Delta k Z_{RG}^2 \sin\phi_0 \exp(-\Delta k Z_{RG}), \qquad (12)$$

for $\Delta k \geq 0$ and $W_G = 0$ for $\Delta k < 0$, where $E_{z0} = 2E_0/kr_0$ is the peak axial field amplitude, $\Delta k = k(\Delta n - 1/2\gamma^2)$, $\alpha = z/Z_{RG}$, $z = 0$ is the focal point, ϕ_0 is the initial phase, and q is the electron charge. The energy gain is maximum for $\sin\phi_0 = 1$ and $\Delta k Z_{RG} = 1$, and is given by

$$W_G[\text{MeV}] \simeq 2.3 P_G^{1/2}[\text{GW}]. \qquad (13)$$

To enhance the energy gain it is necessary to increase the laser propagation distance. In principle, a Gaussian laser beam can be self-guided in a gas by a proper balancing of diffraction, nonlinear self-focusing, and plasma defocusing [4,28].

2. Bessel Beam ICA

For a first-order Bessel beam of the form given by Eq. (2)-(3), the amplitude and phase of the electric field is approximately constant over a total propagation length of $L \simeq 2L_{max}$, where L_{max} is given by Eq. (8). Hence, the maximum energy gain can be estimated by the product of the peak axial field $E_{z0} = k_\perp E_0/k$, Eq. (3), times the total propagation distance, $L = (4/\pi)NZ_{RB}$, i.e., $W_B = 2qE_{z0}L_{max} = 2qaE_0$, where $Z_{RB} = kr_b^2/2$, $k_\perp = \pi/r_b$, $N = a/r_b$ is the number of rings, and a is the aperture radius. Here it has been assumed that the electron remains phase matched to the laser field, i.e., $2\Delta n = k_\perp^2/k^2 + 1/\gamma^2$. In terms of the total Bessel beam power, $P_B \simeq caE_0^2/4\pi k_\perp$, the maximum energy gain is given by

$$W_B[\text{MeV}] \simeq 2.2 N^{1/2} P_B^{1/2} [\text{GW}], \tag{14}$$

in agreement with previous estimates [10]. Hence, by using a Bessel beam with N rings, the energy gain can be enhanced by a factor of $N^{1/2}$ compared to a Gaussian beam with the same total power,

$$W_B/W_G \simeq (NP_B/P_G)^{1/2}. \tag{15}$$

DISCUSSION

Comparisons were made between accelerators driven by Gaussian laser beams and those driven by Bessel laser beams. For equal beam powers, it was shown that the energy gain using a Bessel beam is approximately $N^{1/2}$ times that using a Gaussian beam provide that (i) the accelerating gradient is linearly proportional to the laser field and (ii) the acceleration distance is limited by diffraction and not by phase detuning (or some other mechanism). This is the case for the inverse Cherenkov accelerator (phased-matched acceleration using a J_1 beam in a gas) and for the inverse free electron laser (phase-matched acceleration resulting from the ponderomotive interaction of a J_0 beam and a wiggler magnetic field). For the inverse Cherenkov accelerator, it can be shown that for equal powers and equal peak intensities, the acceleration lengths, peak accelerating field, and energy gain scale as $L_B/L_G \sim 1$, $E_{zB}/E_{zG} \sim N^{1/2}$, and $W_B/W_G \sim N^{1/2}$, respectively.

Scaling laws have been derived for other configurations as well. Direct acceleration by the E_z field in vacuum is limited by phase detuning, hence, there is no advantage in using a Bessel beam (in fact, the energy gain is reduced by $N^{1/2}$, assuming equal beam powers). For cases in which the accelerating gradient is proportional to the square of the laser field (e.g., the laser wakefield accelerator, the plasma beat wave accelerator, and the vacuum beat wave accelerator), the energy gain for both Gaussian and Bessel beam drivers scales as $W \sim P$ (independent of N), assuming an acceleration distance limited by diffraction. For very intense lasers, however, there exists a highly nonlinear regime of the laser wakefield accelerator for which E_{acc} is linearly proportional to the laser field [3,4]; hence, there may be an advantage to using a Bessel beam if the propagation distance is limited by diffraction. In addition, for all laser-plasma accelerators, it appears possible to guide a laser beam large distances (many Rayleigh lengths) using a plasma density channel [3,4,27,28], thus enhancing the acceleration length and the energy gain.

ACKNOWLEDGMENTS

This work was supported by the Department of Energy and the Office of Naval Research. The authors thank M. Baine for his assistance.

REFERENCES

[1] *Advanced Accelerator Concepts*, edited by P. Schoessow, AIP Conf. Proc. **335** (Amer. Inst. Phys., NY, 1995).

[2] *Special Issue on 2nd Generation Plasma Based Accelerators*, edited by T. Katsouleas and R. Bingham, IEEE Trans. Plasma Sci. **PS-24** (1996).

[3] E. Esarey, P. Sprangle, J. Krall, and A. Ting, IEEE Trans. Plasma Sci. **PS-24**, 252 (1996); Phys. Fluids B **5**, 2690 (1993).

[4] P. Sprangle, E. Esarey, and J. Krall, Phys. Plasmas **3**, 2183 (1996); Phys. Rev. E, Oct (1996).

[5] W.B. Mori and T. Katsouleas, in *Advanced Accelerator Concepts*, edited by P. Schoessow, AIP Conf. Proc. **335** (Amer. Inst. Phys., NY, 1995), p. 112.

[6] J.A. Edighoffer and R.H. Pantell, J. Appl. Phys. **50**, 6120 (1979).

[7] E.J. Bochove, G.J. Moore, and M.O. Scully, Phys. Rev. A **46**, 6640 (1992).

[8] E. Esarey, P. Sprangle, and J. Krall, Phys. Rev. E **52**, 5443 (1995); P. Sprangle, E. Esarey, J. Krall, and A. Ting, Opt. Comm. **124**, 69 (1996).

[9] B. Hafizi, A. Ting, E. Esarey, P. Sprangle, and J. Krall, Phys. Rev. E, submitted (1996).

[10] J.R. Fontana and R.H. Pantell, J. Appl. Phys. **54**, 4285 (1983).

[11] W.D. Kimura, G.H. Kim, R.D. Romea, L.C. Steinhauer, I.V. Pogorelsky, K.P. Kusche, R.C. Fernow, X. Wang, and Y. Liu, Phys. Rev. Lett. **74**, 546 (1995).

[12] R.B. Palmer, J. Appl. Phys. **43**, 3014 (1972).

[13] P. Sprangle and C.M. Tang, IEEE Trans. Nucl. Sci. **NS-28**, 3346 (1981).

[14] R.H. Pantell and T.I. Smith, Appl. Phys. Lett. **40**, 753 (1982).

[15] S.Y. Cai, A. Bhattacharjee, and T.C. Marshall, Nucl. Instrum. Meth. A **272**, 481 (1988).

[16] A.E. Siegman, *Lasers* (University Science Books, Mill Valley, CA, 1986); P.W. Milonni and J.H. Eberly, *Lasers* (Wiley, NY, 1988); J.D. Jackson, *Classical Electrodynamics* (Wiley, NY, 1975), Chap. 9.

[17] J. Durnin, J.J. Miceli and J.H. Eberly, Phys. Rev. Lett. **58**, 1499 (1987); J. Durnin, J. Opt. Soc. Am. A **4**, 651 (1987).

[18] P. Sprangle and B. Hafizi, Phys. Rev. Lett. **66**, 837 (1991); B. Hafizi and P. Sprangle, J. Opt. Soc. Am. A **8**, 705 (1991).

[19] Y.Y. Ananev, Opt. Sectrosc. (USSR) **64**, 722 (1988).

[20] M.R. Lapointe, Opt. Laser Tech. **24**, 315 (1992).

[21] J. Turunen, A. Vasara and A.T. Friberg, Appl. Opt. **27**, 3959 (1988).

[22] G. Indebetouw, J. Opt. Soc. Am. A **6**, 150 (1989).

[23] G. Scott and N. McArdle, Opt. Engin. **31**, 2640 (1992).

[24] C. Patterson and R. Smith, Opt. Comm. **124**, 121 (1996).

[25] S.C. Tidwell, G.H. Kim, and W.D. Kimura, Appl. Opt. **32**, 5222 (1993).

[26] I.V. Pogorelsky, W.D. Kimura, and Y. Liu, in *Advanced Accelerator Concepts*, edited by P. Schoessow, AIP Conf. Proc. **335** (Amer. Inst. Phys., NY, 1995), p. 419.

[27] C.G. Durfee III and H.M. Milchberg, Phys. Rev. Lett. **71**, 2409 (1993); C.G. Durfee III, J. Lynch, and H.M. Milchberg, Phys. Rev. E **51**, 2368 (1995).

[28] A. Braun, G. Korn, X. Liu, D. Du, J. Squier, and G. Mourou, Opt. Lett. **20** 73 (1995); X. Liu and D. Umstadter, in *Shortwavelength V: Physics with Intense Laser Pulses*, edited by M.D. Perry and P.B. Corkum (Opt. Soc. Am., Washington DC, 1993), vol. 17, p. 45.

[29] P. Sprangle and E. Esarey, Phys. Fluids B **4**, 2241 (1992); P. Sprangle, E. Esarey, J. Krall, and G. Joyce, Phys. Rev. Lett. **69**, 2200 (1992).

Proof-of-Principle Experiment of the Vacuum Beat Wave Accelerator

A. Ting,[1] B. Hafizi,[2] R. Burris,[1] R. Fischer,[1] C.I. Moore,[3]
K. Krushelnick,[4] E. Esarey,[1] and P. Sprangle[1]

[1] Plasma Physics Division, Naval Research Laboratory, Washington, DC 20375
[2] Icarus Research, Inc., Bethesda, MD, and Omega-P, Inc., New Haven, CT 06520
[3] NRL/NRC Post Doctoral Fellow
[4] Cornell University, Ithaca, NY 14853

The vacuum beat wave accelerator (VBWA) is discussed and design parameters for proof-of-principle experiment are presented. The VBWA utilizes two focused laser beams of differing wavelengths to generate a beat wave that can impart a net acceleration to charged particles. Theory and simulations show that the single-stage energy gain of the VBWA is limited by diffraction of the laser beams, particle slippage in phase and velocity, and radial walk-off. In the simulations the particles are synchronous with the beat wave for a short interval of time and the energy gain has the nature of an impulse delivered near the focal region. Simulations also show that the problem of radial walk-off may be ameliorated by using a converging beam of particles. For terawatt-level laser beams, with wavelengths 1 μm and 0.5 μm, and a 4.5 MeV finite-emittance electron beam the energy can be increased to ~ 12.5 MeV in a non-synchronous interaction over a distance of under 4 mm, with a peak acceleration gradient > 15 GeV/m and an estimated trapping fraction of ~ 1 %.

INTRODUCTION

Advances in the development of high power lasers continue to spur new concepts for accelerators [1]. Laser-based plasma acceleration schemes, however, face a myriad of challenges that must be overcome before any of these schemes can be considered practical. An alternative approach to particle acceleration is to make use of laser beams *in vacuo* in the far field limit, i.e., in regions that are far (compared to the vacuum wavelength) from boundaries [2-10], thus mitigating material breakdown, plasma formation and instability. The vacuum beat wave accelerator (VBWA) [5,6] utilizes a pair of laser beams to accelerate particles. Since the VBWA does not require the proximity of a material medium, nor does it take place in a plasma, it is limited neither by material breakdown (as in an inverse Cerenkov accelerator) nor by pair annihilation of positrons (as in plasma beat wave or plasma wakefield accelerators).

The purpose of this paper is to outline a proof-of-principle experiment of the VBWA. The experiment will be performed at the Naval Research Laboratory (NRL), using a 4.5 MeV RF gun with a photocathode, and a table-top terawatt (T^3) laser [11]. Simulation results, used in the design of the experiment, are reported.

PONDEROMOTIVE BEAT WAVE

In the presence of an electromagnetic field *in vacuo* the change in the relativistic factor γ of an electron of mass m and charge e, moving along the z axis, due to the longitudinal component of the laser field E_z, is given by

$$\Delta \gamma = -\frac{|e|}{mc^2} \int_{-\ell}^{\ell} dt v_z E_z.$$

The usual expression of the Lawson-Woodward theorem states that $\Delta\gamma=0$ in the ultra-relativistic limit ($v_z=c$) provided $\ell \to \infty$ [4-6,12,13]. Since the theorem specifically neglects the **vxB** force, it is natural to consider a configuration wherein this force is significant with a view to escaping the theorem's conclusion. The **vxB** force is also referred to as the ponderomotive force.

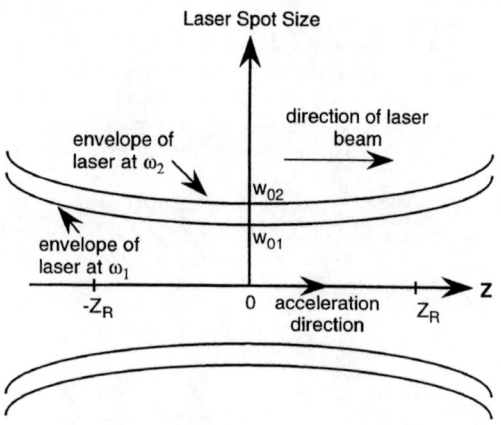

FIGURE 1. Schematic of the Vacuum Beatwave Accelerator.

Consider two laser beams with frequencies ω_1 and ω_2, respectively, that propagate along a common axis and come to a common focus at $z=Z_{focus}$, as shown in Fig. 1. For definiteness circularly polarized laser beams are considered throughout. The vector potential of such a beam is expressible as

$$A_j = \frac{A_{0j} w_{0j}}{w_j} \exp(-r^2 / w_j^2)(e_x \cos \psi_j + e_y \sin \psi_j) + A_{zj} e_z, \tag{1}$$

where the suffix $j=1,2$ identifies the laser beam, $\omega_j = ck_j = 2\pi/\lambda_j$ defines the wavenumber k_j and wavelength λ_j,

$$w_j(z) = w_{0j}\left[1 + (z - Z_{focus})^2 / Z_{Rj}^2\right]^{1/2}$$

is the spot size at z, w_{0j} is the waist (radius), $Z_{Rj} = \pi w_{oj}^2 / \lambda_j$ is the Rayleigh range, A_{0j} is a constant, e_x, e_y and e_z are unit vectors,

$$\psi_j = k_j z - \omega_j t - \tan^{-1}\left[(z - Z_{focus})/Z_{Rj}\right] + r^2 (z - Z_{focus})/(Z_{Rj}\omega_j^2) + \psi_{0j}$$

is the phase [14], and ψ_{0j} is a constant. The expression for the vector potential, Eq. (1), is valid in the paraxial limit, where $w_{oj} \gg \lambda_j$. In Eq. (1) A_{zj} denotes the axial component of the vector potential. The total vector potential is $A = A_1 + A_2$.

Neglecting particle motion in the transverse plane, the equation of motion reduce to

$$\frac{d\gamma}{dz} \approx \frac{(k_2 - k_1)\hat{a}_1 \hat{a}_2}{u_z} \sin(\psi_2 - \psi_1), \qquad (2)$$

and

$$\frac{d(\psi_2 - \psi_1)}{dz} \approx (k_2 - k_1)(\beta_{ph}^{-1} - \beta_z^{-1}), \qquad (3)$$

where

$$a^2 = \hat{a}_1^2 + \hat{a}_2^2 + 2\hat{a}_1\hat{a}_2 \cos(\psi_2 - \psi_1) + a_z^2, \qquad (4)$$

$\hat{a}_j = (a_{0j}w_{0j}/w_j)\exp(-r^2/w_j^2)$, $a_{0j} = |e|A_{0j}/(mc^2)$, the normalized *beat wave* phase velocity is given by

$$\beta_{ph}^{-1} = 1 - \frac{1-(1-\hat{z}_2^2)r^2/w_2^2}{(k_2-k_1)Z_{R2}(1+\hat{z}_2^2)} + \frac{1-(1-\hat{z}_1^2)r^2/w_1^2}{(k_2-k_1)Z_{R1}(1+\hat{z}_1^2)}, \qquad (5)$$

$\hat{z}_j = (z - Z_{focus})/Z_{Rj}$ is the normalized axial coordinate relative to the focal point, and β is the electron velocity normalized to c.

ANALYTICAL ESTIMATE FOR ENERGY GAIN

Equation (3) permits one to identify a (local) slippage distance z_s, given by

$$z_s = \frac{\pi}{|(k_2 - k_1)(\beta_{ph}^{-1} - \beta_z^{-1})|},$$

that is a measure of the distance over which the phase of the particle in the beat wave slips by π. In the limit $z_s < Z_{Rj}$ and in the highly relativistic limit, with $Z_{R1}=Z_{R2}=Z_R$, Eqs. (2) and (3) integrate to [15]

$$\Delta\gamma^2 \sim \frac{4\pi}{3}(k_2 - k_1)Z_R \frac{\gamma_\perp a_{0,1}a_{0,2}}{a}\left\{J_{1/3}\xi + J_{-1/3}\xi\right\}\sin\Phi(Z_{focus}), \qquad (6)$$

where $\xi = 2\pi Z_R \gamma_\perp / (3z_s a)$, the amplitude a is given by Eq. (4), $\gamma_\perp = (1+a^2)^{1/2}$ and $\Phi(Z_{focus})$ is the ponderomotive phase evaluated at $z = Z_{focus}$. Equation (6) is useful in multistage systems, where the relative change in energy $\Delta\gamma/\gamma$ per stage is relatively small.

NUMERICAL RESULTS

The estimate for the energy gain, Eq. (6), ignores beam emittance and radial displacement of particles due to ponderomotive scattering. We now describe the results of numerical simulations which employ a laser beam with wavelength 1 μm and another at 1/2 μm. The laser beam power, determined from the vector potential in Eq. (1), can be written as $P_j[TW]=0.0432(a_{0j}w_{0j}/\lambda_j)^2$ in the paraxial limit. Similarly the peak (on-axis) intensity and electric field are given by $I_j[W/cm^2]=2.75\times10^{18}(a_{0j}/\lambda_j[\mu m])^2$, $E_j[TV/m]=3.11a_{0j}/\lambda_j[mm]$ respectively.

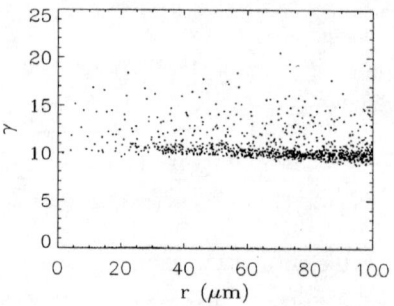

FIGURE 2. Plots of γ-r phase space at end of run. Direction of beam propagation is towards the left.

The simulations employ a finite-emittance convergent beam, consisting of 4000 particles, that is injected upstream, along the z axis. Simulations are performed with equal Rayleigh ranges by appropriate choice of parameters. The simulation parameters are similar to the parameters of the proof-of-principle experiment listed in Table I. In order to approximately match the axial particle velocity with that of the beat wave, and to maximize a_{0j} it was necessary to choose a relatively small waist, $w_{0,1} = 4$ μm.

The computations employ a beam with normalized root mean square (RMS) emittance $\varepsilon_n = 1.2\pi$ mm-mrad that is focused down to an RMS waist on the order of 4 μm, thus achieving a nearly optimal overlap of the optical and particle beams.

Figure 2 shows the γ-r phase space, demonstrating that a small number of the high energy particles are close to the axis. Based on phase space plots it is estimated that the fraction of particles with energy in excess 10 MeV is on the order of 1 %.

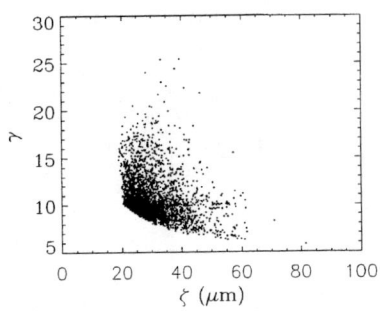

FIGURE 3. Plots of longitudinal phase space (γ-ζ), at end of run. Direction of beam propagation is towards the left.

The lack of significant phase bunching is demonstrated by the plot of the longitudinal phase space (γ-z, where $z = ct - z$) shown in Fig. 3. Note that the particles, initially loaded over a longitudinal distance ~ 1 μm, are now spread over more than 50 μm, and a small fraction of the particles are at the highest energies.

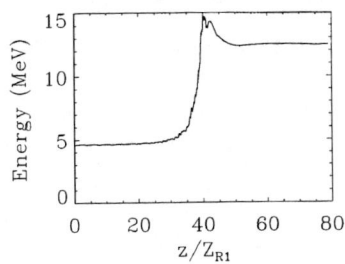

FIGURE 4. Plot of peak energy versus axial distance, with abscissa measured in units of Rayleigh range, Z_{R1}, of laser beam 1.

Figure 4 shows the peak energy of the ensemble of particles, along the interaction length. Observe that the energy rises up to ~15 MeV in the vicinity of the focal point and then drops off, plateauing at ~12.5 MeV. Equation (6) predicts a peak energy of 17 MeV for $\Phi(Z_{focus})=-\pi/2$.

DESIGN OF THE PROOF-OF-PRINCIPLE EXPERIMENT

A schematic of the proof-of-principle experiment is shown in Fig. 5.

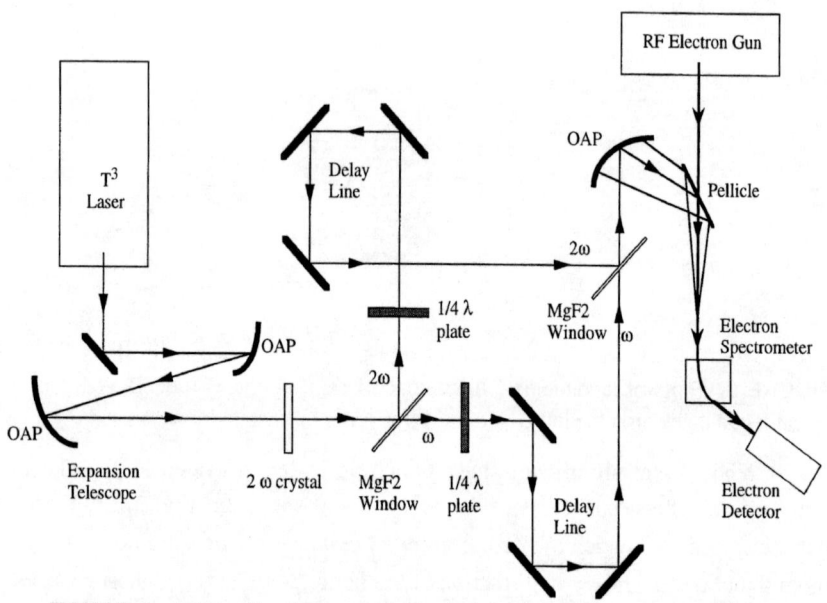

FIGURE 5. Schematic of the proof-of-principle experiment of the VBWA.

Frequency Doubling

Efficient (>50%) frequency doubling of the T^3 laser pulse, to generate high power output at 527 nm, can be achieved by use of KDP crystals. The laser beam is up-collimated to a large beam size (~7 cm) by passing through an expansion telescope that employs two off-axis paraboloids (OAP) at the entrance and exit focusing elements. This allows the peak intensity in the beam to be reduced to prevent possible subsequent damage of optical elements. We use reflective optics to reduce the nonlinear effects that would otherwise occur in refractive optics.

The laser beam, which now consists of two wavelengths with about the same power in each, is split into its two constituent wavelengths by a thin Magnesium Fluoride (MgF_2) window. The window is coated on the front surface to reflect the 2ω green light. The infrared portion of the laser passes through the thin MgF_2 substrate with minimal nonlinear distortion. The reason for splitting the two laser beams is to reduce the walk-off generated between the two beams during frequency doubling in the crystal.

Optical Beamline

To generate the beat wave from the two laser beams, they must have similar polarizations. The T^3 laser generates plane polarized infrared light at 1 μm. During the frequency doubling process, the green light has its polarization rotated by 90°. The quater-wave plates are used to rotate the two laser beams to the same circular polarization. The NRL T^3 laser produces ~3 TW of laser power in the infrared, at 1.05 μm, with the power contained in each of the two laser pulses ~1.5 TW. At such high powers, a gold coated OAP will be damaged at the 527 nm wavelength since gold is not a good reflector at this wavelength. A large bore, high quality, dielectric protected, silver coated OAP will be used for the final focusing of the dual frequency laser pulse. The $f^\#$ of the OAP is chosen to produce an 8 μm laser focal spot size (diameter), as used in the simulations. Equivalent plane imaging diagnostics will be employed to measure the focal spot sizes of the two laser beams. The focal plane of the laser is imaged with a magnifying telescopic optical arrangement onto a CCD array.

Electron Beam Injection

According to the simulations, the electron beam must be injected with a relatively large degree of convergence. Since the injected electron beam is collinear with the laser beam, the OAP would block the incoming electrons if it were used as the last optical element before the final focus. Circumventing this by opening a small hole in the OAP is not practical, because the electron beam has a similar convergence angle as the laser. The combination of laser beam size at the OAP and the $f^\#$ of the OAP is designed to provide a sufficiently long focal length to insert other optical elements that will allow for electron beam injection. A coated thin pellicle reflects the converging laser beams and allow the 4.5 MeV electrons to pass through.

Detection of Accelerated Electrons

The energy spectrum and the spatial distribution of the accelerated electrons exiting the interaction region of the VBWA can be diagnosed with electron spectrometers. Because the electron bunch is longer than the ponderomotive beat wave period, electrons would interact with the beat wave at arbitrary points along the wave and thus experience a range of gradients. A large energy spread (~10 MeV) would be expected regardless of the input electron energy spread. An electron spectrometer is most suitable for analyzing electron beams with large energy spreads due to its large dynamic range. By allowing a collimated electron beam to enter a narrow gap separating two magnetic poles, the electron energy can be determined from the resulting radius of curvature. Direct observation of the electrons in a Wilson cloud chamber is also possible. Because of the angular divergence of the electrons, the electron flux falls off rapidly with distance from the interaction point. The size of the spectrometer therefore has to be small so that it can be placed in close proximity of the interaction point. Conventional electron spectrometers are usually placed outside the vacuum chamber. The large gap of the electromagnet necessitates a large current to bend the electron trajectories. Instead, a small high field electromagnet can be placed inside the evacuated interaction chamber. The electron detector can be a plastic scintillator backed by a photomultiplier tube for large reception area and high gain in the signals.

Photocathode RF Gun

When employing a photocathode, the RF driving circuit of the electron gun must be synchronized with the T^3 laser which drives the beat wave. The NRL T^3 laser utilizes a Ti:sapphire oscillator that incorporates an optical cavity resonating at 76 MHz. This frequency has to be changed to a subharmonic of the S-band RF frequency of the electron gun by suitably varying the cavity length. Piezoelectric mirror mounts will be installed on the end mirrors of the cavity so that the cavity length and therefore the resonant frequency can be slaved to the master oscillator of the klystron that drives the electron gun. An alternate approach is to use the optical oscillator as the master oscillator. The RF signal can be picked up with a photodiode and then frequency multiplied to the S-band frequency of the klystron. [15] Even though the triggering laser has a pulse length < 1 psec, the space charge effect of the high current photo-emitted current is expected to lengthen the electron beam pulse to approximately 2 ps.

Electron Beamline

A photocathode RF electron gun produces a high brightness beam with high current and small emittance. It is necessary to match the emittance phase space of the electron beam to the acceptance phase space of the accelerator. This requires a strong focusing lens to squeeze the horizontal emittance ellipse to increase the overlap of the two phase spaces.

The number of electrons that may be trapped by the beat wave can be estimated as follows. The total number of electrons in a micropulse of the electron beam is ~10^9 using the parameters of the NRL RF gun. From the simulations, the optimal convergence of the injection electron beam is obtained with $f^{\#}$ ~ 6. Using an RMS normalized emittance of 2 π mm mrad for the electron beam from a photocathode, the focal spot radius of the electron beam is estimated to be ~15 μm. This corresponds to a transverse filling factor of ~0.07 which is the square of the ratio of the laser spot size to the electron beam spot size. The temporal filling factor is given by the laser pulse length divided by the electron pulse length, which is ~ 0.25. From the simulations, about 20% of the injected electrons are accelerated to energies higher than 5 MeV. The number of accelerated electrons with energy above 5 MeV is therefore ~3×10^6. The parameters for the experiment are listed in Table I.

TABLE I. Parameters of the p.o.p. VBWA experiment.

Laser Parameters :

λ_1	1.054 μm
λ_2	527 nm
Pulse length	500 fs
Pulse energy/beam	0.75 J
Peak power/beam	1.5 TW
Focal Spot size for λ_1	8 μm
Focal spot size for λ_2	6 μm
Rayleigh length for both beams	50 μm
Peak intensity	~1×10^{18} W/cm^2

Injection electron beam parameters :

Injection energy	4.5 MeV
Beam current	100 A
Pulse length	2 ps
Normalized rms emittance	2 π mm mrad
Focal spot size	30 μm

CONCLUSIONS

Simulations reveal some key characteristics of the VBWA. First, the variation of the phase velocity in the focal region leads to rapid detuning of particles from the beat wave. Second, the interaction is in the nature of an impulse delivered to the particles in the focal region. Third, particles are scattered radially on passing through the focus. As a result, slippage and walk-off tend to limit the acceleration gradient of the VBWA. Simulations demonstrate that radial walkoff can be reduced and the energy gain can be improved by employing a converging particle beam that is focused at the same location as the laser beams. Design parameters and configuration for a proof-of-principle experiment of the VBWA are presented.

ACKNOWLEDGMENTS

The authors have benefited from discussions with Drs. J. Krall, J. L. Hirshfield and A. Ganguly. This work was supported by the DOE and the ONR.

REFERENCES

1. P. Schoessow, ed., *Advanced Accelerator Concepts, AIP Conf. Proc.* **335** (Amer. Inst. Phys., NY, 1995); T. Katsouleas and R. Bingham, ed., *Special issue of 2nd Generation Plasma Accelerators, IEEE Trans. Plasma Sci.* **PS-24** (1996).
2. J. A. Edighoffer and R. H. Pantell, *J. Appl. Phys.* **50**, 6120 (1979).
3. M. O. Scully and M. S. Zubairy, *Phys. Rev. A* **44**, 2656 (1991).
4. E. J. Bochove, G. T. Moore and M. O. Scully, *Phys. Rev. A* **46**, 6640 (1992).
5. P. Sprangle, E. Esarey, J. Krall and A. Ting, *Opt. Commun.* **124**, 69 (1996).
6. E. Esarey, P. Sprangle and J. Krall, *Phys. Rev. E* **52**, 5443 (1995).
7. Y. C. Huang, D. Zheng, W. M. Tulloch and R. L. Byer, *Appl. Phys. Lett.* **68**, 753 (1996).
8. P. Sprangle, E. Esarey and J. Krall, *Phys. Plasmas* **3**, 2183 (1996).
9. F. V. Hartemann, S. N. Fochs, G. P. LeSage, N. C. Luhmann, G. J. Woodworth, M. D. Perry, Y. J. Chen and A. K. Kerman, *Phys. Rev. E* **51**, 4833 (1995).
10. E. Esarey, P. Sprangle, M. Pillof and J. Krall, *J. Opt. Soc. Am. B* **12**, 1695 (1995).
11. M. D. Perry and G. Mourou, *Science* **264**, 927 (1994).
12. J. D. Lawson, *IEEE Trans. Nucl. Sci.* **NS-26**, 4217 (1979).
13. R. B. Palmer, in *Frontiers of Particle Beams*, Lecture Notes in Physics **296**, edited by M. Month and S. Turner (Springer-Verlag, Berlin, 1988), p. 607.
14. A. E. Siegman, *Lasers* (University Science Books, Mill Valley, CA, 1986).
15. S.C. Chen, B.G. Danly, J. Gonichon, C.L. Lin, R.J. Temkin, S.R. Trotz and J.S. Wurtele, "High Power Testing of a 17 GHz Photocathode RF Gun", *IEEE PAC*, Dallas, Texas, May 1995.

GROUP 3

"Structure-Based Accelerator Concepts"—James Rosenzweig (UCLA)

Recent Results & Plans For The Future On SLAC Damped Detuned Structures (DDS)

N.M. Kroll[†‡], R.M. Jones[†‡], C. Adolphsen[†], K.L.F. Bane[†],
W.R. Fowkes[†], K. Ko[†], R.H. Miller[†],
R.D. Ruth[†], M. Seidel[†], and J.W. Wang[†]

[†] Stanford Linear Accelerator Center, M/S 26,
P.O Box 4349, Stanford, CA 94309
[‡] University of California, San Diego,
La Jolla, CA 92093-0319

Abstract. The cells in the SLAC DDS are designed in such a way that the transverse modes excited by the beam are detuned in a Gaussian fashion so that destructive interference causes the wake function to decrease rapidly and smoothly. Moderate damping provided by four waveguide manifolds running along the outer wall of the accelerator is utilised to suppress the reappearance of the wake function at long ranges where the interference becomes constructive again. The newly developed spectral function method, involving a continuum of frequencies, is applied to analyze the wake function of the DDS 1 design and to study the dependence of the wake function on manifold termination. The wake function obtained with the actually realized manifold terminations is presented and compared to wake function measurements recently carried out at the ASSET facility installed in the SLAC LINAC.

1. INTRODUCTION

The wake behind an accelerated bunch in the next linear collider (NLC) design, in which multiple bunches are accelerated within each RF pulse cycle in order to maximize the collider luminosity, has the deleterious effect of beam loading via the longitudinal wake field and a beam break up instability (1) due to the transverse wake field. It is the purpose of this work to investigate minimization of the transverse kick presented to successive bunches in the bunch train. The beam emittance is directly proportional to the mean square of the sum wake function (2) and hence in order to maintain a small beam emittance it is critical to ensure that both the short-range wake field and the long-range wake field are kept within tolerable limits. Here we are concerned with wake fields generated by previous bunches rather than intra-bunch effects, so that short-range and long-range refer to nearby and distant bunches respectively. (Present beam simulations require the wake function be no larger than 1 V/pC/mm/m (3)). In order to reduce the amplitude of the beam-induced wake the cell frequencies are detuned in a Gaussian fashion so that the components of the resulting wake field interfere destructively, causing the wake function to decrease rapidly and smoothly. However, eventually the interference becomes at least in part again constructive leading to reappearance of the wake at an unacceptable amplitude, and one either adopts a method of interleaving a set of multiple structures, each with a slightly different central frequency (to delay reappearance to distance beyond the end of the bunch train) or one utilizes moderate damping to suppress reappearance; the former method entails very tight structure tolerances whereas the latter involves an order of magnitude looser tolerances. The manifold damping scheme provides two additional advantages: significantly enhanced beam diagnostic capability and improved pumping capability for the accelerator structure.

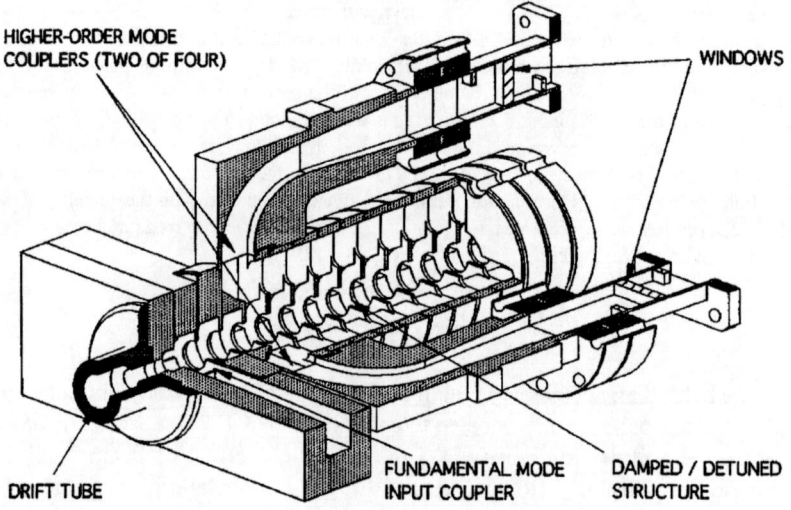

FIGURE 1. Cut-Away View of DDS.

A cross-sectional view of the DDS is shown in fig.1. The features which prove crucial to the reduction of the wake field are clearly illustrated: viz., the tapered cells in cross section and the tapered manifolds, of which two of the four are shown, running parallel to the axis of the accelerator. The figure also shows the way in which the manifolds are coupled to matched loads. The manifold is connected to a rectangular waveguide by means of a 90 degree mitered bend followed by a 90 degree circular H-bend, a taper to WR62 waveguide, a rectangular window, and finally (not shown) a matched load. As will be discussed later the quality of the match to the manifold achieved by this assembly has a very significant effect on the long range wake. Here the concept of "match" applies only to frequencies at which the manifold is propagating. Frequencies for which the manifold is non-propagating at *both* ends will be referred to as stop bands, other frequencies as pass bands.

In the following we report on the computational methods employed, the application of the spectral function method to DDS 1, and our investigation of the sensitivity of the wake function to manifold mismatch. The final section incorporates a new design which results in a considerably reduced short range wake and with prospects for further improvements to the long-range wake function.

2. METHOD OF WAKE FUNCTION CALCULATION

All of our wake function calculations are based upon the equivalent circuit network given in (4), consisting of a sequence of sections each one of which corresponds to a cell and the contiguous section of manifold. The full description of the structure includes a set of 10 "model" parameters for each section. The frequency domain circuit equations are cast in matrix form as (schematically) follows:

$$\begin{pmatrix} \hat{H} & H_x^t \\ H_x & H - GR^{-1}G \end{pmatrix} \begin{pmatrix} \hat{a} \\ a \end{pmatrix} - f^{-2} \begin{pmatrix} \hat{a} \\ a \end{pmatrix} = f^{-2} \begin{pmatrix} B \\ 0 \end{pmatrix} \qquad (2.1)$$

Here each component of the column matrices is an N component vector, and the elements of the square matrix are N by N matrices. The number of sections, N, is taken to be 206 for all calculations reported here. The column matrix on the LHS of (2.1) represents the cell excitation, that on the RHS the drive beam, and f is the frequency. The effect of the manifold is contained in the damping matrix $GR^{-1}G$. There are three possible regimes in which the transverse wake function (i.e. wake potential per unit length) for a particle trailing a distance s behind a velocity c drive bunch (per unit drive bunch charge per unit drive bunch displacement) may be calculated, viz., no damping, weak damping and strong damping:

$$W(s) = 2 \sum_{i=1}^{P} K_p \sin\left(\frac{2\pi f_p}{c} s\right) \theta(s) \qquad (2.2)$$

$$W(s) = 2 \sum_{i=1}^{P} K_p \sin\left(\frac{2\pi f_p}{c} s\right) \exp\left(-\frac{2\pi f_p}{Q_p c} s\right) \theta(s) \qquad (2.3)$$

$$W(s) = \int_{f>0} S(f) \sin\left(\frac{2\pi f}{c} s\right) df \, \theta(s) \qquad (2.4)$$

Here $\theta(s)$ is the unit step function and, K_p, f_p, and Q_p the p-th modal kick factor (5), frequency, and quality factor, respectively. The wake for the first two cases (eqns. 2.2 & 2.3) are expressed as modal sums, while the third (eqn 2.4) encompasses a continuum of modes and is the regime of the spectral function (6). The spectral function S(f) consists of a sum of terms of the form $2 K_p \delta(f-f_p)$ for f in a stop band and a continuous function defined below for f in a pass band. The form (2.4) encompasses the other two as limiting cases.

2.1 Modal Sums and the Perturbation Method

We can obtain the K_p and f_p of (2.2) by setting the RHS and the damping matrix of (2.1) equal to zero, and finding the mode frequencies (real, of course in this case) and amplitude vectors (also real) for which the resultant equation for the amplitudes has non trivial solutions. The K_p are formed from the mode vectors as described in (5), and the undamped wake function (2.2) formed. These results provide the basis for the perturbation method described below. It has turned out to be useful to construct a smoothed spectral function, $S_s(f)$ to represent these results:

$$S_s(f) = 2K_p / (f_{p+1} - f_p), \quad f_p < f < f_{p+1}, \quad \forall p \quad (2.5)$$

The dashed curve of Fig. 6 is an example.

For sufficiently weak damping one can obtain the form eqn. (2.3) by making use of the perturbation formula (7)

$$\delta f_p / f_{p_0} = \tfrac{1}{2} f_{p_0}^2 a_{p_0}^\dagger G R^{-1} G a_{p_0} / (a_{p_0}^\dagger a_{p_0}) \quad (2.6)$$

which is valid to an excellent approximation for small frequency shifts (7).

Here f_{p_0} is the unperturbed p'th modal frequency, G describes the coupling to the manifold, and a_{p_0} is the p'th unperturbed eigenmode. The Q from this equation, together with values obtained from an exact calculation of the complex eigenmodes are shown in figs. 2. The detuning behaves in a similar manner. These results indicate that perturbation theory is unreliable for the DDS 1 parameters. The fact that the absolute value of the perturbation theory complex frequency shift is significantly larger than the mode separation at the center of the band is a strong warning that this will be the case.

The exact eigenmodes and eigenfrequencies can be used to construct a damped modal sum representation of the wake function similar in form to the above (4)(8), and in good agreement with the results obtained by the spectral function method discussed below. The latter is preferred because it is theoretically better founded and far simpler computationally. Indeed

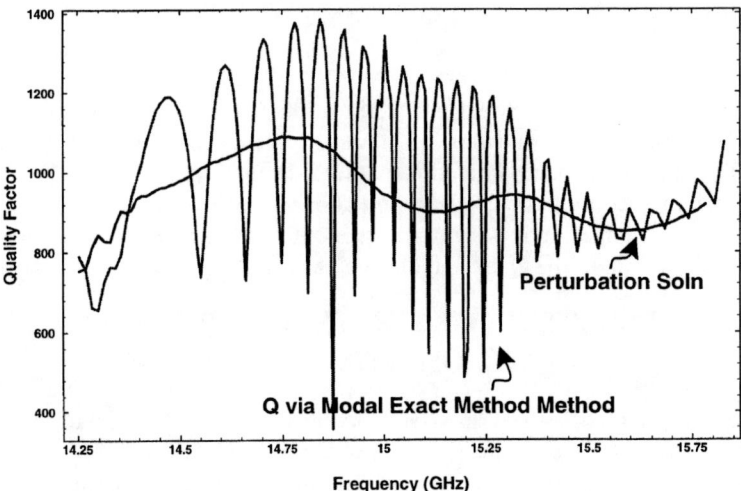

FIGURE 2. Manifold loaded Q. The Q of the DDS is highly oscillatory ranging from 400 to 1400, characteristic of an overcoupled system. The oscillations are centered about the values obtained from peturrbation theory

the numerical difficulties encountered in determining exact eigenmodes and frequencies indicated that it is not a practical method for extensive investigation.

2.2 The Spectral Function Method

In a somewhat more condensed notation eqn. (2.1) may be written:

$$\overline{\overline{H}}\overline{a} - f^{-2}\overline{a} = f^{-2}\overline{B} \tag{2.7}$$

Where the 2N component column vector on the right hand side of eqn. (2.7) represents the drive beam. The wake function is evaluated in terms of the drive beam B, and after some considerable algebraic and matrix manipulations it is evaluated in the form (2.4) where the spectral function is given by:

$$S(f) = 4\pi^{-1} \text{Im}\left\{\sum_{n,m}^{N} \sqrt{K_s^n K_s^m f_s^n f_s^m} \exp[(2\pi jL/c)(f+j\varepsilon)(n-m)]\tilde{H}_{nm}\right\} \tag{2.8}$$

Here L is the cell period, f_s^n and K_s^n, are the synchronous frequency and kick factor respectively, and ε is an infinitesimal quantity (6). The envelope of the wake function is given in terms of the absolute value of the Fourier transform of S(f):

$$W_c(s) = \theta(s) \left| \int_0^\infty S(f) \exp[(2\pi j s / c) f] df \right| \qquad (2.9)$$

In applying this technique S(f) in eqn. (2.8) is sampled in the pass band region for, 2000 or so frequencies and the FFT (Fast Fourier Transform) is evaluated. A typical example of a spectral function in a well damped case is seen in Fig. 6. The individual damped modes appear as oscillations about the smoothed S(f) with the more weakly damped modes being more prominent. The beat-like character of the oscillation echoes the Q oscillations seen in Fig. 2.

3. APPLICATION OF THE THEORY TO DDS 1 AND COMPARISON WITH EXPERIMENT.

The first applications of the spectral function method have been to the DDS 1 structure. The spectral function and wake envelope function for the matched manifold case are given in (6) and (9) The method has been used to study the effect of manifold mismatch on the wake-function. An example is provided by Figs. 3 and 4. Fig. 4 shows that wake quality is much more sensitive to downstream mismatch than to upstream mismatch, and Fig. 3 shows correspondingly that spectral function smoothness is much more degraded for the downstream case.

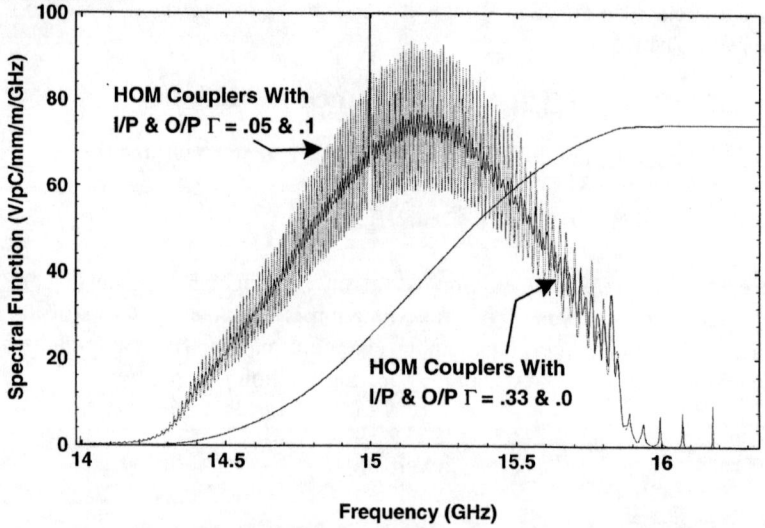

FIGURE 3. Two spectral functions, one incorporating a large mismatch in the input HOM coupler and another including a modest mismatch in the output HOM coupler.

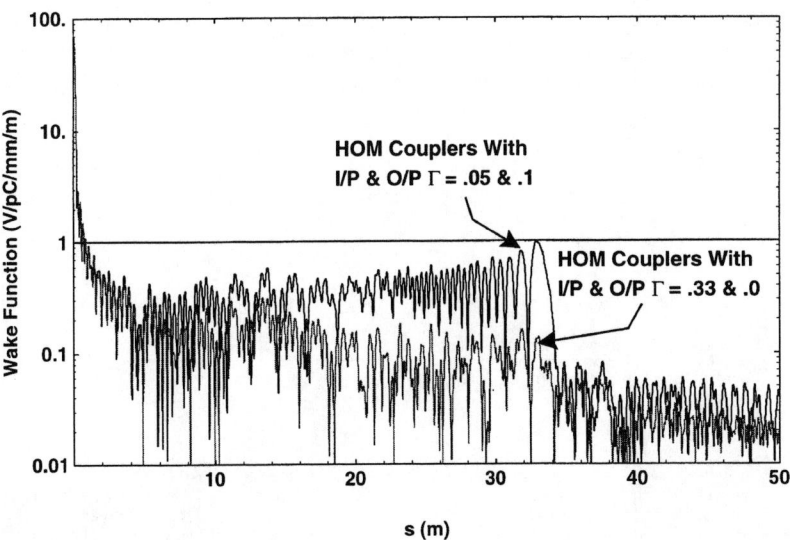

FIGURE 4. Wake function calculated via the spectrum method in which the directional sensitivity of the wake on the input and output HOM couplers is illustrated. The relatively modest mismatch in the output HOM coupler has a significant adverse affect on the wake function (whereas the wake function is relatively insensitive to substantial mismatches in the input HOM coupler)

The manifold terminations actually used in DDS 1 are discussed in (10). The mismatch is substantial and has two principal sources, namely the mitered bend and the rectangular windows (Fig. 1). The downstream manifold is close to cutoff at the low frequency end of the relevant spectral region and is poorly matched there. The windows, which were the only ones available, are poorly matched at the high frequency end.

The resultant spectral function is shown in (6), and the wake envelope function is shown in Fig. 5.

The wake for DDS 1 was measured experimentally in the SLAC LINAC ASSET facility and the results reported in (9). The experimental points shown on the curve in Fig. 5 were obtained from these measurements. The agreement is considered to be reasonably satisfactory given various differences (some planned, some inadvertent) between the DDS model and DDS 1 as fabricated.

4. REDESIGNED DAMPED DETUNED STRUCTURES

In our design of DDS 1 we chose eleven representative sections to obtain frequency-phase pairs from detailed MAFIA simulations and hence obtain ten model parameters (nine circuit parameters plus the cell kick-factors) for each of the eleven sections. Parameters for all sections are subsequently obtained by error function fits and interpolation. A similar procedure may be followed to

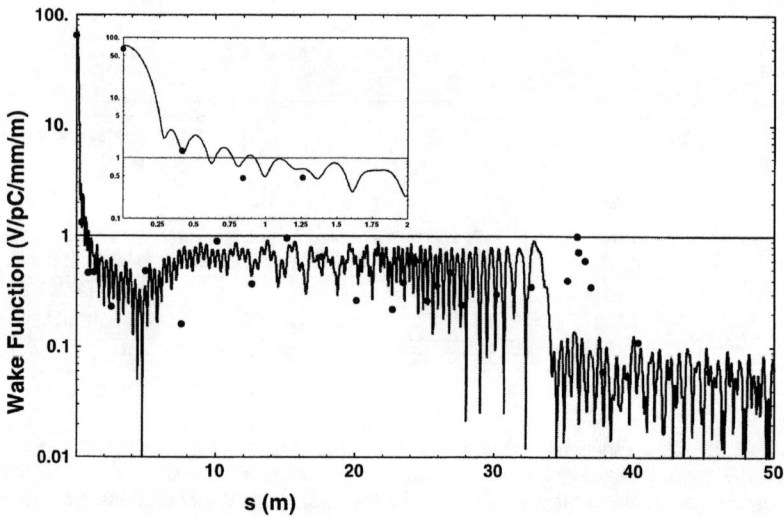

FIGURE 5. Wake function calculated via the spectrum method in which the effect of mismatches in the mitered bend and windows of the higher order mode couplers (HOM) are included. Data determined from the ASSET experiment on the SLC are indicated with dots. The short range wake function is shown inset.

determine the five geometric parameters (i.e., dimensions) (8) for all the sections from those for the original eleven. This is a substantial task for each structure design. However, as we now have all fifteen parameters as a function of synchronous frequency, we can take advantage of this functional dependence to explore new design distributions and to obtain the set of section dimensions which would be needed to realize them. Our new design incorporates a Gaussian kick factor weighted density function with a bandwidth of 5 units of sigma, or 11.25% in units of the central dipole frequency, and the sigma of the Gaussian is 2.25% of the central frequency. Based on our fit parameters we prescribe a smooth uncoupled spectral function, $S_0(f_s)\lambda$ and impose the condition that: $2K(f_s)dn/df = S_0(f_s)\lambda$, where λ is a scale factor to be determined. The upper and lower truncation bounds on the synchronous frequencies are imposed, f_{s1} and f_{sN} and the normalisation condition is obtained:

$$\lambda = N / \int_{f_{s1}}^{f_{sN}} (\tfrac{1}{2} S_0 / K) df_s \qquad (4.1)$$

Then the new synchronous frequencies are determined according to:

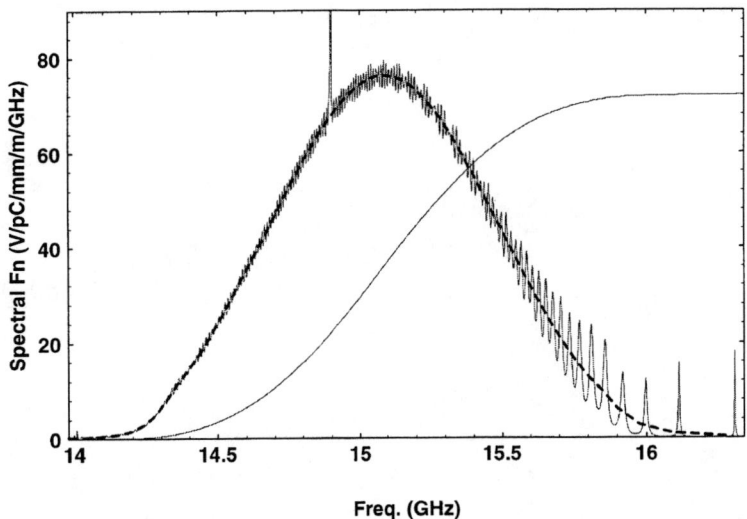

FIGURE 6. Spectral function for proposed DDS with a bandwidth of 5 units of σ width (a frequency bandwidth of 11.25% of the central frequency, and σ is 2.25% of the central frequency). This DDS incorporates perfectly matched HOM couplers terminated in perfectly matched loads. Also shown, dashed and, running through the center of the oscillating spectral function, is the spectral function corresponding to the idealized case (G = 0). The latter spectral function provides a goal in future design strategies.

$$\int_{f_{sn}}^{f_{sn+1}} \tfrac{1}{2K} \lambda S_0 df_s = 1 \tag{4.2}$$

This enables all the cell synchronous frequencies to be determined and hence the new ten parameters are determined. The smoothed spectral function (shown dashed in Fig. 6) is seen to possess the smooth Gaussian form which was the aim of the design procedure.

The spectral function itself (the oscillatory curve in Fig. 6) exhibits the underlying damped mode structure as mentioned previously. The more pronounced oscillations in the upper part of the frequency interval (approximately 15.5 GHz and thereafter) can be reduced by including an enhanced tapered coupling for the last 20 or so cells (11). These more prominent oscillations are responsible for the slight increase of the wake envelope function in the 5 meter region (Fig. 7). The new design has, however, a significantly improved short range wake field, in particular for the first bunch, which sees a wake field an order of magnitude weaker than that seen in DDS 1. An ideally damped design would lead to a spectral function which coincides with the smoothed S(f); the resultant idealized wake envelope function is also shown in Fig. 7.

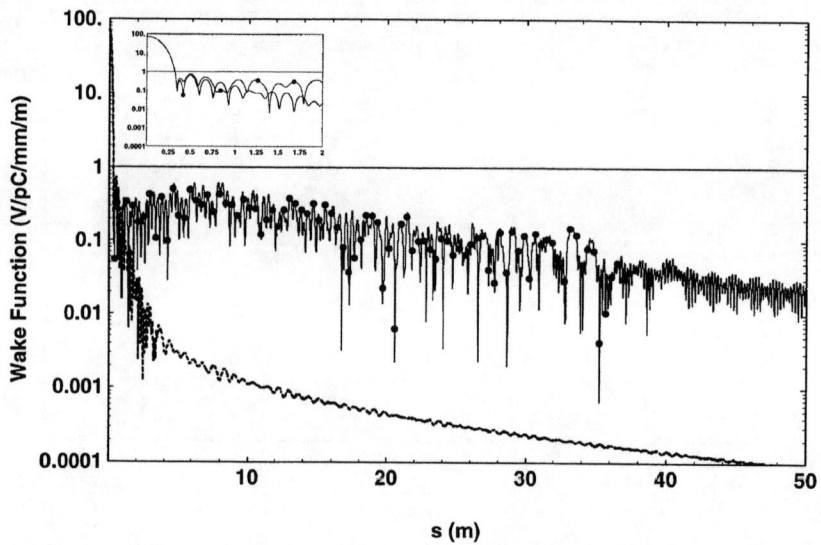

FIGURE 7. Wake envelope function generated by the spectral function of Fig. 6. The dots indicate the positioning of bunches spaced at 1.4ns intervals (not experimental points). The lower curve is the idealized wake function generated by the dashed curve of Fig. 6.

5. ACKNOWLEDGEMENTS

This work is supported by Department of Energy grant number DE-FG03-93ER40759[‡] and DE-AC03-76SF00515[†]. We have benefited greatly from discussions at the weekly structures meeting at SLAC, where these results were first presented and thank all members of the group.

6. REFERENCES

1. Neal, R.B and Panofsky, W.H., *Science*, **152,** 1353 (1966)
2. Bane, K.L.F et al, EPAC94, London, England, 1994; SLAC-PUB- 6581
3. Thompson, K.A., private communication, 1996.
4. Jones, R.M., Ko, K., Kroll, N.M. , Miller, R.H., & Thompson, K.A, EPAC96, Sitges, Spain, June 10-14, 1996; SLAC-PUB-7187 (http://libnext.slac.stanford.edu/slacpubs/7000/slac-pub-7187.ps.Z)
5. Bane, K.L.F, and Gluckstern, *Part. Accel.*, **42,** 123 (1993); SLAC-PUB-5783
6. Jones, R.M., Ko, K., Kroll, N.M., & Miller, R.H., Linac96; SLAC-PUB-7287
7. Jones, R.M., Kroll, N.M. & Miller R.H., To be submitted to *Phys. Rev. E*, 1997
8. Ko, K. et al, PAC95, Dallas, Texas; SLAC-PUB-95-6844
9. Miller, R.H. et al, Linac96, Geneva, Switzerland, 1996; SLAC-PUB -7288
10. Seidel, M., et al, Linac96, Geneva, Switzerland,, 1996; SLAC-PUB-7289
11. Jones, R.M., Kroll, N.M. & Miller, R.H., to be submitted to PAC 97

Analysis And Application Of Manifold Radiation In DDS 1: First Experiences

R.M. Jones[†‡], N.M. Kroll[†‡], M. Seidel[†], C. Adolphsen[†],
K.L.F. Bane[†], W.R. Fowkes[†], K. Ko[†],
R.H. Miller[†], R.D. Ruth[†], and J.W. Wang[†]

*† Stanford Linear Accelerator Center, M/S 26,
P.O Box 4349, Stanford, CA 94309
‡ University of California, San Diego,
La Jolla, CA 92093-0319*

Abstract. The frequency spectrum of the manifold radiation resulting from a uniformly displaced drive beam in DDS 1 has been calculated utilising the spectral function theory and compared with observations in the ASSET experiment. They are found to be in good qualitative agreement. The bulk of the radiation, both measured and computed, is found to emerge from the downstream end of the manifolds. This is in conflict with perturbation theory predictions, and is thus one more indication of the unreliability of perturbation theory for estimating manifold effects for DDS 1. Beam centering as a function of frequency is accomplished by minimizing manifold radiation within a narrow frequency band. This procedure provides information on structure misalignment. While misalignments in DDS 1 degraded the precision of beam centering which could be achieved by minimizing manifold radiation, the results obtained are considered to be encouraging.

1. INTRODUCTION

The damping manifolds of the SLAC Damped Detuned Structure (DDS) (1) are intended to perform two functions in addition to the damping of the long range wakefield. One is to facilitate pumping and the other is to provide information about alignment and beam position by monitoring the radiation which emerges from them. This paper is devoted to preliminary results obtained in connection with the latter function. In the next (second) section we discuss the equivalent circuit theory of manifold radiation and compare the results with microwave measurements carried out during the ASSET experiment. In the third section we discuss the use of manifold radiation to detect structure misalignment and to adjust beam parameters so as to minimize the transverse wake.

2. EQUIVALENT CIRCUIT THEORY CALCULATION OF MANIFOLD RADIATION

The equivalent circuit used to discuss the DDS consists of a sequence of sections each one of which corresponds to a cell and its contiguous section of manifold. For the purpose of discussion we write the equations in matrix form as follows:

$$\begin{pmatrix} \hat{H} & H_x^t \\ H_x & H - GR^{-1}G \end{pmatrix} \begin{pmatrix} \hat{a} \\ a \end{pmatrix} - f^{-2} \begin{pmatrix} \hat{a} \\ a \end{pmatrix} = f^{-2} \begin{pmatrix} xB \\ 0 \end{pmatrix} \quad (2.1)$$

Each component of the column matrices is an N component vector with $\hat{a}$ and a representing the TM and TE components of the cell excitation respectively. The square matrix on the right is a symmetric 2N by 2N matrix requiring nine circuit parameters per section for its description and whose elements are themselves N by N matrices. The N component column vector B on the RHS represents the coupling of the beam to the manifold. It is generally referred to as the cell kick factor and is normally defined "per unit beam displacement". Since we wish to consider cell dependent displacements we have represented the beam displacement at the n'th cell by the n'th diagonal element of the diagonal N by N matrix x. For all of the applications considered here the number of sections N is 206. For further details of the circuit model see (1) and (2).

In order to compute the power spectrum emerging from the manifolds we must first determine the manifold amplitude vector A defined in (2). We compute this quantity as follows: The TE amplitude vector a is obtained by solving Eq. 2.1 for the amplitude vectors in terms of the source vector xB (to which it is evidently linearly related.) A is in turn linearly related to a via the relation (2):

$$A = R^{-1}a \quad (2.2)$$

and thus also to xB. What is actually measured is the power spectrum bled off from the output of the manifolds. One can show (3) that the power spectrum of the radiation emerging from the manifolds is given by:

$$P = (Q\delta x)^2 LF_p \frac{f}{2\pi} \operatorname{Sin}\phi A^\dagger A \quad (2.3)$$

where ϕ is the manifold phase advance function, L the structure length, δx the beam offset, Q the beam charge, F_p the pulse repetition rate and f the frequency component. This quantity, eqn (2.3), is shown in Fig. 1, both for a uniform offset of 1.3 mm for all cells and for a uniform displacement of 1.3 mm confined to cells 19 to 186. The uniform offset curves are in good qualitative agreement with the measurements and exhibit the large asymmetry observed between the upstream and downstream radiation. The sharp fall-offs in the spectra due to the limited range displacement as well as the fact that the central amplitudes for the full and limited range spectra are identical demonstrates a local or quasi local character of the excitation. The synchronous frequencies of cells 19 and 186 are 14.52 GHz and 15.55 GHz respectively and are identified by the dashed vertical lines in Fig. We estimate form Fig. 1. that the excitation frequency is shifted upward from the synchronous frequency by approximately 40 MHz at cell 19 and approximately 100 MHz at cell 186.

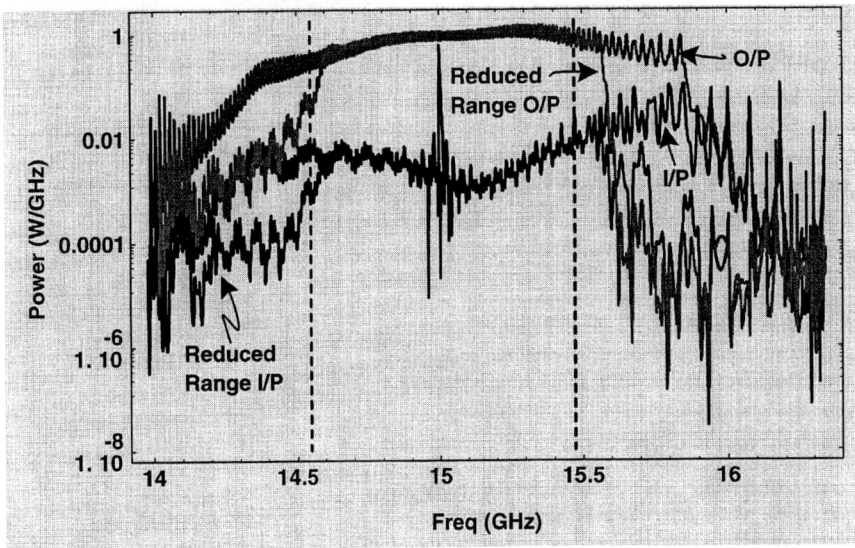

FIGURE 1. Power spectrum generated by a uniformly displaced beam calculated for the manifold input and output HOM couplers. Also shown are the corresponding power spectra when the displacement is limited to cells 19 to 186.

It appears from an examination of the modes themselves that excitation is occurring in the synchronous region of the modes, but that the synchronous points of the detuned modes are shifted from the cells whose synchronous frequency equals the mode frequency. As we shall see in the next section, a localized connection between frequency and excitation point has important practical consequences. The results obtained so far provide strong evidence that such a connection exists and indicate that the methodology applied above can determine it throughout the structure. It appears, however, that there will be a significant shift from the localization relation obtained from the cell synchronous frequencies. Another issue that we plan to investigate is the connection between the width of the localized displacement region and the frequency width of the power spectrum it generates.

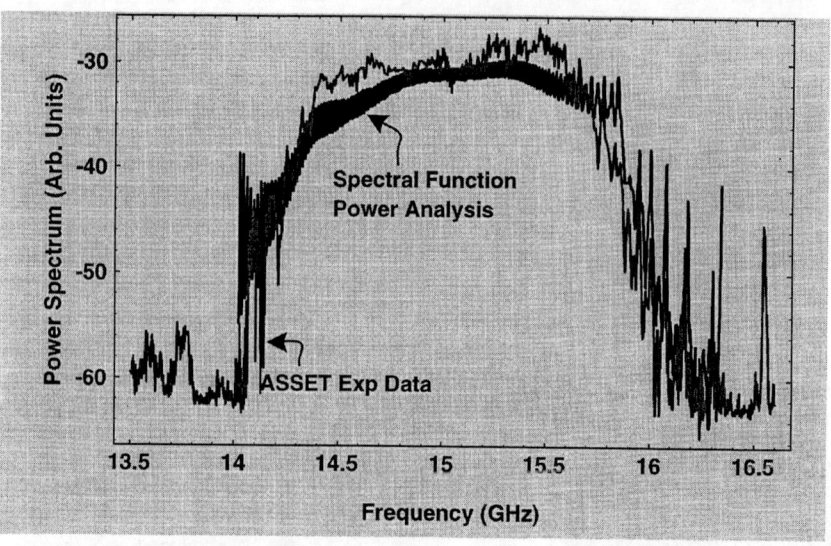

FIGURE 2. Computed and measured downstream power spectra compared for a uniformly displaced beam. Relative amplitudes (comparable absolute values are not known) are adjusted to facilitate comparison.

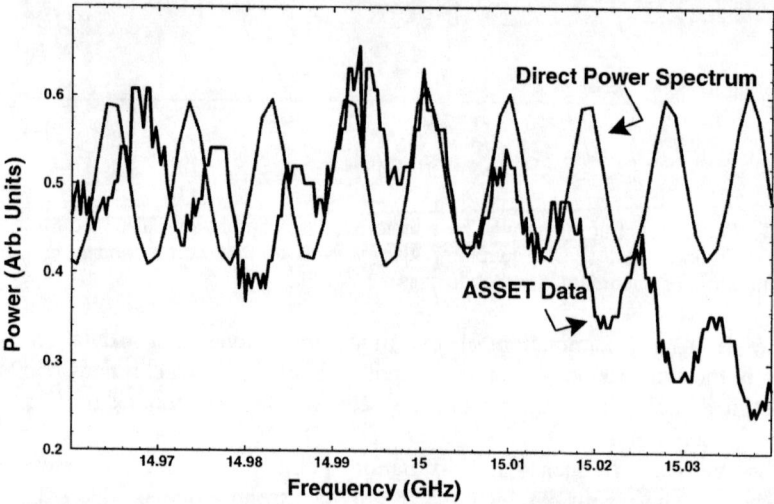

FIGURE 3. Computed and measured downstream power spectra for a uniformly displaced beam compared over a limited frequency band. A 1 MHz spectrum analyzer bandwidth setting facilitates comparison of the oscillatory structure.

Figure 3 shows a comparison between the computed and observed power spectrum (measured a 1 MHz bandwidth) in the region where maximum mode density is expected. One notes that the mean separation of the oscillatory peaks in the observed spectrum is approximately 10% less than that in the computed one. This result can be related to a discrepancy between the observed and computed wakes (ref (1) Fig. 6). Due to the mismatch, the computed wake shows a notable reappearance peak at a distance of 33 m. This reappearance peak was also observed in the measurements but at 36 m. The distance at which this peak is expected is given by c/df where df is a mean minimum mode separation. This relation implies 8.8 mode spacings for the computed manifold spectrum and 9.6 for the observed one in the 0.8 GHz frequency interval covered in the Fig. 3 curves.

3. APPLICATION OF MANIFOLD RADIATION TO THE DETECTION OF STRUCTURE MISALIGNMENT AND TO BEAM POSITION MONITORING

A principle objective of the study of manifold radiation was to determine how well one could minimize the wake by minimizing manifold radiation, and the precision to which one could center the beam in the structure. Unfortunately, a minor accident occurred during the assembly of DDS 1 which resulted in significant misalignment within the structure. This compromised to some extent the wake minimization program, but it provided an opportunity to observe the effect of misalignment on manifold radiation.

Instead of varying the beam offset to minimize total manifold power one can vary the beam offset to minimize manifold power at a particular frequency, thus determining a power minimizing offset $x_0(f)$ as a function of frequency as reported in (4). Mechanical measurement of the misalignment of DDS 1 as a function of cell number provided the opportunity to predict $x_0(f)$ from Eq. (2.2). For this purpose we assume that eqn. (2.1) still holds if the n'th element x_n of the diagonal offset matrix x is written $x_n = X_n + x_0(f)$, where X_n represents the mechanically measured cell offset. Then writing eqn (2.1) in the schematic form

$$P(f) = \sum_{m,n} M_m^*(f) M_n(f)(X_m + x_0(f))(X_n + x_0(f)) \qquad (3.1)$$

one easily sees that P(f) is minimized if

$$x_0(f) = -\sum_{m,n} \text{Re}\{M_m^*(f) M_n(f)\} X_n / \sum_{m,n} M_m^*(f) M_n(f) \qquad (3.2)$$

The experimentally determined and computed $x_0(f)$ are compared in Fig. (4), showing good qualitative agreement. In an attempt to visually relate the mechanical offset data to the computed $x_0(f)$, we plot $-x_0(f_n)$ and the mechanical offset as a function of cell number n in Fig. (5). Here, as suggested by the quasi-locality noted in the previous section, we take fn = fsn + Cn, where f_{sn} is the synchronous frequency of the n'th cell and the constant C is chosen to yield the frequency shifts at cells 19 and 186 previously noted. The agreement is seen to be quite satisfactory. In connection with the poorer agreement shown in Fig. 4 it should be noted that there is some evidence based upon measurements performed at different times that the mechanical offsets shifted in the handling of the structure, so that there is some possibility that the mechanical offsets were different from those shown in Fig. 5 when the Fig. 4 measurements were made. Be that as it may, the gross feature of the mechanical measurement, the 60 μm shift at cell 45, was a constant feature of the measurements and is clearly echoed in both the measured and computed values of $x_0(f)$.

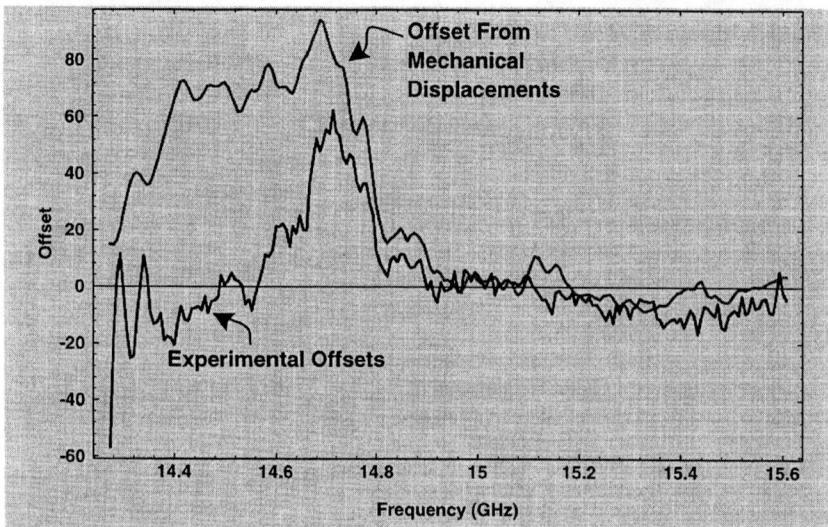

FIGURE 4:. Comparison of measured and computed manifold power spectrum minimizing offsets as a function of frequency.

Given the misalignment discussed above, there is clearly no beam position which zero's the manifold radiation at all frequencies. A number of techniques and criteria that one can use to fix the beam position are discussed in (4) and will not be repeated here. Analysis of the measured residual wake associated with these criteria is in progress and will be reported elsewhere. Because the residual wakes are small, measurements are confined to the wake at short distances, before destructive interference has developed. It is a straightforward matter to compute this wake for the measured offsets and determine the compensating offset needed to negate it. The manifold spectrum associated with this arrangement can also be computed. Such computations may assist in developing beam positioning criteria for minimizing or zeroing the wake.

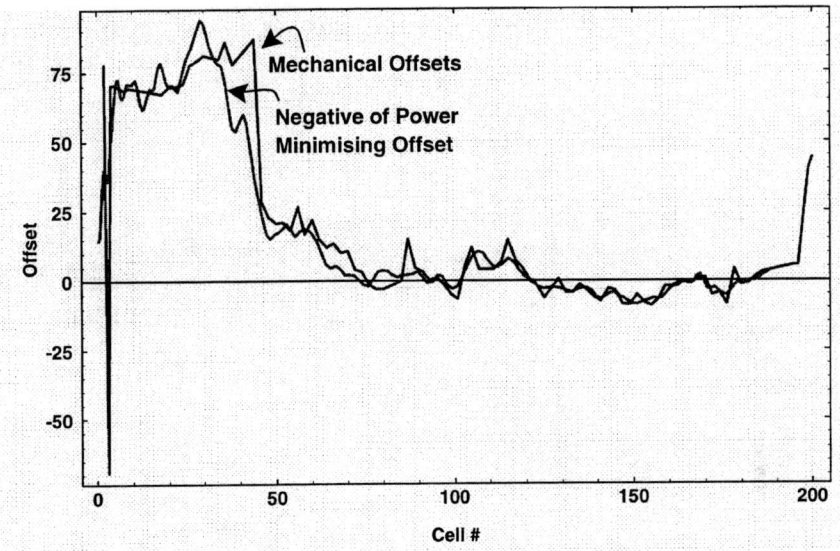

FIGURE 5. Mechanically measured offsets as a function of cell number. The computed power minimizing offsets at a set of cell associated frequencies (as described in the text) are also plotted for comparison.

4. ACKNOWLEDGEMENTS

This work is supported by Department of Energy grant number DE-FG03-93ER40759[‡] and DE-AC03-76SF00515[†]. We have benefited greatly from discussions at the weekly structures meeting at SLAC, where these results were first presented and thank all members of the group.

5. REFERENCES

1 Jones, R.M., Kroll, N.M, Adolphsen, C., Bane, K.L.F., Fowkes, W.R., Ko, K, Miller, R.H., Ruth, R.D, Seidel, M. and. Wang, J.W, "Recent Results & Plans For The Future On SLAC Damped Detuned Structures (DDS)". This issue; also SLAC-Pub-7387

2 Jones, R.M., Ko, K., Kroll, N.M. , Miller, R.H., & Thompson, K.A, "Equivalent Circuit Analysis of the SLAC Damped Detuned Structure", EPAC96, Sitges, Spain, June 10-14, 1996; SLAC-PUB-7187 (http://libnext.slac.stanford.edu/slacpubs/7000/slac-pub-7187.ps.Z)

3 Jones, R.M., Kroll, N.M. & Miller R.H., To be submitted to *Phys. Rev. E*, 1997
4. Seidel, M., et al, "Microwave Analysis of the Damped Detuned Accelerator Structure", Linac96, Geneva, Switzerland,, 1996; SLAC-PUB-7289

Cold Test Results of a Standing Wave Muffin-tin Structure at X-band

P.J. Chou, S.M. Hanna, H. Henke, A. Menegat, R.H. Siemann, and D. Whittum

Stanford Linear Accelerator Center
P.O. Box 4349, Stanford, CA 94309

Abstract A muffin-tin structure is chosen to study high gradient acceleration in the millimeter wavelength range. In order to understand the electromagnetic field characteristics, a standing wave structure operating at a frequency around 11.4 GHz was built. Cold test measurements were performed and results are presented. Comparisons with theoretical predictions based on computer simulation are shown.

1. INTRODUCTION

The muffin-tin structure[1] is chosen for development of high gradient (≥ 1 GV/m) acceleration at SLAC. The operating frequency is tentatively chosen around 32 times SLAC frequency (32× 2.856 GHz). The size of accelerating structures in this frequency range is at the millimeter scale. RF measurements of accelerating structures in this frequency range have never been performed. Since it is important to characterize the electromagnetic field of mm-wave accelerating structures, it is necessary to first perform RF measurements at lower frequencies. Those can provide useful information about the field configurations of muffin-tin structures. A standing wave muffin-tin structure was built at an operating frequency (4× 2.856 GHz) in the X-band. Measurement results for the standing wave muffin-tin structure at X-band are presented. Conceptual descriptions of RF measurements in the mm-wave range are described.

The schematic sketch of the X-band muffin-tin standing wave structure is depicted in Fig. 1. The mechanical design of the muffin-tin slab is given in Fig. 2. There are five full cells and two half cells in the muffin-tin slab. This design is chosen to imitate the field configurations of a $2\pi/3$ mode traveling wave structure[1]. At each end of the muffin-tin slab there is a circular hole of diameter 0.086" drilled through the center. These holes are used to insert RF probes.

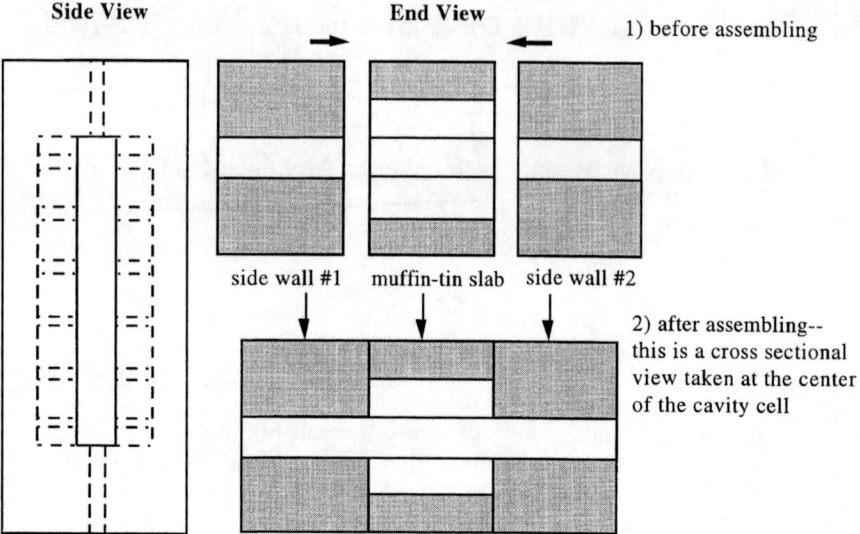

The structure has five full cells and two half cells.

Fig. 1: The schematic sketch of the muffin-tin standing wave structure.

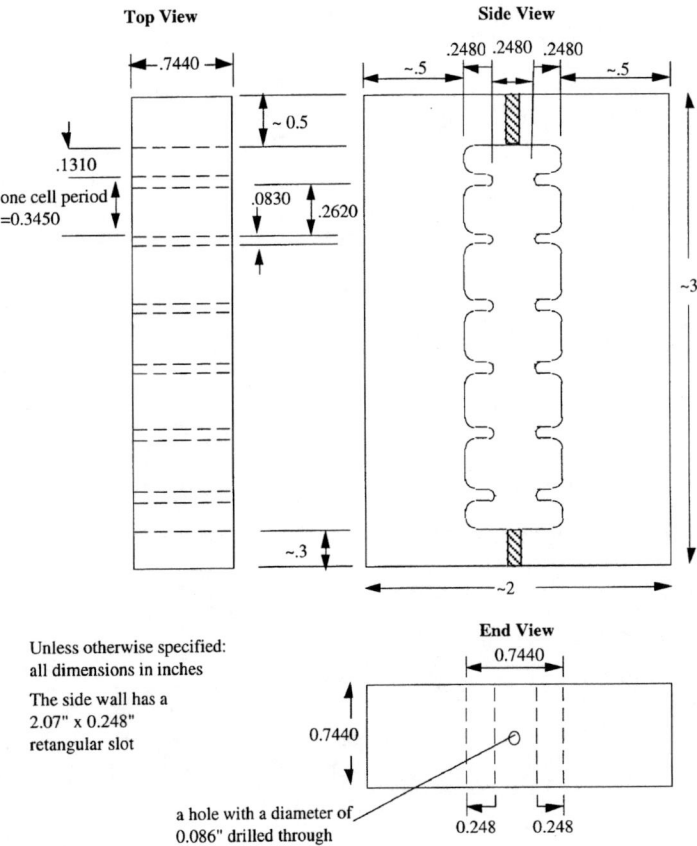

Fig. 2: The mechanical design of the muffin-tin slab as used for low power RF measurements.

2. MEASUREMENTS

An E-probe is inserted through a hole for signal detection as shown in Fig. 3. Reflection measurements were first performed to scan the contents of various modes in the frequency domain.

The reflected signals are given in Figs. 4, 5, and 6. Figure 4 shows the fundamental mode. Each cavity cell can be viewed as an harmonic oscillator. Since there are 7 cells in the structure coupled together through iris openings. Therefore, one expects to observe 7 resonant modes. Each spectral line in Fig. 4 corresponds to one mode. The measured frequency for the $2\pi/3$ mode used for particle acceleration is 11.6475 GHz; the calculated value from MAFIA[2] simulations is 11.5619 GHz.

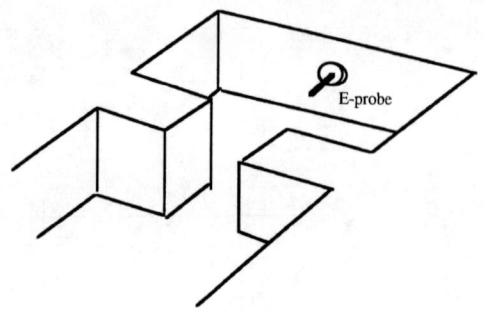

Fig. 3: E-probe placement for signal detection.

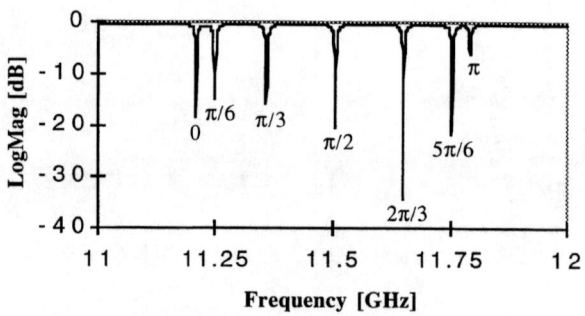

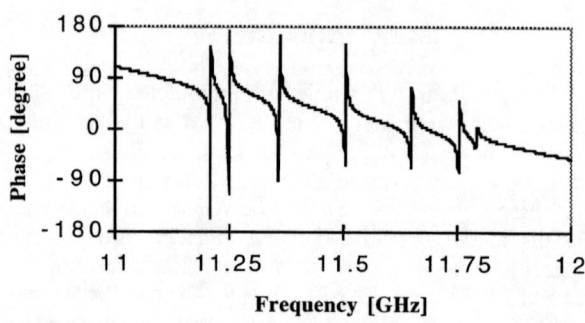

Fig. 4: The measured frequency spectrum of reflected signal from 11 to 12 Ghz.

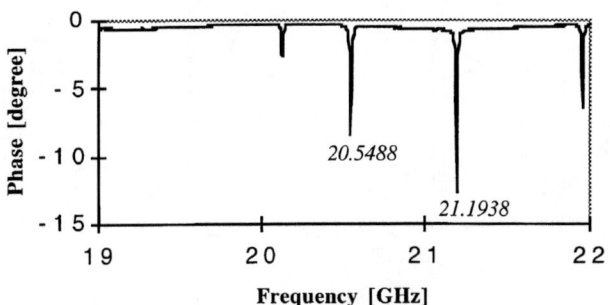

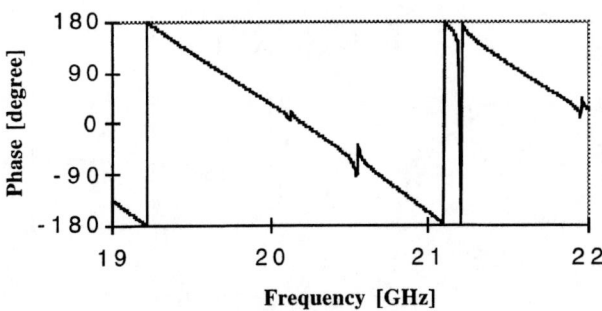

Fig. 5: The measured frequency spectrum of reflected signal from 19 to 22 GHz.

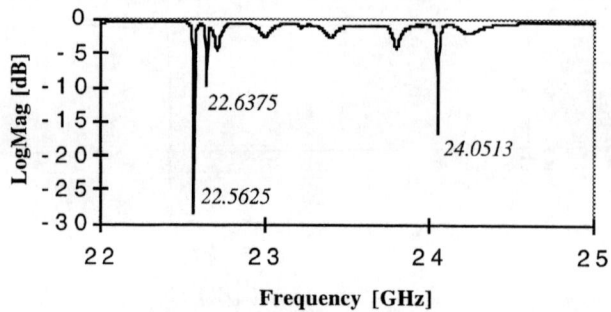

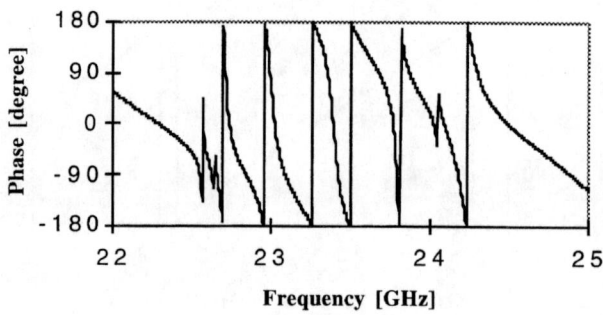

Fig. 6: The measured frequency spectrum of reflected signal from 22 to 25 GHz.

The axial field profile was measured by performing bead pull measurements[3, 4] and recording the shift of resonant frequency for the $2\pi/3$ mode. By using a small cylindrical bead made of metal and moving it along the longitudinal axis without transverse offsets, the normalized amplitude of the accelerating field can be measured. The experimental setup is depicted in Fig. 7. The X-band muffin-tin structure has a longitudinal length of 2.07". The measured frequency shift for the $2\pi/3$ mode as a function of the bead position along the longitudinal axis is given in Fig. 8.

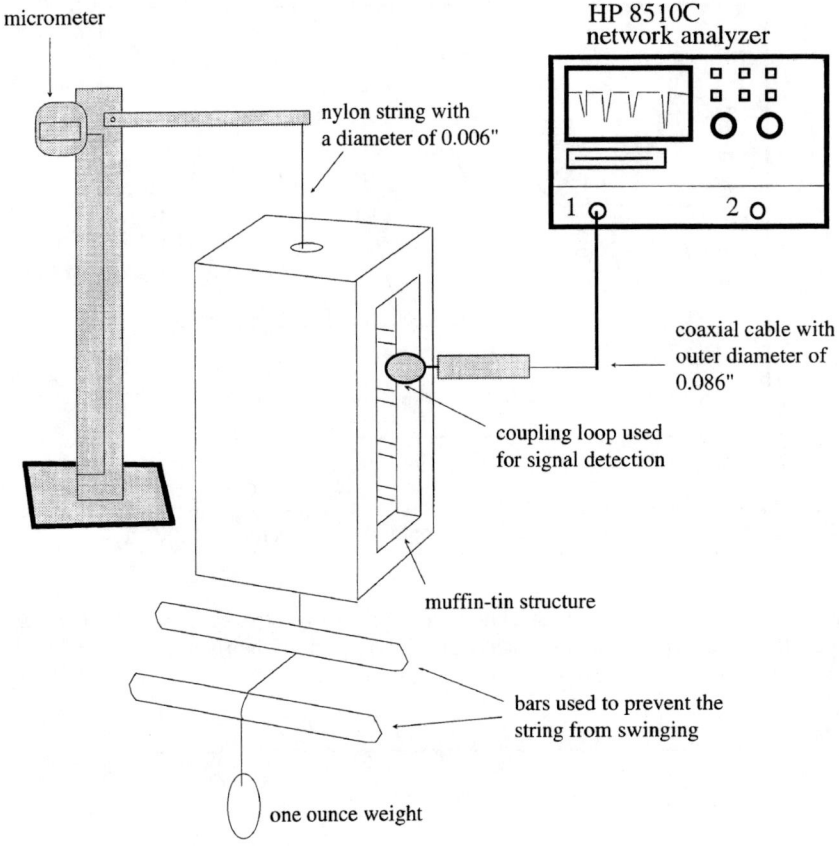

Fig. 7: The experimental setup for bead pull measurements.

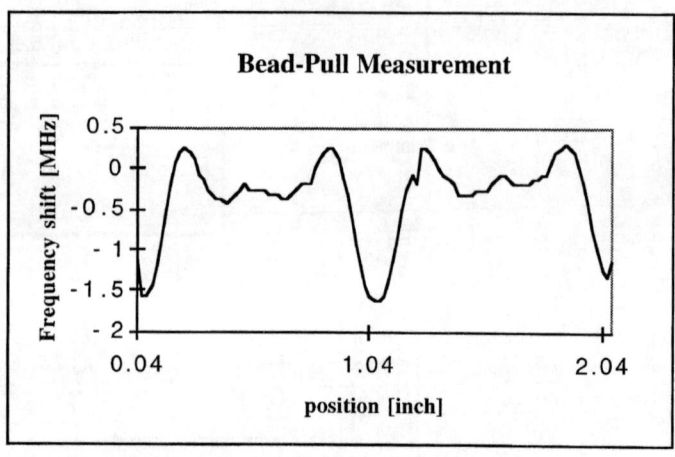

Fig. 8: The measured frequency shift for the $2\pi/3$ mode as a function of the bead position along the longitudinal axis. The length of the structure is 2.07".

For conceptual illustration, a simulated field profile for the $2\pi/3$ mode along the longitudinal axis of a structure with 2 oscillation periods is given in Fig. 9. Only the first three space harmonics are used in the calculation.

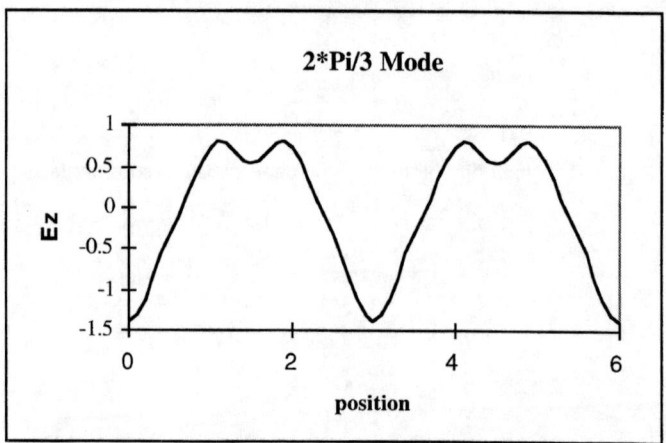

Fig. 9: The simulated field profile along the longitudinal axis of a structure with 2 oscillation periods. The units are arbitrary.

The measured results of loaded Q-factor are given in Table 1 and the calculated results for 2π/3 mode from MAFIA[2] simulation are given in Table 2.

Table 1: The measured results of frequency and Q-factor for various modes.

Frequency [GHz]	Loaded Q-factor
11.6475	1658
22.5625	1336
22.6375	2317
24.0505	1253

Table 2: The calculated parameters for the 2π/3 mode of X-band muffin-tin structure.

Frequency [GHz]	Unloaded Q-factor	R/Q [Ω]	Shunt Impedance [MΩ]
11.5619	6817	127	0.87
22.0781	10103	9.3	0.09
23.6769	11295	7.4	0.08

3. MEASUREMENT METHODS IN THE MM-WAVE RANGE

By using the nonresonant perturbation measurement[5, 6, 7], one can measure the amplitude and phase of the accelerating field. The information needed is the difference between the unperturbed input reflection coefficient S_{11}^u and the perturbed input reflection coefficient S_{11}^p with a bead inside the structure. The mathematical expression of this statement is as follows:

$$\Delta S_{11}(z) = S_{11}^p(z) - S_{11}^u(z) = A|E(z)|^2 e^{-2j\theta_E(z)}$$

where z is the longitudinal coordinate, $\theta_E(z)$ is the phase advance along the longitudinal direction of the structure, E(z) is the accelerating field, and A is a constant which depends on the input power and the bead characteristics. Since a vector network analyzer for millimeter wavelengths is quite expensive, an alternative option depicted in Fig. 10 is being considered for bead pull measurements in the millimeter wavelength range.

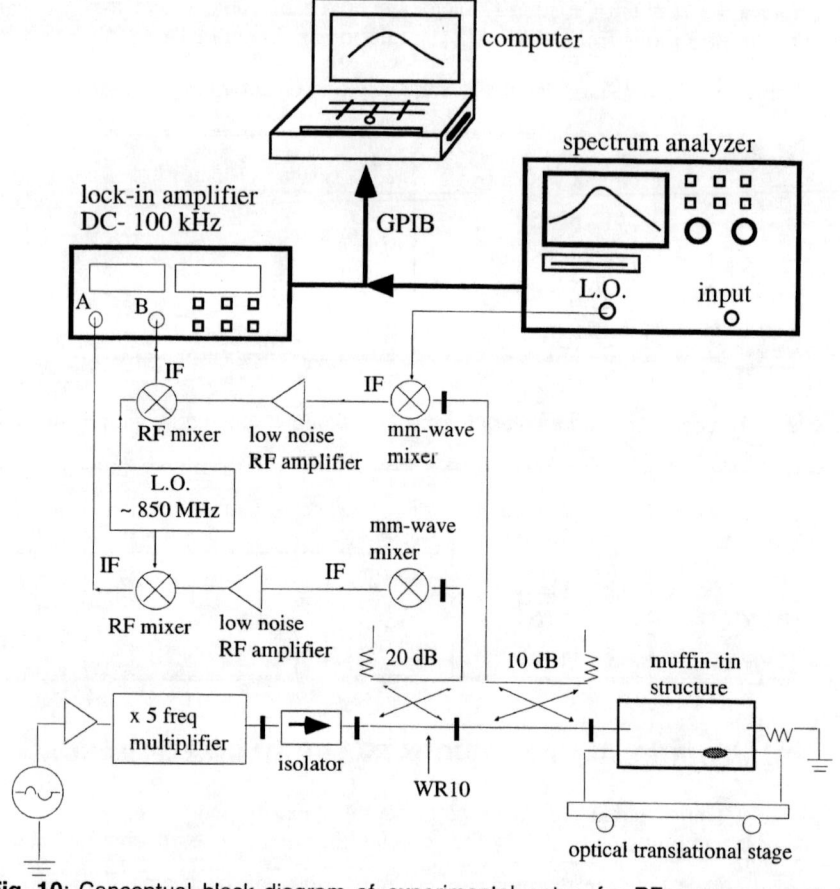

Fig. 10: Conceptual block diagram of experimental setup for RF measurements in the millimeter wavelength range.

The operating frequency of the mm-wave muffin-tin structure is around 92 GHz. A HP8673 frequency synthesizer and a 5 times frequency multiplier are used to generated the 92 GHz source signal. The incident signal is monitored through a 20 dB WR10 waveguide coupler. There are two stages of signal mixing occurring after the WR10 coupler. The signal is mixed down to an intermediate frequency around 850 MHz by a HP11970W mm-wave mixer which uses the 18th harmonic of the spectrum analyzer local oscillator. The average conversion loss of the mixer is about 42 dB. The output signal from the mm-wave mixer, after the first mixing stage, goes through a low noise, narrow-band, RF amplifier (center frequency $\approx$ 850 MHz, bandwidth $\leq$ 20 MHz). After the low noise RF amplifier, the signal goes to the second mixing stage which mixes the signal down to the audio frequency range ($\leq$ 100 kHz). A lock-in amplifier is used to detect the amplitude and phase of the input signal. The reflected signal follows the same procedures for signal processing. The lock-in amplifier takes the input signal from port A as the reference, then it gives the amplitude and phase of the signal from port B with respect to the reference as the output. Therefore, one can use the source signal as the reference and measure the amplitude and phase of reflected signals with and without the perturbing object in the structure.

To perturb the field inside the mm-wave muffin-tin structure, a small bead is needed. Hollow metallic cylinders with diameters ranging from 25 to 127 µm with an approximate length of 500 µm have been fabricated by sputtering aluminum onto silica optical fibers and nylon surgical thread for bead pull measurements[3]. Further miniturization of metallic beads can be achieved by using this sputtering technique. Because the diameter of the string used in bead pull measurements for the mm-wave muffin-tin structure must not be too large ($\leq$ 25 µm), the string will be fixed on two supporting arms which will be mounted on an optical translational stage. An operator can then move the thin string by shifting the translational stage which provides one micron or better accuracy. This arrangement also avoids breaking the thin string used to support the metallic bead.

Since the source signal will contain different harmonics of the frequency multiplier, care needs to be taken in order to remove them in the signal processing. The nearest spurious harmonic is 22 MHz away from the desired signal (Fig. 11).

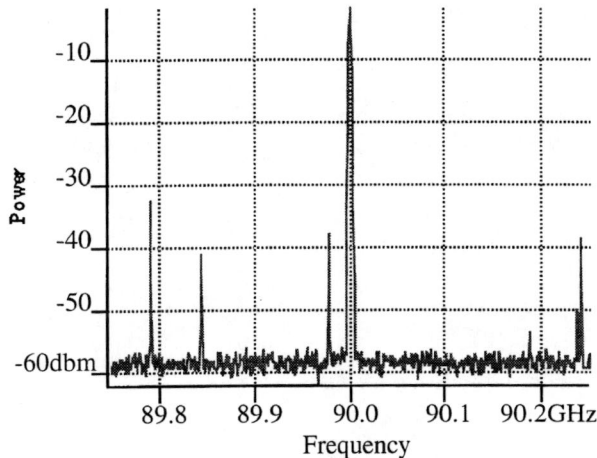

Fig. 11: The spectrum of output signal from the frequency multiplier.

4. SUMMARY

Results of the measurements of the X-band muffin-tin structure have been presented, however more work needs to be done to improve the accuracy of these preliminary experimental results. An alternative scheme for performing RF measurements for the mm-wave muffin-tin structure without the use of a vector network analyzer has been presented. This work will help develop a better understanding of the properties of mm-wave muffin-tin structures.

ACKNOWLEDGMENTS

The authors would like to thank C.G. Ng and R.H. Miller for helpful discussions. This work is supported by DOE contract DE- AC03- 76SF00515.

REFERENCES

[1] R.B. Neal et al. (eds.), *The Stanford Two-Mile Accelerator* (Benjamin, New York, 1967), p. 127.
[2] The MAFIA collaboration, F. Ebeling et al., *MAFIA User Guide*, 1992.
[3] P.J. Matthews et al., *IEEE Trans. Microwave Theory Tech.*, vol. MTT- 44 (1996), p. 1401.
[4] L.C. Maier and J.C. Slater, *J. Appl. Phys.*, vol. 23 (1952), p.68.
[5] C.W. Steele, *IEEE Trans. Microwave Theory Tech.*, vol. MTT- 14 (1966), p. 70.
[6] K.B. Mallory and R.H. Miller, *IEEE Trans. Microwave Theory Tech.*, vol. MTT- 14 (1966), p. 99.
[7] S.M. Hanna et al. in *Proceedings of 1996 European Particle Accelerator Conference* (to be published).

Planar Millimeter-Wave RF Structures

H. Henke
Technische Universitaet, EN-2
D-10587 Berlin, Germany

ABSTRACT

Planar RF-structures have been proposed for operating frequencies between 30 and 240 GHz. The structures are two-dimensional and doublesided (open) and therefore suited for fabrication by lithography. At least two technologies are available: Deep X-ray lithography with plastic resists (thickness < 0.8 mm) and UV-lithography with doped glass resists (thickness up to several mm). The paper presents structure developments and proposals for different applications. The status of fabrication is discussed and first ideas of a full engineering are given.

INTRODUCTION

In the light of the fast development of micromechanic technology millimeter-wave RF structures have been proposed [1-4]. The structures are planar and doublesided and, therefore, ideally suited for fabrication by lithography. The frequency of operation can be in the range between 30 GHz and 240 GHz. The basic structure is a muffin-tin consisting of two metallic slabs supporting match-box like recesses which form cavities when the slabs are facing each other, Fig. 1.

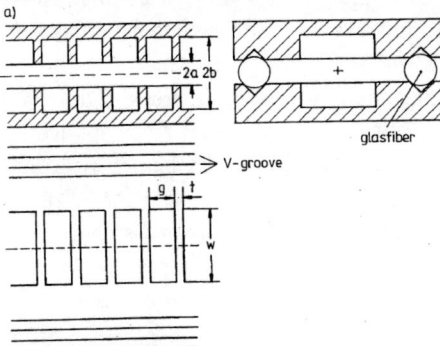

Fig. 1. Basic arrangement of planar structures.

Once the principle of planar structures fabricated by lithography is accepted there is a large variety of different geometries and structures possible. Adding more and more complexity to the structures is no longer limited by cost arguments as usual but rather by the imagination and skill of the designer. The paper presents principal solutions for constant impedance and constant gradient traveling wave structures, for structures with inherent Alternating-Phase-Focusing (APF), for flat-field structures, for choke-mode structures with very low Q higher-order modes (HOM), for short-wavelength RF undulators and for side-coupled standing-wave structures.

Although the relative dimensional tolerances increase with the square root of the frequency, the absolute error is of the order of 1 µm or even below at around 100 GHz. Assuming that tuning of the structure is extremely difficult we obviously need a technology which is able to keep this precision at least within one slab. The overall tuning can then be done by controlling the cooling temperature and/or by squeezing the two slabs together. Up to today the only technology which provides the required precision is deep X-ray lithography with subsequent etching and electroplating (LIGA) [5]. The fabrication of two-dimensional structures with perpendicular walls, good surface quality and submicron precision has been demonstrated. Problem areas are the fotoresist, PMMA plastic, the long exposure time and the achievable depth of the structures which today is limited to ≤ 800 µm, i.e. to structures with operating frequencies ≥ 80 GHz. An alternative solution is UV-lithography (UVL) with fotosensitive glass [6]. It is better suited for deeper structures, that means lower operating frequencies, avoids the problems with organic materials and is appreciably cheaper. On the other hand, due to the graininess of the exposed glass, it has a purer surface quality and it is more difficult to obtain submicron tolerances.

Both technologies are being pursued. The masks for the illumination have been fabricated and first show-pieces are made. In parallel, power couplers are being developed and first efforts are directed towards the alignment of the two slabs and the engineering of a stand-alone structure unit.

Planar structures have more geometric degrees of freedom than axis-symmetric structures. This can be of advantage in particular applications. It can be used to reduce wakefield effects, to accelerate flat beams with large aspect ratio and a very low emittance in one plane, it can be used for RF-focusing, for accelerating sheet beams with several wavelengths width and so on.

Due to the rectangular geometry with plane interfaces between different subregions, planar structures are also easier to calculate. They have been analysed with a mode matching technique [7] where the fields were expanded in series of eigenmodes in the cavity regions and in Fourier integral representations in the aperture region. In that way the RF parameters and HOM's were precisely evaluated. For numerical analysis a finite difference calculus is the obvious choice. MAFIA [8] has been used extensively and recently a new code GdfidL [9] takes advantage of the special geometry in order to reduce memory requirements and CPU time.

TRAVELING-WAVE (TW) CONSTANT-IMPEDANCE (CI) STRUCTURES

This is probably the simplest possible structure and is the one shown in Fig. 1. It can be viewed as an iris loaded groove guide. This basic arrangement has several advantages:
- It is a simple geometry and perfectly suited for lithography.
- It is easy to cool at top and bottom.
- The RF fields decay exponentially in the side-openings thus allowing for mechanical supports after one wavelength.
- The vertically polarized break-up modes are heavily damped.
- The side-openings provide sufficient conduction for vacuum pumping.

The structure has been studied in ref. [1]. The basic RF parameters are repeated in table 1.

Table 1

Geometrical and RF parameters for a $2\pi/3$ traveling wave mode in a 120 GHz muffin-tin.

a = 0.3 mm	b = 0.9 mm	w = 1.8 mm
g = 0.633 mm	t = 0.2 mm	d = 0.8 mm
Q_0 = 2160 for Cu		r_0/Q = 144.9 kΩ/m
r_0 = 312 MΩ/m		k=0.0475
v_g = 0.043 c_0		α = 13.5 m^{-1}

The $2\pi/3$-mode was chosen because it has the highest shunt impedance, a large group velocity and a relatively low error sensitivity.

With the attenuation α, the optimal structure length l, i.e. with the highest energy gain for a given input power, is αl = 1.26 or l = 9.3 cm. However, there are many reasons to go to shorter l and 7 cm were chosen corresponding to N = 84 cells. Then, the input power per structure is 29 kW in order to make 10 MV/m average gradient. The ratio of input to output power dissipation is 6.6 to 1. Depending on the mode of operation pulsed or CM, this may lead to unacceptable thermal stresses or heating of the irises. Certainly, it complicates also the design of the cooling system.

RF FIELD FOCUSING

In cartesian coordinates the synchronous space-harmonic of the E_z-component can be written as

$$E_z = E_0 \cos k_x x \cos k_y y \cdot e^{j\varphi}, \quad \varphi = \omega t - k_z z \quad (1)$$

where

$$k_x \approx \pi/w, \quad k_z = k/\beta, \quad k_y^2 = -k_x^2 - (k/\beta\gamma)^2.$$

Assuming $E_x = 0$ which is very well fulfilled, we can easily derive the Lorentz forces from Maxwell's equations together with (1) and $k_x x$, $k_y y \ll 1$ as

$$F_z = -eE_0 \cos\varphi, \quad F_x = -eE_0 \frac{k_x^2}{k} \beta \sin\varphi \cdot x, \quad F_y = -eE_0 \frac{k_y^2}{k} \beta \sin\varphi \cdot y. \quad (2)$$

As can be seen, the transverse forces are of quadrupole character if the particles are not on the crest of the RF ($\varphi = 0$). For relativistic particles follows from (1) that $k_y 2 \approx -k_x 2$ and the transverse forces have equal magnitude. The peak value at $\varphi = \pi/2$ corresponds to an equivalent magnetic gradient of 42 T/m for $E_0 = 10$ MV/m. For non-relativistic particles the vertical force is larger than the horizontal force by a factor $1 + (2/\beta\gamma)^2$, since $k_x \approx k/\sqrt{2}$.

This particular property of the RF fields is a consequence of the non-axissymmetric structure. One can make use of it by choosing different focusing strategies:

a) External focusing only.

b) External focusing in one plane and RF focusing in the other plane.

c) Alternating gradient (AG) RF-focusing by rotating every other structure by 90° around its axis.

d) Alternating phase focusing (APF) by operating different sections with an RF phase changing
   periodically between plus and minus φ_0.

Case b) is easy to realize. With a negative phase angle $-\varphi_0$ in (2) one has focusing in longitudinal and x-direction. The defocusing in y-direction can be controlled by a one-dimensional magnetic focusing as for instance proposed in [1]. Strategy c), although straightforward on paper, is not very handy in reality. The alignment of rotated structures with micrometer precision may be difficult. Very promising looks the last option d). Since additional structure elements are free of charge when using lithography, one may easily incorporate delay lines as in Fig. 2a or one may modify the cavities itself such that the length changes periodically, Fig. 2 b). The detuning due to the changed cell length is corrected by adjusting the width w. APF has been analysed in ref. [10] and looks very attractive.

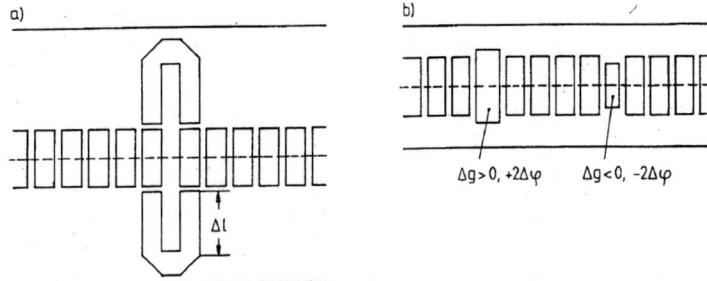

Figure 2. Muffin-tin with a) delay lines, b) periodically changing cavity length.

TRAVELING-WAVE CONSTANT-GRADIENT (CG) STRUCTURES

As pointed out in the chapter on CI-structures they suffer from a very high ratio of input to output power dissipation. The standard solution to the problem is to taper down the group velocity along the structure such that the power dissipation stays constant. With the transported power P(z), the dissipated power per unit length P_d' and the stored energy per unit length W' one gets in first approximation

$$P_d' = -\frac{d}{dz} P(z) = \text{const.} \rightarrow P(z) = P_i \left[1 - \frac{P_i - P_0}{P_i} \frac{z}{l} \right] \quad (3)$$

$$v_g = \frac{P(z)}{W'} = v_{gi} \left[1 - \frac{P_i - P_0}{P_i} \frac{z}{l} \right] \quad (4)$$

for a constant stored energy, i.e. a constant gradient. That is, the group velocity has to change linearly with z and with the same ratio as the power transport. In axis-symmetric structures this is easily achieved by reducing the aperture. The same strategy may be employed for planar structures by machining the surface of the slabs. If this turns out to be too difficult for reasons of tolerances and alignment, one has to choose other options. Looking at the fields in a cavity, Fig. 3a, one identifies regions with electric coupling (k_e) and regions with magnetic coupling (k_m). The group velocity is proportional

$$v_g \sim (k_e - k_m) \text{ with } k_e > k_m. \quad (5)$$

Now it is clear that one can modify the coupling strength and thus choose v_g. Decreasing k_m by inserting metallic corners in the cavity, Fig. 3 b, is not very effective. It increases v_g only up to 50 %. A stronger variation of v_g, reasonably up to a factor of 2.5, results from increasing the iris thickness, Fig. 3d. The most effective option is inserting capacitive loads, Fig. 3c. Thus, v_g can be lowered by a factor of up to 10.

Fig. 4 shows as an example the variation of the group velocity with the thickness d of the capacitive loads. It also shows the required change in cavity width in order to keep it in tune. Obviously, the strong change of the geometry influences badly the shunt impedance. This is the price one has to pay when keeping the structure strictly planar.

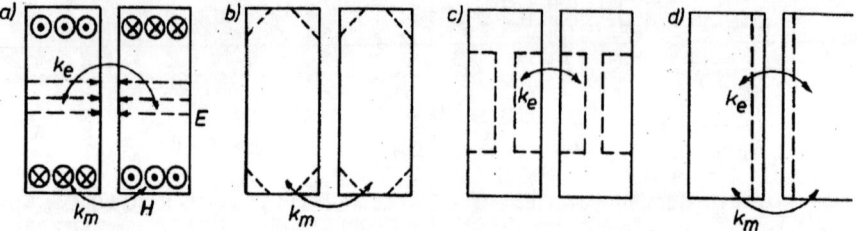

Fig. 3: Top view of two muffin-tin cells; a) accelerating fields with regions of electric k_e and magnetic k_m coupling; metallic inserts modifying b) k_m, c) k_e, d) k_e and k_m.

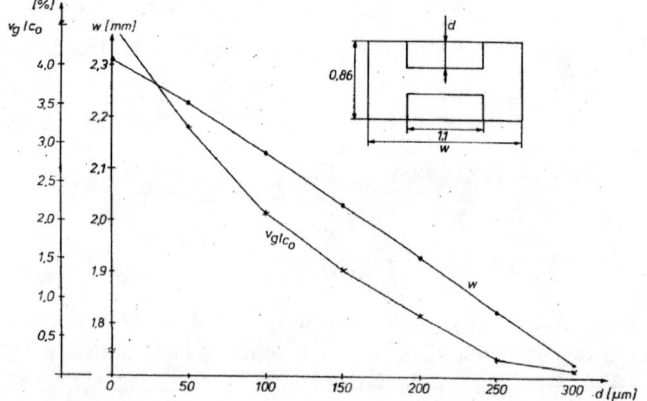

Fig. 4: Group velocity and cavity width versus thickness of capacitive blocks in a 94 GHz muffin-tin structure.

FLAT-FIELD TRAVELING-WAVE STRUCTURES

Some applications may require a field which is independent of the transverse position, at least over a certain fraction of the aperture area. Although planar structures support fields which are intrinsically non-flat (see chapter on RF focusing), they can be modified such that the fields are constant over a wide area. The way to do it is either to introduce a capacitive load, Fig. 5a, or an inductive load, Fig. 5b. As a result the wave number k_x of the space-harmonic, equ. (1), goes to zero and the E_z-component is independent of x and y.

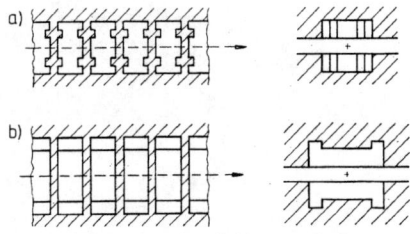

Fig. 5: Muffin-tin structure with flat accelerating fields near the beam axis;
a) increased capacitive and b) increased inductive load at the sides.

The capacitive loading is less effective and is appropriate for applications where only a flattening of the cosine-profile is required. As an example a TW-structure as it may be used in a linac is shown in Fig. 6. The structure was designed for a $2\pi/3$ mode at 34 GHz. The depth of the cavities is 1.6 mm which is probably too deep for fabrication with LIGA. Therefore, it was chosen to apply and study the UVL process.

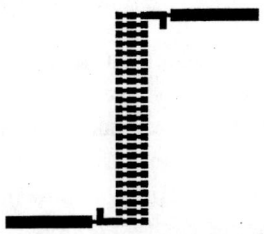

Fig. 6: AUTOCAD drawing for the UV-mask of a flat-field traveling-wave structure with input and output couplers and capacitive cavity loading.

The inductive loading, on the other hand, is appropriate for cavities which are several wavelengths wide. This is easily understood by considering the cavity as a waveguide at cut-off in transverse direction where the inductive load serves as proper termination.

As an example a single-cell flat-field cavity is given in Fig. 7. The cavity is about 3 wavelengths wide and is foreseen for a high-power sheet-beam klystron at 94 GHz. The figure shows the field pattern and the beautiful field flatness over the transverse coordinate. Clearly, the cavity is well suited to accelerate a very wide sheet-beam with an aspect ratio of the order of 30, that means 30 times more current than an axis-symmetric structure with identical current density.

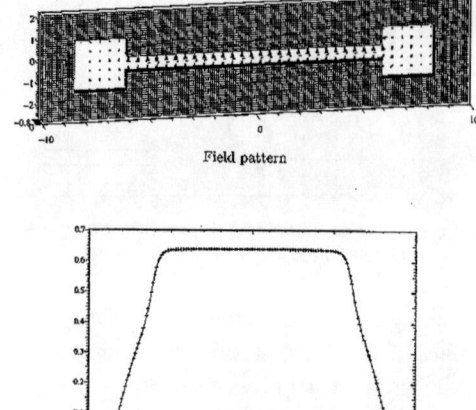

Fig. 7: Flat-field single-cell muffin-tin cavity.

HIGHER-ORDER-MODE (HOM) "FREE" STRUCTURES

Muffin-tin structures support essentially no vertically polarized break-up modes. These modes radiate very strongly into the side openings and can easily be damped by a proper design. In horizontal direction, however, the modes are trapped. To damp the trapped modes two options are available: Firstly, they can be coupled out by lateral waveguides with a cut-off between the fundamental mode frequency and the break-up mode frequency which is typically 60 % higher. Secondly, one can insert stop-band filters into the lateral coupling waveguides where the stop-band is at the fundamental mode frequency and all other modes are coupled out. The principle, called choke mode cavity, has been proposed for axis-symmetric structures [11] and can easily be incorporated in muffin-tin structures, Fig. 8.

The first solution requires coupling waveguides of different depth than the cavities and thus a two-step LIGA process. Although it is in principle possible it certainly will complicate the fabrication and reduce the achievable tolerances. The second solution is perfectly suited for LIGA. The whole device has equal depth and the increased complexity influences only the design process. The trapped fundamental mode is slightly perturbed. Both the r/Q and the Q are decreased by about 10 %. HOM damping is very effective with Q-values in the range of some 10.

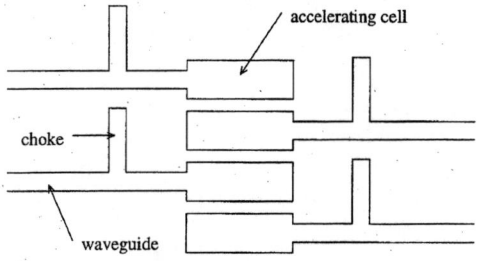

Fig. 8: TW choke-mode structure.

RF UNDULATOR STRUCTURES

The basic mechanism of an electromagnetic undulator is similar to the conventional magnetic undulator. The equivalent magnetic field is given by

$$B_{eq} = \left[1 + \frac{Z_0}{Z_g}\right] \frac{E_0}{c_0} \approx 2 \frac{E_0}{c_0} \quad (6)$$

where Z_0 is the free-space wave impedance, Z_g the waveguide impedance and E_0 is the transverse peak electric field. The undulator period is

$$\lambda_u = \frac{\lambda_0 \lambda_g}{\lambda_0 + \lambda_g} \quad (7)$$

with λ_0 and λ_g being free-space and waveguide wavelength, respectively.

Different possibilities have been investigated to build an RF undulator in planar geometry. Again, the basic muffin-tin arrangement seems to be best suited. One just has to increase slightly the dimensions in order to be well above the cut-off of the TE_{01}-mode, Fig. 9a, which is required because of the transverse electric field. If good field quality is needed, and low losses, then a heavily overmoded structure supporting a hybrid HE_{11}-mode, Fig. 9b, is the appropriate choice. Table 2 shows the parameters for a 120 GHz hybrid mode undulator. The parameters were chosen by a trade-off between low losses, mode density and transverse shunt-impedance.

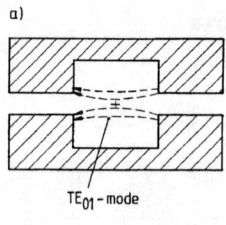 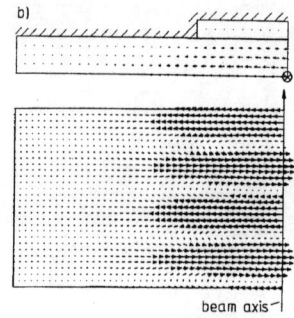

Fig. 9: a) Groove waveguide and b) hybrid HE_{11}-mode in an overmoded muffin-tin.

Table 2
Dimensions and basic RF parameters for a muffin-tin operated in a $2\pi/3$ standing wave mode (120 GHz)

a = 1.0 mm	b = 1.6 mm	w = 5.0 mm
g = 0.633 mm	t = 0.2 mm	
λ = 2.9 mm	Q_0 = 26680	r = 6.9 GΩ/m

An undulator with a slightly different geometry is going to be developed in Argonne National Laboratory [12]. It is foreseen to operate at 108 GHz with 80 periods of 1 mm length and 0.33 T equivalent magnetic field. The Advanced Photon Source will provide an electron beam up to 750 MeV with 1 mA average current and a low enough emittance to drive the undulator. The radiation has been calculated and the first harmonics are expected at 4.0 and 5.2 keV with a brightness around $8 \cdot 10^{11}$ photons/sec/0.1 % band width/mrad^2.

STANDING-WAVE (SW) STRUCTURES

For many applications, typically in CW-operation or in storage rings, an SW-structure is needed. It has the advantage of a constant accelerating field and requires a small number N of coupled cells with well separated modes. Its main disadvantage, however, is a shunt impedance which is only half that of a TW-structure, except for the π-mode. But around the π-mode the mode spacing

$$\Delta\omega = \frac{1}{4} k\omega_0 \left(\frac{\pi}{N}\right)^2 , \text{ k coupling factor} \qquad (8)$$

is a factor $\pi/2N$ smaller than for the $\pi/2$-mode and the structure is very sensitive against dimensional errors and has phase errors from cell to cell. Good mode separation requires the bandwidth to be smaller than half the mode spacing

$$\frac{\omega_0}{Q_0} < \frac{1}{2}\frac{\omega_0}{4} k\left(\frac{\pi}{N}\right)^2. \qquad (9)$$

If one built such a structure with the TW geometry given in table 1, one could barely couple 10 cells together and still have all the stability problems.

A nice trick to overcome these problems is often used in proton machines. The accelerating cells are not coupled directly but via off-axis coupling cells, in such a way that the structure is operated in the $\pi/2$-mode but the effective phase advance from main to main cell is π. This is an effective but very expensive way to fabricate standard structures. In the case of planar structures and fabrication by lithography the solution is given for free.

Several different arrangements of side-coupled muffin-tin structures have been investigated [13]. They all have coalescent pass-bands at π phase advance per cell. The dispersion diagram is symmetric around this point and has a large slope, that means a high group velocity (close to the one of the $\pi/2$-mode) and there are no 1st-order cell-to-cell phase errors. Finally, one geometry has been retained and a full design was made for an operational frequency of 94 GHz [4]. The structure geometry and the RF parameters are repeated here in Fig. 10. and table 3, respectively.

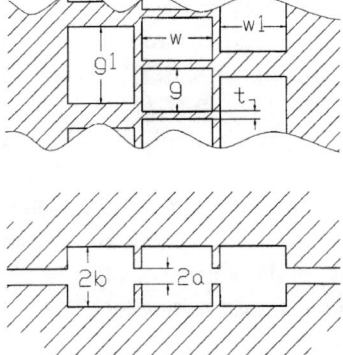

Fig. 10: Geometry of a side-coupled structure:
g1=2.40, w = 2,29, g = 1.34, t = 0.25, w1 = 2.14, 2b = 2.29, 2a = 0.75 (in mm)

Table 3
RF parameters of the 94 GHz side-coupled structure of Fig. 10.

r/Q_0 = 81.5 kΩ/m
Q_0 = 3620 (κ=56*10^6/Ω/m)
r = 295 MΩ/m
v_g/c_0 = 0.0113

$\alpha = 24$ m^{-1} attenuation length
$k_0 = 12*10^3$ V/pCm fundamental mode loss factor

The number of coupled cells was fixed to 21, somewhat lower than the theoretically allowed value. Two structures are positioned on one wafer and powered via a power splitter from a single feed line, Fig. 11. Special care was given to the end-cells. They have to reflect the forward traveling wave and at the same time create a flat π-mode distribution. Both was obtained by numerically optimizing the geometry with the code GdfidL [9].

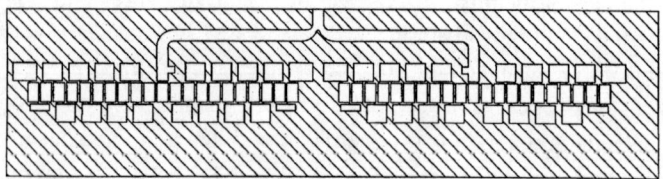

Fig. 11: Two structures with central feed line, power splitter and couplers.

RF POWER COUPLERS

An important and quite delicate task is the design of the power couplers. Two basically different approaches are possible:

1) A device very similar to the standard solution where a waveguide feeds the structure via an iris coupling. In case of a planar structure this translates into a waveguide coming from top or bottom and passing through the corresponding slab. The required matching iris can be located at the surface of the slab. Fig. 12 shows a solution with an incorporated taper as a transition to a heavily overmoded low-loss waveguide which operates in the HE$_{11}$-mode. The coupler has the additional advantage of a contact-free low tolerance joint between the waveguide and taper. Proper optimization of the geometry yields a good transmission of 98 %, ref. [14].

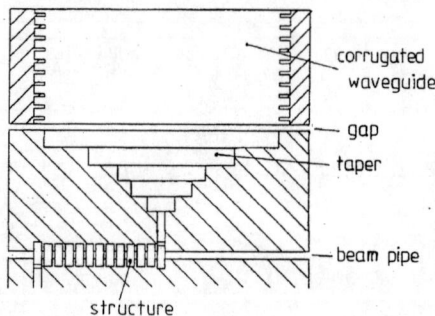

Fig. 12: Power coupler with taper and contact-free joint to waveguide.

2) The second possibility is a completely planar arrangement with a groove waveguide as feeding line coming from the side. As an example a symmetric coupler for a SW side-coupled structure is shown in Fig. 13. It consists of a power splitter, a λ/2-delay line in order to get π-mode excitation and stub-lines which, together with the iris thickness, controls the coupling. At the same time the example shows the complexity of the design which can be incorporated on the slab very easily and free of additional costs. Shape and dimensions of the coupler were found again numerically with the code GdfidL.

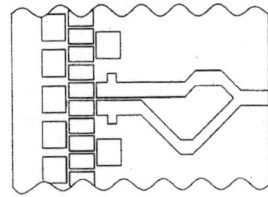

Fig. 13: Power coupler with feed line, power splitter, delay line and matching stub-lines for a side-coupled muffin-tin.

FABRICATION AND ENGINEERING

The simplified LIGA process consists of making an X-ray mask, preparing a substrate and covering it with a fotoresist (PMMA plastic), exposure, developing, etching away the developed part and electroplating the structure. The requirements for high aspect ratio structures are thereby very demanding; good adhesion of the resist, no dissolution of unexposed resist, high mechanical stability, low internal stresses during exposure and development and compatibility with electroplating.

A DXRL mask for an undulator structure was constructed through an intermediate mask at the center for X-ray lithography in Stoughton, ref. [2]. A plating base of Ti/Au with 75/300 A° was written by an e-beam and 3 μm of Au was plated on the intermediate mask. For the final mask 25 μm Au was plated on a structured Si wafer. The substrate was of diamond finished copper with a Ti coating of less than one micron. On the coating a 1 mm thick PMMA film was cast and annealed at temperatures between 100 and 170 °C for one to three hours. The exposure was done at the ALS, Berkeley, and later on at the NSLS, Brookhaven. Different developers are being used. The electroplating process and the subsequent surface finishing is still not fully finished.

In parallel to the work on the undulator structure an intermediate X-ray mask for the standing-wave structure was built at the PMT, Karlsruhe, Germany. It is a 2.5 μm thick Au mask on a 2 μm thick Ti membrane. Next steps will be the fabrication of a 12 μm thick mask and the study of the developing process.

Besides the DXRL work, the fabrication of the 34 GHz traveling-wave structure was started at the Technische Universitaet Ilmenau, Institute for Glass and Ceramics, Germany. The purpose of this development is to prove that the lower frequency structures with a depth of more than 1 mm can be fabricated by UVL. The process is very similar to LIGA. As a fotoresist serves doped Li_2O-Al_2O_3-SiO_2 glass. The part of the glass which will be exposed forms a special crystal phase with a 30 times higher solubility than the unexposed part. Subsequent tempering (developing) and etching creates the microstructured glass plate. In the next step, the structured side of the plate has to be covered by evaporated copper as a starting layer for the electroplating process. Finally, the left-over glass will be etched away and the structure machined to its final state. Five microstructured glass matrices have been fabricated and a first copper structure was made. Problem areas are glass purity and homogeneity, copper evaporation into deep structures and tolerances.

A 10 MV/m accelerating gradient will typically cause some 340 kW dissipation per m structure corresponding roughly to an average power flux of 3.4 W/mm². A pulsed operation with a duty cycle D will lower the heat dissipation by the factor D. Nevertheless, in some applications the cooling will be problematic. Two different techniques are being pursued. A standard pipe cooling with a manifold as shown in Fig. 14a and an advanced microchannel cooling, Fig. 14b.

Standard cooling pipes have surface heat transfer coefficients in the order of $\alpha \approx 1$ W/cm2 °C. Therefore, for the above given power flux and a tolerable temperature gradient of 10 °C, pulsed operation with a few percent duty cycle would be required. Micro-channels have heat transfer coefficients between 10 and 30 W/cm2 °C and the structures could either be operated with higher gradients or with duty cycles between 10 and nearly 100 %.

A heat analysis was done for the SW side-coupled structure shown in Fig. 10 by means of the code GdfidL. With the assumption of a 10 MV/m accelerating gradient a duty cycle of 1 % and a thermal conductivity of 3.84 W/cm °C for a 1 cm thick copper slab, a temperature increase of 7 °C was found on the iris center.

Another major concern is the alignment of the two halves with respect to each other. Circular holes could be produced with lithographic process and could be used to center the two halves by means of small steel balls, see Fig. 14a. The solid mechanical connection then follows from diffusion bonding the side-shoulders. An extensive program is also going on to make use of alignment and bonding techniques developed for micromachined electron microscopes [15]. V-grooves are machined into the wafer, see e.g. Fig.1, and precision glass fibers are placed into the grooves and bonded and clamped in place. Vacuum pumping is provided through the gap between the two halves.

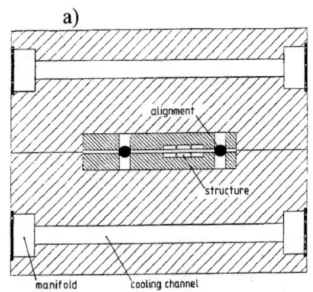 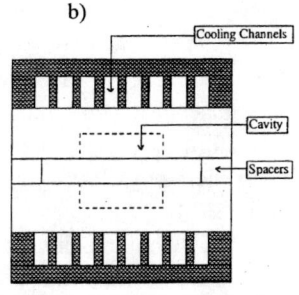

Fig. 14: Cross-section of an RF-Structure with a) cooling pipes and b) microchannel cooling.

CONCLUSIONS

At very high frequencies, around 100 GHz, RF structures have to be fabricated by lithography and, therefore, must be planar. Two technologies are available: Deep X-ray lithography and UV-lithography. Both seem to meet the requirements for an RF structure.

Planar structures have a very different electrical behaviour as compared to axis-symmetric structures. They, also, have a higher geometrical degree of freedom and the costs for fabrication are roughly independent of the complexity of the device. Therefore, many attractive solutions to particular problems are possible. One can adjust the structure geometry to the beam dimensions, modify the transverse dependence of the accelerating field, use the intrinsic RF quadrupole fields or reduce the beam induced fields (wakefields). Possible applications, apart from standard components, are for instance side-coupled structures, APF structures, RF undulators, very wide cavities for sheet-beam klystrons, low HOM structures for linear colliders, integrated couplers, filters, power splitters and many more.

REFERENCES

[1] H. Henke, Y.W. Kang, and R.L. Kustom, "A mm-wave RF structure for relativistic electron acceleration", Argonne National Laboratory, internal report ANL/APS/MMW-1, 1993

[2] H. Henke, "MM-wave linac and wiggler structures", Proceedings of the 4th European Particle Accelerator Conference, London, July 1994, Vol. 1, pp. 322-326.

[3] A. Nassiri et al., "A 50-MeV mm-wave electron linear accelerator system for production of tunable short wavelength synchrotron radiation", IEEE, Intern. Electron Device Conf., Washington D.C., 12/93, pp. 169-172.

[4] W. Bruns, H. Henke and R. Merte, "Design of a 94 GHz accelerating structure", to be published in Proceedings of the 5th European Particle

Accelerator Conference, Barcelona, June 1996.

[5] E.W. Becker, W. Ehrfeld, P. Hagmann, A. Maurer and D. Münchmeyer: Microelectronic Engineering, No. 4 (1986), pp. 35-56.

[6] D. Hülsenberg et. al., "High aspect ratio structures obtained by electroforming in microstructured glass", Proceedings of HARMST 95, Karlsruhe, July 1995, to be published in Micro System Technology, Springer.

[7] M. Filtz, "Coupling impedances of muffin-tin structures with closed and open sides", Proceedings of the 1995 Particle Accelerator Conference, Dallas, May 1995, pp. 2373 - 2375.

[8] R. Klatt et al., Proc. of the 1986 Linear Accelerator Conf., Stanford internal rep. SLAC-Report 303, Sept. 1986.

[9] W. Bruns, "Error sensitivity for side-coupled muffin-tin structures using a finite difference program", Proceedings of the 1995 Particle Accelerator Conference, Dallas, May 1995, pp. 1085 - 1087

[10] F.E. Mills and A. Nassiri, "Alternating phase focusing in mm-wave linear accelerators", Argonne National Laboratory, internal report ANL/APS/MMW-9, 1994.

[11] T. Shintake, "Design of high power model of damped linear accelerating structure using choke mode cavity," in IEEE Proceedings of the 1993 Particle Accelerator Conference, Washington 1993, pp. 1048-1050.

[12] A. Nassiri et al., "Fabrication of mm-wave undulator cavities using deep X-ray lithography", to be published in Rev. Sci. Instrum. 67(8), Sep. 1996.

[13] H. Henke and W. Bruns, "A broad-band side-coupled mm-wave accelerating structure for electrons", Procee-dings of the 1993 IEEE Particle Accelerator Conference, Washington D.C., May 1993, vol. 2, pp. 904-906.

[14] B. Littmann and H. Henke, "Feasibility study of optically coupling RF-power at mm-waves", Proceedings of the 1995 Particle Accelerator Conference, Dallas, May 1995, p. 1593-1595.

[15] A.D. Feinerman et al., J. Vac. Sci. Technology A, 10(4), 1992, pp. 611.

The Fabrication of Millimeter-Wavelength Accelerating Structures

P.J. Chou, G.B. Bowden, M.R. Copeland, A. Farvid, R.E. Kirby,
A. Menegat, C. Pearson, L. Shere, R.H. Siemann, J.E. Spencer and
D.H. Whittum

Stanford Linear Accelerator Center
P.O. Box 4349, Stanford, CA 94309

Abstract There is a growing interest in the research of high gradient ($\geq$ 1GeV/m) accelerating structures. The need for high gradient acceleration based on current microwave technology requires the structures to be operated in the millimeter wavelength. Fabrication of accelerating structures at millimeter scale with sub-micron tolerances poses great challenges. The accelerating structures impose strict requirements on surface smoothness and finish to suppress field emission and multipactor effects. Various fabrication techniques based on conventional machining and micromachining have been evaluated and tested. These will be discussed and measurement results presented.

1. INTRODUCTION

High gradient acceleration based on current microwave technology requires the structure to be operated at higher frequencies in order to reduce power consumption[1]. To determine the limit of particle acceleration based on microwave technology, a millimeter wavelength is chosen. According to frequency scaling[2, 3], the dimensional tolerance required for mm-wave structures ranges from 2 μm to 0.6 μm. The fabrication of accelerating structures at the millimeter scale with sub-micron tolerances poses great challenges. Accelerating structures at millimeter scale are quite small for conventional machining. Another option is to fabricate them by using micromachining techniques such as LIGA[4]. The operating frequency for the test structure is tentatively chosen to be around 32 times SLAC frequency (32 × 2. 856 GHz). Accelerating structures at this scale are too big for most micromachining techniques. The suppression of field emission and multipactor effects places strict requirements on surface roughness and finish. Post machining surface treatment is also important.

The muffin-tin structure[4] is used because it is relatively easy to fabricate by micromachining compared to a cylindrical structure. The schematic sketch of a 6-cell structure is depicted in Fig. 1. Physical dimensions of the muffin-tin slab and two sidewalls are given in Fig. 2.

---

* Work supported by Department of Energy contract DE-AC03-76SF00515. An expanded version of this paper is available from SLAC as SLAC-PUB-7339.

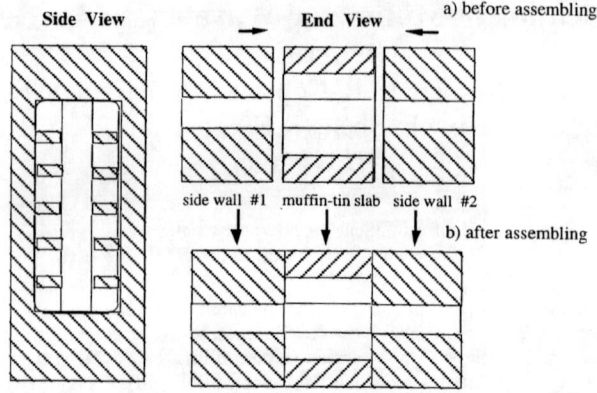

Fig. 1: The schematic sketch of the muffin-tin accelerating structure.

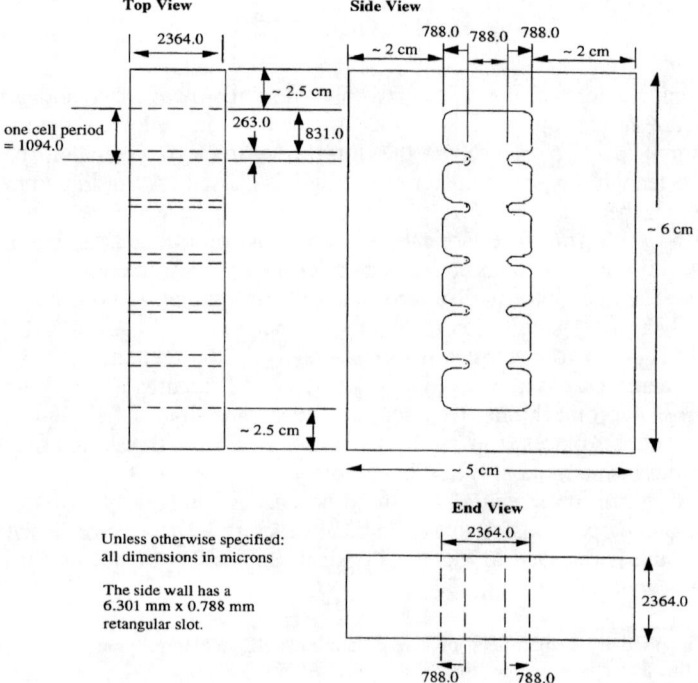

Unless otherwise specified: all dimensions in microns

The side wall has a 6.301 mm x 0.788 mm retangular slot.

Fig. 2: Physical dimensions of muffin-tin slab.

Consider a surface profile at the microscopic scale. The quantity R_a used to characterize the surface roughness is defined below[5]:

$$R_a = \frac{1}{n} \sum_{i=1}^{n} |z_i|$$

where n is the total number of measurements and z_i is the profile amplitude of surface. This definition does not consider cases where average surfaces are tilted or offset with respect to the direction of measurements. R_a is the arithmetic average of the absolute values of all profile amplitudes z_i within the entire measurement range. It should be noted that R_a is just one of several quantities used in surface metrology to characterize a surface profile. It may not be the best way for certain applications.

2. PRECISION MACHINING

I. Wilson gave a thorough review on the fabrication of RF cavities based on conventional machining techniques[6]. Conventional machining techniques are suitable for making large accelerating structures at the scale of a few tens of centimeters or bigger. For mm-wave structures, one also needs to consider micromachining techniques.

2.1. Laser machining

Laser pulses can be used to cut metal with good accuracy. The current dimensional tolerance quoted for this technique is ± 0.0001" (2.5 µm). This is not adequate for mm-wave structures. Too, this technique has other drawbacks. The cutting process will cause burrs and produce a rough surface finish (class 125 or $R_a \approx$ 3.2 µm). It is not a suitable process for copper because of the poor edge definition and 1 mm thick copper sheet is too thick for laser machining. Even for copper sheet of 500 µm, the above problems still exist.

2.2 Hydraulic cutting

A highly pressurized water jet mixed with abrasive particles is used to cut metal. This is an abrasive process and produces rough surface finish. The size of water jet is not small enough to cut small features on the workpiece. Therefore, it is not suitable for the fabrication of mm-wave structures.

2.3 Single-point diamond turning[5]

A cutting tool with a small diamond tip can be used to machine non-ferrous materials such as aluminum and copper to obtain a very fine surface finish. Single-point diamond turning is usually carried out at a turning speed of 2500- 5000 rpm. Typical roughness for ordinary diamond turning may be around 25-50 nm or better.

Compared with conventional cutting, the influence of anisotropy of the material on surface roughness can not be neglected. Single-point diamond turning is best suited to cut axi-symmetric geometry such as mirrors or lenses. For mm-wave structures with complicated 3D geometry, this technique may not be suitable.

One drawback of single-point diamond turning is burrs left along part edges or whenever there is an interrupted cut as depicted in Fig. 3. These burrs are quite pronounced. Sharp points are likely to become field emitters under high power operation. Further surface treatment may be necessary in order to remove burrs.

Because of the concern for surface damage from pulse heating[7, 8], a dispersion strengthened copper called glidcop® might be a better choice as the base material than pure copper. Unfortunately single-point diamond turning does not work well with glidcop®. The cutting tool could be damaged by the alumina particles present in the glidcop®[9].

Single-point diamond turning will also leave machining marks on the surface. Their shape will not be an exact replica of the cutting tool, partly because of the burnishing effect of the cutting tool. Machining marks left by the cutting tool are clearly shown in Fig. 4. The corresponding 3-dimensional surface profile is depicted in Fig. 5. The surface roughness R_a is 0.018 µm.

2.4 Photochemical machining[11, 12]

This technique is suitable for precision machining of thin metal sheets. It offers burr free edges. The design feature is generated on a photomask by using laser plot technology[11]. A thin layer of photosensitive, etchant-resistant polymer called photoresist is coated onto both sides of metal surface. Then the photoresist is exposed and developed. The metal surface is selectively masked by the photoresist. Therefore, the design feature is transferred from the mask to the photoresist. Heated etching solution (acid) is sprayed on both sides of metal. Those regions which are not covered by the photoresist will be etched away. Then the photoresist is removed. When the etching solution impinges on the metal surface, some erosion on the side wall also occurs. This side erosion called undercut is the main factor governing the tolerances that can be held. As the thickness of metal sheet increases, so does the undercut. The tolerances that can be held also increase accordingly. As a general rule of thumb, the minimum feature size can not be less than the thickness of metal sheets used for machining. Typical planar dimensional tolerances are ± 0.0005" within 6" range[12].

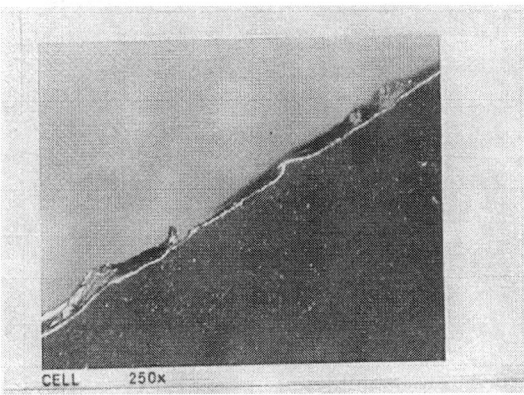

Fig. 3: Burrs left along cutting edges in single-point diamond turning. Picture was taken by optical microscope for a NLC test structure[3].

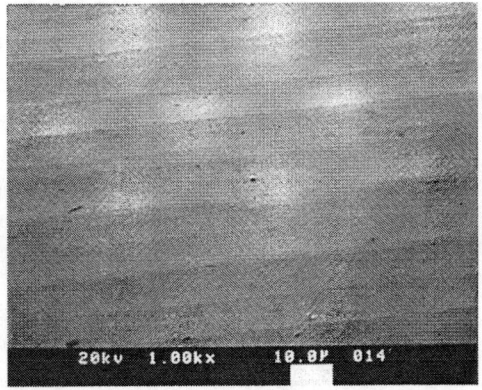

Fig. 4: Machining marks left by the cutting tool in single-point diamond turning. The bar represents 10 μm. From one of NLC test structures. Picture was taken by a scanning electron microscope.

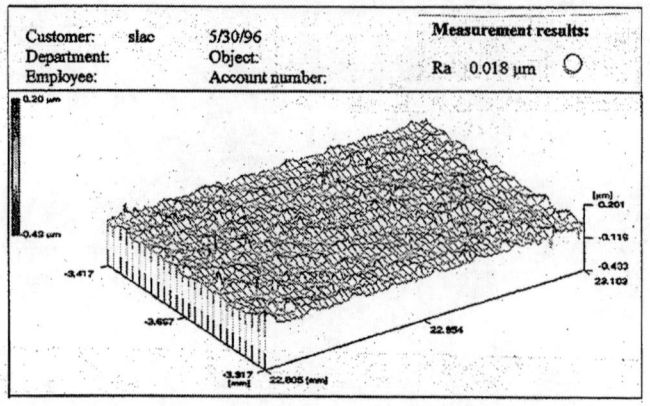

Fig. 5: Three-dimensional surface profile of single-point diamond machined NLC test structure[10].

This technique allows intricate geometries to be made on metal sheets. It is also an inexpensive approach for mass production. In order to achieve tight tolerances required for mm-wave structures, the thickness of metal sheets should not be larger than 0.01" for critical dimensions. Cylindrical structures with high order mode damping slots, pumping ports, input and output couplers can be fabricated into a single piece at the same time. That would require the structure to be built from laminations and probably bonded by diffusion bonding. The flatness of metal sheets and the accuracy of assembling within one micron then become important issues. At this moment, answers to those questions are not clear. Figure 6 depicts the intricate geometries fabricated by photochemical machining. Surface treatments are required to insure a smooth surface for bonding.

Fig. 6: A stainless steel sheet with intricate geometriey made by photochemical machining[11]. The physical dimension of the view area is roughly 2"× 2". Picture was taken by an optical microscope.

2.5 Precision diamond grinding[5, 13]

Grinding wheels are made up of a large number of abrasive grains held in a bonding agent. The abrasive grain can be aluminum oxide, silicon carbide, cubic boron nitride or diamond. The grinding wheel is first formed to the design shape by a dressing tool. Then the rotating grinding wheel is brought to the workpiece to remove metal. The surface roughness has to do with wheel characteristics, e.g. the grain type, grain spacing, grain size, dressing method, wheel balance ... etc. Precision diamond grinders with 0.00001" programmable step are available from commercial companies. Dimensional tolerances are often down to ±0.00005" (± 1.3 µm), and to a 0.000003" surface finish (R_a= 0.08 µm). The drawback of this technique is grain inclusion on the metal surface. The grains can not be removed easily by surface polishing treatments. Since they could become field emitters under high power operation, this is a fatal drawback of the use of grinding techniques for the structure fabrication.

2.6 Electrodischarge machining (EDM)[5, 14]

This method can be used to machine very hard materials. The process is illustrated in Fig. 7.

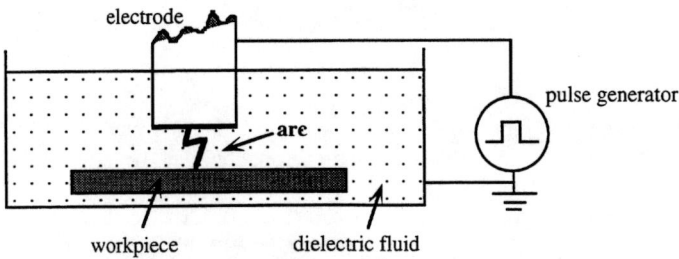

Fig. 7: The basic setup for electrodischarge machining process.

The machining tool (electrode) is placed about 25-50 µm away from the workpiece. The tool electrode is connected to a pulsed voltage source. If the potential difference between the tool and workpiece is high enough, a spark is produced and the dielectric fluid ionized. The dielectric fluid is usually oil or water. If the potential difference is maintained, the spark will developed into an arc which causes the removal of metal. If the potential difference falls the arc decays. The temperature of the arc is the property responsible for removing metal. Typical values are between 5000 and 10000 °C; well above the melting point of most metals. The tool electrode is made of conducting materials, such as copper. For long life and high form accuracy, tungsten carbide is sometimes used. The surface of EDM'ed workpiece looks like shot blasting. There are many craters on the surface, also known as pits in EDM terminology.

A wire EDM machine uses a continuously spooling conducting wire as electrode which moves in prescribed patterns around the workpiece. The typical tolerances can be down to 3-5 μm. A new generation of wire EDM machines began to emerge a couple of years ago. Machine manufacturers claim that a sub-micron accuracy can be achieved. Currently this possibility is being explored. A 3D surface profile of oxygen-free copper machined by wire EDM is shown in Fig. 8. The surface finish produced by wire EDM is inferior to single-point diamond turning. Further surface treatments are definitely needed. Test parts for muffin-tin structures operating around 92 GHz and 11 GHz have been fabricated by wire EDM. Note that the current SLAC operating frequency is 2.856 GHz. Figure 9 depict the 92 GHz muffin-tin slab prior to surface polishing.

Fig. 8: The 3D surface profile of oxygen-free copper machined by wire EDM, $R_a \approx 0.5$ μm[10]. Picture was taken by atomic force microscope.

Fig. 9: A 92 GHz muffin-tin slab fabricated by wire EDM and aligned along a ruler. Each horizontal division is 0.01". Picture was taken by optical microscope.

2.7 Precision electroforming[15, 16]

Electroforming is a process for fabricating a metal part by electrodeposition in a plating bath over a base form or mandrel which is subsequently removed. The advantage of this process is that it faithfully reproduces the form to within one micron[17].The dimensional tolerances of electroformed metal parts are determined by the accuracy of mandrels. Mandrels are prepared by machining techniques with tolerances of ± 0.0002"[17]. Therefore, electroforming can not provide adequate accuracy for mm-wave accelerating structures.

3. MICROMACHINING

Another possibility for fabricating mm-wave accelerating structures is micromachining. For industrial applications machining is normally used to fabricate parts with a size ranging from few millimeters to centimeters and dimensional tolerances ≥ 5 μm. In contrast, micromachining is normally used to fabricate parts with a size ranging from few tens to few hundreds of microns and dimensional tolerances around 5 μm, such as microsensors. Several micromachining techniques have been evaluated and summaries are presented below.

3.1 Optical lithography and wet chemical etching[18]

Optical lithography is an established method and extensively used by the IC industry. The vast majority of lithographic equipment for IC fabrication is optical equipment using ultraviolet (UV) light ($\lambda \approx$ 0.2 to 0.4 μm). The wet chemical etching is used in the lithographic process to produce the design pattern on the silicon substrate. The wet chemical etching is an isotropic process which results in side erosion of the silicon substrate. Optical lithography can also be used to pattern very thick photoresist for the fabrication of microsystems. After the pattern is developed on the thick photoresist, microstructures are made by electroplating. So far the application of optical lithography has been restricted to a resist thickness up to 10 μm due to the limited depth of focus and diffraction[19]. Although patterning of resist layers up to 200 μm using UV light and contact printing was reported[20], it is still not adequate for making mm-wave structures which requires a thickness up to 2 mm with sub-micron tolerances.

3.2 Ion beam etching[18, 21]

The basic procedures are quite similar to optical lithography. The pattern is first developed on the photoresist layer, then an ion beam is used to etch the wafer material instead of wet chemicals. The setup is depicted in Fig. 10. This is a dry etching process. No chemical solution is used as the etchant. Noble gas is used to produce an ion beam for the etching. Typically, argon gas is introduced into the chamber. Upon entering the chamber, the argon is subjected to a stream of

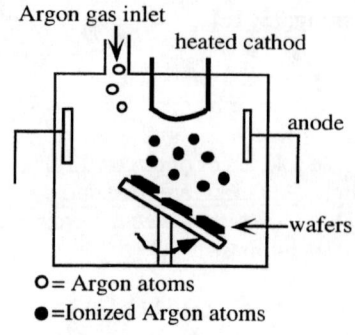

Fig. 10: Schematic of ion beam etching.

energetic electrons between cathode and anode electrodes. The argon atoms are ionized by the electrons into positively charged ions. Wafers are held on a negatively grounded holder which attracts the ionized argon atoms. As argon ions travel toward the wafers, they are accelerated and gain energy. When argon ions impinge on the exposed wafer surface, they literally knock out the atoms on the surface. This is a physical etching process. The etching mechanism is momentum transfer. No chemical reaction takes place between the argon ions and the wafer material. Ion beam etching is also called sputter etching or ion milling. The material removal is highly directional, resulting in good definition on small openings and high spatial resolution. Ion beam etching has very high resolution (about 100 Å) and is suitable for making sub-micron-range structures. Ion beam etching has poor selectivity on material, which means the etching rate is about the same for most materials. For copper the etching rate is 880 Å/min[22]. Silicon wafer of thickness up to 300 µm has been etched successfully by ion beam etching to produce design patterns[23]. At present it is unclear whether ion beam etching is suitable for making mm-wave structures or not.

3.3 Reactive-ion etching[18, 21, 24, 25, 26]

This is a combination of physical and chemical etching processes. The basic apparatus for reactive-ion etching (RIE) is depicted in Fig. 11. The reactive ions generated in the plasma are extracted and accelerated towards wafers. When the reaction between the reactive ions and surface material produces volatile compounds, etching occurs. Those volatile products are subsequently removed by vacuum pump. The ion bombardment helps to make the etching directional which results in almost straight sidewalls (high degree of anisotropy). Different gas composition will affect the selectivity or etch rate. RIE is a highly selective process. For example, using gas mixture of BCl_3 and Cl_2, the relative ratio of etch rate on Al to etch rate on SiO_2 is 25[18]. Using RIE to fabricate microelectrical mechanical systems has been reported for thickness up to few tens of microns[27]. Little work has been done on etching pure copper. Much work needs to be done in order to decide whether RIE is suitable for the fabrication of mm-wave accelerating structures.

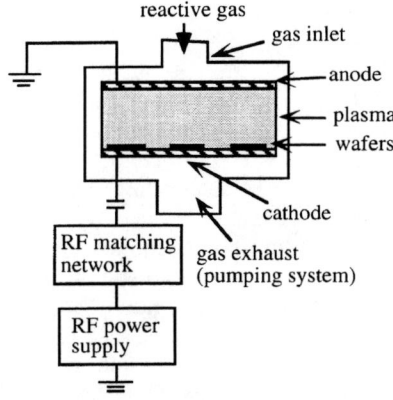

Fig. 11: The schematic sketch of an RIE etch chamber.

3.4 Wet-chemical anisotropic etching of monocrystalline silicon[18, 28]

This technique is a variant of optical lithography with isotropic wet etching. Some chemical etchants dissolve a given crystal plane of a semiconductor much faster than other planes; this results in orientation-dependent etching. Using orientation-dependent etching one can attain V-shaped grooves or straight-walled grooves. Figure 12 depicts the orientation-dependent etching on certain monocrystalline silicon substrates. This technique provides a high degree of anisotropy and selectivity. It is suitable for making micromechanical structures. Micromechanical structures with depths up to 2 mm and dimensional tolerances of 0.0001" can be fabricated[28]. Accelerating structures made by anisotropic etching of silicon with tolerances of 5 µm have been reported[29]. After the silicon substrate is etched, copper is sputtered on to build a metallic layer. Because of pulse heating effects, the copper layer on the silicon substrate may not be able to sustain the fatigue damage. More knowledge from high-power pulse heating experiments is needed to determine whether or not this is a viable candidate for structure fabrication.

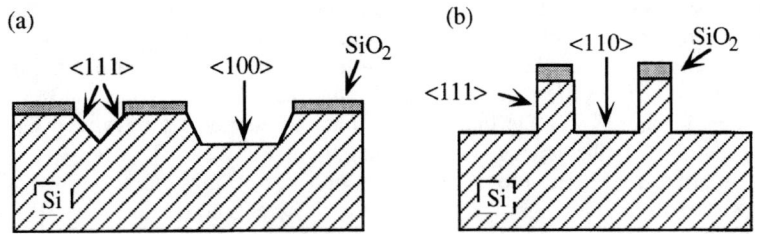

Fig. 12: Orientation-dependent etching (a) through window on <100>-oriented silicon (b) through window patterns on <110>-oriented silicon[18].

3.5 LIGA[30]

LIGA is a German acronym which stands for Lithogrphie, Galvanoformung, and Abformung. This is a combination of deep-etch X-ray lithography, electroforming and plastic molding processes. This is a technique for fabricating microstructures with extreme structural heights. LIGA is an attractive option for fabricating mm-wave accelerating structures. It can provide approximately 2 μm true dimensional tolerances[31]. It is also suitable for mass production of structures. Unfortunately it requires a synchrotron light source to provide the highly collimated X-rays. The preparation of X-ray masks is also more complicated than masks used for optical lithography. At present LIGA remains as an expensive approach for fabricating mm-wave accelerating structures.

4. SURFACE TREATMENT— ELECTROPOLISHING[16, 32, 33, 34]

To prepare smooth surfaces for mm-wave accelerating structures, both chemical cleaning and electropolishing have been investigated. Because of the grain inclusion, mechanical polishing is not being considered. Experimental results show that electropolishing can remove high spots on the surface and incorporated electrode material from EDM[35]. Because we had more success with electropolishing than chemical cleaning at SLAC, most of the work has been devoted to electropolishing with copper and results are presented below.

Electropolishing is an electrochemical process, requiring an electrolyte and DC current. It is similar to electroplating but is the reverse. Acidic chemical solutions are used as the electrolyte and DC current is applied simultaneously. Metal is selectively removed from the surface under controlled conditions. With the application of current, high points of surface roughness and burr areas are higher current density areas than the rest of surface and are dissolved away at a greater rate. This results in smoothing, leveling and deburring. During the process hydrogen is liberated from the cathode and oxygen from the anode. The electropolishing operation requires the proper balance between voltage and current. If the current density is too high, both the valleys and the peaks on the metal surface will be etched, resulting in pitting and gas evolution. When the current density is too low, etching occurs, resulting in non-specific removal of metal. The smoothness of electropolished surface is also dependent on the surface finish prior to the operation. An ideal relationship between current density and voltage is represented in Fig. 13. The key factors which affect the final surface finish are temperature, composition of chemical solutions, voltage, time and solution agitation. Figure 14 depicts the surface finish of copper cut by wire EDM and treated by electropolishing without agitation for 4 minutes. The copper removal rate is about 1 μm per minute. Comparing with Fig. 8, high points on the surface are smoothed out and the roughness is reduced by about a factor of 3 ($R_a \approx 0.2$ μm). After the electropolishing the grain boundary of pure copper can be seen clearly as depicted in Fig. 15.

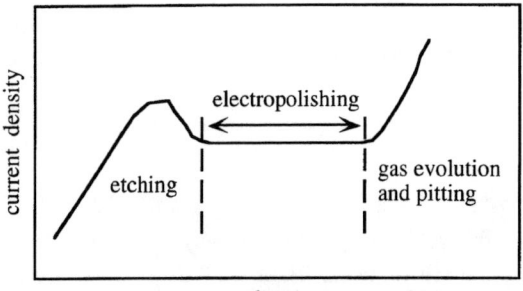

Fig. 13: Ideal relationship between current density and voltage in electropolishing cell.

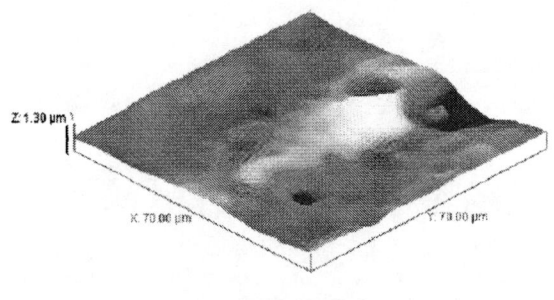

Fig. 14: Surface finish of copper cut by wire EDM and treated by electropolishing for 4 minutes. Picture was taken by an atomic force microscope.

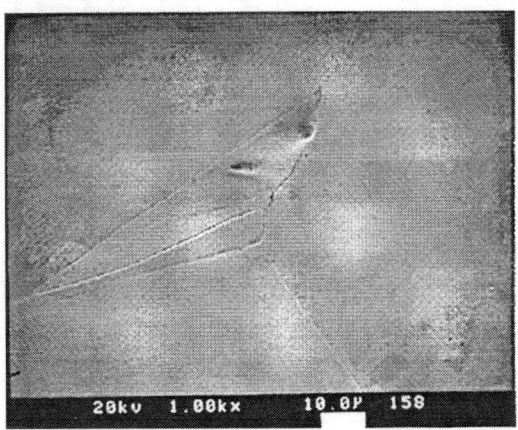

Fig. 15: The surface of copper cut by wire EDM and treated by electropolishing. The white bar represents a length of 10 μm. The crystal grains of pure copper can be seen clearly. Picture was taken by a scanning electron microscope.

When there is no agitation introduced to the bath, the surface waviness begins to develop as the processing time is increased. The surface waviness is depicted in Fig. 16; the processing time is 8 minutes.

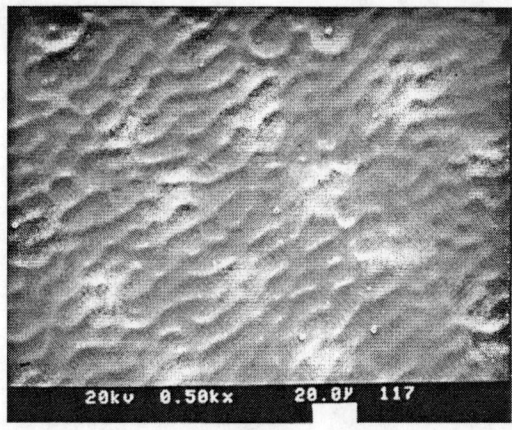

Fig. 16: The surface waviness on a copper block treated by electropolishing for 8 minutes without agitation. Picture was taken by a scanning electron microscope.

5. ENGINEERING DESIGN AND ALIGNMENT ISSUES

In practice an accelerating structure can not be fabricated from a single piece of metal. For the simplest case a structure is made from two identical pieces as depicted below:

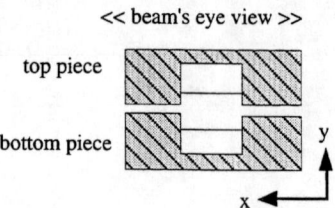

The schematic sketch for one half of a muffin-tin structure is depicted in Fig. 17.

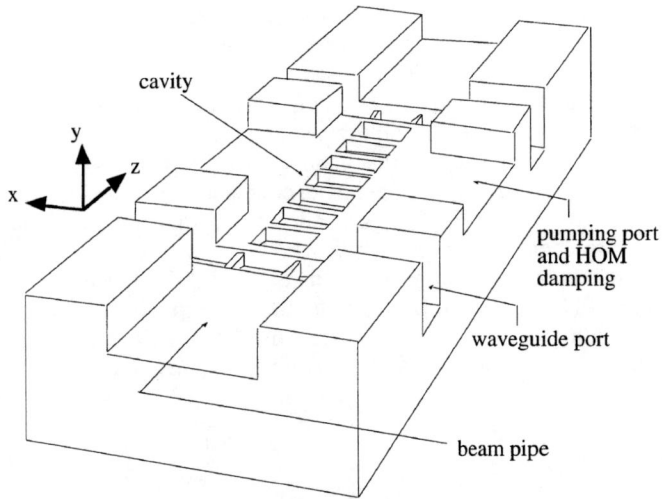

Fig. 17: The schematic sketch for one half of a muffin-tin structure with symmetrical ports for input and output waveguides.

When two pieces are assembled together and bonded, there is always some misalignment in x, y and z directions. An exaggerated sketch for misalignment in x direction is depicted below:

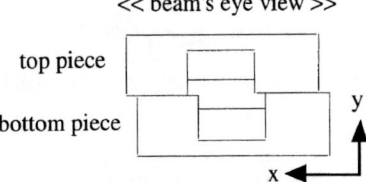

The alignment tolerances required for mm-wave accelerating structures are the same order of magnitude as the dimensional tolerances, i.e. 1 µm or less. One possible solution is to take advantage of the accuracy provided by fabrication technologies (e.g. LIGA can hold 2 µm tolerances or better) and machined a V-shaped gauge groove for structure assembling as depicted below:

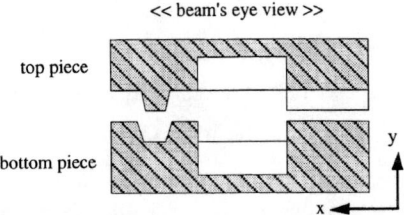

The width of the groove should be slightly larger than the one of the vane by 2-3 µm. This will set an upper limit for misalignments. The top and bottom pieces of the muffin-tin structure, with V-shaped vanes and grooves, are depicted in Figs. 18 and 19. The advantage of using V-shaped gauge grooves is that it is possible to achieve precise alignments in all directions simultaneously. As depicted in Fig. 18, this approach requires the design to have multiple level heights. A multi-level design is a challenge to micromachining techniques such as LIGA. Although microstructures with two level heights fabricated by LIGA have been reported recently[36], it is still uncertain whether the level height can be controlled within an accuracy of one micron or less. There are other alignment techniques being investigated. Alignment tolerances of ± 5 μm in translational direction of beam aperture by using anodic bonding has been reported[29]. Diffusion bonding is being considered as the choice of bonding structural components. The prerequisite for diffusion bonding is to have good surface finishes on all contact planes. This also manifests the importance of surface studies besides the concerns on field emission and RF breakdown.

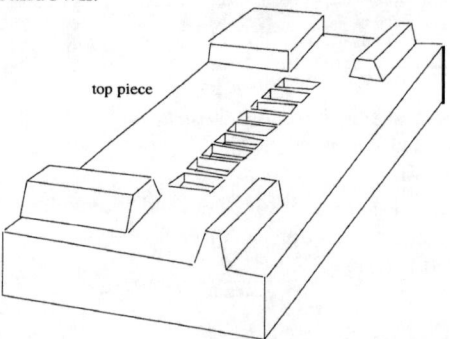

Fig. 18: The top piece of half of a muffin-tin structure with V-shaped vanes.

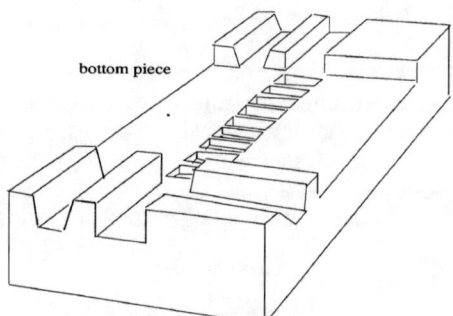

Fig. 19: The bottom piece of half of a muffin-tin structure with V-shaped grooves.

6. SUMMARY

Several fabrication techniques have been surveyed and discussed. The decision on the final choice of design and fabrication technique is quite involved. It will require a good understanding on the relationship between pulse heating and surface damage, properties of transverse wakefield, suppression of field emission and multipactor effects. The fabrication of mm-wave accelerating structures is testing the limit of conventional paradigm for accelerator design based on microwave technology.

ACKNOWLEDGMENTS

The authors would like to thank Eric Lundahl for performing coordinate measurements on test samples and Wayne Shen for helpful discussions.

REFERENCES

1. R.B. Palmer, in *AIP Conference Proceedings No.337, Montauk, NY 1994*, edited by R.C. Fernow (AIP Press, New York, 1995), pp.1- 15.
2. R.B. Neal et al. (eds.), *The Stanford Two-Mile Accelerator* (Benjamin, NY, 1968), p.96.
3. NLC design group, *Zeroth-Order Design Report for the Next Linear Collider*, SL\AC Rpt. 474.
4. H. Henke et al., Argonne National Laboratory Internal Report ANL/APS/MMW-1 (1993).
5. D.J. Whitehouse, *Handbook of Surface Metrology* (IOP Publishing, London, 1994).
6. I. Wilson in *RF Engineering for Particle Accelerators*, CERN 92-03 (1992), p.375.
7. R.H. Siemann (SLAC), private notes.
8. G. Bowden, NLC-ME-Note No. 8-96 (SLAC).
9. Y. Higashi (KEK), private communication.
10. This is performed with a dynamic focus optical scanner which is manufactured by UBM Corp., 544 Weddell Dr., Suite 1, Sunnyvale, CA 94089, U.S.A.
11. Fotofabrication Corp., 3758 Belmont Ave., Chicago, IL 60618, U.S.A.
12. Vaga Industries, http://www.bayrep.com/vaga.
13. Nat Wood, *CNC West* (August/ September 1993), p.16.
14. Ron Witherspoon Inc., 430 Industrial St., Campbell, CA 95008, U.S.A.
15. C.M. Rodia in *Metal Finishing Guidebook and Directory Issue,* edited by M. Murphy et al. (Elsevier Science, New York, 1995), p. 369.
16. C.A. Harper (ed.), *Handbook of Materials and Processes for Electronics* (McGraw-Hill, New York, 1970), p. 10-1.
17. A.J. Tuck Company, P.O. Box 215, 32 Tuck Rd., Brookfield, CT 06804.
18. S.M. Sze, *Semiconductor Devices: Physics and Technology* (Wiley, New York, 1985).
19. B. Loechel et al., *J. Vac. Sci. Technol.* **B13**, 2934 (1995).
20. B. Löchel et al., *Sensors Actuators* **A46**, 98 (1995).
21. P. Van Zant, *Microchip Fabrication*, 2nd ed. (McGraw-Hill, New York, 1995).
22. Ion Beam Milling, Inc., 1000 E. Industrial Park Dr., Manchester, N.H. 03103.
23. John Shott (Stanford University), private communication.
24. H. Mader in *Micro System Technologies 90,* edited by H. Reichl, (Springer-Verlag, Berlin, 1990), pp. 357- 365.
25. D. Bollinger et al., *Solid State Technol.* **27**, No.5 (May 1984), p. 111.
26. D. Bollinger et al., *Solid State Technol.* **27**, No. 6 (June 1984), p. 167.
27. I.W. Rangelow and H. Löschner, *J. Vac. Sci. Technol.* **B13**, 2394 (1995).
28. MicroScape, http://www.microscape1.com.
29. T.L. Willke and A.D. Feinerman, *J. Vac. Sci. Technol.* **B14**, 2524 (1996).
30. W. Ehrfeld in *Micro System Technologies 90,* edited by H. Reichl, (Springer-Verlag, Berlin, 1990), pp. 521- 537.
31. K.H. Jackson (Lawrence Berkeley National Laboratory), private communication.
32. J.F. Jumer in *Metal Finishing Guidebook and Directory Issue*, edited by M. Murphy et al. (Elsevier Science, New York, 1995), p. 420.
33. R.D. Cormia et al., "Electropolishing of Stainless Steel" presented at the *PCMI Annual Meeting*, Anaheim, CA, 1990 (Photo Chemical Machining Institute, Lafayette Hill, PA).
34. H. Diepers et al., *IEEE Trans. Nucl. Sci. Vol.* **20**, No. 3, 68 (1973).
35. E. Hoyt (SLAC), private communication.
36. J. Mohr in *Biomedical Applications of Synchrotron Radiation*, edited by E. Burattini et al. (IOS Press, Oxford, 1996), p. 181.

Recent Progress on Photonic Band Gap Accelerator Cavities

D. R. Smith*, Derun Li*, D. C. Vier*, N. Kroll*†, S. Schultz*

*Department of Physics, University of California, San Diego
9500 Gilman Drive, La Jolla, California 92093-0319
†Stanford Linear Accelerator Center, Stanford, CA 94305

H. Wang

Brookhaven National Laboratory
Upton, New York, 11973

Abstract. We report on the current status of our program to apply Photonic Band Gap (PBG) concepts to produce novel high-energy, high-intensity accelerator cavities. The PBG design on which we have concentrated our inital efforts consists of a square array of metal cylinders, terminated by conducting or superconducting sheets, and surrounded by microwave absorber on the periphery of the structure. A removed cylinder from the center of the array constitutes a site defect where a localized electromagnetic mode can occur. In previous work, we have proposed that this structure could be utilized as an accelerator cavity, with advantageous properties over conventional cavity designs. In the present work, we present further studies, including MAFIA-based numerical calculations and experimental measurements, demonstrating the feasibility of using the proposed structure in a real accelerator application.

INTRODUCTION

A realization of a Photonic Band Gap (PBG) cavity, depicted in Figure 1, consists of a periodic lattice of conducting, or superconducting, cylinders bounded on top and bottom by conducting (or again, superconducting) plates. Removing a cylinder from the center of the array forms a defect site, where a localized electromagnetic accelerating mode can exist over a wide range of lattice parameters. The interior lattice is surrounded by absorbing material at the periphery of the structure. For some applications, it may be desired to substitute dielectric cylinders in place of some or all of the metal cylinders; the use of dielectrics in PBG cavities will be discussed below, and also in a companion paper in these proceedings. In Figure 1 we have presented the absorber geometry as a toothed structure surrounding the periphery; but this is only one of many possible designs that might be used for higher order mode damping.

In a previous paper (1), we introduced the general concept of the PBG cavity, and listed its potential advantages over presently-used cavities. The most fundamental difference between the PBG cavity and a typical metal-walled cavity is with respect to the mode spectrum. A traditional cavity (e. g., a "pill-box" cavity),

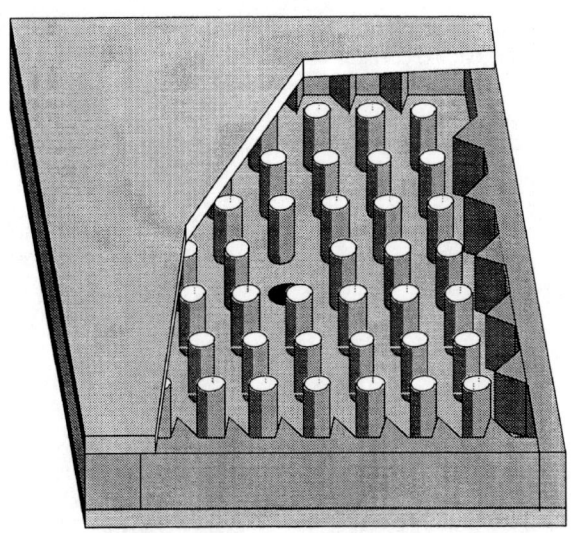

FIGURE 1. A schematic diagram of the components of a metal PBG cavity. The spacing between cylinders is d, the radius of a cylinder is r, and the height of a cylinder is h. Microwave absorber is placed around the periphery of the structure. In our copper test unit, the cylinders were brazed into the bounding metal plates; while in the niobium unit, the cylinders were e-beam welded to the plates.

has a discrete spectrum of modes, beginning with the fundamental and continuing over all frequencies. The type of mode and mode spacing are fixed by the size and geometry of the cavity. It is generally true that over the frequency range of interest for accelerator cavities, there will exist many higher order modes (HOMs) in addition to the accelerating mode, which may be excited by a beam which passes through the cavity. The loss of energy by a beam to the HOMs of an accelerating cavity is an important figure of merit of a cavity or cavity structure, and a significant amount of effort has been expended to design structures that either minimize HOM excitation or detune HOMs (2,3).

In contrast to the traditional cavity, the mode spectrum of a PBG cavity exhibits dense frequency bands of "extended" modes (known as "pass bands" in analogy with solid-state physics), interspersed with frequency bands of very low mode density (known as "stop bands"). In an infinite photonic lattice (no defects or termination), no modes can exist with frequencies in the stop bands. When the lattice contains a defect, such as the removed cylinder in the structures we consider here, "defect" modes can exist within the stop bands with fields localized at the defect site, and decaying exponentially away in all directions. In contrast, modes occuring within the pass bands have fields that are extended throughout the lattice, and can be damped by placing absorber around the periphery of the lattice, as shown in Figure 1. Since defect modes are localized at the center of the lattice, they remain relatively unperturbed by the absorber. Thus, the PBG cavity, which includes a photonic lattice surrounded by absorber, can be expected to have relatively few significant HOMs as compared to a traditional cavity.

In Figure 2 we present schematically the TM mode structures corresponding to a conventional metal-walled cavity, a dielectric-based PBG cavity, and a metal-based PBG cavity. The dielectric structure in general has periodically occuring stop bands that may or may not contain defect modes, depending on the parameters of the lattice. Also, it is a feature of dielectric based structures that the lowest modes (from zero frequency) are extended, and form the first pass band. The third part of the diagram shows schematically the TM modes of the particular metal PBG cavity we have been investigating. There is only one bound mode, which occurs as the lowest-frequency mode of the cavity.

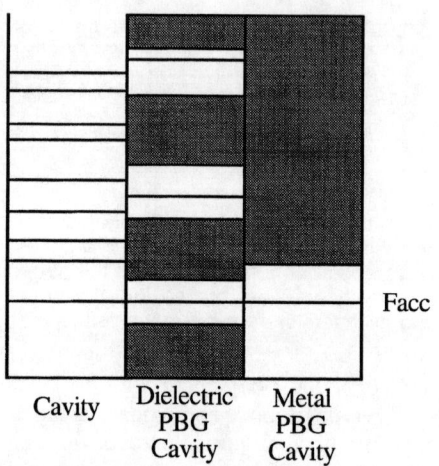

FIGURE 2. A conceptual diagram comparing the TM modes of a conventional cavity, a dielectric PBG cavity, and of a particular metal PBG cavity. For some parameters of a/d the metal PBG will have frequency band gaps in addition to the lowest; however, for the parameters we actually use, the diagram is accurate.

SUMMARY OF ACCELERATING MODE PROPERTIES

Q of the Accelerating Mode

The field pattern of the defect mode of the metal PBG cavity is very similar to that of the TM_{010} mode of a pill-box cavity. The mode has monopole symmetry, with a field maximum at the center of the defect region. The field decays into the surrounding photonic lattice, with an energy (E^2) decay rate of roughly two orders of magnitude per lattice constant; thus, the mode is very well localized. Our initial calculations indicated that a superconducting cavity with a 9 x 9 photonic lattice would have a Q-factor limited by superconductor losses rather than losses at the periphery of the structure. Therefore, absorber placed at the edges of the lattice to damp HOMs should not impact the fundamental.

Since our earlier numerical methods did not have the capabilities to simulate systems with lossy materials in a realistic geometry, we began simulating the configurations with MAFIA, a commercial finite-difference electromagnetic mode solver. With MAFIA, the presence of any losses in the simulated configuration can be solved by the use of MAFIA time-domain code, in which a source excites the cavity and the fields are calculated at successive time increments. To find the Q for a given configuration, we can excite the cavity at a given time t=0 with a current pulse and observe the subsequent time decay of the field at a some point in the cavity. Or, we can excite a continuous sinusoidally varying current density at the cavity's resonant frequency and note the field at which the solution is steady state; this field will be related to the Q of the mode.

For most parameters of interest, however, the Q of the fundamental is so large that it is impractical to run MAFIA until we obtain a either an accurate value for the decay time, or until the steady-state value of field is reached. For Q's of this size, we generally restrict ourselves to calculating the system only during the period where the fields are increasing linearly with time, as in Figure 3. Thus at any point during the time series, we can calculate the stored energy from the field data, the dissipated power in the regions with complex dielectric functions, and therefore the Q corresponding to the configuration. The Q calculated for the PBG cavity simulated in Figure 3 was ~22,000.

We find it convenient to calculate the Q at several places during the time of linear build-up, since the contribution from the excitation source near-fields will

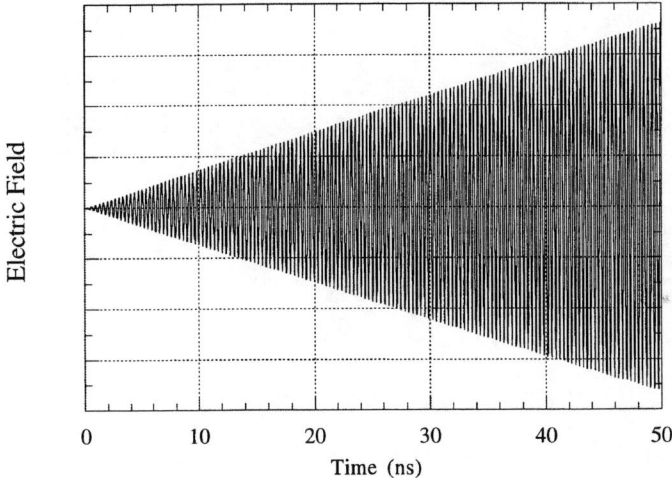

FIGURE 3. Electric field versus time at the center of a 7 x 7 PBG cavity. The cavity was driven by a sinusoidally varying current density at the frequency of the fundamental. The cavity was surrounded by a layer of toothed absorber (as shown schematically in Figure 1), which had a dielectric constant of 22 and conductivity $\sigma=0.3$. The Q at 35 ns was calculated as 22,367, while at 45 ns the Q was calculated as 21,918.

inevitably contribute to the stored energy. These near-fields are bounded in time, however, and thus their relative contribution to the stored energy decreases as a function of time. If we calculate approximately the same value for the Q at several times, then we can be confident that the source fields are not significantly perturbing the result.

Effects of Introducing an Iris

In order to make reasonable predictions about the performance of the accelerating mode in a practical structure, it is necessary to include the effect of a beam hole in the simulations. The presence of the beam hole complicates the numerical problem considerably, due to the loss of translational symmetry along the beam axis. However, MAFIA can accommodate most of the three dimensional geometries of current interest to us. In Figure 4, for example, we present a calculation of the accelerating mode frequency versus phase advance across a cell. The solid points on the curve were computed by MAFIA, for a single cell of a 7 x 7 PBG cavity, while the curve indicates a best fit to the conventional single band dispersion relation.

$$\left(\frac{1}{f}\right)^2 = \left(\frac{1}{F}\right)^2 [1+ \eta \cos \varphi] \tag{1}$$

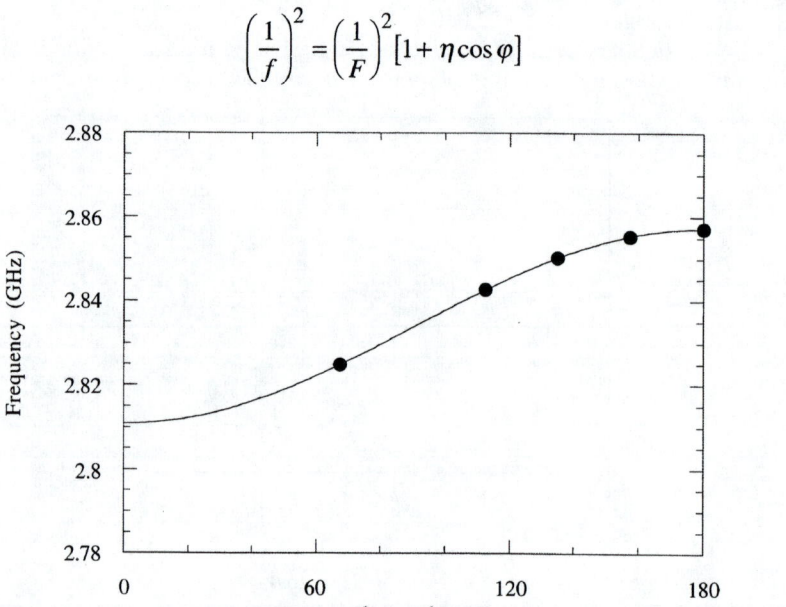

FIGURE 4. Dispersion curve of a 7 x 7 PBG lattice with beam holes. The parameters of the cavity are the same as those described in Figure 5.

The best fit was obtained for the values F=2.83372 GHz and η=0.0163591. The value of η is similar to conventional accelerator cavities, and confirms that we can achieve cell-to-cell coupling of PBG cavities through a beam iris.

In Figure 5 we present a summary of the parameters of the accelerating mode of a 7x7 PBG cavity, calculated using MAFIA frequency-domain code. The spacing between cylinders was 4.2 cm, while the cylinder radii a=0.7875 cm, for an a/d ratio of 0.1875. For the purposes of this calculation, beam holes were included, with radius r_b of 1.5919 cm, centered at the defect site on both the upper and lower plates of the PBG cavity. The beam hole size was chosen arbitrarily, and matches the scaled beam-hole radius of the TESLA cavity. Since both E_{peak}/E_{acc} and the shunt impedance R depend on the characteristics of the beam hole, it should be possible to significantly improve these parameters relative to the TESLA cavity. The PBG cavity length was determined as L=2.6242 cm, with π-phase advance across the cavity.

Effect of Cavity Asymmetry

As evidenced in Figure 5, the parameters of the PBG cavity are very similar to those of presently-used accelerator cavities, which is not surprising considering that the fields are very similar to one another (particularly near the beam axis). However, the unique symmetry and geometry of the PBG suggest that some complications might arise.

For example, since the PBG cavity has a definite four-fold symmetry with respect to rotations, the field of the accelerating mode will have higher multipole components. This is in contrast to the pill-box cavity and other cylindrically symmetric structures. One might anticipate some detrimental effect on a beam which passes through such an asymmetric cavity. In order to obtain some estimate of this effect (without resorting to full beam-dynamics simulations), we can calculate the change in energy of an accelerated test particle as a function of position within the beam hole region.

The energy gain per charge is the integral:

$$|\Delta E| = \left| \int_0^d E_z(r,\varphi,z) e^{i\frac{\omega z}{c}} dz \right| \qquad (2)$$

where E_z is the component of the electric field in the beam direction, and the integral is taken along a line from one end of the cavity to the other, at a radial position within the beam hole. Due to the four-fold symmetry of our structure, ΔE can be expanded in terms of spatial harmonics as

$$\Delta E(r,\varphi) \cong E_0 \left\{ 1 + a_1 \cos(4\varphi) r^4 + a_2 \cos(8\varphi) r^8 \right\} \qquad (3)$$

with only the first three terms of the expansion retained. By taking two values of the field at $\varphi=0°$ and $\varphi=45°$ (constant r), and assuming all other orders are negligible, we can determine the values of the coefficients of a_1 and a_2. Using data from a MAFIA E-code calculation, we find the approximate values a_1=-7.5 x 10^{-5} (1± 0.016) mm^{-4}, and a_2=1.614 x 10^{-8} (1± 0.283) mm^{-8}. As an example, if we take 100 microns for the beam excursion away from the beam axis, then we obtain

MAFIA Simulation Results for a 7x7 PBG Cavity and Comparison with TESLA Cavity

Fundamental Parameter	TESLA (HERA) Cavity	Scaled TESLA Cavity	A 7x7 PBG Cavity
Lattice Width d (mm)	N/A	N/A	41.999
Cylinder Radius a (mm)	N/A	N/A	7.875
Beam Hole Radius r_b (mm)	35.0	15.918	15.919
Nose Radius r_n (mm)	12.0	5.458	5.458
Half Gap Length for π-Mode, L (mm)	57.692	26.238	26.242 (includes two plates; thickness $2 x r_n$)
Fundamental π-Mode Frequency (GHz)	1.299	2.8560	2.8575
Peak Electric Field to Acceleration Field Ratio E_{peak}/E_{acc}	2.39	2.39	2.58
Peak Magnetic Field to Acceleration Field Ratio H_{peak}/E_{acc} [Oe/(MV/m)]	46.36	46.36	104.22
Total Stored Energy U_0/E_{acc}^2 [J/(MV/m)^2]	0.0141	1.335×10^{-3}	1.350×10^{-3}
Q_0 Factor (copper, room temperature)	26,879.6	18,125.0	14,790.8
Shunt Impedance R (MΩ)	3.074	2.073	1.681
R/Q_0 (Ω)	114.4	114.4	113.7
Geometry Factor G	252.7	252.7	206.3
Loss Parameter to Fundamental Mode k_{fu} (Beam Bunch σ=6 mm) (V/pC)	0.2305	0.4811	0.4784

FIGURE 5. Comparison of parameters for the TESLA, Scaled Tesla, and 7x7 PBG cavities. Note that we assume the cavities are made of copper at room temperature.

maximum voltage variations at the 10^{-8} level. This excursion distance is a reasonable beam excursion for the NLC, and thus the asymmetry effect is quite negligible.

Effect of Cylinder Misalignment

Another potential concern is the effect of slight cylinder misalignment on the accelerating mode of the PBG cavity. Typically, we find that the PBG effect is extremely robust, and relatively insensitive to minor random perturbations in the underlying photonic lattice; however, for accelerator applications, even minor results may have significance. In order to assess the effect of randomness on the mode, we simulated (using MAFIA E-code) a 7x7 PBG cavity, using lossless cylinders, and surrounded by copper walls (rather than absorber) on the periphery.

A structure with no cylinder misalignment yields a Q of over 10^{11}; for the structure in which cylinders are randomly perturbed, we expect this Q to be degraded. We also expect a frequency shift. In Table 1 we see that even a sizable amount of disorder produces only minor frequency shifts and Q-degradation of the fundamental mode. The cylinder displacement used in the calculations, ± 1.47 mm, was much larger than the achievable tolerance for fabrication.

It should be pointed out that the fundamental mode depends most strongly on the cylinders closest to the defect region. Since the mode decays exponentially away from the center, equivalent displacements in cylinder positions located farther from the center can be expected to produce smaller perturbations to the localized modes. This further suggests that material and fabrication precision constraints are highest immediately at the center of the structure, and are reduced away from the central group of cylinders. As an illustration of this effect, we have simulated a 9x9 PBG cavity in which we have inserted a cut in the bounding plates around the inner 7x7 photonic lattice. The cut entirely isolates the center defect and (lossless) cylinders from the outer cylinder layer and absorber. Field plots of the defect mode display no significant perturbation after the cut. Furthermore, the Q of the mode is also not significantly reduced. An actual superconducting cavity constructed with a cut would only require the center portion to be cooled to cryogenic temperatures; the thermally isolated outer cylinders, absorber, and walls could be maintained at higher temperatures.

	Δf (MHz)	Q- factor (x 10^{11})
no misalignment	0	2.185
misalignment 1	6.033	1.74269
misalignment 2	8.536	1.841
misalignment 3	58.75	1.117

TABLE 1. Effect of disorder on the frequency and Q of the accelerating mode of the metal PBG cavity. The unperturbed frequency of the mode is 2.920402 GHz. In each case the cylinder spacing was d=3.9272 cm, and cylinder radius was a=0.73635 cm. For the misalignment calculations the cylinders were randomly displaced by ± 1.47 mm in the x and y directions. This translation corresponded to the distance between points on the discretization grid.

SUPERCONDUCTING TEST CAVITIES

Encouraged by early numerical simulation results, as well as initial tests on copper PBG cavities at room temperature, we designed and had fabricated the first set of superconducting (SC) PBG cavities. The SC-PBG cavities consisted of a periodic array of Nb cylinders, forming the underlying photonic lattice, bounded on top and bottm by Nb plates. These first SC cavities were desgned to resonate at ~11 GHz so that they would be small enough for Q measurements and low-power testing in a small liquid-helium dewar. The experience gained in the fabrication and welding techniques, however, are directly applicable to the final high power test structure, which will be at 2.856 GHz in order to be compatible with the available rf power system at the Accelerator Test Facility at Brookhaven National Laboratory.

The foremost consideration in constructing the SC cavity is to determine the optimal method of joining the many cylinders of the photonic lattice to the bounding plates, since the ultimate Q of the cavity in practice may be set by the surface resistance for current flow at these points. The initial approach was to use electron-beam welding to join the Nb cylinders to the Nb plates, which was performed at CEBAF. For a first test, a 3x3 cavity was constructed. The CEBAF shop made a total of 16 weld joints, varying settings on the e-beam welder to optimize welding conditions. As a result of this first test, the settings required to produce clean, reproducible welds were determined.

In order to perform low power testing and Q measurements of the Nb PBG test cavity, a small liquid helium test dewar was assembled, which operates in the frequency range of 1-20 GHz and over the temperature range of pumped helium to room temperature. The testing unit consists of a support structure for mounting the PBG cavity which resides inside of a commercial liquid helium dewar.

After electron beam welding by the CEBAF group, the 7x7 cavity, with no absorber, was tested from room temperature down to 4.2 K, and the cavity Q measured at selected temperatures. At 4.8 K, the measured Q of the 7x7 cavity was found to be at least 1.2×10^6. This Q was determined by analyzing the reflected power spectrum over the frequency range of the resonance, and calculating the FWHM of the resonance curve. The cavity was strongly under-coupled for the measurement, such that the measured Q should be close to the unloaded cavity Q. This measurement provides a lower limit for what the cavity Q actually is, since the resolution of our test equipment places a limit on the Q we can measure. For more accurate determination of such high Q's, an energy decay measurement will be used.

CONCLUSION

We are continuing our numerical exploration of the characteristics of the PBG cavity. The extensive modeling done so far has revealed both the potential advantages and weaknesses of the PBG cavity relative to current and proposed accelerator cavities. We continue to be hopeful about the prospect of producing a PBG cavity with well-damped HOMs; indeed, for room-temperature copper structures, we believe that we have succeeded in that goal. The reduction of these

HOMs indicates the PBG cavity may find use in high-current applications; furthermore, the ability to alter the HOMs without affecting the fundamental suggests that the PBG cavity might also be a potential candidate for a detuned structure, such as the SLAC DDS.

While we remain optimistic with respect to developing high-gradient structures, it is clear from the simulations that it will be necessary to find a geometry in which H_{peak}/E_{acc} can be reduced (see Figure 5). For superconducting applications, this appears to be the most limiting factor for PBG cavities. One potential method of doing this would be to adjust the lattice parameters such that the fundamental would be less localized to the interior, allowing the confining current to be shared throughout more of the cylinders. This would reduce H_{peak}/E_{acc}, but at the expense of also reducing the shunt impedance, and potentially requiring more layers of cylinders. Nevertheless, such design modifications can be made, and the feasibility of the resulting structure will depend on the application and competitive cavities.

Finally, we note that the PBG structure is relatively simple to construct, requiring only rod and flat sheets. We feel that fabrication of these structures could be implemented with current technologies at frequencies as high as 90 GHz. Furthermore, substituting dielectrics for any or all of the metal rods also produces a PBG structure which can have few (possibly one) localized modes. Previous work (4) suggests that suitably treated dielectrics can exhibit very low loss, and are capable of sustaining large gradients; thus, we intend to further investigate both pure dielectric structures, as well as dielectric/metal hybrid structures.

ACKNOWLEDGEMENTS

We thank Ricci Campisi and John Brawley for their asssistance in fabrication of the superconducting PBG cavities, and AccSys Technologies for fabrication of the copper test cavities used in this study. We also thank H. Padamsee for helpful discussions. This research has been supported by the DOE, contracts DE-FG-03-93ER40793 and DE-AC-03-76SF00515.

REFERENCES

1. Schultz S., Smith D. R., and Kroll N., Proc. 1993 Particle Accelerator Conf., (May 1993 Washington, D. C.) IEEE, **4**, 2559 (1994)
2. Thompson K. A., Adolphsen C., Bane K. L. F., Deruyter H., Farkas Z. D., et. al., Particle Acclerators, **47**, 65 (1994)
3. Kroll N., Thompson K., Bane K., Gluckstern R., Ko K., Miller R., and Ruth R., 6th Workshop on Advanced Accelerator Concepts, Lake Geneva, WI, June 12-18 (1994); SLAC-PUB-6660
4. Gai W., Schoessow P., Cole B., Konecny R., Norem J., Rosenzweig J., and Simpson J., Phys. Rev. Lett., **61**, 2756 (1988)

Wake-Field Studies on Photonic Band Gap Accelerator Cavities

Derun Li*, N. Kroll*[†], D. R. Smith* and S. Schultz*

*Department of Physics, University of California, San Diego
9500 Gilman Drive, La Jolla, California 92093-0319

[†]Stanford Linear Accelerator Center
M/S 26, P.O. Box 4349, Stanford, California 94309

Abstract. We have studied the wake-field of several metal Photonic Band Gap (PBG) cavities which consist of either a square or a hexagonal array of metal cylinders, bounded on top and bottom by conducting or superconducting sheets, surrounded by placing microwave absorber at the periphery or by replacing outer rows of metal cylinders with lossy dielectric ones, or by metallic walls. A removed cylinder from the center of the array constitutes a site defect where a localized electromagnetic mode can occur. While both monopole and dipole wake-fields have been studied, we confine our attention here mainly to the dipole case. The dipole wake-field is produced by modes in the propagation bands which tend to fill the entire cavity more or less uniformly and are thus easy to damp selectively. MAFIA time domain simulation of the transverse wake-field has been compared with that of a cylindrical pill-box comparison cavity. Even without damping the wake-field of the metal PBG cavity is substantially smaller than that of the pill-box cavity and may be further reduced by increasing the size of the lattice. By introducing lossy material at the periphery we have been able to produce Q factors for the dipole modes in the 40 to 120 range without significantly degrading the accelerating mode.

INTRODUCTION

One of the more intriguing features of the Photonic Band Gap (PBG) cavity is the unique higher order mode (HOM) spectrum associated with the periodic metal (or dielectric) array of scatterers. As opposed to the modes of conventional metal-walled cavities, the electromagnetic modes of the PBG cavity are mostly extended; only a few of the modes are actually localized to any significant degree within the central defect region of the structure. In the metal PBG structure which we study in this paper, for example, we find that in principle there is only one bound mode, the accelerating mode. All other modes are considerably more extended, and thus will be damped to a greater degree by lossy material (absorber) placed at the boundary of the structure. The possibility of damping these extended modes without seriously perturbing the accelerating mode, led us to conjecture that a PBG cavity could be developed with no HOMs [1]. By this, we suggested that a particle beam passing

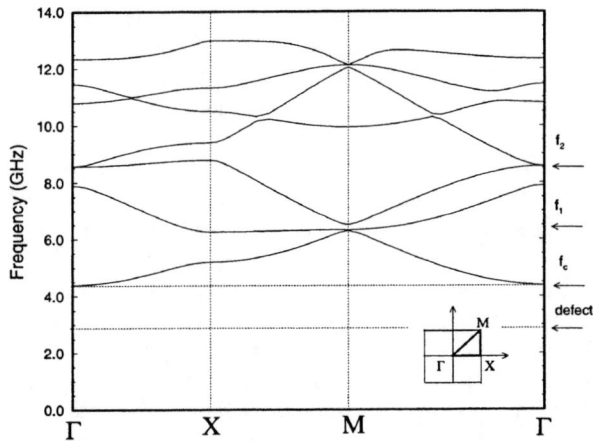

Figure 1: Dispersion relations for a square lattice metal photonic band structure consisting of cylinders of radius a spaced a distance d apart. The calculations were performed for a two-dimensional structure with infinite extent in the lateral dimensions. Periodic boundary conditions are assumed for the calculation and $\frac{a}{d} = 0.1875$

through a damped PBG cavity could have a negligible long-range wake-field, without the necessity for elaborate mode couplers or detuning.

Our initial numerical efforts focused on understanding the dispersion relation, or photonic band structure, associated with the underlying periodic lattice of the PBG cavity. The TM band structure for an infinite metal lattice is shown in Figure 1, over the periphery of a selected octant in k-space; the band structure for k values larger than those of the band edges are folded back into this octant, thus forming the complex series of bands shown. A more detailed discussion of the band structure can be found in a companion paper submitted at this conference [2], or in our previous publications [3, 4]. Here, the only information we require from the band structure is whether or not at a given frequency there are modes associated with the lattice. As can be seen from the Figure 1, the region from zero frequency up to the cut-off frequency f_c contains no modes, while for any frequency above f_c, there is always at least one corresponding mode. Since true localized modes require frequencies to be within a band gap region (e.g. below f_c), the band structure implies that there will be no such localized modes above f_c. The calculation in Figure 1 was performed for an infinite lattice with no defects; removing one cylinder from the array produces a localized defect mode below f_c, and may perturb the modes above f_c such that quasi-bound modes and resonance may occur. We identify two typical types of regions on the band structure where we might expect such behavior, one at f_1 designated in Figure 1 and the other at f_c. Just below f_1, we note that extended modes exist for only a limited region of k-space. Thus, at frequencies such as f_1 there may be a partially bound mode, which we refer to as a resonance. We have studied these resonances in some detail previously [6].

At f_c, we note that the group velocity of modes near the band edge approaches zero. One might anticipate that even an extended mode may have an undesirable large quality factor, Q, since energy may not be transported quickly enough out to the absorber for the mode to be damped. Our studies indicate so far, however, that the band-edge resonances can in fact be effectively damped by absorber at the lattice boundary.

While we find the preliminary results from the band structure calculations encouraging, what is of ultimate importance is the wake-field generated by a drive current in the cavity. The assertion of *no HOMs* only makes sense in this context. In order to carry out the more difficult study of the excitation of cavity modes by a drive beam, we use MAFIA, a commercial finite-difference analysis program. In this paper, we present the results of MAFIA frequency- and time-domain simulations performed to address the issue of wake-fields in single PBG cavities without a beam iris. For comparison purposes we perform similar simulations for a simple cylindrical pill box cavity. Because there is no beam iris and no $v = c$ (speed of light) synchronous mode selection, these are not realistic wake-field calculations for an accelerator, but the comparative amplitudes are expected to be quite similar to those of the realistic case. We expect these simulations to provide useful guidance for the more realistic studies of the wake-field of an iris coupled periodic PBG structure which are now underway.

Figure 2 shows a typical PBG geometry; in this case, we have simulated a 7×7 metal PBG cavity, and obtained the field distribution of a localized mode from a MAFIA frequency-domain simulation. The defect mode (or accelerating mode) is the lowest frequency mode in the structure, and its field pattern resembles very closely the field pattern of the TM_{010} mode of the familiar pill box cavity [2].

HIGHER ORDER MODES AND WAKE-FIELDS STUDIES

Wake-field studies have been carried out by means of both frequency- and time-domain MAFIA simulations. Our attention has been devoted to the simulation of the dipole modes, since coupling of these modes to the beam can result in transverse emittance growth. Modes of other symmetry were studied, but these results will not be presented here.

For all the simulations which follow, we use a square lattice of metal cylinders, with one cylinder removed from the center of the array. The ratio of the cylinder radius to lattice spacing is $\frac{a}{d} = 0.1875$. This ratio determines the form of the band structure shown in Figure 1. The lattice spacing ($d = 39.272$ mm) sets the overall frequency scaling of the modes. The frequency of a given mode varies as $\frac{1}{d}$. The axial boundaries were conducting sheets with a separation distance of 4 mm, intentionally chosen small so that within the limited frequency range of our initial exploration we could restrict our attention to modes with no axial variation. These parameters yielded a defect mode with frequency at 2.92 GHz.

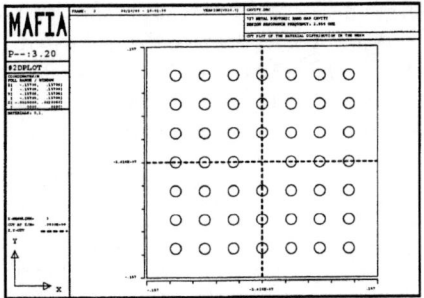
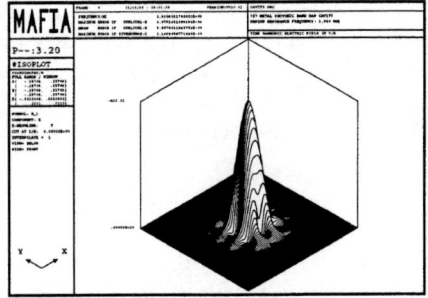

Figure 2: Left graph is a 2-D view of a 7 × 7 square lattice metal PBG cavity with $\frac{a}{d} = 0.1875$, where a and d are the radii of cylinders and lattice constant, the distance between the two centers of two nearest cylinders. The defect is formed by removing a cylinder from the center. The cavity is bounded by a top and bottom conducting sheet and surrounded by conducting walls around the periphery. The resonance frequency of the defect mode is 2.92 GHz obtained from MAFIA frequency- domain simulations with $d = 39.272$ mm. The E_z field distribution of the defect mode is plotted along $z = 0$ cut plane of the PBG cavity (right).

Higher Order Mode Studies in the Frequency-Domain

We initially explored the field patterns of the electromagnetic modes of the PBG cavity, using frequency-domain MAFIA simulations. In all cases one quarter of the structure was simulated with dipole boundary conditions imposed. Because the dipole modes are expected to lie in pass bands, and because the overall linear dimensions of the structure are large, a large density of modes is expected. The first 40 dipole modes of a 7 × 7 structure lay between 4.0 GHz and 8.75 GHz while the first 55 modes of a 9 × 9 structure lay between 4.0 GHz and 8.7 GHz. The increased mode density of the larger structure is of course expected. The field patterns of all modes were visually inspected. Most modes had frequencies in the propagating regions, and the majority of these modes were extended throughout the volume of the cavity. While the mode density of the PBG cavity is greater than that of a conventional cavity, the extended PBG modes have small R/Q due to the large volume of the spatial region which they occupy. Thus, it should be easy to damp them by introducing loss in the outer regions of the structure such that the the accelerating mode remains relatively unperturbed. Several modes exist in the band gap region below 4.2 GHz, corresponding to boundary modes; that is, modes with field patterns greatest between the outer boundary of the cavity and the lattice. These fields decay exponentially as they penetrate the lattice, and consequently have negligible R/Q for beams near the center of the structure, and are particularly easy to damp.

Finally, we find a number of quasi-localized modes. These occur near zero slope portions of the band structure dispersion diagram shown in Figure 1. Within the frequency ranged covered, modes displaying some degree of localization occurred near the band edge at 4.2 GHz and the crossing point at 6.4 GHz, indicated in Figure 1 as f_c and f_1, respectively. These modes have larger R/Q. values more in line with HOMs of conventional cavities, and are more difficult to selectively damp. Subsequent wake-field suppression investigation has been concentrated on modes of this type.

Wake-Field Studies in the Time Domain

Coupling of a drive beam to the modes of a PBG cavity was studied using MAFIA time-domain simulations. Studying this problem in the time-domain provides a more realistic picture of cavity performance, and also allows us to simulate lossy materials. Dipole modes were excited with a $v = c$ Gaussian beam bunch with charge density,

$$\rho(\vec{r},t) = Q\lambda(z-ct)\delta(x)\delta(y-y_0), \qquad \lambda(s) = \frac{1}{\sigma\sqrt{2\pi}}e^{-\frac{(s-s_0)}{2\sigma^2}} \qquad (1)$$

with $\sigma = 8$ mm, $Q = 1$ Coulomb and y coordinate offset y_0 of 5 mm. The resultant wake-field was monitored by recording the x component of the magnetic field at the origin. High frequency mode excitation is suppressed by the width of the Gaussian beam rather than by a beam iris radius (zero for these studies). For comparison purposes, similar calculations were performed on a cylindrical pill box cavity of the same height and with radius chosen to match the lowest monopole mode frequency of the defect mode. Because there is no beam iris and no synchronous mode selection, these are not realistic wake-field calculations for an accelerator; however, the comparative amplitudes are expected to be quite similar to those of the realistic case.

We find that even without damping, the general level of the wake-field for a 7×7 PBG cavity is a factor of three smaller than that of the comparison pill box cavity (see Figure 3). Fourier analysis of the PBG wake-fields shows it to be dominated by a few modes. The characteristics of these modes are then further examined by using a Gaussian drive bunch modulated at the frequency of the mode that one wishes to study. By noting the frequencies of these peaks and inspecting the field patterns which remained after the excitation had died away, we identified the modes below 9 GHz with the quasi-localized modes found in the frequency-domain simulations. Similar modes were found at the higher frequencies of 9.8 GHz and 12.4 GHz; these, in fact, are implied by the dispersion diagram of the band structure in Figure 1. The Gaussian drive bunch used had a larger frequency range than that set by the 40-mode limit used for the frequency-domain simulation. We note that simulations in time-domain are much faster than those in the frequency-domain, and that the time-domain simulations provide a much more effective way of focusing on modes likely to contribute significantly to the wake-field. Apparently only modes with significant R/Q at the excitation point are noticeably excited, and those with the largest R/Q stand out in the Fourier transform.

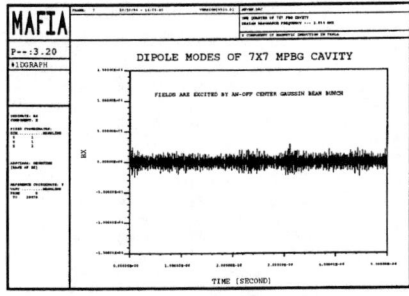
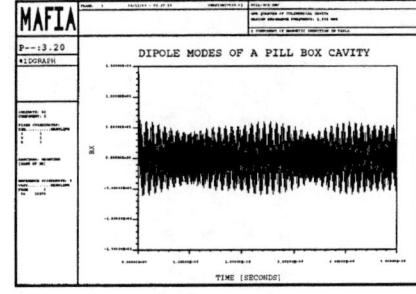
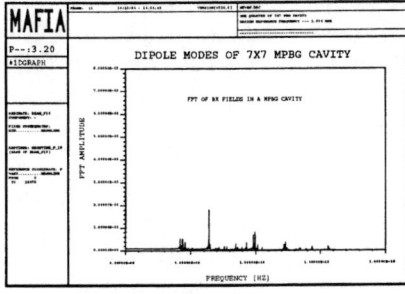
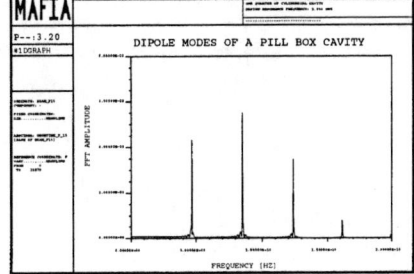

Figure 3: Wake-field $B_x(t)$ at the origin, excited by the y-offset Gaussian bunch, is plotted against time t. For this simulation the total time duration was 50 nano-seconds. Fast Fourier Transformation (FFT) was performed on $B_x(t)$ to obtain the spectrum. The left two graphs show the simulation results for a 7×7 PBG cavity, and the right ones are for that of the comparison cylindrical pill box cavity. For comparison purposes, all the above coordinates are depicted with the same coordinate scaling.

Volume Effects

As most dipole modes are extended, we may expect that R/Q values of these modes could be further reduced with an increase of the cavity volume. Perturbations to the quasi-localized modes should also be anticipated. Such a *volume effect* study was carried out using MAFIA time-domain simulations on PBG cavities of 7×7, 9×9 and 11×11 with the same square lattice geometry parameters given in Figure 2.

The simulations indicated that field strengths of most modes in propagation bands were indeed further reduced with increase in cavity volume. However, the strengths of the particular quasi-localized modes at 6.4 GHz, 9.8 GHz and 12.4 GHz appeared to increase, an effect which is currently being studied.

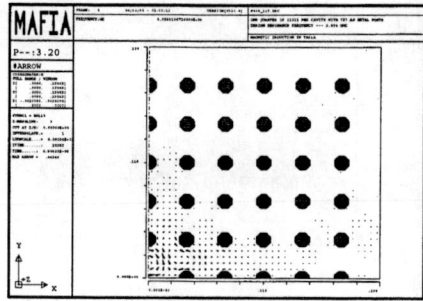 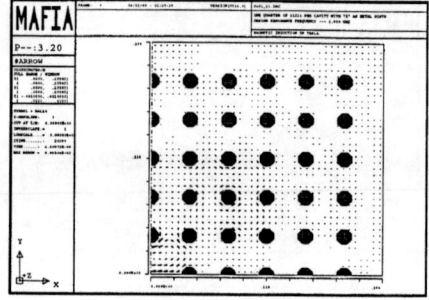

Figure 4: Magnetic fields are depicted within a transverse plane. The simulations were performed on an 11 × 11 metal PBG cavity with the last two rows as lossy cylinders using MAFIA time-domain code; fields were excited by a modulated Gaussian pulse at $f = 6.4$ GHz (left) and $f = 4.2$ GHz (right), respectively. The field patterns were recorded at the end of the simulation time of 50 nano-seconds.

Q FACTOR CALCULATIONS OF HIGHER ORDER MODES

We have developed two methods to calculate the Q factors corresponding to the cavity modes, using MAFIA time-domain simulation results. The first method, in which we measure the the time-decay of an excited field, works best for modes with low Q factors. The second method, in which the Q is calculated at a given time from the stored field patterns of the mode, works best for modes with large Q-factors. A discussion of the latter method is contained in [2] and will not be repeated here. Note that the Qs quoted in the context of lossy materials are the partial Qs associated with losses in the material. Other losses such as copper losses are not included.

The HOMs that we study here all had sufficiently low Q such that we could apply the first method for their analysis. In the time-decay method, we would ideally like to excite a mode at $t = 0$ with a sinusoidal excitation, such that only the mode of interest would grow with time, then switch the current off rapidly and measure the resulting decay time of the field from the time-series. This flexibility with driving functions, however, did not appear to be contained as an option in the version of MAFIA available to us; therefore, we instead utilized the *Gaussin* drive function, which has the mathematical form

$$f(t) = \begin{cases} 0, & \text{for } \tau \leq 0 \\ e^{-((\tau - p_3)\pi p_2)^2} \sin(\omega(\tau - p_3)), & \text{otherwise} \end{cases} \quad (2)$$

with $\tau = t - t_0$, p_2 and p_3 as parameters to define the starting time and band width of the pulse. In choosing the parameters of the *Gaussin* pulse, our goal was to have a pulse width wide enough that only the particular mode of interest was excited, yet narrow enough that we could make an accurate measurement of the decay time after

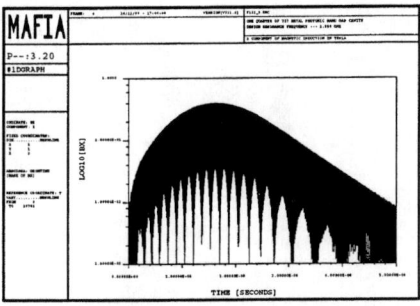
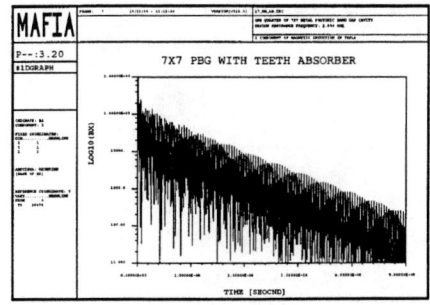

Figure 5: Left graph: Q_0 calculations for a 9×9 PBG cavity with the last row as lossy cylinders. The *Gaussin* drive pulse envelope was down 40 dB at time zero (t_0), peaked approximately at 12.7 ns, and was down 40 dB again at 25.4 ns. The modulation frequency was 12.2 GHz. The extended region of smooth linear decay at later times in the logarithmic plot indicates that only a single mode was excited. The Q determined from the slope and the frequency is 286. Right graph: wake-fields in a 7×7 metal PBG cavity with *teeth* absorber around the periphery as shown in Figure 6, is depicted in a logarithm scale to show the decay pattern of the wake-fields. The wake-fields were driven by the Gaussian beam pulse with an y offset of 5 mm.

the drive pulse had decayed to negligible amplitude. A typical HOM simulation of this type is shown in Figure 5, where it can be seen that the regions of excitation and mode-decay are clearly distinguishable.

EXAMPLES OF PBG CAVITIES WITH DAMPING

Having identified particular modes likely to contribute significantly to the wake-field, we now investigate the possibility of damping these modes relative to the accelerating mode. Initial experimental investigations [6] in which absorber was introduced into a metal PBG cavity indicated that the majority of HOMs could be selectively damped, although certain quasi-localized modes remained with significant Q. Using MAFIA simulations, we can rapidly test numerous absorber configurations and attempt to optimize the resulting wake-field. We present below two examples from the many geometries we have simulated. In each example, we have introduced the absorber in such a way as to minimize the lattice/absorber mismatch, which could potentially contribute to the creation of trapped modes with moderate Q factors.

Tooth-Shaped Lossy Dielectric Material

Since the HOMs tend to be extended, absorber placed at the periphery of the PBG structure can be expected to be effective in reducing the Q of those modes. Such a configuration was modeled, the geometry of which is shown in Figure 6.

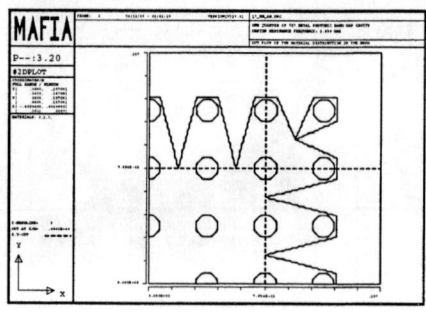
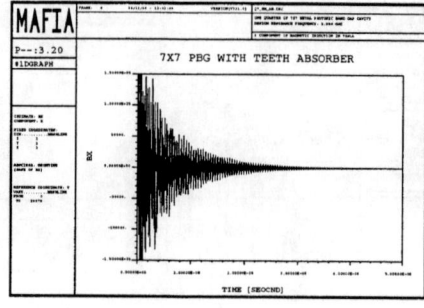

Figure 6: 2-D view of a 7×7 metal PBG cavity (1/4 structure) with *teeth* absorber around the periphery (left). A Gaussian beam bunch was sent to pass through the cavity with an y offset 5 mm. Boundary conditions were imposed to have only dipole modes be excited. The resulting wake-fields are given in the right graph. Note that the ordinate scale here is 10 times smaller than that of the undamped case shown in Figure 3.

The surrounding absorber was composed of triangular teeth, projecting between the cylinders of the last cylinder row to the line centers of the next cylinder row. The dielectric constant of the absorber material is 4, with a conductivity of 0.3 (MKS units). As described above, a MAFIA wake-field simulation was performed to identify significant HOMs, followed by a time-domain Q factor calculation on each mode. The following table lists Q factors for all sizable HOMs within the band width of limited by the excitation pulse. The Q of the accelerating mode was determined by the energy method, and was found to give $Q = 22,000$.

HOM Dipole Modes	4.2 GHz	6.4 GHz	9.8 GHz	12.4 GHz
Q Factors	110	80	46	65

Lossy Dielectric Cylinders

We also studied a number of cases in which absorber was introduced into the PBG cavity by replacing certain of the metal cylinders with high dielectric, lossy cylinders. Our expectation was that due to their high dielectric constant, the dielectric cylinder rows would present a good match to the metal PBG lattice. The dielectric constant and conductivity of the outer cylinders were 22 and 0.3, respectively (CEBAF absorber material [7]). One such example was provided by an 11×11 PBG cavity in which the outer two rows of cylinders were so replaced. The Q for the accelerating mode was found to be 1.4×10^7, limited by the dielectric losses. The Qs for the quasi-localized modes at 6.4 GHz and 4.3 GHz were 282 and 166, respectively; the Qs for all other modes were not computed, but appeared to be smaller.

We note that the quasi-localized mode at 6.4 GHz, shown in Figure 4, couples weakly to another nearby extended mode.

CONCLUSION

These initial wake-field studies have provided us with some guidance for damping the HOMs of a PBG cavity. We are now in the process of extending our investigations to the more realistic case of the synchronous HOMs of an iris-coupled periodic structure. We have performed a preliminary MAFIA frequency-domain calculation on an S-band (fundamental frequency of 2,856 MHz) cavity with $\frac{2\pi}{3}$ phase advance per cell, and iris radius of 18 mm. Inspection of the field plots of HOM's with frequencies up to 6.7 GHz and phase advances from 80 to 180 degrees yielded no modes judged to be quasi-localized. This suggests the possibility that opening the beam iris may facilitate the damping of HOMs. Further investigations and more detailed results for this study will be presented elsewhere.

ACKNOWLEDGMENT

We thank Prof. H. Padamsee at Cornell University for helpful discussions and valuable suggestions, and Mr. H. Wang at BNL for kind help in running MAFIA simulations. The research work is supported by the US Department of Energy, Contracts DOE-DE-FG03-93ER40793 and DOE-DE-AC-03-76SF00515.

References

[1] Schultz, S., Smith, D. R. and Kroll, N. "Photonic Band Gap Resonators for High Energy Accelerators", *Proceedings of the 1993 Particle Accelerator Conference*, IEEE 4, 2559, Washington D. C. (May 1993)

[2] Smith, D. R., Li, Derun, Vier, D. C., Kroll, N. and Schultz, S. "Recent Progress on Photonic Band Gap Accelerator Cavities", *This Proceedings*

[3] Kroll, N., Smith, D. R. and Schultz, S. "Photonic band gap structures: a new approach to accelerator cavities", *Advanced Accelerator Concepts Workshop*, Port Jefferson, New York

[4] Smith, D. R., Schultz, S., Kroll, N., Sigalas, M., Ho, K. M. and Soukoulis, C. M. " Experimental and Theoretical Results for a Two- Dimensional Metal Photonic Band Gap Cavity", *Appl. Phys. Lett.* **65** (5), 645 (1994)

[5] Smith, D. R., Kroll, N. and Schultz, S. "Design Considerations for a 2-D Photonic Band Gap Cavity", *Photonic Band Gap Materials*, Edited by C. M. Soukoulis, NATO AST Series, 315, NY (1996)

[6] Smith, D.R., Kroll, N. and Schultz, S. "Studies on a Metal Photonic Band Gap Cavity", *Proceedings of the Third Workshop on Advanced Accelerator Concepts*, Lake Geneva, Wisconsin, pp. 12-18 June 1994.

[7] Campisi, Isidoro E. and Finger, Keith E. "Artificial Dielectric Ceramics for Microwave Absorption", *private communication*

A Proposed Dielectric-based Laser-driven Electron Accelerator using Crossed Cylindrical Laser Focusing

Y.C. Huang and R.L. Byer

Edward L. Ginzton Laboratory, Stanford University, Stanford, CA 94305-4085

Abstract - we propose a dielectric-based, crossed-laser-beam electron linear accelerator structure operating in a vacuum that is capable of providing 1 TeV electrons in approximately one kilometer. The accelerator structure employs cylindrical laser focusing that allows for simplifying the fabrication process, accelerating more electrons, reducing the electron phase slip, and spreading the structural thermal loading. We present a ~ 0.7 GeV/m average-gradient accelerator structure, repeated every 390 μm, subject to the laser damage fluence 2 J/cm^2 on the optical components for 100 fsec laser pulses.

I. INTRODUCTION

The maximum acceleration gradient of an RF accelerator is limited by the structure breakdown under intense RF fields. In a conventional S-band RF accelerator, field emission on the copper wall occurs when the peak acceleration gradient reaches ~100 MeV/m [1]. The average acceleration gradient of an RF accelerator is thus limited to about 50 MeV/m. To reach the TeV energy level desired for the next generation linear collider, using existing RF schemes requires tens of kilometers of accelerator structure.

With the rapid advance of laser technology, high gradient laser-driven accelerators, primarily the laser wake-field accelerator [2,3] and the boundary-loaded vacuum linear accelerator [4,5], have been proposed in the literature. The laser wake-field accelerator, which accelerates electrons by exciting plasma wake-fields from a high power laser pulse, has experimentally demonstrated a peak gradient of 30 GeV/m over ~ 1 mm [6]. The boundary-loaded vacuum linear accelerator, which is constructed from a series of cascaded individual accelerator cells, provides an average acceleration gradient of ~ 1 GeV/m in theory. The acceleration gradient in a boundary-loaded laser-driven accelerator, like the RF accelerator gradient, is limited by damage. For 100 ~ 200 fsec laser pulses, the laser damage fluence of a dielectric such as glass, fused silica, magnesium fluoride, and sapphire is on the order of 2 J/cm^2 [7, 8]. The corresponding surface damage field and thus the maximum electron acceleration gradient is ~ 10 GeV/m.

Previously we have proposed a 0.7 GeV/m average-gradient, dielectric-based, crossed-laser-beam vacuum linear accelerator [5] with integrated components on a dielectric substrate. That accelerator was axially symmetric and was limited in the ability to accelerator adequate charges for high energy physics applications [9]. The acceleration field and the electron energy gain associated with the accelerator using spherical laser focusing have been reported in Refs. [4, 5]. Here we propose a similar accelerator structure employing cylindrical laser focusing that has the following desirable features:

1. the accelerator structure is constant in one of the transverse directions, and thus it can be fabricated by using current lithographic and etching technology;
2. the transverse wake-field vanishes [10];
3. more electrons can be accelerated by increasing the transverse beam size;
4. the average acceleration gradient can be higher, as will be shown;.
5. thermal loading can be spread over a wider area; and
6. beamstrahlung, the synchrotron radiation loss at the final electron beam collision, can be reduced by increasing the electron beam size in one transverse direction [11].

II. ANALYSIS

Figure 1 shows the proposed crossed-laser-beam accelerator geometry, wherein an electron traverses the focal zone at an angle θ with respect to the two laser beams. The insert in Fig. 1 defines the coordinates used in this paper. The unprimed coordinate system is the laboratory frame which includes the electron velocity axis, z. The primes indicate the rotated coordinates of the laser beams. The two laser beams are derived from a single laser source. They carry equal power, and are phased such that on the z axis the transverse fields in the x cancel and the longitudinal fields in the z add. A crossed-laser-beam accelerator employs the fundamental Gaussian mode that is most common in all types of lasers. The proposed accelerator structure uses repetitive dielectric boundaries over a distance no greater than a π phase slip between the laser field and the electron in a vacuum. Two laser beams are back-coupled from the ± x directions into the *microstage* using two prisms. The total internal reflection (TIR) inside the prisms permits the use of anti-reflective (AR) coatings for beam coupling, which are less complex than high reflective (HR) coatings. Two HR coated flat mirrors provide a secondary reflection and direct the two laser beams into the center of the microstage. For a small angle θ, a minimum prism size (in the z and x directions) of $2w$ is required for coupling ~90% of the laser power into the structure. Beam clipping at the prism sets the geometrical beam coupling condition:

$$3l \times \theta > w, \tag{1}$$

where l is half of the interaction length measured from the focal point, and w is the Gaussian beam electric field $1/e$ radius at A and A′. The structure is constant in the y direction and thus allows cylindrical focusing of the laser beam. The minimum drift space per microstage, where no laser fields exist, is approximately $2w$, and the total length of a microstage is $L_\mu = 2l + 2w$. The average acceleration gradient can be defined as

$$G = \frac{\Delta W}{L_\mu}, \tag{2}$$

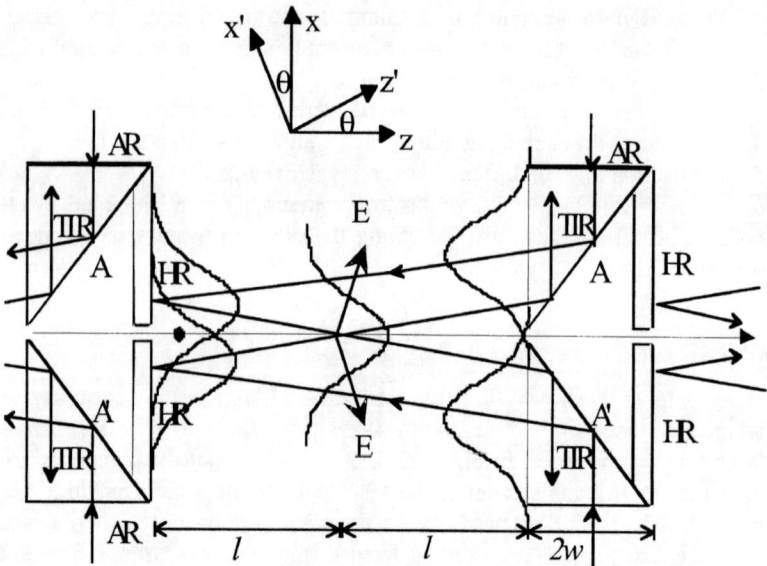

FIGURE 1. The schematic of a crossed-laser-beam accelerator. The electron traverses the focal zone at an angle θ with respect to each of the two beams. The two laser beams are phased such that the longitudinal fields add and the transverse fields cancel. The total-internal-reflection (TIR) prisms couple the laser from both sides of the structure and the high reflectivity (HR) mirrors reflect the two beams toward the center of the accelerator. The structure is constant in y.

where $\Delta W = \int_{-l}^{l} qE_z \cdot dz$ is the single-stage electron energy gain, and q is the electron charge.

With no variation in the y direction, the electrical field in x', $\tilde{E}_{x'}$ in phasor notation, for a fundamental Gaussian mode is given by [12]

$$\tilde{E}_{x'} = \left(\frac{2}{\pi}\right)^{1/4} \sqrt{\frac{2\eta}{w(z')}} P_y \exp[-jkz' + j\frac{1}{2}\Phi(z') - jk\frac{x'^2}{2R(z')} + j\varphi] \times \exp[-\frac{x'^2}{w^2(z')}], \quad (3)$$

where $\eta = 377\ \Omega$ is the wave impedance in vacuum, $w(z') = w_0\sqrt{1+(z'/z_r)^2}$ is the laser field $1/e$ radius in x', z_r is the optical Rayleigh length, P_y is the optical power per unit length in y, $k = \lambda/2\pi$ is the wave propagation constant, $\Phi(z') = \tan^{-1}(z'/z_r)$ is the Guoy phase, $R(z') = z' + z_r^2/z'$ is the radius of curvature of the wavefront, and φ is the electron entrance phase. Compared to the spherical focusing, the Guoy phase term in (3) is reduced by a factor of two [13] due to the removal of the focusing from y'. Therefore the electron phase slip is less severe using cylindrical laser focusing than spherical focusing.

The electrical field component in the z' can be calculated according to $\nabla' \cdot \vec{E} = 0$ in a vacuum. In the paraxial approximation,

$$\tilde{E}_{z'} \approx \frac{-j}{k}\frac{\partial \tilde{E}_{x'}}{\partial x'} = \tilde{E}_{x'} \cdot [\frac{-x'}{R(z')} + 2j\frac{x'}{kw^2(z')}]. \tag{4}$$

Note that even though the radius of curvature is infinite at $z'=0$, $\tilde{E}_{z'} \neq 0$ for a nonzero x', as opposed to the claim in Ref. [14]. The $\tilde{E}_{z'} \neq 0$ at $z'=0$ results from the source-free Maxwell's equation $\nabla \cdot \vec{E} = 0$ that couples the transverse variation of the transverse field to the longitudinal direction in the primed coordinates.

Assume that the electron transit aperture is small and does not noticeably affect the laser fields. The axial electric field component in z can be summed from the two crossed laser beams by

$$\tilde{E}_z = -(\tilde{E}_{x',u} + \tilde{E}_{x',d})\sin\theta + (\tilde{E}_{z',u} - \tilde{E}_{z',d})\cos\theta, \tag{5}$$

where $\tilde{E}_{i,u} = \tilde{E}_i$ with the coordinate transformation $x' = -z\sin\theta$ and $z' = z\cos\theta$, $\tilde{E}_{i,d} = \tilde{E}_i$ with the coordinate transformation $x' = z\sin\theta$ and $z' = z\cos\theta$, and θ is the absolute value of the electron crossing angle with respect to the laser axis. The sign of each field component in (5) has been chosen such that the axial fields add and the transverse fields cancel on the axis. The acceleration field $E_z = \text{Re}(\tilde{E}_z)$ seen by an axial electron becomes

$$E_z = -2\left(\frac{2}{\pi}\right)^{1/4}\sqrt{\frac{2\eta}{w_0}} P_y \cdot \theta \cdot \frac{\exp\left(\frac{-\hat{z}^2\hat{\theta}^2}{1+\hat{z}^2}\right)}{(1+\hat{z}^2)^{3/4}} \cdot \cos\left(\frac{\hat{z}\hat{\theta}^2}{1+\hat{z}^2} + 1.5\tan^{-1}\hat{z}\right), \tag{6}$$

where $\hat{z} \equiv z/z_r$ is the normalized longitudinal coordinate and $\hat{\theta} \equiv \theta/(w_0/z_r)$ is the crossing angle normalized to the far-field diffraction angle w_0/z_r. In deriving (6), a minimum electron injection energy $\gamma \gg 1/\theta$, a small angle $\theta \ll 1$, and an electron entrance phase for maximizing E_z at $z=0$ are assumed.

Laser damage intensity has been measured for spherical laser focusing by previous workers [7, 8]. The damage power per unit length in y, $(P_y)_{max}$, can be related to the measured damage intensity I_{max} by

$$(P_y)_{max} = I_{max}\sqrt{\frac{\pi}{2}}w(z=l), \tag{7}$$

where $z = l$ is the location of the HR mirror. Substituting (7) into (6), one obtains the axial acceleration field under the limit set by structure damage:

$$E_z = -2\sqrt{2\eta I_{max}} \cdot \theta \cdot \frac{(1+\hat{l}^2)^{1/4}}{(1+\hat{z}^2)^{3/4}} \cdot \exp\left(\frac{-\hat{z}^2\hat{\theta}^2}{1+\hat{z}^2}\right) \cdot \cos\left(\frac{\hat{z}\hat{\theta}^2}{1+\hat{z}^2} + 1.5\tan^{-1}\hat{z}\right). \tag{8}$$

where $\hat{l} \equiv l/z_r$ is the location of the high reflectivity mirror in z normalized to the optical Rayleigh length z_r.

As a comparison, the corresponding axial acceleration field for the spherical laser focusing case can be derived from Ref. [4, 5], yielding

$$E_z = -2\sqrt{2\eta I_{max}} \cdot \theta \cdot \frac{(1+\hat{l}^2)^{1/2}}{(1+\hat{z}^2)} \cdot \exp\left(\frac{-\hat{z}^2\hat{\theta}^2}{1+\hat{z}^2}\right) \cdot \cos\left(\frac{\hat{z}\hat{\theta}^2}{1+\hat{z}^2} + 2\tan^{-1}\hat{z}\right). \tag{9}$$

It is seen from Eqs (8 and 9) that the field amplitude and the Guoy phase term are different for the two focusing schemes. The phase slip for the cylindrical focusing case is less severe due to the removal of the 0.5 Guoy phase from the y direction. However, at the focal point $z = 0$, the axial acceleration field for the spherical laser focusing is stronger by a factor of $(1+\hat{l}^2)^{1/4}$.

Figure 2a shows the comparison of the axial field E_z for the two focusing schemes at $\hat{l} = 0.46$ and $\hat{\theta} = 1.37$, where the spherical focusing case gives the maximum electron energy gain [5]. The axial field for the cylindrical focusing is indeed slightly weaker at the focus but decreases more slowly along z due to the smaller phase slip. As can be seen from Fig. 2a, the axial field for the spherical focusing reduces to zero at $|\hat{z}| = 0.46$, corresponding to a π phase shift over the total interaction length, whereas for the cylindrical focusing it remains a finite value. The single-stage energy gain, the area under the two curves in Fig. 2a, is plotted in Fig. 2b. It is evident that the cylindrical focusing provides a higher single-stage energy gain, by ~ 10%, under the same damage fluence assumptions.

The average acceleration gradient, calculated from Eq. (2), can be evaluated as a set of contours for different interaction length in the (θ, w_0) space. These contours, indicating an operation range of (θ, w_0), shrink as the specified average gradient increases or the interaction length increases. A large θ or a small w_0 gives more electron phase slip, while a small θ or a large w_0 (loose focusing) reduces the axial field strength. As a result a large average gradient at a constant interaction length or a long interaction length at a constant gradient, which both require small phase slip and a high axial field, reduces the available (θ, w_0) pairs. Assuming a laser wavelength of $\lambda = 1$ µm at which current solid-state lasers operate efficiently [15], we found that at ~ 0.7 GeV/m acceleration gradient the accelerator stage length can be a few hundred micrometers, and the laser waist and thus the optical components can be a few tens of micrometers.

Figure 3 shows the 0.71 GeV/m gradient contours for an interaction length of $2l = 340$ µm for the two types of laser focusing. The boundary of Eq. (1) is overlaid as a dashed line in the same plot. Only those (θ, w_0) pairs above the dashed line, labeled as a dark solid line, satisfy the geometric coupling condition Eq. (1). It is seen in Fig. 3 that at the specified interaction length and average acceleration gradient the cylindrical focusing scheme is able to provide a range of (θ, w_0) pairs which satisfies the geometrical coupling condition, whereas this particular spherical focusing scheme provides no valid (θ, w_0) pairs for achieving the same acceleration gradient. For example, if "B" in Fig. 3 is chosen to be the operation point using the cylindrical laser focusing, 0.71 GeV/m average acceleration gradient can be obtained with a crossing angle $\theta = 50$ mrad, a laser waist size $w_0 = 25$ µm, a repeat distance $L_\mu = 390$ µm, and the single stage energy gain = 280 keV.

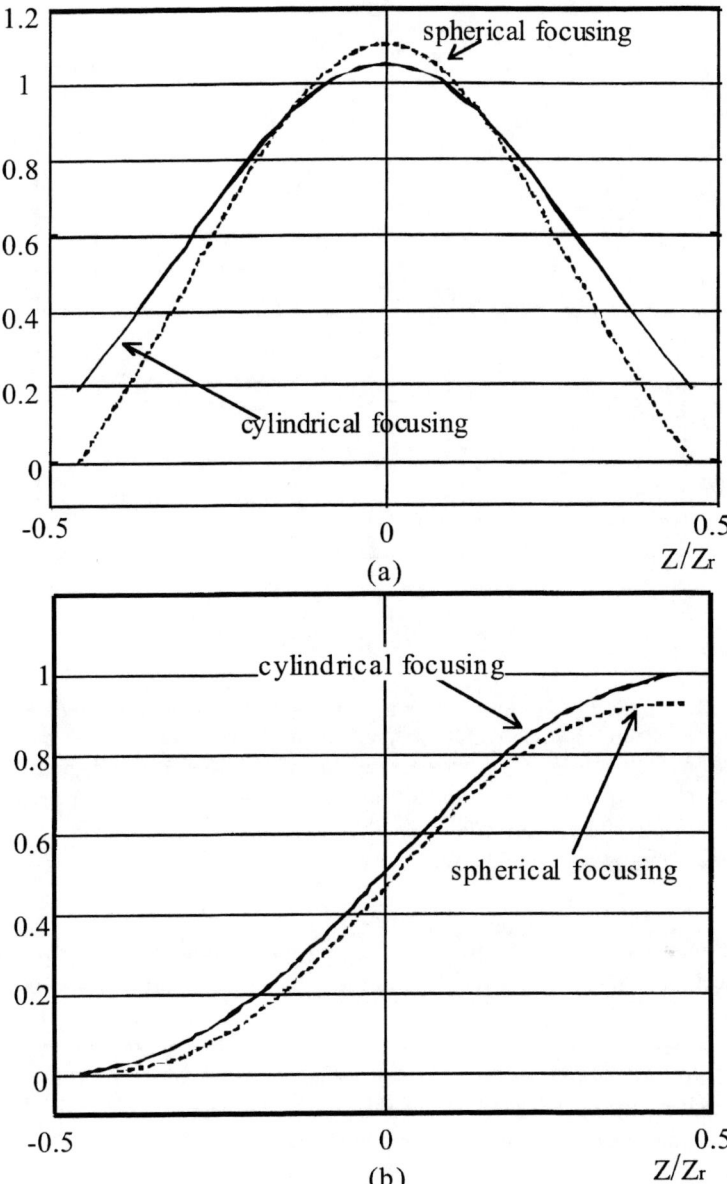

FIGURE 2. (a) The normalized axial acceleration fields for cylindrical laser focusing and spherical focusing for $\hat{l}=0.46$ and $\hat{\theta}=1.37$. The curve for the cylindrical focusing is weaker at the focus but decreases more slowly along z. (b) Electron energy gain integrated from the two curves in (a). The single-stage energy gain for the cylindrical laser focusing scheme is ~10% higher than that for the spherical focusing under the same laser damage limit.

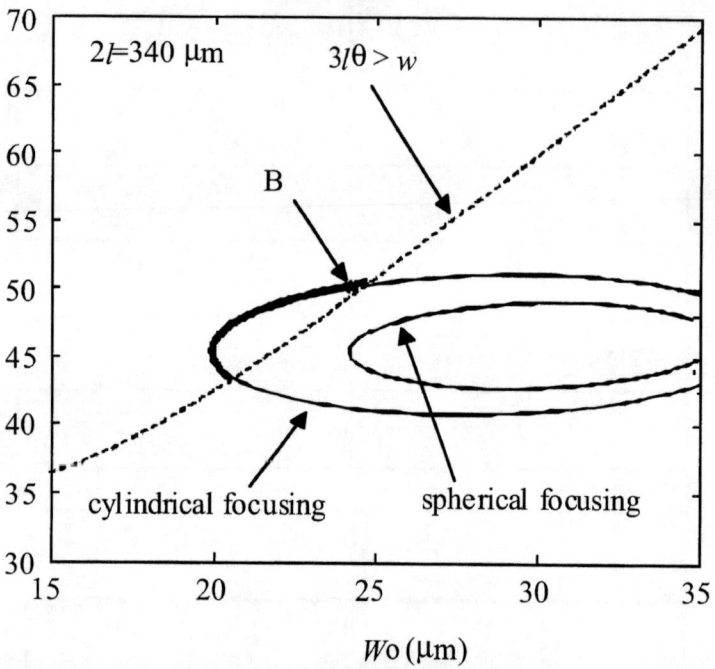

FIGURE 3. 0.7 GeV/m contours for the cylindrical and spherical focusing for $2l$ = 340 μm. Design parameters for (θ, w_0) are allowed above the dashed line, which is governed by Eq. (1). No working (θ, w_0) pairs are obtained for the spherical focusing. Point "B" for the cylindrical focusing gives 0.71 GeV/m average acceleration gradient with θ = 50 mrad, w_0 = 25 μm, L_μ = 390 μm, and single stage energy gain = 280 keV.

Figure 4 illustrates the three-dimensional view of the proposed microaccelerator stage which can be cascaded into a linear array for continuous electron acceleration. In Fig. 4 the two linearly polarized, 1-D focused TEM_{00} laser beams are crossed at the center of the ~ 400 μm long accelerator stage. The phase control of the two laser beams can be achieved by using PZT or electro-optical phase controllers. Modern lithographic and etching technology provides the fabrication precision with a relatively low cost for mass production.

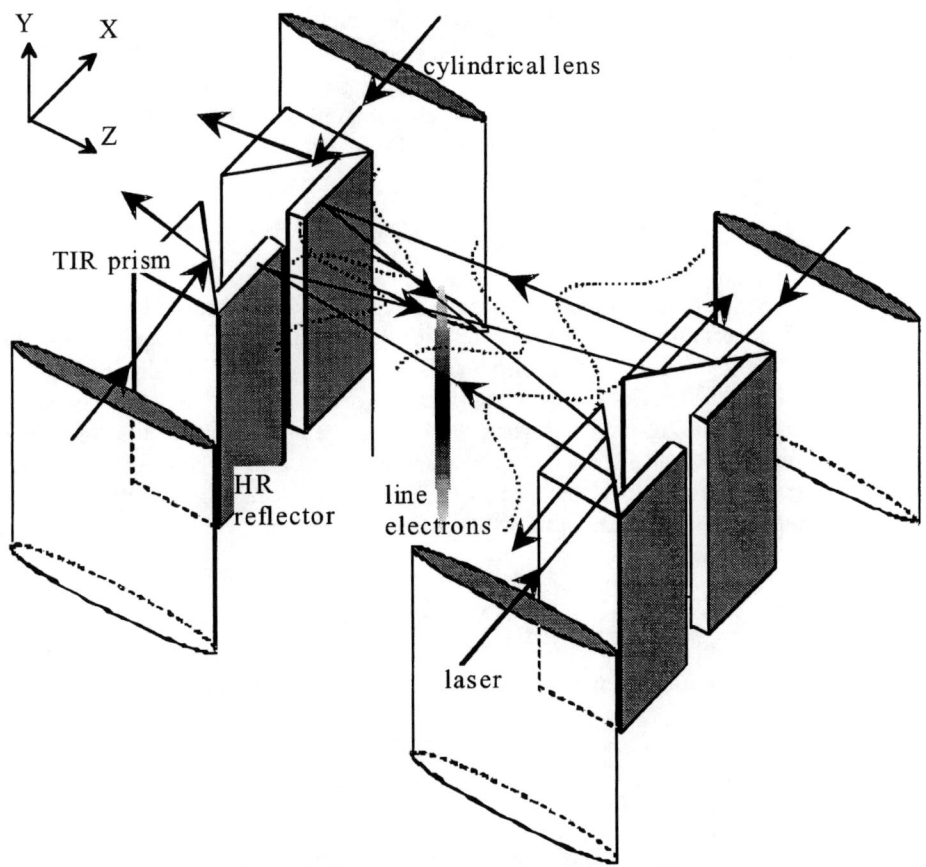

FIGURE 4. The three-dimensional view of the proposed accelerator structure. A linear charge in y is accelerated through two crossed cylindrical laser beams. The cross section of the TIR prism is 50 μm × 50 μm, the height can be a few hundred micrometers or a few millimeters, and the stage length along z, including the distance occupied by the optical components, is 390 μm. Two cylindrically focused TEM_{00} laser pulses are crossed at the accelerator center with a proper phase relationship. The optical components can be implanted or etched on a dielectric substrate.

To accelerate 10^8 electrons per optical cycle in our structure, the required laser peak power is 40 GW per accelerator stage, which gives 3% energy transfer from the laser to the electron beam at a 0.7 GeV/m gradient. With only 3% energy transfer, the accelerator structure can be safely constructed inside a laser resonator without degrading the laser performance. Since all the remaining laser power is recycled, the overall accelerator efficiency can approach the laser wall-plug efficiency, which is about ~20% for diode-pumped solid-state lasers [15].

To transmit the electrons, an electron slit small compared to the laser waist is shown in Fig. 4. Using the numerical technique in Ref. [16], we calculated a 20%

gradient reduction due to the leakage laser field though a 4 μm slit. Following the same calculations in [9], we obtain the maximum electron density $\sim 3 \times 10^7$/mm in y, subject to a 10% radiation loss per accelerator stage and the 10 GV/m damage field for 100 fsec laser pulses. The system parameters used in our calculation are refractive index = 1.5, electron slit width = 4 μm, electron bunch length in z = 0.03 μm, and electron energy ~ 1 TeV.

III. SUMMARY

We have proposed a dielectric-based, cross-laser-beam, vacuum linear accelerator structure that employs cylindrical laser focusing to accelerate electrons. The cylindrical laser focusing scheme is superior in terms of simplifying the fabrication process, achieving a higher average acceleration gradient, spreading structure thermal loading, increasing the beam current, minimizing the transverse wake-field, and reducing the synchrotron radiation loss. A sub-GeV/m average acceleration gradient is achievable in the proposed accelerator structure with a stage repeat distance of 390 μm, wherein the optical components occupy a length of 50 μm per stage.

ACKNOWLEDGMENT

The authors acknowledge valuable opinions from M. M. Fejer.

REFERENCES

[1] Tanabe, E., *Applied Surface Science*, v76-77, 16-20 (1994).
[2] Tajima, T. and Dawson, J.M , *Phy. Rev. Lett.* 43, 267 (1979).
[3] Clayton, C.E., Marsh, K.A., Dyson, A., Everett, M., Lal, A., Leemans, W.P., Williams, R., Joshi, C., *Phys. Rev. Lett.* **70**, 37, (1993).
[4] Sprangle, P., Esarey, E., Krall, J., and Ting, A., *Opt. Comm.* 124, 69-73 (1996).
[5] Huang, Y.C., Zheng, D., Tulloch, W.M., and Byer, R.L., *Appl. Phys. Lett.* **68** (6), 753 (1996).
[6] Nakajima, K. et al., *Phys. Rev. Lett.* **74**, 4288 (1995).
[7] Stuart, B.C., Feit, M.D., Rubenchik, A.M., Shore, B.W., and Perry, M.D., *Phy. Rev. Let.* v 74 n12, 2248-2251 (1995).
[8] von der Linde, D. and Schuler, H., *J. Opt. Soc. Am.*B/Vol. 13, No. 1, 216-222 (1996).
[9] Sprangle, P., Esarey, E., and Krall, J., *Phys. Plasmas* **3** (5), 2183 (1996).
[10] Tremaine, A. and Rosenzweig, J., "Wake-fields in Planar Dielectric-loaded Structure," submitted to *Phys. Rev. Letts.*
[11] Schroeder, D.V., *Beamstrahlung and QED Backgrounds at Future Linear Collider*, Ph.D. dissertation, Stanford University, 1990, p. 12.
[12] Siegman, A.E., *Lasers*, University Science Books, Mill Valley, Ca., 1986. p. 646.
[13] Siegman, A.E., *Lasers*, University Science Books, Mill Valley, Ca., 1986. p. 664.
[14] Haaland, C.M., *Optics Comm.* 124 (1996) 74-78.
[15] Voss, Brauch, U., Wittig, K., Giesen, A., "Efficient High-power-diode-pumped Thin-disk Yb:YAG laser," in *Proceedings of SPIE* v 2426, 1995, p. 501-508.
[16] Edighoffer, J.A., and Pantell, R.H., *J. Appl. Phys.* **50** (10), 6120 (1979).

The Concept of a Linac Driven by a Traveling Laser Focus

A.A. Mikhailichenko
Cornell University, Wilson Laboratory, Ithaca, NY 14850

Here we describe a Linac, driven by a Traveling Laser Focus (TLF), arranged with the help of a special deflecting device. This Linac has an accelerating structure, which is open from one side. The laser radiation is focused from this side onto a spot with minimal transverse dimensions. This spot is moved by a fast deflecting device synchronous with the instantaneous position of the charged particles, thereby providing an accelerating field mostly in the region where the accelerating particles are located. The method described allows a reduction of the power required to create the accelerating gradient, which should be proportional to the number of resolved spots of the deflecting device (typically a hundred). The same number shortens the illumination time for any point on the structure. All this allows for a gradient of up to 100 *GeV/m* with present day techniques.

The systems for the laser beam sweeping, the emittance requirements, the electro-focusing system, the accelerating structure, the final focusing system, the injection system, and some tolerances are considered in this connection. As applicable to the Linear Collider, the resulting luminosity is estimated to be in the range of $10^{33} - 10^{34} cm^{-2}s^{-1}$ at an energy level of about 1*TeV*. This requires a laser flash energy of about 5 *Joules* at a moderate repetition rate of $f=100$ *Hz*. These considerations indicate that a system of this type makes a TLF driven Linac a realistic possibility.

1. INTRODUCTION

A charged particle can be accelerated exclusively by an electrical field, acting at the instantaneous position of the particle. As a rule, an electrical field is localized in the cells, ordered along the particle's trajectory. These cells are excited so that the accelerating field is synchronized with the particle's passage through the individual cell. In a *cm*-range wavelength region, an accelerating structure is filled from its ends. When the accelerating wavelength becomes compatible with the optical one, the structure can be filled with an electromagnetic field only from the transverse side, which causes it to appear more or less like a grating. (We shall use both terms — the grating and the accelerating structure.)

There are many instances describing particle acceleration with a laser field, in which the grating or optical cavity is excited (see for example [1,2]). Typically, the laser burst of high intensity illuminates *the entire structure* along the particle's trajectory. In the transverse direction, the laser light is focused into a spot whose size is the order of a few wavelengths. During the accelerating periods of the laser light, the particles move in the field area. During the decelerating periods, the particles move close to the surface of the grating, with insufficient action from the longitudinal field. In these events, the structure is instantly illuminated on the length L, coinciding with the full particle's trajectory, so the full power of the laser is distributed along the *full* length of the trajectory. The particle, however, is located only in one particular point of the trajectory at every moment of the time. Therefore, the laser power in the place where the particle is located is less than the laser's full power. Again, the illumination of every point of the grating lasts for a time τ, which is the time needed for a particle to pass the structure, i.e., $\tau \cong L/c$, where c is the speed of light.

In contrast, the general idea of the method proposed [3a] is that the laser light is focused onto a spot with a size much smaller than the longitudinal (and transverse) dimension of the grating. This spot is moved in the longitudinal direction by a special sweeping device so that *the focal point follows the particle* in its motion along the accelerating structure. Due to this arrangement, practically all impulse laser power involved in the accelerating field generation is at the position *where the particle is located*. This evidently provides the gain in the accelerating gradient with the same power of a laser. Additionally, as the illumination of every part of the grating lasts an extremely short time in this method, this yields the second advantage of the method.

2. THE METHOD

Figure 1 describes the idea. The pulse of laser radiation lasts for a time τ. So the light beam looks like a needle, with the length $\approx c\tau$. The light beam positions at the different times are numbered 1, 2, and 3 in Fig. 1. The light-sweeping device 4 is positioned on the distance R from the grating 6. Accelerated beam 5 goes close to the surface (or inside) of grating 6 with the velocity V. The grating has the length L in the longitudinal direction (this direction coincides with the direction of the beam accelerated). Device 4 is electrically driven for sweeping the light in the longitudinal direction. Before the accelerated beam arrives at the accelerating structure, the sweeping device deflects the laser beam to the beginning of grating 6 on the left side of Fig. 1. When the beam leaves the grating, the sweeping device deflects the laser beam to the end of the grating. The angle of deflection is $\vartheta \cong L/R$. The deflection lasts the time τ also, which means that the angular velocity of the light spot on the surface of the grating is $\Omega \cong L/(R\tau) \cong V/R$. At the moment when the laser beam arrives at the grating, it has an angle α with respect to the direction of the particle movement. The tangent of this angle is defined as $tan\alpha \cong c\tau/L$, as seen in Fig. 1. The relation between parameters is as follows: $\vartheta \cdot tan\alpha = cL/VR$. This is a basic relation of this method. When the electrons or positrons are accelerated, the velocity V is close to velocity of light, so

$$\vartheta \cdot tan\alpha \cong L/R \rightarrow tan\alpha \cong 1 \rightarrow \alpha \cong \pi/4$$

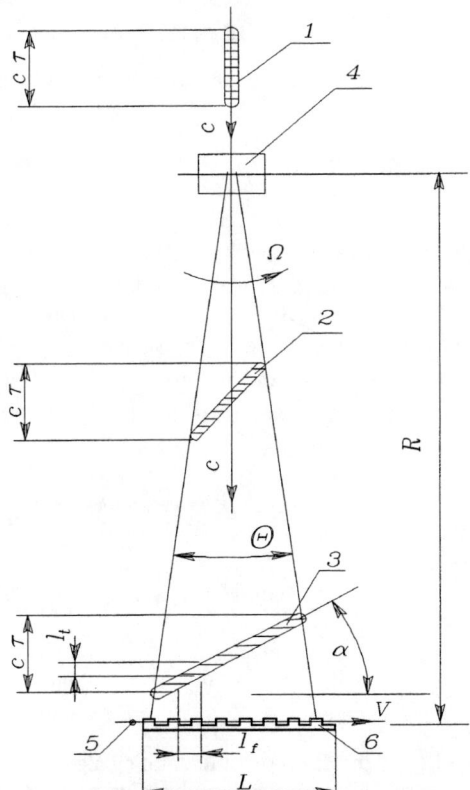

Fig. 1. The principle. The numbers 1, 2, and 3 are the light burst positions in three different moments of time. The length of the laser bust is $c\tau$. 4 is a sweeping device. 5 is the beam which goes through the structure (grating 6). Lines across the light beam indicate the wave front locations.

3. THE FIELD

The field in the wave zone can be represented as integral over aperture of the deflecting device

$$E(t) = -\frac{ik}{2\pi}\frac{\exp(ikR)}{R}\int_{-b/2}^{+b/2}\int_{-a/2}^{a/2} E(x,y)e^{ikSin\vartheta(xCos\varphi+ySin\varphi)}e^{\chi(x,y)}dxdy \quad ,$$

where $E(x,y)$ is the field distribution at the output surface of the device, $a\times b$ is the aperture of the sweeping device, ϑ is an angle in the sweeping plane, φ is an angle in the plane (x, y), $\chi(x,y)$ is a phase distribution over aperture, $k=\omega/c$. As we are interested in the sweeping in only one plane, we can put $\varphi = 0$. By substituting $2x/a=\xi$, with $kaSin\vartheta/2=\alpha_1$ and representing $\chi(x,y)=\delta\cdot x$, i.e., the linear phase variation provided by the sweeping device, we can obtain for uniform distribution $E(x,y)=E_0=$ const,

$$E(t) = -\frac{ikE_0}{2\pi}\frac{\exp(ikR)}{R}e^{i\omega t+\psi}\int_{-1}^{+1}e^{i(\alpha_1-\delta)\xi}d\xi =$$

$$-\frac{ikE_0}{2\pi}\frac{\exp(ikR)}{R}e^{i\omega t+\psi}2\frac{Sin(\alpha_1-\delta)}{\alpha_1-\delta} \quad .$$

The maximum of this expression corresponds to $\alpha_1=\delta$, or $kaSin\vartheta/2 \cong ka\vartheta/2=\delta$. In our case, $\delta=\delta(t)\propto \Omega t$; it means that the maximum of the field distribution in the direction of the grating also moves $\vartheta \propto \Omega t/ka$. The surface of the constant phase is defined by the relation $\omega R/c \pm \Omega t - \omega t + \psi =$ const. More exactly, the light beam looks like a part of opening helix, which has a center in the sweeping device. The points of the helix run from the sweeping center with the velocity of light.

The longitudinal size of the laser focus on the grating in each moment of time is defined as a horizontal cross-section of the light beam, l_f on Fig. 1. The time duration of illumination in each point on the grating is defined by the cross-section in the direction of the sweeping center, l_t in Fig. 1. These dimensions may have the same order of magnitude. The duration of illumination is of the order of $\tau \cong l_f/c$. If l_f is equal to the integer of the wavelength $l_f=n\lambda$, then the structure of the accelerated beam has a sequence of n bunches.

For the focusing and deflecting of the light on the accelerating structure, the phase variation across the aperture of the deflecting device must be $\chi(x,y)=\delta\cdot x+\kappa x^2$. The ordinary optical lens gives the necessary parabolic phase distribution.

4. THE DEFLECTING DEVICE

The fundamental measure of the quality of the deflecting device is the ratio of deflection angle ϑ to the diffraction angle $\vartheta_d \cong \lambda/a$, where a is the aperture of the sweeping device. This basically defines the number of resolved spots $N_R = \vartheta/\vartheta_d$. The deflection angle may be increased by the optical elements, but the number of resolved spots N_R does not vary. Basically, N_R value gives the number for the lowering of the laser power and, also, the number for reducing the degree of the structure heating. Let us consider some possibilities.

4.1 Electro-optical device. For such a sweeping device, an electro-optical crystal can be used. For example, a crystal KDP (KH_2PO_4) is transparent for radiation with $\lambda \cong 0.2 \div 1 \mu m$. Some other crystals, such as Cadmium Telluride (*CdTe*), Cuprum Chloride (*CuCl*), and Gallium Arsenide (*GaAs*), are transparent in the region of wavelengths around $\lambda \approx 10$ µm.

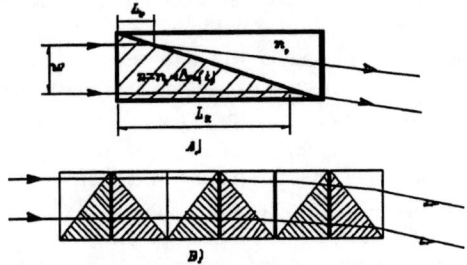

Fig. 2. The deflection device concept. A) elementary cell. B) cells installed in series. Hatched triangles show metallization.

For electrical operation of the sweeping angle, crystal 3 has a triangle metallization 4 (see Figs. 2 and 3). Generally, the device looks like two prisms positioned close to each other. The prism without metallization serves as a compensator of the bending angle when the driving voltage is not applied. When operating voltage is applied to the metallization, the effective refractive index is changed in the transverse direction, so that $n = n_0 + \Delta n(t)$. This yields a deflection of the laser beam. When the refractive index changes by an amount $\Delta n = \Delta n(U(t))$, where $U(t)$ is the voltage applied to metallization, the deflection angle change is $\Delta\vartheta \cong \Delta n \cdot (L_a - L_b)/w$, where w is the width of the incident light beam, L_a and L_b are the distances through which the edges of the light beam traverse the prism [12] (Fig. 2). So the number of resolvable spots N_R can be calculated as

$$N_R \cong \frac{\Delta\vartheta}{\lambda/a} = \frac{\Delta n l a}{\lambda w},$$ where $l = L_a - L_b$. For compensation of the regular deflection angle, the second prism without metallization is used. As the ratio is $a/w \cong 1$, then the number of resolved spots is $N_R \cong \Delta n l / \lambda$. The change of refractive index is equal to $\Delta n \cong (\partial n/\partial E) E(t)$. The typical value of

$\partial n / \partial E \approx n_0^3 r_{63} / 2$, where r_{63} is the electro-optical constant. It can be expressed in terms of half wavelength voltage as $V_\pi = \frac{\pi c}{\omega} \frac{1}{n_0^3 r_{63}} = \frac{\lambda}{2 n_0^3 r_{63}}$, which is the typical value of a few hundred volts. In a KDP (potassium-dihydrogen-phosphate) crystal, $\Delta n \cong 10^{-4}$, and in a KTN (potassium-tantalate-niobat) crystal, $\Delta n \cong 7 \cdot 10^{-3}$ is possible [13,14]. For a segmented crystal with $L \cong 30\,cm$, one can expect $N_R \cong 30 \div 700$. In [4], an applied voltage of about $V = \pm 5\,kV$ provided the deflection angle $\theta = \pm 10^{-3}\,rad$ in the KDP device. The device has 10 KDP crystals with a total length of 20 cm and an approximate 1-cm² input aperture.

4.2 Acousto-optical deflector [12, 22, 23]. Another way to deflect the laser beam is by an acousto-optical deflector. The acoustic wave induced in a medium with a piezooptic effect provides a cyclic change in the index of refraction, with the spatial periodicity comparable to the wavelength of the light wave. The light that is passing through the modulated medium in a transverse direction to the acoustic wave propagation with a small angle, would be deflected in that direction, defined by the Bragg condition. By effectively changing the acoustic wavelength during the light passage through the crystal, one can change the direction of the light that is passed through this device.

The cross-section of the laser beam must be chosen with dimensions such that the power density is below the damage level for this crystal. Also, the sandwich-type arrangement can be used to obtain the heat absorption and for more homogeneous phase distribution in the aperture of the sweeping device.

4.3 Mechanical deflection system. In principle, the mechanical deflection system with a piezoelectric effect can be used as well. The frequency f_m of mechanical oscillations must be around $f_m \cong \Omega / 2\pi \approx 15\,MHz$. Quartz can be used here. To obtain the necessary angle of deflection $\vartheta \cong L / R \cong 10^{-3}$, a quartz plate must also be bent with this angle. For a 1-cm length of plate, the movement must be of the order $10\,\mu m$, which seems high. By using a few deflecting plates, synchronized under the necessary conditions, we can reduce the deflecting angle required for the mechanical deflecting system.

Some light optics also can be used to shorten the distance R from deflector to the cylindrical lens. These optics also can increase the angle of deflection; however, the *number of resolved spots must be kept constant*.

Utilization of the short focusing cylindrical length 7 (Fig.3) reduces *the transverse size* of the spot to a few wavelengths due to its being positioned close to the accelerating structure and the ratio $r / a \approx \pi$, where r is the distance between the lens 7 and structure 8. The size of the laser focus in the region of the second short focusing cylindrical lens 7 is defined by the diffraction angle $\cong \lambda / a$, where a is the aperture of the sweeping device. So it has the order $l_t \approx \lambda \cdot R / a$, where R is the distance between the sweeping device (lens 5 on Fig. 3, see lower) and the lens 7. The maximal aperture of the deflecting system in the longitudinal direction, can be made equal to the sum of the accelerating structure and focusing elements' lengths. For practical purposes, if we accept this figure as $a \cong 3cm$, $R \approx 3m$, then the ratio $R/a \cong 100$. So the diffraction length of the spot in the longitudinal direction can be of the order $l_f \cong 100\lambda$. This value gives the maximal possible value for Q factor of one cell of accelerating structure, $Q \approx 100$.

Each part of the grating structure is illuminated for a duration that is defined by the longitudinal size l_t. For example, if we consider $l_t \cong 100\lambda$, $\lambda = 1\mu m$, then $l_t/c \cong 3 \cdot 10^{-13} sec$. For $\lambda = 10 \mu m$ this value is ten times greater. For a more detailed description of deflecting devices, see [23, 24].

5. A PRACTICAL SCHEME

Let us consider one possible scheme which successfully utilizes this proposed method. In Fig. 3 is represented the source of coherent radiation 1, which provides a ray 2 with polarization along the direction of motion of accelerated particles. Also mounted are a long focusing lens (5) for focusing the laser beam in the longitudinal direction and the sweeping device (3,4). The triangle metallization (4) makes the deflecting prism. After deflection, the laser beam (6) goes through a cylindrical lens (7), which focuses the laser beam on the surface of the structure surface (6) in transverse direction into a spot (9) with a transverse size of a few wavelengths of the laser light. The focal distance of the lens (7) is small enough to reach small cross-sections of the laser focus on the grating and, hence, to gain the electric field strength at the focal point. At this particular moment, the accelerated particles are placed at the focal point. The beam is moving along the trajectory (10) and is focused by the quadrupole lenses (11,12). Each part of the lens (7) is illuminated on short-time duration, so for successful focusing material with low dispersion must be chosen. A parabolic mirror also can be used instead of the lens (7). In that case, the cylindrical surface is displaced along the direction of motion of the particles and the grating is placed in the focus of the parabolic mirror. It is also possible to inverse the sequence of lens (5) and device (3) in order to shorten the distance from the lens to the structure.

The device, represented on Fig. 3, works as follows. The operating voltage in the initial moment deflects the laser beam to the beginning of the accelerating module (left side on Fig.1 and Fig. 3). Starting with that moment of time, synchronized with the phase of the laser light and entrance of the accelerated particles, the sweeping device (3) is supplied by the changing voltage on the metallization (4). This voltage changes the direction of the laser beam (6) so that the focal point (9) follows the beam, on average, at the beam's velocity. Only part of the total radiation generated can be used as the source (1). In this case, a single source can supply several modules, and source (1), utilizing well-known optical techniques, can be therefore used as a device to split the light from the total radiation generated. Due to the possibility of sweeping the laser focus to the limited distance about $2 \div 3cm$, the accelerating device looks like a sequence of 2-cm-long accelerating structures with the focusing elements between them. For an appropriate accelerating structure, the pass holes have a fraction of a wavelength of the accelerating structure.

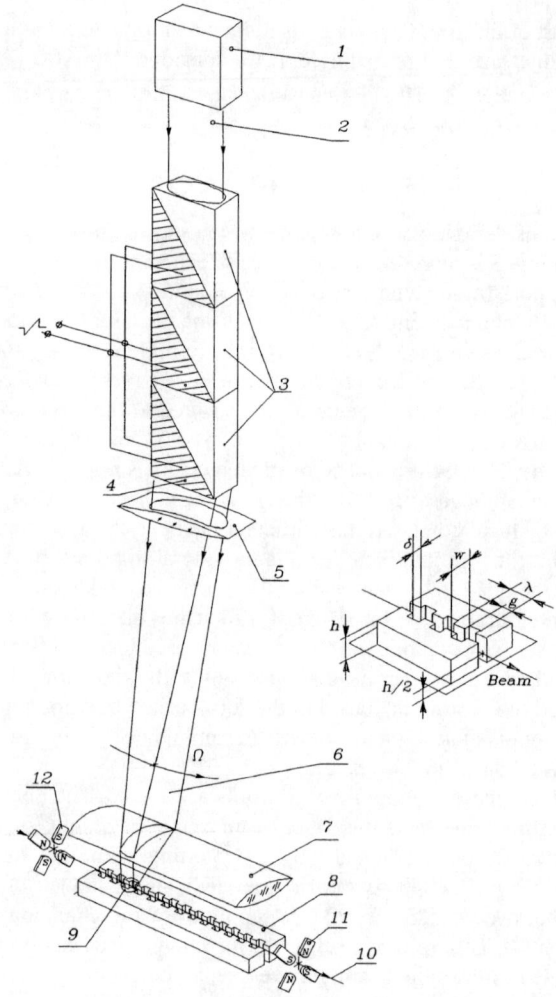

Fig. 3. The Accelerating Device. The accelerating structure is represented to the right.

6. ACCELERATING STRUCTURE

There are proposals for an accelerating structure that could be scaled to the wavelength, corresponding to the laser radiation [1,2,5,18]. We will consider the requirements for the structure described in [4,5] (see Fig.3). This structure is more advanced and can therefore be used to obtain the vacuum, alignment, and cooling required. The most important property, however, is the existence of walls on the sides and a bottom, which allows good positioning for the vector map of the electromagnetic field. The depth of the one cell h is equal approximately to a half of the wavelength. Precise adjustments of the height of the groove is made for the necessary Q-factor value. In the cell the standing wave excited which has only one variation along the y axis. This excludes the transverse kick from the wave. The x axis directed across the grating has a groove of width w in this direction (see Fig. 3) and x has a zero in the middle of the slot. The y is a zero at the groove's bottom. The channels for the passing of the beam have a size $\delta \leq 0.2\lambda$.

The polarization of the electric field in the laser beam is rectangular in each point to the center direction of sweeping device 3. In this case, the radiation comes to the surface almost rectangular to it. This is valid for the central zone of the grating. For a phase correction, which is necessary due to the curvature of the wave front, for example, the cylindrical lens (7) could be made with different thicknesses but with the same focal distance. Let us consider a point on the grating, having a coordinate s, calculated from the center of the grating. The distance r between this point and the sweeping device center can be evaluated as $r = \sqrt{R^2 + s^2} \cong R + s^2/2R \equiv R + \delta r$, which yields a phase variation $\delta\varphi \cong 2\pi\delta r/\lambda = 2\pi s^2/\lambda R$. From the other side, an extra thickness $\Delta(s)$ of the lens provides a phase shift $\delta\varphi \cong 2\pi\Delta(s) \cdot (n-1)/\lambda$, where n is a refractive index of the material of the lens. So, if the thickness variation follows the law $\Delta(s) \cong s^2/(R\cdot(n-1))$ (not dependent of the wavelength), then the phase variation will be compensated. Estimation of this value for the end distance $s \approx 1cm$, $R \cong 3m$, $n-1 \cong 0.5$, will give $\Delta \cong 1/150 cm = 67\mu m$ only. Some phase variation can be used in alternative phase focusing (APF) (see below). If the mirror is used, the analogous correction also can be made with a parabolic profile in the longitudinal direction. This longitudinal profile must have the sweeping device as a focal point.

Synchronization between the particles' motion and the focal spot motion must be made in such a way that most of the particles do not come out of the laser spot. Loss of synchronization yields a decrease in the average accelerating gradient and provides the energy spread inside the accelerated beam. The synchronization of the laser *phase* and position of the particles is more important. So, using the focusing lenses (5,7) and the sweeping device (3), it becomes possible to focus all radiation in a small spot on the structure and to move the focal point along the surface of the grating synchronous with the motion of the particles. Each part of the grating is illuminated by duration, which is defined by the longitudinal size l_t. For example, if we consider $l_t \cong 100\lambda$, $\lambda = 1\,\mu m$, then $l_t/c \cong 3\cdot 10^{-13}\,sec$. For a $\lambda = 10\,\mu m$ this value is ten times more. After the passage of one module, the particles go to the second module, and so on. The short focusing lens (7) is made with material that has a necessary optical bandwidth. Utilization of a polarized filter in the region of the accelerating structure (grating) will prevent illumination of the grating with undesirable field components. This filter can be made on the surface of the lens.

7. THE INJECTION SOURCE

The emittances discussed above on the level of $10^{-10}\,cm\cdot rad$ at the energy about 3 GeV can be obtained *extensively* by scraping the extra particles ejected from the appropriate damping ring [3a,b, 24].

The number of the particles required for the method is four orders of magnitude lower than for the linear collider projects with the same level of luminosity.

7.1 Ordinary systems. A lot of damping rings were considered for Linear Collider schemes. Typical emittances referred to $3 GeV$ are $\varepsilon_x \approx 5\cdot 10^{-8}\,cm\cdot rad$ - radial and $\varepsilon_y \approx 5\cdot 10^{-10}\,cm\cdot rad$ - vertical. The energy spread about $\sigma_\varepsilon \cong 10^{-3}$ and the bunch length $\sigma_z \cong 5\,mm$. The length of the beam after one stage compression is of the order 500 μm and the number of the particles is about $N \approx 10^{10}$. The second stage compresses the beam typically to $100\,\mu m$ at 10 GeV. So if we need only $N \approx 10^6$ we can loose four orders of magnitude in intensity by *scraping* the extra particles ejected from the appropriate damping ring, thereby coming to the necessary figures in the emittance $10^{-10}\,cm\cdot rad$ at 3 GeV. For pre-bunching the FEL mechanism can be used to obtain the bunch of necessary longitudinal dimensions.

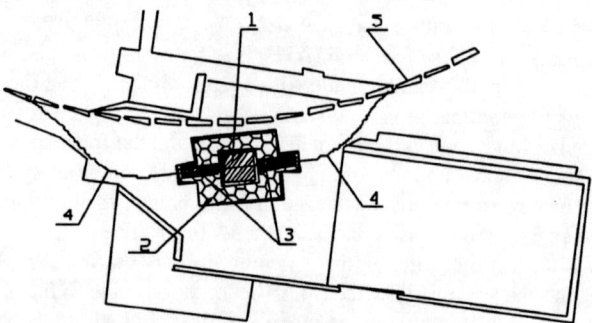

Fig. 4. Sketch of the possible installations of TLF linac in CESR experimental hall. 1- detector, 2- stabilized platform around detector, 3- intermediate platforms for linac, 4- transport lines, including bunching and scrapping, 5- the damping ring.

7.2 Linear Damping System (LDS) [20]. This is another method in which it is possible to obtain the minimal possible emittance. The LDS has the wigglers and accelerating structures displayed one by one along a straight line that provide a decrease of transverse emittances and energy spread. With a length of about 1 km and a beam energy of about 15 GeV, the decrease corresponds to a few damping times in a damping ring. The RF supply is the same as for the main Linac, so the LDS has no dependence on repetition rate connected with preparing the low emittance initial beam.

7.3 Kayak-paddle (Dog's bone) cooler [21]. This is a damping ring that is made of a sequence of wigglers and accelerating structures, installed along a straight line and having bends at the end. The bends can be made to give a small input into the cooling process, so we will neglect its influence. In that sense it is similar to the LDS. In the LDS, as in the damping ring a particle needs to re-radiate its full energy a few times. These methods allow, theoretically, emittances of the order $(\gamma\varepsilon_x) \cong 5 \cdot 10^{-8}\ cm \cdot rad$, $(\gamma\varepsilon_x) \cong 2 \cdot 10^{-9}\ cm \cdot rad$.

7.4 Optical Stochastic Cooling method (OSC) [9]. This is another appropriate method for achieving a beam with an extremely small emittance. This method can give about the same value $\gamma\varepsilon_y \leq 2 \cdot 10^{-9}\ cm\ rad$. An example of an installation of the TLF system is given on Fig. 4. The accelerating system installed on a separate platform (2) is large enough to have a hole in the middle for detector (1) positioning. This platform is stabilized to prevent the influence of the ground motion noise. Two additional platforms (3), positioned on the first one, carry the TLF structures. Stabilization of these two platforms on the first one is made with the help of a laser interferometer system. All of these bunch tailings are made in channel (4), connecting TLF system with the damping ring (5).

8. WAKES AND RESISTIVE INSTABILITY

The beam size must be small enough to pass through the channels. A short wavelength of betatron oscillations helps against the resistive wall instability [15] and reduces the wakefield influence. The longitudinal and transverse wakes *normalized to one cell* are $W_\| \cong -7 kV/pC$ and $W_\perp \cong 2.2 \cdot 10^2\ V/pC/\mu m$ correspondingly for the accelerating structure with $\lambda \cong 10\mu m$, $\delta = 2\mu m$, $w = 7\mu m$ (see Fig. 3) and for the bunch with the longitudinal length $\sigma_l \cong 1\mu m$ (calculated with MAFIA, [16]). However the *relative* variation of the accelerating field is the same as for a typical accelerator due to the huge accelerating field strength.

9. A TRANSVERSE ELECTRON FOCUSING

The focusing system is a combination of a FODO structure arranged with quadrupole lenses of appropriate dimensions and RF focusing of a different nature [3c]. The Final Focusing system is arranged with the help of a bifrequency RF focusing system supplied by a laser radiation of general and doubled frequency (see below). See also [10,11,17].

10. THE FIELD STRENGTH

If we suppose that the full energy of the laser flash is Q, a time duration is t, the number of periods is equal $n = t/T = c\tau/\lambda$, where $T = l/c$ is the period of radiation, then the energy stored in the field of half a period is $W \cong Q/2n$. From the other side $W \cong (1/2)\varepsilon_0 E_m^2 V_{eff}$, where ε_0 is the dielectric permeability of the vacuum, $V_{eff} \cong gwh \cong g\lambda^2$ is the effective volume, where the energy is concentrated. From the expressions for W, one can obtain the maximal field strength $E_m \cong \sqrt{Q/(\varepsilon_0 c\tau\lambda g)}$. For estimation, let us take $Q = 0.01$ J, $\tau \approx 0.1$ ns, $\lambda \cong 1\mu m$, $g \cong \lambda/2 = 0.5\mu m$. This gives the field strength $E_m \cong 270 GeV/m$. This value must be reduced by the factor reflected in the longitudinal length of the illuminated surface of the grating.

In this method the typical flash energy required to supply one structure 2-cm long is ≤ 10 milliJoules. A radiation of $\lambda \cong 1\mu$ m provides ≥ 30 GeV/m (Q factor ≈ 10).

The light flash with the above parameters will feed the accelerating length about $ct \approx 3cm$. And a flash with $Q \approx 1J$ will feed the 3-m total length, providing the energy gain $\approx 300 GeV$.

The level of damage to the grating by a laser light is strongly correlated with the duration of illumination of the grating [6]. There is no experimental data for the damage level, if illumination lasts the time $\tau \cong l_t/c \cong 10^{-13}$ sec. This time is less than the time between electron–electron collisions $\tau \approx l_{free}/v_F \cong 10^{-12}$ sec, where l_{free} is the free path length, and v_F is the electron velocity at Fermi surface [7]. The time of illumination is still, however, longer than the time that corresponds to the reaction of electron plasma in the metal $\tau \cong 2\pi/\omega_p = 2\pi/\sqrt{4\pi n r_0 c^2} \approx 3 \cdot 10^{-16}$ sec, where n is the density of electrons, and r_0 is the classical electron radius. Some scaling of the figures [6] shows, however, the possibility of $10^8 - 10^9$ V/cm without damage to the surface.

11. THE NUMBER OF PARTICLES

The energy accepted from the field by N particles is $W_a \cong eNE_m gI(g)$, where e is the charge of a particle and $I(g)$ is a function of the order of unity, which is an analog of the transit time factor. The share of the energy will be $\eta W \cong \eta \frac{1}{2} Q\lambda/(c\tau) \cong eNgI(g)\sqrt{Q/(\varepsilon_0 c\tau\lambda g)}$. From the last relation it follows that $N \cong \eta \cdot \sqrt{\varepsilon_0 \lambda^3 Q/(ctg)}/(2eI(g))$. With $I(g) = 0.5$, $\eta = 0.1$ (10%), this yields $N \cong 2 \cdot 10^6$ for $\lambda \cong 1\mu m$.

If the final energy of the beam is E, with a repetition rate f, then the total energy carried by these particles per second is $\dot{E}_t \cong eNEf$. Substitute here $f = 100$ Hz, $E = 300$ GeV, and we obtain $\dot{E}_t \cong eNEf \approx 10 J/sec$, or 10 W. As we suggested, for an efficiency of about 10%, the power of the laser must be at least 100 W for a

100 Hz repetition rate. The impulse power of the laser is $P_i \cong Q/\tau$, and we have $P_i \approx 10^8 W$, which is far below the limit achieved.

12. THE FINAL FOCUS

For realization of the high luminosity, the envelope function in the interaction region must be of the order of the bunch length, which is about $0.1 \div 1\lambda$, where λ is the wavelength of the laser radiation.

If we suppose that the beta function value in the interaction region β^* is of the order of the bunch length σ_l, i.e., $\sigma_l \approx \beta^* \cong 0.5$ μm, then the variation of the envelope function from the interaction point at the distance $s = 1$ cm will be $\beta = \beta^* + s^2/\beta^* \cong s^2/\beta^* \approx 1/0.5 \cdot 10^{-4} \cong 2 \cdot 10^4 cm$, where we ignored the focusing arising from the incoming beam. With such an envelope function, the transverse dimension $\sigma_{\perp max}$ will be of the order of $\beta = \beta^* + s^2/\beta^* \cong s^2/\beta^* \approx 1/0.5 \cdot 10^{-4} \cong 2 \cdot 10^4$ cm, or 1.5 micrometers for the transverse emittance value $\varepsilon \cong 10^{-12} cm \cdot rad$.

Preliminarilly let us estimate, the radiation losses in the final lens. As the distance s from IP is of the order of the focal distance of the final lens, $s \approx F = (HR)/GL$, where (HR) is the magnet rigidity of the particle, G is the gradient and L is the length of the lens. So one can conclude that $H_\perp \cong G\sigma_{\perp max} \cong \frac{(HR)}{FL}\sqrt{\frac{\varepsilon}{\beta^*}} \cdot s = \frac{(HR)}{L}\sqrt{\frac{\varepsilon}{\beta^*}}$. This formula reflects the fact that the angle given to the particle by the field of the lens must be of the order of the natural angular spread in the focal point, $\sqrt{\varepsilon/\beta^*} \cong H_\perp L/(HR)$. The bending radius in this magnetic field will be $\rho \cong (HR)/H_\perp = L/\sqrt{\varepsilon/\beta^*}$. The length of formation of radiation is of the order $l_F \cong \rho/\gamma = L/(\gamma \cdot \sqrt{\varepsilon/\beta^*})$, so the number of the radiated photons will be $N_\gamma \cong \alpha L/l_F = \alpha \cdot \gamma \cdot \sqrt{\varepsilon/\beta^*}$, where $\alpha = e^2/\hbar c$. The critical photon energy is $\hbar\omega_{cr} \cong (3/2)\hbar c\gamma^3/\rho$, so the total energy radiated by the particle will be $\Delta E \cong \hbar\omega_{cr} \cdot N_\gamma = (3/2)mc^2\gamma^4 r_0 \varepsilon/(L\beta^*)$, and the energy variation will be $\frac{\Delta E}{E} \cong \frac{3}{2}\gamma^3 \frac{\varepsilon}{\beta^*} \frac{r_0}{L}$. From the other side, variation of the focal distance due to energy variation is $\Delta F \approx -F/(\Delta E/E)$. If we suppose that the focal distance is of the order of the length of the final lens, $F \approx L$, we finally come to the formula $\Delta F \approx -r_0\gamma^3\varepsilon/\beta^*$. For our case $\Delta F \approx -4.5 \cdot 10^{-2} cm$, i.e., much bigger than the length of the bunch. So we come to the fundamental conclusion that the final focus could not be arranged with the help of a single lens (or a doublet) because of the focusing to the β^* that is required. The energy spread (Oide Limit effect) can be estimated as follows

$$\delta(\Delta E) \cong \hbar\omega_{cr} \cdot \sqrt{N_\gamma} \approx \hbar\frac{c}{L}\gamma^3\sqrt{\alpha\gamma} \cdot \left(\sqrt{\varepsilon/\beta^*}\right)^{3/2}.$$

The mostly natural way to arrange the final focusing for our purposes is *a multiplet* of FODO structures with the number of the lenses in multiplet of the order of a few hundred. The gradient in these lenses must vary from the very strong at the side closest to IP, to weak at the opposite side adiabatically. We will call this an *Adiabatic Final Focus* (AFF). Focusing properties of the RF lens, discussed above [3c], can be used here. A laser radiation of general and *multiple* frequency can be used for such focusing.

13. THE LUMINOSITY

For energy $E \cong 300$ GeV, it is possible to reach emittance $\varepsilon \cong 10^{-12}$ cm·rad. This yields the transverse dimension $\sigma_\perp \cong \sqrt{\varepsilon \beta^*} \cong \sqrt{5 \cdot 10^{-5} \cdot 10^{-12}} \cong 0.7 \cdot 10^{-8}$ cm, or $\approx 1 \overset{\circ}{A}$ for $\sigma_l \cong \beta^* \cong 0.5$ μm. For a luminosity, we have the formula $L = N^2 f H / S$, where N is the number of colliding particles, f is the repetition rate, S is the effective cross-section of the beams in the colliding region, and H is the enhancement parameter. Substitute here the previous figures, and we can expect $L \cong 10^{32} \div 10^{33} cm^{-2} s^{-1}$ for $H=2$ and for repetition rate $f=100$ Hz only. Notice here that the total length for acceleration up to 300 GeV can be about few meters, taking into account that a fraction of the accelerating sections is of the order 50%. As the total energy of the full energy accelerator will be around *1 Joule*, the mean power is of the order 100 W.

Operation at a high repetition rate, up to *few tens kHz*, is possible. The beams of electrons and positrons can be polarized, which gives the effective gain in luminosity and reduces the background [19].

14. INTERACTION POINT

Energy losses by synchrotron radiation at the beam interaction point because of the beam's small size is a subject of special consideration. The Y parameter is defined as $Y \equiv 2\hbar\omega_c / 3E = \gamma H / H_c$, where H is the magnetic field of the beam, and $H_c = m^2 c^3 / e\hbar \cong 4.4 \cdot 10^{13}$ Gauss is the Schwinger critical field strength, and ω_c is the critical frequency of classical synchrotron radiation. In our case the magnetic field value in the longitudinal center of the bunch at it boundary H_0 can be estimated $H_0 \cong \dfrac{4\pi}{c} \dfrac{eNc}{2\pi\sigma_\perp \sigma_b} \approx 3.8 \cdot 10^9$ Gs, so for $E = 300$ GeV, $\gamma \cong 6 \cdot 10^5$, $Y_0 \cong 104$, where it was taken into account that the electrical field of the incoming beam doubles the parameter Y_0. Thus, the radiation is in a quantum regime. As the magnetic field value is a function of the particle trajectory inside the incoming beam, the Y parameter also is a function of the particle position. To estimate the energy losses we will follow [8b]. For uniform transverse distribution and Gaussian longitudinal distribution, the Y parameter can be represented $Y = Y_0 \rho_\perp exp(-\varsigma^2/2)$, where Y_0 calculated for the maximal field in the center of the bunch, $\rho_\perp = r/\sigma_\perp$, $\varsigma = z/\sigma_b$ are normalized coordinates. The average fractional energy losses for strong quantum regime $Y_0 \gg 1$ is defined by Sokolov-Ternov formula

$$\varepsilon_E = \frac{\Delta E}{E} = \frac{16 \cdot \Gamma(2/3)}{243} \cdot \frac{\alpha}{\Gamma_0} \cdot (3Y_0)^{2/3} \langle \rho_\perp \rangle^{2/3} \int_{-\infty}^{+\infty} exp(-\varsigma^2/3) d\varsigma ,$$

where $\Gamma_0 = \lambda_C \gamma / \sigma_t$, $\lambda_C = \hbar/mc$, $\langle \rho_\perp \rangle = 1.3$. For the parameters discussed above, $\Gamma_0 \cong 0.46$. By substituting the figures in the formula for losses, one can obtain $\varepsilon_E \cong 0.22$, or 22%, which is somewhat high. One can consider a flat beam for reduction of this effect. For high energy, we can expect to reduce the losses by $\varepsilon_E \approx 1/\gamma^{1/3}$. Correction by changing the regime at some longitudinal distance from the middle of the beam yields an integration of the exponential term with some impact parameter instead of infinity and gives fewer values. We will not represent the results here.

The interaction of the beams with the parameters, $\sigma_t \cong 0.4 \mu m$, $\sigma_\perp \cong 2.5 \overset{\circ}{A}$, $N \cong 1.2 \cdot 10^8$ was considered in [8a]. In this case $Y_0 = 5094$. The quantum beamstrahlung average fractional energy loss was found to be around 10% for the beam of 5 TeV.

15. ALIGNMENT

The ratio of the random component of transverse momentum per stage $\Delta p_\perp$ to the energy gained per stage $\Delta p_\parallel$ must be less if compared with angular dispersion in the beam $\Delta p_\perp / p_\parallel \leq \sqrt{\varepsilon/\beta}$. Substitute here $\varepsilon \cong 10^{-10}$ cm rad, $\beta \cong 10$ cm, and we obtain $\Delta p_\perp / p_\parallel \cong 3 \cdot 10^{-6}$ rad, i.e., the same requirement as for the linear collider. From the other side $\Delta p_\perp / p_\parallel \cong \Delta\theta$, where $\Delta\theta$ is the accuracy of the alignment of the grating. The $\Delta\theta$ can be $\Delta\theta \cong \Delta\phi \lambda / L \cong \Delta\phi \lambda / L \cong \Delta\phi \cdot 10^{-3}/3 = 3 \cdot 10^4 \Delta\phi$, where $\Delta\phi$ is the phase resolution if a diffraction method of alignment is used. The value $\Delta\phi \cong 10^{-2}$ is much greater than the possible registered one. For dynamic alignment the piezoelectric movers also can be used. Alignment is helped by reflected radiation. Optical methods allow the diagnosis within a fraction of the wavelength.

16. PERSPECTIVES

The method proposed gives 50 TeV/km or 500 TeV on 10 km. From the previous estimation of 0.3 J/m, the total energy of the laser beam is 3 kJ per pulse. The gradient, however, is still far below the absolute limit of the acceleration gradient, which is defined by Schwinger field $eE^\infty \cong mc^2/\lambda_C$. $E^\infty \cong 10^9$ GeV/m, or 10^{10} TeV on 10 km.

17. CONCLUSION

Two positive aspects are present in the method proposed. The first aspect is a reduction of illuminating time for every point at the grating. This reduction is approximately equal to the ratio of the laser spot size to the length of the accelerating device (grating). This ratio is defined by the number of resolved spots which are allowed by the deflecting device. The same figure reflects the second positive aspect — lowering the laser power (or the flash energy) for the fixed gradient. With actual parameters, this lowering is about 100 for each aspect.

The general conclusion is that the TLF concept described looks like a realistic model for a Laser Driven Linac.

18. REFERENCES

[1] Y.Takeda, I.Matsui, NI&M, 62(1968), p.306.
[2] R.B. Palmer, Particle Accelerators, vol. 11, p.81, 1980.
[3] A.A. Mikhailichenko:
   a—"The method of acceleration of charged particles," Author's certificate USSR N 1609423, Priority May 1989, Bulletin of Inventions (in Russian), N6, p.220, 1994;
   b—"Excitation of the Grating by Moving Focus of the Laser Beam," EPAC 94, London, Proceedings, p. 802;
   c—"A Beam Focusing System for a Linac Driven by a Traveling Laser Focus," PAC95, Dallas, TX, p. 784.
[4] C.L.Olson, C.A.Frost, E.L.Patterson, J.W.Pokey, IEEE Trans. of Nucl. Sci., vol. NS-30, N4, 1983, p.3189.
[5] R.C.Fernow, J.Claus, "Properties of the Foxhole Accelerating Structure," AIP Conference Proceedings, 279, 1992, 212.
[6] See, for example, AIP Conference N 91, Proceedings, 1982.
[7] C.Kittel, "Introduction to Solid State Physics," Wiley, NY, 1976.
[8] a—T. Himel, J. Siegrest in "Laser Acceleration of Particles," editors C. Joshi and T. Katsouleas, AIP Conf. Proceedings No. 130 (1985).
   b—P. Chen," Quantum beamstrahlung from Gaussian Bunches" in "New Developments in Particle Acceleration Techniques," Proceedings, CERN 87-11, ECFA 87/110, vol.2, p.593 (1987).
[9] A.A. Mikhailichenko, M.S.Zolotorev, "Optical Stochastic Cooling," Phys. Rev. Let. **71**, 4146 (1993).
[10] H.Henke, "mm Wave Linac and Wiggler structure," EPAC 94, London, to be published.
[11] F.E.Mills and A. Nassiri, "Alternating phase focusing in mm-wave linear accelerators," Argonne National Laboratory, internal report ANL/APS/MMW-9, 1994.
[12] P. Das, "Lasers and Optical Engineering," 1990.
[13] F.S.Chen et al., Journ. Applied Physics, Vol.37, N1, 388 (1966).
[14] V.J.Fowler, J.Schlafer, Applied Optics, Vol.5, N10, 1657 (1966)
[15] A.A. Mikhailichenko, V.V. Parkhomchuk, "Resistive wall instability of a single bunch in a Linear Collider," Preprint BINP 91-55, Novosibirsk, 1991.
[16] A. Milich, A private communication, CERN, 1994.
[17] W. Schnell, "Microwave quadrupoles for linear colliders," CLIC Note 34, 1987.
[18] J. Rosenzweig, A. Murokh, C. Pellegrini, "A proposed Dielectric-Loaded Resonant Laser Accelerator", Phys. Rev. Let.**74**, 2467(1995).

[19] A.A. Mikhailichenko, "A polarized positron sources," Sources-94, International Workshop on e^+, e^- Sources and Pre-Accelerators for Linear Colliders, Schwerin, Germany, Sept.29–Oct. 4, 1994, Proceedings, edited by Rainer Wanzenberg, 1994, 634p., p. 61–75.
[20] N.S. Dikansky, A.A. Mikhailichenko, "Linear Damping System for obtaining Beams of $e^{\pm}$ with minimal emittance," International Conf., EPAC 92, Berlin, 1992, Proceedings, p.898.
[21] A.A. Mikhailichenko, "Damping Ring For VLEPP Linear Collider," III international Workshop on Linear Colliders LC91, Protvino, September 17–27, 1991. Proceedings, Edited by V.E. Balakin, S.Lepshokov, N.A. Solyak, Serpukhov, (IFVE). 1991. Serpukhov, USSR: BINP (1991).
[22] "Optical scanning," Edited by G. Marshall, Marcel Dekker, Inc., 1991.
[23] "Selected Papers on Laser Scanning and Recording," SPIE Vol. 378, Editor L. Beiser, 1985.
[24] A.A.Mikhailichenko, "Damping Ring For Testing The Optical Stochastic Cooling," EPAC 94, London, June 27–July 1, 1994, Proceedings, p. 1214..

An Inverse Cherenkov Accelerator Using a Dielectric Channelled Waveguide

Wei Gai and J. Simpson

High Energy Physics Division
Argonne National Laboratory, Argonne, IL 60440

Abstract: A cylindrical dielectric structure based inverse Cherenkov, laser driven accelerator scheme is proposed. The scheme uses the inverse process of a charged particle radiating inside a partially filled dielectric wave guide. Due to the efficient coupling of the laser beam to the accelerating fields, a very modest amount of laser power can produce GV/m scale gradients. Numerical examples are given for several cases. One particular parameter set demonstrates that a net acceleration of 1 GeV can be achieved using a 200 kW, 3 ns, radially polarized 10.6 µm laser drive beam. A proof of principle prototype experiment is also discussed.

New methods of particle acceleration have been a subject of intense study in the past decade. One of the more appealing possibilities is the use of a laser beam to efficiently accelerate electrons. Laser-plasma interaction based techniques such as laser wake-field [1,2] and laser plasma beatwave [3.4] acceleration have already been successfully demonstrated. In principle, any reverse process of a charged particle radiation can also be used for particle acceleration. Recently, there is increasing interest in non-plasma based laser particle acceleration schemes [5,6]. For example, Fabry-Perot type resonance acceleration structures have been proposed [7], and W. Kimura *et al.* have successfully demonstrated inverse Cherenkov acceleration in a high pressure H_2 gas by using a radially polarized CO_2 laser beam [8]. In this paper we propose a new scheme which uses a hollow dielectric fiber in which light propagates as a travelling TM_{0n} wave. Advantages of our proposed scheme are: 1) efficient coupling of the laser power to the electron beam; 2) it overcomes the problem of electron beam scattering in the gas; and 3) the acceleration distance can be sustained due to the nature of the travelling wave structures (hollow fiber).

The concept of charged particle acceleration by the inverse Cherenkov effect is not new. Pantell and Fontana proposed a scheme using pressurized gas as the medium to support an acceleration field [9] and a subsequent experiment has been carried out [8]. The Pantell et al. scheme is based on the creation of an accelerating electric field component due to the wavefront tilt when a radially polarized laser beam is focused by an axicon onto the direction of the charged particle beam. They also discussed the possibility of using an infinite medium with a small hole to accelerate the charged particles based on the same principle. In their proposal, the total acceleration distance is limited due to the finite transverse size of the driving laser beam. Therefore, sustained high gradient acceleration cannot be achieved unless the laser beam can propagate in the same direction as the electron beam. The required laser power for the original ICA scheme is high (>> GW) in order to achieve the high gradient. In this paper, we propose a new method using the inverse process of Cherenkov radiation as proposed in the reference [9], in which the supporting structure is a cylindrically symmetric hollow dielectric channeled waveguide with a total reflective outer wall.

The Cherenkov radiation of a charged particle travelling inside the dielectric channel has been studied in detail [10]. The radiation pattern is dominated by TM_{0n} (n is the radial mode number), which has a strong longitudinal, E_z, component. Previously we proposed using Cherenkov radiation emitted by an intense electron beam in a dielectric structure as a means of particle acceleration, a dielectric wakefield accelerator (DWA) [11]. The electromagnetic wave frequency in the DWA scheme is in the GHz range rather than THz as would be inverse Cherenkov acceleration using a CO_2 laser, but the fundamental physics is identical. If dimensions of the dielectric channeled waveguide are chosen properly, a particular TM_{0n} wavelength λ can be matched to a laser wavelength. Therefore, a laser beam with the same wavelength as TM_{0n} mode will propagate inside the dielectric channel and be used as the power source to accelerate charged particles (inverse Cherenkov process). An attractive way to power such a dielectric channeled waveguide would be to use a radially polarized laser beam as proposed by Pantell et al. [9], the reason being that the azimuthal magnetic and radial electric field in such a beam are similar to the corresponding TM_{0n} fields in the dielectric channeled waveguide. This permits the laser power to be efficiently coupled into the dielectric channeled waveguide. Once coupled, the laser beam will propagate in the TM$_{0n}$ mode. We can therefore treat this problem as a travelling wave linear accelerator made of dielectric channeled waveguide with a laser as the power source. Coupling of the laser beam into the dielectric tube can be done using a tapered transition section, a technique commonly used in the optical fiber technology.

The acceleration structure under the study has inner vacuum radius a and a layer of dielectric (ϵ) with radius b. A conductor tube is used to guide the laser beam. The radially polarized laser beam propagates through a focusing lens and then transition section in which the beam interferes with itself to satisfy proper boundary condition and thus establishes the correct mode for synchronous particle acceleration in the acceleration section. This proposed scheme is shown in the Figure 1.

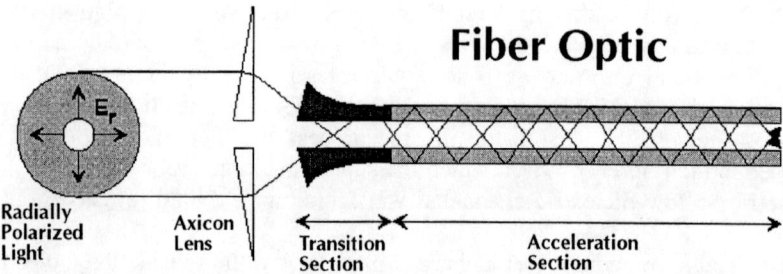

Figure 1. Schematic diagram of a dielectric based "inverse Cherenkov Accelerator". A radially polarized laser beam propagates through an axicon type lens and is subsequently focus into a hollow dielectric structures. The laser beam interferes with itself and the resulting patterns are TM$_{On}$ modes with strong longitudinal electrical fields. The structure has inner radius a and outer radius b partially filled with dielectric material. A transition section serves the purpose of matching the laser beam into the waveguide with minimum reflection.

Let us first review some important characteristics of Cherenkov radiation in dielectric structures. A charged particle passing through a medium of refractive index $\sqrt{\epsilon}$ with velocity v will radiate photons when the Cherenkov radiation condition is satisfied ($v > c/\sqrt{\epsilon}$) where c is velocity of the light in vacuum. The intensity of the radiation per unit length in an infinite medium is

$$\frac{dW}{dl} = \frac{e^2}{c^2} \int_{\beta\sqrt{\epsilon}>1} \left(1 - \frac{1}{\epsilon\beta^2}\right) \omega d\omega \qquad (1)$$

which is proportional to $\sin^2(\theta_c) = (1 - 1/\epsilon\beta^2)$, and θ_c is commonly known as the "Cherenkov angle." Because the dielectric constant of a pressurized gas system is very close to 1, the radiated power is very small compared to the radiation in a solid medium. Cherenkov radiation inside a dielectric loaded waveguide has some additional and very important characteristics which distinguish it from that in infinite media. First, whereas in infinite media the radiation spectrum is continuous, radiation inside a dielectric loaded waveguide is constrained to accessible eigenmodes of the guide which are, for an axially moving charge, pure transverse magnetic (TM). Furthermore, the effective Cherenkov angle can be quite large due to the relatively large value of the dielectric constant of the lining compared to gaseous media such as those used in "traditional" ICA experiments. Together, these result in there being strong coupling between the excited laser fields and a charge particle to be accelerated. Numerical examples verify this point and are presented later in this paper.

The axial electric fields inside the acceleration structure shown in Figure 1 can be described by

$$E_z^{(1)}(\omega, r, z, t) = E_0 I_0(kr) e^{i(k_z z - \omega t)}$$
$$E_z^{(2)}(\omega, r, z, t) = [B_1 J_0(s_1 r) + D_1 N(s_1 r)] e^{i(k_z z - \omega t)} \qquad (2)$$

where E_0, B_1 and D_1 are the field amplitudes in the region 0 (vacuum) and 1 (dielectric) respectively and are related by boundary conditions, and

$$k^2 = \frac{\omega^2}{v^2}(1 - \beta^2)$$
$$s^2 = \frac{\omega^2}{v^2}(\beta^2 \epsilon - 1) \qquad (3)$$

where $\beta c = v = \frac{\omega}{k_z}$ is phase velocity of the wave travelling inside the tube. Therefore β determines the synchronism of the wave and the accelerated particles. The transverse electric field can be written as [12]

$$E_r = i\frac{\omega/v}{\omega^2/v^2(\mathrm{B}^2 \in -1)} \frac{\partial E_z}{\partial r} \tag{4}$$

and magnetic field $H_\phi = \in E_r$ everywhere inside the tube. By matching the boundary conditions at a and b (E_z and D_r continuous), all the components in the field can be expressed in terms of E_0. The electric field inside the hole described by equations 2 and 4 have very interesting characteristics. When $k \to 0$, i.e., the phase velocity of the wave is c, E_z is constant across the vacuum hole. When we apply the Panofsky and Wenzel theorem [13] to this case, which relates the longitudinal and transverse forces exerted on a charged particle as,

$$F_r(r,z) = e\int \frac{\partial E_z(r,z)}{\partial r} dz \tag{5}$$

there are no focusing and de-focusing forces. Although it is possible for the laser beam to couple the energy into other deflecting modes, such as HEM_{1n}, it can be easily shown that the phase velocities of those modes are not synchronized with the accelerated beam.

The stored energy per unit length defined as U in the tube is the sum of contribution from both regions, and can be expressed as,

$$\begin{aligned} U &= \sum_{0,1} 2\pi \tfrac{1}{2} \int_s \left(\epsilon_0 \overline{E}^2 + \mu_0 \overline{B}^2\right) r dr \\ &= E_0^2 u \end{aligned} \tag{6}$$

where u is a geometric factor which solely depends on the structure geometry and dielectric constant. For a given laser beam power, the axial electric field in the center region of the tube can be expressed as,

$$E_0 = \left[\frac{P}{u\beta_g c}\right]^{\frac{1}{2}} \tag{7}$$

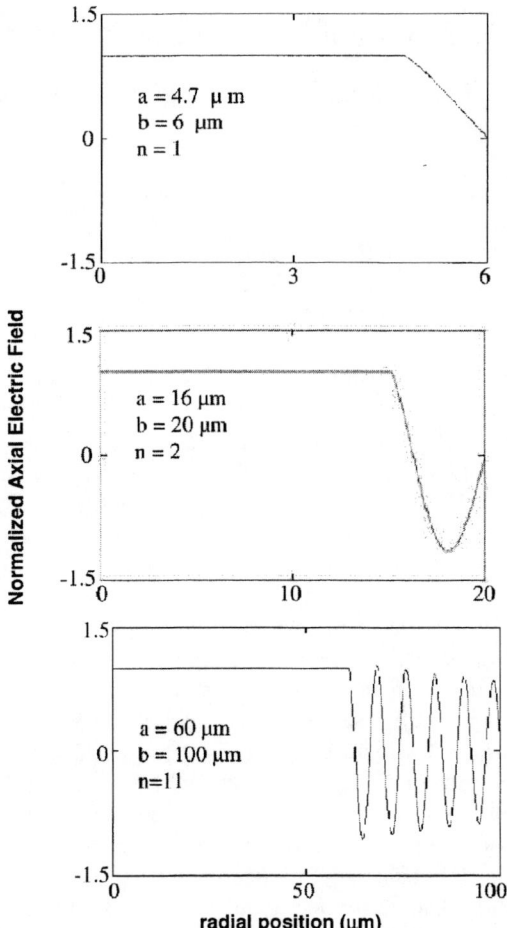

Figure 2 Normalized axial electric field. Longitudinal electric field distribution vs transverse positions. Where a is inner radius of the vacuum region and b is the radius of dielectric layer. n is a mode number corresponding the phase velocity = 1. The longitudinal field in the vacuum region is constant. Field inside the dielectric layer varies in the form of zero`s order bessel function.

where β_g is the group velocity. The total acceleration distance $L = \dfrac{v\tau\beta_g}{1-\beta_g}$ depends on the laser pulse length. Where τ is the laser pulse duration and v is the velocity of the accelerated particles.

We numerically calculated the field distributions and energy densities for the boundary value problem discussed above [equations (2) (3) (4) (5) and (6)] for various geometries. For simplicity and ease of comparison, we fix the dielectric constant at 3 in these calculations. Other values yield similar results provided the geometry is adjusted accordingly.

The ideal way to accelerate a charged particle is in the lowest radial TM_{01} mode. However, at the optical frequencies being considered, the size of the structure would be prohibitively small. By using a higher order mode TM_{0n} ($n > 1$), we can still have a uniform axial electric field in the center region and the field inside the dielectric layer would still governed be by equation 2. Figure 2 shows the E_z vs the radius for different radial modes. Hollow glass fibers with inner diameters of 10 microns can be fabricated and in fact, have been used for laser-guided atom experiments [14].

The following table gives several examples of cases that illustrate that a very strong longitudinal field can be established by a relatively modest laser power. Here we are using a CO_2 laser ($\lambda = 10.6\ \mu m$) as the power source. Example 1 shows that for $n = 1$, a structure with $a = 4.7\ \mu m$ and $b = 6\ \mu m$ is required for a dielectric of constant of 3, and less than *200 kW* power is required to achieve *1 GV/m* acceleration. The group velocity in this case is *0.522 c*. The power required is considerably less than that needed to achieve a correspondingly high gradient in the gas ICA scheme. In order to have a gain of *1 GeV* net acceleration, a total acceleration distance of *1 m* is required. Thus, at least a 3 ns laser pulse length is needed. Example 2 gives the result for the radial mode $n = 2$. The structure has a *15 μm* inner radius and *20 μm* outer radius. This is significantly larger than example 1. In this case, *10 MW* laser power will produce *1 GV/m* gradient. One can even operate at still higher order modes. Example 3 shows that for $n = 11$, a structure size of ~100 μm can be used. Approximately *1.3 GW* of laser power is necessary to produce *1 GV/m* in this case.

Example	Power	a(μm)	b(μm)	E_z	radial modes	β_g
1	0.2 MW	4.7	6	1.1 GV/m	1	0.5 c
2	10 MW	15	20	1.0 GV/m	2	0.78 c
3	1.3 GW	60	100	1.0 GV/m	11	0.7

We would like to point out that a small, gas filled gap between the dielectric fiber and the outer conducting wall could be used to fine tune the phase velocity by adjusting the gas pressure (serving in a sense as do the tuning stubs in some conventional linac structures). Fields inside the gap have interesting properties though longitudinal electric field is very small, the transverse electric and magnetic fields are large. Therefore, a large portion of the laser power will flow through this vacuum section. In a case similar to that of example 3, except at 50 μm tuning gap, the axial electrical field is lowered by 20%, although group velocity and geometries remain very close to example 3.

A key question which must be addressed for all structure supported laser accelerator schemes is the laser damage threshold. For materials commonly used in CO_2 laser applications, the damage threshold is typically about *0.5 J/cm²* with laser pulse length in the range 10 ps – 1 ns [15]. Continuing research on high damage threshold materials may find a much better material to support the higher gradient which is much more desirable for our applications. For a proof of principle experiment currently considered at the Accelerator Test Facility of Brookhaven National Laboratory, using the geometry of example 2 (above), this would impose a limit of 1 μJ. For a 10 ps, 0.1 MW laser beam, an acceleration gradient of ~100 MV/m would be established over an acceleration distance of 13 mm due to the high group velocity of the wave. This would be a feasible and interesting demonstration.

The transport of a charged particle beam through these small dielectric structures presents a major technical challenge. However, Wang *et al.* [16] have studied the Brookhaven ATF low emittance beam line and have shown that in an emittance of 10^{-10}*m-rad* the electron beam contains 10^6 particles, and can be focused to <1μm and transported through a distance of a few tens of centimeter. In principle, this should be sufficient for a proof of principle experiment.

Next we would like to discuss collective beam effects, i.e., wakefields and beam break-up. As in any slow wave structure, the longitudinal wake function and frequency strongly depend on the size of structure. For the structures considered here (to be used at optical frequencies), the wake function can be very large. Calculation of the longitudinal wake functions for the structure considered here have been made. For radial mode $n = 1$ structures, the wake field is on the order of TV/m (assuming 1 nC beam and 1 µm rms bunch length). If the ratio of wake field amplitude/acceleration field is to be kept below 10% to control the momentum spread of the beam at 1 GV/m acceleration gradient, the maximum charge which could be accelerated is about 0.1 pC which corresponds to a wake field amplitude of 100 MV/m. However, for a 100 µm structure such as example 3 above, the wake field would be on the order of GV/m for 1 nC beam; therefore, 10–100 pC beam can be accelerated through the dielectric channeled waveguide.

As in any accelerating structure, a misaligned beam travelling in the dielectric structures will excite transverse wake-fields, with possible serious head-tail instabilities. As stated above, in the relativistic limit, there is no direct focusing effect by the laser beam. External focusing by solenoids or quadrupoles with BNS damping [17] must be used. Design and fabrication of miniature strong magnets are already a subject of much current study [18], and more work is needed to explore the various options for controlling the beam break up effects.

In conclusion, we propose a new laser driven accelerator technique which promises an efficient way to couple laser energy to a charged particle beam. The scheme has all the properties of a travelling wave linac structure. Numerical examples show that a very modest amount of laser energy is required to achieve high gradient acceleration, much less than that in gas based inverse Cherenkov acceleration in a pressured gas system. Beam loading has been calculated and beam break-up issues have been discussed. An outline of a proof of principle experiment has been proposed. The Accelerator Test Facility (ATF) at Brookhaven National Laboratory may be a desirable place to do such an experiment because a suitable, radially polarized laser beam and low emittance e⁻ beam are already available.

ACKNOWLEDGMENTS

The authors would like to acknowledge helpful discussions with P. Schoessow. This work is supported by DOE, Division of High Energy Physics, Contract No. W-31-109-ENG-38.

REFERENCES

[1] A. Ogata, "Laser Wakefield Acceleration Experiments a ~1ps 10 TW Nd:Glass Laser," in the Proceedings of 6th Advanced Acceleration Concept, Edited by P. Schoessow, June 12–18, 1994, Fontana, WI.

[2] A.E. Dangor *et al.*, "Electron Acceleration to 44 MeV in Plasma Waves by the Rutherford 35 TW Single-Frequency Laser," in the Proceedings of 1995 Particle Accelerator Conference and International Conference on High Energy Accelerators (to be published), May 1–5, 1995, Dallas, TX.

[3] C.E. Clayton, K.A. Marsh, A. Dyson, M. Everet, A. Lal, W.P. Leemans, R. Williams and C. Joshi, Phys. Rev. Lett. **70**, 37 (1993).

[4] F. Amiranoff *et al.*, Phys. Rev. Lett. **74**, 5220 (1995).

[5] D. Zhang and R. Byer, "Proposed Waveguide Structure for Laser Driven Electron Acceleration," in the Proceedings of 6th Advanced Acceleration Concept, Edited by P. Schoessow, AIP, June 12–18, 1994, Fontana, WI.

[6] C.M. Haaland, Optics Comm. **114**, 280 (1995).

[7] J. Rosenzweig, A. Murokh, and C. Pellegrini, Phys. Rev. Lett. **74**, No. 13, 2467, 1995.

[8] W.D. Kimura, G.H. Kim, R.D. Romea, L.C. Steinhauer, I.V. Pogorelsky, K.P. Kusche, R.C. Fernow, X. Wang, and Y. Liu, Phys. Rev. Lett. **74**, 546 (1995).

[9] J. Fontana and R. Pantell, Journal of Applied Physics, **54**, 4285–4288 (1983).

[10] B.M. Bolotovskii, Usp. Fiz. Nuak. 75, 295 (1961) [Sov. Phys. Usp. 4, 781, (1962)].

[11] R. Keinigs, M. Jones and W. Gai, Particle Accelerator, **24**, 223 (1989), also see, W. Gai *et al.*, Phys. Rev. Lett. **61**, 2756 (1989).

[12] J.D. Jackson, *Classical Electrodynamics*, page 242, 1st Edition (John Wiley & Sons, 1962).

[13] Panofsky and W. Wenzel, Rev. Sci. Instrum., **27**, 967 (1956).

[14] M.J. Renn, D. Montgomery, O. Vdovin, D. Anderson, C. Wieman, and E. Cornell, Phys. Rev. Lett. **75**, No. 18, 3253 (1995).

[15] I. Pogorelsky et al., "Approach to Compact Terawatt CO_2 Laser System for Particle Acceleration" in Proceedings of 6th Advanced Acceleration Concept, Edited by P. Schoessow, June 12–18, 1994, Fontana, WI.

[16] X. Wang and H. Kirk, in Proceedings of 1993 Particle Accelerator Conference, San Francisco, 1991.

[17] K. Bane, AIP Conference Proceedings **153** (AIP, New York) p. 971, 1987.

[18] L. Turner, A. Nassiri, F. Mills and S. Kim, "A Micro-Undulator Fabricated by LIGA Process," in the Proceedings of the 14th International Conference on Magnet Technology, June 11–16, 1995, Tampere, Finland.

Theory and Simulation of an Inverse Free-electron Laser Experiment

S. K. Gou and A. Bhattacharjee*
Department of Physics and Astronomy, University of Iowa, Iowa city, IA 52242

J.-M. Fang and T. C. Marshall
Department of Applied Physics, Columbia University, New York, NY 10027

An experimental demonstration of the acceleration of electrons using a high-power CO_2 laser interacting with a relativistic electron beam moving along a wiggler has been carried out at the Accelerator Test Facility of the Brookhaven National Laboratory [Phys. Rev. Lett. **77**, 2690 (1996)]. The data generated by this inverse free-electron-laser (IFEL) experiment are studied by means of theory and simulation. Included in the simulations are such effects as: a low-loss metallic waveguide with a dielectric coating on the walls; multi-mode coupling due to self-consistent interaction between the electrons and the optical wave; space charge; energy spread of the electrons; and arbitrary wiggler-field profile. Two types of wiggler profile are considered: a linear taper of the period, and a step-taper of the period. (The period of the wiggler is ~3cm, its magnetic field is ~1T, and the wiggler length is 0.47m.) The energy increment of the electrons (~1-2%) is analyzed in detail as a function of laser power, wiggler parameters, and the initial beam energy (~40 MeV). At a laser power level ~ 0.5 Gw, the simulation results on energy gain are in reasonable agreement with the experimental results. Preliminary results on the electron energy distribution at the end of the IFEL are presented. Whereas the experiment produces a near-monotone distribution of electron energies with the peak shifted to higher energy, the simulation shows a more structured and non-monotonic distribution at the end of the wiggler. Effects that may help reconcile these differences are considered.

I. INTRODUCTION

In an inverse free-electron laser (IFEL) [1-3], energy is transferred from an intense laser beam to a relativistic electron beam in the presence of the periodic magnetic field of an undulator, using the principle of stimulated absorption of radiation. In order to interact with the electron beam effectively, the laser field should have a wavelength satisfying the resonance condition

* Corresponding author, email: amitava@iowa.physics.uiowa.edu

$$\lambda_r = \frac{\lambda_w}{2\gamma_0^2}\left(1 + \frac{e^2 B_w^2 \lambda_w^2}{8\pi m^2 c^4}\right), \quad (1.1)$$

for a planar wiggler. This formula implies that in the frame of reference moving with the mean motion of the electrons the frequency of oscillation of the electrons is the same as that of the laser field.

A proof-of-principle experiment that verifies essential aspects of the IFEL concept was first reported in [4]. An experimental demonstration of energy gain, using a high power CO_2 laser at the Accelerator Test Facility (ATF) of the Brookhaven National Laboratory (BNL) has been reported recently [5]. The main goal of this paper is to carry out a theoretical treatment and simulations of this experiment.

The following is a plan of this paper. In Section II, we derive the basic equations describing IFEL dynamics in a waveguide. Section III presents the simulation results for the parameters of the ATF-BNL experiment, as well as comparisons with the experiment data. We conclude with a summary and the implications of our results in section IV.

II. THE IFEL MODEL AND BASIC EQUATIONS

The IFEL described here uses a tapered planar wiggler specified by the vector

$$\vec{A}_w = \vec{e}_x A_w(z)\sin\left(\int_0^z K_w(z')dz'\right), \quad (2.1)$$

where the amplitude $A_w(z)$ and wave number $K_w(z)$ of the wiggler are given functions of the axial coordinate z. The polarization of the laser radiation is the same as that of the wiggler and denoted by $\vec{A}_r$. The laser field is injected into a coated rectangular waveguide [6] in which the x components of the TE_x modes vanish; so only the TM_x modes can interact with electrons effectively. Although this type of waveguide is expected to operate in a chosen mode, multi-mode effects are included in our formulation. The vector potential for the TM_x mode is

$$\vec{A}_r(\vec{r},t) = \vec{e}_x \sum_{mn}\{a_{mn} e^{i(K_{//}^{mn}z-\omega_r t)} + c.c.\}\cos\frac{m\pi}{a}(x+\frac{a}{2})\sin\frac{n\pi}{b}(y+\frac{b}{2})$$
$$+\vec{e}_x \sum_{mn}\{b_{mn} e^{i(K_{//}^{mn}z-\omega_r t)} + c.c.\}\sin\frac{m\pi}{a}(x+\frac{a}{2})\sin\frac{n\pi}{b}(y+\frac{b}{2}), \quad (2.2)$$

where a and b are the widths of the rectangular waveguide, ω_r is the laser frequency, $K_{//}^{mn} = \sqrt{(\omega_r/c)^2 - (m\pi/a)^2 - (n\pi/b)^2}$, and $a_{mn}(z)$ and $b_{mn}(z)$ are amplitudes for the $TM_x^{\gamma=0,\pi}$ and $TM_x^{\gamma=\pi/2}$ modes satisfying the

inequalities $|\partial a_{mn}/\partial z| \ll |K_{//}^{mn} a_{mn}|$ and $|\partial b_{mn}/\partial z| \ll |K_{//}^{mn} b_{mn}|$, respectively.
The axial component of the radiation field is neglected in (2.2). The attenuation parameters for the laser field are given in [6].

The current distribution, which is the source term in Maxwell's equations, is obtained by adding the contributions of all electrons. With the assumption that the strength of the laser field is small with respect to that of the wiggler field, we obtain the following equations of motion for a single electron:

$$\frac{d}{dz}\psi_{mn} = K_w^{mn}(z)\{1 - \frac{\gamma_{rmn}^2}{\gamma^2}\}$$

$$+ i\frac{eK_{//}^{mn}K(z)}{\sqrt{2}m\beta_z c^2 \gamma^2} \sum_{m'n'} \{J_0(\xi_{m'n'}) - J_1(\xi_{m'n'})\}\{a_{m'n'} e^{i\psi_{m'n'}} - c.c.\}$$

$$\cdot \cos\frac{m'\pi}{a}(x+\frac{a}{2})\sin\frac{n'\pi}{b}(y+\frac{b}{2})$$

$$+ i\frac{eK_{//}^{mn}K(z)}{\sqrt{2}m\beta_z c^2 \gamma^2} \sum_{m'n'} \{J_0(\xi_{m'n'}) - J_1(\xi_{m'n'})\}\{b_{m'n'} e^{i\psi_{m'n'}} - c.c.\}$$

$$\cdot \sin\frac{m'\pi}{a}(x+\frac{a}{2})\sin\frac{n'\pi}{b}(y+\frac{b}{2}), \qquad (2.3)$$

$$\frac{d}{dz}\gamma = -\frac{e\omega_r K(z)}{\sqrt{2}m\beta_z\gamma c^3} \sum_{mn} \{J_0(\xi_{mn}) - J_1(\xi_{mn})\}\{a_{mn} e^{i\psi_{mn}} + c.c.\}$$

$$\cdot \cos\frac{m\pi}{a}(x+\frac{a}{2})\sin\frac{n\pi}{b}(y+\frac{b}{2})$$

$$- \frac{e\omega_r K(z)}{\sqrt{2}m\beta_z\gamma c^3} \sum_{mn} \{J_0(\xi_{mn}) - J_1(\xi_{mn})\}\{b_{mn} e^{i\psi_{mn}} + c.c.\}$$

$$\cdot \sin\frac{m\pi}{a}(x+\frac{a}{2})\sin\frac{n\pi}{b}(y+\frac{b}{2}), \qquad (2.4)$$

where $\beta_z = v_z/c$, $K(z) = eA_w(z)/\sqrt{2}mc^2$, $K_w^{mn}(z) = K_w(z) + \{K_{//}^{mn} - K_r\}/\beta_z$,

$\xi_{mn} = K^2(z)\{K_{//}^{mn} + K_w(z)\}/4\beta_z K_w(z)\gamma^2$,

$\gamma_{rmn}^2 = K_{//}^{mn}\{1 + K^2(z)\}/2\beta_z K_w^{mn}(z)$ is the square of the resonant energy of

electron beam, $\psi_{mn}(z,t) = K_{//}^{mn} z + \int_0^z K_w(z')dz' - \omega_r t$ is the relative phase,

and J_0 and J_1 are Bessel functions.

The laser field is governed by Maxwell's equation which gives

$$\cos\frac{m\pi}{a}(x+\frac{a}{2})\sin\frac{n\pi}{b}(y+\frac{b}{2})\{K_{//}^{mn}\frac{\partial}{\partial z}+\frac{\omega_r}{c^2}\frac{\partial}{\partial t}\}a_{mn}$$

$$+\sin\frac{m\pi}{a}(x+\frac{a}{2})\sin\frac{n\pi}{b}(y+\frac{b}{2})\{K_{//}^{mn}\frac{\partial}{\partial z}+\frac{\omega_r}{c^2}\frac{\partial}{\partial t}\}b_{mn}$$

$$=\frac{\pi e^2 A_w(z)}{mc^2}\sum_{j=1}^{N_e}\frac{\delta(\vec{r}-\vec{r}_j)}{\gamma_j}\{J_0(\xi_{mn})-J_1(\xi_{mn})\}e^{-i\psi_{mn}}, \qquad (2.5)$$

where N_e is the number of electrons. Equations (2.3-2.5) contain transverse variables which make the problem three-dimensional. In a simplified model, we can eliminate the transverse variables from these equations by averaging over the transverse motion of the electrons. Under the assumption that the transverse profile of the electron density can be described by a Gaussian,

$$n(\vec{r}) = 2\pi n_0 R_x R_y \sigma(x,y), \qquad (2.6)$$

we can make the replacement

$$(...) \to \int_{-a/2}^{a/2} dx \int_{-b/2}^{b/2} dy \sigma(x,y)(...), \qquad (2.7)$$

where n_0 is the maximum density, R_x and R_y are the radii of the electron beam in the x and y directions, and the electron beam transverse profile is

$$\sigma(x,y) = \frac{1}{2\pi R_x R_y}\exp(-\frac{x^2}{R_x^2}-\frac{y^2}{R_y^2}). \qquad (2.8)$$

We assume that the electron density profile is longitudinally uniform. Thus, we obtain

$$\sum_{j=1}^{N_e}\delta(\vec{r}-\vec{r}_j) \to n_0 \exp(-\frac{x^2}{2R_x^2}-\frac{y^2}{2R_y^2})\frac{1}{N_e}\sum_{j=1}^{N_e}. \qquad (2.9)$$

Finally, averaging over the transverse dimension of the electron beam and a ponderomotive wavelength, we obtain the system of equations

$$\frac{d}{dz}\psi_{mnj} = K_w^{mn}(z)\{1-\frac{\gamma_{rmn}^2}{\gamma_j^2}\}$$

$$+i\frac{eK_{//}^{mn}K(z)}{\sqrt{2}m\beta_z c^2\gamma_j^2}\sum_{m'n'}I_{m'}(a,R_x)I_{n'}(b,R_y)\{J_0(\zeta_{m'n'j})-J_1(\zeta_{m'n'j})\}$$

$$\cdot\{a_{m'n'}e^{i\psi_{m'n'j}}-c.c.\}\cos\frac{m'\pi}{2}\sin\frac{n'\pi}{2}$$

$$+i\frac{eK_{//}^{mn}K(z)}{\sqrt{2}m\beta_z c^2\gamma_j^2}\sum_{m'n'}I_{m'}(a,R_x)I_{n'}(b,R_y)\{J_0(\xi_{m'n'j})-J_1(\xi_{m'n'j})\}$$

$$\cdot\{b_{m'n'}e^{i\psi_{m'n'j}}-c.c.\}\sin\frac{m'\pi}{2}\sin\frac{n'\pi}{2}, \qquad (2.10)$$

$$\frac{d}{dz}\gamma_j = -\frac{eK_r K(z)}{\sqrt{2}m\beta_z c^2 \gamma_j} \sum_{mn} I_m(a,R_x)I_n(b,R_y)\{J_0(\xi_{mnj}) - J_1(\xi_{mnj})\}$$

$$\cdot \{a_{mn} e^{i\psi_{mnj}} + c.c.\} \cos\frac{m\pi}{2}\sin\frac{n\pi}{2}$$

$$-\frac{eK_r K(z)}{\sqrt{2}m\beta_z c^2 \gamma_j} \sum_{mn} I_m(a,R_x)I_n(b,R_y)\{J_0(\xi_{mnj}) - J_1(\xi_{mnj})\}$$

$$\cdot \{b_{mn} e^{i\psi_{mnj}} + c.c.\} \sin\frac{m\pi}{2}\sin\frac{n\pi}{2}, \qquad (2.11)$$

$$\frac{d}{dz}a_{mn} = \frac{8\pi^2 R_x R_y A_w(z)e^2 n_0}{mc^2 K_{//}^{mn} ab} I_m(a,R_x)I_n(b,R_y)\cos\frac{m\pi}{2}\sin\frac{n\pi}{2}$$

$$\cdot \frac{1}{N_e}\sum_{j=1}^{N_e} \frac{\{J_0(\zeta_{mnj}) - J_1(\zeta_{mnj})\}}{\gamma_j} \exp(-i\psi_{mnj}), \qquad (2.12)$$

$$\frac{d}{dz}b_{mn} = \frac{8\pi^2 R_x R_y A_w(z)e^2 n_0}{mc^2 K_{//}^{mn} ab} I_m(a,R_x)I_n(b,R_y)\sin\frac{m\pi}{2}\sin\frac{n\pi}{2}$$

$$\cdot \frac{1}{N_e}\sum_{j=1}^{N_e} \frac{\{J_0(\zeta_{mnj}) - J_1(\zeta_{mnj})\}}{\gamma_j} \exp(-i\psi_{mnj}), \qquad (2.13)$$

where

$$I_m(a,R_x) = \int_{-a/2}^{a/2} dx \frac{\exp(-\frac{x^2}{2R_x^2})}{\sqrt{2\pi}R_x}\cos\frac{m\pi x}{a}. \qquad (2.14)$$

It can be shown that the system (2.10-2.13) admits the energy conservation law

$$\{a_{mn}a_{mn}^* + b_{mn}b_{mn}^*\} + \frac{16mc^2\pi^2 R_x R_y n_0}{abK_r^2}\frac{1}{N_e}\sum_{j=1}^{N_e}\gamma_j = \text{constant}, \qquad (2.15)$$

which makes it evident that electrons can be accelerated by transferring the energy of the laser field to the electron beam.

When the electron beam is bunched, because of its interaction with the wiggler and the laser field, the non-uniform longitudinal charge distribution produces a longitudinal electric field. This introduces an additional term in the equation for the electron energy equation. The space-charge term associated with a charge distribution modulated at the laser wavelength can be represented as

$$(\frac{d}{dz}\gamma_j)_{sc} = -i\frac{\beta_z \omega_p^2}{c\omega_r}\sum_{mn}\{\langle e^{-i\psi_{mn}}\rangle e^{i\psi_{mnj}} - \langle e^{i\psi_{mn}}\rangle e^{-i\psi_{mnj}}\}, \qquad (2.16)$$

where $\langle e^{\pm i\psi_{mn}}\rangle = \frac{1}{N_e}\sum_{j=1}^{N_e} e^{\pm i\psi_{mnj}}$ describes the bunching of the electron beam.

It is easy to check that the space-charge effect will vanish if the electron phase is homogeneously distributed in such a way that the electron bunching is zero. Equations (2.10)--(2.13) (augmented by equation (2.16) when relevant) are the basic equations underlying the simulation.

III NUMERICAL SIMULATIONS AND COMPARISON WITH THE ATF-BNL EXPERIMENT

The simulations are performed for the parameters given in Table I which are representative of the BNL-ATF experiment. The waveguide employed is the type described by Zakowicz [6].

Table I

Injection energy γ_0 (Mev)	35	40
Nominal current (mA)	5	5
Maximum current (A)	30	30
$\Delta\varepsilon/\varepsilon$	3×10^{-3}	3×10^{-3}
Beam radius R_x, R_y (mm)	0.3	0.3
Wiggler strength B_w (KG)	8.5-8.70	10-10.24
Wiggler period λ_w (mm)	28.57-31.38	28.79-31.27
Wiggler length L_w (cm)	47	47
Field-free gap length (mm)	4	4
Laser wavelength λ_r (μm)	10.6	10.6
Laser power (Gw)	0.4-1	0.4-1
Photon beam radius w (mm)	0.7	0.7
Waveguide width a,b (mm)	2.8	2.8

For attenuation parameters of the TM_x modes, see [6].

For a given Gaussian laser beam at input, with the maximum electric field strength E_0 and waist-size w, the initial amplitudes for the waveguide modes are

$$a_{mn}(0) = -i\frac{8\pi c E_0 w^2}{ab\omega_r}\cos\frac{m\pi}{2}\sin\frac{n\pi}{2}\cdot I_m(a/\sqrt{2}w)\cdot I_n(b/\sqrt{2}w), \quad (3.1)$$

$$b_{mn}(0) = -i\frac{8\pi c E_0 w^2}{ab\omega_r}\sin\frac{m\pi}{2}\sin\frac{n\pi}{2}\cdot I_m(a/\sqrt{2}w)\cdot I_n(b/\sqrt{2}w). \quad (3.2)$$

In our simulations, only the modes with $m,n(\leq 3)$ are considered. We note that the basic equations are periodic in ψ; so it is sufficient to consider the dynamics in one period of ψ. In what follows, we use $N_e = 1060$ in one period of ψ.

The energy increase ratio $\eta(=\Delta E/E)$ of accelerated electrons is measured in the ATF-BNL experiment and is investigated in detail in our simulation. Two types of wiggler configurations are considered: a linearly tapered wiggler and a step-tapered wiggler with a field-free gap of equal length.

The dependence of η on laser power is given in Fig. 1. It is shown that a step-tapered wiggler is generally more effective in accelerating electrons than a linearly tapered wiggler. The slopes of the curves depend on the wiggler parameters and the electron injection energy. Moreover, the lower the initial electron energy $\gamma_0 mc^2$, the higher the ratio η.

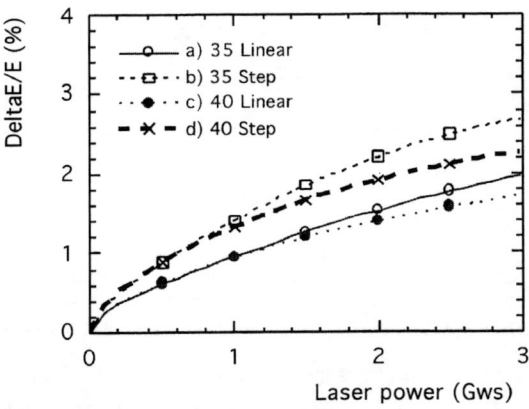

Fig. 1 The energy increase of accelerated electrons vs. laser power
   a) linearly tapered wiggler with 35 Mev initial electron beam energy
   b) step-tapered 4-section wiggler with 35 Mev initial electron energy
   c) linearly tapered wiggler with 40 Mev initial electron beam energy
   d) step-tapered 4-section wiggler with 40 Mev initial electron energy

There is no direct measurement of the actual amount of laser power available for acceleration, but it is estimated in Ref. 5 that the power is 1 Gw. In the simulations that follow we vary the power level over the range (0.4-1) Gw.

Fig. 2 gives the dependence of η on laser wavelength for both types of taper. The maximum value of η is obtained at the wavelengths 12.25μm for linearly tapered and 11.50μm for step-tapered configurations. The peak of η is bracketed reasonably well by the relation (1.1) which has been derived under the assumption of a constant taper and predicts two wavelengths when we substitute the magnitude of the wiggler field at the beginning and end of the taper. The numerically observed value lies in the range between these two predictions.

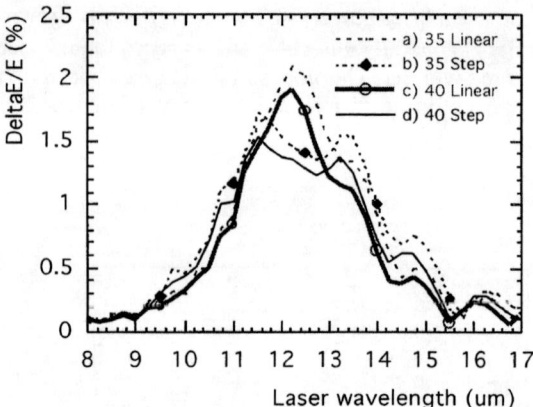

Fig. 2 The energy increase of accelerated electrons vs. laser wavelength for 0.5 Gw laser power.
- a) linearly tapered wiggler with initial electron beam energy 35 Mev
- b) step-tapered 4-section wiggler with initial electron beam energy 35 MeV
- c) linearly tapered wiggler with initial electron beam energy 40 MeV
- d) step-tapered 4-section wiggler with initial electron beam energy 40 MeV

Fig. 3 gives η as a function of the untapered strength of wiggler. We see once again that the relation (1.1) works reasonably well as a predictor for the position of the peak. The peak value of η for the linearly tapered wiggler is a little higher than that for the step-tapered wiggler. The simulation results are compared with experimental data.

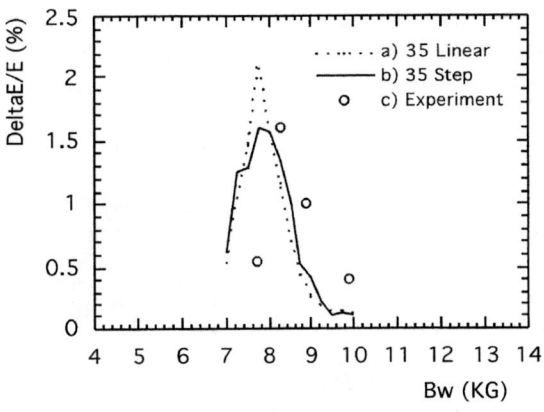

Fig. 3.1

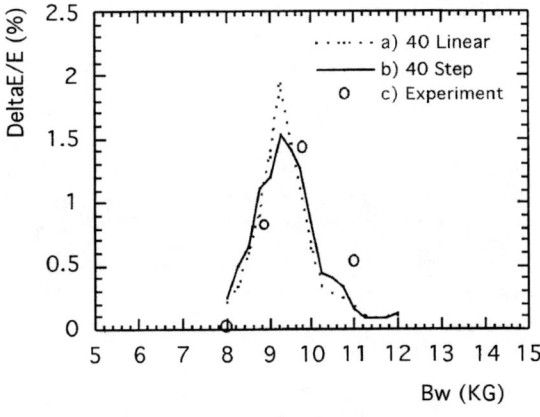

Fig. 3.2

Fig. 3.1 The energy increase of accelerated electrons vs. wiggler strength obtained for 0.5 Gw laser power and the parameters for the 35 Mev electron beam (Column 1, Table I).
    a) linearly tapered wiggler
    b) step-tapered wiggler
    c) experimental result

Fig. 3.2 The energy increase of accelerated electrons vs. wiggler strength obtained for 0.5 Gw laser power and the parameters for the 40 Mev electron beam (Column 2, Table I).
    a) linearly tapered wiggler
    b) step-tapered wiggler
    c) experimental result

Fig. 4 indicates the dependence of η on the initial electron beam energy. Four simulation curves (a-d) are given for the parameters of the 35 Mev and 40 Mev injected electrons in the case of constant wiggler strength (B_w =9.5 KG) in linearly and step-tapered wiggler configurations. These simulation results compare well with experimental data (e).

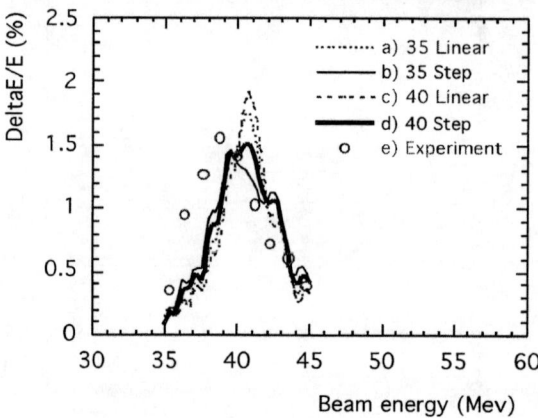

Fig. 4 The energy increase of accelerated electrons vs. the initial electron beam energy
 a) linearly tapered wiggler with the parameters of 35 Mev beam
 b) step-tapered 4-section wiggler with the parameters of 35 Mev beam
 c) linearly tapered wiggler with the parameters of 40 Mev beam
 d) step-tapered 4 section wiggler with the parameter of 40 Mev beam
 e) experimental result

Figs. 5 give the energy histograms of an accelerated electron beam with 40 Mev injection energy just after it emerges from a linearly tapered wiggler. For low (~0.5 Gw) as well as high (~10 Gw) laser power, the fine structure of the distribution function is sensitive to space-charge effects. Roughly speaking, when space-charge effects are included, the distribution function of accelerated electrons has two principal peaks for lower laser power (Fig. 5.2) and multiple peaks (Fig. 5.4) for higher laser power. The simulation shows that the space-charge effect tends to pull the peaks in together at lower laser power. Note that all of these distribution functions differ from the observed distribution function of Fig. 6 which appears to be monotone.

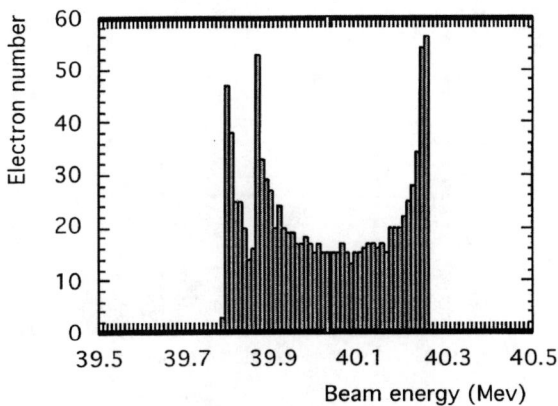

Fig. 5.1

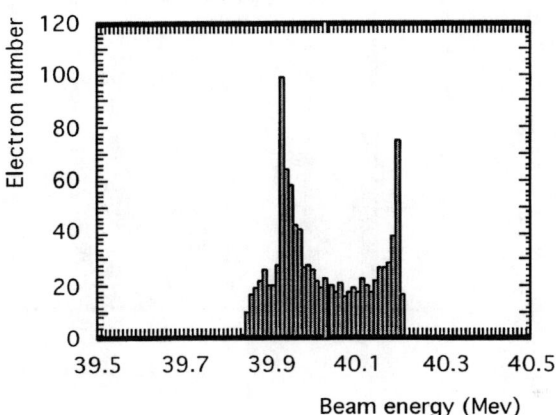

Fig. 5.2

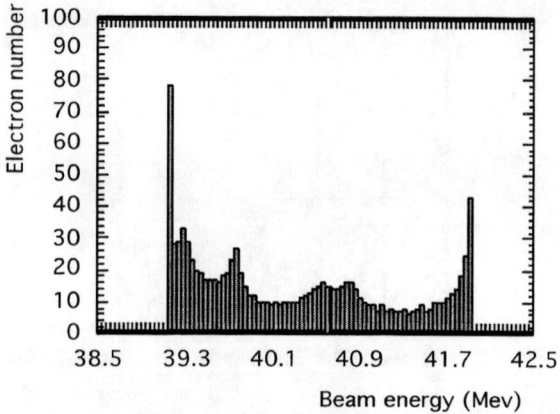

Fig. 5.3

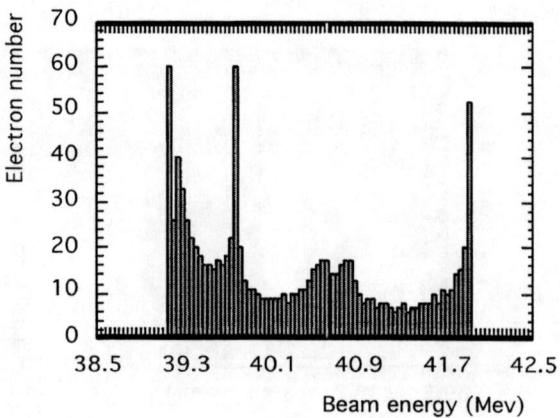

Fig. 5.4

Fig. 5 The distribution function of accelerated electrons with 40 Mev injection energy just after the electrons emerge from a linearly tapered wiggler.
    5.1) for 0.5 Gw laser power, without space charge
    5.2) for 0.5 Gw laser power, with space charge
    5.3) for high laser power, without space charge
    5.4) for high laser power, with space charge

In the experiment, the electron distribution is measured not at the end of the wiggler, but at a distance of $8m$ downstream. One might expect that the details of the distribution would erode as the beam moves some distance along the drift tube before reaching the spectrometer. In an effort to reproduce the observed distribution from our simulation, we have carried out a one dimensional time-dependent IFEL simulation [7,8], including the effects of slippage and space charge, and calculated the evolution of the electron distribution at the end of the wiggler [Fig. 7.1] and at the spectrometer [Fig. 7.2]. The parameters used are: 0.2 Gw laser power, 10 KG wiggler field strength, 40 Mev input electron beam energy with 0.2% energy spread. The distribution at the spectrometer shows a smoother distribution on the higher energy side, compared to the one at the end of the wiggler.

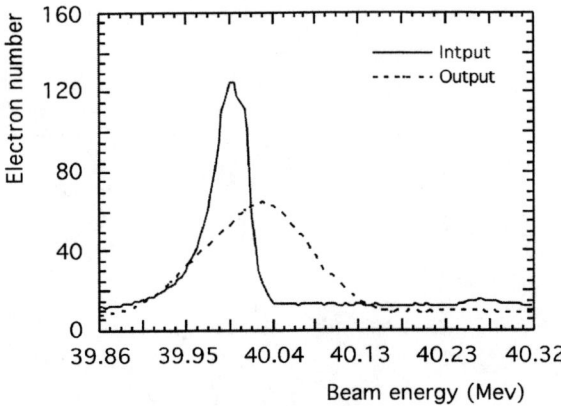

Fig. 6 The experimentally observed distribution function of the electron beam. The solid line is the distribution without the laser, and the dotted line is the distribution function in the presence of the laser.

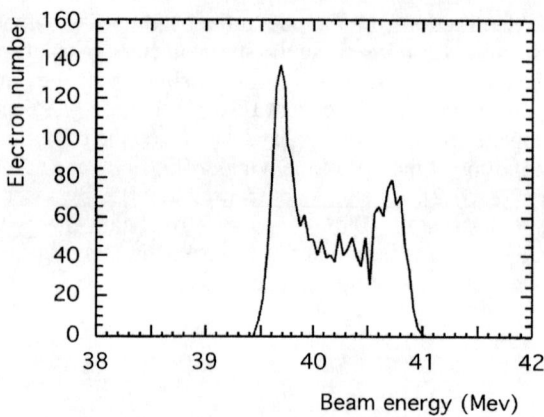

Fig. 7.1

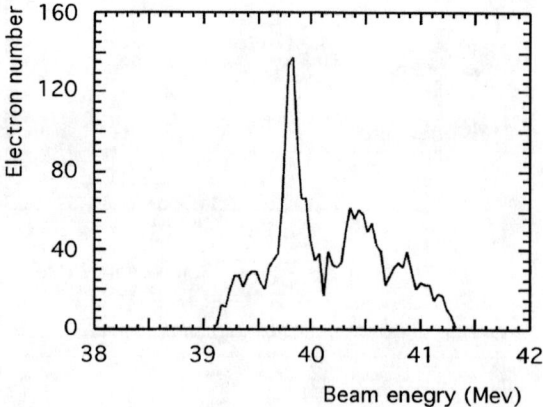

Fig. 7.2

Fig. 7 The distribution function from the one-dimensional simulation
   7.1) at the end of the wiggler
   7.2) at 8 meter downstream from the wiggler (where the
         spectrometer is located)

We have also calculated the bunching (in radians) produced by the IFEL at 25 cm and 100 cm downstream from the wiggler [Fig. 8]. Further downstream, a progressive erosion of the bunching occurs, caused by the drift and space-charge effects. These calculations underscore the potential usefulness of the IFEL as a "beam-buncher" (on the 10 micron wavelength scale) for injection into another laser accelerator.

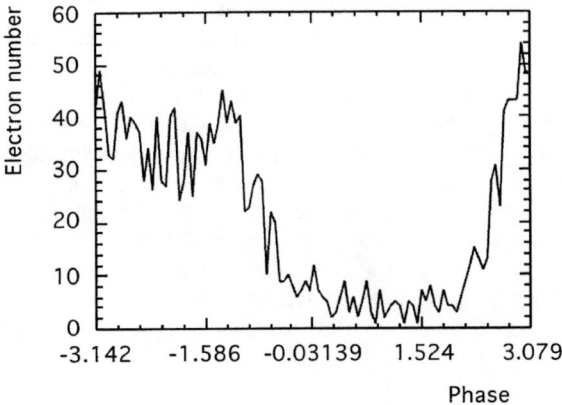

Fig. 8.1

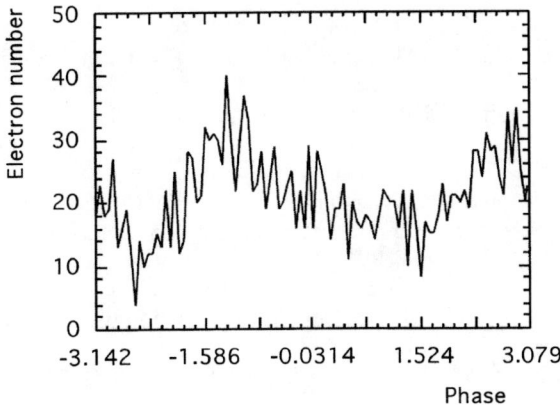

Fig. 8.2

Fig. 8 The bunching (in radians) of the electron beam with the same parameters as Fig. 7
       8.1) 25 cm downstream from the wiggler
       8.2) 100 cm downstream from the wiggler

IV CONCLUSIONS

We have presented simulations of the IFEL accelerator at ATF-BNL. The energy increase ratio is investigated in detail as a function of laser power, laser wavelength, wiggler strength, and initial electron energy. The simulations for ~0.5 Gw laser power are in good accord with experimental results on the energy increase of the electron beam, measured as functions of the wiggler field and the initial electron beam energy. There remains, however, a qualitative discrepancy between the simulation and the experiment for the energy histogram of the accelerated electron beam. Whereas the experiment yields a monotonic, one-peak histogram for accelerated electrons, the simulation generally predicts multi-peak histograms at the end of the wiggler as well as at distances downstream. We have also presented some numerical results on the bunching characteristics of the electron beam after IFEL interaction to indicate the potential usefulness of the IFEL as a beam-buncher.

ACKNOWLEDGMENTS

This research is supported by the U. S. Department of Energy Grant No. DE-FG02-91-ER40669.

REFERENCES

[1] R. B. Palmer, J. Appl. Phys. **43**, 3014 (1972).
[2] P. Sprangle and C. M. Tang, IEEE Trans. Nucl. Sci. **NS-28**, 3346 (1981).
[3] E. D. Courant, C. Pellegrini, and W. Zakowicz, Phys. Rev. A **32**, 2813 (1985)
[4] I. Wernick and T. C. Marshall, Phys. Rev. A **46**, 3566 (1992).
[5] A. van Steenbergen, J. Gallardo, J. Sandweiss, and J.-M. Fang, Phys. Rev. Lett. **77**, 2690 (1996).
[6] W. Zakowicz, J. Appl. Phys. **55**, 3421 (1984).
[7] N. M. Kroll, P. L. Morton, and M. N. Rosenbluth, IEEE J. Quantum Electron. **QE-17**, 1436 (1981).
[8] T. B. Zhang and T. C. Marshall, Phys. Rev. Lett. **74**, 916 (1995).

Status of the BNL IFEL Accelerator

A. van Steenbergen, J. Gallardo, M. Babzien, K. Kusche,
R. Malone, I. Pogorelsky, T. Romano, J. Sheehan, J. Skaritka,
X-J Wang

*Brookhaven National Laboratory
Upton, New York 11973-5000*

J. Sandweiss

*Physics Department
Yale University
New Haven, CT 06511*

J.-M Fang

*Department of Applied Physics
Columbia University
New York, NY 10027*

X. Qiu

*Physics Department
State University of New York
Stony Brook, NY 11794*

Abstract. A 40 MeV electron beam, using the inverse free-electron laser interaction, has been accelerated by $\Delta E/E = 2.5\%$ over a distance of 0.47 m. The electrons interact with a 1-2 GW CO_2 laser beam bounded by a 2.8 mm ID sapphire circular waveguide in the presence of a tapered wiggler with $B_{max} \approx 1\,\text{T}$ and a period $2.89\,\text{cm} < \lambda_w < 3.14\,\text{cm}$. The experimental results of $\Delta E/E$ as a function of electron energy E, peak magnetic field B_w and laser power W_l compare well with analytical and 1-D numerical simulations and permit scaling to higher laser power and electron energy. The present status of the IFEL accelerator and planned near term development are indicated.

INTRODUCTION

The study of the Inverse-Free Electron-Laser (IFEL) as a potential mode of electron acceleration has been pursued at Brookhaven National Laboratory (BNL) for a number of years. [1–4] Recent studies have focused on the development of a low energy, high gradient, IFEL accelerator [5] as a first step toward a multi-module electron accelerator of maximum operating energy of a few GeV. Experimental verification of the IFEL accelerator concept was obtained [6] in 1992, using a radiation wave length of $\lambda = 1.65$mm, and more recently [7] using a wavelength of 10.6 μm. In this report further experimental evidence of the IFEL interaction ($\lambda = 10.6\,\mu$m) is presented. The experiment used a 50 MeV electron beam, a 1- 5 GW CO_2 laser beam provided by BNL's Accelerator Test Facility (ATF) and a uniquely designed period length tapered wiggler.

The wiggler is a fast excitation electromagnet with stackable, geometrically and magnetically alternating substacks of Vanadium Permendur (VaP) ferromagnetic laminations, periodically interspersed with conductive (Cu), non-magnetic laminations, which act as eddy current induced field reflectors [8,9]. Four current conducting rods, parallel to the wiggler axis, are connected at the ends of the assembly, constituting the excitation loop that drives the wiggler. The overall wiggler stack is easily assembled, is compressed by simple tie rods, and readily permits wiggler period (λ_w) variation. Configured as a constant period wiggler, $\lambda_w = 3.75$ cm and $B_{max} = 1$ T, the system has shown [10] an rms pole-to-pole field variation of approximately 0.2 %.

The CO_2 laser beam is brought into the IFEL interaction region by a low loss dielectric (Al_2O_3, sapphire) circular waveguide which evidenced very good transmission properties [11] of the high power CO_2 laser beam. Extensive studies were carried out to establish optimum coupling into the guide and to measure the transmission loss of the long (1.0 m) extruded single crystal sapphire guides. Also, because of the overmoded guide configuration (ID = 2.8 mm), attempts were made to determine the transverse mode spectrum. To this end various wave guide configurations were tested at low laser beam power with the beam focused to a Gaussian waist with adjustable radius at the entrance of the waveguide. The beam profile was measured using a pyroelectric vidicon TV camera combined with digital frame grabber. For the 2.8 mm. ID sapphire dielectric guide a laser power attenuation factor of 0.2 dB/m was measured. The laser beam profile within the guide was inferred by measuring the beam diameter at the guide exit for various guide lengths. The results show that, commensurate with the near constant beam profile within the guide, the mode structure is dominated by the guide fundamental mode only. This is in accord with the absence of mode mixing reported in Ref. 11 for filamentary sapphire guides for CO_2 laser radiation transport.

The laser power must be efficiently coupled into the desired mode (H_{11}). To determine the transition region over which the mode becomes established, a series of scalar diffraction calculations were performed to find the fields propagating from the coupling aperture. It was found that the mode pattern transformed from the input Gaussian to a stable field distribution over a distance comparable to Z_R, the Rayleigh length. For the waist sizes employed here, after the mode has stabilized, the amplitude typically fluctuates by $\pm 5\,\%$ and the phase by ± 0.05 radians. These calculations suggest a $90\,\%$ coupling efficiency into the desired mode, consistent with the experimental observations reported below.

In the IFEL accelerator, the electron beam is accelerated by the interaction with the laser radiation wave in the medium of a periodic wiggler field. The theoretical description of the interaction has been given by a number of authors [3,12]. Approximate analytical expressions derived in Ref. [3] were used to parameterize a single acceleration stage. Subsequently, 1-D and 3-D simulation programs were written solving the self consistent system of Lorentz equations for the electrons and the wave equations for the input laser field as discussed in Ref. [12]. The 1-D program has been used to determine the self- consistent wiggler period length and its taper for given values of electron beam energy and laser power and to calculate the bucket acceptance and bucket leakage for a single or multi module accelerator. The 3-D code has been used to study beam walk-off, transverse phase space distributions and emittance growth.

EXPERIMENTAL ARRANGEMENT AND RESULTS

Extensive IFEL simulation studies were carried out both for a single IFEL accelerator module and for a sequence of IFEL modules. The objective of the present experiment was a proof of principle performance of a single IFEL unit incorporated in beam line II of the ATF [13,14]. A schematic layout, specific to the IFEL experiment only, is shown in Fig. 1. Beam transport downstream from the nominal 50 MeV Linac is so dimensioned as to yield a dispersion free IFEL interaction region. The electron beam, at the IFEL location, is matched vertically to the natural wiggler betatron amplitude $\beta_y = 0.17\,m$, $\alpha_y = 0.0$ and to a horizontal amplitude $\beta_x = 0.3\,m$, $\alpha_x = 0.0$. Downstream of the IFEL interaction region the optical system is configured as a momentum spectrometer with adjustable dispersion magnitude ($0.0 < \eta_p < 3.0\,m$) at a diagnostic end station; there, the beam momentum dispersion is measured by means of a phosphor screen-vidicon TV camera-Spiracon frame grabber. Also shown schematically in Fig. 1 is the CO_2 laser beam entry into the interaction region vacuum envelope through a ZnSe window, and its propagation

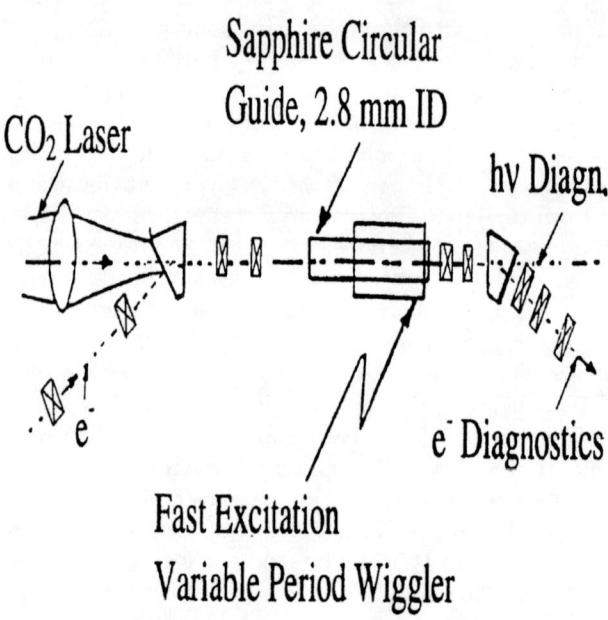

FIGURE 1. Schematic of the experimental configuration.

as a free-space mode, to the sapphire dielectric waveguide entry. With deliberation, the dielectric guide was taken to be 0.6 m in length, whereas the accelerator module length (wiggler length) was set at 0.47 m. This was done to approximate a mode matching section, enhancing thereby the mode purity in the IFEL module proper.

The design parameters used in this IFEL accelerator experiment are listed in Table 1. With optimized overlap of the electron and CO_2 laser beams, both spatially and time wise, and the interleaving of the lower repetition rate CO_2 laser pulses with the higher repetition rate electron beam pulses, the IFEL electron beam acceleration could readily be established. Electron acceleration was measured with the spectrometer at the diagnostic screen. An example of the momentum spectrum of the unaccelerated and accelerated electrons is given in Fig. 2, where the beam intensity distribution is shown versus $\sqrt{\beta_x \epsilon_x} + \eta_p \Delta p/p$, with the spectrometer optics adjusted so that $\eta_p \Delta p/p \gg \sqrt{\beta_x \epsilon_x}$. Optimization of the IFEL effect and exploration of parameter space, with variation of the electron beam injection energy, CO_2 laser power and wiggler maximum magnetic field magnitude was carried out in several consecutive runs, the results of which established the unambiguous signature of the IFEL acceleration. This is illustrated in Figs. 3, 4 and 5 where $(\Delta E/E)_{\text{IFEL}}$ is shown

TABLE 1. IFEL Experiment, first phase

e^- beam	Injection Energy	40.0	MeV
	Exit Energy	42.3	MeV
	$<Accel.Field>$	4.9	MV/m
	Current, nominal	5	mA
	N(bunch)	10^9	e^-
	I(max.)	30	A
	$\Delta E/E(1\sigma)$	$\pm 3.0 \times 10^{-3}$	
	Emittance (1σ)	7×10^{-8}	m rad
	Beam radius	0.3	mm
Wiggler	Wiggler Length	0.47	m
	Section Length	0.6	m
	Period Length, λ_w	2.9–3.1	cm
	Wiggler Gap	4	mm
	Field max.	10	kG
	Beam oscill., $a_{1/2}$	0.16–0.2	mm
CO_2 Laser	Power, W_l(Laser)	10^9	Watts
	Wave Length, λ	10.6	μm
	Max.Field, E_o	0.78×10^3	MV/m
	Guide Loss, α	0.05	m^{-1}
	Field Attenuation	0.26	dB/sec
	Pulse, (fwhm)	220	psec
	A_o	1.53×10^3	
	$r_o(L_w/2)$	1.0	mm

both as given by the 1-D model simulations and as obtained experimentally. Fig. 3 shows the relative energy gain for B_w and W_l constant; in Fig. 4 the plot $(\Delta E/E)_{\text{IFEL}}$ vs. B_w is given and in Fig. 5 the relative energy gain versus laser power W_l is plotted.

The approximate IFEL design equations [3] are:

$$d\gamma/dz = \frac{AKf(K)}{\gamma}\sin\psi \qquad (1)$$

where $\psi = (k + k_w)z - kct$ is the electron phase relative to the laser phase; the normalized laser electric field is $A = (\frac{e}{mc^2})\frac{1}{R_o}\sqrt{(\pi W_l Z_o)}$, $K = (eB_w\lambda_w)/(2\pi mc) \approx 2.7$ is the wiggler parameter, $f(K) \approx 0.38$ is a correction factor due to the linear polarization of the wiggler, $Z_o = 377\,\Omega$ is the vacuum impedance, R_o is the waveguide radius and k, k_w are the radiation and wiggler wavevectors, respectively. The resonance condition leads to:

$$\lambda = \frac{\lambda_w}{2\gamma^2}(1 + \frac{1}{2}K^2) \qquad (2)$$

The relative energy gain of the electron beam in a wiggler of length L_w is:

$$\Delta\gamma/\gamma = (\Delta p/p)_{\text{IFEL}} = A(K/\gamma^2)f(K)\sin\psi_r L_w \qquad (3)$$

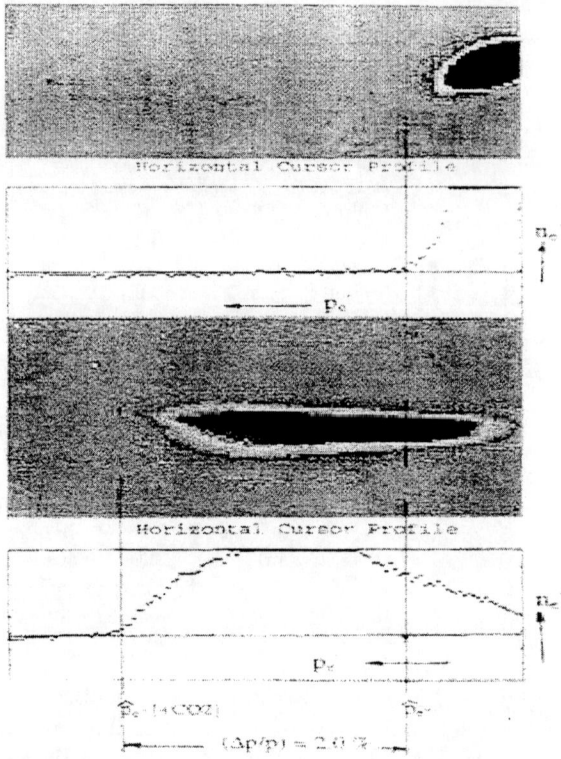

FIGURE 2. Momentum spectrum of the unaccelerated and IFEL accelerated electron beam. $E_l = 40$ MeV, $B_w = 10$ kG, $l_w = 2.9$ - 3.1 cm, $W_l = 1$ GW.

where ψ_r is the resonance phase (45° for optimum bucket size).

In Fig. 3 the solid line shows the results of the numerical simulations with laser power $W_l = 1$ GW and $B_w = 10$ kG normalized to the maximum experimental value. The agreement of the simulations with the experimental results are good. Similarly, in Fig.4 experimental results are compared with the simulations for 35 MeV and 40 MeV, in both cases the agreement is good. The maximum $(\Delta p/p)_{\text{IFEL}}$ for initial electron energy of 35 MeV leads to a value of the magnetic field $B_w = 8.35$ kG, to be compared with the experimental value of 8.44 kG, and for $E = 40$ MeV, the calculated B_w is 9.98 kG and the experimental value was $B_w = 9.96$ kG. Fig.5 shows the relative energy gain as function of the square root of the laser power; the scattering of data points reflect the typical laser power pulse to pulse variation; as a consequence, every set of experimental data needs to be normalized to $\sqrt{W_l}$. With the present spectrometer, the energy gain could be measured with good accuracy due to

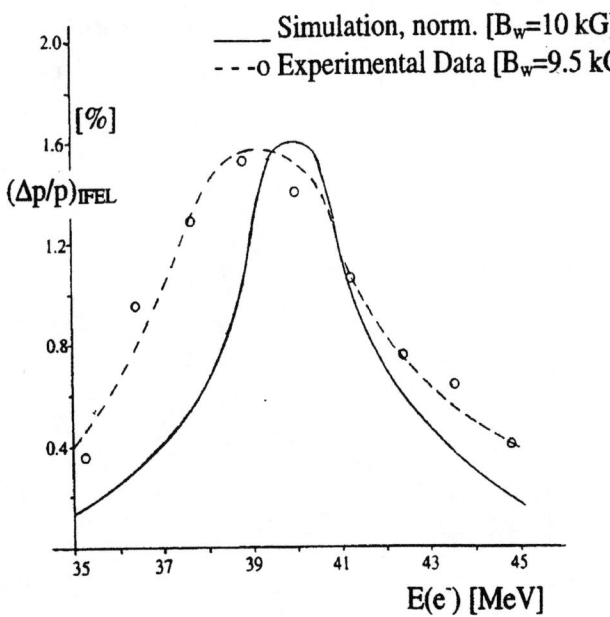

FIGURE 3. Relative energy gain $\Delta E/E$ vs E with B_w, W_l constant.

the sharp intensity fall-off of the high energy edge of the non-accelerated particles. A quantitative intensity ratio of the accelerated to unaccelerated beam could not be obtained due to the extended low energy edge of the unaccelerated beam. This limited the ability to measure the bucket size and leakage for comparison with model predictions and therefore, the value of the synchronous phase angle ψ_r could not be unambiguously established. Analytically, ψ_r and $\Delta\gamma/\gamma$ as a function of laser power W_l and wiggler parameters are given by:

$$\sin\psi_r = (3/16)\frac{k}{k_w}K(Af(K)L_w)^{-1}[(\frac{\lambda_w(L_w)}{\lambda_w(0)})^2 - 1] \quad (4)$$

$$\Delta\gamma/\gamma = 2\sqrt{[\frac{Kf(K)A}{k(1+\frac{1}{2}K^2)}]}\Gamma(\psi_r) \quad (5)$$

Eq.4 permits to calculate the moving bucket [12] parameter $\Gamma(\psi_r)$ and its maximum energy extent $\Delta\gamma/\gamma$. For the experimental value $\Delta\gamma/\gamma = 2.5\%$, we find : $\psi_r = 34°$ in reasonable agreement with the optimal 45° and a laser power of $W_l = 2.7$ GW which is larger than the 1 GW estimated experimentally.

In conclusion, the IFEL acceleration of a 40 MeV electron beam by $\Delta E/E = 2.5\%$ with a 1 GW CO_2 laser and a tapered wiggler with peak field on axis of 10 kG has been confirmed. Agreement with the model predictions is satisfactory, permitting the scaling of anticipated results to higher laser power.

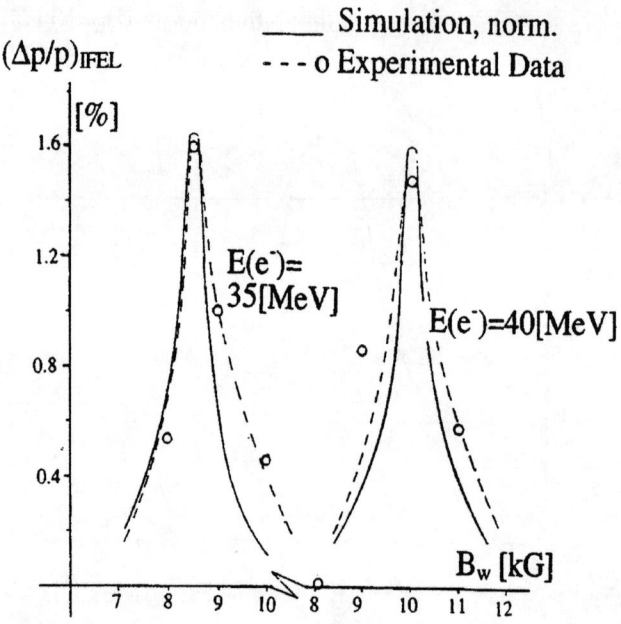

FIGURE 4. Relative energy gain $\Delta E/E$ vs B_w with E and W_l constant.

Present IFEL operation is limited to a maximum laser power of $\leq 2\,\mathrm{GW}$. With the enhanced vacuum pump-out capability of the IFEL interaction region and modified entry cone to the sapphire guide configuration, as presently incorporated, operation at a laser power of 5 GW is anticipated in the near term, which would enable close to 10% energy gain. With the upgrading of the ATF CO_2 laser to the 1 TW level as presently underway, an IFEL mean acceleration gradient of 100 MeV/m might become achievable. With regards to this, limitations on energy gain may arise from at least two sources [15]: firstly, the damage threshold of the sapphire waveguide and the consequent potential decrease of the power transmission; secondly, the decrease of the electron beam intensity associated with the self-field interaction due to the small aperture of the guide. We will address these problems in future experimental work.

Near term development of the IFEL accelerator concept will incorporate two approaches: first, the construction of a second VaP fast excitation wiggler - sapphire guide IFEL interaction region, for incorporation into a two accelerator modules IFEL accelerator, to test realistically a synchronized multi-module IFEL accelerator sequence and aim, with the above cited CO_2 laser developments, at a 100 MeV IFEL Linac ; second, in a joint developmental approach with the STI Inverse Čerenkov Accelerator (ICA) experiment [16], use of the

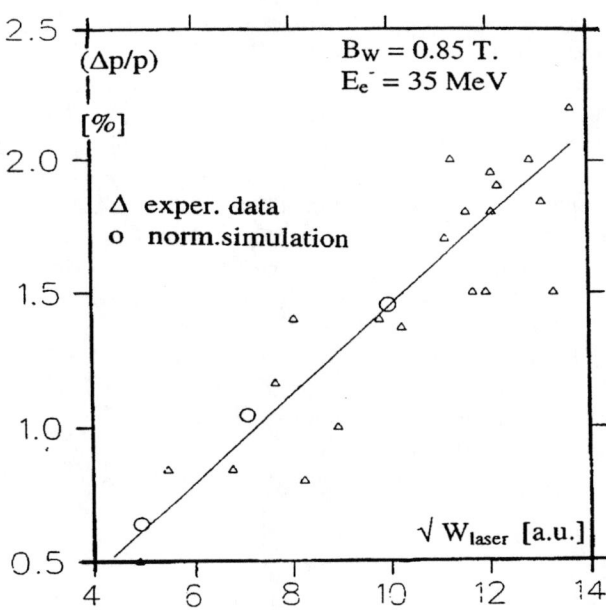

FIGURE 5. Relative energy gain $\Delta E/E$ vs W_l with E and B_w constant.

IFEL accelerator as a synchronized prebuncher for the IC accelerator in an IFEL - ICA buncher - accelerator sequence.

ACKNOWLEDGMENTS

The authors wish to acknowledge the continued support of I. Ben-Zvi and the technical staff of the ATF. This work was supported by the Advanced Technology R & D Branch, Division of High Energy Physics, U.S.Department of Energy, DE-AC02-76CH00016.

REFERENCES

1. R. Palmer, J. Appl. Phys. 43, 3014 ,1972.
2. C. Pellegrini, P. Sprangle,W. Zakowicz, Proc. of the XIII Int. Conf. on High Energy Accelerators, p.473, 1983.
3. E. Courant,C. Pellegrini,W. Zakowicz, Phys. Rev. **A32**, 2813, 1985.
4. A. Fisher, J. Gallardo, J. Sandweiss, A.van Steenbergen, *Inverse Free Electron Laser Accelerator*, Proc. Adv. Accel. Concepts, Port Jefferson, NY, AIP 279, p.299 ,1993.

5. A. Fisher, J. Gallardo, A. van Steenbergen, J. Sandweiss, *IFEL Accelerator Development*, Nucl. Instr. Meth. A341, 1994.
6. I. Wernick and T. C. Marshall, Phys. Rev. **A46**, 3566 ,1992.
7. A. van Steenbergen, J. Gallardo, J. Sandweiss, J. Fang, M. Babzien, K. Batchelor, A. Fisher, K. Kusche, R. Malone, I. Pogorelsky, J. Qiu, T. Romano, J. Sheehan, J. Skaritka, T. Srinivasan-Rao, X J Wang, *Inverse Free Electron Laser Single Module e^- Acceleration*, Proceedings BNL CAP/ATF Users Meeting, Dec., 1995
8. A. van Steenbergen, Patent Application 368618, June 1989 (Issued Aug. 1990)
9. A. van Steenbergen, J. Gallardo, T. Romano, M. Woodle, *Fast Excitation Wiggler*, Proc. PAC San Francisco, IEEE NS, May 1991
10. A. Fisher, J. Gallardo, A. van Steenbergen, J. Sandweiss, J. Fang, *IFEL Development*, Proc. Adv. Accel. Concepts, Fontana, WI 1994
11. J. Harrington, C. Gregory, Optics Letters 15, (1990)
12. N. Kroll, P. Morton, M. Rosenbluth, IEEE, QE 7, 89, 1979
13. I. Ben-Zvi, Proc. Adv. Accel. Concepts, Port Jefferson, NY, AIP 279, 590 (1993)
14. I. Pogorelsky, Proc. Adv. Accel. Concepts, Port Jefferson, NY, AIP 279 608 (1993)
15. P. Sprangle, E. Esarey, J. Krall, Phys. of Plasmas 3, 2183 (1996)
16. W. Kimura, I. Pogorelsky, Y. Liu, K. Kusche, A. van Steenbergen, J. Gallardo, J. Sandweiss, D. Quimby, M. Babzien *Inverse Čerenkov Acceleration using an IFEL Prebuncher*, Proceedings this Workshop.

Observation of 10 μm Smith-Purcell Radiation From 45 MeV/c Electrons

R. C. Fernow and Harold G. Kirk

Physics Department, Bldg. 901-A
Brookhaven National Laboratory
Upton, New York 11973-5000

S. H. Robertson

Department of Physics
University of Colorado
Boulder, Colorado 80309

J. H. Brownell and John E. Walsh

Department of Physics and Astronomy
Dartmouth College
Hanover, New Hampshire 03755

Abstract.
Using the high-brightness, high-energy electron beam at the Brookhaven Accelerator Test Facility we observe forward directed Smith-Purcell radiation in the mid-infrared spectral regime. This radiation can prove useful as a source of infrared radiation for other scientific studies as well as a providing a precursor investigation of the inverse process, namely the acceleration of electrons by means of the coupling of laser light with electrons via micro-structures.

INTRODUCTION

Motivated by a scarcity of radiation sources in the 10 to 100 μm regime, we are studying the production of Smith-Purcell radiation using the high-brightness electron beam of the Brookhaven National Laboratory's Accelerator

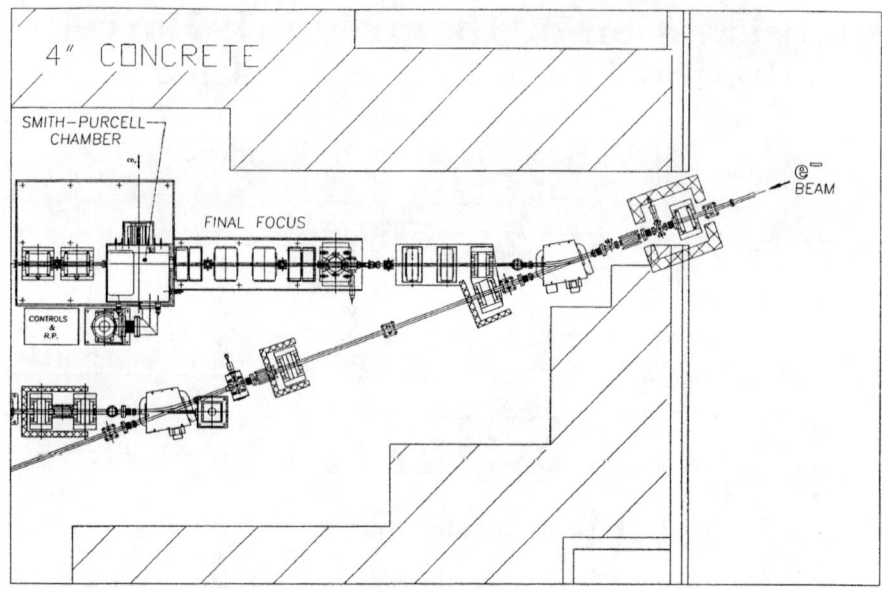

FIGURE 1. The ATF beam line immediately preceding the experimental chamber.

Test Facility. The production of Smith-Purcell radiation in the mm regime using a low energy 3 MeV beam has already been reported [1].

EXPERIMENTAL LAYOUT

The experiment is installed in the Brookhaven Accelerator Test Facility [2] high-energy beam line. Immediately preceding the experimental chamber (Figure 1) is a final-focus system [3] designed to reduce a nearly parallel beam to a small focused spot within the experimental chamber.

We show, in Figure 2, the layout of the experimental apparatus within the experimental chamber. The scale of the apparatus can be inferred by the 12" x 12" breadboard upon which the mirrors and grating are placed. The entire breadboard is mounted on a remotely controlled linear stage with a 2" stroke capability. The radiation is guided out of the chamber through a 5/8" clear aperture ZnSe window at the output port by means of a collection mirror mounted on a 360° remotely controlled rotating stage and two fixed angle mirrors. All three mirrors are 2" plane mirrors. The angular acceptance of the radiation from the grating extends from 5° to 11°, which, using the well known formula [4] relating wavelength (λ) to grating period (d) and emission

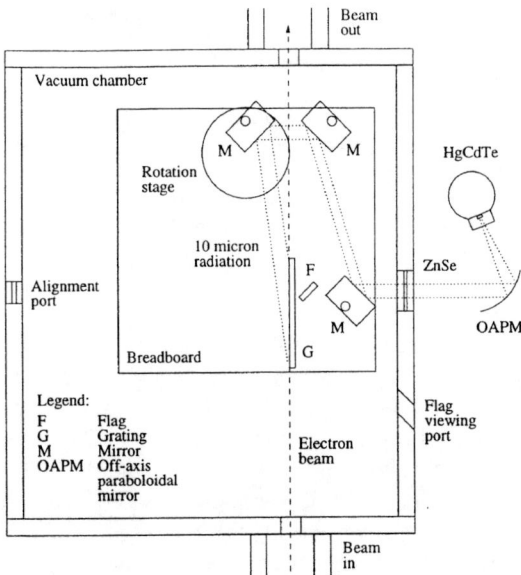

FIGURE 2. Layout of the Experimental chamber and the radiation detector.

angle (θ),

$$\lambda = d(\frac{1}{\beta} - \cos\theta),$$

corresponds to expected radiation wavelengths of 4 to 18 μm. The ZnSe output window is broadband AR-coated giving it an 70 % transmission spectrum extending from 7 to 12 μm.

Outside the experimental chamber the radiation is 000 onto a detector using a 10 cm focal length, 3" diameter off-axis paraboloidal mirror. The detector used for the generated radiation is a HgCdTe detector from InfraRed Associates, Inc. It's 40 % spectral response curve extends from 9 to 20 μm with a d* of 8 x 10^9 cm-$Hz^{\frac{1}{2}}$/W. During operation, the detector is cooled to liquid N_2 temperature. The detector is connected to a preamplifier, which when run with a bias voltage of 12V, has a gain of 40 db. At 10 μm the detector responsivity and risetime were 100 V/W and 27 ns, respectively. Detector output was observed on a 100 MHz bandwidth oscilloscope. In order to reduce background from x-ray radiation, the detector was enclosed in a wrap of 3/16" lead, leaving only the detector window exposed. This resulted in background signals being reduced from 10 mV to less than 1 mV.

Alignment of the grating was achieved by first placing a special alignment tee into the chamber before the experimental platform was installed. This tee was set upon the experimental chamber and mechanically aligned with the surveyed beam direction. A mirror, attached to the bottom of the tee, was suspended into the chamber to the position where the grating was eventually

TABLE 1. Typical Beam Parameters

Beam Energy [MeV]	45
dE/E [%]	0.3 - 0.5
Emittance [π mm-mrad]	1-4
Beam Charge [nC]	0.1 - 1.0
Micro-pulse length [ps]	10
Micro-pulse separation [ns]	25
Beam spot at Focus, σ_x [μm]	120
Beam spot at Focus, σ_y [μm]	180

installed. The mechanical tolerances of the tee were such that the plane of the mirror was parallel to the surveyed beam path to within 30 micro-radians. A laser beam from a Hamar laser alignment system was projected into the experimental chamber through the alignment port. The reflected beam was monitored with a total path length of 11.6 m from the laser source to the monitoring point. After removal of the alignment tee, the experimental platform was installed with a mirror placed within the grating holder. The grating mount was then mechanically aligned so that the reflected laser light coincided with the original reflection on the monitoring point. The two reflected laser spots overlapped to within 0.5 mm insuring that the grating mount is aligned to the alignment tee to within 50 micro-radians.

The grating used for this study is installed within the grating holder so that the grating face is vertical. The grating is constructed of aluminum, has a length of 126 mm, a 1 mm period, and a blaze angle of 5 degrees (set so the electron bunch footprint travels up the long slope of the grating). After installation, the grating pitch angle was surveyed and found to be 4 mrad with respect to the beam direction (a pitch angle of 0° corresponds to the nominal beam path being perpendicular to the grating lines).

The electron beam sent into the experimental chamber can be delivered in either of two modes, single- or multi-pulsed. The length of each micro-pulse can vary from 3 ps to 15 ps while the total charge for each micro-pulse varies from 0.1 nC to 0.5 nC depending on the mode of operation and the operating phase of the rf gun [5]. Typical beam parameters encountered during this experiment are given in Table 1.

We show in Figure 3 the beam profile of the electron beam operating in single-pulse mode at 45 MeV/c. The horizontal projection is the beam profile normal to the grating while the vertical projection is the beam profile parallel to the surface of the grating. In this case the total beam charge was 0.2 nC with a normal beam profile of $\sigma_x = 115$ microns. In order to ensure that the beam centroid deviates by less than one sigma in displacement height while traversing the length of the grating it is necessary for the beam direction to be parallel with the grating surface to within 1 mrad. This is an important issue in running the experiment in that it is noticed that if the beam is not steered

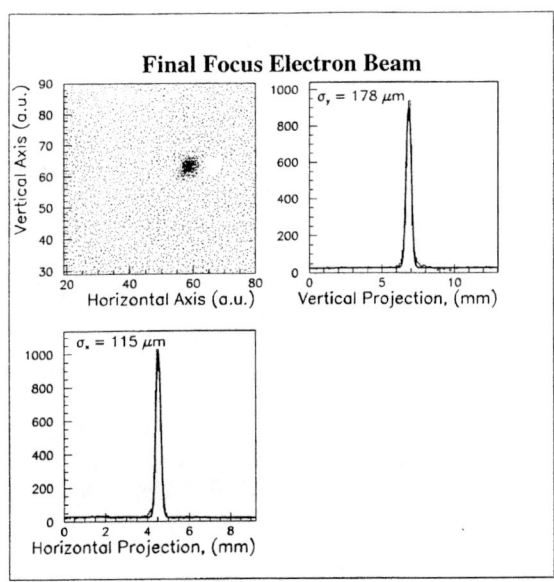

FIGURE 3. The beam profile at the middle of the grating. The result is for the ATF 45 MeV/c electron beam operating in the single-pulse mode.

along the magnetic axis of the final-focus quadrupoles, beam directional deviations of as much as 10 mrad can occur.

EXPERIMENTAL RESULTS

In order to scan for Smith-Purcell radiation, both the impact parameter of the beam relative to the grating and the emission angle relative to the grating surface are remotely varied. A signal was obtained when the electron beam was operating in a multi-pulse mode so that 15 micro-pulses were contained within the macro-pulse. The beam spot was roughly twice the size as that obtained for the single-pulse case (Figure 3), namely 260 μm in the direction normal to the grating surface and 590 μm parallel to the surface.

Initially the collection mirror was remotely set to collect radiation emitted at 8° while the grating was moved relative to the beam. The results are shown in Figure 4. Next the grating was remotely set to a position near the peak of the radiation signal and the collection mirror was then rotated. This mirror scan is shown in Figure 5. In order to verify that the signal did not result from x-ray emissions due to bremsstrahlung within the experimental chamber, we repeated the mirror scan with the ZnSe window blocked with a card. This scan is also shown in Figure 5.

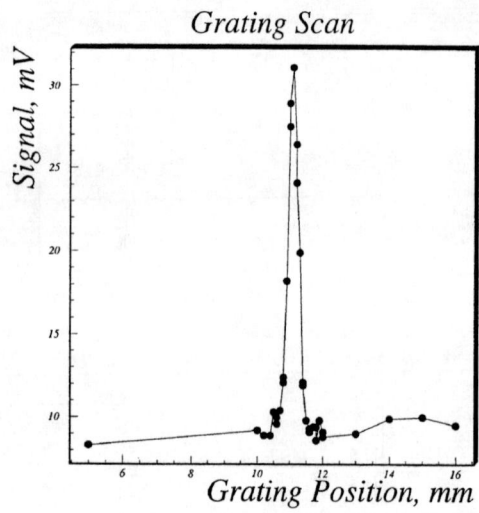

FIGURE 4. Radiation signal as a function of grating position. The electron beam is fixed.

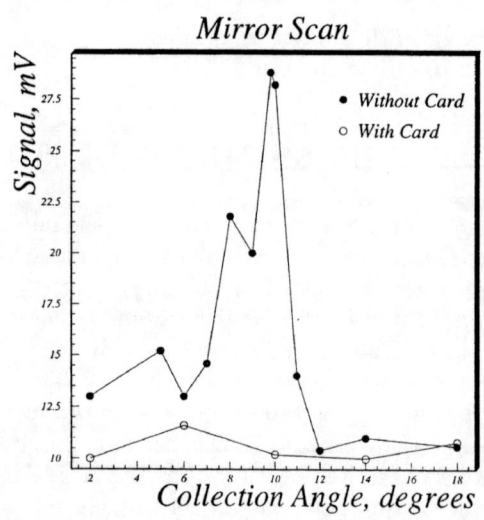

FIGURE 5. Radiation signal as a function of collection angle.

Given the detector characteristics, we find that the peak signal corresponds to about 5 pJ within each micro-pulse which are on the order of 3 to 10 ps long. Noting further that the clear angular acceptance of the ZnSe output window is 30 mrad and that the beam area is of the order of 1.5 cm^2 we determine that the radiation intensity at the source is of the order of 1 kW/cm^2-sr. Theoretical analysis [1], based on a consideration of profiles for both the grating and electron beam, places this value at about a factor 50 higher than that expected by spontaneous emission alone. We attribute this increased power to be due to a coherence enhancement of the radiation.

CONCLUSIONS

The spontaneous emission from the grating is remarkably bright and sufficiently strong enough to warrant using it as a source in some types of spectroscopic investigations. The possibility of using gratings in place of undulators for a free-electron laser is apparent. The demonstrated coupling of the electron beam with the grating in the $10\mu m$ regime has obvious applications when the inverse process is considered, namely the acceleration of electrons with the light from a CO_2 laser.

ACKNOWLEDGMENTS

The authors wish to thank N. Kurnit for providing helpful suggestions as well as the infrared detector used for this study. We also thank M.F. Kimmitt for many useful discussions. This research was supported by the U.S. Department of Energy under Contract Nos. DE-ACO2-76-CH00016 and DE-FG02-95ER40926. Support of U.S. ARO grant DAAH04-95-1-0640 is also gratefully acknowledged.

REFERENCES

1. K.J. Woods, J.E. Walsh, R.E. Stoner, H.G. Kirk, and R.C. Fernow, *Physical Review Letters* **74** 3808-3811 (1995).
2. K. Batchelor, et al, The Brookhaven Accelerator Test Facility, *Proceedings of the 1988 Linear Acceleration Conference*, CEBAF report 89-001, 540 (1988).
3. X.J. Wang and H.G. Kirk, The Brookhaven ATF Low-Emittance Beam Line, *Proceedings of the 1991 Particle Accelerator Conference*, San Francisco, California, 604 (1991).
4. S.J. Smith and E.M. Purcell, *Physical Review* **92** 1069 (1953).
5. X.J. Wang, X. Qiu and I. Ben-Zvi, *Physical Review E* **54** R3121 (1996).

INVERSE CERENKOV ACCELERATION USING AN IFEL PREBUNCHER

W. D. Kimura[*], I. V. Pogorelsky[†], Y. Liu[‡], K. P. Kusche[*†], A. van Steenbergen[†], J. C. Gallardo[†], J. Sandweiss[§], D. B. Cline[‡], D. C. Quimby[*], and M. Babzien[†]

[*]STI Optronics, Inc. Bellevue, WA 98004-1495
[†]Brookhaven National Laboratory, Upton, NY 11973-5000
[‡]University of California at Los Angeles, Los Angeles, CA 90024-1594
[§]Yale University, New Haven, CT 06511-8167

ABSTRACT

The BNL IFEL will be used to optically prebunch the *e*-beam before sending it into an inverse Cerenkov acceleration (ICA) stage. Prebunching the beam will greatly improve the efficiency of the ICA process. The basic experimental design and preliminary model predictions for the combined ICA/IFEL experiment are discussed. Near-term goals are to demonstrate optical prebunching, rephasing of the prebunched beam with the optical field, and more efficient acceleration. Long-term goals are to demonstrate 100 MeV net acceleration using an ICA accelerator.

I. INTRODUCTION

In inverse Cerenkov acceleration (ICA), a gas is used to slow the phase velocity of the laser light wave. By intersecting the laser light at the Cerenkov angle θ_C, where $\theta_C = \cos^{-1}(1/n\beta)$, n is the index of refraction of the gas and β is the ratio of the velocity of the electron to the velocity of light, the electrons will stay in phase with the light wave and experience net energy exchange.

The particular laser focusing geometry used in the present ICA experiment on the Brookhaven National Laboratory (BNL) Accelerator Test Facility (ATF) is depicted in Fig. 1. A radially-polarized laser beam is focused by an axicon onto the *e*-beam at an angle θ_C inside a cell filled with hydrogen gas at typically ~2 atm pressure [1].

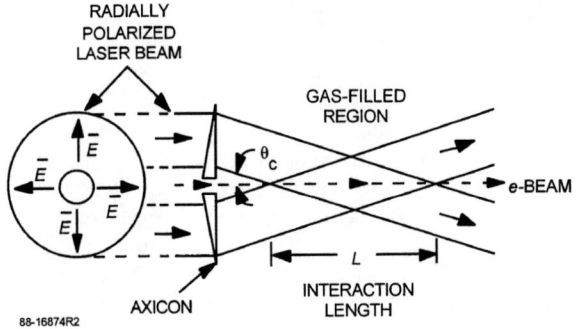

Figure 1. Basic focusing geometry used in present ICA experiments [taken from Ref. 1].

During the experiment, the *e*-beam energy spectrum is measured using a spectrometer whose output is detected using a CCD camera. Figure 2 depicts raw video images from the spectrometer camera. Energy dispersion is in the vertical direction as shown. When there is no laser present [Fig. 2(a)], the spectrum is narrow corresponding to an intrinsic *e*-beam spread of ~±0.1 MeV (~0.2% of the 56-MeV *e*-beam). When the laser is present electrons are either accelerated or decelerated depending upon their relative phase to the laser light. [The *e*-beam pulse length (~10 ps) is much greater than the laser wavelength.] Hence, the energy spectrum widens dramatically as shown in Fig. 2(b). [Note, the spectrum in Fig. 2(b) actually goes well beyond the top and bottom of the photo; however, due to

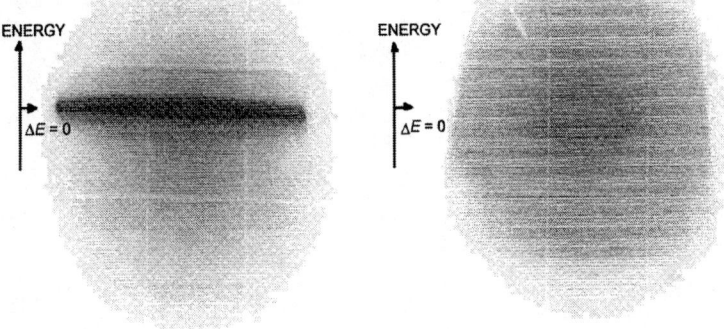

Figure 2. Raw video image of electron energy spectrum. (a) With laser not present in ICA interaction region. (b) With laser present in ICA interaction region.

output acceptance limitations of the spectrometer, the image decreases in intensity and becomes clipped at the edges.]

Peak energy gains of ~3.5 MeV for a delivered CO_2 laser power of 580 MW have been observed corresponding to an acceleration gradient of ~31 MeV/m [2].

The next goal of this experiment is to demonstrate 100-MeV net acceleration [3]. As part of this effort the electrons will need to be prebunched to a fraction of the optical wavelength in size. When these prebunched electrons intersect the laser light wave at the optimum phase for acceleration, the bunch will be accelerated as a group, thereby increasing the efficiency of the process. Hence, instead of yielding a modulated spectrum as shown in Fig. 2(b), the spectrum will contain primarily only accelerated electrons.

Prebunching electrons to optical dimensions is a natural consequence of modulating the *e*-beam energy with the laser light. If the electrons in Fig. 2(b) are allowed to drift, the accelerated particles will catch up with the decelerated ones, thereby creating bunches which are only a few microns in longitudinal direction and with a bunch separation equal to the laser wavelength (i.e., 10.6 μm).

Although the ICA process can, in fact, also be used as an optical prebuncher, a much better prebuncher for this particular experiment is an inverse free electron laser (IFEL). This is primarily because the relatively low-γ of the ATF *e*-beam leads to a higher sensitivity to the effects of gas scattering during the ICA process, which can disrupt the quality of the bunching. Such an IFEL has already been demonstrated on the ATF [4]. This paper discusses plans for using this BNL IFEL as the prebuncher for the ICA experiment.

II. COMBINED ICA/IFEL EXPERIMENT

A. BASIC APPROACH

The conceptual layout for the 100-MeV ICA Demonstration Experiment is illustrated in Fig. 3.

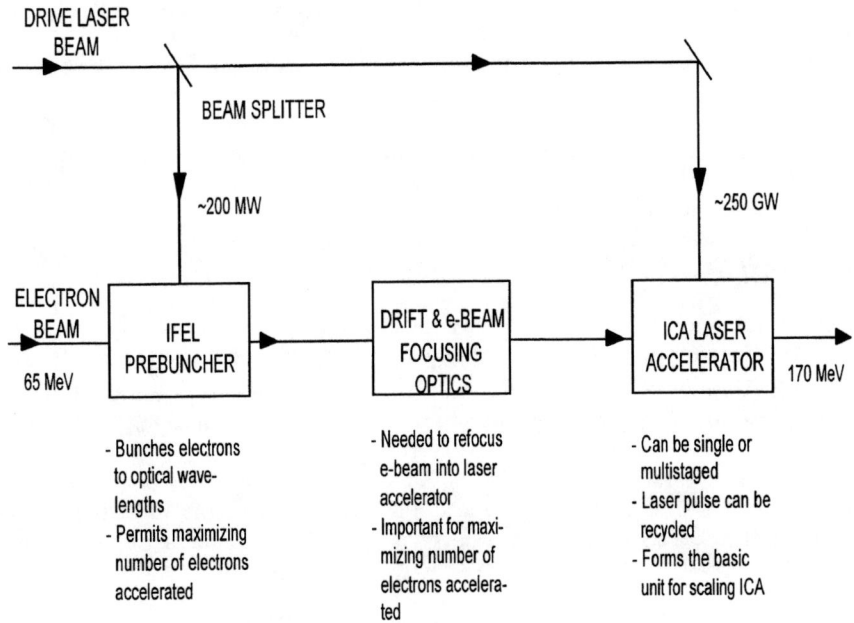

Figure 3. Conceptual Layout for 100-MeV ICA Laser Accelerator.

The IFEL will modulate the energy of the electron beam such that after drifting a certain distance the electrons will longitudinally bunch. For a given wiggler configuration (i.e., spacing and magnetic field strength), *e*-beam energy, and laser wavelength, this bunching distance L_D is controlled by the amount of energy modulation imparted by the laser beam. This distance scales as

$$L_D = \frac{\gamma^2 \lambda_l}{2(\Delta\gamma/\gamma)_{p-p}}, \qquad (1)$$

where λ_l is the laser wavelength and $(\Delta\gamma/\gamma)_{p-p}$ is the peak-to-peak energy spread induced by the IFEL. This expression assumes that the IFEL induces an approximately sinusoidal modulation of the energy in (γ,ψ) phase space, i.e., it is not saturated. Thus, during the experiment the IFEL prebunching characteristics can be "tuned" by adjusting the amount of laser power sent into the device.

Between the IFEL and the ICA gas cell is a drift space in which quadrupoles will be located to focus the e-beam into the ICA gas cell. Next comes the ICA laser accelerator stage where most of the laser energy will be directed. The entrance of this cell will be located at the optimum bunching distance. (Or, equivalently, the laser power driving the IFEL will be adjusted so that optimum bunching occurs at the entrance to the cell.) The primary technical issue here is properly rephasing the optically bunched beam with the laser light wave in the ICA cell so that the bunch is accelerated.

We estimate that roughly 200 MW of laser power will be needed to drive the IFEL. To achieve 100 MeV net acceleration, previous analysis predicted that ~250 GW of peak power is needed to drive the ICA cell [3].

B. BNL IFEL

Before presenting more details of the combined experiment, we briefly review the specifications for the BNL IFEL. This experiment has demonstrated acceleration of 40-MeV electrons by $\Delta E/E = 2.5\%$ over a distance of 47 cm using ~1 GW of laser power [4]. Other relevant parameters for the wiggler are listed in **Table 1**.

These wiggler parameters, in particular the amount of taper, may not be optimum for the combined ICA/IFEL experiment. Fortunately, one of the attractive features of the BNL IFEL is that it can be easily reconfigured. For example, it can be made into an untapered device, which may be more desirable for this experiment.

Table 1. IFEL Wiggler Parameters During Recent Experiments.

Parameter	Value
L_W (m)	0.47
Section Length (m)	0.6
Period λ_W (cm)	2.89 - 3.14
Gap (mm)	4
B_w^{max} (T)	1.0 - 1.024

Another noteworthy feature of the BNL IFEL is the use of a 2.8-mm ID, 60-cm long circular sapphire waveguide for channeling the laser radiation down the center of the wiggler. Approximately 90% coupling of the laser energy into the desired mode (HE_{11}) inside the waveguide has been demonstrated [4]. Previous tests measured an attenuation factor for the laser beam traveling through the guide of 0.2 dB/m. This same waveguide would be used during the combined ICA/IFEL experiments.

C. DESCRIPTION OF EXPERIMENTAL SYSTEM

Figure 4 is a preliminary plan-view of the combined ICA/IFEL experiment. The experiment will be located on Beamline #1 of the ATF where the ICA experiment is currently positioned. The distance of separation between the IFEL and the ICA interaction region (H_2 gas cell) is ~170 cm; hence, this corresponds to the desired bunching distance. Note that the *e*-beam direction of travel in Fig. 4 is opposite that shown in Fig. 3.

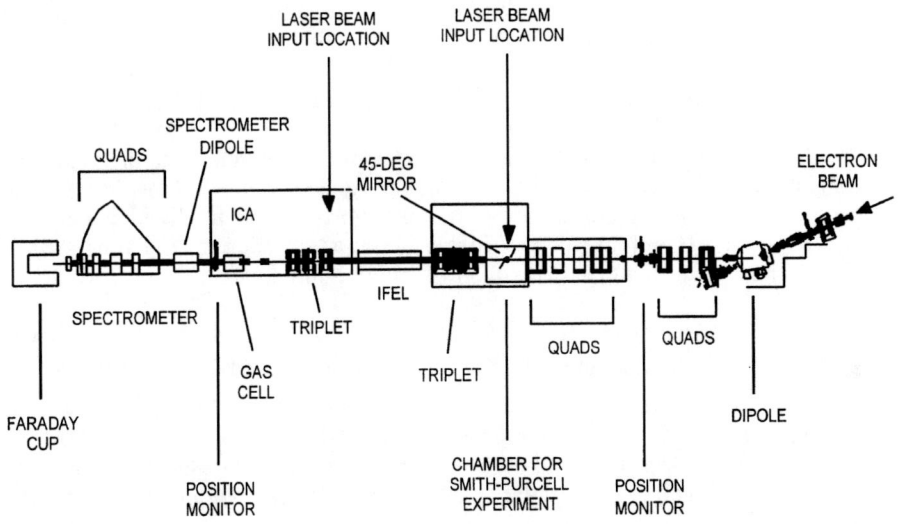

Figure 4. Plan view of preliminary ICA/IFEL experiment located on Beamline #1 of the ATF.

Triplets before the IFEL and in between the IFEL and ICA gas cell will be used to focus the *e*-beam into the respective devices. Introducing the laser beam into the IFEL will be accomplished by using a 45°-mirror positioned inside a chamber already on Beamline #1. The laser beam reflects off the mirror and is focused into the entrance of the sapphire waveguide. A small hole through the center of the 45°-mirror allows passage of the *e*-beam. Unlike the previous IFEL experiments, the laser driving the IFEL will be annular in shape to avoid striking the hole in the 45°-mirror. Although further analytical studies are needed, we do not expect that this will significantly affect the coupling efficiency of the laser beam into the waveguide.

The laser beam source for driving both the IFEL and ICA devices will be the ATF psec CO_2 laser that has been used in the previous experiments for both programs. Its linearly-polarized output beam will be sent through an axicon telescope, which converts the Gaussian-shaped beam into an annular one. A beamsplitter will then divide this beam into two beams, where most of the laser power will be sent to the ICA experiment. The rest of the laser power will be sent through an adjustable attenuator and then to the IFEL. This attenuator will permit adjusting the laser power delivered to the IFEL.

Not shown in Fig 4 is the optical delay system that will be part of the laser beam transport to the IFEL. Due to pathlength differences it will be necessary to delay the IFEL laser beam with respect to the ICA laser beam. This delay will be an optical "trombone" and can also be used to fine tune the phase delay between the two laser pulses sent to the IFEL and ICA cell.

III. PRELIMINARY MODEL PREDICTIONS

The ICA Monte Carlo code has been used to perform preliminary prediction for the combined ICA/IFEL experiment. Although the code was not designed to explicitly model an IFEL, the energy modulation of the e-beam is essentially identical to the ICA process. Therefore, for this first analysis attempt, the model used an ICA prebuncher, but with all gas and window scattering turned off.

Electron beam parameters for the simulations are: E = 40 MeV, intrinsic energy spread $\Delta E/E$ = 0.2%, and an assumed uniformly filled phase space with a normalized emittance ε_n = 3 π mm-mrad. In the simulation the distance between the exit of the wiggler and the entrance window of the ICA gas cell is 167 cm. The model includes a triplet quadrupole positioned between the wiggler and the gas cell, 84 cm from the exit of the wiggler. Laser power into the pseudo-IFEL prebuncher was adjusted so that optimum bunching occurs around 170 cm from the wiggler.

Figure 5 shows the model predictions for the electron number distribution N over phase ϕ and the change of energy ΔE over phase space at the optimum bunching point. As can be seen, the bunch length (FWHM) is 1.8 µm and 45% of the electrons are within the bunch length. The parameters for the ICA gas cell were: interaction length = 4 cm, Cerenkov angle = 20 mrad, gas pressure = 2.22 atm.

Figure 6 shows the predicted results for a laser power of 2 GW delivered to the interaction region (note the ordinate scale is 10× that of Fig. 5). The bunch length (FWHM) is 2.2 µm and 49% of the electrons are trapped within the bunch length. A peak acceleration of nearly 4 MeV is predicted corresponding to an acceleration gradient of 100 MeV/m for these electrons. Obviously, in Fig. 6(b) there is some evidence of bunch smearing in energy. Further studies with the model are necessary to determine the proper conditions to minimize this. While only preliminary simulations, these results are very promising and indicate that the IFEL does indeed make an excellent prebuncher for the ICA experiment.

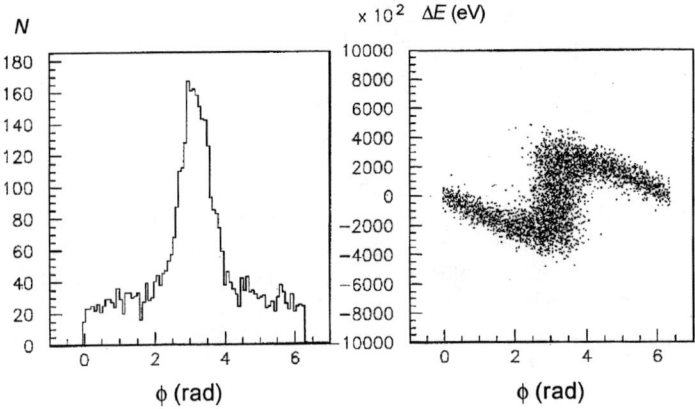

Figure 5. Model predictions for *e*-beam characteristics at entrance to ICA gas cell. (a) Number of electrons versus phase. (b) Change in energy versus phase.

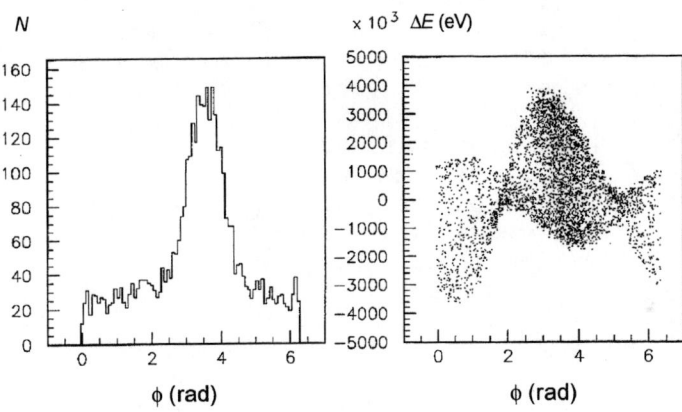

Figure 6. Model predictions for *e*-beam characteristics at exit of ICA gas cell for 2-GW laser power. (a) Number of electrons versus phase. (b) Change in energy versus phase.

The previous simulation assumed a delivered laser power of only 2 GW. Energy gain scales as the square-root of the laser peak power [1]; hence, much higher gains are possible using a more powerful laser beam. The ATF is currently upgrading the psec CO_2 laser [5]. The goal is to eventually reach the terawatt level with several hundred gigawatts available initially. This will enable the combined ICA/IFEL experiment to reach its goal of 100 MeV net energy gain.

IV. FUTURE PLANS

We are in the process of refining the design for the combined ICA/IFEL experiment, including optimizing the IFEL wiggler configuration for the experiment. Other experimental issues still need to be examined. For example, in order to ensure the bunched electrons reintersect the laser light wave in the ICA cell at the optimum phase for acceleration, it will be necessary to control the pathlength difference between the laser beams going to the IFEL and the ICA cell to within a few microns. This can be easily done using a piezoelectric-driven mirror, which is capable of submicron accuracy, mounted on the optical delay trombone. However, once this optimum pathlength has been found, it is important that the length stay constant within better than a few microns. This implies the need to have an optical mounting system that is stable to this degree. While standard mounts are quite stable, any temperature gradients or vibrations between different parts of the optical system could introduce micron errors in pathlengths.

Near-term plans are to finish the separate ICA and IFEL experiments that are currently being performed on the ATF. This includes characterization of a coherent transition radiation (CTR) diagnostic [6] that will be used to analyze the optically prebunched electrons created by the present IFEL. This same diagnostic or ones based upon it will be used during the combined ICA/IFEL experiment to monitor the bunching characteristics of the *e*-beam along the beamline.

Currently, the IFEL is located on a different beamline at the ATF. At the next shutdown of the facility, plans are to relocate the IFEL upstream from the ICA experiment. This will also require moving the ICA experiment slightly further downstream on the beamline. Once these moves are complete and additional hardware has been installed (e.g., triplets, beamsplitting optics), each of these devices will be separately recommissioned. At that point the combined experiment can begin.

One of the first objectives of the combined experiment will be demonstrating rephasing of the prebunched electrons with the laser light wave inside the ICA gas cell. This optimum phase point can be found by monitoring the spectrometer output as a function of the optical delay trombone mentioned earlier. At the optimum phase point the video image, as shown earlier in Fig. 2, will display a group of accelerated electrons. The number of electrons within this group and the amount of electrons distributed over the rest of the spectrum will give an indication of how well the electrons were trapped during the ICA interaction.

V. CONCLUSIONS

An experiment is being planned to use the BNL IFEL as a prebuncher for the ICA experiment. This will provide high quality bunches, which can be trapped and accelerated by the ICA interaction. Using prebunched electrons will significantly improve the efficiency of this process. This combined experiment would be one of the first to demonstrate rephasing of optically bunched electrons, and efficient trapping and acceleration of the bunch.

VI. ACKNOWLEDGMENTS

The authors wish to acknowledge the helpful advice and support of J. R. Fontana, R. H. Pantell, L. C. Steinhauer, and R. C. Fernow. This work was supported by the U.S. Department of Energy Grant Nos. DE-FG06-93ER40803 and DE-AC02-76CH00016.

REFERENCES

1. J. R. Fontana and R. H. Pantell, J. Appl. Phys. **54**, 4285 (1983).
2. W. D. Kimura, G. H. Kim, R. D. Romea, L. C. Steinhauer, I. V. Pogorelsky, K. P. Kusche, R. C. Fernow, X. Wang, and Y. Liu, Phys. Rev. Lett. **74**, 546-549 (1995).
3. R. D. Romea, W. D. Kimura, and L. C. Steinhauer, in *Advanced Accelerator Concepts*, Fontana, WI, AIP Conference Proceedings No. 335, P. Schoessow, Ed., (American Institute of Physics, New York, 1995), p. 390-404.
4. A. van Steenbergen, J. Gallardo, J. Sandweiss, M. Babzien, J-M Fang, X. Qiu, J. Skaritka, and X-J. Wang, Phys. Rev. Lett., **77**, 2690, 1996.
5. I. V. Pogorelsky, I. Ben-Zvi, A. van Steenbergen, R. Fernow, W. D. Kimura, and S. V. Bulanov, "CO_2 Laser Technology for Advanced Particle Accelerators," in these proceedings.
6. Y. Liu and D. B. Cline, "Micro-Bunching Diagnostics for the IFEL by Coherent Transition Radiation," in these proceedings.

Microwave Inverse Cerenkov Accelerator

T. B. Zhang[1], T. C. Marshall[1,2], M. A. LaPointe[1], and
J. L. Hirshfield[1,3]

[1]*Omega-P, Inc., 202008 Yale Station, New Haven, CT 06520*
[2]*Department of Applied Physics, Columbia University, New York, NY 10027*
[3]*Physics Department, Yale University, New Haven, CT 06520-8120*

A Microwave Inverse Cerenkov Accelerator (MICA) is currently under construction at the Yale Beam Physics Laboratory. The accelerating structure in MICA consists of an axisymmetric dielectrically lined waveguide. For the injection of 6 MeV microbunches from a 2.856 GHz RF gun, and subsequent acceleration by the TM_{01} fields, particle simulation studies predict that an acceleration gradient of 6.3 MV/m can be achieved with a traveling-wave power of 15 MW applied to the structure. Synchronous injection into a narrow phase window is shown to allow trapping of all injected particles. The RF fields of the accelerating structure are shown to provide radial focusing, so that longitudinal and transverse emittance growth during acceleration is small, and that no external magnetic fields are required for focusing. For 0.16 nC, 5 psec microbunches, the normalized emittance of the accelerated beam is predicted to be less than 5π mm-mrad. Experiments on sample alumina tubes have been conducted that verify the theoretical dispersion relation for the TM_{01} mode over a two-to-one range in frequency. No excitation of axisymmetric or non-axisymmetric competing waveguide modes was observed. High power tests showed that tangential electric fields at the inner surface of an uncoated sample of alumina pipe could be sustained up to at least 8.4 MV/m without breakdown. These considerations suggest that a MICA test accelerator can be built to examine these predictions using an available RF power source, 6 MeV RF gun and associated beam line.

I. INTRODUCTION

The stimulated Cerenkov effect is a well-understood mechanism for generating coherent radiation from an energetic electron beam [1-3]. The radiating electrons move at speeds greater than that of the velocity of light in the structure (hence the name "Cerenkov"). Although there are several ways to slow light waves, as a general rule the term is used when the slowing is caused by a dielectric element. When one does a linearized treatment of the fields and the self-consistent motion of the particles, a dispersion relation is obtained for growth or decay of radiation in the system. One of the three roots obtained corresponds to a damped wave; this we identify with the mechanism of stimulated absorption, whereby an electron will gain energy at the expense of the RF field. In the discussion which follows, we consider the application of stimulated absorption in the nonlinear regime of particle trapping, which applies to an electron accelerator. This we refer to as a Microwave Inverse Cerenkov Accelerator ("MICA") [4,5].

Acceleration of the electrons is achieved by appropriate phasing of a 6 MeV electron bunch which is emitted from a thermionic cathode RF gun, so that a continuous accelerating force is applied to all the electrons which move synchronously with the slow RF wave. Variation of the wave speed, if necessary, can be done by using a small taper in the filling factor of the dielectric element. Thus the device resembles an RF linac, but without the periodic loading structures in the waveguide. As the MICA is smooth-bore and the motion of the particles is essentially one-dimensional, the quality of the electron beam produced can be expected to be good. The MICA under consideration will use a SLAC klystron source of microwave power at 2.856 GHz. With a bunch length of only 5 psec compared with the RF period of 350 psec, we can expect excellent trapping and acceleration of a monoenergetic bunch of electrons. Another approach [6-7], the ICA(Inverse Cerenkov Accelerator) experiment at Brookhaven National Laboratory, uses a CO_2 laser and an axicon to accelerate a 40 MeV electron beam, and the light wave is slowed by introducing hydrogen gas into the beam line. The gas contributes to some electron scattering, and the main disadvantage

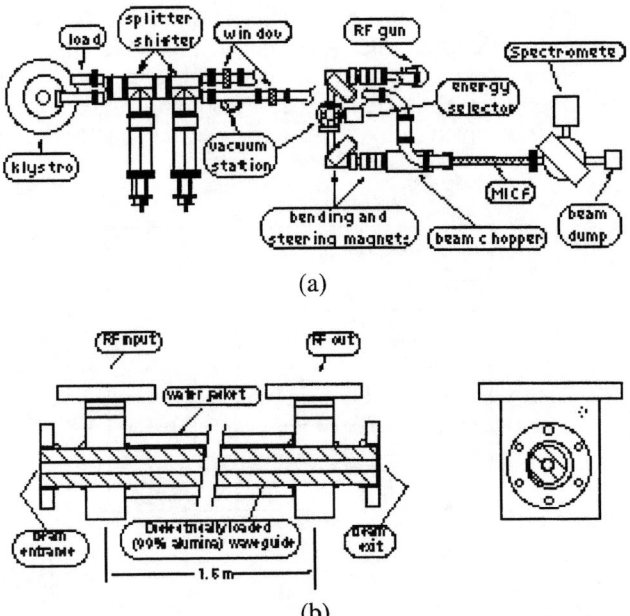

FIGURE 1. (a) Schematic diagram of MICA layout. A pre-accelerated short bunch of ~6 MeV electrons from an RF gun is injected in the appropriate accelerating phase into MICA, so that trapping and further acceleration of the entire bunch can occur.(b) MICA accelerating section.

of the short laser wavelength is that electrons interact with the wave over the full range (2π) of phase; that is, the bunch length is long compared with the RF wavelength. In MICA, the electrons move down a 1 cm diameter hole in an alumina dielectric liner as a filamentary beam of under 1 mm diameter. The main limitation here is that of the maximum axial field gradient (120-160 kV/cm [8]) along the dielectric surface. Shown in Fig. 1 is a schematic layout of the MICA.

II. EIGENMODES OF THE DIELECTRIC-LOADED WAVEGUIDE

The MICA configuration is a circular waveguide loaded by high ε dielectric material, with a small hole on axis for passage of the beam as shown in Fig. 1(b). The axisymmetric modes of this cylindrical dielectric-loaded system are either TE or TM, and we consider the TM_{0n}-like mode of this system, i.e. the mode with finite axial electric field on axis, no azimuthal variations and one radial maximum for the axial electric field. The use of a high ε annulus inside a circular waveguide maintains a large uniform E_z field inside the hole which is ideal for an accelerator. Using the appropriate boundary conditions at the interface of two differing media as well as at the outer metallic conducting wall, we solve Maxwell's equations using standard procedures [9], and arrive at a dispersion relation of the system for TM_{0n} eigenmodes. Accordingly, the normalized phase velocity v_{ph}/c can be obtained from the respective eigenvalue [5]. As an example,

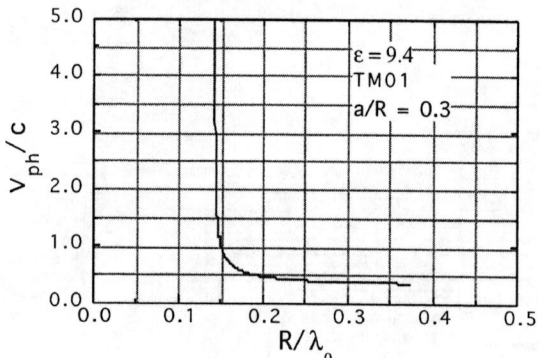

FIGURE 2. Normalized phase velocity v_{ph}/c vs. normalized outer radius R/λ_0 for TM_{01} mode waveguide with an alumina liner when the radius of the vacuum hole $a/R = 0.3$, Note: $v_{ph} \leq c$ when $R \geq 0.15 \lambda_0$.

Fig. 2 shows the normalized phase velocity of the TM_{01} mode as a function of the ratio R/λ_0, the dielectric constant ε is taken to be 9.4, and the ratio of hole radius to outer dielectric radius a/R is taken to be 0.3. One finds from Fig. 2 that a phase velocity of c and below obtains when $R/\lambda_0 > 0.15$, from which one finds the design values of outer waveguide radius and hole radius in terms of the RF vacuum wavelength λ_0 for a given desired phase velocity.

For operation at 2.856 GHz, the frequency of the RF source available for the MICA, one has λ_0 = 10.504 cm. Since the injected 6 MeV electrons have velocities of about 0.997c, the desired phase velocity is very close to c. From the curve in Fig. 2 with a/R = 0.30, one selects R = 1.59 cm and a = 0.48 cm for $v_{ph}/c \approx 1.0$. The cut-off frequency of the TM_{01} mode is 2.69 GHz, and for the TM_{02} mode, it is 6.72 GHz. By neglecting the hole, one can estimated cut-off frequencies for the TE_{11}-like and TE_{01} modes to be 1.80 GHz and 3.76 GHz, respectively. The dimensions selected for operation at 2.856 GHz thus appear to guarantee freedom from interference by other nearby modes, since only the TE_{11}-like mode with nearly orthogonal polarization is not cutoff at 2.856 GHz.

III. PARTICLE ACCELERATION

Particle simulation studies have been carried out for MICA. The simulation parameters are shown in Table 1. At the beginning of MICA, the relative phase of beam particles with respect to the field is very close to π, therefore they are accelerated by the maximum axial field while suffering little influence from the transverse field components, because the intense axial field E_z dominates the other field components near the axis. The electron energy increases almost linearly when the particles move down the waveguide as shown in Fig. 3(a). Particles with good phase matching will experience the maximum axial field, as we illustrated in Fig. 3 (b). Also shown in Fig. 3 (in dotted lines) is the result when a 10 percent RF power depletion is taken into account. As is seen, small power deletion does not drop the particle energy significantly.

The group velocity of TM_{01} mode in the loaded waveguide is much slower than the phase velocity. Their relation is very close to $v_{ph}v_g = c^2/\varepsilon$, the result for completely-filled waveguide. MICA operation with $v_g/c \approx 0.10$ would be similar to that of conventional RF linac operation, where low group velocities are also employed. In the MICA case, the bulk of the energy is stored in the high-dielectric-constant material, while in the conventional linac the bulk of the energy is stored in periodic structures that act similar to cavities. Low group velocity implies that the energy fill time is much longer than the microbunch transit time

TABLE 1. Simulation parameters of MICA

Electron beam parameters		
Initial electron energy	$\gamma_0 = 13$	
Maximum initial transverse velocity	$\beta_\perp = 2.60 \times 10^{-3}$	
Initial axial velocity (6 MeV)	$\beta_z = 0.9970$	
Beam radius	$r_b = 0.05$	(cm)
	$r_b/R = 0.032$	
Waveguide parameters		
Waveguide radius	$R = 1.59$	(cm)
Radius of vacuum hole	$a = 0.48$	(cm)
Ratio of two radii	$a/R = 0.30$	
Dielectric constant (alumina)	$\varepsilon = 9.4$	
Waveguide length	$z = 150$	(cm)
Waveguide mode	TM_{01}	
Radiofrequency wave		
Field power	$P = 15$	(MW)
Maximum field strength	$E_{zmax} = 6.29$	(MV/m)
Frequency	$f_0 = 2.856$	(GHz)
Normalized phase velocity	$V_{ph}/c = 0.9943$	
Free space wavelength	$\lambda_0 = 10.50$	(cm)
Waveguide wavelength	$\lambda_g = 10.46$	(cm)

along an accelerator section, so that significant energy depletion, if any, would cause late-following bunches to experience less acceleration than early-leading bunches. After several fill times, a steady-state can be reached, but beam loading will reduce the field amplitudes, bringing about less net acceleration than in the absence of beam loading. This situation is undesirable when the accelerator is designed for high energy gain and narrow energy resolution. Therefore one must

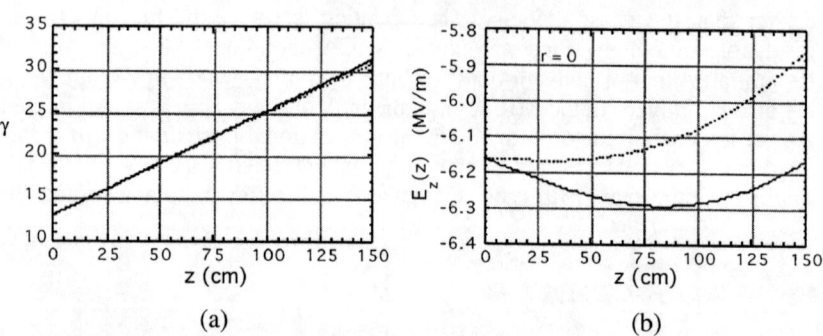

FIGURE 3. (a) Electron energy as a function of the axial distance. (b) The axial accelerating field seen by one particle as it moves down the waveguide. The dotted lines are the case when a 10 percent power depletion is taken into account.

insure that the energy carried away by the beam during a sequence of microbunches (i.e. during one fill time $\tau = L/v_g$) is much less than the stored energy. Our beam loading computation shows that for an average current of $I_0 = 0.143$ A, which corresponds to peak current of 10A with 5 ps bunch averaged over the 350 ps RF cycle established in the MICA, there is only a slight drop of 2.3 % from the maximum no-load energy gain 9.06 MV.

Due to the small difference between the electron velocity and the wave phase velocity, one may expect that the electron will gradually slip from the maximum acceleration position, forward or backward depending on whether the beam is going faster or slower. In the current simulation, we find a phase slippage of $\Delta\phi \approx 24°$ in 1.5 m with the electrons moving ahead of the RF field, corresponding to a slippage interval of $\Delta\tau_0 \approx 23$ ps. For a RF gun with a beam bunch length of only $\Delta\tau_0 = 5$ ps, we can expect excellent trapping and acceleration of electrons during the entire propagation along the waveguide, without a taper of the dielectric element.

When the electrons are located in a small "phase window" of acceleration, the radial component of the field E_r will prevent the electrons from spreading out even though the particles have an initial transverse velocity distribution. Our simulation shows that electrons remain well confined inside the hole in the dielectric and the transverse velocity spread shrinks. Beam spreading becomes serious only when electrons are out of acceleration phase.

IV. SELF FIELD EFFECT AND EMITTANCE EVOLUTION

The self field effect due to finite electron current has been investigated by running the PARMELA accelerator code [10], a versatile multiparticle electron linac code widely used in accelerator community [11]. In PARMELA the electron beam, represented by a collection of macroparticles, may be transformed through a linac and/or transport system. The self field effects (both electric and magnetic) are automatically taken into account in the simulation. Since the code usually applies to a periodic loading structure, it was necessary to modify the code so that it has the capability of modeling traveling wave acceleration in the smooth-bore MICA structure. In Table 2, we summarize some parameters of main interest for a bunch at the entrance (z=0) and the exit (z=150 cm) of the waveguide. Entrance conditions are taken from the RF gun and beam line computation, while the output is determined by the simulation results. The initial parameters in the PARMELA simulation are the same as we used in the single particle dynamics run shown in Table 1, except a finite bunch charge of q = 0.05 nC is now included.

TABLE 2. PARMELA results for beam emittance and energy spread at the entrance and exit of waveguide

z (cm)	a_0 (mm)	E_{beam} (MeV)	$4\sigma_\phi$ (deg.)	$4\sigma_E$ (keV)	ε_{nxrms} (πmm mrad)	ε_{nyrms} (πmm mrad)	ε_{nzrms} (πdeg-keV)
0	0.5	6.0	5.78	68.9	3.05	2.90	24.9
150	3.0	15.6	5.83	80.2	3.05	2.91	25.8

Comparing the PARMELA output results with the single particle results, we find that in both simulations the acceleration gradients, the final particle energy, the beam cross-section and the particle velocity evolution are all in excellent agreement [5]. The 1000 particles used in PARMELA simulation are all "good" particles, meaning that there is no particle loss in the MICA. PARMELA shows very clearly that the transverse emittance ε_{nxrms} and ε_{nyrms} are constant throughout the acceleration, even though the longitudinal emittance ε_{nzrms} has a very slight change because of the minute longitudinal bunching which makes the particle energy spectrum more narrow. This PARMELA simulation is also compared to a test run where the net charge is set to zero: we observe only trivial differences. This shows that the self field effects are not significant for 10 A peak current. However, when the beam current is increased, the self field effects do affect the ultimate beam quality. For PARMELA runs with 20 A peak micropulse current a noticeable growth in normalized transverse and longitudinal emittance is found, while for 200 A the growth is substantial. These results suggest that achievement of the goal of a normalized transverse emittance of 5π mm-mrad for a 0.16 nC (10^{19} particles), 5 psec bunch is realistic.

V. ATTEMPT TO MEASURE THE DIELECTRIC BREAKDOWN LIMIT

In order to determine the breakdown limits at 2.856 GHz, we have designed and constructed a cavity resonator with an alumina liner. Thus it is necessary to determine the resonance frequency and quality factor Q for a TM_{01n} resonator constructed with a section of dielectric-lined waveguide with conducting plates closing the ends. Since the exact value of the alumina dielectric constant of the sample we used was not accurately known, we began with a low power test to determine the resonant modes of a simple cavity incorporating an alumina annulus with outer metallic surfaces, which is coupled to a signal generator and a detector as shown in Figure 4.

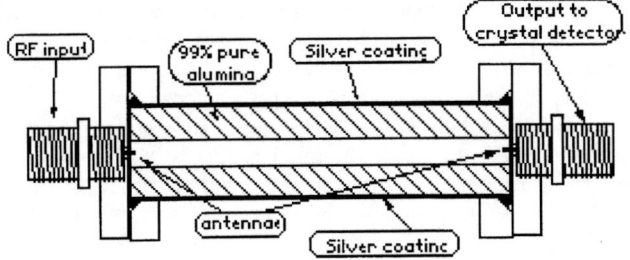

FIGURE 4. Sketch of the test resonator fabricated from a short section of alumina pipe coated on its exterior with silver paint. A low power RF input was used.

The design parameters of an ideal cavity resonator, based on the sample available, are listed in Table 3. The cavity operates in the TM_{012} mode; that is, the length of the cavity is one guide wavelength of TM_{01} mode. The cavity has a moderate quality factor $Q = 4620$. The relation between the maximum axial field in the cavity and the power coupled in is also given in Table 3 in terms of the parameter $E_{zmax} / P^{1/2}$ where P is the total power lost in both walls and dielectric.

TABLE 3. Simulation Parameters of Dielectric-loaded Cavity Resonator

Cavity parameters:		
Radius of empty hole	$a = 0.508$	(cm)
Radius of cylindrical cavity	$R = 1.429$	(cm)
	$a/R = 0.356$	
Length of cavity	$d = 12.11$	(cm)
Real part of the dielectric constant	$\varepsilon_r = 9.62$	
Imaginary part of the dielectric constant	$\varepsilon_i/\varepsilon_r = 9.4 \times 10^{-5}$	
RF wave parameters:		
Cut-off frequency	$f_c = 3.118$	(GHz)
Resonance frequency	$f_0 = 3.210$	(GHz)
Transverse wave number in the hole	$k_{1r} = 0.4281$	(1/cm)
Transverse wave number in the dielectric	$k_2 = 2.021$	(1/cm)
Wavelength in the cavity resonator	$\lambda_g = 12.11$	(cm)
Cavity mode	TM_{012}	
Quality factor of the cavity		
Q for the conducting walls	$Q_w = 7929$	
Q for the dielectric	$Q_d = 11070$	
Q of the cavity	$Q = 4620$	
Power loss ratio (MV/m)/(MW)$^{1/2}$	$E_{max}/P^{1/2} = 20.32$	

Measurements were conducted with a resonator fabricated from a short section of alumina pipe coated on its exterior with silver (Fig. 4). The alumina samples, supplied by LSP Ceramics, Inc., had inner and outer radii of 1.429 cm and 0.508 cm, respectively. There are some differences in dimension between the proposed waveguide and the test cavity (the outer radius is 10% smaller than the required value of 1.5875 cm, while the inner radius is 7% larger). Nevertheless, measurements on the samples available still provide a good test of theory. Raw data for the observed RF transmission by the cavity is shown in Fig. 5(a), over a frequency range between 3 and 6 GHz. Fig. 5(b) shows a plot of the square of the 12 observed resonance frequencies in Fig. 5(a) *versus* the square of the resonance index. From theoretical analysis we know that the slope of this line should be the reciprocal of the relative dielectric constant; for the data in Fig. 5(b), this reciprocal slope is 9.62. This differs from 9.4, the canonical value taken in the analysis given above, but 9.62 is well within the range quoted for good purity alumina. It is highly significant that no other resonance for this structure between 3 and 6 GHz were found that did not fit on the line shown in Fig. 5(b), despite attempts having been made to excite non-axisymmetric modes using a

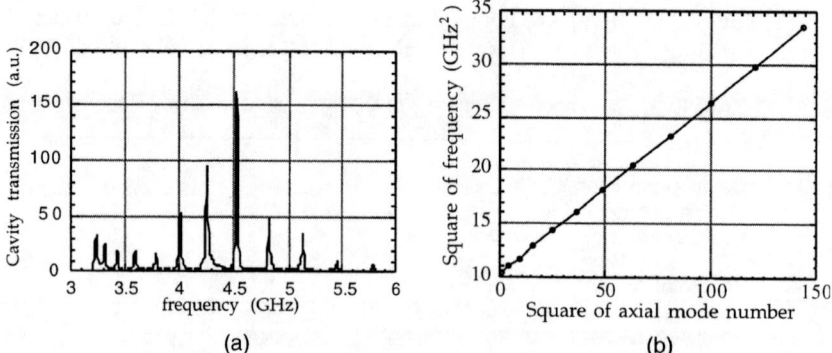

FIGURE 5. (a) Mode spectrum of the resonator obtained in measurement cavity transmission is in arbitrary units; (b) Square of mode resonance frequency *versus* square of axial mode number from measurements on the resonator shown in (a). The reciprocal of the slope of the line is 9.62, which can be inferred to be the dielectric permittivity ε.

non-axisymmetric antenna. One can conclude from this observation that potentially disruptive non-axisymmetric modes of the dielectric pipe were not excited. Calculation of the properties of non-axisymmetric modes would be a formidable task, one which these experimental tests appear to render unnecessary.

The vertical intercept for the line in Fig. 5(b) should be the square of the TM_{01}-mode waveguide cutoff frequency, which in this case is observed to be 3.216 GHz. For $\varepsilon = 9.62$, $R = 1.429$ cm and $a = 0.508$ cm, the calculated value is 3.118 GHz, a value 3.05% lower than the measurement. This discrepancy is not unreasonable, considering the added circuit reactance of the coupling antennas, and the incomplete closure of the end walls. Typical Q values for the observed cavity resonance were in the range of 400-500, much lower than the calculated unloaded value of 4620. This is also not too surprising, considering the open ends of the beam hole, and the strong external loading that was required to make accurate resonance measurements on all 12 modes. But this exercise emphasizes the need to carefully test alumina samples prior to acceptance, and prior to selection of the final parameters for the 150 cm accelerating sections. In particular, an accurate advance measurement of dielectric constant, phase velocity and loss tangent must be made from samples taken from the alumina batch to be used for the final accelerating sections.

Measurements using high power microwaves applied to the alumina samples were also carried out to determine RF breakdown limits. Since the cavity described above has resonance above 3.2 GHz, an alternative experimental arrangement was devised to subject the alumina surfaces to high tangential RF electric fields at 2.856 GHz (obtained from a XK-5 klystron). A standing-wave resonance was established in WR-284 rectangular waveguide using inductive irises. Measurements with the alumina sample in place showed this arrangement to give an effective gain of over 11 dB, as deduced from signals on the sample probe with and without the irises. Under these conditions, the peak tangential RF electric field at the inner alumina surface is calculated to be 33.6 $P^{1/2}$ V/cm, where P is the incident power in watts. This indicates that a field of 63 kV/cm would be applied when P = 3.52 MW. In the experiments, the RF power level was increased over a ~12 hour period to provide gradual RF processing of the structure, without allowing the background pressure to exceed 2×10^{-6} Torr. It was found that this procedure could be continued up to a power level of 6.25 MW, without evidence of arcing at the alumina surface. This corresponds to a tangential field of 84 kV/cm. These observations suggest that acceleration gradients of at least 8.4 MV/m should be achievable in MICA, where a design with superior vacuum integrity and coating of the alumina is planned.

V. CONCLUSIONS

We have studied a Microwave Inverse Cerenkov Accelerator, which has an acceleration mechanism similar to that of a conventional RF linac. However, the accelerating structure, which comprises a continuous coated ceramic pipe, should be less expensive to fabricate than that of the linac. In the absence of any periodic loading structures in the waveguide, wakefield generation that can lead to emittance growth and beam breakup should be minimized. Thus MICA's advantages of a relatively compact structure, smooth-bore design and no need of magnetic focusing make it a very competitive facility as a simple, low cost electron accelerator.

In this paper, we have discussed briefly the eigenmode, field profile, particle dynamics, beam loading and space charge effects. Experimentally, we failed to observe dielectric breakdown of alumina at fields up to 8.4 MV/m. We find that a thick liner with a high dielectric constant is very helpful not only to store high RF energy but also to maintain an intense and uniform axial accelerating field in the central hole. The particle motion in the waveguide is nearly one dimensional with all input particles being accelerated and no interception by the dielectric. There is no beam breakup, and the beam bunches have good stability even if they are slightly off-axis. For the beam current under consideration, the initial low normalized emittance--less then 3 π mm mrad--is constant throughout the acceleration. The acceleration gradient in the simulation is 6.3 MV/m in which case the electron energy increases from 6 to 16 MeV in 150 cm. However, without exceeding the breakdown limit measured by experiment (>8.4 MV/m), and using higher microwave power and/or a higher Q structure, the electron energy could increase even more, perhaps in the range of 10-15 MeV/m if techniques for improving the dielectric breakdown[8] on the surface using Ti or TiN evaporated coatings can be used successfully.

Challenging technical issues must be overcome. These include precision grinding such as the finish of the waveguide liner, since the phase velocity of a RF wave in the vicinity of c is very sensitive to the radius of the vacuum hole and tube, and careful design of the matching for the power feeding system and the accelerator waveguide because of the substantial difference of the wave group velocity (or impedance) in these two sections.

ACKNOWLEDGMENT

This work is sponsored by the DOE, Division of High Energy Physics. The authors thank Dr. Hongxiu Liu of CEBAF for his assistance in running PARMELA for space charge effect simulation.

REFERENCES

[1] J.E. Walsh, T.C. Marshall, and S.P. Schlesinger, Phys. Fluids 20, 709(1977)
[2] J.E. Walsh and E. Fisch, Nucl. Instrum. Meth. A318 ,772 (1992)
[3] W. Main, R. Cherry, and E. Garate, Appl. Phys. Lett. 55, 1498 (1989)
[4] T. B. Zhang and T. C. Marshall, Nucl. Instrum. Meth. A375 , 614(1996)
[5] T. B. Zhang, T. C. Marshall, M. A. Lapointe, J. L. Hirshfield and Amiram Ron, Phys. Rev. E54 1918(1996)
[6] A. Van Steenbergen, J. Gallado, J. Sandweiss, and J.-M. Fang, Phys. Rev. Lett. 77,2690(1996)
[7] W.D. Kimura et al, Phys. Rev. Lett. 74, 546 (1995)
[8] Y. Saito, "Breakdown Phenomena in RF Windows", paper presented at Montauk, October 1994 (to be published)
[9] M. Shoucri, Phys. Fluids 26 ,2271 (1983)
[10] H. Liu, Private communication.
[11] K.R. Crandall and L. Young, Las Alamos Natl. Lab. Rep. La-UR-90-1766 (May 1990)

Simulation Results and Experimental Design for the Microwave Inverse FEL Accelerator

R. B. Yoder,[1] T. B. Zhang,[2] T. C. Marshall,[3] and J. L. Hirshfield[1,2]

[1]*Physics Dept., Yale University, New Haven, CT 06520-8120*
[2]*Omega-P, Inc., 202008 Yale Station, New Haven, CT 06520-2008*
[3]*Dept. of Applied Physics, Columbia University, New York, NY 10027*

A microwave inverse free-electron-laser accelerator (MIFELA) is currently under construction at the Yale Beam Physics Laboratory. MIFELA is an accelerator based on stimulated absorption of microwave energy by electrons moving in a magnetic wiggler field and axial guiding field; both fields are tapered to maintain near-resonance during acceleration. The acceleration structure is a simple, smooth-bore cavity, surrounded by a bifilar helical wiggler. A 2-1/2 cell RF gun provides 5 ps bunches of 6 MeV electrons, and ~4 MW of RF energy at 11.4 GHz is taken from the output of the Yale gyroharmonic converter. 3D simulation results are presented for electron acceleration from 6 to 11 MeV in 1.5 m, which show a high trapping fraction (78%) and a final FWHM energy spread of 0.9%. Details of the experimental parameters and current design issues are discussed.

INTRODUCTION

The idea of accelerating charged particles through an inverse free-electron laser (IFEL) mechanism, in which electromagnetic energy is taken up by an undulating electron beam, was first proposed by Palmer in 1972 (1). In an IFEL device, electrons are trapped in moving potential buckets formed by the beating of a propagating electromagnetic field with an external periodic magnetic field. The strengths of the IFEL scheme—strong trapping and high theoretical acceleration gradients attainable in a simple smooth-walled structure—have engendered continued interest in the topic, especially since the development of advanced lasers in the past two decades.

Since 1972, a substantial body of theoretical work on IFEL acceleration has been produced (2), and in the last few years there have been some experimental efforts to demonstrate energy gain in such a configuration. To date, experimental proof of the principle has been reported by researchers at Columbia University (3), where a two-stage FEL was reconfigured as an autoaccelerator, with a small fraction of electrons reaching energies of ~1 MeV, and at Brookhaven (4), where a group using the CO_2 laser and linac beam of the ATF with a planar wiggler has recently announced energy gain of 2.5% at 40 MeV (5). We report on current work on the development and construction of a prototype IFEL accelerator, operating at microwave wavelengths, which is designed to demonstrate the IFEL principle in a practical, efficient and small-scale device.

OVERVIEW OF THE EXPERIMENT

The Yale/Omega-P accelerator, known as the Microwave Inverse Free-Electron-Laser Accelerator (MIFELA), consists of a cylindrical microwave cavity surrounded by both a helical wiggler and a series of solenoids which produce an axial magnetic field (Fig. 1). Use of a resonant cavity, rather than a waveguide, provides higher acceleration fields; the axial guiding field is added to preserve orbital stability as the electrons are accelerated. RF power at 11.424 GHz is to be provided by the output of Yale's gyroharmonic converter, which uses a gyrating electron beam to convert the output of a SLAC klystron at 2.856 GHz to its fourth harmonic and is expected to produce up to 5 MW of available RF (14). The injected electron beam is provided by a 2-1/2 cell RF gun which produces 5 ps bunches of 6 MeV electrons with an energy spread less than 1%. The wiggler, a bifilar helical winding, will be tapered in both pitch and radius to maintain optimum acceleration throughout the interaction region. Design parameters for the MIFELA, as used for the simulations reported here, are given in Table 1.

TABLE 1. Simulation parameters for the MIFELA.

Entry region (0 < z < 50 cm)

Electron beam energy	$\gamma = 13$
Electron beam radius	$r_b = 0.7$ mm
Wiggler magnetic field	$B_W = 0$–4.1 kG, sine-squared ramp
Axial magnetic field	$B_z = 0$–2.4 kG, linear ramp
Wiggler period	$\lambda_W = 10.0$ cm, constant
Wiggler radius	$r_W = 3.0$ cm, constant

Acceleration region (50 cm < z < 200 cm)

Cavity mode	TE_{11n} (n ~ 120)
Cavity radius	$R = 1.45$ cm
Electron beam peak current	$I_b \sim 0.1$ A
Wiggler period	$\lambda_W = 10.0$–12.3 cm, linear ramp
Wiggler coil radius	$r_W = 3.0$–2.8 cm, linear ramp
Wiggler current	$I_W = 42$ kA
Wiggler magnetic field strength	$B_W = 4.1$–5.1 kG
Axial magnetic field	$B_z = 0.6$–2.6 kG
Normalized RF field strength	$a_S = 0.17$, circularly polarized
RF wavelength	$\lambda_S = 2.626$ cm

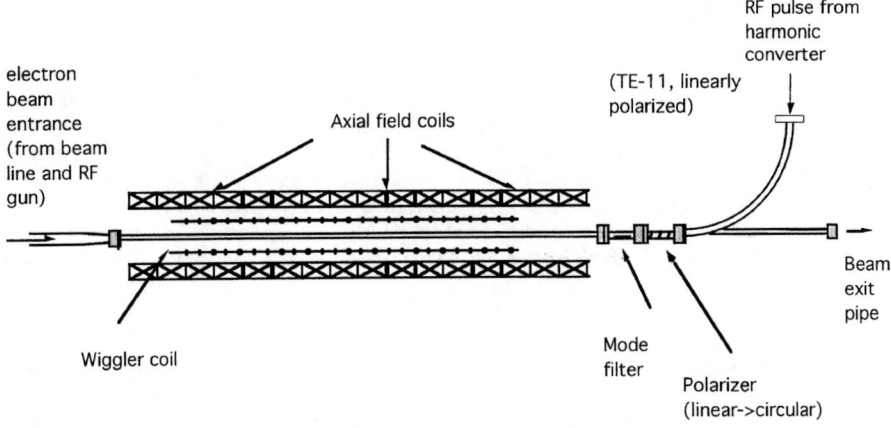

FIGURE 1. Schematic of the MIFELA components and layout.

An approximate energy gain relation for the IFEL interaction has been well-established (2) and is given by

$$\frac{d\gamma}{dz} = -\frac{\pi a_s a_w}{\lambda \gamma} \sin \varphi \qquad (1)$$

where a_s and a_w are the normalized vector potentials of the RF drive field and helical wiggler field respectively, λ is the drive wavelength, and φ is the phase of the driven electron motion with respect to the beat wave. The phase is controlled by the wiggler pitch and amplitude; negative values of φ give increasing γ (acceleration), while positive values correspond to electron deceleration—i.e., the FEL amplifier. For useful acceleration gradients to be achieved, a_w must be on the order of 6–8, which leads to large electron orbits, on the order of several millimeters in radius. Because of this, the MIFELA is divided into three separate regions, in which the electron beam is injected and spun up, accelerated, and extracted, respectively. Initial simulation studies (6) using a one-dimensional particle code and 2D fields showed that orbital stability could be maintained over an overall cavity length of 2 m using this arrangement.

In the entry region, electrons are injected from the RF gun along the z-axis of the cavity, and the wiggler and guide fields are brought up from zero over five wiggler periods, causing the electron orbits to spiral outwards. There follows an acceleration region, of length ~10 wiggler periods, in which the electron energy γ increases smoothly. A final exit region reduces the orbit radius to a negligible value by linearly tapering the wiggler and guide fields to zero, allowing a thin electron beam to be extracted for analysis. (Of course, should a gyrating electron beam be desired, the guide field could be continued.)

PARTICLE DYNAMICS AND SIMULATION RESULTS

To refine the experimental design for the MIFELA, extensive simulation studies of the interaction have been carried out using a modified version of the fully nonlinear, three-dimensional FEL code ARACHNE (7). This code has been in use for some time and has been extensively benchmarked against FEL experiments (8); it represents a slow-time-scale description of a steady state FEL amplifier (or IFEL accelerator) in three dimensions, assuming a single propagating frequency. The propagating electromagnetic field is taken to be a superposition of vacuum TE and TM modes, and the exact formula for the off-axis magnetic field of a helical wiggler is used, as derived by Park et al. (9).

In the current experiment, a single cavity TE_{11n} mode is used, where $n \sim 120$, and the wiggler amplitude and guide magnetic field are varied adiabatically over the length of the structure. The RF wavelength used is 2.63 cm, with an initial wiggler period of 10 cm, so the electrons undergo about 20 undulation cycles along the 2 m structure. In the entry region, with length 50 cm, it was found most advantageous to taper the axial guide field linearly from zero to its full value, but the wiggler field amplitude follows a sine-squared profile [i.e., $B_w(z) = B_w\sin^2(\pi z/50)$]. The variation of the wiggler field amplitude and guide field over the length of the entry and acceleration regions is shown in Fig. 3(a).

It is clear from Eqn. 1 that it is critical to inject the electron beam in the correct phase—or, more precisely, that trapping depends on the choice of a correct "phase window" for beam injection (10). In the simulation results presented here, the phase window used was $\pi/4$ radians, with good results; in practice, the 5 ps bunch length expected to be produced by the RF gun is equivalent to a phase spread of approximately $\pi/8$ radians, well within this restriction.

A plot of the average electron energy over the length of the MIFELA is shown in Fig. 2. Note that in the acceleration region (from 50 cm to 200 cm) energy growth is almost linear, with a gradient of 3.3 MeV/m. During the entry region, axial momentum is transferred to rotational momentum with little change in total energy, as is shown in Fig. 3(b), which plots the axial and transverse values of β in the MIFELA.

Figure 4 shows the energy cross-section at the exit of MIFELA for this simulation run. The injected beam was made up of 1000 particles, spread uniformly in initial azimuthal angle and radius, with no initial energy spread. 801 particles were trapped and accelerated, and 730 particles remain within a 2% spread in γ at the end of the acceleration region, with a FWHM of 0.9% in the γ distribution. Since energy spread in the injected MIFELA beam is expected to be ≤1%, we anticipate experimental results of a similar order. Examination of the longitudinal phase space [Fig. 5(a)] and beam spot [Fig. 5(b)] at the end of MIFELA confirms that the particles remain strongly bunched during acceleration and that the beam size remains small, with a 1 mm spread in radius containing 89% of the accelerated particles—in other words, 71% of the accelerated electrons have remained within the phase and beam radius windows that they had upon injection.

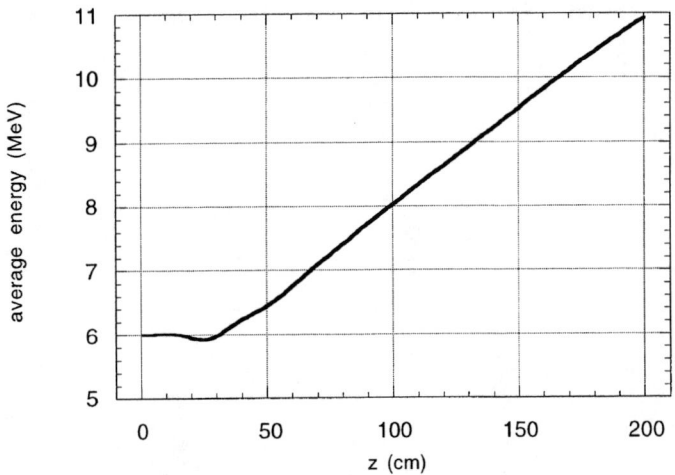

FIGURE 2. Average electron energy in MIFELA as a function of distance. Parameters for this simulation are as reported in Table 1, with a peak electric field strength $E_p = 41$ MV/m, corresponding to an RF pump parameter $a_s = 0.17$, and magnetic fields as shown in Figure 3(a).

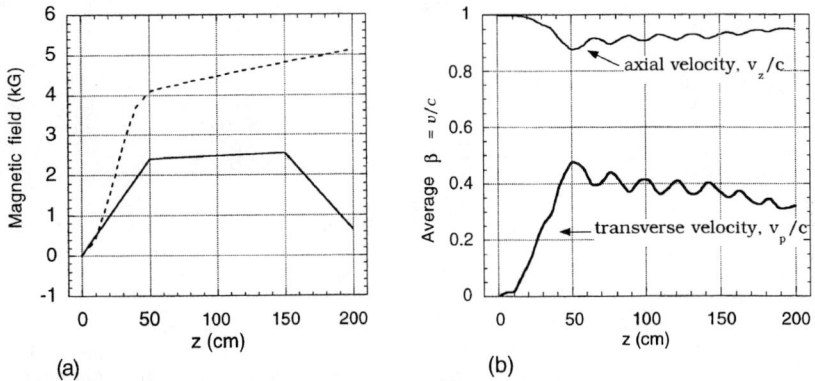

FIGURE 3. (a) Variation of axial guide field (solid line) and wiggler field amplitude (dashed line) in the MIFELA. The wiggler field taper is accomplished by linearly varying the wiggler pitch from 10.0 to 12.3 cm and the radius from 3.0 to 2.8 cm during the acceleration region. (b) Average normalized axial (top) and transverse (bottom) velocity components in the MIFELA.

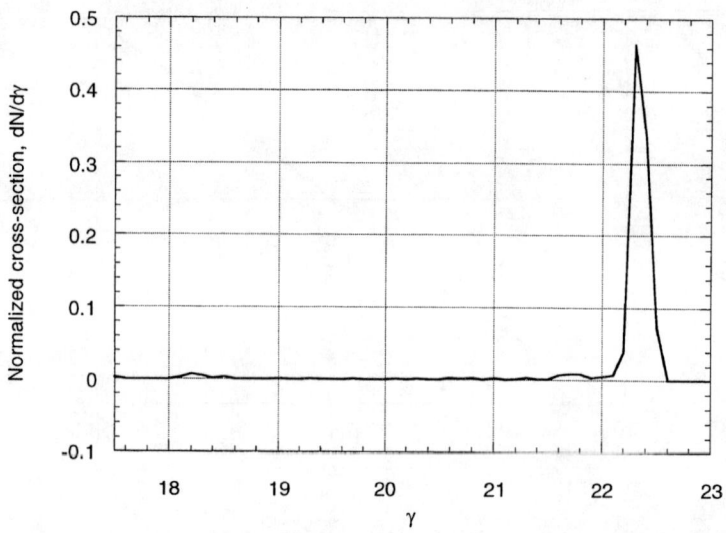

FIGURE 4. Normalized energy cross-section dN/dγ at the exit of MIFELA. The handful of particles outside the main peak are electrons which have become detrapped.

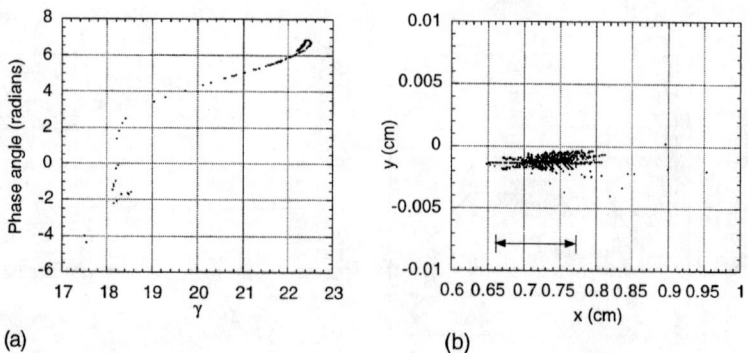

FIGURE 5. (a) Longitudinal phase space at the exit of MIFELA. 801 particles of 1000 in the simulation run were trapped; 730 particles are within a 2% γ-spread in the upper right corner. (b) Beam spot at the exit of MIFELA. Note the different x and y scales. The arrow along the x-axis marks 1 mm, containing 710 particles.

If an axial beam is desired for extraction, and the gyrating electrons are "spun down" by reducing the wiggler and guide fields, the particles pass out of resonance and little energy change occurs. The exit phase space is not significantly altered, and the beam spot has a radius on the order of 1 mm.

EXPERIMENTAL DESIGN AND CONSTRUCTION

Currently, construction and testing of the beam line and microwave circuit for the MIFELA is well underway, with final design work being done on the wiggler and acceleration cavity.

A single microwave source, the SLAC XK-5 klystron mentioned previously, serves as the driver for both the RF gun and the gyroharmonic converter. This klystron produces up to 24 MW of output at 2.856 GHz—the operating frequency of the gun and the input frequency of the converter, which produces 11.424 GHz output with an estimated efficiency in the range of 40–50% (14). Power is shared between the devices via a power splitter/phase shifter, with the difference in frequencies requiring that electrons be injected into every fourth RF cycle in the MIFELA.

The 2-1/2 cell RF gun was designed by M. Borland of Argonne National Laboratory and built by AET Associates, and is based on a 1-1/2 cell design currently in use at the Stanford Synchrotron Radiation Laboratory. It is designed to produce 5 ps bunches of 6 MeV electrons, with 10^9 particles per bunch, during a macropulse of up to 2 μs. The gun has been extensively simulated using the PARMELA and RFGUN codes, and the expected energy spread at the gun output is ~3% (11). A beamline has been constructed which will serve to focus the beam into the MIFELA cavity, as well as reducing the energy spread by means of an achromat and slit, so that only about 10^8 electrons per bunch will be injected into MIFELA with an energy spread of better than 1%. As the pump power becomes depleted during the macropulse, beam loading effects are expected. The beamline will also include a switching magnet, so that the beam can be used either in MIFELA or in the microwave inverse Cerenkov accelerator, a second novel device which is being built parallel to the MIFELA and which is described in another paper at this Conference (12). Diagnostic devices on the beamline include a synchrotron radiation probe, which is designed to provide an accurate measure of the bunch length by spectroscopic analysis of the bending magnet radiation.

Several prototypes of the wiggler coil have been tested, and a current driver for the wiggler has been constructed, which uses a five-segment pulse-forming network to produce a current pulse of up to 50 kA and about 40 μs in duration, with less than 1% ripple. To produce the fields used in the simulations described here, about 42 kA will be required. In order to create the required smoothly up-tapered wiggler field amplitude in the entry region, a series of resistive shunts will be used to partially short the wiggler windings and decrease the effective current in that portion of the wiggler. About 40 such shunts, placed according to an algorithm which we have derived (13), are sufficient to give a smooth amplitude variation of the correct shape in the entry region without changing the other field parameters.

The axial guide magnetic field is provided by a series of 16 solenoids, which are independently controlled by computer so as to automatically generate any given field profile. Such a system has been used successfully for several years in the Yale Beam Physics Laboratory's gyroharmonic conversion experiments.

The acceleration cavity itself is a section of circular waveguide, of radius 1.45 cm and length about 2 m. The electron beam will enter the cavity through a small aperture, well below cutoff for the microwave frequencies in MIFELA, while the cavity is filled with RF from the opposite end (see Fig. 1) through a coupling iris of appropriate size. A system of polarizers and mode filters on the waveguide line from the microwave source (the gyroharmonic converter) will ensure the presence of the single rotating TE_{11} mode required for operation of the MIFELA. A beam exit port will allow the beam to be extracted for spectrum analysis and other diagnostics.

The Q-value of the acceleration cavity determines the field strength within the cavity and hence the acceleration gradient. For a high-purity copper waveguide, the expected ohmic Q in a high-axial-mode-number TE_{11p} mode can be calculated to be greater than 30,000 (14). Using critical coupling to give a loaded Q of 15,000, we have calculated that an input power level of 5 MW will give a peak RF electric field inside the cavity of 29 MV/m, corresponding to an RF pump parameter (normalized vector potential) of 0.12, with a fill time of order 1 μsec. With such a Q in the cavity, the pump parameter of 0.17 used in the simulations would be obtained with an input power level of 10.5 MW.

CONCLUSION

The MIFELA represents a potentially attractive charged-particle accelerator with particularly simple fabrication and component requirements. Our design study indicates that reasonable acceleration gradients can be achieved with strong trapping and minimal spread in phase and energy. While a gyrating beam is generated in MIFELA, an extraction section could produce an axial beam for multistage devices. Components for a prototype accelerator which would demonstrate this principle efficiently are currently being manufactured and assembled at the Beam Physics Laboratory, and experimental operation is expected to begin in the next six months.

ACKNOWLEDGEMENTS

The authors acknowledge the assistance and collaboration of Mei Wang (Yale), Y.-H. Liu (Columbia), and M. A. LaPointe (Omega-P), and helpful discussions with A. K. Ganguly. This work is supported by the US Department of Energy.

REFERENCES

1. Palmer, R. B., *J. Appl. Phys.* **43**, 3014–3023 (1972); also see Sprangle, P., and Tang, C. M., *IEEE Trans. Nucl. Sci.* **NS-28**, 3346–3348 (1981).

2. Ting, A. C., and Sprangle, P. A., *Part. Accel.* **22**, 149–160 (1987), and references therein.

3. Wernick, I., and Marshall, T. C., *Phys. Rev. A* **46**, 3566–3568 (1992).

4. Fisher, A., Gallardo, J., Sandweiss, J., and van Steenbergen, A., "Inverse Free Electron Laser Accelerator," in *Proceedings of the 3rd Advanced Accelerator Concepts Workshop*, Port Jefferson, New York, 1992, pp. 299–318.

5. Van Steenbergen, A., Gallardo, J., Sandweiss, J., and Fang, J.-M., *Phys. Rev. Lett.* **77**, 2690–2693 (1996).

6. Zhang, T. B., and Marshall, T. C., *Nucl. Instr. Meth. Phys. Res. A* **375**, 515–518 (1996).

7. Ganguly, A. K., and Freund., H. P., *Phys. Rev. A* **32**, 2275–2286 (1985); Freund, H. P., and Ganguly, A. K., *Phys. Rev. A* **34**, 1242–1246 (1986).

8. Ganguly, A. K., and Freund, H. P., *Phys. Fluids* **31**, 387–393 (1988); Freund, H. P., and Ganguly, A. K., *IEEE Trans. Plasma Sci.* **PS-20**, 245–255 (1992) and references therein.

9. Park, S. Y., Baird, J. M., Smith, R. A., and Hirshfield, J. L., *J. Appl. Phys.* **53**, 1320–1325 (1982).

10. Zhang, T. B., and Marshall, T. C., *Phys. Rev. E* **50**, 1491–1495 (1994).

11. Borland, M., unpublished report (1995).

12. Zhang, T. B., Marshall, T. C., LaPointe, M. A., and Hirshfield, J. L., "Microwave Inverse Cerenkov Accelerator," presented at this Workshop.

13. Zhang, T. B., and Marshall, T. C., "Shunt Resistors in Wiggler Entrance," unpublished report (1996).

14. LaPointe, M. A., Yoder, R. B., Wang, M., Ganguly, A. K., Wang, C., Hafizi, B., and Hirshfield, J. L., "High-Power Microwave Production by Gyroharmonic Conversion and Co-Generation," presented at this Workshop.

Ionization Effects in Inverse Cherenkov Accelerators

Phillip Sprangle, Bahman Hafizi,[+] and Richard F. Hubbard

Beam Physics Branch, Plasma Physics Division
Naval Research Laboratory, Washington, DC 20375

[+] *Icarus Research Inc., P. O. Box 30780, Bethesda, MD 20824-0780*

ABSTRACT

Ionization processes limit the accelerating gradient and pulse duration of the electromagnetic driver in the Inverse Cherenkov Accelerator (ICA). Two configurations for the ICA are considered in which the electromagnetic driver is propagated in a waveguide that is, i) lined with a dielectric material or ii) filled with a neutral gas. The intensity of the driver in the ICA, and therefore the acceleration gradient, is limited by tunneling and collisional ionization effects. Ionization of the dielectric liner or gas can lead to significant modification of the optical properties of the waveguide, altering the phase velocity of the accelerating field and causing particle slippage, thus disrupting the acceleration process. To avoid the effects of ionization short pulse durations are necessary, and therefore pulse lethargy, i.e., electrons outrunning the driving pulse, can present an additional limitation. Limitations on the driver pulse duration and accelerating gradient, due to tunneling and collisional ionization, are obtained for the two ICA configurations. Maximum accelerating gradients and pulse durations are presented for a 10 μm and a 1 mm wavelength electromagnetic driver.

I. INTRODUCTION

Although lasers are capable of providing extraordinarily high electric fields for the acceleration of particles, numerous fundamental and technological issues must be resolved before practical high gradient accelerators can be realized [1-4]. There are three fundamental issues that must be addressed in any type of high gradient laser driven acceleration mechanism. These are: i) laser beam guiding over extended distances, ii) phase coherence (slippage) between the accelerated particles and laser field, and iii) material breakdown arising from either the laser fields or from the accelerated particles.

The electromagnetic driver in the ICA can be an intense laser or longer wavelength pulse. In free space and in the absence of boundaries the phase velocity of the accelerating field exceeds the vacuum speed of light, and particles continually slip relative to the field and acceleration ceases. In the ICA the phase velocity of the accelerating field is reduced by introducing a gas in the accelerating region or by lining the interior of a waveguide with a dielectric material [2,3,5-9].

Using a dielectrically lined or gas-filled waveguide in the ICA configuration, i) avoids diffraction of the driving laser beam, ii) overcomes electron slippage, and iii) the acceleration is accomplished in a vacuum. However, the large electric fields associated with short, intense electromagnetic pulses can readily ionize the dielectric material or gas[10] Ionization will affect the optical characteristics of the waveguide which may cause phase slippage and disrupt the acceleration process. However, if the pulse is too short, group velocity slippage, i.e., pulse lethargy, may impose a further limitation on the pulse length.

II. FIELDS AND DISPERSION RELATION

A dielectrically lined waveguide can support an axially symmetric transverse-magnetic (TM) mode [9]. The cross-sectional view of the dielectrically-lined waveguide is shown in Fig. 1, where the inner surface of the dielectric is at $r = a$ and the conducting surface is at $r = b$. The particular TM mode under consideration consists of a radial and an axial electric field and an azimuthal magnetic field. The electric field components within the central vacuum region ($0 < r < a$) are

$$E_r = E_0 \left(J_1(k_\perp r) / J_1(k_\perp a) \right) f(z,t) + c.c., \quad (1a)$$

$$E_z = i(k_\perp / k) E_0 \left(J_0(k_\perp r) / J_1(k_\perp a) \right) f(z,t) + c.c., \quad (1b)$$

and within the outer dielectric region ($a \leq r \leq b$),

$$E_r = \left(A J_1(\alpha r) + B Y_1(\alpha r) \right) f(z,t) + c.c., \quad (2a)$$

$$E_z = i(\alpha / k) \left(A J_0(\alpha r) + B Y_0(\alpha r) \right) f(z,t) + c.c \quad (2b)$$

where ω is the frequency, k is the axial wavenumber, $f(z,t) = (1/2)\exp(i(kz - \omega t))$, $k_\perp = (\omega^2 / c^2 - k^2)^{1/2}$ is the transverse wavenumber within the inner region, c is the vacuum speed of light, $\alpha = (\varepsilon \omega^2 / c^2 - k^2)^{1/2}$ is the transverse wavenumber within the outer region, ε is the dielectric constant, E_0, A, B are the constant amplitudes, $J_n(x)$ is the ordinary Bessel function of first kind of order n and c.c. denotes the complex conjugate. The accelerating axial electric field given by Eq. (1b) is peaked along the z-axis. For phase velocities equal to c, i.e., $k_\perp = 0$, the axial electric field is independent of radial position,

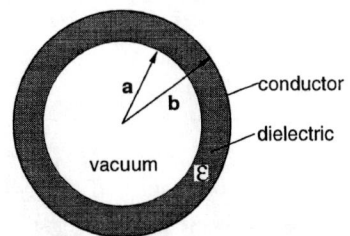

FIGURE 1. Dielectrically-lined optical waveguide schematic. The dielectric material with dielectric constant ε lies between the vacuum ($0 \leq r \leq a$) region and perfect conductor at $r = b$.

$E_z = i(\lambda/\pi a)E_0 f(z,t) + c.c.$, where $\lambda = 2\pi c/\omega$ is the wavelength. Applying the appropriate boundary conditions at $r = a$ and $r = b$ we obtain the following dispersion relation for phase velocities near c

$$\omega^2/c^2 - k^2 = 8/a^2 - F(\alpha_0, a, b), \tag{3}$$

where $F(\alpha_0,a,b) = 4(\alpha_0/\varepsilon a)\tan(\alpha_0(b-a))$ and it has been assumed that $\alpha_0 a \gg 1$, i.e., $2\pi(\varepsilon-1)^{1/2} a/\lambda \gg 1$. In the limit $\lambda/a \ll \pi(\varepsilon-1)^{1/2}/\varepsilon$, the wavelength associated with mode ℓ is $\lambda/a = 2[(b-a)/a](\varepsilon-1)^{1/2}/\ell$, where ℓ denotes the number of nodes in the liner. In the thick dielectric limit, $\ell \gg 1$, the dominant electric field component in the dielectric is given by

$$E_z = E_0(a/r)^{1/2}(-1)^\ell[(\varepsilon-1)/\varepsilon]\sin(\alpha_0(b-r))f(z,t) + c.c., \tag{4}$$

III. CRITICAL AVERAGE PLASMA DENSITY

The large fields needed for acceleration can ionize the dielectric material and change the optical properties of the waveguide. If the optical properties change sufficiently electron slippage will occur and the acceleration process will cease. As ionization takes place the dielectric constant changes to $\varepsilon + \delta\varepsilon$, where $\delta\varepsilon = -\omega_p^2/\omega^2$ is the contribution to the dielectric constant due to ionization, $\omega_p = (4\pi q^2 \langle n_p \rangle/m)^{1/2}$ is the average plasma frequency, q is the charge and m is the mass of an electron and $\langle n_p \rangle$ is the average plasma density within the optical waveguide. When the change in $F(\alpha_0,a,b)$ in Eq. (3) due to ionization equals ~ $8/a^2$, the optical properties of the guide change significantly and slippage disrupts the acceleration process. Ionization in the dielectric material will not disrupt the acceleration process if $|\delta\varepsilon| = \omega_p^2/\omega^2 \ll (8/a^2)|\partial F/\partial\varepsilon|^{-1}$. This inequality defines a critical average plasma density within the dielectric, n_{crit}, given by

$$n_{crit} = \frac{2}{\pi r_e}\left(\frac{\omega/c}{a}\right)^2 |\partial F/\partial\varepsilon|^{-1}, \tag{5}$$

where $r_e = q^2/mc^2 = 2.8 \times 10^{-13}$ cm is the classical electron radius. If the average plasma density $\langle n_p \rangle$ generated within the dielectric exceeds n_{crit} the acceleration process is disrupted because of slippage.

In the limiting case discussed in connection with the dispersion relation,

$$n_{crit} = \frac{\varepsilon}{\pi} \frac{1}{a^2 r_e} \frac{a}{b-a}. \tag{6}$$

As an illustration we take $\varepsilon = 2$, $a = 100$ μm, $a/(b-a) = 2$ and find that the critical density for slippage is $n_{crit} \cong 5 \times 10^{16}$ cm^{-3}, independent of wavelength.

IV. IONIZATION IN SOLIDS

The rate of change of plasma density n_p is given by

$$\frac{\partial n_p}{\partial t} = S + W n_p, \tag{7}$$

where S is the photo-ionization source due to electron tunneling and W is the collisional ionization rate. The photo-ionization source acts as an initial source of free electrons which are further increased in number by collisional processes, i.e., by electron avalanche. The solution of Eq. (7) is

$$n_p = (S/W)(\exp(Wt) - 1) + n_{p0} \exp(Wt), \tag{8}$$

where n_{p0} is the initial density of plasma (free) electrons and it is assumed that the plasma density remains small compared to the neutral density.

A. Collisional Ionization

The collisional ionization rate can be estimated by using a classical free electron model for the electron energy gain [11,12]

$$\frac{dU}{dt} = \frac{q^2 E^2 v_m}{2m(\omega^2 + v_m^2)} - \frac{2m}{M} v_m U, \tag{9}$$

where U is the electron energy, E is the peak electric field amplitude, v_m is the momentum transfer frequency, q is the electron charge, m is the electron mass and M is the mass of the neutral atoms. Solving Eq. (9), we obtain

$$U = U_{max}(1 - \exp(-v_0 t)), \tag{10}$$

where $U_{max} = (v_m / v_0)\left(1 + v_m^2 / \omega^2\right)^{-1} U_{0s}$ is the maximum electron energy, $U_{0s} = m v_{0s}^2 / 2$ is the electron oscillation energy, $v_{0s} = qE/m\omega$ is the oscillation velocity and $v_0 = 2 m v_m / M$ is the energy damping frequency. Typically, in solids $v_m \sim 10^{15}$ sec^{-1} [13] and $v_0 \sim 10^{11}$ sec^{-1}. When the electron energy reaches the ionization energy U_I, the electron collisionally ionizes the atom. The time for this to occur is the ionization time W^{-1}. Setting $U = U_I$ and $t = W^{-1}$ in Eq. (10) determines W,

$$W = -\nu_0 / \log_e(1 - U_I / U_{max}), \quad (11)$$

for $U_I/U_{max} < 1$ and $W = 0$ for $U_I/U_{max} > 1$. For $U_I/U_{max} \ll 1$, Eq. (11) becomes $W \cong (U_{0s}/U_I)\nu_m / (\nu_m^2/\omega^2 + 1)$. Note that $(U_I/U_{0s})^{1/2} = \gamma_k$ is the Keldysh parameter [14]. In the tunneling ionization regime $\gamma_k < 1$ and therefore $U_{0s}/U_I > 1$. Taking $\nu_m \cong 10^{15}$ sec^{-1} for solids we find that the collisional ionization rate for the case where $U_I/U_{max} \ll 1$ is $W[\text{sec}^{-1}] \cong 3.4 \times 10^{13}(U_{0s}/U_I)$, for 10 μm wavelength radiation.

B. Tunneling Ionization Source

The ionization source in solids, averaged over times much longer than the field period $2\pi/\omega$, in the tunneling regime $\gamma_k = (U_I/U_{0s})^{1/2} < 1$, given by [15]

$$S = A_s \tilde{E}^{5/2} \exp[-\beta_s / \tilde{E}], \quad (12)$$

where $A_s = 1.5 \times 10^{40} \tilde{U}_I^{-5/4}$, $\beta_s = 0.56 \tilde{U}_I^{3/2}$. In Eq. (12) A_s has units of sec^{-1} cm^{-3}, $\tilde{U}_I = U_I / U_H$ is the atomic ionization energy U_I normalized to that of hydrogen, $U_H = 13.6$ eV, and $\tilde{E} = E / E_H$ is peak applied electric field normalized to the hydrogenic electric field $E_H = 5.2$ GV / cm.

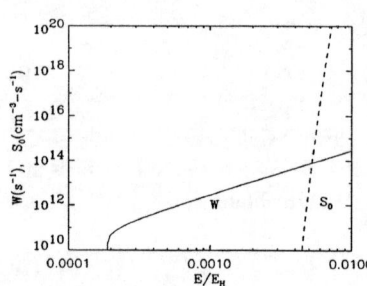

FIGURE 2. Plot of tunneling ionization source function S and collisional ionization rate W as a function of normalized electric field $\tilde{E} = E / E_H$. In this plot $U_I = 8$ eV, $\lambda = 10$ μm, $\nu_m = 10^{15}$ sec^{-1}, and $\nu_0 = 10^{11}$ sec^{-1}.

Figure 2 shows the tunneling ionization source term and the collisional ionization rate as functions of the normalized electric field. For pulses shorter than a few picoseconds, collisional ionization is negligible for $W < 10^{10}$ sec^{-1}. The precipitous drop in W below $E/E_H = 2 \times 10^{-4}$ occurs as U_{max} approaches U_I in Eq. (11). The tunneling term S does not lead to significant ionization for $E/E_H < 5 \times 10^{-3}$, but ionization increases very rapidly for higher fields.

V. LIMITATION ON LASER PULSE DURATION AND ACCELERATING GRADIENT

To avoid modifying the optical properties of the dielectric waveguide the average induced plasma density must be less than the average critical density n_{crit}, i.e., $\langle n_p \rangle \ll n_{crit}$. Using Eq. (8) for n_p we find that

$$\langle n_p \rangle = \langle (S/W)(\exp(W\tau_L) - 1) \rangle + n_{p0} \langle \exp(W\tau_L) \rangle \ll n_{crit}, \qquad (13)$$

where τ_L is the laser pulse duration. In performing the cross-sectional average we make use of the fact that in the tunneling ionization part, the source function S is a highly sensitive function of the field amplitude through the exponential term [see Eq. (12)]. The cross-sectional average of the contribution from the initial plasma electron density can be written as $\langle \exp(W\tau_L) \rangle \cong f \exp(W_m \tau_L)$ where f is a filling factor, due to the spatial variation of the field within the dielectric. Employing these approximations, Eq. (13) becomes

$$W_m^{-1}(\exp(W_m \tau_L) - 1)\langle S \rangle + f n_{p0} \exp(W_m \tau_L) \ll n_{crit}, \qquad (14)$$

where W_m is the maximum value of W, i.e., W is evaluated at the peak electric field E_m in the dielectric. The filling factor f is given by $f \cong (\pi/8)^{1/2} (W_m \tau_L)^{-1/2}$, for $W_m \tau_L \gg 1$. In terms of the laser pulse duration we obtain the inequality

$$\tau_L \ll \tau_{crit} = W_m^{-1} \log_e \left[(\tau_0 W_m + 1) / (1 + f n_{p0} W_m / \langle S \rangle) \right], \qquad (15)$$

where τ_{crit} is the critical pulse duration and $\tau_0 = n_{crit}/\langle S \rangle$. The inequality in Eq. (15) places a limit on the laser pulse duration which is a function of the accelerating gradient, E_z. For $\ell \gg 1$, $E_z = [\lambda \varepsilon / a\pi(\varepsilon - 1)^{1/2}] E_m$.

The average ionization source term is $\langle S \rangle \cong 2\pi \int_a^b S r dr / (\pi b^2)$, where the radial variation in S arises from that of the electric field. Using the electric field given in Eq. (4) and expanding about the ℓ peaks we obtain

$$\langle S \rangle = \frac{a^2}{2b^2} S_m \pi^{1/2} \frac{(b-a)}{a} \left(\tilde{E}_m / \beta_s \right)^{1/2}, \qquad (16)$$

where S_m is S evaluated at $\tilde{E}_m = E_m / E_H$, i.e., $S_m = A_s \tilde{E}_m^{5/2} \exp\left[-\beta_s / \tilde{E}_m\right]$. Using Eq. (16) for $\langle S \rangle$ and Eq. (6) for n_{crit}, we obtain

$$\tau_0 = n_{crit} / \langle S \rangle = \frac{2\varepsilon}{a^2 r_e} \left(\frac{a}{b-a}\right)^2 \frac{b^2}{a^2 S_m} \pi^{-3/2} \left(\beta_s / \tilde{E}_m\right)^{1/2}. \tag{17}$$

VI. GAS-FILLED INVERSE CHERENKOV ACCELERATOR

Another configuration for the inverse Cherenkov accelerator consists of a waveguide that is filled with a neutral gas [2-5]. For sufficiently low gas densities, the collisional scattering of the accelerated particles can be neglected [16].

For the TM mode the fields are given by Eq. (1), with $k_\perp = (\varepsilon\omega^2/c^2 - k^2)^{1/2}$, where $\varepsilon \geq 1$ is the dielectric constant of the gas. The dispersion relation for the TM mode is given by $k_\perp a = p_{0n}$, or $\varepsilon\omega^2/c^2 - k^2 = (p_{0n}/a)^2$, where a is the waveguide radius, and p_{0n} is the nth zero of J_0. Writing $\varepsilon = 1 + \Delta\varepsilon$, where $\Delta\varepsilon > 0$ is the contribution of the neutral gas to the dielectric constant, we find that for the phase velocity to equal the vacuum speed of light the wavelength is given by $\lambda/a = (2\pi/p_{0n})\Delta\varepsilon^{1/2}$. Since $\Delta\varepsilon \ll 1$ for gases at moderate pressures, i.e., a few atmospheres, the dispersion relation implies that $\lambda/a \ll 1$. Since $E_z = (\lambda/\pi a)E_0$ on axis, the accelerating field is small compared to the transverse field E_0.

To determine the critical average plasma (electron) density in the gas we proceed as in Sec. III. Accounting for the ionization of the ambient gas, we write $\varepsilon = 1 + \Delta\varepsilon - \omega_p^2/\omega^2$, where the critical plasma density is given by

$$n_{crit} = \pi\Delta\varepsilon / (r_e \lambda^2) = p_{0n}^2 / (4\pi a^2 r_e). \tag{18}$$

Typically $\Delta\varepsilon \sim 10^{-4}$ for a gas at atmospheric pressure and for a $\lambda = 10$ μm laser-driven gas-filled ICA the critical density is found to be $n_{crit} \cong 10^{15}$ cm^3.

The development of plasma density in the gas, undergoing ionization due to tunneling and electron avalanche, is given by Eq. (7). The ionization source term in Eq. (7) for a gas is given by [14]

$$S = A_g \tilde{E}^{-1/2} \exp(-\beta_g / \tilde{E}), \tag{19}$$

where $A_g = 1.6 \times 10^{17} \tilde{U}_I^{7/4} n_{n0}$, in units of sec^{-1} cm^{-3}, $\beta_g = 0.67 \tilde{U}_I^{3/2}$, n_{n0} is the neutral gas density in units of cm^{-3}. The collisional ionization rate W is given by Eq. (11). Similarly, the rate of change of electron energy is given by an Eq. (9), where the momentum transfer frequency ν_m in a gas is proportional to the pressure and is on the order of 10^{12} sec^{-1} at atmospheric pressure [11,12]. The limitation of the pulse duration and accelerating gradient in the gas-filled ICA is given by Eq. (15). For a gas-filled ICA, $\tau_0 = n_{crit}/\langle S \rangle$, where n_{crit} is given by Eq. (18), and

$$\langle S \rangle \cong S_{g0}(2\pi\tilde{E}_0 / \beta_g)^{1/2}, \tag{20}$$

where S_{g0} is the value of S evaluated at peak electric field E_0, i.e., $\tilde{E}_0 = E_0 / E_H$. The filling factor f in Eq. (15) for the gas-filled waveguide ICA is $f \cong (\pi/2)^{1/2}$ $(W_m\tau_L)^{-1/2}$, where $W_m\tau_L \gg 1$.

VII. RESULTS

Figures 3 through 6 show τ_{crit} versus the accelerating gradient $E_z = \lambda E_0/\pi a$ for the dielectric liner and gas-filled ICA configurations. In the dielectric-lined ICA plots, E_0 is the radial electric field amplitude at the inner (vacuum) surface of the dielectric, while in the gas-filled ICA, E_0 is the peak field within the waveguide.

Figure 3 is a plot of the critical time as a function of the accelerating gradient, for $\lambda = 10$ μm, as the inner radius of the waveguide is varied. The dielectric layer thickness b - a = 50 μm in all three cases, with a = 30, 50, and 100 μm for the three curves. Figure 4 shows the variation of τ_{crit} with E_z for $\lambda = 1$ mm,

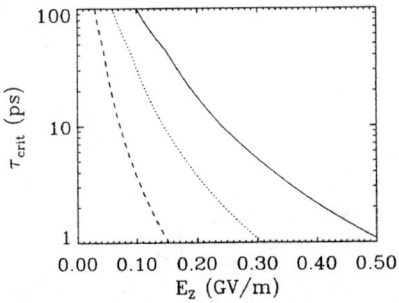

FIGURE 3. Plot of critical pulse time τ_{crit} versus the accelerating gradient E_z for a laser of wavelength $\lambda = 10$ μm, inner radius a = 30 μm (solid curve), 50 μm (dotted curve) and 100 μm (dashed curve), b - a = 50 μm. In addition, $\varepsilon = 3$, $U_i = 8$ eV, $n_{po} = 0$, $v_m = 10^{15}$ sec^{-1} and $v_o = 10^{11}$ sec^{-1}.

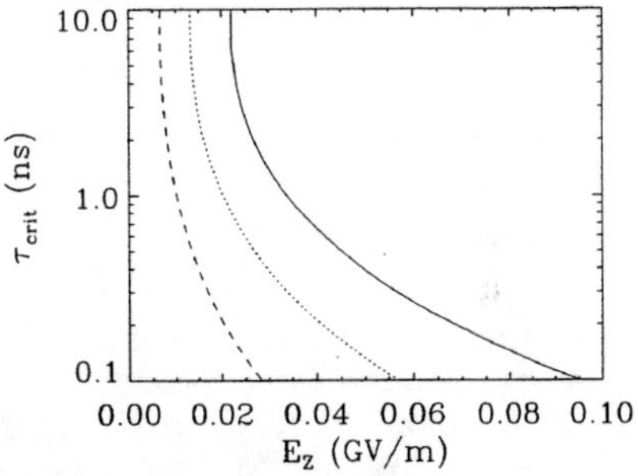

FIGURE 4. Plot of critical pulse time $\tau_{\chi crit}$ versus the accelerating gradient E_z for a millimeter wave driver, $\lambda = 1$ mm. The inner radius a = 3 mm (solid curve), 5 mm (dotted curve) and 10 mm (dashed curve), where b - a = 5 mm. In addition, $\varepsilon = 3$, $U_i = 8$ eV, $n_{po} = 0$, $v_m = 10^{15}$ sec^{-1} and $v_o = 10^{11}$ sec^{-1}.

as the radius of the inner wall is varied. The scaling here is similar to the $\lambda = 10$ μm case shown in Fig. 3, but the allowable accelerating gradient is lower. The longer pulse lengths, compared to the 10 μm wavelength example, are the principle reason for the lower accelerating gradients.

Figure 5 shows the critical pulse duration versus the accelerating gradient for a helium gas-filled ICA. The variation of the critical time τ_{crit} with the accelerating gradient E_z at $\lambda = 10$ μm, is shown for gas pressures P_g of 3, 10, and 30 atmospheres. Since $\Delta\varepsilon = 7 \times 10^{-5}$ at 1 atm for helium, the dispersion relation for the lowest order mode, $p_{0n} = 2.4$, gives a = 265, 145, and 84 μm for the three pressures. This gas-filled case has a lower accelerating gradient than the dielectric liner case shown in Fig. 3 because the dispersion relation requires the wall radius to be so large that $E_0 = E_r(r = a) \gg E_z$. In dielectric liner cases, E_z can be the dominant component.

Finally, Fig. 6 is a plot of τ_{crit} versus E_z for a $\lambda = 1$ mm source with the same three helium gas pressures shown in Fig. 5. The wall radius a = 26.5, 14.5, and 8.4 mm for gas pressures of 3, 10, and 30 atmospheres, respectively. The allowable accelerating gradient is only a few MV/m. This is due again in part to the large a/λ ratio required to satisfy the dispersion relation. In addition, the avalanche time W^{-1} even at these modest fields is much less than the typical pulse duration.

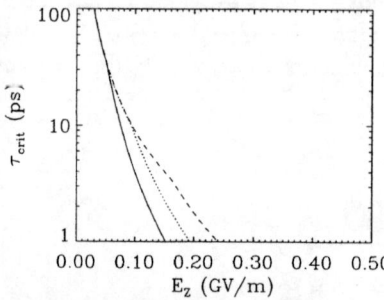
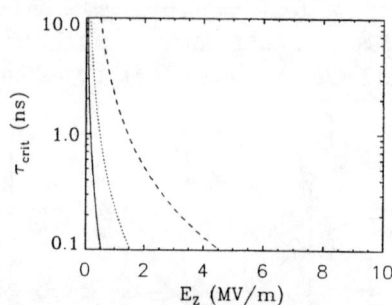

FIGURE 5. Plot of critical pulse time τ_{crit} versus the accelerating gradient E_z in a gas (helium)-filled ICA with laser wavelength $\lambda = 10$ μm. The gas pressures and corresponding wall radii are 3 atm and 265 μm (solid curve), 10 atm and 145 μm (dotted curve), and 30 atm and 84 μm (dashed curve), with $U_i = 24.6$ eV, $v_m/P_g = 10^{12}$ sec^{-1}/atm and $v_d/P_g = (2m/M)v_m/P_a = 3 \times 10^8$ sec^{-1}/atm.

FIGURE 6. Plot of critical pulse time τ_{crit} versus the accelerating gradient E_z in a gas (helium)-filled ICA with wavelength $\lambda = 1$ mm. The gas pressures and corresponding wall radii are 3 atm and 26.5 mm (solid curve), 10 atm and 14.5 mm (dotted curve), and 30 atm and 8.4 mm (dashed curve), with $U_i = 24.6$ eV, $v_m/P_g = 10^{12}$ sec^{-1}/atm and $v_d/P_g = (2m/M)v_m/P_a = 3 \times 10^8$ sec^{-1}/atm.

VIII. SUMMARY AND DISCUSSION

Ionization processes limit the accelerating gradient and pulse duration of the electromagnetic driver in the ICA. Using a dielectrically-lined or gas-filled waveguide, i) avoids diffraction of the driving laser or millimeter wave beam, ii) overcomes electron slippage, and iii) the acceleration is accomplished in a vacuum. The intensity of the driver in the ICA, and therefore the acceleration gradient, is limited by tunneling and collisional ionization effects. Partial ionization of the dielectric liner or gas can lead to significant modification of the optical properties of the waveguide, altering the phase velocity of the accelerating field and causing particle slippage, thus disrupting the acceleration process.

We obtained an expression for the limit on the electromagnetic pulse duration in terms of the accelerating gradient. To avoid the effects of ionization, the pulse duration of the electromagnetic driver in the ICA must be less than a critical pulse duration time τ_{crit} given by Eq. (15). The value of τ_{crit} is a sensitive function of the peak electric field and thus the accelerating gradient. For a wide range of parameters, the accelerating gradient is limited to less than ~0.5 GV/m (100 MV/m) for a driver wavelength of $\lambda = 10$ μm ($\lambda = 1$ mm).

Acknowledgments

This work was supported by DOE Contract DE-A102-93ERR40797 and by ONR.

References

1. *Advanced Accelerator Concepts*, edited by P. Schoessow, AIP Conf. Proc. **335** (American Institute of Physics, New York, 1995).
2. Sprangle, P., Esarey, E., and Krall, J., *Phys. Plasmas* **3**, 2183 (1996).
3. Sprangle, P., Esarey, E., and Krall, J., to appear in *Phys. Rev. E*.
4. Hafizi, B., Esarey, E., and Sprangle, P., *Phys. Rev. E* (submitted).
5. Edighoffer, J.A., Kimura, W.D., Pantell, R.H., Piestrup, M.A., Wang, D.Y., *Phys. Rev. A* **23**, 1848 (1981).
6. Fontana, J.R., and Pantell, R.H., *J. Appl. Phys.* **54**, 4285 (1983).
7. Romea, R.D., and Kimura, W.D., *Phys. Rev. D* **42**, 1807 (1990).
8. Kimura, W.D., Kim, G.H., Romea, R.D., Steinhauer, L.C., Pogorelsky, I.V., Kusche, K.P., Fernow, R.C., Wang, X., and Liu, Y., *Phys. Rev. Lett.* **74**, 546 (1995.
9. Zhang, T.B., Marshall, T.C., LaPointe, M.L., Hirshfield, J.L., and Ron, A., *Phys. Rev. E* **54**, 1918 (1996).
10. Stuart, B.C., Feit, M.D., Rubinchick, A.M., Shore, B.W., and Perry, M.D., *Phys. Rev. Lett.* **74**, 2248 (1995).
11. Raizer, Y.P., *Gas Discharge Physics* (Springer-Verlag, Berlin, 1991), Chap. 2.
12. Zeldovich, Y.B., and Raizer, Y.P., *Physics of Shock Waves and High-Temperature Hydrodynamic Phenomena,* (Academic Press, NY, 1966), Chap. VI.
13. Shen, Y.R., *The Principles of Nonlinear Optics* (Wiley, New York, 1984).
14. Keldysh, L.V., *Sov. Phys. JETP* **20**, 1307 (1965) [*Zh. Eksp. Teor. Fiz.* **47**, 1945 (1964)].
15. Keldysh, L.V., *Sov. Phys. JETP* **6**, 763 (1958) [*Zh. Eksp. Teor. Fiz.* **33**, 994 (1957)].
16. Hafizi, B., Sprangle, P. and Ting, A., in Ref. 1, p. 480.

GROUP 4

"Beam Monitoring, Conditioning, and Control at High Frequencies and Ultrafast Timescales"—Xijie Wang (BNL)

Temporal characterization of a self-modulated laser wakefield

S. P. Le Blanc[†], M. C. Downer[†], R. Wagner,

S.-Y. Chen, A. Maksimchuk, G. Mourou, and D. Umstadter

Center for Ultrafast Optical Science, University of Michigan, Ann Arbor, MI 48109

[†]Department of Physics, The University of Texas at Austin, Austin, Texas 78712

Abstract

The temporal envelope of plasma density oscillations in the wake of an intense (I $\sim 4 \times 10^{18}$ W/cm^2, $\lambda = 1$ μm) laser pulse (400 fs) is measured using forward Thomson scattering from a copropagating, frequency doubled probe pulse. The wakefield oscillations in a fully ionized helium plasma ($n_e = 3 \times 10^{19}$ cm^{-3}) are observed to reach maximum amplitude ($\delta n_e/n_e \sim 0.1$) 300 fs after the pump pulse. The wakefield growth (4 ps^{-1}) and decay (1.9 ps^{-1}) rates are consistent with the forward Raman scattering instability and beam loading, respectively.

1 Introduction

Because electrostatic fields in a plasma wave ($E \geq 100$ GV/m) can exceed by three orders of magnitude those in conventional RF linacs, plasma based accelerators can potentially offer a compact method for accelerating high energy electron pulses for the next generation of collider experiments [1]. However, particle beams accelerated by plasma based accelerators must meet challenging requirements, including high luminosity, low emittance, and high beam loading efficiency. Of the several methods for driving large amplitude plasma waves, the laser wakefield accelerator (LWFA) [1] and its variant, the self-modulated LWFA [1, 2, 3, 4, 5, 6], have recently received considerable attention because the the terawatt class laser systems [7] needed to drive the wakefield have been significantly reduced in size. In the LWFA, the amplitude $\alpha = \delta n_e/n_e$ of the plasma wave can be resonantly excited by the ponderomotive force of the laser pulse if the laser pulse duration is approximately half of the plasma wave period $\tau_p = 2\pi/\omega_p$, where $\omega_p = \sqrt{4\pi e^2 n_e/m_e}$ is the electron plasma frequency and n_e is the plasma density.

For the self-modulated LWFA, the plasma density is chosen to be much larger than for the standard LWFA so that the forward Raman scattering (FRS) instability can grow. The FRS instability is the conversion of an electromagnetic wave (ω_o, k_o) into a plasma wave (ω_p, k_p) and Stokes $(\omega_o - \omega_p, k_o - k_p)$ and anti-Stokes $(\omega_o + \omega_p, k_o + k_p)$ electromagnetic side bands [8]. Electron density perturbations in the plasma cause local variations in the group velocity $v_g = c(1 - \omega_p^2/w_o^2)^{1/2}$ of the laser pulse. As a result, light that propagates near a density maximum (minimum) will slow down (speed up). Eventually, the light is bunched to positions where $\delta n_e = 0$. Because the plasma density perturbation and the bunching of the light are $\pi/2$ out of phase, the ponderomotive force of the bunched light will reinforce the original density perturbation. Since the maximum longitudinal electric field E_z scales as $E_z \propto \alpha\sqrt{n_e}$, the self-modulated wakefield can produce a much larger accelerating field than the standard LWFA. If the plasma wave is driven to wavebreaking amplitudes, then background electrons can be trapped and accelerated in the propagation direction of the laser pulse [3, 4, 5, 6].

In these proceedings, we report time resolved measurements of the amplitude of the self-modulated laser wakefield obtained using forward Thomson scattering from a co-propagating, frequency doubled probe pulse [9]. In addition to measuring the growth and decay rate of the wakefield, we report that the onset of FRS is consistent with a plasma density perturbation driven by the ionization front of the laser pulse. These observations are important for testing recently developed 2D particle-in-cell (PIC) simulations of laser plasma interactions [10, 11, 12] and for the design of plasma based accelerators. Although future plasma based accelerators are likely to be based on the standard LFWA scheme, the present investigation of a self modulated LFWA nevertheless provides a useful testbed to investigate plasma instabilities and beam loading issues for atmospheric density plasmas.

2 Experimental setup

In the present experiments, a hybrid Ti:Sapphire - Nd:Glass laser system capable of delivering 3 J, 400 fs laser pulses was used to drive the self modulated LWFA. The 43 mm diameter beam was focused by an f/4 off-axis parabolic mirror to a spot size (e^{-2} intensity) of $r_o = 8.9$ μm, giving a maximum vacuum intensity of $I = 6 \times 10^{18}$ W/cm². The laser was focused onto a supersonic helium gas jet whose neutral density varied linearly with backing pressure [6]. The neutral gas density produced by the nozzle was measured to have a flat top shape with a width of 500 μm and 200 μm gradients on each side. To probe the lifetime of the plasma wave, a small portion (20%) of the infrared laser pulse was split off, frequency doubled in a 4 mm Type I KDP crystal, and then made to co-propagate with the infrared pump pulse. The temporal overlap between the IR pump pulse and orthogonally polarized green probe pulse was measured with a resolution of ±100 fs by frequency domain interferometry. The probe pulse had a maximum energy of 15 mJ and could be focused to a spot size of 6.4 μm by the harmonically coated parabolic mirror. Forward scattered light from the probe pulse was collected on axis, passed through a polarizer to suppress scattered pump light, and measured with a prism spectrometer which has a resolution of $\lambda/\Delta\lambda = 600$ at $\lambda = 1.053$ μm.

3 Forward Raman scattering

When the peak power of the IR pump pulse (P ≥ 1 TW) is near the critical power for relativistic self-focusing $P_c = 17(\omega_o^2/\omega_p^2)$ GW, the forward scattered light from the pump pulse shows the appearance of three anti-Stokes Raman shifted side bands separated by the plasma frequency ($\omega_p \sim 3 \times 10^{14}$ s^{-1}). [6]. Numerical simulations indicate that the appearance of multiple side bands is clear evidence of FRS. From the relative amplitude of the Raman satellites, the plasma wave amplitude is estimated to be $\delta n/n$=0.08-0.4, depending on the pump power and plasma density [6]. Under these conditions, a collimated beam of 2 MeV electrons with a transverse emittance of 1 mm mrad is emitted in the laser propagation direction [6].

In one dimension, the spatio-temporal growth rate for the FRS instability starting from a uniform noise source δn_s is given by [10]:

$$\delta n = \begin{cases} \delta n_s \cosh(\gamma_o \tau) & \psi \geq \tau c \\ \delta n_s \sum (\frac{\psi/c}{\tau - \psi/c})^n I_{2n}(2\gamma_o \sqrt{(\tau - \psi/c)\psi/c}) & \tau c \geq \psi \end{cases} \quad (1)$$

where I_{2n} is the modified Bessel function of the first kind, ψ is the distance from the leading edge of the pulse, τ is the propagation time, a_o is the normalized vector potential of the laser ($a_o = 8.5 \times 10^{-10} \lambda(\mu m)(I(W/cm^2))^{1/2}$), and

$$\gamma_o = \frac{\omega_p^2}{\sqrt{8}\omega_o} \frac{a_o}{(1+a_o^2/2)^{1/2}} \tag{2}$$

is the the temporal growth rate. Because the laser pulse changes shape as the instability grows with distance, FRS for relativistic intensity laser pulses is a highly nonlinear process.

When the green probe pulse propagates through the plasma, collective Thomson scattering from the wakefield plasma wave causes multiple side bands to appear in the spectrum of the forward scattered probe light. For P = 3 TW and a backing pressure of 100 psi, Fig. 1 shows the appearance of first and second order Thomson scattered satellites which are separated by the plasma frequency $\omega_p = 2.7 \times 10^{14}$ s^{-1}. The amplitude of the plasma wave can be determined from both the absolute and relative scattering efficiency of the Thomson side bands. For collective Thomson scattering, the absolute scattering efficiency is given by [13]:

$$\frac{P_s}{P_o} = \frac{1}{4}(\delta n_e)^2 r_o^2 \lambda_o^2 L^2 \frac{\sin^2(\Delta k L)}{(\Delta k L)^2} \tag{3}$$

where r_o is the classical electron radius, λ_o is the wavelength of the incident light, L is the interaction length, $\Delta k = k_o - k_s \pm k_p$ is the wavevector mismatch, $k_p = \omega_p/v_p$, and the phase velocity v_p of the plasma wave is equal to v_g of the pump pulse. For direct forward scatter of the first anti-Stokes sideband ($\Delta k = 8 \times 10^3$ m^{-1}) and L equal to the confocal beam parameter (L=430 μm), the phase mismatch factor F = $\sin^2(\Delta k L)/(\Delta k L)^2 = 0.02$ and the amplitude of the plasma wave is determined to be $\alpha = 0.08$. Similarly, from the first Stokes line, α=0.06. Applying the same analysis to the second order satellites, the amplitude of the second harmonic of the plasma wave is $\delta n_2/n$=0.01. From harmonic wave analysis [14], the fundamental amplitude of the plasma wave is related to its second harmonic by $\alpha \sim (\delta n_2/n)^{1/2}$, or $\alpha = 0.1$ for the present conditions at 100 psi backing pressure. Using harmonic analysis and Eq. 2, the relative amplitude of the first and second order Thomson satellites can be used to determine α without specifying L: $\alpha \sim (P_2/P_1)^{1/2}$, or α=0.1. Thus, the absolute and relative scattering efficiencies yield plasma wave amplitudes that are in good agreement. For a Gaussian laser focus, the peak amplitude is estimated to be α_p=0.2.

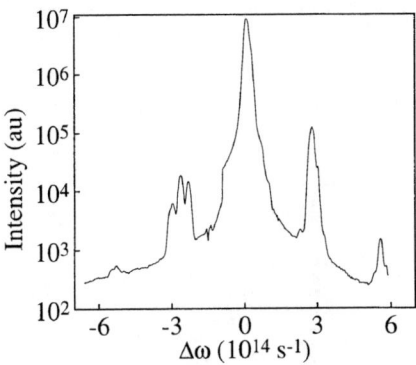

Figure 1: Spectrum of the Thomson scattered probe light for a helium backing pressure of 100 psi, P_{pump}= 3 TW, and Δt=0.

4 Wakefield temporal characterization

By measuring Thomson scattering from the probe pulse as a function of the delay between the pump and probe, the temporal envelope of the wakefield oscillations can be recorded. Fig. 2a shows the change in the spectrum of the probe pulse as the delay (Δt) between a 600 mJ (1.5 TW) pump and a 15 mJ probe pulse is varied from - 1 to 3 ps. Only the first order Thomson scattered satellites are shown. When the Thomson scattered satellites first become observable ($\Delta t \sim -700$ fs), their frequency shift is $\Delta \omega = 3 \times 10^{14}$ s^{-1}. The frequency shift then increases gradually and becomes fixed at $\Delta \omega = 3.3 \times 10^{14}$ s^{-1} for $\Delta t \geq 0$, because it takes a finite time for the helium to become fully ionized. Simultaneously, a blue shifted wing appears on the green probe pulse because of the rapid increase in the electron density [15]. This temporal position of the ionization front is consistent with a calculation of field ionization of helium for our pump pulse parameters.

At most delay times, the amplitude of the anti-Stokes satellite is larger than that of the Stokes due to more favorable phase matching of the anti-Stokes for direct forward scatter. Using the scattering efficiencies of the Stokes and anti-Stokes sidebands and Eq. 3, Fig. 2b shows the plasma wave amplitude as function of the probe delay time. The plasma wave is measured to have a peak amplitude $\alpha = 0.1$, which corresponds to a maximum longitudinal field of $E_z = 56$ GV/m for a cold, nonrelativistic fluid. Under the current tight focusing conditions, the maximum radial electric field is $E_r = 2E_z/k_p r_o = 0.2 E_z$. Large radial fields can

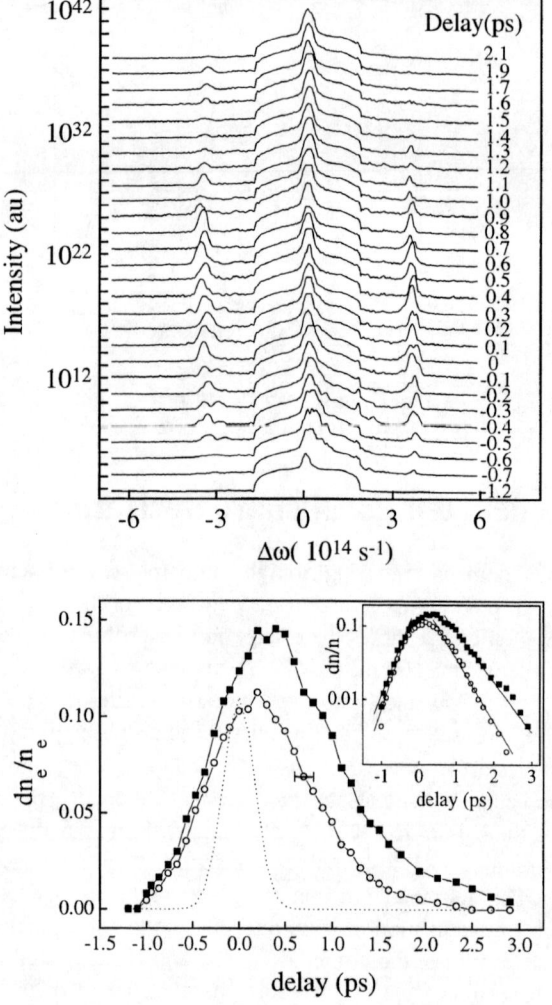

Figure 2: a) Forward scattered probe spectra at a helium backing pressure of 180 psi as a function of the delay between the 1.5 TW pump. b) Plasma wave amplitude determined from the scattering efficiency of the Stokes (filled squares) and anti-Stokes (open circles) satellites as a function of delay. The dotted line indicates the 400 fs pump pulse. The inset shows the exponential fits for the wakefield growth and decay.

give rise to an increase in the transverse emittance of the accelerated electron beam. Fig. 2b shows that the wakefield amplitude maximizes at the end of the pump pulse ($\Delta t = 300 \pm 100 fs$) and lasts for approximately 2 ps longer. The leading edge of the wakefield rises sharply due to the exponential growth rate of FRS. Note that the probe pulse duration (~ 300 fs) is much longer than the plasma period ($\tau_p = 21$ fs for $n_e = 3 \times 10^{19}$ cm^{-3}) in the present experiment. Therefore, we do not resolve individual wakefield oscillations [16].

4.1 Wakefield growth

As shown in Fig. 2b, the onset of the plasma wave occurs ~ 1 ps before the peak of the pump pulse. This observation is contrary to recent 2D simulations which show the plasma wave growing closer to the peak of the pulse [10, 12]. In the simulations, the onset of FRS occurs when the leading edge of the pulse is steepened by backward Raman scattering and pump depletion. In the present experiments, FRS starts near the position of the ionization front. The ponderomotive force from an ionization front can create a large amplitude noise source to seed FRS [17, 18]. The ionization front travels at the phase velocity of the plasma wave and creates a noise source that scales as $\delta n_{si}/n \sim a_o^2(\psi_i)/4$, where ψ_i is the position in the pulse where the ionization threshold occurs [19]. For single and double ionization of helium, $\delta n_{si}/n = 10^{-4}$ and 10^{-3}, respectively, which is larger than the noise source due to intensity gradients of the laser pulse ($\delta n_s/n \sim 10^{-6}$) [10]. Consequently, future numerical simulations should include ionization effects.

From the measurements in Fig. 2b, the wakefield growth and decay rates can be determined. Fitting an exponential to the growth of the plasma wave, the deconvolved growth rate is determined to be 3.5 ± 0.3 ps^{-1} from the anti-Stokes Thomson signal, and 3.3 ± 0.3 ps^{-1} from the Stokes signal. For the current experimental parameters ($\omega_p = 3.3 \times 10^{14}$ s^{-1}, $a_o = 1$), the temporal growth rate of the forward Raman scattering instability ($\gamma_o = 18$ ps^{-1}) is over estimated if the peak intensity of the laser pulse is substituted in the expression for γ_o, since the theoretical derivation for γ_o assumes a constant amplitude pulse. In reality, exponential growth actually takes place during the leading edge of the pulse where $a_o \ll 1$. For the value of a_o that is 400 fs before the pump pulse maximum, $\gamma_o = 8$ ps^{-1}. Since the Rayleigh time $\tau_r = z_r/c$ is longer than the pulse width ψ/c in Eq. 1, the growth of FRS is approximately given by $\delta n = \delta n_s I_o(2\gamma_o\sqrt{\psi\tau/c})$. Evaluation of the Bessel function gives a rise time of 9 ps^{-1} when a_o is allowed to follow the temporal pulse shape and the interaction length is equal to $2z_r$. A number of factors can cause the measured growth rate to be slightly lower than the calculated growth of four wave resonant FRS. Because FRS is a convective instability, the actual growth rate changes at each delay time (or position ψ). Furthermore, FRS evolves into different regimes - four wave nonresonant, three wave, and self-modulation - as time progresses, each with its own growth rate [19]. For example, the ratio for 1D growth of FRS (Γ_{1D}) to that for 3D self modulation (Γ_{3D}) is given by : $\Gamma_{1D}/\Gamma_{3D} = (k_p^2 r_o^2/2k_o^2)^{1/3}$ [4]. In the present experiments, this ratio is unity, so 3D instabilities will compete with 1D FRS. Since the plasma wavelength ($\lambda_p \sim 6\,\mu m$) is comparable to r_o, the plasma wave is near the 3D limit where it has been shown that 1D FRS cannot be resonantly driven because of the complicated three dimensional shape of the plasma wave. [10, 12].

To further investigate the growth of the wakefield, Thomson scattering was measured as a function of the gas jet backing pressure and the peak power of the pump pulse. Fig. 3a shows variation in the Thomson spectra for a fixed laser power (P=3 TW) as the helium backing pressure varies from 30 to 180 psi. As expected, the frequency separation between the satellites, given by ω_p, increases as the square root of the backing pressure. For 3 TW pump power, the first anti-Stokes Thomson satellite appears for a backing pressure of 40 psi. From the measured frequency shift of the satellite, the critical power for self focusing at this backing pressure is $P_c = 1.4$ TW ,and thus $P/P_c = 2.1$ and $\gamma_o = 9$ ps^{-1}. In a similar pressure scan conducted at P=1.7 TW, the first Thomson satellite appeared at a backing pressure of 80 psi where P/P_c=1.9 and $\gamma_o = 10$ ps^{-1}. As the peak power of the pump pulse increases, Fig. 3b shows the change in the Thomson spectra for a fixed backing pressure (180 psi) and at a fixed delay time (Δt=0). The first Thomson scattered satellite appears at P=0.78 TW, or where P/P_c=1.7 and $\gamma_o = 14$ ps^{-1}. At the threshold for Thomson scattering, the ratio P/P_c decreases as the backing pressure (and hence electron density) increases. This trend is evident from Eq. 1 which indicates that the threshold for FRS depends on ω_p/ω_o and P/P_c [8]. For both the pressure and power scan in Fig. 3, $\gamma_o \sim 10$ ps^{-1} even though P/P_c and the backing pressures are different at threshold. The growth rate of the plasma wave as determined by the FRS threshold measurements is in reasonable agreement with the rate obtained from the delay scan in Fig 2b. These observations are also in agreement with those of Ref. [6] which showed that Raman satellites from the pump pulse first appear at P/P_c= 0.5 and $\gamma_o = 6$ ps^{-1} for a helium backing pressure of 150 psi. This indicates that our experiment is in the multidimensional regime of self-modulation [21, 22], as distinguished from previous experiments [3, 4, 5] which were interpreted within the context of 1D FRS.

As the laser power increases, Fig. 3b shows that the width of the sidebands

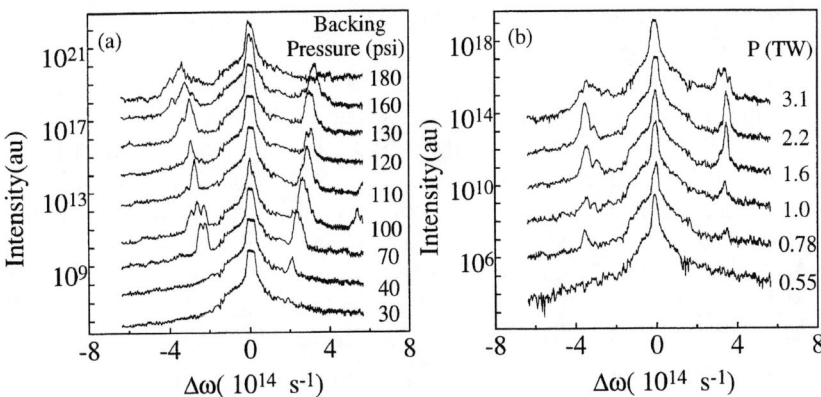

Figure 3: a) Pressure dependence of the Thomson scattered probe light for P=3 TW and $\Delta t = 0$. b) Variation in Thomson scattering with pump pulse power for a fixed backing pressure of 180 psi and $\Delta t = 0$.

increases by a factor of 2, indicating a loss of coherence of the plasma wave. Although sideband broadening has been reported and attributed to wavebreaking [5], the present observations and Ref. [6] are the first showing broadening and spectral modulation of both Thomson and Raman side bands. A number of processes could give rise to the modulations observed on the scattered sidebands. The high frequency modulations may come from phase matching effects in the scattering efficiency. However, the frequency modulation due the sinc^2 factor in Eq. 3 ($\delta\omega_{sinc} \sim 10^{12}$ s^{-1}) is smaller than the observed modulation frequency $\delta\omega = 3 \times 10^{13}$ s^{-1}. Similar modulation has been observed in backscattered Raman spectra and attributed to ion plasma wave fluctuations [20]. Lastly, simulations show that relativistic self phase modulation can produce spectral modulations with the experimentally observed frequency modulation. However, experimental evidence for this mechanism is inconclusive thus far.

4.2 Wakefield decay

The decay of the wakefield is caused by the conversion of collective plasma wave energy into particle energy. In Fig. 2b, the exponential decay rate of the wakefield is 1.6 ± 0.1 ps^{-1} and 1.9 ± 0.2 ps^{-1} as determined from the Stokes and anti-Stokes Thomson signals, respectively. If the observed damping of the wakefield is due to beam loading [23], then the energy in the plasma wave should be transferred to the accelerated electrons. The energy of the plasma wave is approximately given by : $W_{wave} = \varepsilon_o E_z^2 AL/2$, where $A = \pi r_o^2$ and $L = 2z_r$. For $\alpha = 0.1$, $W_{wave}=$ 1.6 mJ. Since this compares well to the energy of 10^9 electrons being accelerated to 2 MeV, or $W_{beam} = 1$ mJ, it is likely that beam loading and particle acceleration contribute to wakefield decay.

A complementary way of discussing beam loading is by the concept of Landau damping of trapped electrons. Such trapped electrons come from hot electrons generated by Raman back-and side-scatter, and above threshold ionization (ATI). For $a_o = 1$, ATI plasma heating is estimated to yield a plasma with $T_e \sim$ 80 keV [24], while Raman back- and side scattering generate hot electrons with $T_e \sim 4$ keV [25] and $T_e \geq 10$ keV [26], respectively. If the damping of the plasma wave is dominated by the acceleration of trapped electrons, the rate of decrease of W_{wave} must be equal and opposite to the rate of increase of W_{beam}. From linear theory [27], the Landau damping rate for trapped particles is approximately equal to $\gamma_L = L_d/c$, where L_d is the dephasing length. Under the present conditions, $\gamma_L = 2$ ps^{-1}, which is consistent with the measured wakefield decay rate of 1.9 ps^{-1}. Thus, the wakefield decays on the time scale that trapped particles exchange energy with the plasma wave. Damping of relativistic plasma waves due to nonlinear Landau damping has also been observed in simulations of the resonant laser plasma accelerator [28].

Beam loading can also be analyzed by considering the superposition of the laser induced wakefield and the wakefield generated by the accelerated particle bunch itself [23]. The maximum number of particles that can be accelerated is approximately given by the number of electrons required for the electron bunch generated wakefield to cancel the laser induced wakefield. For a short electron bunch, the maximum number of particles N_{max} is given by [23]:

$$N_{max} = 5 \times 10^5 \alpha A \sqrt{(n_o)} \qquad (4)$$

where n_o is the electron density in cm^{-3} and A is the effective cross sectional area of the beam in cm^2. Under the present conditions of $n_o = 3.3 \times 10^{19}$ cm^{-3}, $A = \pi r_o^2$, and $\alpha = 0.15$, $N_{max} = 10^9$, which is in good agreement with the measured value of 2×10^9 particles. If N_{max} particles are loaded into a given acceleration phase of the wakefield, all subsequent wakefield oscillations will be cancelled since all of the wave energy is absorbed by the particles. Experimentally, the wakefield is observed to damp over many wakefield oscillation periods ($\tau_p \sim 20 fs$). Consequently, the laser induced wakefield is diminished by contributions from many accelerated particle bunches, each of which contains a number of particles N_b less than N_{max}. Since the number of acceleration buckets b occupied during the decay of the wakefield is $b \sim 2(ps)/\tau_p \sim 100$, the number of accelerated particles in each bunch is approximately $N_b = N_{max}/b = 10^7$ particles if it is assumed that each particle bunch contributes equally to the wakefield decay.

5 Summary

In summary, we have determined that the rise time of the self-modulated wakefield is in agreement with that expected from the growth of the FRS instability and that the decay of the wakefield is consistent with beam loading or Landau damping of the plasma wave. The onset of FRS is consistent with an ionization front induced plasma wakefield, and the detection of a large amplitude plasma wave by Thomson scattering occurs near the critical power for relativistic self-focusing. Channeling of the laser pulse due to relativistic self-focusing was not conclusively observed in the present experiments since the plasma length (defined by the width of the gas jet) was nearly equal to the confocal parameter of the laser beam. Using a gas jet that produced a longer interaction length, channeling of the laser pulse was recently observed over 750 μm [29].

Acknowledgements

This work was supported by DOE Grant No. DEFG03-96-ER-40954, DOE/LLNL subcontract No. B307953, and the NSF STC PHY 8920108. M. C. D. acknowledges a Faculty Research Assignment from the University of Texas.

References

[1] T. Tajima and J. Dawson, Phys. Rev. Lett **43**, 267 (1979)

[2] P. Sprangle *et al.*, Phys. Rev. Lett **69**, 2200 (1992); T. M. Antonsen and P. Mora, Phys. Rev. Lett **69**, 2204 (1992); N. E. Andreev *et al.*, JETP Lett., **55**, 571 (1992).

[3] C. A. Coverdale *et al.*, Phys. Rev. Lett **74**, 4659 (1995).

[4] K. Nakajima *et al.*, Phys. Rev. Lett **74**, 4428 (1995)

[5] A. Modena *et al.*, Nature **377**, 606 (1995).

[6] D. Umstadter *et al.*, Science **273**, 472 (1996).

[7] P. Maine *et al.*, IEEE J. Quantum Electron. **24**, 398 (1988).

[8] W. B. Mori *et al.*, Phys. Rev. Lett **72**, 1482 (1994).

[9] C. I. Moore *et al.*, Bullet. Amer. Phys. Soc. **40**, 1797 (1995); A. Ting *et al.*, these proceedings.

[10] C. D. Decker *et al.*, IEEE Trans. Plasma Sci. **24**, 379 (1996); C. D. Decker *et al.*, Phy. Rev. E. **50**, R3338 (1994).

[11] S. V. Bulanov *et al.* Phys. Rev. Lett **74**, 710 (1995).

[12] K.-C. Tzeng *et al.*, Phys. Rev. Lett **76**, 3332 (1996).

[13] R. E. Slusher and C. M. Surko,Phys. Fluids **23**, 472 (1980).

[14] D. Umstadter *et al.*, Phys. Rev. Lett **59** 292, (1987).

[15] W. M. Wood *et al.*, Phys. Rev. Lett **67**, 3523 (1991).

[16] C. W. Siders *et al.*, Phys. Rev. Lett **76**, 3570 (1996); J. R. Marques *et al.*, ibid., 3566 (1996).

[17] W. B. Mori and T. Katsouleas, Phys. Rev. Lett **69** 3495 (1992).

[18] T. Tajima and D. Fisher, Phy. Rev. E. **53**, 1844 1996).

[19] C. D. Decker *et al.*, Phys. Plasmas **3**, 1360 (1996).

[20] A. Ting *et al.* , Opt. Lett. **21**, 1096 (1996).

[21] E. Esarey *et al.*, Phys. Rev. Lett **72**, 2887 (1994).

[22] N. E. Andreev *et al.*, Phys. Plasmas **2**, 2573 (1995).

[23] T. Katsouleas *et al.*, Particle Accelerators **22**, 81 (1987).

[24] B. M. Penetrante and J. N. Bardsley, Phys. Rev. A **43** 3100 (1991).

[25] S. C. Wilks *et al.*,Phys. Plasmas **2**, 264 (1995).

[26] Y. Kishimoto *et al.*, to be published.

[27] D. R. Nicholson, *Introduction to Plasma Theory* (Krieger, Florida, 1992).

[28] D. Umstadter *et al.*, "Resonantly driven laser plasma electron accelerators," in Proceedings of the 1994 Workshop on Advanced Accelerator Concepts, P. Schoessow, ed. (AIP No. 335) (American Institute of Physics, New York, 1995).

[29] S.-Y. Chen *et al.* , these proceedings.

Micro-bunching Diagnostics for the IFEL by Coherent Transition Radiation

Y. Liu and D.B. Cline

Center for Advanced Accelerators,
University of California, Los Angeles, CA 90095

X.J. Wang, M. Babzien, J.M. Fang and V. Yakimenko

Brookhaven National Laboratory, Upton, NY 11973

Abstract. Here, we propose an effective method for detecting micro-bunching effects (10-fs bunch length) produced by the IFEL interaction, by measuring the CTR spectrum. The pre-bunching of an initially energy-modulated e^- beam passing through a wiggler (IFEL interaction) is studied. Simulation shows that more than 40% of electrons are pre-bunched in the micro-bunches. The longitudinal distribution of an optically pre-bunched beam is Fourier analyzed to find the dominant harmonics contributing to the CTR. The CTR spectrum is calculated analytically for the IFEL situation. A detection system has been built to demonstrate this technique.

1. INTRODUCTION

The Inverse Free Electron Laser (IFEL) acceleration[1,2] has been observed at the Accelerator Test Facility (ATF) at Brookhaven National Laboratory. Recently it was decided to study the bunching by the IFEL for the possible use as an injector for a high energy Inverse Cerenkov Accelerator (ICA).[3] Since the IFEL operates in a vacumm environment, it is possible to provide a high quality, pre-bunched e^- beam for the next acceleration stage. As we know, ICA operates in a phase matching medium with diamond windows which are used to isolate the medium from the vacumm environment. These additional materials will scatter electrons and change e^- beam emittance.

A basic approach for such a two-stage system is to send an e^- beam into a laser-driven pre-buncher (IFEL wiggler), whose output feeds directly into the laser accelerator (ICA gas cell). The IFEL stage is needed to maximize the number of electrons maintaining a specific phase relationship with the optical mode required for optimum acceleration. An optical mirror will be mounted in the middle of the ICA gas cell which could be inserted into the interaction region by remote control. It is a diagnostic element using Coherent Transition Radiation (CTR) to monitor micro-bunchingin a planned IFEL pre-buncher, ICA accelerator experiment.

We have previously studied the use of a CTR to predict the bunching of the ICA.[4] In this paper the IFEL self-bunching process (without the space-charge effect) has been simulated. Photon production from the CTR is estimated according to the current IFEL setup at the ATF. Furthermore, a feasible CTR diagnostic system to detect and measure the micro bunches is presented.

2. OPTIMUM SELF-BUNCHING DISTANCE AND ELECTRON DISTRIBUTION

The e^- beam passing through the wiggler and interacting with the high intensity CO_2 laser leaves the interaction region with a distinct velocity distribution pattern which, propagated over a certain distance, results in self-bunching of the e^- beam. In our bunching experiment, this characteristic distance is optimized to be about 45-50 cm, and we expect to observe strong bunching on a scale of several microns. Since there is no external electromagnetic field present in the drift space, the effects of the space charge within the e^- beam will dominate and they should be considered. However, previous study using PARMELA to simulate phase-space evolution in the drift space for ICA shows that particles at the head of the bunch tend to gain energy, while particles from the tail are decelerated.[4] In effect, the energy-phase space is rotated in the counter-clockwise direction. The amount of this local rotation is weighted by the longitudinal charge distribution gradient. Therefore, the strongest effect occurs near the center of the distribution (smearing). Under strong space-charge conditions (1-nC bunch) the effect of self-bunching is still present. The bunching peak is slightly 'washed out' by the space-charge defocusing and smearing — at about the 10% level (FWHM) compared with the case of no space charge.

In the current setup the maximum charge delivered to IFEL wiggler at ATF is up to several-hundred pC. It is far less than our simulation charges. Thus, we should not worry about space-charge effect too much in the IFEL self-bunching study. By optimizing laser power and magnetic field at 40 MeV e^- beam energy we predict that more than 40% of electrons are bunched within several microns at the optimum

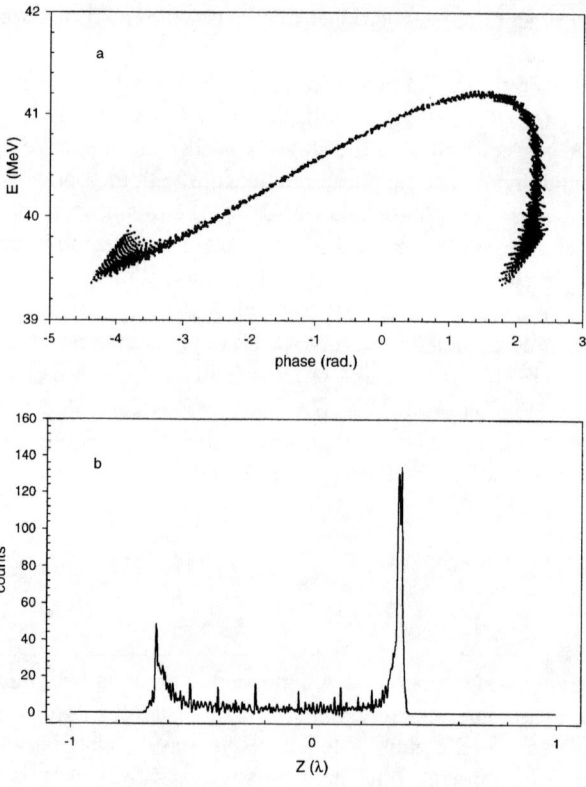

Figure 1: (a) The electron distribution at predicted optimum bunching distance (50 cm) in energy-phase space. $\lambda = 10.6 \mu m$ (b) The electron distribution in real space.

bunching distance. Figure 1b shows the electron distribution in the longitudinal direction at predicted optimum bunching distance. Figure 1a shows the electron distribution in energy-phase space. The code we used to simulate the electron energy modulation is a 1D IFEL Fortran code written by J. Gallardo at BNL.[5]

3. COHERENT TRANSITION RADIATION

Backward transition radiation generated by a particle crossing the interface of the two media at an oblique angle, θ, is characterized by the following distribution:[6]

$$\frac{d^2u}{d\omega d\Omega} = \frac{e^2}{4\pi^2 c} \frac{|rt|^2 \sin^2\theta}{(1-\beta\cos\theta)^2}, \qquad (1)$$

where $r\prime$ is the Fresnel reflection coefficient for the mirror. However, CTR is a collective effect produced by a large ensemble of electrons being in phase with each other. The total number of photons radiated is highly enhanced and the total radiation distribution becomes[7]

$$\frac{d^2U}{d\omega d\Omega} = [N + N(N-1)F(\omega,\theta)]\frac{d^2u}{d\omega d\Omega}, \qquad (2)$$

where

$$F(\omega,\theta) = |f(\omega,\theta)|^2 = \left|\int\int\int f(r,z)\exp(-i\vec{k}\cdot\vec{x})d^3x\right|^2 \qquad (3)$$

is a bunching factor, containing information about the electron distribution. The coherent effect scales like N^2 compare to the incoherent part, which scales linearly with the electron number, N. Since N is about 10^9 at least in our case, therefore Eq.(2) is simplified as

$$\frac{d^2U}{d\omega d\Omega} \approx N^2 F(\omega,\theta)\frac{d^2u}{d\omega d\Omega}, \qquad (4)$$

This technique has been used at several synchrotron light sources to measure a single bunch.[8,9,10] The multiple bunch case for fundamental wavelength (first harmonic wavelength) also has been estimated.[4,11]

In multiple bunches the total coherent intensity depends on electron distribution in each bunch and number of bunches. To estimate the backward CTR coming from a 45^0 mirror for the IFEL experiment, we assume the following set of conditions describing the experimental setup at the ATF. (1) The initial e^- beam distribution (before entering the IFEL wiggler, where the electrons are modulated by the CO_2 laser) is assumed to be a smooth bi-Gaussian in r and z. (2) The energy modulation imposed in the interaction region is strong enough and the affected electrons should self-bunch at a certain distance outside the wiggler. Under such assumptions we should expect that the modulated electron distribution evolves in a drift space (z) according to the following formula[4,11]:

$$\begin{aligned} f(r,z) &= g(r)h(z) \\ &= \left[\frac{\exp(-r^2/2\sigma_r^2)}{2\pi\sigma_r^2}\right]\left[\frac{\exp(-z^2/2\sigma_z^2)}{(2\pi)^{1/2}\sigma_z^2}\right]\left[1+\sum_{n=1}^{\infty}b_n\cos(nk_rz)\right], \end{aligned} \qquad (5)$$

where σ_z and σ_r present the longitudinal and the vertical e^- beam size, k_r is the wave-number of the CO_2 laser modulation and b_n are the Fourier coefficients of the longitudinal electron density distribution.

By combining Eq. (3) with Eq. (5) the transverse part F_T and the longitudinal part F_L are given by the following expressions:

$$\begin{aligned} F_T(\omega,\theta) &= \left|\int\int g(r)\exp(-ikr\sin\theta\cos\phi)rdrd\phi\right|^2 \\ &= \exp(-(k\sigma_r\sin\theta)^2), \end{aligned} \qquad (6)$$

$$F_L(\omega,\theta) = \left|\int h(z)\exp(-ikz\cos\theta)dz\right|^2$$

$$= \left|\exp\left(-\frac{(k\sigma_z\cos\theta)^2}{2}\right) + \sum_{n=1}^{\infty}\left(\frac{b_n}{2}\right)\left[\exp\left(-\frac{\sigma_z^2}{2}[k\cos\theta - nk_r]^2\right)\right.\right.$$
$$\left.\left. + \exp\left(-\frac{\sigma_z^2}{2}[k\cos\theta + nk_r]^2\right)\right]\right|^2. \tag{7}$$

Since the micro-bunch peaks are narrow compared to their separation distance ($k_r\sigma_z \gg 1$), only the contributions from near each harmonic wavelength could be significant. Thus, we can simplify Eq. (7) referring to each harmonic wavelength as follows:

$$F_L(\omega,\theta) \approx \left(\frac{b_n}{2}\right)^2 \left[\exp\left(-\sigma_z^2(k\cos\theta - nk_r)^2\right)\right]. \tag{8}$$

The total number of the CTR photons in the narrow band around the harmonic wavelength is given by the following expression:

$$\frac{d^2N_{ph}}{dkd\theta} \approx \frac{\alpha|rt|^2}{2\pi}\left(\frac{Nb_n}{2}\right)^2 \frac{\sin^3\theta}{(1-\beta\cos\theta)^2}\frac{1}{k}\exp\left[-(k\sigma_r\sin\theta)^2 - \sigma_z^2(k\cos\theta - nk_r)^2\right], \tag{9}$$

where $\alpha = e^2/\hbar c$. To find the angular distribution of the CTR photons in the narrow band (approximately one percent bandwidth) around this frequency, we should integrate Eq. (9) over the described frequency range. Due to a very narrow frequency band, the value of k, in the fourth term of Eq. (9), could be set to nk_r and then the integration limits could be extended from $-\infty$ to ∞. Finally, we get the angular distribution of the CTR photons around each harmonic frequency, in the following compact form:

$$\frac{dN_{ph}}{d\theta} \approx \frac{\alpha|rt|^2}{2\pi}\left(\frac{Nb_n}{2}\right)^2 \frac{1}{k_r\sigma_z}\frac{\sin^3\theta}{(1-\beta\cos\theta)^2}\left(\frac{1}{n}\right) \times$$
$$\left[\frac{\pi}{((\sigma_r/\sigma_z)\sin\theta)^2 + \cos^2\theta}\right]^{1/2} \exp\left[-\frac{(nk_r\sigma_r\sin\theta)^2}{((\sigma_r/\sigma_z)\sin\theta)^2 + \cos^2\theta}\right]. \tag{10}$$

In general $\sigma_r \ll \sigma_z$ and $\theta \ll 1$. Using this approximation, Eq. (10) reduces to the following simple formula:

$$\frac{dN_{ph}}{d\theta} \approx \frac{\alpha|rt|^2}{8\sqrt{\pi^3}}\left(\frac{k_r\sigma_z}{\pi}\right)\left(\pi\frac{Nb_n}{k_r\sigma_z}\right)^2\left(\frac{1}{n}\right)\frac{\sin^3\theta}{(1-\beta\cos\theta)^2}\exp\left[-(nk_r\sigma_r\sin\theta)^2\right]. \tag{11}$$

The physical meaning of Eq. (11) is quite clear. The second term represents the full length of the macro-bunch in units of the modulation wavelength λ_r. The third term is the square of the total number of electrons within the micro-bunch (the laser modulation 'chops' the macro-bunch into a number of micro-bunches). The fourth term describes the contribution of the single-electron transition radiation and the last term reflects the transverse distribution of electrons. n is a harmonic number. The coherent contribution from higher harmonic will be significantly weaker than fundamental wavelength. By Fourier analysis of the CTR spectrum at the optimum bunching distance, the coefficients b_n are easily found. Since the micro-bunches are very narrow the contribution from the higher harmonic components may be detectable in our experiment. Here, we estimate the number of photons generated by each harmonic, separately.

For the IFEL experimental conditions, a 40-MeV e^- beam pulse ($2\sigma_z \sim 3000$ μm) is about 10 ps long. Its σ_r at the focal waist is about 300 μm. After passing through 50 cm of the drift space the beam size will grow and reach the transverse size given by $\sigma_r \sim 390$ μm. The total number of electrons in a single macro-bunch is assumed to be $N \sim 1.25 \times 10^9$, which is equivalent to 200 pC of total charge. From the spectrum analysis, the Fourier coefficients are evaluated as follows: $b_1 = 1.134$, $b_2 = 0.978$, $b_3 = 0.867$ and $b_4 = 0.778$. Since the total photons at a certain wavelength will significantly decrease with increasing harmonic number (see Eq. (11)), therefore we confine our consideration to the first four harmonics only. The total number of photons at these wavelengths is evaluated as follows: $N_1 \sim 5.90 \times 10^{11}$, $N_2 \sim 1.79 \times 10^{10}$, $N_3 \sim 1.97 \times 10^9$ and $N_4 \sim 3.86 \times 10^8$. These values imply that the IFEL CTR signal is quite strong. The possibility of detecting the first, second, even third or fourth harmonic frequencies with a narrow bandwidth detector gives more flexibility to our measurement. Theoretically predicted angular distributions for the first four harmonics are shown in Figure 2. As a comparison, incoherent transition radiation distribution is shown in Figure 2 also.

4. DIAGNOSTIC SETUP FOR THE IFEL

At the ATF, beam-line 2 is provided for the IFEL experiment. The detection system has been assembled and is located downstream near the wiggler exit. The basic experimental setup can be described as follows. A 45^0 mirror (CTR mirror) held by a 6-way cross is inserted in the e^- beam path to generate backward CTR and transport the CTR light through a ZnSe window into the detecting system. The CTR light will be focused by a 3" diameter remote-controlled parabolic mirror into a cooled detector (InSb) with 1×1 mm^2 sensitive area. The CTR mirror can be moved in and out easily by an actuator. The 6-way cross connects with a pair of bellows which allow the mirror to travel back and forth for 40 cm along the e^- beam axis without disturbing the vacumm environment. The whole system including the 6-way cross sits on a movable table driven by a digitized remote-controlled stepping motor. A strip-line detector is placed between the upstream bellows and the exit of wiggler to provide the e^- beam charge information. The schematic of

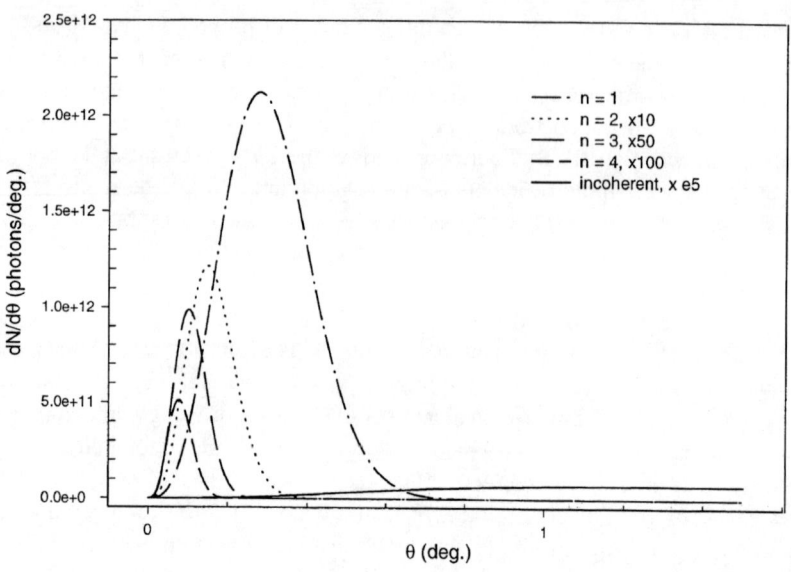

Figure 2: Angular distribution of the CTR photons within 1% bandwidth. (a) $n = 1$, first harmonic frequency, peak at $\theta_m = 0.28^0$; (b) $n = 2$, second harmonic frequency, peak at $\theta_m = 0.15^0$; (c) $n = 3$, third harmonic frequency, $\theta_m = 0.10^0$; (d) $n = 4$, fourth harmonic frequency, $\theta_m = 0.06^0$; and (e) incoherent, $\theta_m = 1.27^0$.

the complete detection system is shown in Figure 3. Control and data collection will be performed at the ATF control room.

The disadvantage of measuring the backward CTR is the fact that the intensity may be slightly weaker than for the forward CTR, diminished by the factor $|rt|^2$. Due to IFEL setup CO_2 laser will directly hit the CTR mirror and be transported into the detection system. This implies that the first harmonic (10.6μm) CTR will not be useful to identify bunching effect. A filter should be placed between ZnSe window and parabolic mirror to block the CO_2 laser light. Only second and higher harmonic frequencies CTR should be detected. In principle it is not necessary to use the filter since the detector is not sensitive to this wavelength. But the laser intensity is too strong without attenuation, and damage the detector.

By moving the CTR mirror back and forth the optimum bunching distance could be located by looking at the micro-bunching effect variation. The adjustment of the input laser power at the fixed e^- beam energy and magnetic field is an alternative way of finding the optimum bunching distance. However, it may affect the bunching

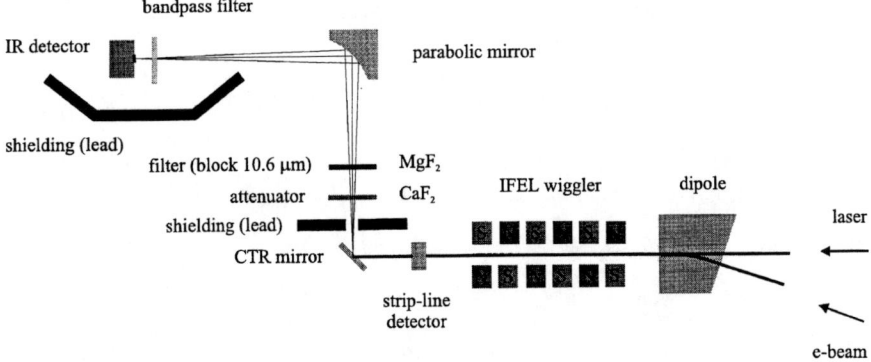

Figure 3: Schematic of the CTR diagnostic setup for the IFEL experiment.

characteristics significantly, which will lead to a complex calibration procedure.

In conclusion, this is a flexible and easy operated detection system. It could help us optimize phase matching condition quickly in planned IFEL/ICA joint experiment. The detection system has been built to demonstrate this technique.

5. ACKNOWLEDGMENTS

This work was supported by the U.S. Department of Energy, Grant No. DE-FG03-92ER40695. We thank I. Ben-Zvi, E.B. Blum, J. Gallardo, W. Kimura, I. Pogorelsky, J. Sandweiss and A. van Steenbergen for their help and discussions. Also we thank J. Skaritka for the engineering design work and B. Cahill, B. Harrington, M. Montemagno and K. Kusche for their technical support.

REFERENCES

1. A. van Steenbergen, J. Gallardo, J. Sandweiss, J.M. Fang, M. Babzien, K. Batchelor, A. Fisher, K. Kusche, R. Malone, I. Pogorelsky, J. Qiu, T. Romoano, J. Sheehan, J. Skaritka, T. Srinivasan-Rao, X.J. Wang, "Inverse Free Electron Laser e^- Acceleration", Proc. BNL/ATF Users Meeting, December, 1995

2. A. van Steenbergen, J. Gallardo, J. Sandweiss, J.M. Fang, "Observation of Energy Gain at the BNL Inverse Free Electron Laser Accelerator", to be pub., PRL

3. W.D. Kimura, et al, *Phys. Rev. Lett.* **74**(4), 546 (1995).

4. Y. Liu, et al., *AIP Conference Proceedings 367*, AIP Press, New York, 445,

5. Private communication.

6. D.W. Rule, et al, *Nucl. Instr. Methods in Phys. Res.* **A296**, 739 (1990).

7. E.B. Blum, U. Happek and A.J. Sievers, *Nucl. Instr. Methods in Phys. Res.* **A307**, 568 (1991).

8. T. Nakazato, et al, *Phys. Rev. Lett.* **63**(12), 1245 (1989).

9. Y. Shibata, et al., *Phys. Rev. E*, **50**(2), 1479 (1994).

10. R. Lai and A.J. Sievers, *Phys. Rev. E*, **50**(5), R3342 (1994).

11. J. Rosenzweig, G. Travish, and A. Tremaine, *Nucl. Instr. Methods in Phys. Res.* **A365**, 255 (1995).

INVERSE TRANSITION RADIATION

L.C. Steinhauer, R.D. Romea, and W.D. Kimura
STI Optronics, Inc., 2755 Northup Way, Bellevue, WA 98004

ABSTRACT

A new method for laser acceleration is proposed based upon the inverse process of transition radiation. The laser beam intersects an electron-beam traveling between two thin foils. The principle of this acceleration method is explored in terms of its classical and quantum bases and its inverse process. A closely related concept based on the inverse of diffraction radiation is also presented: this concept has the significant advantage that apertures are used to allow free passage of the electron beam. These concepts can produce net acceleration because they do not satisfy the conditions in which the Lawson-Woodward theorem applies (no net acceleration in an unbounded vacuum). Finally, practical aspects such as damage limits at optics are employed to find an optimized set of parameters. For reasonable assumptions an acceleration gradient of 200 MeV/m requiring a laser power of less than 1 GW is projected. An interesting approach to multi-staging the acceleration sections is also presented.

I. INTRODUCTION

Transition radiation (TR) is emitted when electrons pass suddenly from one medium into another having a different index of refraction [1,2]. TR arises because the electron self-fields (which are affected by the surrounding medium) must change in passing from one medium to another. The continuity of electromagnetic fields at the interface requires additional fields. These additional fields generally appear as waves (the TR) that propagate away from the interface in the upstream and downstream directions. Diffraction radiation (DR) is a related phenomena except that the modification of the self fields is caused by the presence of a small aperture in a screen through which the electron passes. Here, continuity at the plane of the screen requires additional fields, which propagate as waves away from the screen (the DR).

Both TR and DR extract energy from the electron, causing it to decelerate. In principle the inverse of these processes can be exploited to accelerate electrons [3]. In the inverse process the emitted radiation is replaced by the electromagnetic field of a laser focused onto the electron path. Here the inverses of TR and DR are called inverse-transition acceleration (ITA) and inverse-diffraction acceleration (IDA). An essential feature of ITA/IDA is that the electron-light interaction is bounded between two optical surfaces separated by a characteristic distance.

The properties of the forward and inverse processes are closely related: this is illustrated by the correspondence between their characteristic length scales. Consider TR/DR emission at a particular wavelength, λ. Near the interface, the various angular components of that emission interfere destructively. However beyond a characteristic distance from the interface, the superimposed components organize coherently [1]. This distance, the formation length, is $L_f = \lambda \gamma^2/\pi$, where γ is the relativistic parameter of the particle. The characteristic length scale of ITA/IDA is the slip distance, the distance over which the particle slips by π in the phase of the laser wave that intersects the particle path at the angle θ_L [4,5]:

$$L_\pi = (\lambda/2)/(1/\beta - \cos\theta) \approx \lambda/(\theta^2 + 1/\gamma^2) \qquad (1)$$

where the approximation applies for $\gamma \gg 1$ and $\theta \ll 1$. The correspondence between the forward and inverse processes is evident, since $\pi L_f = L_\pi(\theta_L = 0)$. The close relationship between forward and inverse processes means that one can explore the properties of TR/DR to learn about ITA/IDA.

The ITA/IDA processes offer an attractive method of high-gradient electron acceleration because of the high electric fields that can be produced at the focus of a laser beam. However a powerful principle, the Lawson-Woodward theorem [6,7], rules out laser-particle acceleration in a vacuum for a broad range of conditions. The theorem can be stated as follows. In the far field, any optical beam is the superposition of propagating plane waves. Thus the overall effect of an optical beam on a particle is the superposition of the effects of individual plane waves. Since in vacuum the particle velocity is less than the speed of light, the particle slips relative to the phase of a plane wave. Finally, since the electric field of the plane wave has a periodic dependence on phase, there can be no net laser-electron energy exchange in an unbounded interaction.

A possible way to escape the pessimistic conclusion of the Lawson-Woodward theorem is to impose spatial bounds on the optical beam-particle interaction [8]. If the electron-light interaction in vacuum is limited to distances of the order of L_π then net acceleration should be possible. This is called the "double-boundary" arrangement because it limits the interaction to the region between two surfaces.

In this paper we present the ITA/IDA concept from the perspective both of its basic physics and of its practical embodiment. Sections II-IV establish the principle of the concept from three perspectives. Section II presents the quantum picture of TR and ITA. Here a quantum-mechanical basis for the Lawson-Woodward theorem is established, and the conditions under which the theorem *doesn't* apply are determined, i.e. a bounded interaction. Section III applies a classical analysis to the double-boundary TR case, extending the usual single-boundary case. Here a correspondence with the inverse process, ITA, is established. Section IV treats the double-boundary DR and IDA cases, applying results found elsewhere in the literature. Here the connection with TR/ITA is established, as well as the conditions in which the interaction between light and the *e*-beam is little diminished compared with TR/ITA. Section V changes the emphasis from the principle of ITA/IDA to its practical embodiment. An optimized set of parameters for maximum acceleration gradient is derived subject to reasonable constraints. An interesting approach to multi-staging the acceleration sections is also presented.

II. QUANTUM APPROACH TO TR AND ITA

The quantum picture of TR and its inverse treats the emission (absorption) of a single photon by a free electron. This interaction conserves both momentum and energy. It can be viewed in terms of electron and photon dispersion relations [9]. These form a surface in four-dimensional space (p_x, p_y, p_z, W) that relate the energies to the momenta. In order for a transition to conserve both momentum and energy, the surfaces must intersect at two or more points. On the electron surface, a pair of intersection points represent the initial and final electron states. The electron dispersion relation is $W_e^2 = p_e^2 c^2 + m_e^2 c^4$ where W_e, $p_e = |\mathbf{p}_e|$, and m_e are the energy, momentum, and rest mass respectively, and c is the speed of light in vacuum. If only two momentum components are considered (the "plane" of the interaction, *e.g.* p_{ex}, p_{ez}) then the electron dispersion surface is a hyperboloid of revolution (single sheet) in the three-dimensional space (p_{ex}, p_{ez}, W_e). The photon dispersion relation is a cone (single sheet) with angle $dW_\omega/dp_\omega = v_\omega$ where W_ω and p_ω are the photon energy and momentum, and v_ω is the velocity of the photon. These dispersion surfaces are illustrated in Fig. 1. In a medium with index of refraction exceeding unity, $v_\omega < c$, the two surfaces can intersect at two or more points: this is the case of Cerenkov emission (or inverse Cerenkov acceleration). However in a vacuum, the local "slope" of the hyperboloid dW_e/dp_e is always less than c: *i.e.* the two surfaces cannot intersect at two points. This is a quantum-mechanical statement of the Lawson-Woodward theorem.

The non-intersection of the two dispersion surfaces in Fig. 1 represents a momentum discrepancy in an energy-conserving interaction. This *momentum gap* is illustrated in Fig. 2 for photon absorption by an electron. The large dot represents the initial momentum (electron + momentum), and the large circle represents the locus of final electron momenta with the same energy. For small θ (angle between the photon and initial electron paths) and large γ the *longitudinal momentum gap* (*i.e.* final electron state "A" in Fig. 2) is

$$\Delta p_\shortparallel \approx \frac{\hbar\omega}{2c}\left(\theta^2 + \frac{1}{\gamma^2}\right) \tag{2}$$

where $\hbar$ is Planck's constant and ω is the frequency of the photon. The point "B" in Fig. 2 corresponds to a *lateral* momentum gap, which is somewhat larger.

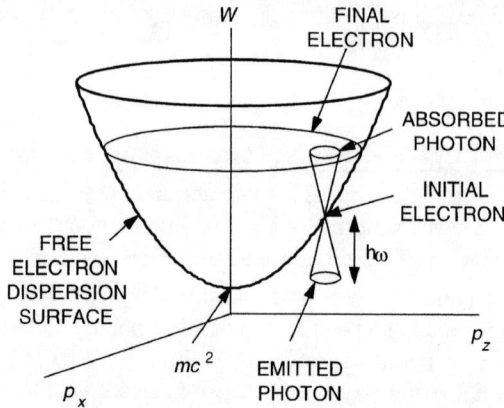

Fig. 1. Dispersion surface perspective of the interaction between a free electron and a photon. The figure illustrates the case of photon absorption.

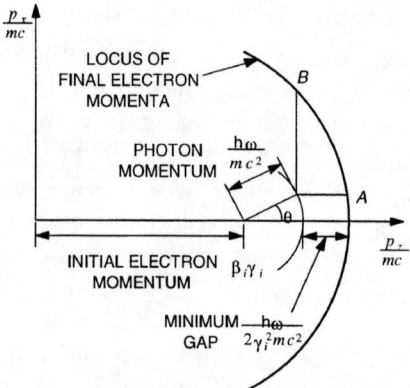

Fig. 2. Illustration of the momentum gap in photon absorption by a free electron. The interaction takes place in (p_x, p_z) momentum space (transverse and longitudinal directions, respectively, with respect to the initial electron motion).

Despite the momentum gap, an electron-photon interaction is possible if a length scale is introduced by material structures. Then the missing momentum is supplied by a virtual photon, representing the interaction of the electron self fields with the bounding surfaces. In effect the surfaces absorb the extra momentum as a recoil. This can be viewed in terms of the uncertainty principle as suggested by Pantell and Piestrup [10]. Identify the uncertainty, δp, with the momentum gap (Eq. 2); then $\Delta p \sim \eta/\lambda$ where λ is the mirror separation and therefore the interaction length. In particular, if $\Delta p = \pi \eta/\lambda$ then $\lambda = L_\pi$, *i.e.* the phase slip distance (Eq. 1). If λ much exceeds L_π then an interaction is still possible but with a much reduced probability. For $\lambda \to \infty$ the probability goes to zero: this is a quantum-mechanical confirmation of the Lawson-Woodward theorem. Note that the agreement between the characteristic length scale in the quantum picture with that in the classical picture is a demonstration of the correspondence principle, *i.e.* the agreement between the classical and quantum-mechanical viewpoints. A detailed quantum-mechanical analysis that confirms this approximate treatment is given elsewhere [11].

The quantum analysis of TR/ITA can be summarized under three points. (1) Using the uncertainty principle, it was shown that the interaction is maximized when that distance is comparable to the classical slip distance, L_π: this verifies the correspondence principle between the quantum and classical TR/ITA processes. (2) The conservation of momentum and energy is satisfied when the virtual photon arising in the interaction with the mirror surfaces is taken into account. (3) The Lawson-Woodward theorem is verified, as well as the conditions when it does not apply, *viz.* when the interaction is bounded.

III. CLASSICAL ANALYSIS OF TR

An analysis of TR in the double-boundary case is performed in order to demonstrate corresponding properties of ITA. In the original single-boundary analysis of TR, Frank and Ginsburg [12] used the following artifice. As an electron moving in vacuum approaches a mirror boundary the field in the vacuum is the sum of that produced of the electron and that of its mirror image moving towards it. After crossing the boundary both the electron and its image cease to exist from the point of view of the field in the vacuum. Thus the crossing of the boundary thus gives rise to exactly the same radiation as that due to a sudden stop at the same point of the electron and its image. This artifice has also been used in later treatments of TR [13]. A natural extension of this artifice can be used to represent the TR emitted by an electron passing through the vacuum space between two mirrors [14]. In this case the TR equals to the radiation generated by *three* electrons: the first moves from $z = -\infty$ and stops abruptly at $z = 0$; the second starts at $z = 0$ and stops at $z = \lambda$; and the third starts at $z = \lambda$ and continues to $z = +\infty$. While in motion all three electrons move at constant speed $v = \beta c$ along the z-axis (β is the relativistic parameter).

The two-boundary case can be analyzed as follows. An electron moving with arbitrary path $\mathbf{r}(t)$ and velocity $\mathbf{v}(t)$ emits radiation in the solid angle $d\Omega$ and frequency range $d\omega$ [15]:

$$\frac{d^2 U_{double}}{d\omega d\Omega} = \frac{e^2}{c}\left|\frac{1}{2\pi\varepsilon}\int_{-\infty}^{\infty} dt\, \mathbf{k}\times\mathbf{v}\,\exp[i(\mathbf{k}\cdot\mathbf{r}-\omega t)]\right|^2 \qquad (3)$$

where e is the electron charge, ε is the permittivity of the medium, $\mathbf{k}$ is the wave vector ($k = |\mathbf{k}| = \omega/c$ in vacuum), and t is time. The squared quantity is the far-field approximation of the vector potential of the emitted radiation. The first and third electrons move in regions (inside the mirrors) where no electromagnetic radiation can propagate since $\varepsilon \to \infty$. In this limit matching at the two boundaries can be ignored, just as it was in the single-boundary TR analysis. Thus in the two-boundary case, one need only consider the region $(0, \lambda)$ and an electron moving along the z axis with the position $z = \beta ct$ within the time range $0 < t < \lambda/\beta c$. Then the radiated energy is

$$\frac{d^2 U_{double}}{dk d\theta} = \frac{2e^2}{\pi}\sin^3\theta \sin^2\left[\frac{k\lambda}{2}\left(\frac{1}{\beta}-\cos\theta\right)\right]\bigg/\left(\frac{1}{\beta}-\cos\theta\right)^2 \qquad (4)$$

where θ is the angle between the wave vector and the electron path. Here $0 \leq \theta \leq \pi$, i.e. wave propagation is possible in both the $+z$ and $-z$ directions. This result can be placed in a form resembling the single-boundary case multiplied by a correction factor. This is done by lumping together the forward ($0 \leq \theta \leq \pi/2$) and backward ($\pi/2 \leq \theta \leq \pi$) moving waves using the change of variable, $\theta \to \pi - \theta$, for the backward waves. Then

$$\frac{d^2 U_{double}}{dk d\theta} = \frac{2e^2 \sin^3\theta}{\pi\beta^2}\left(\frac{1}{\beta^2}-\cos^2\theta\right)^{-2}\cdot\left\{(1+\beta\cos\theta)^2\sin^2\left[\frac{k\lambda}{2}\left(\frac{1}{\beta}-\cos\theta\right)\right]+\right.$$
$$\left.+(1-\beta\cos\theta)^2\sin^2\left[\frac{k\lambda}{2}\left(\frac{1}{\beta}+\cos\theta\right)\right]\right\} \qquad (5)$$

where the range is $0 \leq \theta \leq \pi/2$, and the factor in curly brackets is the correction for two boundaries. For each θ in this range, Eq. 5 is the sum of the forward (θ) and backward ($\pi-\theta$) components.

Figure 3 shows the angular distribution of TR for both single- and double-boundary cases. The double-boundary case resembles the single-boundary case multiplied by an oscillatory interference factor. The modification factor (curly brackets in Eq. 5) is dominated by the first term, which for θ << 1 is approximately $(1+\beta)^2 sin^2[(k\lambda/2)(1/\beta-cos\theta)]$. The average of the lobed sin^2 factor is ~1/2 so that for β ≈ 1 the "average" of the modification factor is {...} ≈ 2. Where this averaging is appropriate double-boundary TR is roughly twice single-boundary TR.

The spectral emission for all angles is shown in Fig. 4 as a function of $k\lambda$, a dimensionless measure of the mirror separation. The double-surface case approaches a constant for $k\lambda > 2\gamma^2$ where the TR is approximately twice the single-surface case (shown as a horizontal bar). This threshold corresponds exactly to $\lambda > L_f$, the formation distance. The fall-off for lower values of $k\lambda$ corresponds to incomplete development of coherent TR.

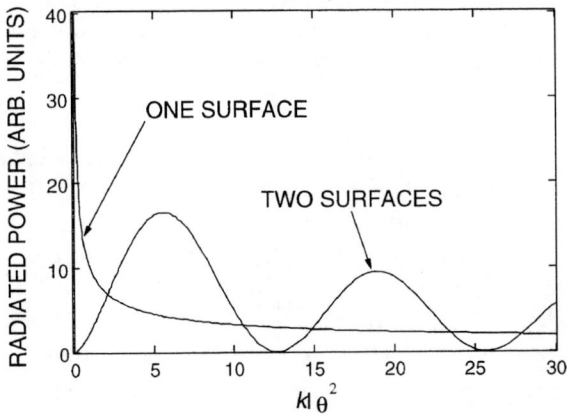

Fig.3. Angular distribution of TR for both single- and double-boundary cases as a function of a dimensionless parameter representing the spacing between two mirrors. This example assumes γ = 1000.

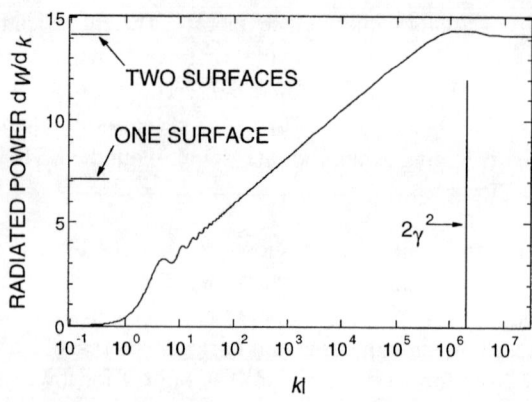

Fig.4. Total TR emission (all angles) for two surfaces as a function of another dimensionless parameter representing the spacing between two mirrors. This example assumes $\gamma = 1000$.

In the inverse process, ITA, a laser of wavelength λ is directed upon the *e*-beam at an angle θ_L. The optimum laser angle is nominally the value of θ at the peak of the first lobe in Fig. 3 (given fixed λ). This occurs when the argument of the dominant first sin^2 in the curly brackets of Eq. 5 is about $\pi/2$, *i.e.*,

$$\frac{k\lambda}{2}\left(\frac{1}{\beta}-\cos\theta_l\right) \approx \frac{\pi}{2} \qquad (6)$$

By inspection of Eq. 1 this corresponds precisely to $\lambda \approx L_\pi$, the classical slip distance. Thus the angle of the strongest double-boundary TR emission matches the laser angle in the ITA case with a phase slip of π.

The classical analysis of the double-boundary TR case can be summarized under three points. (1) Double-surface TR exhibits an interference pattern not found in the single-surface case: it arises because propagating waves move both forward and backward in the space between the mirrors. (2) For $L > L_f$ the double-surface TR is roughly twice the single-surface case. (3) The angle of maximum TR in the double-surface case corresponds to the laser angle in ITA for which $\lambda = L_\pi$, manifesting a correspondence between the forward and inverse processes.

IV. CLASSICAL ANALYSIS OF DR and IDA

In a practical acceleration system, the mirrors must have small holes for free passage of the *e*-beam. Otherwise significant scattering and beam degradation will arise unless extremely thin membranes can be used for the mirrors. Introducing an aperture changes the laser-*e*-beam interaction in some respects. In particular, since the electron travels in an unbounded vacuum, questions re-emerge about whether the Lawson-Woodward theorem applies, which would rule out net acceleration. Here the forward "emission" process is called diffraction radiation DR. The single-boundary DR case has been analyzed for an infinitely-thin conducting screen with a circular aperture [16]: the presence of the aperture reduces the emission (compared with TR) by the factor

$$(F_{DR})_{screen} = [J_0(\theta ka)]^2 \left[\frac{ka}{\beta\gamma} K_1\left(\frac{ka}{\beta\gamma}\right)\right]^2 \qquad (7)$$

where J_0, K_1 are Bessel functions and a is the aperture radius. This expression assumes $\gamma \gg 1$ and that the electron path is normal to the screen and passing through the center of the aperture. The J_0^2 factor represents the diffraction pattern of the aperture. The second factor $\sim [xK_1(x)]^2$ represents the radial fall-off of the electron self fields, which fall exponentially for radii exceeding $\beta\gamma/k$. Its exponential smallness for larger apertures arises because a large large aperture causes little disturbance to the self fields and thus produces little DR. For small apertures, the reduction factor F_{DR} goes to unity, since $J_0(x) \to 1$ and $xK_1(x) \to 1$ for $x \to 0$. In this limit the TR result is recovered even though the electron travels entirely in a vacuum along an unbounded path. That is, the conditions for the Lawson-Woodward theorem are violated if material boundaries can be placed suitably close to an electron path which is entirely in vacuum.

In another work [17] double-boundary DR has been analyzed in an arrangement where the aperture opens into an infinitely long tube of constant radius. This "tube-aperture" case experiences less diminishment: $(F_{DR})_{tube} \approx [(F_{DR})_{screen}]^{1/2}$. The inverse process, IDA, has also been analyzed in the double-boundary, tube-aperture arrangement [17]. The reduction factor relative to ITA is approximately

$$(F_{IDA})_{tube} \approx J_0(\theta_L ka)\exp(-0.43\, ka/\beta\gamma) \qquad (8)$$

The classical analysis of the double-boundary DR/IDA case can be summarized under two points. (1) for small enough apertures DR gives the same result as TR. (2) this implies that even for an electron whose unbounded path lies entirely in vacuum, the Lawson-Woodward theorem does not apply if material surfaces are sufficiently close to the electron path.

V. PRACTICAL EMBODIMENT OF IDA

We have established the principle of ITA/IDA from both classical and quantum perspectives, and have identified them as the inverse processes of TR/DR. However the attractiveness of these concepts depends on how high an acceleration gradient can be achieved subject to such real limitations as imposed by optical damage, available lasers, and fabricability of small-scale structures. Therefore the next step is to consider the practical embodiment of the concept and uncover its principle design constraints. These two issues are treated in reverse order here since the objective of achieving high-gradient acceleration subject to constraints somewhat drives the system embodiment. The focus is on IDA, viewed as the more practical of the two concepts since the *e*-beam travels only in vacuum.

A. Optimum Acceleration Subject to Engineering Constraints

Three obvious limitations on the IDA arrangement are considered: (1) damage at optical surfaces; (2) the available laser power; and (3) the size of optical structures that can be fabricated (*viz.* how small). In the first two cases specific limitations (damage threshold, maximum laser power) can be applied using known results or achievements. The third relates to the system dimensions and is less clear. For this analysis the interaction is assumed to take place between two more-or-less planar optical surfaces. The laser beam crosses the *e*-beam path at a fixed angle θ_L. There are three length scales in the problem: the optical surfaces are separated by a distance λ; each contains an aperture of radius a for passage of the electron beam; and the nominal laser beam radius at the optics r_L (the laser beam is assumed to be clipped at the radius $2r_L$). Two intuitive choices that link these parameters are adopted: (1) $2r_L = \theta_L \lambda$, *i.e.* the outermost laser ray from the upstream optic crosses the electron path just as the *e*-beam passes through the downstream optic; and (2) λ is set equal to the slip distance L_π.

The basic equations for the optimization are the following. (1) The relationship between the accelerating electric field (in the *e*-beam path) and the maximum electric field at the optics where the damage threshold is applied (Eq. 9 to follow). In the absence of any reducing factors the former is the acceleration gradient. (2) The acceleration reduction from phase slippage and the finite aperture (Eq. 10 to follow). (3) The relationship between the laser power and the maximum electric field (Eq. 13 to follow). (4) A less well characterized limitation is imposed by the minimum size for practical fabrication (Eq. 11 to follow).

Consider a Bessel function laser profile, $E_r \approx \mathrm{const}\, J_1(\theta_L k r)$, $E_z \approx \mathrm{const}\, \theta_L J_0(\theta_L k r)$ where $k = 2\pi/\lambda$ is the wave number. The ratio of the nominal accelerating field $E_{acc} \approx E_z$ (ignoring the complex phase) to the maximum electric field is

$$\frac{E_{acc}}{E_{max}} \approx 4.6 \theta_L^{3/2} (a/\lambda)^{1/2} \qquad (9)$$

Here E_{max} is approximated using the maximum values of J_1 in the large argument approximation (valid for $\theta_L a/\lambda > 0.1$). The effective accelerating field is reduced from E_{acc} by two effects, phase slippage, and finite aperture. Since $\lambda = L_\pi$ is assumed, the phase slips by π and the phase slippage reduction factor is $2/\pi$, i.e. the average of one lobe of a sine function. Finite-aperture size leads to reduced effectiveness because of stray electric fields that leak through the aperture. Based on the reduction factor found in diffraction radiation analyses, this factor is $\lambda/\pi^2 \theta_L a$ (again assuming $\theta_L a/\lambda > 0.1$). Including these two factors in Eq. 9 the acceleration gradient is

$$W' = 0.3(\theta_L \lambda/a)^{1/2} eE_{max} \qquad (10)$$

The fabricability requirement is less well characterized. Suppose for simplicity that it imposes a minimum on the interaction length, $\lambda > \lambda_{min}$. This corresponds to an upper bound on θ_L which (using $L_\pi = \lambda$ and $\gamma >> 1/\theta_L$) is

$$\theta_L < (\theta_L)_{max} = (\lambda/\lambda_{min})^{1/2} \qquad (11)$$

Then the maximum acceleration gradient ($\lambda = \lambda_{min}$) is

$$W' = 0.3 \frac{\lambda}{(L_{min} a)^{1/2}} eE_{max} \qquad (12)$$

The laser power assuming truncation at $r = 2r_L$ is

$$P_L \approx 6.8(\lambda a/\theta_L) I_{max} \qquad (13)$$

where $I_{max} = \varepsilon_0 c E_{max}^2/2$ is the maximum intensity at the optic. Here $\theta_L > 1/\gamma^2$, $\theta_L a/\lambda > 0.1$, and the large-argument form of J_1, were assumed.

Table I shows an interesting example from these results. Here $\gamma >> 1/\theta_L$ was assumed, i.e. $\gamma >> 5$, and the damage threshold is typical for a 1 psec laser pulse [18,19]. Moderately high accelerating gradients (~200 MeV/m) are projected. Interestingly, the required laser energy is relatively modest, less than 1 GW. This example assumes a particular aperture size and interaction length: in view of Eq. 12, reduction of either increases the acceleration gradient. Also, $W' \propto \lambda$ implying that larger gradients are achieved for longer wavelengths.

The fact that the required laser energy is so low is related to a curious feature of the optimization procedure. Here the optimization assumed constrained interaction length and damage limit. If, alternately, the constraints were laser power and damage limit, and a state-of-the-art laser power (e.g. 10 TW) were assumed as elsewhere [20], then a much less attractive result follows. At such high power with fixed fluence, the scaling relations are forced into a regime in which different expressions than Eqs. 12 and 13 apply. In this case the energy gain is a reasonable 60 MeV but over a very long interaction length of $\lambda = 27$ m. Consequently the acceleration gradient is an uninteresting 2 MeV/m.

Table I
Example of Optimum Acceleration in an IDA

Assumed quantities		Resulting parameters	
wavelength:	$\lambda = 10.6$ μm	acceleration gradient:	W' = 195 MeV/m
damage flux:	$I_{max} = 5$ TW/cm^2	laser crossing angle:	$\theta_L = 230$ mrad
interaction length:	$\lambda = 0.2$ mm (200 μm)	energy gain:	$\Delta W = W'\lambda = 0.04$ MeV
aperture diameter:	$2a = 0.1$ mm (100 μm)	laser power:	$P_L = 0.78$ GW

B. Practical embodiment of an IDA

Since the optimum IDA requires relatively short interaction sections, a practical accelerator must have multiple stages with many sections stacked end to end. A single laser pulse should be used for many sections without requiring unreasonable optics. A practical approach proposed by Fontana [21] is illustrated in Fig. 5. The sections are separated by a series of thin, transparent windows, each with a centered aperture for passage of the *e*-beam. A laser beam introduced at one end would then propagate through the series of thin windows and interaction sections. The windows could be slightly concave to continuously refocus the laser, but the same might be accomplished more simply by placing the system inside a circular waveguide as shown.

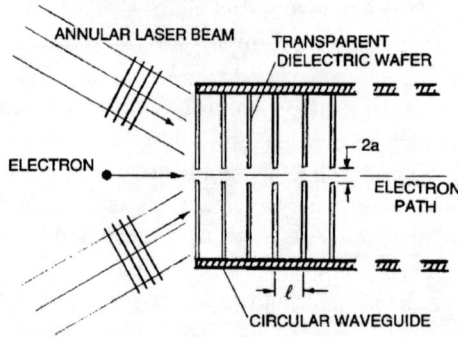

Fig. 5. A possible embodiment for a multi-stage IDA with short-length interaction sections.

ACKNOWLEDGEMENT

The authors wish to acknowledge the helpful comments of J.R. Fontana, R.H. Pantell, K.T. McDonald, and R.B. Fiorito. This work was supported by U.S. Department of Energy, Grant No. DE-FG06-93ER40803.

REFERENCES

1. J.D. Jackson, Classical Electrodynamics, 2nd. Ed. (John Wiley, New York, 1975), p. 685.
2. L.D. Landau and E.M. Lifshitz, *Electrodynamics of Continuous Media, 2nd Ed.*, (Pergamon, Oxford, 1984), p. 409-412.
3. J.R. Fontana, University of California at Santa Barbara, private communication.
4. L.C. Steinhauer and W.D. Kimura, J. Appl. Phys. **72**, 3237-3245 (1992).
5. L.C. Steinhauer and W.D. Kimura, J. Appl. Phys. **74**, 4813 (1994).
6. R.B. Palmer, Part. Accel. **11**, 81 (1980).
7. R.B. Palmer, in *Advanced Accelerator Concepts*, AIP Conference Proceedings No. 335, P. Schoessow, Ed. (American Institute of Physics, New York, 1995), p. 90.
8. J.A. Edighoffer and R.H. Pantell, J. Appl. Phys. **50**, 6120 (1979).
9. R.H. Pantell, in *Physics of Quantum Electronics*, Vol. 7, (Addison Wesley, Reading, 1979), pp. 1-14.
10. R.H. Pantell and M.A. Piestrup, Appl. Phys. Lett. **32**, 781 (1978).
11. L.C. Steinhauer, R.D. Romea, and W.D. Kimura, "Energy exchange between electrons and photons within a bounded vacuum", submitted to J. Appl. Phys. (1996).
12. I. Frank and V. Ginsburg, J. Phys. **9**, 353 (1945).
13. M.L. Ter-Mikaelian, *High-Energy Electromagnetic Processes in Condensed Media*, Wiley-Interscience, New York, 1972, p. 230.
14. R.D. Romea, L.C. Steinhauer, and W.D. Kimura, "Inverse transition accleration," submitted to Phys. Rev. Lett. (1996).
15. M.L. Ter-Mikaelian, *op. cit.*, p. 199.
16. Yu.N. Dnestrovkii, D.P. Kostomarov, Sov. Phys. Dokl. **4**, 132 (1959); **4**, 159 (1959). The expression found here is more general than Eq. 31.15 of M.L. Ter-Mikaelian, *op. cit.*, pp. 376-390 which assumes $\gamma \to \infty$.
17. L.C. Steinhauer, W.D. Kimura, and R.D. Romea, "Inverse diffraction acceleration of particles," submitted to J. Appl. Phys. (1996).
18. D. Du, X. Liu, G. Korn, J. Squier, and G. Mourou, Appl. Phys. Lett. **64**, 3071 (1994).
19. B.C. Stuart, M.D. Feit, A.M. Rubenchik, B.W. Shore, and M.D. Perry, Phys. Rev. Lett. **74**, 2248 (1995).

20. P. Sprangle, E. Esarey, and J. Krall, "Laser driven electron acceleration in vacuum, gases and plasmas," Naval Research Laboratory, Report NRL/MR/6790--96-7816 (1996).
21. J.R. Fontana, private communication, 1994.

Generation, Measurement and Application of Synchronized 700 fs Electron, 100 fs T³ Laser and Picoseconds X-ray Single Pulses

M.Uesaka, T.Ueda, T.Watanabe, M.Kando*[1], K.Nakajima*[2]*[3], H.Kotaki*[3]m and A.Ogata*[2]

Nuclear Engineering Research Laboratory, University of Tokyo,
2-22 Shirakata-Shirane, Tokai-mura, Naka-gun, Ibaraki 319-11, Japan
[1] Institute for Chemical Research, Kyoto University, Japan
[2] National Laboratory for High Energy Physics, Japan
[3] Japan Atomic Energy Research Institute, Japan

Abstract. A subpicosecond (700 femtosecond at FWHM) electron pulse from the S-band (2.856GHz) linear accelerator (linac) of the NERL(Nuclear Engineering Research Laboratory) was synchronized with a femtosecond (100 femtosecond at FWHM) laser pulse from a T³ (Table-Top Terawatts) laser with a time jitter whose standard deviation is 3.7 picoseconds. Then we generated a picosecond characteristic X-ray pulse by irradiating the electron pulse to a Cu target (Ka, 8.1keV, 1.54Å) and obstained the Bragg diffraction from an NaCl ionic monocrystal using a high sensitivity X-ray imaging plate. Further, we discuss its applications to observe lattice vibration of the monocrystal by using the synchronized laser (pump) and X-ray (probe), which is named as the pulse-snapshot method.

INTRODUCTION

Recently, there have been remarkable progresses in producing ultrashort pulses by lasers and other particle beam accelerators. Now high-powered 100 femtosecond laser pulses are available by table-top lasers and a 1kA subpicosecond electron pulse by our linac (1). Ultrashort synchronized pulses lead to an ultrafast time-resolved dynamic spectroscopy, which is capable of observing new ultrafast phenomena in quantum beam-material interaction. Thus, the precise synchronization of these pulses is of recent much interest. In the NERL of University of Tokyo, the synchronization system is to be applied to several new experiments; the laser wakefield acceleration (2-4), the T³ laser, femtosecond X-rays generation and application to ultrafast dynamic microspectroscopy. It is necessary for the above applications that the subpicosecond electron beam and the femtosecond laser pulse are precisely synchronized with the time resolution of less than 1ps. In this paper, the synchronization system, the measured evidences and its application are presented.

SYNCHRONIZATION SYSTEM

Many synchronized trigger pulses are necessary for the generation of an ultrafast electron and laser pulses. For example, we have to send the synchronized pre- and main trigger pulses to the grid pulser, the sub-harmonic bunchers (SHB), the RF(Radio Frequncy) power source for the accelerating tubes. The T³ laser consists of a Ti-Sapphire oscillator, a stretcher, a regenerative amplifier, a pulse selector and a multipass amplifier. The trigger pulses are also sent to the timing stabilizer, the YAG laser, the pulse selector and other pre-trigger pulses for the T³ laser. At first we determine 119 MHz as the main RF frequncy. Then we generate higher harmonics (476MHz, 2.856GHz, etc.) via frequency multipliers. Then, the trigger pulses synchronized with a specified phase of 79.3 MHz is generated with a time jitter of a few picoseconds. The repetition rate of the laser oscillator is precisely fixed to 79.3 MHz by the timing stabilizer (or so-called mode locker). The RF synchronized trigger pulses trigger the electron gun of the linac and also select the laser pulse at 10Hz. Further, this trigger pulse are used to run several diagnosis devices including a femtosecond streak camera. The selected laser pulse is transported to the compressor through a vacuum pipe and compressed from 300ps to 100fs at FWHM to be 2.5TW as a peak power.

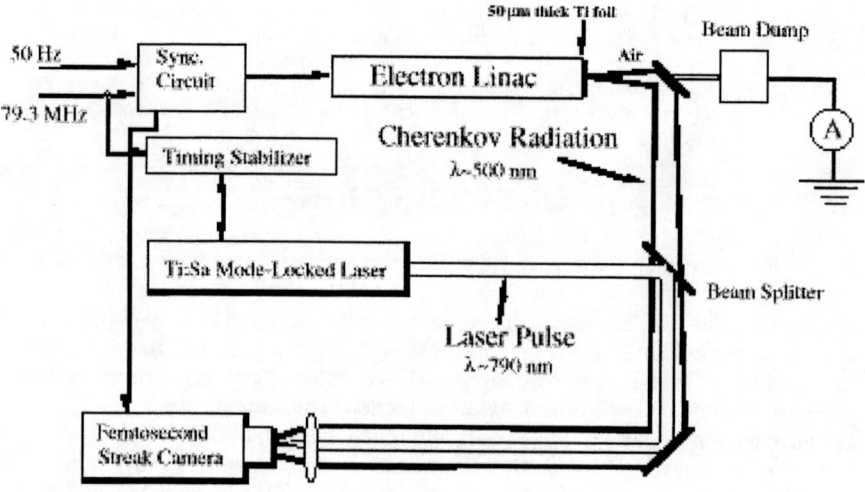

Figure 1. Measurement setup of the synchronization.

The measurement setup is schematically shown in Fig.1. Cherenkov radiation pulse emitted by the electron pulse in air and the laser pulse are introduced to the streak camera. The time interval between the electron and laser pulses is adjustable by several time-delay units in the synchronization system. The synchronization was confirmed and those pulse shapes were obtained by the femtosecond streak camera whose time resolution is 180fs at FWHM (HAMAMATSU FESCA-200)(5). We can measure them on-line by a single shot. The measured streak image and pulse shape are shown in Fig.2. We can generate the 700 fs electron single bunch(1) and the 100 fs laser single bunch and multi-bunches. The reason why we have such longer pulse than 1 ps in the Fig.2 is that we did not use very narrow band-pass optical filter in order to enhance their brightness on the phospher screen in the streak camera. The time jitter between the two pulses were 3.7 picoseconds at the standard deviation among about 100 shots. We are improving the synchronization circuit in order to have the time jittor of less than 1 picosecond at FWHM.

Light Intensity

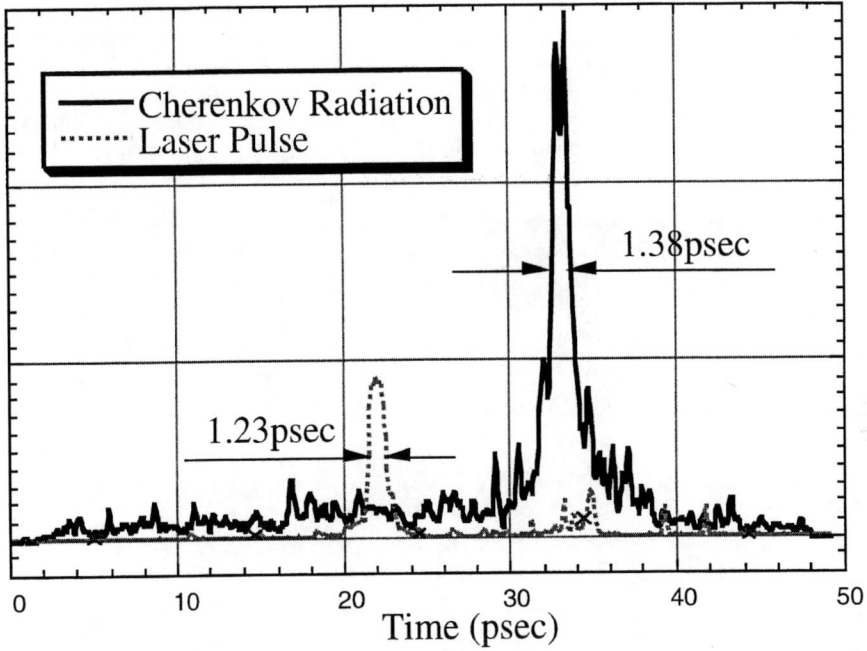

FIGURE 2. Synchronized laser pulse and elctron single pulse (via Cherenkov radiation in air) obtained by the femtosecond streak camera.

APPLICATIONS

Pulse-snapshot X-ray Diffraction

Here we propose a new method of femtosecond time-resolved dynamic X-ray diffraction, which enables us to observe ultrafast quantum phenomena such as lattice vibration.

Now they can determine the microscopic structure of macropolimers or proteins via X-ray diffraction using high brightness synchrotron radiation. On the other hand, the pulseradiolysis opened the window to picosecond time-resolved researches in radiation chemistry (6,7). However, the former is still the static analysis and the latter gives us only morochromatic data of light emission ord absorption as a function of time.

The new femtosecond time-resolved dynamic microspectroscopy, named by us as "the pulse-snapshot method", is the combination of the above two techniques. We can have a three-dimensional (3D) snapshot of microscopic motion of atoms. Here we need to have the two synchronized femtosecond laser pulse (as a pump pulse in femtosecond laser spectroscopy) and X-ray pulse (probe). We irradiate a matter by the 100fs laser and induces lattice vibration, and then the succeeding X-ray pulse with a specified delay-time gives us an X-ray diffraction image on an X-ray imaging plate. This process is repeated until we get a clear diffraction image with sufficient signal-to-noise ratio. The lattice structure is reconstructed by introducing a certain robust inverse analysis. Next, we change the delay-times and get several snapshots of the lattice vibration at several time-steps. Computer graphics technique enables us to obtain the animation of the lattice vibration. The concept is schematically explained in Fig. 3.

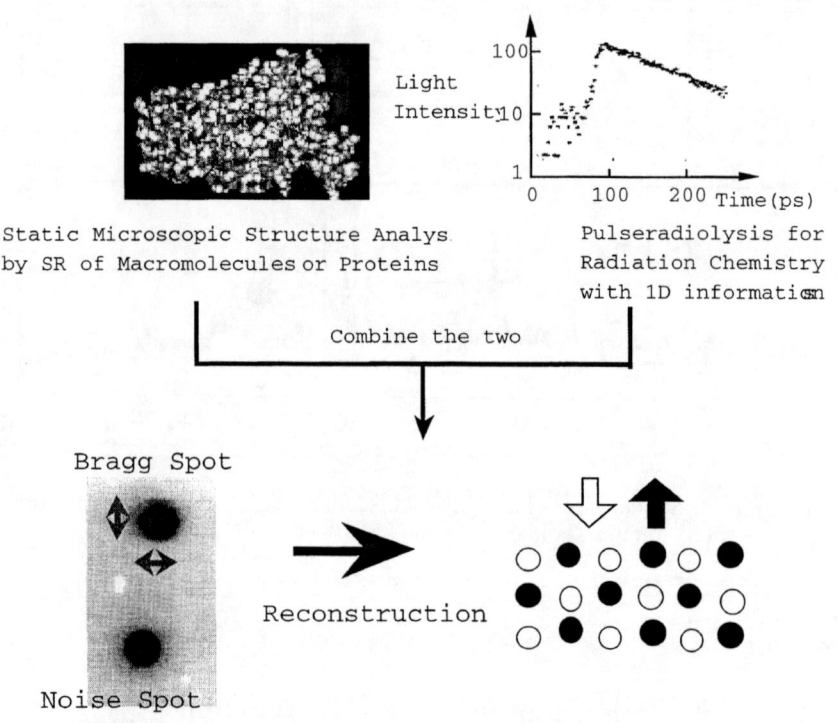

FIGURE 3. Schematic drawing of the pulse-snapshot method.

As the first step, we generated the picoseconds Kα characteristic X-ray (8.1keV; 1.54Å) from a solid Cu target by irradiating it by the 10 ps electron single pulse. Here we used a NaCl ionic monocrystal because it has the simplest cubic lattice structure. The first experiment setup is shown in Fig.4. The static Bragg Diffraction spot due to its lattice structure has been obtained so far by using the measurement system as shown in the left bottom of Fig.3. Since we did not use any X-ray monochrometer there, we had both the Bragg diffraction spot due to the cubic structure and the noise spot due to the X-rays with other wavelengths.

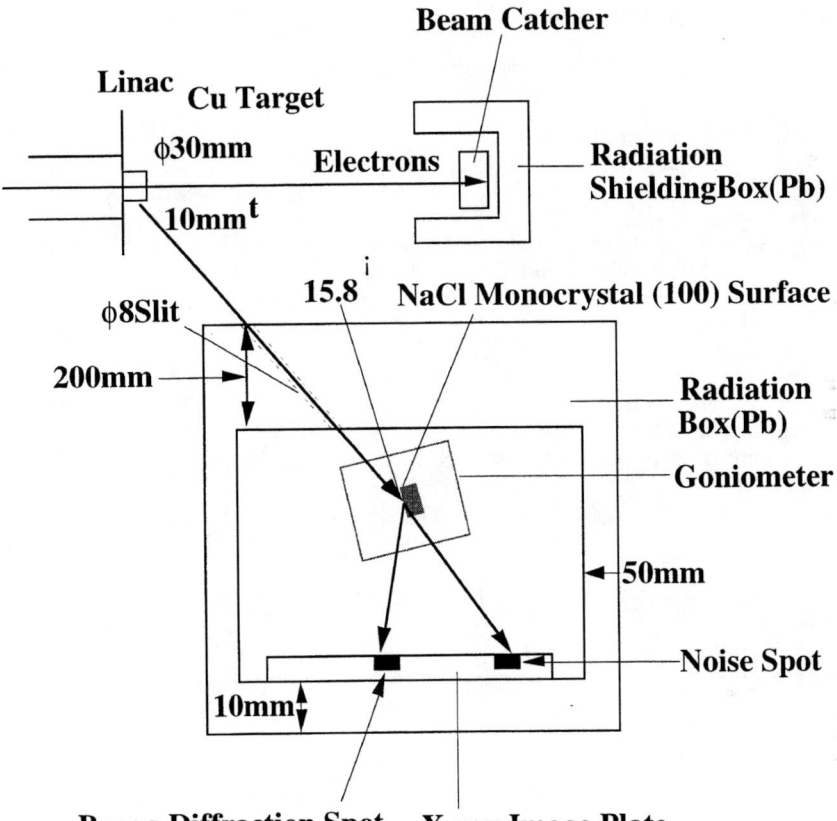

FIGURE 4. Picosecond X-ray diffraction measurement configuration

We are going to improve both the synchronization system and the X-ray diffraction system to generate a subpicosecond X-ray pulse and to get a 3D snapshots of the lattice vibration in the next year. Pulse shape measurement of the subpicosecond X-ray pulse is also planned by using a subpicosecond X-ray streak camera.

Thomson Scattering X-rays Generation

There have been several other methods for generating femtosecond X-rays such as Thomson scattering and laser plasma X-ray emission. Thomson scattering is one of the promising methods which are highly directional (~0.6 degree divergence) and can be tuned in energy(8,9). This method has a high potential for the application to the pulse snapshot X-ray diffraction. The femtosecond X-rays are produced by the Thomson scattering between the ultrashort terawatt laser pulse and the 38MeV electron beam at 90 degree(8). Here the laser pulse must be rigorously synchronized with the electron pulse in both the time domain and space.

CONCLUSION

In this work, we constructed the synchronization system by which we could synchronize the subpicosecond electron beam with the femtosecond laser pulse with the time resolution of 3.7 ps standard deviation. Then the experiment of the laser wakefield acceleration is now being done by using the system. Furthermore, the experiment of the pulse-snapshot X-ray diffraction to visualize lattice vibration in matter is under way simultaneously. We had already got the static view the cubic lattice structure of the NaCl ionic monocrystal using the picosecond X-rays.

REFERENCES

1. M.Uesaka et al., Phys. Rev. E, Vol. 50, No. 4 (1994), p. 3068.
2. A.Ogata et al., Physica Scripta, Vol. T52 (1994), p. 69.
3. H.Nakanishi et al., Nucl.Instrum. & Meth. in Phys. Res., A 328 (1993), p. 596.
4. K.Nakajima et al., Phys. Rev. Lett., Vol. 74, No. 22 (1995), p. 4428.
5. A.Takahashi et al., SPIE(The Inernational Society of Optical Engineering), Proc.(1994), Vol.2116,p.275.
6. C.D.Jonah, Rev. Sci. Instrum. Vol. 46 (1975), p. 62.
7. Y.Tabata, Radiat. Phys. Chem. Vol. 18 (1981), p. 43.
8. K.-J.Kim et al., Nucl. Instrum. & Meth. in Phys. Res. A, 341 (1994), p. 351.
9. R.W.Schoenlein et al., Science, Vol.274(1996), p.236.

GROUP 5

"Particle Beam Sources"—Luca Serafini (INFN, Milano/UCLA) and Anahid Dian Yeremian (SLAC)

Initial Commissioning Results of the Next Generation Photoinjector*

D. T. Palmer*, X. J. Wang[†], R. H. Miller*
M. Babzien[†], I. Ben-Zvi[†], C. Pellegrini[**], J. Sheehan[†],
J. Skaritka[†], H. Winick*, M. Woodle[†], V. Yakimenko[†]

*Stanford Linear Accelerator Center
Stanford University, Stanford, CA 94309

[†]Brookhaven National Laboratory
Accelerator Test Facility
Upton, NY 11973

[**]University of California Los Angeles
Department of Physics
Los Angeles, CA 90095

Abstract

The BNL/SLAC/UCLA symmetrized 1.6 Cell S-band emittance-compensated photoinjector has been installed at the Brookhaven Accelerator Test Facility(ATF). The commissioning results and performance of the photocathode injector are presented. This emittance-compensated photoinjector consists of the symmetrized BNL/SLAC/UCLA 1.6 cell S-band photocathode radio-frequency (rf) gun and a single solenoidal magnet for transverse emittance compensation [1]. The highest acceleration field achieved on the cathode is 150 $\frac{MV}{m}$, and the normal operating field is 130 $\frac{MV}{m}$. The quantum efficiency of the copper cathode was measured to be 4.5×10^{-5}. The transverse emittance and bunch length of the photoelectron beam were measured. The optimized rms normalized emittance for a charge of 300 pC is 0.7 π mm-mrad. The bunch length dependency of photoelectron beam on the rf gun phase and acceleration fields were experimentally investigated.

---

*Work supported by Department of Energy contract DE–AC03–76SF00515.

1 Introduction

The BNL/SLAC/UCLA symmetrized 1.6 Cell S-band emittance-compensated photoinjector has been installed at the Brookhaven Accelerator Test Facility(ATF) as the electron source for beam dynamics studies, laser acceleration and free-electron laser experiments. The emittance-compensated photoinjector consists of the symmetrized BNL/SLAC/UCLA 1.6 cell S-band photocathode rf gun, powered by an XK-5 klystron and a single emittance-compensation solenoidal magnet. There is a short drift space between the photoinjector and the input to the first of two SLAC three meter travelling wave accelerating sections. This low-energy drift space contains a copper mirror that can be used in either transition-radiation studies or laser alignment. There is also a beam-profile monitor/faraday plate located 66.4 cm from the cathode plane.

The ATF consists of two SLAC travelling-wave linacs powered by a single XK-5 klystron. The high-energy beam-transport system consists of nine quadrupole magnets, an energy spectrometer, an energy-selection slit and a high-energy faraday cup. Diagnostics located in the high-energy transport consist of beam-profile monitors and strip lines. The strip lines are used as an on-line laser/rf phase-stability monitor.

The drive laser is a Nd:YAG master oscillator/power-amplifier system. A diode-pumped oscillator mode-locked at 81.6 MHz produces 21 psec FWHM pulses with 100 mW of average power. Gated pulses seed two flashlamp-pumped multi-pass amplifiers and are subsequently frequency quadrupled. This nonlinear process leads to a factor of two reduction in the laser pulse length. The 266 nm beam is transported to the rf gun area via a 20 meter-long evacuated pipe. The laser-beam transport system near the injector includes a set of telescoping lenses and a limiting aperture. This limiting aperture is imaged onto the cathode with a spherical lens and a pair of Littrow prisms which compensate for the anamorphic magnification introduced by the 72° incidence on the cathode. The time slew across the cathode caused by this oblique incidence is corrected by using a diffraction grating. The relay imaging technique used throughout the optical transport improves the beam-pointing stability. Since the laser beam overfills the limiting aperture, the transverse profile of the beam is a truncated gaussian. The spot size of the laser beam on the cathode is 2 mm diameter, edge to edge.

2 Injector Design

The BNL/SLAC/UCLA symmetrized 1.6 Cell S-band emittance-compensated photoinjector consists of the symmetrized BNL/SLAC/UCLA 1.6 cell S-band photocathode rf gun mated to a single emittance-compensation solenoidal magnet.

The 1.6 cell rf gun differs from the original BNL 1.5 cell rf gun [2] in that the half cell has been lengthened to decrease the rf field levels on the cell to cell coupling iris and also to provide more rf focusing in the iris region. The 1.6 cell rf gun is not a side-coupled 0-mode suppressed rf gun, as in the previous BNL type rf guns. High power rf is coupled only into the full cell. The enlarged beam iris diameter increases the cell to cell coupling, which provides a mode separation between the π and 0-modes of 3.225 MHz for a balanced field configuration. The half cell is fully symmetrized with two 72° oblique incidence laser ports. The cathode plate is removable using a single helico flex seal for both the vacuum and rf seals. This removable cathode plate eliminates the multipactoring problem common to choke-joint cathodes. The removable cathode facilities the use of different cathode materials such as Cu and Mg. The Cu cathode results are presented in this paper. The full cell has two symmetrized plunger-type tuners with a total tuning range of ±2 MHz. The rf coupling slot is symmetrized by an identical coupling slot that provides additional vacuum pumping [3]. The 1.6 cell gun uses resistive heating to maintain the resonant frequency.

The single emittance-compensation solenoidal magnet was specifically designed to be used with the 1.6 cell gun, utilizing POISSON [4] field maps into PARMELA [5] to study the beam dynamics with different magnet designs. In previous emittance-compensation system designs a bucking coil is positioned upstream of the cathode plane to null the magnetic field at the cathode. This is unnecessary in the present design because the single solenoid magnet produces less than 9 gauss at the cathode when the peak solenoidal field is 3 Kgauss.

After rf conditioning the 1.6 cell rf gun operates at 5×10^{-9} torr with a field gradient of 125 $\frac{MV}{m}$ and in the quiescent state the vacuum is 1×10^{-9} torr.

3 Gun Energy / Dark Current

The BNL/SLAC/UCLA 1.6 cell S-band photocathode rf gun is designed to attain a field level at the cathode and at the middle of the full cell of up to 150 $\frac{MV}{m}$, and to operate with rf pulse widths up to 3.5 μs. Calibration of the field levels in the gun were verified by measurements of the beam energy using a $\cos(\theta)$ deflection magnet located inside the bore of the emittance-compensation solenoidal magnet. The results of these energy measurements are shown in figure 1 where Φ is the laser injection phase. $\Phi = 0$ and $\Phi = 90$ are the zero crossing and crest of the rf respectively.

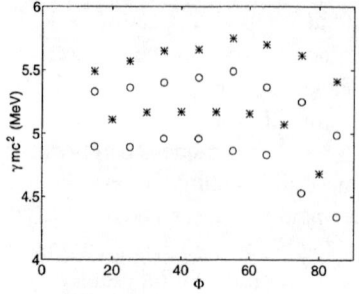

Figure 1: Energy versus cathode field levels of 123, 117, 110 and 105 $\frac{MV}{m}$ respectively.

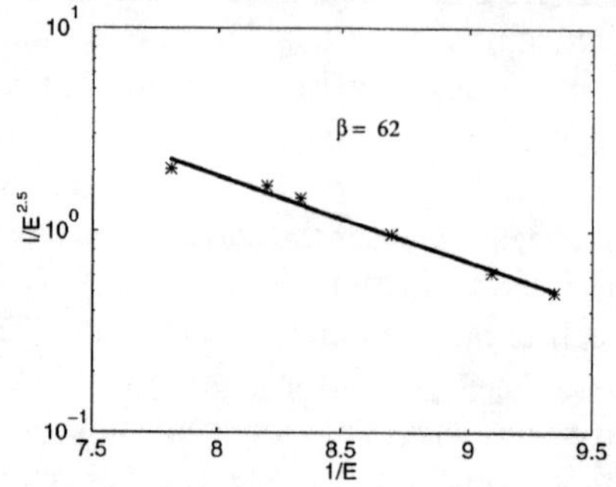

Figure 2: Fowler-Nordheim plot for the Cu cathode.

Standard machining and Cu wool polishing techniques were used to fabricate both the full and half cells. The cathode plate was prepared using published procedures [6]. In combination these techniques produce a field enhancement factor $\beta = 58$ as can be seen in the Fowler-Nordheim Plot of figure 2.

4 Multi-Pole Fields

Multi-pole field effects were studied by decreasing the laser spot size to 400 μm and setting the laser injection phase to the Schottky peak. This laser injection phase causes an effective electron bunch-lengthening and a large energy-spread tail. By adjusting the laser spot position we were able to eliminate this energy spread tail. This alignment minimizes the integrated higher-order-mode contribution to the beam distortion. Analysis indicates that the symmetrized BNL/SLAC/UCLA 1.6 cell photocathode rf gun's electrical and geometric center are within 170 μm of each other, which is within the laser spot alignment error. Compared to similar experimental results of the 1.5 cell BNL gun whose electrical and geometric centers differ by 1.0 mm [7], the 1.6 cell gun has fulfilled the symmetrization criteria.

Future work with custom laser masks to study the field patterns at larger diameters are planned [8].

5 Quantum Efficiency versus Polarization

The laser time-slew correction has the drawback of decreasing the available laser energy at the cathode by 50%. The available charge was measured as a function of polarization. In figure 3 it can be seen that the charge is maximized at a polarizer angle of $56°$ which corresponds to P polarized light on the cathode.

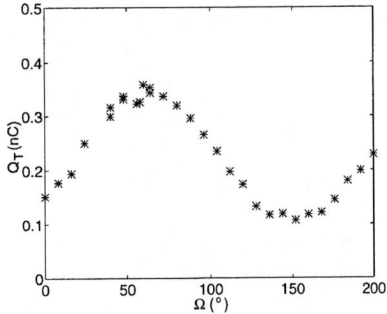 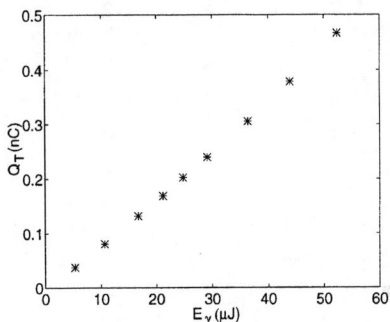

Figure 3: Electron bunch charge versus polarizer angle for the Cu cathode.

Figure 4: Electron bunch charge versus laser power for the Cu cathode.

The measured value of the Cu cathode's quantum efficiency is $QE = 4.4 \times 10^{-5}$, which is calculated from figure 4. These studies were conducted at a laser injection phase of $90°$ which utilized the Schottky effect to increase the available charge.

6 Transverse Phase Space

The normalized rms emittance, $\varepsilon_{n,rms}$, measurements were taken using a variation of the three screen method. Two screens were utilized while insuring that a beam waist was located at the down stream profile screen. The two-screen method is compared to the standard quadrupole-scan technique in figure 5 and the results from these two methods are compared in table 1 for a total charge of $Q = 250$ pC.

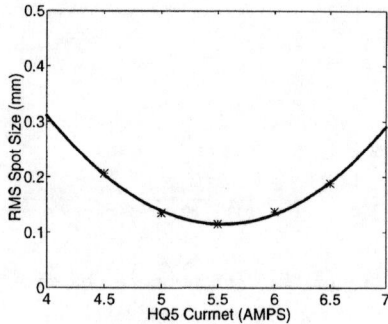

Figure 5: Quadrupole scan RMS emittance results.

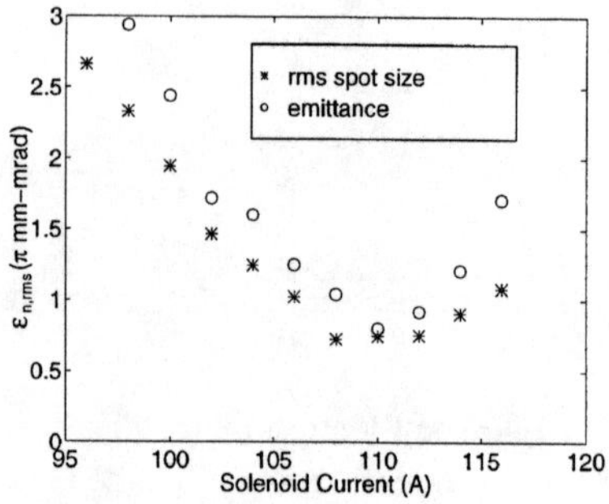

Figure 6: Emittance and RMS beam size versus solenoidal current

PARMELA was used to simulate the emittance compensation process and the subsequent acceleration to 40 MeV [9]. A correlation of the minimum spot size with an emittance minimum was noted during these simulations. This was experimentally verified during the commissioning of the 1.6 cell rf gun, using the beam profile monitor located at the output of the second linac section, as can be seen in figure 6. These results are consistent with similar results of the BNL 1.5 cell rf gun [7].

$\varepsilon_{n,rms}$ Quad Scan	2.29 π mm-mrad
$\varepsilon_{n,rms}$ Two Screen	2.42 π mm-mrad

Table 1: Quadrupole-scan and two-screen method results for $Q_T = 250$ pC.

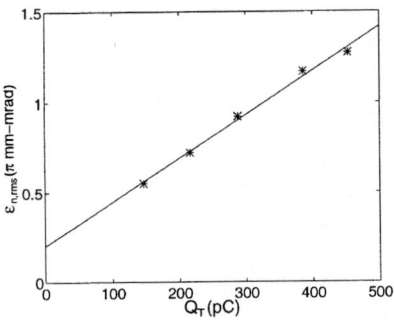

Figure 7: Emittance versus electron bunch charge.

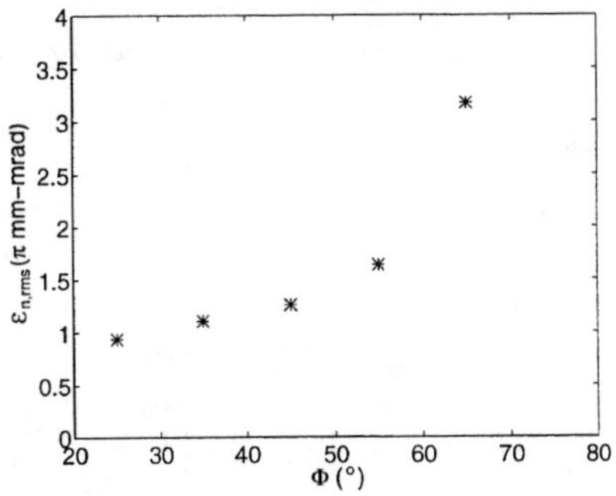

Figure 8: Emittance versus laser injection phase.

There are four emittance terms that contribute to the total $\varepsilon_{n,rms}$; these are the space charge, rf, thermal and magnetic terms. The last term is due to the small, but finite, magnetic field at the cathode.

Studying the dependence of transverse emittance on the bunch charge in figure 7, we have noted that at $Q = 0$ there is a residual emittance term of 0.2 π mm-mrad. This term is possibly due to ε_{rf}, $\varepsilon_{thermal}$, ε_{mag} and measurement errors. If we neglect the magnetic term, which is reasonable due to the initial cathode spot-size and the small magnetic field at the cathode we can estimate the thermal and rf emittance terms to be less than 0.2 π mm-mrad. Since the measured $\varepsilon_{n,rms}$ is a factor of three less than the space-charge emittance term, ε_{sc}, that Kim's theory [10] predicated, we are confident that we have produced an emittance-compensated beam.

Due to laser power limitation, rf gun bunch compression and the Schottky effect it is not possible to keep the peak current constant for different laser injection phases. Therefore in figure 8 the plot is not for a constant current but for a decreasing charge from a maximum of 400 pC to a minimum of 178 pC. The functional dependence of this plot has been verified by comparison with Kim's theory.

7 Longitudinal Phase Space

When measuring the bunch length and energy spread of the electron bunch the rf system is adjusted such that the bunch initially has a minimum energy-spread. This is accomplished by adjusting the overall linac phase with respect to the laser injection-phase by means of the low level rf system. The $\delta\phi$ between the two linac sections is adjusted by means of a high power rf phase shifter, such that the energy spread of the beam is minimized. The beam energy is set to 40 MeV by adjusting the low level rf drive to the linac klystrons.

The energy spread is estimated by measuring the beam size on a phosphor screen in the dispersion region of the high-energy spectrometer. The dispersion in this region is 5.4 $\frac{mm}{\%}$. Figure 9 is a plot of the energy spread of the electron bunch as a function of the phase difference between the linacs.

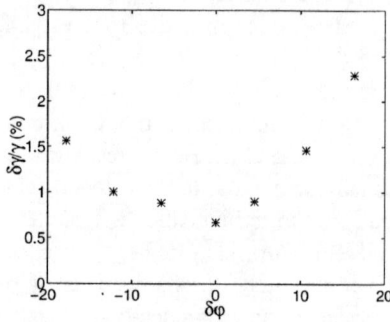
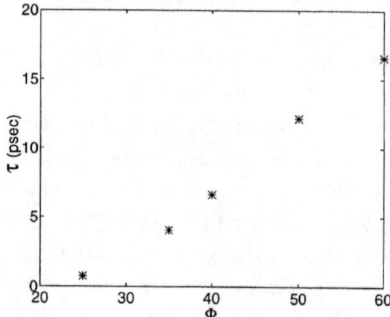

Figure 9: FWHM energy spread versus linac phase.

Figure 10: 95% bunch length versus laser injection phase.

Electron bunch length was measured by dephasing the second linac section such that a linear energy chirp is produced along the bunch. This allows the bunch length to be correlated to the energy spread. Using the technique discussed previously to measure the energy spread, the bunch length is measured as a function of laser injection phase. Figure 10 is experimental verification of bunch compression in the 1.6 cell rf gun. Bunch compression in rf guns has been experimentally demonstrated [11].

8 Conclusions

We have experimentally studied the six dimensional phase space of the electron beam that is produced by the BNL/SLAC/UCLA 1.6 cell S-band photocathode rf gun. We have experimentally verified longitudinal bunch-compression, electron-bunch energy and transverse emittance as a function of injection phase, solenoidal field and charge for peak fields in the rf gun of $127\frac{MV}{m}$. The optimized rms normalized emittance for a charge of 300 pC is 0.7 π mm-mrad.

Future work includes studies of the multi-pole fields that this gun is designed to suppress and cathode magnetic field effects along with slice emittance and inverse RADON transforms that will elucidate the electron beams transverse phase space. Emittance measurements for a bunch charge of 1 nC are also planned.

9 Acknowledgments

The authors would like to thank the technical staff at UCLA, the Stanford Linear Accelerator Center and at Brookhaven Accelerator Test Facility for all their dedicated work on this project. We would also like to thank Mr. James N. Weaver from SSRL for all the technical discussions and help that he has provided.

References

[1] B. E. Carlsten, *NIM*, **A285**, 313 (1989)

[2] K. Batchelor *et al.*, Proc. of 1990 EPAC p. 541

[3] D. T. Palmer *et al.*, Proc. 1995 Part. Accel. Conf. (1995) p. 982

[4] K. Halbach and R. F. Holsinger, *Particle Accelerators* **7**, 213 (1976)

[5] L. M. Young, private communications

[6] T. Srinivasan-Rao *et al.*, BNL-62626

[7] X. J. Wang *et al.*, Proc. 1995 Part. Accel. Conf. (1995) p. 890

[8] Z. Li, private communications

[9] D. T. Palmer *et al.*, Proc. 1995 Part. Accel. Conf. (1995) p. 2432

[10] K. J. Kim, *NIM*, **A275**, 201 (1989)

[11] X. J. Wang *et al.*, Phys. Rev. E, Oct 1989

The Design and Fabrication of an X-Band RF Gun

C.H. Ho, W.K. Lau, T.T. Yang, J.Y. Hwang, S.Y. Hsu, Y.C. Liu

Synchrotron Radiation Research Center, Hsinchu, Taiwan

G.P. Le Sage

University of California at Los Angeles, Electrical Engineering Department, Los Angeles, CA 90095, USA

F.V. Hartemann, and N.C. Luhmann, Jr.

University of California at Davis, Department of Applied Science, Davis, CA 95616, USA

Abstract. A recently proposed [1] high brightness, high repetition rate, multibunch photoinjector project has reached the high power construction stage. The accelerator structure consists of a 1-1/2 cell, side wall coupled, X-Band (8.548 GHz) standing wave cavity, driven by a 20 MW SLAC Klystron, and a GHz repetition rate (burst mode) rf modelocked AlGaAs laser diode oscillator and Chirped Pulse Amplification (CPA) Ti:Al_2O_3 multipass amplifier. The photocathode gun will be used to accelerate a train of one hundred, 0.1 - 1 nC electron bunches to an energy in the range of 5 MeV. A joint collaboration between the UC Davis Department of Applied Science (DAS), and the Synchrotron Radiation Research Center (SRRC) has been established to expedite the construction and characterization of the accelerator structure. A prototype copper cavity has been fabricated and characterized. The results of the low power rf measurements are presented, as well as a description of the high power cavity design. The solenoid focusing system design and construction is also described.

1. INTRODUCTION

The requirements of high brightness and high intensity for relativistic electron beams for Free Electron Laser (FEL) and laser scattering and acceleration experiments have made the photoinjector an attractive source. Since the invention of the photoinjector in the mid eighties [2,3], variations of the concept [4] have been used for FEL experiments, high luminosity colliders, sources of high charge pulses for wakefield accelerators [5], and low emittance beams for general high energy physics research.

Recently, a multibunch photoinjector project was proposed at 8.548 GHz, which produces a compact structure with higher accelerating gradients, and thus lower emittance characteristics than the existing rf guns whose frequencies typically range from 100 MHz to 3 GHz. The X-Band frequency is also low enough to avoid limiting tolerance requirements, as well as deleterious wakefield effects associated with RF structures designed for very high frequencies.

A collaborative research effort for the rapid development of the X-Band photoinjector system has been formed between UC Davis DAS, and SRRC. A prototype X-Band structure has been fabricated by SRRC, and continues to be characterized by UC Davis DAS at the Lawrence Livermore National Laboratory (LLNL) site. Both the design and fabrication procedures have included separate verification from both groups. The high power, ultra-high vacuum, brazed structure will also be constructed at SRRC. The gun will be installed and fully characterized at LLNL by the UC Davis DAS personnel as well as the SRRC staff.

2. PROTOTYPE FABRICATION

The initial prototype RF cavity is a scaled version of the S-Band Brookhaven gun [6]. Extensive simulation runs using the codes SUPERFISH and PARMELA predict an energy of 4.3 MeV with a peak gradient of 150 MV/m, corresponding to a drive power of 9 MW [7]. With a microbunch charge of 0.1 nC, the predicted transverse, rms, invariant emittance is less than 1π mm-mrad at the gun exit. The energy spread for this case is 0.31 %. A prototype copper cavity was fabricated in a CNC machine shop near the SRRC facility. The measured accuracy of the machining tolerance is +/- 0.05 mm. A supporting structure made of Aluminum was also built for the purposes of bead pull cavity perturbation measurements.

The central frame, which is used to press together the cavity components, is detachable from the supporting stand, and can be used to test a variety of cavity structures. The experimental arrangement for the bead pull measurements is shown in Fig. 1. A machineable ceramic (Aluminum Oxide) bead with a 1.2 mm diameter was used as an axial cavity perturbation. The bead was drilled using an Nd:Yag laser, and a nylon string with diameter 0.05 mm was used to pull the bead down the axis of the cavity. A 0.69 mm length, 0.49 mm diameter metal cylinder, made from a section of a hypodermic needle has also been used to check the convergence of the bead pull data.

A steel alignment pin was placed in a groove along the outer edge of the cavity body to avoid rotation of the separate sections of the cavity. For the purposes of the cold test, the waveguide lateral position was not fixed, but free to be moved and optimized. The waveguide can be moved over a distance +/- 7 mm.

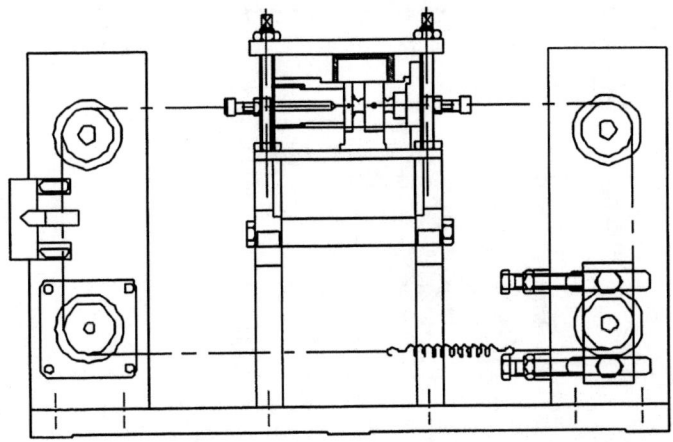

Figure 1: Assembly drawing for bead-pull measurement

The large diameter of the cathode (20 mm) has the purpose of avoiding the well known problems of rf breakdown and field emission leading to dark current near the gap between the cathode plug and cavity wall by moving the gap to a low field region. A 32 UNC threaded cathode plug was originally used to make the fine adjustments of the cathode position. A BAL SEAL coil spring is placed in a groove near the cathode surface to provide a short path for the wall currents which leak into the gap region. The contact of the clean, tight tolerance copper surfaces of the cathode plugs was found to cause galling problems initially with the coil spring arrangement. A choke type cathode was built to alleviate this problem, as shown in Fig. 2. The choke type plunger is based on the principle of a quarter-wave transformer [8]. This type of cathode seems to produce similar results to that of the coil spring arrangement. The galling problem eventually persisted, even with the choke type cathode plug. By using a micrometer and avoiding the twisting motion of the plug inside the cavity, the problem of galling seems to have been eliminated to the point where we can use a combination of a coil spring and a choke design.

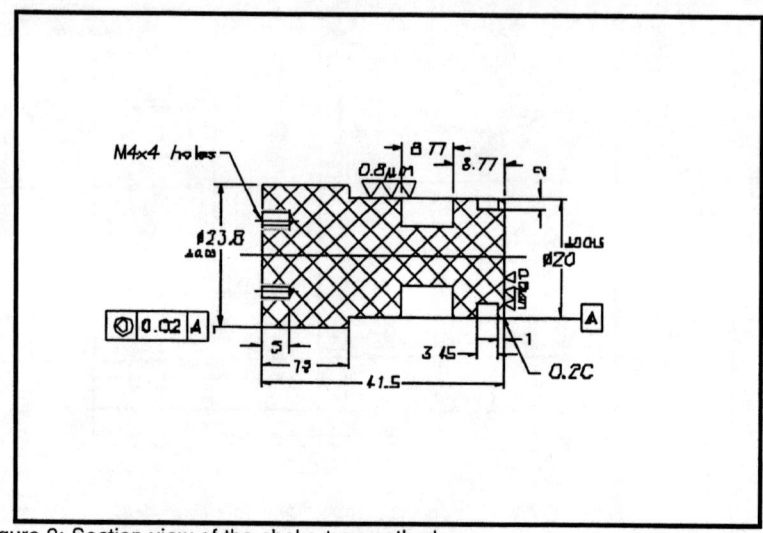

Figure 2: Section view of the choke-type cathode

Coupling to the accelerator structure is accomplished through two apertures in the broad wall of a TE_{10} waveguide, and side wall apertures in the full and half cell cavities. The coupling occurs through the ϕ component of the TM_{010} mode magnetic fields in the cavities, and the axial magnetic field component in the waveguide. This waveguide axial field is phased 180 degrees apart on either side of the broad wall center line, and leads to the dominant excitation of the π-mode in the cavities [9]. The wall currents are also useful for visualizing the coupling between the fundamental modes in the waveguide and cavities, and also demonstrates the π-mode dominance. Since the width of the WR-112 waveguide originally used to couple the RF energy to the accelerator structure was much greater than the length of the coupling slot, the lateral position of the waveguide strongly effected the relative coupling of the cells. This phenomenon was due to the fact that the magnetic field reaches a null in the center of the waveguide. When the half cell aperture is near the edge of the broad wall of the waveguide, the resonance peak becomes significantly deeper than that of the full cell, when these resonances are detuned. The case where the full cell aperture is near the outer waveguide edge produces an even more pronounced effect, since the aperture in the full cell is longer than that of the half cell. Unfortunately, the situation is even more complicated when the resonance peaks are close together, since a balanced field can be achieved even when the separate resonances initially have greatly disparate amplitudes. This is due in part to coupling through the aperture between the cells.

Many of these complications were eliminated when a waveguide taper was employed to reduce the width of the coupling waveguide to the size of standard WR-90 rectangular waveguide. The WR-112 waveguide is used in the rest of the RF system since the losses are reduced at 8.548 GHz. The width of the WR-90 waveguide closely matches the total length of the 1-1/2 cell structure, and allows the half cell and full cell coupling holes to simultaneously occupy the opposite edges of the waveguide broadwall, maximizing the coupling to both cells. Critical coupling is easily achieved with a balanced field, even when the apertures are reduced in size from their original dimensions used for the WR-112 waveguide coupling. Smaller coupling apertures will introduce less azimuthal dependence in the structure, and reduce the higher order modes. The addition of the taper to the accelerator structure was the idea of one of our design team members (Greg Le Sage). A high power taper was designed for the full power structure, and is shown in Fig. 3.

The sizes of the coupling holes were optimized empirically, in small iterative steps, until critical coupling was achieved. The coupling for the original structure using the WR-112 waveguide was accomplished using a lateral slot with rounded ends, and started with dimensions 18 mm $\times 2$ mm. The coupling using the WR-90 waveguide will be accomplished using equally sized round holes at the outer edges of the half and full cells. Each cavity is also equipped with a field tuner and monitor. The tuners are rounded copper plungers which radially penetrate the outer wall of the cavity. The double choke design is used for the field tuners to present a short circuit at the cavity penetration, as in the case of the cathode plug. In the case of the tuners, however, the shorting plane is fixed at the inner wall of the RF cavity independent of the field tuner position. A fixed short circuit at the cavity penetration point avoids RF leakage through the small gap around the tuner head.

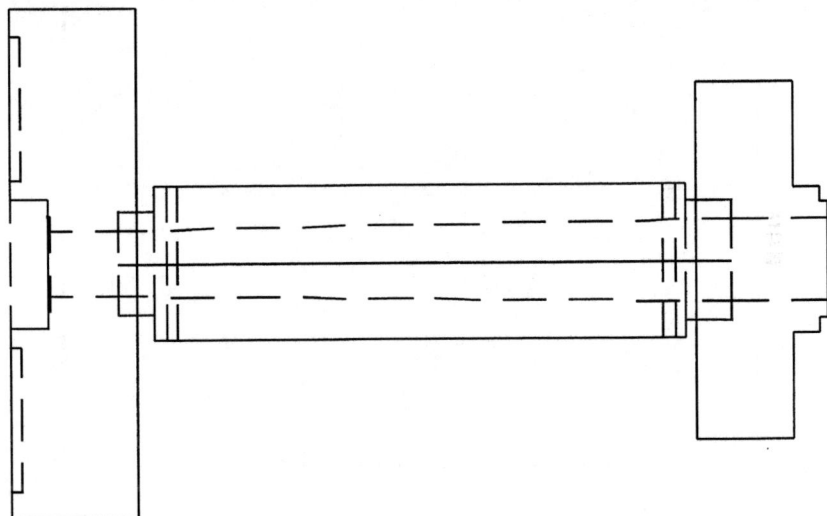

Figure 3: High power waveguide taper design

The field monitors consist of recessed loops which couple to the ϕ component of the TM_{010} mode magnetic field. The field monitors are adjusted by rotation, since the coupling loop polarization strongly effects the interception of the azimuthal magnetic fields. The cross sectional view of the tuner and monitor assemblies are shown in Figs. 4 and 5.

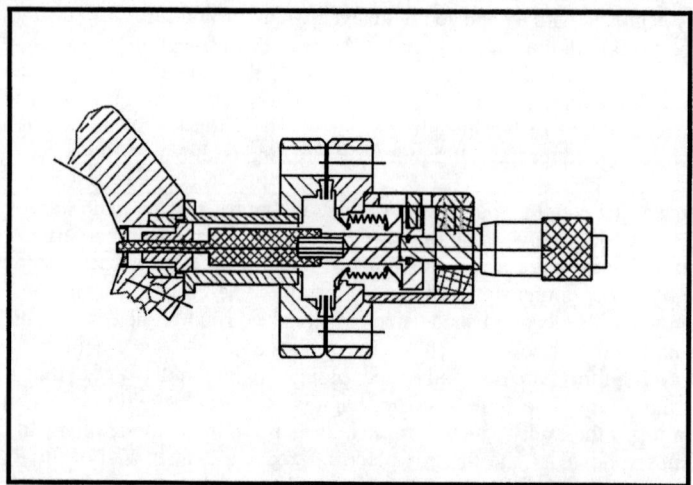

Figure 4: Section view of the field tuner in the high power cavity design

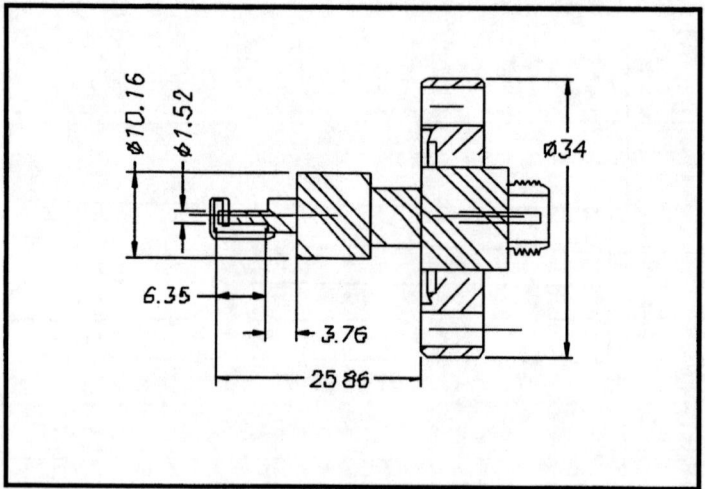

Figure 5: Sectional view of the field monitor in the high power cavity design

A 22.5 cm long, 200 turn, 3.400 kG solenoid is installed at the exit of the rf gun for focusing the diverging beam. An identical bucking solenoid is located at an equal distance behind the cathode plane, providing zero field on the cathode surface where the electron beam is photoemitted. The assembly drawing of the solenoid is shown in Fig. 6. Note that an alternating crossover design was employed to minimize transverse field errors through the solenoid magnet. Low carbon steel yokes with 2 cm thickness were used to reduce stray fields, and to increase the peak field on axis.

The field profile at the maximum drive current of 256 amps is shown in Fig. 7. A plot of the magnetic field lines produced by a POISSON simulation is shown in Fig. 8.

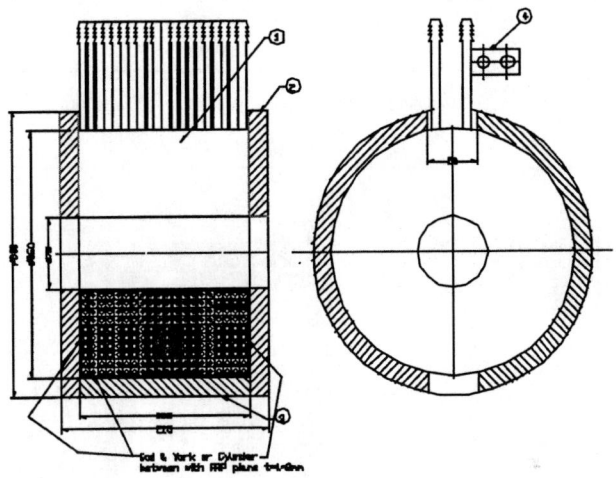

Figure 6: Assembly drawing of solenoid

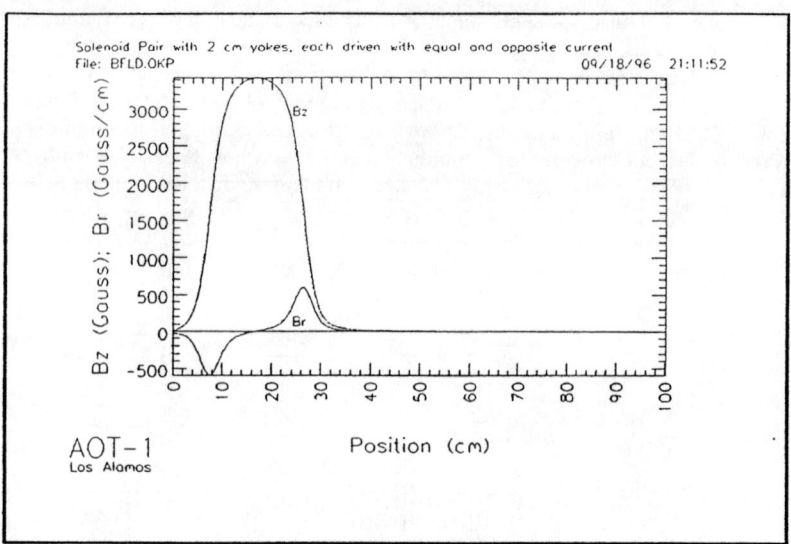

Figure 7: Solenoid magnetic field profile

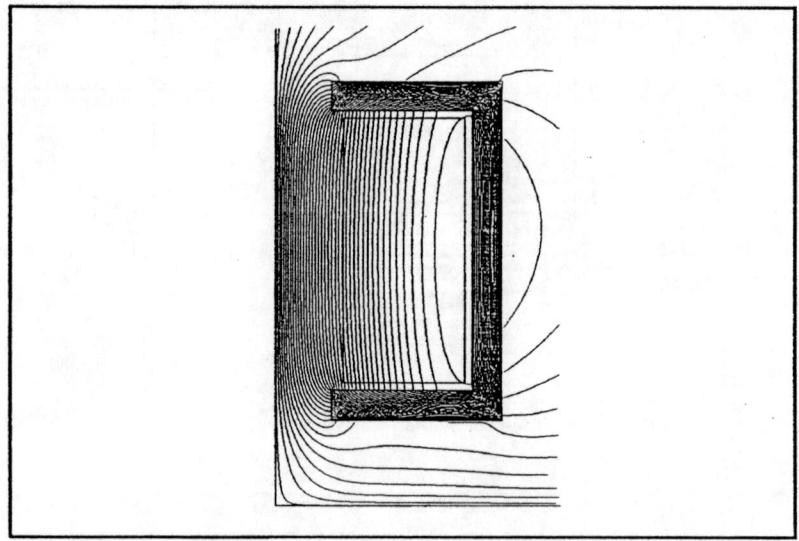

Figure 8: Solenoid magnetic field lines

3. COLD TEST RESULTS

The cold test cavity optimization procedure is as follows. A sliding short plate at one end of the feed waveguide is adjusted to a distance of $N\lambda_g + 3/4\ \lambda_g$ from the center of the coupling hole. The standing wave peak in the guide is thus maximized over the coupling apertures. The frequencies of the half cell and full cell are initially detuned so that their resonances do not overlap. The waveguide lateral position is then adjusted until the magnitude of the reflection coefficient for the half cell and full cell are roughly equal. This assures that rf power is coupled into the separate cells with similar amplitude. The field tuners are then used to merge the resonances of the half cell and full cell. A bead pull measurement is performed to check the balance between the field amplitudes in the half cell and full cell. The phase from each field monitor is also measured to look for 180 degrees phase difference, corresponding to the π-mode. The monitors form a second port for the network analyzer in this case. This procedure is iterated with small changes to the size of the coupling holes.

The effect of cathode tuning is drastically larger than that of the field tuners, obviously due to its larger diameter. The cathode plug will be tightly locked in position very nearly flush to the rear cavity wall for high power operation. The waveguide shorting plane will also be fixed for the high power structure.

An example bead pull measurement is shown in Fig. 9 for a roughly balanced field case. We also show the consistency of the measured data with an URMEL calculation of the simulated field profile. A similar measurement performed at UCD DAS is shown in Fig. 10, and compared to a SUPERFISH simulation of the cavity field profile.

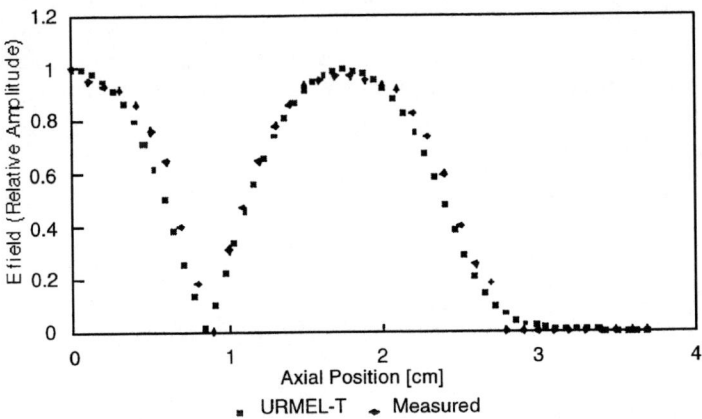

Figure 9: E-field distribution on axis

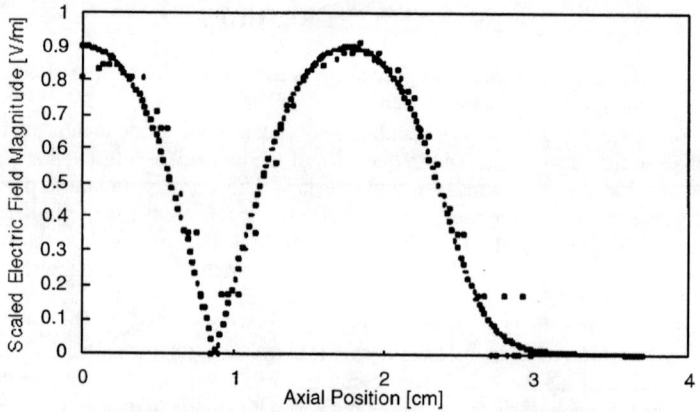

Figure 10: Bead pull result comparison to SUPERFISH

4. HIGH POWER CAVITY DESIGN

After the cold test results had confirmed the resonance, electric field structure, and quality factor of the accelerator structure, and the effectiveness of the field tuners and monitors had been demonstrated, the high power cavity design was completed using the same inner dimensions, but taking into account the need for very good vacuum at high electric field gradients. Since temperature adjustment of the high power structure will ultimately be used to fine tune the resonance of the overall structure, cooling channels were also included to allow careful control of the cavity temperature. Our foremost concern for the high power structure is to maintain 10^{-9} Torr or better vacuum in the accelerator structure during high power operation, since any degradation of vacuum in the cavity can cause arcing problems. The assembly drawing of the high power cavity now under construction is shown in Fig. 11.

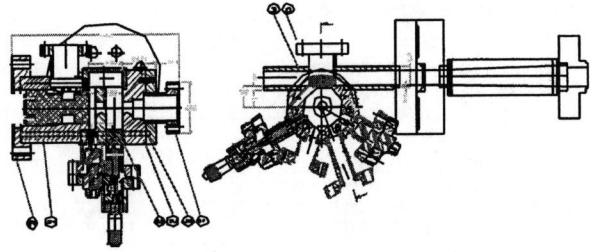

Figure 11: High power cavity assembly drawing

5. SUMMARY

A prototype X-Band rf accelerator cavity has been fabricated at SRRC, and has been characterized and optimized at UC Davis DAS. Many unexpected, and very important developments have been revealed through experimentation with this test apparatus. The high power OFE copper cavity is currently being machined, and will shortly be used to measure some basic characteristics including dark current, VSWR, maximum obtainable gradient, beam energy and charge, as well as the profile of the beam as a function of solenoid focusing. Careful measurement of emittance and energy spread will follow these initial experiments.

6. ACKNOWLEDGMENTS

We would like to thank Dr. G. Westenskow (LLNL) and Dr. Y.J. Chen (LLNL) for many useful discussions and suggestions. We also would like to thank Dr. Roger Miller (SLAC) for useful insight directly prior to cavity machining, and Dr. Dian Yeremian (SLAC) for beamline diagnostic advice. This work is supported in part by the National Science Council (Taiwan), as well as by MURI (USA) under contract F49620-95-1-0253, by ATRI (USA) under contract F30602-94-2-0001, and by DoE (USA) under contract DE-FG03-95ER54295.

REFERENCES

[1] G.P. Le Sage, et al., "2.142 Ghz Repetition Rate High Brightness X-Band Photoinjector", in Proc. 1995 IEEE Particle Accelerator Conf. (Dallas, Texas, USA).

[2] G. A. Westenskow and J.M.J. Madey, "Microwave Electron Gun", Laser and Particle Beams (1984), Vol. 2, Part 2, p. 223.

[3] J.S. Fraser, et al., "High-Brightness Photoemitter for Electron Accelerators", Proc. 1985 IEEE Particle Accelerator Conf. (Vancouver, BC), p.1791.

[4] C. Travier, Nucl. Instr. Methods Phys. Res. A, A304, 285 (1991).

[5] C.H. Ho, "A High Current, Short Pulse Electron Source for Wakefield Accelerators", Ph. D. Dissertation (UCLA), 1992.

[6] K.T. McDonald, "Design of the Laser-Driven RF Electron Gun for the BNL Accelerator Test Facility", IEEE Trans. on Electron Devices, Vol. 35, No. 11, November 1988.

[7] G.P. Le Sage, et al., "A High-Brightness X-Band Photoinjector", in this proceeding.

[8] R.E. Collin, "Foundations for Microwave Engineering", 2nd ed., McGraw-Hill, Inc., 1992, p. 394-397.

[9] L.C.-L. Lin, et al., "Waveguide Side-Wall Coupling for RF Guns", Proc. 4th European Particle Accelerator Conf. (London, 1994), p. 1471.

High Power Operation of a 17 GHz Photocathode RF Gun

S. Trotz, W. J. Brown, B. G. Danly, J.-P. Hogge, M. Khusid, K. E. Kreischer, M. Shapiro and R. J. Temkin

Plasma Fusion Center
Massachusetts Institute of Technology
Cambridge, MA 02139 USA

ABSTRACT

We report the first operation of a 17 GHz RF photocathode electron gun. This is the first photocathode electron gun to operate at a frequency above 2.856 GHz. Such electron guns have the potential for achieving record high values of electron beam quality. The $1\frac{1}{2}$ cell, π-mode, copper cavity was tested with 5-10 MW, 100 ns, 17.145 GHz pulses from a 24 MW Haimson Research Corp. klystron amplifier. Klystron power is stable to within ±5% up to 15 MW. Conditioning resulted in a maximum surface field of 250 MV/m, corresponding to an on-axis gradient of 150 MV/m. Dark current of 0.5 mA was observed at 175 MV/m, consistent with Fowler-Nordheim field emission theory if a field enhancement factor of about 100 is assumed. Electron bunches were generated by a regenerative laser amplifier that produces 2 ps, 1.9 mJ pulses at 800 nm with ±10 % energy stability. These pulses were frequency tripled to 46 µJ of UV, and then focused on the wall of the cavity. Preliminary beam measurements indicate 0.12 nC bunches were produced with a kinetic energy of about 1 MeV. This corresponds to a peak current of about 100 A, and a density at the cathode of 2 kA/cm². Both single and multiple pulse laser induced beam emission were observed.

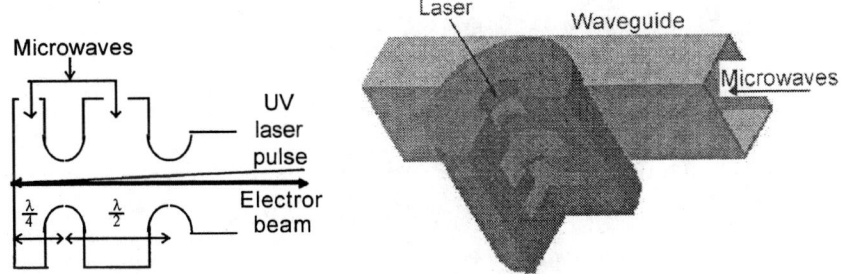

Figure 1. side view of rf gun structure   **Figure 2.** rendered view of rf gun

INTRODUCTION

The goal of the 17 GHz photocathode experiment is to construct an ultra high-brightness source of electrons which can be used for free electron lasers or future linear colliders. The photocathode RF gun is a novel electron beam source intended to meet the requirements set by future high-energy linear colliders and next generation free electron lasers[1]. A coupled pair of pillbox TM_{010}-like resonators is excited by sidewall coupled microwaves at 17 GHz. The axisymmetric structure is shown in cross section in Figure 1. Note that the axial length of the structure is determined by the microwave wavelength. A three dimensional view of the RF gun structure and coupling waveguide is shown in Figure 2. A picosecond ultraviolet laser pulse illuminates one wall of the structure at the axis of symmetry. Electrons are released by the photoelectric effect and are accelerated by the electric field of the microwaves in the cavity.

According to photoinjector scaling laws[2,3], operating at high frequency will allow production of extremely high brightness beams. The high frequency of operation raises the RF breakdown limit allowing strong electric fields to be used. In turn, the intense fields result in rapid acceleration of the electrons to relativistic speeds and reduced space charge induced emittance growth. The variation of rms normalized emittance with frequency is shown in Figure 3 according to scaling laws including the assumption of emittance compensation (solid curve). Data points indicate experimental results. Note that it is possible to achieve lower emittance than the scaling law suggests at the cost of lower charge in an electron bunch. Figure 4 shows the theoretical variation in beam brightness with frequency[4].

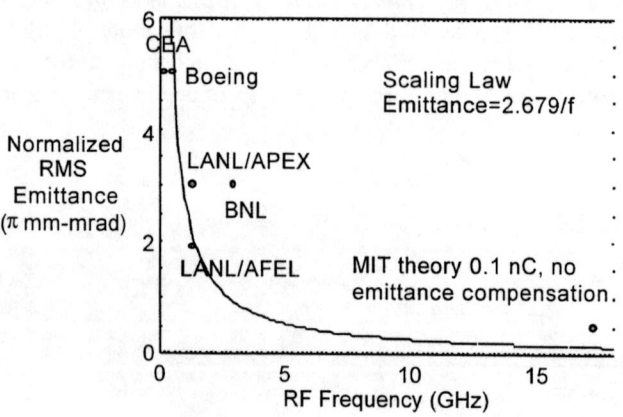

Figure 3. emittance frequency scaling

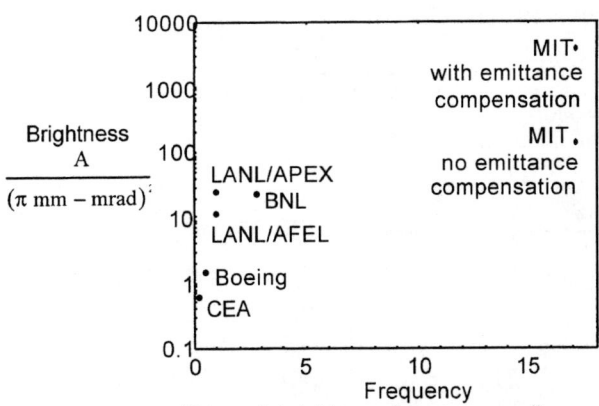

Figure 4. brightness frequency scaling

RF CAVITY AND TRANSPORT LINE

As shown in Fig. 1, the MIT 17 GHz RF gun is a traditional B.N.L.-style sidewall coupled 1.5 cell structure[5]. The RF gun is clamped, not brazed, and has been cold tested using a vector network analyzer. The measured parameters are given in Table 1. Note that the peak electric field is not measured directly but inferred from the energy balance equation.

Table 1 - rf gun measured parameters

Frequency	17.145 GHz (6*SLAC frequency)
Length	1.31 cm (1.5 cells)
Inner Diameter of Cavities	0.538"
Unloaded Quality Factor	1790
Loaded Quality Factor	808
Coupling Coefficient	1.22
Ratio of Peak to Average field	1.21
Material	OFHC Copper
Peak Electric Field at Cathode	250 MV/m (7.2 MW pwr into gun)

The RF gun requires a 7.2 MW, 50 ns pulse to reach the design level field strength of 250 MV/m on axis. The experiment utilizes a 17 GHz relativistic klystron amplifier constructed by Haimson Research Corporation[6]. The klystron is driven by a 580 kV, 1 μs flattop modulator pulse. A Thomson CSF gun produces a 100 A pulse. The beam is space-charge limited and the gun perveance is 0.27 μperv. The amplifier chain includes a TWTA to provide about 10 W to the klystron. The klystron gain is approximately 60 dB. A block diagram of the system components is shown in Figure 5.

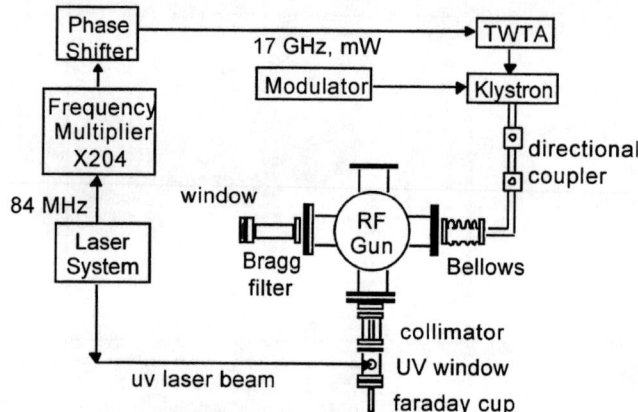

Figure 5. system block diagram

The klystron amplifier has been found to generate power in two spurious modes when connected to a mismatched load. Microwaves associated with these modes are slightly higher in frequency than the operating frequency of the klystron. To eliminate these modes which reduce the stability and gain of the amplifier, a Bragg filter has been designed and installed on the RF gun coupling waveguide. This filter is reflective for frequencies in the range 16.7 to 17.1 GHz and is transparent for higher frequencies. The filter has eliminated one spurious mode and variation of magnetic optics has shifted the second mode so that it occurs after the microwave output pulse. The klystron power stability is better than ±5%. The Bragg filter structure is shown in Figures 6 and 7. As shown in Figure 5, the microwaves transmitted through the Bragg filter and exit the RF gun vacuum vessel via a window. Note that the filter is not completely reflective at the operating frequency reducing the power coupled into the RF gun.

Figure 6. rendered Bragg filter   **Figure 7.** Bragg filter cross section

The phase stability of the klystron output is better than 5° measured using a phase discriminator. However, this measurement was limited by noise and the actual phase stability figure may be lower.

LASER

The illumination for the RF gun photocathode is produced by a Ti:Sapphire laser system. The laser system produces 2 ps, 1.9 mJ pulses at 800 nm. The pulse duration of 2 ps at 800 nm was verified using a single shot autocorrelator. Pulse-to-pulse laser energy fluctuations are approximately ±10% at 800 nm. The laser system parameters are summarized in Table 2.

Table 2. laser system parameters

Parameter	Design	Measured
Initial Wavelength	800 nm	800 nm
Tripled Wavelength	267 nm	267 nm
Repetition Rate	10 Hz	10 Hz
IR Energy per pulse	1.5 mJ	1.9 mJ
UV Energy per pulse	10 µJ	46 µJ
Energy Fluctuation (IR)	± 10 %	± 10 %
Pulse Length	2 ps	1.9 ps
Phase Jitter	< 1 ps	unknown
Timing Jitter	< 1 ns	< 1 ns
Polarization	> 99 %	> 99 %
Beam divergence	0.5 mrad	0.2 mrad
Laser Spot Radius	0.5 mm	1 mm
Beam pointing error	< 50 µrad	< 50 µrad
Modelock frequency	84 MHz (tunable)	84 MHz tunable

The pulses are frequency converted using the nonlinear optical properties of a 10 mm cube of KDP crystal for second harmonic generation to blue, 400 nm, light and a 5 mm cube of BBO crystal for sum frequency generation into the ultraviolet, 267 nm. The collinear arrangement of crystals shown in Figure 8 is possible since group velocity dispersion is not significant for 2 ps pulses in crystals of these sizes and walk-off is not significant.

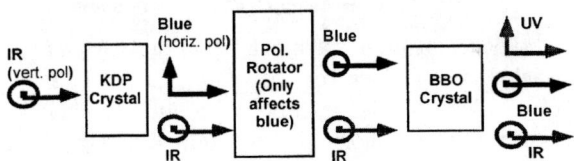

Figure 8. harmonic generation system

The UV pulses each contain up to 46 µJ of energy which is more than sufficient to create 0.1 nC from a copper cathode under the strong microwave field enhancement of the photoelectric effect. Equation 1 gives the dependence of quantum efficiency on field strength and wavelength for copper[7].

$$\eta = A\left(h\nu - \phi + \sqrt{\frac{eE}{4\pi\varepsilon_0}}\right)^2 \quad (1)$$

$A = 13 \cdot 10^{-3} eV^{-2}, \phi = 4.65 eV$ (Cu), $h\nu = 4.69 eV$ (267 nm)

The laser is located in a lab down the hall from the accelerator due to space constraints and requires stable beam transport over about 40 m to the RF gun. The beam is focused using a 5:1 reduction telescope to 2 mm diameter near the nonlinear optic crystals. The UV laser pulse is injected into the RF gun by means of a prism mounted inside a Faraday cup which faces the exit hole of the RF gun.

TIMING

In order to produce a high quality electron beam, emission of the electron bunch via the photoelectric effect must be synchronized with the electromagnetic field inside the RF gun. Synchronization of the UV laser pulse with the microwaves in the RF gun has two components, amplitude and phase. Amplitude corresponds to cavity filling on a 50 ns timescale. Phase corresponds to the phase of the accelerating field on a 1 ps timescale. (A picosecond is 6 degrees of RF phase at 17 GHz.) These two physical constraints are illustrated in Figure 9.

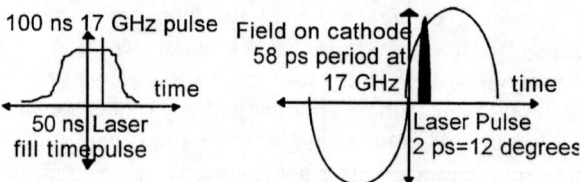

Figure 9. synchronization timescales

Phase synchronization is achieved via a somewhat novel approach. There is no external master clock or oscillator. The Ti:Sapphire oscillator is regeneratively modelocked. As laser pulses bounce back and forth in the cavity, a photodiode samples a small portion of the light. This signal is filtered, amplified, and used to drive an acousto-optic modulator. As a result, the frequency of modelocking is determined by the optical length of the cavity. $f_{Modelocking} = \frac{c}{2L}$. The length of the optical cavity is made stable by use of an INVAR metal tube which has a low thermal expansion coefficient. Typically, the frequency is 84 MHz plus/minus 0.1 ppm over hours of operation. The stability of the laser and the presence of the photodiode signal permit the use of the laser as the master clock. The 84 MHz photodiode signal is multiplied by 204 using solid-state electronics to 17 GHz for use as a milliwatt level input to the microwave amplifier chain.

Experimentally, the photodiode signal is taken out of the laser, amplified at the 84 MHz frequency, and transported to the RF gun control room via BNC cable. Next, the 84 MHz signal is converted to 17 GHz, phase shifted, and attenuated. The low power 17 GHz signal is then sent to the TWTA via WR-62 waveguide. As shown in Figure 10, spectral analysis of the derived 17 GHz signal is a narrowband, high quality signal with weak sidebands and wings.

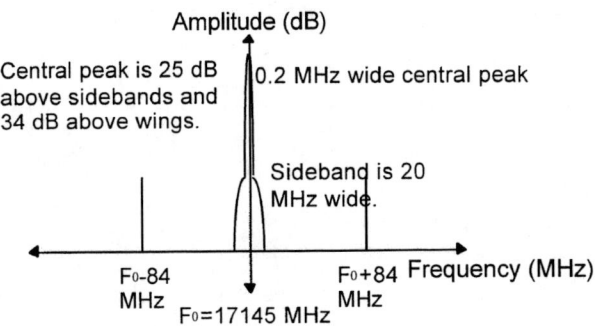

Figure 10. spectral analysis of derived 17 GHz

The synchronization of the laser pulse arrival to the correct amplitude of the electric field in the RF gun involves timescales from 1 second to 1 nanosecond. The sequence of events is shown in Figure 11. The klystron is operated at 1 Hz. The laser fires at 10 Hz. The buildup of the voltage in the high voltage modulator for the klystron requires approximately 150 ms. The buildup of energy in the YAG resonator requires about 140 µs after the time the lamps energizing the rod are fired. The risetime of the current pulse from the modulator is about 2 µs from the time the modulator fire command is sent. The laser pulses are 12 ns apart (84 MHz) and require about 500 ns to reach the RF gun from the time they are injected into the regenerative amplifier.

In order to synchronize these time scales, the laser is used as the master trigger. The laser produces two timing signals, lamp and Q-switch. The lamp signal occurs synchronously with the firing of the YAG lamps and therefore precedes the arrival of the laser pulse by 140 µs ± 12 ns. The variation of the delay comes from the spacing between laser pulses. The lamp signal fires at 10 Hz. A divide-by-ten box is used to send the charge/fire command pair to the klystron modulator. This permits the coarse synchronization of the klystron gain to the laser pulse arrival (to within 12 ns). The second timing signal from the laser is the Q-switch signal. The laser electronics phase lock to the 84 MHz photodiode signal in order to select a pulse to inject into the regenerative amplifier. As a result, the Q-switch signal precedes the laser pulse arrival at the RF gun by 500 ns ± 1 ns (perhaps less jitter, unobservable using existing oscilloscopes.) The Q-switch signal is used to turn on and off the microwave drive for the klystron by modulating the TWTA voltage grid and a microwave pin diode switch. Using this scheme, we have observed < 1 ns timing jitter between laser pulse arrival at the gun and the microwave input to the klystron. Laser synchronization is verified using an infrared photodiode located near the RF gun which detects scattered light from a turning mirror.

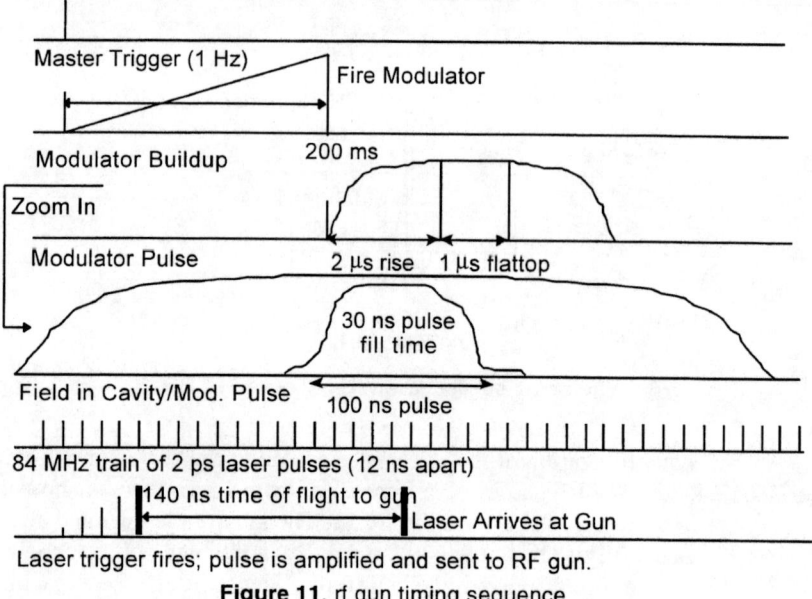

Figure 11. rf gun timing sequence

BEAM DIAGNOSTICS

Existing electron beam diagnostics are limited to a high speed Faraday cup shown in Figure 12. The Faraday cup is designed as a tapered 50 Ω transmission line with a type N connector[8]. The hole in the central conductor serves to reduce secondary emission noise and allows passage of the UV laser beam into the RF gun. It has been found that the Faraday cup produces anomalous signals about 1 V in magnitude unrelated to the presence of electrons, when high power microwaves are present. Therefore, the signal to noise ratio of the Faraday cup was dramatically improved when isolated from microwaves and other electronic noise using a narrow metal collimator acting as a cutoff waveguide. With a 1.95" long, 0.2" i.d. collimator in place, the noise level on the Faraday cup was about 15 mV.

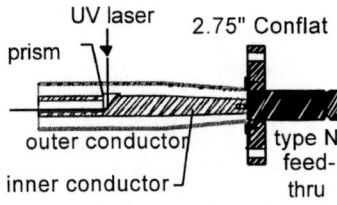

Figure 12. faraday cup

HIGH POWER OPERATION

High power tests of the RF gun with the klystron indicate consistent filling of the RF cavity and the generation of field gradients up to 180 MV/m without breakdown. Occasional breakdown of the RF gun is observed at higher powers corresponding to field gradients up to 250 MV/m.

The laser photodiode signal has been frequency multiplied to 17 GHz and used to drive the klystron. Using the laser derived 17 GHz signal, over 10 MW stable power output has been demonstrated. The RF gun cavity has demonstrated stable filling using the laser derived megawatt microwave pulses. The laser pulse arrival has been synchronized with the RF gun cavity filling at the sub-nanosecond level. This time scale corresponds to the amplitude of the electric field in the RF gun as explained previously. The IR laser beam has been transported to the experimental area, converted to UV, and injected into the RF gun.

MEASUREMENT OF DARK CURRENT EMISSION

The reduced noise level on the Faraday cup diagnostic made it possible to observe the variation of dark current emission from the RF gun with field strength in accordance with the Fowler-Nordheim law. See Figure 13.

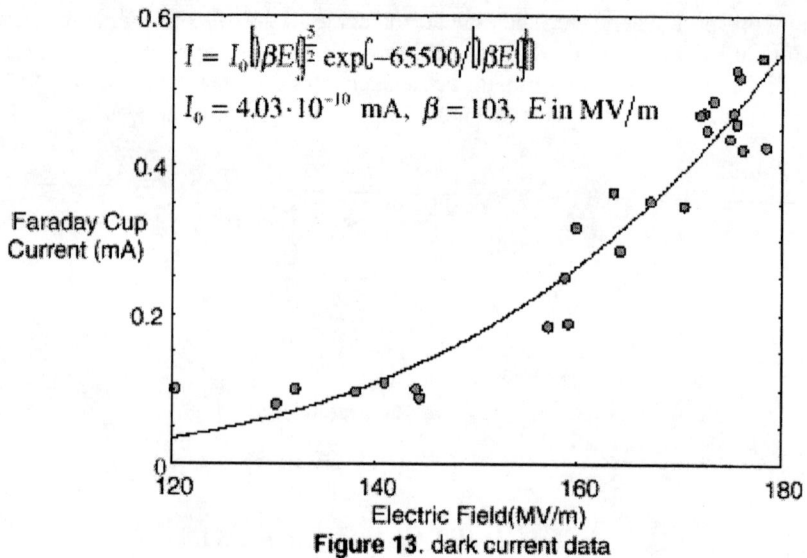

Figure 13. dark current data

EXPERIMENTAL RESULTS ON LASER-INDUCED ELECTRON BEAM EMISSION

The most important achievement in the 17 GHz RF gun experiment has been the demonstration of laser induced emission. Production of an electron beam from the 17 GHz RF gun at M.I.T. marks the first successful operation of a photocathode RF gun at a frequency higher than 3 GHz. The goal of laser induced emission was expected to be signaled by the appearance of an extremely brief Faraday cup signal whose width would be dictated by the capacitance of the measurement system, a few ns, and whose area would correspond to the total charge emitted from the RF gun. This signal would appear different from dark current signals whose duration would be commensurate with RF drive pulse lengths. Also, breakdown signals on the Faraday cup would be wider, appear in the absence of the laser, and would be accompanied by a rise in pressure.

Pulses matching the expected signature were observed the first time the system was operated using the collimator on the Faraday cup. As shown in Figure 14, an extremely narrow, about 4 ns long, negative Faraday cup signal follows the positive photodiode signal by about 6 ns. This delay is consistent with the physical location of the photodiode detector relative to the RF gun and Faraday cup. UV pulses were measured to be about 20 μJ. The integrated Faraday cup signal indicates 0.12 nC of charge, consistent with theoretical expectations. The calculated electric field for this shot was 130 MV/m at the time of laser injection.

These Faraday cup signal spikes are present only when the laser beam is injected into the RF gun.

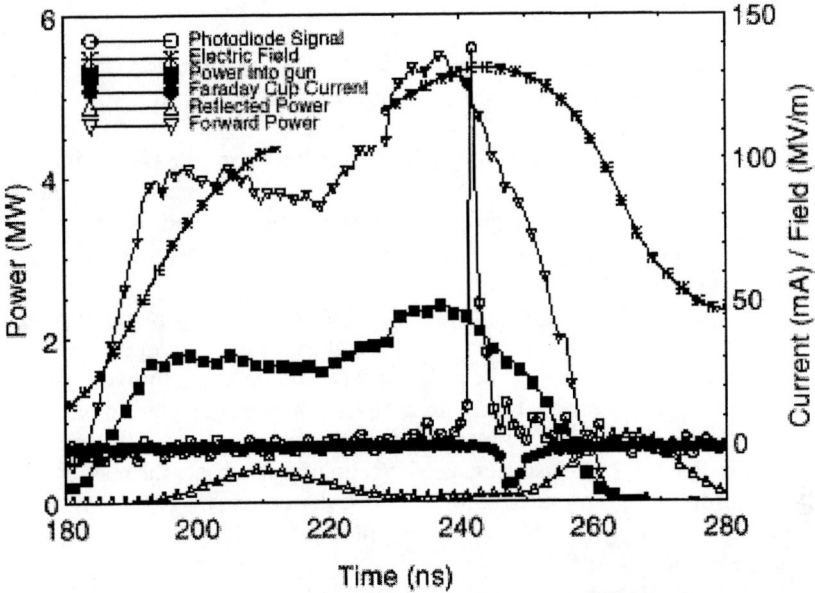

Figure 14. observation of laser induced electron beam emission (Faraday cup signal)

Further evidence for the identity of the Faraday cup spikes is given by multiple pulse emission. Imperfect dumping of the regenerative amplifier cavity occasionally produces multiple UV laser pulses of reduced energy about 10 ns apart in time. These multiple pulses are observed on the photodiode diagnostic and in some cases produce multiple Faraday cup spikes with the same temporal structure.

Table 3. beam properties

Parameter	Design	Measured
Bunch Charge	0.1 nC	0.12 nC
Kinetic Energy	2 MeV	≈ 1 MeV
Bunch Length	0.5 ps	-
RF Injection Phase	12°	-
Current Density at Cathode	6.7 kA/cm^2	≈ 2 kA/cm^2
Emittance (normalized rms)	0.47 π mm-mrad	-
Energy Spread	0.18 %	-
Peak Current	210 A	≈ 100 A
Brightness	9.9E13 A/(m-rad)^2	-

FUTURE PLANS

Short term plans focus on phase locking and beam quality measurement. Existing beam measurements are summarized in **Table 3**. To date, phase locking of the laser to the microwave field in the RF gun has not been demonstrated. Approximately 1 in 6 shots demonstrate laser emission regardless of the setting of the phase shifter in the microwave chain. This percentage is consistent with random phasing of the laser relative to the microwaves. Sources of phase jitter are being investigated.

A Browne-Buechner magnetic energy spectrometer employing a Faraday cup as a detector has been constructed and will be installed on the RF gun beamline. The use of a pepper pot or beam position monitor is being considered for emittance measurements. Pulse length measurement is planned using an RF kicker cavity.

Long term plans include modifications to improve the performance of the 17 GHz RF gun. While scaling laws indicate very high beam quality at 17 GHz without magnetic focusing, beam quality can be further improved (see Figure 2 theory data points for MIT experiment) with the implementation of a solenoidal magnetic field[9]. Typically a bucking coil to zero the field at the cathode and a second solenoid for focusing are used. Scaling laws indicate the need for about 1 T strength field for emittance compensation. Permanent magnets may be suitable considering the small field volume required. Iron pole pieces will be needed to concentrate magnetic flux.

It would be useful to implement a brazed RF gun to reduce power consumption and increase the maximum attainable accelerating gradient. In order to predict the microwave properties of a brazed RF gun from mechanical dimensions to sufficient accuracy, theoretical work is being done on the coupling of the RF gun to the WR-62 waveguide via holes in a finite thickness wall. The axisymmetric code, URMEL, is used to model the unperturbed 1.5 cell resonances and field structure. This data, combined with the quasi-analytical theory for the coupling hole polarizability accurately reproduces the measured S_{11} data for the RF gun. Figure 15 illustrates the theoretical prediction of the RF gun reflection coefficient. Despite careful modelling, a brazed gun will probably require tuners since a 25 μ m error in the radius of a cavity will produce approximately 30 MHz shift in the 17 GHz resonance. Symmetrization of RF fields is being investigated by the adoption of coupling holes on two sides of a racetrack geometry[10].

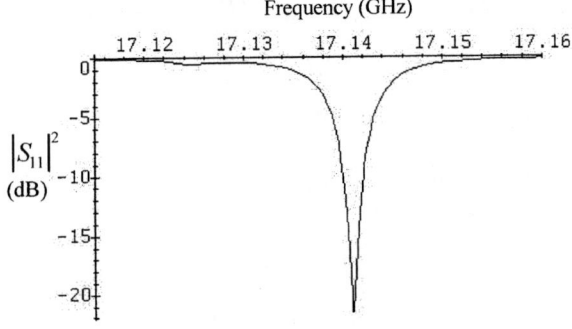

Figure 15. reflection coefficient for rf gun

ACKNOWLEDGMENTS

This work is supported by D.O.E. H.E.P. contract number DE-FG02-91ER40648.

REFERENCES

[1] C. Travier, RF Guns: bright injectors for FEL, Nucl. Instr. And Meth. **A304**, 285 (1991).
[2] Leon Lin, S. C. Chen, and J. S. Wurtele, On the Frequency Scalings of RF Guns, *Proceedings of the Sixth Advanced Accelerator Concepts Workshop*, 1994.
[3] Colby and Rosenzweig, Charge and Wavelength Scaling of RF Photoinjector Designs, *Proceedings of the Sixth Advanced Accelerator Concepts Workshop*, 1994.
[4] Brightness calculated using emittance data from above and published typical current values
[5] K. T. McDonald. Design of the laser-driven RF electron gun for the BNL accelerator test facility. *DOE/ER/3072-43. Princeton University*, 1988.
[6] J. Haimson, B. Mecklenburg, and B. G. Danly. Initial Performance of a high gain, high efficiency 17 GHz traveling wave relativistic klystron for high gradient accelerator research. *In Pulsed RF Sources for Linear Colliders, Montauk, L. I.*, 1994.
[7] J. Fischer and T. Srinivasan-Rao. UV Photoemission studies of metal photocathodes for particle accelerators. *Brookhaven National Laboratory Report*, 42151, 1988.
[8] J. Gonichon. *Theoretical and Experimental Study of a High Frequency Gun Driven by Laser Functioning at 17 GHz*. Doctoral Thesis. University of Paris 7, 1995.
[9] B. E. Carlsten, Nucl. Instr. And Meth. **A285**, 313 (1989).
[10] Jake Haimson. Private communication.

Table Top, Pulsed, Relativistic Electron Gun with GV/m Gradient

T. Srinivasan-Rao, and J. Smedley

Brookhaven National Laboratory, Upton, NY, 11973

ABSTRACT

We present the design and performance characteristics of a compact, high voltage pulser with 150 ps rise time, 0.2 to 2 ns adjustable flat top and up to 1 MV amplitude on a 80 Ohm load or up to 0.5 MV on a 20 Ohm load, at 1 Hz repetition rate. Combination of a laser triggered SF_6 and a liquid gap is used to form the fast rising pulse and maintain a low jitter between the laser, external trigger, and the high voltage pulsed output. The dark current and breakdown studies with this pulse applied between the electrodes of a diode indicate that fields up to 1 GV/m could be supported by stainless steel and copper cathodes without breaking down. The dark current from a conditioned cathode in a background pressure of 10^{-7} Torr is below the detection limit of 0.5 mA of our system. Photoemission studies had been conducted with 300 kV applied between copper cathode and stainless steel anode seperated by 2 mm. KrF laser of 5 eV photon energy and 20 ns FWHM was used to irradiate the cathode. In these preliminary measuremements, 3 nC charge and corresponding quantum efficiency of 3.5×10^{-4} have been obtained. Future plans include increasing the gradient to GV/m range, decreasing the laser pulse duration to ps and subps range and increasing the electron energy to a few MeV.

INTRODUCTION

In the past decade, there has been extensive research [1] in the development of low emittance, high brightness electron injectors for linear collider and free electron laser applications. RF injectors with a few nC charge in a few ps, with an emittance of ~1-5 π mm mrad are operational in a number of facilities [2-4]. The frequency of the RF and the design of the cavity are chosen to minimize the RF and space charge effects on the electron bunch so that low emittance, high brightness electron beam could be generated. Minimization of RF effects on emittance growth require a low RF frequency while minimizing the space charge effects require high field and hence high RF frequency. The design is hence a compromise between these two conflicting requirements. Some of these limitations could be overcome by using a large pulsed electric field at the cathode rather than a RF field. An added advantage of these high fields on metal surface is the lowering of the work function due to Schottky effect. The change in the work function is given by [5]

$$\Delta\phi = \sqrt{\frac{e}{4\pi\varepsilon_0}\beta E} \quad (1)$$

where e is the charge of the electron, ε_0 is the dielectric constant of free space, E is the applied field and β is the field enhancement.

For a field of 1 GV/m, and a field enhancement factor of 3 [5], the change in the work function can be calculated to be ~2 eV. This opens up the possibility of using either the infrared or visible radiation to overcome the work function and to extract the electrons from low workfunction metals such as yttrium or magnesium. The complexity and cost of the laser system associated with the photocathode is then significantly reduced.

DESIGN OF THE PULSER

The breakdown studies conducted by Juttner et. al. [7] and Mesyats et. al. [8] indicate that metals could withstand field gradients of a few GV/m if the duration of the field is ~ few ns. For HV pulses with pulse duration less than 10 ns, both cathode initiated and anode initiated breakdowns become less probable, the breakdown voltage become relatively insensitive to the cathode and anode materials, formation of microscopic craters due to explosive emission become less frequent and the erosion of the electrodes decreases significantly. Hence uniform field gradients of ~1 GV/m on macroscopic surfaces would necessitate a HV pulse with voltage amplitude ~ 1 MV, pulse duration of a few ns with subns rise and fall times. In addition, to extract photoelectrons, this pulse should be synchronizable to a laser beam. HV pulses synchronizable to the laser pulse within 150 ps had been achieved [9] by laser triggering high pressure gas closure switches. In this section, we describe the design of such a high voltage pulser constructed to meet these requirements.

A photograph of the pulser is presented in Fig. 1. The pulser consists of three major sections: the low voltage pulse generator, the transformer and the high voltage transmission line with pulse sharpeners.

Fig. 1 Photograph of 1 MV pulser including the SF_6 gap, transmission line, diode and associated detection system and vacuum cell.

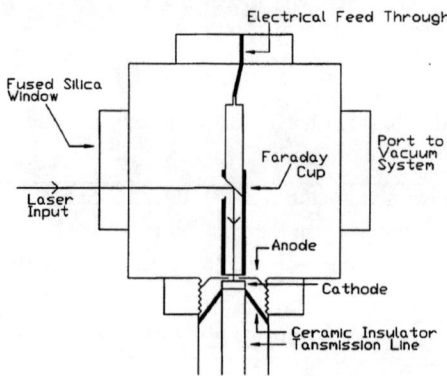

Fig. 2 Schematic of the vacuum cell containing the cathode, anode, and Faraday cup.

The low voltage pulse generator section is made up of a D. C. voltage source of 10 - 25 kV, a trigger generator and a low voltage switch. The trigger generator derives its trigger pulse from an external pulse generator that controls the timing of the rest of the system. The low voltage switch is a three electrode spark gap, with the central electrode connected to the trigger generator. The output of this low voltage system is fed to the primary of the high voltage transformer. the voltage regulator of the DC source can be used to vary the final voltage from the pulser.

The high voltage transformer is a Tesla transformer with a copper sheet as the primary and a 80 turn copper winding as the secondary, with the secondary immersed in transformer oil. The primary and the secondary windings of the transformer are designed to meet the resonant conditions. This enables a doubled voltage with polarity opposite to that of the primary to be applied to the diode at the end of the transmission line. The capacitance of the transformer is 1.5 nF, resulting in a rise time of ~500 ns for the high voltage pulse. The duration and the amplitude of the transformer output can be measured using a built in resistive divider with an attenuation coefficient of 18000.

The high voltage from the transformer is transmitted along a pulse forming line with two switches, a high pressure SF_6 gap and a liquid gap, to sharpen the rise and fall times of the HV pulse. The SF_6 gap consists of a hemispherical electrode connected to the HV transformer and a flat electrode connected to the transmission line with an electrically isolated stainless steel wire in the middle of the gap to aid in laser triggering. Two CaF_2 windows centered on the wire permit the laser to irradiate the wire normal to the gap. The gas pressure in this gap can be varied up to 150 psi, although most of the measurements were done at 90 psi. Since the voltage hold off increases with the pressure, the gas pressure could be used to vary the voltage delivered to the gun. Additional pulse sharpening is achieved by a self discharging liquid switch with 0.8-1.5 mm adjustable gap at the end of the pulse forming line. The duration of the pulse can be varied from 200 ps to 2 ns by changing the length of the pulse forming line.

A tapered line transformer, terminated at the cathode, doubles the voltage to 1 MV. The system could be used either with or without this tapered line depending on the required voltage range. Three calibrated capacitive dividers help measure the amplitude and duration of the voltage pulse along the transmission line, as well as to optimize the gap spacing for the required performance. They are located as follows: one is positioned before the liquid switch, one at the beginning of the transmission/transformer line and the third at the end of this line,. Both the transmission line and the capacitive divider lines were designed to accommodate the high frequencies encountered in this system.

The cathode at the termination of the line and the grounded anode held parallel to the cathode act as the diode for generating the electron bunch (Fig 2). Both the electrodes are removable and hence the performance of the diode for different electrode material, and geometry can be investigated without changing the characteristics of the applied voltage significantly. Alternately, for a given electrode material and geometry, the performance of the diode can be studied for various shapes and amplitudes of the voltage pulse, by changing the SF_6 gas pressure, the amplitude of the low voltage and the length of the pulse forming line, without breaking the vacuum. The pulser and the diode with its diagnostics are housed in an enclosure designed to filter the electromagnetic noise associated with such a system. This enables us to minimize the length of the cables to be used and possible distortion associated with the cables.

PERFORMANCE OF THE PULSER

As shown in the timing diagram Fig.3, a pulse generator (Stanford Research System DG 535) acts as the master clock and two suitably delayed trigger pulses trigger the laser and the HV pulser. The laser beam, derived from a KrF laser, was focused between the stainless steel wire and the high voltage electrode of the SF_6 gap such that the beam nearly fills the gap between the two.

The typical laser parameters were, energy: 75 mJ, spot size: 2 mm x 0.2 mm, pulse duration: 20 s and laser wavelength: 248 nm. As shown in Fig 4, with 90 psi of SF_6, the optimum delay between the arrival of the laser at the gap and the breakdown of the gap was ~30 ns, with jitter of 0.7 ns. The maximum voltage amplitude measured at the last probe was 0.9 MV. A 5:2 mix of argon:SF_6 at this pressure reduces the voltage by 10% and jitter to 0.5 ns and the delay to 5 ns. Since this jitter value includes the fluctuations in the laser energy and hence the trigger time, the real jitter between the laser and the HV pulse is expected to be much smaller than the measured 0.5 ns. Jitter values of ~150 ps have been reported [9] with similar arrangements. The pulse shape, shown in Fig 5., indicates that the duration of the pulse is ~0.7 ns, rise and fall times are ~150 ps, and the fluctuation in the voltage amplitude in the flat region is negligibly small.

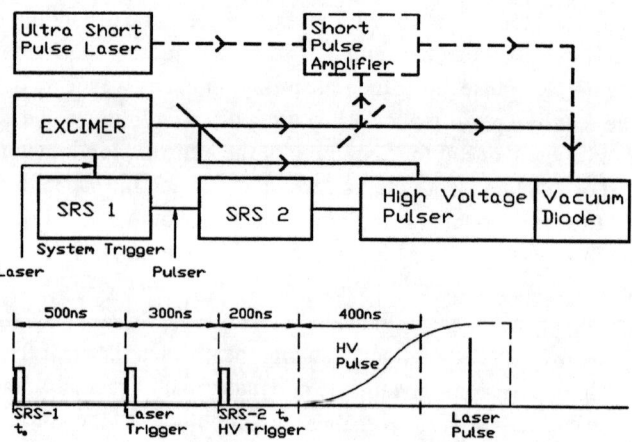

Fig. 3 Timing diagram of the system. SRS 1 is the master clock. The laser is triggered by a signal from SRS 1 suitably delayed and the pulser is triggered by SRS 2 driven by SRS 1

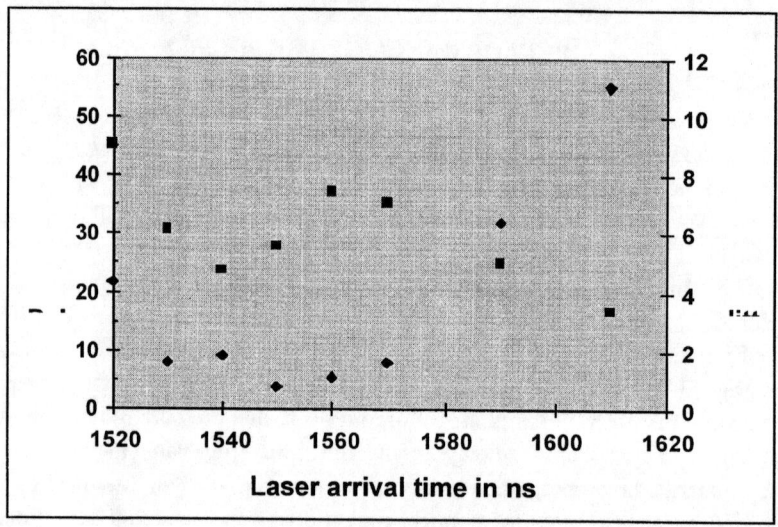

Fig. 4 Delay (squares) and jitter (diamonds) of the HV pulse with reference to the laser pulse for 90 psi of SF_6

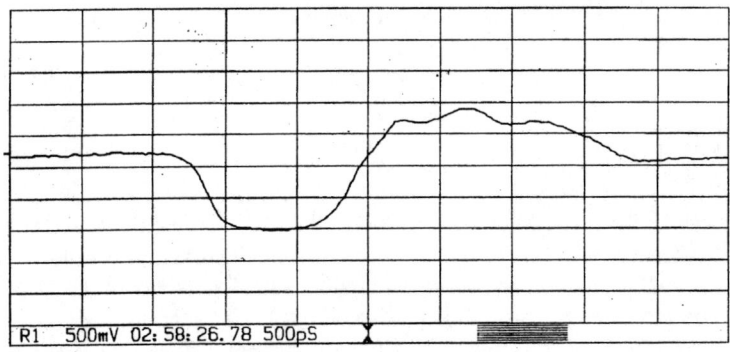

Fig. 5. Shape of the high voltage pulse. The maximum amplitude is 0.9 MV and deconvoluted rise and fall times are ~150 ps.

DARK CURRENT STUDIES

The dark current measurements were done with the pulser in the self trigger mode, without the laser irradiating the SF_6 gap. This enabled us to apply the maximum field on the diode. two different cathode materials, stailess steel and copper, were tested for their voltage hold off properties and dark current emission. Fields exceeding 1 GV/m could be applied to both the cathodes. As can be expected, the dark current depended criticall on the background pressure between the electrodes. The dark current from stainless steel after conditioning, in a field gradient of 0.7 MV/m, was below the detection limit of our our system even at background pressures of ~10^{-3} Torr. Copper on the other hand yielded a dark current of 200 A under similar conditions. When the background pressure was reduced to 10^{-7} Torr, the dark current fell below the detection limit of our system, 0.5 mA.

PHOTOEMISSION STUDIES

The cathode used for these measurements was OFC II copper. The surface preparation is described in detail elsewhere [10]. The anode is a stainless steel cup with a 2 mm diameter hole in the middle for laser irradiation and extraction of electrons. The electrode gap for these measurements were 2 mm. The charge is collected at the Faraday cup consisting of a 3/8" copper rod and copper sleeve fitted tightly over it. The end of the copper rod was cut at 45° and polished to mirror finish and was used as the reflector for introducing the laser beam onto the cathode. A small cut out in the copper sleeve facilitates the transport of the laser beam. The signal from the Faraday cup was measured by the 7 GHz oscilloscope. The entire arrangement was maintained in 10^{-7} Torr pressure.

The laser beam irradiating the SF_6 gap was split into two parts by passing it through a CaF_2 window. The beam transmitted through the window was used to irradiate the gap, while the reflected fraction was used to irradiate the photocathode. The arrival time of the laser at the gap was adjusted for maximum voltage and minimum jitter. The gap pressure was adjusted such that the voltage at the cathode was 300 kV. The photocathode beam was optically delayed so that both the laser and the HV pulse arrive at the cathode synchronously. The beam was then focussed by a lens such that the smallest spot size is at the anode hole to reduce beam loss. The diameter of laser spot on the cathode was ~1 mm. The photoelectric signal from the Faraday cup is shown in Fig.4. The integrated charge arriving at the Faraday cup was 3 nC. The refelctivity of the copper mirror at this wavelength was measured to be 25%. If one assumes that the electrons emitted in the high field alone arrive at the Faraday cup, the energy of the laser beam intercepted by the cathode during this time is ~ 42 µJ. The quantum efficiency of the cathode can then be calculated to be 3.5×10^{-4}, comparable to the value obtained at the ATF [11] with copper cathode in a field gradient of 120 MV/m. Measurement of photocurrent in GV/m gradient is currently underway.

COMPUTER SIMULATIONS

Results of computer simulations using MAFIA to predict the gun performance under different diode geometries are shown in Table 1. The parameters for the simulations are: parallel plate diode geometry, 1 nC charge, flat top charge distribution in both r and z with 10 ps bunch length and 100 µm diameter, and 1 MV on the cathode. The emittance was monitored in three locations along the beam path, in front of the anode (ε_1), immediately after the anode (ε_2) and after a drift distance of 7 mm (ε_3). As can be seen from Table 1, the lowest value of 0.5 mm mrad for ε_3 is obtained for a gap spacing of 1 mm and a hole size of 2 mm. This emittance may further be reduced if the diode geometry, bunch length, and charge distributions are optimized.

Table 1

Gap mm	Hole Radius mm	ε_1 mm-mrad	ε_2 mm-mrad	ε_3 mm-mrad
1	.5	0.3	0.4	0.7
1	1	0.4	0.4	0.6
1	2	0.4	0.4	0.5
2	1	0.4	0.5	0.6
2	2	0.4	0.4	0.7
3	2	0.5	0.5	0.7
4	2	0.5	0.6	0.6

CONCLUSION

In conclusion, a high voltage pulser capable of delivering up to 1 MV with a duration of ~1 ns and rise and fall times of ~0.15 ns has been constructed and tested. Copper and stainless steel cathodes have been tested to withstand 1 GV/m gradient. Under background pressures of 10^{-7} Torr, the dark current is below the detection limit of our system. Preliminary photoemission studies with 5 eV photons of 20 ns duration and field gradients of 150 MV/m have yielded up to 3 nC and a quantum efficiency of 3.5×10^{-4}, in agreement with previous results. Computer simulations indicate that electron beams with 1 nC charge, ps bunch length and 0.5 mm mmrad emittance can be obtained with this diode. Future plans include increasing the field gradient to GV/m range, decreasing the laser pulse duration to ps and subps ranges and increasing the energy of the electron beam to a few MeV.

Acknowledgments

The authors would like to thank Drs. V. Radeka, R. Palmer, W. Willis, and I. Ben-Zvi for their support. The authors would also like to thank X. J. Wang for the fruitful discussions on the detection system, H. Kirk for the valuable help in setting up and running MAFIA, and J. Schill for expert assistance. This work was supported by DE-AC02-76CH00016.

REFERENCES

1. C. Travier, Nucl. Instr. and Meth in Phys. Res. A 340, (1994), 26.

2. B. Carlsten, IEEE J. Quant. Electron. QE 27, (1991), 2580.

3. K. Batchelor et.al., Nucl. Instr. and Meth. in Phys. Res. A 318, (1992), 372.

4. R. Sheffield et.al. in Proc. 1992 Linear Acceleration Conf., Ottawa, Canada, Aug. 24-28, 1992.

5. T. Srinivasan-Rao, J. Fischer, and T. Tsang, J. Opt. Soc. Am. B8, (1991), 294.

6. G. A. Mesyats and D. I. Proskurovsky, Pulsed Electrical Discharge in Vacuum, (Springer-verlag, 1989), p. 109.

7. B. Juttner, V. F. Puchkarev, and W. Rohrbeck, preprint ZIE 75-3 (Akad. Wiss., Berlin, GDR, 1975), B. Juttner, w. Rohrbeck,and H. Wolff, Proc. iii Int'l Symp. on Discharge and Electrical Insulation in Vacuum, Paris, France, 1968, p. 209.

8. G. A. Mesyats, and D. I. Proskurovsky, Pulsed Electrical Discharge in Vacuum, (Springer- Verlag, 1989), Chap 1-5.

9. A. H. Guenther, J. R. Bettis, R. E. Anderson and R. V. Wick, IEEE J. Quant. Electron. QE 6, (1970), 492.

10. T. Srinivasan-Rao, J. Fischer, and T. Tsang, J. Appl. Phys. 69, (1991), 3291.

11. X. J. Wang et.al., Nucl. Instr. and Meth. in Phys. Res. A375, (1996), 82.

Acceleration of Kiloampere Current at 2.65 GV/m

Francesco Villa
Frantel, Inc., 13847 Skyline Blvd., Los Gatos, CA, 95030

Introduction

It is unlikely that the rate of growth of high energy physics machines could be sustained at a rate similar to the rate in the past, unless some way of reducing (substantially) the cost per MeV is found. The Livingstone plot (energy in the center of mass versus time) is a straight line in a log-log scale from the experiments of Mme. Curie (a few Mev at the beginning of this century fit the plot rather well) to a time close to the present. However, the main lesson of the Livingstone plot is not an inexorable growth in the center of mass energy, but the fact that this growth is the result of ever-changing techniques, or accelerator types, that evolve following the "straight line" of the plot. After some time, every machine reaches saturation and is replaced by new ideas and more efficient (technically and economically) ways to increase the energy.

The length and to some extent the cost of a linear accelerator is controlled by the accelerating gradient. Until the mid-eighties, the gradient of future linear colliders was advertised as 200/300 MV/m. .[1]. Slowly, reality set in: there is no RF power source efficient and economical enough to drive a 200 MV/m machine, and the measured high fields at breakdown were not usable in practice. By 1991 the gradient had dropped to 100 MV/m while 50-100 MV/m is the range advertised at present, followed by vague statements promising higher gradients at a later stage of development. The demise of the SSC reinforces the perception that we are in a period of slow or "flat" energy growth. Even though the next linear collider incorporates the latest advances in funding strategies and advertising, it is basically a machine of the late fifties, in the multibillion dollars class.

The original motivation for the high gradient work was for a linear collider. From existing data, it was known that high fields (GV/m) can be sustained for pulse duration less than 1 ns, in vacuum, without breakdown of the stressed volume. Fast pulse (aka switched power) machines would solve the problem of breakdown/dark currents and would allow ultrahigh gradients, of the order of a few Gv/m. The work along these lines was abandoned, ostensibly because a good, efficient switch did not appear to be available or technically feasible. For energies well below ~1000 MeV, there are applications of high gradient machines with considerable commercial value i.e. infrared and short wavelength FELs operating in the SASE mode, production of monochromatic x rays for various uses [2],[3], compact injectors for x-ray lithography storage rings, etc. In view of the potential market of high brilliance, compact machines, private funds were made available to produce a proof of principle experiment. The questions to be answered were: can these fields be used for acceleration? Can we extract electrons from a highly stressed cathode without destroying the electrodes? Another problem was the diagnostic for such an experiment. The generation of a megavolt pulse lasting much less than one nanosecond is not trivial. The direct measurement of such pulse is still an unresolved, extremely difficult problem. The description of this experiment will be published elsewhere [4], and we report briefly the results. Work along the same lines is in progress at Brookhaven [5], and it will be reported at this workshop.

Experimental Apparatus

The high voltage pulse was generated by a conventional Marx circuit compressed by an intermediate stage; the final pulse shaping was done by a gas avalanche switch [6, 7] in a Blumlein line. A photograph of the apparatus is shown in Figure 1, and an artist's rendition of the complete setup is shown in Figure 2.

The high voltage pulse was transferred by a coaxial line and applied to a 1 mm anode/cathode gap. The duration of the pulse was of the order of 120 ps. This value is controlled by the geometry of the Blumlein line pulse forming network. The amplitude of the high voltage pulse was determined by the measurement of the energy of the accelerated electrons, by deflecting them with a small dipole made of a permanent magnet. The electrons were extracted from the cathode by a Nitrogen laser pulse impinging on the cathode before the application of the high voltage pulse.

Figure 1. Photograph of the apparatus on a table top.

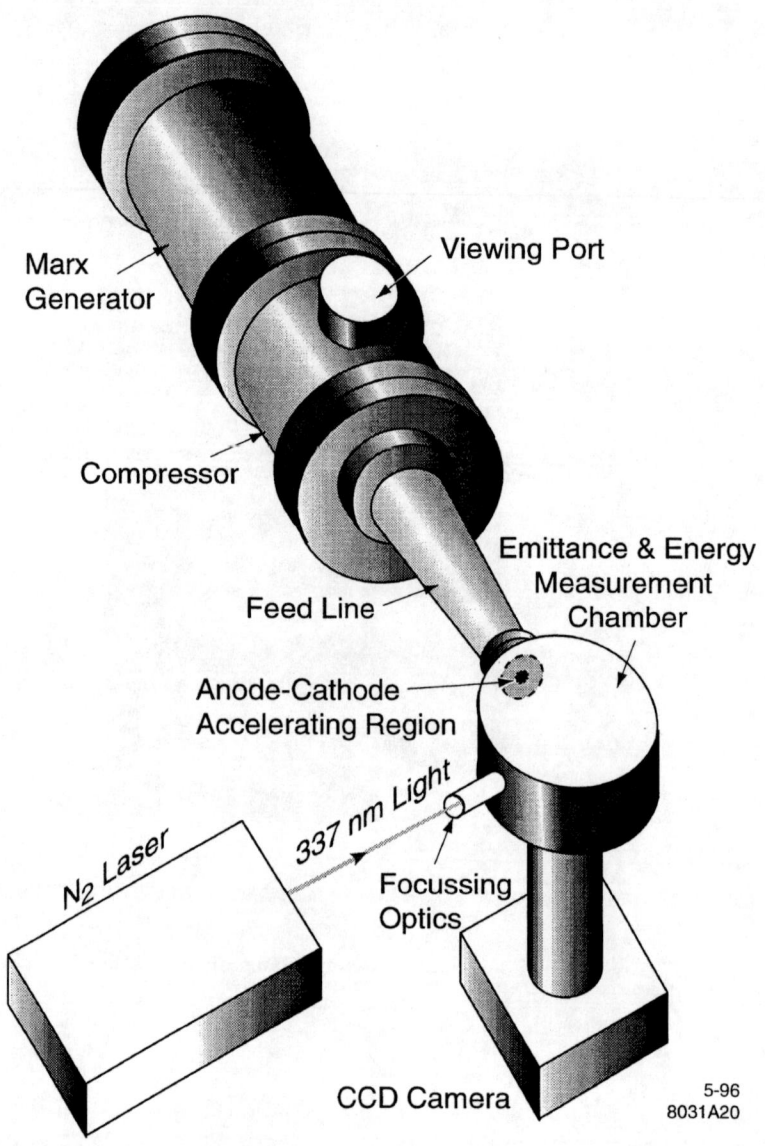

Figure 2. Artist's rendition of the apparatus, including the energy and emittance measurement chamber.

When the energy of the laser pulse exceeds a certain threshold, electrons are emitted by an "anomalous emission" mechanism [8], probably a thermo-assisted emission, or by emission from a plasma This effect has a "memory" in the sense that the cathode will continue to emit electrons for some time (at least more than 15 ns) after the application of the laser pulse, therefore relaxing the requirement of synchronization of the laser pulse with the high voltage pulse. The emittance of the extracted bunch was measured with thepepper pot method [9]. By varying the total extracted charge q and electric field E, we found that indeed the transverse emittance scales as q/E, consistent with the prediction of K..J. Kim [10]. The electric field was varied from 2.65 GV/m to 0.4 GV/m, and the charge extracted varied from a minimum of 0.8 nanocoulomb to a maximum of 170 nanocoulomb. Measured emittance data are plotted in Figs. 3 and 4.

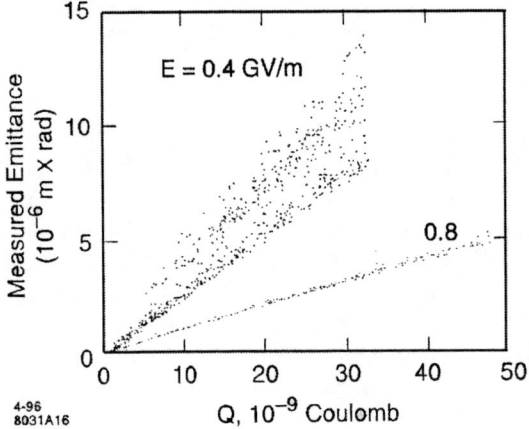

Figure 3. Measured transverse emittance for constant field (0.4 GV/m and 0.8 GV/m) as a function of the charge in the extracted bunch.

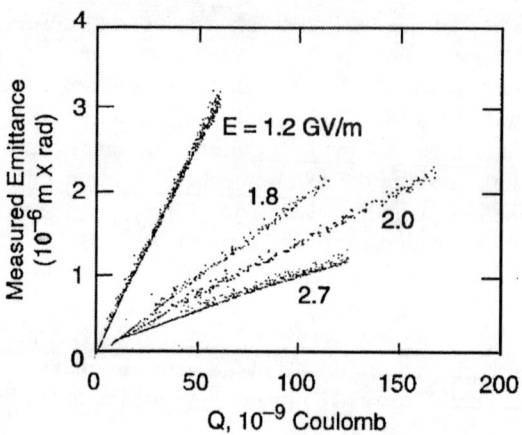

Figure 4. Measured transverse emittance, for constant field (1.2 , 1.8, 2.0 , 2.7 GV/m) as a function of the charge in the extracted bunch.

The measured emittance is plotted versus the charge in the bunch, for constant electric field. From these data we extract the plot in Figure 5, where the invariant emittance is plotted versus the electric field, for constant value of the charge. The maximum electric field was limited by some instability of the Marx generator, not by breakdown in the anode/cathode gap. Therefore the value of 2.65 GV/m is not the upper limit of the field for pulse duration of the order of 120 ps.

There are several problems to be solved in order to use short pulse acceleration for high energy machines, the most critical of all being the synchronization of multiple switches at the picosecond level. For low energy (below ~100 MeV), the problem could be solved by having a single switch. For higher energy machines, multiple switches must be synchronized. Laser "seeding" of the gas avalanche switch could be a possibility. Application to a future collider is precluded by the single shot nature of switched power acceleration, unless bunch neutralization becomes a viable technique to minimize beam beam interaction.

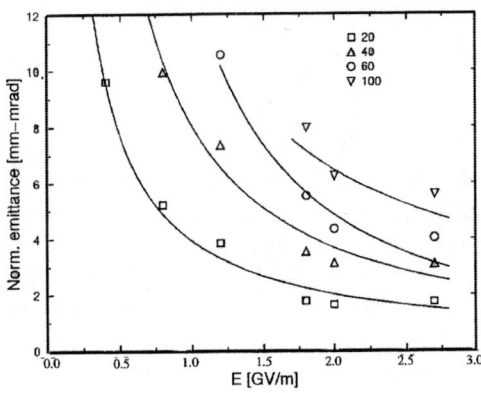

Figure 5. Normalized emittance for constant charge versus field.

Future Plans

The experiment described above shows a possible way to high gradients for a "single bunch" accelerator. High voltage modulators in the hundreds of pulses per second are certainly feasible, but if the final switching (to generate subnanosecond pulses) is done with gas avalanche switches, it is unlikely that the repetition rate could exceed a few hundred/sec. The problem is not the erosion of the switch's electrodes, but its recovery time.

We have demonstrated a current of the order of 1 Kiloampere, for a charge of 120 nC, with a transverse emittance of 1 mm × mrad (Figure 4). The results must be considered an upper limit, since there all sources of error in our apparatus will produce an emittance larger than the actual value. It is likely that this emittance value can be reduced considerably, by studying the dynamics of the gun and by refining the electrode's geometry. An appropriate longitudinal magnetic field, and the rejection of some of the charge may also contribute to a lower value of emittance. The energy spread of the bunch (due to the fact that the electrons are emitted continuously during the application of the high voltage pulse) can be improved by rejecting "off energy" electrons.

The first application of this technology will be a table top FEL in the far infrared, (wavelength longer than 10 microns), if it is found that a substantial market for such a device exists. The FEL will operate in the SASE mode, to be used as a laboratory instrument for infrared spectroscopy, plasma diagnostics, combustion studies etc. The beam energy will be between 5 and 10 MeV, and a set of different undulators will allow the FEL to cover the spectrum from 20 microns up to a fraction of millimeter.

Conclusions

Experimental data show that short pulse acceleration is capable of generating gradients comparable, if not higher, than any other technique proposed so far. The high gradient can be obtained in presence of very large currents without breakdown in the anode/cathode gap.

The transverse emittance of an electron bunch extracted from a cathode scales as q/E (total charge/electric field).

It has been shown that high fields can assist significantly photoelectric emission with the result that metallic photocathodes may achieve a respectable quantum efficiency.

Gas avalanche switches are not limited by electrode erosion if operated in a regime where actual sparking does not occur. Although limited in repetition rate to a few hundred pulses/sec, they appear to have a life expectancy comparable to magnetic compressors driven by thyratrons.

Acknowledgments

This experiment was made possible by the financial support of Harris Blake, Inc. K. Cook, M. Artusy, G. Bowden gave invaluable assistance during different phases. We are particularly grateful to T. Srinivasan and E. Ben-Zvi for their continued interest and helpful comments. This work was evaluated as non productive and not relevant by SLAC management.

References

1. B. Richter, Invited talk presented at the ICFA Seminar on Future Perspectives in High Energy Physics, Upton, N.Y. (October 5-10, 1987)
2. F. Villa, et al. *Microsystem Technologies* 2 (1996) 79-82
3. F.C. Carroll, *Journal of X-Ray Science and Tech.* 4, 323 (1994)
4. F. Villa, A. Luccio, Laser and Particle Beams, to be published
5. T. Srinivasan, paper presented at Advanced Accelerator Concepts Workshop, 1996 , and private communication.
6. F. Villa, R.E. Cassell, SLAC-PUB-4858, (1989)
7. Pincosy, et al., *1992 Proc. of the SPIE* 1631, 277
8. X.J. Wang, J. *Appl. Phys.*72 (3), 888-894 (1992)
9. S.C. Hartman, et al., IEEE Part. Acc. Conference, 1993a , p.561
10. K.J. Kim, *Nucl. Instr. and Methods*, A275, 201, (1989)

Thin Film Photoemission Experiments

R. Brogle, P. Muggli, and C. Joshi

Electrical Engineering Department
University of California Los Angeles, CA 90024

Abstract

The "promptness" of the photoemission process is an important issue in the production of ultra-short electron bunches from photoinjector guns. One would ideally hope to obtain electron emission whose duration faithfully mimics the photon pulse for photon pulses on the order 100 fs for many advanced accelerator applications. If all the photoelectrons were emitted at the surface, the electron current would indeed mimic the photon pulse. However, it is important to experimentally determine the relative contributions of the surface versus the bulk of the material to the net photoelectron current. It is the purpose of these experiments to address this issue.

PHOTOELECTRON DYNAMICS

The standard model of volume photoemission consists of a three step process: (1) the electron absorbs a photon of energy $h\nu$ which elevates it to an excited energy state, (2) the excited electron travels to the surface, and (3) the electron crosses the surface barrier to escape the material. The height of the surface barrier above the Fermi level is $e\Phi$, where e is the magnitude of the electron charge and Φ is the work function of the material. The total emitted charge will depend on the depth into which the incident light can penetrate and the depth from which a photoexcited electron can reach the surface before losing its escape energy.

For optically excited electrons in metals, the primary mechanism for energy loss will be inelastic collisions with the conduction band electrons [1]. A single such collision will on average result in a significant loss of energy for the excited electron because the conduction electrons with which it collides have much lower energies. After the collision the electron most likely will no longer have sufficient energy to overcome the surface barrier and will not be emitted ($h\nu - e\Phi < e\Phi$ for the wavelengths in this experiment). Thus the maximum electron escape depth is dependent on the range l that the photoexcited electrons can travel before suffering a collision.

In general, electrons may also suffer elastic collisions with lattice phonons and may lose energy through plasmon excitation. However, for visible photon excitation energies the electron-phonon mean free path is much larger than the electron-electron mean free path. Therefore, electron-phonon collisions can be

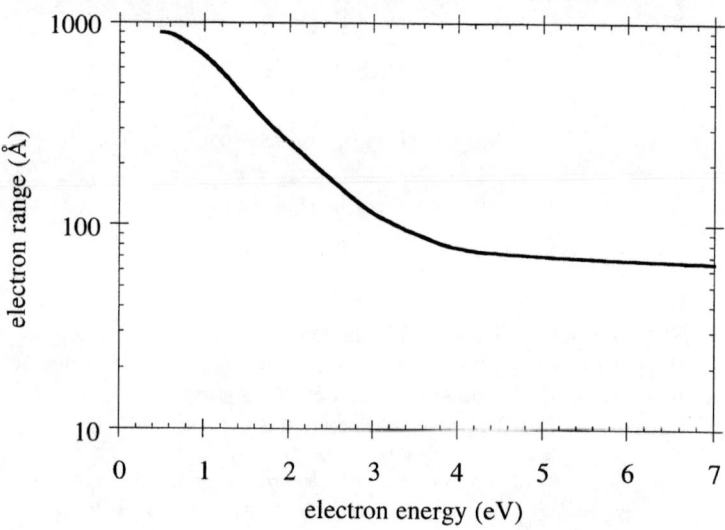

Figure 1: Empirical curve for electron range vs. excitation energy (with respect to the Fermi level) as measured in gold films (from *Sze* et al.)

ignored. Also, the photon energies are too small to create photoelectrons capable of exciting plasma waves ($\hbar\omega_p \approx 11$ eV for copper)—this effect can be ignored as well. The electron range l will then be equal to the electron-electron mean free path, and any electrons which are emitted will have traveled ballistically to the surface.

At visible wavelengths, l is a decreasing function of the excitation energy [2]. This can be explained in terms of the Pauli exclusion principle for the conduction electrons. A photoelectron that has been excited to an energy ε (measured with respect to the the Fermi level ϵ_F) can interact only with those conduction electrons having energy between ϵ_F and $\epsilon_F - \varepsilon$. Conduction electrons below this energy cannot be excited above the Fermi level because the photoelectron energy is less than this energy difference, and there are no unoccupied energy states below the Fermi level. Thus no energy exchange between the conduction electrons below $\epsilon_F - \varepsilon$ and the excited electron can take place. The greater the excitation energy ε, the larger the number of conduction electrons that the photoelectron can interact with, and therefore the shorter the electron range. Fig. 1 shows an empirical curve taken from [1] showing this behavior of the electron range l in gold films.

Linear photoemission from thin metal films can be investigated using a simple one-dimensional analysis [3]. This approach is sufficient because the

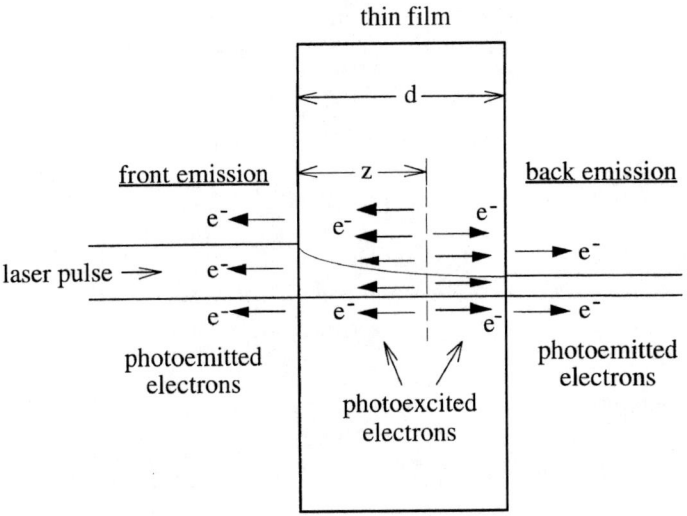

Figure 2: Front and back side photoemission from a film of thickness d.

isotropy of the electron velocities makes it possible to study motion in the $\hat{z}$ direction (the direction normal to the surface) independently from the electron motion in the $\hat{x}$ or $\hat{y}$ directions. In addition, for optically excited electrons most of the kinetic energy must be in the $\hat{z}$ direction in order to overcome the metal surface barrier, thus the emitted electrons will have had primarily z motion in the metal.

Fig. 2 depicts a laser pulse incident on a thin metal film of thickness d. Let us look at electrons at a depth z from the front surface (the surface onto which the laser pulse is incident). As the pulse propagates through the metal, its intensity decreases exponentially:

$$I(z) = I_0 e^{-z/\delta} \qquad (1)$$

where δ is the optical absorption depth at the laser wavelength. The probability of an electron at depth z absorbing one photon is proportional to this intensity:

$$P_\gamma(z) \propto I_0 e^{-z/\delta} \qquad (2)$$

An electron which absorbs a photon must then travel to the front surface without suffering a collision in order to have a chance to escape. The probability P_{nc} of no collision occuring over this distance z is

$$P_{nc}(z) \propto e^{-z/l} \qquad (3)$$

Thus the total probability for the front surface emission of an electron from a depth z is

$$P_{front}(z) = P_\gamma(z) \cdot P_{nc}(z) \propto I_0 e^{-z/\delta} e^{-z/l} \qquad (4)$$

For a film of thickness d, the electrons which lie a distance z from the front surface will be at a distance $d - z$ from the back surface. The probability for emission from the back surface is then

$$P_{back}(z) = P_\gamma(z) \cdot P_{nc}(d - z) \propto I_0 e^{-z/\delta} e^{-(d-z)/l} \tag{5}$$

To find the total electron yield from each surface, we must integrate the emission probabilities over the thickness of the film. For a given incident light intensity the emitted charge from the front is

$$Q_{front} \propto \int_0^d I_0 e^{-z/\delta} e^{-z/l} dz = \frac{I_0}{\frac{1}{\delta} + \frac{1}{l}} \left[1 - e^{-(\frac{1}{\delta} + \frac{1}{l})d}\right] \tag{6}$$

and the charge emitted from the back is

$$Q_{back} \propto \int_0^d I_0 e^{-z/\delta} e^{-(d-z)/l} dz = \frac{I_0}{\frac{1}{\delta} - \frac{1}{l}} \left[e^{-d/l} - e^{-d/\delta}\right] \tag{7}$$

For a laser pulse with a gaussian spatial and temporal profiles the total energy E in the pulse is related to the peak intensity I_0 by

$$E = \sqrt{\pi} \tau r_0^2 I_0$$

where τ and r_0 are $1/e$ "widths" of the temporal and spatial profiles. Therefore we can express the total integrated charge from a one-photon emission process in terms of the total integrated energy of the gaussian laser pulse:

$$Q = bE \tag{8}$$

where the constant b is the electron yield in units of pC/μJ. For the front and the back side emission from a thin film the electron yields will be

$$b_{front} = \frac{K}{\frac{1}{\delta} + \frac{1}{l}} \left[1 - e^{-(\frac{1}{\delta} + \frac{1}{l})d}\right] \tag{9}$$

$$b_{back} = \frac{K}{\frac{1}{\delta} - \frac{1}{l}} \left[e^{-d/l} - e^{-d/\delta}\right] \tag{10}$$

where K is the constant of proportionality having units of pC/(μJ·Å). For large thicknesses b_{front}

$$\lim_{d \to \infty} b_{front} = b_{bulk} \tag{11}$$

and therefore we can determine the value of K from the front emission properties:

$$K = \left(\frac{1}{\delta} + \frac{1}{l}\right) b_{bulk} \tag{12}$$

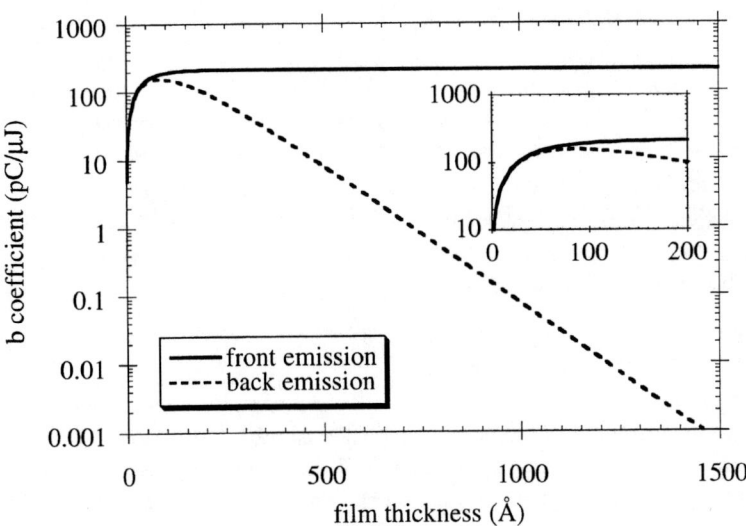

Figure 3: The theoretical behavior of the electron yield (as measured by the b coefficient in Eqs. 9 and 10) vs. film thickness d for front and back illumination of copper by 217 nm light. The optical absorption depth is taken to be $\delta_{217} = 105$ Å and the electron range is $l = 70$ Å.

Fig. 3 shows the behavior of Eqs. 9 and 10 as a function of film thickness d for 217 nm incident light on copper. We used the published optical absorption depth value for copper of $\delta_{217} = 105$ Å [4] and the electron range value suggested by Fig. 1 of $l = 70$ Å for a 5.7 eV excitation energy. For small film thicknesses ($d < l$) the front and back electron yields are equal and increase as the thickness becomes larger. For greater thicknesses ($d > l$) the front yield reaches the bulk b value after about 200 Å and remains constant, while the back yield turns around and decays exponentially, dropping by 5 orders of magnitude over 1500 Å.

FRONT AND BACK ILLUMINATION EXPERIMENT

The setup for the thin film illumination experiment is shown in Fig. 4. The samples were $1'' \times 2''$ fused silica slides with steps of various thicknesses of copper film evaporated onto one side. Two samples were used, one with film thicknesses ranging from 50–500 Å and the other with thicknesses 500–1400 Å. The hollow anode was mounted on a rotation stage with an external control so that it could be positioned under vacuum on either side of the sample for the front illumination (Fig. 4a) and back illumination (Fig. 4b) configurations. A mechanical feedthrough was used to move the sample to illuminate each thickness and also to flip the slide so that the copper side would face the anode for each configuration. The transmission of the laser pulse through each thickness of copper film was measured using a photodiode placed behind the sample outside of the vacuum chamber.

The electron yield b was determined for each film thickness by measuring the emitted charge Q as a function of incident laser pulse energy E. Both the front and back side emission demonstrated the expected proportionality given in Eq. 8. Each 217 nm laser pulse had a duration of 400 fs, and the pulses were varied in energy from 0.001 to 1 μJ. A plot of the front and back electron yield as a function of film thickness is shown in Fig. 5.

From the plot, we see that the front b coefficient increases as the films become thicker until it reaches a bulk value of $b_{bulk} = 190$ pC/μJ at 250 Å. This increase is more gradual than predicted by the theory. As the films become thicker than this value we observe no additional emitted charge, thus we take 250 Å to be the maximum depth (d_{max}) from which electrons can escape the metal for this incident wavelength. Note that the front emission from thicknesses $d > 250$ Å is a factor of five greater than the emission from the thinnest 50 Å sample. This indicates that the majority of the emitted electrons originated in the bulk of the metal and not on the surface. The back b coefficient initially decays exponentially, but then reaches a constant value of ~0.2 pC/μJ at 1000 Å. However, this charge was found to be present with no sample in place—therefore it is background charge caused by scattered 217 nm light.

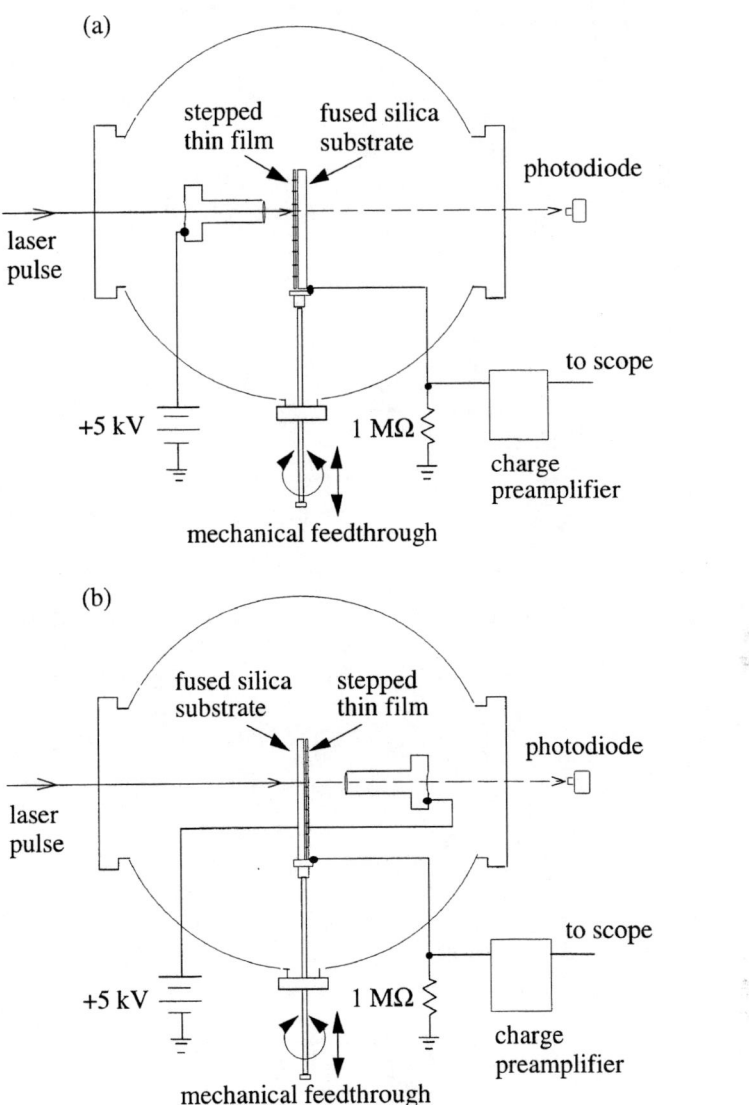

Figure 4: Setup for the multiphoton thin film illumination experiment: (a) front illumination configuration (b) back illumination configuration.

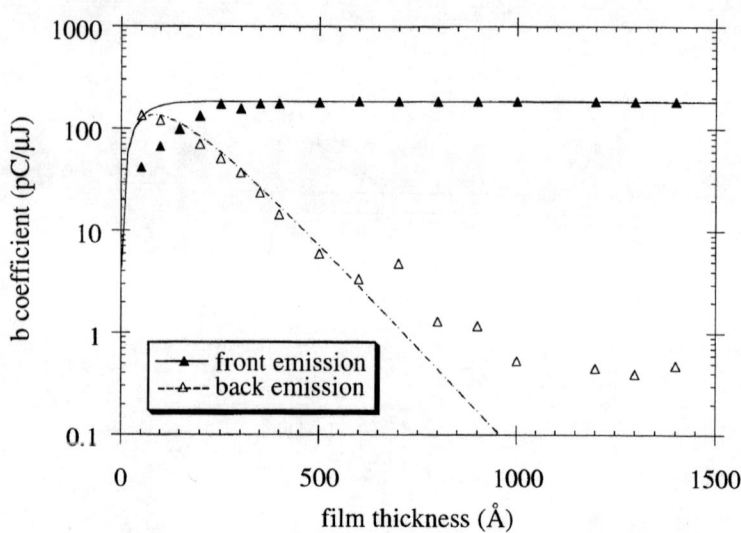

Figure 5: The measured electron yield (b coefficients) vs. copper film thickness for front and back illumination by 217 nm laser pulses, plotted with the theoretical curves. The maximum emission depth is 250 Å.

An unexpected result shown in this plot is that for small thicknesses (50 and 100 Å) the back electron yield is greater than the front yield. However, this may be due to the presence of copper oxide or other contaminants on the metal surface. In the front illumination configuration the laser pulses are incident on the bare copper side; any surface contaminants may absorb some of the light and thereby reduce the energy available to the conduction electrons. In the back illumination configuration, on the other hand, the laser pulses are incident on the surface which has been deposited on the fused silica substrate. This surface is protected by the substrate and thus has no contaminants which may absorb the light. Therefore, for small thicknesses the back emission can be greater because the laser energy incident on the conduction electrons is higher.

EMISSION PROMPTNESS

The measurements of the maximum emission depth d_{max} can be used to obtain an upper limit on the time delay between an incident laser pulse and the resulting emitted electron bunch. Because the photoemitted electrons traveled ballistically to the metal surface, the maximum time delay will be the transit time for electrons from the maximum emission depth to reach the surface:

$$\Delta t_{max} = \frac{d_{max}}{v_e} \qquad (13)$$

The minimum electron velocity v_e for escape will be the Fermi velocity v_F plus the velocity required to overcome the work function $v_{e\Phi}$:

$$v_e = v_F + v_{e\Phi} = \sqrt{\frac{2\epsilon_F}{m_e}} + \sqrt{\frac{2e\Phi}{m_e}} \qquad (14)$$

where m_e is the electron mass. For copper $\epsilon_F = 7.0$ eV and $e\Phi = 4.6$ eV, thus the minimum escape velocity is

$$v_e = 2.0 \times 10^6 \text{ m/s}$$

For the maximum escape depth $d_{max} = 250$ Å measured for the 217 nm emission, the delay time is

$$\Delta t_{max} = 12 \text{ fs}$$

This delay time is relatively small, and will not produce significant electron bunch broadening for incident laser pulses > 100 fs.

Another result from this work is that the electron yield from the back side of a thin copper film can be comparable to the yield from the front side of bulk copper. For example, the yield from the back emission of a 150 Å film is only a factor of two less than the front bulk yield (Fig. 5). Thus a back illuminated photocathode could be used in DC guns coupled to a high frequency RF linac where laser beam access from the front side may be difficult or impossible. Such a photocathode would allow for synchronization to RF cycles or other fast events down the beam line (such as a beat-wave accelerator) which is not possible with conventional thermionic cathodes.

References

[1] S.M. Sze, J. L. Moll, and T. Sugano. *Solid-State Electronics,* volume 7, pages 509-23. Permagon Press, 1964.

[2] J. J. Quinn and R. A. Ferrell. Electron self-energy approach to correlation in a degenerate electron gas. *Phys. Rev.*, 112(3):812-27, Nov. 1958.

[3] C.R. Cromwell, W.G. Spitzer, L.E. Howarth, and E.E. LaBate. Attenuation length measurements of hot electrons in metal films. *Phys. Rev.*, 127(6):2006-15, Sep 1962.

[4] R.C. Weast, ed. *Handbook of Chemistry and Physics.* CRC Press, Boca Raton, Florida, 1984.

Measurements of the Argonne Wakefield Accelerator's Low Charge, 4 MeV RF Photocathode Witness Beam

J. Power, E. Chojnacki[*], M. Conde, W. Gai, R. Konecny,
P. Schoessow, J.Simpson

Argonne National Laboratory
9700 S. Cass Ave Argonne IL, 60439

[*]Nuclear Science, Cornell
Ithaca, NY 14853

ABSTRACT

The Argonne Wakefield Accelerator's (AWA) witness RF photocathode gun produced its first electron beam in April of 1996. We have characterized the charge, energy, emittance and bunch length of the witness beam over the last several months. The emittance was measured by both a quad scan that fitted for space charge using an in house developed Mathematica routine and a pepper pot technique. The bunch length was measured by imaging Cherenkov light from a quartz plate to a Hamamatsu streak camera with 2 psec resolution. A beam energy of 3.9 MeV was measured with a 6 inch round pole spectrometer while a beam charge was measured with both an ICT and a Faraday Cup. Although the gun will normally be run at 100 pC it has produced charges from 10 pC to 4 nC. All results of the measurements to date are presented here.

I. INTRODUCTION

The central purpose of the AWA [Schoessow, 95] is to study high gradient acceleration schemes in dielectric structures [Rosing, 90] and in plasmas [Barov, 95]. An extremely high charge (100 nC) and relatively short electron bunch (FWHM less than 50 ps) is necessary to produce gradients that surpass today's conventional gradients. This high charge bunch is called the drive beam [Ho, 92]. Once the high gradient electromagnetic wave is excited in the structure, a low charge (100 pC) electron bunch is injected into the structure to witness the fields. This low charge bunch is called the witness beam [Power, 95].

In this paper the results of the characterization of the witness beam are presented. The characterization done here includes measurements of charge, energy, bunch length, and emittance.

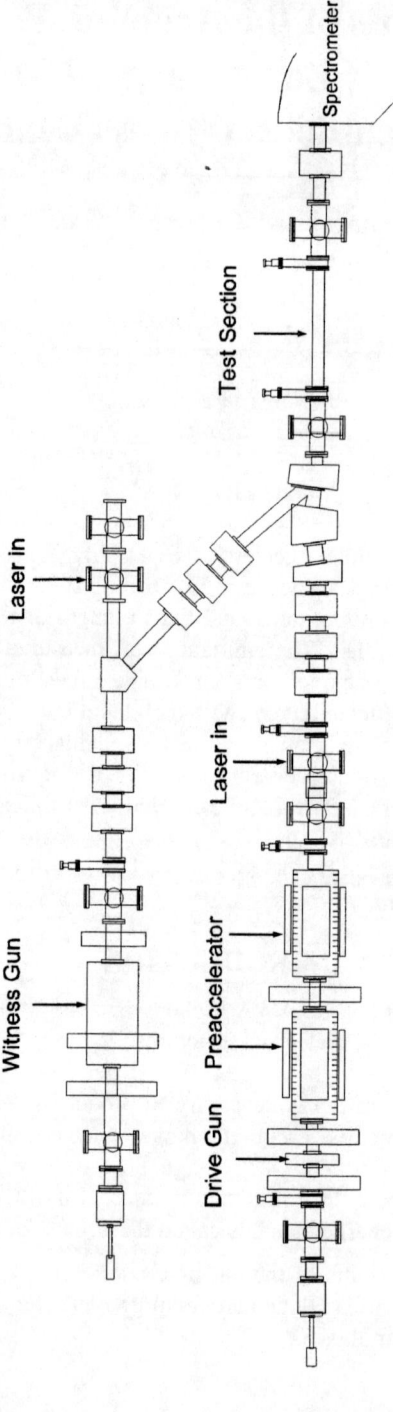

Figure 1. Plan View of the Drive and Witness Beam Lines

II. AWA FACILITY

The AWA employs two RF photocathode guns to create the drive and witness beams. A schematic diagram of the AWA facility is given in Figure 1 showing the drive and witness beam photocathode gun. A complete facility discussion can be found in the reference [Schoessow, 95].

The drive and witness beams are individually steered and focused into the test section (Figure 1) where either a dielectric tube or plasma cell resides. The laser pulse enters the vacuum system through a UV quality quartz window and is reflected toward the PC gun by internally mounted UV mirrors situated off-axis. The drive beam exits the gun at an energy of about 2 MeV and is immediately injected into the drive linac preaccelerator [Chojnacki, 93] where it is further accelerated to 16 MeV. A 4 MeV beam is ejected out of the witness gun into an achromatic and isochronous transport line. The two beams are combined by a bending magnet from which they emerge collinearly for propagation into the test section. Finally, the drive beam is directed into a beam dump and the energy of the witness beam is resolved in a spectrometer immediately following the test section.

III. BEAM GOALS

Witness beam parameters were chosen by considering what type of beam would make the best probe for wakefield measurements. A complete beam parameter set, which includes the charge, energy, emittance, bunch length, and energy spread of the witness beam is briefly discussed here. A more thorough discussion of this procedure is given in the reference [Power, 96].

The witness beam energy of 4 MeV was chosen for two reasons. First, an energy near 4 MeV is easy to resolve in the spectrometer since it is well separated from the 16 MeV drive beam. Second, the beam has $\beta = .994$ which has relatively little phase slippage compared to the drive beam's $\beta = .9995$.

In general, a good probe should not significantly change the parameter being measured, while still being easy to observe. A charge of 100 pC is sufficiently low so as not to significantly perturb the wakefield being measured while still being easy to observe on the phosphor screen beyond the spectrometer.

The emittance need only be small enough to allow the beam to transport from the gun to the test section with most of the charge out of the gun. The other consideration is that the beam transport through the same test section that the drive beam passes through. Since the drive beam has a normalized rms emittance of 400 mm mrad there is very little need to keep the beam emittance low. An emittance on the order of 1 mm mrad is more than adequate.

We would like the bunch length of the test charge to be much less than the wavelength of the excited wakefield. At the speed of light a 1.0 mm bunch spans 3.3 ps of the RF wave which for a 10 GHz wave is 12 degrees of its RF cycle. This is about the shortest bunch that can be easily produced from our RF photocathode gun at 100 pC, and we must target for this number. The reason we are able to tolerate this long bunch (12°) is that we are interested primarily in measuring the peak gradient of the RF wave which can be measured by studying the witness beam centroid.

Wakefield measurements are taken by measuring a change in energy of the witness beam as described in Chapter 7. At first thought, this would seem to imply that the resolution of the wakefield measurement is determined by the energy spread of the witness bunch. However, this is not strictly true since the wakefield measurement depends on the beam energy centroid, which can be determined much more accurately than the full energy distribution of the bunch. Thus energy spread is not a primary limitation. Under 1% rms is adequate.

IV. GUN DESIGN

The AWA witness gun chosen was a six cell, $\pi/2$ mode, iris-loaded, RF photocathode gun. The gun can be described as a copper cylinder with six copper washers placed inside along its axis. Figure 2 is a side view of the cavity showing the five full cells surrounded by the two half cells. In the left most half-cell (the upstream half-cell) is an extractable photocathode plug where the laser strikes to produce photoelectrons. In the right-most half cell (the downstream half cell) is the hole that allows the laser beam to enter and electrons to exit.

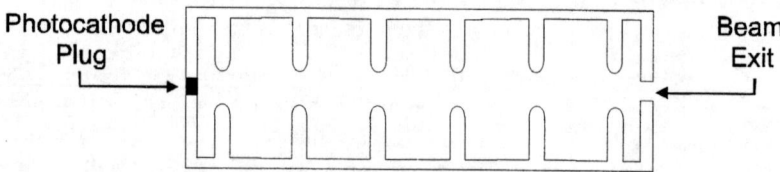

Figure 2. The Witness Gun for the AWA. A Side View of the Entire Gun. The Plug at the Beginning and the Exit Hole at the End of the Cavity.

PARMELA simulations showed [Power, 95] that this $\pi/2$ gun meets the design requirements outlined above. For 3.8 MW of RF power the witness gun produces a 4 MeV, 100 pC beam with a FW energy spread of 0.75% and an emittance of 1.9 mm mrad.

V. MEASUREMENTS

The charge is measured immediately after the witness gun (see Figure 1) with an integrating current transformer (ICT) from Bergoz. The ICT is an on-line, non-destructive current monitor that integrates the total charge passing through its center, with a rise time of about 10 ns. Since we are operating at 30 Hz, we can use the ICT as a single shot charge monitor with a fast digitizing scope to integrate the area under the voltage curve. The ICT was calibrated with a Wavetek function generator and was found to produce a peak output voltage signal of 72 mV/nC. The witness beam charge has been measured over a range of 10 pC to 4 nC.

The energy and energy spread measurements were made with a magnetic spectrometer at the end of the beam line as shown in Figure 3. The spectrometer is a D = 6 inch diameter, round-dipole magnet and vacuum chamber with ports located at the $\theta_0 = 60°$ and $\theta_0 = 0°$ (the undeflected position). Energy

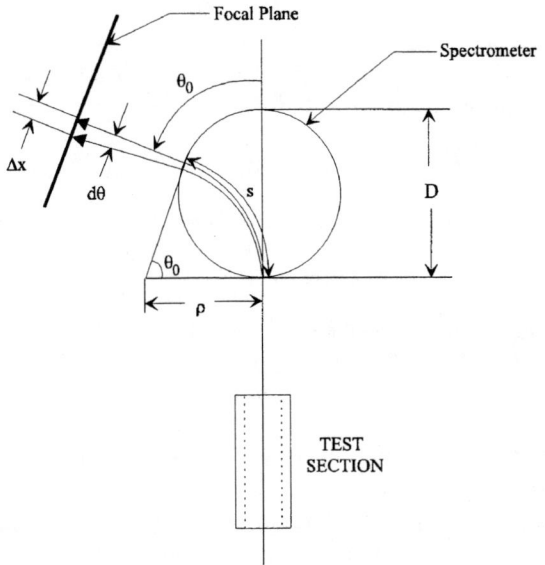

Figure 3. The Energy Spectrometer Following the Test Section Used to Measure Energy and Energy Spread

measurements are taken by bending the beam through an angle, θ_0, with a known magnetic field, B, and calculating the momentum according to

$$\theta(\deg) = 1.71\left(\frac{BL[G \cdot m]}{p[MeV/c]}\right) \quad (1)$$

where B is the magnetic field strength, p is the momentum, and L is the effective length of the dipole. (An effective length must be used here since the magnetic field does not drop to zero at the pole edge.) The effective length for the $\theta_0 = 60°$ port is 17.0 cm. The witness beam was centered on the 60° port for a magnetic field, B = 895 Gauss; corresponding to a momentum, p = 3.81 MeV/c and a kinetic energy, T = 3.85 MeV.

The energy spread is best measured in a region of high dispersion. In the case of the witness beam a spot size measurement is made on the focal plane shown in Figure 3 where the beam width is indicated by Δx in the figure. The dispersion function, D, and its derivative, D', increase in a dipole according to [Wiedemann, 93]

$$D(s) = \rho_0 \left(1 - \cos\frac{s}{\rho_0}\right) \qquad (2)$$

$$D'(s) = \sin\frac{s}{\rho_0} \qquad (3)$$

where s is the path length and ρ_0 is the bending radius. Outside the dipole, the dispersion grows according to the formula

$$D(s) = D(s=0) + D'(s=0)\Delta s \qquad (4)$$

The bending radius of the dipole at the $60°$ port is calculated to be $\rho = 13.1$ cm and the distance from the pole edge to the focal plane is 120 cm. Using Equations 2-4 we find the dispersion to be 60 cm at the focal plane.

In the case of the energy spread measurement for the witness beam we can only give a highly overestimated upper limit. This is because we have violated the requirement that the betatron term is negligible compared to the dispersion term. The betatron function is normally made small at the focal plane by placing a horizontally focusing quad directly in front of the spectrometer and then focusing the beam to a minimum (i.e. small β_x), so that the width that remains is due to the energy spread, not the emittance. At the time the measurements were taken, there was no quad in front of the spectrometer. In our case we measured a spot size of $\Delta x = 1.0$ cm which gives an upper limit to the energy spread, $\delta < 1.6\%$.

The emittance was measured by a quad scan technique with a screen in a diagnostic chamber beyond the bend magnet at the top of Figure 1 and by varying the first quad after the witness gun in Figure 1. A new quad scan technique [Power, 96b] is used that includes space charge.

The envelope of a round beam, r = R (the maximum r value of a particle in the bunch) is a surface that surrounds the all the particles in the beam in the transverse direction. The beam envelope [Reiser, 94] evolves along the s axis according to

$$R''(s) = \frac{\varepsilon^2}{R^3(s)} + \frac{K}{R(s)} \tag{5}$$

where the derivative is taken with respect to s, and K is the perveance [Lawson, 88] in vacuum

$$K = \frac{I}{I_0} \frac{2}{\beta^3 \gamma^3} \tag{6}$$

where I is the beam current and I_0 is the Alfven current equal to 17 kA. As opposed to most quad scans, we keep the space charge term in Equation 5 for out fit of the data. The resulting fit involves a non-linear least squares fit and is done with a Mathematica routine written by this author. The normalized emittance measured by this new quad scan technique is 3.4 mm mrad. The geometrical emittance is $3.4/\beta\gamma = 0.41$ mm mrad as shown in Figure 4 (where $\beta\gamma$ is equal to 8.6).

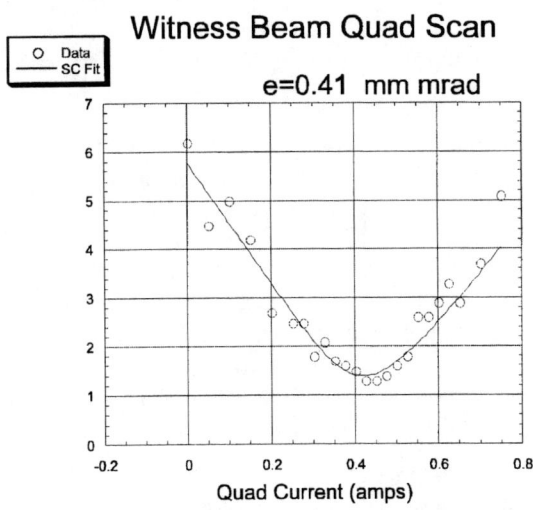

Figure 4. Quad Scan for the Witness Beam

The bunch length is measured by sending the witness beam through a piece of quartz plate and imaging the subsequent Cherenkov light into a Hamamatsu streak camera.

The light from the quartz plate is imaged onto the entrance slit of the streak camera set to a slit width 25 μm. Using the fastest sweep rate on the streak camera we have a measured resolution of 3 ps.

A streak camera measurement of the witness beam is shown in Figure 5. For this measurement, the FWHM bunch length is 10 ps.

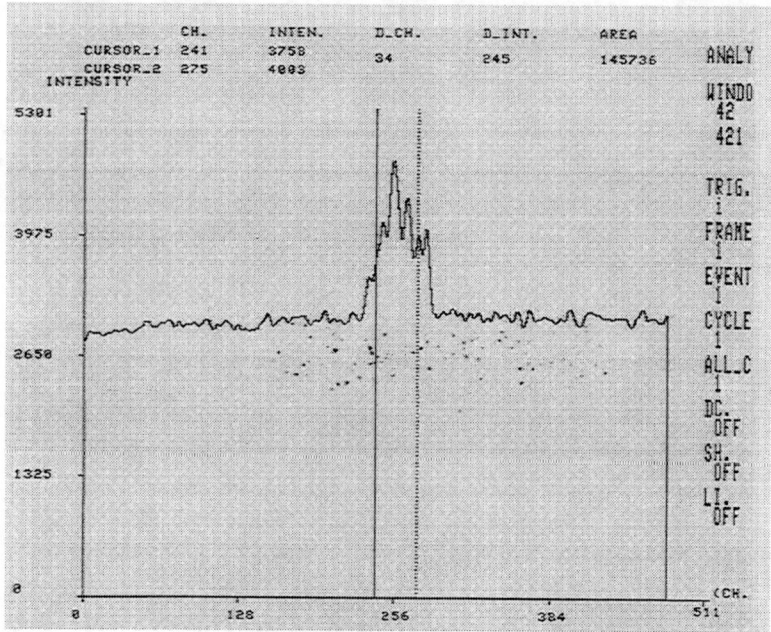

Figure 5. Streak Camera Image of the Witness Beam. Δz = 10 ps FWHM

VI. CONCLUSIONS

The end result of this work is a functioning, wakefield measurement system. This system has been used to measure the first wakefields produced at the AWA with these early measurements at 7 MV/m. The wakefield measurement system consists of a 6 cell, $\pi/2$, RF photocathode gun, transport lines to deliver both the drive and witness beams to the test section, and a synchronization system responsible for the timing between the drive and witness beams.

The RF photocathode gun used to generate the witness beam has produced an electron beam with an energy of 4 MeV, a charge of 100 pC, a normalized transverse emittance of 3.4 π mm mrad, and a FWHM bunch length of 10 ps. The beam was characterized using a spectrometer for energy, an integrating current transformer for charge, a new quad scan technique for emittance, and a streak camera for bunch length. The beam characterization is found to agree well with the predictions of the PARMELA simulations.

VI. ACKNOWLEDGMENTS

This work was supported by DOE grant W-31-109-Eng-38.

VII. REFERENCES

[Barov, 95] Barov, N., "Measurements of Plasma Wake-Fields in the Blow-Out Regime," Particle Accelerator Conference, Dallas, 1995.

[Chojnacki, 93] Chojnacki, E., "Drive Linac for the Argonne Wakefield Accelerator, " in Proceedings of the 1993 Particle Accelerator Conference, Washington D.C., 1993.

[Ho, 92] Ho, C.H., A High Current, Short Pulse Electron Source for Wakefield Accelerators, Ph.D. Thesis, UCLA, 1992.

[Lawson, 88] Lawson, J.D., The Physics of Charged-Particle Beams, Second Edition, Clarendon Press, Oxford, 1988.

[Power, 95] Power, J., et al., "Witness Gun for the Argonne Wakefield Accelerator," Particle Accelerator Conference, Dallas, 1995.

[Power, 96a] Power, J., "A 4 MeV Photocathode Beam for Measurements of the Dielectric Wakefield Accelerator: An Experimental Thesis", (unpublished), 1996.

[Power, 96b] Power, J., "A Quad Scan with Space Charge," (unpublished), 1996.

[Reiser, 94] Reiser, M., Theory and Design of Charged Particle Beams, John Wiley & Sons, Inc., New York, 1994.

[Rosing, 90] Rosing, M. and Gai, W., "Longitudinal-and-Transverse-Wakefield Effects in Dielectric Structures," Physical Review D, **42**, No. 5, pp. 1829-1834, 1990.

[Schoessow, 95] Schoessow, P., et al., "The Argonne Wakefield Accelerator High Current Photocathode Gun and Drive LINAC," Particle Accelerator Conference and International Conference on High-Energy Accelerators, Dallas, 1995.

[Wiedemann, 93] Wiedemann, H., Particle Accelerator Physics, Springer-Verlag, Berlin, 1993.

Microbunching and Coherent Acceleration of Electrons by Subcycle Laser Pulses

Bernhard Rau, T. Tajima, and H. Hojo[†]

Physics Department
The University of Texas at Austin
Austin, Tx 78712

[†]Plasma Research Center
University of Tsukuba
Tsukuba 305, Japan

I. ABSTRACT

The pick up and acceleration of all plasma electrons irradiated by an intense, subcyclic laser pulse is demonstrated via analytical and numerical calculations. It is shown that the initial low emittance of the plasma electrons is conserved during the process of acceleration, leading to an extremely cold, bunched electron beam. Compression of the electron bunch along the longitudinal coordinate is naturally achieved due to the interaction of electrons and laser pulse. In this paper, we find the localized solutions to Maxwell's equations of a subcyclic laser pulse and use these to determine the acceleration of charged particles and we suggest future application for this acceleration mechanism as low energy particle injector and as electron source for coherent x-ray generation.

II. INTRODUCTION

The issue of particle bunching and pre-acceleration has always been a crucial point in accelerator physics. A low emittance source of tightly bunched particles is needed for almost all practical application of high energy particles. But while there is a great effort to replace the rf-driven accelerator schemes with high gradient linear accelerators such as the laser wakefield accelerator, plasma wakefield accelerator, plasma beat wave accelerator, and others, less work has been done to replace electron guns with higher technology standards. Photo cathode work has been proposed and developed and recently, optical generation of pre-accelerated electron bunches has been proposed [1]. As beam loading and emittance issues become more important for a small scale high energy accelerator, researchers are in need of a cold source of bunched particles.

In this paper, we investigate the collective acceleration of plasma electrons by a properly shaped high intensity laser pulse. Those investigations are done analytically as well as with the help of a 1-$\frac{2}{2}$ PIC (particle in cell) code. We study the electron motion in the focal region of an ultrashort laser pulse and determine the electron bunch parameters such as emittance, bunch length, and beam energy.

In the next section, we analytically derive the 3-dimensional solution for the electromagnetic fields of a laser pulse of arbitrary length present inside the focal region. We use these fields to determine the acceleration of plasma electrons due to their interaction with a strong, ultrashort laser pulse in section IV. Section V will contain the analytical and numerical results for the acceleration parameters and in section VI, we will show applications of such electron bunches for high brilliance x-ray sources. We will conclude this paper in section VII.

III. 3-DIMENSIONAL FIELDS OF A SUBCYCLIC LASER PULSE

Electromagnetic waves of finite radial extent take a 3-dimensional expression. While other authors have derived expressions valid for long laser pulses [3, 2], their analysis fail to describe pulses whose envelope changes on a length scale comparable to the laser wavelength. Here, we derive solutions for a pulse of arbitrary length (and thus valid even for subcyclic pulses).

Assuming the *x*-polarized laser pulse to propagate along the *z*-direction far away from any boundaries, we can solve the vacuum wave equation for the *x* component of the electric field $E_x(r,z,t)$ in Fourier space $(\nabla^2 + k^2)\tilde{E}_x(r,z;k = \omega/c) = 0$ to determine this field in a closed form expression

$$\tilde{E}_x(r,z;k) = \frac{1}{2\pi} \int dp \int dq A(p,q;k) \exp[ik(px+qy+\sqrt{1-p^2-q^2}z)], \quad (1)$$

where the borders of integration must be chosen in a way that the resulting modes will be bounded for all x, y, and z, i.e. p, q, and $\sqrt{1-p^2-q^2}$ must be real. The amplitude $A(p,q;k)$ is determined as

$$A(p,q;k) = \tilde{E}_0(k)\frac{k^2 w_0^2}{2}\exp[-k^2\frac{(p^2+q^2)w_0^2}{4}], \qquad (2)$$

where we assumed $\tilde{E}_x(r,z=0;k) = \tilde{E}_0(k)\exp[-r^2/w_0^2]$. Choosing E_x to be cylindrically symmetric, the double integral in (1) can be reduced to an integral over the dimensionless variable $b = \sqrt{1-p^2-q^2}$. Assuming further that E_x has the same Fourier spectrum as its 1-dimensional counterpart, we find for the Fourier coefficients $\tilde{E}_0(k)$

$$\tilde{E}_0(k) = \frac{1}{\sqrt{2\pi}}\int_{-\infty}^{\infty} E_{1-D}(z=0,t)\exp[ikct]c\,dt$$

$$= \frac{E_0}{\sqrt{2\pi}}\int_{-\infty}^{\infty}\exp[-\frac{c^2 t^2}{2\sigma^2}+ikct]\cos[-k_0 ct]c\,dt$$

$$= E_0\sigma\exp[-\frac{k_0^2+k^2}{2}\sigma^2]\cosh[kk_0\sigma^2], \qquad (3)$$

which, together with (1) and (2) gives the integral expression

$$E_x(r,z,t) = \frac{E_0 w_0^2 \sigma}{2\sqrt{2\pi}}\exp[-\frac{k_0^2\sigma^2}{2}]\int_{-\infty}^{\infty} dk\, k^2 \cosh[kk_0\sigma^2]\int_0^1 db\, b \times$$

$$\times \exp[-k^2(\frac{\sigma^2}{2}+\frac{w_0^2(1-b^2)}{4})+ik(zb-ct)]J_0(kr\sqrt{1-b^2}). \qquad (4)$$

Here, w_0 is the spotsize of the laser pulse at the focal plane, k_0 is the k-number affiliated with the center frequency of the pulse, σ characterizes the pulse length, and E_0 is the pulse amplitude. With (4) and the assumption that E_y is 0 for all r, z, and t, we use the vacuum equations $\vec{\nabla}\cdot\vec{E} = 0$ and $-\partial \vec{B}/(c\,\partial t) = \vec{\nabla}\times\vec{E}$ to find the other components of the electric and magnetic fields as

$$E_z(x,y,z,t) = \frac{\partial}{\partial x}\left\{\frac{E_0 w_0^2 \sigma}{2\sqrt{2\pi}}\exp[-\frac{k_0^2\sigma^2}{2}]\int_{-\infty}^{\infty} dk\, ik\cosh[kk_0\sigma^2]\int_0^1 db \times\right.$$

$$\left.\times \exp[-k^2(\frac{\sigma^2}{2}+\frac{w_0^2(1-b^2)}{4})+ik(zb-ct)]J_0(kr\sqrt{1-b^2})\right\}, \qquad (5)$$

$$B_x(x,y,z,t) = \frac{\partial^2}{\partial x\,\partial y}\left\{\frac{E_0 w_0^2 \sigma}{2\sqrt{2\pi}}\exp[-\frac{k_0^2\sigma^2}{2}]\int_{-\infty}^{\infty} dk \cosh[kk_0\sigma^2]\int_0^1 db \times\right.$$

$$\left.\times \exp[-k^2(\frac{\sigma^2}{2}+\frac{w_0^2(1-b^2)}{4})+ik(zb-ct)]J_0(kr\sqrt{1-b^2})\right\}, \qquad (6)$$

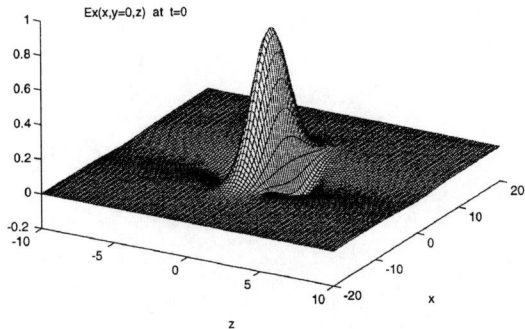

Figure 1: $E_x(x, y=0, z, t=0)$ [see Eq. (4)] for $k_0\sigma = 1$ and $w_0 = 5\sigma$. The units of the x and z axis are in σ.

$$B_y(x,y,z,t) = \frac{E_0 w_0^2 \sigma}{2\sqrt{2\pi}} \exp[-\frac{k_0^2\sigma^2}{2}] \int_{-\infty}^{\infty} dk \cosh[kk_0\sigma^2] \int_0^1 db \left\{ k^2 b^2 - \frac{\partial^2}{\partial x^2} \right\} \times$$
$$\times \exp[-k^2(\frac{\sigma^2}{2} + \frac{w_0^2(1-b^2)}{4}) + ik(zb - ct)] J_0(kr\sqrt{1-b^2}), \qquad (7)$$

and

$$B_z(x,y,z,t) = \frac{\partial}{\partial y} \left\{ \frac{E_0 w_0^2 \sigma}{2\sqrt{2\pi}} \exp[-\frac{k_0^2\sigma^2}{2}] \int_{-\infty}^{\infty} dk\, ik \cosh[kk_0\sigma^2] \int_0^1 db\, b \times \right.$$
$$\left. \times \exp[-k^2(\frac{\sigma^2}{2} + \frac{w_0^2(1-b^2)}{4}) + ik(zb - ct)] J_0(kr\sqrt{1-b^2}) \right\}. \qquad (8)$$

Equations (4) - (8) of course only represent one special localized solution to Maxwell's equations with the given frequency spectrum (3). A general solutions involves a super composition of different fields of that kind.

Other authors [3] have argued that the integral over b can be solved to a good approximation for large values of kw_0. While this is correct for mono-frequent (and therefore long) laser pulses, the approximation fails for short (large bandwidth) pulses as they always include frequencies with $kw_0 \ll 1$.

On axis, equations (5),(6), and (8) vanish so that we are left with E_x and B_y only. Furthermore, E_x can be expressed in closed form for $z = 0$ or for large values of z:

$$E_y(r=0,z,t) = E_z(r=0,z,t) = B_x(r=0,z,t) = B_z(r=0,z,t) = 0, \qquad (9)$$

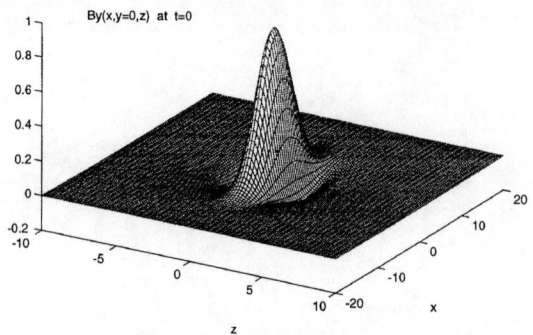

Figure 2: $B_y(x, y=0, z, t=0)$ [see Eq. (7)] for $k_0\sigma = 1$ and $w_0 = 5\sigma$. The units of the x and z axis are in σ. For $w_0 \gg \sigma$, E_x (see Fig. 1) and B_y are very similar.

$$E_x(r=0, z=0, t) = E_0 \exp[-\frac{c^2 t^2}{2\sigma^2}] \cos[k_0 ct]$$
$$-\frac{E_0}{\sqrt{1+\rho^2}} \exp[-\frac{k_0^2 \sigma^2}{4(1+\rho^2)}] \exp[-\frac{c^2 t^2}{2\sigma^2(1+\rho^2)}] \cos[\frac{k_0 ct}{1+\rho^2}], \quad (10)$$

and

$$E_x(r=0, |z| > \sigma \text{ and } |z| > \frac{w_0^2}{\sigma}, t) = \frac{E_0}{2} \exp[-\frac{(z-ct)^2}{2\sigma^2}] \times$$
$$\times \left\{ \exp[ik_0(z-ct)] \frac{z - ct - ik_0\sigma^2}{z - ct - ik_0\sigma^2 + z/\rho^2} + c.c. \right\}, \quad (11)$$

where we defined ρ as a ratio of transverse to longitudinal extent of the laser pulse to be $\rho^2 = w_0^2/(2\sigma^2)$. Equations (9) - (11) approach to correct limits for a 1-dimensional pulse of arbitrary length ($\rho \to \infty$) and for a 3-dimensional long pulse ($\rho \to 0$). Those fields will be used in the next section to determine the forces acting on charged particles around the focal point of a laser pulse.

IV. ELECTRON ACCELERATION BY SUBCYCLIC LASER PULSES

Assuming the fields (4)-(8) in the focal region of a large-spotsize pulse ($r \ll w_0, |z| \ll w_0, \rho = w_0/(\sqrt{2}\sigma) \gg 1$), we can use the 1-dimensional fields to determine the electron motion due to its interaction with the electromagnetic radiation. With

$$E_x(r,z,t) = B_y(r,z,t) \simeq E_0 \exp[-\frac{(z-ct)^2}{2\sigma^2}] \cos[k_0(z-ct)]$$
$$E_y(r,z,t) = E_z(r,z,t) = B_x(r,z,t) = B_z(r,z,t) \simeq 0, \qquad (12)$$

we can use the z component of the relativistic Lorentz equation along with the time dependence of the power

$$m_0 c \frac{d\gamma}{dt} - m_0 c \frac{d(\gamma \beta_z)}{dt} = -e(\vec{\beta} \cdot \vec{E}) + e(E_z + \beta_x B_y - \beta_y B_x) \simeq 0 \qquad (13)$$

to find the approximate constant $\gamma(1-\beta_z)$ of the electron motion. The motion in the y direction is determined by the Lorentz force to become

$$m_0 c \frac{d(\gamma \beta_y)}{dt} = -e(E_y + \beta_z B_x - \beta_x B_z) \simeq 0. \qquad (14)$$

Thus, inside the focal region we can neglect the motion along the y-direction. In this parameter regime, we can follow the analysis of Scheid & Hora [4]: For electron initially at rest ($\vec{v}_{init} = 0$), we integrate the x-component of the Lorentz equation with respect to the relative coordinate u to obtain the final normalized momentum in the x direction as

$$(\gamma \beta_x)_{final} = \frac{-e}{m_0 c} \int_{-\infty}^{+\infty} (E_x + \beta_y B_z - \beta_z B_y) dt$$
$$\simeq \frac{-e}{m_0 c} \int_{-\infty}^{+\infty} ((1-\beta_z) E_x) dt = \frac{e}{m_0 c^2} \int_{-\infty}^{+\infty} E_x \, du \equiv A. \qquad (15)$$

Using (13), the final relativistic Lorentz factor γ_{final} can then be expressed in terms of the electric field E as $\gamma_{final} = 1 + A^2/2$. In particular, we find that a net energy gain due to the interaction of electrons and laser pulse is possible as long as A does not vanish, i.e. as long as the electric field E has a "zero-frequency" or DC component. For wave trains with more than a few oscillations, A vanishes in accordance with the Lawson-Woodward theorem [5].

Introducing the normalized vector potential $a_0 \equiv (eE_0)/(m_0 k_0 c^2)$, the electron gain for a wave packet of the form (4) - (8) becomes

$$\Delta E = a_0^2 m_0 c^2 \pi (k_0 \sigma)^2 \exp[-(k_0 \sigma)^2]. \tag{16}$$

Maximum gain for a given field amplitude E_0 is thus obtained for $k_0\sigma = 1$, which entails the optimum pulse width being shorter than one wavelength (see Fig. 1). Such pulses have been successfully generated in the microwave [6], far infrared [7], and even in the femtosecond regime [8]. However, their amplification to higher intensities proves to be complicated [9] as the gain media normally used for amplification do not support such large bandwidths.

In the regime of $\beta_y^2 \ll \beta_x^2 + \beta_z^2$ (i.e. inside the focal region of a loose focus), we can find the angle φ between final electron motion and z axis to be $\varphi = \arctan[\beta_x/\beta_z]$. But, since $\gamma = 1 + A^2/2$ and $\gamma\beta_x = A$, φ is determined by

$$\varphi = \arctan \frac{2}{A} = \arctan \sqrt{\frac{2}{\gamma - 1}}. \tag{17}$$

Hence, we expect the electrons to be ejected at an angle which depends on the electron energy. For electrons with kinetic energies of or exceeding 1MeV, the main part of their momentum however will be along the z axis, while for highly energetic particles ($\gamma \gg 1$), the angle approaches $\varphi \simeq \sqrt{2}\gamma^{-1/2}$.

So far, we have only addressed the pickup of a single electron by a subcyclic laser pulse in vacuum. In the next section, we will address the coherent acceleration of all the plasma electrons on the spot using those electromagnetic pulses. The relevant injector properties will be discussed.

V. MICROBUNCHING AND COHERENT ELECTRON ACCELERATION

For the generation of tightly bunched electron pulses, we propose to inject a subcyclic laser pulse into a thin layer of plasma. While the heavy ions remain inertial during such a short time scale, the plasma electrons will be expelled from the initial plasma region and accelerated. Since this acceleration to relativistic velocities takes place on a time scale of the pulse length, space charge forces between the electrons are practically absent. However, an ambipolar electrostatic field E_{amb} will be generated due to the separation of plasma electrons and ions which requires the laser pulse to be powerful enough such that the accelerated electrons overcome this repulsive potential. Using (16), this condition can be rewritten as $\Delta E > \int_0^\infty eE_{amb}(z,t)\,dz$. The ambipolar field however will decay both along the longitudinal axis (due to a finite spotsize) as well as in time (due to the electron response of the plasma remaining outside the transverse focal region), so that we can find the lower limit for the laser pulse intensity to be $m_0c^2\pi(k_0\sigma)^2a_0^2\exp[-(k_0\sigma)^2] > \int_0^{L_p} 4\pi n_0 e^2 z\,dz$, or, assuming $k_0\sigma = 1$,

$$a_0^2 > \frac{\exp[1]}{2\pi}\left(\frac{L_p}{c/\omega_p}\right)^2. \tag{18}$$

Here n_0 is the electron density of the initial plasma, L_p is its longitudinal extent, and ω_p is the plasma frequency. Eq. (18) shows that even plasma length of the order $O(c/\omega_p)$ require relativistically strong field amplitudes ($a_0 > 1$). Very much the same way, we can estimate the number of electrons accelerated per shot:

$$N/S = n_0 L_p < a_0^2/(2r_e L_p \exp[1]), \tag{19}$$

where $r_e = e^2/(m_0c^2)$ and S is the transverse area over which the intensity of the wave packet is nearly constant (typically $O(1/k_0^2)$). For plasma lengths of the order of 1 μm, the upper limit for the number of electrons per unit area is $N/S[\text{mm}^{-2}] < a_0^2 \cdot O(10^{13})$, which even for moderate values of a_0 seems to be sufficient to fulfill the needs of particles per bunch in modern accelerators/injectors.

Finally, we can estimate the emittance of the electron bunch by calculating the initial longitudinal and transverse emittance. The normalized emittance ε_N is defined as the product of spatial and normalized momentum spread by $\varepsilon_N = \sqrt{<\Delta z^2><\Delta p_z^2> - <\Delta z \Delta p_z>^2}/(m_0c)$, with the usual definition of the variance $<\Delta A^2> = <A^2> - <A>^2$ and $<A> = \sum_{i=1}^N A_i/N$. For a plasma of longitudinal extend L_p in thermal equilibrium with a plasma electron temperature T_e, we find $<\Delta z^2> = L_p^2/12$ and $<\Delta p_z^2> = m_0 T_e$ (and $<\Delta x \Delta p_x>^2 = 0$), so that the initial longitudinal emittance becomes

$$\varepsilon_{N,long}^{init} = \sqrt{\frac{L_p^2}{12}\frac{T_e}{m_0c^2}}, \tag{20}$$

or about $4.0 \cdot 10^{-4} L_p[\mu\text{m}]\sqrt{T_e[\text{eV}]}$ mm mrad. (Or $\varepsilon_{N,long}^{init} \approx 6.9 \cdot 10^{-10} L_p[\mu\text{m}]\sqrt{T_e[\text{eV}]}$ eV-sec when we express $\varepsilon_{N,long}^{init}$ in eV-sec.) The initial transverse emittance for a circular area S becomes with $<\Delta r^2> = S/(2\pi)$ and $<\Delta p_r^2> = 2m_0 T_e$

$$\varepsilon_{N,trans}^{init} = \sqrt{\frac{S}{\pi}\frac{T_e}{m_0c^2}}, \tag{21}$$

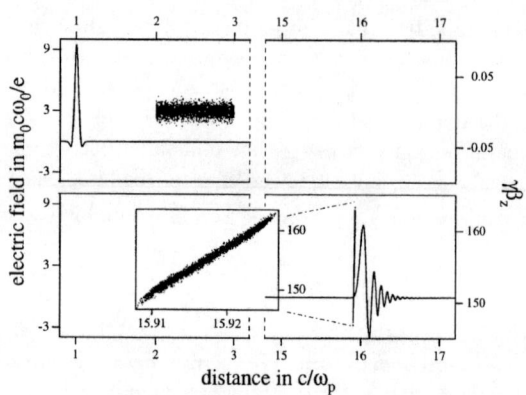

Figure 3: Initial and final transversal electric field and phase space for $L_p = 1c/\omega_p$, $\omega_0/\omega_p = 30$, and $a_0 = 10$. Inset: blow up of the final longitudinal phase space

or ca. $7.9 \cdot 10^{-1} \sqrt{S[\text{mm}^2] T_e[\text{eV}]}$ mm mrad. With the above threshold ionization of atoms in the ultrashort pulse regime, the expected electron temperature T_e is only a few eV. Thus for a plasma volume $L_p \times S \sim 1\mu\text{m} \times 3 \cdot 10^{-4} \text{mm}^2$, $\varepsilon_{N,long}^{init}$ is expected to be of the order $1 \cdot 10^{-3}$ mm mrad and $\varepsilon_{N,trans}^{init}$ of the order of $1.5 \cdot 10^{-2}$ mm mrad.

As mentioned above, space charge effects are practically absent due to the rapid acceleration of the electrons to relativistic velocities and do therefore not increase the emittances. However, the three dimensional nature of the electromagnetic fields may lead to unstable regions of betatron oscillations that will affect the electron momentum spread so that the final emittance might grow in the process of acceleration. Additional growth due to the transverse accelerating structure is possible, but both growth rates can be minimized for a geometry with a loose focus ($\rho = w_0/(\sqrt{2}\sigma) \gg 1$).

We simulated the interaction of a subcyclic laser pulse with a plasma slab of finite extent using a $1-\frac{2}{2}$ dimensional PIC code. The laser pulse was initialized in the vacuum region outside the plasma and evolved into the layer (see Fig. 2). After its propagation through the plasma layer, the pulse enters the vacuum region behind the slab, carrying with it the expelled electrons, while the heavy ions are left behind. For this simulation, we used a $2^{11}\Delta$ large simulation box with vacuum boundary conditions. About 8000 macro particles made up the quasi neu-

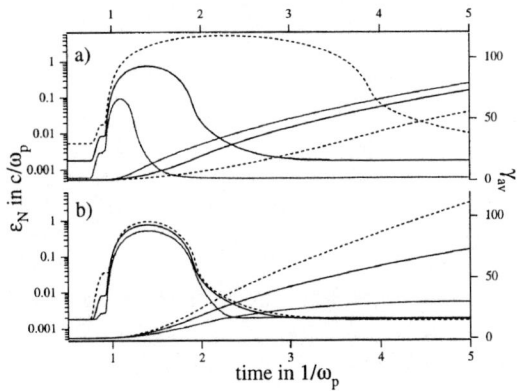

Figure 4: Normalized phase space volume and γ_{av} as a function of time for a) $a_0 = 10$, $L_p = 1/3$(dotted), 1(solid), and 3(dashed line) c/ω_p and b) $a_0 = 5$(dotted), 10(solid), and 20(dashed line) $L_p = 1\ c/\omega_p$. The initial growth of the phase space volume is due to fact that the electron are not accelerated simultaneously. Once The pulse passes the plasma slab, all the electron experience the acceleration force and the longitudinal phase space volume decays to its initial value.

tral plasma initially in thermal equilibrium with an electron temperature varied between 1 and 1,000 eV. Since the ambipolar field E_{amb} does no decay in a 1-D simulation, we stopped the run at a point where the laser pulse just passed the electrons. As mentioned above, multidimensional effects let the potential decrease with both increasing distance and time. Such effects where manually implemented into the 1-D code but did not lead to any qualitatively different results. A rigorous investigation however requires the help of a 2 or 3-D simulation and has not been conducted yet.

Due to the interaction with the laser pulse, the expelled electron bunch becomes compressed along its longitudinal axis (see inset of Fig. 2) and spread in momentum space. The normalized longitudinal emittance ε_N^{long} however remains approximately constant (see Fig. 3 (a) and (b)). Typically, a spatial compression of a factor 100 is naturally achieved within the laser-electron interaction. This seems to be due to the fact that the pulse loses its "zero-frequency" or DC component energy to the electrons. Particles in the trailing part of the beam extract this energy more efficiently, since the pulse, traveling with the speed of light, passes the bunch from the behind and supplies undepleted energy to those electrons being located in the back of the bunch. Thus, those electrons are accelerated forward until they line up with the leading part. Ideally, all the electrons gather in a plane in the reference frame of the particle bunch and coherently extract the pulse energy.

The parameters used for the results displayed in Fig. 2 are $a_0 = 10$, $\omega_0/\omega_p = 30$, and $L_p = c/\omega_p$. From Eq. (16), we would expect an average relativistic Lorentz factor $\gamma_{av}^{pred.} \simeq 116.6$. Comparing this with the simulation (see inset of Fig. 2), the average Lorentz factor is well above 150. This finds its explanation in the fact that the laser pulse group velocity slows down as the particle-wave interaction takes place and thus extents the time for the energy exchange. In particular, the phase velocity of the DC component slows down while the component itself becomes depleted, leading to a chirp of the laser pulse.

VI. APPLICATIONS OF ELECTRON MICROBUNCHES

The parameters found in the previous section can now be evaluated in few of possible future applications of this particle accelerating mechanism. So could a Ti:Sapphire laser capable of generating subcyclic pulses and operating at an intensity of about $3.6 \cdot 10^{18} \text{W/cm}^2$ ($a_0 \approx 1.3$, see Fig.4 for similar setup) be used as a table top particle injector. From the irradiation of a plasma layer of about 1-2 μm thickness at a plasma electron density of about $2 \cdot 10^{18}/\text{cm}^3$ we could expect about $3 \cdot 10^{12}$ electrons/mm^2 at a kinetic energy of 1 MeV. The longitudinal emittance $\varepsilon_{N,long}$ should be as low as $2 \cdot 10^{-3}$ mm mrad (assuming a temperature of 10 eV). The initial transverse emittance $\varepsilon_{N,trans}^{init}$ is then (assuming the irradiated area of constant intensity to be about $c^2/\omega_p^2 \approx 14 \mu m^2$) ca. $9 \cdot 10^{-3}$ mm mrad.

Another future application of this acceleration mechanism might be the use as a bright x-ray source. Since the synchrotron radiation of bunched charged particles becomes enhanced by a factor of N_b, the number of particles per bunch, if the wavelength λ of the synchrotron radiation is less than the bunch length λ_b [11], we can expect highly coherent and enhanced radiation from bunches with a bunch length λ_b smaller than the critical wavelength $\lambda_c = (4\pi m_0 c^2)/(3\gamma^2 eB)$, where B is the magnetic field strength of the synchrotron magnets. Since the longitudinal particle bunching is small to begin with ($O(c/\omega_p)$) and becomes compressed by a factor of about 100 in the process of acceleration (see Fig. 2), we might be able to achieve electron bunch lengths of the order of Å's. To estimate the brilliance

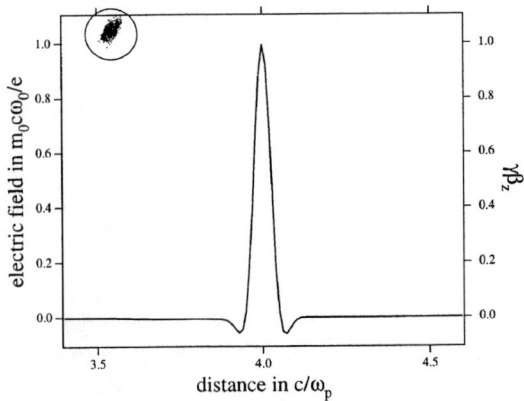

Figure 5: Final electric field and phase space for $a_0 = 1$. The plasma was initialized as shown in Fig. 2 (between $2 \leq \frac{c}{\omega_p} \leq 3$). The particles inside the circle represent the entire final phase space.

b of such a particle bunch, we use the frequency and angular dependence of the radiated energy E_{rad} [12] on axis for a single electron

$$\frac{d^2 E_{rad}}{d\omega\, d\Omega}\bigg|_{\text{on axis}} = \frac{3e^2\gamma^2}{4\pi^2 c}\left(\frac{\omega}{\omega_c}\right)^2 K_{2/3}^2\left(\frac{\omega}{2\omega_c}\right) \qquad (22)$$

and express the number of photons per electron radiated in the frequency bandwidth $\Delta\omega/\omega$ as

$$\frac{d\,\#\text{ of photons}}{d\Omega} = \frac{3\alpha\gamma^2}{4\pi^2}\left(\frac{\omega}{\omega_c}\right)^2 K_{2/3}^2\left(\frac{\omega}{2\omega_c}\right)\frac{\Delta\omega}{\omega}, \qquad (23)$$

which leads to a spectral brilliance of

$$b = 10^{-9}\frac{3\alpha\gamma^2}{4\pi^2}\left(\frac{\omega}{\omega_c}\right)^2 K_{2/3}^2\left(\frac{\omega}{2\omega_c}\right)\rho[\text{mm}^{-2}]\nu[\text{s}^{-1}]. \qquad (24)$$

Here b is the brilliance in $\frac{\#\text{ of photons}}{\text{sec.\ mrad}^2\ \text{mm}^2\ 0.1\%\frac{\Delta\omega}{\omega}}$, $K_{2/3}$ is the modified Bessel function of the second kind, $\alpha \simeq 1/137$ is the fine structure constant, ρ is the particle density N/S for the incoherent and N^2/S for the coherent synchrotron radiation, ν is the repetition rate (for the average brilliance) or the inverse radiation time for a single bunch ($\approx c/\lambda_b$ for the peak brilliance), and $\omega_c = 2\pi c/\lambda_c$ is the critical frequency. Due to the angular divergence of the photon packet, the spectral brilliance at the target however will be smaller by a factor $(r_0\gamma/(r_0\gamma+d))^2$ with r_0 being the radius of the irradiated area and d the distance between metallic film and target.

Using a metallic film ($n_0 \sim O(10^{23}/\text{cm}^3)$) of several ten atomic layers for the initial plasma slab, we could produce bunch lengths in or even below the angstrom regime, which would lead to the coherent production of x-ray radiation. Of particular interest is the generation of soft x-ray radiation at the "water-window" spectral region ($\lambda = 2.3 - 4.4$ nm) [13]. Assuming the electron bunch production from a $L_p = 0.1 c/\omega_p$ thick metal film at $n_0 = 10^{23}/\text{cm}^3$ by this mechanism, the final bunch length is expected to be in the angstrom regime. Assuming further that the bunch could be accelerated to electrons energies of 2 GeV while preserving this bunch length, a magnetic field of $B = 0.24$T would be sufficient to generate synchrotron radiation with a critical wavelength λ_c of 2nm. With $N/S \simeq 1.68 \cdot 10^{14}/\text{mm}^2$ and $S \simeq 10^{-3}\text{mm}^2$, Eq. (24) leads to an average brilliance of coherent synchrotron radiation at $\omega \approx \omega_c$ exceeding $3 \cdot 10^{21} \frac{\text{photons}}{\text{sec. mrad}^2 \text{ mm}^2 \, 0.1\% \frac{\Delta \omega}{\omega}}$ at a repetition rate ν of 10Hz. Assuming the bunch length λ_b to be 1 Å, the peak intensity at $\omega \approx \omega_c$ would be greater than $10^{39} \frac{\text{photons}}{\text{sec. mrad}^2 \text{ mm}^2 \, 0.1\% \frac{\Delta \omega}{\omega}}$. The value for the average brilliance would be of the same order as the one proposed for the LINAC coherent light source at the Stanford Linear Accelerator Center [14] and the one proposed for TESLA-FEL at DESY, Hamburg [15], while the value for the peak brilliance would exceed those proposed for SLAC and DESY by several orders of magnitude.

Extreme short electron bunches created by this mechanism might also be of interest for exciting a wakefield in a crystal. Short electron bunch lengths and high electron densities within the bunch may be used to resonantly excite electrostatic waves along a nano [17] or meso hole structure [18] of a crystal. In [17], pores of radii of several lattice spacings are etched through finite volumes of a single crystal. Electron scattering off valence- and conduction-band electrons (and nuclei) is drastically reduced along those channels while a coherent wakefield structure over many lattice spacings is still possible. The length scale of the acceleration structure ($\sim c/\omega_p$) along a nanohole (background electron density $\sim 10^{24}/\text{cm}^3$) is about ≈ 50 Å and thus of the same order as the hole radius. Since the particle density of the driving beam should be on the order of (or even higher than) the ambient electron density, the desired electron bunches could be produced using the a metallic film as discussed above. In this case, the thickness of the film should be several hundreds Å which would require a laser pulse $a_0 > 3$ (see Eq.(18)). Since the transverse dimension of the electron bunch is expected to be much larger than the nanohole diameter, a honeycomb-like hole structure in the material would be necessary to reduce the number of "unused" electrons in the bunch. For the length scale of the wakefield wavelength to be of the order of the transverse dimension of a microhole, a material with an ambient plasma electron density of about $10^{20}/cm^3$ could be used. Here, the electron bunch could be produced from a $\sim 5\mu m$ thin plasma layer like the one mentioned before in this section. Once again, Eq. (18) requires a laser strength for this case of $a_0 > 2$.

The energy gain of the electrons due to their acceleration from rest by subcyclic pulses scales was found to scale proportional to a_0^2, as could be expected from (16). Comparing this to the energy gain of electrons in the laser wakefield accelerator scheme (which is proportional to $\omega_0^2/\omega_p^2)a_0^2$, [10]), we may not suggest this mechanism to be used for acceleration of particles to very high energies. Furthermore, the energy gain for highly relativistic particles per stage is only proportional to $a_0^{2/3}$ as will be shown in the following: The equation of motion for the z component of a highly relativistic particle ($\gamma \gg 1$, $\beta_z \approx 1$) becomes

$$m_0c\frac{d\gamma}{dt} \approx m_0c\frac{d\gamma\beta_z}{dt} = -e(E_z + \beta_x B_y - \beta_y B_x) \approx -e\beta_x E_x \approx -e\sqrt{2}\gamma^{-1/2}E_x, \quad (25)$$

since $\beta_x/\beta_z = \sqrt{2/(\gamma-1)}$. Time integrating this, we find $\gamma(t+\Delta t) = [\gamma(t)^{(3/2)} + (3E_x\Delta t)/(\sqrt{2}m_0c)]^{(2/3)}$. Since the interaction length $c\Delta t$ is limited to about twice the vacuum diffraction length Z_R, iteration over n acceleration stages leads to a γ-factor of

$$\gamma(n\Delta t) \simeq \left(\gamma(0)^{\frac{3}{2}} + n3\sqrt{2}a_0k_0Z_R\right)^{2/3} \simeq \left(n3\sqrt{2}a_0k_0Z_R\right)^{2/3} \quad (26)$$

Assuming the "long pulse" diffraction length $Z_R = k_0w_0^2$, (26) can be rewritten as $\gamma \approx (n3\sqrt{2}a_0k_0^2w_0^2)^{2/3}$ in the highly relativistic limit. [1] Thus, a multi-stage electron collider at 1 TeV with a center wavelength $\lambda_0 \simeq 1\mu m$, $a_0 \simeq 1$, and $w_0 \simeq 10\lambda_0$ (which amounts to a laser power of about 13 PW per stage) would then require about 160,000 of such stages.

Finally, we should mention that this acceleration mechanism might have astrophysical applications as well: Magnetic field lines wrapped tightly around rotating

---

[1] This result is somewhat misleading as one could conclude that the energy gain can be made infinitely big just by increasing the spotsize w_0 while keeping the laser power $P \propto a_0^2w_0^2$ constant. However, for equation (26) we assumed that the electron does not dephase significantly with the pulse center during the interaction. This becomes wrong in the 1-dimensional limit ($w_0 \to \infty$) and (26) must be replaced by $\gamma(t+\Delta t) \approx (2+\pi a_0^2/\exp[1])/(2\gamma(t)(1-\beta_z))$ [4].

discs, as it is believed to be the case for binary stars or super-massive black holes [16], may generate strong, whip-like motions of Alfvén waves due to magnetic reconnection. Those Alfvén waves can convert into subcyclic electromagnetic pulses which then causes particle acceleration and γ-bursts.

VII. CONCLUSION

In conclusion, we proposed a new method for the generation of cold and tightly bunched electrons. Electron bunches produced by the irradiation of thin plasma layers with subcyclic laser pulses are expected to have extremely low emittance while still supporting a sufficient number of electrons per bunch to meet future accelerator requirements. The longitudinal extent of those bunches can be so small that highly intense coherent synchrotron radiation at x-ray frequencies could be generated. In addition to this, these particle bunches could be used as drivers for a crystal plasma wakefield accelerator. We also pointed out that this acceleration mechanism could be of astrophysical interest as well.

ACKNOWLEDGMENTS

This work was supported by the US Department of Energy.

References

[1] D. Umstadter, J. K. Kim, and E. Dodd, Phys. Rev. Lett. **76**, 2073 (1996).

[2] M. Lax, W. H. Louisell, and W. B. McKnight, Phys. Rev. A **11**, 1365 (1975); L. W. Davis, Phys. Rev. A **19**, 1177 (1979).

[3] G. P. Agrawal and D. N. Pattanayak, J. Opt. Soc. **69**, 575 (1979); L. Cicchitelli, H. Hora, and R. Postle, Phys. Rev. A **41**, 3727, (1990).

[4] W. Scheid and H. Hora, Laser Part. Beams **7**, 315 (1989).

[5] J. D. Lawson, IEEE Trans. Nucl. Sc. **NS-26**, 4217 (1979), P. M. Woodward, Journal IEE **93**, Part III A, 1554 (1947).

[6] C. W. Domier, N. C. Luhmann, A. E. Chou, W-M. Zhang, and A. J. Romanowsky, Rev. Sci. Instrum. **66**, 339 (1995).

[7] C. Raman, C. W. S. Conover, C. I. Sukenik, and P. H. Bucksbaum, Phys. Rev. Lett. **76**, 2436 (1996).

[8] A. Bonvalet, M. Joffre, J. L. Martin, A. Migus, Appl. Phys. Lett. **67**, 2907 (1995).

[9] G. Mourou and C. P. J. Barty, private communication.

[10] T. Tajima and J. M. Dawson, Phys. Rev. Lett. **43**, 267 (1979), E. Esarey and M. Pilloff, AIP Conf. Proc. **335**, 574 (1995).

[11] J. K. Koga, T. Tajima, and Y. Kishimoto, "The Future of Accelerator Physics", ed. T. Tajima (AIP Conf. Proc., NY 1996), pg. 424, F. C. Michel, Phys. Rev. Lett. **48**, 580 (1982).

[12] J. D. Jackson, "Classical Electrodynamics", 2nd ed., J. Wiley & Sons (1975).

[13] I. C. E. Turcu et. al., J. Appl. Phys. **73**, 8081 (1993), R. A. London, M. D. Rosen, and J. E. Trebes, Appl. Opt. **15**, 3397 (1989).

[14] G. Materlik, Phys. Bl. **51**, 286 (1995).

[15] J. Roßbach, Phys. Bl. **51**, 283 (1995).

[16] C. A. Haswell, T. Tajima, J.-I. Sakai, Astrophys. J. **401**, 495 (1992).

[17] B. Newberger, T. Tajima, F. R. Huson, W. Mackay, B. C. Convington, J. R. Payne, Z. G. Zou, N. K. Mahale, and S. Ohnuma, "Proc. of the 1989 IEEE Particle Accelerator Conference", ed. F. Bennett and J. Kopta, pg. 630 (1989).

[18] B. Meerson and T. Tajima, Opt. Comm **86**, 283 (1991).

NONLINEAR SPACE-CHARGE EFFECTS IN HIGH-BRIGHTNESS BEAMS

Y. Fink and C. Chen

Plasma Fusion Center, Massachusetts Institute of Technology, Cambridge, MA 02139

W. P. Marable

Department of Mathematics, Hampton University, Hampton, VA 23668

The dynamics of continuous, high-brightness, space-charge-dominated beams propagating through a periodic solenoidal focusing channel is studied. It is shown that nonlinearities in the self fields induce chaotic particle motion and beam halo formation for beams that are root-mean-square (rms) matched into the focusing channel but have nonuniform density profiles transverse to the direction of beam propagation. In particular, two parabolic density profiles are considered. For beams with hollow density profiles, it is found that excessive space charge at the edge of the beam induces two pairs of stable and unstable period-one orbits in the vicinity of the beam core envelope, and that the chaotic layer associated the unstable period-one orbits allows particles to escape from the core to form a halo. On the other hand, for beams with hump density profiles (i.e., with high densities on the beam axis and low densities at the beam edge), it is found that excessive space-charge on the beam axis induces an unstable fixed point on the axis and two stable period-one orbits off the axis inside the beam, and that the chaotic layer associated with the unstable fixed point is responsible for halo formation. In both cases, the halo is found to be bounded by a Kolmogorov-Arnold-Moser (KAM) surface. The ratio of halo to beam core envelope is determined numerically.

I. INTRODUCTION

Beam halo formation is an important issue in the design and development of next generation high-power particle accelerators and high-power microwave and millimeter wave tubes for a wide range of applications such as high energy and nuclear physics research, accelerator production of tritium, heavy ion fusion, and high-power, high-resolution radar [1]. Depending upon the application, beam halos, if not controlled, can lead to intolerable beam losses, radio-frequency (rf) breakdown, radioactivity buildup in the accelerator, and emittance growth, to mention a few examples. It has been recognized recently [2-9] that for space-charge-dominated beams, halo formation is due to chaotic beam dynamics induced by nonlinear space-charge effects. Chaotic particle orbits not only are *sensitive to initial conditions*, but also occupy a larger region in phase space than regular particle orbits, resulting in beam halo formation and growth in the total (edge) emittance.

In this paper, we explore the mechanisms of chaotic behavior and halo formation in continuous, space-charge-dominated beams propagating through a periodic solenoidal focusing channel with well matched root-mean-square (rms) beam envelopes. For a periodic solenoidal focusing channel with the periodicity length S and the vacuum phase advance σ_0, a *space-charge-dominated beam* satisfies the condition [9]

$$\frac{SK}{4\sigma_0 \varepsilon} > 1,$$

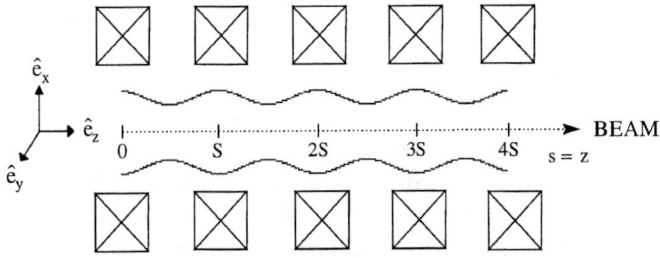

Fig. 1 Schematic of charged-particle beam propagation through a periodic solenoidal focusing channel, where the oscillatory curves illustrate the envelope for the rms-matched beam in the focusing channel.

whereas an *emittance-dominated beam* satisfies the condition

$$\frac{SK}{4\sigma_0 \varepsilon} \ll 1.$$

Here, $K = 2\nu/\gamma_b^3 \beta_b^2$ is the normalized beam perveance, ε is the unnormalized rms emittance of the beam [10], ν is the Budker parameter, and $\beta_b c$ and γ_b are the (average) velocity and relativistic mass factor of the particles, respectively. For an electron beam,

$$\frac{SK}{4\sigma_0 \varepsilon} = 2.9 \times 10^{-5} \frac{1}{\sigma_0} \left(\frac{S}{\varepsilon_n}\right) \frac{I_b}{\gamma_b^2 \beta_b^2},$$

where I_b is the electron beam current in amperes, $\varepsilon_n = \gamma_b \beta_b \varepsilon$ is the normalized rms emittance in meter-rad, and S is in meters. For an ion beam,

$$\frac{SK}{4\sigma_0 \varepsilon} = 1.6 \times 10^{-8} \frac{1}{\sigma_0 A} \left(\frac{q}{e}\right) \left(\frac{S}{\varepsilon_n}\right) \frac{I_b}{\gamma_b^2 \beta_b^2},$$

where A and q/e are the atomic mass and magnitude of the charge state of the ion, respectively, I_b is the ion beam current in amperes, $\varepsilon_n = \gamma_b \beta_b \varepsilon$ is the normalized rms emittance in meter-rad, and S is in meters.

II. DESCRIPTION OF THE MODEL

We consider an intense, continuous charged-particle beam propagating at axial velocity $\beta_b c \hat{e}_z$ through a periodic solenoidal focusing channel, as shown schematically in Fig. 1. In the thin-beam approximation, the applied magnetic field for the focusing channel is given by

$$\vec{B}^{ext}(x, y, s) = B_z(s)\hat{e}_z - \frac{1}{2} B'_z(s)(x\hat{e}_x + y\hat{e}_y) \tag{1}$$

and

$$\vec{B}^{ext}(x, y, s) = \vec{B}^{ext}(x, y, s + S), \tag{2}$$

where $s=z$ is the axial coordintate, S is the fundamental periodicity length of the focusing field, and the prime denotes derivative with respect to s.

A. Beam Self Fields

To derive the transverse equations of motion for individual test particles, we make the paraxial approximation which implies (a) the Budker parameter is small compared with unity, i.e., $q^2 N/mc^2 \ll 1$, (b) the beam is thin compared with the lattice period S, and (c) the transverse kinetic energy is small compared with the axial kinetic energy, i.e., $v_x^2 + v_y^2 \ll v_z^2 \cong \beta_b^2 c^2$. Here, N is the number of particles per unit axial length, m and q are the particle rest mass and charge, respectively, c is the speed of light in vacuo, and $\vec{v}$ is the particle velocity. Furthermore, we assume that the beam is rms matched into the focusing channel and has the following density profile:

$$n_b(r,s) = \begin{cases} \hat{n}_b(s) + \delta\hat{n}_b(s)\left[1 - \dfrac{2r^2}{r_b^2(s)}\right], & \text{for } r < r_b(s), \\ 0, & \text{for } r > r_b(s), \end{cases} \quad (3)$$

where $r = (x^2 + y^2)^{1/2}$ is the radial coordinate, $r_b(s) = r_b(s+S)$ is the radius (core envelope) for the rms-matched beam, $\hat{n}_b(s) = N/\pi r_b^2(s)$, and $\delta\hat{n}_b(s) = \delta N/\pi r_b^2(s)$ is a measure of nonuniformity in the beam density profile. It is readily shown that the beam radius $r_b(s)$ is related to the rms beam radius $\langle r^2(s)\rangle^{1/2}$ by

$$\langle r^2(s)\rangle = N^{-1}\int dx\,dy\,n(r,s)\,r^2(s) = \frac{r_b^2(s)}{2g}, \quad (4)$$

where the geometric factor g is defined by

$$g = (1 - \delta\hat{n}_b/3\hat{n}_b)^{-1}. \quad (5)$$

For a beam with a uniform density profile, $\delta\hat{n}_b(s) = 0$, which corresponds to the Kapchinskij-Vladimirskij (KV) equilibrium [11].

The self-electric and self-magnetic fields associated with the beam space-charge and current are expressed as

$$\vec{E}^{(s)}(x,y,s) = -\left(\hat{e}_x \frac{\partial}{\partial x} + \hat{e}_y \frac{\partial}{\partial y}\right)\Phi^{(s)}(x,y,s), \quad (6)$$

$$\vec{B}^{(s)}(x,y,s) = \left(\hat{e}_x \frac{\partial}{\partial y} - \hat{e}_y \frac{\partial}{\partial x}\right)A_z^{(s)}(x,y,s), \quad (7)$$

where $\partial/\partial s \cong 0$ in the paraxial approximation, the scalar potential for the self-electric field is obtained by integrating Poisson's equation

$$\left(\frac{\partial^2}{\partial x^2} + \frac{\partial^2}{\partial y^2}\right)\Phi^{(s)} = -4\pi q n_b(r,s), \quad (8)$$

and the vector potential for the self-magnetic field is defined by

$$\vec{A}^{(s)}(x,y,s) = \beta_b \Phi^{(s)}(x,y,s)\hat{e}_z. \quad (9)$$

The solution to Poisson's equation (8) is

$$\Phi^{(s)}(r,s) = \begin{cases} -q(N+\delta N)r^2/r_b^2(s) + q\delta Nr^4/2r_b^4(s), & \text{for } r \le r_b(s), \\ -q(N+\delta N/2) - 2qN \ln[r/r_b(s)], & \text{for } r > r_b(s). \end{cases} \quad (10)$$

B. Determination of the RMS-Matched Envelope

The radius for the rms-matched beam is determined from the envelope equation

$$\frac{d^2 r_b}{ds^2} + \kappa_z(s) r_b - \frac{gK}{r_b} - \frac{(4g\varepsilon)^2}{r_b^3} = 0, \quad (11)$$

which is derived following the analysis by Sacherer [10]. In Eq. (11), the geometric factor is defined in Eq. (5); the focusing parameter is defined by

$$\kappa_z(s) = \left[\frac{qB_z(s)}{2\gamma_b \beta_b mc^2}\right]^2 = \kappa_z(s+S), \quad (12)$$

where $\gamma_b = (1-\beta_b^2)^{-1/2}$; the normalized beam perveance is defined by

$$K = \frac{2q^2 N}{\gamma_b^3 \beta_b^2 mc^2}; \quad (13)$$

and the rms emittance ε is assumed to be constant and is defined by $\varepsilon = \varepsilon_{\tilde{x}} = \varepsilon_{\tilde{y}}$ and [10]

$$\varepsilon_{\tilde{x}} = \left(\langle \tilde{x}^2 \rangle \langle \tilde{x}'^2 \rangle - \langle \tilde{x}\tilde{x}' \rangle^2 \right)^{1/2}, \quad (14a)$$

$$\varepsilon_{\tilde{y}} = \left(\langle \tilde{y}^2 \rangle \langle \tilde{y}'^2 \rangle - \langle \tilde{y}\tilde{y}' \rangle^2 \right)^{1/2}. \quad (14b)$$

Here, $\langle \ \rangle$ represents the ensemble average over the beam particle distribution, and the particle transverse displacement in the Larmor frame of reference, $(\tilde{x}, \tilde{y})$, is related to that in the laboratory frame of reference, (x, y), by

$$\tilde{x}(s) = x(s)\cos[\phi(s)] - y(s)\sin[\phi(s)], \quad (15a)$$

$$\tilde{y}(s) = x(s)\sin[\phi(s)] + y(s)\cos[\phi(s)], \quad (15b)$$

with $\phi(s) = \int_{s_0}^{s} \sqrt{k_z(s)} ds$.

In general, the solutions to the envelope equation (11) can exhibit both regular and chaotic behavior [3,8]. The present model describes the dynamics of an rms-matched beam whose radius corresponds to a periodic solution to the envelope equation (11). When the strength of the focusing field is moderate, Eq. (11) has a unique periodic solution with $r_b(s) = r_b(s+S)$ [3].

For the case of an even focusing lattice with $\kappa_z(s) = \kappa_z(-s)$, it can be shown [3] that Eq. (11) is invariant under the time reversal transformation $(s, r_b) \to (-s, r_b)$, and that the periodic solution $r_b(s) = r_b(s+S)$ has the property $r_b'(0) = 0$. In this case, the rms-matched beam envelope can be determined numerically using a shooting method.

Figure 2 shows the periodic envelope for an rms-matched beam propagating through a periodically interrupted solenoidal focusing channel with the focusing parameter defined by the following periodic step function:

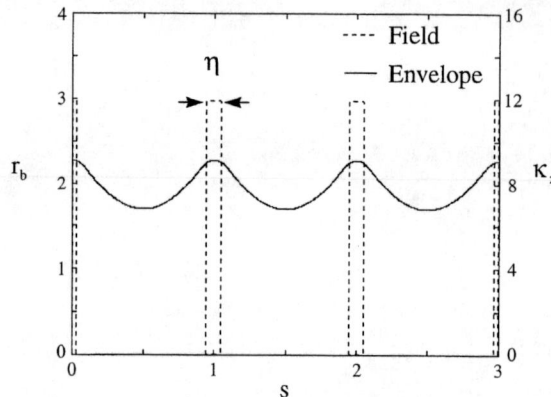

Fig. 2 Beam radius as a function of propagation distance s for an rms matched beam propagating through a step-function lattice defined by Eq. (16). Here, the choice of system parameters corresponds to: $\eta = 0.2$, $S^2\kappa_{z0} = 12.0$ $(\sigma_0 = 88.8°)$, and $SK/4\varepsilon = 10$. The horizontal and vertical axes s, r_b, and κ_z are scaled by the multiplication factors S^{-1}, $(4g\varepsilon S)^{-1/2}$, and S^2, respectively.

$$\kappa_z(s) = \begin{cases} \kappa_{z0}, & \text{for } -\eta/2 \leq s/S < \eta/2, \\ 0, & \text{for } \eta/2 \leq s/S < 1-\eta/2, \end{cases} \quad (16)$$

where η is the filling factor. The vacuum phase advance for the particle motion in this lattice is given approximately by

$$\sigma_0 = \left[S \int_0^S \kappa_z(s)ds \right]^{1/2} = (\eta S^2 \kappa_{z0})^{1/2}, \quad (17)$$

which is a measure of the strength of the average focusing field. The choice of system parameters in Fig. 2 corresponds to: $\eta = 0.2$, $S^2\kappa_{z0} = 12.0$ $(\sigma_0 = 88.8°)$, and $SK/4\varepsilon = 10$. It is evident in Fig. 2 that $r_b'(0) = 0$, as expected for $\kappa_z(s) = \kappa_z(-s)$. Note also that the results shown in Fig. 2 are independent of g in terms of the scaled variables defined by s/S, $S^2\kappa_z$, and $(4g\varepsilon S)^{-1/2}r_b$.

C. Transverse Equations of Motion

It can be shown that in the Larmor frame of reference, the transverse equations of motion for a test particle in the combined periodic solenoidal and self fields are expressed as

$$\frac{d^2x}{ds^2} + \kappa_z(s)x + \frac{q}{\gamma_b^3 \beta_b^2 mc^2} \frac{\partial}{\partial x} \Phi^{(s)}(x,y,s) = 0, \quad (18)$$

$$\frac{d^2y}{ds^2} + \kappa_z(s)y + \frac{q}{\gamma_b^3\beta_b^2 mc^2}\frac{\partial}{\partial y}\Phi^{(s)}(x,y,s) = 0, \qquad (19)$$

where $\Phi^{(s)}(x,y,s)$ is defined in Eq. (10), and the tilde over the variables x and y has been omitted. Heretofore, the variables x and y should be understood as the variables $\tilde{x}$ and $\tilde{y}$, respectively.

For a uniform-density beam with $\delta N = 0$, the equations of motion (18) and (19) are linear for the test particle in the beam interior with $r \le r_b(s)$ but become nonlinear for the test particle outside the beam with $r > r_b(s)$. It is important to point out that for nonuniform-density beams, however, the equations of motion are always nonlinear, regardless of whether the test particle is inside or outside the beam. It will be shown in Sec. III that for beam propagation through a periodic solenoidal focusing channel, Eqs. (18) and (19) are generally nonintegrable and support chaotic solutions.

In the limit of a uniform solenoidal focusing channel with $\kappa_z(s) = $ const., the rms-matched beam radius is constant. As a result, the equations of motion (18) and (19) are integrable. In this case, test particles have regular orbits and are always confined inside the beam envelope.

D. The Initial Distribution

In the present test-particle model, an initial distribution function corresponding to the parabolic density profile defined in Eq. (3) has been derived and is expressed as

$$f_b(x,y,x',y',s) = \begin{cases} \dfrac{N-\delta N}{16\pi^2\varepsilon_m^2}\delta(W-1) + \dfrac{\delta N}{8\pi^2\varepsilon_m^2}H(W), & 0 \le \delta\hat{n}_b \le \hat{n}_b, \\[2mm] \dfrac{N-\delta N}{16\pi^2\varepsilon_m^2}\delta(W-1) + \dfrac{2\delta N}{\pi r_b^2(s)}R^2H(R^2)\delta(x')\delta(y'), & -\hat{n}_b < \delta\hat{n}_b \le 0, \end{cases} \qquad (20)$$

where $R^2 = (x^2+y^2)/r_b^2$,

$$W(x,y,x',y',s=s_0) = \frac{r^2}{r_b^2} + \frac{1}{16\varepsilon_m^2}\left[(r_bx' - xr_b')^2 + (r_by' - yr_b')^2\right], \qquad (21)$$

$$H(x) = \begin{cases} 1, & 0 \le x \le 1, \\ 0, & \text{otherwise}, \end{cases} \qquad (22)$$

and the maximum emittance ε_m is defined by

$$\frac{\varepsilon}{\varepsilon_m} = \begin{cases} 1 - \dfrac{\delta\hat{n}_b}{3\hat{n}_b}, & \text{for } 0 \le \delta\hat{n}_b \le \hat{n}_b, \\[2mm] \left(1 + \dfrac{2\delta\hat{n}_b}{3\hat{n}_b} - \dfrac{\delta\hat{n}_b^2}{3\hat{n}_b^2}\right), & \text{for } -\hat{n}_b < \delta\hat{n}_b < 0. \end{cases} \qquad (23)$$

It is readily verified that $n_b(x,y,s=s_0) = \int f_b(x,y,x',y',s=s_0)dx'dy'$. Moreover, the distribution function f_b has the property that it approaches the KV equilibrium distribution [11] continuously as $\delta\hat{n}_b \to 0$. Therefore, the beams under the present investigation are perturbed directly from the KV equilibrium which is the only known Vlasov equilibrium for periodically focused intense charged-particle beams.

III. NUMERICAL RESULTS

In this section, we discuss briefly results of a numerical study of the beam dynamics for the case of the step-function lattice described by Eq. (16). Detailed results are presented in [13]. In the numerical study, the envelope equation (11) and the particle equations (18) and (19) are solved simultaneously using a fourth-order Runga Kutta integrator. The initial conditions for the envelope equation (11) are chosen such that they yield the periodic beam envelope as described in Sec. II. Because $y = 0 = y'$ is invariant, we choose the initials conditions $y(0) = 0 = y'(0)$ in all of the analyses discussed in this section. Moreover, for all of the results presented in Figs. 3-5, the phase space variables are scaled according to:

$$s \to \frac{s}{S}, \quad x \to \frac{x}{r_b(0)}, \quad y \to \frac{y}{r_b(0)}, \quad x' \to \frac{r_b(0)x'}{4\varepsilon_m}, \quad \text{and} \quad y' \to \frac{r_b(0)y'}{4\varepsilon_m}, \qquad (24)$$

where ε_m is the maximum emittance defined in Eq. (23).

A. Uniform-Density Profile

Although the equations of motion (18) and (19) have a simple form, the transverse beam dynamics exhibits rich behavior whenever space-charge effects become significant. This is illustrated in Fig. 3, where Poincare surface-of-section plots [12] are shown in the phase space (x, x') for both emittance- and space-charge-dominated uniform-density beams. The choice of the system parameters in Fig. 3 corresponds to: $\eta = 0.2$, $S^2\kappa_{z0} = 10.0$ $(\sigma_0 = 81.0°)$, $g = 1.0$ $(\delta\hat{n}_b = 0)$, and $SK/4\varepsilon = 0.5$ for the case of an emittance-dominated beam in (a) and $SK/4\varepsilon = 6.0$ for the case of a space-charge-dominated beam in (b). For each case shown in Fig. 3, 41 particles are loaded initially at $s = 0$ uniformly along the x-axis from $x = -2.0$ to 2.0, and the initial conditions are indicated by the crosses. The Poincare map [12] is generated here by plotting the positions and momenta of the test particles as they pass through the lattice points $s = 1$, $2, \ldots, 2000$.

Figure 3(a) shows a simple and regular phase space structure for the case of an emittance-dominated beam. By contrast, Fig. 3(b) shows a rather complicated phase space structure for the case of a space-charge-dominated beam, containing a mixture of regular orbits, nonlinear resonances, and chaotic layers. In Fig. 3(b), all of the test particles loaded initially inside the beam envelope have regular orbits, and these particles correspond to those in the KV distribution. However, because Eqs. (18) and (19) are *nonlinear* for $r > r_b(s)$ *and* because the strength of the nonlinearity is proportional to $SK/4\varepsilon$, the orbits of some of the test particles that cross the beam envelope become chaotic, i.e., *sensitive to initial conditions*. The chaotic particle orbits lie in the chaotic layers bounded by the invariant tori known as Kolmogorov-Arnold-Moser (KAM) surfaces [12].

It should be emphasized that all of the test particles in the KV distribution will remain inside the beam envelope, despite the fact that the underlying equations of motion (18) and (19) are nonintegrable and support chaotic solutions for $r > r_b(s)$. As far as beam halo formation is concerned, it is important to identify the mechanisms by which the test particles initially in the perturbed KV distribution f_b defined in Eq. (20)

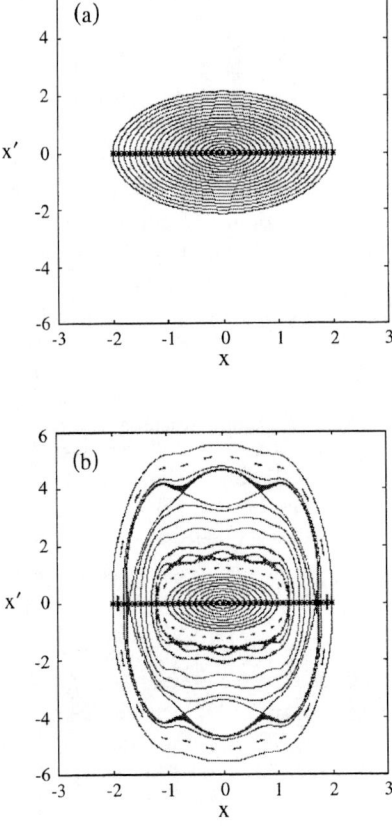

Fig. 3 Poincare surface-of-section plots in the phase space (x,x') for emittance- and spcae-charge-doinated beams propagating through 2000 lattice periods with uniform density profiles. Here, the choice of the system parameters corresponds to: $\eta = 0.2$, $S^2\kappa_{z0} = 10.0$ $(\sigma_0 = 81.0°)$, $g = 1.0$ $(\delta\hat{n}_b = 0)$, and $SK/4\varepsilon = 0.5$ for the case of an emittance-dominated beam in (a) and $SK/4\varepsilon = 6.0$ for the case of a space-charge-dominated beam in (b).

enter the chaotic layer. This is the subject matter discussed in the remainder of this section.

B. Nonuniform Density Profiles

Figure 4 shows Poincare surface-of-section plots in the phase space (x,x') for a beam with a hollow density profile. The system parameters in Fig. 4 are:

$\eta = 0.2$, $S^2 \kappa_{z0} = 12$ $(\sigma_0 = 88.8°)$, $SK/4\varepsilon = 10$, and $\delta\hat{n}_b/\hat{n}_b = -0.57$ $(g = 0.84)$. In Fig. 4, 21 test particles are loaded initially on a circle defined by $W(x,x',0,0) = 1$ in the phase space. Note that $W = 1$ is the maximum value achieved by any particle in the perturbed KV distribution f_b. The Poincare surface-of-section plots are generated here in the same way as in Fig. 3.

In Fig. 4, there is a pair of stable and unstable fixed points at the edge of the beam, i.e., at $(x,x') \approx (1,0)$ in the phase space. The unstable fixed point is located inside the beam, whereas the stable fixed point and associated island are located outside of the beam. Because of the symmetry in the underlying equations of motion (18) and (19), there is another pair of stable and unstable fixed points at $(x,x') \approx (-1,0)$. These fixed points, which correspond to periodic solutions of the equations of motion (18) and (19), are induced by excessive space-charge at the edge of the hollow beam. Associated with the two unstable fixed points is a thin chaotic layer (separatrix) which occupies both the region with $W < 1$ and the region with $W > 1$ in the phase space. Particles in this thin chaotic layer can cross the beam envelope, forming a halo around a dense core of beam determined by $W \leq 1$ in the phase space. The chaotic layer extends to $x \approx \pm 1.7$ along the x-axis where a KAM surface is located and shown as a solid curve. Therefore, the halo size in both examples shown in Fig. 4 is about 70% larger than the beam core radius.

Figure 5 shows Poincare surface-of-section plots in the phase space (x,x') for a beam with a hump density profile. The system parameters in Fig. 5 are: $\eta = 0.2$, $S^2 \kappa_{z0} = 12$ $(\sigma_0 = 88.8°)$, $SK/4\varepsilon = 10$, and $\delta\hat{n}_b/\hat{n}_b = 0.4$ $(g = 1.24)$. In Fig. 5, 41 particles are loaded initially on a circle defined by $W(x,x',0,0) = 1$ in the phase space, and the initially conditions are indicated by the crosses in Fig. 5. There are two stable fixed points at $(x,x') = (\pm 0.85, 0)$ and an unstable fixed point at the origin $(x,x') = (0,0)$. Particles initialized near the unstable fixed point assume chaotic motion which results in the formation of a halo. In this case, the halo extends to about 1.15 times the beam radius.

IV. CONCLUSIONS

The dynamics of continuous space-charge-dominated beams propagating through a periodic solenoidal focusing channel has been studied using a test-particle model. The studies were carried out in the regime where the beam is assumed to be root-mean-square (rms) matched into the focusing channel but have a nonuniform density profile transverse to the direction of beam propagation. It was shown that nonlinearities in the self fields induce chaotic particle motion and beam halo formation.

For beams with hollow density profiles (i.e., with low densities on the beam axis and high densities at the beam edge), it was found that excessive space charge at the edge of the beam induces two pairs of stable and unstable period-one orbits (i.e., two pairs of stable and unstable fixed points of the Poincare map) in the vicinity of the beam core envelope, and that the chaotic layer associated the unstable period-one orbits allows particles to escape from the core to form a halo. The halo was found to be bounded by a Kolmogorov-Arnold-Moser (KAM) surface. The ratio of halo to beam core envelope, which, depending on system parameters, can be up to a value of 1.7, was determined numerically.

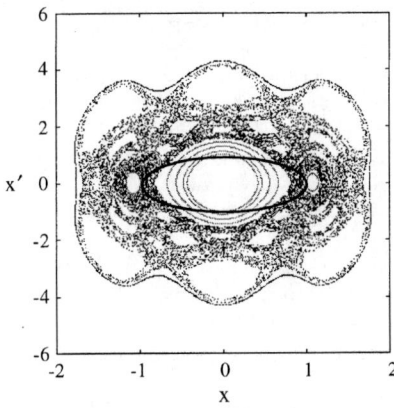

Fig. 4 Poincare surface-of-section plots in the phase space (x, x') for a beam propagating through 2000 lattice periods with a hollow density profile. Here, the choice of system parameters corresponds to: $\eta = 0.2$, $S^2 \kappa_{z0} = 12.0$ $(\sigma_0 = 88.8°)$, $SK/4\varepsilon = 10$, and $\delta \hat{n}_b / \hat{n}_b = -0.57$ $(g = 0.84)$.

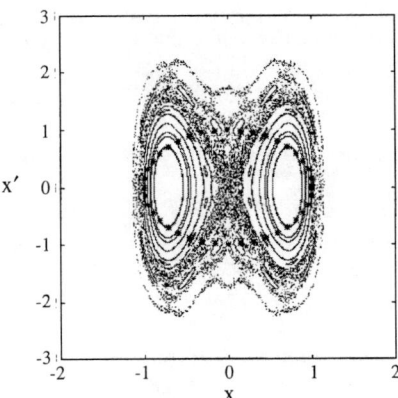

Fig. 5 Poincare surface-of-section plot in the phase space (x, x') for a beam propagating through 2000 lattice periods with a hump density profile. Here, the choice of systems parameters corresponds to: $\eta = 0.2$, $S^2 \kappa_{z0} = 12.0$ $(\sigma_0 = 88.8°)$, $SK/4\varepsilon = 10$, and $\delta \hat{n}_b / \hat{n}_b = 0.4$ $(g = 1.24)$.

On the other hand, for beams with hump density profiles (i.e., with high densities on the beam axis and low densities at the beam edge), it was found that excessive space-charge on the beam axis induces an unstable fixed point on the axis and two stable period-one orbits (i.e., two stable fixed points of the Poincare map) off the axis inside the beam. In this case, the mechanism of beam halo formation was identified with the chaotic layer associated with the unstable fixed point on the beam axis. The ratio of halo to beam core envelope for a beam with a hump density profile was found to be less than that for a beam with a hollow density profile, for, otherwise, the same choice of system parameters.

It should be emphasized that for rms-matched beams propagating through a uniform solenoidal focusing channel, test particles do *not* exhibit either chaotic behavior or beam halo formation, because the equations of motion are integrable for an arbitrary density profile with axisymmetry.

ACKNOWLEDGMENTS

This work was supported by the Department of Energy, Office of High Energy and Nuclear Physics, Grant No. DE-FG02-95ER-40919 and by the Air Force Office of Scientific Research, Grant No. F49620-94-1-0374.

REFERENCES

1. *Space Charge Dominated Beams and Applications of High Brightness Beams*, edited by S. Y. Lee (AIP Conf. Proc. **377**, 1996).
2. J. S. O'Connell, T. P. Wangler, R. S. Mills, and K. R. Crandall, Proc. 1993 Particle Accel. Conf. (IEEE Service Center, Piscataway, New Jersey, 1993), Vol. 5, p. 3567.
3. C. Chen and R. C. Davidson, Phys. Rev. Lett. **72**, 2195 (1994); Phys. Rev. **E49**, 5679 (1995).
4. Q. Qian, R. C. Davidson, and C. Chen, Phys. Plasmas **1**, 3104 (1994); Phys. Plasmas **2**, 2674 (1995).
5. L. M. Lagniel, Nucl. Instrum. Methods Phys. Res. **A345**, 1516 (1994).
6. R. L. Gluckstern, Phys. Rev. Lett. **73**, 1247 (1994).
7. C. Chen and R. A. Jameson, Phys. Rev. **E52**, 3074 (1995).
8. S. Y. Lee and A. Riabko, Phys. Rev. **E51**, 1609 (1995).
9. C. Chen, AIP Conf. Proc. **377**, 169 (1996).
10. F. J. Sacherer, IEEE Trans. Nucl. Sci. **NS-18**, 1105 (1971).
11. I. M. Kapchinskij and V.V. Vladimirskij, Proc. Int. Conf. High Energy Accelerators (CERN, Geneva, 1959), p. 274.
12. A. J. Lichtenberg and M. A. Lieberman, *Regular and Chaotic Dynamics*, Second Edition (Springer-Verlag, New York, 1992).
13. Y. Fink, C. Chen, W. P. Marable, ``Halo formation and chaos in root-mean-square matched beams propagating through a periodic solenoidal focusing channel,'' Phys. Rev. E, submitted for publication (1996).

The Space Charge Limits of Longitudinal Emittance in RF Photoinjectors

D. H. Dowell, S. Joly and A. Loulergue

Commissariat a l'Energie Atomique
B.P. 12, 91650 Bruyeres-le-Chatel, France

Abstract

The excellent electron beam quality of radiofrequency (RF) photocathode injectors is well-established, and has been verified by numerous measurements of the transverse emittance. However, there have been few experimental determinations of the longitudinal phase space. This paper describes measurements of the electron distribution produced by a 144 MHz photocathode RF gun at the ELSA Free Electron Laser facility in Bruyeres-le-Chatel, France. Phase space distributions were found by analysis of beam energy spectra and pulse shapes at 17.5 MeV for micropulse charges between 0.5 and 5 nC. A simple ray tracing model was used to transform a parameterized phase space distribution from the injector exit, through the accelerator and around a 180 degree, three dipole, non-isochronous bend. The RF phase of the accelerator which accelerates the beam from 2 to 17.5 MeV is varied to obtain the longitudinal emittance in a method analogous to the quadrupole scan technique for transverse emittance. The phase space parameters at the injector exit are obtained by fitting the data at 17.5 MeV. This work shows the longitudinal emittance increases linearly with cathode surface charge density below the space charge limit. Above this limit, the longitudinal phase space distribution either elongates or fragments into sub-bunches. The consequences these effects have upon pulse compression at higher beam energies is discussed.

INTRODUCTION.

This paper describes the measurement and analysis of electron beam energy spectra and pulse shapes to determine the electron distribution in longitudinal phase space at the photocathode injector RF cavity exit, and discusses the application of these results to pulse compression. The distribution in longitudinal phase space from the photoinjector is important in establishing the degree an electron bunch can be compressed to achieve the high peak currents needed for the Next Linear Collider(1), short wavelength Self-Amplified Spontaneous Emission

(SASE) experiments(2), free electron laser (FEL) amplifiers and high peak current applications in general.

The paper begins with a brief review of the experimental technique and the data for micropulse charges from 0.5 to 5 nC. This is followed by a discussion of the method used to determine the experimental phase space parameters at the exit of the injector cavity. The dependence of the uncorrelated energy spread and bunch length upon the surface charge density at the cathode is presented. The paper ends with the application of these results to pulse compression experiments.

DESCRIPTION OF THE EXPERIMENT AND THE DATA.

These experiments were performed at the ELSA free electron laser facility in Bruyeres-le-Chatel, France(3). The accelerator consists of a single 144 MHz photocathode radiofrequency (RF) injector followed by three 433 MHz accelerator sections. The beam energy at the 144 MHz cavity exit is 1.8 to 2 MeV and after the 433 MHz sections is 17.5 to 18 MeV. The principal beamline component is a three-dipole, doubly-achromatic, non-isochronous, 180 degree bend referred to as the demi-tour(4). In the FEL experiments, the demi-tour is used to bend the electron beam onto the wiggler axis.

In this experiment the electron beam energy spectrum is measured at the point of largest energy dispersion occurring in the middle dipole of the demi-tour, and the beam pulse shape is measured after the demi-tour using an optical transition radiation view screen viewed by a streak camera. Self-consistent analysis of these data is performed using a parameterized electron distribution in longitudinal phase space at the exit of the photoinjector cavity.

Measurements were made at 17.5 MeV for micropulse charges of 0.5, 1, 2, 3, 4 and 5 nC with all the RF phases of the accelerator adjusted to obtain the minimum energy spread for the electron beam at each micropulse charge. Additional data taken at other settings of the 433 MHz accelerator RF phase are also used in the analysis. A standard ray-tracing technique is used to calculate energy spectra and pulse shapes at the six beam charges. The spectra and bunch shapes represent projections of this electron distribution in longitudinal phase space.

Figure 1 shows measurements of the electron beam pulse shapes for micropulse charges of 0.5, 1, 2, 3, 4 and 5 nC at a beam energy of 17.5 MeV. The drive laser pulse length is 25 ps full width half maximum (FWHM) in all cases. The electron beam bunch length is nearly unchanged at 23 to 26 ps (FWHM) for charges below 2 nC. However, the electron bunch shape changes significantly when the micropulse charge is increased beyond 2 nC. At these higher micropulse charges, the bunch evolves an interesting multipulse structure. For 3 nC, there is some pulse length elongation and indication of a small pulse approximately 100 ps behind the main pulse. This hint of multipulse structure grows to three distinct, but unequal pulses at 4 nC. For a charge of 5 nC the electron bunch dissociates into three nearly equally spaced and equally charged bunches. Similar effects have been observed in another photocathode RF injector(5).

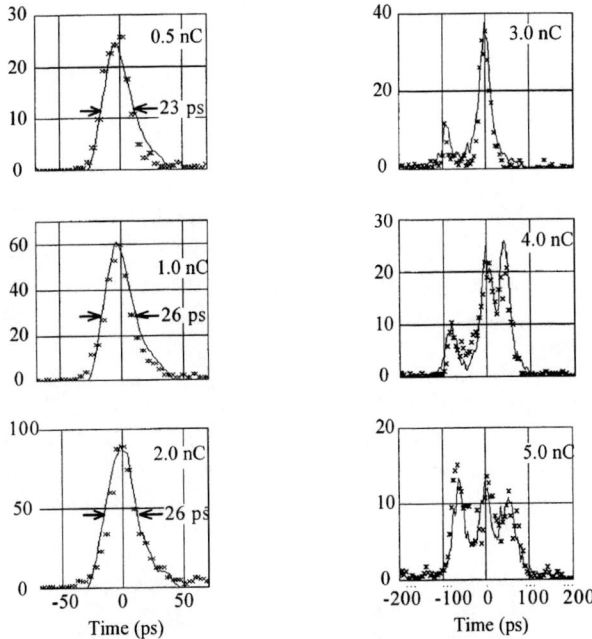

FIGURE 1. Streak camera measurements of electron beam pulse shapes at 17.5 MeV after the 180 degree demi-tour bend.

In contrast, the energy spectra (not shown) with the RF sections phased for minimum energy spread at 17.5 MeV are observed to be nearly gaussian in shape for all six micropulse charges. This energy spread increases from 26 keV (FWHM) at 0.5 nC to 41 keV at 2 nC, and is 53 keV, 41 keV and 48 keV at 3, 4 and 5 nC, respectively.

THE LONGITUDINAL PHASE SPACE PARAMETERS.

The emittance data for electron beam experiments is usually assumed to be described by an ellipse in the phase space plane of energy and time. In the case of the longitudinal phase space, each electron's conjugate variables, ΔE and Δt, are assume to fall within the boundaries of an ellipse defined by

$$\gamma \Delta t^2 + 2\alpha \Delta t \Delta E + \beta \Delta E^2 = \frac{\varepsilon_\ell}{\pi} \qquad (1)$$

Here α, β and γ are the Twiss parameters for the longitudinal phase space ellipse and ε_ℓ is the longitudinal emittance.

The longitudinal beam matrix, τ, is defined in terms of the ellipse parameters in a manner analogous to the transverse phase space ellipse, i.e.,

$$\tau = \begin{pmatrix} \tau_{11} & \tau_{12} \\ \tau_{12} & \tau_{22} \end{pmatrix} = \varepsilon_\ell \begin{pmatrix} \beta & \alpha \\ \alpha & \gamma \end{pmatrix} \qquad (2)$$

In terms of the beam matrix elements, the pulse length is given by $\sqrt{\tau_{11}}$ and the energy spread is by . The units of the τ-matrix have been chosen to be picoseconds (ps) and keV, and represent rms quantities in this paper.

The analysis of transverse emittance data is then a matter of determining the parameters of the beam ellipse. However, in general the longitudinal phase space is non-linear, and cannot be represented by a simple phase space ellipse. Therefore the approach taken here is to distort the ellipse in a power series although other methods are equally valid(6). In particular, we begin with the ellipse defined in Equation 1, solve for the energy spread, and add quadratic and cubic terms to introduce the distortion. The distorted ellipse boundary is then described by,

$$\Delta E = -\frac{\alpha}{\beta}\Delta t \pm \sqrt{\left(\frac{\alpha}{\beta}\Delta t\right)^2 - \frac{\gamma \Delta t^2 - \frac{\varepsilon_\ell}{\pi}}{\beta}} + a\Delta t^2 + b\Delta t^3 \qquad (3)$$

Therefore the analysis of longitudinal data determines not only the three ellipse parameters but also the additional coefficients, a and b, at the exit of the injector cavity.

The experiment consists of measuring the beam energy spectrum and bunch shape at various accelerator RF phases, in a manner analogous to measuring the transverse emittance using the quadrupole scan technique, except that in the longitudinal case, it is necessary to determine these five independent parameters by fitting the shapes of distributions, instead of just fitting the width of a distribution. The accelerated beam energy and time spectra are measured after the non-isochronous bend. The data is analyzed by assuming the following transformation of the energy and time variables for a group of macroparticles from the exit of the 144 MHz injector cavity to the entrance of the demi-tour,

$$\Delta E_1 = \Delta E_0 + E_{433}(\cos(\phi_{433} + 2\pi f_{rf}\Delta t_0) - \cos(\phi_{433})) ; \qquad \Delta t_1 = \Delta t_0$$
(4)

The temporal transformation assumes the beam is relativistic and no RF bunching is present. These assumptions are valid once the electrons are fully relativistic, as they are after the injector cavity.

The transport around the demi-tour is given very simply in terms of its non-isochronicity, R_{56} (in units of ps/keV),

$$\Delta E_2 = \Delta E_1; \qquad \Delta t_2 = \Delta t_1 + R_{56}\Delta E_1 \tag{5}$$

These relations are used with a ray tracing model to propagate the initial phase space distributions at the exit of the injector cavity through acceleration by the 433 MHz sections and around the demi-tour. The energy and time projections of the phase space distribution give the calculated energy and time spectra for comparison with the experimental spectra. The three phase space ellipse and two distortion parameters are varied to obtain the best fits to the low charge data.

At micropulse charges greater than 2 nC it is necessary to use a separate five parameter phase space distribution for each sub-bunch. Therefore for the three sub-bunches observed at 5 nC, there are five parameters for each phase space distribution, in addition to the energy and time offsets of the head and tail phase spaces relative to the center bunch. This gives a total of 3x5+4 or 19 parameters used to fit the energy spectra and bunch shapes measured at various RF phases of the 433 MHz sections.

DISCUSSION OF THE EXPERIMENTAL RESULTS.

Applying the analysis method described in the previous section to the bunch shape and energy spectrum data one obtains the phase space distributions shown in Figure 2. The solid curves drawn with the data in Figure 1 are the time projections of these phase space distributions and illustrate the excellence of the fits to the data. The energy projections compare equally well with the measured energy spectra.

Similar to the data, the phase space distributions also exhibit the separation of the single bunch into first two and then three parts with increasing charge, but with more detail. The phase space analysis indicates that the separated bunches are not only separated by 55 to 60 ps but the middle bunch is approximately 200 keV lower in energy relative to the head and tail pulses. This energy distribution (as well as the bunch separation) is a consequence of the local depression in electric potential produced by the electrons and their image charges inside the photocathode-vacuum boundary. According to classical diode theory, under certain conditions the electron bunch can oscillate in this potential well during

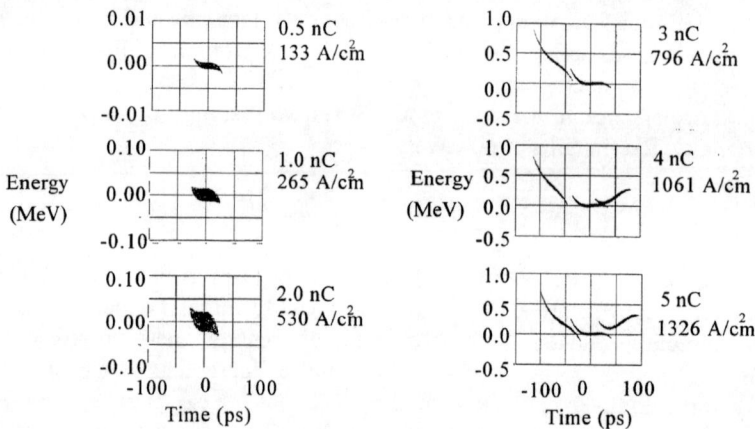

FIGURE 2. The deduced longitudinal phase space distributions at the exit of the RF photocathode injector for micropulse charges from 0.5 to 5.0 nC.

acceleration. This effect is referred to as the 'virtual cathode', and occurs when the electric field produced by the electron bunch exceeds the applied electric field, reflecting electrons back toward the photocathode(7). This transition to space charge limited emission, in analogy to the planar diode, is defined to occur when the field generated by the electron surface charge density at the photocathode equals the applied RF field. In this experiment, the peak field at the crest of the RF is 25 MV/m, however since the electron bunch is launched 20 to 30 degrees before the RF field peaks, the cathode field is slightly lower at 22 to 23 MV/m.

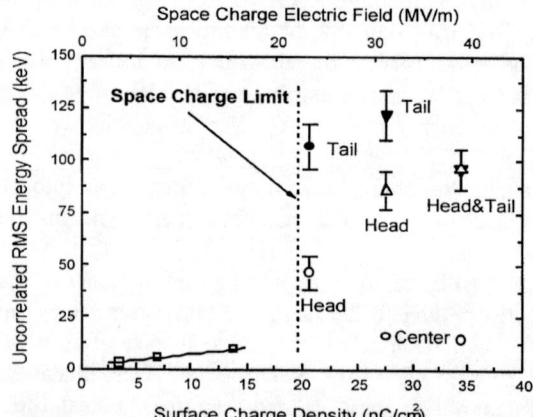

Figure 3. The uncorrelated energy spread at the exit of the photocathode injector as a function of the surface charge density at the cathode. Above the space charge limit, the notations: 'Head', 'Tail' and 'Center' denote the energy spread of each sub-bunch.

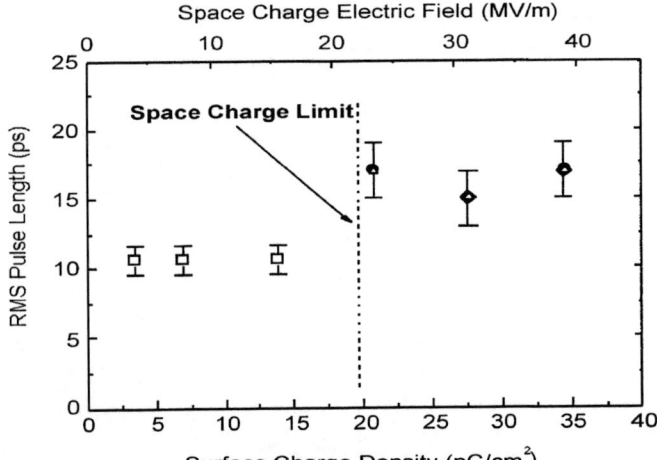

Figure 4. Bunch length as a function of the surface charge density. Above the space charge limit, the length of each sub-bunch is the same within the uncertainties of the measurements.

Above the space charge limit, the electrons' space charge forces dominate the beam dynamics, causing not only bunch breakup but also growth in the bunch energy spread and length. Figures 3 and 4 show the uncorrelated energy spreads and the bunch lengths below and above space charge limited emission for the individual phase space distributions given in Figure 2. Here the uncorrelated rms energy spread is given by $\sqrt{\tau_{22}}$ and the bunch length by $\sqrt{\tau_{11}}$ as defined in Equation 2 for each phase space ellipse. Below the space charge limit, the rms energy spread is small and increases linearly with surface charge density. The bunch length remains constant and approximately equal to the initial drive laser pulse length. However above the space charge limit the behavior becomes radically different with the formation of first two and then three bunches at the higher surface charge densities. In the two bunch case, the energy spread of the head bunch is slightly less than half of the 110 keV tail energy spread. For three bunches, the head energy spread grows to be equal to that of the tail (approximately 100 keV), while the center bunch has an energy spread not much larger than the below space charge limit values.

The center bunch electron appears undisturbed by the space charge driven expansion of the head and tail electrons, consisting of the bunch core electrons which are unaffected due to the cancellation of the opposing space charge forces of the head and tail electrons. Even though the total bunch charge extracted from the cathode is as high as 5 nC, the charge in each of the sub-bunches is less than 5/3 or 1.7 nC with a surface charger density of 11.9 nC/cm^2, well below the space charge limit. Therefore, however many electrons one attempts to extract from the cathode above the space charge limit, the resulting bunch will either elongate or dissociate into sub-bunch densities less than the space charge limited density.

THE LONGITUDINAL EMITTANCE AND PULSE COMPRESSION.

The uncorrelated longitudinal emittance is defined as,

$$\varepsilon_\ell = \pi\sqrt{\det \tau} \qquad (6)$$

and is computed using the beam matrices for the phase space distributions shown in Figure 2. This emittance does not include the correlated distortions of the phase space ellipse and therefore results by assuming the quadratic and cubic terms defined in Equation. 3 are zero. The emittances deduced at the exit of the injector cavity for micropulse charges from 0.5 to 5 nC are plotted in Figure 5.

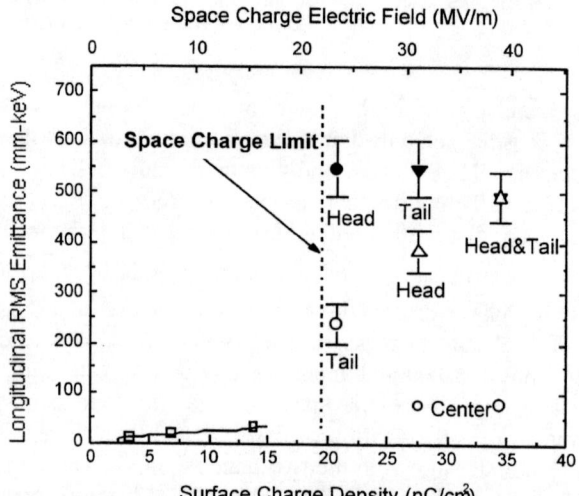

Figure 5. The experimentally deduced longitudinal emittance at the exit of the injector as it depends upon the surface charge density. This emittance growth is mainly caused by the increasing uncorrelated energy spread with charge.

Comparison of Figures 3, 4 and 5 indicate the longitudinal emittance is dominated by the uncorrelated energy spread. Therefore it is insufficient to only measure the electron bunch length as a confirmation the longitudinal emittance has not increased with charge. It is also necessary to measure the energy spread, which is seen to increase more rapidly than bunch length.

At surface charge densities below the space charge limit, the longitudinal rms. emittance has the following dependence upon the surface charge density, σ,

$$\varepsilon_\ell^{rms} = (2.6 + 2.1\sigma(nC/cm^2))\pi \text{ mm-keV}. \tag{7}$$

The uncertainty of the zero charge intercept is significant at +/-1.9 π mm-keV, while the slope is better determined with an uncertainly of +/-0.3 π mm-keV/(nC/cm^2).

An estimate of the bunch compression achievable with these emittances can be obtained by assuming the phase space ellipse is oriented at a waist at the entrance to the accelerator and is maximally compressed in the non-isochronous compressor. In this case, it can be shown that the compressed bunch length can be written as

$$\tau_{11}^{compressed} = R_{56}^2 \tau_{22}^{injector} \text{ or } \Delta t_{compressed} = R_{56} \Delta e_{injector} \tag{8}$$

where R_{56} is the non-isochronicity of the buncher and $\Delta e_{injector}$ is the uncorrelated energy spread of the beam out of the injector. Still assuming the longitudinal beam ellipse is at a waist, the corresponding compressed energy spread, $\Delta e_{compressed}$, is,

$$\tau_{22}^{compressed} = \frac{\varepsilon_\ell^2}{\pi^2 R_{56}^2 \tau_{22}^{injector}} \text{ or } \Delta e_{compressed} = \frac{\varepsilon_\ell}{\pi R_{56} \Delta e_{injector}}. \tag{9}$$

Combining these relations with those for the energy spread and emittance given above, one obtains the following relation between the compressed bunch length and the photocathode surface charge density,

$$\Delta t_{compressed}^{rms} = R_{56}(ps/keV)[0.81 + 0.66\sigma(nC/cm^2)] \text{ ps}. \tag{10}$$

Similarly the compressed rms energy spread is

$$\Delta e_{compressed}^{rms} = \frac{2.6 + 2.1\sigma(nC/cm^2)}{0.3R_{56}(ps/keV)[0.81 + 0.66\sigma(nC/cm^2)]} \text{ keV} \tag{11}$$

Consider the case of the demi-tour compressor used in the present experiment which has a R_{56} of 0.25 ps/keV, then in the zero charge limit the compressed rms bunch length is 0.20 ps and the rms energy spread is 43 keV. Just below the space charge limit ($\sigma=15$ nC/cm^2), the compressed rms bunch length is 2.7 ps and the corresponding rms energy spread remains 42 keV.

It is important to remember these estimates assume the longitudinal phase space has no distortion and is perfectly represented by an ellipse. In general, this is not a valid assumption since the phase space distribution is distorted by space charge forces in the injector, wake fields and the RF waveform of the accelerator. In addition, the dependence of the uncorrelated energy spread and the emittance upon the surface charge density found here pertains specifically to an injector with 25 MV/m of peak RF field on the photocathode. Based on the simple definition of space charge limited emission used in this paper, one expects the space charge limit to scale linearly with the RF field. Therefore, higher charge densities with lower longitudinal emittances should be observed in higher field injectors.

ACKNOWLEDGMENTS.

This paper has benefited greatly from discussions with T.D. Hayward, A. Vetter, J. Adamski, C. Parazzoli (Boeing), and P. O'Shea (Duke University). The author wishes to thank the research staff at Bruyeres-le-Chatel for the generous hospitality shown him during his visits.

REFERENCES.

1. Zimmermann, F. and Raubenheimer, T.O., "Compensation of Longitudinal Nonlinearities in the NLC Bunch Compressor," presented at Micro Bunch: A Workshop on Production, Measurement and Applications of Short Bunches of Electrons and Positrons in Linacs and Storage Rings, Upton L.I. New York, September 28-30, 1995.
2. Tatchyn, R. et al., "Research and Development Toward a 4.5-1.5 Angstrom Linac Coherent Light Source (LCLS) at SLAC," presented at the 17th International Free Electron Laser Conference (FEL95), New York, New York, August 21-25, 1995.
3. Dei-Cas, R. et al., Nuclear Instruments and Methods **A296**(1990)209.
4. Auturier, J., de Brion, J.P. and Leboute, H., Nuclear Instruments and Methods **A304**(1991)305.
5. Travier, C., Devanz, G., Leblond, B., and Mouton, B., "Experimental Characterization of Candela Photo-Injector", presented at the 18th International Free Electron Laser Conference (FEL96), Rome, Italy, August 26-30, 1996.
6. Krafft, G.A., "Correcting M56 and T566 to Obtain Very Short Bunches at CEBAF," presented at Micro Bunch: A Workshop on Production, Measurement and Applications of Short Bunches of Electrons and Positrons in Linacs and Storage Rings, Upton L.I. New York, September 28-30, 1995.
7. Birdsall, Charles K. and Bridges, William B., *Electron Dynamics of Diode Regions*, New York, Academic Press, 1966, ch. 3.

GROUP 6

"Microwave Sources"—Glen Westenskow (LLNL/LBNL)

THE NEXT LINEAR COLLIDER TEST ACCELERATOR'S RF PULSE COMPRESSION AND TRANSMISSION SYSTEMS

S. G. Tantawi, A.E.Vlieks, K. Fant, T. Lavine, R. J. Loewen, C. Pearson, R. Pope, J. Rifkin, and R. D. Ruth

Stanford Linear Accelerator Center, 2575 Sand Hill Rd., Menlo Park, CA,94025, U.S.A

Abstract. The overmoded rf transmission and pulsed power compression system for SLAC's Next Linear Collider (NLC) program requires a high degree of transmission efficiency and mode purity to be economically feasible. To this end, a number of new, high power components and systems have been developed at X-band, which transmit rf power in the low loss, circular TE_{01} mode with negligible mode conversion. In addition, a highly efficient SLED-II [1] pulse compressor has been developed and successfully tested at high power at two accelerator test facilities at SLAC. The systems produced a 200 MW pulse with a near-perfect flat-top with pulse widths ranging from 150-245 ns. In this paper we describe the design and test results of a rectangular-to-circular mode converter and the components/transmission systems based on them, as well as the design and measurements of the high power pulse compression systems using SLED-II. We will also describe how these components are being used to efficiently provide high power rf in the NLC Test Accelerator (NLCTA) program at SLAC.

1 INTRODUCTION

The NLC rf systems use low loss highly over-moded circular waveguides operating at the TE_{01} mode. The efficiency of the systems is sensitive to the mode purity of the mode excited inside these guides. We used the so- called flower petal mode transducer [2] to excite the TE_{01} mode. This type of mode transducers is efficient, compact and capable of handling high levels of power.

To make more efficient systems, we modified this device by adding several mode selective chokes to act as a mode purifiers. To manipulate the rf signals we used these modified mode converters to convert back and forth between over-moded circular waveguides and single-moded WR90 rectangular waveguides. In general, we used the relatively simple rectangular waveguide components to do the actual manipulation of rf signals. For example, two mode transducers and a mitered rectangular waveguide bend comprise a 90 degree bend. Also, a magic tee and four mode transducers would comprise a four-port-hybrid, etc. We will discuss the efficiency of an rf transport system based on the above methodology.

We also used this methodology in building the SLED-II pulse compression system.. In this paper we describe the SLED-II system constructed at SLAC. We report the experimental procedures used to measure the performance of the system.

2 IMPLEMENTATION OF SLED II

Figure 1 shows the pulse compression system. It uses two 35.35-meter long cylindrical copper waveguides as delay lines, each 12.065 cm in diameter and operating in the TE_{01} mode. In theory, these over-moded delay lines can form a storage cavity with a quality factor $Q > 1 \times 10^6$. Each of the delay lines is terminated by a shorting plate whose axial position is controllable to within ±12 μm by a DC motor with a position-monitor feed back system. The input of the line is tapered down to a 4.737 cm diameter waveguide at which the mode TE_{02} is cut-off; hence, the circular irises that determine the coupling to the lines do not excite higher order modes provided that they are perfectly concentric with the waveguide axis.

A flower petal choked mode converter excites the TE_{01} mode just before each iris. Both mode converters are connected to the coplanar arms of a high-power WR90 magic tee. The arms differ in length by a quarter wavelength at the operating frequency of 11.424 GHz. Therefore, the reflection from the lines exits through the H-arm when the input to the lines enters from the E-arm. The distance from the irises to the center of the magic tee has been adjusted to within ±13 μm to maximize this transmission.

3 CHOKED MODE CONVERTER

The flower petal mode converter as used in the current SLED-II systems and associated transport lines to convert rf power from/to the rectangular TE_{10} mode to/from the circular TE_{01} low loss mode and is described in detail in [2]. This device also has the property that if two units are connected together at their rectangular ports, a circular 90 degree bend results. We made use of this fact to design a very compact 90 degree bend of high mode purity with losses equal to that of two mode converters.

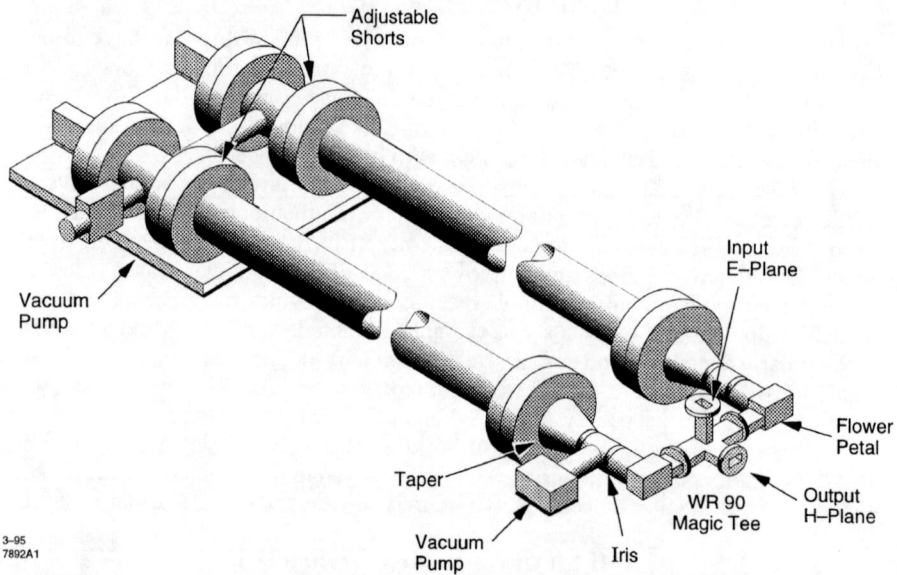

FIGURE 1. SLED-II layout showing the two 12.065 cm diameter waveguide delay lines. Guide length is 35.35 m (223 ns round trip time).

The measured level of power in spurious TE mode is ~ 0.5%. Simulations show similar levels of contamination due to TM modes. Although these levels of contamination are small, measured losses, observed in a transmission line composed of a 7.442 cm diameter waveguide surrounded by two mode converters, can be high.

These losses are due to the mode conversion-reconversion phenomenon observed in over-moded waveguides used in communication systems[3]. Since the mode converter is efficient, the circular waveguide with the two mode converters represents a high quality-factor cavity for a spurious mode. Because the line is large compared to the wavelength, there is a large number of resonating modes in the frequency band of interest. These resonant modes are coupled to the input signal because of the small mode contamination produced by the mode converters.

We, basically, have two choices in order to reduce the mode conversion losses. We can either reduce the coupling to spurious modes by improving the mode converter or reduce the quality factors of these resonating modes by inserting mode filters. Vacuum compatible mode filters are expensive and hard to make. We, therefore, tried to improve the performance of the mode converter by using mode selective chokes.

According to [2] the modes TE_{11} TE_{41} and TM_{11} represent the main components of the mode contamination produced by the flower petal. The output waveguide of the mode converter has a diameter of 4.737 cm. Except for the TE_{41} mode this guide size will not allow TE_{4n} modes to propagate. Hence, a single choke designed to reflect TE_{41} mode at the operating frequency of 11.424 GHz will greatly reduce the contamination from this mode. Also, if the choke width along the waveguide axis is less than the free space half wavelength it will not affect the TE_{01} mode. Since the modes TE_{11}, TM_{11}, and TE_{12} have the same azimuthal variations, any choke, designed to reflect any of them, will produce some mode conversion to the other modes. However, several chokes, properly spaced, can act as an effective reflector for all modes.

We added three chokes to the flower petal mode converter, one for the TE_{41} and two for the TM_{11} and TE_{11} modes. This reduced the number of resonance modes and the depth of these resonating modes.

4 MEASUREMENTS

All low level rf measurements were performed as shown in figure 2 using an HP8510C network analyzer with the results examined in the time domain using a PC.

The frequency domain measurements were transferred to the PC via a GPIB link and multiplied by the FFT of a maximally flat pulse modulating an 11.424 GHz signal. This pulse has a limited frequency response that is smaller than the measurement's frequency span. This is shown in figure 3 where the Fourier transform of a 10 ns risetime pulse is shown over the bandwidth used by the Network Analyzer.

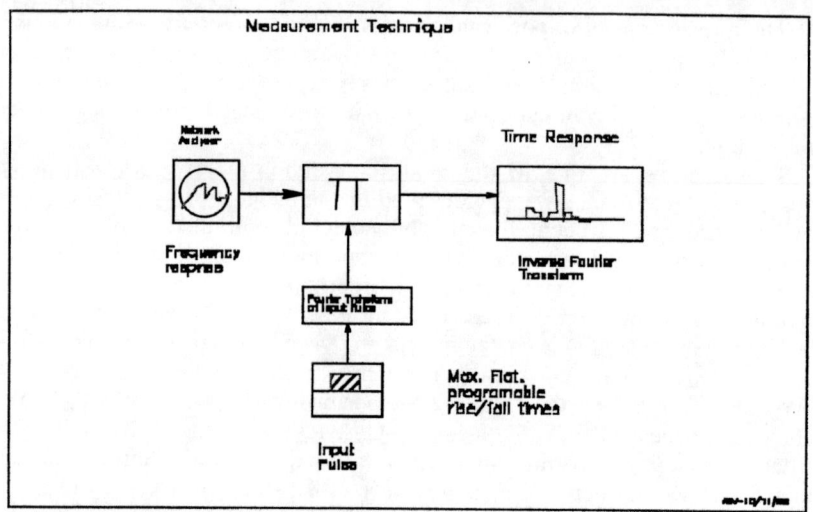

FIGURE 2. Schematic of low power rf measurement technique

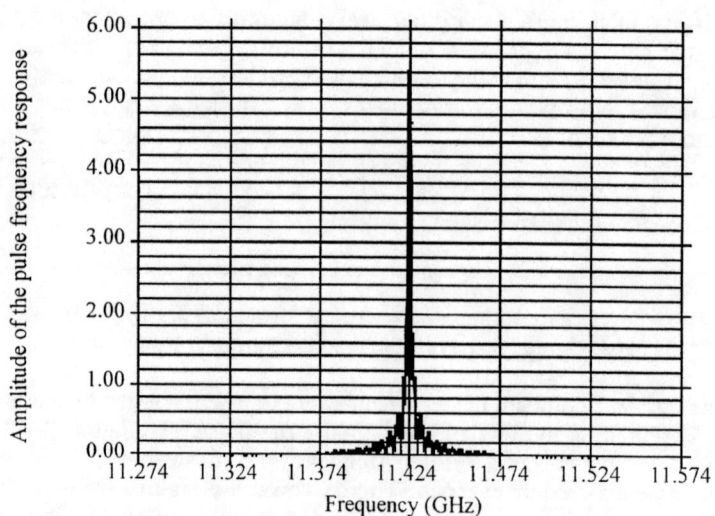

FIGURE 3. Frequency response of a 10 ns risetime maximally flat pulse

The actual time domain output is produced by taking the IFFT of this frequency domain product. Note that once we obtain the frequency characteristics of the system from the network analyzer, we can calculate the time domain response for any arbitrary input pulse.

It should be noted that using this method of rf component measurement the accuracy of the measurement is determined by the accuracy of the network analyzer, a precision device, and not by ancillary equipment having their own sources of error. We can also synthesize the phase reversed pulse required for SLED II by linear addition of two maximally flat pulses.

5 EXPERIMENTAL RESULTS

5.1 Transmission Line Measurements

The transmission line that connects the output of the first pulse compressor, in the NLCTA, to the injector section of the accelerator is 19.86 meter long. It contains two 90 degree bends, and two mode converters, one at each end. To measure the performance of that line one end was shorted. We measured the response of this system with the technique described in the above section. Figure 4 shows the reflected pulse after a round trip through the line. The one way *theoretical* power loss through the line due to the circular waveguide, which has a 7.442 cm diameter, is 1.4%. Then from the measurements, the one way line losses is 5.82%. Hence, because the system contains 6 mode converters (a 90 degree bend to be equivalent to two mode converters), an upper limit for the losses per mode converter can be set to 0.74%.

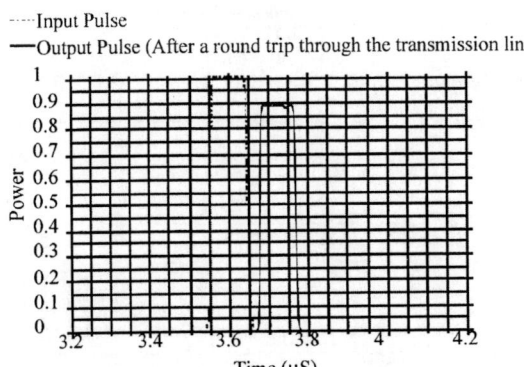

FIGURE 4. Performance of the transmission line that connects the pulse compressor to the accelerator.

5.2 SLED II Low Power Measurements

Figure 5 shows the response of the system to a 1.338 µS pulse. The last 223 ns of the pulse (A compression ratio of 6) has a 180 degree phase shift. The measured efficiency of the system is 67.5%. with a power gain of 4.08.

One can determine the values of the key parameters (i.e. iris reflection coefficient, delay line losses and external losses) in the SLED-II system by performing a measurement of this type for several different pulse compression factors.

A least-squares fitting of these measurements to the theoretical response [4] is shown in Figure 6. The round trip power loss was found to be 1.51%, indicating an intrinsic Q for the lines of 1.05×10^6. The theoretical value for the round trip losses is 1.15%. This low level of loss in the lines indicates that an extremely pure TE_{01} mode is being excited in these lines. The external losses are 6.89%, and the iris reflection coefficient is 0.70. The iris was designed using a mode matching code to have a reflection coefficient of 0.685. This is the optimum value for a compression ratio of 6.

FIGURE 5. Measured Output of the NLCTA's First RF Pulse Compressor.

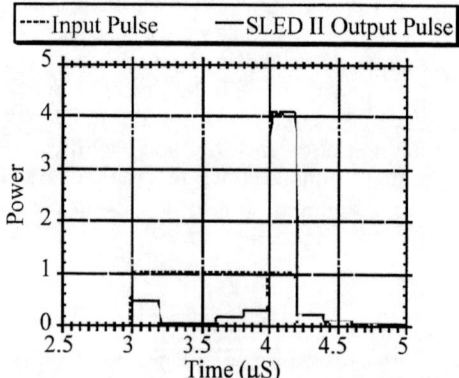

	R_c	0.70 (Design Value=0.685)
	Round Trip Losses	1.51% (Theoretical Value=1.15%)
	External Components Losses	6.89% (Theoretical Value=4%)

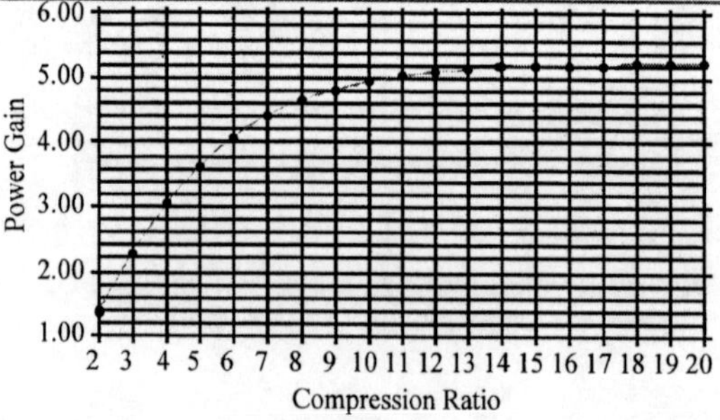

FIGURE 6 The points are measured power gains. The above table shows the fitting parameters.

5.3 SLED-II High Power Measurements

High Power Measurements were performed at the NLCTA and compared with the low power response. An example of one measurement is shown in figure 7 for a output power level of 160 MW.

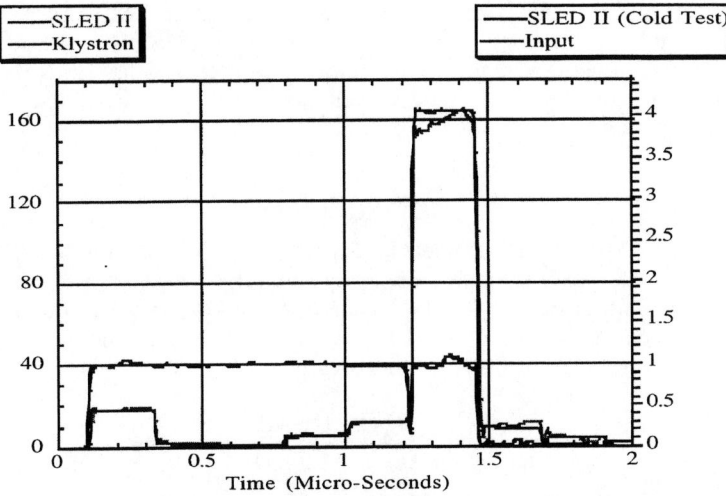

FIGURE 7. Comparison of high power and low power test measurements

Table 1. Comparison of SLED-II pulse compression systems at SLAC.

	Test Lab SLED II	NLCTA Station #0	NLCTA Station #1	NLCTA Station#2
Compression Ratio	8	6	6	6
Pulse width (nS)	150	223	245	245
Delay Line Losses/100nS (%)	1.63	0.68	0.52	0.59
Intrinsic Q	4.3×10^5	1.05×10^6	1.37×10^6	1.21×10^6
Total Delay Line Losses (%)	5.05	3.63	2.34	2.39
External Losses (%)	4.85	6.89	6.36	5.26
Total Efficiency (%)	58.6	67.5	68.6	69.3

One sees that the high power response closely matches that of the lower power results. The deviations in the pulse shape and magnitude of the high power compressed pulse are understood as resulting from the klystron input pulse which has deviations in phase and amplitude across the pulse. In addition, the rf phase reversal at the klystron input is not exactly 180 degrees and has a settling time of several nanoseconds. A compilation of high power SLED-II measurements performed at the NLCTA and the Accelerator Structure Test Area (ASTA) is shown in table 1. The component improvements in the NLCTA SLED-II systems are evident by the higher intrinsic Q values and the total delay line losses. One cannot compare the efficiency values however because of the different compression ratios.

6 CONCLUSION

We have demonstrated efficient transport and pulse compression rf systems suitable for the NLC program. These systems are based on an efficient and compact mode transducer. The implemented systems approached the theoretical design values. We also, demonstrated measurement techniques capable of measuring rf systems with very small losses.

ACKNOWLEDGMENTS

This work is supported by the US Department of Energy under contract DE-AC03-76F00515.

REFERENCES

[1] P. B. Wilson, Z. D. Farkas, and R. D. Ruth, *"SLED II: A New Method of RF Pulse Compression,"* Linear Accelerator Conference, Albuquerque, NM, September 1990; SLAC-PUB-5330.

[2] S. G. Tantawi et al., *"Numerical Design and Analysis of a Compact TE10 to TE01 Mode Transducer,"* Conference on Computational Accelerator Physics, Los Alamos, NM, 1993, AIP Conference Proceedings 297, pp. 99-106.

[3] S. E. Miller, *"Waveguide as a Communication Medium,"* The Bell System Technical Journal, Nov. 1954, pp. 1209-1265.

[4] S. G. Tantawi et al, *" Active Radio Frequency Pulse Compression Using Switched Resonant Delay Lines,"* Nuclear Instruments & Methods in Physics Research Section A, Vol. 370 (1996) pp. 297-302.

Active High Power RF Pulse Compression Using Optically Switched Resonant Delay Lines

Sami G. Tantawi[*], Ronald D. Ruth, Arnold E. Vlieks, and Max Zolotorev[**]

Stanford Linear Accelerator Center, Stanford University, Stanford, CA 94309, USA
[*]Also with Electrical Communications and Electronics Department, Cairo University, Giza, Egypt
[**] Lawrence Berkeley Lab. Berkeley, CA., USA

Abstract. We present the design and a proof of principle experimental results of an optically controlled high power rf pulse compression system. The design should, in principle, handle few hundreds of Megawatts of power at X-band. The system is based on the switched resonant delay line theory (1). It employs resonant delay lines as a means of storing rf energy. The coupling to the lines is optimized for maximum energy storage during the charging phase. To discharge the lines, a high power microwave switch increases the coupling to the lines just before the start of the output pulse. The high power microwave switch, required for this system, is realized using optical excitation of an electron-hole plasma layer on the surface of a pure silicon wafer. The switch is designed to operate in the TE_{01} mode in a circular waveguide to avoid the edge effects present at the interface between the silicon wafer and the supporting waveguide; thus, enhancing its power handling capability.

1 INTRODUCTION

During the past few years high power rf pulse compression systems have developed considerably. These systems provide a method for enhancing the peak power capability of high power rf sources. One important application is driving accelerator structures. In particular, future linear colliders, such as the proposed NLC (2), require peak rf powers that can not be generated by the current state of the art microwave tubes(3).

The SLED Pulse compression system (4) was implemented to enhance the performance of the two mile accelerator structure at Stanford Linear Accelerator Center (SLAC). One drawback of SLED is that it produces an exponentially decaying pulse. To produce a flat pulse and to improve the efficiency, the Binary Pulse Compression (BPC) system (5) was invented. The BPC system has the advantage of 100% intrinsic efficiency and a flat output pulse. Also, if one accepts some efficiency degradation, it can be driven by a single power source (6). However, The implementation of the BPC(7) requires a large assembly of over-moded waveguides, making it expensive and extremely large in size. The SLED II pulse compression system is a variation of SLED that gives a flat output pulse(8). The SLED II intrinsic efficiency is higher than SLED (9), but not as good as BPC. However, from the compactness point of view SLED II is far superior to BPC. In this paper we present a variation on SLED II that will enhance its intrinsic efficiency without increasing its physical size.

2 ACTIVE PULSE COMPRESSION USING SINGLE EVENT SWITCHED RESONANT DELAY LINES

The theory of active pulse compression with several time events is detailed in (1). Here, we state briefly the special case of a *single event* switched pulse compression system.

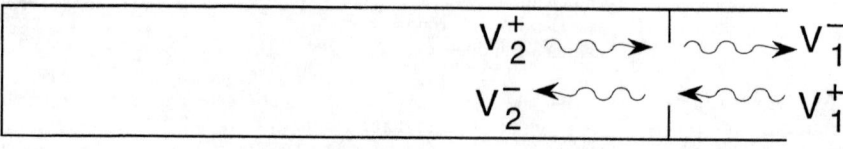

Figure 1. Resonant delay line.

Consider the waveguide delay line with a coupling iris shown in Figure 1. The *lossless* scattering matrix representing the iris is unitary. At a certain reference plane the matrix takes the following form :

$$\underline{S} = \begin{pmatrix} -R_0 & -j(1-R_0^2)^{1/2} \\ -j(1-R_0^2)^{1/2} & -R_0 \end{pmatrix}. \qquad (1)$$

In writing Eq. (1) we assumed a symmetrical structure for the iris two port network. During the charging phase we assume a constant input, i.e., $V_1^+(t) = Vin$ which equals a constant value. We, also, assume that all the voltages are equal to zero at time $t < 0$. After the energy has been stored in the line one may dump the energy in a time interval τ by flipping the phase of the incoming signal just after a time interval $(n-1)\tau$, and changing the iris reflection coefficient; i.e.

$$V_1^+(t) = \begin{cases} Vin & 0 \le t < (n-1)\tau \\ -Vin & (n-1)\tau \le t < n\tau \ ; \\ 0 & otherwise. \end{cases} \qquad (2)$$

The reflection coefficient during the discharging phase, which dumps all the energy stored in the line, is then

$$R_d = \cos\left[\tan^{-1}\left(\frac{1-(R_0 p)^{n-1}}{1-R_0 p}(1-R_0^2)^{1/2} p\right)\right]. \qquad (3)$$

This new reflection coefficient is greater than zero and the switch need only change the iris between R_0 and R_d. The output, then, reduces to

$$Vout = R_d \left[1 + \left(\frac{1-(R_0 p)^{n-1}}{1-R_0 p} \right)^2 (1-R_0^2)p^2 \right] Vin . \qquad (4)$$

The compressed pulse takes place in the interval $(n-1)\tau \le t < n\tau$. The optimum value of R_0 is such that it fills the system with maximum possible amount of energy in the time interval $(n-1)\tau$.

One can calculate the maximum power gain to be

$$Maximum\ Power\ Gain = \frac{p^2}{1-p^2} ; \qquad (5)$$

which occurs at

$$R_0 = p . \qquad (6)$$

Unlike the passive system, the maximum power gain has no intrinsic limit. It is only limited by the amount of losses in the storage line. In this case the gain can be much higher than 9, which is the limit of the passive system.

3 MICROWAVE CONTROL USING A SYMMETRIC THREE PORT NETWORK

Consider the three port device shown in Figure 3. The device is composed of a basic *lossless* three port device with two similar ports, namely, port 1 and port 2. The third port is terminated so that all the scattered power from that port is completely reflected. However, the phase of the reflected signal from the third port can be changed actively. For any lossless and reciprocal 3-port network the scattering matrix is unitary and symmetric. By imposing these two conditions on the scattering matrix $\underline{S}$ of our device and at the same time taking into account the symmetry between port one and port two, at some reference planes, one can write:

$$\underline{S} = \begin{pmatrix} \dfrac{e^{j\phi}-\cos\theta}{2} & \dfrac{-e^{j\phi}-\cos\theta}{2} & \dfrac{\sin\theta}{\sqrt{2}} \\ \dfrac{-e^{j\phi}-\cos\theta}{2} & \dfrac{e^{j\phi}-\cos\theta}{2} & \dfrac{\sin\theta}{\sqrt{2}} \\ \dfrac{\sin\theta}{\sqrt{2}} & \dfrac{\sin\theta}{\sqrt{2}} & \cos\theta \end{pmatrix} ; \qquad (7)$$

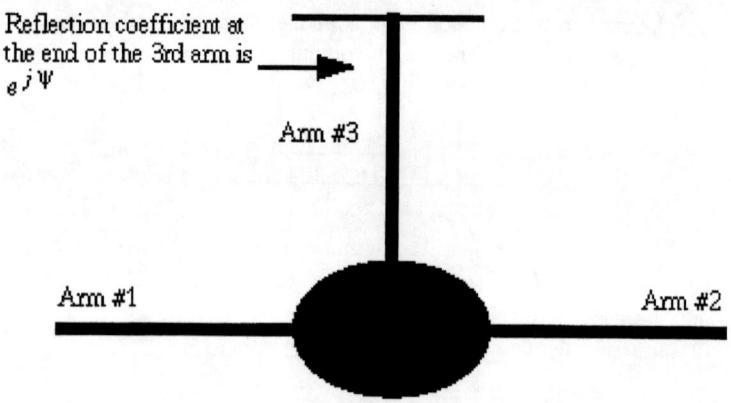

Figure 2. A symmetric three port network. The third arm is terminated with a short circuit.

Indeed, with the proper choice of the reference planes, this expression is quite general for any symetric three port network. The scattering matrix properties are determined completely with only two parameters: θ and ϕ. We terminate the third port so that all the scattered power from that port is completely reflected; i.e.,

$$V_3^+ = V_3^- e^{j\psi}. \tag{8}$$

The resultant, symmetric, two port network, then, has the following scattering matrix parameters:

$$S_{11} = S_{22} = \frac{\left(e^{j\psi} + e^{j\phi}\right) - \left(1 + e^{j(\phi+\psi)}\right)\cos\theta}{2\left(1 - \cos\theta\, e^{j\psi}\right)}, \tag{9}$$

$$S_{12} = S_{21} = \frac{\left(e^{j\psi} - e^{j\phi}\right) - \left(1 - e^{j(\phi+\psi)}\right)\cos\theta}{2\left(1 - \cos\theta\, e^{j\psi}\right)}. \tag{10}$$

By changing the angle ψ of the third port terminator, the coupling between the first and the second ports can vary from 0 to 1.

4 THE OPTICAL SWITCH

A. Device Physics

To actively change the angle of the reflection coefficient at the third port we place a piece of semiconductor material in the third arm. An external stimulus such as a laser light can induce an electron-hole plasma layer at the surface of the semiconductor, thus changing its dielectric constant. Therefore, the propagation constant of rf signals through the active arm changes; and consequently the coupling between the other two ports also changes.

For the pulse compression system application associated with the NLC (2), for which we choose a compression ratio of 8, it is required to change the reflection coefficient at the first arm between two fixed values. The device should remain in one state for approximately 1.75μsec, and in the other state for 250 nsec. Since silicon has a carrier lifetime that can extend from 1 μsec to 1 msec it seems like a natural choice for this application. One can excite the plasma layer with a very short pulse from the external stimulus (~5nsec) and the device will stay in its new status longer than the duration of the rf signal. Since repetition rate for this pulse compression system is 180 pulse/sec there is sufficient time between pulses for the switch to completely recover.

To be useful, this switch needs to have a very small amount of loss. One can show that the dielectric constant is given by the classical relation

$$\varepsilon = \varepsilon_0 \varepsilon_r \left(1 - j \frac{\sigma}{\omega \varepsilon_0 \varepsilon_r}\right); \quad (11)$$

where

$$\sigma = e \sum_i \mu_i N_i, \quad (12)$$

which is the conductivity of the semiconductor.

At a carrier density of $10^{19}/cm^3$, silicon has a conductivity of ~3.3×10^3 mho/cm. This is two orders of magnitude smaller than that of copper. However, it is high enough to make an effective reflector. The skin depth of an rf signal at the NLC frequency at this conductivity level is ~8μm. In choosing the laser wavelength to produce the photo-induced carriers, light penetration depth should be comparable to this skin depth.

B. Design Methodology

During the charging time, we choose the angle $\psi_c = \pi$. The angle ψ_c is the angle of the reflection coefficient of the third arm during the charging time. That determines the angle ϕ completely;

$$R_0^2 = \sin^2\frac{\phi}{2}. \qquad (13)$$

During the discharging time the angle ψ would change from π to the new value ψ_d, Hence the active layer, i.e. silicon wafer will be placed at a point which has a reduced electric field by a factor of $\sin \psi_d$. One then, writes an expression for the maximum field seen by the silicon wafer during the charging time:

$$E_{max} = 2\left|\tan\frac{\theta}{2}\cos\frac{\psi_d}{2}\right|\left|j\frac{1-(R_0 p)^{C_r-2}}{1-R_0 p}(1-R_0^2)^{1/2}p+1\right|\left(\frac{P_{in}Z_3}{A_3 G_3}\right)^{1/2}; \qquad (14)$$

where P_{in} is the constant level input power, Z_3 is the wave impedance of the mode excited in the waveguide that forms the third arm, A_3 is the cross sectional area of that guide, and G_3 is a geometrical factors that depends on the mode and the waveguide shape of the third arm. The angle ψ_d should satisfy:

$$|S_{11}|^2 = R_d^2 = \frac{\left(\cos(\frac{\phi+\psi_d}{2}-\theta)+\cos(\frac{\phi+\psi_d}{2}+\theta)-2\cos(\frac{\phi-\psi_d}{2})\right)^2}{4-8\cos(\psi_d)\cos(\theta)+4\cos^2(\theta)}; \qquad (15)$$

where R_d is given by Eq. (3). Finally, we write an expression for the amount of losses in the silicon wafer during the discharging time, P_l:

$$P_l = \left(\frac{2\sin\theta}{(3-4\cos\theta\cos\psi_d+\cos 2\theta)^{1/2}}\right)^2\left|j\frac{1-(R_0 p)^{G-1}}{1-R_0 p}(1-R_0^2)^{1/2}p-1\right|^2\frac{R_s}{Z_3}P_{in}; \qquad (16)$$

where R_s is the surface resistance and is given by

$$R_s = \left(\frac{\omega\mu_0}{2\sigma}\right)^{1/2}. \qquad (17)$$

The value of the conductivity σ is given by Eq. (12). Clearly one wants to use as much laser power as possible to maximize σ.

Equations (13), (14), (15), and (16) are the design equations. The goal of the design is to reduce the electric field below 100 kV/cm during the charging time; which is the estimated breakdown field for a silicon wafer with a relatively large size. At the same time one should keep the losses in the silicon wafer below a certain limit so that the temperature of the wafer does not rise above a certain temperature, say 70 C°. If this temperature is exceeded, a risk of thermal runaway exists; as the silicon wafer gets hotter the losses, during discharging time increase, causing the temperature rise further until the silicon wafer becomes conductive because of thermal effects alone.

The calculations of the switching time of this system is governed by the filling time of the third arm. To calculate this time accurately one must know how all the system components behave with frequency. This is beyond the scope of the current analysis. However, one can have a conservative estimate for that time by considering only the third arm at resonance. If this arm has an approximate length of one-half wavelength, and couples to the outside world with an iris that has a reflection coefficient equal to $\cos\theta$ ($S_{33}=\cos\theta$; see Eq. (7)) the filling time T_f then can readily shown to be :

$$T_f = \frac{-\left(1-\left(\frac{f_c}{f}\right)\right)^{-1/2}}{f \ln(\cos\theta)} ; \qquad (18)$$

where f is the operating frequency. This equation assumes that the third port is at resonance. However, in the real operation of the switch the third arm is never brought to resonance. Hence, the expression puts an upper limit on the switching time.

5 PROOF OF PRINCIPLE EXPERIMENT

Figure 3 shows the schematic diagram of an active pulse compression experiment. A flower petal mode transducer(16) and a long circular waveguide act as the storage delay line. The waveguide is excited at the TE_{01} mode. A matched magic tee, terminated with a short circuit at the E arm acts as the three port network. The TE_{01} mode switching arm (third arm) is connected to the H arm of the magic tee with a special side coupled mode converter(17). The circular guide representing the third arm is terminated from one side by a short circuit plate and a 250 micron thick, 6000 ohm cm silicon wafer is placed between the shorting plate and the mode converter. From the other side of the mode converter, a TE_{01} choke act as a terminator for this circular guide, while allowing the laser light to reach the silicon wafer. We start our tuning of this switch by adjusting the shorting plate until the field in the circular arm reaches a maximum. The field is observed by a small H probe placed near the choke during the cold test adjustments. This makes the angle $\psi_c = \pi$. Then the circular guide is connected to H arm of the magic tee. The movable short, which is connected to the E arm of the magic tee, is tuned until the reflection coefficient reaches R_0. Then, the laser is fired and the silicon wafer position is adjusted to get a reflection coefficient equals to R_d.

Figure 4 shows the output of this system at a compression ratio of 8. The system has a gain of 6. The passive pulse compression system, SLED II, has a theoretical gain of 5.1, and if one assumes similar losses in the delay line SLED II gain would drop to 4.2. Figure 4, also, shows the output of the system for a compression ratio of 32. The system has a gain of 11. SLED II has a theoretical gain of 7.4, and if one assume similar losses in the delay line SLED II gain would drop to ~5. Indeed, a gain of 11 is much more than the theoretical gain of any passive pulse compression system. These have a maximum gain of 9 as the compression ratio goes to infinity.

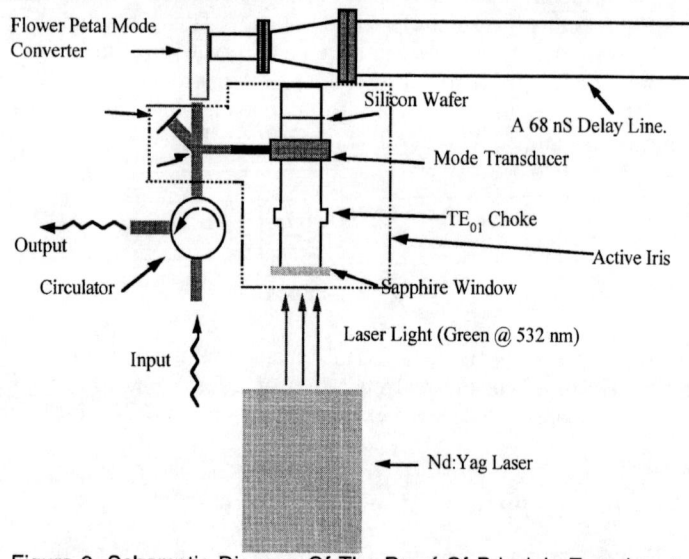

Figure 3. Schematic Diagram Of The Proof Of Principle Experiment.

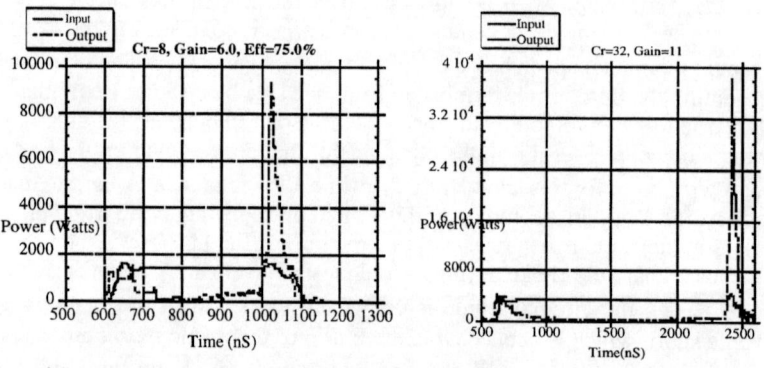

Figure 4. Experimental Output of the System at a Compression Ratio of 8, and 32.

6 CONCLUSION

We have developed the theory for a single-time-switched resonant delay line pulse-compression system. We gave a design example for an active iris operating at the TE_{01} mode. We, finally, demonstrated the operation of such a switch. The system achieved a power gain of 11 at a compression ratio of 32. This is more than the maximum theoretical gain of SLED II even as the compression ratio goes to infinity.

7 REFERENCES

(1) S. G. Tantawi, et. al, "Active RF Pulse Compression Using Switched Resonant Delay Lines," Nuc. Inst. and Meth, A, Vol. 370 (1996), pp. 297-302; SLAC-PUB 6748.

(2) R. D. Ruth et. al., "The Next Linear Collider Test Accelerator," Proc. of the IEEE Particle Accelerator Conference, Washington DC, May 1993, p. 543.

(3) Proc. Conference on Pulsed RF Sources for Linear Colliders, Montauk, Long Island New York, October 2-7, 1994

(4) Z. D. Farkas et. al., "SLED: A Method of Doubling SLAC's Energy," Proc. of the 9th Int Conf. on High Energy Accelerators, 1976, p. 576.

(5) Z. D. Farkas, "Binary Peak Power Multiplier and its Application to Linear Accelerator Design," IEEE Trans. MTT-34, 1986.

(6) P. E. Latham, "The Use of a Single source to Drive a Binary Peak Power Multiplier," Linear Accelerator Conference, Williamsburg, Virginia, 1988, CEBAF-R-89-001, pp. 623-624.

(7) Z. D. Farkas, et . al., "Two-Klystron Binary Pulse Compression at SLAC," Proc. of the IEEE Particle Accelerator Conference, Washington DC, May 1993, p. 1208.

(8) P. B. Wilson, Z. D. Farkas, and R. D. Ruth, "SLED II: A New Method of RF Pulse Compression," Linear Accl. Conf., Albuquerque, NM, September 1990; SLAC-PUB-5330

(9) Z. D. Farkas, et. al., " Radio frequency pulse compression experiments at SLAC," SPIE's Symposium on High Power Lasers, Los Angeles, CA, January, 1991, also SLAC-PUB-5409.

(10) Lee, H. Chi, Editor, "Picosecond optoelectronic devices, " Academic Press Inc., Orlando, 1984.

(11) W.R. Fowkes,et. al. "Reduced Field TE_{01} X-Band Traveling Wave Window;" Proc. of the 16th IEEE Particle Accelerator Conference (PAC 95) and International Conference on High-energy Accelerators (IUPAP), Dallas, Texas, 1-5 May 1995; SLAC-PUB-6777, Mar 1995.

(12) Wait, J. R. "Electromagnetics and plasmas," Holt, Rinehart and Winston, Inc. New York, 1968

(13) SZE S. M. "Physics of semiconductor devices, "John Wiley & Sons, Inc., New York, 1969.

(14) D. Sprehn, et. Al. "PPM Focused X-Band Klystron Development At The Stanford Linear Accelerator Center," Proc. of the 3rd International Workshop on RF Pulsed Power Sources for Linear Colliders (RF 96), Hayama, Japan, 8-12 Apr 1996; SLAC-PUB-7231, Jul 1996

(15) S. G. Tantawi, et. al., "Design of a Multi-Megawatt X-Band Solid State Microwave Switch" Presented in the IEEE int. Conf. on Plasma Sci., Wisconsin, June, 1995.

(16) S. G. Tantawi et. al., "Numerical Design and Analysis of a Compact TE_{01} to TE_{11} Mode Transducer," Conference on computational Accelerator Physics, Los Alamos, NM, 1993, AIP Conference Proceedings 297, pp. 99-106.

(17) S. G. Tantawi et. al., "Compact TE_{10} (rectangular) to TE_{01} (Circular) Mode Converter for Over-Moded Waveguides," Patent disclosure to DOE April 30, 1996, DOE Patent Docket No. S 85,926 (RL-13545).

Pulse Compressor Based on Electrically Switched Bragg Reflectors

M. I. Petelin,[1] A. L. Vikharev,[1] and J. L. Hirshfield[2]

[1] *Institute for Applied Physics (IAP), Russian Academy of Sciences, 46 Ulyanov St., 603600 Nizhny Novgorod, Russia*
[2] *Yale University, Physics Dept., New Haven, CT 06520-8120, and Omega-P Inc., 202008 Yale Station, New Haven, CT 06520-2008*

A novel switched energy storage (SES) pulse compressor is described with the apparent capability for high efficiency compression of high power 11.4 GHz pulses in the pulse energy range of interest for future electron-positron collider applications.

1. INTRODUCTION

Methods of microwave pulse compression in the time domain to obtain high peak power have been under development for over two decades. The results have been described in a large number of papers [e.g., see (1)]. The main methods used for high-power pulse compression are identified by the labels switched energy storage (SES), SLAC Energy Development (SLED), and binary power multiplication (BPM). For comparing these methods, we introduce the well-known standard definitions:

$$M = P_1/P_0, \quad C = t_0/t_1, \quad \eta_c = M/C$$

where the initial pulse is characterized by power and pulse width P_0 and t_0, the final pulse is characterized by P_1 and t_1, and η_c is the compression efficiency. Sample results obtained by these methods are presented in Table 1 [taken from (1)].

The most widely studied system is SES (2–6). It represents a microwave pulse compressor based on storage of microwave energy in cavities and the subsequent quick extraction of the energy to a load by a rapid change in the coupling to the load by the use of gas switches. The best recent results obtained by SES are given in Table 2.

TABLE 1. Comparison of typical results for each of three main methods of pulse compression.

Method	Power gain (M)	Time gain (C)	Efficiency ($\eta_c = M/C$)	P_{peak} MW
SES	~ 160	~ 200	0.8	~ 200
SLED	2.7	4.3	0.62	160
BPM	4.8	12.4	0.38	120

TABLE 2. Best results for pulse compression by SES.

Frequency GHz	Power gain (M)	Output power MW	Pulse duration ns	Efficiency η_c	Reference
3	44	70	15	0.3	(3)
3	70	100	20	–	(6)
2.85	23	160	20	0.35	(2)
10	50	2.5	8	0.4	(6)

These results were obtained with cavities built using waveguides of standard cross-section or on oversized waveguides. The switches are based on an H-plane junction. The most thoroughly investigated switched energy storage device is shown in Fig. 1.

It is seen from this figure that the cavity which stores the TE mode energy is connected to an interference switch in the form of a hybrid H-plane tee. A standing wave is produced in the energy storage mode in the H-plane junction. The node of this standing wave is situated at the input cross-section of the other arm and therefore the energy does not enter the load. For release of the stored energy, a plasma is produced at a distance of $\lambda/4$ from the short-circuited end of the H-plane junction and the energy is fed to the load. Several switching mechanisms have been studied, namely spontaneous breakdown of a gas in the waveguide, high voltage plasma discharge, vacuum arc, and electron beam injection. The most widely used switch is the plasma discharge tube.

It is seen from Table 2 that using SES systems, we obtained compressed microwave pulses at long wavelength ($f = 3$ GHz) owing to the successful design of the energy output based on the H-plane junction. Unfortunately, at shorter wavelengths such a commutator operates less efficiently and is not capable of sustaining very high RF powers. Primarily, this is because the single-mode T-junction has a small cross-section and therefore limits the extracted power. For example, at room temperature and atmospheric pressure the power withdrawn from the SES system at X-band amounts only to a few megawatts (6).

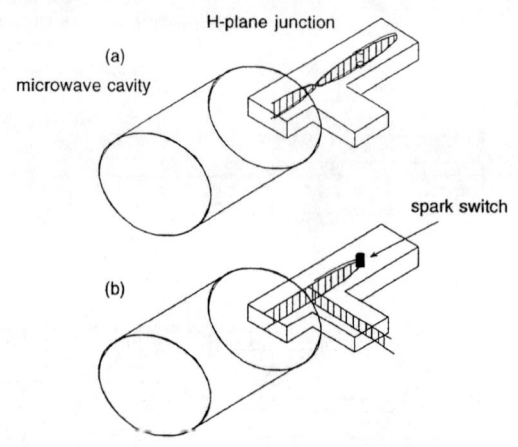

FIGURE 1. Switched energy storage and spatial distribution of the electric field during the energy storage cycle (a) and the energy extraction cycle (b).

In the SLED method of pulse compression, two high-Q cavities store energy from a klystron. Release of the stored energy is triggered by a reversal in the phase of the klystron pulse. In SLED systems, about the same output powers ($\simeq$160 MW) have been reached, but at higher frequency: $f = 11.4$ GHz. This result was obtained for a moderate power gain ($M = 2.67$) and moderate efficiency $\eta_c = 0.62$.

Comparison of the different methods of microwave pulse compression shows that SES systems possess higher efficiency (80%) than SLED systems. However, at higher frequencies (e.g., X-band), one requires radically new designs of energy extraction from the cavity in order to operate at high peak powers.

2. CONCEPT OF A COMPRESSOR USING BRAGG REFLECTORS

We propose a novel SES system enabling one to overcome the above-mentioned difficulties when operating in X-band (or even at higher frequencies). Our idea consists of using Bragg reflectors in oversized cylindrical cavities to store the microwave energy. A schematic diagram of the proposed SES is shown in Figs. 2 and 3. The compressor is a cylindrical cavity defined by Bragg

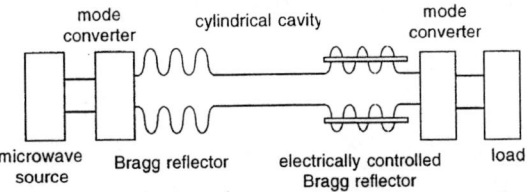

FIGURE 2. Schematic layout of microwave pulse compressor with Bragg reflectors.

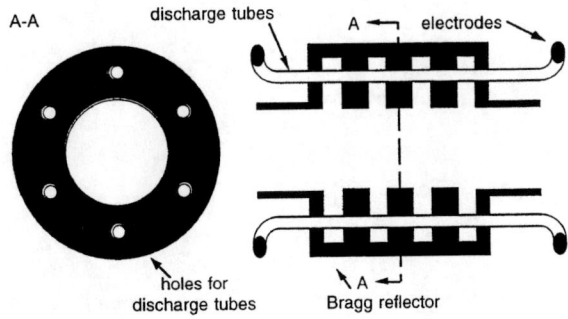

FIGURE 3. Electrically controlled Bragg reflector.

reflectors at each end. The microwave radiation from the generator (klystron) is fed into the cavity through one of the Bragg reflectors (left side) and is extracted through the other (right side). The plasma switch placed into the right-hand Bragg reflector is used to switch from the regime of energy accumulation and storage to the regime of energy extraction. Since the compressor will operate in the "breakdown-proof" mode TE_{0n}, the primary microwave generator will be coupled to the compressor using a mode converter.

For a high Q in the cavity ($Q = 10^4$–10^5), the transmission coefficients of the left and right Bragg reflectors are $\alpha_1 \simeq 10^{-2}$ and $\alpha_2 \simeq 10^{-3}$. The length of the Bragg reflectors is rather small, about 0.5 m. The total length of the SES system is estimated to be several meters. For reaching an output pulse duration of $\simeq 100$ ns with a cylindrical cavity of length $\simeq 1$ m, the transmission coefficient of the second reflector in the energy extraction regime would be switched to $\alpha \simeq 0.1$ by igniting the discharge tubes.

3. THE THEORY OF BRAGG REFLECTORS

We consider a cylindrical waveguide and perturb its walls with small inhomogeneities. We introduce into the waveguide one of its unperturbed eigenmodes. Any inhomogeneity will scatter the incident wave, in particular, in the backward direction. The sum of back-scattered waves is clearly maximum if the phase shift between elementary waves scattered by any two neighboring inhomogeneities is equal to 2π. This condition is satisfied, if the distance d between the two scatterers is equal to the half wavelength in the waveguide:

$$d = \lambda_i/2, \tag{1}$$

For a periodic inhomogeniety, the wave corrugation constant $h_c = 2\pi/d$ is twice the wave propagation constant $h_i = 2\pi/\lambda_i$:

$$h_c = 2h_i \tag{2}$$

This equation represents a particular case of the general combination equation (1)

$$h_s - h_i = h_c \tag{3}$$

where h_i and h_s are propagation constants of incident and a scattered waves.

Under condition (2), if the waveguide wall corrugation is strictly periodic, the incident and scattered waves are coupled into an evanescent mode with exponential decay of the RF field along the waveguide.

The Bragg scattering effect (in particular, reflection as described above) takes place not only in corrugated waveguides, but in a more general class of periodic media, being the foundation for a wide range of physical phenomena and numerous applications with lasers, microwave techniques, plasma physics, etc. In particular, waveguide Bragg reflectors are essential components of cavities (7) used in high power microwave generators driven by relativistic electron beams (8,9).

Note that the Bragg back scattering takes place not only at exact resonance, but within a non-zero interval of mismatch Δ around condition (2):

$$|2h_i - h_c| < \Delta/2 \tag{4}$$

the value of Δ depending on the corrugation profile.

For the pulse compressor under discussion, the parameter Δ should be small enough so that the space resonance condition (Eq. 4) would be disturbed by a very small variation of the wave propagation constant h_i. The latter can be changed by altering the effective radius of the waveguide, which can be realized by the use of gas discharge switches.

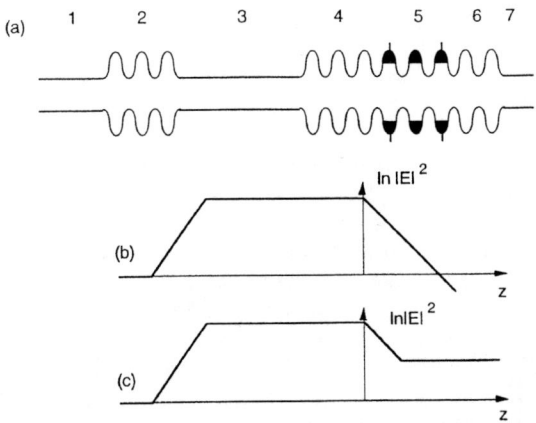

FIGURE 4. Microwave pulse compressor with Bragg reflectors. (a) 1–input waveguide, 2–input Bragg reflector, 3–storage cavity, 4...6–output Bragg reflector (composed of 4–passive section, 5–section with electrically controlled distributed gas discharge switches, 6–section with self-breakdown switches), and 7–output waveguide. (b) Longitudinal distribution of RF fields during energy storage cycle. (c) Same distribution during output cycle.

The microwave pulse compressor can be envisaged (Fig. 4) as a section of an oversized waveguide with Bragg reflectors at its ends.

The input reflector is passive: it is purely metallic and does not include any non-linear or electrically controlled elements.

The output reflector consists of passive, electrically controlled (e.g., externally triggered), and self-breakdown sections. The output passive section is similar to, but shorter than, the input section. The trigger section is the shortest, but most complicated, because it includes electrically controlled gas discharge elements. The self-breakdown section includes tubes filled with a gas which, if illuminated by a sufficiently strong microwave field, becomes ionized. The total length of the output section is considerably greater than that of the input section.

In the microwave energy accumulation phase [Fig. 4 (b)] the input Bragg reflector can be regarded as semi-transparent, and the output reflector, being unactivated, as totally reflecting; the cavity Q-factor is determined by the coupling to the input waveguide.

In the microwave energy radiation phase [Fig. 4 (c)] the control voltage pulse breaks down gas in the tubes in the trigger section, the wave propagation constant in this section shifts out of the evanescence zone, and the section is filled by the formerly reflected wave. In the self-breakdown section, the enlarged RF field produces plasma in the gas-filled tubes and this section transmits the reflected wave as well.

Thus, of the several output sections, the only functioning one is the passive one. But because it is short, the accumulated microwave power can be transmitted quickly to the load.

To provide a high power capability, the pulse compressor should have a broad cross section and the RF field at the compressor walls should be small. From this viewpoint the most attractive system appears to be one with axial symmetry, operating in TE modes. The electric field of such modes has a ring-like structure and falls to zero at the metallic walls, including those of the Bragg reflectors. (We note in passing that according to the combination relation (3) it is possible to operate at a combination of TE_{0m} and $TE_{0m'}$ modes possessing different radial indices m and m', but the advantages of such combinations are not quite clear at this stage.)

Of course, at the location of the gas-filled tubes, the RF electric field will be non-zero to provide wave switching, but still small enough to exclude breakdown during the energy accumulation phase. To satisfy the last condition, the tubes are protected by the output passive section and are hidden in the grooves of the trigger section. Finally the trigger section design should be based on a compromise, and parameters of the gas switches are of primary importance for the power obtainable with the Bragg compressor.

Assuming the gas switch parameters to be optimized, the only remaining possibility to increase the microwave power capability of the structure is by expansion of the compressor cross-section, i.e., by the use of TE modes with high radial indices m. Use of a higher radial index will reduce ohmic wall losses as well. The main practical limitation on this approach is the appearance of parasitic resonances, because any real system cannot be ideally axisymmetric. One method to filter out the parasitic modes is the use of a middle section in the form of a quasi-optical waveguide. The method is very effective, but nevertheless has its own limits. So the maximum power limit of the Bragg compressor remains an open question.

The use of high-order modes implies the use of mode converters to match a Bragg compressor with a power source and with a load. Presently there are several types of mode converters able to meet any reasonable specification of the sort.

4. DESIGN AND OPERATION OF A PLASMA SWITCH

The function of a plasma switch is to ensure the switching of the system from the regime of energy storage to the regime of energy extraction. The plasma switch is placed inside the right-hand Bragg reflector. It consists of a set of discharge tubes. A schematic diagram of the Bragg reflector with a plasma switch is shown in Fig. 3.

Gas-discharge tubes are placed along the Bragg reflector. They are inserted inside the corrugation. The tubes are filled with gas (air) under subatmospheric pressure. The plasma in the tubes is produced by applying a high-voltage pulse to one electrode of the tube. The total number of tubes for SES in the TE_{01} mode, for example, can be 6-8. The tubes are not switched on in the energy storage regime. When a plasma is produced simultaneously in all tubes, the wave constant in the Bragg reflector is changed, the Bragg reflection conditions are not satisfied, and the reflector becomes transparent.

It should be noted that for obtaining an output pulse duration of 100 ns when the total length of the SES system is a few meters, the transmission coefficient of the right-hand reflector must be $\alpha_2 = 0.1$. Such a value of α_2 can be achieved by placing the gas-discharge tubes only along part of the length of the Bragg reflector.

Preliminary laboratory results show that the time of switching of such a plasma is determined by the propagation velocity of the ionization front in the gas-discharge tube, which, in turn, depends on the high voltage supplied to the tube, the voltage rise-time, and the gas pressure in the tubes (10). Experiments with gas-discharge tubes show (10) that under voltage $U = 30-50$ kV and gas pressure $p = 10-100$ Torr the velocity of the ionization front in the tubes can reach $V = 5 \times 10^9 - 10^{10}$ cm/s. Therefore, the time for plasma production in a switch of length $l \sim 50$ cm can amount to 5-10 ns.

5. CONCLUSION

The above-described design of an SES system possesses at least two attractive properties which provide significant advantages over previous systems. It is proposed to use the breakdown-proof TE_{0n} mode of a circular waveguide for storing a large amount of microwave energy. The energy will be extracted with the aid of an oversized waveguide enabling one to raise the output power appreciably. In this SES design, the output power will be determined by spontaneous breakdown in the gas-discharge tubes, while the remaining volume of

SES is evacuated. The tubes are located in a weak field region at the periphery of the waveguide; thus, rather high output power can be achieved.

Estimates show that if the diameter of the right-hand Bragg reflector is 10 cm, and a high-voltage power supply of 30–50 kV is used for production of plasma in the tubes, then such an SES system can have an output power of at least 100 MW. The output power can be increased by building up the pressure in the discharge tubes and increasing the amplitude of the high-voltage pulses in the plasma switch.

In this compressor, energy losses occur as ohmic dissipation in the device walls and in the plasma switch tubes. We have estimated the compression efficiency (see Section 1 for definition) to be in the range $\eta_c = 0.7$–0.8. The output pulse shape will have a rapidly rising leading edge and a slowly drooping following edge, perhaps as long as 100 nsec. It is possible that pulse-shaping can be effected by gentle radial tapering of the energy storage cavity.

Demonstration of the microwave pulse compression method presented in this paper is currently being undertaken in a collaboration between IAP, Omega-P, Inc., and Yale University. At IAP, participants in the work include O. A. Ivanov, V. A. Isaev, N. F. Kovalev and S. Kuzikov, at Omega-P participants include M. A. LaPointe, and at Yale participants include R. B. Yoder and M. Wang. This work was supported in part by the US Department of Energy.

6. REFERENCES

1. Benford, J., and Swegle, J., *High-Power Microwaves*, Boston: Artech House, 1991.

2. Alvarez, R. A., Birx, D. L., Byrne, D. P., Lauer, E. J., and Scalapino, D. J., *Particle Accel.* **11**, 125–130 (1981).

3. Devyatkov, N. D., Didenko, A. N., Zamyatina, L. Y., Razin, S. V., and Yushkov, Y. G., *Radiotekh. Elektron.* **25**, 1227–1230 (1980) [*Radio Eng. Electron. (USSR)* **25**, 87–90 (1980)].

4. Alvarez, R. A., Byrne, D. P., and Johnson, R. M., *Rev. Sci. Instrum.* **57**, 2475–2480 (1986).

5. Alvarez, R. A., *Rev. Sci. Instrum.* **57**, 2481–2488 (1986).

6. Avgustinovich, V. A., Novikov, S. A., Razin, S. V., and Yushkov, Y. G., *Isvest. Vyssh. Ucheb. Zav. Radiofizika* **28**, 1347–1348 (1985) [*Radiophys. Quant. Electron. (USSR)* **28**, 1347 (1985)].

7. Kovalev, N. F., Orlova, I. M., and Petelin, M. I., *Radiophys. Quant. Electron. (USSR)* **11**, 784 (1968).

8. Kovalev, N. F., Pankratova, T. B., and Shestakov, D. I., *Radiotekh. Elektron.* **19**, 2205–2206 (1974).

9. Bratman, V. L., Denisov, G. G., Ginzburg, N. S., and Petelin, M. I., *IEEE J. Quant. Electron.* **QE-19**, 282–296 (1983).

10. Asinovsky, E. I., Vasilyak, L. M., and Markovets, V. V., *Teplofiz. Vis. Temp.* **21**, 577 (1983) [*High Temperature (USSR)* **21**, 293–305 (1983)].

Development of an X-Band Magnicon Amplifier for the Next Linear Collider

Steven H. Gold and Arne W. Fliflet
Beam Physics Branch, Plasma Physics Division,
Naval Research Laboratory, Washington, DC 20375-5346

Allen K. Kinkead
Sachs/Freeman Associates, Inc., Landover, MD 20785

B. Hafizi
Icarus Research, Inc., P.O. Box 30780, Bethesda, MD 20824-0780

Oleg A. Nezhevenko,[†] Viacheslav P. Yakovlev,[†] and Jay L. Hirshfield
Omega-P, Inc., 202008 Yale Station, New Haven, CT 06520
[†]*Permanent Address: Budker Institute of Nuclear Physics, Novosibirsk, Russia 630090*

Richard True
Litton Systems, Inc., Electron Devices Division, San Carlos, CA 94070

ABSTRACT

The magnicon is a scanning-beam microwave amplifier that is being developed as a high power, highly efficient microwave source for use in powering the next generation of high gradient electron linear accelerators. In this paper, we first discuss the results from a cold cathode magnicon experiment at 11.12 GHz, driven by a single-shot Marx generator. We then present the design of a new thermionic magnicon experiment to produce more than 50 MW at 11.4 GHz, using a 210 A, 500 kV beam from an ultrahigh convergence thermionic electron gun driven by a rep-rated modulator. This new design has a predicted efficiency in excess of 60%.

INTRODUCTION

The magnicon [1-3] is a scanning-beam microwave amplifier tube that is under development as a radio-frequency (RF) source to power future high gradient electron accelerators such as the proposed TeV Next Linear Collider (NLC). The magnicon uses rotating TM modes to spin up a pencil electron beam to high transverse momentum in a series of deflection cavities, the first externally driven, and then a synchronously rotating RF mode of the output cavity is used to extract the transverse momentum as microwave power at the drive frequency or one of its harmonics.

The Naval Research Laboratory (NRL) has been investigating magnicon physics for the past five years with the goal of building a high power 11.4 GHz magnicon tube [4–12]. We published a design study for a 50 MW, 50% efficient frequency-doubling 11.4 GHz magnicon employing a 500 kV, ~170 A, 2-mm-diameter electron beam from an advanced thermionic electron gun [7], and then designed and built an initial magnicon experiment using a cold-cathode diode on the NRL Long Pulse Accelerator Facility [13]. However, this experiment employed a 5.5-mm-diameter beam, for which the predicted efficiency falls to ~20%. In preliminary results, high power magnicon operation was demonstrated (14 MW ±3 dB), but a plasma loading problem restricted operation to parameters for which the final deflection cavity (the penultimate cavity) was unstable [10].

The Budker Institute of Nuclear Physics (INP) is presently carrying out thermionic magnicon experiments at 7 GHz. They recently reported 30 MW at 35% efficiency, with this relatively low efficiency attributed to a slightly oversized electron beam that was mismatched into the magnetic field [14]. Some members of the INP magnicon team are collaborators in the present work.

In this paper, we present a summary of results from the cold-cathode magnicon experiment, including work performed after Ref. [10] was completed. We then present the results of a design study for the new thermionic magnicon amplifier that is currently being constructed at NRL. The new design has been optimized for the parameters of a new ultrahigh convergence 500 kV, 210 A thermionic electron gun, and incorporates improvements that should raise the efficiency while reducing the possibility of RF breakdown. The predicted efficiency is in excess of 60%.

COLD–CATHODE MAGNICON EXPERIMENT

Apparatus

Figure 1 shows a schematic of the NRL cold-cathode magnicon experiment, and its design parameters. The electron beam is produced by a single-shot Marx generator driving a smooth metal cathode located in the fringing fields of the main solenoidal magnet. The 5.5-mm diameter beam size is set by an aperture located at the end of a graphite down-taper in the uniform field of the magnet. The typical output voltage waveform consists of a ~150 nsec risetime, a ~250 nsec voltage flat-top, and a ~800 nsec falltime that can be eliminated by a triggered divertor switch. The complete circuit includes four 5.560 GHz, TM_{110} deflection cavities (a drive cavity, two gain cavities, and a two-section π-mode penultimate cavity), followed by an

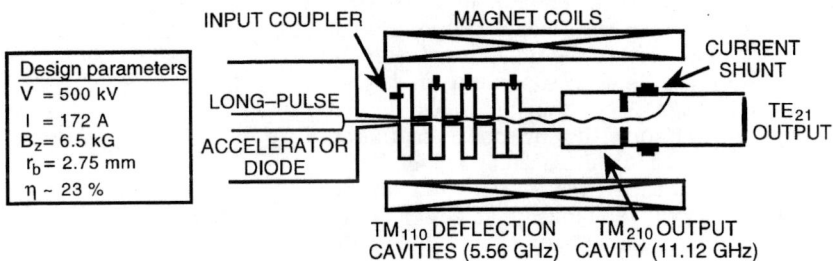

FIGURE 1. Schematic diagram of the NRL X-band cold-cathode magnicon experiment.

11.120 GHz TM_{210} output cavity with a quality factor Q~250. The cavities are assembled using bolts and O-rings, with copper gaskets to provide RF seals. The first cavity is driven by a tunable C-band magnetron. Each of the deflection cavities has a calibrated RF pickup. The power from the output cavity is coupled into a 3.5-cm-diameter output waveguide that is cutoff to the TM_{21} mode, resulting in TE_{21} output. This output is measured using waveguide pickups in the far field of a lucite vacuum window. The various RF signals are measured using calibrated attenuators and crystal detectors. The beam current is monitored by a resistive current shunt.

In the design simulations, externally driven fields in the first cavity (~700 A/m) induce progressively larger fields in the gain cavities, reaching ~80 kA/m in the penultimate cavity. The electron beam is progressively spun up, but the penultimate cavity induces ~80% of the transverse beam momentum used to drive the output cavity [7]. Generating high fields in the penultimate cavity is thus a prerequisite to high power operation of the output cavity.

Results

Initial tests of the magnicon circuit were carried out at the design voltage, current, and magnetic field (500 kV, ~170 A, ~6.5 kG). In the design calculations, ~40 dB of RF gain occurs between the drive cavity and the penultimate cavity. In the experiment, this gain could only be observed at very low levels of drive power (<<1 W), with the signal in each of the deflection cavities saturating at ~1-10 kW as the drive power was progressively increased. This gain saturation effect was caused by plasma loading of the deflection cavities at very low RF field levels. The presence of plasma was confirmed by observation of light emission from the cavities when the saturation effect occurred. The plasma generation is related to the vacuum and surface conditions of a typical pulsed power vacuum (~10^{-5} Torr), and to the lack of high-temperature bakeout and RF conditioning. Under these conditions, only small signals (<100 kW) were seen from the output cavity. High pass filters determined that the emission frequency was <9.5 GHz.

Deflection cavity gain increases both with current and with the ratio B/γ, where B is the axial magnetic field and γ is the relativistic factor. It was discovered that by operating at higher currents, voltages, and magnetic fields, the RF gain could overcome the low power saturation caused by plasma loading, leading to high fields in the deflection cavities, in particular in the penultimate cavity, and thus to the possibility of high power operation in the magnicon mode. At this point, a heterodyne diagnostic was used to search for the predicted magnicon output frequency of 11.12 GHz. This diagnostic mixes

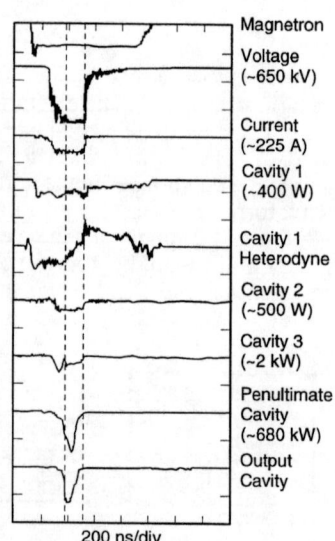

FIGURE 2. Oscilloscope traces for magnicon operation at 650 kV, 7.9 kG.

the output signal with one from a tunable local oscillator, and then performs an FFT (fast Fourier transform) on the resulting signal, yielding $|\Delta f| = |f - f_{LO}| < 1$ GHz to a precision of a few MHz, where f is a frequency from the experiment and f_{LO} is the frequency of a local oscillator.

Using this diagnostic, magnicon output was found at 650 kV, ~225 A, and magnetic fields ranging from 6.7 to 8.2 kG. Figure 2 shows a set of traces corresponding to a typical discharge at 7.9 kG. A 5.560 GHz magnetron signal was injected into the first cavity. The signals in the first three cavities are ~400 W, ~500 W, and ~2 kW, exhibiting gain saturation. However, the penultimate cavity ramps up to ~680 kW in a narrow gain spike before breaking down, driving a large signal from the output cavity.

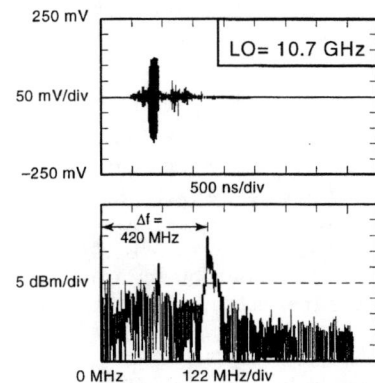

FIGURE 3. Fast Fourier transform of a heterodyne signal, showing the 11.120 GHz line.

The penultimate cavity is clearly operating close to instability, even though a drive signal was required. Under these conditions, the 11.120 GHz line is the dominant spectral feature, and the only strong line at $f > 9.5$ GHz. A typical FFT is shown in Fig. 3. A comparison of the penultimate and output cavity frequencies was carried out by means of two simultaneous FFTs employing a single local oscillator and separate mixing crystals to generate $f_1 = |f_{penult} - f_{LO}|$ and $f_2 = |f_{out} - 2f_{LO}|$. This test clearly demonstrated that $f_2 = 2f_1$, indicating a synchronous interaction. However, it was determined that the drive frequency could not control the penultimate cavity frequency.

An anechoic far-field measurement setup was used to measure the output radiation. It used two K_u-band waveguide-to-coaxial adapters (9.5 GHz cutoff frequency) as microwave pickups, one stationary, and one that swept either in the horizontal plane or azimuthally about the axis of symmetry at a distance of ~81 cm from the output window. Figure 4 shows a planar scan of E_θ and E_r, and the calculated far-field radiation pattern for a rotating TE_{21} mode (with the height of the E_θ curve normalized to the data). They are in reasonable agreement. Here, each plotted point is the average of three experimental values, each normalized to the signal at the stationary pickup. (The data extend only from −75° to +40°, because the stationary pickup was at +50°.) No unusual periodic structure was found in an azimuthal scan, in

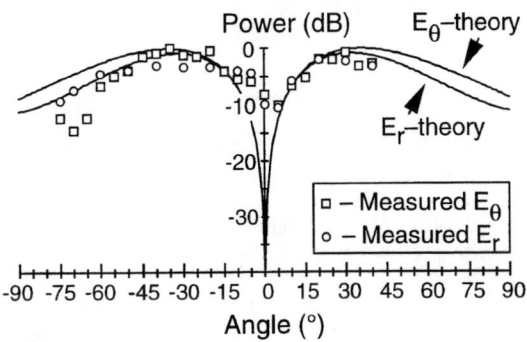

FIGURE 4. Measured far-field antenna pattern and predicted TE_{21} mode pattern.

835

agreement with the assumption of a rotating mode. Using Fig. 4, one can calculate the total power radiated into 2π. The largest measured signal at the peak of the antenna pattern was 16.0±0.5 dBm at a magnetic field of 7.4 kG. Summing attenuation factors, integrating over 2π, and adding the power radiated into E_r, yields a total radiated power of ~101.4 dBm, or 14 MW, at an efficiency of ~10%. The error bar is estimated to be ±3 dB.

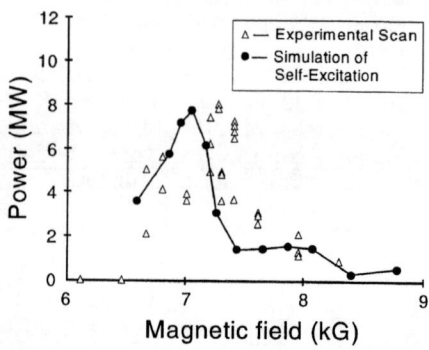

FIGURE 5. Output power versus magnetic field. Open triangles are experimental data, and closed circles are from a time-dependent simulation of self-excitation.

Figure 5 shows a scan of output power versus magnetic field, using a pickup at the angular maximum of the antenna pattern, and results from the time-dependent magnicon simulation code [14,15]. The power peaks at approximately 8 MW in the vicinity of 7.3 kG, and vanishes for magnetic fields less than 6.5 kG or greater than 8.5 kG. (The 14 MW signal discussed previously did not occur in this data set.) The loss of output below 6.5 kG occurs because plasma loading clamps the penultimate cavity RF fields at a low level. However, the output also falls off above 7.5 kG, even though large penultimate cavity signals persist. The magnicon simulation code was used to model this behavior. The process in the penultimate cavity can be modeled either as self-excitation or as phase-locked oscillation. In the case of self-excitation, with <1 µW of drive power, the penultimate cavity reaches powers >750 kW for B>7 kG. Figure 5 shows the predicted output power in this regime. The agreement with the data is excellent, with only a slight shift in the optimum magnetic field (~250 G) that may be due to calibration error. However, the apparently larger signals mentioned in the previous paragraph, ~14 MW±3 dB, cannot be reproduced. On the other hand, if the penultimate cavity is modeled as a phase-locked oscillator, we find output powers in the range of 12-13.5 MW at magnetic fields of 6.5-6.7 kG, which lie below the experimental value of 7.4 kG.

Summary of cold-cathode experiment

Operating at its design parameters, the cold-cathode magnicon experiment exhibited low field saturation of the deflection cavities due to plasma loading attributed to inadequate vacuum and surface conditions. This effect was overcome by increasing the current, voltage, and magnetic field to levels that resulted in instability, or at least near instability, of the penultimate cavity. Experimental parameters were found that produced a large penultimate cavity signal at 5.560 GHz, and a simultaneous output pulse at 11.120 GHz. The two frequencies differed by exactly 2×, indicating a synchronous interaction. However, the drive cavity signal did not control the penultimate cavity frequency, indicating a lack of true amplifier operation. The dependence of output power on magnetic field was in good agreement with simulations modeling self-excitation of the penultimate cavity, while the highest experimental powers could only be matched (at somewhat lower magnetic fields) by modeling the penultimate cavity as a phase-locked oscillator. Our original design study had previously demonstrated the necessity of a small beam size (<2.5-mm diameter) to achieve high (>50%) efficiency [7]. For these reason, it became clear that a completely new experimental configuration was required in order to demonstrate the feasibility of efficient, long-pulse, high-duty-factor magnicon amplifiers at 11.4 GHz for linear accelerator applications. In the remainder of this paper, we discuss the design of this new experiment and its present status.

THERMIONIC MAGNICON EXPERIMENT

The essential components of a high power thermionic magnicon experiment are a high power modulator, an ultrahigh convergence electron gun, a matched magnet system, and a set of cavities designed for high vacuum, high temperature bakeout, and high efficiency operation. In this section, we discuss the design of a novel high convergence electron gun to produce a 500 kV, 210 A, <2-mm-diameter beam, and the redesign of the magnicon circuit to optimize the efficiency, to lower the peak surface RF fields, and to side-couple the microwave power from the output cavity into X-band waveguide. The electron gun will be driven by a 1.5 μsec, 10 Hz modulator. The new magnicon circuit consists of a drive cavity, three gain cavities, a penultimate cavity, and an output cavity, as illustrated in Fig. 6. The drive and gain cavities are identical to those in the previous NRL design [7]. In the following sections, we discuss the redesign of the penultimate and output cavities, the design of the electron gun, and the simulation of the entire circuit.

The penultimate cavity

The penultimate cavity is an iris-coupled TM_{110} π-mode two-section cavity that carries out the final stage of beam spin up. As a result, it contains very high RF fields, creating the danger of RF breakdown. The objective of the new design was to optimize the beam spin-up, while minimizing the surface RF fields. The redesign involved adjusting the beam pipe diameter, rounding of the beam tunnel and the iris, and reducing the length of the first cavity section. For the new penultimate cavity, the combination of reduced field enhancements, and of more effective beam spin-up due to shortening the first cavity section, reduced the maximum surface electric field from 670 kV/cm in the original design to 572 kV/cm, comparable to values employed in high power X-band klystrons at SLAC [16]. The maximum electric field on the iris aperture is 472 kV/cm.

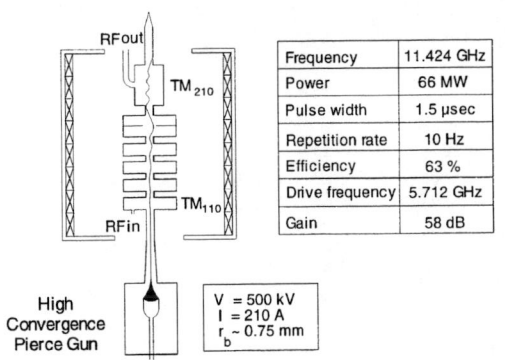

Frequency	11.424 GHz
Power	66 MW
Pulse width	1.5 μsec
Repetition rate	10 Hz
Efficiency	63 %
Drive frequency	5.712 GHz
Gain	58 dB

V = 500 kV
I = 210 A
r_b ~ 0.75 mm

FIGURE 6. Schematic of the high efficiency thermionic magnicon amplifier.

The output cavity

In order to improve the efficiency and lower the maximum surface electric fields, the output cavity was also redesigned, adjusting beam tunnel and cavity dimensions and increasing the rounding of the beam tunnel apertures. The maximum surface RF field in the output cavity was reduced from 830 kV/cm to 520 kV/cm. This is discussed in greater detail in Ref. [17]. The second goal was to redesign the cavity for side coupling rather than end coupling. For purposes of analyzing the output coupling, the rotating TM_{210} mode of the output cavity is modeled as two degenerate non-rotating modes separated by 45° of rotation about the cavity axis and by a $\pi/2$ phase difference. The INP 7 GHz magnicon employs two rectangular waveguides separated by 135° to couple out both non-rotating components. However, these waveguides break the quadrupole symmetry and can cause mode conversion. A recent INP publication has suggested a design with eight equally-spaced protrusions to minimize this problem [14]. We have taken this approach, and analyzed it using the two-dimensional code CFISH [18]. Due to reflection symmetry, only one half of the cavity need be modeled. By choosing the boundary conditions on the midplane, one can force CFISH to find either of the two non-rotating modes. Figure 7 shows CFISH plots of the cavity H' and H'' field lines for the two modes, and the values of $|H|$ and H_x along the radius extending through the waveguide midplane. Here, H' and H'' are the real and the imaginary parts of the RF magnetic field H, H_x denotes the magnetic field along the radius, and $|H|$ is the magnitude of H. To mock up an outgoing-wave boundary condition, a lossy dielectric was placed at the end of the rectangular waveguide with dielectric constants chosen to minimize reflections. The frequency of the two modes are 11.3856 GHz with Q=624 and 11.3819 GHz with Q=627. These frequencies are well within their 1/Q bandwidths, ensuring that both can be simultaneously excited to generate a rotating mode. Additionally, the field plots clearly display quadrupole structure, with no noticeable dipole distortion. However, it is important to remember that the effect of beam tunnels and the finite axial extent of the waveguides and protrusions will alter

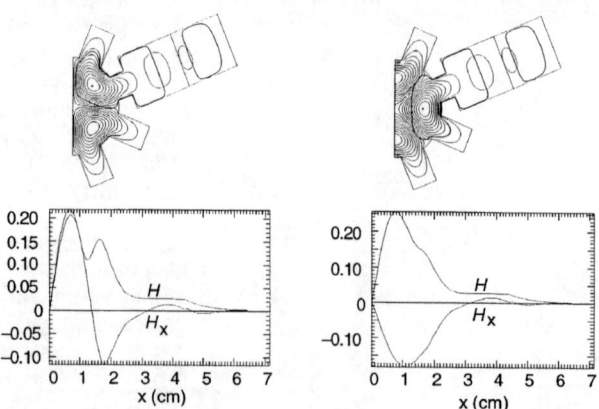

FIGURE 7. CFISH simulation of the TM_{210} modes of the output cavity: (top) RF magnetic field (*H*) lines, (bottom) plot of $|H|$ and H_x measured along the radius leading from the origin through the X-band waveguide.

the actual cavity frequency and loading. Therefore, the CFISH design must be further refined, either by three-dimensional simulation or cold test.

Discussion of the electron gun and final magnicon circuit

The point design for the 11.424 GHz magnicon amplifier uses of a 500 kV, 210 A, 1.5-mm diameter electron beam, with a 1.5 µsec pulse width and a 10 Hz repetition rate. (The Brillouin diameter is ~1.2 mm at 6.5 kG.) The electron gun is a novel high convergence relativistic Pierce gun using a dispenser cathode. The gun design and beam matching into the magnetic field are complete (see Fig. 8). Beam compression in the gun must necessarily be high in order to limit the field gradient at the focus electrode and to limit cathode loading. A 7.5-cm diameter, 30°-half-angle cathode was used, implying an overall compression ratio of 2500:1 and a mean cathode loading of 4.5 A/cm^2. The maximum focus electrode gradient is <190 kV/cm, which should allow pulse lengths >2 µsec, and the gun is designed for high duty factor operation. A key feature of the gun is an electrically-isolated focus electrode next to the cathode, which is biased negatively by a few hundred volts with respect to the cathode. This is used to control the beam edge discharge angle and to eliminate emission from the outer cathode cylinder, both of which can give rise to unwanted halo electrons. This gun is being fabricated by Litton Systems, Inc., and a matching magnet is being fabricated by the INP.

Figure 9 shows a cross section of the new magnicon circuit, and shows the magnet coils and iron yoke. The new circuit consists of a drive cavity, three simple gain cavities, and a redesigned penultimate cavity, all operating in the TM$_{110}$ mode at

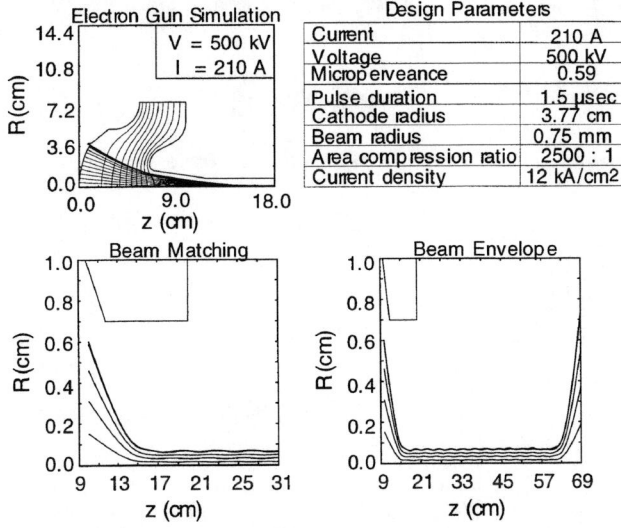

FIGURE 8. Design for the 500 kV, 210A magnicon electron gun, showing electrode geometry, selected electron trajectories, equipotentials, and beam envelope with magnetic field matching [19,20].

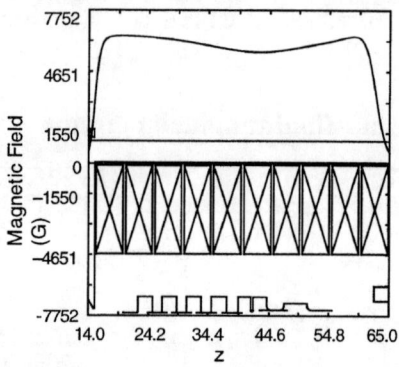

FIGURE 9. Cross-section of the magnicon circuit, showing the axial magnetic field.

5.712 GHz, and a new output cavity operating in the TM_{210} mode at 11.424 GHz. The addition of a gain cavity has increased the gain to ~58 dB, reducing the drive power for a 66 MW device to ~100 W at 5.712 GHz. This gain is consistent with values used at SLAC in the design of X-band klystrons for the NLC [16]. Figure 9 also shows the calculated on-axis magnetic field. However, note that the simulations discussed below employed a preliminary version of the magnet.

Figure 10 shows steady-state and time-dependent simulations of the complete magnicon circuit for a 1.5-mm-diameter electron beam. The time-dependent simulation shows the evolution of RF fields in the three gain cavities, the penultimate cavity, and the output cavity. The field amplitudes and the output cavity phase reach stable steady-state values. The steady-state simulation follows the trajectories and the spatial evolution of the electron energy through the asymptotic RF fields of the time-dependent simulation. The transverse dimension is stretched by ~4× in order to see details of the orbits. The beam scalloping that is evident results from simulating a zero canonical momentum beam without space charge. Notice that energy spreads and large energy excursions occur in the region of the penultimate cavity, but that the energy loss in the output cavity is remarkably uniform. The result is 66 MW at 63% efficiency. Our simulations show only a few percent loss in efficiency at beam diameters up to 2.5 mm.

SUMMARY

We have designed an electron gun and circuit for a high gain, high efficiency

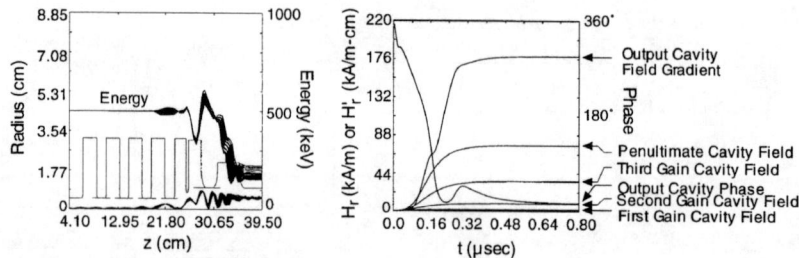

FIGURE 10. Simulations of the thermionic magnicon design for a 1.5-mm-diam. beam: (left) Steady-state simulation, showing electron trajectories and energy versus z; (right) Time-dependent simulation, showing the amplitude of the on-axis fields in the deflection cavities and the gradient and phase of the on-axis field in the output cavity.

11.424 GHz frequency-doubling magnicon amplifier. The beam parameters are 500 kV, 210 A, and 1.5-mm beam diameter. The electron gun will be driven by a 1.5 μsec, 10 Hz modulator. The predicted power is 66 MW at 63% efficiency. The maximum surface electric field is ~570 kV/cm, a value that we believe is consistent with 1.5 μsec pulse widths. This design forms the basis of a new thermionic magnicon experiment, which we plan to assemble and test in 1997.

ACKNOWLEDGMENTS

This work was supported by the U.S. Department of Energy (DoE) under Interagency Agreement DE-AI02-94ER40861, and by the Office of Naval Research. A portion of this work was also supported by a DoE Small Business Innovation Research (SBIR) grant to Omega-P, Inc. and carried out under a Cooperative Research and Development Agreement (CRADA) between NRL and Omega-P.

REFERENCES

1. Karliner, M., et al., *Nucl. Instrum. Methods Phys. Res. A* **269**, 459–473 (1988).
2. Manheimer, W. M., *IEEE Trans. Plasma Sci.* **18**, 632–645 (1990).
3. Nezhevenko, O. A., *IEEE Trans. Plasma Sci.* **22**, 765–772 (1994).
4. Hafizi, B., Seo, Y., Gold, S. H., Manheimer, W. M., and Sprangle, P., *IEEE Trans. Plasma Sci.* **20**, 232–239 (1992).
5. Gold, S. H., Sullivan, C. A., Hafizi, B., and Manheimer, W. M., *IEEE Trans. Plasma Sci.* **21**, 383–387 (1993).
6. Hafizi, B., Gold, S. H., Manheimer, W. M., and Sprangle, P., *Phys. Fluids* B **5**, 3045–3055 (1993).
7. Nezhevenko, O. A., Yakovlev, V. P., Gold, S. H., and Hafizi, B., *IEEE Trans. Plasma Sci.* **22**, 785–795 (1994).
8. Hafizi, B., and Gold, S. H., *Phys. Plasmas* **2**, 902–914 (1995).
9. Fliflet, A. W., and Gold, S. H., *Phys. Plasmas* **2**, 1760–1765 (1995).
10. Gold, S. H., Kinkead, A. K., Fliflet, A. W., Hafizi, B., and Manheimer, W. M., *IEEE Trans. Plasma Sci.* **24**, 947–956 (1996).
11. Fliflet, A. W., and Gold, S. H., *IEEE Trans. Plasma Sci.* **24**, 957–963 (1996).
12. Hafizi, B., and Gold, S. H., *IEEE Trans. Plasma Sci.*, to be published.
13. Jaitly, N. C., et al., in *Digest of Technical Papers—Eighth IEEE Int. Pulsed Power Conf.* (IEEE, New York, 1991), p. 161–165.
14. Kozyrev, E. V., et al., *Part. Accel.* **52**, 55–64 (1996).
15. Yakovlev, V., Danilov, O., Nezhevenko, O., and Tarnetsky, V., in *Proc. 1995 Particle Accelerator Conf.* (IEEE, Piscataway, NJ, 1995), vol. III, p. 1569–1571.
16. Wright, E., et al., in *Pulsed RF Sources for Linear Colliders—AIP Conf. Proc.* **337**, ed. R. C. Fernow (AIP Press, New York, 1995), p. 58–66.
17. Gold, S. H., Nezhevenko, O. A., Yakovlev, V. P., and Hafizi, B., *NRL Memo Report 7857* (1996).
18. Billen, J. H., and Young, L. M., "POISSON SUPERFISH," Los Alamos National Laboratory Report LA-UR-96-1834 (1996).
19. Myakishev, D. G., and Yakovlev, V. P., *Int. J. Mod. Phys. A* **2** (Proc. supp.), 915–917 (1993).
20. Tiunov, M. A., Fomel, B. M., and Yakovlev, V. P., in *Proc. 13th Int. Conf. on High Energy Accelerators* [NAUKA (Siberian Branch), 1987], vol. 1, p. 353.

RK-TBA Studies at the RTA Test Facility*

S. Lidia[a], D. Anderson[a], S. Eylon[a], E. Henestroza[b], T. Houck[c],
L. Reginato[a], D. Vanecek[a], G. Westenskow[c], and S. Yu[a]

[a]*Lawrence Berkeley National Laboratory, 1 Cyclotron Road, Berkeley, CA 94720 USA,*
[b]*Department of Applied Science, University of California, Davis, CA 95616 USA,*
[c]*Lawrence Livermore National Laboratory, P.O. Box 808, Livermore, CA 94550 USA*

Abstract. Construction of a prototype RF power source based on the RK-TBA concept, called the RTA, has commenced at the Lawrence Berkeley National Laboratory. This prototype will be used to study physics, engineering, and costing issues involved in the application of the RK-TBA concept to linear colliders. The status of the prototype is presented, specifically the 1-MV, 1.2-kA induction electron gun and the pulsed power system that are in assembly. The RTA program theoretical effort, in addition to supporting the development of the prototype, has been studying optimization parameters for the application of the RK-TBA concept to higher-energy linear colliders. An overview of this work is presented.

INTRODUCTION

A Lawrence Berkeley National Laboratory (LBNL) and Lawrence Livermore National Laboratory (LLNL) collaboration has been studying RF power sources based on the RK-TBA concept for several years [1, 2, 3]. The collaboration prepared a preliminary design study for a RF power source suitable for the NLC [4] last year. This RF power source, referred to as the TBNLC, is comprised of subunits, each approximately 340 m in length with 150 extraction structures generating 360 MW per structure. The number of subunits is dependent on the power requirement for the collider, e.g. 76 subunits are required for a 1.5-TeV collider. A test facility has been established at LBNL to verify the analysis used in the design study. The primary effort of the facility is the construction of a prototype TBNLC subunit that will permit the study of technical issues, system efficiencies, and costing.

Relativistic klystrons are not limited to X-band frequencies. We have formed a collaboration with the CERN Compact Linear Collider (CLIC) group to study a 30 GHz RK power source for CLIC. We refer to this RF power source design as the RK-CLIC [5]. Many of the design features are shared between the TBNLC and RK-CLIC. More recently we have attempted to optimize the RK-TBA concept relative to the required IP physics of the collider instead of matching the RK's to existing collider designs.

---

*The work was performed under the auspices of the U.S. Department of Energy by LLNL under contract W-7405-ENG-48, LBNL under contract AC03-76SF00098, and FAR under SBIR Grant DE-FG03-95ER81974.

In the first section of this paper, we will discuss the development of the RTA injector cathode and pulsed-power system, which has recently been constructed and is now undergoing testing. We will shift focus in the second section in order to discuss some recent work on extending the RK-TBA concept to higher energy and higher frequency colliders.

RTA: PROTOTYPE RF POWER SOURCE

Construction of the RTA, a prototype of the proposed TBNLC RF power source subunits [6], has started at LBNL. Details of the RTA design has been presented elsewhere [7, 8]. Here, we describe the 1-MV, 1.2-kA induction electron gun and the pulsed power system for the gun.

Induction Electron Gun

An illustration of the 1-MV, 1.2-kA induction electron source, referred to as the gun, is shown in Figure 1. The cores are segmented radially to reduce the individual aspect ($\Delta r/\Delta z$) ratios with each driven separately at about 14 kV. The lower aspect ratio reduces the variation in core impedance during the voltage pulse simplifying the pulse forming network (PFN) design. We chose a constant radius design for the cathode-side cells. This design increased the METGLAS® core volume by about 10%, but the added cost was recovered in reduced insulator and fabrication costs. Figure 2 is a photograph of the completed cathode-half of the gun undergoing initial pulsed power tests. Currently, the cathode-half gun is being used to test various insulator configurations. The test results will be incorporated into the RTA's induction accelerator design.

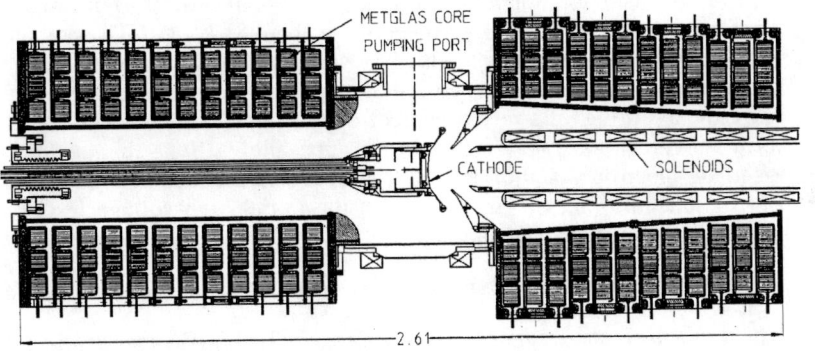

FIGURE 1. Illustration of the RTA gun, a 1.2-kA, 1-MeV induction electron source.

FIGURE 2. Photograph of the assembled cathode-half of gun. The cathode will be located within the pumping spool. Flange for the anode-half of gun is blanked off.

A novel feature of the gun design is the insulator. We are doing high voltage testing with a single, 30 cm ID, PYREX® tube for the insulator with no intermediate electrodes. Average gradient along the insulator at the operating voltage of 500 kV is about 5.1 kV/cm. Maximum fields at the triple points, intersection of insulator, vacuum, and metal, is less than 3.5 kV/cm. Maximum surface fields in the cathode half of the gun are about 85 kV/cm. The rationale for using PYREX® is to explore methods of reducing the costs of induction injectors. PYREX® is significantly less expensive than ceramic, and additional savings are realized by avoiding intermediate electrodes. Since there is additional risk associated with this approach, our design allows for the addition of intermediate electrodes and/or substitution of a ceramic insulator with minimal impact to schedule or expense. However, the initial high-voltage tests on the cathode-side insulator are encouraging.

Pulsed Power System

The pulsed power system will consist of a 20-kV Energy Storage Bank Charging Power Supply, 3-kJ Energy Storage Bank, two Command Resonant Charging Chassis, 24 Switched Pulse Forming Networks, and four Induction Core Reset Pulsers. A photograph of one PFN is shown in Figure 3. Each PFN will drive a single 3-core induction cell. A sample pulse is shown in Figure 4.

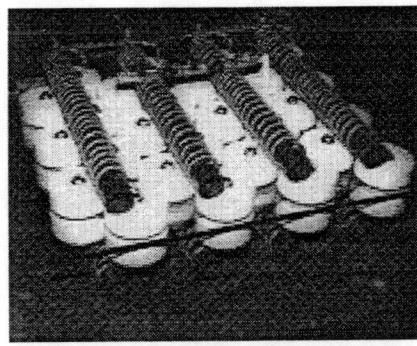 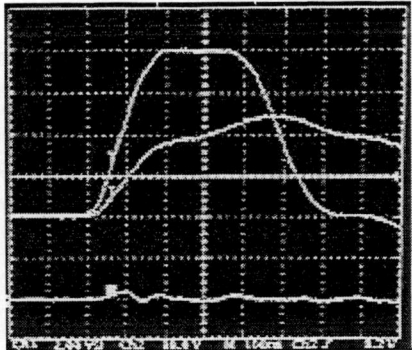

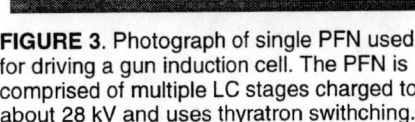

FIGURE 3. Photograph of single PFN used for driving a gun induction cell. The PFN is comprised of multiple LC stages charged to about 28 kV and uses thyratron swithching.

FIGURE 4. Oscilloscope trace of pulsed power pulse applied to induction cell. Top trace is voltage (10 kV/div) and middle trace is current (4 kA/div). Time scale : 100 ns/div.

Segmenting the core in the induction cell and driving the individual core segments avoids a high-voltage step-up transformer. This reduces the developmental effort needed to achieve a "good" flattop pulse (minimal energy variation) and improves the efficiency of the overall pulsed power system. Our system of low-voltage PFNs driving multiple core induction cells is similar to the system envisioned for the extraction section in the TBNLC design. For the core material, we choose METGLAS® alloy 2605SC instead of the 2714AS, the preferred material for the TBNLC, due to the larger inner diameter gun cores. In the RTA gun configuration, the larger flux swing of 2605SC was of greater importance than the lower loss per unit volume of 2714AS. The RTA extraction section will use 2714AS to permit an accurate measurement of the pulsed power system efficiency expected for the TBNLC.

An area of concern is the consistency of the METGLAS® cores. Several core materials were tested at the RTA Test Facility [9] to establish a data base for design studies. However, this testing did not address the issue of consistency between cores of the same material. We now have a data base including the 38 cores of METGLAS® alloy 2605SC used in the construction of the cathode-half of the gun. Figure 5 shows the energy loss per unit volume for these cores at a flux swing rate, dB/dt, of 5 T/µs. The cores used 20 µm thick 2605SC layers with mylar insulation and achieved an average packing fraction of 76%, minimum of 72% and maximum of 78%. The cores had a radial thickness of 5.8 cm with an inner radius of 19 cm, 27 cm, or 35 cm. The small, medium, and large cores in Figure 5 refer to the different inner radii. The three horizontal lines represent the average loss per volume for the respective core sizes. The smaller the core radius, the higher the loss per volume, as shown in the figure. However, total loss per core for the 38 cores was approximately the same with no significant dependence on core size.

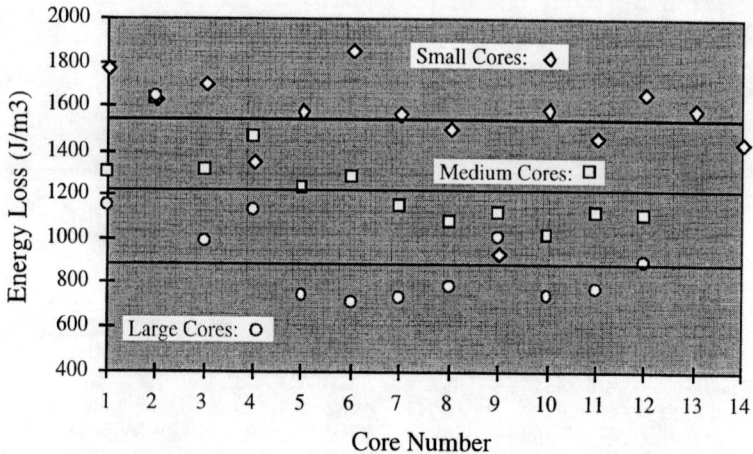

FIGURE 5. Test results for the METGLAS® alloy 2605SC cores.

The standard deviation in loss per volume for the small and medium cores was 14% and the large core was 29%. However, by matching the cores, the standard deviation for a three core cell was reduced to 4%. If the core losses vary sufficiently, it becomes necessary to tailor individual PFNs to adjust for the different cell loads. For a large relativistic klystron, matching cell cores should permit acceptable loss variation.

HIGHER-ENERGY, HIGHER-FREQUENCY COLLIDERS

We are currently evaluating possible designs for a 5-TeV-scale collider [10], based on operating frequencies higher than 11.424 GHz. We propose the use of high current, high power beams in the main collider linacs, while loosening some of the stringent parameters in the final focus section. The accelerating structure in the main linacs must of necessity be heavily damped, as well as detuned to allow for fast roll-off of the short-range wakefields. This damped, detuned two-beam accelerator we dub the DD-TBA.

Interaction Point Physics

Requirements of high average luminosity, a usable level of beamstrahlung induced energy spread, and a low background of high energy photons lead to tradeoffs between beam power and beam quality. The definition of terms and a comprehensive review of the relevant IP physics in a linear collider can be found in Wilson's article [11], Palmer [12], and Irwin [13].

The NLC [14] klystron-based collider designs have exhibited overall wall plug to beam efficiencies around 10%. In order to hold down total power consumption, a heavy burden is usually placed on generating and maintaining higher quality beams, keeping the beam power at lower levels. In this DD-TBA collider design, the net efficiency can be 50% or more. In this scheme, we choose instead to operate with much higher beam power in order to relax some of the constraints and challenges at the final focus. Various proposed schemes, and their IP parameter sets are listed in Table I. The parameters of the 1-TeV NLC case are included for comparison.

For the DD-TBA design we have allowed for both a larger beam spot size and normalized emittance, while keeping Υ and δ_B at moderate values. The range of Υ considered in the various designs spans an order of magnitude. The physics of high ($\gg 1$) Υ interactions is still not understood, so placing any upper limit is somewhat premature. Also, the issue of energy resolution in the detector systems must be addressed before an upper limit on δ_B can be imposed as a design constraint. However, in a reasonable 5-TeV collider design, it is very difficult to achieve an energy spread below 10%.

Table 1. Comparison of linear collider IP parameters.

	Palmer [12]	Irwin [13]	Wilson [11]	DD-TBA**	NLC
E_{cm} (TeV)	5.0	5.0	5.0	5.0	1.0
$L(10^{35}cm^{-2}s^{-1})$	2.5	2.5	2.5	2.5	0.11
$N(10^{10})$	0.31	0.03	0.44	0.25	1.1
$f_{rep}n_b$(kHz)#	12.7	330	5.6	71	12.6
σ_y(nm)	0.2	0.1	0.1	0.4	5.1
R	136	156*	700	180	49
σ_z(mm)	20	27*	20	20	150
ε_{ny}(nm)	1.5	3.3	1	4	110
D_y	7.3	7	7	2	7.6
H_D	2	1.4*	1.1*	2	1.4
Υ	21	4*	28	7.8	0.29
δ_B(%)	27	10*	20	15.5	12.6
P_{beam} (MW)	54	40	18	72	7.9

* These parameters are not given explicitly by the authors, but have been derived from scaling relationships.
** We have used a value of A_y equal to 0.10.
# f_{rep} is the pulse train repetition frequency; n_b is the number of bunches per train.

This loosening of beam quality does not come without its price. The RK power source is most efficient when generating long RF pulses (100's to 1000's of ns). Efficient use of that pulse means that we must use bunch trains that span it. To achieve the required luminosity, we must also pack the bunches tightly together. The current DD-TBA design uses trains of 4761 0.4-nC bunches with a separation of 2 RF wavelengths. This gives a large DC current of 6.01 A during the pulse, which has a repetition rate of 10 Hz.

High Gradient Structures

The transport and acceleration of such large current beams necessitates a hard study of the high gradient structures. The introduction by Wilson [15] provides an in-depth discussion of the pertinent physics. Once an average current is chosen, the structure design becomes a tradeoff between accelerating gradient and RF to beam power conversion efficiency. We adopt an approach that uses heavy beam loading to boost efficiency, while maintaining relatively high loaded accelerating gradients. The linac structures are designed to have high efficiency in transfer of RF power to beam power (~80%), with high input RF power (400 MW/structure). The structure parameters are listed in Table 2.

The transverse wakefields in this structure are quite severe due to the large current. By using heavily damped structures we can produce designs with low dipole mode Q's. This can significantly damp wakefield levels generated by a bunch at a given point in the structure by the time the next bunch arrives.

Table 2. Linac structure parameters.

Frequency	30 GHz	Idc	6.01 A
βgroup	0.10	Input power	400 MW
a/λ	0.214	Peak gradient	244 MV/m
r/Q	23.7 kΩ/m	Average gradient	126 MV/m
Q	4425	Power into beam	316 MW
Fill time	14 ns	Power into walls	80 MW
τ	0.298	Structures per m	2

Relativistic Klystron Source

The relativistic klystron power source design is similar to the proposed TBNLC. For this design, each unit would power 600 high gradient structures, so that each linac arm would require 79 DD-TBA units.

Each DD-TBA consists of a 3.5-kA, 5.0-MeV injector, a beam modulation unit, an adiabatic capture section to bunch and accelerate the beam, the main RF extraction section, and an afterburner section to extract power from the beam while decelerating it prior to the dump. At the entrance to the main extraction section, the beam has an average energy of 25 MeV and carries 3150 A of RF current with 1750 A of DC current. Each relativistic klystron has 300 extraction sections to power 600 high gradient structures. The ultimate efficiency of the relativistic klystron is limited by the number of extraction sections the beam can pass through before succumbing to beam breakup (BBU) instabilities. Careful attention must then be paid to transport and stability.

Transport and Beam Stability

Permanent magnet quadrupoles are employed to provide a magnetic FODO lattice. The lattice has a 0.33 m period with a 60° phase advance per period, giving a 2 m betatron period. The quadrupole magnets are ferrites with an 800 G poletip field, 1.0 cm bore radius, and 0.48 occupancy factor. For a normalized edge emittance of 2000π mm-mrad, the equilibrium beam edge radius will be about 2.0 mm.

Two severe transverse instabilities have been identified in the RK-TBA. One is a low frequency mode associated with the induction modules, and the other is a high frequency mode due to the RF extraction structures. Similar instabilities will exist in this design, but at higher frequencies. Simple scaling arguments [16] imply that the high frequency instability growth rate in this design could be a factor of 4 higher than in the TBNLC design, while the low frequency instability rate could be 10 times higher, if left uncorrected.

Beam energy spread should result in effective Landau damping to counter the low frequency instability. Transport of the beam depends upon the ferrite permanent magnet quadrupoles. Increasing the poletip field of the magnets will also increase the quadrupole gradient. Alternatively, we can increase the bore of the beam pipe as well as induction gaps, while increasing the poletip field at fixed beam energy, and maintain the same betatron period. Thus, we can decrease the transverse impedance due to the induction gaps, and hence the low frequency instability growth rate.

The higher frequency mode is more severe. Our solution is to place the extraction structures at half-betatron wavelengths, on the nodes. The growth rate should be similarly depressed as in the betatron node scheme for the TBNLC [3]. Field error tolerances in the quadrupoles become an issue, since this instability is sensitive to the details of the focusing lattice with respect to the positions of the RF output structures.

Another beam dynamic issue related to the induction cell is the extraction of RF power from the modulated beam. Power is absorbed by various materials in the cell and reduces efficiency. Techniques for lowering the longitudinal impedance of the cell at 30 GHz, therefore minimizing power loss in the output structures, is an active area of study.

The idler cavities in the adiabatic capture section and the extraction structures in the main section are detuned from synchronism at 30 GHz. This compensates for bunch lengthening effects, and provides longitudinal focusing. The synchrotron oscillation, induced by the power extraction and reacceleration, has a period of 91 m.

Induction Modules

We have designed a system to provide 155 kV per induction cell, to replace the beam energy lost in the RF output structures. For our long pulse (300 ns), and assuming that we drive the core to saturation, the 2714AS material has the lowest losses, and hence the largest efficiency. For a DC current of 1750 A and voltage of 155 kV/cell, the net core efficiency is ~91%.

Travelling Wave Output Structures

We obtain a zero-order design for the 30-GHz output structures by scaling the physical dimensions from our 11.424 GHz design. The structure is initially designed to operate in the $2\pi/3$ mode, but is then detuned by 30° so that it will actually resonate in the $\pi/2$ mode when driven at 30 GHz. The structure parameters are listed in Table 3.

Table 3. Travelling wave output structure parameters.

Frequency	30 GHz	R/Q	19 W/cell
Mode	$2\pi/3$ *	Pout	801 MW
βgroup	0.65	Max. field	344 MV/m
a/λ	0.62		

* Detuned by 30° - resonant travelling mode is $\pi/2$.

System Efficiencies

The pulse power system suitable for this design would utilize a DC power supply, a Command Resonant Charging (CRC) chassis, and thyratron switching, like the earlier TBNLC proposal. We can make predictions of the efficiency of the pulse power system based on our previous work. These estimates are listed below in Table 4. Here the drive beam fall time has been included to account for losses at the end of the voltage pulse that are dissipated in the induction cores. Drive beam to RF losses account for the beam losses at the front end of the relativistic klystron, and for beam power lost at the dump. Auxiliary power accounts for cooling and vacuum systems, etc. We include the RF to beam efficiency of the high gradient structures, and calculate the net efficiency of the RK-TBA to be ~52%.

Table 4. Power source efficiencies.

DC Power	0.93	Drive Beam Fall Time	0.94
Command Resonant Charging	0.96	Drive Beam to RF	0.93
Modulator (thyratron)	0.94	Auxiliary Power	0.98
Induction Cells	0.91	RF to High Energy Beam	0.79
Net Wall Plug to Beam			0.52

Future Collider and Power Source Studies

We are exploring techniques to modulate beams and extract power at much higher frequencies (30 GHz - 120 GHz or higher). At these frequencies, free electron lasers become good candidates for modulating beams [17]. We are currently designing inductively detuned RF extraction cavities at 30-35 GHz, to take advantage of some available beam sources. W-band (90-120 GHz) systems are also being considered in support of high-gradient structure research occurring elsewhere.

ACKNOWLEDGMENTS

We thank Andy Sessler and Swapan Chattopadhyay for their support and guidance. Yu-Jiuan Chen and George Caporaso provided valuable assistance with the induction accelerator design. Ming Xie provided helpful advice with our initial collider modelling. Wayne Greenway and Bob Candelario we thank for their excellent technical support.

REFERENCES

1. Sessler, A.M. and Yu, S.S., "Relativistic Klystron Two-Beam Accelerator," *Phys. Rev. Lett.* **54**, 889 (1987).
2. Westenskow, G.A., and Houck, T.L., "Relativistic Klystron Two-Beam Accelerator," *IEEE Trans. on Plasma Sci.*, **22**, 750 (1994).
3. Giordano, G., et al., "Beam Dynamic Issues in an Extended Relativistic Klystron," in *Proc. of the 1995 IEEE Particle Accelerator Conf.*, 1995, pp. 740–742.
4. Loew, G.A., and Weiland, T., *International Linear Collider Technical Review Committee Report 1995*, Stanford, Stanford University, 1995, pp. 61–65, 83, 84.
5. Houck, T.L., et al., "Scaling the TBNLC Collider to Higher Frequencies," in *Proceedings of XVIII International Linac Conference (LINAC96)*, 1996.
6. Houck, T.L., and Westenskow, G.A., "Prototype Microwave Source for a Relativistic Klystron Two-Beam Accelerator" *IEEE Trans. on Plasma Sci.*, **24**, 938 (1996).
7. Houck, T., et al., "RK-TBA Prototype RF Source, in *Proceedings of 3rd ICFA Workshop on Pulsed RF Sources for Linear Colliders (RF96)*, 1996.
8. Westenskow, G.A., et al., "Relativistic Klystron Two-Beam Accelerator Studies at the RTA Test Facility," in *Proceedings of XVIII International Linac Conference (LINAC96)*, 1996.
9. Reginato, L., et al., "Engineering Conceptual Design of the Relativistic Klystron Two-Beam Accelerator Based Power Source for 1-TeV Next Linear Collider," in *Proc. of the 1995 IEEE Particle Accelerator Conf.*, 1995, pp. 743–745.
10. Lidia, S.M., et. al., "Relativistic Klystron Two-Beam Accelerator Approach to Multi-TeV e^+e^- Linear Colliders," contributed to the APS New Directions for High Energy Physics Workshop, Snowmass, CO, USA, July 1996.
11. Gaponov-Grekhov, A.V., and Granatstein, V.L., eds., *Applications of High Power Microwaves*, Boston: Artech House, 1994, ch. 7.
12. Palmer, R.B., Ann. Rev. Nucl. Sci. 40, 529-92 (1990).
13. Irwin, J., "Bird's IP View of Limits of Conventional e+e- Linear Collider Technology," presented at the 6th Workshop on Advanced Accelerator Concepts, Lake Geneva, Wisconsin, June 12-18, 1994.
14. The NLC Design Group, *Zeroth-Order Design Report for the Next Linear Collider*, SLAC Report 474, Stanford University, Stanford, CA, May 1996.
15. Wilson, P., SLAC-PUB-2884 (rev.) (1991).
16. Chao, A.W., *Physics of Collective Instabilities in High Energy Accelerators*, New York: John Wiley & Sons, 1993.
17. Gardelle, J., et. al., "Analysis of the beam bunching produced by a free electron laser," submitted for publication in Physics of Plasmas, June 1996.

Progress in High Power, High Efficiency Relativistic Traveling Wave Tube Amplifiers

J. A. Nation, S. A. Naqvi, G. S. Kerslick and L. Schächter*

Laboratory of Plasma Studies & School of Electrical Engineering
Cornell University, Ithaca, NY 14853, USA
* Department of Electrical Engineering, Technion-Israel Institute
of Technology, Haifa 32000, Israel

ABSTRACT

We present an overview of recent research at Cornell University on the use of relativistic traveling wave tube amplifiers for high power microwave generation. We consider three topics namely the dependence of the amplifier gain on the beam energy, axial energy extraction using a TM to TEM mode converter, and techniques for enhancing the efficiency of the amplifier to at least 50%.

INTRODUCTION

A number of high power microwave sources are based on the interaction between a relativistic electron beam and axial electric fields. Within this category the klystron [1] and the traveling wave tube amplifier (TWT) are perhaps the most common devices when control of phase and frequency are essential requirements. In both devices a beam is bunched as a result beam-wave interactions in either discrete cavities or in a periodic traveling wave structure. In this paper we review recent work at Cornell University using traveling wave tube amplifiers aimed at the production of very high power microwave pulses at X band frequencies.

We discuss three topics:

1. The dependence of system performance on the electron beam energy

2. Experimental performance and theoretical design of a coaxial output section for the TWTA and,

3. Techniques to increase the beam-to-microwave conversion efficiency to $\geq 50\%$

The first topic presents a summary of an analysis which shows that design and operation of relativistic amplifiers does not depend critically on the matching of the wave phase velocity to the electron beam energy. This arises since the coupling between the beam and structure waves scales inversely with the beam energy, hence the growth rate for the interaction is greater at lower beam energy and this, in zero order, compensates for the phase slip between the bunched electrons and the wave, which scales with the beam velocity.

In the second topic we discuss the design and performance of an output coupler for the TWTA. In high current relativistic amplifiers the gain of the structure is high so it is necessary to carefully taper the ends of the amplifier to prevent oscillation. As the output of slow wave structure is tapered the wave phase velocity increases so that the electrons slip to a wave phase such that energy is returned from the amplified rf wave to the beam electrons. As a result the rf power drops from its saturation level and the extraction efficiency, defined by $\eta_{extract}$ = rf power extracted / rf power at saturation, becomes low. We report on the use of a mode converter from the TM_{01} amplifier mode to a coaxial TEM mode. The electron beam is dumped in the center conductor of the coaxial converter, which penetrates through the tapered region into the amplifier. As a consequence the beam electrons do not see the accelerating wave and the microwave signal is frozen at the higher level found at the location of the coaxial dump, i.e. at the saturation level of the amplifier. This effect alone can increase the extraction efficiency by a factor of order five. In order for this extraction scheme to be useful it is essential that the mode converter be non-reflecting at the wave and at adjacent frequencies. Simulation data using the MAGIC code, and experiments, indicate that the reflection coefficient may be very small over a range of several hundred MHz and that the scheme is viable.

Finally we outline theoretical and simulation studies at present in progress in which we examine ways of increasing the amplifier conversion efficiency, defined as η_{conv} = rf power at saturation / beam power. Typical efficiencies peak at 25-30% in amplifiers of the type discussed in item 2. It is straightforward to show that the efficiency in a TWT (and also in a klystron) depends on the degree of bunching of the beam. Initially half of the electrons are in the accelerating phase of the structure wave and half are in the decelerating phase. Subsequent velocity modulation leads to bunching in the decelerating phase which in turn is opposed by the space charge associated with the bunched electrons. The best amplifier performance is achieved when the output section of the amplifier starts at a location where the bunching is significant but not maximum. Unfortunately in the bunching process the electrons slip with respect to the wave so that the best bunching occurs when the slowed bunches significantly lag the peak in the decelerating phase of the structure wave. This effect limits the efficiency achievable in the amplifier. In addition the use of a sever to separate the first and second stages of the amplifier also leads to debunching of the beam due to its space charge. We overcome these limitations by the use of a two stage amplifier in which the first stage is only used to weakly modulate the beam. The output of this stage is a sever followed by the main amplifier. The first part of the amplifier is used to tightly bunch the beam. At the location where the bunching is close to its maximum and the rf current is perhaps 5-6 times the dc beam current, the amplifier structure is changed to one where the wave velocity at the bunching frequency is less than the beam velocity. The shift in the wave phase causes the bunched beam to slip back into the decelerating phase. If this section is properly designed the $\mathbf{J} \cdot \mathbf{E}$ conversion of beam energy to wave energy is maximized and the interaction efficiency is increased. Simulation results, not yet optimized, show that conversion efficiencies of over 50% are achievable.

REVIEW OF TWT CHARACTERISTICS

In this section we review the topics outlined above. The first two topics have been published in the last few months and the reader is referred to these papers for further detail [2,3].

Beam Energy Effects on Amplifier Efficiency

As indicated in the introduction the coupling coefficient between the beam and the structure wave scales inversely with the beam velocity and energy ($\sim 1/\beta\gamma$). The growth rate α_0 is related to the coupling coefficient by $\alpha_0 = Im(k)/k_0 \approx \sqrt{3}C/2$, where C is given by $C^3 = (Ie\omega/2\gamma mc^2 k_0 \beta c \gamma^2 \beta^2) Z_{int}$. Since the electron velocity is relatively insensitive to the beam energy a small fractional decrease in the relativistic beam velocity produces a large increase in the coupling coefficient. This relativistic effect compensates for the increased slip factor caused by the change in beam velocity. Thus a substantial interaction may be achieved over a broader range of electron beam energies when compared to non-relativistic amplifiers.

We have examined analytically the case of a dielectric loaded waveguide amplifier and the results presented were obtained specifically for this case. Similar results have been obtained using simulation codes for periodically loaded disk TWT amplifiers. We consider the performance of three similar amplifiers designed to operate at a wave frequency of 9 GHz. The three structures are designed so that the structure wave is synchronous with the beam electrons at energies of 10 keV, 100 keV and 1 MeV respectively. For adequate comparison with the other two cases (100 keV and 10 keV) we ensure that (i) the normalized growth rate at

Table 1: PARAMETERS AT RESONANCE: ($\omega = k_0\beta_0 c$)

V_0	Radius, a, (mm)	I_0 (A)	α_0^P	α_0
1 MV	20	500	0.0295	
100 kV	10.1	11.22	0.0290	0.0275
10 kV	2.65	0.88	0.0285	
1 MV	15	500	0.0407	
100 kV	7.46	11.22	0.0402	0.0385
10 kV	1.89	0.88	0.0395	
1 MV	10	500	0.0613	
100 kV	4.84	11.22	0.0605	0.0586
10 kV	1.15	0.88	0.0600	

resonance calculated from dispersion relation is kept the same and (ii) the beam term $I/\gamma_0^3\beta_0^2$ in the Pierce coupling parameter C is kept constant so that when the first condition is satisfied, the interaction impedance $Z_{int} \equiv 1/2\langle E_z^2\rangle/k_0^2 P$ is approximately the same for each of the three cases. The second condition determines the current and the former, combined with the requirement for resonance, sets the radius and dielectric constant for the 10 kV and 100 kV cases. The scaling results are summarized in Table 1.

In the table the radius a is that of the guide, I_o the beam current, $\alpha_0^P = \frac{Im(k)}{k_0} \approx \frac{\sqrt{3}}{2}C$, where C is the Pierce parameter, and α_0 is the exact growth rate calculated from the dispersion relation. The results are shown graphically in fig 1. which plots the values of α versus the beam velocity. In particular we note that although the growth rate at resonance is identical at all three energies the peak growth rate occurs at a substantially lower electron velocity than the synchronous value of 0.93 c. In fact the growth rate maximizes at an energy of about 300 keV.

In conclusion, we observed that (i) in the relativistic regime the maximum growth rate does not occur at resonance but at a beam velocity lower than the phase velocity of the wave in the cold structure, (ii) this maximum growth rate can be much larger than the growth rate at resonance, and (iii) relativistic devices are expected to be less sensitive to beam voltage fluctuations than their non-relativistic counterparts. This relative insensitivity is expected to be significant, for example, in the design of modulators used to produce the electron beams which drive high power traveling wave amplifiers. In addition we note that in both the relativistic

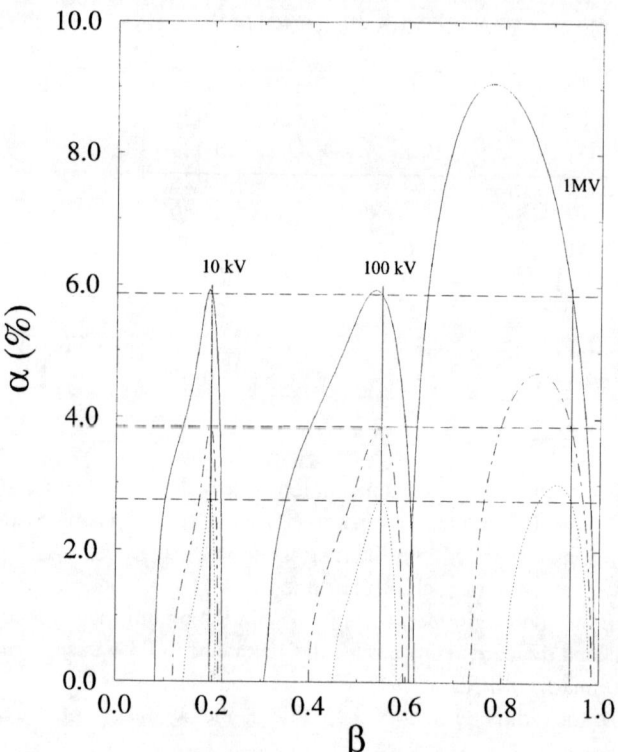

Figure 1: Normalized growth rate α variation with beam velocity β for 10 kV, 100 kV and 1 MV systems. The solid curve represents systems which have the value of growth rate at resonance $\alpha_0 = 5.86\ \%$; the dotted-dashed curve represents $\alpha_0 = 3.85\ \%$;and the dotted-curve represents $\alpha_0 = 2.75\ \%$. The vertical solid line indicate β_0 the resonant value of β. The dashed horizontal lines indicate α_0, the corresponding value of α at resonance.

and non relativistic regimes it is the product $\gamma\alpha$ that peaks at resonance [2].

Axial Extraction Technique for High Power Microwave Amplifiers

We present results from theoretical and experimental studies describing the coupling from a high power traveling wave tube amplifier, operating in the TM_{01} circular mode, to a TEM mode in a coaxial waveguide while the beam is dumped in the hollow inner conductor of the coaxial guide. It is found that this configuration exhibits advantages over that used in existing devices, namely: (i.) The mode conversion efficiency of the extractor is insensitive to the dimensions of the inner conductor and is relatively broad-band. (ii.) The inner and outer conductors are parallel to the confining magnetic field and hence the output section is magnetically insulated.

The coaxial converter also allows one to eliminate the return of energy from the rf wave to the beam electrons following saturation or wave speed up in tapered output sections. This enhances the amplifier extraction efficiency by a factor of up to five compared to extraction into a cylindrical waveguide.

The coaxial extraction section of the TWT used in this work is shown in fig. 2. The amplifier consists of a two-stage TWT with a phase advance of $\frac{\pi}{2}$ per cell designed to work with an 800 kV, 500 A pencil electron beam. The periodic structure has an inner (iris) radius of 0.9 cm, an outer radius of 1.56 cm, a periodic length of 0.77 cm and has a 12 period taper to the outer tube radius at each end. The output of the amplifier is fed to a coaxial section in which the radius of the inner conductor has been varied between 0.7 cm and 1.1 cm., and power extracted with various penetration lengths of the inner conductor into the tapered section.

The axial converter design arose from noting that the TM_{01} circular mode can be very effectively coupled into a TEM coaxial waveguide mode in a guide with the same outer radius if the inner radius is selected correctly. Analytic calculations, supported by simulation data indicate that for a tube radius of 1.56 cm a range of inner conductor radii from 1.0 to 1.1 cm have power reflection coefficients below 0.5%. Simulation data indicate that similar results may be obtained within the tapered section of the TWT provided that the end of the inner coaxial tube is located midway between two irises. In fig. 3 we show simulation data for the energy reflected when an incident 300 MHz wide Gaussian pulse enters the coaxial mode converter. Data is presented with the mode converter at 3 axial locations, and is given as a function of the radius of the inner conductor. The data points are obtained from a modal analysis of the junction between a uniform circular waveguide and a coaxial waveguide at 9 GHz. Note that the reflected energy

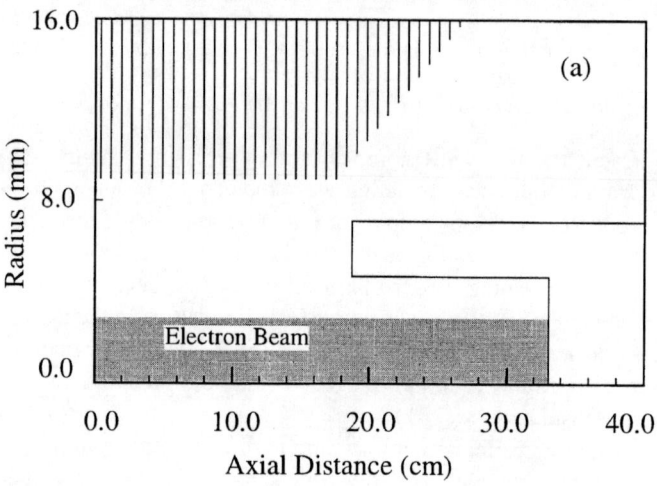

Figure 2: Schematic of coaxial extraction geometry with beam dump located between the 1st and 2nd iris of tapers (position (a) in fig. 3)

is less than 0.5% approximately 1.5 periods from the end of the uniform amplifier for an inner conductor radius of 0.7 cm.

The most important feature of the axial converter is the shielding of the electron beam from the periodic structure by the inner conductor. Fig. 4 shows the total Poynting flux, including both rf and beam components as a function of distance, for both a uniform cylindrical output (trace (b)) and for the coaxial output case (trace (a)). It can be seen that the microwave signal grows in amplitude throughout the amplifier and decays as the wave enters the tapered section. The reduction in power level occurs as energy is extracted by bunches rapidly slipping out of the decelerating phase and entering the accelerating phase of the wave (in the absence of the inner conductor the wave phase-velocity gradually increases from $0.9c$ at the start of the tapers to $1.7c$ at the end). The DC level shifts at $z = 0$, 19, and 33 cm correspond to changes in the Poynting flux associated with the beam as the return conductor geometry is changed. The RF power output in the two cases shown is 25 MW in the absence of the inner conductor and 90 MW with the inner conductor. In this figure the mode converter is located at $z = 19$ cm and the beam is dumped in the collector at 33 cm. The inner conductor decouples the

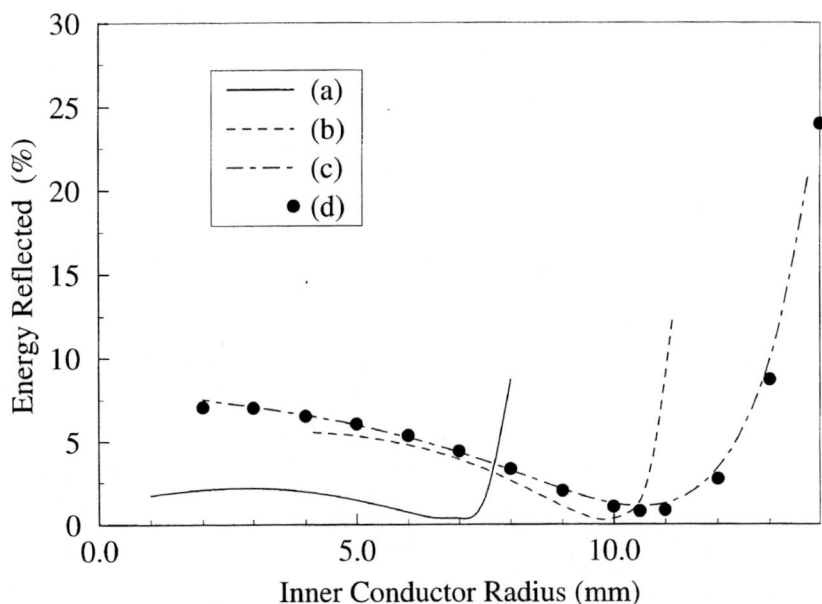

Figure 3: RF pulse energy reflected from the junction of a tapered slow wave structure and a coaxial waveguide as a function of the inner conductor radius (outer radius is 15.6 mm). The three axial locations are: (a) between 1st and 2nd iris of tapers; (b) between 6th and 7th iris; and (c) 5mm beyond the tapered section. Data points (d) are from modal analysis of the junction between a circular waveguide and a coaxial waveguide at 9 GHz.

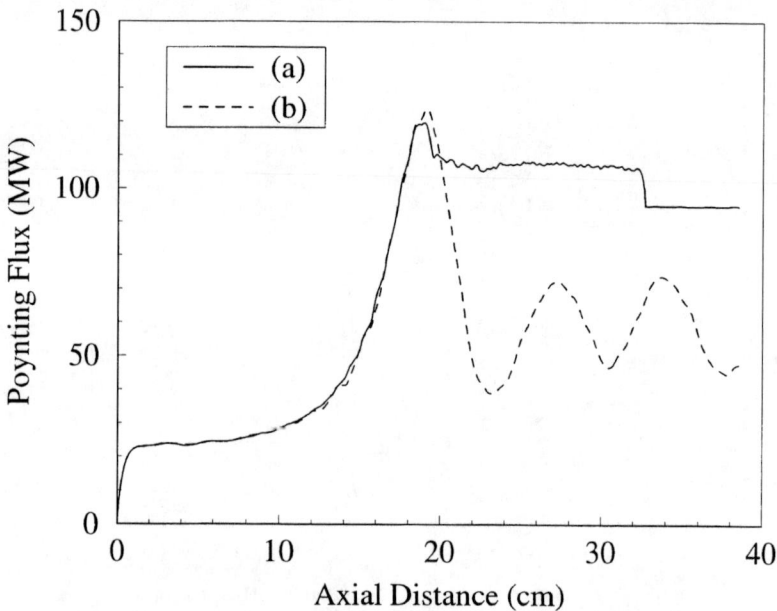

Figure 4: MAGIC code output showing the time averaged Poynting flux as a function of axial distance. Results are shown for the coaxial extraction geometry (a) and for a cylinder without a center conductor present (b)

wave and the beam when the beam enters the inner conductor.

The optimal design for coaxial extraction is a compromise between (a) maximizing power extracted relative to saturation level (b) minimizing reflections from the extractor and (c) minimizing surface fields. The power extracted will be maximized when the inner conductor is close to the start of the tapers, but this also reduces the radial gap between the coaxial conductors and thus increases the probability of RF breakdown. A smaller inner conductor radius may slightly increase the reflections (see fig. 3) but will increase the gap and hence reduce the breakdown probability.

We have tested the axial extraction system experimentally with a 100 ns, 700 kV, 500 A beam generated using a field emission diode. It should be noted that the beam energy is 100 kV less than the design figure, but based on earlier results [2] we expect good amplifier gain even if the beam and slow wave structure parameters do not satisfy the classical resonance condition exactly. The beam is 0.6 cm in diameter and is guided by a 10kG magnetic field. After passing through the amplifier the beam is dumped into the hollow center conductor. Data are shown

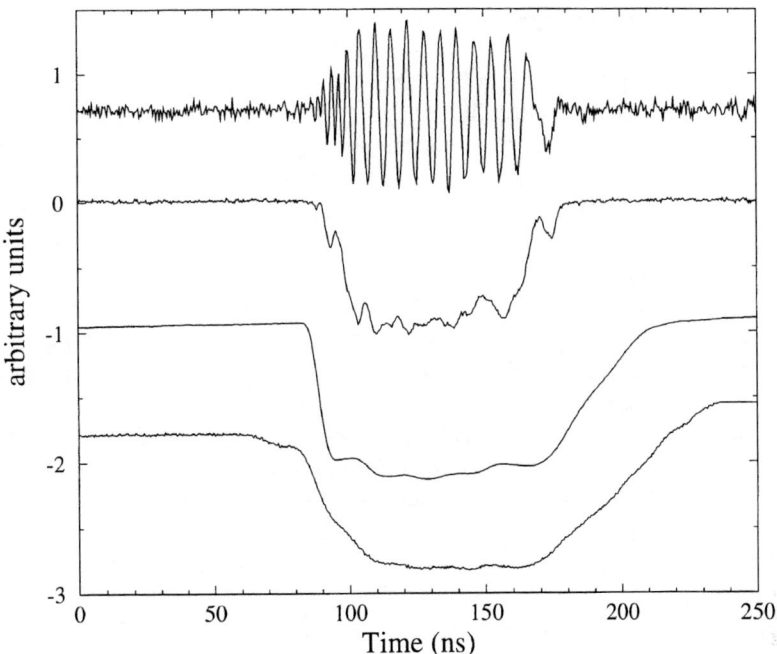

Figure 5: Waveforms showing, from the top: Output signal heterodyned with local oscillator signal; Output power, measured by Er in coaxial extractor; Beam Current; and Diode Voltage.

in fig. 5 which presents results for a 7.1 mm radius inner conductor located 3.5 periods into the taper from the end of the uniform section of the amplifier. The microwave envelope, as detected by a calibrated radial electric field probe mounted in the coaxial section, and a mixer output signal are shown. The microwave output is relatively constant for approximately 70 ns matching the stable portion of the beam profile, and the signal has the correct frequency as indicated from a FFT of the mixed signal. Peak output signal levels exceed 70 MW, and are 3 times that measured when the signal is detected in the uniform pipe at the end of the taper. The smoothness of the output pulse indicates that the reflection from the mode converter is small. The fluctuations observed are believed to be largely due to the relatively poor beam quality associated with the field emission diode used.

Optimization of Microwave Conversion Efficiency

In the previous sections we have outlined recent results concerning the design of the structure to eliminate reflections, to extract the microwave signal close to its saturation level, and discussed electron energy effects. We now examine possible techniques for maximizing the rf conversion efficiency.

In our TWT's we typically use a two stage amplifier. In the first stage the beam is weakly bunched in a low gain amplifier. This precludes the possibility of oscillation, and reduces the probability of sideband development, in the first stage where the reflection coefficient at the input to the amplifier may be relatively large. The output from the first stage is taken through a resistive sever made of SiC, hence eliminating the rf on the beam while allowing the modulated beam to propagate through the sever to the input of the second stage. In this stage the rf wave is reconstructed and grows to saturation at the end of the amplifier. To eliminate reflections the output is either fed into a tapered section, or better, into a coaxial extraction section.

In the late stages of amplification the bunched beam slips in phase with respect to the axial electric field and the conversion efficiency is limited since the power generated per unit volume $\mathbf{J}\cdot\mathbf{E}$ is not maximized. An approach to this problem is to taper the phase velocity of the slow wave structure bringing the rf current back into phase with the axial electric field. We have found that it is better to change the wave velocity in the slow wave structure before wave saturation. In this manner we attempt to maximize the bunching in the decelerating phase and maintain the bunches in this phase for as long as possible before spreading occurs due to the space charge. Using this approach and we have obtained, in simulation, overall system efficiencies of up to 52%. This figure compares favorably with the 25-30% efficiency obtained in uniform amplifiers with output tapers and is comparable to the efficiencies found in klystrons at similar frequencies.

In fig. 6 we show output from a simulation run illustrating the results obtained using this technique. Reading from the top of the figure we show plots of the beam bunching; J_z, the rf beam current; E_z, the axial electric field; the axial momentum and the total Poynting flux (Including the dc beam flux). The steady and progressive slip in phase of the beam bunches as the wave grows is evident, until between $z = 0.4$ and 0.45 m the phase slip is corrected by a change in the structure design. For the 800 keV, 500A beam used the output rf power is about 200 MW and the overall efficiency 50%. To achieve this result we used a structure designed to have a wave phase velocity at the operating frequency of c and in the extraction part of the TWTA a phase a velocity of $0.8c$.

We are currently exploring variations in system performance with beam and structure parameters to determine the best overall configuration. Our experience with the experimental system performance compares well with the code predic-

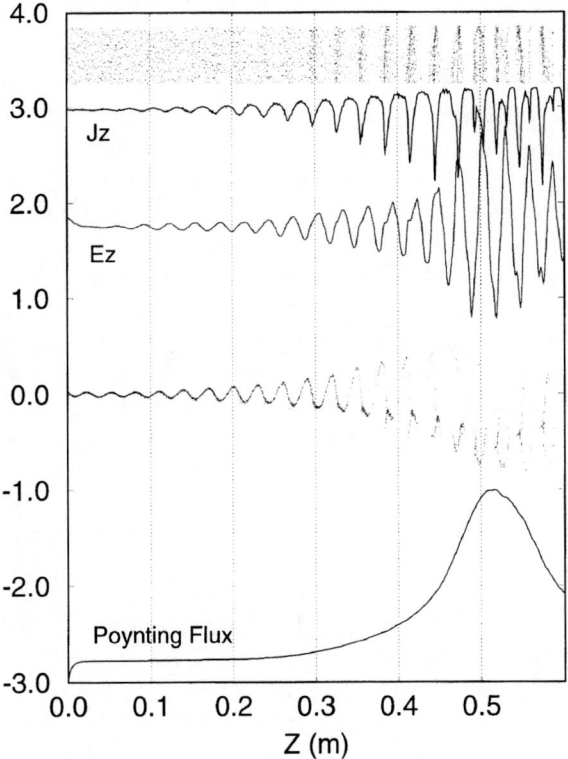

Figure 6: MAGIC simulation results from a TWT with the phase velocity tapered from 1.0c to 0.8c between z=0.4m and z=0.45m

tions so we have confidence that high efficiencies are attainable.

CONCLUSIONS

We have presented in this paper analysis and experimental data illustrating three features of TWT amplifier design. In the first section we showed that the structure wave should probably not be synchronous with the beam electrons due to the relativistic dependence of the growth rate. An important consequence of this is that the beam modulator design does not require the same degree of control needed in non relativistic devices. In the second section we showed both experimentally and theoretically that the beam should be decoupled from the structure wave at saturation and that this may be accomplished by the use of a TM to TEM mode converter in which the beam is dumped in the inner conductor of the coaxial output section. The system has the added advantages that the rf fields are relatively weak in the output region, that the structure is not very sensitive to variations in the dimensions of the guide, and that the output section is magnetically insulated. The output section design may well be relevant to klystron output design. Finally we showed data from simulations illustrating that efficiencies of over 50% are achievable in TWTA's by use of non adiabatic transitions in the structure such that the phase of the rf current and the axial electric field lead to maximum amplifier conversion efficiency.

ACKNOWLEDGEMENTS

This work was supported by the Department of Energy, and by the AFOSR under the MURI program. The MAGIC code was provided by MRC.

References

[1] G. Caryotakis, *IEEE Trans. Plasma Sci.*, **22**, 683, (1994).

[2] S. Naqvi, G. S. Kerslick, J. A. Nation and L. Schächter, *Phys. Rev. E*, **53**, 4229, (1996).

[3] S. A. Naqvi, G. S. Kerslick, J. A. Nation and L. Schächter, *Appl. Phys. Lett.*, **69**, 1550, (1996).

High Power Gyroklystron Development for Advanced Accelerator Applications

W. Lawson, J. Anderson, J. P. Calame, J. Cheng, M. Castle,
V. L. Granatstein, B. Hogan, M. Reiser, and G. P. Saraph

*Institute for Plasma Physics and Electrical Engineering Department
University of Maryland, College Park, MD 20742, USA*

Abstract

In this paper we present the design details and report on the current status of both a first harmonic two-cavity coaxial gyroklystron and a second harmonic three-cavity amplifier tube, each of which is designed to produce over 100 MW of output power. Both tubes utilize a fundamental mode TE_{011} input cavity which is driven by a 150 kW magnetron at 8.56 GHz. The former tube also has an 8.56 GHz TE_{011} output cavity while the latter system has a buncher cavity and an output cavity that resonate at twice the drive frequency in the TE_{021} mode. We present details of all system aspects, including the test bed modifications required to produce the enhanced beam characteristics, simulated beam properties, and simulated circuit interactions. Cold test results of both tubes are detailed and our preliminary HFSS modeling efforts are described. Results to date indicate that both systems should be at least 40% efficient.

INTRODUCTION

At the University of Maryland, we have been running a comprehensive program to study the suitability of gyroklystrons as drivers for linear collider applications (1). We have previously reported a variety of experimental results, all of which were achieved on a test bed which produced a small-orbit beam with a nominal voltage and current of 450 kV and 200 A, respectively. Published accounts of our effort include an amplified power level of 27 MW at 32% efficiency in a three-cavity first harmonic gyroklystron near 10 GHz (2); 32 MW at 29% efficiency in a two-cavity second harmonic gyroklystron near 20 GHz (3); 28 MW in a second harmonic coaxial gyroklystron (4); and 2 MW at 2% efficiency in a two-cavity third harmonic gyroklystron near 30 GHz (5). We have also produced 22 MW at 22% efficiency in a fundamental gyrotwystron (6), and 12 MW at 11% efficiency in a second harmonic gyrotwystron (7).

In this paper we present the design details of two coaxial gyroklystron tubes which are predicted to produce at least 100 MW of output power with an efficiency of nearly 40%. These tubes utilize a fundamental mode TE_{011} input cavity which is driven by a 150 kW magnetron at 8.568 GHz. The first tube also has an 8.568 GHz TE_{011} output cavity, whereas the second tube has 17.136 GHz TE_{021} buncher and output cavities. We present details of all system aspects, including the test bed modifications required to produce the enhanced beam characteristics, simulated beam properties, and simulated circuit interactions. Cold test results of both cavities are discussed. In the next section we describe the test bed and in the following section we present the results of our simulations. The cold-test results and HFSS simulations are described in the fourth section. The project status and a description of our future plans are summarized in the final section.

TEST BED MODIFICATIONS

We have just completed an upgrade of our facility which should enable us to produce amplified microwave powers in excess of 100 MW (see Fig. 1). Our modulator voltage and current have been increased to 500 kV and 800 A, respectively. We have designed, installed, and completed acceptance testing of a single-anode Magnetron Injection Gun (MIG) which is capable of producing a 480 - 720 A rotating electron beam at the nominal beam voltage with an axial velocity spread less than 7%. The simulated space-charge-limited perveance of 5.5 µP was in good agreement with the measured result. The maximum current produced in the acceptance test was limited by our modulator (due to an applied voltage which was lower than the nominal operating voltage) to 670 A.

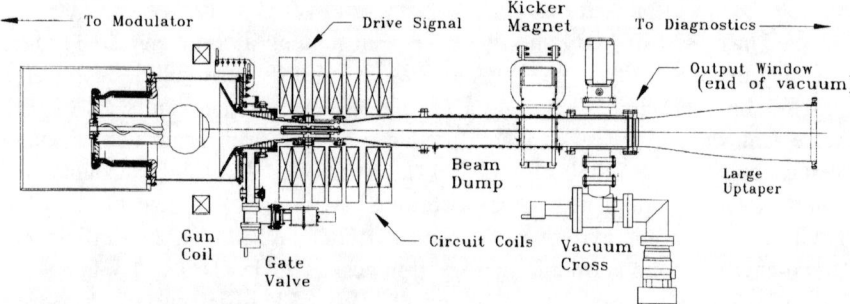

Fig. 1. The gyroklystron test bed.

The original water-cooled magnets have been used, but a larger power supply for the gun coil was required because of a decrease in the magnetic compression. We reduced our drive frequency from 10 GHz to exactly three times the current SLAC frequency, so a new coaxial magnetron and a modified input waveguide were required. The output waveguide (uptapers, beam dump, window, kicker magnet, pumping cross) was totally rebuilt to accommodate the expected larger peak powers. The anechoic chamber was modified to accommodate the new output waveguide and the directional coupler diagnostic was completely redesigned.

THEORETICAL CIRCUIT PERFORMANCE

A detailed design analysis has been carried out on a number of coaxial, two- and three-cavity gyroklystron systems with the aid of our partially self-consistent nonlinear code. The input cavity in all tubes is in resonance with the signal frequency at 8.568 GHz and the output cavity is resonant with either the first (8.568 GHz) or the second harmonic (17.136 GHz) frequency. In a three-cavity system, an additional buncher cavity is introduced which is resonant at either the first or second harmonic frequency. In the following sections, we describe only the two tubes which are scheduled to undergo hot testing in the near future.

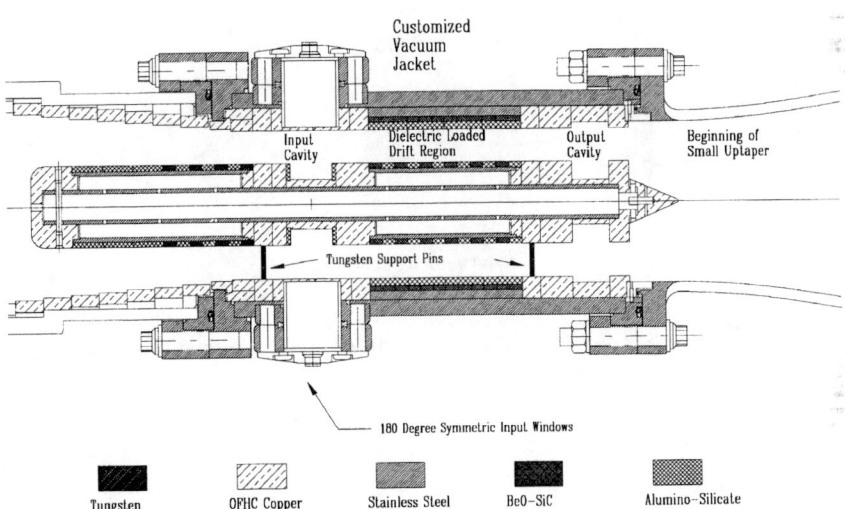

Fig. 2. The first harmonic two-cavity tube.

Two-Cavity First Harmonic Tube

The first tube that we will hot test is the two-cavity first harmonic tube which is shown in Fig. 2 and consists of an input cavity and an output cavity separated by a drift section. The input cavity is defined by a decrease in the inner conductor radius only and the quality factor is brought down to Q ≈ 50-65 by loading the cavity with two thin rings of carbonized aluminum-silicate placed at either end of the cavity. The inner radius is 1.05 cm and the length is 2.29 cm. Power is injected through two radial coupling ports which are separated by 180° and excited in phase. Our start-oscillation code predicts that the input cavity is completely stable up to a current of 800 A.

The drift section has inner and outer radii of 1.825 cm and 3.325 cm, respectively. The inner conductor is required so that the drift tube is cutoff to the operating mode. The regions adjacent to each cavity are made of copper, but lossy ceramics line the majority of the drift tube to eliminate spurious modes. The total length of the drift region is 9.1 cm. Lossy ceramics are also used in the downtaper between the gun and the input cavity.

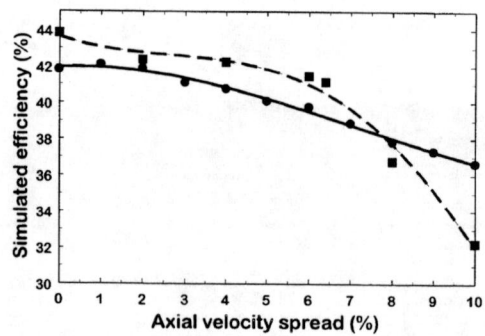

Fig. 3. Efficiency of the first (solid line) and second harmonic (dashed line) tubes vs. velocity spread.

The output cavity is defined by changes in both radii and has a length of 1.70 cm. Power is extracted axially into the output waveguide via a coupling aperture. The aperture has the same radii as the drift tube and has a length of 0.9 cm. The diffractive quality factor is about 122. The start-oscillation code also predicts the output cavity to be stable at the nominal current, which is given in the middle column of Table 1 along with the other operating parameters. The efficiency is nearly 40% and the output power is about 95 MW. The dependence of tube efficiency on axial velocity spread is plotted in Fig. 3 with the solid line. The simulated velocity spread of the electron gun is 6.4 % at the nominal current. The curve shows a slow but steady decrease in efficiency with increasing spread and indicates that an efficiency of 37% is still possible if the spread is as high as 10%. A simplified schematic of the tube dimensions along with the optimal axial magnetic field profile is indicated in Fig. 4.

TABLE 1. Comparison of the 1st and 2nd harmonic designs.

Parameters	1st harmonic	2nd harmonic
Voltage	500 kV	500 kV
Current	480 A	770 A
Velocity ratio	1.508	1.508
Input Cavity Q	50	50
Buncher Cavity Q	-	389
Output Cavity Q	122	320
Gain	21 dB	49 dB
Efficiency	39.4%	41.1 %
Output Power	94.6 MW	158.2 MW

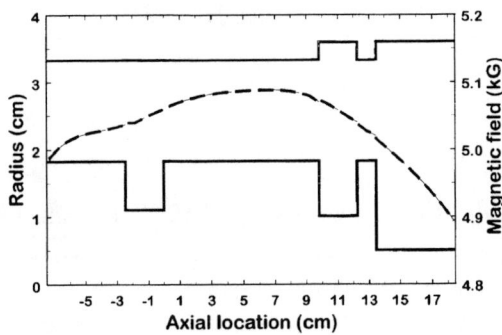

Fig. 4. The first-harmonic two-cavity tube and the optimal magnetic field profile.

Three-Cavity Second Harmonic Tube

The second harmonic design which we intend to test is a three-cavity system. A simplified schematic with the tube along with a plot of the optimal magnetic field profile is given in Fig. 5. The buncher cavity operates at the second harmonic and is formed with non-adiabatic radial wall transitions. Mode conversion from the TE_{02} mode to the TE_{01} is estimated to be about -40 dB. Dielectric loading of the cavity is used in order to obtain a Q of 389 and is achieved by reducing the thickness of the copper sections that separate the cavity from the drift tube dielectrics. The linear start oscillation code indicates that the buncher cavity is stable to beam currents below about 1000 A at the design value of the magnetic field ($B_0 = 4.81$

kG). The output cavity is also designed with non-adiabatic radial wall transitions. The scattering matrix code estimates the purity of the TE_{02} operation in the output cavity to be 97%. The ratio of the power flowing into the drift tube to the power flowing into the output waveguide is better than -24 dB. Furthermore, the lossy dielectric loading in the drift tube, which will have a minimum effect on the Q-value of the operating mode, will suppress the excitation of the spurious modes and reduce the cross-talk.

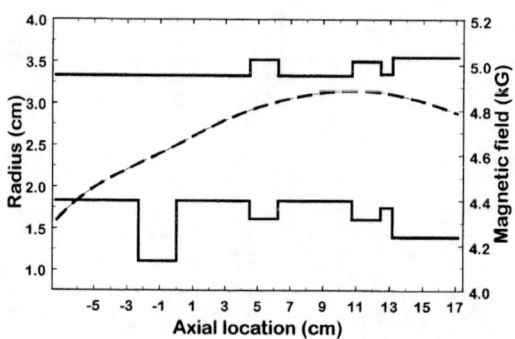

Fig. 5. The second-harmonic three-cavity tube and the optimal field profile.

The output cavity is highly overmoded and is linearly stable only up to a current of 400-450 A for the operating regime from 4.8 kG to 5.0 kG. The beam can excite various other modes at higher current levels. In the actual system the signal injected in the input cavity modulates the beam. The length of the drift section is chosen such that the beam is tightly bunched (in gyro-phase) when it enters the output cavity. The well-bunched beam at 8.568 GHz leads to forced excitation of the operating mode (TE_{021}). The operating mode grows in amplitude first. Then, in the presence of the large amplitude operating mode, the gain of the other modes is suppressed. Nonlinear gain calculations show that the cavity is stable under the operating conditions given in Table 1.

The simulated results at the nominal operating point are indicated in the final column of Table 1. The optimal current according to the simulations is 770 A and the estimated peak output power is over 150 MW. The corresponding gain and efficiency are 49 dB and 41%, respectively. The dependence of efficiency on velocity spread is shown as the dashed line in Fig. 2. Note that the efficiency begins to drop off fairly rapidly for spreads above 7%. However, these simulations are not re-optimized with respect to magnetic field profile, etc., at each point, and additional investigations indicate that higher efficiencies can be achieved if the velocity spread is higher than expected.

COLD-TEST RESULTS AND HFSS SIMULATIONS

The construction and cold testing of the first experimental tube has been completed. Cold-testing yielded the final dimensions of the input cavity required to achieve the frequency of 8.568 GHz and a quality factor of 55. They are quite near the theoretical estimates given in the previous section. The lossy ceramic ring dimensions have also been finalized. Cold-test drift tube attenuation measurements have indicated adequate isolation.

The output cavity has twelve separate metal pieces and has been completely fabricated and cold-tested. The cavity's outer radial wall extends to 3.59 cm while the inner radial wall dips to 1.007 cm. As indicated in Fig. 2, a fairly short taper of the inner conductor radius follows immediately after the diffractive lip to convert the coaxial waveguide to a circular waveguide. Cold testing of the output cavity (and adjacent drift tube region) was performed with a symmetric injection scheme and the resonant frequency and quality factor of the operating TE_{011} mode were found to be 8.565 GHz and 134, respectively.

An aluminum mock-up of the second harmonic buncher cavity has been constructed and cold-tested. Preliminary results have indicated that the required quality factor and frequency can be achieved for this design.

We have recently started using the commercial High-Frequency Structure Simulator software package (HFSS). We have been using it to model the drive cavity in order to optimize the design of the coupling apertures. The results indicate that sufficient coupling through our dual input waveguide configuration can be achieved. We have also begun to model the second harmonic buncher cavity. All HFSS results to date have been in good agreement with cold test results in terms of both resonant frequency and Q.

PROJECT STATUS AND FUTURE PLANS

The upgrade of our facility is complete. We have designs of first and second harmonic tubes which promise to produce peak powers of 100 MW or more with efficiencies of at least 40%. Upon completion of a few minor details, the pump-down of the first configuration with the two-cavity fundamental tube will begin. Several variations on the first harmonic tube may be tested, depending on the measured performance characteristics. A buncher cavity may also be added if it is found that tube performance is gain-limited.

A mock-up of the second harmonic output cavity is currently under construction, along with the mode converters and non-linear waveguide transitions that will be required to perform cold-testing. When cold testing has determined the final dimensions of the buncher and output cavities, the vacuum compatible version of the tube will be constructed. When the hot testing of the first two-cavity system is completed, we will begin testing the three-cavity second harmonic configuration. We expect this to occur early in the spring of 1997.

We continue to work on improving our simulation capabilities. Time-dependent capability has been added to our nonlinear (single-mode) code by researchers from the Naval Research Laboratory and initial results have confirmed the steady-state code predictions. We hope in the future to add multi-mode capability to our time dependent code.

We are also looking at advanced cavity concepts for future tubes. For the input cavity, we are using HFSS to simulate a single waveguide injection scheme which couples to the input cavity via an outer coaxial cavity. Furthermore, we are investigating an output cavity which couples through the inner radial wall to a circular waveguide, thereby decoupling the microwaves and the beam beyond the output cavity and enabling the use of additional tube stabilization schemes. Preliminary HFSS simulations of this output cavity scheme have been successful. Both concepts promise to improve performance of the gyroklystron tubes beyond the current predictions.

Finally, we are beginning to look at the application of gyroklystron circuits to Ka-band and beyond.

References

[1] Granatstein, V. L. and W. Lawson, "Gyro-Amplifiers as Candidate RF Drivers for TeV Linear Colliders," *IEEE Trans. Plasma Sci.* **24**, 648 - 665 (1996).
[2] Tantawi, S., W. Main, P. E. Latham, G. Nusinovich, W. Lawson, C. D. Striffler, and V. L. Granatstein, "High Power X-Band Amplification from an Overmoded Three-Cavity Gyroklystron with a Tunable Penultimate Cavity," *IEEE Trans. Plasma Sci.* **20**, 205 - 215 (1992).
[3] Matthews, H. W., W. Lawson, J. P. Calame, M. K. E. Flaherty, B. Hogan, J. Cheng, and P. E. Latham, "Experimental Studies of Stability and Amplification in a Two-Cavity Second Harmonic Gyroklystron," *IEEE Trans. Plasma Sci.* **22**, 825 - 833 (1994).
[4] Flaherty, M. K. E., W. Lawson, B. Hogan, H. W. Matthews, and J. P. Calame, "Operation of a K-Band Second Harmonic Coaxial Gyroklystron," *J. Appl. Phys.* **76**, 4393 - 4398 (1994).
[5] Lawson, W., B. Hogan, M. K. E. Flaherty, and H. Metz, "Design and Operation of a Two-Cavity Third Harmonic Ka-Band Gyroklystron," *Appl. Phys. Lett.* **69**, 1849-1851 (1996).
[6] Latham, P. E., W. Lawson, V. Irwin, B. Hogan, G. S. Nusinovich, H. W. Matthews, and M. K. E. Flaherty, "High Power Operation of an X-Band Gyrotwistron," *Phys. Rev. Lett.* **72**, 3730 - 3733 (1994).

[7] Lawson, W., P. E. Latham, J. P. Calame, J. Cheng, B. Hogan, G. S. Nusinovich, V. L. Granatstein, and M. Reiser, "High Power Operation of First and Second Harmonic Gyrotwystrons," J. Appl. Phys. 78, 550 - 559 (1995).

Gyroklystrons for Driving Linear Colliders at 35 GHz

V. L. Granatstein, G. P. Saraph, G. S. Nusinovich, A. Singh, and W. Lawson

Institute for Plasma Research
University of Maryland, College Park, MD 20742

INTRODUCTION

Various high power microwave sources in the frequency range from 8 to 35 GHz are being developed for driving linear colliders. Research efforts are in progress to develop efficient microwave sources for this application with typically 50-100 MW power level and 1 μs pulse-length. Klystrons [1], [2], intense beam traveling-wave-tubes [3], magnicons [4], CARMs [5], ubitrons (FELs) [6], and gyroklystrons [7] are the main microwave sources being considered to fulfill these requirements. By choosing a relatively high microwave drive frequency, the accelerating gradient can be larger and the overall length of the collider can be minimized; there has been considerable interest both in Europe and in the U.S. in colliders operating in the 30-35 GHz range [8], [9]. The gyroklystron is a microwave amplifier type that is especially well configured to handle high power at wavelength of about 1 cm ($f = 30$ GHz) or less.

This paper presents initial design studies of gyroklystron amplifiers in the 30-35 GHz frequency range. In gyro-devices, the relativistic dependence of the electron cyclotron frequency on electron energy leads to cyclotron maser instability which causes bunching in gyro-phases. This bunching in gyroklystrons proceeds in a way similar to the electron ballistic bunching in conventional klystrons. However, the frequency selectivity of the cyclotron resonance interaction enables one to use large, overmoded cavities and drift regions in gyroklystrons. This has two advantages over conventional klystrons; first, the device is less susceptible to breakdown at high power levels and second, it can operate at higher frequencies. The high power levels required for the future accelerators can be reached by using relativistic electron beams in the device.

The development of relativistic gyroklystrons for the accelerator applications is being carried out at the University of Maryland. The early investigations have demonstrated power levels up to 27-32 MW and efficiency of 28 % in 10 GHz and 20 GHz experiments at the fundamental and second cyclotron harmonics, respectively [10]-[12]. Ongoing experiments are based on 100-150 MW designs in 8.57 GHz and 17.14 GHz presented in Ref. [13]. Simulations predict over 40

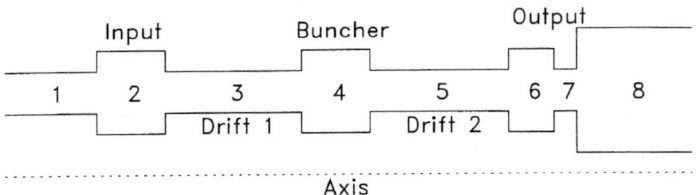

Figure 1: Schematic diagram of a three-cavity, coaxial gyroklystron system

% efficiency and over 45 dB gain for these experiments. This paper extends the design studies presented in Ref. 13 to develop sources at 35 GHz.

Designs of two relativistic gyroklystron systems are presented here, both employing three-cavity, coaxial microwave circuits. A coaxial microwave circuit has been chosen to alleviate the problems associated with the highly overmoded waveguides, especially self-excitation in the drift regions. It also serves to reduce the potential depression due to the space charge in the beam. A schematic diagram of a coaxial, three-cavity gyroklystron system is shown in Fig. 1. For both systems, the input cavities are resonant at the input signal frequency of 17.5 GHz and the penultimate (buncher) and output cavities are resonant at 35 GHz. We refer to each system by the specific cyclotron harmonic in each cavity, e.g., 1-2-2 system is a three-cavity system with the input cavity at the fundamental frequency and the buncher and output cavities at the second harmonic of the cyclotron frequency. The first design is of 1-2-2 type system which requires approximately 10 kG magnetic field that could be supplied either by a water-cooled solenoidal electromagnets or by a superconducting magnet. The second design is of 2-4-4 system with the output at the fourth cyclotron harmonic. It requires only 5 kG magnetic field that could be supplied by permanent magnets which would be required to reduce power consumption and/or costs in a large collider. We have also investigated 1-2-4 scheme [14], but 2-4-4 scheme results in higher efficiencies.

SIMULATION STUDIES

The design analysis is carried out using a set of numerical codes developed at the University of Maryland for the relativistic gyroklystron systems. This set includes a scattering matrix code for cold-cavity fields [15], [16], a linear start-oscillation code for stability, and a nonlinear gyroklystron code [17] for optimizing parameters to get maximum efficiency and gain.

It is important to note that in a relativistic gyroklystron system, the distance traveled by an electron in one cyclotron period is on the order of the cavity length. For such short cavities the equations of particle motion cannot be simplified by averaging over the cyclotron period. Also the cyclotron resonance is broad such that more than one cyclotron harmonic can interact at a given frequency of operation. A detailed formulation of this problem is presented in Ref. 17.

The short cavities also imply that the inner and outer radii vary rapidly as compared to the wavelength, producing linear mode conversion. We model each cavity as a series of straight, uniform sections with abrupt radial discontinuities. The electromagnetic fields in each coaxial region are expanded in terms of its eigenmodes and matched at the boundary using a scattering matrix formulation. The cold cavity resonant frequencies and quality factors, Q's, are determined for each cavity. We assume that the Q is sufficiently high as compared to its diffraction limit so that the field pattern is not altered significantly by introducing the electron beam.

Optimizing a design is an iterative procedure and involves modifying the cavity shapes, magnetic field profile, and operating parameters to achieve maximum efficiency and gain. The cavities are designed to have high mode purity, good intercavity isolation, strong beam coupling and desired value of Q. Simulations are carried out for a 500 kV beam with a current up to 700-800 A and a pitch-angle (velocity ratio $v_\perp/v_z$) of 1.5. The beam radius is 1.8 cm (at 10 kG) for the 1-2-2 design and 2.6 cm (at 5kG) for the 2-4-4 design. Specific details of the two designs are given below.

Design of 1-2-2 System

The input cavity is designed to operate in TE_{01} mode at the input signal frequency (17.5 GHz) and at the fundamental cyclotron resonance. The buncher and the output cavities operate in the TE_{02} mode at 35 GHz at the second harmonic resonance. The drift sections are cutoff to the respective operating cavity modes at these frequencies. The dimensions of each section are tabulated in Table 1.

The input and buncher cavities have infinitely large diffractive quality factor due to the cutoff drift sections on either side. The Q-value is brought down by loading the cavity with lossy dielectric materials to make it stable to self-excitation. The dielectric-loaded quality factor is adjusted experimentally to the desired value to ensure stability and efficient operation. The output cavity is formed with non-adiabatic radial wall transitions. A small lip at the end of the output cavity is used to confine the field energy and get the Q-value to the desired level (Q = 435). In circular cavities, mode conversion from TE_{02} to TE_{01} mode

Cavity or Section	No.	1-2-2 Design			2-4-4 Design		
		Inner Radius (cm)	Outer Radius (cm)	Length (cm)	Inner Radius (cm)	Outer Radius (cm)	Length (cm)
Inlet	1	1.420	2.180	4.000	1.825	3.325	4.000
Input	2	1.320	2.280	1.270	1.620	3.530	1.410
Drift 1	3	1.420	2.180	4.000	1.825	3.325	2.000
Buncher	4	1.350	2.250	1.060	1.680	3.470	1.245
Drift 2	5	1.420	2.180	8.000	1.825	3.325	3.000
Output	6	1.350	2.250	1.050	1.685	3.465	1.268
Output Lip	7	1.415	2.185	0.300	1.790	3.360	0.300
Outlet	8	1.300	2.300	2.000	1.600	3.550	3.000

Table 1: Dimensions of the 1-2-2 and 2-4-4 designs

would occur at these transitions. Since the TE_{01} mode is above cut-off at the operating frequency (35 GHz), this converted power would flow back into the drift tube and could potentially cause cross-talk, instability, and heating problems. In a coaxial cavity, however, it is possible to choose inner and outer radial dimensions to suppress this mode conversion. Our output cavity has been designed with these considerations in mind. The scattering matrix code estimates the purity of the TE_{02} operation in the output cavity to be 97 % and a left-to-right power ratio (ratio of the power flowing backward to the power flowing forward) of -24 dB.

Using the nonlinear simulation code the efficiency of this design is estimated to be 42 % for an ideal beam. For a beam with 6 % RMS spread in axial velocity the efficiency is estimated to be 38 %. The total device efficiency can be boosted futher by employing energy recovery from the spent electron beam which will be discussed in a later section of this paper.

Design of 2-4-4 System

In this design the input cavity is resonant at the second cyclotron harmonic in the TE_{02} mode. The buncher and output cavities are resonant at the fourth harmonic in the TE_{04} mode. The output cavity is designed to have a Q value of 1100. The dimensions of this system are tabulated in Table 1. The efficiency of this system is estimated to be 22 % for an ideal beam. The operation at the fourth harmonic is more susceptible to the spread in axial velocity. The efficiency value drops to 16 % for 6 % RMS spread in the beam. Energy recovery can of course be more effective in raising overall efficiency when intrinsic efficiency is smaller

Parameters	Design 1-2-2	Design 2-4-4
Voltage	500 kV	500 kV
Current	700 A	700 A
Pitch-Angle	1.50	1.50
Magnetic Field	10 kG	5 kG
Beam Radius	1.80 cm	2.57 cm
Q-Output Cavity	435	1100
Gain	48 dB	43 dB
Efficiency (ideal)	42 %	22 %
Efficiency (w/spread)	38 %	16 %
Output Power	133 MW	56 MW

Table 2: Operating parameters for the 1-2-2 and 2-4-4 systems

The main advantage of this 2-4-4 scheme operation at the magnetic field strength of only 5 kG. The operating parameters for these two schemes are tabulated in Table 2.

ENHANCEMENT OF DEVICE EFFICIENCY WITH ENERGY RECOVERY

Initial Single-stage Depressed Collector Design

A significant portion of the initial electron energy is still left in the spent electron beam after it exits the interaction region. Depressed collectors can be used to recover this energy from the spent beam by decelerating it by an external electric field. Such energy recovery is commonly used in traveling-wave tubes, klystrons, and other microwave amplifiers [18], [19]. In gyro-devices a large component of the beam energy is in transverse motion. This orbital component of the energy should be converted into the longitudinal component before the energy recovery [20], [21].

Multi-stage depressed collectors are being used for energy recovery in low to medium power (up to 100's of kW), CW microwave sources. However, the present gyroklystron amplifiers are very high power, pulsed devices. It is very challenging to couple the recovered energy back to the power supply of such devices. In order to minimize the difficulties in implementing the recovery scheme we consider only a single electrode depressed collector system in this section. Multi-stage depressed collectors which will be more difficult to implement but which will potentially result in higher overall efficiency will be discussed in the next section. The spent beam can be extracted either radially through a gap in the wall or axially.

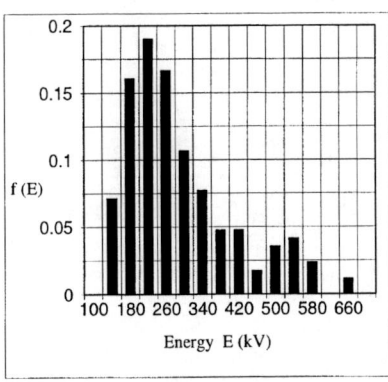

Figure 2: Distribution function of the spent beam as a function of the total energy of each particle

In our case, the Larmor radius of the beam is very large (1/4 times guiding center radius) and hence the gap required for radial extraction is very large (about 4 cm); the electromagnetic power loss through this gap is unacceptable. Therefore we have chosen a simple geometry in which the cylindrical beam-dump is at a depressed potential.

We have taken the spent beam from the 1-2-2 design operating at the maximum efficiency to design the energy recovery system. The distribution function of the spent beam as a function of the total energy of each particle is shown in Fig. 2. The lowest energy of the particles is about 140 keV. The distribution of the pitch angle as a function of the total energy is shown in Fig. 3. Notice that the pitch angle of the lowest energy electrons is only 0.7, indicating that only about one third of their energy is in the orbital component. Further, the beam enters a tapered region in which the axial magnetic field is gradually reduced. For strictly adiabatic variation, a factor of 7 reduction in axial magnetic field would lead to a factor of $\sqrt{7}$ in radial expansion of the beam and a factor of 7 reduction in orbital energy. Thus, most of the orbital energy would be converted into the axial component.

The axial distance required for adiabatic reduction of orbital energy is very large. Therefore we have chosen magnetic field pattern which is non-adiabatic to get the best energy recovery possible in a limited distance. We have introduced an additional coil and iron shield to get the required magnetic field pattern in the depressed collector region. Using the EGUN code we have traced the particle trajectories through this magnetic field from the end of the RF structure as shown in Fig. 4. For 100 % collection of the beam at the depressed collector, the maximum

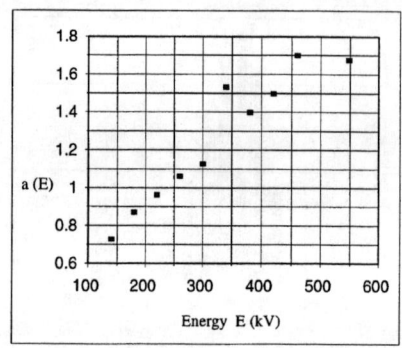

Figure 3: Variation of the average pitch angle in each energy group as a function of the particle energy in the spent beam

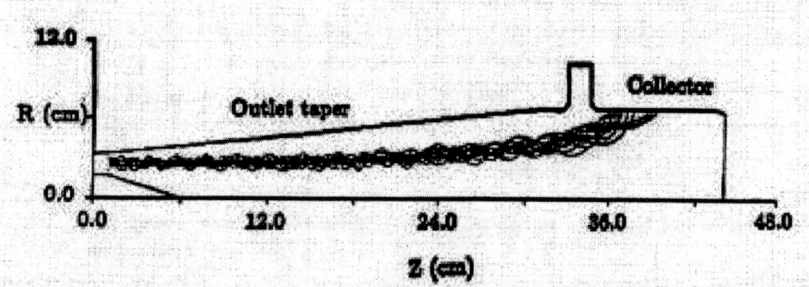

Figure 4: Traces of electron trajectories in the outlet and depressed collector region using the EGUN code

allowable potential difference between the anode and the collector is 125 kV. The power supply requirements can therefore be lowered from 500 kV to 375 kV. Thus the device efficiency can be improved from 42 % to 56 % for an ideal beam. For a more realistic beam with 6 % spread in the initial axial velocity the device efficiency improves from 38 % to 51 %. There will be some additional power loss through the gap between the taper and the collector (1.6 cm in length). It can be minimized by designing a resonant choke flange.

Similar analysis can be performed for the 2-4-4 system. Since the elctronic efficiency is only 22 % (or 16 % with spread), the spent beam carries a relatively large portion of the initial energy. The minimum energy of the electron distribution is above 300 kV. In this case a depressed collector at 275 kV can collect the entire beam. Thus, the total efficiency of the 2-4-4 system can be improved to 48 % (or 35 % with 6 % spread).

In order to implement the energy recovery scheme the system would require two power supplies. A high current power supply connected between the cathode and the depressed collector would provide the entire beam current. A low current biasing supply would provide voltage between the anode and the depressed collector. This also implies that the potential for the depressed collector is likely to be fixed beforehand and cannot be changed based on the operating conditions. In the case of the 1-2-2 system, the high current supply would require only 375 kV voltage. This reduces the power supply requirements by a factor of 4/3. In the case of 2-4-4 system the voltage required is reduced to 225 kV, which means reduction by a factor of about 2.2. An implementation scheme for energy using a double-anode MIG-gun is being considered. A full discussion of this scheme will be presented in a later publication.

PERSPECTIVE ON SINGLE AND MULTI-STAGE DEPRESSED COLLECTORS

Single stage depressed collectors have the advantage of simplicity in the collector geometry and the power supply. However where higher values of collector efficiency are desired, one needs to go to multi-stage depressed collectors. A multi-stage collector requires precise tailoring of electric and magnetic field configurations and the electrode geometry in order to ensure sorting of beamlets of different energy values so as be collected on depressed potentials appropriate to their energy. This requirement is also coupled with the need for avoiding return of electrons before collection. The prime reason for that is the residual rotational energy in the beamlets, which cannot be recovered by the electrostatic field. Where

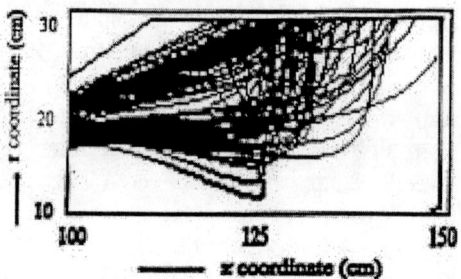

Figure 5: Electron trajectories in the collector region of a single-stage depressed collector for 170 GHz gyrotron

the energy of forward motion goes to zero at any location in the collector region the electrons turn back from there. The magnetic field has to have a blend of adiabatic and non adiabatic transitions. The former is used to reduce the energy of rotational motion, though at the cost of increasing the size of the collector. In this case the beamlets of all energy values follow the lines of magnetic flux. There is no energy sorting, and also the beam is collected on a relatively restricted area of the collector. This increases the heat dissipation density. In a non adiabatic transition, the flux lines are bending with sharper curves. The beamlets of lower energy tend to follow them more closely, than their more energetic counterparts. However beamlets also tend to cross the lines of magnetic flux, resulting in transfer of energy from forward motion to transverse motion. Thus, too abrupt a change in the magnetic field value can result in turning back of electrons before collection.

Computer-aided-design of Depressed Collectors

We have used the concept of effective potential as an aid to design of depressed collectors. It takes into account the electric as well as the magnetic field, and the motion of the electrons as influenced by the latter [22]. The contours of effective potential define the areas in the collector region which are accessible to electrons of different energy. A library of codes has been developed to assist in various stages of depressed collector design [23]. A code, that we have called ProfilEM, allows us to obtain diagrams of contours of effective potential and of lines of magnetic flux against the backdrop of the electrode geometry. The effect on these contours of changing the control parameters, such as coil currents and electrode potentials, can be seen on-line. This narrows the parameter space to be explored. In this way the magnetic field configuration can be designed and the

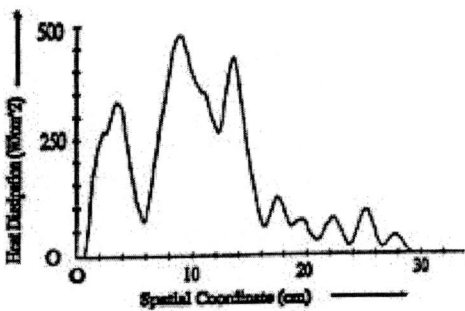

Figure 6: Profile of heat dissipation density on the collector surface of a single-stage depressed collector for 170 GHz gyrotron

electrode potentials chosen to give optimum collection and energy separation of the beamlets of different energy values. The final results are obtained by simulation of electron trajectories. Three dimensional representation of the distribution of effective potential can be obtained, which provides insights into the behaviour of trajectories of electrons. Besides ProfilEM other codes have been developed to evaluate the heat density profile at the surface of the collectors, the collector efficiency, and the distribution of current among different collectors.

An example of the electron trajectories in a single-stage depressed collector for a 170 GHz gyrotron, is shown in Figure 5, and that of the heat dissipation profile for the same case in Figure 6 [24]. It is seen that the presence of non adiabaticity in the transition through the magnetic field has resulted in a substantial spread of the beam over the collector surface. In turn the peak heat dissipation density is only 480 W/cm^2, for a tube with a power output of 1 MW CW. In the case of the gyroklystrons under consideration, the pulse power in the beam is of the order of 350 MW. For a pulse width of 1.5 μsec at a pulse repetition rate of 120 Hz, the average power in the beam is of the order of 60 kW. In this case the average heat dissipation density is not likely to be a problem. However because of the high peak power in the beam, the transient rise in the temperature would be just as important. In this case too it would thus be an advantage to have a larger spread of the beam on the collector surface, so as to avoid hot spots during the pulse.

Using these techniques and codes, two-stage depressed collectors have been designed for gyrotrons to operate at one megawatt output power level at 170 and 110 GHz It is of interest to note that at 110 GHz a two-stage depressed collector design gave an estimated collector efficiency of 68 %, where the energy range in the spectrum of the spent beam is of the order of 5:1. . For a case where the spent beam has an energy spread of 3:1, a collector efficiency of 72 % could be achieved. For a gyroklystron of 1-2-2 design, a collector efficiency of 68 % could raise the electronic efficiency of 38 % to an overall efficiency of 66 %. As indicated in the previous section, the energy range in the gyroklystron spent beam is narrower where the electronic efficiency is low. For such a case, it would be reasonable to assume that a collector efficiency of 72 % would be feasible. Thus for the case of 2-4-4 design a two-stage depressed collector could raise an electronic efficiency of 16 % to 40 %.

CONCLUSIONS

In conclusion, gyroklystron amplifiers operating at the second and fourth harmonics appear to be feasible candidates for driving linear colliders at 35 GHz. In initial design studies, we find that the second harmonic system (1-2-2) has 38 % efficiency which can be improved to 51 % using a single-stage depressed collector, while the fourth harmonic system (2-4-4) has lower efficiency of 16 % which can be improved to 35 % with a single-stage depressed collector. With a multi-stage depressed collector, further improvement in efficiency would be expected. For the fourth harmonic system, the external magnetic field requirement is 5 kG which could be supplied by permanent magnets. Operation at the second harmonic would require a superconducting solenoid which would have acceptably low power consumption; however, capital costs would need to be carefully evaluated.

ACKNOWLEDGMENTS

This work is supported by a research grant from the U.S. Department of Energy, High Energy Physics Division.

REFERENCES

[1] M. A. Allen, J. K. Boyd, and R. S. Callin, *Phys. Rev. Lett.*, vol. 63, pp. 2472-2475, 1989.

[2] G. Caryotakis, *IEEE Trans. on Plasma Science*, vol. 22, pp. 683-691, 1994.

[3] D. Shiffler, J. A. Nation, L. Schachter, J. D. Ivers, and G. Kerslick, *J. Appl. Phys.*, vol. 70, pp. 106-113, 1991.

[4] O. A. Nezhevenko, *IEEE Trans. on Plasma Science*, vol. 22, pp. 756-772, 1994.

[5] W. L. Menninger, *et al.*, *Proc. 1991 Part. Accel. Conf.*, pp. 754-756, 1991.

[6] R. Phillips, *Proc. of Workshop on Pulsed RF Sources for Linear Colliders* ed. by R. C. Fernow, AIP Conf. Proc. 337, Montauk, New York, Oct. 1994

[7] V. L. Granatstein, P. Vitello, K. R. Chu, K. Ko, P. E. Latham, W. Lawson, C. D. Striffler, and A. Drobot, *IEEE Trans. Nucl. Sci.*, vol. NS-32, p. 2957, 1985.

[8] L. Thorndahl, *Proc. of Workshop on Pulsed RF Sources for Linear Colliders*, ed. by R. C. Fernow, AIP Conf. Proc. 337, Montauk, New York, Oct. 1994.

[9] P. B. Wilson, *Proc. 1996 DPF/DPB Summer Study of New Directions for High Energy Physics*, Snowmass, 1996 (to be published).

[10] J. P. Calame, W. Lawson, *et.al.*, *J. Appl. Phys.*, vol. 70, pp. 2423-2434, 1991.

[11] S. G. Tantawi, W. T. Main, *et.al.*, *IEEE Trans. on Plasma Science*, vol. 20, pp. 205-215, 1992.

[12] W. Lawson, H. W. Matthews, *et.al.*, "High-power operation of a K-band second harmonic gyroklystron," *Phys. Rev. Lett.*, vol. 71, pp. 456-459, 1993.

[13] G. P. Saraph, W. Lawson, M. Castle, J. Cheng, J. P. Calame, and G. S. Nusinovich, *IEEE Trans. on Plasma Science*, vol. 24, pp. 671-677, 1996.

[14] G. S. Nusinovich and O. Dumbrajs, *Phys. Plasmas*, vol. 2, pp. 568-577, 1995.

[15] J. M. Neilson, P. E. Latham, M. Caplan, and W. G. Lawson, *IEEE Trans. Microwave Theory Tech.*, vol. 37, pp. 1165-1170, 1989.

[16] W. Lawson and P. E. Latham, *IEEE Trans. Microwave Theory Tech.*, vol. 37, pp. 1165-1170, 1989.

[17] P. E. Latham, W. Lawson, and V. Irwin, *IEEE Trans. on Plasma Science*, vol. 22, pp. 804-817, 1994.

[18] H. G. Kosmahl, *Proc. IEEE Electron Devices*, vol. 70, no. 11, p. 1325, 1982.

[19] W. Neugebauer and T. G. Mihran, *IEEE Trans. Electron Devices*, vol. ED-19, no. 1, p. 111, 1972.

[20] M. E. Read, W. Lawson, A. J. Dudas, and A. Singh, *IEEE Trans. Electron Devices*, vol. ED-37, no. 6, pp. 1579-1589, 1990.

[21] A. Singh, G. Hazel, V. L. Granatstein, and G. P. Saraph, *Int. J. Electronics*, vol. 72, nos. 5 and 6, pp. 1153-1163, 1992.

[22] A. Singh, D. S. Weile, S. Rajapatirana and V. L. Granatstein, (to be published), 1996.

[23] A. Singh, D. S. Weile, S. Rajapatirana and V. L. Granatstein, *Conf. Record - IEEE Int. Conf. on Plasma Science*, Boston, June 3-5, 1996.

[24] A. Singh, *Final Report of US Gyrotron Team*, pp. 17-19, May 15, 1994.

High-Power Microwave Production by Gyroharmonic Conversion and Co-Generation

M. A. LaPointe,[*] R. B. Yoder,[§] Mei Wang,[§] A. K. Ganguly,[*]
Changbiao Wang,[§] B. Hafizi[*] and J. L. Hirshfield[*§]

[*]*Omega-P, Inc., 202008 Yale Station, New Haven, CT 06520*
[§]*Physics Department, Yale University, New Haven, CT 06511*

An rf accelerator that adds significant gyration energy to a relativistic electron beam, and mechanisms for extracting coherent radiation from the beam, are described. The accelerator is a cyclotron autoresonance accelerator (CARA), underlying theory and experimental tests of which are reviewed. The measurements illustrate the utility of CARA in preparing beams for high harmonic gyro interactions. Examples of preparation of gyrating axis-encircling beams of ~400 kV, 25 A with $1 < a < 2$ using a 2.856 GHz CARA are discussed. Generation of MW-level harmonic power emanating from a beam prepared in CARA into an output cavity structure is predicted by theory. First measurements of intense superradiant 2nd through 6th harmonic emission from a CARA beam are described. Gyroharmonic conversion (GHC) at MW power levels into an appropriate resonator can be anticipated, in view of the results described here. Another radiation mechanism, closely related to GHC, is also described. This mechanism, dubbed "co-generation", is based on the fact that the lowest TE_{sm} mode in a cylindrical waveguide at frequency sw with group velocity nearly identical to group velocity for the TE_{11} mode at frequency w is that with $s = 7$, $m = 2$. This allows coherent radiation to be generated at the 7th harmonic co-existent with CARA and in the self-same rf structure. Conditions are found where co-generation of 7th harmonic power at 20 GHz is possible with overall efficiency greater than 80%. It is shown that operation of a cw co-generator can take place without need of a power supply for the gun. Efficiency for a multi-MW 20 GHz co-generator is predicted to be high enough to compete with other sources, even after taking into account the finite efficiency of the rf driver required for CARA.

INTRODUCTION

Considerable interest exists in new concepts for producing high peak power (>150 MW), narrow band, low duty-cycle, low pulse repetition frequency (~1 kHz) pulsed rf amplifiers operating at 1.0 GHz or higher for use in large electron/positron linear colliders. For these devices, electrical efficiencies greater than 45% are considered essential. As a result, effort is underway for development of rf sources to meet this need, including klystrons [1], relativistic klystrons [2], cluster klystrons [3], gyroklystrons [4], magnicons [5] and gyro-harmonic converters [6]. Much of the klystron development is at 11.424 GHz, where the Next Linear Collider Test Accelerator (NLCTA) is to be operated at SLAC. A new magnicon design at 11.424 GHz has recently been perfected [6], in which higher power and efficiency are predicted than have been achieved with klystrons. A relativistic klystron [2] and second harmonic gyroklystron [4] are being investigated for operation at 17.136 GHz. But little effort has gone into study of sources at higher frequencies, even though it has been pointed out [7] that high

CP398, *Advanced Accelerator Concepts*, edited by S. Chattopadhyay, J. McCullough, and P. Dahl
AIP Press, New York © 1997

energy colliders (>1 TeV) will probably have to operate at 20 GHz or above. For example, a study for a 5 TeV collider shows that operation at 34 GHz could allow a loaded acceleration gradient of 150 MV/m to be sustained, leading to an active accelerator length of "only" 17 km [8]. It will not be argued either way in this paper whether or not construction of a machine of this magnitude will be practical in the forseeable future. But it is clear that demonstration of an acceptable rf source at a frequency of 20 GHz or higher is essential for the rational planning of such a machine.

This paper describes recent progress on experimental and theoretical understanding of gyroharmonic conversion (GHC) and co-generation. These processes could have the potential for efficient production of high-power cm-wavelength radiation suitable for collider applications. In GHC, low frequency rf drive power is used to energize an electron beam in a magnetic field, and the beam is then allowed to selectively emit coherent radiation at a harmonic of the drive frequency [6]. As with the magnicon, no bunching is induced on the beam in GHC, and all electrons can experience nearly identical forces from the rf fields. This can lead to high conversion efficiencies, computed to be above 70% at the 5th harmonic under ideal conditions [6]. With efficient energy recovery from the spent beam using a depressed collector, the electrical efficiency might be as high as 85%. If a 150 MW rf drive pulse at 2.9 GHz were used, such as that produced with DESY klystrons [9], one finds 100 MW and 300 J/pulse to be produced at the harmonic. This is sufficient energy/pulse to drive ~11 m of a 34 GHz multi-TeV collider. If the efficiency of the rf drive source is 53%, one finds that GHC can have an overall efficiency of 45%, as required. This crude estimate defines the goals that must be achieved for GHC to be a serious contender for future collider applications. Herein, a progress report is presented on work to date on GHC, with first experiemtal results of 2nd through 6th harmonic superradiant emission at 5.7 through 17.1 GHz from a beam prepared using gyroresonance. In addition, a new approach for efficient utilization of a such a beam for 7th harmonic co-generation at 20 GHz is described, wherein efficiencies exceeding 80% are predicted.

CYCLOTRON AUTORESONANCE ACCELERATION (CARA)

A sketch of the Yale/Omega-P GHC experimental device is given in Fig. 1. The device has three major parts: (i) a Pierce-type diode electron gun that injects a pencil beam; (ii) a cyclotron autoresonance accelerator (CARA) that energizes the beam and produces axis-encircling gyrating orbits that rotate in time at the CARA drive frequency; and (iii) a converter waveguide or cavity in which coherent radiation is generated by the beam at harmonic of the CARA drive frequency. In the Yale/Omega-P apparatus, CARA is driven at 2.856 GHz, using a former SLAC XK-5 klystron that produces up to 24 MW, 3 μsec pulses at up to a 10 Hz repetition rate. The 1.0 μPerv injector gun operates up to 100 kV, with a current of up to 31 A. The guide magnetic field is produced by up to 20 individually-energized coils whose power supplies are externally-programmed to produce desired field profiles. Here, measurements of CARA performance are described in which a beam collector with a built-in calorimeter is substiuted for the converter section shown in the bottom half of Fig. 1.

Prior theory and simulation studies for CARA predict that high efficiency transfer of rf power to beam power can occur, provided that the injected beam does not have too large an axial velocity spread, and provided the magnetic field profile is judiciously adjusted [10,11]. To date, ~25 A beams of up to ~10 MW beam power have been produced, with efficiency values of between 90% and 96%. Higher beam powers might seem possible, since 24 MW of rf power is available, but this is not the case, since the upper energy limit in CARA for 100 kV injection is about 435 keV for beam matching into GHC at grazing [11]. A summary of the data for CARA operation is given in Fig. 2, where total input power (beam + rf) is the independent variable, with output beam power and efficiency are the dependent variables. Details of this experiment are found in two recent publications [12,13], and will not be restated here. Significant points established by the measurements include (a) confirmation of theory regards high rf power transfer efficiency; (b) observed resilience to detrapping when the actual magnetic field profile deviates from the ideal profile; and (c) an achievement of final beam energies in accordance with theoretical predictions.

These measurements add evidence to earlier contentions [6] that CARA is well suited for preparation of beams for injection into GHC output structures for generation of harmonic output power. In planned GHC experiments, the rf power levels are limited, since the maximum CARA beam power achieved is about 10 MW. By comparison, injection of a 350 kV, 200 A beam into an upgraded CARA, driven with 140 MW of rf from a DESY klystron, would allow a 1.0 MeV, 200 MW beam to be produced and injected at grazing into a GHC.

GYROHARMONIC CONVERSION (GHC)

Theoretical and computational studies have shown that the efficiency of gyroharmonic conversion (GHC) is sensitive to the quality of the injected beam, and to the strength of competing rf modes in the output structure [6]. In the Yale/Omega-P apparatus, the injected beam (up to 100 kV, 31 A) is computed to have an rms axial velocity spread of 0.10%, based on simulation studies using the relativistic DEMEOS code [14]. The axial velocity spread is magnified in CARA, and upon entrance to the rf output structure it can reach rms values in the range 0.5% - 1.0%, depending upon the magnitude of the magnetic guide field up-taper between CARA and the output section. The growth of axial velocity spread can be controlled to some extent, by use of a small detuning away from exact resonance of the magnetic field profile in CARA [15]. For an axis-encircling cold beam at grazing, theory predicts that no output radiation is coupled into TM modes, and that radiation at the s-th harmonic is coupled only into TE_{sm} modes, where $m = 1,2,3....$ The most dangerous competing mode in a waveguide output section has been shown to be the one lower than the design mode, if it is not cutoff; otherwise, it is at the next higher harmonic. In a cavity output section designed to operate in the TE_{411} mode, for example, 4th harmonic competition could be anticipated from the TE_{412} and TE_{413} modes, if these are so closely spaced so as to be within the beam's gain spectrum.

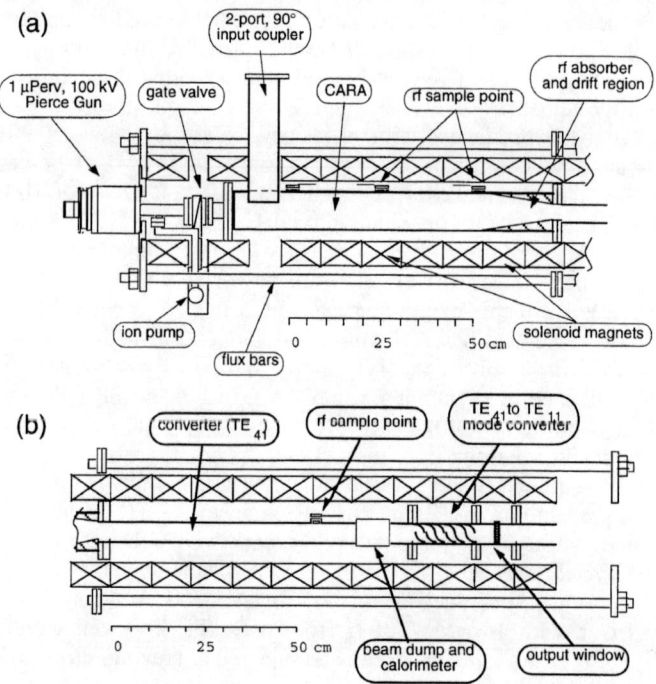

FIGURE 1. Schematic diagram of the Yale/Omega-P cyclotron autoresonance accelerator (a) and gyroharmonic converter (b). During measurements, the GHC is terminated either with a radiating horn or with a calorimeter.

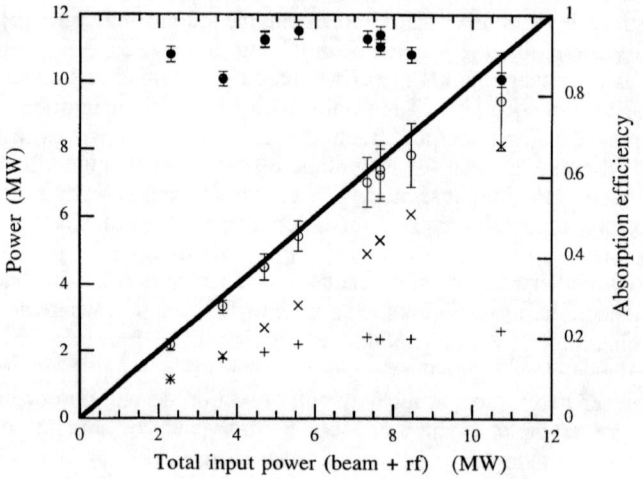

FIGURE 2. Beam power and efficiency for the Yale/Omega-P CARA. Key to the symbols: (+) initial beam power; (×) initial rf power; (o) final beam power; (•) efficiency.

Experiments have been conducted with a variety of travelling wave and cavity output sections, including cases with the output radiation passing through a mode converter designed to transform the TE_{41} rotating mode at 11.424 GHz into a TE_{11} rotating mode. The radiation also passed a single-disk alumina window and was either broadcast by a horn and detected in an anechoic chamber, or absorbed in a calorimeter. For the experiments to be describe here, the output structure consisted of a copper cavity of radius 2.205 cm and length 10 cm, followed by a 1.34 deg, 5-cm up-taper to a copper output waveguide of radius 2.322 cm. The cavity and up-taper having eight azimuthally-symmetric 2.4-mm wide axial slots cut through the wall, designed to supress any mode without four-fold azimuthal symmetry. The surrounding stainless-steel vacuum jacket had irregular grooves cut into its inner surface to scatter any radiation that leaks out through the slots. An eigenvalue solver was used to find the TE_{411} mode eigenfrequency and Q for this system (absent slots); the results were 11.577 GHz and 183. This frequency is noted to be 133 MHz displaced from the 4th harmonic 11.424 GHz; 133 MHz is well in excess of half the cavity response width, namely 32 MHz, and much larger than the 400 kHz natural line width of the resonance (see Fig. 3). As a result, there is little reason to expect cavity excitations at any of the harmonics of the CARA drive frequency in this experiment. Nevertheless, dramatic sharp spectral features were observed at harmonics 2 through 6, i.e, near 5.7, 8.5, 11.4, 14.3 and 17.1 GHz. These features were 15-40 db above background radiation, although the frequency response of the receiving antenna, transmission waveguide and spectrum analyzer have not been calibrated. Small adjustments in the guide magnetic field profile near the injection gun had a significant influence on the sharp features; a much broader emission spectrum resulted from detuning of the field profile. Radiation near 2.856 GHz cannot propagate through the ~20 m transmission waveguide, which is cutoff below 3.5 GHz.

Examples of the spectral distribution of 4th harmonic radiation observed is shown in Fig. 3. Two spectra are shown, corresponding to two frequencies of the XK-5 klystron that energized CARA, with a separation of 1.000 MHz. The spectra are vertically displaced arbitrarily in the figure for clarity. Each spectrum is obtained by accumulating data from about 2,500 pulses. It is seen that the radiation peaks are ~25 db above the background, and that the peaks for the two spectra are separated by 4.00 MHz, as would be expected for radiation precisely at the 4th harmonic of the drive frequency. Small differences that are evident between the spectra may be attributable to a ~1 db variation in the klystron output power as the drive frequency was changed. The quasi-periodic variations on the spectra are probably due to reflections in the transmission system. A detailed measurement of the FWHM for the spectra gives values of 380 kHz for the bottom case and 406 kHz for the upper case; both of these values compare closely with the Fourier limit FWHM of 442 kHz for a 2 μsec rectangular pulse. Similar spectra are found at the 2nd, 3rd, 5th and 6th harmonics.

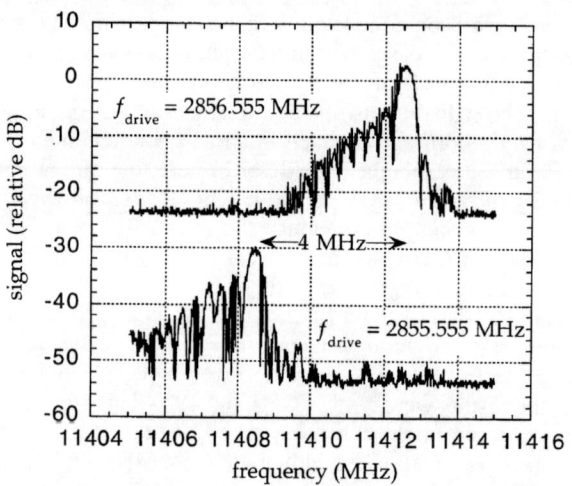

FIGURE 3. 4th harmonic superradiant emission from CARA beam.

These unusual emission spectra do not appear to be associated with resonances of the output structure, as is evident by the harmonic relationship of each feature to the klystron frequency, and by the frequency tracking described above. The tentative explanation offered here is that superradiant emission from the beam is being observed which, because of the apparent high degree of phase coherence between the gyrating particles, can reach the high power levels necessary for detection in this experiment. If so, this would be the first observation of superradiant emission up to the 6th harmonic from a phase coherent gyrating beam. It can be expected that adjustment of the resonance frequency of the TE_{411} cavity down from 11.577 GHz will allow MW-level 4th harmonic power to be coupled from the beam.

SEVENTH-HARMONIC CO-GENERATION

It has not generally been noted that certain $TE_{s\ell}$ electromagnetic modes of a cylindrical waveguide at frequency $s\omega$ exhibit coincidental near matching between their group velocities and those of the TE_{11} mode at frequency ω. This basic characteristic of the simplest closed guided wave structure may allow the excitation of strong s-th harmonic radiation when a fundamental injected wave interacts with a nonlinear medium in the waveguide. Group velocity matching can allow classical parametric interactions to occur [17] but, as will be shown here, it may also allow coupling in which the nonlinear medium supplies and receives significant power to and from the waves. The nonlinear medium for the work presented here is a relativistic gyrating electron beam; the fundamental TE_{11} mode excites cyclotron autoresonance acceleration (CARA) of the beam [2]; and the lowest mode for which a near match in group velocity obtains is the $TE_{7,2}$ at the seventh harmonic. It will be shown that high efficiency production of multi-megawatt power at 20

GHz is possible through this "co-generation" mechanism, when a beam is energized using 2.856 GHz power. Other high power microwave TE-mode fast-wave harmonic interactions may also be enhanced by this matching phenomenon, including the gyro traveling-wave amplifier, gyroklystron and waveguide free-electron laser [18]. Similar matching phenomema for TM-modes may enhance other types of harmonic interactions, such as in the frequency-doubling magnicon [5].

The term "co-generation" is used here to signify that harmonic power is generated in the self-same structure where CARA acceleration (and deceleration) of the beam occurs. Since the interaction to be discussed occurs in a traveling-wave structure, injected rf power that does not become transformed into the harmonic wave can be recovered, recycled and perhaps used in a subsequent co-generator stage. Energy in the spent electron beam emerging from the interaction can also be recovered with high efficiency, since the energy spread on the beam induced in the acceleration and generation processes can be small. It is these unusual features that permit overall 7th harmonic co-generation efficiencies greater than 80% to be predicted, as is illustrated below.

The requirement for group velocity matching for efficient power transfer between two guided waves that interact with an electron beam near gyroresonance and its harmonics can be easily appreciated. Below are written the resonance conditions that must be satisfied for simultaneous acceleration due to the CARA mechanism, and for deceleration due to the generation of radiation near the s-th gyroharmonic, namely $\omega = \frac{\Omega}{\gamma} + ck_{z,11}\beta_z$, and $s\omega = \frac{s\Omega}{\gamma} + ck_{z,s\ell}\beta_z$. Here, ω is the radian frequency at which CARA is operated, $\Omega = eB_o/m$ is the rest gyrofrequency for electrons of mass m and charge e in a magnetic field B_o, γ and $c\beta_z$ are the relativistic energy factor and the axial velocity for the beam electrons, and $k_{z,s\ell} = (s\omega/c)(v_{g,s\ell}/c) = (s\omega/c)n_{s\ell}$ is the axial wavenumber, itself proportional to the normalized group velocity $n_{s\ell}$ for each wave; for CARA, $s = \ell = 1$. Clearly, these cannot be satisfield simultaneously unless $\frac{k_{z,s\ell}}{s} = k_{z,11}$ or $n_{s\ell} = n_{11}$ namely equal group velocities for the two modes, when the frequency of the $TE_{s\ell}$ mode is s times that of the TE_{11} mode. This is consistent with the selection rule for gyroharmonic conversion with an axis-encircling beam in a cylindrical waveguide, namely that power at the s-th harmonic can flow cumulatively from the beam only into $TE_{s\ell}$ modes [6]. It can be shown that group velocity matching to within about 1% occurs for the $TE_{7,2}$, $TE_{13,3}$, $TE_{24,5}$ and $TE_{30,6}$ modes. This suggests that one might find conditions where a prescribed magnetic field profile allows efficient CARA acceleration, and simultaneously allows strong radiation at one or more of the 7th, 13th, 24th and 30th harmonics. Since succesively higher beam energies are required to obtain significant coupling as s increases, there is reason to expect that conditions might be found where only one of these modes will be excited at a time. In a rectangular waveguide, group velocity matching is much more prevalent, occuring (for example) between the TE_{01} and all TE_{0s} modes at the s-th harmonics. This situation could lead to serious mode competition at high harmonics; moreover, the aforementioned selection rule for axis-encircling beams does not apply for rectangular waveguides. Thus, only cylindrical waveguides are appropriate for co-generation.

Here we examine co-generation only at the 7th harmonic, but competing modes are also considered. We take the rf source frequency to be 2.856 GHz, corresponding to the frequency of the 65 MW klystrons (type 5045) that drive the Stanford Linear Collider (SLC), so that the 7th harmonic is at 19.992 GHz. A calculation could be made for any rf source frequency, but we purposely have chosen 2.856 GHz in anticipation of conducting experiments using an existing rf source, and in conceiving of the rf driver system for a future multi-TeV electron-positron collider being built using existing mature rf technology, including type 5045 klystrons. Thus to examine in detail the co-generation of harmonic radiation within CARA, particle simulation studies were carried out with rf and injected beam powers corresponding to Yale/Omega-P experimental capabilities described above, namely for an rf source power level at 2.856 GHz of 10.0 MW, and an injected beam of 100 kV, 25 A.

Results to illustrate the principle of co-generation are shown in Fig. 4 for an ideal "cold" beam, i.e., a beam with no initial axial momentum spread or guiding center spread. Fig. 4 shows the evolution along the axis in CARA of rf power in the TE_{11} and $TE_{7,2}$ modes at 2.856 GHz and 19.99 GHz in a waveguide of radius 3.65 cm (n_{11} = 0.5374 and $n_{7,2}$ = 0.5329); Fig. 4 also shows the variation in imposed resonant axial magnetic field. A pronounced recurrence phenomenon is observed, with rf power shifting back and forth between the fundamental and the 7th harmonic waves. At z = 84.0 cm, where the 7th harmonic power has reached 3.32 MW, the fundamental power is 6.22 MW, and the sum of the two is only 5% less than the 10.0 MW fun fundamental power injected initially. The beam energy at z = 84.0 cm is found to be 100.4 kV, so the beam is seen to have acted essentially as a catalyst by exchanging power with the rf modes up to this point, but returning nearly to its original state. At z = 167.2 cm, the fundamental rf power level has risen to 9.45 MW, and only 0.31 MW resides in the 7th harmonic wave; again the beam energy is back to 99.9 kV. In the course of this one full reccurence cycle, the magnetic field varies quasi-sinusoidally between 0.86 and 1.55 kG. The maximum total of 6th and 8th harmonic power is found to be 0.13 MW, the 13th harmonic power is found to reach 0.427 MW at z = 84.0 cm, and 0.949 MW at z = 251 cm. The fact that the $TE_{13,3}$ mode is the only serious competitor to the $TE_{7,2}$ enforces the claim made above that group velocity matching is a major factor in co-generation, since $n_{13,3}$ = 0.5438. More power appears at the 13th harmonic during the second recurrence cycle because of injection of finite power from the first cycle. Were the device to be terminated at z = 84.0 cm, rf power could be extracted with about one-third at the 7th harmonic and two-thirds at the fundamental; the beam power could in principle be largely recovered using a depressed collector. The 6.22 MW of extracted fundamental power could, in principle, be used (with an additional 3.78 MW) to drive a second CARA identical to the first, and the sequence continued indefinitely. Such a system would have a nearly ideal conversion efficiency, the net effect of which is to convert rf power from the fundamental to the 7th harmonic. Losses in this system arise from finite conductivity in the waveguide walls, from imperfect beam power recovery at a depressed collector, and from power converted to competing harmonics. But these can each be small fractions of the total 12.5 MW in circulation. Of course, an injected beam with zero axial velocity spread is an idealization that cannot be met in practice.

Results of a computation for an injected beam with a finite rms axial velocity spread of 0.10% are shown in Fig. 6, with the various quantities displayed as in Fig. 5. The basis for employing this value of axial velocity spread is a design study using the DEMEOS code [14] for the moderate convergence, perveance $K = 1.0 \times 10^{-6}$ A-V$^{-3/2}$ Pierce-type electron gun that is installed on the Yale/Omega-P CARA. Two significant additional features are present in this example, namely a detuning of the axial magnetic field from exact resonance, and a finite amplitude injected signal at the 7th harmonic [16]. Injection at the CARA input of 0.10 MW of 7th harmonic power in the $TE_{7,2}$ mode was found to yield a 7th harmonic power level at $z = 82$ cm of 1.99 MW, as shown in Fig. 3a. If the injected 7th harmonic power level is increased from 0.1 to 1.0 MW, the 7th harmonic output increases to 3.9 MW, for gain of an additional 1.0 MW. Injection can be accomplished by recirculating to the input a portion of the 7th harmonic output power.

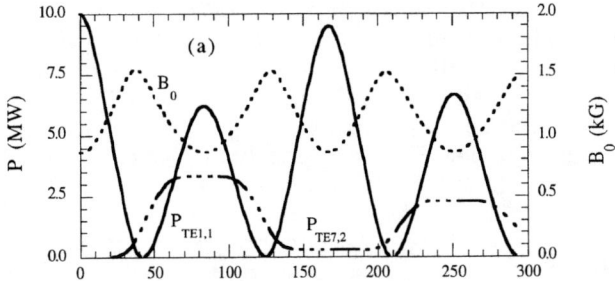

FIGURE 4. Parameters for 7th harmonic co-generation for an initially cold beam. Parameters are described in text.

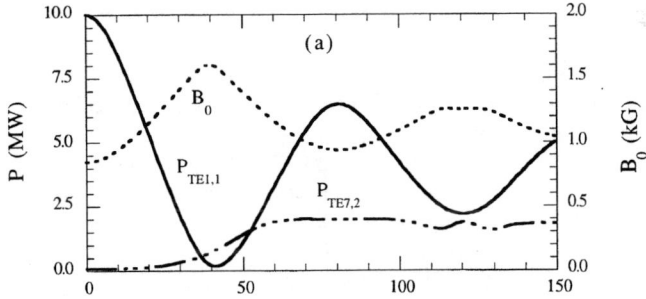

FIGURE 5. Same as Fig. 4, except for a beam with an initial rms axial velocity spread of 0.10%.

If a device based on the results shown in Fig. 5 were terminated at $z = 82$ cm, one could extract 1.99 MW of 7th harmonic power, 6.47 MW of fundamental harmonic power, and 0.098 MW of power in competing harmonics. The efficiency for utilization of rf power then follows as $\eta_{rf} = (1.99 + 6.47)/10.10 = 83.8\%$. If energy in the spent electron beam is included (3.86 MW), one would have $\eta_{total} = (8.46 + 3.86\,\eta_{rec})/12.6$, where η_{rec} is the efficiency for recovery of beam power using a depressed collector. One can define $\eta_{rec} = (\gamma_r - 1)/(\langle\gamma_s\rangle - 1)$ as an ideal recovery efficiency for a single-stage depressed collector, where $\gamma_r = 1 + e|V_r|/mc^2$ is the energy factor corresponding to the collector retarding potential V_r, and where $\langle\gamma_s\rangle$ is the average energy factor of the spent beam.
particles. To avoid undesirable beam reflection at the collector, γ_r is taken to be the lowest value of beam energy factor in the spent beam. For the beam at $z = 82$ cm in Fig. 5, computations have been performed that yield $\eta_{rec} = 63.0\%$. Using this gives $\eta_{total} = 86.4\%$. This figure embodies several idealizations, but it seems high enough to provide strong motivation for further study of co-generation.

DISCUSSION

Experiments on CARA show that gyration energy can be imparted to a high power electron beam with efficiency values that exceed those for other fast-wave interactions. A CARA efficiency of 96% has been measured, a value that exceeds that of any prior high power rf/beam interaction, and which also exceeds the efficiency of industrial rf accelerators.

Significant emission from energetic beams prepared in CARA has been observed, at the 2nd through 6th harmonics of the CARA drive frequency; this emission is tentatively identified as superradient emission, not previously observed from gyrating beams. Superradiant emission could not arise on a gyrating beam unless the particles maintained good phase coherence. A measure of the coherence time is given by the FWHM of the spectra, shown to be about 400 kHz in Fig. 3. This indicates that good coherence is maintained throughout the 2 μsec duration of the beam pulse. Since the data shown in Fig. 3 are accumulated over more than 2,500 shots (~40 minutes), it can also be concluded that good stability is possible with such a beam.

Gyroharmonic co-generation described here as a new radiation mechanism suggests that it may be not unreasonable to conceive of high power 20 GHz co-generators with a total power efficiency of greater than 80%. These could be driven using existing SLC klystrons and associated modulators. It does not seem unreasonable for each co-generator in such a system to provide 50-60 J output pulses at 20 GHz with an overall rf system efficiency, including the klystron drivers, that would exceed 50%. This system could constitute the sought-after rf source for a future multi-TeV electron-positron collider.

ACKNOWLEDGEMENTS

This research was sponsored in part by the US Department of Energy and in part by the US Office of Naval Research.

REFERENCES

1. Caryotakis, G, *IEEE Trans. Plasma Science* **22**, 756 (1994).
2. Haimson, J., Mecklenburg, B., and Danly, B. G., in *Pulsed RF Sources for Linear Colliders*, AIP Conf. Proc. **337**,146, R. C. Fernow, ed. (Amer. Inst. of Phys., New York, 1995).
3. Palmer, R. B., *et al*, ibid., p. 94; Wang, H., *et al*, ibid., p. 103.
4. Calame, J. P., *et al*, ibid., p. 195.
5. Nezhevenko, O. A., ibid., p. 172; Gold, S. H., Hafizi, B., and Fliflet, A. W., ibid., p. 184.
6. Ganguly, A. K., and Hirshfield, J. L., *Phys. Rev.* E **47**, 4364 (1993).
7. Nusinovich, G. S., and Granatstein, V. L., ibid., p. 16.
8. Wilson, P. B., ibid., p. 293.
9. Sprehn, D., Phillips, R. M., and Caryotakis, G., ibid., p. 43.
10. Hafizi, B., Sprangle, P., and Hirshfield, J. L., *Phys. Rev.* E **50**, 3077 (1994).
11. Wang C., and Hirshfield, J. L., *Phys. Rev.* E **51**, 2456 (1995).
12. LaPointe, M. A., Yoder, R. B., Wang, C., Ganguly, A. K., and Hirshfield, J. L., *Phys. Rev. Lett.* **76**, 2718 (1996).
13. Hirshfield, J. L., LaPointe, M. A., Ganguly, A. K., Yoder, R. B., and Wang, C., *Phys. Plasmas* **3**, 2163 (1996).
14. True, R. (private communication).
15. Hirshfield, J. L., Ganguly, A. K. and Wang, C., in *Pulsed RF Sources for Linear Colliders*, AIP Conf. Proc. **337**, 200; R. C. Fernow, ed. (Amer. Inst. of Phys., New York, 1995).
16. Wang, C., Hirshfield, J. L., and Ganguly, A. K., *Phys. Rev. Lett.* **77** (1996).
17. See, for example, Yariv, A., *Quantum Electronics* (New York: Wiley, 1967).
18. Liu, Y.-H., and Marshall, T. C., *Proc. 11th Intl. Conf. on High-Power ParticleBeams*, Prague, June 1996 (to be published).

A Racetrack Geometry to Avoid Undesirable Azimuthal Variations of the Electric Field Gradient in High Power Coupling Cavities for TW Structures[*]

J. Haimson, B. Mecklenburg and E. L. Wright

*Haimson Research Corporation
3350 Scott Blvd., Building 60, Santa Clara, CA 95054-3104*

Abstract. The high levels of peak RF input power required for linear collider TW accelerator structures, and the need to prevent undesirable transverse momentum contributions being imparted to the beam during traversal of the linac coupler cavities, have led to the use of symmetric dual feed cylindrical cavities having diametrically opposed side wall coupling apertures. Analyses of these cavities have shown that the coupling apertures in the wall of the cavity cause azimuthal variations of the longitudinal electric field that can enhance the maximum surface gradient (and secondary electron emission) and reduce the peak power rating of the structure by 10 to 20 percent. These field variations also produce transverse gradients of longitudinal electric field amplitude and phase that can seriously degrade the beam emittance due to quadrupolar transverse momentum contributions. The racetrack shaped coupler cavity, described herein, avoids many of these undesirable features while still allowing the correct cavity match and tuning condition to be readily achieved.

INTRODUCTION

Previous studies of single feed RF linac coupler cavities have examined the effects of field asymmetries introduced by the large coupling aperture in the sidewall of the cavity, and have shown that a transverse gradient of the axial electric field amplitude causes transverse spreading of a bunch located at the crest of the accelerating wave and that a phase variation in the transverse plane causes a net deflection of the accelerating bunch [1,2,3]. These transverse momentum contributions have been partially compensated in the past by using various design offset couplers, by arranging the input and output coupler feeds on constant gradient accelerator sections to be on the same side of the structure, and by feeding successive linac sections from alternate sides [3].

---

[*] Work supported by U.S. Department of Energy SBIR Grant No. DE-FG03-93ER81487.

For phase velocity of light, S-band linac coupler cavities having the same diameter beam aperture (2a) in both the cavity end wall and the contiguously located disc, the transverse gradients of the axial electric field amplitude and phase are typically 3 to 4% per cm and 1° per cm, respectively, for single feed (uncompensated) couplers. It is possible to reduce the field amplitude transverse gradient by two orders of magnitude by offsetting the center of the cylindrical cavity a distance Ω from the beam centerline, on the side remote from the coupling aperture; but the undesirable phase asymmetry remains essentially unaltered. (For the DESY Hamburg, NIKHEF Amsterdam and MIT Middleton linacs [4], the $2a/\lambda_0$ values ranged from 0.20 to 0.29, and the associated Ω/λ_0 values varied from .013 to .022.)

A desire to avoid the phase transverse gradient while maintaining amplitude symmetry has led to the use of dual feed cavities having diametrically opposed coupling apertures. This is especially important for high gradient accelerator applications since the transverse momentum imparted to the electron bunch due to phase asymmetry in the transverse plane is directly proportional to the product of this asymmetry and the accelerating gradient.

The microwave characteristics of 11.4 GHz symmetric double input couplers having cylindrical cavities have recently been evaluated [5] using the MAFIA code [6]. These studies confirmed that double input couplers can successfully avoid bipolar asymmetries, by preventing the presence of transverse RF field gradients at the beam axis, and also revealed that quadrupolar field asymmetries, introduced by the dual feeds, resulted in an azimuthal variation and enhancement of the electric field amplitude.

With the need to establish an input coupler design for the high gradient 17 GHz linac project [7], the opportunity was taken to investigate means of preventing electric field enhancement in dual feed coupler cavities and to minimize the quadrupolar transverse momentum contributions experienced by the beam during traversal of the coupler cavity.

HIGH GRADIENT 17 GHz TW LINAC STRUCTURE

The waveguide section for the 17 GHz linac comprises a 94 cell, $2\pi/3$ mode structure designed to have a 200 mA loaded beam energy of 25 MeV with an RF input power of 20 MW. The injected beam phase orbit studies were conducted in parallel with the structure design work to ensure convergence of parameters so that the impedance required to ensure the correct field characteristics for electron capture and bunching at the input of the structure would be consistent with the subsequent variation of group velocity and impedance necessary to satisfy the desired quasi-constant gradient condition. The investigations were based on two different injection energy regimes, namely, 400 to 600 keV, and approximately

2 MeV, so that beam injection using either a pulsed HV or an RF electron gun could be evaluated.

The phase orbit studies established the injection and acceptance characteristics for the 17 GHz high gradient TW structure. These studies also confirmed that the RF fringe field at entry to the structure, and the presence of a standing wave (SW) domain in the input coupler [8], had a dominating influence on both the capture and bunching process, and on the asymptotic bunch location of the accelerated beam.

A combination of high injection voltage, short operating wavelength and high accelerating gradient enabled the desired field characteristics to be achieved without the conventional use of reduced phase velocity cavities. The input section of the accelerator waveguide comprises a dual feed input coupler followed by several uniform impedance, phase velocity of light cavities designed to provide a maximum instantaneous field (at the midplane of the cavities) of 70 MV/m, and an average accelerating gradient of 54 MV/m (amplitude of the synchronous component of the total field). To achieve this gradient with an RF input power level of 20 MW, the product of the circuit loss per unit length and the TW shunt impedance per unit length must equal 72 Np.MΩ/m^2, and this requirement uniquely defined the diameter of the beam aperture ($2a_0$) at the input of the $2\pi/3$ mode waveguide structure. After applying an empirically determined correction [9] to allow for the enhanced skin losses anticipated at 17 GHz, a final value of $2a_0 = 0.325\lambda_0$ was established for the input section of the accelerator waveguide. (The beam aperture diameter chosen for the output section of the accelerator waveguide was $0.2468\lambda_0$ [7].)

EVALUATION OF DUAL FEED INPUT COUPLERS

Having determined the diameter of the beam aperture, the input section of the 17.136 GHz linac structure was modeled using a four cell TW circuit comprising two identically matched and tuned dual feed coupler cavities and two, phase velocity of light, $2\pi/3$ mode accelerator cavities, as illustrated in Figure 1. The longitudinal (z) dimensions were maintained constant for each of the cavities such that $t = (\lambda_0/3)$ - d and d = 0.2777 λ_0; and the cavity transverse dimensions were chosen to establish the correct phase advance for the TW circuit, i.e., a total of 360°, comprising zero phase shift in the drift tube terminated SW region of the matched coupler cavities, 60° through the remaining region of the coupler cavities, and 120° through each of the accelerator cavities. [This phase distribution is discussed further in a later section; for example, refer to Figure 8.]

The transverse distribution of the longitudinal electric field was investigated for a variety of dual feed coupler configurations by evaluating the field amplitude and phase gradients at three contiguous cross sections through the coupler cavity, as defined by the (xy) planes shown in Figure 1. The four cell structure is shown

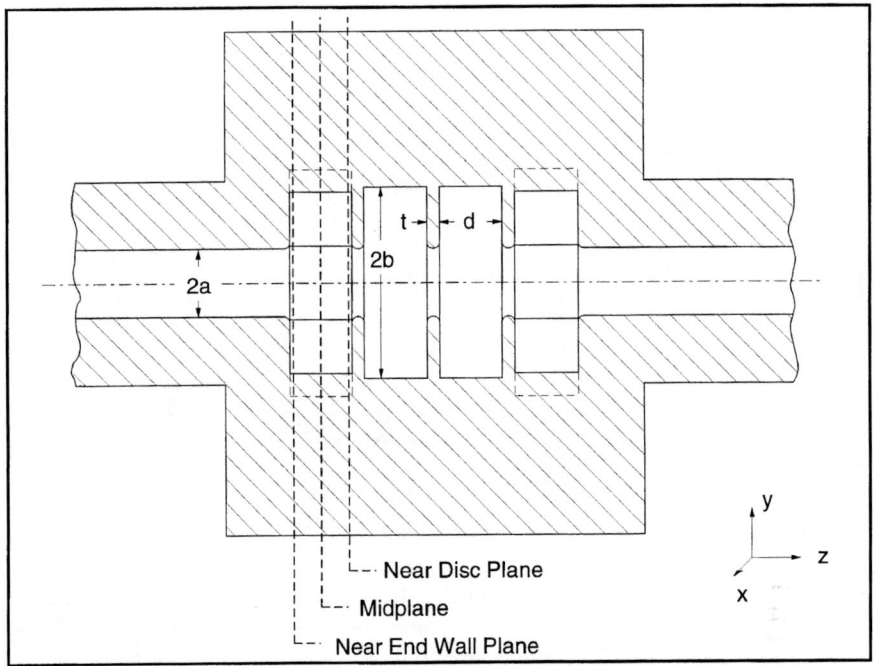

FIGURE 1. Geometry of the modeled 17 GHz, 2π/3 mode structure showing the transverse planes at which the field asymmetries were evaluated when using accurately matched and tuned, dual feed input and output coupler cavities and flat terminating loads.

oriented with the rectangular waveguide feeds entering and exiting the end cells along the x coordinate, and the side wall coupling apertures are shown in the (yz) plane.

The cross-sectional views in Figure 2 show the geometric features of the dual feed, cylindrical and racetrack shaped cavity couplers that were studied during this investigation. The racetrack cavity geometry was formed by generating two semi-cylindrical shaped walls with radii R having centers offset by a distance Ω on diametrically opposite sides of the cavity centerline, as indicated in Figure 2 (b); and the semi-cylindrical sections of the cavity were connected with straight parallel sidewalls of length 2Ω. The coupling apertures between the cavity and the rectangular waveguide feeds were cut through the elongated sides of the racetrack cavity, and since these apertures extended for the full length (d) of the cavity and had widths (A_w) considerably larger than 2Ω, nothing remained of the original straight sidewalls, as indicated in Figure 2 (b).

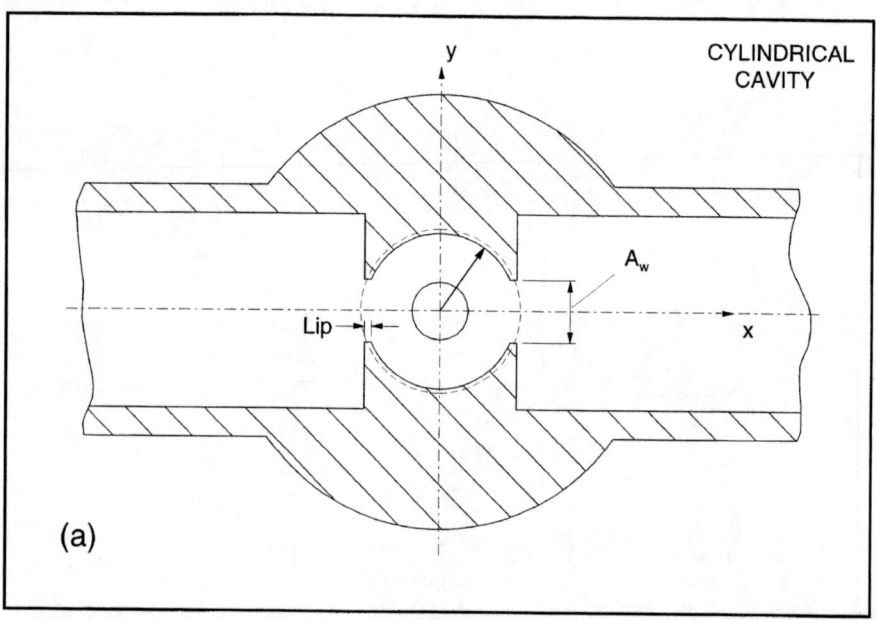

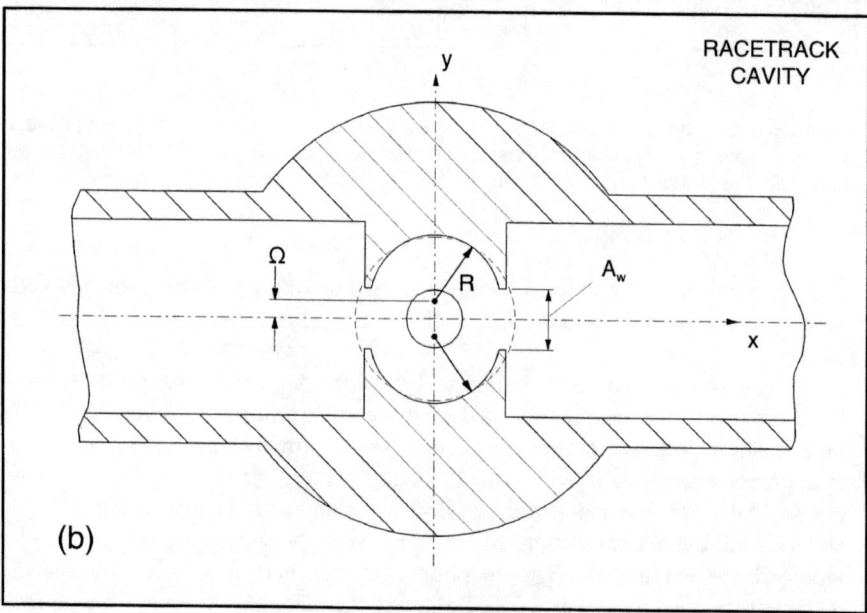

FIGURE 2. Comparison of cylindrical (conventional) and racetrack (compensated) dual feed coupler cavities for high power traveling wave structures.

The results of MAFIA simulations are shown in the Figure 3 comparison of the electric field transverse distributions for dual feed cylindrical and racetrack shaped coupling cavities. These results were obtained using matched and tuned cavities having identical dimensions for 2a, d, t and the lip thickness of the coupling apertures.

For the dual feed cylindrical cavity, the correct tune and match were readily achieved by iterative adjustment of the cavity diameter and the coupling apertures (A_w); but the coupler fields were not rotationally symmetric about the axis, as indicated by the divergence of the total electric field distributions, E(x) and E(y), shown in Figure 3 (a) for two transverse planes through the cavity. The field plots have been normalized to the on-axis value at the cavity midplane and at the peak of the accelerating field.

The parameter, Ω, defining the racetrack geometry [refer Figure 2 (b)] provides a means of minimizing or eliminating the above field asymmetry because adjustment of the geometric ratio $R/(R + \Omega)$ enables the E(x) and E(y) distributions to be equalized (within the cylindrical volume bounded by the beam apertures) while still retaining the capability of readily matching and tuning the coupler by adjustment of A_w and the transverse dimensions of the cavity; i.e., the cavity tune is dependent on the absolute values of R and Ω, while the field symmetry $E(x)/E(y) = 1$ is dependent on the factor $R/(R + \Omega)$. The near-symmetrized field distributions of the dual feed racetrack coupler cavity are shown plotted in Figure 3 (b), for $R/(R + \Omega) = 0.790$.

The transverse distributions of the longitudinal component of the electric field, $E_z(x)$ and $E_z(y)$, for the cylindrical cavity dual feed coupler are shown expanded in Figure 4 for all three planes identified in Figure 1. These distributions show the field enhancement caused by the x-y asymmetry, and indicate that the maximum gradient occurs on the surface of the disc in radial alignment with the sidewall coupling apertures (along the x coordinate at y = 0) and in close proximity to the beam aperture iris. The plots also indicate that the field asymmetry is present in all three transverse planes, and that the beam will experience quadrupolar forces during the full traversal of the coupler cavity. The resulting transverse momentum impulses will be proportional to $\int_0^d \partial E_z/\partial x, \partial E_z/\partial y \, dz$, and sinusoidally dependent on the phase location of the particle with respect to the crest of the TW field; i.e., particles ahead and behind the crest will experience quadrupolar forces acting in opposite directions, and the magnitude of these focusing and defocusing forces will be dependent on the bunch width.

Also, the transverse distributions of phase, $\Theta(x)$ and $\Theta(y)$, shown plotted in Figure 4 at the cavity midplane, indicate the presence of phase gradients in each quadrant of the beam cross section; and this will result in additional quadrupolar transverse momentum contributions being imparted to the beam. These phase gradient related contributions will be proportional to $E \int_0^d \partial\Theta/\partial x, \partial\Theta/\partial y \, dz$, and cosinusoidally dependent on the particle phase position with respect to the crest of the TW accelerating field.

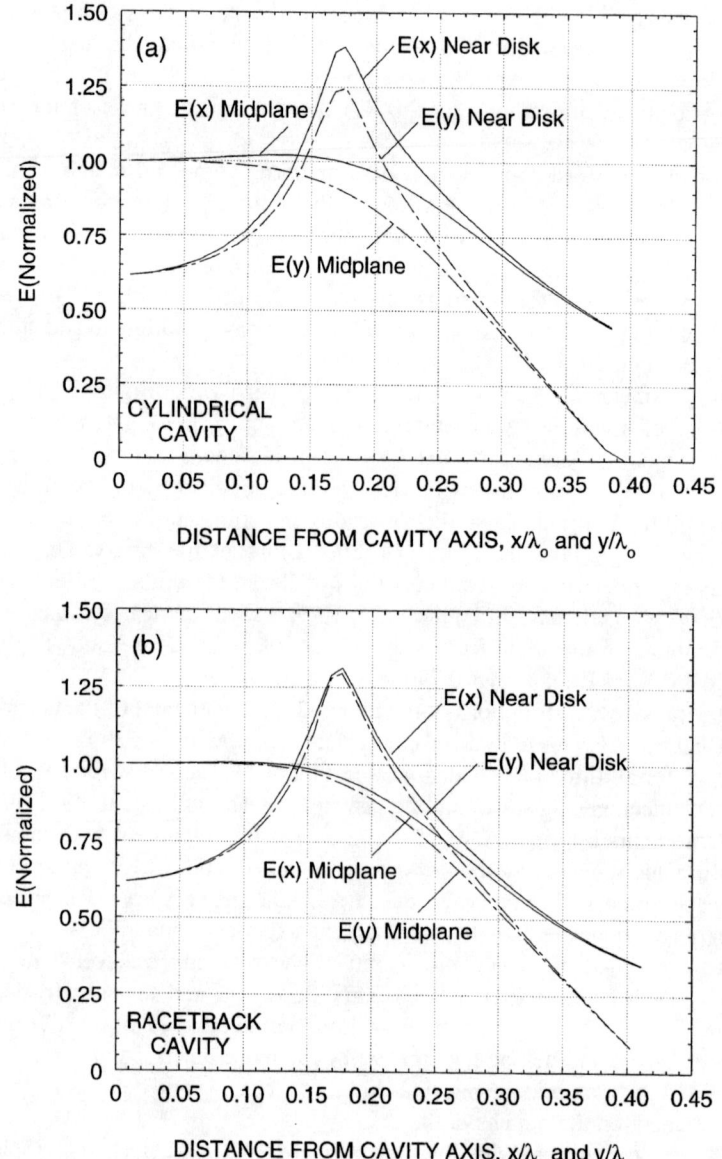

FIGURE 3. Comparison of transverse variation of total electric field in cylindrical and racetrack dual feed coupling cavities.

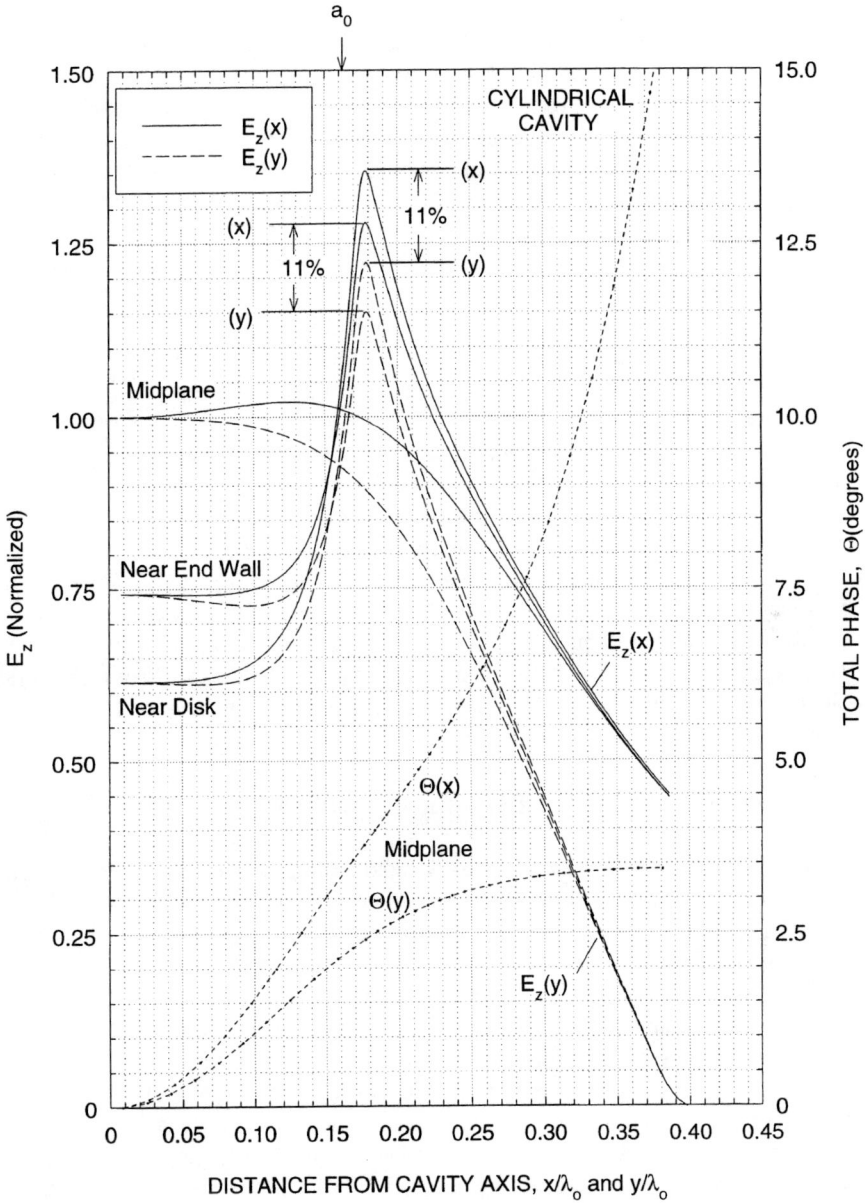

FIGURE 4. Longitudinal electric field distributions, $E_z(x)$ and $E_z(y)$, in three parallel transverse planes through the cylindrical cavity dual feed coupler, showing considerable amplitude variation between orthogonal components in each of the three transverse planes.

The longitudinal electric field distributions for the 17 GHz dual feed racetrack shaped coupler cavity, with the transverse dimensions optimized to more accurately symmetrize the fields [using $R/(R + \Omega) = 0.772$] are shown plotted in Figure 5 at three transverse planes through the matched and tuned cavity. The equality of the $E_z(x)$ and $E_z(y)$ distributions, especially within the cylindrical volume bounded by the beam apertures, indicates the high degree of field symmetry that can be achieved with the racetrack geometry.

Expanded plots of the radial and azimuthal variations of the longitudinal electric field in the axial region of the cavity are compared in Figures 6 and 7, respectively, for both types of dual feed coupler design. For example, at a radius of $0.05\lambda_o$, Figure 7 shows that the field amplitude in the racetrack coupler cavity has an azimuthal variation of only 0.045%, a factor of approximately 20 less than that of the cylindrical cavity coupler.

By avoiding field asymmetries, the racetrack geometry offers a means of minimizing the growth of beam emittance and of preventing the enhancement of surface gradient. For the structure dimensions typically used in TW linac applications, the surface gradient variations due to field asymmetry in dual feed cylindrical cavity couplers is approximately 10% (refer Figure 4). However, this surface gradient asymmetry can exceed 25% in the large beam aperture, reduced phase velocity cavities typically used in TW output circuits for high power RF tubes [10]. Applications of this nature can derive considerable benefit from a racetrack shaped output coupler because of the substantial reduction in peak surface gradient.

REDUCTION TO PRACTICE AND FUTURE PLANS

The 17 GHz linac dual feed racetrack input coupler was cold test modeled prior to fabrication of the final coupler subassembly. The results obtained from the MAFIA simulations were found to be in good agreement with the RF cold test measurements, including the predicted sensitivity of the VSWR and the cavity phase shift to small changes in the coupling apertures.

A typical example of the 17 GHz dual feed coupler response to a small change in A_w is illustrated by the simulated steady state axial electric field and phase distributions in Figure 8. These distributions show that for the field symmetrized, dual feed racetrack coupler characterized in Figure 8 (a), increasing the coupling aperture dimension A_w by 0.1 mm (keeping all other parameters unchanged) will cause the VSWR to be increased from 1.01 to 1.40, and the cavity phase shift to be reduced by approximately 2°, as indicated in Figure 8 (b). The simulated distributions show that, for this geometry, the peak axial electric field and the TW phase origin are displaced approximately 3° from the coupler cavity midplane, and that a clearly defined phase dwell occurs at the midplanes of the $2\pi/3$ mode accelerator cavities. Figure 8(b) also indicates that for this over-coupled condition,

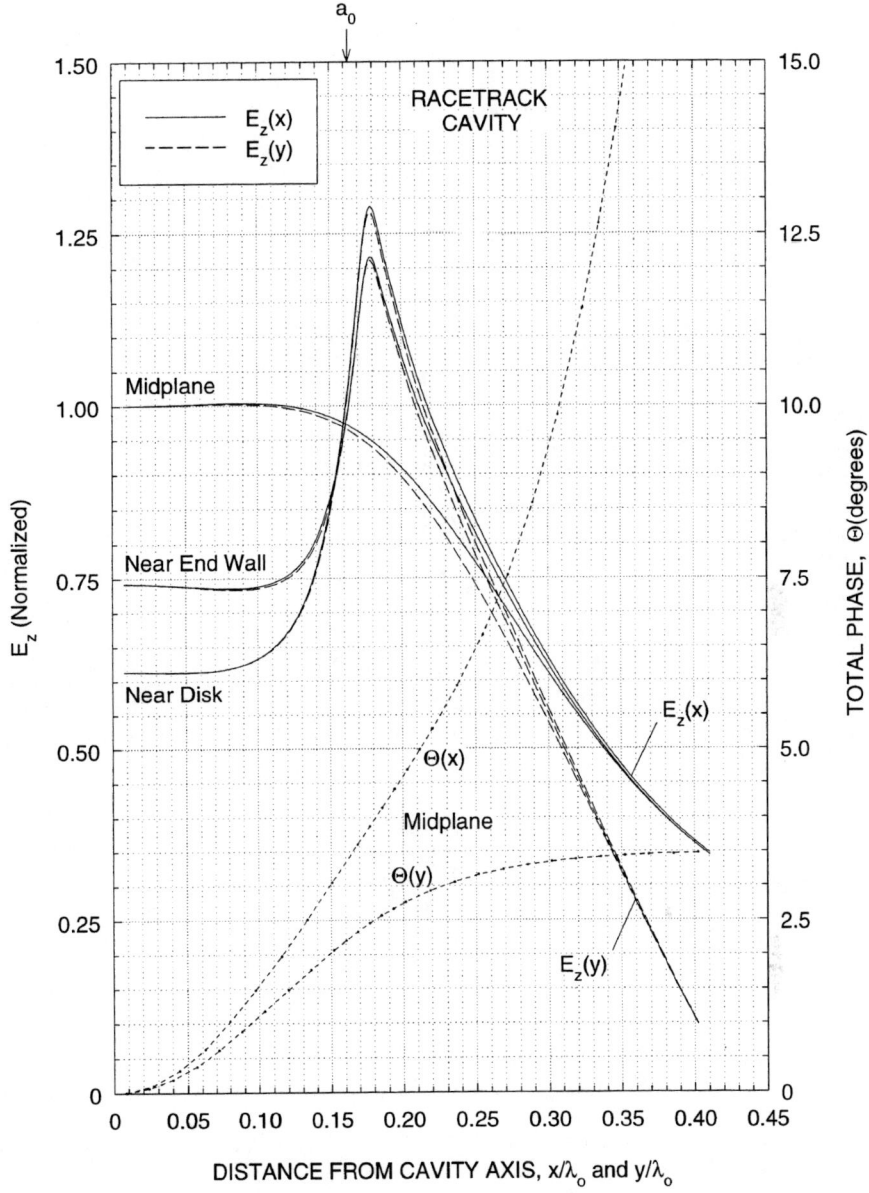

FIGURE 5. Longitudinal electric field distributions, $E_z(x)$ and $E_z(y)$, in three parallel transverse planes through the racetrack cavity dual feed coupler, showing the orthogonal components symmetrized in all three transverse planes.

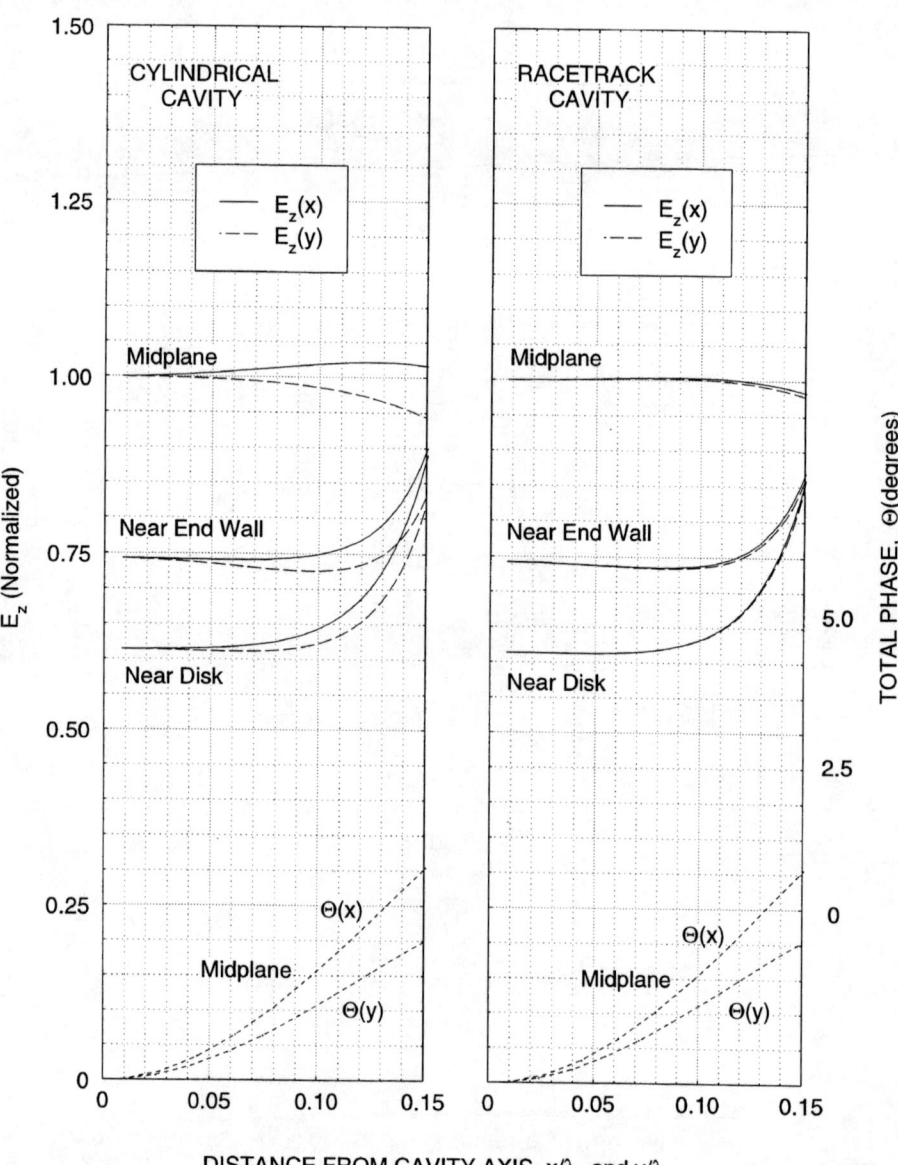

FIGURE 6. Near axis field distributions of the dual feed coupler cavities comparing the divergent $E_z(x)$ and $E_z(y)$ distributions of the cylindrical cavity (azimuthal asymmetry) with the symmetrized distributions given by the racetrack cavity.

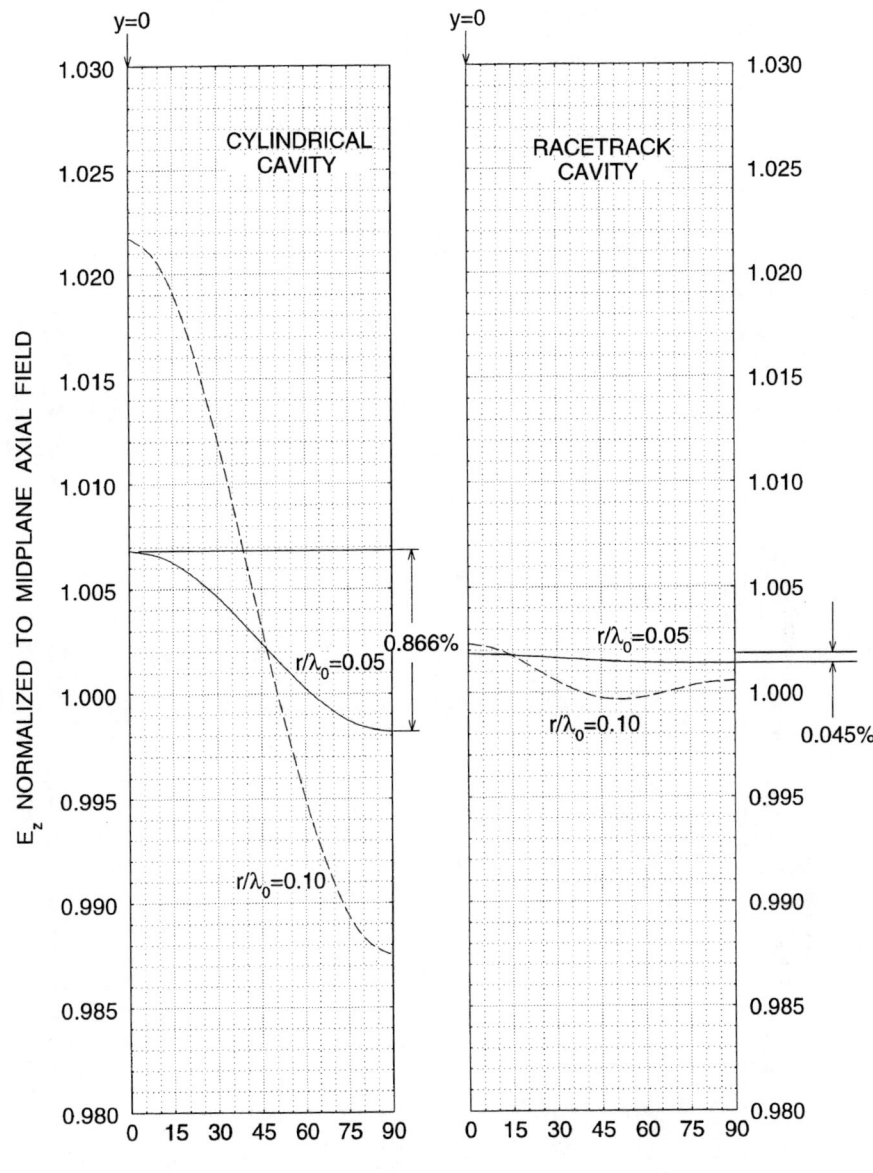

FIGURE 7. Azimuthal variation of normalized longitudinal electric field in the midplane of cylindrical and racetrack dual feed coupler cavities, at constant radii of $0.05\lambda_0$ and $0.10\lambda_0$.

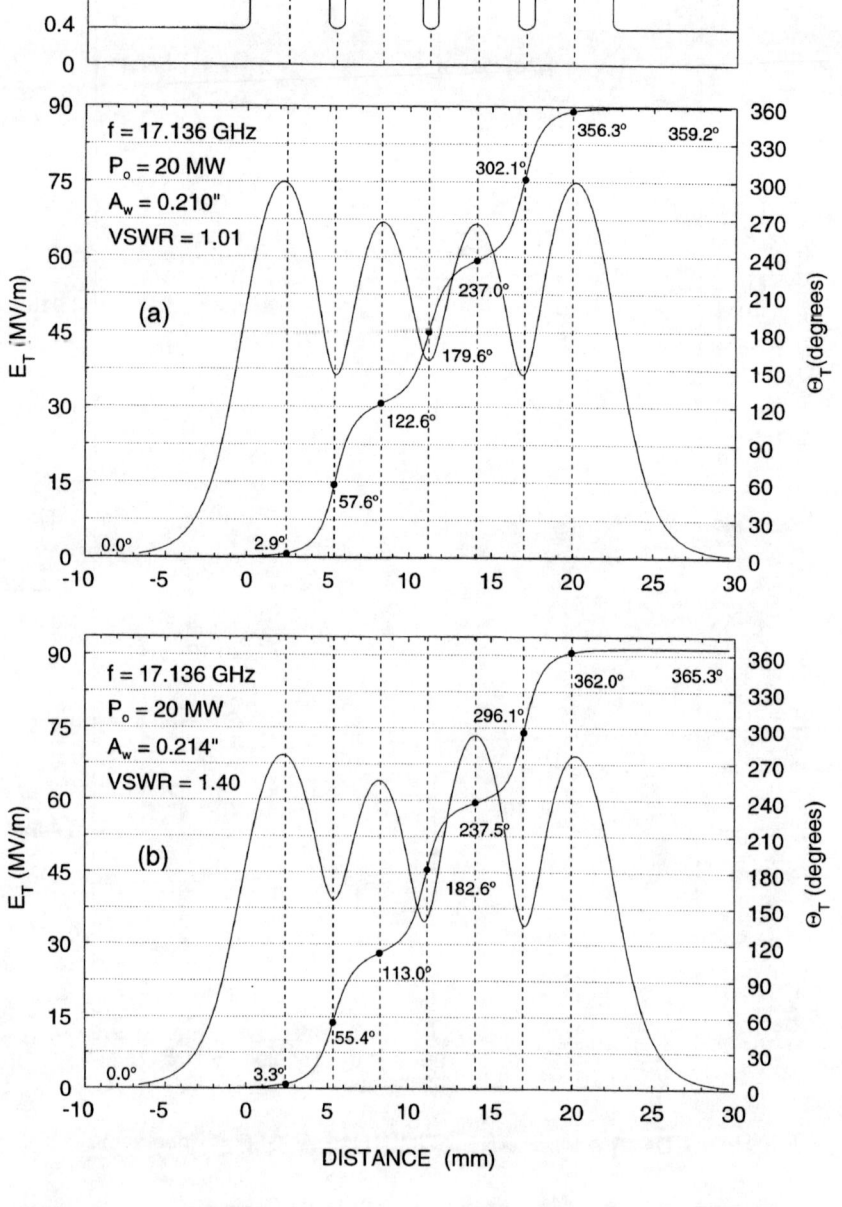

FIGURE 8. Comparison of steady state axial electric field and phase distributions for a 4-cell 2π/3 mode circuit having (a) critically coupled and (b) over coupled dual feed racetrack terminating cavities.

the field envelope becomes distorted, and the coupler peak fields are reduced by approximately 10%.

Future plans include (a) the design and fabrication of a large beam aperture, dual feed racetrack coupler to be incorporated in a new TW output structure for a high peak power 17 GHz relativistic klystron and (b) the investigation of racetrack shaped coupling cavities to minimize beam emittance growth caused by field asymmetries in photo cathode RF gun structures.

ACKNOWLEDGMENTS

The authors wish to acknowledge C.-K. Ng, K. Ko and W. Herrmannsfeldt for their helpful suggestions and interest in this investigation.

REFERENCES

1. Helm, R. H., *A Note on Coupler Asymmetries in Long Linear Accelerators*, Report No. M–167, Stanford Linear Accelerator Center, Stanford University, Stanford, California, 1960.
2. Helm, R. H., *Effects of Stray Magnetic Fields and RF Coupler Asymmetry in the Two-Mile Accelerator with Sector Focusing*, Report No. SLAC-20, Stanford Linear Accelerator Center, Stanford University, Stanford, California, October 1963.
3. Loew, G. A. and Neal, R. B., eds. P. M. Lapostolle and A. L. Septier, *Linear Accelerators*, Amsterdam: North Holland Pub. Co., 1970, ch. B.1.1, pp. 97–99, 143–144.
4. Haimson, J., eds. P.M. Lapostolle and A. L. Septier, *Linear Accelerators*, Amsterdam: North Holland Pub. Co., 1970, ch. B.3.2, pp. 421–423.
5. Ng, C.-K. and Ko, K., "Numerical Simulation of Input and Output Couplers for Linear Accelerator Structures," presented at the Computational Accelerator Conference (CAP 93), Pleasanton, California, February, 1993.
6. The MAFIA Collaboration, F. Eberling et al., MAFIA User Guide, 1992.
7. Haimson, J. and Mecklenburg, B., "HV Injection Phase Orbit Characteristics for Sub-picosecond Bunch Operation with a High Gradient 17 GHz Linac," in *Proceedings of the 1995 IEEE Particle Accelerator Conference*, **95CH35843**, vol. **2**, 1995, pp. 755–757.
8. Haimson, J., "Electron Bunching in Traveling Wave Linear Accelerators," in *Nuclear Instruments & Methods*, **39**, 1966, pp. 13–34.
9. Haimson, J. and Mecklenburg, B., "Design and Construction of a 33 GHz Brazed Accelerator Waveguide for High Gradient Operation," in *Proceedings of the 1987 IEEE Particle Accelerator Conference*, **87CH2387-9**, vol. **3**, 1987, pp. 1928–1930.
10. Haimson, J., Mecklenburg, B. and B. G. Danly, "Initial Performance of a High Gain, High Efficiency 17 GHz Traveling Wave Relativistic Klystron for High Gradient Accelerator Research," in *Pulsed RF Sources for Linear Colliders*, ed. R.C. Fernow, AIP Conference Proceedings, **337**, New York: AIP Press, 1995, pp. 146–159.

The Results of the 7 GHz Pulsed Magnicon Investigation

O.A.Nezhevenko, E.V.Kozyrev, I.G.Makarov, A.A.Nikiforov,
G.N.Ostreiko, B.Z.Persov, G.V.Serdobintsev, V.V.Tarnetsky,
S.V.Shchelkunoff, V.P.Yakovlev, and I.A.Zapryagaev

Budker Institute of Nuclear Physics, 630090 Novosibirsk, Russia

Abstract. The paper presents the latest results of research of a new RF amplifier – magnicon, in which the beam is modulated by a deflecting RF magnetic field (no bunching at that). This device was invented and developed at Budker INP as a prototype of a microwave power source for linear electron-positron colliders. The theoretical and experimental studies of physical mechanisms of an output power and efficiency limitation, that allowed to develop the operational version of the magnicon, are described in the paper. In 1995, an output power of 30 MW at an efficiency of 35 % and gain of 55 dB was obtained during the tests of this magnicon version. At the end of 1995, improvements of the electron gun were made. The present RF fields non-symmetry in the output cavity will be eliminated in the version that is under way now. The tests of the magnicon version with the above improvements are scheduled to be started in 1996.

INTRODUCTION

The magnicon [1–4] belongs to a new class of microwave amplifiers — deflection-modulated devices. The first magnicon was built and tested in the 1980's in INP [2].

This paper presents the results of testing the advanced version of magnicon, which was described in detail at RF'94 workshop [5] and developed in INP as a prototype of the microwave power source for linear colliders.

A schematic diagram of the device is shown in Fig.1. The magnicon consists of the following basic units: an electron source, RF system, magnetic system and a collector. RF system consists of two parts: the deflecting system for beam modulation and the output cavity for conversion of the beam energy into the RF energy. A magnetic system provides a long-term interaction between beam electrons and RF fields in the cavities as well as beam focusing. The tube is an amplifier operating at frequency of 7 GHz in frequency-doubling mode.

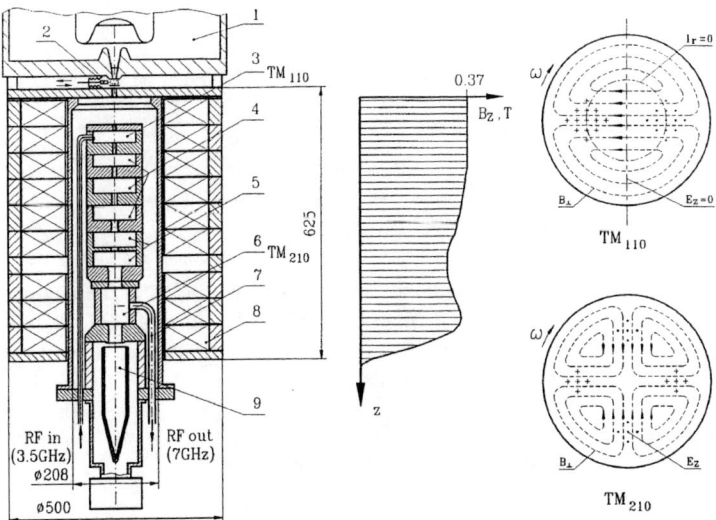

FIGURE 1. Schematic layout of the magnicon: 1 — electron source; 2 — vacuum valve; 3 — drive cavity; 4 — gain cavities; 5 — penultimate cavity; 6 — output cavity; 7 — waveguide (×2); 8 — solenoid; 9 — collector

In all deflection system cavities the circularly-polarized TM_{110} mode (Fig.1) oscillations are excited. The drive cavity 3 is excited by the drive generator. The passive (gain) cavities 4 and 5 are excited by a deflected beam. The penultimate cavity 5 consists of two coupled cavities in which the beam excites the opposite-phase (π-mode) oscillations, thereby enabling the realization of the deflection angle "summing" mode of operation [1]. This enables one to attain a deflection angle $\alpha > 50°$ at a cavity surface RF field not more than $E \approx 250$ kV/cm. In the output cavity 6 the modulated beam excites the TM_{210} mode (Fig.1) with a frequency of 7 GHz, which is two times higher than the drive frequency (3.5 GHz). All the cavities are located inside the solenoid 8, which produces a longitudinal magnetic field with the required field distribution. For an effective beam interaction in the output cavity it is necessary for the cyclotron frequency of electron rotation in the solenoid field to be 1.5–2 times higher than the drive frequency [2,4,7].

The output cavity is more than 8 cm long that provides the maximum surface electric field $E < 250$ kV/cm.

The magnicon design parameters are listed in Table 1.

The results of the beam behavior simulations in the process of deflection and deceleration are shown in Fig.2.

The results were obtained using the computer codes for both steady state and time dependent simulations, that were created by our team.

The physical model for simulations considers a beam of finite transverse size,

TABLE 1. Desingned parameters of the magnicon.

Operating frequency	7 GHz	Drive frequency	3.5 GHz
Output power	55 MW	Gain	53 dB
Pulse duration	1.5 μs	Beam voltage	420 kV
Repetition rate	5 pps	Beam current	240 A
Efficiency	56 %		

real space distribution of DC magnetic field and real RF fields of the cavities [8]. Those fields were calculated by SAM and SuperLANS2 codes [9,10]. We did not take into account space charge effects and finite beam emittance. The numerical model is based on macro particle methods.

EXPERIMENTAL STUDIES

1. The parameters obtained at the first tests of the described version are listed in Table 2.

The oscillograms presented in Fig.3(a) are: beam voltage (U), signal from the penultimate cavity (PC4), output signal (OUT1), and signal from the drive cavity (IN1).

The output signal (peak power) calibration was carried out by the calorimetric measurements of average RF power.

The measured dependence between the output power and drive signal (Fig.3b) is in quite good agreement with the simulation results.

The magnicon cavities consist of separated copper parts connected with one another by indium seals. This design allows to replace the RF system parts operatively but does not allow to bake-out the cavities up to high temperatures. This leads to long RF conditioning times for the cavities.

During the deflecting cavities conditioning the self-excitation was observed at different frequencies. After dismantling of the tube autographs of electric discharges were found almost in all the cavities. The discharges autographs indicate the self-excitation of various modes (symmetrical and non-symmetrical). However,

TABLE 2. Measured parameters of the magnicon.

Frequency	7.006 GHz	Drive frequency	3.503 GHz
Power	30 MW	Gain	55 dB
Pulse width	0.7 μs	Beam voltage	401 kV
Repetition rate	3 pps	Beam current	210 A
Efficiency	35 %		

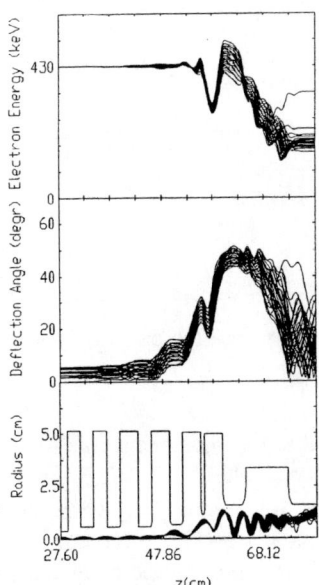

FIGURE 2. Simulation of the magnicon for a 3 mm diameter beam.

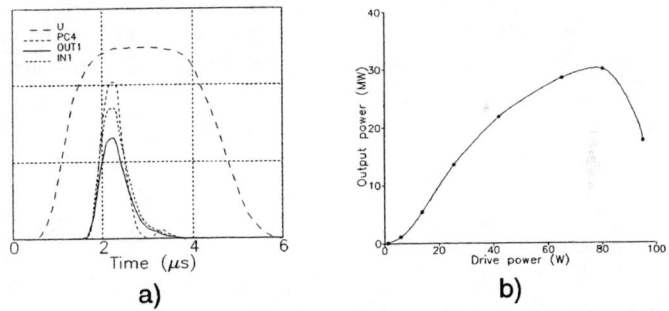

FIGURE 3. a) – the oscillograms; b) – the output power versus the drive signal.

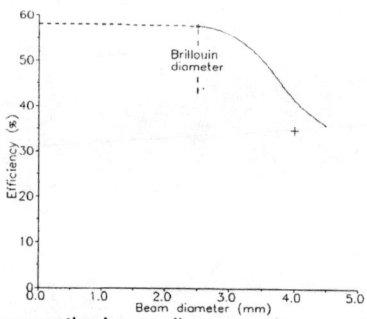

FIGURE 4. The efficiency versus the beam diameter; the experimental point is marked with +.

these self-excitations disappeared during conditioning and were not observed at the operating range of drive signals.

The main problems were concerned with conditioning of the output cavity, waveguides and loads, that are a single vacuum chamber (there are no ceramic windows). During conditioning self-excitation in the output cavity appeared and after a time disappeared at various frequencies (11.8 GHz, 5.92 GHz, and 12.04 GHz, one after another). The first two frequencies disappeared during conditioning, however the self-excitation at the frequency 12.04 GHz is still here and limits the pulse width at a power over 20 MW (at lower power the output signal duration is 1.5 μs). Results of simulation and measurements at atmosphere proved the presence of a resonance at this frequency. Oscillation presents also in the output cavity and penultimate cavity. The oscillation mode has a field distribution with a quadrupole nature.

2. The main causes of decreasing efficiency with respect to the designed value are a larger (4 mm) than the calculated one (2.8 mm) beam diameter, a non-optimal loaded Q-factor, and distortion of an electromagnetic field distribution in the output cavity.

For an effective magnicon operation the solenoid magnetic field should be about 0.37 T. At present time the electron gun designed for magnetic field value of 0.45 T is used [6]. This magnetic field decreasing by 20 % led to the beam diameter increasing up to d_{max}=4 mm. The calculated efficiency value versus beam diameter d_{max} is shown in Fig.4.

It is clear from Fig.4, that at d_{max}=4 mm efficiency cannot exceed 42 % and for the projected value of 56 % it is necessary to have the beam with d_{max} <3 mm. Moreover, the coupling between the output cavity and load was equal to 200 and was have been chosen optimal for the efficiency of 56 %. In the case of the beam with a large cross-size the efficiency decreases and it takes a higher loaded Q-factor value to obtain a maximal output power.

The efficiency obtained with the real beam is 0.8 from the calculated value. A new focusing electrode has been produced and installed to improve the situation, and the distance between the gun and solenoid has been increased to better match

Present design   Improved version

FIGURE 5. Orthogonal TM_{110} modes field maps in the output cavity: 1 — waveguides, 2 — protrusions.

the beam into the magnetic field. During testing with the accompanying magnetic field of 0.37 T the maximal measured beam diameter less than $d_{max} < 3$ mm was obtained.

Another cause leading to decreasing efficiency is the major RF fields distribution distortion in the output cavity due to presence of coupling apertures with waveguides.

The field maps (2D-simulation [10]) for orthogonal TM_{110} modes, superposition of which defines the RF fields distribution in output cavity, are presented in Fig.5. For compensation of the coupling apertures with the waveguides 1 effect there are two protrusions 2 in the present design output cavity, however, their effect is inadequate. One can see that field distributions of these two modes differ sufficiently. The loaded Q-factors of these modes also somewhat differ (180 and 220). As a result the interaction with the beam is found to be irregular along the azimuth that leads to the decreasing efficiency.

This problem can be solved by increasing the number of protrusions (see the field map on Fig.5) [1].

However, the simulation and model testing bring out, that this method of RF field azimuth nonuniformity eliminating leads to increasing of the longitudinal field disturbance as compared with the cavity without holes. Results of 2D-simulations for several special cases are presented in Fig.6. Presence of holes is taken into account by introducing the equivalent local increasing of the cavity diameter which has been chosen so that calculated TM_{210} mode frequency is equal to the measured value.

Compensating cavity diameter increasings near the upper and lower cavity faces have been made to eliminate RF disturbanse and do not give rise to azimuthal in-

---

[1] The measured loaded Q-factors for the modes in a cavity model are equal within the measuring error limits

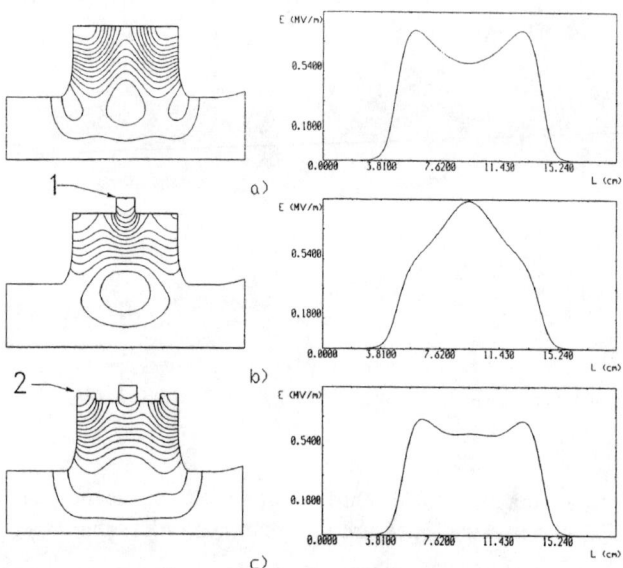

FIGURE 6. Magnetic field map and longitudinal electric field component versus attitude: an output cavity a) without holes; b) with coupling windows and protrusions (simulated by the equivalent local increasing of the cavity diameter 1); c) with the additional local cavity diameter increasing 2 near faces.

homogeneity because they are homogeneous along the azimuth.

3. It is traditional (beginning with a gyrocon) for the oscillations with circular polarization obtaining that the deflection cavity is driven by two signals of equal amplitude through two power inputs separated in azimuth by 90° [1,11]. These signals must also be shifted in phase by 90°. In magnicon the beam is magnetized and its gyrotropic properties lead to the circular deflection "self-stabilization" effect, i.e. if oscillations with an elliptical polarization are excited in the cavity, the ellipticity is reduced in the presence of the beam [1,2]. The experimental tests have verified that in the present magnicon version this wholesome effect shows itself so strongly that one can drive the deflection cavity by one signal (like a klystron) without a loss in output power and efficiency.

SUMMARY

During investigations many effects interfering with normal device operating were revealed [5,12,13] and methods to eliminate them were found. The improved version of the output cavity has been designed and is manufactured now. Moreover, provision is made for the addition vacuum pumping of waveguides and matched waveguide loads in the further experiments. It will allow to decrease the device conditioning time significantly. All these improvements should allow to achieve the designed parameters of the magnicon.

REFERENCES

1. O.A. Nezhevenko, *IEEE Trans. of Plasma Science* **22**, No. 5, 756–772 (1983).
2. M.M. Karliner et al., *NIM-A* **A 269**, No. 3, 459–473 (1988).
3. V.E. Akimov et al., "High Power X-Band Pulse Magnicon", in *Proceedings of Europ. Part. Acc. Conf.*, Nice, 1990, vol. 1, pp. 1000-1002, World Scientific, 1992.
4. O.A. Nezhevenko, "The Magnicon: a New Power Source for Accelerators", in *Proceedings of IEEE Part. Accel. Conf.*, San Francisco, 1991, vol. 5, pp. 2933–2942.
5. O. Nezhyagenko et al., "7 GHz Pulsed Magnicon", in *AIP Conf. Proc. 337 (RF94 Workshop)*, Montauk, NY, 1994, pp. 174–183.
6. Y.V. Baryshev et al., *NIM-A* **A 340**, 241–258 (1994).
7. M.M. Karliner et al., *"An Approximate Theory of the Magnicon"*, Preprint INP 88–66, Novosibirsk, 1988 (in Russian).
8. V. Yakovlev et al., "Numerical Simulation of Magnicon Amplifier", in *Proceedings. of IEEE Part. Accel. Conf.*, Dallas, 1995, (to be published).
9. B. Fomel, M. Tiunov, and V. Yakovlev, "Computer-Aided Electron Gun Design," in *Proceedings of XIII Int. Conf. on High-Energy Acc.*, vol. 1, 1987, pp. 353–355.
10. D. Myakishev, and V. Yakovlev, "The New Possibilities of SuperLANS Code," in *Proceedings. of IEEE Part. Accel. Conf.*, Dallas, 1995, (to be published).
11. M.M. Karliner et al., *"Circular Deflection of Electron Beam in the Gyrocon"*, Preprint INP 82–147, Novosibirsk, 1982 (in Russian).
12. O.A. Nezhevenko et al., "First Test of the X-Band Pulsed Magnicon", in *Proceedings. of IEEE Part. Acc. Conf.*, Piscataway, NJ, vol. 4, 1993, pp. 2650–2652.
13. I. Zapryagaev et al., "Status of the X-Band Pulsed Magnicon", in *Proceedings. of Europ. Part. Acc. Conf.*, vol. 3, 1994, pp. 1927–1929.

GROUP 7

"Laser Sources For Particle Acceleration"—Michael Downer (UT Austin) and Craig Siders (LANL)

CO_2 LASER TECHNOLOGY FOR ADVANCED PARTICLE ACCELERATORS

I.V. Pogorelsky[1], A. Van Steenbergen[1], R. Fernow[1], W.D. Kimura[2], S.V. Bulanov[3]

[1]*Accelerator Test Facility, Brookhaven National Laboratory, 725C, Upton, NY 11973*
[2]*STI Optronics, 2755 Nortup Way, Bellevue, WA 98004*
[3]*Plasma Physics Laboratory, General Physics Institute, Moscow, Russia*

Abstract. Short-pulse, high-power CO_2 lasers open new prospects for development of high-gradient laser-driven electron accelerators. The advantages of $\lambda=10$ µm CO_2 laser radiation over the more widely exploited solid state lasers with $\lambda \approx 1$ µm are based on a λ^2-proportional ponderomotive potential, λ-proportional phase slippage distance, and λ-proportional scaling of the laser accelerator structures. We show how a picosecond terawatt CO_2 laser that is under construction at the Brookhaven Accelerator Test Facility may benefit the ATF's experimental program of testing far-field, near-field, and plasma accelerator schemes.

I. INTRODUCTION

Lasers are the sources of the most intense electromagnetic radiation and strongest electric and magnetic fields available for laboratory research. For example, focusing of a terawatt laser beam into a 10-µm spot results in intensity of 10^{18} W/cm^2 and, associated with it, an electric field of 30 GV/cm that exceeds by five orders of magnitude fields attainable in conventional particle accelerators. Such capability stimulates a new high-energy physics discipline to emerge: laser-driven high-gradient particle accelerators. The primary practical goal is to find an alternative to conventional accelerators in order to build, in the future, more economical high-energy (~TeV) machines, or compact moderate-energy (~GeV) accelerators.

Presently, a variety of acceleration methods are under consideration and study. All laser accelerator schemes proposed so far may be split into three major categories: far electromagnetic (EM) field, near EM field, and plasma accelerators. How to define the border between the first two methods? EM field may be presented as a sum of propagating EM waves

$$\vec{E}(\vec{r},t) = \sum_j A_j \exp[i(\vec{k}_j \times \vec{r}_j - \omega_j t)]. \quad (1)$$

When all wave vectors $\vec{k}_j$ are real, we talk about far field accelerators. For these schemes it is essential that any distances from the source of the field or from any boundary surface are $\gg \lambda$. Otherwise, near field effects come into play.

Fields with imaginary $\bar{k}_j$ are called near fields. Actually, in this case, we talk about evanescent EM fields vanishing within a λ-thick layer above the surface.

In the third group of methods, particles are accelerated not by EM fields directly but by electrostatic fields due to the charge separation in laser-induced plasma waves. So far, the record of ~100 MeV over the 0.5 cm distance laser acceleration has been achieved using laser wakefield acceleration (LWFA) in plasma[1].

In general, particle acceleration by a fast-oscillating electromagnetic field becomes possible when a relativistic particle moves in synchronism with the phase of the driving field. The long wavelength of a CO_2 laser helps to meet this requirement. This feature is of particular importance for far-field acceleration schemes, examples of which are the Inverse Cherenkov Accelerator (ICA)[2,3] and Inverse Free Electron Laser (IFEL) accelerator[4,5].

Proposed near-field accelerator schemes, Grating Linac[6] and Resonant Accelerator[7], are based on the accelerating action of evanescent fields developed near the periodically shaped surfaces under laser irradiation. Since the spatial scale of such structures is comparable with the laser wavelength, these schemes look practical when a CO_2 or longer-wavelength laser is used as the accelerator driver.

Among known plasma-based laser acceleration techniques, self-modulated LWFA[8] looks presently the most promising. This method requires a so called "relativistically strong" laser beam that satisfies conditions for relativistic self-focusing in plasma. With the CO_2 laser, this condition may be satisfied at a 100 times lower plasma density than with a 1-μm laser of the same power. Another advantage of CO_2 lasers for plasma accelerators is based on a λ^2-proportional ponderomotive potential that controls the intensity of the laser-induced plasma wake. Simulations demonstrate a possibility to accelerate electrons by 250 MeV over a 4 cm distance using properly shaped 5 TW CO_2 laser pulses propagating inside a low- density plasma channel[9].

The approach to utilize the long-wavelength laser radiation for particle acceleration study is pursued at the ATF where the first terawatt picosecond CO_2 laser is under construction to study electron acceleration using far-field, near-field, and plasma acceleration schemes. We discuss here how all these schemes can benefit from using the CO_2 laser.

Description of the ATF, its current experiment program, and design of the present and upgraded laser system can be found elsewhere[10,11]. Table 1 summarizes the performance characteristics of the ATF primary components: photocathode RF linac and CO_2 laser system.

TABLE 1. ATF CO_2 Laser and e-Beam Parameters

CO_2 Laser (Present)	
Pulse Duration [ps]	100
Output Energy [J]	1
Output Peak Power [GW]	10
Repetition Rate [Hz]	0.1
CO_2 Laser (Upgrade)	
Pulse Duration [ps]	3
Output Energy [J]	~15
Output Peak Power [TW]	~5
Repetition Rate [Hz]	0.1
LINAC	
Bunch Duration FWHM [ps]	0.4-10
Electron Energy [MeV]	50
Peak Current [A]	170-50
Electron Energy Spread [%]	0.2
Normalized Emittance [mm.mrad]	0.5-2
Repetition Rate [Hz]	3

II. FAR FIELD ACCELERATORS

II.1. Inverse Cherenkov Accelerator

In any laser acceleration scheme, the key question is how to maintain synchronism between particles and oscillating electric fields over an appreciable distance. One of the possibilities is when the particle, traveling in medium with velocity βc, is intersected by the EM wavevector at the Cherenkov angle, θ, which is described by the condition

$$\cos\theta = \beta n^{-1}. \qquad (2)$$

Here, the inclination of the wavevector is responsible for developing a longitudinal accelerating field component, while the medium is chosen to produce retardation of the phase velocity of the wave to match the speed of electrons. Inverse Cherenkov acceleration is the only example of a first order, far field acceleration process where the particle interacts with a single EM wave.

In the first ICA demonstration at Stanford University[12], a linear polarized, focused Nd laser beam crossed the path of electron beam in the interaction cell filled with hydrogen. The observed energy shift was 50 keV over the 7 cm interaction length.

In a modified scheme[2] (see Fig.1) which is under test at the ATF, we start with a radially polarized beam. By an axicon, the laser beam is converged to the e-beam axis, z, producing a cylindrically symmetrical interference pattern. An amplitude of the longitudinal component of the electric field, which is responsible for electron acceleration, is described analytically by Bessel function of the first kind of the order 0:

$$E_z(r,z) = tg\theta \times E_0(z) J_0(2\pi\theta r / \lambda), \qquad (3)$$

where $E_0(z)$ is a field amplitude that depends upon the laser intensity distribution at the axicon surface, $W(R)$,

$$E_0(z) = 2\pi\theta \sqrt{\frac{2zW(R)}{\varepsilon_0 c\lambda}} \qquad (4)$$

with $z = R/\theta$. The distribution described by Eq.(3) has the maximum along the z axis that defines the acceleration gradient attainable under the phase matching condition, Eq.(2). The radial position of the first minimum in the distribution Eq (3) is observed at

$$r_{min} = 0.38\lambda/\theta; \qquad (5)$$

for $\lambda=10.6$ μm and $\theta=20$ mrad, $r_{min} \approx 200$ μm. In a practice, the parameter r_{min} shall be chosen according to the realistic size of the e-beam, $r_{min} \approx r_e$, that propagates along the axicon axis. Then, as follows from Eq.(5), the longer λ permits to choose the proportionally bigger angle θ. Combining this condition with Eqs.(3) and (4), we come to the conclusion: $E_z \sim \lambda^{3/2}$, due to stronger inclination of the laser wavefront to the e-beam propagation.

Another advantage of using the relatively long-wavelength CO_2 laser radiation for the ICA scheme is due to the increase of the "phase slippage" distance, L_{slip}, where the accelerating particle stays in partial synchronism with the

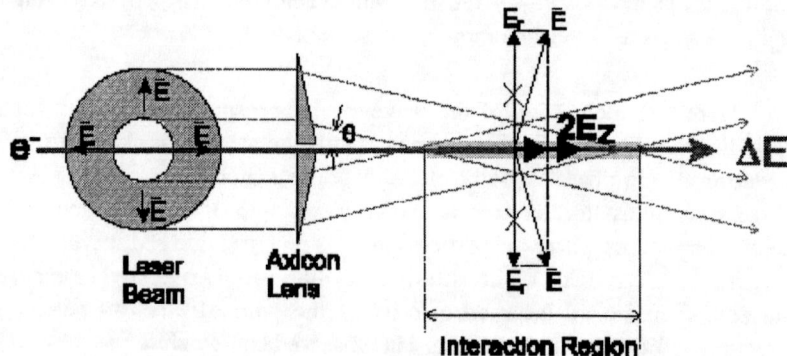

FIGURE 1. Axicon focusing of radially polarized laser beam in the inverse Cherenkov accelerator

driving EM field. The expression for L_{slip} comes from the condition $L_{slip}\Delta\beta = \lambda/2$, where $\Delta\beta$ is the electron velocity increase above the β-number defined by Eq.(2). Using $\beta = \sqrt{1-1/\gamma^2}$, we derive

$$L_{slip} = \frac{\lambda\gamma^3}{2\Delta\gamma}. \qquad (6)$$

For example, for the conditions of the ATF ICA experiment, L_{slip}=20 cm at $\Delta\gamma$=25 that corresponds to the 12.5 MeV electron acceleration. It becomes evident that the 1-μm laser would require much stronger acceleration gradient in order to attain similar net acceleration over the 10 times reduced distance. However, that requirement would be difficult to satisfy considering the above mentioned proportion $E_z\sim\lambda^{3/2}$.

So far, maximum 3.7 MeV acceleration has been measured when ~1 GW CO_2 laser power was delivered to the interaction region. After the completion of the ongoing ATF CO_2 laser upgrade to a 3 ps pulse duration, up to 200 GW laser peak power may be delivered into the hydrogen cell prior to gas breakdown or optics damage. Monte-Carlo computer simulation of the ICA process shows a possibility of a 100 MeV acceleration demonstration over the 30 cm long interaction range.[13,14] To avoid phase slippage, gradual or step-wise change of the Cherenkov angle will be introduced over the interaction distance. To produce monochromatic accelerated electrons, the short electron bunches should be phased with the peak accelerating field. For this purpose, periodical electron prebunching with the spatial interval equal to the laser wavelength will be produced in the IFEL accelerator section placed upstream the high-power ICA accelerator cell.

II.2 Inverse FEL Accelerator

The IFEL scheme is an example of a second order, far field laser acceleration process. In this case, a second field of a wiggler magnet is used to bring the relativistic particles into a transverse oscillating motion. Thus, transverse EM laser field has a projection of its electrical component along the local direction of the e-beam propagation (see Fig.2). Hence, electric forces may produce an additional kick to the electrons in the direction of their propagation, provided the laser field is in phase with the electron wiggling.

In vacuum, the oscillating electron can not propagate with a phase velocity of light along the direction of the laser beam. Now, synchronism means that the electron should slip exactly one period (or integer number of periods) of the EM wave while traveling a wiggler period, λ_w. The synchronism condition at a small angle limit takes the form

$$\lambda = \frac{\lambda_w}{2\gamma^2}(1+K^2) \qquad (7)$$

where K is a dimensionless wiggler parameter equal to $K = \frac{eB_w \lambda_w}{2\pi mc^2}$ and B_w - wiggler magnetic field. Hence, the condition Eq.(7) may be satisfied over a long acceleration distance by adjusting wiggler field and period.

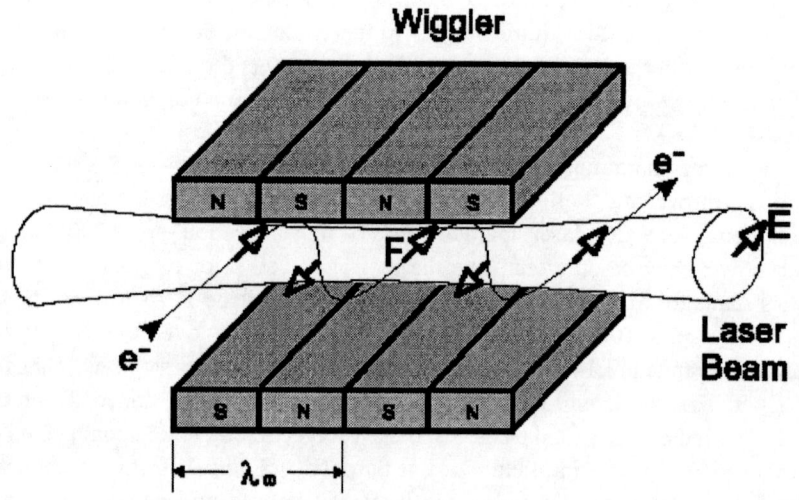

FIGURE 2. Principal diagram of the IFEL accelerator

First proposed in 1972 [4], IFEL acceleration has been demonstrated using FEL[15] and a moderate-power CO_2 laser[16] as drivers.

The goal of the ATF IFEL experiment is further optimization of the accelerator parameters at a higher CO_2 laser power. The laser beam is guided inside a low-loss sapphire waveguide of a 2.8 mm internal diameter mounted inside the 0.5 m-long wiggler.

Electron spectra obtained during the ATF IFEL experiment show that practically all the electrons are trapped and accelerated. Because of problems with vacuum degradation when the laser is delivered inside the guide, the laser power has been kept below 0.5 GW. Observed acceleration is 2.2%. Further optimization is under way.

Near term plans call for increasing the CO_2 laser power to 200 GW. This should result in the accelerating gradient of ~100 MeV/m.[5]

IFEL is also expected to be a good buncher. In the mentioned above oncoming ATF experiment, electrons bunched in the IFEL to the period of λ will be sent to the ICA interaction cell to demonstrate quasi-monochromatic acceleration. A challenge in this experiment is to produce, deliver through the e-beam transport and focusing system, and maintain during the ICA process tiny electron bunches sized to the fraction of λ. To achieve this goal with the long-wavelength CO_2 laser is a problem. But with the λ≈1 μm, the same may be hardly feasible.

III. NEAR FIELD ACCELERATORS

III.1 Grating Linac

Near-field laser accelerator scheme illustrated by Fig.3 is based on excitation of an evanescent field when a laser beam is cylindrically focused onto a periodic structure, e.g., diffraction grating. Electrons injected parallel to the surface will be accelerated when moving in phase with standing wave oscillations. This approach is similar to the RF linac, and is therefore known as a Grating Linac[6]. For a relativistic e-beam to satisfy the synchronism conditions, the structure period is nearly equal to the laser wavelength. Thus, a long-wavelength CO_2 laser radiation helps to use reasonably "macroscopic" structures that may be produced by a conventional lithographic etching technique. Taking into account that the evanescent accelerating field is observed within one-wavelength distance from the surface, the requirement to the electron beam dimensions are also not as severe as would be with the 1-μm laser driver. In addition, the reduced phase slippage at a long wavelength is relevant for this scheme as well.

When a 1 GW CO_2 laser beam is focused to the 5×0.03 mm strip at the "foxhole" structure, 1 GeV/m acceleration is predicted[17]. Using a short, picosecond laser pulse is a way to avoid the optical damage of the structure. To ensure the interaction of the e-bunch, directed perpendicular to the laser beam, over the appreciable acceleration distance (much longer than the laser pulse length), a linear delay shall be introduced across the laser intensity front. It may be done by reflecting the laser beam from the diffraction grating prior to cylindrical lensing.

III.2 Dielectric-Loaded Resonant Laser Accelerator

In another proposed near-field electron accelerator scheme[7], linearly polarized laser radiation penetrates through the periodically modulated dielectric structure filling the gap of the Fabry-Perot interferometer as shown in Fig.4. At

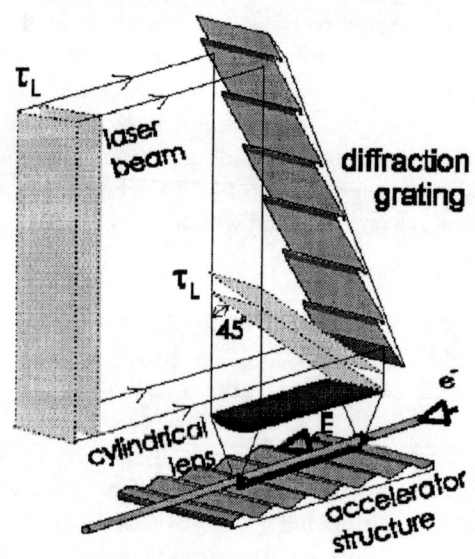

FIGURE 3. Principal diagram of the grating linac

the high quality factor of the interferometer cavity, Q, equal to the number of the optical double passes required to reach the field saturation in the cavity, a large stored EM energy and field amplitude can be obtained at the relatively low input laser power. Periodic modulation of the permittivity, ε, within the dielectric masks sandwitched between the interferometer mirrors

$$\varepsilon(z) = \varepsilon_0 + \Delta\varepsilon \cos(2\pi z / \lambda) \qquad (8)$$

ensures a space-periodical modulation of the field phase in the central vacuum gap, provided the gap width is of the order of the laser wavelength. The electron beam, propagating inside the vacuum gap in the direction of the laser beam polarization, experiences acceleration due to the synchronously oscillating electric field. The average accelerating field is

$$\overline{E}_z \cong \sqrt{\frac{4\pi Q I}{(1+\Delta\varepsilon^2/2)}}, \qquad (9)$$

where I is the incident laser intensity. The accelerating field in excess of 1 GV/m has been estimated for the realistic experimental parameters.

Similar to Grating Linac, this scheme looks more practical with the relatively long-wavelength CO_2 laser, provided the problem of laser damage of the structure is solved.

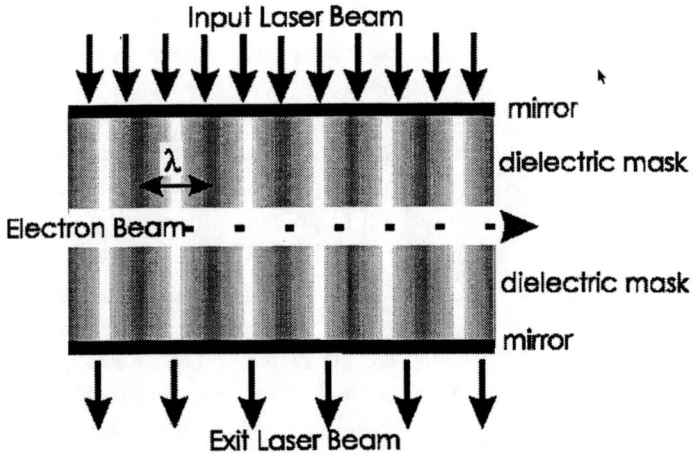

FIGURE 4. Schematic of the dielectric loaded laser resonant accelerator structure

IV. LASER-DRIVEN PLASMA ACCELERATORS

It looks logical to use a high-power CO_2 laser, which is a strong ionizer, in the schemes where such ionization and related effects are not problems but are desirable, as it happens in plasma accelerators.

An EM wave packet propagating in plasma ponderomotively separates charges initiating their oscillation at the plasma frequency, ω_p, that depends upon the electron density, N_e, by

$$\omega_p = 2e(\pi N_e / m)^{1/2}. \tag{10}$$

Plasma wave follows the laser pulse with a phase velocity equal to the group velocity of the laser pulse, $v_{ph} = v_{gr}^L = cn$, where $n = (1 - \omega_p^2/\omega^2)^{1/2}$ is a plasma refractive index.

A relativistic particle propagating together with plasma wave will experience acceleration until it slips out of synchronism over the distance

$$l_a \approx \lambda_p (\omega/\omega_p)^2. \tag{11}$$

Initiated via oscillation of free plasma electrons in the laser field, amplitude of a plasma wave is proportional to the energy of the electron oscillatory motion called ponderomotive potential

$$W_{osc} = e^2 E_L^2 / 2m\omega^2, \tag{12}$$

where E_L is the laser field amplitude. This quadratic dependence of the ponderomotive potential upon the laser wavelength makes CO_2 laser an attractive candidate to drive plasma accelerator.

Amplitude of the accelerating field due to the longitudinal charge separation depends upon the particular method of the plasma wave excitation.

In the LWFA scheme, plasma waves are initiated by an instant "shock" that is produced with a short laser pulse optimally equal in duration to the half-period of the plasma wave, $\tau_L = \lambda_p / 2c$. Note, that after developing a high-power picosecond CO_2 laser, there will be the first-time opportunity to use CO_2 laser in this scheme in a practically meaningful range of the plasma density, $N_e = 10^{14} - 10^{15}$ cm^{-3}. The advantage of using CO_2 laser in the LWFA scheme stems from the proportionality of the accelerating field to the laser wavelength that is the result of the strong ponderomotive potential in the CO_2 laser field and, in particular, follows from the expression[8]

$$E_a = \frac{\pi^2 mc^2 a^2}{4\lambda_p e\sqrt{1+a^2/2}}, \quad (13)$$

where
$$a \equiv eE_L / m\omega c \quad (14)$$
is the dimensionless laser strength parameter.

In the laser beatwave accelerator (LBWA) scheme we choose periodical force that matches the plasma frequency and resonantly enhances plasma oscillations. Such laser intensity modulation is produced by mixing two laser beams of different frequencies that satisfy a condition $\omega_1 - \omega_2 = \omega_p$.

The accelerating field in the LBWA reaches an amplitude of

$$E_a = (5a_1 a_2)^{1/3} N_e^{1/2}. \quad (15)$$

Thus, due to the proportion of $E_a \propto \lambda^{2/3}$ (see Eqs.(14) and (15)), CO_2 laser can produce 6 times stronger acceleration than a 1-µm laser of the equal intensity.

Exploiting this feature, up to 30 MeV acceleration over a 1 cm interaction distance has been demonstrated using subnanosecond, multi-gigawatt, dual-wavelength CO_2 lasers.[18,19] The expected enhancement of the acceleration gradient with a picosecond terawatt CO_2 laser is due to higher strength parameter entering Eq.(15) and higher stability of the fast-excited wakefield.

The third, self-modulated (SM) LWFA scheme, looks the most promising. It comes to the scene when the laser pulse is relativistically strong. Relativistically strong means that laser power, P, satisfies the condition of relativistic self-focusing

$$P \geq 17(\omega/\omega_p)^2 [GW]. \quad (16)$$

In this case, initially small plasma density oscillations cause modulation at the plasma frequency of the laser beam envelope and its intensity. Then, the pulse resonantly enhances the plasma oscillation similar to the LBWA scheme matched automatically to the local plasma density. It is understood that the laser pulse length in this case shall extend over several plasma periods, $\tau_L \gg 2\pi/\omega_p$.

It has been shown by simulations that, if the laser pulse is just strong and long, it is still not enough to produce an intense and regular wake[9]. The wake needs an efficient initiation, similar to that provided in the LBWA or LWFA schemes. One of the possibilities is when a steep leading front, with $\tau_{fr} < \lambda_p/c$, serves as a good initiator for a plasma wave. Simulations done for a 5-TW, 1.5-ps, appropriately shaped CO_2 laser pulse propagating in a plasma channel[9] predict electron acceleration to 250 MeV.

Finally, let us address the question of how to compare potential performance of 10-μm and 1-μm lasers in the SMLWFA configuration. An accelerating field attainable at the plasma wave breaking limit is

$$E_a \propto \omega a^2 / \sqrt{1 + a^2/2}. \qquad (17)$$

According to Eq.(14), a is proportional to λ. In spite of this fact, the expression for the net acceleration

$$\gamma_{max} \approx a(\omega/\omega_p)^3, \qquad (18)$$

obtained from Eqs.(11) and (17), still seems beneficial for shorter wavelength lasers. However this first impression is misleading. The thing is that, for the SMLWFA scheme, the choice for parameters entering Eq.(18) is not arbitrary. The values of ω and ω_p are linked together through the self-focusing condition, Eq.(16).

Assume for our wavelength comparison two laser beams of the equal power but different wavelength, both close to the relativistic self-focusing condition, and focused to the equal spot size. Then, the ratio ω/ω_p should be chosen equal for any laser wavelength. The result of it is that the maximum acceleration is proportional to λ, $\gamma_{max} \propto \lambda\sqrt{P}$.

When λ=1 μm, we need to channel the laser beam and, hence, e-beam in a 10-μm waveguide in order to obtain similar acceleration as with the CO_2 laser beam inside a 100-μm waveguide. Doing it, we may encounter severe problems with the electron beam scattering. Indeed, when we increase ω and ω_p 10 times, the plasma density will be increased 100 times, according to Eq.(10), with the increase of the multiple scattering in the gas according to the formula[20]

$$\Delta\theta_{1/e} = (W_s/W)(z/L_R)^{1/2}[1 + 0.1\log_{10}(z/L_R)], \qquad (19)$$

where $\Delta\theta_{1/e}$ is the angular spread of the e-beam, W is the mean electron energy, W_s=19.7 MeV is a multiple scattering constant, z is the path length through the gas traversed by the electron, and L_R is the radiation length of the medium, which for hydrogen gas is L_R=7×10^5 [cm]/p[atm]. The result of gas scattering is the e-beam emittance growth and reduced acceleration efficiency due to a poor overlap of the diverging e-beam with the narrow channeled laser beam. A simplified formula for the e-beam radius growth due to the multiple scattering

$$\Delta r(z) = \frac{2}{3}\frac{W_s}{W\sqrt{L_R}}z^{3/2} \qquad (20)$$

indicates that the e-beam expands by Δr=10 μm over the distance of z=2 cm in a 0.25 atm of H_2.

It follows that the CO_2 laser, that permits similar acceleration at 10 times wider waveguide and e-beam diameters and a 100 times lower pressure, looks more attractive for a prospective high-energy plasma accelerator.

Considering the feasibility of conducting a sub-GeV plasma acceleration experiment at the ATF, we should remember that not just a terawatt CO_2 laser will be available for this purpose, but also one of the world's brightest e-beams that may be fitted inside a 100-μm wide channel. Fig.5 shows the principal schematic of the ATF LWFA experiment where a plasma channel is produced via gas ionization by axicon-focused linear or radial polarized laser beam [21]. A split fraction of the drive CO_2 laser beam may be used for this purpose. Parameters of the hypotetic LWFA experiment are presented in Table 2.

Dielectric capillary tubes may offer another possibility for guiding high-power laser beams in LWFA. This opportunity is based on a finite velocity of the plasma front that is originated at a capillary wall due to the laser ablation. At a typical plasma implosion velocity of several km/sec, it should take several nanoseconds for the overdense plasma to fill a capillary. Thus, the picosecond laser pulse will not be effected by the capillary ablation. Test with a picosecond Nd:YAG laser[22] shows the viability of this approach.

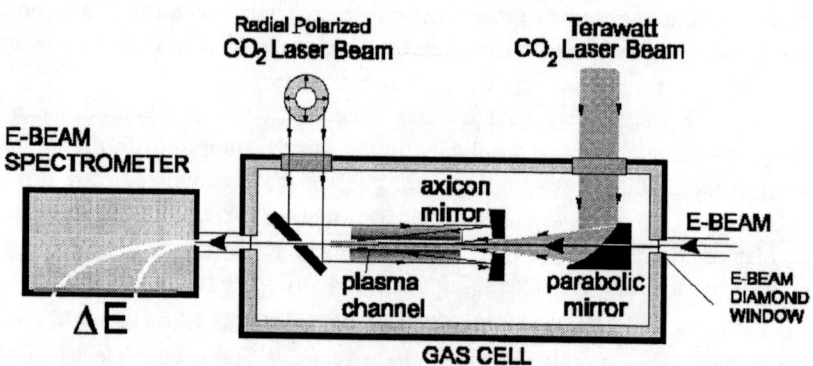

FIGURE 5. Principal diagram of the LWFA experiment

TABLE 2. Design Parameters of the LWFA Experiment with CO_2 Laser Driver

Seed Electron Beam	
Electron Energy [MeV]	50-70
Bunch Duration FWHM [ps]	0.4-10
Electron Energy Spread [%]	0.2
Normalized Emittance [mm.mrad]	0.5-2
Peak Current [A]	170-50
Plasma Channel by Axicon-Focused CO_2 Laser	
Laser Energy [J]	1-3
Channel Radius [μm]	60
Axicon Angle [mrad]	50
Channel Length [cm]	4
Plasma Density Inside the Channel [cm^{-3}]	3.5×10^{16}
Plasma Density at $r = r_{ch}$ [cm^{-3}]	8×10^{16}
Laser Accelerator	
Laser Peak Power [TW]	5
Laser Pulse Duration [ps]	1-3
Laser Radius at Focus [μm]	60
Laser Peak Intensity [TW/cm^2]	4×10^4
Laser Field [GV/m]	500
Laser Strength	1
Plasma Wavelength [μm]	170
Acceleration Gradient [GV/m]	6
Pump Depletion Length [cm]	4
Phase Detuning Length [cm]	4
Rayleigh Length [cm]	0.3
Interaction Length [cm]	4
Energy Gain with Guiding [MeV]	250

V. CONCLUSIONS

The first terawatt picosecond CO_2 laser is under construction at the ATF. There are several reasons why we are interested in using a CO_2 laser for particle acceleration: With far-field accelerators, we may benefit from a slow phase slippage. For near-field accelerators, the CO_2 laser helps to use macroscopically-sized accelerator structures and e-beams. With plasma accelerators, we capitalize on the strong ponderomotive potential of electron, oscillating in the laser field, and all outcomes of it such as: strong plasma wave formation, relativistic self-focusing at a low plasma density, etc.

We intend to utilize all above-listed features of the picosecond terawatt CO_2 laser to benefit the laser acceleration experiments at the ATF. So far, high-brightness, 10-ps, 50-MeV electron bunches have been accelerated to several MeV at the ATF with the ~1 GW CO_2 laser pulse using far-field accelerator schemes. Due to the expected acceleration scaling as a square root of the laser power, 100 MeV ICA and IFEL demonstration is possible after the laser upgrade.

Combined IFEL-ICA experiment is under preparation to enhance the acceleration efficiency by the electron periodical prebunching at the laser wavelength. A laser linac near-field accelerator experiment is scheduled for tests. By the SMLWFA method, 250 MeV acceleration over a 4-cm distance is feasible when using the properly shaped 5-TW laser pulse guided in a plasma channel.

ACKNOWLEDGMENTS

The success of the ATF laser accelerator experiments, reviewed in this paper, is due to the team work of scientific and technical personal of the ATF and associates.

The work is supported by the US Department of Energy.

REFERENCES

1. A. Madena, Z. Najmudin, A.E. Dangor, et.al., Nature, **377**, 606 (1995); T. Katsouleas, "Laser Acceleration of Electrons in Plasmas", *Joint Meeting of APS and AAPT*, May 2-5, 1996, Indianapolis, IN
2. J.R. Fontana and R.H. Pantell, *J. Appl. Phys.*, **54**, 4285 (1983)
3. W.D. Kimura, G.H. Kim, R.D. Romea, et al., *Phys. Rev. Lett.*, **74**, 546 (1995)
4. R.B Palmer, *J. Appl. Phys.*, **43**, 3014 (1972)
5. A. Fisher, J. Gallardo, J. Sandweiss, and A. vanSteenbergen, *Proc. Advanced Accelerator Concepts*, Port Jefferson, NY, 1992, AIP **279**, 299 (1993)
6. 1992)R.B. Palmer, Proc. *Laser Acceleration of Particles*, AIP **91**, 179 (1982)
7. J. Rosenzweig, A. Murokh, and C. Pelligrini, *Phys. Rev. Lett.*, **74**, 2467 (1995)
8. P. Sprangle, E. Esarey, J. Krall, and G. Joice, *Phys. Rev. Leii.*, **69**, 2200 (
9. S.V.Bulanov, T.J. Esirkepov, N.M. Naumova, et al., *IEEE Trans. on Plasma Sci.*, **24**, 393 (1996); see also S.V. Bulanov, G.I. Dudnikova, T.Zh. Esirkepov, N.M. Naumova, F. Pegoraro, I.V. Pogorelsky, A.M. Pukhov, V.A.Vshivkov, "Controlled Wakefield Acceleration via Laser Pulse Shaping", Proc. of Int. Conf. on Plasma Phys., Paper 12Q02, Nagoia, Japan, Sept. 9-13, 1996; S.V. Bulanov, G.I. Dudnikova, N.M. Naumova, F. Pegoraro, I.V. Pogorelsky, V.A.Vshivkov, "Charged Particle Acceleration in Nonuniform Plasmas", in these Proceedings.
10. I. Ben-Zvi, *Advanced Accelerator Concepts, Port Jefferson*, NY, 1992, AIP **279**, 590 (1993).
11. I.V. Pogorelsky, J. Fischer, K. Kusche, et al., *IEEE J. Quant. Electron.*, **31**, 556 (1995); see also I.V. Pogorelsky, I. Ben-Zvi, J. Skaritka, et al., "The First Terawatt Picosecond CO_2 Laser for Advanced Accelerator Study at the Brookhaven ATF", these Proceedings
12. J.A. Edighoffer, W.D. Kimura, R.H. Pantell, et al., *Phys. Rev.*, **A23**, 1848 (1981)
13. R.D. Romea and W.D. Kimura, *Phys.Rev.*, **42D**, 1807 (1990)
14. W.D. Kimura, R.D. Romea, and L.C. Steinhauer, *Particle World*, **4**, #3, 22 (1995)
15. I. Wernick and T.C. Marshall, Proc. *Advanced Accelerator Concepts*, Port Jefferson, NY, 1992, AIP **279**, 292 (1993)
16. A.T. Amatuni et al, *Part. Accel.*, **32**, 221 (1990)
17. R.C. Fernow and J. Claus, *Proc. Advanced Accelerator Concepts*, Port Jefferson, NY, 1992, AIP 279, 212 (1993)
18. M. Everett, et al., *Nature*, **368**, 527 (1994)
19. N.A. Ebrahim, *J. Appl. Phys.*, **76**, 7645 (1994)
20. V.L. Highland, *Nuclear Instruments and Methods*, **129**, 497 (1975)
21. I.V. Pogorelsky, W.D. Kimura, and Y. Liu, *Proc. Advanced Accelerator Concepts*, Fontana, WI, 1994, AIP **335**, 419 (1995)
22. S. Jackel, , et al., *Opt. Lett.*, **20**, 1086 (1995)

THE FIRST TERAWATT PICOSECOND CO₂ LASER FOR ADVANCED ACCELERATOR STUDIES AT THE BROOKHAVEN ATF

I.V. Pogorelsky, I. Ben-Zvi, J. Skaritka, Z. Segalov, M. Babzien, K. Kusche
Accelerator Test Facility, Brookhaven National Laboratory, 725C, Upton, NY 11973

I.K. Meskovsky, V.A. Lekomtsev, A.A. Dublov, Yu.A. Boloshin
Optoel Co., St. Petersburg, Russia

G.A. Baranov
Efremov Inst, NIEFA, St. Petersburg, Russia

Abstract

The first terawatt picosecond CO_2 laser system is under development at the Brookhaven Accelerator Test Facility. Presently operational 1-Joule 100-ps ATF laser will be upgraded with a 10 atm amplifier capable of delivery ~15 Joules of laser energy in a 3-ps pulse. We describe the design of the x-ray preionized 10-atm amplifier of a 10-liter active volume energized by a 1-MV, 200 kA transverse electric discharge. The amplifier, equipped with internal optics, permits the accommodation of a regenerative stage and a multi-pass booster in a relatively compact single discharge volume. The ATF terawatt CO_2 laser shall become operational in 1997 to serve for laser acceleration, x-ray generation and other strong-field physics experiments.

I. INTRODUCTION

Lasers are the sources of the most intense electromagnetic radiation and strongest electric and magnetic fields available for laboratory research. For example, focusing of a terawatt laser beam into a 30-mm spot results in an intensity of 10^{17} W/cm² and, associated with it, an electric field of 10 GV/cm that exceeds by four orders of magnitude fields attainable in conventional particle accelerators. Such capability stimulates a new high-energy physics discipline to emerge: laser-driven high-gradient particle accelerators. Experimental activity in this field has been vitalized by recent development of, so-called, T^3 (table-top terawatt) solid state lasers operating at wavelengths of l≈1 mm.

Another novel driver source for advanced particle accelerators may be provided by the emerging picosecond terawatt CO_2 (psTW-CO_2) laser technology. There are a number of considerations that favor such an application of long-wavelength (l≈10 mm) psTW-CO_2 lasers.

In general, particle acceleration by a fast-oscillating electromagnetic field becomes possible when a relativistic particle moves in synchronism with the phase of the driving field. The long wavelength of a CO_2 laser helps to meet this requirement. This feature is of particular importance for far-field acceleration schemes, examples of which are the inverse Cherenkov and inverse FEL accelerators.

Proposed near-field accelerator schemes are based on the accelerating action of evanescent fields developed near the periodically shaped surfaces under laser irradiation. Since the spatial scale of such structures is comparable with the laser wavelength, these schemes look practical when a CO_2 or longer-wavelength laser is used as the accelerator driver.

The advantage of slow-oscillating fields for particle acceleration in plasma stems from the fact that the energy of oscillatory motion acquired by the electron from an electromagnetic wave is quadratically proportional to the wavelength. Hence, any process where the field-induced electron oscillation is paramount is dramatically enhanced. The examples of such processes are: relativistic self-focusing, avalanche and tunneling ionization, and plasma wave excitation, which are especially relevant for electron acceleration in a plasma.

Conventional subnanosecond, multigigawatt CO_2 lasers, intensity-modulated with the period of a plasma wave, have been used successfully in laser beatwave accelerator (LBWA) experiments[1,2] with up to 30 MeV electron acceleration over a 1 cm interaction distance demonstrated. The efficiency of plasma acceleration schemes that, in addition to LBWA, include laser wakefield accelerator [3] would be greatly enhanced by laser power increase and pulse shortening provided by the psTW-CO_2 laser. The detailed analysis of advantages of the psTW-CO_2 lasers for particle acceleration studies is done in [4].

The laser synchrotron x-ray source described in [5] serves as another example of how the psTW-CO_2 laser may benefit strong-physics applications. Here, we gain from a l-proportional photon flux and l-proportional laser strength parameter that defines the efficiency of high harmonics generation via nonlinear Compton scattering.

The approach to a picosecond, high peak power CO_2 laser is being pursued at the ATF, where a 10-GW, 100-ps table-top CO_2 laser system is in operation to test several laser acceleration schemes and the upgraded psTW version of this laser is under construction.

The problems that hindered thus far the development of psTW-CO_2 lasers are related to the relatively narrow rotational structure typical for molecular gas spectra ($\sim 10^{10}$ Hz at atmospheric pressure). That is why picosecond pulse formation via a mode-locking technique and subsequent amplification have not been as successfully obtained with CO_2 lasers as with solid state lasers, which have wide crystal-host broadening of ion spectral lines (10^{12}-10^{13} Hz). However, at high gas pressure, the individual rotational CO_2 lines can be collisionally broadened to a quasi-continuum of a 10^{12} Hz width. Also, alternative ways to produce CO_2 pulses with picosecond time scales have been developed including: optical parametric oscillation and semiconductor switching. The last method is based on modulating the reflective and transmissive properties of a semiconductor by optically controlling the free-carrier charge density. Subpicosecond IR pulses have been demonstrated by this method[6].

These principles constitute the approach of the ATF CO_2 laser upgrade to the terawatt level which has been presented previously[7]. The key technical issue that needs to be resolved to make a psTW-CO_2 laser possible is construction of a high-pressure, big-aperture CO_2 laser amplifier. We provide an update on the progress in design and construction of a 10-atm, 10-l amplifier discharge module that shall boost the output power of the ATF CO_2 laser to the several terawatt level.

II. EMERGING PICOSECOND TERAWATT CO2 LASER TECHNOLOGY

In solid state lasers, radiation transitions within the outer electron shells of active ions exhibit broadening to 5-50 THz due to the perturbation action of a host matrix. Such a broad gain spectrum makes possible the generation and amplification of picosecond and even femtosecond laser pulses. On the contrary, it has been realized that to build a picosecond CO_2 laser is a problem because the spectral gain in the gas discharge is periodically modulated by a molecular rotational structure. Due to the discrete spectrum, and for other technical reasons, mode-locking techniques do not work for CO_2 lasers as well as for solid state lasers. Such a modulated spectrum also impedes amplification of picosecond pulses.

Two alternative methods have been proposed to produce picosecond and sub-ps CO_2 laser pulses. Both of them require a short-wavelength, short-pulse laser. By frequency mixing in a nonlinear crystal, the difference frequency at 10 mm may be generated in a parametric oscillator. At the ATF we use another method, semiconductor switching, to generate picosecond CO_2 laser pulses of a variable duration.

The semiconductor optical switching method is based on the modulation of the reflective and transmissive properties of a semiconductor slab, placed into the CO_2 laser beam, by optically controlling the free-carrier charge density. A short-wavelength picosecond laser pulse with a photon energy above the band gap of the semiconductor creates a highly reflective electron-hole plasma in the surface layer of a semiconductor, such as germanium, which is normally transparent to 10-mm radiation. To define the trailing edge of the pulse, shortening it to a few picoseconds, the complement to reflection switching, transmission switching, may be used for a second stage. An optically delayed control pulse cuts off the trailing edge of the transient pulse by initiating reflection and absorption. The resulting "sliced" transmitted pulse has a variable length defined by optical delay adjustment of the control radiation before the transmission switch.

Instead of the transmission switch in a double semiconductor slicer configuration, the transient response of a thin etalon can be exploited[8]. The etalon, tuned to minimum reflectivity, serves as a differentiator transmitting radiation with a constant or slow-varying intensity while reflecting at intensity gradients shorter than the double optical thickness of the etalon. For instance, a 100-mm thick Ge etalon placed after the reflection switch may serve for efficient differentiating of the transient pulse with a steep leading front, producing ~ 3-ps high-contrast IR pulses.

Another way to eliminate the tail in the reflected pulse is to use a semiconductor material with previously introduced radiation damage to the lattice structure. For such materials, subpicosecond electron-hole recombination times have been measured[9].

Let us address now the problem of amplification of short picosecond laser pulses in the active medium of the CO_2 laser. If the input laser pulse is shorter than 18 ps, its spectrum covers several discrete rotational transition lines. The electric field of such an input pulse excites a polarization in CO_2 molecules, which are in various rotational states. Since molecules in different states are characterized by different frequencies, these polarization components eventually become dephased. As a result, the spectral and time structure of the induced radiation will not remain equal to those of the initial pulse. At a low, ~1 atm, gas pressure the discrete gain spectrum transforms the spectrum of the input pulse from continuous to discrete, and its inverse Fourier transform corresponds to a pulse train with an 18 ps period. At higher pressure, the broadening effect smoothes the discrete gain spectrum. As a result, the pulse splitting is reduced and ultimately disappears at an amplifier pressure of 15 atm.

An alternative to achieving gain smoothing is by reduction of the spectrum modulation period using a multi-isotope gas mixture. Replacement of one of the oxygen nuclei by that of a different isotope destroys the symmetry of the CO_2 molecule. This means that twice as many radiation transitions are allowed and the gain spectrum becomes twice as dense as with a regular CO_2 molecule. If we consider a mixture of $^{12}C^{16}O_2 : ^{12}C^{16}O^{18}O : ^{12}C^{18}O_2 = 1:2:1$, then, due to isotopic shifts, the combined spectrum will have in overlap regions an approximately 4-times denser rotational line structure than with a regular CO_2 molecule. Computer modeling[10] shows that the reduction in spectral line interval, together with pressure broadening, results in considerably less short-pulse distortion during amplification. The technical advantage of this approach is related to the greater ease of establishing a large-aperture stable discharge at 4-5 atm in comparison to 10-15 atm.

At the total bandwidth of the 10P CO_2 branch, $\Delta \nu \approx 1$ THz, the fundamental limit for the laser pulse duration, as defined by the ratio $\Delta \nu \times \tau \approx 0.5$, is $t^a 0.5$ ps. As short as 0.6 ps CO_2 laser pulses have been demonstrated from a high-pressure regenerative amplifier[11].

When estimating laser amplifier efficiency, the following two physical parameters play the major role: small signal gain $g_0 = sN^*$, where s is the gain cross-section, and N^* - population inversion at the laser levels; and saturation fluence $E_s = hn/2s$.

Parameters g_0 and E_s regulate the energy amplification process described by the Franz-Nodvik equation

$$E_{out} = E_s \ln\{1 + \exp(g_0 l)[\exp(E_{in}/E_s) - 1]\}. \tag{1}$$

The product of these parameters gives also an estimate of the energy potentially extractable from the amplifier in a single pass in a strongly saturated regime:

$$E_{max} = g_0 E_s l S, \tag{2}$$

where l and S are, correspondingly, the length and aperture of the amplifier.

The ratio E_{max}/t characterizes the peak laser power from the amplifier. We know already that the pressure increase helps to reduce the pulse duration, t. Now, we need to understand how E_{max} depends upon the pressure.

Due to pressure broadening of the gain spectrum, there is a linear proportionality of E_s to the pressure via the parameter s. The small signal gain, g_0, is inversely proportional to s. However, g_0 depends also upon N^* which is subject to the electric discharge conditions. The electron-molecule collision frequency and, hence, the pump rate both rise proportionally to the pressure. If the discharge is faster than collisional quenching of the inversion, then we may consider $N^* \sim P$ and g_0 invariant with pressure. Ultimately, we come to the conclusion that $E_{max} \sim P$.

Computer simulation[10] for a $t0=3$ ps pulse propagating in a 10-atm amplifier gives $E_s^a 500$ mJ/cm^2 and, at the typical $g_0 = 4\%/cm$, the extractable specific energy is $E_{max}/lS^a 20$ mJ/cm^3. Taking into account that the total discharge volume may exceed 10 l, extraction of as high as 100 J of energy in a few-picosecond pulse from a single, reasonably compact CO_2 laser amplifier looks possible. However, the limiting factor to the high energy extraction will be the damage threshold of the output window, $E_{th}^a 500$ mJ/cm^2.[10] For an optical window of a 10¥10 cm^2 size, the extractable energy is 30-50 J which corresponds to 10-15 TW peak power at a 3-ps laser pulse duration. These estimates make psTW-CO_2 lasers quite competitive with the T^3 solid state lasers.

The main physical parameters of solid state and CO_2 lasers are compiled in Table 1. It is interesting that the gain cross-section per ion or molecule is comparable for both. However, about ten times higher concentration of active ions in solid state than CO_2 molecules in a gas makes gain in solid state lasers about ten times higher. About ten times higher photon energy makes the specific stored energy in solid state about hundred times higher than in gas. However, much bigger volume of gas amplifiers makes the total stored energy per CO_2 amplifier stage similar or higher than for a big-size slab solid state amplifier. Because of the ease of the heat removal by fast gas exchange in the CO_2 amplifier, it is potentially capable of high repetition rates that are difficult to attain with solid optical elements. This may be important for future advanced particle accelerators.

Table 1. Typical Parameters of Solid State and CO_2 Lasers

PARAMETER	Solid State	10-atm CO_2
Bandwidth (THz)	5-50	1
Cross section ($\times 10^{-20}$ cm^2)	1-30	5
Gain (%/cm)	~50	3-4
Saturation energy (J/cm^2)	1-20	0.5
Breakdown threshold (J/cm^2)	1	3
Stored energy (J/cm^3)	1	0.01
Active volume (cm^3)	10-100	10,000
Gain relaxation time (ms)	>1	0.2
Average power short-term limit (W)	1-10	100-1000

III. TERAWATT CO_2 LASER PROJECT AT THE ATF

The psTW-CO_2 laser system is the upgrade version of the presently operational 10-GW ATF CO_2 laser[10]. As long as a number of basic principles and elements of the present ATF laser will be preserved after the upgrade, it would be relevant to give its brief overview.

The ATF CO_2 laser system includes: a hybrid TEA oscillator, picosecond semiconductor switch, and a UV-preionized multipass TE amplifier.

In the laser oscillator, a diffraction grating tunes the laser wavelength stepwise between the individual rotational lines in the gain spectrum of the CO_2 molecules, which are vibrationally excited in the electric discharge. In addition to the 1-atm discharge-cell, the oscillator also includes an auxiliary low-pressure discharge cell. The narrow spectral line of the low-pressure discharge selects the particular longitudinal eigenmode to build-up inside the laser cavity. Piezo-tuning of the cavity length matches a mode spectral position to the gain peak. The output single-mode laser pulse has a smooth envelope, free from the stochastic mode-beat spikes otherwise typical for free-running TEA CO_2 lasers.

The semiconductor double-stage switching method is used to slice a 100-ns oscillator pulse to the desired picosecond width. A 10-ps Nd:YAG laser, that serves as a photocathode driver for the ATF linac, also supplies a pulse for slicing. Using the same initiator for the linac and for the CO_2 pulse slicing ensures the desired picosecond synchronization of the electron bunch and laser pulse at their interaction region (see Fig.1).

To reach a power level needed for laser accelerator studies, the switched picosecond pulse is transmitted through the 8-pass CO_2 amplifier that features a 120-cm long, 3-atm, UV-preionized, transverse electrical discharge energized by a 150-kV pulse. The limited spectral bandwidth of the amplifier defines ~100 ps minimum duration of the output laser pulse. Thus, at the amplifier output energy of 1 J, the available peak power is ~10 GW.

The design concept for the CO_2 laser upgrade presumes slicing and then amplification of a short (~3 ps) laser pulse. We also expand the amplifier cross-section, thus allowing a high energy extraction through the large-aperture output window. Both pulse shortening and energy increase should permit an increase of the peak power from several GW to the several TW level. Fig.2 presents a principal optical diagram for the modified CO_2 laser system. The presently operational oscillator and semiconductor switch will supply a picosecond seed pulse into a regenerative preamplifier which will share a portion of the active discharge region in a large-aperture multi-isotope amplifier. After the regenerative amplifier, four or five additional passes, with the laser beam expanded to ~80 cm^2, will boost the output power to the 5 TW level in a ~3-ps pulse.

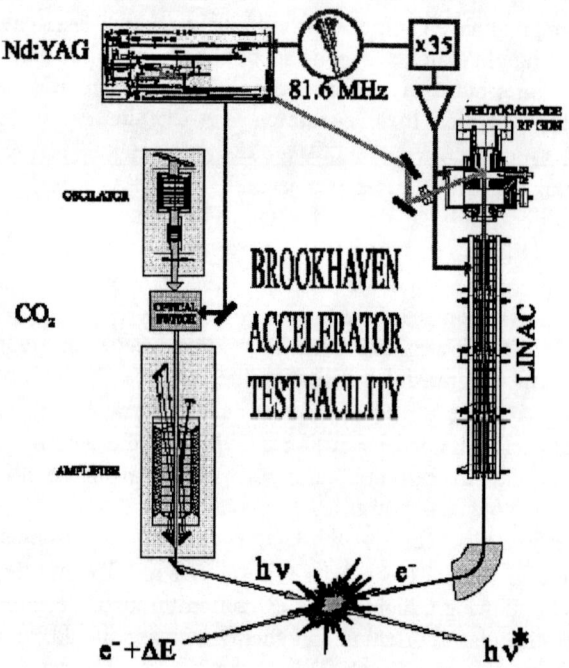

Fig.1 Principal diagram of the ATF laser system and linac synchronized to 1 ps for e⁻-hn interaction experiments

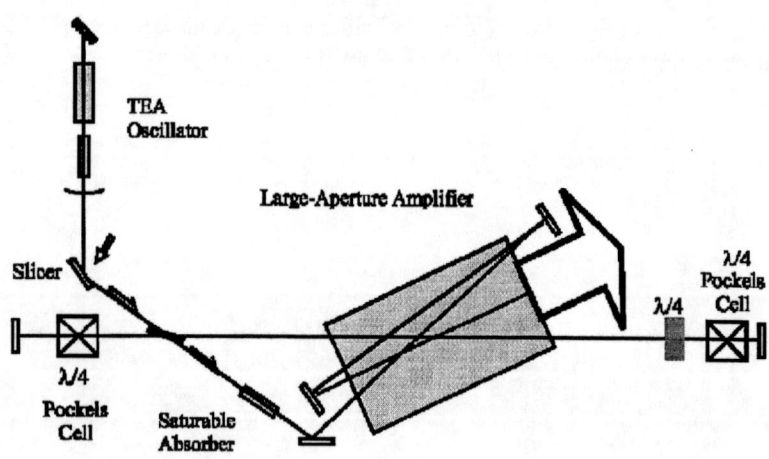

Fig.2. Principal optical diagram of the psTW-CO_2 laser system

Essential for the amplifier design is the discharge mechanism to create uniform gas excitation in a ~10 l volume under high pressure. A self-sustained glow discharge can exist only at $dP £ 0.1$ cm.atm, where d is the interelectrode distance and P - the gas pressure. Above this point, an external ionization source should be applied to prevent discharge from arcing. The simplest corona and UV-preionization methods work well up to $dP^a 15\text{-}20$ cm.atm. Above that level, more intense and volume-penetrating x-ray and e-beam preionizers are used. High ionization efficiency of e-beams makes it possible even to reduce the discharge voltage two times below the self-glow level (so-called "e-beam sustained" discharge). It helps to maintain the discharge at the optimum normalized field strength, V/dP, resulting in high efficiency of the upper laser level excitation. However, a high probability of e-beam window failure (usually thin metal foils), especially at a high pressure, makes e-guns inconvenient for this application.

X-ray preionization, while somewhat more complex than the UV-preionization which is used in the present 3-atm ATF CO_2 amplifier, has advantages for large-aperture high-pressure discharge applications. Because of strong absorption of UV radiation by CO_2 molecules, it is difficult to implement even for $dP>25$ cm.atm (e.g., $P=5$ atm and $d=5$ cm). The penetration range for >30keV x-rays is much larger. Another important advantage of x-ray preionization is the elimination of spark discharges associated with the UV preionization method which contribute significantly to the dissociation of CO_2, thus shortening gas lifetime.

With relatively simple corona-cathode e-guns, collimated "sheets" of x-rays with the cross-section of 0.1 m^2 or more can be readily produced. At applied ~100 kV cathode voltage, the e-beam with the current density of ~1 A/cm^2 maintained during ~1 ms is adequate to generate in the interelectrode space the initial photoelectron concentration necessary for starting a uniform volumetric discharge at $dP^a 50\text{-}100$ cm.atm.

After the choice of the preionization mechanism is done, the next key decision would be regarding the discharge parameters.

An electrical energy deposition into the discharge of ~120 J/l.atm is needed to attain a typical gain of ~2.5%/cm. For the 10-atm, 10-l discharge, that results in $E_{dep}^a 10$ kJ.

The high-voltage pulsed power supply for the discharge shall comply with requirements of the breakdown and sustain potentials set by parameters d, P, and x, where x is the proportion of the molecular components in the CO_2:N_2:He mixture. Based on available data, we can draw a semi-empirical law for the reduced sustain voltage:

$$V_s/dP[kV/cm.atm]=2.5(1+0.1x[\%]).$$

(3)

As follows from Eq.(3), a technically feasible pulse generator with the output voltage of 1 MV is capable of sustaining a discharge in x=15% mixture at dP=80 cm.atm. The discharge voltage and energy define the capacitance of the high-voltage generator bank that should be ~20 nF. Together with inductance and active resistance of the discharge circuit, the storage capacitance defines the discharge duration that can be 300-500 ns in our case.

There is at least one extra requirement that may further restrict our choice of the discharge duration. Glow discharge is a transient phenomenon at $P\geq0.1$ atm, and contracts into streamer channels during the time interval that decreases inversely proportional to the gas pressure. For a 10 atm pressure, that condition requires the discharge duration of <300 ns that is difficult to provide using a Marx-type generator with the specified above parameters. That is why the discharge circuit of the ATF psTW-CO_2 laser includes also a pulse forming line built as a tunable water capacitor connected to the Marx generator. This design permits variation of the pump pulse duration between 100-300 ns at a peak current of up to 200 kA.

A cross-sectional diagram of the high-pressure x-ray preionized amplifier designed for the ATF terawatt CO_2 laser system by Optoel Co. (St. Petersburg, Russia) is shown in Fig.3. X-rays penetrate into the active volume of 100¥10¥10 cm^3 through the mesh ground electrode and Be window that separates the vacuum x-ray tube from the high-pressure discharge volume. The big, 10¥10 cm^2, optical aperture of the amplifier helps to accommodate the regenerative amplifier and subsequent multipass amplification in one discharge cell. This, together with the multipass mirror set-up inside the discharge volume, permits a reasonably compact design of the terawatt laser system.

The layout of the main components of the high-power CO_2 laser amplifier is shown in Fig.4. It includes: a high-pressure discharge vessel, x-ray preionizer with a pulsed power supply, Marx generator, and a water capacitor. The drawing in Fig.4 does not show a gas circulation and recovery system, external optics and other auxiliary components that will be assembled around the amplifier.

The psTW-CO_2 laser system, with parameters summarized in Table 2, will become operational in 1997. The collimated laser beam will be transported to the experimental hall and interact with picosecond 50-MeV, 1-nC electron bunches to test several laser acceleration schemes, nonlinear Compton scattering and other prospective applications

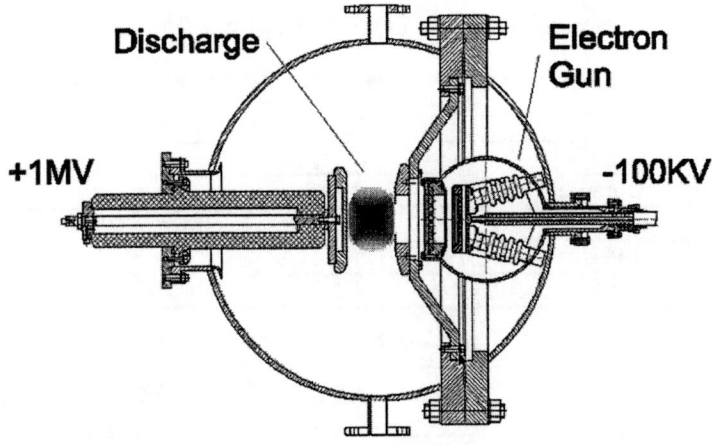

Fig.3 Cross-sectional diagram of the x-ray preionized CO_2 laser amplifier

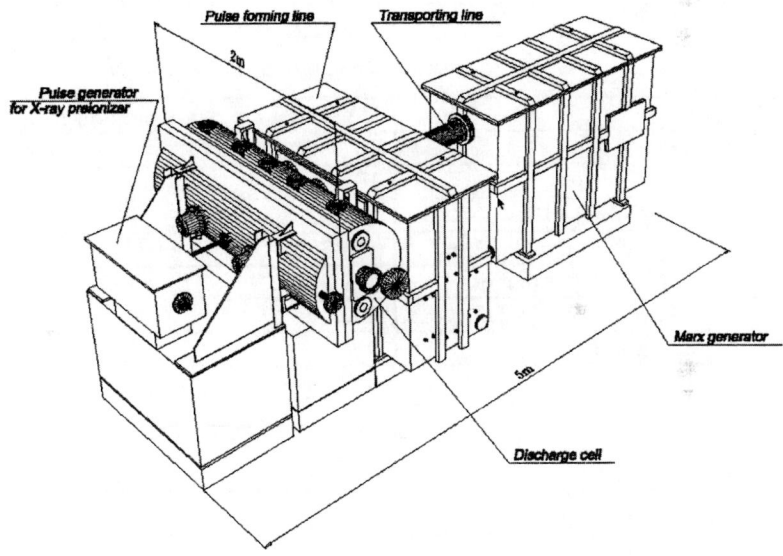

Fig.4 Layout of the main components of the high-power CO_2 laser amplifier

Table 2. Design Parameters of the ATF psTW-CO_2 Laser System

Oscillator	
Type	Hybrid, TEA
Pulse Duration [ns]	100
Output Energy [mJ]	100
Output Peak Power [MW]	1
Max Repetition Rate [Hz]	0.3
Picosecond Slicer	
Type	Semiconductor switching
Control Laser	Nd:YAG
Control Laser Energy [mJ]	10
Control Pulse Duration [ps]	10
Sliced Pulse Duration [ps]	3
Sliced Energy [mJ]	1-2
Output Peak Power [MW]	~0.5
Max Repetition Rate [Hz]	3
Amplifier (electro-physical)	
Type	High-pressure, x-ray preionized
Pressure [atm]	10
Active Volume [l]	10
Discharge Voltage [MV]	1
Discharge Pulse Duration [ns]	100-300
Discharge Peak Current [kA]	150-200
Stored Electric Energy [kJ]	10
Repetition Rate [Hz]	0.1
Amplifier (optical)	
Regenerative Stage Output [mJ]	30
Final Output Aperture [cm^2]	80
Final Output Energy [J]	~15
Pulse Duration [ps]	3
Output Peak Power [TW]	~5

IV. CONCLUSIONS

The first terawatt picosecond CO_2 laser is under construction at the ATF. It is an upgrade version of the 10-GW, 100-ps laser that is presently used at the ATF for laser acceleration studies. The power boost to several terawatt will be attained via laser pulse shortening to 3 ps and energy increase from 1 J to ~15 J. This is possible due to installation of a new x-ray preionized amplifier of 10 atm pressure and 10 l volume. The high pressure and the possibility of using a multi-isotope gas mixture expand the amplifier bandwidth to 1 THz, thus permitting amplification of short picosecond pulses. The overall dimensions of the ATF laser will not exceed the foot-print size of a typical T^3 solid state laser.

The new ATF laser may become a valuable instrument for strong-field physics study. For example, there are several reasons why a CO_2 laser, with its wavelength ten times longer than that of conventional solid state lasers, may be attractive for advanced laser acceleration study. That includes:
- slow phase slippage of accelerated particles from a relatively shallow electromagnetic wave crest;
- macroscopically-sized accelerator structures and e-beams for near-field accelerators;
- strong plasma wave formation and relativistic self-focusing at a low plasma density useful for plasma accelerators.

All these features are discussed in more detail in [4]. Similarly, the psTW-CO_2 laser opens new prospects for development of high-intensity laser synchrotron x-ray and gamma sources as discussed in [5].

Expected for commissioning in 1997, the first psTW-CO_2 laser laser will become available to the ATF users to explore the proposed above and other strong-field physics phenomena in mid-IR region.

Acknowledgments

The authors wish to thank everybody who provided their help at various stages of design and manufacturing of the high-power CO_2 amplifier, with special thanks to L. Smith, A. Feldman, P. LeDoux, D. Lynch, R. Bauman, T. Srinivasan-Rao, L. DiMauro, C.H. Fisher, N.A. Kurnit and A. Kuchinsky.

The work is supported by the US Department of Energy.

References

1. M. Everett, A. Lal, D. Gordon, C. Clayton, K. Marsh, C Joshi, *Nature*, **368**, 527 (1994)
2. N.A. Ebrahim, *J. Appl. Phys.*, **76**, 7645 (1994)
3. T. Tajima and J.M. Dawson, *Phys.Rev. Lett.*, **43**, 267 (1979)
4. I.V. Pogorelsky, A. Van Steenbergen, R. Fernow, W.D. Kimura, S.V. Bulanov, "CO_2 Laser Technology for Advanced Particle Accelerators", in these Proceedings
5. I.V. Pogorelsky, "Prospects for Compact High-Intensity Laser Synchrotron X-Ray and Gamma Sources", in these Proceedings
6. C. Rolland, P.B. Corkum, *J.Opt.Soc.Am.B*, **3**, 1625 (1986)
7. I.V. Pogorelsky, W.D. Kimura, C.H. Fisher, F. Kannari, and N.A. Kurnit, *6th Workshop on Advanced Accelerator Concepts*, June 12-18, 1994, Fontana, WI, AIP Conference Proceedings, **335**, 405 (1995)
8. P.B. Corkum and D. Keith, *J. Opt. Soc. Am.*, **B2**, 1873 (1985)
9. A.Y. Elezzabi, J. Meyer, M.K.Y. Hughes, and S.R. Johnson, *Opt. Lett.*, **19**, 898 (1994)
10. I.V. Pogorelsky, J. Fischer, K. Kusche, M. Babzien, N.A. Kurnit, I.J. Bijio, R.F. Harrison, and T. Shimada, *IEEE J. Quant. Electron.*, **31**, 556 (1995)
11. P.B. Corkum, *IEEE J. Quant. Electron.*, **QE-21**, 216 (1985)

PROSPECTS FOR COMPACT HIGH-INTENSITY LASER SYNCHROTRON X-RAY AND GAMMA SOURCES

I.V. Pogorelsky

Accelerator Test Facility, Brookhaven National Laboratory, 725C, Upton, NY 11973

Abstract. A laser interacting with a relativistic electron beam behaves like a virtual wiggler of an extremely short period equal to half of the laser wavelength. This approach opens a route to relatively compact, high-brightness x-ray sources alternative or complementary to conventional synchrotron light sources. Although not new, the laser synchrotron source (LSS) concept is still waiting for a convincing demonstration. Available at the BNL Accelerator Test Facility (ATF), a high-brightness electron beam and the high-power CO_2 laser may be used for prototype LSS demonstration. In a feasible demonstration experiment, 10-GW, 100-ps CO_2 laser beam will be brought to a head-on collision with a 10-ps, 0.5-nC, 50 MeV electron bunch. Flashes of collimated 4.7 keV (2.6 Å) x-rays of 10-ps pulse duration, with a flux of ~10^{19} photons/sec, will be produced via linear Compton backscattering. The x-ray spectrum is tunable proportionally to the e-beam energy. A rational short-term extension of the proposed experiment would be further enhancement of the x-ray flux to the 10^{22} photons/sec level, after the ongoing ATF CO_2 laser upgrade to 5 TW peak power and electron bunch shortening to 3 ps is realized.

In the future, exploiting the promising approach of a high-gradient laser wake field accelerator, a compact "table-top" LSS of monochromatic gamma radiation may become feasible.

I. INTRODUCTION

Generation and application of high-brightness, quasi-monochromatic x-rays and γ-rays is a fast developing area of science and technology. Contemporary synchrotrons equipped with wiggler magnets are the sources of the most intense x-ray fluxes, approaching the level of 10^{18} photon/sec within the spectral window of 0.1% of the photon energy. Diverse demands of multidisciplinary research, industrial, and medical applications drive the search for even more intense (as well as compact) x-ray sources. Meeting these requirements, the next generation synchrotron light sources (SLS) are under development based on the principles of the free-electron laser.

Another approach to a relatively compact x-ray LSS is based on using contemporary picosecond lasers of ultra-high peak power. The laser beam acts on relativistic electrons as an electromagnetic wiggler but has a period 10^4-10^5 times shorter than the conventional undulator. Thus, LSS permits generation of proportionally heavier photons at the same e-beam energy as a conventional SLS. Similarly, LSS permits the use of ~100 times less energetic electrons to generate x-rays of a particular wavelength, significantly downsizing the required electron accelerator.

Limited by the repetition rate of the laser driver, LSS does not compete with conventional synchrotron sources in average radiation power. However, it has the potential of delivering extremely intense x-ray flashes within picosecond and femtosecond time intervals. X-ray beams with 10^{10}-10^{11} photons/pulse may be produced via relativistic Compton scattering (which is a quantum-mechanical foundation of the LSS) using state-of-the-art terawatt picosecond lasers and high-brightness RF linacs that operate with picosecond photocathode injectors. Such pulses are of particular interest for a number of applications.

Why have these attractive features of the LSS not been utilized so far? Maybe because the appropriate laser and linac have yet to shake hands within one facility. We believe that at the ATF [1] we have such a unique opportunity. Indeed, the ATF's linac, the world's-brightest, delivers up to 0.5 nC, 10-0.4 ps, 50-MeV electron bunches with a peak current of ~100 A, momentum spread of 0.2%, and normalized emittance down to 0.5 mm.mrad [2]. That permits the high concentration of relativistic electrons necessary for efficient photon scattering. In addition, the ATF-operated picosecond CO_2 laser, which delivers ~10 times more photons than more common solid state picosecond lasers of a similar energy, permits the proportional increase of x-ray flux.

In Section II, we discuss the efficiency of x-ray generation via Compton backscattering of the CO_2 laser photons on a counter-propagating relativistic electron beam. As an example, in Section III we consider a feasible demonstration experiment at the ATF where the 10-GW pulses of 100-ps duration (from the presently operational CO_2 laser) will be brought into a head-on collision with the 0.5 nC, 10-ps bunches of the 50-MeV electron beam. The 4.7 keV (2.6 Å) x-rays of 10-ps pulse duration with a peak flux of 2×10^{19} photons/s will be obtained within the 0.4% bandwidth window. At the second stage of the ATF experiment, after the ongoing laser upgrade to the ~5 TW power level and the electron bunch shortening to 3 ps, x-ray flashes with peak flux of up to 2×10^{22} photons/sec may be produced.

As the next step to even more compact and economical x-ray and gamma sources, we consider substitution of a conventional electron accelerator with the projected ultra-high gradient laser accelerator. The enthusiasm that drives research in the area of laser acceleration is based on ultra-high electric fields, up to 10^{11} V/cm, attainable upon the tight focusing of terawatt laser beams. Being used for particle acceleration, such fields may permit reduction of accelerator dimensions by many orders of magnitude, putting SLS on a table top. After demonstration of a ~100 GeV/m acceleration gradient over a few millimeters of interaction region in the recent laser wakefield accelerator (LWFA) experiment [3], the next goal for the laser accelerator development is to extend the interaction region over a longer distance and to ensure the production of a good quality electron beam. The primary approach to longer interaction distances is based on laser waveguiding in dielectric or plasma channels. In Section IV, we consider the possibility of using the channeled LWFA to produce an electron beam with parameters required for a high-brightness LSS operating in the gamma-region and discuss considerations for choosing a particular type of laser to drive such a facility.

So far, the prime attention of researchers in the area of strong-field laser applications was attracted to the fast advancing technology of table-top terawatt (T^3) solid state lasers that operate at wavelengths of $\lambda \approx 1$ μm. A picosecond terawatt CO_2 (psTW-CO_2) laser may be suggested as another candidate for the LSS driver. The first laser of this kind is under development at the ATF [4]. In the present paper, we show how the LWFA-LSS concept may benefit from the relatively long wavelength ($\lambda \approx 10$ μm) CO_2 laser driver. Up to 250 MeV electron acceleration over a 4 cm interaction region with a plasma-channeled 5 TW CO_2 laser beam is predicted [5]. A feasible high-intensity LSS is based on using split beams of the psTW-CO_2 laser to accelerate copropagating electrons and, then, to generate up to 1.2 MeV gamma-rays at a peak flux above 10^{23} photons/sec in head-on collision with electrons within the same compact interaction chamber.

II. LINEAR COMPTON SCATTERING OF CO_2 LASER RADIATION FROM A RELATIVISTIC ELECTRON BEAM

The advantage of using a laser as a virtual wiggler in the x-ray LSS stems from the equation for the wavelength of photons backscattered from a relativistic electron beam,

$$\lambda_X = \lambda/4\gamma^2, \qquad (1)$$

where γ is the relativistic Lorentz factor. Short laser wavelength, λ, which is normally in the range of 0.3-10 μm, permits relatively easy access to the x-ray and gamma-ray region.

A comprehensive theoretical analysis of Compton photon scattering from relativistic electron beams may be found elsewhere [6,7]. Here, we compile and derive the expressions that help us to understand how the efficiency of x-ray Compton scattering depends upon the laser wavelength and other laser and e-beam parameters.

An electron, oscillating in an external electromagnetic (EM) field, emits radiation at a power of

$$P_e = 2e^2 \dot{V}^2 / 3c^2, \qquad (2)$$

where $\dot{V}$ is the acceleration vector (averaged over the oscillation period) defined by the acting oscillatory electric field, $|E_L|$, of the EM wave,

$$\dot{v} = e|E_L|/m, \quad (3)$$

and $|E_L|$ is related to the laser intensity $I_L = P_L / S_L$ via

$$|E_L|^2 = 2I_L / \varepsilon_0 c. \quad (4)$$

Here, P_L is the laser beam power, $S_L = \pi r_L^2$ is the effective cross-section for the laser beam, and r_L is the laser beam radius. Combining Eqs.(2) through (4), we obtain:

$$P_e = \frac{4c}{3\pi\varepsilon_0} P_L \left(\frac{r_e}{r_L}\right)^2, \quad (5)$$

where $r_e = e^2/mc^2 = 2.82 \times 10^{-13}$ cm is the classical electron radius.

The expression for P_e, corrected for the relativistic electron motion in a Gaussian laser beam, is [6]

$$P_e = \frac{64}{3} P_L \gamma^2 \left(\frac{r_e}{r_L}\right)^2. \quad (6)$$

Considering for simplicity a rectangular laser pulse of duration τ_L and a uniform electron distribution with a density n_e over the e-bunch duration τ_b, the power radiated by the electron ensemble within the immediate interaction region $L(t)$ is equal to

$$P_X(t) = P_e n_e L(t) S_L, \quad (7)$$

where $L(t)$ at any moment t is equal to or smaller than $\min\{c\tau_L, c\tau_b\}$. Here, we consider counter-propagating electron and laser beams because this geometry ensures the maximum number of generated x-ray photons for given laser and e-beam parameters. The radiation process will continue during the time interval while the laser pulse and the electron bunch are at least in a partial overlap. It is presumed that the e-beam radius r_b matches, or is smaller than, r_L. That ensures that the total electron bunch charge is used efficiently. Then, the total backward scattered radiation energy is

$$\mathbf{E}_X = \int_0^{(\tau_L+\tau_b)/2} P_X(t)\partial t = P_e n_e S_L c \left[\min\{\tau_L, \tau_b\}\left(\frac{1}{2}\min\{\tau_L, \tau_b\} + \left|\frac{\tau_L - \tau_b}{2}\right|\right)\right] =$$

$$= P_e n_e S_L c \frac{\tau_L \tau_b}{2}. \quad (8)$$

Note that the actual duration of the produced backscattered pulse is not equal to the total laser-electron interaction time interval, $\tau_{int} = (\tau_L + \tau_b)/2$. Since x-rays originate from electrons in the direction of the electron propagation, it turns out that the x-ray pulse length is defined primarily by the electron bunch duration. On top of it is the x-ray pulse stretching due to the x-ray and electron speed mismatch accumulated over the interaction time,

$$\Delta\tau = (1-\beta)\frac{\tau_L + \tau_b}{2} \approx \frac{\tau_L + \tau_b}{4\gamma^2}. \quad (9)$$

Hence, the total x-ray pulse duration is

$$\tau_X = \tau_b + \Delta\tau = \frac{1-\beta}{2}\tau_L + \frac{3-\beta}{2}\tau_b \approx \tau_b + \frac{\tau_L}{4\gamma^2}. \quad (10)$$

However, for the practically important case when $\tau_L \ll 4\gamma^2\tau_b$, we can consider $\tau_X \approx \tau_b$. It follows that relatively long laser pulses may be tolerated without a noticeable increase in the x-ray pulse duration above τ_b. Combining Eqs.(6), (8) and (10), the expression for x-ray power is

$$P_X = \frac{\mathbf{E}_X}{\tau_X} \approx \frac{32\pi}{3}\mathbf{E}_L n_e c\gamma^2 \mathbf{r}_e^2 \approx \frac{10\mathbf{E}_L Q \gamma^2 r_e^2}{e\tau_b r_L^2}, \quad (11)$$

where $\mathbf{E}_L = P_L \tau_L$ is the laser pulse energy, Q is the electron bunch charge and r_b is the e-beam radius in the interaction region (we consider $r_b \leq r_L$, so that no electrons are lost for interaction). Corresponding to Eq.(11), the engineering formula is

$$P_X[W] \approx \frac{0.5\mathbf{E}_L[J]Q[nC]\gamma^2}{\tau_b[ps]r_L^2[mm]} \quad (12)$$

The conclusion drawn from Eq.(11) is that the intensity of the produced x-rays does not depend directly upon the laser pulse duration. The high total laser pulse energy, its tight focusing, and short-duration of the relativistic electron bunch become the most important parameters that define the x-ray peak power. In addition, in order to be in a good overlap, the electron beam shall be focused to the same or smaller spot size as the laser beam.

Because of the natural divergence of the electron and laser beams, they remain tightly focused only within a limited distance. For the Gaussian optical beam, this distance is related to the Rayleigh length, z_0, measured from the focal point to the point where the laser beam expands two times in a cross-sectional area,

$$L_L \approx \pi z_0 = \pi^2 r_L^2 / \lambda. \quad (13)$$

A similar condition is applicable to electron beams with their divergence characterized by geometric emittance ε/γ in the same way as diffraction-limited laser beams are characterized by wavelength,

$$L_b = \pi^2 r_b^2 \gamma / \varepsilon. \quad (14)$$

The comparison of λ with ε/γ tells which one, e-beam or laser beam, is the limiting factor for making the interaction region narrow. Typical low-emittance electron beams provided by RF linacs permit much tighter filamentation than focused laser beams. Thus, we consider that limitations to the interaction length are imposed by the laser beam.

For the efficient use of the laser pulse in the Compton process, the laser beam waist length, L_L, shall extend over the overlap distance defined by the electron and laser pulse duration,

$$L_L \geq c(\tau_L + \tau_b)/2. \tag{15}$$

If the interaction distance is below the limit defined by Eq.(15), it may result in the proportional drop of the laser energy coupled to the electron bunch. Thus, overly-tight laser focusing would not help to gain any additional x-ray power. From Eqs.(13) and (15), it follows that, at $\tau_L \geq \tau_b$, laser parameters shall also satisfy the condition

$$r_L^2 \geq \frac{c\tau_L \lambda}{2\pi^2}. \tag{16}$$

Combining Eqs.(12) and (16), we obtain the estimate for the maximum attainable x-ray power, P_X^{max}, when the laser beam divergence is the limiting factor for the length and cross-section of the interaction area,

$$P_X^{max}[kW] \approx \frac{30 P_L[TW] Q_{[nC]} \gamma^2}{\tau_{b[ps]} \lambda_{[\mu m]}}. \tag{17}$$

Note, that Eq.(12) is still valid when realistic r_b is above the limit set by Eq.(16).

Power of the x-ray beam is just one of parameters entering into the expression for spectral brightness.

$$B = \frac{P_X}{(2\pi R \theta_{\Delta v})^2 \Delta v_X}. \tag{18}$$

Here, $\theta_{\Delta v}$ is the angular divergence of the radiation within the spectral bandwidth Δv_X. The total angular divergence of the backscattered radiation is defined by γ,

$$\theta_o = 1/\gamma. \tag{19}$$

Within this cone, the scattered radiation frequency drops off axis according to

$$\Delta v_X / v_X \approx \gamma^2 \Delta \theta^2, \tag{20}$$

thus, $\Delta v_X / v_X \approx 1$ over $\Delta \theta = \theta_o$.

It follows from Eq.(1) that the bandwidth of the backscattered x-rays is directly related to the momentum spread or the "temperature" of the e-beam as

$$\Delta v_X / v_X = 2\Delta \gamma / \gamma. \tag{21}$$

This narrow bandwidth radiation is observed within the cone with the opening angle

$$\theta_{\Delta v} \approx (\Delta v_X / v_X)^{1/2} / \gamma. \tag{22}$$

From the above expressions for P_X, Δv_X, and $\theta_{\Delta v}$ we can conclude that, in order to produce high-brightness x-rays for a particular wavelength, it is beneficial to choose possibly high values for both parameters, γ and λ. To illustrate this conclusion, let us compare CO_2 ($\lambda = 10$ μm) and Nd:YAG ($\lambda = 1$ μ

m) lasers of equal power. As long as $\lambda_x = \lambda/4\gamma^2$ is considered as an invariant, then choosing the CO_2 laser, with its wavelength 10 times longer than the solid state laser wavelength, requires a $\sqrt{10}$ times more energetic e-beam. These leads to the improvement of the angular divergence of the produced x-rays which is equal to $1/\gamma$, and reduction of spectral bandwidth defined by $\Delta\gamma/\gamma$. According to $P_X \sim E_L \gamma^2$ (see Eq.(12)), the backscattered x-ray power will rise 10 times due to the higher γ. This actually stems from the facts that the x-ray flux is proportional to the delivered laser photon flux, and, at the fixed laser energy, the last parameter is proportional to λ. When combining together all of the mentioned above, we come to the conclusion that using a CO_2 laser as the LSS driver opens the prospect for up to 300 times increase in the spectral brightness of the produced x-rays of a particular wavelength compared with using a 1-μm laser of the same energy.

The above remark on the advantage of using a long-wavelength CO_2 laser is valid provided other parameters entering to Eq.(12) are equal. In a situation when the cross-section of the interaction region is limited by the diffraction divergence of the laser beam, which is proportional to λ, P_X does not depend any more upon λ. This conclusion follows also from Eq.(17) under the assumed above condition of $\gamma \sim \sqrt{\lambda}$. However, it is most likely that the radius of the interaction region will not be reduced much below 30 μm due to such practical considerations as alignment and pointing stability of the laser and e-beam. Focusing to r_L=30 μm may be obtained with the CO_2 laser as well. Under this condition, the wavelength power scaling discussed above is still valid.

III. DESIGN PARAMETERS FOR THE ATF LASER SYNCHROTRON SOURCE

LSS, based on linear relativistic Compton backscattering of picosecond CO_2 laser pulses, may be realized at the ATF with its presently available 10 GW, 100-ps CO_2 laser and 10-ps, 0.5 nC, 50 MeV (γ=100) electron beam. For the electron beam, we assume the normalized emittance of ε =2 mm.mrad and momentum spread of $\delta p/p$=0.2%, as already demonstrated at the ATF. Quantitative estimates for the parameters of the feasible ATF LSS experiment are done in Table1 (Stage 1).

By Eq.(1), the wavelength of the backscattered radiation is $\lambda_x = \lambda/4\gamma^2$=2.6 Å. That corresponds to the photon energy of $h\nu_x[eV]=1.25\times10^4/\lambda_x[Å]$=4.7 keV.

Spectral tuning of the generated x-rays is possible via γ adjustment. The x-ray power, Eq.(12), will vary proportionally to γ^2, and the x-ray brightness,

Eq.(18), proportionally to γ^5. Another way for fine tuning the x-ray wavelength is by tuning the CO_2 laser wavelength within the CO_2 gain spectrum (9.2-11.8 µm).

According to Eq.(15), the length of the interaction region, where the focused laser beam shall match in a cross-section the counter-propagating e-beam, is L_{int}=17 mm.

As discussed above, the comparison of λ with ε/γ tells which one, e-beam or laser beam, is the principal limiting factor for making the interaction region narrow. The demonstrated at the ATF geometric emittance, $\varepsilon/\gamma = 2\times10^{-2}$ µm.rad, being orders of magnitude smaller than the CO_2 laser wavelength, λ =10.6 µm, indicates that the CO_2 beam diffraction divergence will ultimately define the effective radius of the interaction region. Using Eq.(13), we calculate the waist radius of the diffraction limited CO_2 laser beam to be $r_L = \sqrt{L_{int}\lambda/\pi} \approx 140$ µm. Hence, the e-beam shall be focused to the same size, r_b=140 µm

At the specified laser and e-beam parameters, the x-ray flux is 2×10^{19} photons/sec and the total number of the x-ray photons per pulse is equal to 2×10^8 photons/pulse.

Utilizing prospective features of the ATF that are under development, we may project an advancement to the Stage 2 of the ATF LSS experiment. For the upgraded LSS version, consider a 70 MeV, 3-ps electron bunch, attained via compression of the photocathode UV laser driver pulse in a nonlinear crystal [8], and a 5-TW 3-ps CO_2 laser [4].

Note, that with a short, 3-ps laser pulse, the x-ray intensity increase will be attained, according to Eq.(12), via shortening of the interaction region that permits tighter focusing of the laser and electron beams. With the short interaction length L_{int}=1 mm (defined by the short laser and electron pulses, both 3 ps), the laser beam waist may be brought to r_L=30 µm. Assuming the corresponding focusing of the e-beam, the 9.36 keV x-ray peak flux will be increased to 2×10^{22} photons/sec. The total x-ray photon number in this case is 6×10^{10} photons/pulse. At the short, 3-ps laser pulse, the x-ray spectrum is defined not just by the e-beam temperature, as according to Eq.(21), but will broaden to

$$\Delta\lambda'_x/\lambda_x = 1/N, \qquad (23)$$

where N is the total number of laser wavelengths during the electron-laser interaction, that is 100 for τ_L=3 ps, thus, $\Delta\lambda'_x/\lambda_x \approx 1\%$. The angle-averaged brightness is $B(1\% \text{ b.w.}) \approx 2\times10^{21}$ photons/sec.mm^{2}mrad^2 or $B(0.1\% \text{ b.w.}) \approx 2\times10^{20}$ photons/sec. mm^{2}mrad^2.

It becomes evident that, in order to maintain high x-ray monochromaticity, the CO_2 laser pulse duration shall be relatively long, $\tau_L \geq 30$ ps. The contradiction involved in the desire to have a long interaction distance with a narrow cross-section may be resolved by channeling the laser beam. The idea of laser beam channeling, relevant also for x-ray lasers, laser acceleration in plasma, and laser fusion, is presently under extensive study. Two primary approaches look feasible and have been demonstrated experimentally: dielectric waveguide [9] and preformed plasma channel [10].

With a 5-TW CO_2 laser beam focused to a 30 µm diameter spot, as high as 10^{17} W/cm^2 intensity will be attained. Such intensity corresponds to the unitless laser strength of $a=3$ defined by the expression

$$a = 0.85 \times 10^{-9} \lambda [\mu m] I^{1/2} [W/cm^2] = 0.3 E_L [TV/m] \lambda [\mu m]. \quad (24)$$

It is known that at $a>1$, the nonlinear Compton scattering effect comes to the scene [6]. With the maximum order of generated harmonics equal to $n=a^3$, as short as $\lambda = 0.05$ Å radiation may be produced when a CO_2 laser beam is backscattered by the 70-MeV e-beam. The maximum of the intensity distribution shifts to $n=11-13$, with these components ~3 times more intense than the fundamental $n=1$ component [6].

Other interesting features of harmonics are that their angular divergence and spectral bandwidth are inversely proportional to n, offering prospects for considerable increase in a spectral brightness. Study of nonlinear Compton scattering may be carried out as the next stage of the ATF LSLS experiment.

The parameters of the advanced LSS summarized in Table 1 (Stage 2) are calculated for linear Compton scattering only. In addition, a multitude of high-brightness harmonics will be observed as discussed above.

Table 1. Design Parameters of the ATF X-Ray LSS Experiment

Experiment Staging:	Stage 1	Stage 2
ELECTRON BEAM		
Energy [MeV]	50	70
Bunch Charge [nC]	0.5	0.5
Bunch Duration FWHM [ps]	10	3
Radius at Focus [µm]	140	30
Waist Length [mm]	>100	>10
Normalized Emittance [mm.mrad]	2	2
Electron Momentum Spread [%]	0.2	~0.05
CO_2 LASER		
Pulse Duration [ps]	100	3
Peak Power [GW]	10	5000
Laser Energy [J]	1	15
Radius at Focus [µm]	140	30
Waist Length [mm]	17	1
X RAYS		
Wavelength [Å]	2.6	1.35
Pulse Duration [ps]	10	3
Angular Spread [mrad]	10	7
Spectral Bandwidth [%]	0.4	1
Photons per Pulse	2×10^8	6×10^{10}
Peak Flux [photons/s]	2×10^{19}	2×10^{22}
Peak Brightness (/0.1%) [photon/sec.mm^{2}mrad^2]	10^{16}	2×10^{20}

IV. TABLE-TOP LASER WAKEFIELD γ-RAY SYNCHROTRON SOURCE

The idea of a laser wakefield electron accelerator (LWFA) [11,12], that is currently under intense study, leads to record accelerating gradients, hundreds times higher than with conventional linacs [3]. A combination of plasma accelerator with LSS may open a route to the extremely compact, table-top wakefield x-ray and gamma LSS.

Another feature important for the LSS is the high quality of the produced e-beam that includes: a small (close to 0.1%) energy spread, and a low emittance. In order to generate a high-quality e-beam in the LWFA, we might want to start with a similar or better quality seed e-beam and preserve it in the course of the laser acceleration in plasma. The requirement for small divergence of the Compton backscattered photons puts a limit on the e-beam normalized emittance: ε[mm.mrad]<<r_L [μm]. The requirement for the small energy spread presumes that the seed electron bunches should be much shorter than the plasma wake period

$$\tau_b \ll \lambda_p/c, \qquad (25)$$

where plasma wavelength, λ_p, is related to the plasma density N_e via

$$N_e[cm^{-3}] = 10^{21} \lambda_p^{-2}[\mu m]. \qquad (26)$$

The 0.1% energy spread calls for a monochromatic electron bunch of the $\tau_b \leq \lambda_p/30c$ duration. It follows from Eq.(26) that the long wavelength of the CO_2 laser permits a 100 times plasma density reduction in comparison with the more common solid state lasers that have λ≈1 μm. The reduced electron scattering in a low density plasma will help to preserve a low e-beam emittance.

Contemporary photocathode RF guns may serve as injectors of electrons with the required parameters. In particular, 370 fs electron bunches of 5 MeV energy, 40 pC charge (2.5×10^8 electrons), 0.15% energy spread, and 0.5 mm.mrad normalized emittance have been demonstrated at the ATF by proper phasing of photoelectrons with RF field [2]. As short as 100 fs bunches may be expected after further development of this technique.

Let us consider a prospective LWFA driven by a 5 TW CO_2 laser. For the standard LWFA scheme [11], the duration of the driver laser pulse shall satisfy the requirement

$$\tau_L \approx \lambda_p/2c. \qquad (27)$$

It is understood that, due to the natural diffraction of the laser beam, the interaction length may be limited to a few millimeters. Thus, some sort of laser beam channeling is required. The condition for plasma waveguiding is [13]

$$\Delta N_e \left[cm^{-3} \right] = 1.1 \times 10^{20} / r_L^2 [\mu m]. \tag{28}$$

Here, ΔN_e is the height of a wall in a plasma channel that is required to guide a laser beam.

Thus, we see that τ_b and τ_L define λ_p through Eqs.(25) and (27), and N_e via Eq.(26). At the same time, the chosen plasma density limits the laser focal spot size to the radius r_L defined by Eq.(28), thus limiting the attained laser field, E_L, to the value

$$E_L[TV/m] = 2.6 \times 10^{-9} I_L^{1/2} \left[W/cm^2\right] = 15 \times P_L^{1/2} [TW] r_L [\mu m]. \tag{29}$$

N_e is also one of the factors that defines the efficiency of conversion of the laser EM field into the accelerating wakefield, E_a,

$$\eta \equiv E_a / E_L = 1.6 \times 10^{-11} N_e^{1/2} \left[cm^{-3}\right] \lambda [\mu m] a_L / \left(1 + a_L^2 / 2\right)^{1/2}. \tag{30}$$

Another parameter that enters Eq.(30) is unitless laser strength, a, defined by Eq.(24).

This completes the self-consistent set of equations that defines the only solution for plasma and waveguide parameters and, ultimately, the acceleration gradient for the given input laser and e-bunch parameters.

With the theoretically shortest CO_2 laser pulse duration of $\tau_L=1$ ps, that has been demonstrated experimentally [14], we obtain wakefield parameters presented in Table 2 (Standard LWFA). With the calculated rather moderate acceleration gradient of $E_a=0.5$ GV/m, there is still an opportunity to obtain a 50-MeV electron beam after passing a 10 cm long plasma waveguide.

Conventional linacs are capable to produce up to 100 MV/m accelerating fields. To demonstrate practical advantages of the new technology for compact LSS in a more pronounced way, the prototype laser accelerator should be capable of producing at least several hundred MeV electrons in a compact device.

Acceleration gradients much higher than with the standard LWFA, have been demonstrated for, so called, self-modulated (SM) LWFA regime [12] when laser power, P_L, satisfies the condition for relativistic self-focusing (RSF):

$$P_L \geq 17 \left(\omega/\omega_p\right)^2 [GW]. \tag{34}$$

With a CO_2 laser of a $P_L=5$ TW power, the condition in Eq.(34) is satisfied at $\lambda_p \leq 170$ μm ($N_e \geq 3.5 \times 10^{16}$ cm^{-3}). The corresponding wake period is 500 fs, which is still much longer than the projected electron bunch duration (~100 fs). Note, that a 1-μm laser of similar power requires 10 times smaller λ_p and, accordingly, 100 times higher plasma density [15]. This will adversely impact both e-beam divergence, due to the multiple scattering in the gas, and energy spread, due to the high proportion of $\tau_b c / \lambda_p$.

The complexity of the processes involved in the generation of a wake field with a self-modulated relativistically strong laser beam guided in a plasma channel does not allow analytical solution. Simulations performed close to the conditions specified in Table 2 (Self-Modulated LWFA) predict 250 MeV acceleration over a 4-cm interaction distance [5]. Even higher, up to 450 MeV, acceleration has been simulated for longer interaction distances [16]. It has been shown also that the net acceleration does not depend much upon the initial electron energy, and 5-MeV seed electrons may be used with the same efficiency as 50-MeV electrons. Based on these results, we assume 250 MeV SMLWFA accelerating stage for a table-top LSS in Table 2.

Further simulations are required to determine the quality of the produced e-beam. Preliminary estimates of the electron energy spread are based on the proportion between the e-bunch length and wake period. Normalized emittance of the electron beam is considered not to be effected by the plasma wake.

Linear and nonlinear Compton scattering of laser photons on 250 MeV electrons will produce high-intensity monochromatic hard x-rays and gamma rays in a table-top device that may offer unique capabilities for multidisciplinary research, ultra-high resolution lithography and microscopy.

Parameters in Table 2 are calculated for conditions of laser guiding in a plasma channel. Dielectric capillary tubes may offer another possibility for guiding high-power laser beams in LWFA. This opportunity is based on the finite velocity of the plasma front that is originated at the capillary wall due to the laser ablation. At a typical plasma implosion velocity of several km/sec, it should take several nanoseconds for the overdense plasma to fill a capillary. Thus, the picosecond laser pulse will not be effected by capillary ablation. Tests with a picosecond Nd:YAG laser [9] show the viability of this approach.

The principal schematic of the table-top x-ray LSS is presented in Fig.1. A 5-MeV photocathode electron gun serves as the injector of a high-brightness electron beam. A quadrupole triplet or solenoid focuses the e-beam into the 120 µm wide laser waveguide. The interaction cell is filled with H_2 at ~1 torr and is separated from the gun by the ≤1 µm thick diamond window. The CO_2 laser beam is split into two beams that serve to drive both LWFA and LSS. The delay between the split beams is adjusted optically to make both beams synchronized with the femtosecond electron bunch. The LWFA driver laser beam is focused at the entrance of the waveguide with a parabolic Cu mirror drilled in the center to transmit the input e-beam. Via tunnel ionization of H_2, the laser beam produces plasma of 3.5×10^{16} cm^{-3} electron density and generates a strong plasma wave. Another parabolic mirror with a central hole serves to focus the LSS driver laser beam at the exit of the waveguide. At this exit, electrons accelerated in the plasma wake field interact with a split portion of the CO_2 laser beam generating backscattered x- or γ-rays. A downstream dipole serves to separate photons from electrons which are sent to a dump to reduce the x-ray noise. X-rays or gamma quanta are extracted through the output foil window.

Note, that for simplicity of the drawing, we presume a LWFA with a dielectric laser waveguide. In the case of the plasma channel, the accelerator stage will be modified as illustrated in Fig.5 in the reference [15] of these Proceedings.

Table 2. Design Parameters for Compact X-Ray and Gamma LSS

Laser Accelerator Type:	Standard LWFA	Self-Modulated LWFA
INPUT ELECTRON BEAM		
Energy [MeV]	5	5
Bunch Charge [nC]	0.5	0.5
Bunch Duration FWHM [fs]	100	100
Radius at Focus [μm]	30	30
Waist Length [cm]	16	16
Normalized Emittance [mm.mrad]	0.5	0.5
Electron Momentum Spread [%]	0.2	0.2
PLASMA ACCELERATOR		
Laser Peak Power [TW]	5	5
Laser Pulse Duration [ps]	1	3
Laser Focus Radius [μm]	220	60
Laser Waist Length [cm]	4	0.3
Plasma Density [cm^{-3}]	2.3×10^{15}	3.5×10^{16}
Plasma Wavelength [μm]	660	170
Channel Radius [μm]	220	60
Channel Length [cm]	10	4
Acceleration gradient [GV/m]	0.5	6
Energy Gain with Channeling [MeV]	50	250
10-th HARMONIC IN COMPTON BACKSCATTERING		
Laser Peak Power [TW]	5	5
Laser Pulse Duration [ps]	1	3
Laser Focus Radius [μm]	30	30
Laser Waist Length [cm]	0.1	0.1
Photon Energy [keV]	47	1200
Pulse Duration [fs]	100	100
Angular Spread [mrad]	1	0.2
Spectral Bandwidth [%]	0.2	0.1
Photons per Pulse	2×10^{10}	6×10^{10}
Peak Flux [photons/s]	2×10^{23}	6×10^{23}

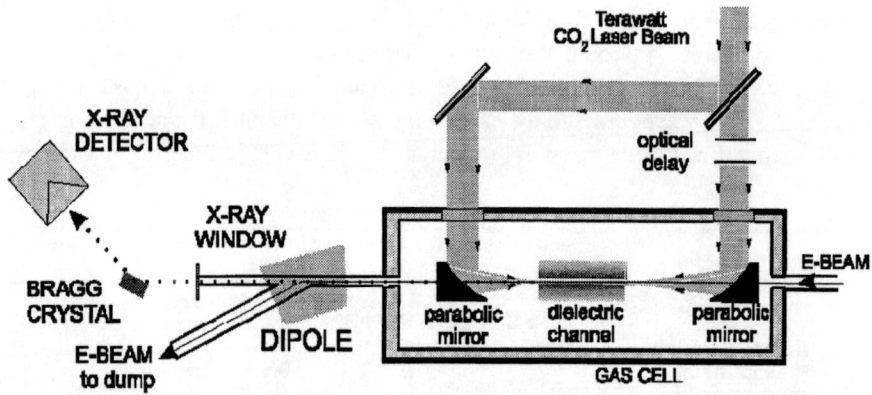

FIGURE 1. Principal diagram of the compact LSS based on laser wakefield acceleration in a waveguide

Then, before interacting with the LSS driver laser beam, the accelerated electrons penetrate the hole in the 45° mirror that serves to direct the third CO_2 laser beam to the axicon mirror.

V. CONCLUSIONS

At the ATF, the world's brightest 50-MeV linac and a high-power picosecond CO_2 laser may serve as brick stones for the demonstration Laser Synchrotron Source experiment. After the ongoing laser upgrade to the terawatt power level, monochromatic x-ray flashes with peak flux of up to 10^{22} photons/sec may be produced, orders of magnitude above the numbers achieved with conventional synchrotron light sources. With a terawatt CO_2 laser, the expected nonlinear Compton scattering effect become efficient and may be used for significant expansion of the x-ray spectrum to benefit the potential applications.

As a next step to even more compact and economical x-ray and gamma sources, we consider substitution of a conventional electron accelerator with a high-gradient laser accelerator. We discuss laser wavelength impact on the efficiency of the laser wakefield electron acceleration, quality of the accelerated e-beam, and parameters of the produced gamma radiation, demonstrating advantages of CO_2 lasers for such an application.

250 MeV electrons and 1.2 MeV gamma quanta at a flux of 6×10^{23} photons/sec may be produced in a table-top device using a conventional photocathode RF gun and 5-TW CO_2 laser that will be commissioned at the ATF in 1997-1998. The previously discussed feasible LSS may offer unique

capabilities for multidisciplinary research, ultra-high resolution lithography, and microscopy.

ACKNOWLEDGMENTS

The author wish to thank I. Ben-Zvi for encouragement of this study and fruitful discussions, T. Srinivasan-Rao, K. McDonald, P. Siddons, J. Hastings, X-J. Wang, J. Skaritka, and K.Kusche for discussions and help in preparation of the upcoming tests.
The work is supported by the US Department of Energy.

REFERENCES

1. www.bnl.gov/atf
2. X.J. Wang, X. Qui, and I. Ben-Zvi, "Experimental Observation of High-Brightness Micro-Bunching in a Photocathode RF Electron Gun", *Phys. Rev. E*, **54**, R3121 (1996)
3. A. Madena, Z. Najmudin, A.E. Dangor, et.al., *Nature*, **377**, 606 (1995); T. Katsouleas, "Laser Acceleration of Electrons in Plasmas", *Joint Meeting of APS and AAPT*, May 2-5, 1996, Indianapolis, IN
4. I.V. Pogorelsky, I. Ben-Zvi, J. Skaritka, Z. Segalov, M. Babzien, K. Kusche, I.K. Meskovsky, V.A. Lekomtsev, A.A. Dublov, Yu.A. Boloshin, and G.A. Baranov, "The First Terawatt Picosecond CO_2 Laser for Advanced Accelerator Study at the Brookhaven ATF", *these Proceedings*
5. S.V.Bulanov, T.J. Esirkepov, N.M. Naumova, F. Pegoraro, I.V. Pogorelsky, and A.M. Pukhov, *IEEE Trans. on Plasma Sci.*, **24**, 393 (1996)
6. E. Esarey, P. Sprangle, and S.K. Ride, "Nonlinear Thomson Scattering of Intense Laser Pulses from Beams and Plasmas", *Preprint NRL/MR/6790-93-7365* (1993)
7. S.K. Ride, E. Esarey, and M. Baine, "Thompson Scattering of Intense Lasers from Electron Beams at Arbitrary Interaction Angles", *Phys. Rev. E*, **52**, #5 November (1995)
8. A. Umbrasas, J-C. Diels, J. Jacob, G. Valiulis, A. Piskarkas, *Opt. Lett.*, **20**, 2228 (1995)
9. S. Jackel, R. Burris, J. Grun, A. Ting, C. Manka, K. Evans, and J. Kosakowskii, *Opt. Lett.*, **20**, 1086 (1995)
10. C.G. Durfee III and H.M. Milchberg, *Phys. Rev. Lett.*, **71**, 2409 (1993)
11. T. Tajima and J.M. Dawson, *Phys. Rev. Lett.*, **43**, 267 (1979)
12. P. Sprangle, E. Esarey, J. Krall, and G. Joice, *Phys. Rev. Lett.*, **69**, 2200 (1992)
13. E. Esarey, P. Sprangle, J. Krall, and A. Ting, *IEEE Trans. On Plasma Sci.*, **24**, 252 (1996)
14. P.B. Corkum, *IEEE J. Quant. Electron.*, **QE-21**, 216 (1985)
15. I.V. Pogorelsky, I. Ben-Zvi, A. Van Steenbergen, R. Fernow, W.D. Kimura, and S.V. Bulanov, "CO_2 Laser Technology for Advanced Particle Accelerators", *these Proceedings*
16. S.V. Bulanov, G.I. Dudnikova, N.M. Naumova, F. Pegoraro, I.V. Pogorelsky, V.A.Vshivkov, "Charged Particle Acceleration in Nonuniform Plasmas", *these Proceedings*

LIST OF PARTICIPANTS

Alvis, Rosa M.	Lawrence Berkeley National Laboratory
Assmann, Ralph W.	Stanford Linear Accelerator Center
Bane, Karl	Stanford Linear Accelerator Center
Barov, Nikolai	Argonne National Laboratory
Barty, Chris	University of California, San Diego
Ben-Zvi, Ilan	Brookhaven National Laboratory
Bernard, Denis	Ecole Polytechnique
Bhawalkar, Dilip	Center for Advanced Technology, India
Bobin, J. Louis	Univ. Pierre et Marie Curie
Carlsten, Bruce E.	Los Alamos National Laboratory
Carrigan, Jr., Richard A.	Fermi National Accelerator Lab.
Caryotakis, George	Stanford Linear Accelerator Center
Chattopadhyay, Swapan	Lawrence Berkeley National Laboratory
Chen, Chiping	Massachusetts Inst. of Technology
Chen, Pisin	Stanford Linear Accelerator Center
Chen, Szu-yuan	University of Michigan
Chiou, Tzeng-Chih	University of Southern California
Chou, Ping J.	Stanford Linear Accelerator Center
Clayton, Chris	University of California, Los Angeles
Colby, Eric	UCLA/FNAL
Colestock, Patrick	Fermi National Accelerator Lab.
Conde, Manoel	Argonne National Laboratory
Corlett, John N.	Lawrence Berkeley National Laboratory
Corsini, Roberto	CERN, Switzerland
Cowan, Tom	Lawrence Livermore National Lab.
Dahl, Per	Lawrence Berkeley National Laboratory
Debenham, Philip H.	U.S. Department of Energy
Delayen, Jean	Jefferson Lab.
Dodd, Evan S.	University of Michigan
Dowell, David H.	Boeing Physical Sci. Res. Center
Downer, Michael	University of Texas at Austin
Esarey, Eric	Naval Research Laboratory
Fang, Jyan-min	Columbia University
Fazio, Michael V.	Los Alamos National Laboratory
Fernow, Richard C.	Brookhaven National Laboratory
Finley, David A.	Fermi National Accelerator Lab.
Fiorito, Ralph	Naval Research Laboratory
Fisch, Nat	Princeton Plasma Physics Lab.
Flechtner, Donald D.	Cornell University
Fontana, Jorge R.	University of California, Santa Barbara
Freeman, Richard R.	Lawrence Livermore National Lab.
Gai, Wei	Argonne National Laboratory
Gallardo, Juan C.	Brookhaven National Laboratory
Gold, Steven H.	Naval Research Laboratory
Gordon, Dan	University of California, Los Angeles
Gou, Sankui	University of Iowa
Govil, Richa	Lawrence Berkeley National Laboratory
Granatstein, Victor L.	University of Maryland

Haimson, Jacob	Haimson Research Corp.
Hartemann, Frederic V.	UC Davis/LLNL
Hemker, Roy	University of California, Los Angeles
Henke, Heino	Technical University, Berlin, Germany
Hirshfield, Jay L.	Yale University
Ho, Ching-Hung	SRRC, Taiwan
Houck, Timothy L.	Lawrence Livermore National Lab.
Hsu, Jui-Lung	University of Southern California
Huang, Yen-Chieh	Stanford University
Huang, Zhirong	Stanford Linear Accelerator Center
Hubbard, Richard F.	Naval Research Laboratory
Jones, Roger	Stanford Linear Accelerator Center
Joshi, Chan	University of California, Los Angeles
Kando, Masaki	Japan Atomic Energy Research Inst.
Katsouleas, Thomas C.	University of Southern California
Kerman, Arthur	Massachusettes Inst. of Technology
Kim, Jin-soo	Fusion and Accelerator Research
Kim, Kwang-Je	Lawrence Berkeley National Laboratory
Kimura, Wayne	STI Optronics
Kirk, Harold G.	Brookhaven National Laboratory
Kreischer, Kenneth E.	Massachusetts Inst. of Technology
Krishnagopal, Srinivas	Center for Advanced Technology, India
Kroll, Norman M.	University of California, San Diego
Kusche, Karl P.	Brookhaven National Laboratory
Lawson, Wesley G.	University of Maryland
LeBlanc, Stephen Paul	University of Texas at Austin
LeSage, Gregory P.	LLNL/UC Davis
Leemans, Wim P.	Lawrence Berkeley National Laboratory
Li, Derun	University of California, San Diego
Lidia, Steve M.	Lawrence Berkeley National Laboratory
Liu, Yabo	UCLA/BNL
Luhmann, Jr., Neville G.	UC Davis/LLNL
Mako, Frederick M.	FM Technologies, Inc.
Marquardt, Niels	Univ. Dortmund, Germany
Marques, Jean-Raphael	Ecole Polytechnique, France
Marsh, Ken	University of California, Los Angeles
Mikhailichenko, Alexander	Cornell University
Milchberg, Howard M.	University of Maryland
Mills, Fred	Fermi National Accelerator Lab.
Mine, Philippe	Ecole Polytechnique, France
Moore, Christopher I.	Naval Research Laboratory
Moretti, Alfred	Fermi National Accelerator Lab.
Mori, Warren B.	University of California, Los Angeles
Mourou, Gerard	University of Michigan
Nakajima, Kazuhisa	KEK, Japan
Narang, Ritesh	University of California, Los Angeles
Nation, John A.	Cornell University
Nezhevenko, Oleg	Omega-P Inc.
Noble, Robert J.	Fermi National Accelerator Lab.
Ogden, Viki C.	University of California, Los Angeles
Palmer, Dennis T.	Stanford Linear Acclerator Center

Pantell, Richard H.	Stanford University
Pellegrini, Claudio	University of California, Los Angeles
Peters, Gerald J.	U.S. Department of Energy
Phillips, Robert M.	Stanford Linear Accelerator Center
Pogorelsky, Igor V.	Brookhaven National Laboratory
Ponce, David M.	Lawrence Berkeley National Laboratory
Power, John	Argonne National Laboratory
Quimby, David	STI Optronics
Rau, Bernhard R.	University of Texas at Austin
Schoessow, Paul	Argonne National Laboratory-East
Schroeder, Carl B.	UCB/LBNL
Schultz, Sheldon	Stanford Linear Accelerator Center
Serafini, Luca	INFN, Italy
Sessler, Andrew	Lawrence Berkeley National Laboratory
Shadwick, Brad	UCB/LBNL
Siemann, Robert	Stanford Linear Accelerator Center
Shvets, Gennady	Princeton University
Siders, Craig	Los Alamos National Laboratory
Simpson, James D.	Argonne National Laboratory-East
Skvortsov, Vladimir A.	High Energy Density Res. Center, Russia
Smedley, John M.	Brookhaven National Laboratory
Smith, David R.	University of California, San Diego
Song, Joshua	Argonne National Laboratory
Spitkovsky, Anatoly	SLAC/UCB
Sprangle, Phillip	Naval Research Laboratory
Srinivasan-Rao, Triveni	Brookhaven National Laboratory
Steinhauer, Loren C.	University of Washington, Redmond
Sutter, David F.	U.S. Department of Energy
Tajima, Toshiki	University of Texas at Austin
Tang, Huan	Stanford University
Tani, Keiji	Japan Atomic Energy Research Inst.
Tantawi, Sami	Stanford Linear Accelerator Center
Temkin, Richard J.	Massachusetts Inst. of Technology
Ting, Antonio	Naval Research Laboratory
Tremaine, Aaron M.	University of California, Los Angeles
Trotz, Seth R.	Massachusetts Inst. of Technology
Tzeng, Kuo-Cheng	University of California, Los Angeles
Uesaka, Mitsuru	University of Tokyo, Japan
Van Bibber, Karl A.	Lawrence Livermore National Lab.
Van Steenbergen, Arie	Brookhaven National Laboratory
Vanecek, David	Lawrence Berkeley National Laboratory
Villa, Francesco	Stanford Linear Accelerator Center
Vlieks, Arnold	Stanford Linear Accelerator Center
Vogel, Nadja I.	Lawrence Berkeley National Laboratory
Volfbeyn, Pavel	Massachusetts Inst. of Tech./LBNL
Wang, Haipeng	Brookhaven National Laboratory
Wang, Xijie	Brookhaven National Laboratory
Westenskow, Glen A.	Lawrence Livermore National Lab.
Wheeler, Susan	Lawrence Berkeley National Laboratory
Whittum, David	Stanford Linear Accelerator Center
Williams, Ronald	Florida A&M University

Winick, Herman	Stanford Linear Accelerator Center
Wurtele, Jonathan	Lawrence Berkeley National Laboratory
Xie, Ming	Lawrence Berkeley National Laboratory
Xie, Zu-Qi	Lawrence Berkeley National Laboratory
Yeremian, Anahid Dian	Stanford Linear Accelerator Center
Yoder, Rodney B.	Yale University
Yu, Simon	Lawrence Berkeley National Laboratory
Zhao, Yongxiang	Brookhaven National Laboratory
Zholents, Alexander	Lawrence Berkeley National Laboratory
Zolotorev, Max	Lawrence Berkeley National Laboratory

Author Index

A

Adolphsen, C., 455, 465
Ahn, H., 83, 390
Anderson, D., 842
Anderson, J., 865
Arinaga, M., 390

B

Babine, A., 372
Babzien, M., 591, 608, 664, 695, 937
Baine, M., 337, 400
Bane, K. L. F., 455, 465
Baranov, G. A., 937
Barov, N., 116
Ben-Zvi, I., 40, 695, 937
Bhattacharjee, A., 575
Biswal, S., 68
Bobin, J. L., 417
Bogacz, S. A., 286
Boloshin, Y.-A., 937
Bowden, G. B., 501
Brogle, R., 747
Brown, W. J., 717
Brownell, J. H., 601
Bulanov, S. V., 422, 923
Burris, H. R., 400
Burris, R., 443
Byer, R. L., 538

C

Calame, J. P., 865
Carrigan, R. A., Jr., 146
Castle, M., 865
Chattopadhyay, S., 233
Chen, C., 782
Chen, P., 273
Chen, S.-Y., 408, 651
Cheng, J., 865
Chiou, T. C., 357
Chojnacki, E., 757
Chou, P. J., 473, 501
Chu, T. S., 309

Clark, T. R., 76
Clayton, C. E., 13
Cline, D. B., 286, 608, 664
Conde, M., 116, 757
Copeland, M. R., 501
Corsini, R., 126
Cox, G., 116

D

Danly, B. G., 717
Dodd, E., 106
Dowell, D. H., 793
Downer, M. C., 214, 372, 651
Dublov, A. A., 937
Dudnikova, G. I., 422

E

Esarey, E., 96, 337, 400, 433, 443
Eylon, S., 842

F

Fang, J. M., 664
Fang, J.-M., 575, 591
Fant, K., 805
Farvid, A., 501
Fernow, R., 923
Fernow, R. C., 601
Fink, Y., 782
Fisch, N. J., 344
Fischer, R., 400, 443
Fisher, D., 372
Fliflet, A. W., 832
Fowkes, W. R., 455, 465

G

Gai, W., 116, 564, 757
Gallardo, J., 591
Gallardo, J. C., 608
Ganguly, A. K., 887

Gold, S. H., 832
Gou, S. K., 575
Granatstein, V. L., 865, 874

H

Hafizi, B., 96, 433, 443, 638, 832, 887
Haimson, J., 898
Hanna, S. M., 473
Hartemann, F. V., 309, 705
Henestroza, E., 842
Henke, H., 473, 485
Hirshfield, J. L., 618, 629, 822, 832, 887
Ho, C. H., 705
Hogan, B., 865
Hogge, J.-P., 717
Hojo, H., 766
Houck, T., 842
Hsu, S. Y., 705
Huang, Y. C., 538
Huang, Z., 254
Hubbard, R., 96, 400
Hubbard, R. F., 638
Hwang, J. Y., 705

J

Joly, S., 793
Jones, R. M., 455, 465
Joshi, C., 747

K

Kando, M., 83, 390, 687
Katsouleas, T., 175, 357
Kawakubo, T., 83, 390
Kerman, A. K., 309
Kerslick, G. S., 852
Khusid, M., 717
Kim, J. K., 106
Kim, K.-J., 243
Kimura, W. D., 608, 673, 923
Kinkead, A. K., 832
Kirby, R. E., 501
Kirk, H. G., 601
Kishimoto, Y., 390
Ko, K., 455, 465

Koga, J., 390
Konecny, R., 116, 757
Kotaki, H., 83, 390, 687
Kozyrev, E. V., 912
Krall, J., 96
Kreischer, K. E., 717
Krishnagopal, S., 263
Kroll, N., 518, 528
Kroll, N. M., 455, 465
Krushelnick, K., 337, 400, 443
Kusche, K., 591, 937
Kusche, K. P., 608

L

LaPointe, M. A., 618, 887
Lau, W. K., 705
Lavine, T., 805
Lawson, W., 55, 865, 874
Le Blanc, S. P., 651
LeBlanc, S. P., 372
Leemans, W. P., 23
Lekomtsev, V. A., 937
Le Sage, G. P., 705
Li, D., 518, 528
Lidia, S., 842
Liu, Y., 608, 664
Liu, Y. C., 705
Loewen, R. J., 805
Loulergue, A., 793
Luhmann, N. C., Jr., 309, 705

M

Makarov, I. G., 912
Maksimchuk, A., 408, 651
Malone, R., 591
Manka, C., 400
Marable, W. P., 782
Marshall, T. C., 575, 618, 629
Mecklenburg, B., 898
Menegat, A., 473, 501
Meskovsky, I. K., 937
Mikhailichenko, A. A., 294, 547
Milchberg, H. M., 76
Miller, R. H., 455, 465, 695
Milyutin, P. V., 301
Moore, C., 337

Moore, C. I., 400, 443
Mori, W. B., 357
Mourou, G., 68, 651
Muggli, P., 747

N

Nakajima, K., 83, 390, 687
Nakanishi, H., 83, 390
Naqvi, S. A., 852
Nation, J. A., 852
Naumova, N. M., 422
Nees, J., 68
Nezhevenko, O. A., 832, 912
Nikiforov, A. A., 912
Nikitin, S. P., 76
Noble, R. J., 273
Nusinovich, G. S., 874

O

Ogata, A., 83, 390, 687
Ostreiko, G. N., 912

P

Palmer, D. T., 695
Pearson, C., 501, 805
Pegoraro, F., 422
Pellegrini, C., 695
Persov, B. Z., 912
Petelin, M. I., 822
Pogorelsky, I., 591
Pogorelsky, I. V., 422, 608, 923, 937, 951
Pope, R., 805
Power, J., 116, 757

Q

Qiu, X., 591
Quimby, D. C., 608

R

Rau, B., 372, 766
Reginato, L., 842
Reiser, M., 865
Rifkin, J., 805
Robertson, S. H., 601
Romano, T., 591
Romea, R. D., 673
Rosenzweig, J. B., 181
Ruth, R. D., 254, 455, 465, 805, 813

S

Sanders, D. A., 286
Sandweiss, J., 591, 608
Saraph, G. P., 865, 874
Schächter, L., 326, 852
Schoessow, P., 116, 757
Schultz, S., 518, 528
Segalov, Z., 937
Seidel, M., 455, 465
Serafini, L., 196
Serdobintsev, G. V., 912
Sergeev, A., 372
Shapiro, M., 717
Shchelkunoff, S. V., 912
Sheehan, J., 591, 695
Shere, L., 501
Shvets, G., 344, 357
Siders, C. W., 214, 372
Siemann, R. H., 3, 473, 501
Simpson, J., 116, 564, 757
Singh, A., 874
Skaritka, J., 591, 695, 937
Skvortsov, V. A., 301
Smedley, J., 730
Smith, D. R., 518, 528
Spencer, J. E., 501
Sprangle, P., 96, 337, 400, 433, 443, 638
Srinivasan-Rao, T., 730
Steinhauer, L. C., 673
Stepanov, A., 372

T

Tajima, T., 233, 372, 766
Tani, K., 83, 390

Tantawi, S. G., 805, 813
Tarnetsky, V. V., 912
Temkin, R. J., 717
Ting, A., 96, 337, 400, 443
Troha, A. L., 309
Trotz, S., 717
True, R., 832

U

Ueda, T., 83, 390, 687
Uesaka, M., 83, 390, 687
Umstadter, D., 106, 408, 651

V

Vanecek, D., 842
Van Meter, J. R., 309
Van Steenbergen, A., 923
van Steenbergen, A., 591, 608
Vier, D. C., 518
Vikharev, A. L., 822
Villa, F., 739
Vlieks, A. E., 805, 813
Vogel, N., 378
Vshivkov, V. A., 422

W

Wagner, R., 408, 651
Walsh, J. E., 601
Wang, C., 887
Wang, H., 518
Wang, J. W., 455, 465

Wang, M., 887
Wang, X. J., 187, 664, 695
Wang, X.-J., 591
Watanabe, H., 390
Watanabe, T., 83, 390, 687
Westenskow, G., 842
Westenskow, G. A., 207
Whittum, D., 473
Whittum, D. H., 501
Winick, H., 695
Woodle, M., 695
Wright, E. L., 898
Wurtele, J. S., 167

X

Xie, M., 233

Y

Yakimenko, V., 664, 695
Yakovlev, V. P., 832, 912
Yang, T. T., 705
Yeremian, A. D., 196
Yoder, R. B., 629, 887
Yokoya, K., 233
Yu, S., 842

Z

Zapryagaev, I. A., 912
Zhang, T. B., 618, 629
Zolotorev, M., 813

AIP Conference Proceedings

	Title	L.C. Number	ISBN
No. 336	Dark Matter (College Park, MD 1994)	95-76538	1-56396-438-4
No. 337	Pulsed RF Sources for Linear Colliders (Montauk, NY 1994)	95-76814	1-56396-408-2
No. 338	Intersections Between Particle and Nuclear Physics 5th Conference (St. Petersburg, FL 1994)	95-77076	1-56396-335-3
No. 339	Polarization Phenomena in Nuclear Physics Eighth International Symposium (Bloomington, IN 1994)	95-77216	1-56396-482-1
No. 340	Strangeness in Hadronic Matter (Tucson, AZ 1995)	95-77477	1-56396-489-9
No. 341	Volatiles in the Earth and Solar System (Pasadena, CA 1994)	95-77911	1-56396-409-0
No. 342	CAM -94 Physics Meeting (Cacun, Mexico 1994)	95-77851	1-56396-491-0
No. 343	High Energy Spin Physics Eleventh International Symposium (Bloomington, IN 1994)	95-78431	1-56396-374-4
No. 344	Nonlinear Dynamics in Particle Accelerators: Theory and Experiments (Arcidosso, Italy 1994)	95-78135	1-56396-446-5
No. 345	International Conference on Plasma Physics ICPP 1994 (Foz do Iguaçu, Brazil 1994)	95-78438	1-56396-496-1
No. 346	International Conference on Accelerator-Driven Transmutation Technologies and Applications (Las Vegas, NV 1994)	95-78691	1-56396-505-4
No. 347	Atomic Collisions: A Symposium in Honor of Christopher Bottcher (1945-1993) (Oak Ridge, TN 1994)	95-78689	1-56396-322-1
No. 348	Unveiling the Cosmic Infrared Background (College Park, MD, 1995)	95-83477	1-56396-508-9
No. 349	Workshop on the Tau/Charm Factory (Argonne, IL, 1995)	95-81467	1-56396-523-2
No. 350	International Symposium on Vector Boson Self-Interactions (Los Angeles, CA 1995)	95-79865	1-56396-520-8
No. 351	The Physics of Beams Andrew Sessler Symposium (Los Angeles, CA 1993)	95-80479	1-56396-376-0

	Title	L.C. Number	ISBN
No. 352	Physics Potential and Development of $\mu^+\mu^-$ Colliders: Second Workshop (Sausalito, CA 1994)	95-81413	1-56396-506-2
No. 353	13th NREL Photovoltaic Program Review (Lakewood, CO 1995)	95-80662	1-56396-510-0
No. 354	Organic Coatings (Paris, France, 1995)	96-83019	1-56396-535-6
No. 355	Eleventh Topical Conference on Radio Frequency Power in Plasmas (Palm Springs, CA 1995)	95-80867	1-56396-536-4
No. 356	The Future of Accelerator Physics (Austin, TX 1994)	96-83292	1-56396-541-0
No. 357	10th Topical Workshop on Proton-Antiproton Collider Physics (Batavia, IL 1995)	95-83078	1-56396-543-7
No. 358	The Second NREL Conference on Thermophotovoltaic Generation of Electricity	95-83335	1-56396-509-7
No. 359	Workshops and Particles and Fields and Phenomenology of Fundamental Interactions (Puebla, Mexico 1995)	96-85996	1-56396-548-8
No. 360	The Physics of Electronic and Atomic Collisions XIX International Conference (Whistler, Canada, 1995)	95-83671	1-56396-440-6
No. 361	Space Technology and Applications International Forum (Albuquerque, NM 1996)	95-83440	1-56396-568-2
No. 362	Two-Center Effects in Ion-Atom Collisions (Lincoln, NE 1994)	96-83379	1-56396-342-6
No. 363	Phenomena in Ionized Gases XXII ICPIG (Hoboken, NJ, 1995)	96-83294	1-56396-550-X
No. 364	Fast Elementary Processes in Chemical and Biological Systems (Villeneuve d'Ascq, France, 1995)	96-83624	1-56396-564-X
No. 365	Latin-American School of Physics XXX ELAF Group Theory and Its Applications (México City, México, 1995)	96-83489	1-56396-567-4
No. 366	High Velocity Neutron Stars and Gamma-Ray Bursts (La Jolla, CA 1995)	96-84067	1-56396-593-3
No. 367	Micro Bunches Workshop (Upton, NY, 1995)	96-83482	1-56396-555-0

No.	Title	L.C. Number	ISBN
No. 368	Acoustic Particle Velocity Sensors: Design, Performance and Applications (Mystic, CT, 1995)	96-83548	1-56396-549-6
No. 369	Laser Interaction and Related Plasma Phenomena (Osaka, Japan 1995)	96-85009	1-56396-445-7
No. 370	Shock Compression of Condensed Matter-1995 (Seattle, WA 1995)	96-84595	1-56396-566-6
No. 371	Sixth Quantum 1/f Noise and Other Low Frequency Fluctuations in Electronic Devices Symposium (St. Louis, MO, 1994)	96-84200	1-56396-410-4
No. 372	Beam Dynamics and Technology Issues for + - Colliders 9th Advanced ICFA Beam Dynamics Workshop (Montauk, NY, 1995)	96-84189	1-56396-554-2
No. 373	Stress-Induced Phenomena in Metallization (Palo Alto, CA 1995)	96-84949	1-56396-439-2
No. 374	High Energy Solar Physics (Greenbelt, MD 1995)	96-84513	1-56396-542-9
No. 375	Chaotic, Fractal, and Nonlinear Signal Processing (Mystic, CT 1995)	96-85356	1-56396-443-0
No. 376	Chaos and the Changing Nature of Science and Medicine: An Introduction (Mobile, AL 1995)	96-85220	1-56396-442-2
No. 377	Space Charge Dominated Beams and Applications of High Brightness Beams (Bloomington, IN 1995)	96-85165	1-56396-625-7
No. 378	Surfaces, Vacuum, and Their Applications (Cancun, Mexico 1994)	96-85594	1-56396-418-X
No. 379	Physical Origin of Homochirality in Life (Santa Monica, CA 1995)	96-86631	1-56396-507-0
No. 380	Production and Neutralization of Negative Ions and Beams / Production and Application of Light Negative Ions (Upton, NY 1995)	96-86435	1-56396-565-8
No. 381	Atomic Processes in Plasmas (San Francisco, CA 1996)	96-86304	1-56396-552-6
No. 382	Solar Wind Eight (Dana Point, CA 1995)	96-86447	1-56396-551-8
No. 383	Workshop on the Earth's Trapped Particle Environment (Taos, NM 1994)	96-86619	1-56396-540-2
No. 384	Gamma-Ray Bursts (Huntsville, AL 1995)	96-79458	1-56396-685-9
No. 385	Robotic Exploration Close to the Sun: Scientific Basis (Marlboro, MA 1996)	96-79560	1-56396-618-2

	Title	L.C. Number	ISBN
No. 386	Spectral Line Shapes, Volume 9 13th ICSLS (Firenze, Italy 1996)		1-56396-656-5
No. 387	Space Technology and Applications International Forum (Albuquerque, NM 1997)	96-80254	1-56396-679-4 (Case set) 1-56396-691-3 (Paper set)
No. 388	Resonance Ionization Spectroscopy 1996 Eighth International Symposium (State College, PA 1996)	96-80324	1-56396-611-5
No. 389	X-Ray and Inner-Shell Processes 17th International Conference (Hamburg, Germany 1996)	96-80388	1-56396-563-1
No. 390	Beam Instrumentation Proceedings of the Seventh Workshop (Argonne, IL 1996)	97-70568	1-56396-612-3
No. 391	Computational Accelerator Physics (Williamsburg, VA 1996)	97-70181	1-56396-671-9
No. 392	Applications of Accelerators in Research and Industry: Proceedings of the Fourteenth International Conference (Denton, TX 1996)	97-71846	1-56396-652-2
No. 393	Star Formation Near and Far Seventh Astrophysics Conference (College Park, MD 1996)	97-71978	1-56396-678-6
No. 394	NREL/SNL Photovoltaics Program Review Proceedings of the 14th Conference— A Joint Meeting (Lakewood, CO 1996)	97-72645	1-56396-687-5
No. 395	Nonlinear and Collective Phenomena in Beam Physics (Arcidosso, Italy 1996)	97-72970	1-56396-668-9
No. 396	New Modes of Particle Acceleration— Techniques and Sources (Santa Barbara, CA 1996)	97-72977	1-56396-728-6
No. 397	Future High Energy Colliders (Santa Barbara, CA 1997)	97-73333	1-56396-729-4
No. 398	Advanced Accelerator Colliders Seventh Workshop (Lake Tahoe, CA 1996)	97-72788	1-56396-697-2 (set) 1-56396-727-8 (cloth) 1-56396-726-X (CD-Rom)